INSTRUMENT COMMERCIAL MANUAL

JEPPESEN.
Sanderson Training Products

55 Inverness Drive East, Englewood, CO 80112-5498
International Standard Book Number 0-88487-274-2
Printed in Canada

JS314520—002

To Paul Sanderson, whose pioneering efforts in aviation training and education inspired generations of pilots and aviation enthusiasts. Since 1956, Paul has made numerous contributions to the development of innovative new products and training courses. His devoted service is greatly appreciated and continues to benefit all of us in aviation.

Acknowledgments

This manual could not have been produced without the tireless commitment of the Guided Flight Discovery (GFD) team members listed below. Additional personnel in Jeppesen's Aviation Courseware Development department deserve special thanks for assuming responsibility for other projects so the GFD team members could focus their efforts on the creation of this text.

Managing Editor
Pat Willits

Associate Editors
Mike Abbott
Liz Kailey
Jim Mowery

Primary Writers
Julie Boatman
Jon Hiles
George McCray
Dave Schoeman
Ken Shockley
Chuck Stout
Anthony Werner

Technical Support
Jerry Farrell
Michelle Gable
Judi Glenn
Tanya Letts
Chad Pomering
Matt Ruwe

Media Productions Manager
Richard Hahn

Lead Artist
Pat Brogan

Graphic Artists
Jennifer Crowe
Paul Gallaway
Kerrie Hughes
Lee Korak
Dean McBournie
Lawrence Montano
David Navarro
Richard E. Patterson
Karen Sadenwater
Scott Saunders
Jay Weets

Photographers
Dave Chance
Jake Christian
Gary Kennedy
Virgil Poleschook

Welcome to Guided Flight Discovery

Guided Flight Discovery is a new concept in pilot training designed to make your professional training exciting and enjoyable. This revolutionary system is comprehensive, application-oriented, and it leads you logically through essential aeronautical knowledge areas. The program exposes you to a variety of useful, interesting information which will enhance and expand your understanding of the professional world of aviation.

While each element of the Guided Flight Discovery Pilot Training System may be used separately, the effectiveness of the materials can be maximized by using all of the individual components in a systems approach. To help you efficiently organize your studies and get the most out of your training, Guided Flight Discovery incorporates cross-references which are used to direct you to related Guided Flight Discovery study materials. The main components of the Instrument/Commercial Program are described below.

Instrument/Commercial Manual

The *Instrument/Commercial Manual* is your primary source for initial study and review. The text contains complete and concise explanations of the advanced concepts and ideas that every professional pilot needs to know. The subjects are organized in a logical manner to build upon previously introduced topics. Subjects are often expanded upon through the use of Discovery Insets which are strategically placed throughout the chapters. Periodically, human factors principles are presented in Human Factors Insets to help you understand how your mind and body function while you fly. Throughout the manual, concepts which directly relate to FAA test questions are highlighted by FAA Question Insets. Additionally, you can evaluate your understanding of material introduced in a particular section by completing the associated review questions. Finally, Chapter 14 — Commercial Maneuvers uses colorful graphics and step-by-step procedure descriptions to help you visualize and understand each maneuver. Performance standards from the Commercial PTS also are included. A more detailed explanation of the text and how to use its unique features is contained in the Preface starting on page vii.

Instrument/Commercial Syllabus

The syllabus permits concurrent or separate enrollments in the instrument rating course and the commercial pilot certification course. The instrument segment provides a variety of training options for operators and students including a PC-based aviation training device (PCATD), a flight training device (FTD), and the airplane. The commercial segment includes single-engine complex and multi-engine airplane training. Ground and flight lessons are coordinated to ensure that your training progresses smoothly and you are consistently introduced to topics during ground training prior to application in the airplane.

Support Materials

A variety of support materials are available to further enhance your understanding of subject matter for professional pilot training. A brief description of these resources is provided below.

Instrument Rating and Commercial Pilot Airmen Knowledge Study Guides

These valuable study tools provide you with all the FAA questions which may be included on the Instrument Rating and Commercial Pilot computerized test. Answers and explanations for each question are provided to allow you to instantly check your understanding of required material.

Instrument and Commercial Pilot Test Preparation Software

The test preparation software contains the same information as the Instrument Rating and Commercial Pilot Study Guides with the added features of question search, simulated test taking, and performance tracking. Cross-references to Jeppesen Sanderson and FAA material are also included for every question.

Flight School Support Materials

Flight schools which use the Guided Flight Discovery Pilot Training System may provide a variety of additional resources and instructional support materials. Designed specifically to provide you with a well administered, quality training program, Guided Flight Discovery fight school support materials help foster an environment which maximizes your potential for understanding and comprehension on your way to becoming a professional pilot. Some of these resources are described below.

Instrument/Commercial Videos

The Instrument/Commercial videos present instrument concepts, airplane systems, and commercial maneuvers. The dynamic videos use state-of-the-art graphics and animation, as well as dramatic aerial photography to help easily explain complex ideas.

PC-Based Aviation Training Device (PCATD)

Flight schools may also provide access to Jeppesen Sanderson's PC-based aviation training device. The PCATD is designed specifically for instrument training and skill enhancement, and can be a great tool for giving you a head start on certain instrument maneuvers and procedures.

Instructors Guide

The *Instructors Guide* is available for use by flight instructors and flight school operators. The Instrument/Commercial Insert for the *Instructors Guide* helps flight training professionals effectively implement the Guided Flight Discovery advanced courses.

Preface

The purpose of the *Instrument/Commercial Manual* is to provide you with the most complete explanations of aeronautical concepts for professional pilots through the use of colorful illustrations, full-color photos, and a variety of innovative design techniques. The *Instrument/Commercial Manual* and other Guided Flight Discovery materials are closely coordinated to make learning fun and effective. To help you organize your study, the *Instrument/Commercial Manual* is divided into four parts:

Part I — Discovering New Horizons

The information needed to begin your professional pilot journey is introduced in this part. The first chapter, Building Professional Experience, answers many of your questions about the training process and introduces advanced human factor concepts. Principles of Instrument Flight, Chapter 2, is the foundation for becoming an instrument-rated pilot. Chapter 3 — The Flight Environment introduces air traffic control for instrument flight and emphasizes ATC clearances.

Part II — Instrument Charts and Procedures

Chapters 4 through 7 of Part II provide a broad introduction to instrument charts in the same sequence of use as an instrument flight — departure, enroute, arrival, and approach. Procedural sections follow each chart presentation topic so you are immediately exposed to operational factors that require interpretation of IFR chart symbology and other cartographic features. Chapter 8 — Instrument Approaches combine approach chart formats with an operationally oriented narrative designed to show you how to precisely fly various types of instrument procedures.

Part III — Aviation Weather and IFR Flight Operations

In Part III, you will become familiar with aviation weather from the perspective of conducting IFR flight operations. In Chapter 9 — Meteorology you will learn how to minimize your exposure to hazardous weather phenomena by using weather reports, forecasts, and charts in the planning phase. Chapter 10 — IFR Flight Consideration provides a framework for coping with the complex options presented by IFR flight operation, including emergencies.

Part IV — Commercial Pilot Operations

Specialized information for commercial pilots characterizes the content of Part IV. Chapter 11 — Advanced Systems provides operational insight for complex and high performance aircraft. Chapter 12 — Aerodynamics and Performance Limitations helps you define the flight envelope and prepares you for predicting performance and controlling weight and balance as you move up to higher performance aircraft. Chapter 13 and 14 — Commercial Flight Considerations and Commercial Maneuvers conclude this part with in-depth analyses of emergency procedures, commercial decision making, and performance of precision flight maneuvers.

Table of Contents

PART III — AVIATION WEATHER AND IFR FLIGHT OPERATIONS

PART IV — COMMERCIAL PILOT OPERATIONS

APPENDICES

Using the Manual

The *Instrument/Commercial Manual* is structured to highlight important topics and concepts and promote an effective and efficient study/review method of learning. To get the most out of your manual, as well as the entire Guided Flight Discovery Pilot Training System, you may find it beneficial to review the major design elements incorporated in this text.

PART I

A new era in aviation began in 1929 when Jimmy Doolittle flew for the first time from takeoff to landing solely by reference to the flight instruments. Doolittle expanded aviation's horizons with his accomplishments and you will continue to benefit from his efforts as you embark on training for an instrument rating or a commercial certificate. Your own new horizons lay ahead as you increase your knowledge of flight operations, master new skills, and discover unique facets of your role as pilot in command. You begin in Chapter 1 by exploring instrument and commercial training and the opportunities available to you as you gain professional experience. You also will be introduced to human factors concepts that are essential to operating safely in the instrument/commercial environment. As you delve into Chapter 2, you will gain an understanding of flight instrument systems, as well as examine the principles of attitude instrument flying and instrument navigation. Chapter 3 provides a review of the flight environment and offers new insight into ATC services and clearances.

Learning Objectives

Learning objectives are provided at the beginning of each part to help you focus on important concepts.

CHAPTER 3

THE FLIGHT ENVIRONMENT

Instrument/Commercial
Part I, Chapter 3 — The Flight Environment 

Cross-Reference Icon

A cross-reference icon is included at the beginning of each chapter to direct you to the corresponding video which supports and expands on introduced concepts and ideas.

CHAPTER 2 — PRINCIPLES OF INSTRUMENT FLIGHT

Several important values are not marked on the airspeed indicator. During gusty or turbulent conditions, you should slow the aircraft below the design maneuvering speed, V_A, to ensure the load factor is within safe limits. At or below V_A, the airplane will stall before excessive G-forces can occur. Design maneuvering speed, which decreases with the total weight of the aircraft, appears in your POH or on a placard. V_{LE}, the maximum speed with the landing gear extended on a retractable gear airplane, and V_{LO}, the maximum speed for extending or retracting the landing gear, also appear in the POH but not on the airspeed indicator.

Design maneuvering speed is one important value not shown by the color coding of an airspeed indicator. [Figure 2-21] If you encounter severe turbulence during an IFR flight, slow the airplane below this speed. This decreases the amount of excess load that can be imposed on the wing.

INSTRUMENT CHECK

Unless the airplane is facing into a strong wind, the airspeed indicator should read zero before you taxi. If it indicates some value due to wind, verify the indicator drops to zero when you turn the airplane away from the wind. As you accelerate during the takeoff roll, make sure the airspeed indicator comes alive and increases at an appropriate rate. If not, discontinue the takeoff.

ALTIMETER

An altimeter is required for both VFR and IFR flight. For IFR, you need a **sensitive altimeter**, adjustable for barometric pressure. Obviously an accurate altimeter is essential when you are operating in IFR conditions. In addition to helping you maintain terrain clearance, minimum IFR altitudes, and separation from other aircraft, the altimeter helps you maintain aircraft control. An understanding of the operation and limitations of the altimeter will enable you to interpret its indications correctly. [Figure 2-22]

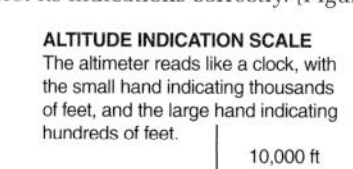

Figure 2-22. This altimeter indicates 2,800 feet MSL. The barometric scale in the window is at 29.92 inches of mercury (in. Hg.).

Now that you will be flying under IFR, your ability to accurately read and adjust the altimeter is far more important than it was under VFR. See Figure 2-22.

2-18

Full-Color Graphics

The full-color graphics used throughout the text are carefully designed to make difficult concepts easy to understand.

ARRIVAL CHARTS — SECTION A

scription, and these charts are filed with the airport's approach charts. NOS includes STARs at the front of each *Terminal Procedures Publication* regional booklet.

To illustrate how STARs can be used to simplify a complex clearance and reduce frequency congestion, consider the following arrival clearance issued to a pilot flying to Seattle, Washington. *"Cessna 32G, cleared to the Seattle/Tacoma International Airport as filed. Maintain 12,000. At the Ephrata VOR intercept the 221° radial to CHINS Intersection. Intercept the 284° of the Yakima VOR to SNOMY Intersection. Cross SNOMY at 10,000. Continue via the Yakima 281° radial to AUBRN Intersection. Expect radar vectors to the final approach course."*

Now consider how this same clearance is issued when a STAR exists for this terminal area. *"Cessna 32G, cleared to Seattle/Tacoma International Airport as filed, then CHINS TWO ARRIVAL, Ephrata Transition. Maintain 10,000 feet."* A shorter transmission conveys the same information. A **transition** is one of several routes that bring traffic from different directions into one STAR.

STARs are established to simplify clearance delivery procedures.

REDUCING PILOT/CONTROLLER WORKLOAD

Safety is enhanced when both pilots and controllers know what to expect. Being able to rehearse a proposed route or clearance in advance means that the pilot can pay more attention to situational awareness during the corresponding portion of a flight. Effective communication increases with the reduction of repetitive clearances, freeing up congested control frequencies.

To accomplish this, STARs are developed according to the following criteria:

- STARs must be simple, easily understood and, if possible, limited to one page.
- A STAR transition should be able to accommodate as many different types of aircraft as possible. That way, both jets and single-engine piston aircraft may be able to use the same chart for arrival.
- VORTACs are used wherever possible, so that military and civilian aircraft can use the same arrival.
- DME arcs within a STAR should be avoided as not all aircraft in the IFR environment are so equipped.
- Altitude crossing and airspeed restrictions are included when they are assigned by ATC a majority of the time.

6-3

FAA Question Insets

Information which relates directly to FAA test questions appears in tan insets. In addition to highlighting important concepts, the FAA Question Insets provide a good review tool when preparing for the Instrument and Commercial computerized tests.

Human Element Insets

Human Element Insets are presented in Chapter 2 through Chapter 13 to introduce the human factors aspect of flight. The topics, which can range from physiology to decision making, are presented with emphasis on flight related applications.

AIRPORTS, AIRSPACE, AND FLIGHT INFORMATION | SECTION A

Runways used for instrument operations have additional markings. A nonprecision instrument runway is used with an instrument approach that does not have an electronic glide slope for approach glide path information. This type of runway has the visual runway markings, plus the threshold and aiming point markings.

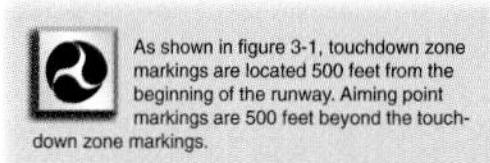

As shown in figure 3-1, touchdown zone markings are located 500 feet from the beginning of the runway. Aiming point markings are 500 feet beyond the touchdown zone markings.

An Airport Project That Didn't Stay Afloat

In the 1920's, the economic range of a commercial airplane was approximately 500 to 600 miles. To fly farther than that nonstop, an airplane had to carry so much weight as fuel that it was difficult for companies to turn a profit. To make transatlantic air travel successful, Edward Armstrong, an engineer, had an idea to construct seadromes — floating airports strung across the ocean.

Armstrong designed a structure which consisted of a deck, approximately 1,500 feet long and 300 feet wide, placed on 28 steel columns, each 170 feet long. The seadrome was designed to float with the airfield deck 70 feet above the water's surface. A buoyancy chamber would be built into each column 100 feet below the deck, and 60 feet below that would be a ballast chamber filled with iron ore. The seadrome would be attached to a buoy connected by two 18,000-foot steel cables to a concrete mushroom anchor. This system reduced the great strain that would have been caused by connecting the anchor line directly to the structure, and ensured that the seadrome was free to swing about into the wind. Propellers would allow the seadrome to maneuver while anchored and to navigate in an emergency.

Beneath the deck of the seadrome, Armstrong envisioned hangars, which would use a large elevator to transport airplanes to and from the flight deck. In addition, there were plans for repair shops, a radio station, and a hotel. The 50-room hotel would feature a gymnasium, swimming pool, miniature golf course, billiard room, movie theater, bowling alley, and several tennis courts.

The seadrome project had enthusiastic approval from famous aviators such as Charles Lindbergh. However, it was discontinued by the U.S. government when concern arose about the difficulty of maintaining adequate protection of seadromes built over international waters.

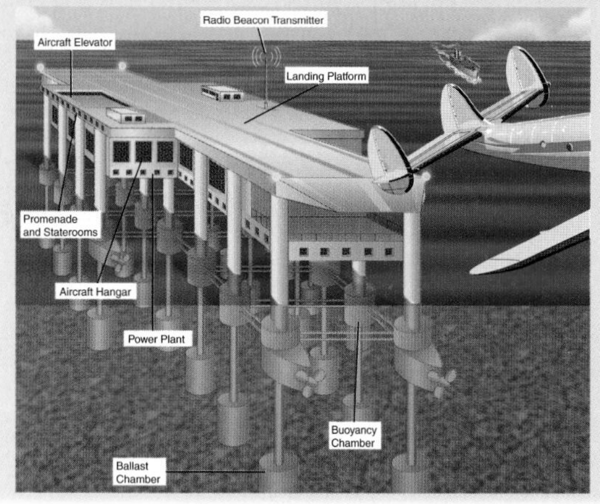

Discovery Insets

Discovery Insets are included throughout the text beginning with Chapter 2 to expand on ideas and concepts presented in the accompanying material. The information presented in each Discovery Inset varies, but is designed to enhance your understanding of the world of aviation. Examples include references to National Transportation Safety Board investigations, aviation history, and thought-provoking questions and answers.

INSTRUMENT/COMMERCIAL TRAINING AND OPPORTUNITIES | SECTION A

Common Carriage is any operation for compensation or hire in which the operator holds itself out, by advertising or any other means, as willing to furnish transportation for any member of the public. Private carriage does not involve holding out.

Figure 1-15. FAR Parts 121, 125, and 135 govern operations ranging from scheduled air carriers to on-demand charters.

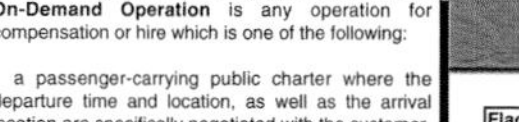

On-Demand Operation is any operation for compensation or hire which is one of the following:

a passenger-carrying public charter where the departure time and location, as well as the arrival location are specifically negotiated with the customer.

a common-carriage operation using an airplane (including turbojet-powered) having a passenger-seat configuration of 30 seats or less and a payload capacity of 7,500 pounds or less.

a private carriage operation conducted with an airplane having a passenger-seat configuration of less than 20 seats or a payload capacity of less than 6,000 pounds.

• a scheduled operation with a frequency of less than 5 round trips per week conducted with a nonturbojet-powered airplane with a maximum of 9 passenger seats and a maximum payload capacity of 7,500 pounds.

a cargo operation conducted with an airplane having a payload capacity of 7,500 pounds or less.

Flag Operation is any scheduled operation to any point outside the U.S. conducted in a turbojet-powered airplane or an airplane which has a passenger-seat configuration of more than 9 seats or a payload of more than 7,500 pounds.

Commuter Operation is any scheduled operation conducted in a nonturbojet-powered airplane having a maximum passenger-seat configuration of 9 seats or less and a maximum payload capacity of 7,500 pounds or less. A commuter operation must be scheduled with a frequency of at least 5 round trips per week on at least 1 route according to published flight schedules.

FARs GOVERNING SPECIFIC COMMERCIAL OPERATIONS

Airplane Size/Weight	Part 121 Domestic (Scheduled)	Part 121 Flag (Scheduled)	Part 121 Supplemental (Not Scheduled)	Part 135 Commuter (Scheduled)	Part 135 On-Demand (Not Scheduled)	Part 125 (Not Scheduled)
Common Carriage:						
≤9 seats and ≤7,500 lbs	No[1]	No[1]	No[2]	Yes[1]	Yes[2]	No
10-30 seats and ≤7,500 lbs	Yes	Yes	No[2]	No	Yes[2]	No
>30 seats or >7,500 lbs	Yes	Yes	Yes	No	No	No
Common Carriage Is Not Involved:						
<20 seats or <6,000 lbs	No	No	No	No	Yes	No
≥20 seats or ≥6,000 lbs	No	No	No	No	No	Yes

[1]Turbojet-powered airplanes used in scheduled passenger-carrying operations must comply with Part 121 regardless of passenger seating or payload capacity.

[2]If turbojet-powered airplanes and other airplanes with 10-30 passenger-seat configurations are used for Part 121 domestic or flag operations, non-scheduled or charter operations with that airplane shall be conducted under Part 121 supplemental rules.

1-23

Color Photographs

Color photographs are included to enhance learning and improve understanding.

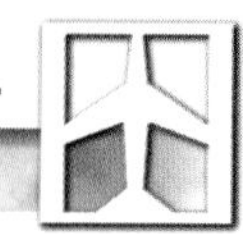

ELECTRONIC FLIGHT PUBLICATIONS

You can access a variety of flight information by using a modem-equipped personal computer and the appropriate software or Internet service. The government uses electronic bulletin boards and world wide web sites to make flight publications and information updates more readily accessible to the general aviation community. For example, the FAA Home Page contains information on various aviation subjects and links to other home pages, such as FedWorld Information Network. Through FedWorld, you can access and order government publications, such as advisory circulars, FARs, practical test standards, and the *Federal Register.* [Figure 3-28]

Figure 3-28. You can access the FAA Home Page at **http://www.faa.gov**.

SUMMARY CHECKLIST

- ✓ A visual runway normally is marked only with the runway number and a dashed white centerline. It may include additional markings for specific operations. A runway used for instrument approaches has additional markings, such as threshold markings, touchdown zone markings, aiming point markings, and side stripes.
- ✓ Additional markings for displaced thresholds include taxi, takeoff, and rollout areas, as well as blastpad/stopway areas. Closed runways and taxiways may be marked by Xs at some airports.
- ✓ Hold lines keep aircraft clear of the runways, and at controlled airports, serve as the point that separates the responsibilities of ground control from those of the tower.
- ✓ There are six basic types of airport signs — mandatory, location, direction, destination, information, and runway distance remaining.
- ✓ A variety of lighting systems, including approach light systems, sequenced flashing lights, runway alignment indicator lights, and runway end identifier lights are used at airports to aid the pilot in identifying the airport environment, particularly at night and in low visibility conditions.
- ✓ Various visual glide slope indicators, such as the visual approach slope indicator (VASI), precision approach slope indicator (PAPI), and tri-color VASI help pilots establish and maintain a safe descent path to the runway.
- ✓ Runway and taxiway lighting are installed at some airports to assist you in landing and taxiing at night or during low visibility conditions. This lighting can consist of runway edge lights, threshold lights, displaced threshold lights, touchdown zone

Summary Checklists

Summary Checklists are included at the end of each section to help you identify and review the major points introduced in the section.

KEY TERMS

Low Altitude Enroute Charts
Victor Airways
High Altitude Enroute Charts
Jet Routes
Mileage Break Point
Intersections
Noncompulsory Reporting Point
Compulsory Reporting Point
Minimum Enroute Altitude (MEA)
Minimum Obstruction Clearance Altitude (MOCA)
Maximum Authorized Altitude (MAA)
Minimum Reception Altitude (MRA)
Minimum Crossing Altitude (MCA)
Changeover Point (COP)
Remote Communications Outlet (RCO)
Area Charts

QUESTIONS

1. True/False. Low altitude enroute charts generally depict localizers that have only approach functions.

Match the following navaid symbols with the appropriate facility names.

2. NDB
3. TACAN
4. Localizer
5. VORTAC
6. VOR

A.

B.

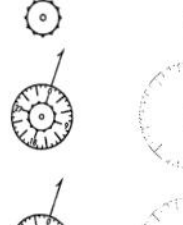

C.

D.

E.

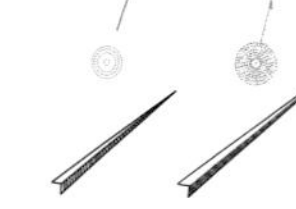

Key Terms

For ease of recognition and quick review, key terms are highlighted in red type when they are first introduced and defined. A list of key terms is included at the end of each section.

Questions

Questions are provided at the end of each section beginning with Chapter 2 to help you evaluate your understanding of the concepts which were presented in the accompanying section. Several question formats are provided including completion, matching, true/false, and essay. Perforated answer sheets, which are organized by chapter, are included at the back of the text.

PART I
Discovering New Horizons

Flying by instruments soon outgrew the early experimental phase. It became a practical reality, and aviation entered a new era. I was grateful for the opportunity to participate in the initial experiments. This work was, I believe, my most significant contribution to aviation.
— James H. "Jimmy" Doolittle

PART I

A new era in aviation began in 1929 when Jimmy Doolittle flew for the first time from takeoff to landing solely by reference to the flight instruments. Doolittle expanded aviation's horizons with his accomplishments and you will continue to benefit from his efforts as you embark on training for an instrument rating or a commercial certificate. Your own new horizons lay ahead as you increase your knowledge of flight operations, master new skills, and discover unique facets of your role as pilot in command. You begin in Chapter 1 by exploring instrument and commercial training and the opportunities available to you as you gain professional experience. You also will be introduced to human factors concepts that are essential to operating safely in the instrument/commercial environment. As you delve into Chapter 2, you will gain an understanding of flight instrument systems, as well as examine the principles of attitude instrument flying and instrument navigation. Chapter 3 provides a review of the flight environment and offers new insight into ATC services and clearances.

CHAPTER 1

BUILDING PROFESSIONAL EXPERIENCE

SECTION A

INSTRUMENT/ COMMERCIAL TRAINING AND OPPORTUNITIES

If I have seen further it is by standing upon the shoulders of Giants.
— Sir Isaac Newton

I believe that simple flight at least is possible to man and the experiments and investigations of a large number of independent workers will result in the accumulation of information and knowledge and skill which will finally lead to accomplished flight . . . I wish to avail myself of all that is already known and then if possible add my mite to help on the future worker who will attain final success.
— Wilbur Wright

The future is constructed by pioneers who build upon the knowledge of those who came before. Sir Isaac Newton discovered basic principles of motion. Wilbur and Orville Wright used those principles to construct a flying machine. To design innovative aircraft, create revolutionary technology, and discover new flying techniques each aviation pioneer sees further by standing on the shoulders of such giants as Leonardo daVinci, Otto Lilienthal, the Wright Brothers, and Charles Lindbergh.

As you begin training to earn an instrument rating or commercial pilot certificate, you are poised to take full advantage of the wisdom of those who have blazed the aviation trail. Continue to follow their flight path, and build upon the knowledge and skills that you have already gained as a pilot. Look back into aviation history and then direct your gaze forward to your future. It is time now for *you* to see further.

January 1, 1914 — The first scheduled passenger-service airline in the United States is born. The fare for the trip from Tampa to St. Petersburg, Florida is five dollars.

May 15, 1918 — The first standard airmail route in the United States is established between New York City and Washington, D.C.

1918

July 1, 1927 — Having successfully bid for the transcontinental airmail route from Chicago to San Francisco, the Boeing Airplane Company begins service flying the B-40, a new Boeing-produced airplane powered by the latest air-cooled Pratt and Whitney 400-horsepower Wasp radial engines.

Courtesy of United Technologies Archive

1928

1928 — Edwin A. Link develops the Link Trainer flight simulator as a means of providing affordable flight training to pilots without ever leaving the ground.

1929

September 24, 1929 — Jimmy Doolittle makes the first totally blind flight from takeoff to landing in the NY-2 Husky biplane.

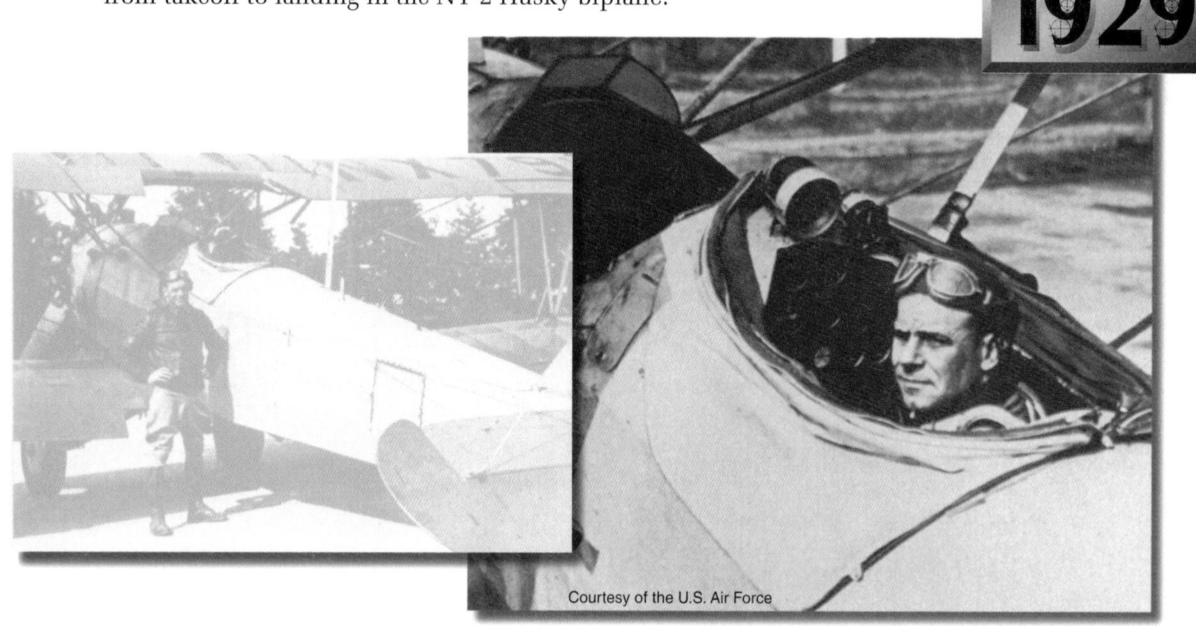

Courtesy of the U.S. Air Force

1933

1933 — Boeing develops the Model 247, the first passenger airliner with an autopilot, pneumatically operated de-icing equipment, a variable-pitch propeller, and retractable landing gear. With 7 stops, the Model 247 completes the trip between New York and Los Angeles in 20 hours, 7 1/2 hours less than the best previous airliner time.

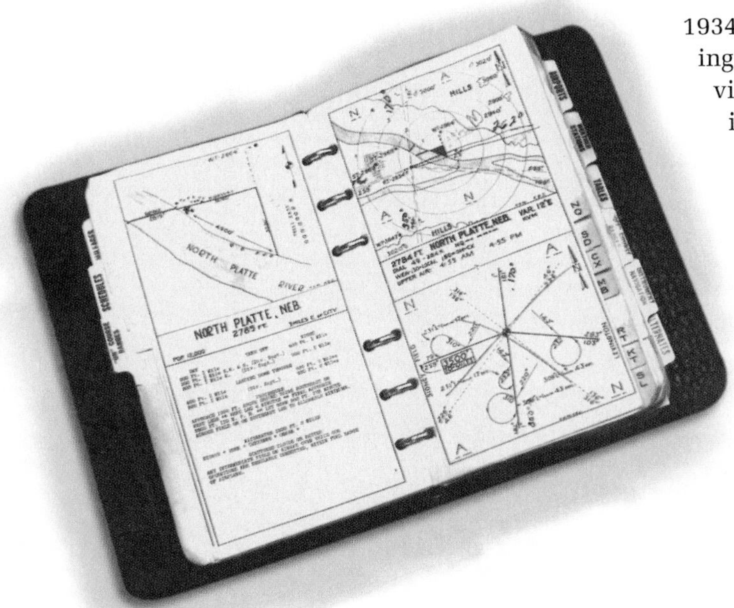

1934 — Elrey B. Jeppesen begins publishing the first airway manuals which provide pilots with valuable navigation information, airport elevations, heights of obstructions, and runway lengths. Each manual sells for ten dollars.

1934

March 15, 1935 — Wiley Post is the first pilot to take advantage of the high winds of the jet stream. Wearing a pressure suit he designed with BFGoodrich, Post covers 2,035 miles in his Lockheed Vega, the Winnie Mae, at an average groundspeed of 279 miles per hour, 100 miles per hour faster than the airplane's normal speed.

Courtesy of United Technologies Archive

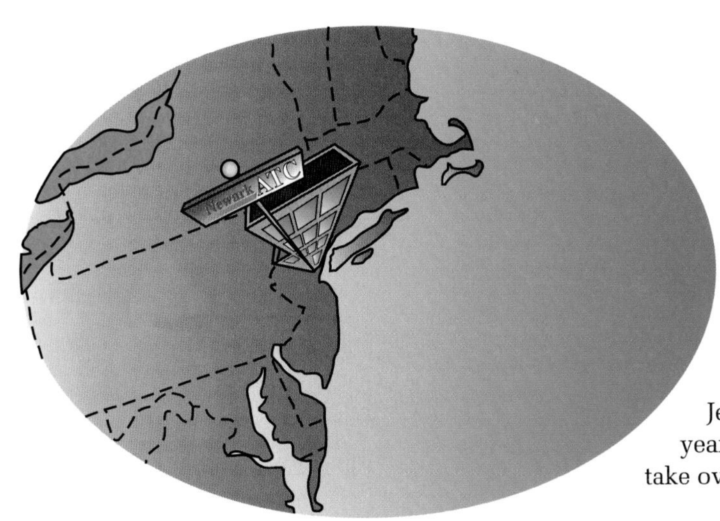

1935

December 1, 1935 — The first air traffic control center is established at Newark, New Jersey. Initially run by the airlines, less than a year later the Bureau of Air Commerce arranges to take over air traffic control.

July, 1949 — Beginning a new era in commercial aviation, the world's first jet airliner, the deHavilland Comet takes its maiden flight. In 1952, the Comet enters service with the British Overseas Airways Corporation but tragically, in 1954, two Comets are ripped apart in mid-air due to explosive decompression. All of the aircraft are grounded.

1955

1955 — The Boeing 707, the first jet airliner built in the United States, is purchased by Pan American for passenger service beginning in 1958. The B-707 cuts intercontinental travel time almost in half.

People thought we were crazy. — Juan Tripp, Pan American president regarding the first order of B-707s

1958

August 23, 1958 — Prompted by recent midair collisions, the Federal Aviation Act is passed. This action creates the Federal Aviation Agency, an independent government organization which has the sole responsibility for developing and maintaining a common civil-military system of air navigation and air traffic control. In 1966, the agency becomes the Federal Aviation Administration within the Department of Transportation.

Courtesy of Boeing Archives

1969

February 9, 1969 — The first jumbo jet, the Boeing 747 takes to the air for the first time, opening up the world to the traveling masses. The original B-747 is 225 feet long, has a tail as tall as a 6-story building, and can carry 3,400 pieces of baggage in its cargo hold. The airplane contains a ton of air when pressurized.

October 1, 1969 — On its 45th test flight, the French-assembled Concorde prototype 001 exceeds Mach 1 for the first time. It holds Mach 1.05 for 9 minutes at an altitude of 36,000 feet.

1978

1978 — The Airline Deregulation Act frees airlines to compete on routes, pricing, and service.

1993

1993 — The global positioning system (GPS) reaches initial operational capability with a full constellation of 24 satellites.

We had all learned that if we got inside the clouds for any length of time we became confused and fell. We tried things like bobs on the end of a string. I remember one fellow even had a half-full milk bottle that he thought he could use as a level. Of course none of these things worked . . . we were pretty well convinced that you just couldn't fly blind for any length of time. However, we all learned ways to prolong such flying. We got so that we could climb up through clouds by feel, by bracing our feet and feeling the wind on our cheeks, and things like that. Eventually, even then, we would get confused and the plane would stall and spin down.

— Dean Smith, one of the best-known early airmail pilots, as quoted in *The American Heritage History of Flight*

INSTRUMENT FLIGHT

Early airmail planes were not equipped with the proper instruments and navigation equipment to allow pilots to safely fly in clouds or low visibility conditions and weather information often was unavailable. Of the first 40 airmen hired to fly the mail, 31 were killed. A record of this nature was not likely to aid in the development of a reliable, safe air transportation system. As a result, numerous efforts were made to improve the ability of airmail pilots to reach their destinations, day or night, during a wide variety of weather conditions.

In 1927, the U.S. government began installing the first radio navigation system, the LF/MF four-course radio ranges. This arrangement of multiple radio beacons allowed four courses, called ranges, to be transmitted from one facility. To intercept and follow these ranges, pilots listened to the signals on headsets or speakers. The volume of the radio signal and whether it was a constant hum or series of Morse code dots and dashes enabled pilots to determine whether they were approaching or leaving a station and whether they were *on the beam.* [Figure 1-1] However, this system did not allow pilots to fly safely in the clouds. Flying blind with precision required far more sophisticated instruments than any that existed through the 1920s.

In 1928, the Daniel Guggenheim Fund for the Promotion of Aeronautics installed a Full Flight Laboratory at Mitchell Field on Long Island. Jimmy Doolittle was assigned to the program and given the task of solving the problems inherent in flying solely by instrument reference, or flying blind. He enlisted the help of Elmer A. Sperry, founder of the Sperry Gyroscope Company, who had harnessed the principles of gyros to build flight instruments.

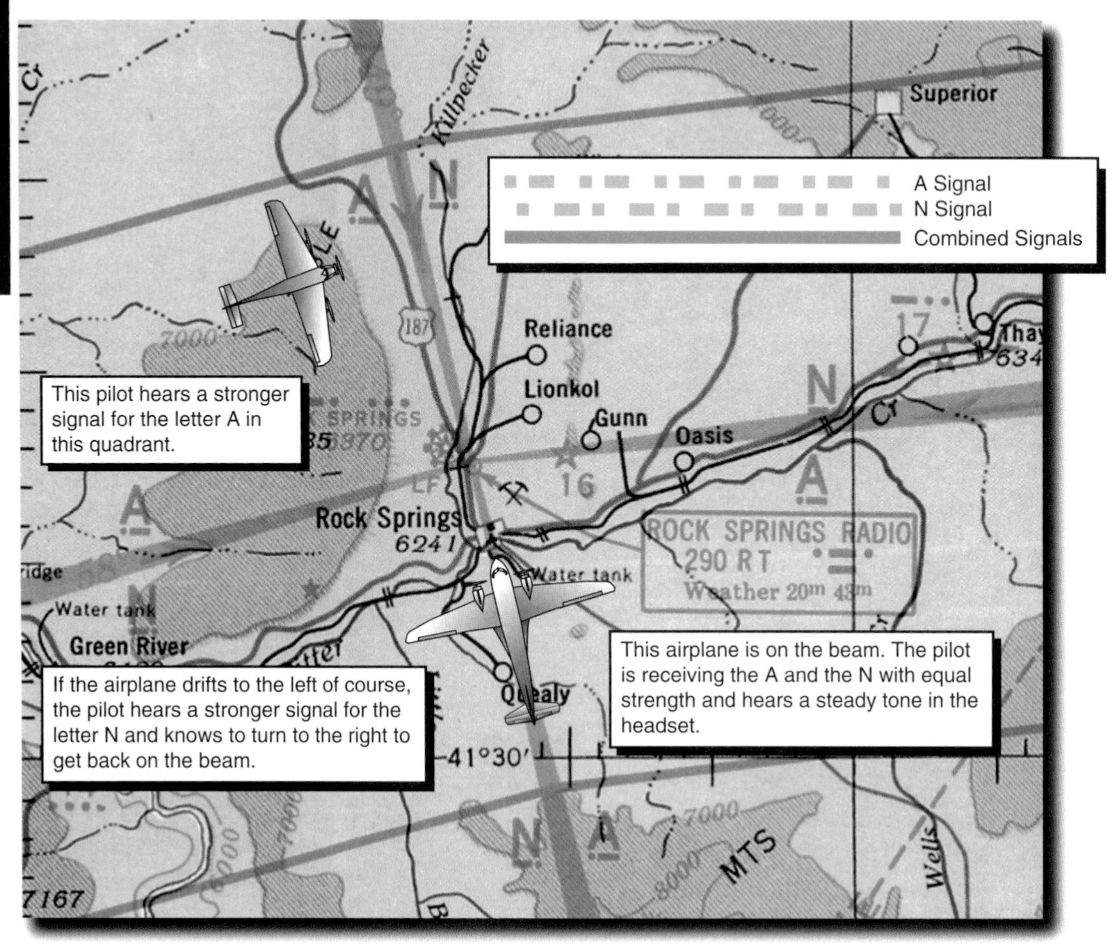

Figure 1-1. Each station broadcast the Morse code letter A (• —) in two quadrants and the letter N (— •) in the other quadrants. The areas where the signals overlapped formed the four legs of the range.

Courtesy of Honeywell

Courtesy of the U.S. Air Force

Figure 1-2. In the cockpit of Jimmy Doolittle's NY-2, instrument flight was born.

Sperry developed an artificial horizon, which provided a pictorial example of the airplane's attitude, as well as a directional gyro, which could be set to the magnetic compass.

For a blind landing, an altimeter that was much more precise than any yet available was needed. Doolittle turned to Paul Kollsman who, with the help of a Swiss watch-making firm, developed an altimeter that was accurate to within a few feet.

On September 24, 1929, Jimmy Doolittle, with Ben Kelsey as his safety pilot in the front cockpit of their NY-2 Husky biplane, completed the first flight solely by instrument reference. After a blind takeoff, the flight lasted 15 minutes and included two 180° turns. Doolittle navigated on the beam for part of the exercise prior to setting up for the approach, then successfully made a blind landing. [Figure 1-2]

Jimmy has more gifts than any one man has a right to be blessed with. — newspaper man Ernie Pyle, who often accompanied Jimmy Doolittle on stints around the military bases

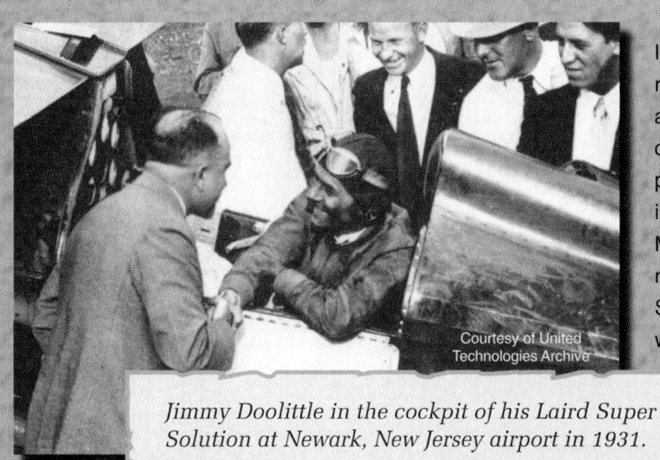

Courtesy of United Technologies Archive

Jimmy Doolittle in the cockpit of his Laird Super Solution at Newark, New Jersey airport in 1931.

In addition to his instrument flight research, Jimmy Doolittle was a top aeronautical engineer and one of the country's best pursuit and acrobatic pilots. Doolittle won almost every honor in civil aviation including the coveted Mackay Trophy given annually to the most outstanding flier. Flying the Laird Super Solution, he was also the first to win the Bendix Trophy Race from Los Angeles to Cleveland.

In 1940 at age 44, Doolittle was made president of the Institute of Aeronautical Science. Soon after, Major Doolittle became America's first World War II hero as he led 16 planes in a raid that wreaked destruction on Tokyo, Yokohama, Osaka, Kobe, and Nagoya. He was promoted to general and received the Medal of Honor on April 18, 1942.

In addition to improving flight safety by developing instrument systems, the 1920s and 30s realized the genesis of aviation charting. With no aeronautical charts available, many pilots used road maps for navigation. When visibility was limited, early aviators often followed the railroad tracks, which they called *hugging the UP*, or Union Pacific. At times, deteriorating weather conditions forced pilots to land in emergency fields to sit and wait for the skies to improve.

I got the information I needed any place I could — city and county engineers, surveyors, farmers. I drove all the way from Chicago to Oakland, California, and checked out the emergency fields and the obstructions around them, different ways to get in, how far they were from the railroad track and the highway. When the radio ranges came in, I used to take the chief pilot's airplane and go work out a procedure. I guess I devised 80% of the letdown procedures between Oakland and Chicago . . . I'd come in from a United trip at, say, two or three a.m. and go to the plant where they'd have a stack of letdown procedures, or approach plates, as we started calling them. I'd go all the way through it — I called it dry flying — and check the flight, and initial it. Not one chart went out of there without my initial.

— Elrey B. Jeppesen

In 1930, Elrey B. Jeppesen signed on with Varney Airlines and later with Boeing Air Transport as an airmail pilot to fly the Salt Lake City-Cheyenne/Salt Lake City-Oakland routes. During the winters of 1930 and 1931, Jeppesen experienced the loss of many of his fellow pilots due partly to the lack of published aeronautical information. To improve safety, he began recording pertinent information about flight routes and airports in a 10 cent black notebook. Jeppesen included field lengths, slopes, drainage patterns, information on lights and obstacles, drawings which profiled terrain and airport layouts, as well as phone numbers of local farmers who could provide weather reports. Equipped with an altimeter to record accurate elevations, Jeppesen climbed hills, smokestacks, and water towers on his days off. He flew each leg of the radio ranges and jotted down safe letdown procedures for airports.

Word soon began to circulate that Jeppesen had an amazing record of flight completions and that one of the principal reasons was his secret little black book on airports and landing procedures. Pilots began asking for copies and Jeppesen made them for his friends. However, the demand became so great that he entered the chart publishing business in 1934. [Figure 1-3]

Figure 1-3. Starting in 1934 with his 10 dollar airway manual, the company Elrey B. Jeppesen founded continues to produce charts for airlines and pilots worldwide.

WHY AN INSTRUMENT RATING?

The addition of an instrument rating to your private pilot certificate allows you to fly in a wider range of weather conditions than you can as a VFR pilot. When you are operating under **IFR (instrument flight rules)**, you can fly in the clouds with no reference to the ground or horizon. This is sometimes referred to as flying in **IMC (instrument meteorological conditions)**. The terms IFR, IMC, VFR, and VMC are used frequently in several different ways. Operating under **VFR (visual flight rules)**, you are governed by specific regulations which include minimum cloud clearance and visibility requirements. Instrument flight rules (IFR) govern flight operations in weather conditions below VFR minimums. [Figure 1-4] When referring to weather conditions, the terms IFR and IMC often are used interchangeably, as are the terms **VMC (visual meteorological conditions)** and VFR. In addition, the terms VFR and IFR can define the type of flight plan under which you are operating. [Figure 1-5]

Figure 1-4. Instrument and visual flight rules are contained in the Federal Aviation Regulations (FARs). The FARs are identified by a specific title number (Aeronautics and Space Title 14) within the larger group or rules contained in the Code of Federal Regulations (CFR). The FARs are divided into numbered parts (FAR Part 61, FAR Part 91, etc.) and each regulation typically is identified by the part number, followed by the specific regulation number, for example: FAR 61.65.

Figure 1-5. The *Aeronautical Information Manual* (AIM) defines the terms VFR, VMC, IFR, and IMC in the Pilot/Controller Glossary. It is possible for you to be operating on an IFR flight plan in VFR weather conditions.

VFR (Visual Flight Rules) are rules that govern the procedures for conducting flight under visual conditions. The term VFR is also used in the U.S. to indicate weather conditions that are equal to or greater than minimum VFR requirements. In addition, it is used by pilots and controllers to indicate type of flight plan.

IFR (Instrument Flight Rules) are rules governing the procedures for conducting instrument flight. This is also a term used by pilots and controllers to indicate type of flight plan. The International Civil Aviation Organization (ICAO) defines IFR as a set of rules governing the conduct of flight under instrument meteorological conditions.

VMC (Visual Meteorological Conditions) are meteorological conditions expressed in terms of visibility, distance from cloud, and ceiling equal to or better than specified minima.

IMC (Instrument Meteorological Conditions) are meteorological conditions expressed in terms of visibility, distance from cloud, and ceiling less than the minima specified for visual meteorological conditions.

While you are provided with more options regarding weather, perhaps the greatest benefit an instrument rating provides is the increase in safety. Instrument training enhances your skill at precisely controlling the aircraft, improves your ability to operate in the complex ATC system, and increases your confidence level. A study of aircraft accidents over an 11-year period showed that continuing, and initiating VFR flight into IMC without an instrument rating were the first and second most prevalent causes of weather-related general aviation accidents. Statistics have shown the risk of a weather-related accident declines as pilots gain instrument flying experience. Pilots with less than 50 hours of instrument time were involved in 58% of all weather accidents, and 47% of fatal weather accidents. As pilots gain more experience (50 to 100 hours of instrument flying time) their risk decreases by more than 80% to a level slightly below 9% of all accidents. [Figure 1-6]

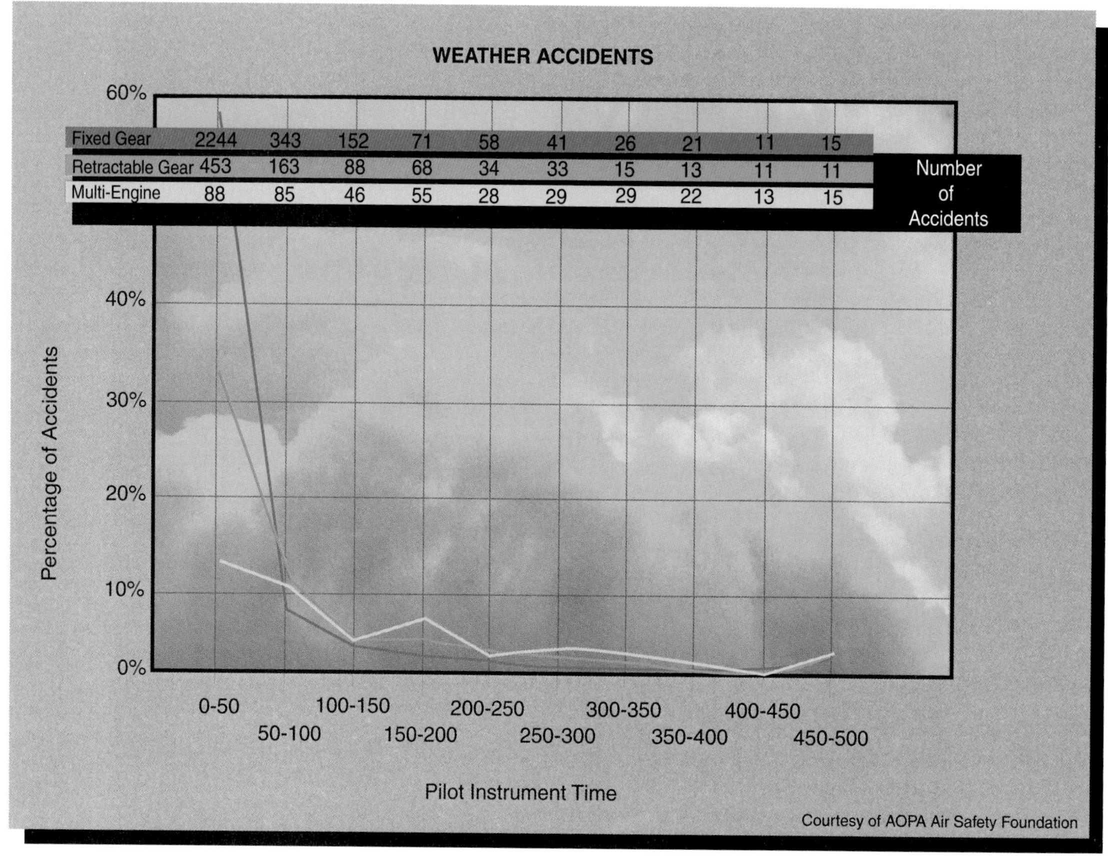

Figure 1-6. As instrument time progresses beyond 100 hours, the accident risk factor decreases to a statistically insignificant level.

INSTRUMENT TRAINING

To be eligible for an **instrument rating**, you must possess at least a private pilot certificate with an aircraft rating appropriate to the instrument rating sought, be able to read, write, speak, and understand the English language, and complete specific training and flight time requirements described in the FARs. You also must pass a knowledge test and successfully complete a practical test which consists of oral quizzing, performing pilot operations, and executing instrument procedures in the airplane. The FARs

require that you have received instruction in specific flight operations and maneuvers, as well as ground instruction in certain knowledge areas. [Figure 1-7]

During instrument training you will

learn how to accurately control the airplane during attitude instrument flying.

increase your knowledge about weather and use of aviation weather reports and forecasts.

interpret instrument charts and increase your skill at radio navigation.

follow ATC clearances and procedures while operating in the IFR environment.

perform instrument operations such as holding and approach procedures.

develop your aeronautical decision-making skills and judgment.

practice emergency procedures.

Figure 1-7. During your instrument training, you will become proficient in controlling the airplane solely with reference to instruments.

You must meet minimum flight hour requirements to apply for an instrument rating. According to FAR Part 61, you must have at least 50 hours of cross-country time as pilot in command (PIC) and 40 hours of actual or simulated instrument time in the areas of operation specified in the regulations. This includes at least 15 hours of instrument flight training from an authorized instructor in the airplane. Some of your instrument time may be conducted with a safety pilot who is appropriately rated for the airplane. If your training is accomplished under FAR Part 141, you must have 35 hours of instrument training from an authorized instructor in the areas specified in Appendix C, FAR Part 141 and need not comply with the 50-hour PIC cross-country requirement. [Figure 1-8]

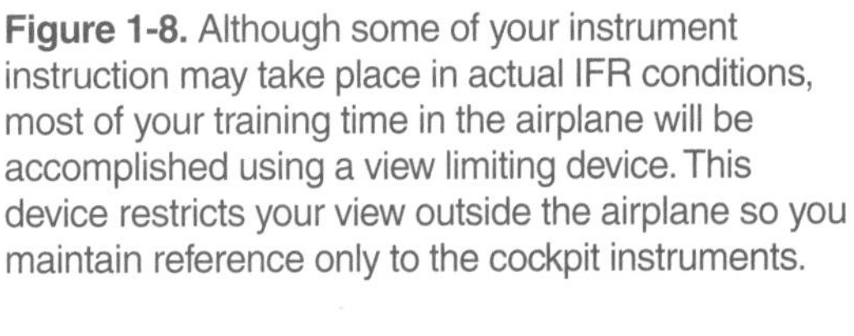

Figure 1-8. Although some of your instrument instruction may take place in actual IFR conditions, most of your training time in the airplane will be accomplished using a view limiting device. This device restricts your view outside the airplane so you maintain reference only to the cockpit instruments.

FLYING ON THE GROUND

Part of your instrument training may be accomplished in a **flight simulator flight training device**, or a **personal computer-based aviation training device (PCATD)**. To count the time toward the instrument rating requirements, an authorized instructor must be present during the simulated flight. The amount of time which may be credited toward

your instrument rating depends on the type of device that you are using and on your training curriculum. The FARs specify the maximum hours which can be used in either a flight simulator or flight training device. Advisory Circular (AC) 61-126 provides the requirements that a PCATD must meet for FAA certification. An AC-compliant PCATD may be used for up to 10 hours of the flight training time required for the instrument rating. [Figure 1-9]

FLIGHT SIMULATOR
A flight simulator is a full-size cockpit replica of a specific type of aircraft, or make, model and series of aircraft which uses a force (motion) cueing system and visual system.

PCATD
A PCATD must replicate a type of airplane or family of airplanes and meet the virtual control requirements specified in AC 61-126.

FLIGHT TRAINING DEVICE
A flight training device is a full-size replica of the instruments, equipment, panels, and controls of an aircraft, or set of aircraft, in an open flight deck area or in an enclosed cockpit. A force (motion) cueing system or visual system is not required.

Figure 1-9. Each type of simulator or training device must have the hardware and software necessary to represent the aircraft in ground and flight operations and must be evaluated, qualified, and approved by the FAA.

Equipment such as a simulator or training device is a very valuable tool for developing your instrument scan and for practicing procedures such as holding and approaches. The simulation can be placed on hold so your position is frozen while you discuss the procedure with your instructor. Exercises can be repeated as many times as necessary and most equipment allows you to position the airplane at any point on a procedure. In addition, a training device provides a lesson opportunity regardless of the weather. Although you should take full advantage of training in actual IFR conditions, at times icing or thunderstorms may prevent practice in the airplane.

It (the Link Trainer) is a box set on a pedestal and cleverly designed to resemble a real airplane. On the inside the deception is quite complete, even to the sound of slip stream and engines. All of the usual controls and instruments are duplicated within the cockpit, and once under way the sensation of actual flight becomes so genuine that it is often a surprise to open the top of the box and discover you are in the same locality.

—Ernest K Gann, Fate is the Hunter

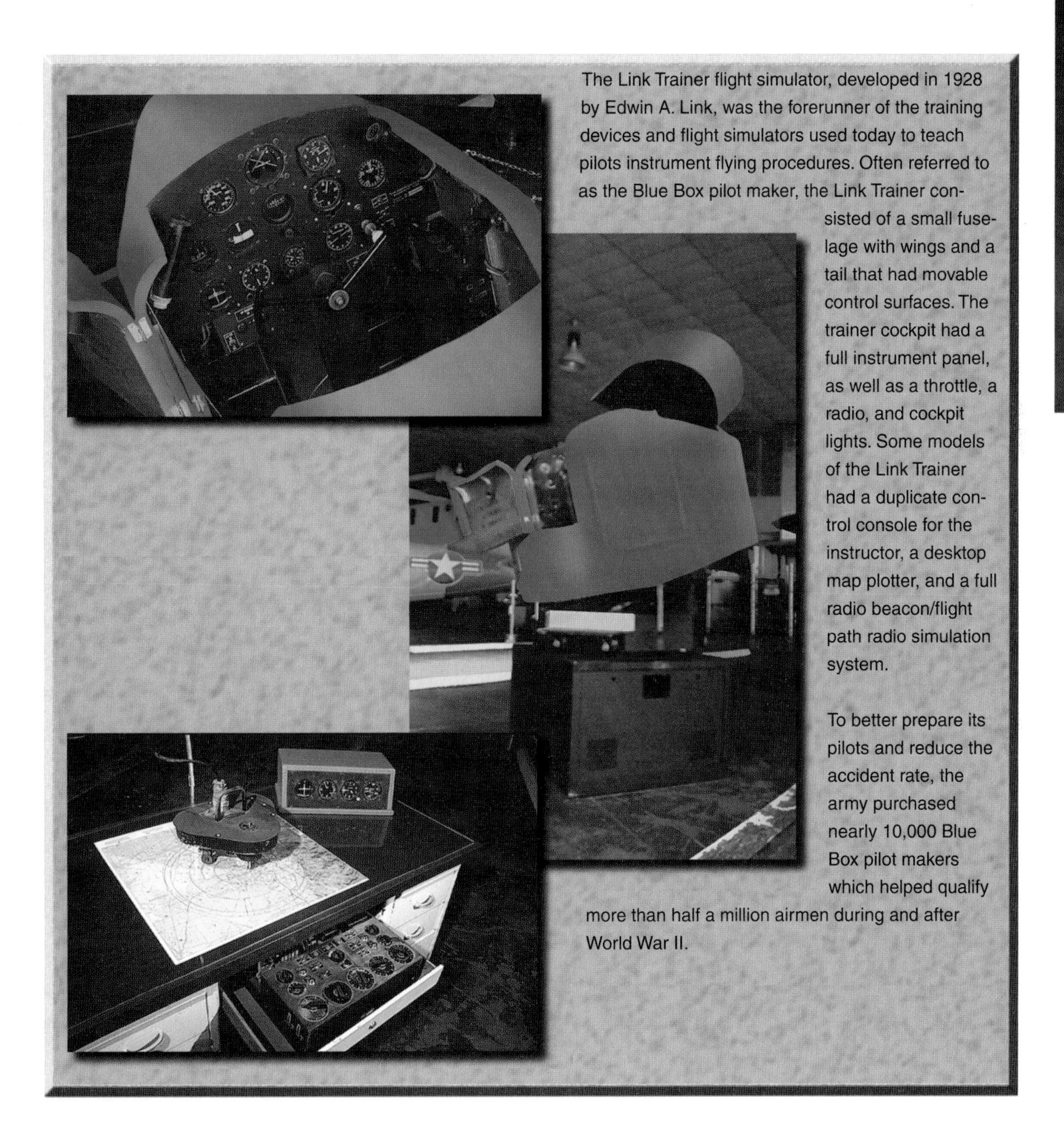

The Link Trainer flight simulator, developed in 1928 by Edwin A. Link, was the forerunner of the training devices and flight simulators used today to teach pilots instrument flying procedures. Often referred to as the Blue Box pilot maker, the Link Trainer consisted of a small fuselage with wings and a tail that had movable control surfaces. The trainer cockpit had a full instrument panel, as well as a throttle, a radio, and cockpit lights. Some models of the Link Trainer had a duplicate control console for the instructor, a desktop map plotter, and a full radio beacon/flight path radio simulation system.

To better prepare its pilots and reduce the accident rate, the army purchased nearly 10,000 Blue Box pilot makers which helped qualify more than half a million airmen during and after World War II.

CURRENCY FOR THE CLOUDS

Once you hold an instrument rating, you must meet certain recency of experience requirements to act as PIC under IFR or in weather conditions which are less than the minimums prescribed for VFR operations. Within the preceding six calendar months, you must have intercepted and tracked courses through the use of navigation systems, performed holding procedures, and flown at least six instrument approaches. These instrument procedures must be accomplished under actual or simulated instrument conditions in flight, in a flight simulator, or in a flight training device.

The location and type of each instrument approach and the name of the safety pilot, if required, must be recorded in your logbook. A flight simulator or flight training device may be used to log instrument time provided an authorized instructor is present during the simulated flight. If you do not meet the instrument experience requirements within the six calendar months or within six calendar months after that, you must pass an **instrument proficiency check** which consists of a representative number of tasks required by the instrument rating practical test. [Figure 1-10]

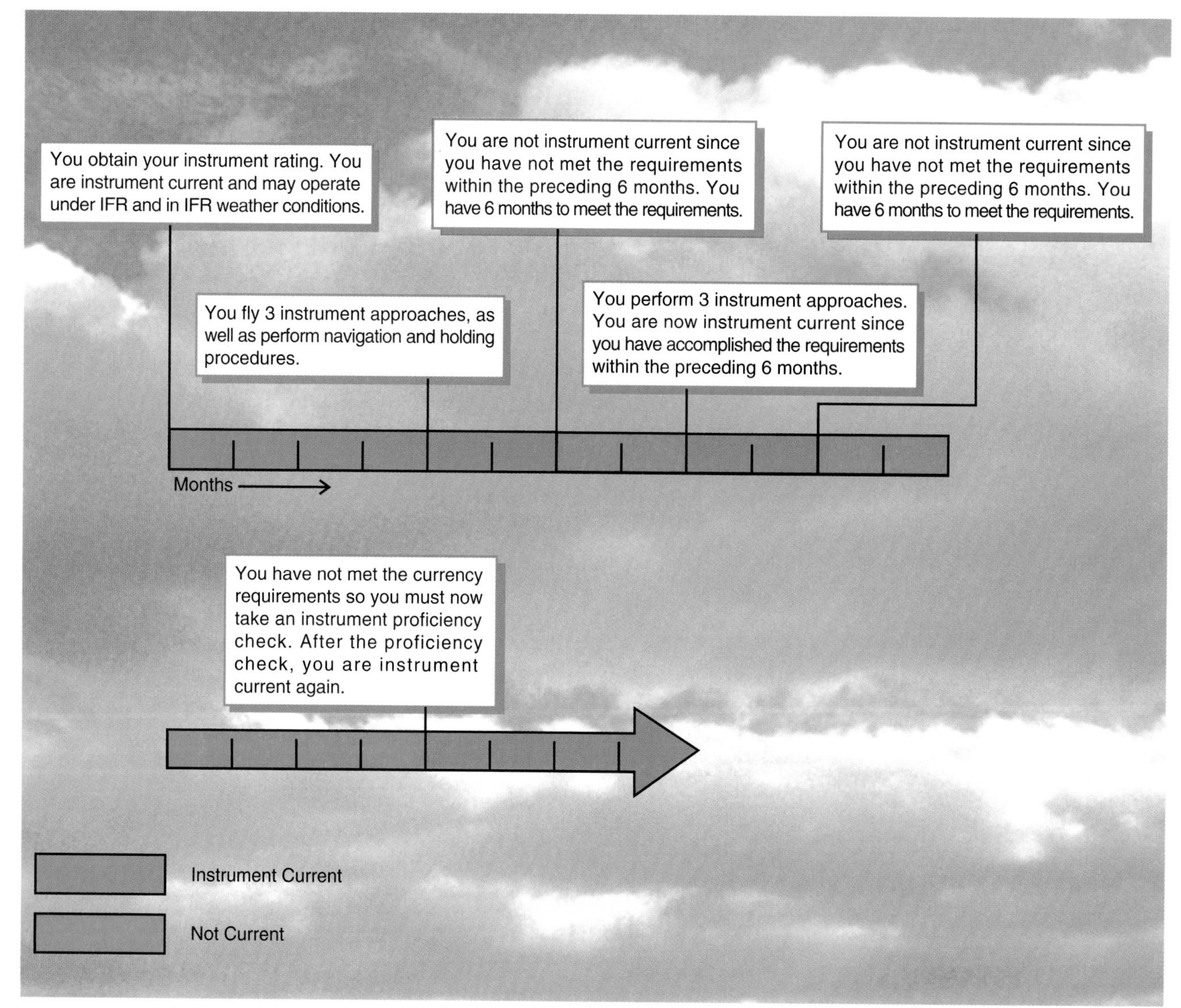

Figure 1-10. You may log instrument time only for that flight time when you operate the airplane solely by reference to instruments under actual or simulated instrument flight conditions.

THE COMMERCIAL PILOT CERTIFICATE

One of the unique aspects of flying is that you can transform what you have a passion for as a hobby into an exciting and rewarding career. The desire to fly for a living was so strong in the early days of aviation that aviators were willing to pursue dangerous and unstable employment as barnstormers or airmail pilots. You do not have to take the same type of risks today, but you must be willing to face the many challenges that lay ahead on the road to becoming a professional pilot. [Figure 1-11]

For you to be eligible for a **commercial pilot certificate**, you must be at least 18 years of age and hold a third-class medical certificate. However, a second-class medical certificate is required for you to exercise the privileges of your commercial pilot certificate. You must be able to read, write, speak, and understand the English language, and meet specific training and flight time requirements described in the FARs. In addition, you must pass an aeronautical knowledge test and a practical test.

You also must possess at least a private pilot certificate and, under FAR Part 141, hold an instrument rating or be concurrently enrolled in an instrument rating course. If you are training under Part 61 regulations and you do not hold an instrument rating in the same aircraft category and class, you will be issued a commercial pilot certificate that contains the limitation, "The carriage of passengers for hire on cross-country flights in excess of 50 nautical miles or at night is prohibited." The experience requirements in Part 61 of the FARs include 250 hours of flight time as a pilot including training from an instructor in various areas of operation specified within that part. Under Part 141 rules, you must receive 120 hours of training which includes at least 55 hours of instruction in operations specified in Appendix D, FAR Part 141. [Figure 1-12]

Figure 1-12. Mastering commercial maneuvers such as the lazy eight requires a high level of proficiency in advanced planning, accuracy, and control coordination.

. . . flying in heavy rain on one occasion, one of the blades of my wooden propeller started to come apart. The fabric coating, which covered one of these blades, had worn through, and the first thing I knew there was a terrific bang and very heavy vibration. I had to come down on the beach at Le Toquet. We were short of daylight, but my two passengers were very nice. I explained what the problem was, and one of them said, "I've got a penknife on me. I'll have a go at fixing this thing." So he helped me, and we hacked this loose piece of wood and fabric off the blade and got it started up again . . . In those days the passengers took part in the whole thing in very good spirit.

— Alan Campbell Orde, a British pilot recalling the startup of European commercial airlines in 1919, as quoted in *The American Heritage History of Flight*

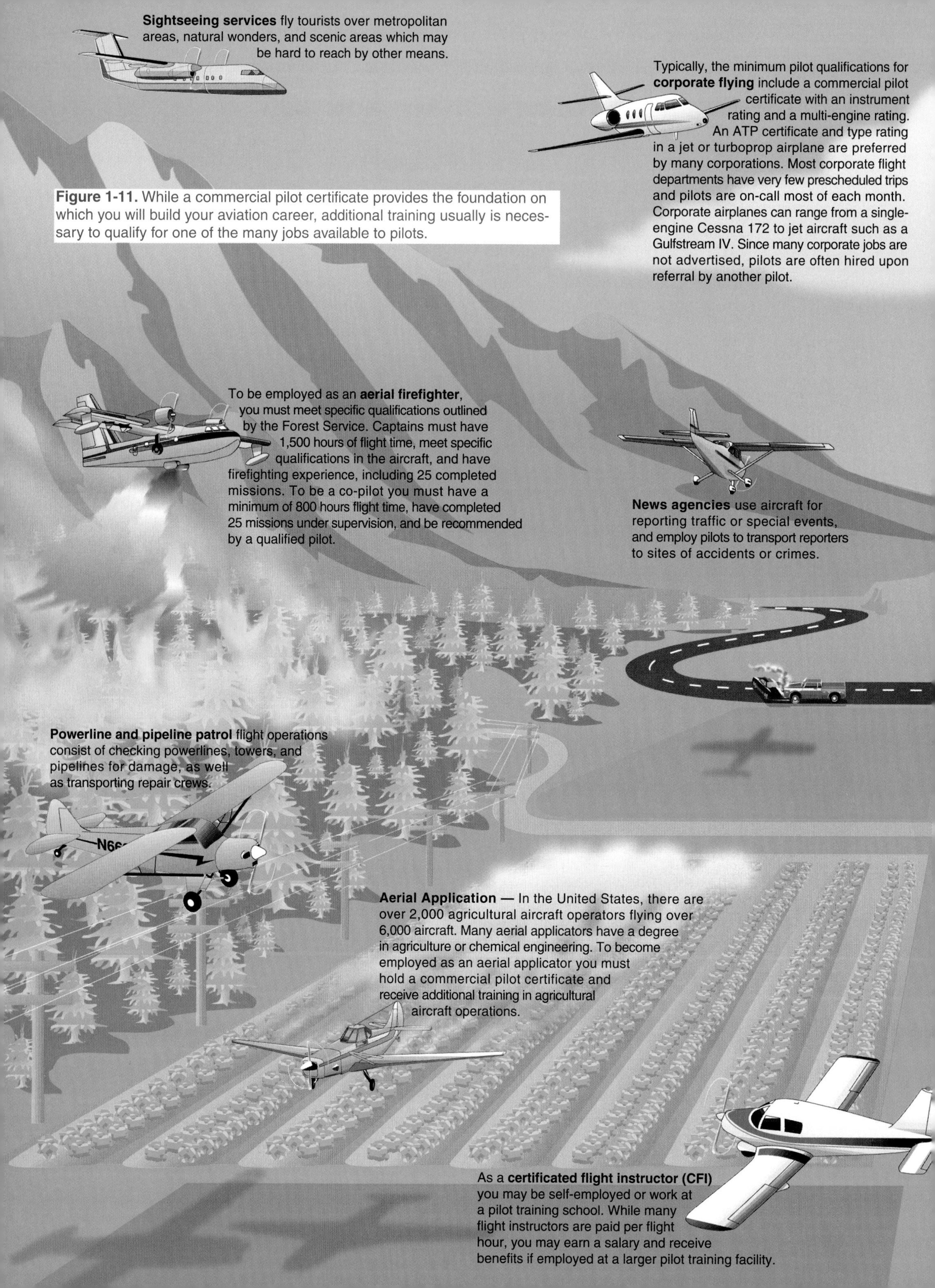

Sightseeing services fly tourists over metropolitan areas, natural wonders, and scenic areas which may be hard to reach by other means.

Typically, the minimum pilot qualifications for **corporate flying** include a commercial pilot certificate with an instrument rating and a multi-engine rating. An ATP certificate and type rating in a jet or turboprop airplane are preferred by many corporations. Most corporate flight departments have very few prescheduled trips and pilots are on-call most of each month. Corporate airplanes can range from a single-engine Cessna 172 to jet aircraft such as a Gulfstream IV. Since many corporate jobs are not advertised, pilots are often hired upon referral by another pilot.

Figure 1-11. While a commercial pilot certificate provides the foundation on which you will build your aviation career, additional training usually is necessary to qualify for one of the many jobs available to pilots.

To be employed as an **aerial firefighter**, you must meet specific qualifications outlined by the Forest Service. Captains must have 1,500 hours of flight time, meet specific qualifications in the aircraft, and have firefighting experience, including 25 completed missions. To be a co-pilot you must have a minimum of 800 hours flight time, have completed 25 missions under supervision, and be recommended by a qualified pilot.

News agencies use aircraft for reporting traffic or special events, and employ pilots to transport reporters to sites of accidents or crimes.

Powerline and pipeline patrol flight operations consist of checking powerlines, towers, and pipelines for damage, as well as transporting repair crews.

Aerial Application — In the United States, there are over 2,000 agricultural aircraft operators flying over 6,000 aircraft. Many aerial applicators have a degree in agriculture or chemical engineering. To become employed as an aerial applicator you must hold a commercial pilot certificate and receive additional training in agricultural aircraft operations.

As a **certificated flight instructor (CFI)** you may be self-employed or work at a pilot training school. While many flight instructors are paid per flight hour, you may earn a salary and receive benefits if employed at a larger pilot training facility.

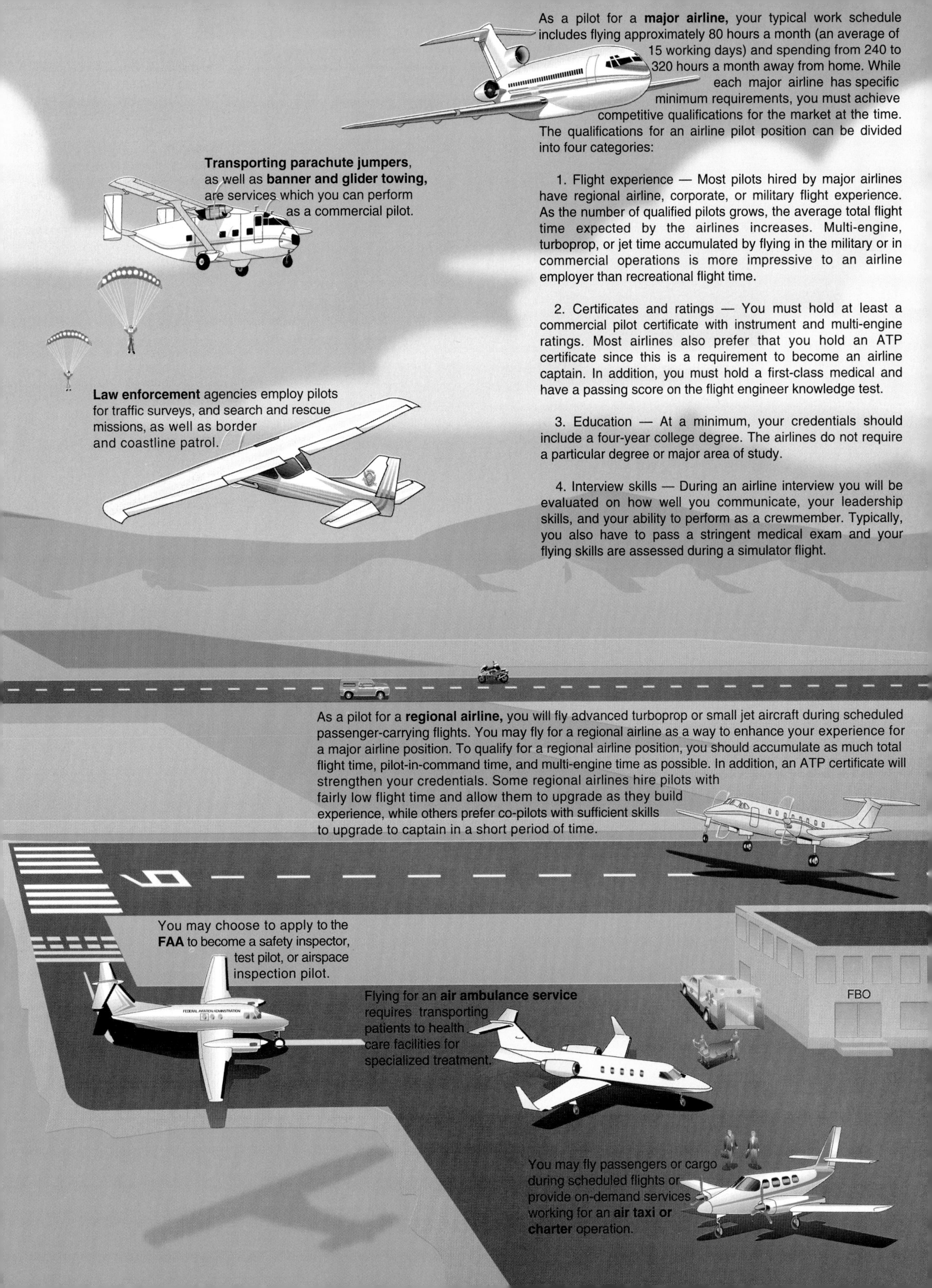

As a pilot for a **major airline,** your typical work schedule includes flying approximately 80 hours a month (an average of 15 working days) and spending from 240 to 320 hours a month away from home. While each major airline has specific minimum requirements, you must achieve competitive qualifications for the market at the time. The qualifications for an airline pilot position can be divided into four categories:

1. Flight experience — Most pilots hired by major airlines have regional airline, corporate, or military flight experience. As the number of qualified pilots grows, the average total flight time expected by the airlines increases. Multi-engine, turboprop, or jet time accumulated by flying in the military or in commercial operations is more impressive to an airline employer than recreational flight time.

2. Certificates and ratings — You must hold at least a commercial pilot certificate with instrument and multi-engine ratings. Most airlines also prefer that you hold an ATP certificate since this is a requirement to become an airline captain. In addition, you must hold a first-class medical and have a passing score on the flight engineer knowledge test.

3. Education — At a minimum, your credentials should include a four-year college degree. The airlines do not require a particular degree or major area of study.

4. Interview skills — During an airline interview you will be evaluated on how well you communicate, your leadership skills, and your ability to perform as a crewmember. Typically, you also have to pass a stringent medical exam and your flying skills are assessed during a simulator flight.

Transporting parachute jumpers, as well as **banner and glider towing,** are services which you can perform as a commercial pilot.

Law enforcement agencies employ pilots for traffic surveys, and search and rescue missions, as well as border and coastline patrol.

As a pilot for a **regional airline,** you will fly advanced turboprop or small jet aircraft during scheduled passenger-carrying flights. You may fly for a regional airline as a way to enhance your experience for a major airline position. To qualify for a regional airline position, you should accumulate as much total flight time, pilot-in-command time, and multi-engine time as possible. In addition, an ATP certificate will strengthen your credentials. Some regional airlines hire pilots with fairly low flight time and allow them to upgrade as they build experience, while others prefer co-pilots with sufficient skills to upgrade to captain in a short period of time.

You may choose to apply to the **FAA** to become a safety inspector, test pilot, or airspace inspection pilot.

Flying for an **air ambulance service** requires transporting patients to health care facilities for specialized treatment.

You may fly passengers or cargo during scheduled flights or provide on-demand services working for an **air taxi or charter** operation.

Wiley Post climbs out of the Winnie Mae after his solo flight around the world in 1933.

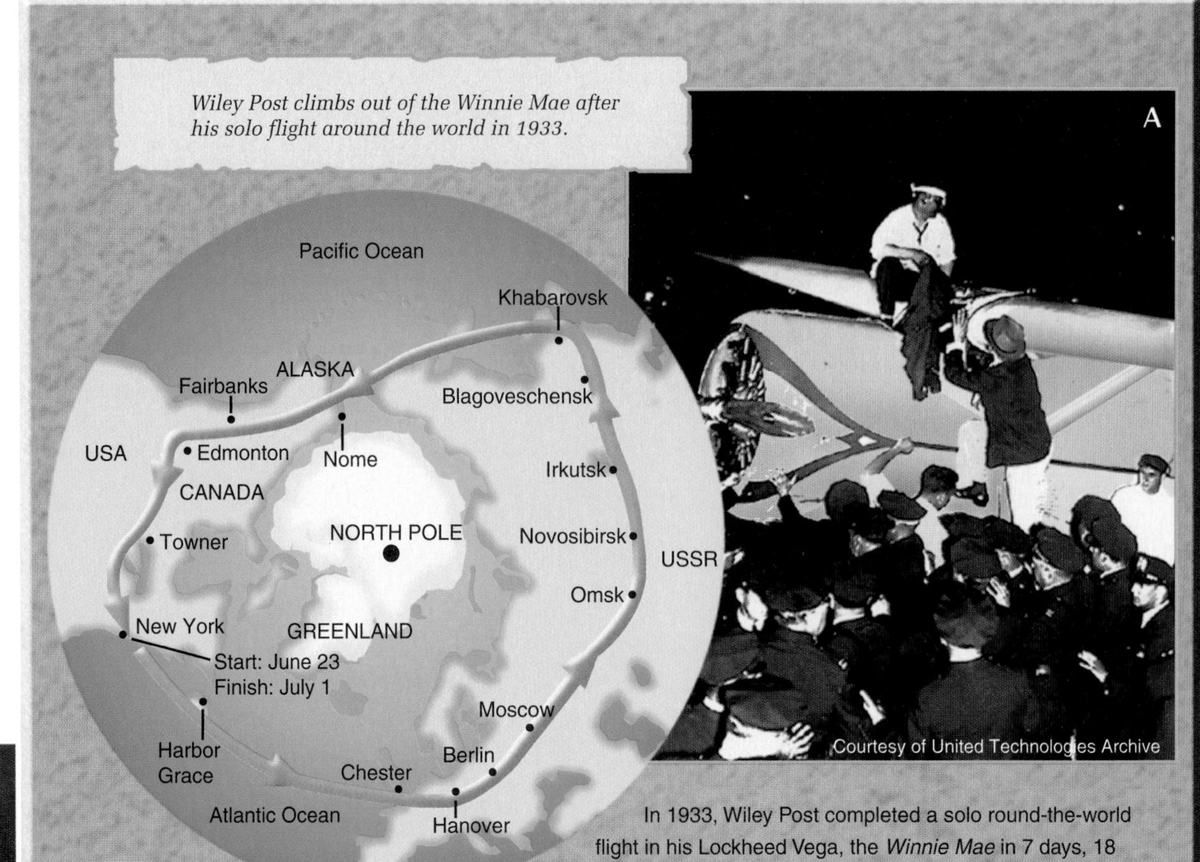

In 1933, Wiley Post completed a solo round-the-world flight in his Lockheed Vega, the *Winnie Mae* in 7 days, 18 hours, and 49 1/2 minutes. [Figure A] His goal was to prove that flying could be safer and more precise with the benefits of new instruments and equipment, such as the automatic pilot and radio direction finder. Having accomplished his objective, Post was determined to take aviation technology even further. Upon his return, he announced that if man wanted to fly long distances safely and faster, he would have to fly higher — into the stratosphere, where the powerful jet stream winds blew.

Since Post could not pressurize the *Winnie Mae's* cabin, he enlisted the help of the BFGoodrich Company to develop a full-pressure suit he could wear while flying the airplane. Post's suit, consisted of three layers (long underwear, an inner black rubber air pressure bladder, and an outer cloth contoured suit), and a helmet (containing a special oxygen breathing system and outlets for earphones and a throat microphone). [Figure B]

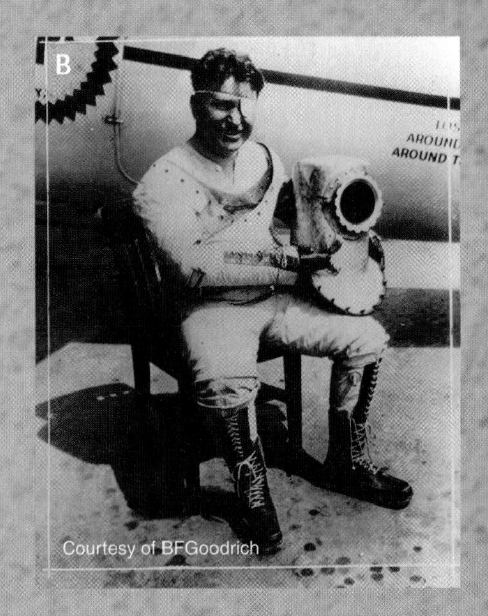

On February 22, 1935, Post made his first attempt to fly across the country at an altitude of more than 30,000 feet MSL. Only 31 minutes into the flight, he was forced to make an emergency landing in the Mojave Desert after his engine began throwing oil. Post approached a man near the landing sight to ask for assistance in removing his helmet. The man nearly fainted from fright when he saw Post lumbering toward him in his pressure suit.

It was determined that Post's airplane had been sabotaged by a jealous competitor. This did not deter Post from his mission and 3 weeks later, on March 15, 1935, he embarked on a second transcontinental record attempt. Although Post had to turn back when he ran out of oxygen, he had covered 2,035 miles in 7 hours and 9 minutes. This meant that the *Winnie Mae's* groundspeed averaged 279 miles per hour, over 100 miles per hour faster than the airplane's normal speed. At times the airplane had reached groundspeeds of up to 340 miles per hour. Wiley Post and his *Winnie Mae* had been in the jet stream.

Within a quarter of a century of Post's high altitude flights, men, women, and children would be hurtling through the stratosphere at almost the speed of sound in the comfortable pressurized cabins of jetliners, wholly ignorant of the frustrating labors of 1934 and 1935, unmindful of the man who met the difficulties in their rudest shapes. Yet every time a contrail runs its white chalkline across the blue, it deserves recollection that it was Wiley Post who pointed the way to putting it there.

— Stanley R. Mohler, M.D. and Bobby H Johnson, Ph.D., written for a National Air and Space Museum Smithsonian Annals of Flight monograph

The FARs specify that you must receive 10 hours of flight training in an airplane with retractable landing gear, flaps, and a controllable pitch propeller. To operate as pilot in command of an airplane meeting this criteria you need training and an endorsement from your instructor for **complex airplanes**. It also is possible that you will be introduced to **high performance airplanes** during your flight training. A high performance airplane is defined as an airplane having an engine of more than 200 horsepower. The training required to receive a high performance or complex airplane endorsement will focus on the operation of advanced airplane systems. [Figure 1-13]

Figure 1-13. To act as pilot in command of a high performance or complex airplane, you must receive specific training outlined in the FARs, as well as an instructor's endorsement in your logbook.

Designing retractable landing gear was first attempted as early as 1911, but the idea was not widely popular until the size of airplanes increased, resulting in a need to reduce airframe drag. The first fully retractable gear was introduced in 1920. The Dayton Wright R.B. Racer created a sensation as its wheels folded into its fuselage during the Gordon Bennett Cup Race in France. The R.B. Racer was built with balsa wood and plywood, including the movable control surfaces. Linen was glued to the plywood and then several coats of varnish were applied.

Retractable landing gear was first demonstrated in the United States during the Pulitzer Trophy Race of 1922. Two Verville-Sperry R-3s were procured by the army for the race while the Navy competed with the Bee-Line Racer. The Verville-Sperry retracted its gear into its wings, however, unnecessary drag was created since the airplane did not have wheel-well covers. The landing gear of the Bee-Line had full skirts so when the wheels retracted into the wings, the wheel wells were completely covered. While pilots operated this early landing gear by use of cranks, as wood and fabric were replaced with stressed-skin aluminum fuselages, retractable landing gear evolved from manually operated to hydraulically actuated.

Bee-Line Racer

Dayton Wright R.B. Racer

Verville Sperry R-3

COMMERCIAL PILOT PRIVILEGES

FAR 61.133 states that as a commercial pilot, you may act as pilot in command of an aircraft for compensation or hire and you may carry persons or property for compensation or hire provided you meet the qualifications which apply to the specific operation. While some commercial operations are governed by FAR Part 91, many others must meet additional requirements described in FAR Parts 119, 121, 125, 129, 135, and 137. [Figure 1-14] Terms which help explain various types of commercial operations are defined in FAR Part 119. [Figure 1-15]

FAR PART 119 — Certification: Air Carriers and Commercial Operators

FAR PART 121 — Operating Requirements: Domestic, Flag, and Supplemental

FAR PART 125 — Certification and Operation: Airplanes Having a Seating Capacity of 20 or More Passengers or a Maximum Payload Capacity of 6,000 Pounds or Greater

FAR PART 129 — Operations: Foreign Air Carriers and Foreign Operators of U.S.-Registered Aircraft Engaged in Common Carriage

FAR PART 135 — Operating Requirements: Commuter and On-Demand Operations

FAR PART 137 — Agricultural Aircraft Operations

Figure 1-14. The FAR parts shown here may be applicable to you as you pursue a career as a professional pilot, however, you normally will not study these regulations for commercial pilot certification.

Common Carriage is any operation for compensation or hire in which the operator holds itself out, by advertising or any other means, as willing to furnish transportation for any member of the public. Private carriage does not involve holding out.

Courtesy of Chicago Dept. of Aviation

Figure 1-15. FAR Parts 121, 125, and 135 govern operations ranging from scheduled air carriers to on-demand charters.

On-Demand Operation is any operation for compensation or hire which is one of the following:

- a passenger-carrying public charter where the departure time and location, as well as the arrival location are specifically negotiated with the customer.
- a common-carriage operation using an airplane (including turbojet-powered) having a passenger-seat configuration of 30 seats or less and a payload capacity of 7,500 pounds or less.
- a private carriage operation conducted with an airplane having a passenger-seat configuration of less than 20 seats or a payload capacity of less than 6,000 pounds.
- a scheduled operation with a frequency of less than 5 round trips per week conducted with a nonturbojet-powered airplane with a maximum of 9 passenger seats and a maximum payload capacity of 7,500 pounds.
- a cargo operation conducted with an airplane having a payload capacity of 7,500 pounds or less.

Flag Operation is any scheduled operation to any point outside the U.S. conducted in a turbojet-powered airplane or an airplane which has a passenger-seat configuration of more than 9 seats or a payload of more than 7,500 pounds.

Commuter Operation is any scheduled operation conducted in a nonturbojet-powered airplane having a maximum passenger-seat configuration of 9 seats or less and a maximum payload capacity of 7,500 pounds or less. A commuter operation must be scheduled with a frequency of at least 5 round trips per week on at least 1 route according to published flight schedules.

FARs GOVERNING SPECIFIC COMMERCIAL OPERATIONS

Airplane Size/Weight	Part 121 Domestic (Scheduled)	Part 121 Flag (Scheduled)	Part 121 Supplemental (Not Scheduled)	Part 135 Commuter (Scheduled)	Part 135 On-Demand (Not Scheduled)	Part 125 (Not Scheduled)
Common Carriage:						
≤9 seats and ≤7,500 lbs	No[1]	No[1]	No[2]	Yes[1]	Yes[2]	No
10-30 seats and ≤7,500 lbs	Yes	Yes	No[2]	No	Yes[2]	No
>30 seats or >7,500 lbs	Yes	Yes	Yes	No	No	No
Common Carriage is Not Involved:						
<20 seats or <6,000 lbs	No	No	No	No	Yes	No
≥20 seats or ≥6,000 lbs	No	No	No	No	No	Yes

[1]Turbojet-powered airplanes used in scheduled passenger-carrying operations must comply with Part 121 regardless of passenger seating or payload capacity.

[2]If turbojet-powered airplanes and other airplanes with 10-30 passenger-seat configurations are used for Part 121 domestic or flag operations, non-scheduled or charter operations with that airplane shall be conducted under Part 121 supplemental rules.

Under FAR Part 91, you may not engage in **common carriage**, which involves holding out, or advertising your services to furnish transportation for any member of the public. FAR Part 119 lists specific activities not governed by FAR Parts, 121, 125, or 135 for which the holder of a commercial pilot certificate may be paid. [Figure 1-16]

Figure 1-16. Student instruction, certain nonstop sightseeing flights within limited areas, crop dusting, banner towing, aerial photography or survey, firefighting, and some corporate flights are examples of operations which are not governed by FAR Parts 121, 125, and 135.

ADDITIONAL CERTIFICATES AND RATINGS

Although the path to a flying career can vary, many jobs, such as corporate or airline pilot positions, require additional pilot certificates and ratings, as well as flight experience. A multi-engine rating is an essential requirement for most flying jobs, and you must hold an airline transport pilot certificate to operate as an airline captain. In addition, you may choose a rewarding career as a certificated flight instructor, or use flight instruction as a step to gain experience and enhance your professional qualifications.

MULTI-ENGINE RATING

FAR Part 61 does not specify a minimum number of ground or flight instruction hours required for the addition of a **multi-engine rating** to a pilot certificate, but you will have to pass a practical test. Under FAR Part 141, a multi-engine rating course must include the ground and flight instruction hours in accordance with the applicable Part 141 appendices. Typically, the training can be completed in a short period of time, but most aircraft insurance policies require that you obtain a substantial amount of multi-engine flight time before operating the airplane as pilot in command. To accumulate the necessary experience, you may be able to share flight time and expenses with a qualified pilot who meets the insurance requirements. In general, a multi-engine rating is considered an addition to your private or commercial pilot certificate and the training will be conducted separately. However, it is possible in some curriculums for the multi-engine training to be incorporated within the commercial pilot training requirements. [Figure 1-17]

Figure 1-17. Learning the procedures for flying a multi-engine airplane after an engine failure is perhaps the most challenging aspect of multi-engine training.

CERTIFICATED FLIGHT INSTRUCTOR

Under FAR Part 61, a specific number of ground or flight instruction hours is not required to become a **certificated flight instructor (CFI)**, however, you are required to pass two knowledge exams and a practical test. Under FAR Part 141, a flight instructor course must include the ground and flight instruction hours specified in Appendix F, Part 141. Your CFI training will focus on aspects of teaching which include the learning process, student evaluation, and lesson planning. Additional ratings may be added to your flight instructor certificate, such as an instrument instructor, or multi-engine instructor rating.

AIRLINE TRANSPORT PILOT CERTIFICATE

To apply for an **airline transport pilot (ATP)** certificate, you must be at least 23 years of age and hold a first-class medical. The flight time requirements to obtain an ATP certificate are demanding: a total of 1,500 hours of flight time including 250 hours of pilot-in-command time, 500 hours of cross-country time, 100 hours of night flight, and 75 hours of instrument experience. Refer to FAR Part 61, Subpart G and Part 141, Appendix E for specific training requirements. The ATP knowledge test emphasizes subjects such as navigation, meteorology, aircraft performance, and air carrier flight procedures. During the practical test, your instrument skills will be evaluated, as well as your ability to correctly perform emergency procedures.

At first some of the pilots took the whole idea of stewardesses as kind of a joke. Then they realized that they didn't have to hand out box lunches and take care of sick passengers any more . . . Refueling was sometimes interesting. Sometimes we had to land at an emergency landing field and then we had the gas in two-and-a-half or five-gallon cans. They would form a sort of fire brigade, handing the cans from one to the other, including the stewardess and some of the passengers. Then, if we were some place where there was no crew on the field, somebody had to go out on the left wing to the engine to do something there. The pilot and the copilot were busy inside, so the third member of the crew had to go out on the wing—and that was the stewardess. We did it without a murmur because of the argument that when a third person was needed for something like that the third person should be a man.

— Ellen Church, who became the first airline stewardess when she was hired by Boeing Air Transport in 1930, as quoted in *The American Heritage History of Flight*

I feel we are on the brink of an era of expansion of knowledge about ourselves and our surroundings that is beyond description or comprehension at this time. Our efforts today and what we've done so far are but small building blocks on a very huge pyramid to come . . . Knowledge begets knowledge. The more I see, the more impressed I am not with how much we know but with how tremendous the areas are that are as yet unexplored.
— Lieutenant Colonel John H. Glenn Jr. in a speech given to a joint session of congress six days after he orbited the earth in a Mercury space capsule

All Photos Courtesy of NASA

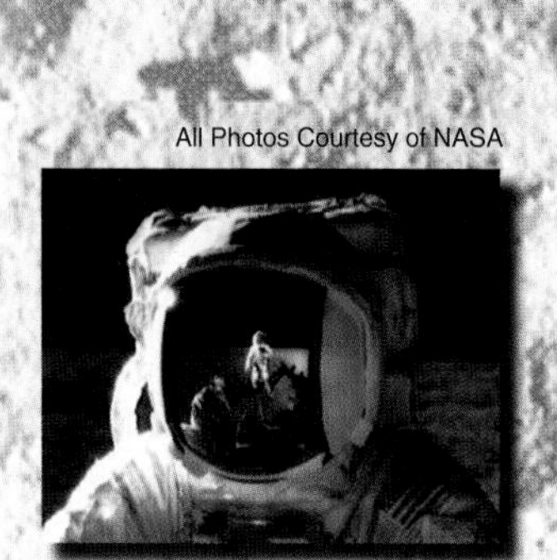

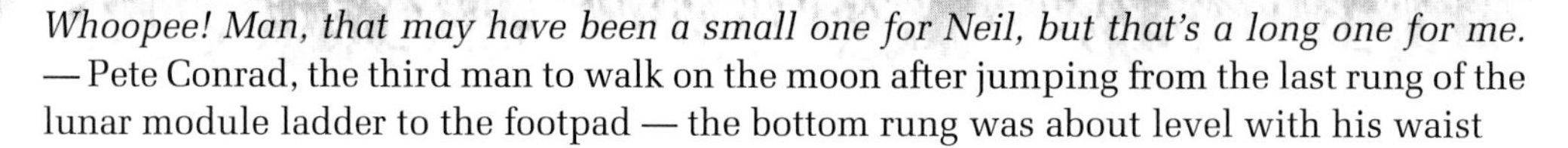

Whoopee! Man, that may have been a small one for Neil, but that's a long one for me.
— Pete Conrad, the third man to walk on the moon after jumping from the last rung of the lunar module ladder to the footpad — the bottom rung was about level with his waist

John Glenn's statement about the future of knowledge is just as true today as it was in 1962. Although aviation and space technology has come a long way since Jimmy Doolittle climbed into the cockpit of his NY-2 Husky biplane for the first blind flight and Wiley Post donned his space suit to cruise the stratosphere, there is still so much more yet to explore. If we approach the future with the same kind of unbridled enthusiasm as Pete Conrad, anything is possible. Conrad is chairman of Universal Space Lines, a venture he hopes will become the first commercial space airline, opening up a new world of discovery and opportunity.

Whether your goal is to earn an instrument rating, embark on a career as a commercial pilot, or reach for the stars as a spacecraft commander carrying passengers to the moon and beyond, you are a part of an industry with unlimited possibilities. You are one of a unique group of people who see further, dare to dream, and then have the passion, courage, and commitment to take the steps, large or small, to see their visions become reality. As you continue your aviation training and set your own dreams in motion, do not forget — you are standing on the shoulders of Giants.

SUMMARY CHECKLIST

✓ The addition of an instrument rating to your private pilot certificate allows you to fly under IFR (instrument flight rules). These regulations govern flight operations in weather conditions below VFR minimums.

✓ When referring to weather conditions, the terms IFR and IMC (instrument meteorological conditions) are often used interchangeably, as are the terms VFR and VMC (visual meteorological conditions). In addition, the terms IFR and VFR can define the type of flight plan under which you are operating.

✓ Statistics have shown the risk of a weather-related accident declines as a pilot gains instrument flying experience.

✓ To be eligible for an instrument rating, you must hold a private pilot certificate, be able to read, write, speak, and understand the English language, and complete specific training and flight time requirements described in the FARs, as well as pass a knowledge and practical test.

✓ Part of your instrument training may be provided by an authorized instructor in a flight simulator, flight training device, or a personal computer-based aviation training device (PCATD).

✓ To meet recency of experience requirements for instrument flight, you must have intercepted and tracked courses through the use of navigation systems, performed holding procedures, and flown at least six instrument approaches within the preceding six calendar months.

✓ If you do not meet the instrument currency requirements within six calendar months or within six calendar months after that, you must pass an instrument proficiency check.

✓ For you to be eligible for a commercial pilot certificate, you must be at least 18 years of age, hold a private pilot certificate, be able to read, write, speak, and understand the English language, and meet specific training and flight time requirements described in the FARs, as well as pass a knowledge and practical test. Under FAR Part 141, you must hold an instrument rating or be concurrently enrolled in an instrument rating course.

✓ Although you need at least a third-class medical certificate to be eligible for a commercial pilot certificate, you must have a second-class medical certificate to exercise commercial pilot privileges.

✓ As part of the commercial pilot training requirements, you must receive 10 hours of flight training in a complex airplane—an airplane with retractable landing gear, flaps, and a controllable pitch propeller.

✓ A high performance airplane is defined as an airplane having an engine of more than 200 horsepower.

- ✓ FAR Parts 119, 121, 125, and 135 govern operations ranging from scheduled air carriers to on-demand charters.

- ✓ Under FAR Part 91, you may not engage in common carriage which involves holding out, or advertising your services to furnish transportation for any member of the public.

- ✓ Student instruction, certain nonstop sightseeing flights within limited areas, crop dusting, banner towing, aerial photography or survey, firefighting, and some types of corporate flights are examples of operations which are not governed by FAR Parts 121, 125, and 135.

- ✓ The addition of a multi-engine rating to your private or commercial certificate does not require a minimum number of ground or flight instruction hours under FAR Part 61. Under FAR Part 141, a multi-engine rating course must include the ground and flight instruction hours in accordance with the applicable Part 141 appendices.

- ✓ To become a certificated flight instructor (CFI), you must pass two knowledge exams and a practical test. Under FAR Part 61, a specific number of ground or flight instruction hours is not required for CFI training. An FAR Part 141 flight instructor course must include the ground and flight instruction hours specified in Appendix F, Part 141.

- ✓ To apply for an airline transport pilot (ATP) certificate, you must be at least 23 years of age and hold a first-class medical. A total of 1,500 hours of flight time is required including 250 hours of pilot-in-command time, 500 hours of cross-country time, 100 hours of night flight, and 75 hours of instrument experience.

KEY TERMS

IFR (Instrument Flight Rules)

IMC (Instrument Meteorological Conditions)

VFR (Visual Flight Rules)

VMC (Visual Meteorological Conditions)

Instrument Rating

Flight Simulator

Flight Training Device

Personal Computer-Based Aviation Training Device (PCATD)

Instrument Proficiency Check

Commercial Pilot Certificate

Complex Airplane

High Performance Airplane

Common Carriage

Multi-Engine Rating

Certificated Flight Instructor (CFI)

Airline Transport Pilot (ATP)

SECTION B
ADVANCED HUMAN FACTORS CONCEPTS

The thing we call luck is merely professionalism and attention to detail, it's your awareness of everything that is going on around you, it's how well you know and understand your airplane and your own limitations. Luck is the sum total of your abilities as an aviator. — Stephen Coonts, *The Intruders*

From your previous training and pilot experience, you know that flying requires a continuous series of decisions. If you are at your best, you should be able to plan effectively and safely handle most situations that occur during a flight. You have a foundation in decision making within the visual environment, and you have assumed responsibility for yourself, and any passengers that have flown with you since your private pilot training.

Now, you enter a different realm. As an instrument pilot, you often will rely exclusively on avionics, instrumentation, charts, and air traffic control for aircraft guidance. You will learn to process far more information, and you must consider the instructions and suggestions of others during the course of each flight. As a commercial pilot, you will fly faster and more complex aircraft, and, much of the time, you will be required to fly these aircraft in the IFR environment.

Figure 1-18. The demands of instrument and commercial flight require you to be highly organized in the cockpit, use available resources effectively, and maintain a professional attitude as you work with others.

If you used the Guided Flight Discovery *Private Pilot Manual* during your initial training, you are familiar with many of the topics explored here. You may wish to refer to that text to review basic human factors concepts. The discussion in this section is tailored to address both the instrument and commercial environments since many operations in each overlap. You are stepping up to a new level of flying, one where you must frequently make more complex decisions and utilize additional resources through application of human factors principles. You will gain more insight into crew relationships, managing demanding workloads, and working with ATC.

Human Element Insets provided within each chapter explore in detail the human factors concepts that apply to each phase of flight, and deepen your understanding of the principles reviewed and introduced in this section. Chapter 10, Section B — IFR Decision

Making more closely examines decision making as it relates to instrument flight operations. Chapter 13, Section B — Commercial Decision Making applies human factors principles to common scenarios in commercial operations.

AERONAUTICAL DECISION MAKING

Flying has come a long way since aviation pioneers first figured out how to use crude instruments to find their way out of the clouds. Even the simplest training aircraft has far more instrumentation than the panels of yesterday, and safety has increased greatly as a result. Although aircraft and the flight environment have become more sophisicated, pilots today are still challenged by the same types of decisions that early aviators faced.

Aeronautical decision making (ADM) is a systematic approach to the mental process used by aircraft pilots to consistently determine the best course of action in response to a given set of circumstances. *Pilot error* is the description given to an action or decision made by a pilot that caused, or contributed substantially to an accident. However, the phrase *human factors-related* more accurately describes these accidents. Rarely does a single action or event cause the inopportune termination of a flight, but rather a chain of events. Human factors-related causes are responsible for approximately 75% of all aviation accidents. Sources of pilot error in the IFR and commercial environments include misinterpretation of a chart, failure to understand a clearance, inability to use equipment properly, and lack of coordination among crewmembers. It should be noted that pilot error also can be the result of poor ergonomics, which is the integration of people and the equipment they use.

(The history of aviation safety in the United States) is somewhat analogous to golf. It's fairly easy to get under 100 if you take some lessons and play once in a while. With a little more investment and a little more practice, you can get to 90. With still more effort, maybe you can get to 80. But if you are ever going to reach 72, you have to work very hard. It's shaving those last five or ten strokes off your game that takes a disproportionate investment. — David Hinson, former FAA Administrator

CREW RESOURCE MANAGEMENT

Human factors-related accidents motivated the airline industry to implement crew resource management (CRM) training for flight crews. The focus of CRM programs is the effective use of all available resources; human resources, hardware, and information. Human resources include all groups routinely working with the cockpit crew (or pilot) who are involved in decisions which are required to operate a flight safely. These groups include, but are not limited to: dispatchers, cabin crewmembers, maintenance personnel, and air traffic controllers. Although the CRM concept originated as airlines developed ways of facilitating crew cooperation, CRM principles such as workload management, situational awareness, communication, the leadership role of the captain, and crewmember coordination have direct application to the general aviation cockpit.

Though the single pilot can employ CRM skills , the flight environment changes when more than one pilot shares the cockpit. Your first experience working in a cockpit crew may occur during the training for an instrument rating or commercial certificate. A measure of redundancy is provided in the crew environment since one pilot may notice something that escapes the attention of another. However, this information must be shared for it to be useful. CRM training helps flight crews understand the limitations of human performance, especially under stressful situations, and makes them aware of the importance of crew coordination to combat error.

A typical CRM program includes classes which provide background in group dynamics, the nature of human error, and the elements of people working with machines. In addition, flight crews may be asked to review accident reports that highlight the importance of crew coordination. [Figure 1-19]

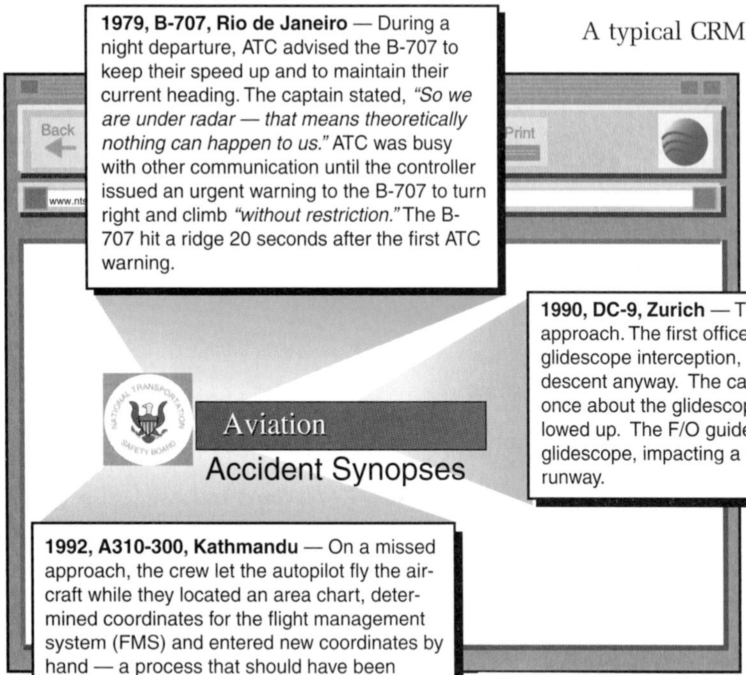

Figure 1-19. These three reports illustrate how the breakdown of CRM can contribute to or be the primary cause of an accident.

CRM training also involves the use of flight simulators and training devices. While these devices were once used primarily to teach flying skills, they now enable crews to test their abilities to handle complex problems, such as severe weather situations and mechanical failures which require more than simply following a procedures checklist. The airlines refer to CRM training in the simulator as line-oriented flight training (LOFT). [Figure 1-20] Many universities, colleges, and flight schools recognize that CRM training produces better pilots, whether those pilots go on to work as flight instructors, for corporate flight departments, the airlines, or flight schools after graduation.

Figure 1-20. A LOFT session normally allows a crew to conduct a complete flight, including filling out the necessary paperwork and performing crew briefings, while an instructor directs the session and analyzes the group behavior. A post-simulation debriefing, in which video recordings of the simulator session are reviewed, is an important part of LOFT training.

THE DECISION-MAKING PROCESS

Before you can make an effective decision, you need the motivation to select and follow through on a reasonable course of action. You can elect to either act or do nothing, but your choice must be within societal norms in order for it to be deemed suitable. If you are leaving for work and it is raining outside, you can either choose to wear a raincoat or dash to your car, but you probably would not use your shower curtain as a substitute for proper rain gear. An example of this type of decision making in the flight environment is the choice to use proper phraseology rather than regional slang while communicating with ATC. However, flying involves decisions that are far more involved than most daily

concerns. Since instrument and commercial flight demand complex decision making, your motivation and attitude as you make decisions are critical. [Figure 1-21]

Figure 1-21. The decision-making process normally involves several steps in which you make choices based on a variety of factors, some beyond your control. By recognizing the elements you can control, you will improve your ability to make wise and effective decisions.

(C/L/R) was developed so that as flight crews work together, they work effectively, they are able to use effective communication with each other; that if they need to advocate or to speak up, they can speak up. If there are things that are unclear, they can inquire . . . if there is conflict, the conflict can be addressed in an effective way. Many of the (C/L/R) principles . . . are equally relevant to a one-person crew as they are to a two-person or three-person crew. It's great to be a good stick, but it's better to be a safe pilot. — Cal Hutchings, Director of United Airlines Command/Leadership/Resource Management (C/L/R) program, an extensive human factors training course which teaches flight crews CRM principles

PILOT-IN-COMMAND RESPONSIBILITY

As pilot in command, you are the final authority in the airplane you are flying. When only one pilot is in the cockpit, the PIC is obvious, but when two pilots are present, each pilot's responsibilities must be defined before the flight. Within the cockpit, one person is pilot in command, and the other serves to assist the PIC. An important **pilot-in-command responsibility** is to establish an atmosphere of open communication in the cockpit and ensure that the suggestions and concerns of the co-pilot are validated and considered carefully.

It is a truism that the most important thing in flying is to learn your limitations . . . I had realized that the pilot who flew within his limitations would probably live to a ripe old age, whereas the pilot who flew beyond them would not. I also knew that different pilots had different limitations. — Jimmy Doolittle with Carroll V. Glines in Doolittle's autobiography, *I Could Never Be So Lucky Again*

Understanding your limitations is essential to acting as an effective pilot in command. Your performance during a flight is affected by many factors, such as your health, recency of experience, knowledge, skill level, and attitude. Studies have identified five **hazardous attitudes** which can interfere with a pilot's ability to make sound decisions and exercise authority properly. [Figure 1-22]

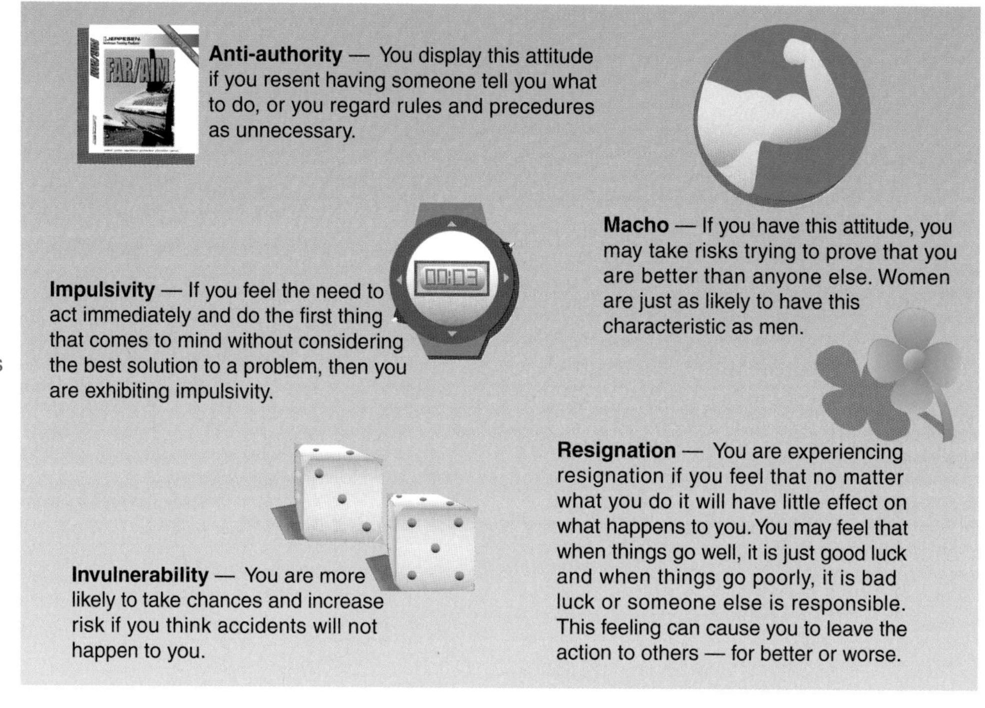

Figure 1-22. As pilot in command, you must examine your decisions carefully to ensure that your choices have not been influenced by a hazardous attitude.

They gave us cotton wool to stuff in our ears, the Tin Goose [Ford Trimotor] *was so noisy. The thing vibrated so much it shook the eye glasses right off your nose. In order to talk to the guy across the aisle, you had to shout at the top of your lungs.*— Arthur Raymond, designer, Douglas Aircraft

By this time, you are familiar with maintaining currency by flying and refreshing your skills on a regular basis. To fulfill recency of experience requirements as an instrument pilot, you need to log at least six instrument approaches, as well as holding procedures, and intercepting and tracking courses through the use of navigation systems. Prior to a flight in IFR weather conditions, you should be certain that your instrument flying skills are sharp and that you are fully prepared to operate in the IFR environment.

COMMUNICATION

The nature of the IFR environment requires that you have more contact with ATC than you may have previously experienced. Perhaps you communicated with ATC for flight following, weather updates, and other services during VFR cross-country flights, or you

have operated within a high density terminal area. If so, you will be better prepared for the increase in communication workload that you will experience when flying on an instrument flight plan. As the amount of communication with ATC increases, so does the chance of error. Readback of ATC clearances is crucial in the IFR environment. You should not assume controller silence after a readback is verification of your transmission. If you are unsure that the controller has understood your communication, ask for a verbal confirmation. [Figure 1-23]

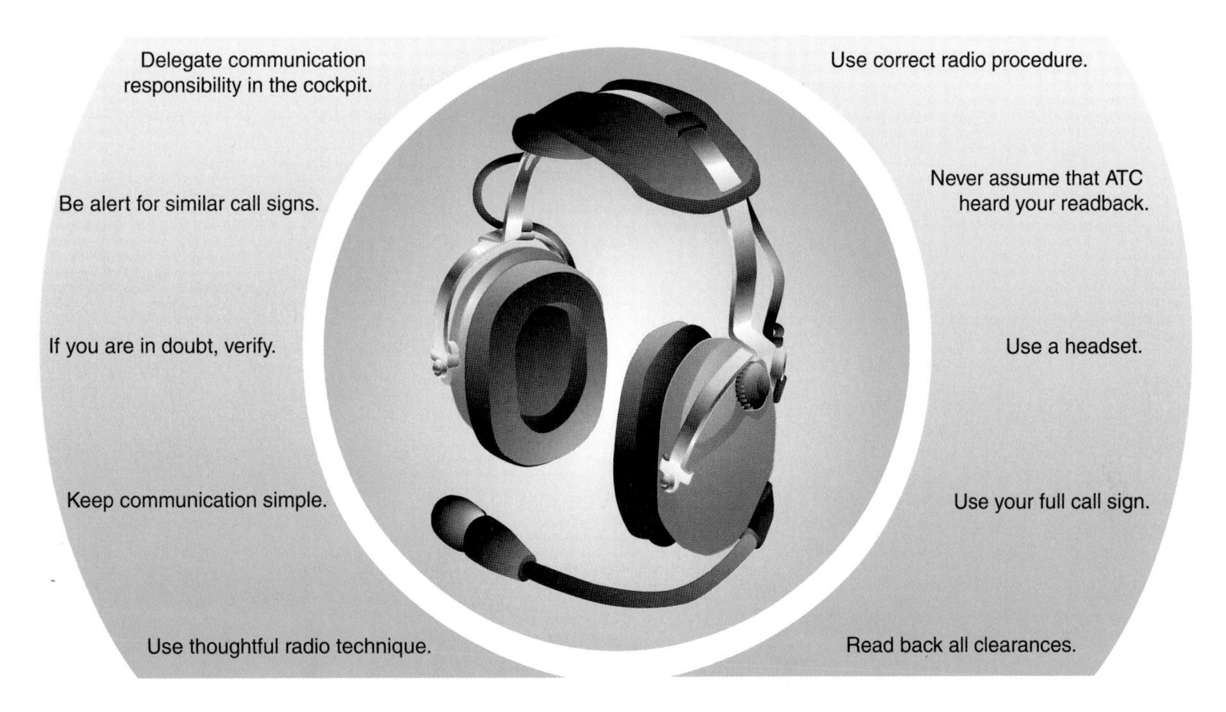

Figure 1-23. There are several procedures which you can follow to reduce the risk of miscommunication with air traffic controllers.

Another form of communication comes into play when you fly with another pilot. If you do not use standard terminology and verify that your meaning is understood, miscommunication can occur. A breakdown in communication can cause friction and frustration, detracting from important tasks, or lead to a hazardous situation where one pilot believes the other is controlling the airplane, but in reality, neither pilot has control.

RESOURCE USE

Resource use is an important part of human factors training. A flight in the IFR environment requires you to rely on additional resources beyond those used in flight under VFR. Planning an instrument flight involves gathering charts for enroute navigation, departures, arrivals, and instrument approaches. Effective use of all your paper resources requires some advance planning. If you thought that the cockpit became easily cluttered on a VFR cross-country flight, you will be amazed at how quickly you may become disorganized when you add the myriad of instrument charts that you will take on every IFR flight. You also remain in closer contact with the larger resource of air traffic control

while on an IFR flight plan. With radar services, weather information, and local knowledge, ATC can assist you in almost any situation. However, you must be vigilant for inconsistencies on the part of air traffic controllers and not allow ATC to make decisions for you. [Figure 1-24]

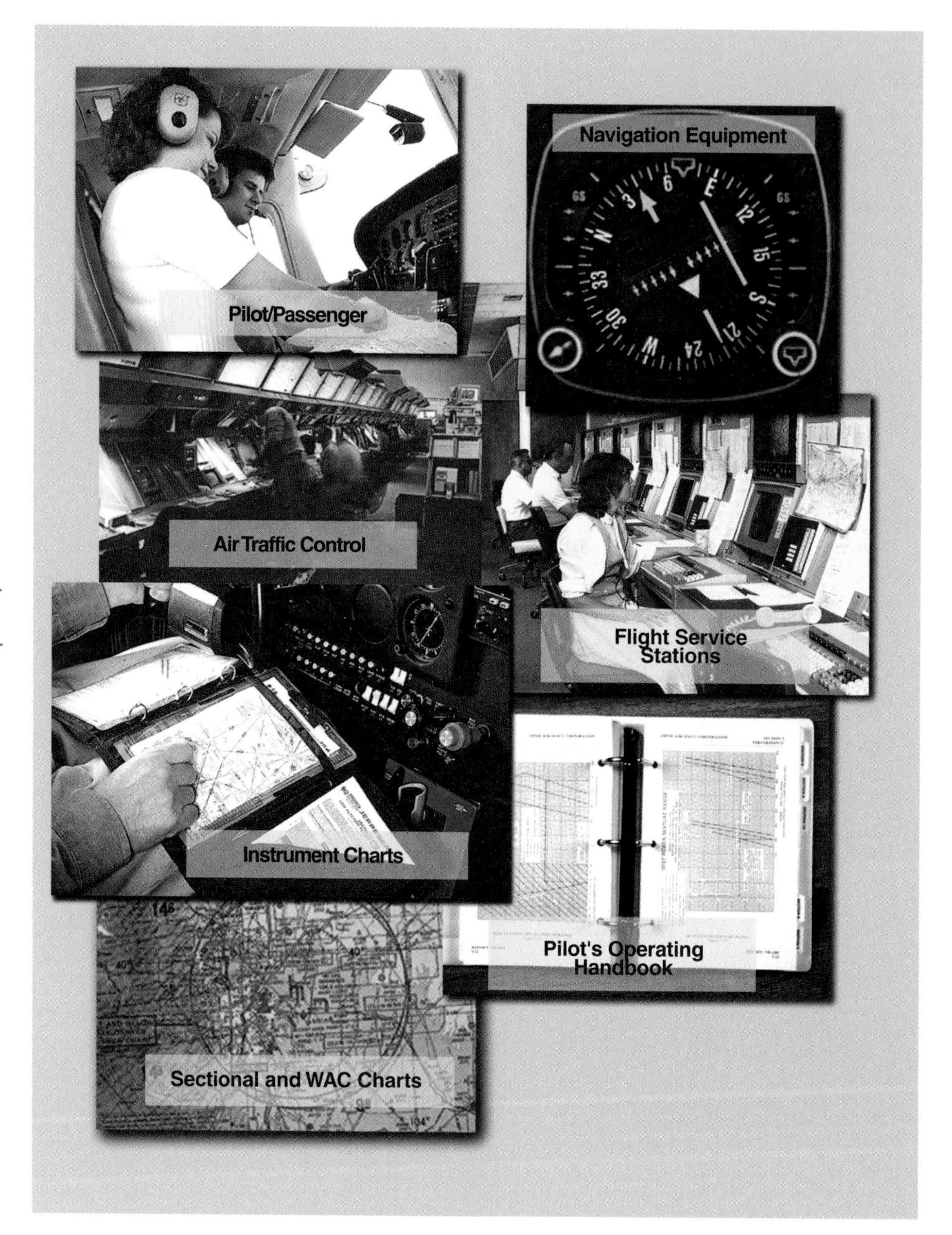

Figure 1-24. Many resources are available to the instrument/commercial pilot. Balancing the amount of information that you can access during the flight, while remaining in control, is a critical skill you must develop.

Your cockpit resources will increase as you fly more complex aircraft with advanced systems. An autopilot, on-board oxygen or pressurization systems, and advanced navigation equipment, such as GPS, are valuable resources. However, if you are not thoroughly familiar with the equipment in your aircraft or you rely on it so much that you become complacent, flight safety is compromised. Automation in airliners and corporate aircraft has heightened the need for crews to communicate more effectively to maintain situational awareness.

Competing For Air Time

Nonaviation technology can have a direct impact on flying safely. Cellular phones are now inexpensive, and the average pilot can have access to basic service at reasonable rates. A cell phone may be invaluable if you land in a remote area, only to find the pay phone out of service, or your wallet out of quarters. Carrying your cell phone can make the difference between life and death in crash survival, helping rescuers to locate your party, in case of ELT or cockpit radio failure.

You may have noticed that on airline flights, the use of cellular phones is prohibited. FAR 91.21 states that no person may operate a portable electronic device on any U.S. air carrier or during IFR flight. Use of cellular phones in flight may interfere with navigation systems and electronic cockpit displays, resulting in erroneous heading information and, at times, blacking out entire displays. The FCC also prohibits the use of cell phones in flight, and can revoke your service under 47 CFR 22.925 and 22.901. In addition to problems with interference, cell phones operate on a roaming procedure that uses any and all available stations to send or receive a call. When you are airborne, you have access to many times the number of stations that you would on the ground. This blanketing effect ties up stations across an area, rendering the sites unusable to others operating cell phones.

WORKLOAD MANAGEMENT

In the cockpit environment, your tasks are not evenly distributed over time. Through advance planning and prioritizing, you can make use of the periods in which relatively few duties demand your immediate attention. Effective **workload management** directly impacts safety by ensuring that you are prepared for the busiest segments of the flight through proper use of down time. The workload during an IFR flight can be demanding, however, as you gain experience you will realize what tasks can be accomplished ahead of time, and which need to be left until the moment. Items such as organizing charts in the order of use, setting radio frequencies, and writing down expected altitudes and route clearances will help you visualize and mentally prepare for what comes next. [Figure 1-25]

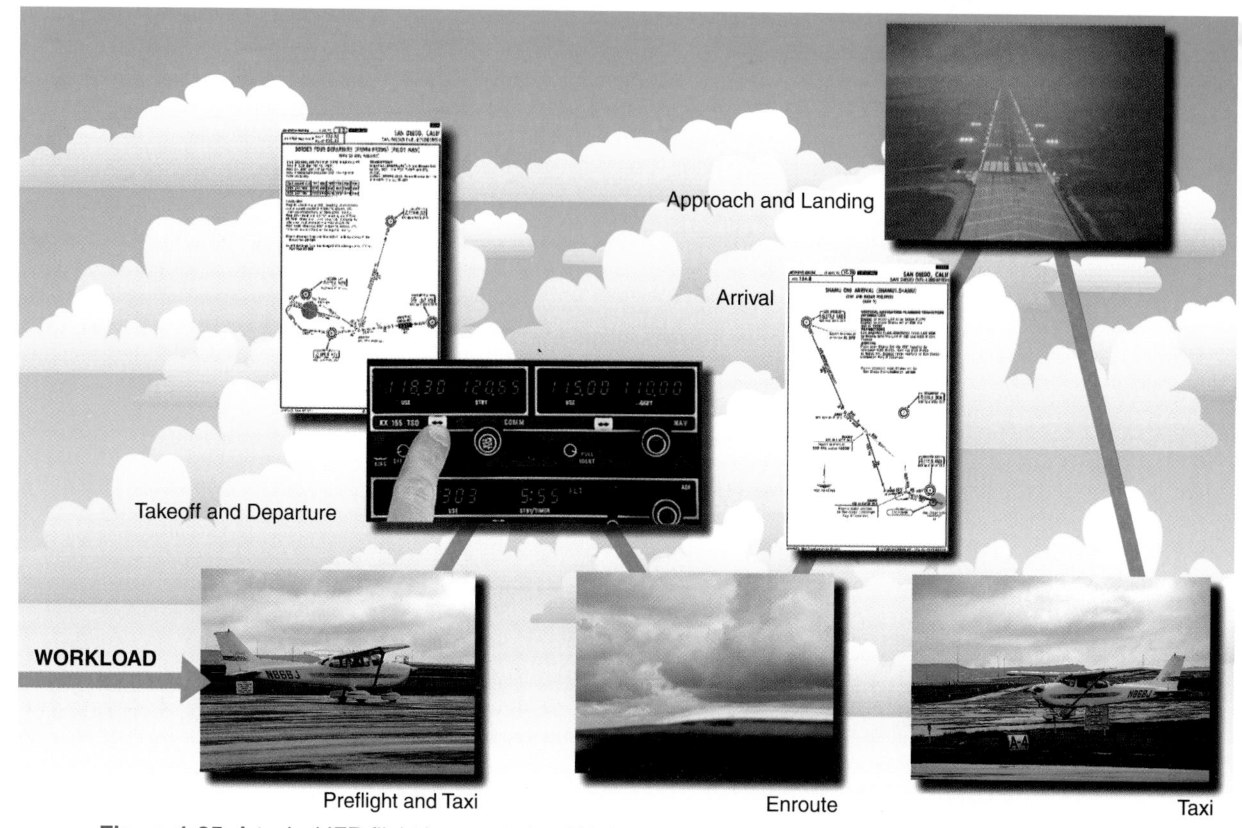

Figure 1-25. A typical IFR flight has periods of high and low workload. Plan what you can accomplish during the least busy times to enhance your preparation and safety.

SITUATIONAL AWARENESS

Maintaining **situational awareness** with no outside visual reference is perhaps the largest challenge to the instrument pilot. From the newly-rated instrument pilot to the most senior airline captain, a certain number of pilots fail to meet this challenge, as accident records attest. Good situational awareness occurs when you have a solid mental picture of the flight, from your own fitness and that of passengers to the operating conditions of the airplane, weather trends, and ATC instructions. Fatigue, stress, emergencies and other distractions can cause you to focus on one aspect of the flight, and omit others from your attention. Situational awareness can erode to the point where it may take serious effort on your part to regain your grasp of what is going on around you, especially during heavy workload periods. You also are prone to losing situational awareness when cockpit workload is low, as the ease of the flight may contribute to complacency.

If the pilot flies blind, has bad luck correcting his drift, is dubious about his position, that peak begins to stir with a strange life and its threat fills the breadth of the night sky in the same way as a single mine, drifting at the will of the current, can render the whole of the ocean a danger. — Antoine de Saint-Exupery

Many professional flight departments have made **controlled flight into terrain (CFIT)**, and the breakdown in situational awareness that leads to it, their top training priority. CFIT occurs when an aircraft is flown into terrain or water with no prior awareness on the part of the crew that the crash is imminent. For example, CFIT can happen through the crew's poor interpretation of charts or through misunderstood or misleading clearances from ATC. [Figure 1-26]

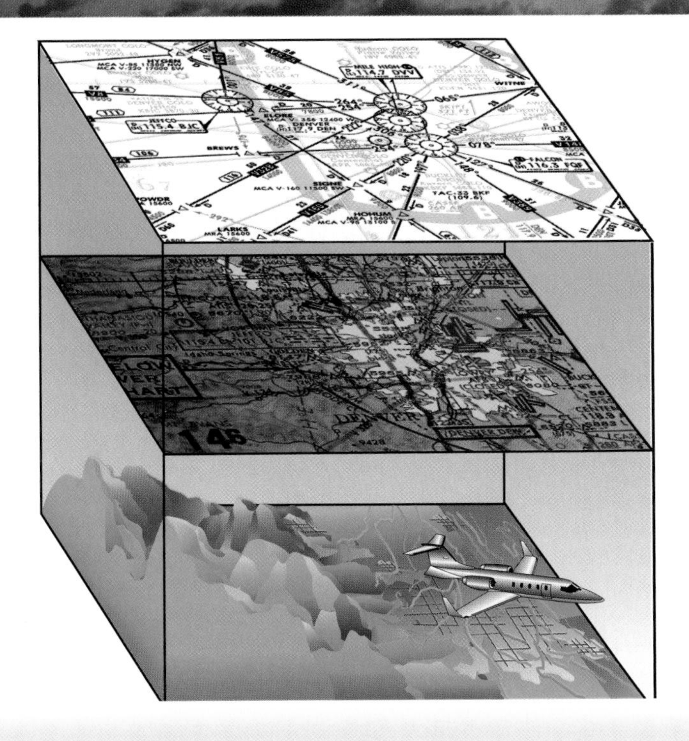

Figure 1-26. Part of maintaining IFR situational awareness involves picturing the terrain over which you are flying. Lack of a good mental picture can be misleading and dangerous. Looking at the appropriate low altitude enroute chart may not give you the full picture. A VFR chart, such as a WAC or sectional, may give you the best terrain information and improve your situational awareness.

Another Set of Eyes

Controlled flight into terrain (CFIT) accidents claim an average of 5 large commercial jets every year, and up to 25 commuter airline, corporate and air taxi aircraft. CFIT accidents have declined significantly since the mandatory installation of Ground Proximity Warning Systems (GPWS) in all large turboprop and jet aircraft began in 1974. The GPWS equipment alerts the crew when the aircraft is flown too close to surrounding terrain by providing instructions for evasive maneuvers, such as: *"Terrain! Terrain! Pull up! . . ."* along with an annunciator and a warning horn. Limitations to early GPWS equipment included a significant number of false signals which led some crews to disable the device. In addition, at times, signals came too late for the crew to react appropriately. Little time was left for the pilots to think critically about the warning.

CFIT risk was reduced by about 20 times when the original GPWS equipment was installed in aircraft, and subsequent equipment improvements have reduced the risk by approximately 50 times. The next generation of equipment, the Enhanced Ground Proximity Warning System (EGPWS) promises to again lower the accident risk significantly. The augmented system provides a visual display to indicate an aircraft's relative postion with respect to terrain. [Figure A] The EGPWS uses sophisticated flight path algorithms that predict the aircraft's future location and scan the area 30° in either direction. Wind shear warnings and customized altitude callouts also are included. In addition, the EGPWS gives as much as a 60-second advance warning, allowing pilots to take effective evasive action. Hopefully, these improvements will relegate CFIT accidents into the annals of commercial aviation history, along with routine engine failures.

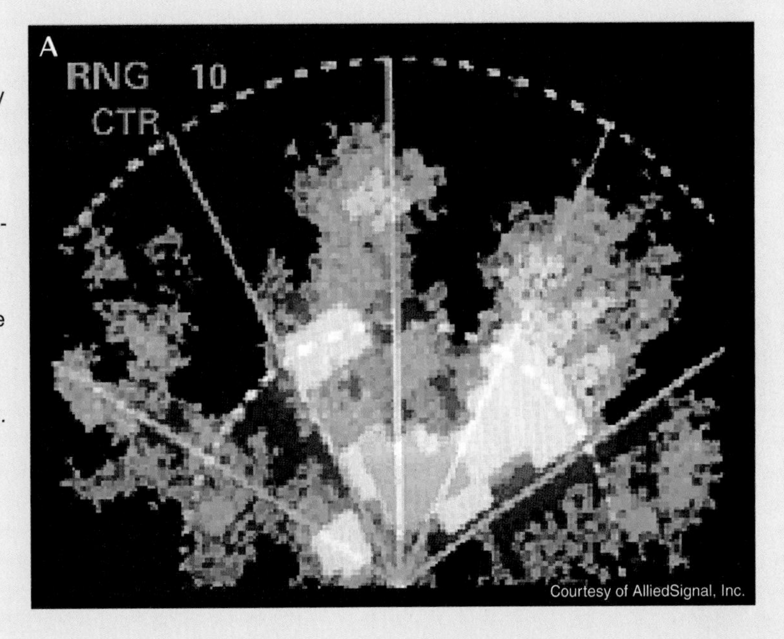

AVIATION PHYSIOLOGY

The study of aviation physiology is an important part of human factors training. How you feel physically has a direct impact on how well you fly. As you gain advanced certificates and ratings, you become accustomed to elements of flight, such as motion and pressure differences. However, flying within the instrument and commercial environments also means that you will be exposed to situations you may not have previously encountered. You should be aware of the effects that these situations have on you physically and mentally as you continue your professional training. This information also will help you understand the reactions your passengers may experience, so you can better prepare them for flight.

DISORIENTATION

Awareness of your body's position in relation to your environment is a result of input from three primary sources: vision, the vestibular system located in your inner ear, and your kinesthetic sense. During flight, you may experience disorientation if your brain receives conflicting messages from your senses. **Kinesthetic sense** is the term used to describe an awareness of position obtained from the nerves in your skin, joints, and muscles. Kinesthetic sense is unreliable, however, because the brain cannot tell the difference between input caused by gravity and that of maneuvering G-loads.

In good weather and daylight, you obtain your orientation primarily through your vision. In IFR conditions or at night, there are fewer visual cues, and your body relies upon the vestibular and kinesthetic senses to supplement your vision. Since these senses can provide false cues about your orientation, the probability of disorientation occurring in IFR weather is quite high. Fatigue, anxiety, heavy pilot workloads, and the intake of alcohol or other drugs increase your susceptibility to disorientation and visual illusions. These factors increase response times, inhibit decision-making abilities, cause a breakdown in scanning techniques and impair night vision. To alleviate symptoms of disorientation, you must rely on and properly interpret the indications of the flight instruments. Reducing your workload with the use of an autopilot or flight director, and improving your cockpit management skills can help prevent overload and the possibility of disorientation.

To prevent or overcome spatial disorientation in IFR conditions, you must rely on and properly interpret the indication of the flight instruments.

You are more subject to disorientation if you use body signals to interpret flight attitude.

Lightheadedness, dizziness, the feeling of instability, and the sensation of spinning are common symptoms of disorientation. Experiencing these sensations during flight may either be the result of spatial or vestibular disorientation. Although the term spatial disorientation often is used to describe vestibular disorientation, the two terms have different meanings.

SPATIAL DISORIENTATION

Spatial disorientation occurs when there is a conflict between the signals relayed by your central vision and information provided by your peripheral vision. Spatial disorientation is more likely when you are in IFR conditions, as your peripheral vision has practically none of the references needed to establish orientation. The movement of rain or snow seen out the window by your peripheral vision also may lead to a misinterpretation of your own movement and position in space. This is similar to the illusion of motion that you experience when an airplane next to yours begins to taxi away from its parking space. Your peripheral vision can misinterpret this visual cue and lead you to believe that your stationary airplane is in motion.

VESTIBULAR DISORIENTATION

When subjected to the different forces of flight during instrument maneuvers, the vestibular system may send misleading signals to the brain resulting in vestibular disorientation. The vestibular system, located in your inner ear, consists of the vestibule and three semicircular canals. The utricle and saccule organs within the vestibule are responsible for the perception of gravity and linear acceleration. A gelatinous substance within the utricle and saccule is coated with a layer of tiny grains of limestone called otoliths. Movement of the vestibule causes the otoliths to shift, which in turn causes hair cells to send out nerve impulses to the brain for interpretation.

The semicircular canals are oriented in three planes to sense yaw, pitch, and roll. The canals are filled with a fluid and a gelatinous structure, called the cupula. When the body changes position, the canals move but the fluid lags behind, causing the cupula to lean away from the movement. Movement of the cupula results in the deflection of hair cells which stimulate the vestibular nerve. This nerve transmits impulses to the brain which interprets the signals as motion about an axis. [Figures 1-27 and 1-28]

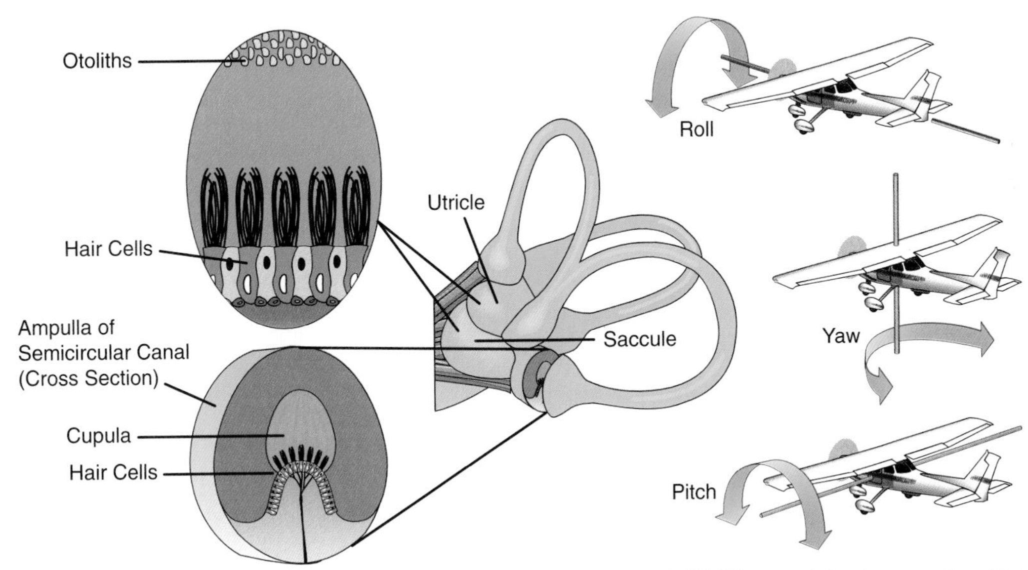

Figure 1-27. The semicircular canals, which lie in three planes, sense the motions of roll, pitch, and yaw. The vestibular nerve transmits impulses to the brain to interpret the motion.

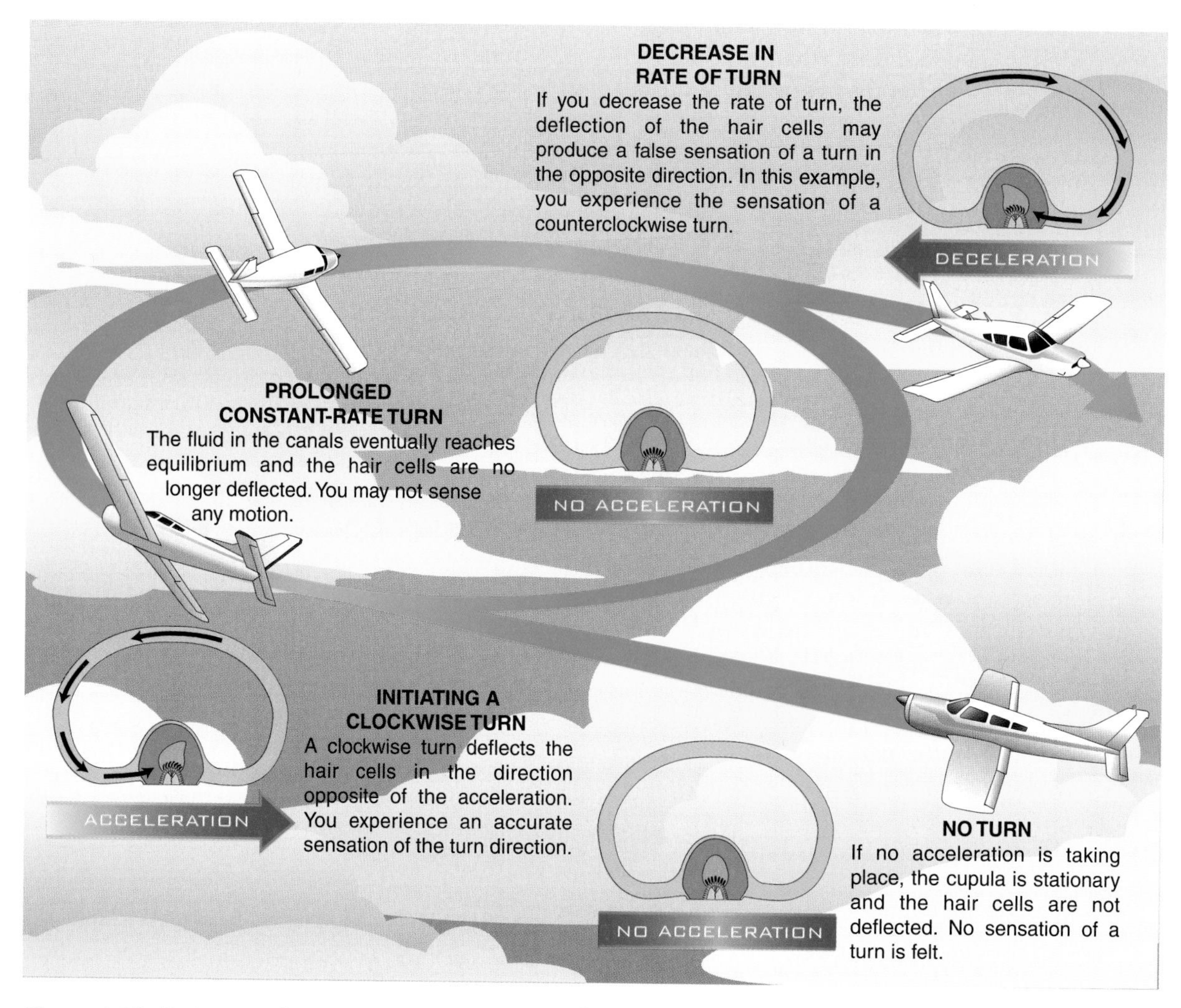

Figure 1-28. During a prolonged, constant-rate turn, you may not sense any motion since the fluid in the semicircular canals eventually reaches equilibrium and the hair cells are no longer deflected.

The majority of the illusions which lead to vestibular disorientation occur when visibility is restricted. Awareness of these illusions will aid you in coping with them in flight. It takes many hours of training and experience before you are competent to fly an aircraft solely by reference to instruments. [Figure 1-29]

A rapid acceleration during takeoff can create the illusion of being in a nose-up attitude, and an abrupt change from climb to straight-and-level flight can create the illusion of tumbling backward. See figure 1-29.

If you move your head abruptly during a prolonged constant-rate turn in IFR conditions, you may become disoriented. See figure 1-29.

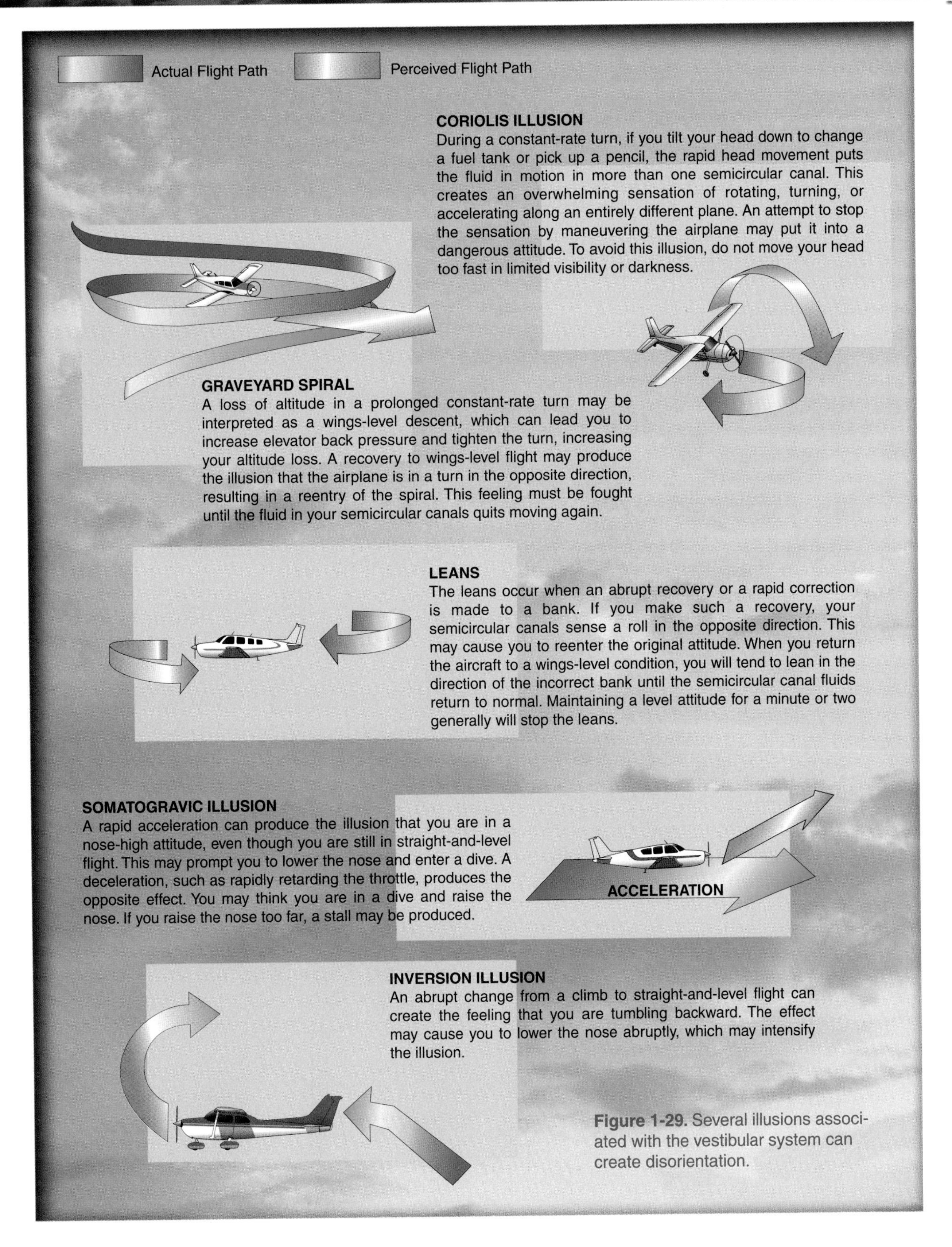

CORIOLIS ILLUSION
During a constant-rate turn, if you tilt your head down to change a fuel tank or pick up a pencil, the rapid head movement puts the fluid in motion in more than one semicircular canal. This creates an overwhelming sensation of rotating, turning, or accelerating along an entirely different plane. An attempt to stop the sensation by maneuvering the airplane may put it into a dangerous attitude. To avoid this illusion, do not move your head too fast in limited visibility or darkness.

GRAVEYARD SPIRAL
A loss of altitude in a prolonged constant-rate turn may be interpreted as a wings-level descent, which can lead you to increase elevator back pressure and tighten the turn, increasing your altitude loss. A recovery to wings-level flight may produce the illusion that the airplane is in a turn in the opposite direction, resulting in a reentry of the spiral. This feeling must be fought until the fluid in your semicircular canals quits moving again.

LEANS
The leans occur when an abrupt recovery or a rapid correction is made to a bank. If you make such a recovery, your semicircular canals sense a roll in the opposite direction. This may cause you to reenter the original attitude. When you return the aircraft to a wings-level condition, you will tend to lean in the direction of the incorrect bank until the semicircular canal fluids return to normal. Maintaining a level attitude for a minute or two generally will stop the leans.

SOMATOGRAVIC ILLUSION
A rapid acceleration can produce the illusion that you are in a nose-high attitude, even though you are still in straight-and-level flight. This may prompt you to lower the nose and enter a dive. A deceleration, such as rapidly retarding the throttle, produces the opposite effect. You may think you are in a dive and raise the nose. If you raise the nose too far, a stall may be produced.

INVERSION ILLUSION
An abrupt change from a climb to straight-and-level flight can create the feeling that you are tumbling backward. The effect may cause you to lower the nose abruptly, which may intensify the illusion.

Figure 1-29. Several illusions associated with the vestibular system can create disorientation.

MOTION SICKNESS

Nausea, sweating, dizziness, and vomiting are some of the symptoms of motion sickness which often is caused by vestibular disorientation. During visual flight, you overcome motion sickness by focusing your eyes on the outside horizon. However, in the clouds, precipitation, fog, or haze that constitute IFR conditions, this becomes impossible. To overcome motion sickness without outside visual references, you should focus on the instrument panel, since it is your only source of accurate position information.

Lost in the Air

A figure skater executes a jump, and follows an arcing flight path over the ice. Like a pilot, the skater must remain aware of his or her position so that a smooth landing can be made. The skater uses a visual reference during the takeoff portion of the jump to maintain consistency in the rotation and landing. This reference helps the skater balance and know when to open up, stop the rotation, and reach for the ice with the landing foot. If the visual reference is lost, the skater risks becoming *lost in the air* and unsure of his or her position on the ice.

During flight, much like the skater, you must use a visual reference to maintain orientation. Instrument flight is made more challenging by the lack of an outside visual reference. If you become unsure of your position in space, you must learn to rely on the instruments to avoid becoming lost in the air.

HYPOXIA

Hypoxia occurs when the tissues in the body do not receive enough oxygen. Hypoxia can be caused by several factors including an insufficient supply of oxygen, inadequate transportation of oxygen, or the inability of the body tissues to use oxygen. An insidious characteristic of hypoxia is that its early symptoms include euphoria which may prevent you from recognizing a potentially hazardous situation. You should remain alert for the other symptoms of hypoxia such as headache, increased response time, impaired judgment, drowsiness, dizziness, tingling fingers and toes, numbness, blue fingernails and lips (cyanosis), and limp muscles. The forms of hypoxia are divided into four major groups based on their causes: hypoxic hypoxia, hypemic hypoxia, stagnant hypoxia, and histotoxic hypoxia.

Symptoms of hypoxia may be difficult to recognize before your reactions are affected.

HYPOXIC HYPOXIA

As you progress to airplanes which are capable of flying at high altitudes, you must have a good understanding of the adverse physiological effects that can occur due to changes in atmospheric pressure. [Figure 1-30] Although the percentage of oxygen in the atmosphere is constant, its partial pressure decreases proportionately as atmospheric pressure decreases. For example, at 18,000 feet, atmospheric pressure decreases to approximately one-half of sea level pressure. Hypoxic hypoxia occurs when there are not enough molecules of oxygen available at sufficient pressure to pass between the membranes in your respiratory system.

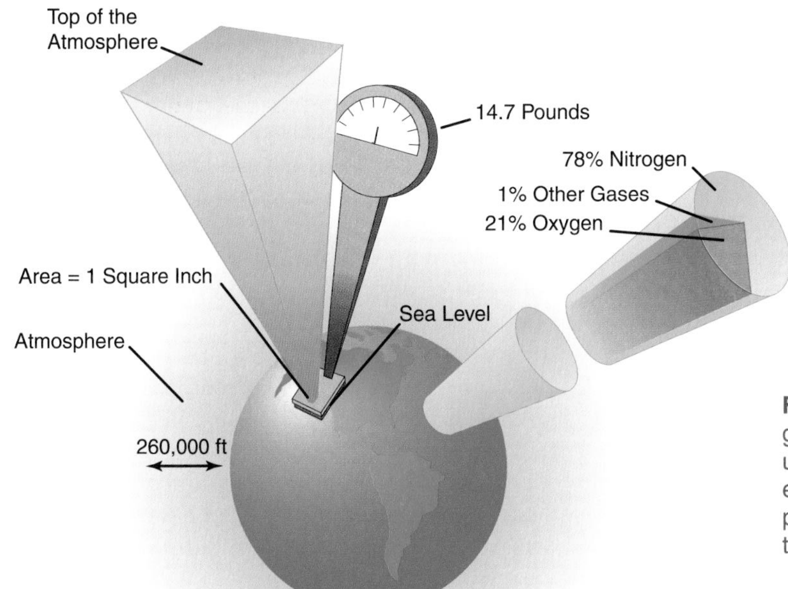

Figure 1-30. The atmosphere is a mixture of gases that exists in fairly uniform proportions up to approximately 260,000 feet above the earth. Typically, air exerts about 14.7 pounds per square inch (lb/in^2) at sea level. As altitude increases, pressure steadily decreases.

As you ascend, the hemoglobin that carries the oxygen molecules throughout your body receives a lower oxygen saturation. For example, at 10,000 feet MSL, the oxygen saturation level of the hemoglobin is approximately 85%. When the saturation value is 85% or lower, your body's functions start to degrade, becoming worse with increased altitude.

Hypoxia is the result of the brain and body tissue not receiving a sufficient supply of oxygen.

Hypoxic hypoxia can occur very suddenly at high altitudes during rapid decompression, or it can occur slowly at lower altitudes when you are exposed to insufficient oxygen over an extended period of time. The **time of useful consciousness** is the maximum time you have to make a rational, life-saving decision and carry it out following a lack of oxygen at a given altitude. [Figure 1-31]

Altitude	Time of Useful Consciousness
45,000 feet MSL	9 to 15 seconds
40,000 feet MSL	15 to 20 seconds
35,000 feet MSL	30 to 60 seconds
30,000 feet MSL	1 to 2 minutes
28,000 feet MSL	2 1/2 to 3 minutes
25,000 feet MSL	3 to 5 minutes
22,000 feet MSL	5 to 10 minutes
20,000 feet MSL	30 minutes or more

Figure 1-31. Hypoxic hypoxia is considered to be the most lethal of all physiological causes of accidents. Beyond the time of useful consciousness, you may not be able to complete even the simplest task, such as speaking on the radio.

OTHER FORMS OF HYPOXIA

Hypemic hypoxia occurs when your blood is not able to carry a sufficient amount of oxygen to your body's cells. This type of hypoxia can be caused by any condition which results in a reduced number of healthy blood cells such as anemia, disease, blood loss, or deformed blood cells. In addition, hypemic hypoxia can be caused by any factor, such as carbon monoxide poisoning, which interferes with the attachment of oxygen to the blood's hemoglobin. Since it attaches itself to the hemoglobin about 200 times more easily than does oxygen, carbon monoxide (CO) prevents the blood from carrying sufficient oxygen. As the hemoglobin becomes progressively saturated with CO, the body tissues are deprived of oxygen, eventually producing physiological symptoms similar to those encountered with hypoxic hypoxia.

Hypemic hypoxia can be encountered at any altitude; however, hypoxia susceptibility due to the inhalation of CO increases as altitude increases. Carbon monoxide can enter an airplane through a faulty cabin heater system. With most single-engine airplanes, ventilation air is heated by flowing through a shrouded cavity which surrounds the exterior surface of the exhaust muffler or manifold. If the exhaust system develops a crack or hole within the cavity, carbon monoxide can combine with the ventilation air and subsequently be delivered to the cabin. During flight, if you suspect that carbon monoxide is entering the cabin, you should immediately shut off the cabin heat, open the fresh air vents or windows, and land as soon as possible. As an added precaution, if supplemental breathing oxygen is available, you and your passengers should use it until the landing is made. Even after you have eliminated the CO exposure, it can take up to 48 hours for your body to dispose of the carbon monoxide. [Figure 1-32]

Hypoxia susceptibility due to the inhalation of CO increases as altitude increases.

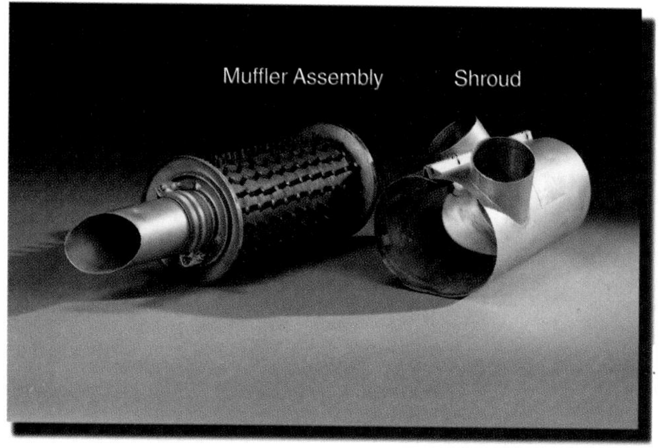

Figure 1-32. To aid in preventing carbon monoxide poisoning, the cabin heater shroud should be frequently removed from the exhaust system by an appropriately certificated aircraft technician. With the shroud removed, a thorough inspection can be accomplished to determine the condition of the exhaust and heater system.

Frequent inspections should be made of aircraft exhaust manifold-type heating systems to minimize the possibility of exhaust gases leaking into the cockpit.

If you are a smoker, you most likely contain a certain level of CO in your bloodstream at any time. Approximately 2.5% of the volume of cigarette smoke is carbon monoxide. A blood saturation of 4% carbon monoxide may result from inhaling the smoke of 3 cigarettes at sea level. This causes a reduction in visual acuity and dark adaptation similar to the mild hypoxia encountered at 8,000 feet MSL. Heavy smokers can have carbon monoxide blood saturation as high as 8%.

Stagnant hypoxia is an oxygen deficiency in the body due to the poor circulation of the blood. During flight, stagnant hypoxia can be the result of pulling excessive positive Gs, or cold temperatures can decrease the blood supply to the extremities. The inability of the cells to effectively use oxygen is defined as histotoxic hypoxia. This impairment of cellular respiration can be caused by alcohol and other drugs such as narcotics and poisons. [Figure 1-33]

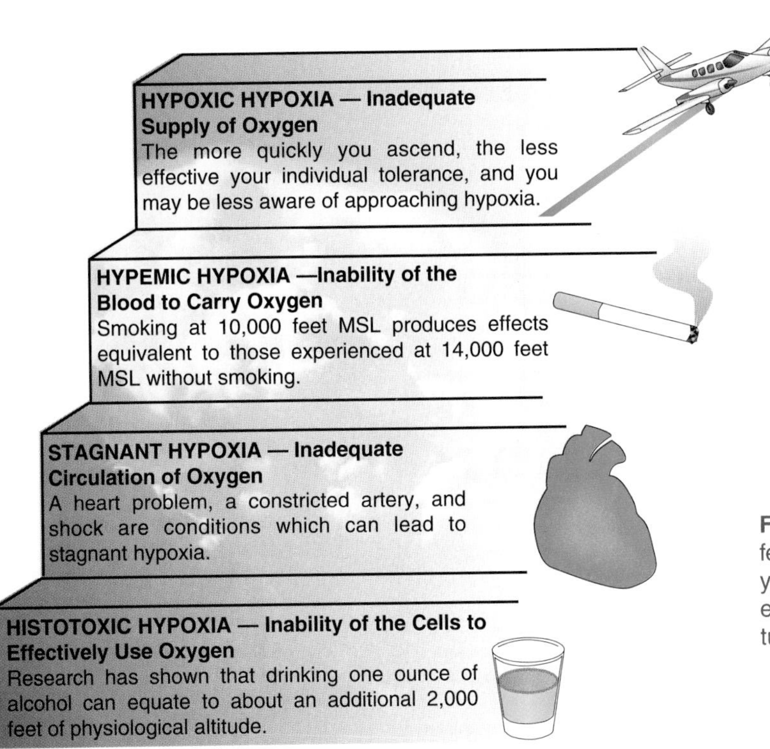

Figure 1-33. A combination of different types of hypoxia affecting your body may cause you to experience symptoms at much lower altitudes than expected.

PREVENTION OF HYPOXIA

Even if you learn the early symptoms of hypoxia, do not assume that you will be able to take corrective action whenever they occur. Since judgment and rationality deteriorate when you are suffering from hypoxia, prevention is the best approach. Your susceptibility to hypoxia is related to many factors, many of which you can control. You can increase your tolerance to hypoxia by maintaining good physical condition, eating a nutritious diet, and by avoiding alcohol and smoking. If you live at a high altitude and have become acclimated, you normally have an increased tolerance to the conditions that would lead to hypoxia compared to a person living at a lower altitude.

Your body requires more oxygen during physical activity. For example, you risk becoming hypoxic more readily when you are flying the aircraft manually in turbulent IFR conditions than during a smooth VFR flight. In addition, as your body copes with temperature extremes in the cockpit, you are using energy, which is comparable to increased activity.

SUPPLEMENTAL OXYGEN

If you are planning a flight with a cruise altitude over 12,500 feet MSL, you should review FAR Part 91 for the requirements regarding supplemental oxygen. [Figure 1-34] As a general rule, the FAA recommends that you begin using supplemental oxygen whenever you are flying at cabin pressure altitudes greater than 10,000 feet MSL during the day. However, since night vision acuity is highly affected by the partial pressure of oxygen, the FAA recommends the use of supplemental oxygen above a cabin pressure altitude of 5,000 feet while flying at night. Chapter 11, Section B — Environmental and Ice Control Systems describes the various supplemental oxygen systems used in aircraft.

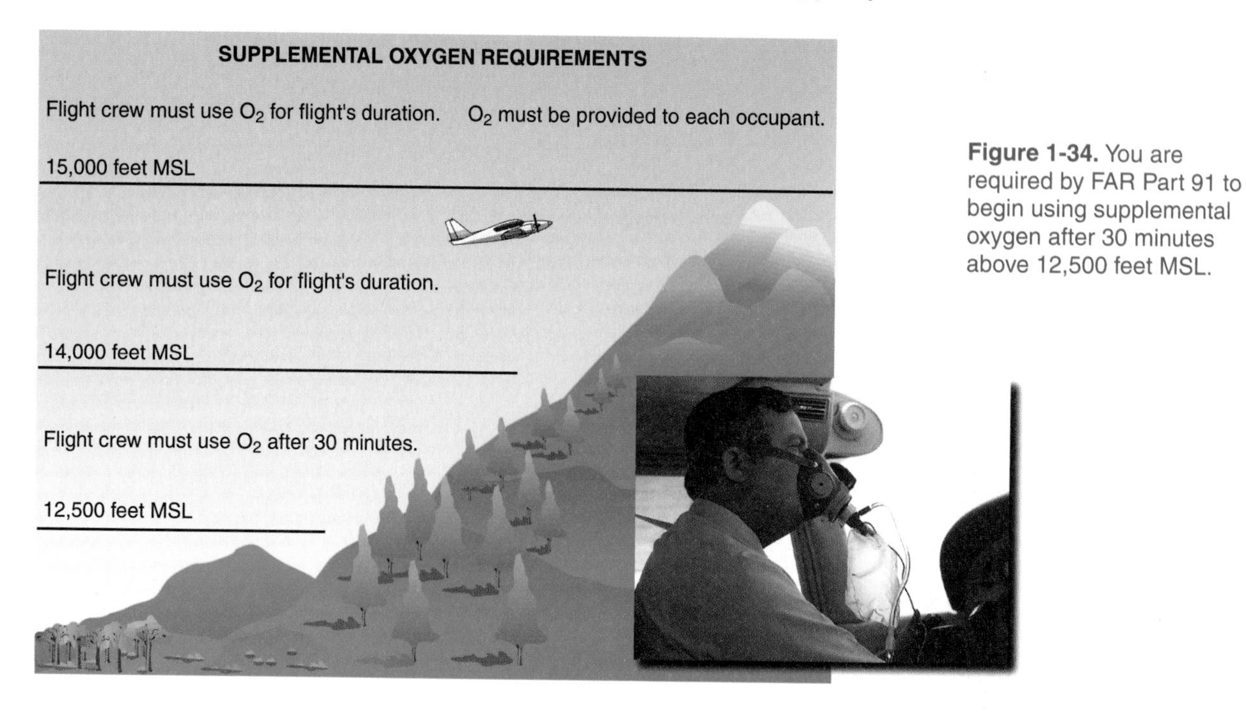

Figure 1-34. You are required by FAR Part 91 to begin using supplemental oxygen after 30 minutes above 12,500 feet MSL.

HIGH-ALTITUDE TRAINING

Aircraft cabin pressurization is the maintenance of a cabin altitude lower than the actual flight altitude by a system which compresses air. Pressurization systems reduce some of the physiological problems experienced at high altitudes. However, prior to operating a pressurized aircraft, with a service ceiling or maximum operating altitude higher than 25,000 feet MSL, you must complete high-altitude training. This training consists of ground instruction on high-altitude aerodynamics and meteorology, respiration, hypoxia, use of supplemental oxygen, and other physiological aspects of high-altitude flight. More information on high-altitude operations is provided in Chapter 11, Section B — Environmental and Ice Control Systems. [Figure 1-35]

Figure 1-35. Several single-engine and light twin-engine aircraft are pressurized, and you must have a high-altitude logbook endorsement before you can operate them as pilot in command.

DECOMPRESSION SICKNESS

Decompression sickness (DCS) is a condition caused by a rapid reduction in the ambient pressure surrounding the body. When decompression occurs, nitrogen and other inert gases which are normally dissolved in body tissue and fluid, expand to form bubbles that rise out of solution, much like uncapping a bottle of soda. These bubbles produce a variety of symptoms, that range from pain in the large joints of the body, such as elbows, shoulders, hips, wrists, knees and ankles, to seizures and unconsciousness. These symptoms tend to increase in severity depending on the rate and amount of pressure change and, if severe enough, can result in death. When DCS develops as a result of the decreased pressure at high altitude, it is commonly referred to as high-altitude sickness or altitude-induced DCS.

Pressure changes resulting from underwater activities such as scuba diving can cause DCS. For example, when diving from sea level to a depth of 33 feet, the resulting water pressure can subject a body to twice the sea level air pressure. When combined with breathing high pressure air from diving equipment, there is a significant increase in the amount of nitrogen dissolved in the body. If the nitrogen level does not stabilize during ascent to the water surface, DCS may occur. Even after a dive, it will take a period of time for the body to completely eliminate the excess nitrogen. If the nitrogen is not removed, a flight at high cabin altitudes may produce severe symptoms of altitude-induced DCS. [Figure 1-36]

Figure 1-36. If you or a passenger plan to fly after scuba diving, it is important that enough time is allowed for the body to rid itself of excess nitrogen absorbed during diving.

The recommended waiting time before ascending to 8,000 feet MSL is at least 12 hours after a dive which has not required a controlled ascent (nondecompression stop diving), and at least 24 hours after a dive which has required a controlled ascent (decompression stop diving). The waiting time before ascending to flight altitudes above 8,000 feet MSL should be at least 24 hours after any scuba dive.

Normally, altitude-induced DCS is not experienced below 29,000 feet unless the ascent has been extremely rapid or there has been exposure to a high ambient pressure environment before flying. However, a situation that can significantly increase its potential is encountered when a pressurized airplane rapidly decompresses while at altitude. Although rapid decompressions are rare, if you encounter one, you should be extremely alert to DCS symptoms. Even after the flight, you should continue to watch for symptoms for a couple of days. In addition, if the decompression occurred with passengers, you should brief them on the possible delayed effects so they will also continue to monitor their own condition. If symptoms do occur, supplemental breathing oxygen should be used until proper medical treatment can be obtained, since oxygen tends to help flush the nitrogen from the body.

HYPERVENTILATION

Hyperventilation is a physiological disorder that develops when too much carbon dioxide (CO_2) has been eliminated from the body, usually caused by breathing too rapidly or too deeply. Without a sufficient quantity of CO_2, normal respiration is disturbed, producing symptoms that resemble hypoxia. If you are hyperventilating, you may experience drowsiness, dizziness, shortness of breath, and feelings of suffocation. In addition, hyperventilation may produce a pale, clammy appearance and muscle spasms compared to the cyanosis and limp muscles associated with hypoxia. An excessive loss of CO_2 from your body can lead to unconsciousness due to the respiratory system's overriding mechanism to regain control of breathing. After becoming unconscious, your breathing rate will be exceedingly low until enough CO_2 is produced to stimulate the respiratory center.

Hyperventilation may cause you to experience drowsiness, dizziness, shortness of breath, and feelings of suffocation. The condition is usually caused by an insufficient supply of carbon dioxide, produced by breathing too rapidly or too deeply.

Hyperventilation can be triggered by tension, fear, or anxiety. Slowing your breathing rate, talking aloud, or breathing into a paper bag normally restores the body's proper carbon dioxide level. While you may experience hyperventilation in a stressful situation, you should be especially alert to the symptoms of hyperventilation in passengers who may feel anxious about flying.

STRESS

Stress is the body's reaction to the physical and psychological demands placed upon it. You may typically think of stress as negative, such as the work- or family-induced stress that may confront you in daily living. However, stress can be positive as well. When your body is placed under stress, chemical hormones are released into the blood, and your metabolism speeds up. Heart rate, respiration, blood pressure, and perspiration all increase. A small amount of stress is good since it helps keep you alert and aware. When stress builds, however, it interferes with your ability to focus and cope with a given situation.

To overcome the symptoms of hyperventilation, you should slow your breathing rate.

For example, a normal flight contains a certain amount of stress, depending on the weather, your familiarity with the route and destination, and the condition of the airplane. Since stress is cumulative, you also bring into the flight varying degrees of stress left over from the other areas of your life. Trouble with a spouse or parent, tension on the job and major life changes can cause large amounts of stress that affect your ability to fly. [Figure 1-37]

FATIGUE

Fatigue deserves special mention when considering the instrument and commercial flight environments. As an instrument pilot, your level of concentration may need to be highest at the end of a flight, when you are most likely to be tired. If you operate as a commercial pilot for hire, you may be asked to fly many times in one day, several days in a row, and you may feel tired as you adapt to the schedule. Since both realms of flight are typically more demanding than a personal flight in VFR conditions, any residual

fatigue you have, from lack of sleep to excess physical work, may significantly affect your ability to operate safely. Many accidents attributed to pilot error occur at the end of a long duty day, when cockpit crews are tired and pilot performance suffers.

Figure 1-37. Stress comes from many different sources, and its cumulative effects can interfere with your capacity to operate an aircraft safely.

The cockpit environment also adds to your fatigue level, due to noise and vibration. If you do not already own one, you will find that purchasing a headset is one of the best investments you can make as a pilot. Make sure the headset fits snugly, and is at least noise-attenuating. Low-cost earplugs are also beneficial, and you should carry some for passenger use during flight.

No amount of training will allow you to overcome the effects of fatigue. Getting adequate rest on a regular basis is the only way to perform at your best. A drug or remedy touted as a panacea for those who are chronically tired should be treated with caution. It is wise not to rely on any drug to help you offset fatigue, since there is no substitute for a good night's sleep.

The Key to Success is Knowing When to Quit

On June 4, 1935, two brothers, Fred and Algene Key, took off in a Curtiss Robin named *Ole Miss* from the airport at Meridian, Mississippi. The *Ole Miss* did not touch down until July 1 after remaining aloft 653 hours and 34 minutes — a total of 27 days. The airplane was refueled and the brothers received supplies in flight from another airplane 432 times.

A metal catwalk had been constructed around the front of the fuselage to enable the Keys to lubricate the engine and conduct emergency repairs. Since Fred was the smaller of the two brothers, he was tasked with climbing out on the catwalk when necessary. The Key brothers endured thunderstorms, an electrical fire in the cabin, and a close call with turbulence while Fred was on the catwalk, but perhaps their greatest obstacle was fatigue.

By June 30 the stress and fatigue of the grueling flight had taken its toll and it was apparent the brothers would not be able to stay aloft until July 4, their original target date.

The constant vibration was causing two types of fatigue: metal fatigue in which wires and braces were threatening to weaken and mental fatigue which created disorientation in the pilots. Nerve shock and shear weariness caused the Key who was resting to have to take as many as five minutes before he could wake and get his bearings. Under such circumstances, emergencies could not be dealt with easily. — The Flying Key Brothers and Their Flight to Remember, by Stephen Owen

Upon landing, a crowd of 35,000 to 40,000 spectators flocked to pay tribute to the exhausted aviation heroes. The mob lifted the weary heroes from the plane after the *Ole Miss* touched down on the newly christened Key Field.

ALCOHOL AND DRUGS

Anytime you are ill enough to require medication, you should closely examine your plans to conduct a flight. Many drugs that are used to alleviate symptoms of illness and disease also have side effects that interfere with your ability to fly safely. Prior to flying you should consult an aviation medical examiner about any medication you are using.

Alcohol and other depressants impair the body's functioning in several critical areas, causing decreased mental processing and slow motor and reaction responses. Due to the high level of performance required by instrument and commercial flight operations, using depressants while acting as pilot in command severely increases your risk of an accident. The FARs specifically state that you should not fly within 8 hours of using alcohol, or when you have a blood alcohol level of .04% or greater. However, the regulations also state that anytime your ability is impaired by alcohol or any drug, you are unfit for flight. Most commercial flight departments require 12 or 24 hours to pass before allowing their pilots to fly after using alcohol. This is a policy that all pilots should find beneficial to flight safety since judgment and decision-making abilities can be adversely affected by even small amounts of alcohol.

Judgment and decision-making abilities can be adversely affected by even small amounts of alcohol.

Commercial airlines and most flight departments conduct random and pre-hire drug tests to reinforce the idea that drugs should stay out of the cockpit. Illicit drugs can cause hallucinations and other withdrawal effects that last long after the drugs have been taken. [Figure 1-38]

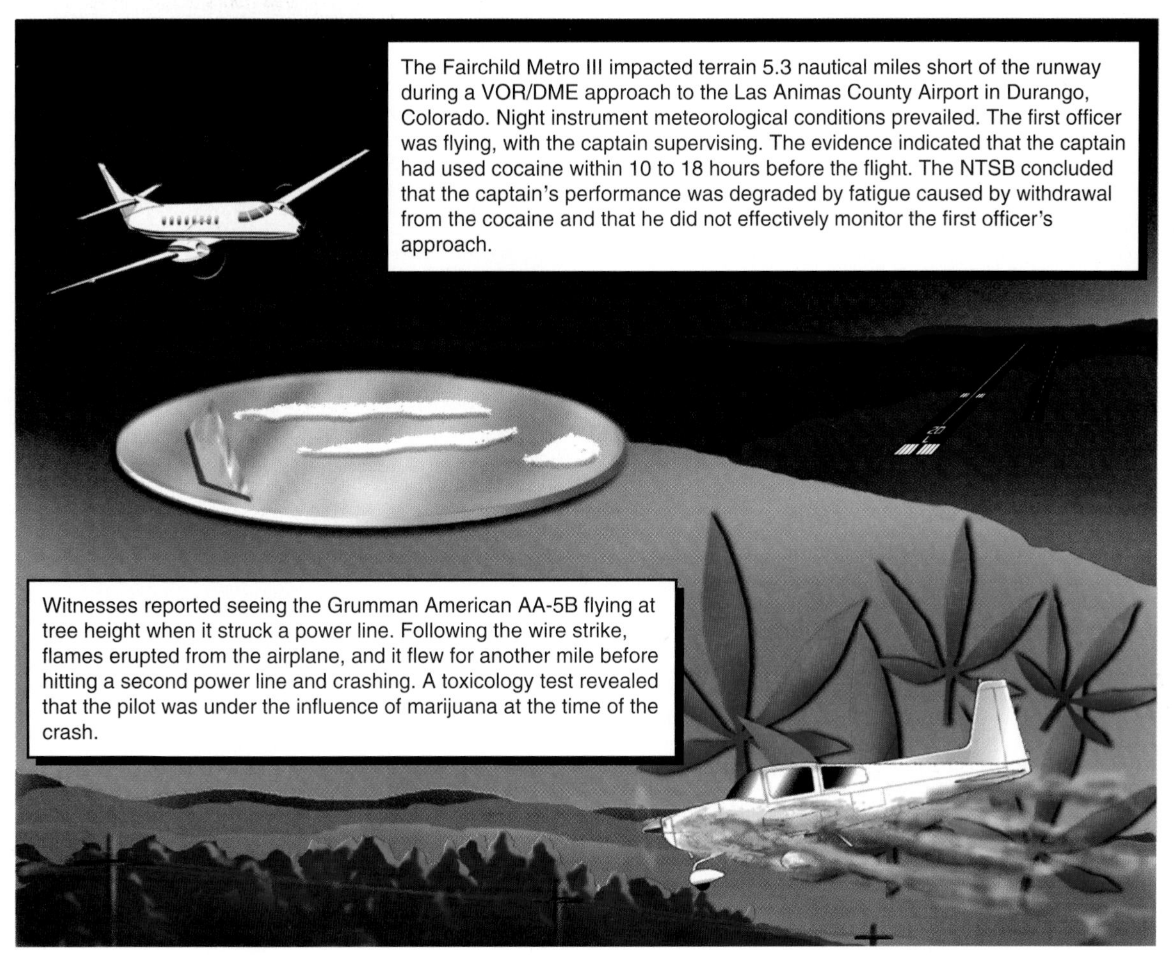

Figure 1-38. These accidents graphically illustrate why drugs and flying do not go together.

FITNESS FOR FLIGHT

Your overall health has a large impact on how you fly. A general program of exercise and a balanced diet will improve your mental clarity and energy level, and your piloting skills will benefit. [Figure 1-39]

Figure 1-39. An exercise regimen can be as little as 30 minutes of aerobic exercise, such as walking, jogging or cycling, performed at least 3 times a week. This amount of exercise increases your stamina and makes you less prone to a heart attack or stroke.

Physiology of the Final Frontier

The International Space Station (ISS) is one of the greatest international scientific and technological endeavors ever undertaken. [Figure A] Thirteen nations around the world are uniting to create this permanent laboratory where gravity, temperature, and pressure can be manipulated for a variety of scientific and engineering pursuits in ways that are impossible in ground-based laboratories.

To better prepare crews for missions aboard the ISS, NASA has a wide variety of programs in place to study the physiological effects of living in space. As humans travel to the low-gravity environment of Earth orbit, virtually every system in the body, from bones and muscles to the immune system, is affected. The physiological effects of weightlessness include bone loss, atrophy of muscles, and motion sickness. One of the objectives of the Biomedical Research and Countermeasures Program, managed by the NASA Johnson Space Center, is to study the problems associated with extended periods of flight that will be characteristic of ISS missions and of missions to explore the solar system. Researchers are working to develop methods that will allow humans to live and work in microgravity for durations of over a year and to minimize the risks in readapting to Earth's gravity.

Another research program, the COIS (Canal and Otolith Integration Studies) Investigation is designed to study changes in the coordination of head and eye movements associated with adaptation to microgravity, and to examine how vestibular and visual information is processed in the absence of a gravitational reference.

Assembly of the ISS will require hundreds of hours of space walks, or extravehicular activities (EVAs). The goal of the Neutral Buoyancy Lab (NBL) located at the Johnson Space Center is to prepare for space missions involving EVAs. NASA team members utilize the NBL to develop flight procedures, verify hardware compatibility, train EVA astronauts, and refine EVA operations. [Figure B] The ability to successfully and predictably perform assembly and maintenance operations in orbit is critical to the success of future space endeavors.

The history of aviation has involved not only improvements in aircraft design and equipment, but also an increased understanding of how our minds and bodies function in flight. Both endeavors are crucial to flight safety. Now, as we develop new technology to venture out and investigate the solar system, human factors come to the fore. In order to explore the universe, we must first explore ourselves.

We used to joke about canned men, putting people in a can and seeing how far you can send them and bring them back. That's not the purpose of this program . . . Space is a laboratory, and we go into it to work and learn the new.
— John H. Glenn Jr.

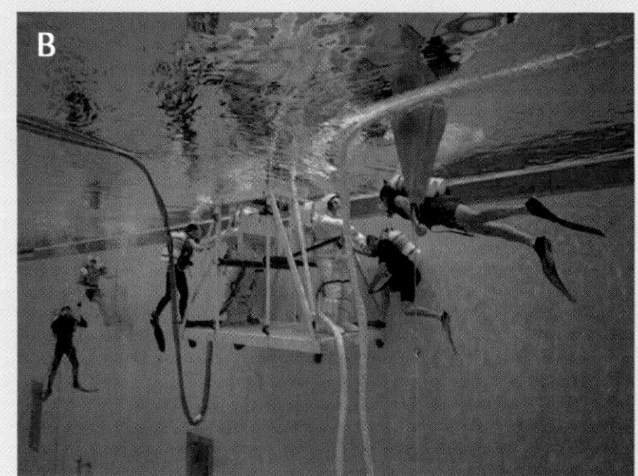

Photos and emblem courtesy of NASA

Before each flight, a run-through of the **I'm Safe Checklist** will aid you in determining your fitness for flight. In broad terms, ask yourself about any reservations you have concerning the flight. Am I ill, or taking any drugs that might affect my safety as a pilot? Have I had enough rest? Did I eat a good breakfast? Are my issues at work going to interfere with my concentration level in the airplane? If you have any reservations about your ability to make the flight, save the trip for another time. Do not let the pressures of returning home, or impressing friends, or proving your worth as a pilot disrupt your honest evaluation of your fitness to fly. None of these reasons have anything to do with your skill, but they have everything to do with good judgment. [Figure 1-40]

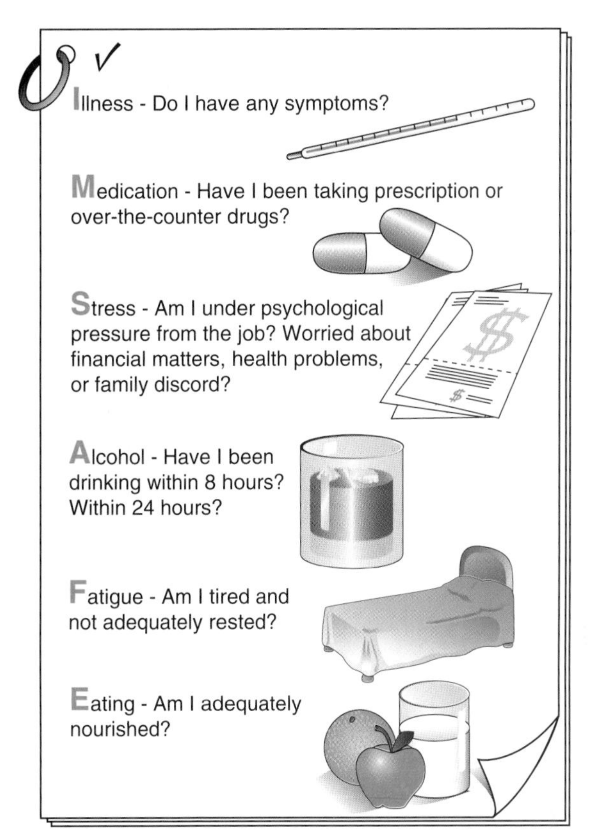

Figure 1-40. The I'm Safe Checklist is a good tool to help you evaluate your fitness for flight.

SUMMARY CHECKLIST

- ✓ Aeronautical decision making is a systematic approach to the mental process used by aircraft pilots to consistently determine the best course of action in response to a given set of circumstances.
- ✓ Approximately 75% of all aviation accidents are attributed to human factors-related causes.
- ✓ The focus of CRM programs is the effective use of all available resources: human resources, hardware, and information.
- ✓ When two pilots share the cockpit, each pilot's responsibilities must be defined before the flight.
- ✓ Studies have identified five hazardous attitudes which can interfere with a pilot's ability to make sound decisions and exercise authority properly.

- ✓ Readback of ATC clearances is crucial in the IFR environment. You should not assume controller silence after a readback is verification of your transmission.

- ✓ Your cockpit resources will increase as you fly more complex aircraft with advanced systems. If you are not thoroughly familiar with the equipment in your aircraft or you rely on it so much that you become complacent, flight safety is compromised.

- ✓ Since your duties in the cockpit are not evenly distributed over time, you must use prioritizing and planning to effectively manage your workload.

- ✓ Maintaining situational awareness is perhaps the largest challenge to an instrument pilot, and requires you to have a solid mental picture of the flight, from your own fitness and that of passengers to the operating conditions of the airplane, weather trends, and ATC instructions.

- ✓ Controlled flight into terrain (CFIT) occurs when an aircraft is flown into terrain or water with no prior awareness on the part of the crew that the crash is imminent.

- ✓ When there is a conflict between the information relayed by your central vision and your peripheral vision, you may suffer from spatial disorientation.

- ✓ When subjected to the various forces of flight, the vestibular system can send misleading signals to the brain resulting in vestibular disorientation.

- ✓ Hypoxia occurs when the tissues in the body do not receive enough oxygen. Hypoxia can be caused by several factors including an insufficient supply of oxygen, inadequate transportation of oxygen, or the inability of the body tissues to use oxygen.

- ✓ Hypoxic hypoxia occurs when there are not enough molecules of oxygen available at sufficient pressure to pass between the membranes in your respiratory system.

- ✓ Hypemic hypoxia occurs when your blood is not able to carry a sufficient amount of oxygen to your body's cells.

- ✓ Since it attaches itself to the hemoglobin about 200 times more easily than does oxygen, carbon monoxide (CO) prevents the blood from carrying sufficient oxygen.

- ✓ Stagnant hypoxia is an oxygen deficiency in the body due to the poor circulation of the blood. During flight, stagnant hypoxia can be the result of pulling excessive positive Gs.

- ✓ The inability of the cells to effectively use oxygen is defined as histotoxic hypoxia. This impairment of cellular respiration can be caused by alcohol and other drugs such as narcotics and poisons.

- ✓ If you are planning a flight with a cruise altitude over 12,500 feet MSL, you should review FAR Part 91 for the requirements regarding supplemental oxygen.

- ✓ Prior to operating a pressurized aircraft with a service ceiling or maximum operating altitude higher than 25,000 feet MSL, you must complete high-altitude training.

- ✓ Hyperventilation is a physiological disorder that develops when too much carbon dioxide (CO_2) has been eliminated from the body, usually caused by breathing too rapidly or too deeply.

- ✓ Decompression sickness (DCS) is a condition caused by a rapid reduction in the ambient pressure surrounding the body causing nitrogen and other inert gases which are normally dissolved in body tissue and fluid to expand and form bubbles. These bubbles produce a variety of symptoms that range from pain in the large joints of the body to seizures and unconsciousness.
- ✓ If you or a passenger plan to fly after scuba diving, it is important that enough time is allowed for the body to rid itself of excess nitrogen absorbed during diving.
- ✓ Stress is the body's reaction to the physical and psychological demands placed upon it, and it can adversely affect your ability to fly safely.
- ✓ When you are fatigued, you are more prone to error in the cockpit. No amount of training will allow you to overcome the effects of fatigue. Getting adequate rest on a regular basis is the only way to perform at your best.
- ✓ Prior to flying you should consult an aviation medical examiner about any medication you are using.
- ✓ Improving your overall fitness can have a positive effect on your performance as a pilot.
- ✓ Preflight use of the I'm Safe Checklist will help ensure you are fit for flight.

KEY TERMS

Aeronautical Decision Making (ADM)

Crew Resource Management (CRM)

Pilot-In-Command Responsibility

Hazardous Attitudes

Communication

Resource Use

Workload Management

Situational Awareness

Controlled Flight Into Terrain (CFIT)

Kinesthetic Sense

Spatial Disorientation

Vestibular Disorientation

Motion Sickness

Hypoxia

Hypoxic Hypoxia

Time of Useful Consciousness

Hypemic Hypoxia

Stagnant Hypoxia

Histotoxic Hypoxia

Supplemental Oxygen

High-Altitude Training

Decompression Sickness (DCS)

Hyperventilation

Stress

Fatigue

I'm Safe Checklist

CHAPTER 2

PRINCIPLES OF INSTRUMENT FLIGHT

Instrument/Commercial
Part I, Chapter 2—Principles of Instrument Flight

SECTION A
FLIGHT INSTRUMENT SYSTEMS

To fly under Instrument Flight Rules (IFR), you need to control an aircraft while reading charts, tuning radios, and performing a variety of other complex tasks. During your training, you will study and practice attitude instrument flying until it becomes second nature. One essential skill for successful instrument flying is instrument interpretation. You need a good working knowledge of each of the instruments before you can interpret them consistently and accurately.

The instruments which provide information about the airplane's attitude, direction, altitude, and speed are collectively referred to as the flight instruments. These instruments are categorized according to their method of operation. Gyroscopic instruments, a convenience for VFR flight, are an absolute necessity for IFR flight. The pitot-static instruments also are essential when flying IFR. In the IFR environment, proper instrument interpretation is the basis for aircraft control. Knowing these instruments and systems will help you determine quickly what the instruments are telling you and translate it into an appropriate control response. This will enhance your safety under IFR.

The FARs require certain instruments for IFR flight. [Figure 2-1] The altimeter and static system, as well as the transponder, must have been inspected within the preceding 24 calendar months. It is your responsibility as pilot in command to determine that each system has been checked and found to meet FAR requirements for instrument flight.

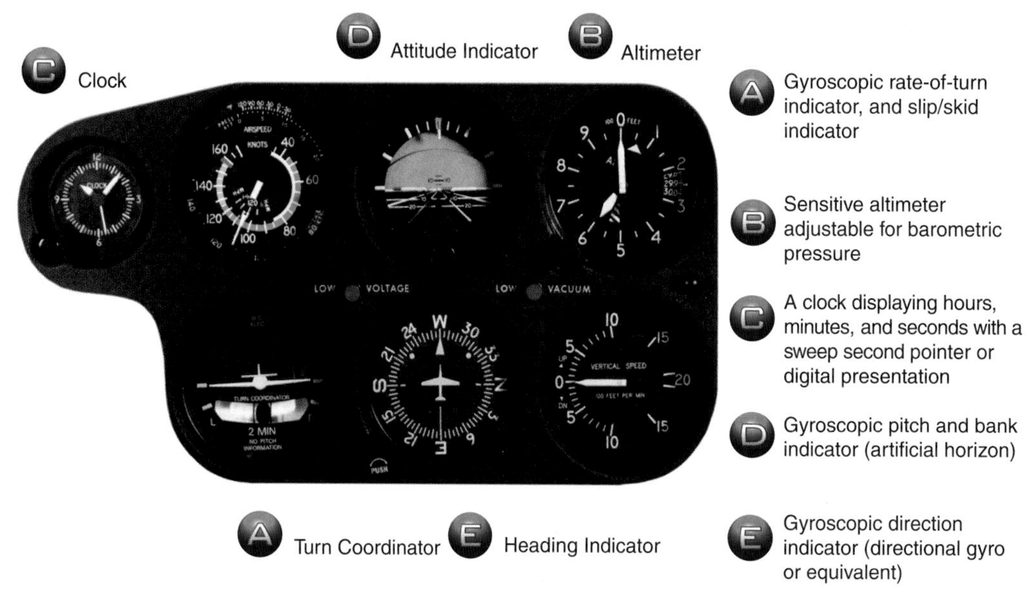

Figure 2-1. In addition to the instruments required for VFR flight, these instruments are required for flight under IFR.

GYROSCOPIC FLIGHT INSTRUMENTS

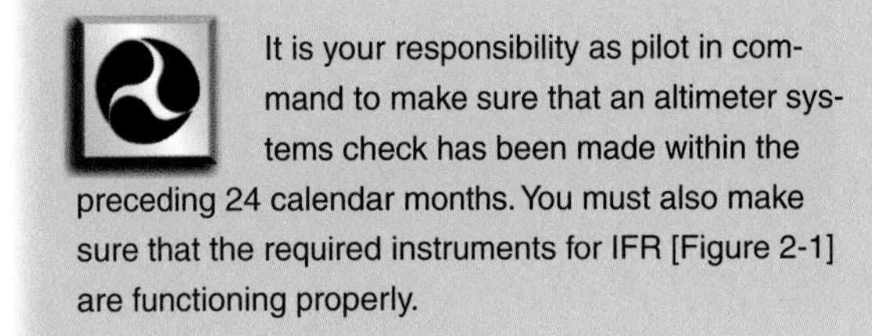

It is your responsibility as pilot in command to make sure that an altimeter systems check has been made within the preceding 24 calendar months. You must also make sure that the required instruments for IFR [Figure 2-1] are functioning properly.

The three gyroscopic instruments in your aircraft are the attitude indicator, heading indicator, and turn coordinator. On most small airplanes, the vacuum system powers the attitude and heading indicators [Figure 2-2], while the electrical system powers the turn coordinator. Gyroscopic instrument operation is based on two fundamental concepts that apply to gyroscopes — rigidity in space and precession.

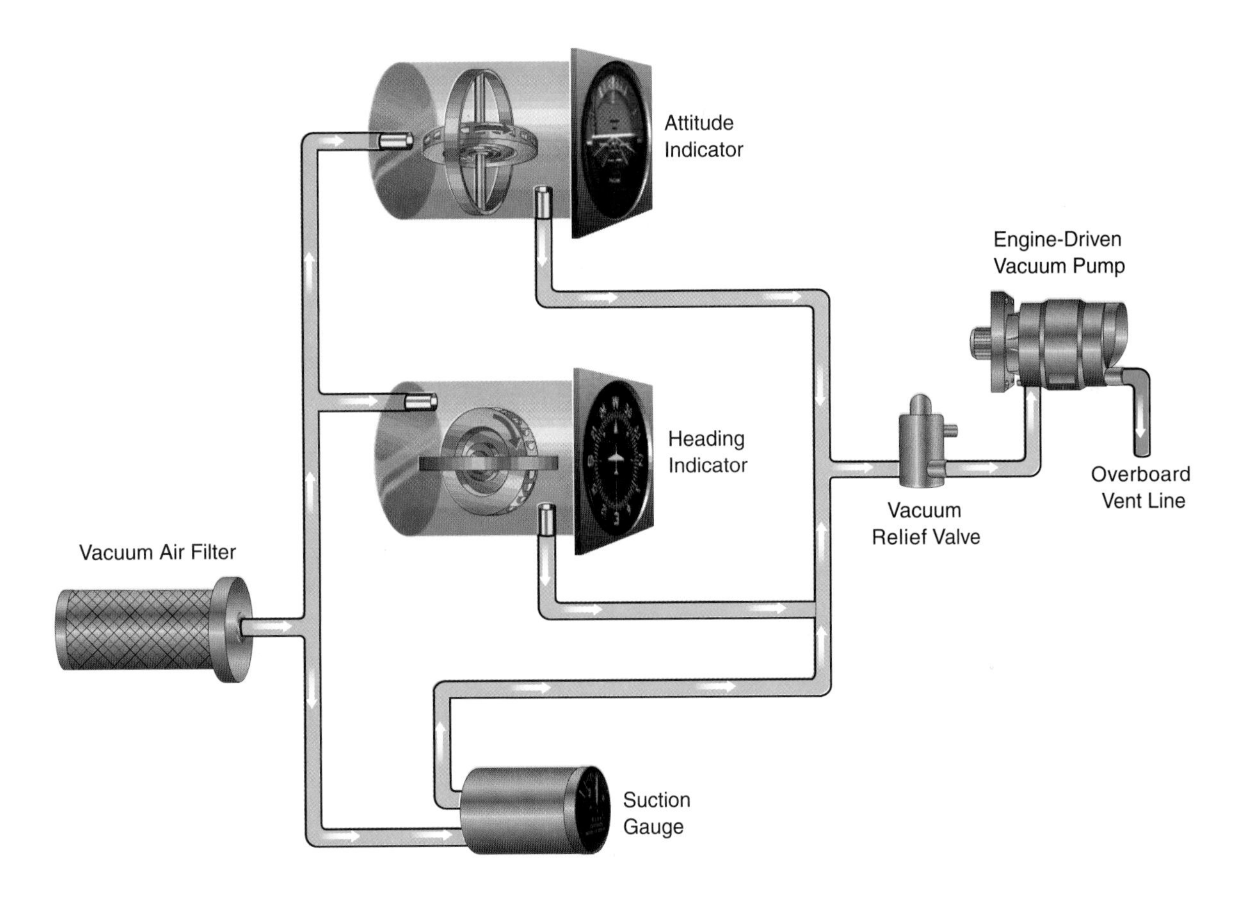

Figure 2-2. The vacuum system draws air in through a filter assembly. The air then moves through turbines in the attitude and heading indicators where it causes the gyros to spin at up to 18,000 r.p.m. The airflow continues on to the engine-driven vacuum pump where it is expelled. A relief valve prevents the vacuum pressure from exceeding prescribed limits.

One advantage of an electric turn coordinator is that it serves as a backup in case of vacuum system failure.

A gyro depends upon the resistance to deflection of its internal, spinning disc for proper operation.

RIGIDITY IN SPACE

Rigidity in space refers to the principle that a wheel with a heavily weighted rim spun rapidly tends to remain fixed in the plane in which it is spinning. By mounting this wheel, or gyroscope, on a set of **gimbals**, the gyro is able to rotate freely in any plane. If the gimbals' base tilts, twists, or otherwise moves, the gyro remains in the plane in which it was originally spinning. [Figure 2-3] This principle allows a gyroscope to be used to measure changes in the attitude or direction of an airplane.

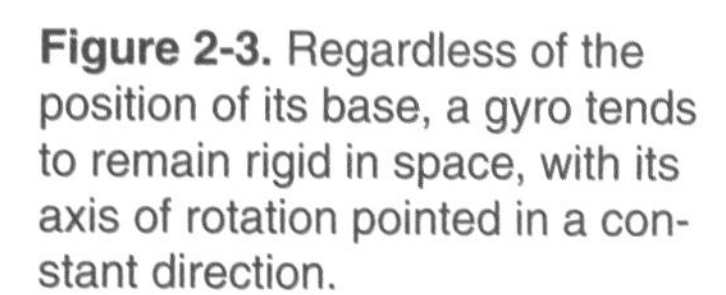

Figure 2-3. Regardless of the position of its base, a gyro tends to remain rigid in space, with its axis of rotation pointed in a constant direction.

Seat of the Pants?

Today we take gyroscopic flight instruments for granted. It was not always apparent that the unique properties of spinning gyros offered the solution to the problem of controlling an aircraft in instrument meteorological conditions.

Some early aviators scoffed at the notion of using delicate flight instruments, preferring to rely on their senses. Their "instruments" consisted of weighted strings hanging from windscreens and silk stockings tied to wing struts. The wiser aviators quickly learned these tools were inadequate when flying in the mist, where even the most talented seat-of-the-pants flyers could be in a spin or inverted without even knowing it. An important invention was needed before safe flight in the clouds was possible.

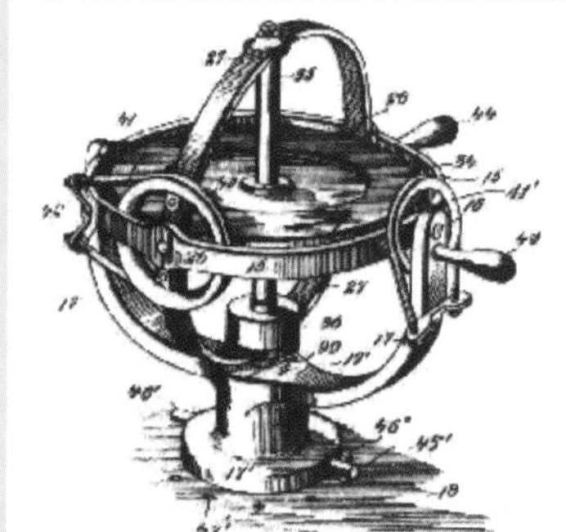

Courtesy of Sperry Marine, Inc.

Elmer A. Sperry founded the Sperry Gyroscope Company in 1910, after playing with a child's spinning top and realizing the potential for navigation. Sperry first invented a stabilizer system for ships and then determined gyroscopic principles could be applied to instruments of flight. He invented the bank and turn indicator in 1918, followed by the gyrocompass and the artificial horizon. These three basic gyro instruments are still in use today.

PRECESSION

When an outside force tries to tilt a spinning gyro, the gyro responds as if the force had been applied at a point 90° further around in the direction of rotation. This effect is called **precession**, because the cause precedes the effect by 90°. [Figure 2-4] Unwanted precession is caused by friction in the gimbals and bearings of the instrument, causing a slow drifting in the heading indicator and occasional small errors in the attitude indicator.

Figure 2-4. When a force (including friction) acts to tilt a spinning gyro, the effect of that force is felt in the direction of rotation 90° from where the force is applied.

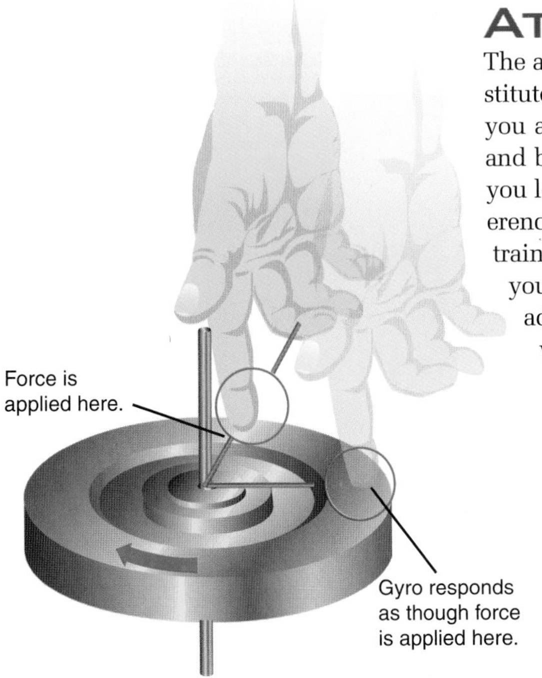

ATTITUDE INDICATOR

The attitude indicator, or artificial horizon, is a mechanical substitute for the natural horizon. It is the only instrument that gives you an immediate and direct indication of the airplane's pitch and bank attitude. During your integrated private pilot training you learned to use the attitude indicator with outside visual references to control the aircraft precisely. During your instrument training, the attitude indicator will become the central part of your scan; you will use the attitude indicator to make precise adjustments to the aircraft's pitch and bank without outside visual reference.

HOW IT WORKS

The heart of the attitude indicator is a gyro that spins in the horizontal plane, mounted on dual gimbals that allow it to remain in that plane regardless of aircraft movement. Before the gyro can spin in the horizontal plane, it must erect itself. While the aircraft is taxiing, gravity provides the force to level the gyro. On a vacuum-driven attitude indicator, this is accomplished through the action of **pendulous vanes**. [Figure 2-5]

Figure 2-5. Air exits the gyro assembly through four ports at right angles to each other located near the base of the assembly. These ports can be individually blocked and opened by pendulous vanes which swing in front of them.

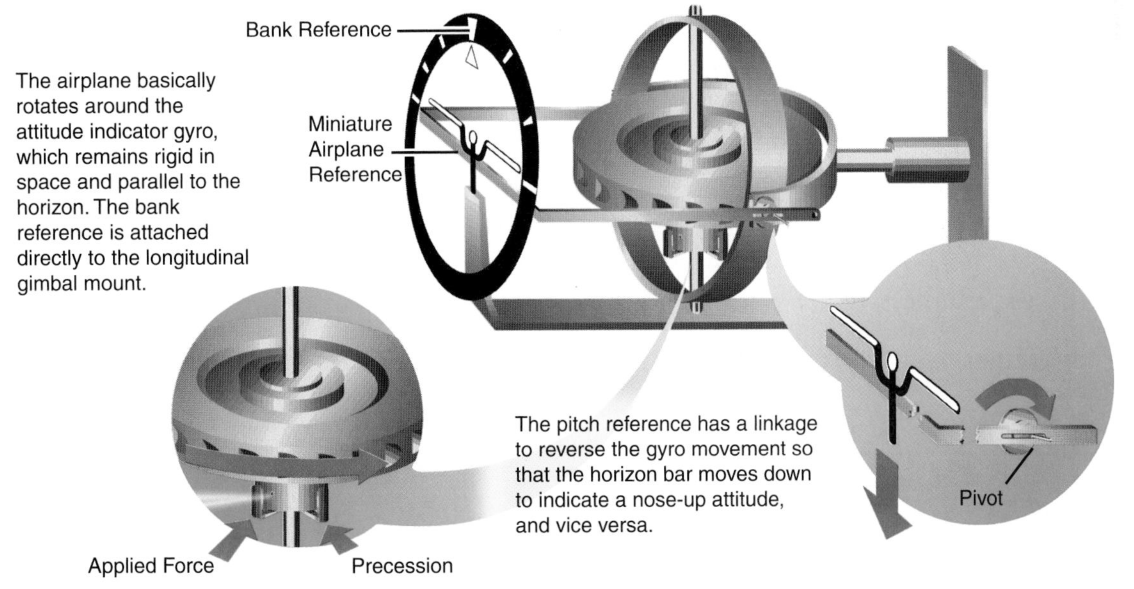

If the gyro tilts in such a way that it is not parallel to the ground or horizon, gravity swings some of the pendulous vanes open and others closed. The differential thrust from the air exiting the ports acts to tilt the gyro back to level. Because of precession, the effect of the differential thrust is felt 90° in the direction of gyro rotation.

ERRORS

Errors in modern attitude indicators are usually very minor, resulting in less than 5 ° of bank error and 1 bar width of pitch error in a 180° turn. These errors occur because the pendulous vanes act on the attitude indicator's gyro in an undesirable way during turns. Since the gyro is parallel to the horizon, and the G-force inside the airplane is toward the bottom of the airplane (not directly down the gyro's axis of rotation), the vanes open asymmetrically and cause the gyro to precess. The same action which erects the gyro during taxi tries to line up the gyro with the G-force from the turn. Errors in both pitch and bank indication are usually at a maximum as the aircraft rolls out of a 180° turn, and cancel after 360° of turn. [Figure 2-6]

Errors in both pitch and bank indication are usually at a maximum as the aircraft rolls out of a 180° turn. Other errors occur during acceleration and deceleration. See Figure 2-6.

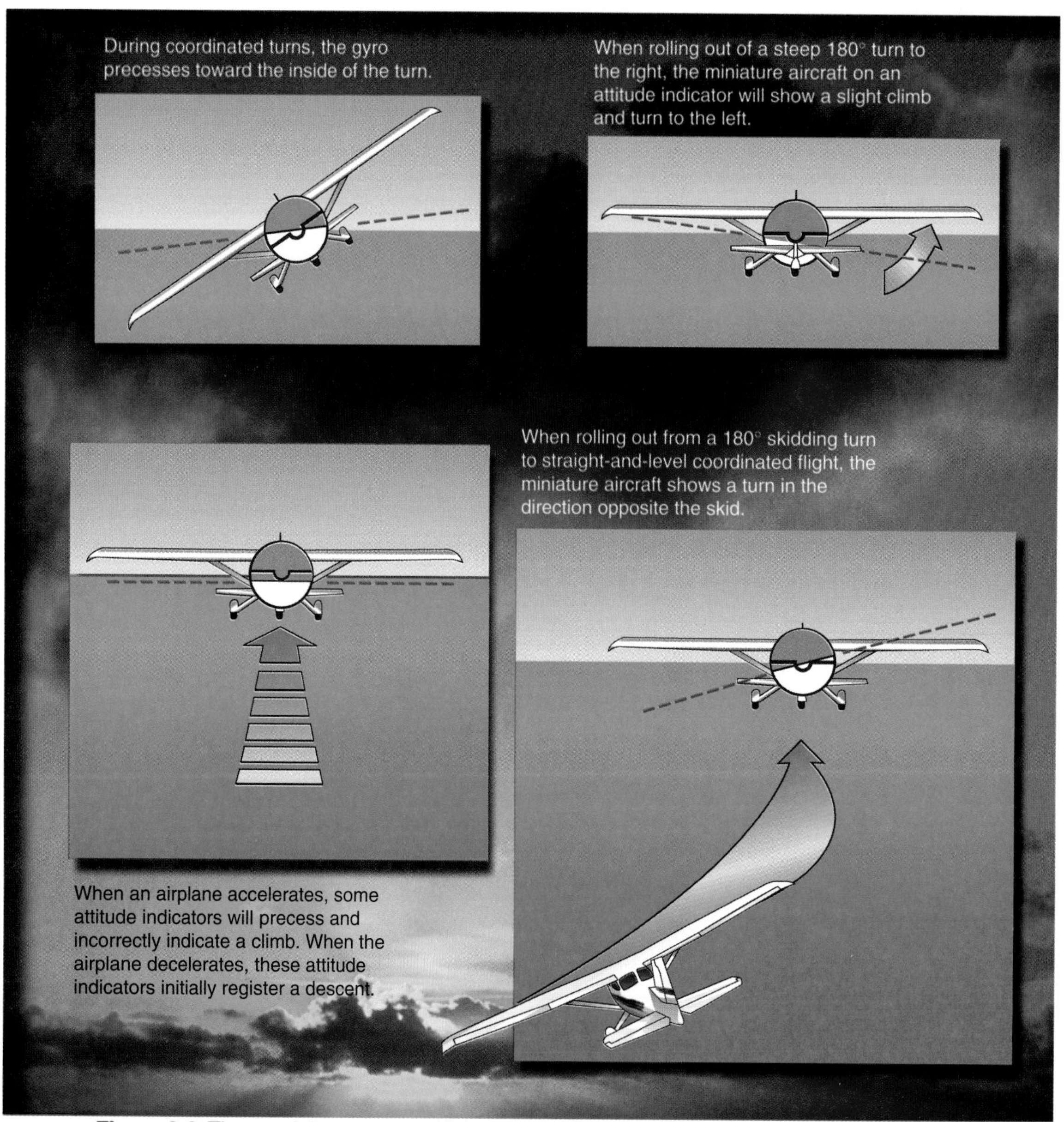

Figure 2-6. The pendulous vanes, which work to erect the gyro in a vacuum-driven attitude indicator, are adversely influenced by centrifugal force during turns.

Acceleration and deceleration also may induce precession errors, depending on the amount and duration of the force applied. During acceleration, the horizon bar moves down, indicating a climb, which, unfortunately reinforces the same illusion a pilot can experience during acceleration. That illusion, called somatogravic illusion (see Chapter 1) falsely makes a pilot feel the aircraft is in a nose-high attitude. If you apply control pressure to correct this indication, it will result in a lower pitch attitude than the instrument shows. The danger is that your angle of climb when accelerating for a missed approach may be too shallow to clear obstacles.

INSTRUMENT TUMBLING

As long as sufficient vacuum is maintained, modern attitude indicators usually are very reliable instruments. Some are designed to function properly during 360° of roll or 85° of pitch. When the gimbals in older indicators hit their limits, the gyros precess rapidly, or tumble, from their plane of rotation. This generally occurs beyond approximately 100° of bank or beyond 60° of pitch. A tumbled instrument is unusable and may take several minutes to re-erect itself. Some of these instruments employ caging devices to prevent the gyro from tumbling or to stabilize the spin axis after it has tumbled.

HEADING INDICATOR

A gyroscopic heading indicator is required for IFR flight. When properly set, this instrument is your primary source of heading information. Because changes in heading during coordinated flight imply the wings are not level, the heading indicator also indirectly indicates bank.

HOW IT WORKS

The heading indicator usually is vacuum powered and senses rotation about the aircraft's vertical axis. In most training airplanes, the heading indicators contain free (as opposed to slaved) gyros. This means they have no automatic, north-seeking system built into them. For the heading indicator to display the correct heading, you must align it with the magnetic compass before flight and recheck it periodically during flight. [Figure 2-7]

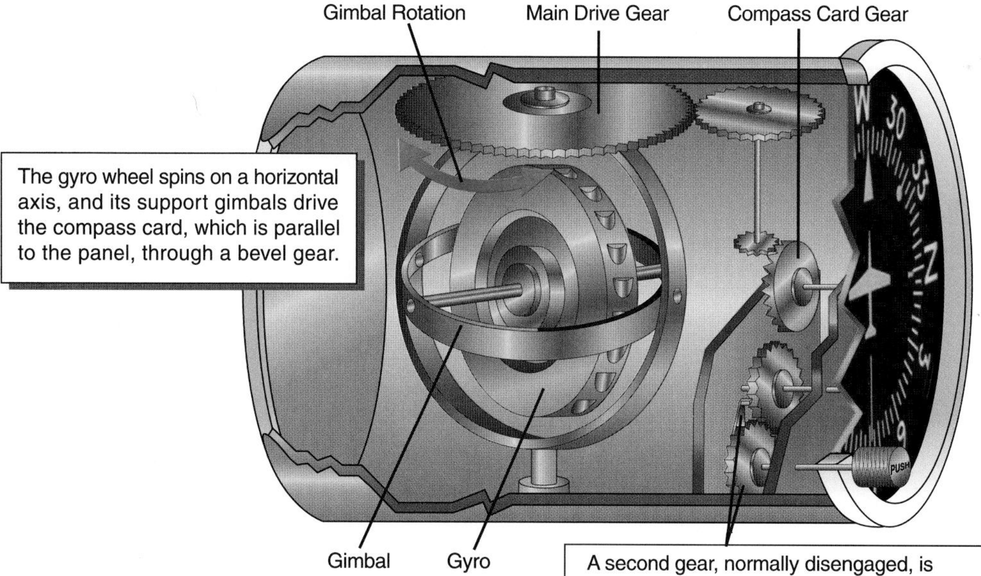

Figure 2-7. The heading indicator gyro spins in a vertical plane. Rigidity in space keeps the gyro pointing in the same direction as the aircraft turns about its vertical axis.

ERRORS

If the airplane were never pitched or banked, the heading indicator's gyro could turn freely within a single gimbal with negligible error. Because an airplane does more than yaw, an additional gimbal is needed to allow free rotation of the gyro. Precession can cause the heading to drift from the proper setting, so you must check the heading indicator against the magnetic compass at approximately 15-minute intervals during flight. When you reset the heading indicator, make sure you are in straight-and-level, unaccelerated flight to ensure an accurate magnetic compass indication. Like the attitude indicator, the heading indicator may tumble during excessive pitch and roll conditions. If the indicator has tumbled, you must realign it with a known magnetic heading or with a stabilized indication from the magnetic compass.

TURN INDICATORS

The turn indicator allows you to establish and maintain constant rate turns. A **standard-rate turn** is a turn at a rate of 3° per second. At this rate you complete a 360° turn in 2 minutes. The bank required to maintain a specific rate of turn increases with true airspeed (TAS). You can calculate the approximate required bank for a standard-rate turn in light training aircraft using the following formula:

Angle of Bank = [True Airspeed in Knots ÷ 10] + 5

At 100 knots, you must bank the airplane approximately 15° to make a standard-rate turn. To avoid the need for excessive angles of bank, the turn indicators on high-speed airplanes are calibrated for half-standard-rate turns. [Figure 2-8]

A standard-rate turn is 3° per second. It takes 60 seconds to turn 180°. A half-standard-rate turn is 1-1/2° per second. It takes 4 minutes to turn 360°.

Figure 2-8. Turbine aircraft often use 4 minute turn indicators to avoid the steeper banks associated with higher speeds. Your instrument training aircraft probably has a 2 minute turn coordinator.

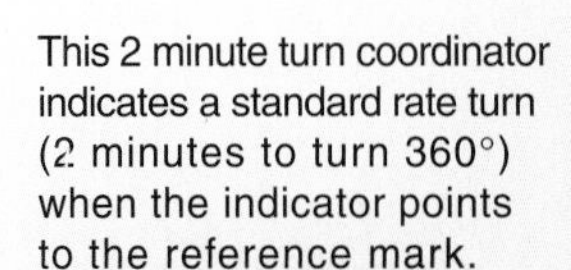

This 4 minute turn-and-slip indicator also depicts a standard rate turn (2 minutes to turn 360°) when the needle points to the reference mark.

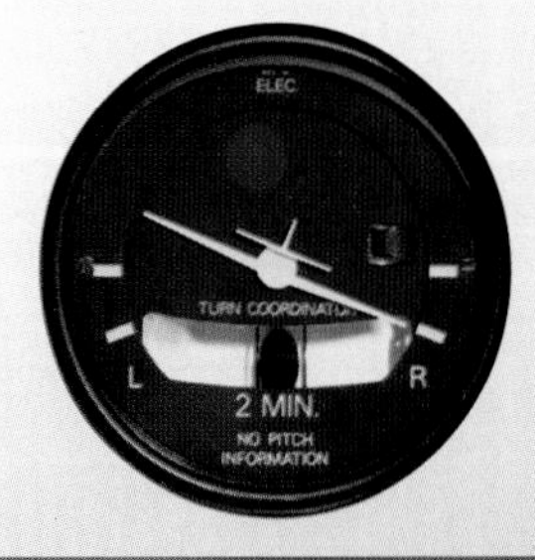

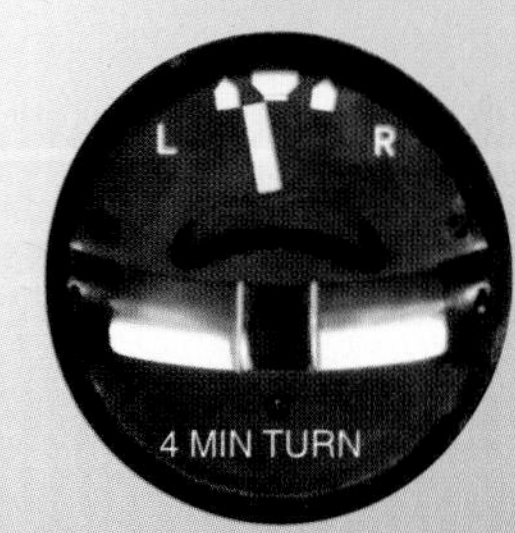

Unlike the 2 minute turn indicator, this instrument is calibrated so a one-needle-width deflection precisely indicates a half-standard-rate turn (4 minutes to turn 360°).

During a constant-bank level turn, an increase in airspeed results in a decreased rate of turn, and an increased turn radius.

There are two types of turn indicators; the older **turn-and-slip indicator** and the **turn coordinator**. These instruments have different appearances and work a little differently. Both instruments indicate rate of turn, but because of the improved design of the turn coordinator, this instrument also indicates rate of roll as you enter a turn. [Figure 2-9] Since a coordinated turn requires the aircraft to be banked, both the turn-and-slip indicator and the turn coordinator give you an indirect indication of bank. If other bank instruments fail, it is easier to control the aircraft with the turn coordinator because of the additional information it provides. Because the turn coordinator has largely replaced the turn-and-slip indicator in modern training aircraft, this book will mostly refer to that instrument.

The **inclinometer** is the part of the turn coordinator that contains the fluid and the ball. The position of the ball indicates whether you are using the correct angle of bank for the rate of turn. In a **slip**, the rate of turn is too slow for the angle of bank, and the ball moves

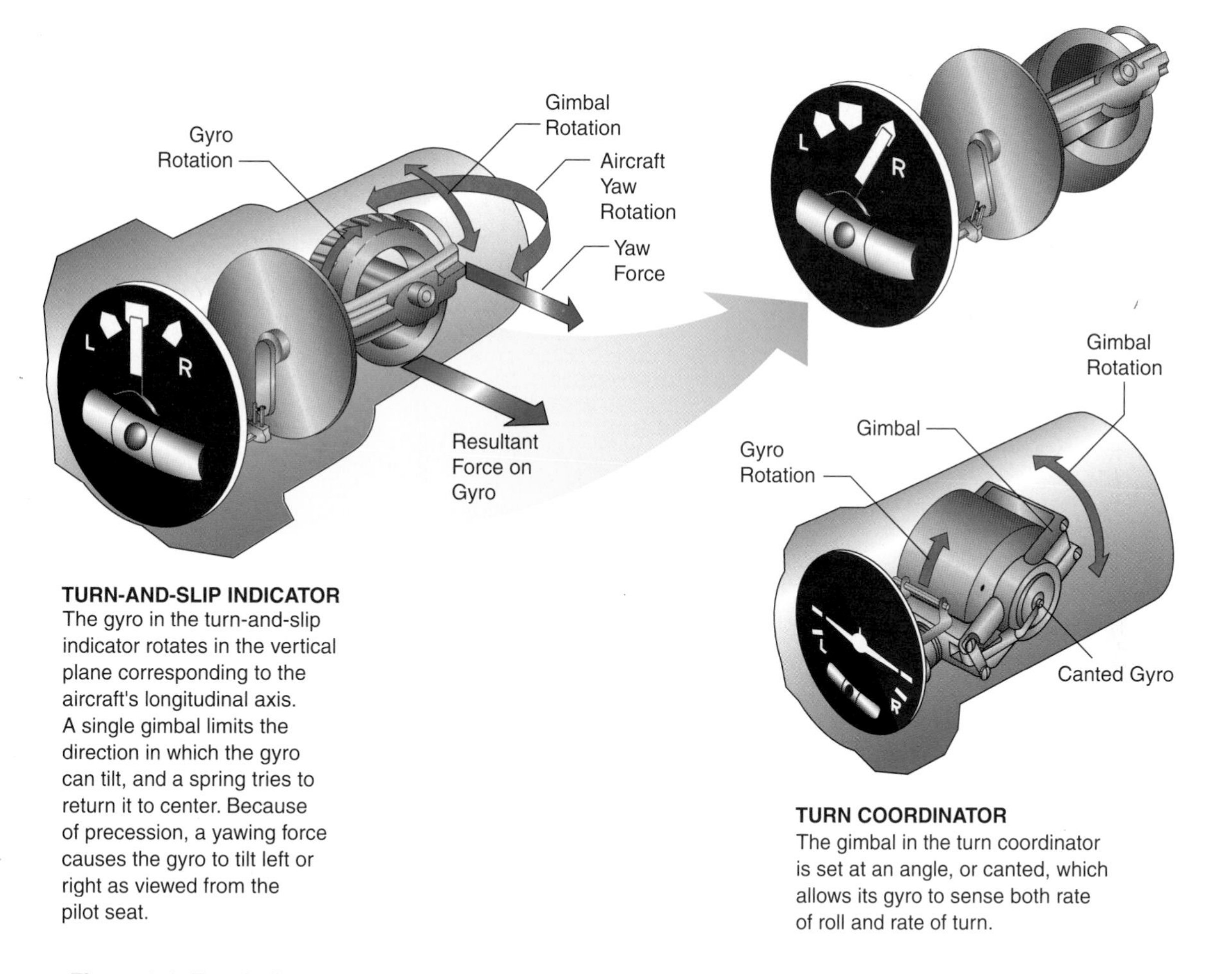

Figure 2-9. Turn indicators rely on controlled precession for their operation.

The miniature aircraft of the turn coordinator directly displays rate of roll and rate of turn information; the turn-and-slip indicator only gives the rate of turn. Both instruments indirectly indicate the bank attitude; the needle displacement increases as angle of bank increases.

to the inside of the turn. In a **skid**, the rate of turn is too great for the angle of bank, and the ball moves to the outside of the turn. Step on the ball, or apply rudder pressure on the side the ball is deflected, to correct an uncoordinated flight condition. [Figure 2-10]

Figure 2-10. The inclinometer helps you coordinate a turn by measuring the balance between centrifugal force (CF) and horizontal component of lift (HCL).

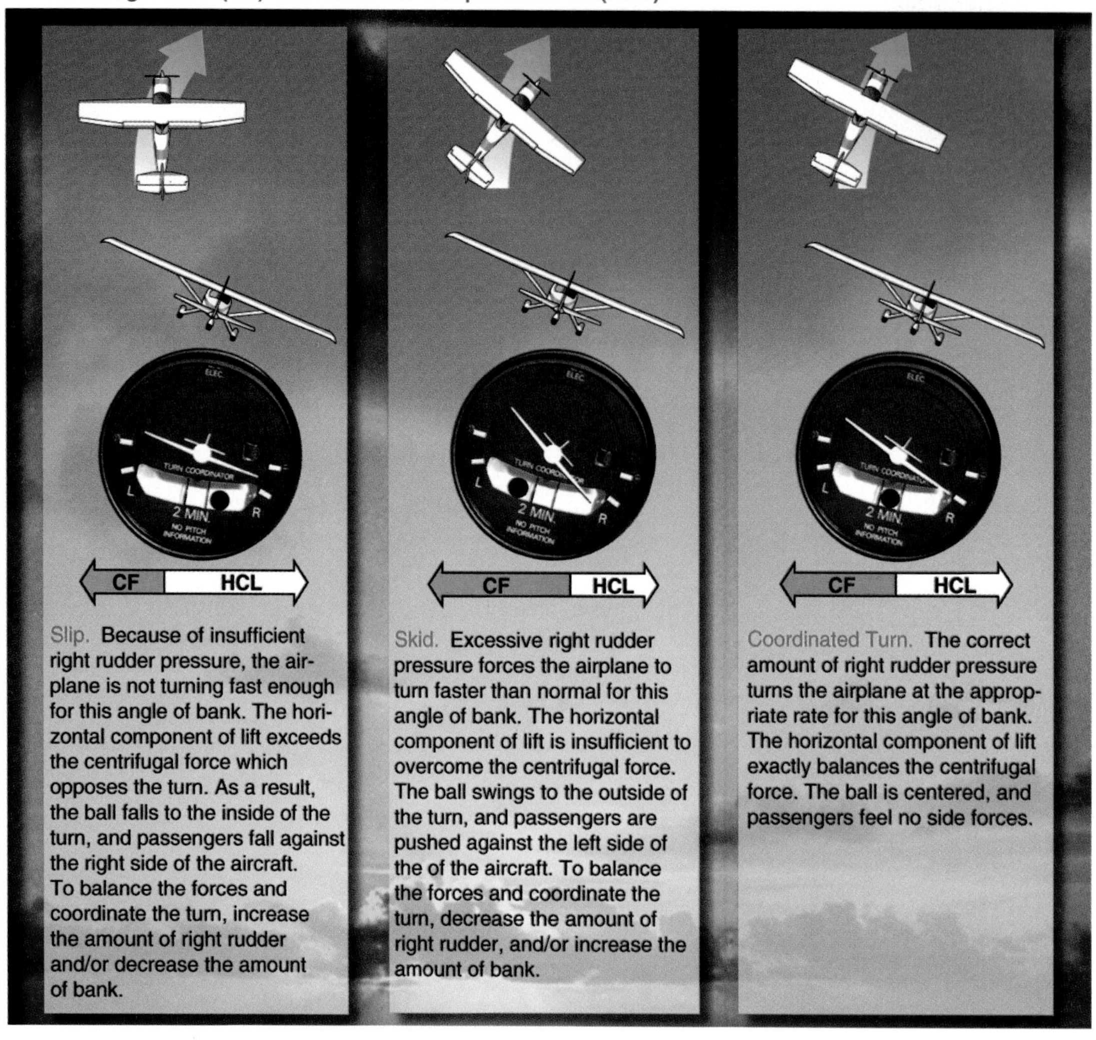

Slipping or skidding also alters the normal load factor you experience in turns. This happens because the wings must generate enough lift to support the weight of the airplane plus overcome centrifugal force. Since a skid generates a higher-than-normal centrifugal force, load factor is increased. In a slip, load factor decreases because centrifugal force is lower than normal.

The ball of the turn coordinator indicates the quality of the turn. The horizontal lift component causes an airplane to turn. In a skidding turn, the load factor is increased because of the excess centrifugal force. See Figure 2-10.

INSTRUMENT CHECKS

The preflight check of the gyroscopic flight instruments and their power sources is particularly important if departing under IFR. Some instruments display warning flags when they lose their source of vacuum or electric power. Before turning on the master switch or starting the engine, make sure the instruments that have these warning flags are displaying OFF indications. [Figure 2-11]

Prior to engine start, check the turn-and-slip indicator to determine if the needle is approximately centered and the tube is full of fluid. During taxi, the ball should move to the outside of the turn, and the needle should deflect in the direction of the turn.

Figure 2-11. Before turning on the master switch, verify the operation of all instrument failure indicators.

The inclinometer should be full of fluid, with the ball resting at its lowest point. When you turn on the master switch, listen to the electrically driven gyro(s). There should be no abnormal noises, such as grinding sounds, that would indicate an impending failure.

Before you start the engine, listen for any unusual mechanical noise. Noise might indicate a problem in the electric gyro instruments.

Check the ammeter immediately after starting the engine for a positive charging rate, or if your aircraft is equipped with a load meter, make sure it indicates a normal value. Listen for the vacuum-driven gyros. If they are malfunctioning, you may hear them over the noise of the engine. If you suspect something abnormal, shut down the engine and listen to the gyros spin down. The gyros should reach full operating speed in approximately five minutes. Until that time, it is common to see some vibration in the instruments. When the gyros stabilize, the miniature airplane in the turn coordinator and the horizon bar in the attitude indicator should be level while the airplane is stopped or taxiing straight ahead.

During turns, the turn coordinator and heading indicator should display a turn in the correct direction. The ball in the inclinometer should swing to the outside of the turn; since you do not bank the airplane on the ground, taxi turns are basically skids. Align the heading indicator with the magnetic compass. Then, recheck it prior to takeoff to ensure it has not precessed significantly. A precession error of 3° or less in 15 minutes is acceptable for normal operations.

Give the vacuum-driven heading indicator and attitude indicator 5 minutes to spin up. Make sure that the horizon bar on the attitude indicator tilts no more than 5° during taxi turns. After setting the heading indicator to the magnetic heading, verify that it maintains proper alignment with the magnetic compass during taxi turns.

When making a left taxiing turn, the miniature aircraft on the turn coordinator shows a turn to the left and the ball moves to the right.

You should include the ammeter and suction gauge in your pretakeoff check to ensure the gyro instruments are receiving adequate power. The ammeter should not show a discharge during runup, even with lights and pitot heat turned on. If the vacuum is outside its normal range, the vacuum-driven indicators become unreliable. Some airplanes have vacuum warning lights, as well as low and high voltage warning lights. [Figure 2-12]

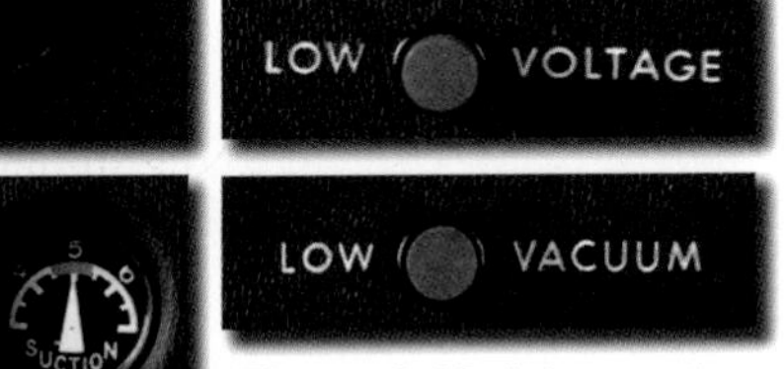

Figure 2-12. It is very important to verify the proper operation of your electrical and vacuum systems before departing under IFR.

MAGNETIC COMPASS

The magnetic compass is the only direction-seeking instrument in most light airplanes. It is a self-contained unit not requiring electrical or suction power. To determine direction, the compass uses simple bar magnets suspended in a fluid so it can pivot freely and align itself with the earth's magnetic field.

ERRORS

Because the magnetic compass is sensitive to in-flight turbulence, FAA regulations require a stable, gyroscopic heading indicator for IFR flight. The heading indicator must be synchronized with the compass regularly to give accurate information, and you need to know how to effectively navigate with the compass in the event the heading indicator fails during flight. In light turbulence, you may be able to use the compass by averaging the readings. Other errors and limitations you must consider are magnetic variation, compass deviation, and magnetic dip.

VARIATION

As an IFR pilot, you do not concern yourself with variation as much as VFR pilots because all courses on IFR charts are published as magnetic. However, you do need to convert the true winds aloft direction to magnetic before factoring winds into your flight planning. Variation is the angular difference between the true and magnetic north poles. The amount of variation depends on where you are located in relation to these poles. [Figure 2-13]

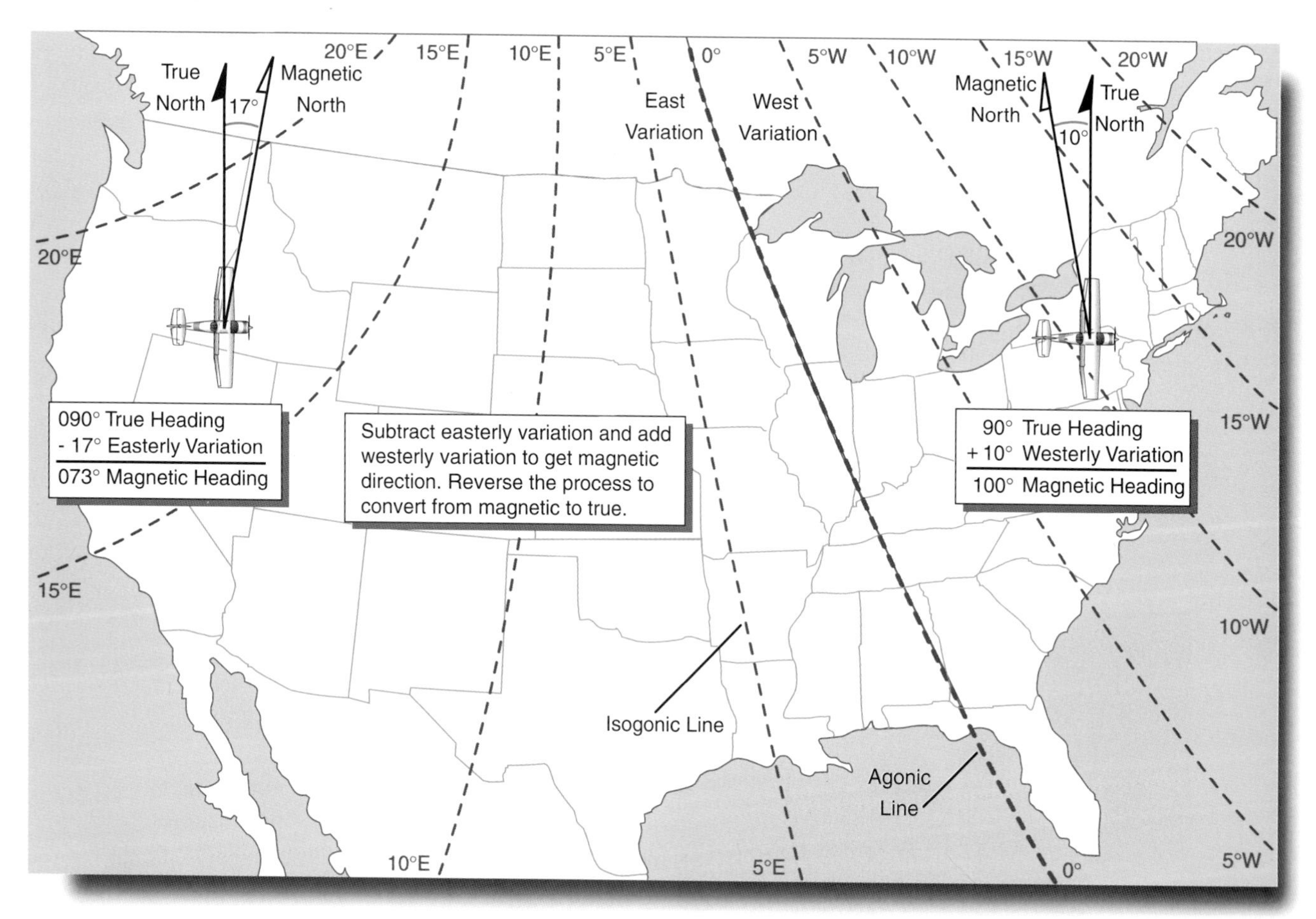

Figure 2-13. Isogonic lines connect points where variation is equal. The agonic line connects points where variation is zero.

DEVIATION

Deviation is error due to magnetic interference with metal components in the aircraft, as well as magnetic fields from aircraft electrical equipment. Compensating magnets within the compass housing can reduce, but not eliminate, deviation. These magnets are usually adjusted with the engine and all electrical equipment operating in a procedure called swinging the compass. Any remaining errors are then recorded on a compass correction card. [Figure 2-14]

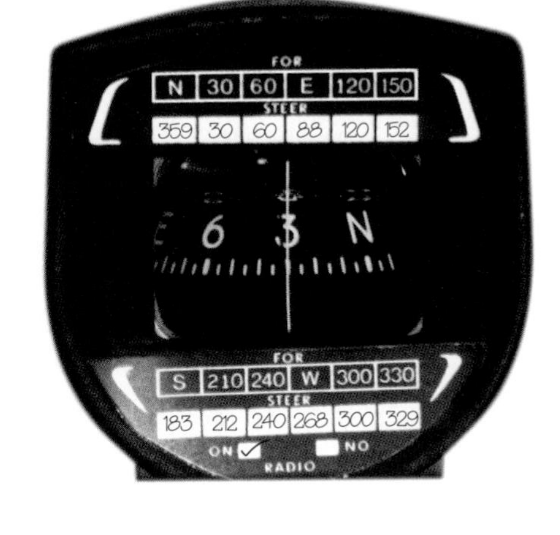

Figure 2-14. A correction card mounted on or near the compass tells you what direction to steer to get specific headings. If you use this compass to fly a magnetic heading (MH) of 180°, you must steer a compass heading (CH) of 183°.

Magnetic deviation varies for different headings of the same aircraft.

MAGNETIC DIP

Magnetic dip is responsible for the most significant compass errors. This phenomenon makes it difficult to get an accurate compass indication when maneuvering on a north or south heading. Magnetic dip exists because the magnet in the compass tries to point three dimensionally toward the earth's magnetic north pole, which is at a point deep inside the earth. [Figure 2-15] Magnetic dip is responsible for compass errors during turns and during acceleration. This section

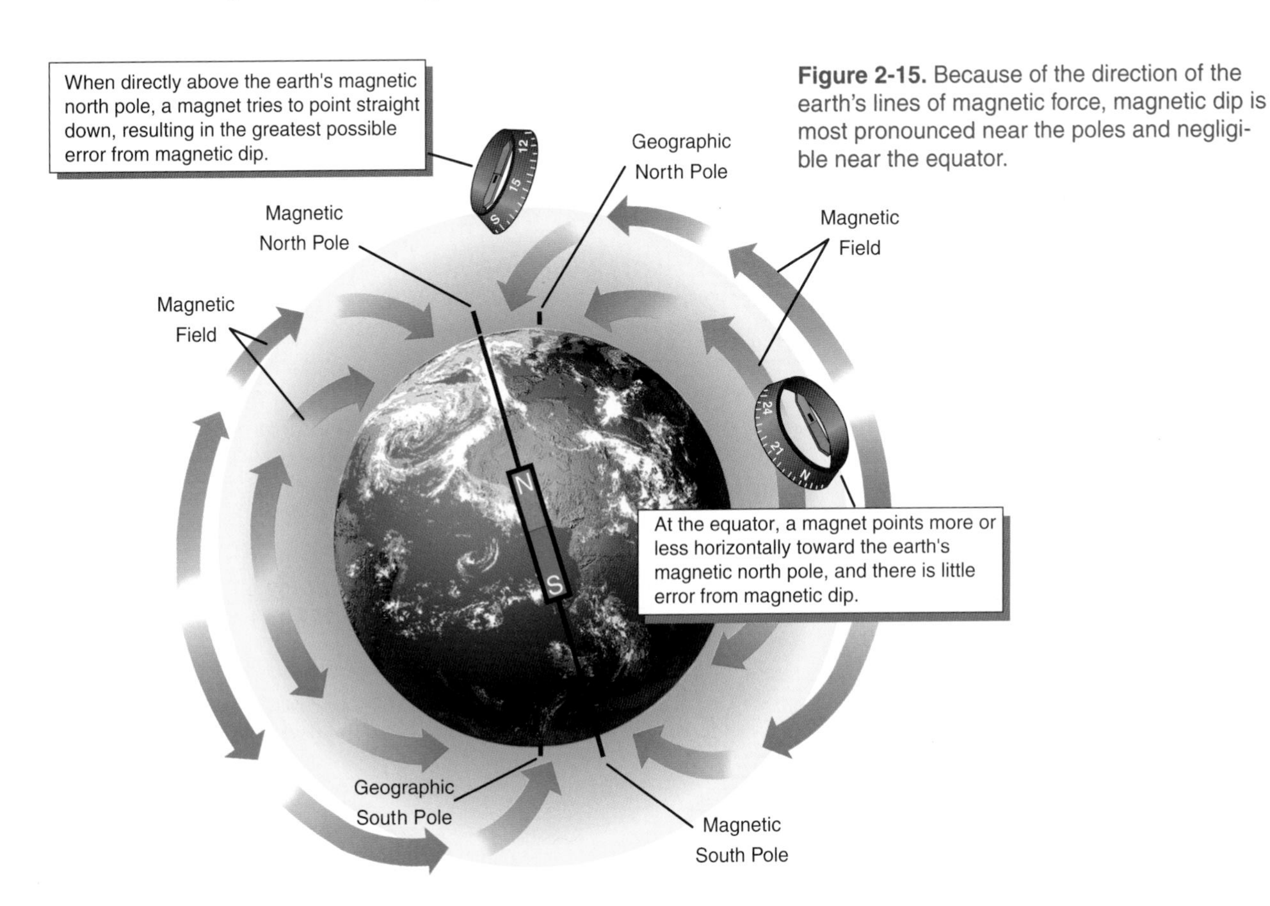

Figure 2-15. Because of the direction of the earth's lines of magnetic force, magnetic dip is most pronounced near the poles and negligible near the equator.

describes the errors that occur in the northern hemisphere due to magnetic dip; in the southern hemisphere, compass behavior is exactly opposite to what is described here.

Turning error occurs when you are turning to or from a heading of north or south. This **northerly turning error** is most apparent at the poles and disappears as you approach the Equator. When rolling into a turn from a northerly heading in the northern hemisphere, the compass swings in the opposite direction of the turn. As you proceed with the turn, the compass card reverses and moves in the correct direction, catching up with your actual heading as you reach an east or west heading. When rolling into a turn from a southerly heading, the compass card swings in the correct direction, but leads the actual heading. As you proceed with the turn, the compass card slows down, matching your actual heading as you reach east or west. [Figure 2-16]

Figure 2-16. A magnetic compass will indicate correctly entering a turn from an east or west heading, but will experience turning error when entering or completing a turn on a north or south heading. To accurately compensate for compass errors, it is essential to use standard-rate turns.

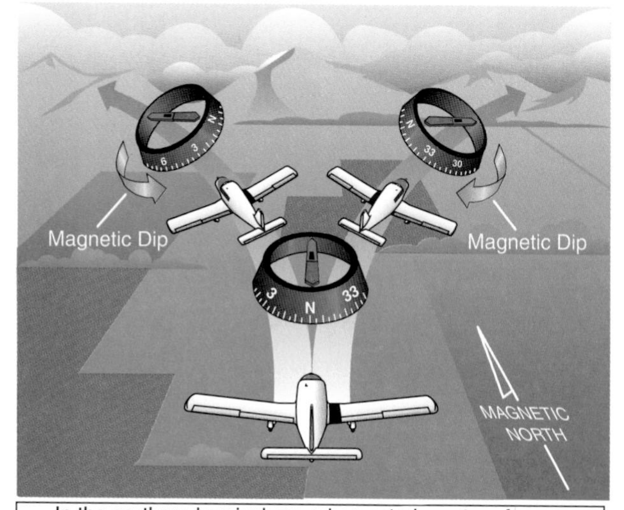

In the northern hemisphere, when entering a turn from a north heading, the compass will initially indicate a turn in the opposite direction.

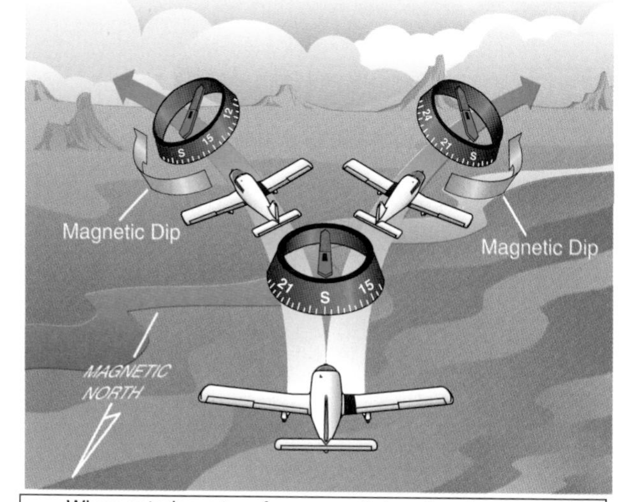

When entering a turn from a south heading, the compass will turn in the proper direction, but will lead the turn until on a heading of east or west.

It is essential you use standard-rate turns if relying on the magnetic compass. When performing a compass turn to a northerly heading, you must roll out of the turn before the compass reaches the desired heading. When turning to a southerly heading, you must delay the roll-out until the compass card swings past the desired heading. When determining whether to lag or lead the desired heading on roll-out, remember the acronym, OSUN (Overshoot South, Undershoot North). The amount of correction depends on your latitude and angle of bank. With 15° to 18° bank (a standard-rate turn in a typical piston-powered airplane), the amount of lag or lead approximately matches your latitude, plus the one-half angle of bank you lead the roll-out on any turn. For example, at 35° N latitude and a 16° bank, a right turn to north requires a roll-out point of 317° (360 – 35 – 8). A right turn to south requires a roll-out point of 207° (180 + 35 – 8). When turning left to a north heading the roll-out point is 43° (360 + 35 + 8). When turning left to a south heading it is 153° (180 – 35 + 8).

Northerly turning error in a magnetic compass is caused by magnetic dip. See Figure 2-16.

Magnetic dip also causes acceleration and deceleration errors. In the northern hemisphere, the compass swings toward the north during acceleration and toward the south during deceleration. When the speed stabilizes, the compass returns to an accurate indication. This error is greatest on east and west headings and decreases to zero on north and south headings. Remember the acronym, ANDS (Accelerate North, Decelerate South) to describe this error in the northern hemisphere. In the southern hemisphere, the error occurs in the opposite direction (Accelerate South, Decelerate North). Since rapid changes in airspeed are infrequent when flying an airplane, acceleration error is not nearly as troublesome as turning error. Nonetheless, you want to be aware of this error if using your compass for navigation when adding power to increase speed, or when reducing power to slow airspeed.

INSTRUMENT CHECK

Since a magnetic compass is required equipment for VFR or IFR flight, you should never take off without verifying its proper operation. Although the compass does have errors, it is predictable and reliable. Simply make sure the compass is full of fluid, and during taxi, verify the compass swings freely and indicates known headings.

PITOT-STATIC INSTRUMENTS

The pitot-static instruments (airspeed indicator, altimeter, and vertical speed indicator) rely on air pressure differences to measure speed and altitude. Pitot pressure, also called impact, ram, or dynamic pressure, is connected only to the airspeed indicator, while static pressure, or ambient pressure, is connected to all three instruments. [Figure 2-17]

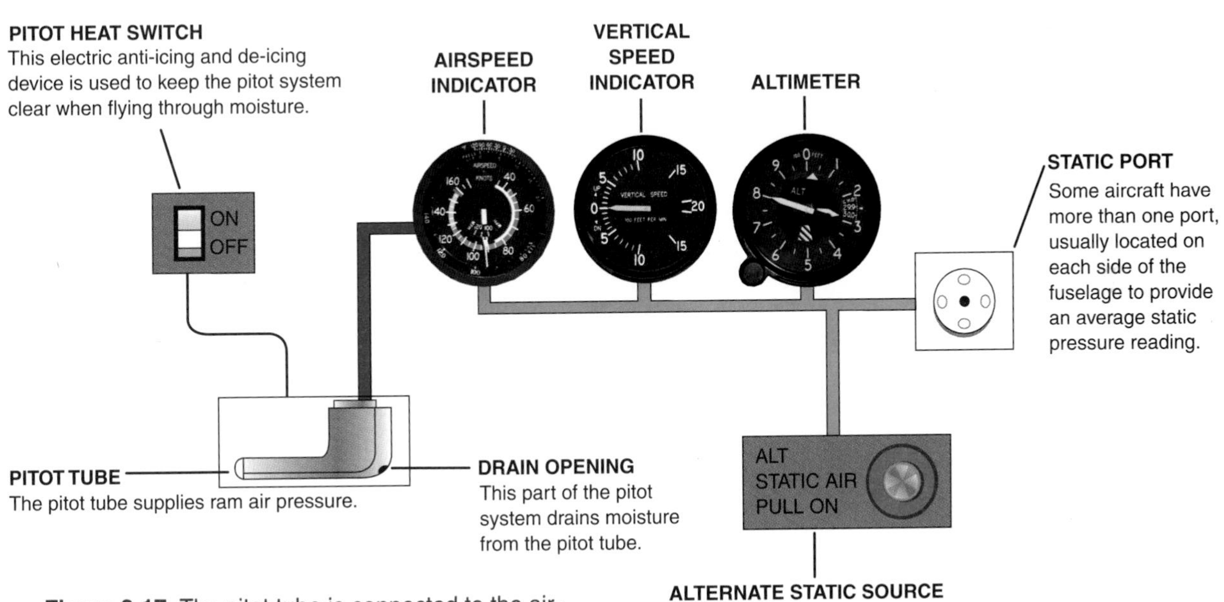

Figure 2-17. The pitot tube is connected to the airspeed indicator only, while the static port is connected to all three pitot-static instruments.

AIRSPEED INDICATOR

The airspeed indicator displays the speed of your airplane by comparing ram air pressure with static air pressure — the faster the aircraft moves through the air, the greater the pressure differential measured by this instrument. [Figure 2-18] Manufacturers use **indicated airspeed** (IAS) as the basis for determining aircraft performance. Takeoff, landing, and stall speeds listed in the POH are indicated airspeeds and do not normally vary with altitude or temperature. This is because changes in air density affect the aerodynamics of the airframe and the airspeed indicator equally.

At higher elevation airports, the indicated airspeed for approach and landing remains unchanged, but the corresponding groundspeed is faster.

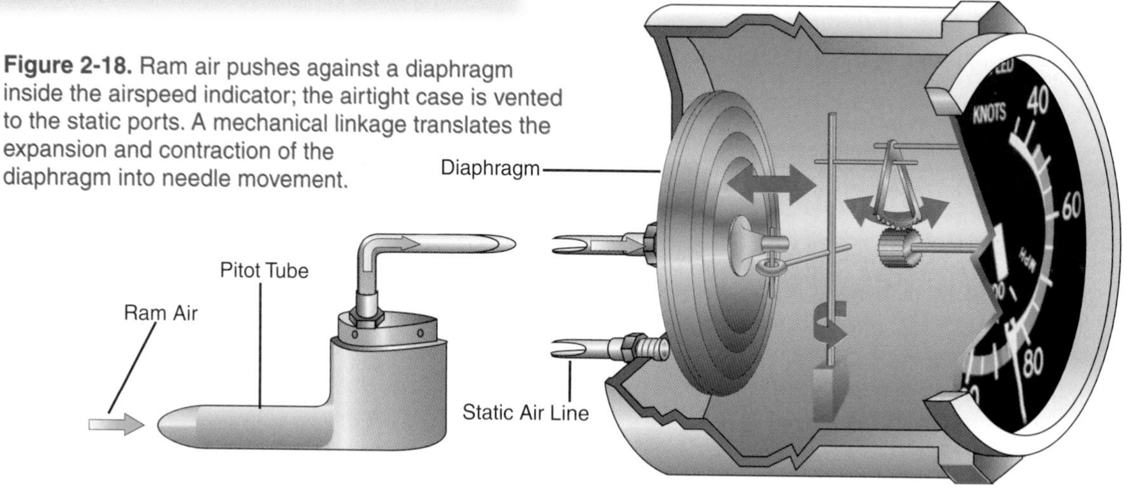

Figure 2-18. Ram air pushes against a diaphragm inside the airspeed indicator; the airtight case is vented to the static ports. A mechanical linkage translates the expansion and contraction of the diaphragm into needle movement.

AIRSPEEDS

As an instrument pilot, you need to understand the different types of airspeed. **Calibrated airspeed** (CAS) is indicated airspeed corrected for installation and instrument errors. Most of the discrepancy occurs because, at high angles of attack, the pitot tube does not point straight into the relative wind. This tends to make the airspeed indicator indicate lower-than-normal at low airspeeds. [Figure 2-19] Manufacturers compensate for this as best they can, but some errors are inevitable. The difference between indicated and calibrated airspeed is minimal at cruise speeds. You can find the corrections in the pilot's operating handbook (POH). As a practical matter, you normally use specific indicated airspeeds for various operations, and only concern yourself with CAS when you need to convert to true airspeed.

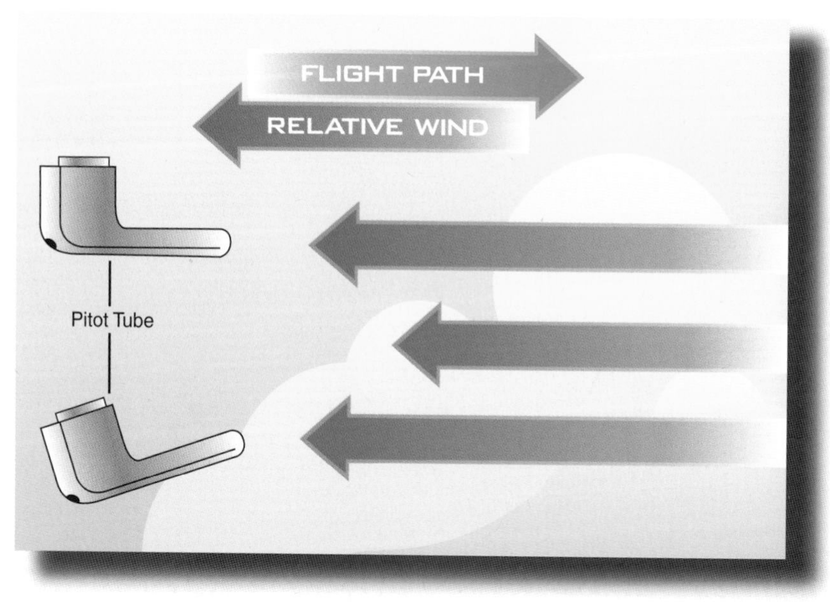

Figure 2-19. At high angles of attack, the relative wind does not strike the pitot tube straight on. This results in lower-than-normal indicated airspeed.

Equivalent airspeed (EAS) is calibrated airspeed corrected for adiabatic compressible flow at a particular altitude. At airspeeds above 200 KIAS and altitudes above 20,000 feet, air is compressed in front of an aircraft as it passes through the air. Compressibility causes abnormally high airspeed indications, so EAS is lower than CAS. Many electronic and mechanical flight computers are designed to compensate for this error. It is significant to pilots of high speed aircraft, but relatively unimportant to the average light airplane pilot.

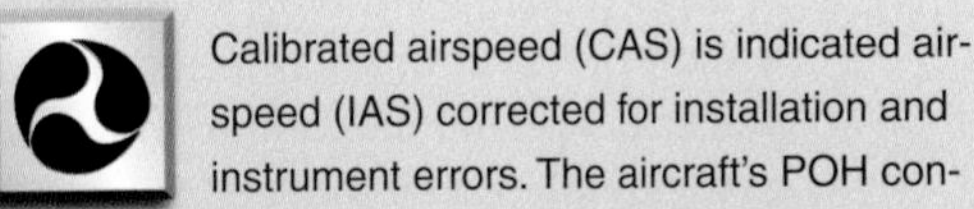

Calibrated airspeed (CAS) is indicated airspeed (IAS) corrected for installation and instrument errors. The aircraft's POH contains a chart allowing you to convert IAS to CAS. True airspeed (TAS) is CAS corrected for nonstandard temperature and pressure.

True airspeed (TAS) is the actual speed your airplane moves through undisturbed air. At sea level on a standard day, CAS (or EAS, as appropriate) equals TAS. As density altitude increases, true airspeed increases for a given CAS, or for a given amount of power. You can calculate TAS from CAS (or EAS), pressure altitude and temperature using your flight computer. Assuming conditions close to standard temperature, you can get an approximate true airspeed by adding 2% of the indicated airspeed for each 1,000-foot increase in altitude.

During a flight at constant power and at a constant indicated altitude, true airspeed increases as outside air temperature increases.

Many high performance aircraft have a Mach indicator incorporated with the airspeed indicator. [Figure 2-20] **Mach** is the ratio of the aircraft's true airspeed to the speed of sound. A speed of Mach 0.85 means the aircraft is flying at 85% of the speed of sound at that temperature. When computing true airspeed from a conventional airspeed indicator, you must factor in air density, which requires a correction for temperature and altitude. These corrections are unnecessary with a Mach indicator because the temperature determines the speed of sound. Thus, Mach is a more valid index to the speed of the aircraft.

Figure 2-20. A Mach Indicator provides a more meaningful speed index for high performance aircraft.

V-SPEEDS AND COLOR CODES

The color codes on an airspeed indicator provide important information on the operation of the aircraft. These markings actually reflect the airplane's performance envelope. [Figure 2-21]

A Mach meter presents the ratio of the aircraft's true airspeed to the speed of sound.

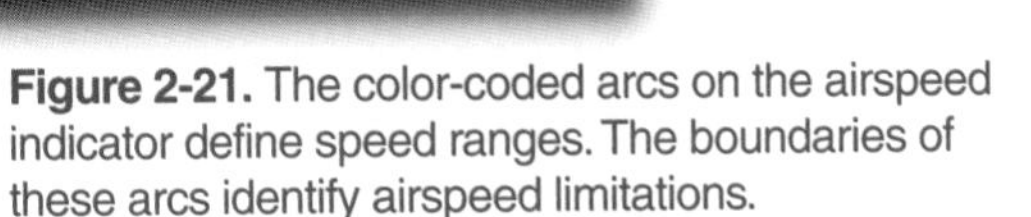

Figure 2-21. The color-coded arcs on the airspeed indicator define speed ranges. The boundaries of these arcs identify airspeed limitations.

V_{S0} is the stalling speed, or minimum steady flight speed, in the landing configuration, at the maximum landing weight.

V_{NE} is the never-exceed speed.

YELLOW ARC
You can operate in this caution range only in smooth air.

V_{NO} is the maximum structural cruising speed.

GREEN ARC
The green arc is the normal operating range.

V_{S1} is the stalling speed, or minimum steady flight speed, obtained in a specified configuration. For light airplanes, this is normally power-off stall speed at maximum takeoff weight in a clean configuration (gear and flaps up).

WHITE ARC
The white arc is the full-flap operating range.

V_{FE} is the maximum speed with the flaps fully extended. Some aircraft allow partial flap extensions above this speed for approach operations.

Several important values are not marked on the airspeed indicator. During gusty or turbulent conditions, you should slow the aircraft below the design maneuvering speed, V_A, to ensure the load factor is within safe limits. At or below V_A, the airplane will stall before excessive G-forces can occur. Design maneuvering speed, which decreases with the total weight of the aircraft, appears in your POH or on a placard. V_{LE}, the maximum speed with the landing gear extended on a retractable gear airplane, and V_{LO}, the maximum speed for extending or retracting the landing gear, also appear in the POH but not on the airspeed indicator.

Design maneuvering speed is one important value not shown by the color coding of an airspeed indicator. [Figure 2-21] If you encounter severe turbulence during an IFR flight, slow the airplane below this speed. This decreases the amount of excess load that can be imposed on the wing.

INSTRUMENT CHECK

Unless the airplane is facing into a strong wind, the airspeed indicator should read zero before you taxi. If it indicates some value due to wind, verify the indicator drops to zero when you turn the airplane away from the wind. As you accelerate during the takeoff roll, make sure the airspeed indicator comes alive and increases at an appropriate rate. If not, discontinue the takeoff.

ALTIMETER

An altimeter is required for both VFR and IFR flight. For IFR, you need a **sensitive altimeter**, adjustable for barometric pressure. Obviously an accurate altimeter is essential when you are operating in IFR conditions. In addition to helping you maintain terrain clearance, minimum IFR altitudes, and separation from other aircraft, the altimeter helps you maintain aircraft control. An understanding of the operation and limitations of the altimeter will enable you to interpret its indications correctly. [Figure 2-22]

ANEROID WAFERS
The main component of the altimeter is a stack of sealed aneroid wafers that expand and contract as atmospheric pressure from the static source changes. A mechanical linkage translates these changes into pointer movements on the indicator.

ALTITUDE INDICATION SCALE
The altimeter reads like a clock, with the small hand indicating thousands of feet, and the large hand indicating hundreds of feet.

ALTIMETER SETTING WINDOW
When the setting on an altimeter is changed, the indication changes in the same direction by approximately1,000 feet for each inch of pressure.

100 ft Pointer

10,000 ft Pointer

1,000 ft Pointer

CROSS HATCH FLAG
A cross-hatched area appears on some altimeters when displaying an altitude below 10,000 feet MSL.

Altimeter Setting Adjustment Knob

Static Port

Figure 2-22. This altimeter indicates 2,800 feet MSL. The barometric scale in the window is at 29.92 inches of mercury (in. Hg.).

Now that you will be flying under IFR, your ability to accurately read and adjust the altimeter is far more important than it was under VFR. See Figure 2-22.

TYPES OF ALTITUDE

The altimeter measures the vertical elevation of an object above a given reference point. The reference may be the surface of the earth, mean sea level (MSL), or some other point. There are several different types of altitude, depending on the reference point used.

The altimeter indicates height in feet above the barometric pressure level set in the altimeter window. For example, if the altimeter is set to 30.00, it would indicate the height of the airplane above the pressure level of 30.00 in. Hg. If this is the correct local altimeter setting, then the 30.00 in. Hg. pressure level would be at sea level and the altimeter would indicate the true altitude above sea level (assuming standard temperature). **Indicated altitude** is what you read on the altimeter when it is correctly adjusted to show your approximate height above mean sea level (MSL). Use indicated altitude, when operating IFR below 18,000 feet MSL.

Pressure altitude is displayed on the altimeter when it is set to the standard sea level pressure of 29.92 in. Hg. It is the vertical distance above a theoretical plane, or **standard datum plane**, where atmospheric pressure is equal to 29.92 in. Hg. Regulations require that you set the altimeter to 29.92 when operating at or above 18,000 feet MSL. These high altitudes are referred to as flight levels (FL); 18,000 feet above the standard datum plane is FL180.

Density altitude is pressure altitude corrected for nonstandard temperature. It is a theoretical value used to determine airplane performance. When density altitude is high (temperatures above standard), aircraft performance suffers. Most aircraft documentation gives you performance information based on pressure altitude and temperature, rather than explicitly using density altitude.

Set your altimeter to 29.92 in. Hg. in order to obtain pressure altitude below 18,000 feet MSL. And use pressure altitude when operating at or above 18,000 feet MSL. Always set the altimeter to 29.92 in. Hg. upon climbing to 18,000 feet MSL.

True altitude is the actual height of an object above mean sea level. On aeronautical charts, the elevations of such objects as airports, towers, and TV antennas are true altitudes. Unfortunately, your altimeter displays true altitude in flight only under standard conditions. Nonstandard temperature and pressure cause your indicated altitude to differ from true altitude. The true altitude computations you make with a flight computer assume that pressure and temperature lapse rates match a perfectly standard atmosphere, which is rarely the case. [Figure 2-23]

Pressure altitude is the altitude read on the altimeter when the instrument is set to indicate height above the standard datum plane. It is the same as density altitude at standard temperature, and is equal to true altitude under standard atmospheric conditions.

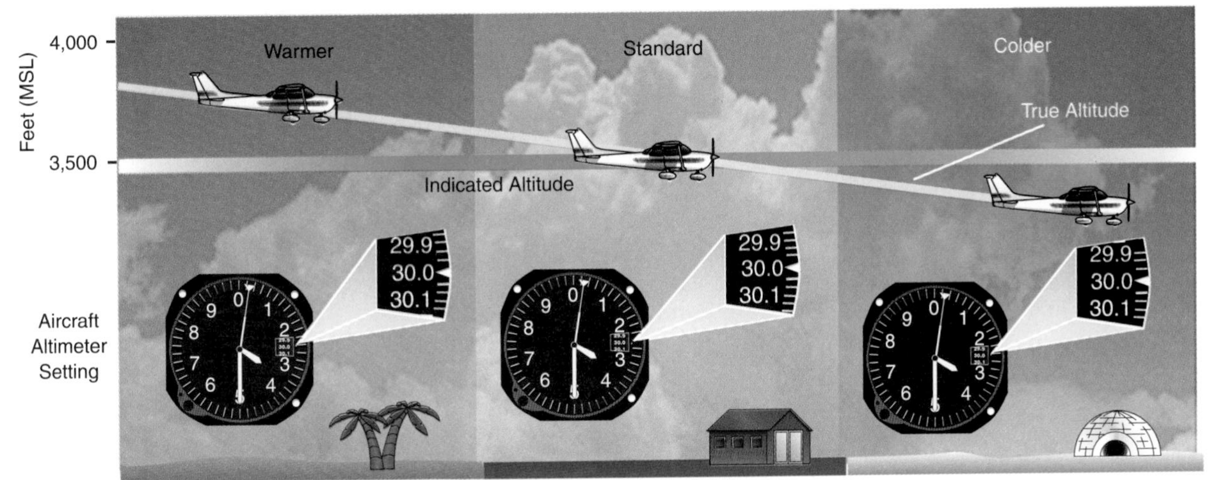

Figure 2-23. When the air temperature is warmer than standard, the altimeter will indicate a lower altitude than that actually flown. When the temperature is colder than standard, the altimeter will indicate too high, that is, true altitude will be lower than indicated altitude.

True and indicated altitude are equal when you are flying with the correct altimeter setting and temperature conditions match **International Standard Atmospheric (ISA)** values. However, if the temperature is 10°C colder than standard, true altitude is about 4% lower than indicated altitude. This is an error of 500 feet at 12,000 feet MSL; a significant discrepancy if flying over mountainous terrain on a cold day. True altitude also equals indicated altitude when you are sitting on the airport ramp with the altimeter set to the local altimeter setting where it indicates the field elevation.

When flying from hot to cold, look out below. See Figure 2-23.

When you set the scale of the pressure altimeter to the local altimeter setting, it will indicate the true altitude at field elevation. If an altimeter setting is unavailable, set the altimeter to the field elevation.

Absolute altitude is the actual height of the aircraft above the earth's surface. Some airplanes are equipped with radar altimeters that measure this height above ground level (AGL) directly. During instrument approaches, absolute altitude is used to define the height above the airport (HAA), height above the touchdown zone (HAT), and the threshold crossing height (TCH).

ALTIMETER SETTING

The most common altimeter error is also the easiest to correct. It occurs when you fail to keep the altimeter set to the local altimeter setting. When flying from an area of high pressure to an area of low pressure without resetting your altimeter, the instrument interprets the lower pressure as a higher altitude. Since you will lower the nose of the airplane to maintain the same indicated altitude, you will end up at a lower true altitude. This is why, when flying from high to low pressure, look out below. [Figure 2-24]

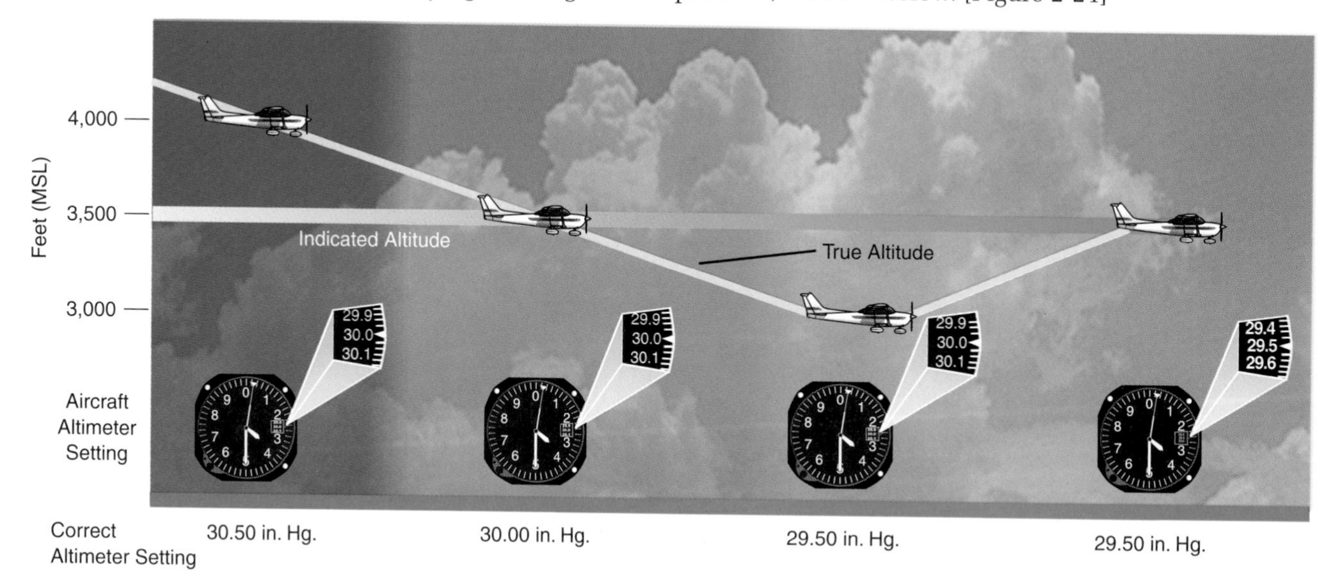

Figure 2-24. If you fly from an area of higher pressure to an area of lower pressure without resetting your altimeter, you may fly at a lower altitude than you had intended. If you reset the altimeter to the correct setting, you can maintain the desired altitude.

In the event you are departing an airport where you cannot obtain a current altimeter setting, you should set the altimeter to the airport elevation. After departure, obtain the current altimeter setting as soon as possible from the appropriate ATC facility.

The local altimeter setting should be used by all pilots primarily to provide for better vertical separation of aircraft. ATC periodically advises the pilot of the proper altimeter setting.

INSTRUMENT CHECK

In addition to the required static system check, you should make sure the altimeter is indicating accurately during the IFR preflight check. To accomplish this, set the altimeter to the current altimeter setting. If it indicates within 75 feet of the actual elevation of that location, it is acceptable for IFR flight.

Before an IFR flight, set the altimeter to the current altimeter setting. For acceptable accuracy, the indication should be within 75 feet of the actual elevation.

VERTICAL SPEED INDICATOR

The vertical speed indicator (VSI), sometimes called a vertical velocity indicator (VVI) or rate-of-climb indicator, measures how fast the static (ambient) pressure increases or decreases as the airplane climbs or descends. It then displays this pressure change as a rate of climb or descent in feet per minute. Unlike the altimeter, the VSI is not affected by air temperature since it measures only *changes* in air pressure. [Figure 2-25]

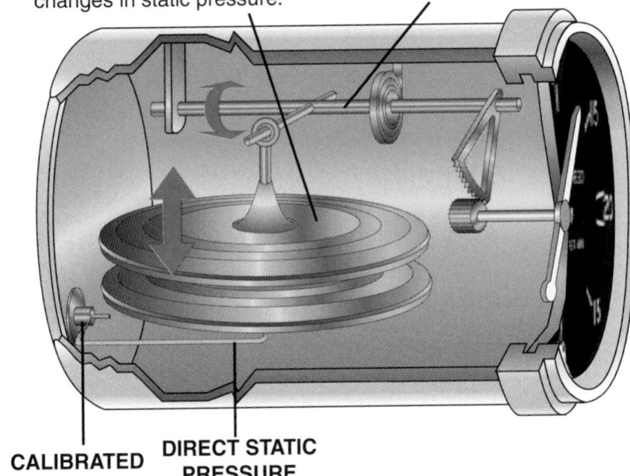

Figure 2-25. When the static pressure is constant, it remains equal inside and outside of the diaphragm. Any change in pressure is felt immediately inside the diaphragm but delayed outside the diaphragm, and it expands or contracts due to the momentary pressure differential. When the aircraft levels off, the pressure outside the diaphragm equalizes with the pressure inside the diaphragm, and the needle indicates zero.

The VSI displays **rate information** and **trend information**. Although the instrument is designed to display rate of climb or descent, it can take six to nine seconds of lag for the needle to stabilize on an accurate vertical speed indication after you change power or pitch. Even though it takes a few moments to indicate the exact vertical speed, the VSI is valuable because it instantaneously indicates changes in vertical speed, or trend information. When making a steep turn, the VSI usually is the first instrument to tell you a small correction in pitch is needed, since the attitude indicator does not give a precise enough indication of pitch. As you study attitude instrument flying in the next section and practice it in the airplane, you will appreciate the VSI's early warning of deviations from the desired pitch. This instrument provides even more essential pitch information in the event of a failure of the gyroscopic attitude indicator.

Since the VSI is not designed to instantaneously indicate the rate of climb or descent, it will not give a clear indication during turbulence or when applying abrupt control inputs. You may be able to average the erratic readings during turbulence to determine whether you are climbing or descending. Because the VSI uses static air pressure, this instrument will not function if the static port is clogged.

In some advanced aircraft, you may find an instantaneous vertical speed indicator (IVSI). This device incorporates acceleration pumps to compensate for the limitations of the calibrated leak, eliminating the lag found in the typical VSI.

Instrument Check

If, during taxi, you notice that the VSI indicates a descent or climb, you may use that value as a zero indication.

The VSI, while not legally required for instrument flight, is an extremely useful instrument. Some pilots will not take off into low IFR conditions unless this instrument is operating properly. Before starting the aircraft engine, check to see that the VSI indicates zero. If you wait until after engine start to check the VSI, you may see needle fluctuations due to the propeller slipstream. Some VSIs have an adjustment screw to zero the instrument. If yours does not, simply make a mental note of where "zero" is, and compensate for that during flight.

PITOT TUBE ICING LEADS TO IN-FLIGHT BREAKUP

From the files of the NTSB. . .

Aircraft: *Piper PA-46-350P*

Injuries: *2 Fatal*

Narrative: *Before takeoff, the pilot was advised of IFR conditions along the first part of the route, with flight precautions for occasional moderate turbulence below 15,000 feet, and mixed icing from freezing level (6,000 feet) to 18,000 feet. He filed an IFR flight plan with a cruise altitude of 11,000 feet. During departure, the pilot was cleared to climb to 9,000 feet, and told to expect clearance to 11,000 feet 5 minutes later.*

Radar data showed the aircraft climbed at about 1,500 feet per minute and 100 knots slowing slightly above 8,000 feet. At about 9,000 feet the aircraft started to level and accelerate. It then climbed momentarily, deviated laterally from course, and entered a steep descent. In-flight breakup occurred and wreckage was scattered over a 4,100 foot area. A trajectory study showed breakup occurred between 4,500 and 6,500 feet as the aircraft was in a steep descent in excess of 266 knots. Metallurgical exam of wings and stabilizers revealed features typical of overstress separation; no preexisting cracks or defects were found.

The probable cause, according to the NTSB, was the pilot's failure to activate the pitot heat before flying at and above the freezing level in instrument meteorological conditions (IMC), followed by his improper response to erroneous airspeed indications that resulted from blockage of the pitot tube by atmospheric icing. Spatial disorientation of the pilot was listed as a contributing factor.

The chain of events which led to this accident most likely began with the pilot's lack of experience and training in IFR emergencies, followed by poor judgement regarding the weather conditions. In addition, the pilot failed to effectively manage his workload and was unable to maintain situational awareness during the flight.

System Errors

The pitot-static instruments usually are very reliable. Gross errors almost always indicate blockage of the pitot tube, the static port, or both. Blockage may be caused by moisture (including ice), dirt, or even insects. During preflight, always check the pitot tube for blockage. If you do this, you also will remember to remove the pitot tube cover. Always check the static port openings as well. If the pitot or static ports are clogged, have them cleaned by a certificated mechanic. It is also possible for the pitot tube to become blocked by visible moisture during flight when temperatures are near the freezing level. If you are flying in visible moisture and your airplane is equipped with pitot heat, it should be on to prevent pitot tube icing.

PITOT BLOCKAGE

The airspeed indicator is the only instrument affected by a pitot tube blockage. There are two types of pitot blockage that can occur. If the ram air inlet clogs while the drain hole remains open, the pressure in the line to the airspeed indicator will vent out the drain hole, causing the airspeed indicator to drop to zero. This typically occurs when ice forms over the ram air inlet. [Figure 2-26]

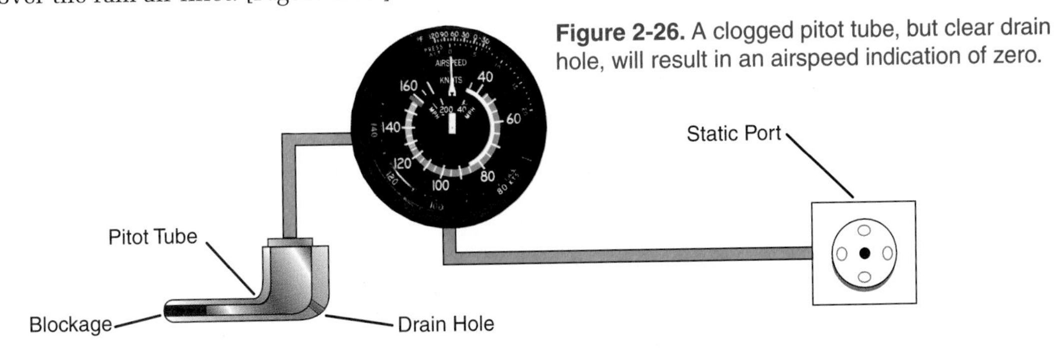

Figure 2-26. A clogged pitot tube, but clear drain hole, will result in an airspeed indication of zero.

The second situation occurs when both the ram air inlet and drain hole become clogged, trapping the air pressure in the line. In level flight, the airspeed indicator typically remains at its present indication, but no longer indicates changes in airspeed. If the static port remains open, the indicator will react as an altimeter, showing an increase in airspeed when climbing and a decrease in speed when descending. This is opposite the normal way the airspeed indicator behaves, and can result in inappropriate control inputs because you will observe runaway airspeed as you climb and extremely low airspeeds in a descent. This type of failure can be very hazardous because it is not at all obvious when it occurs. [Figure 2-27]

If the pitot tube's ram air input and drain hole are blocked, the airspeed indicator will act as an altimeter with indicated airspeed increasing as altitude increases. See figure 2-27.

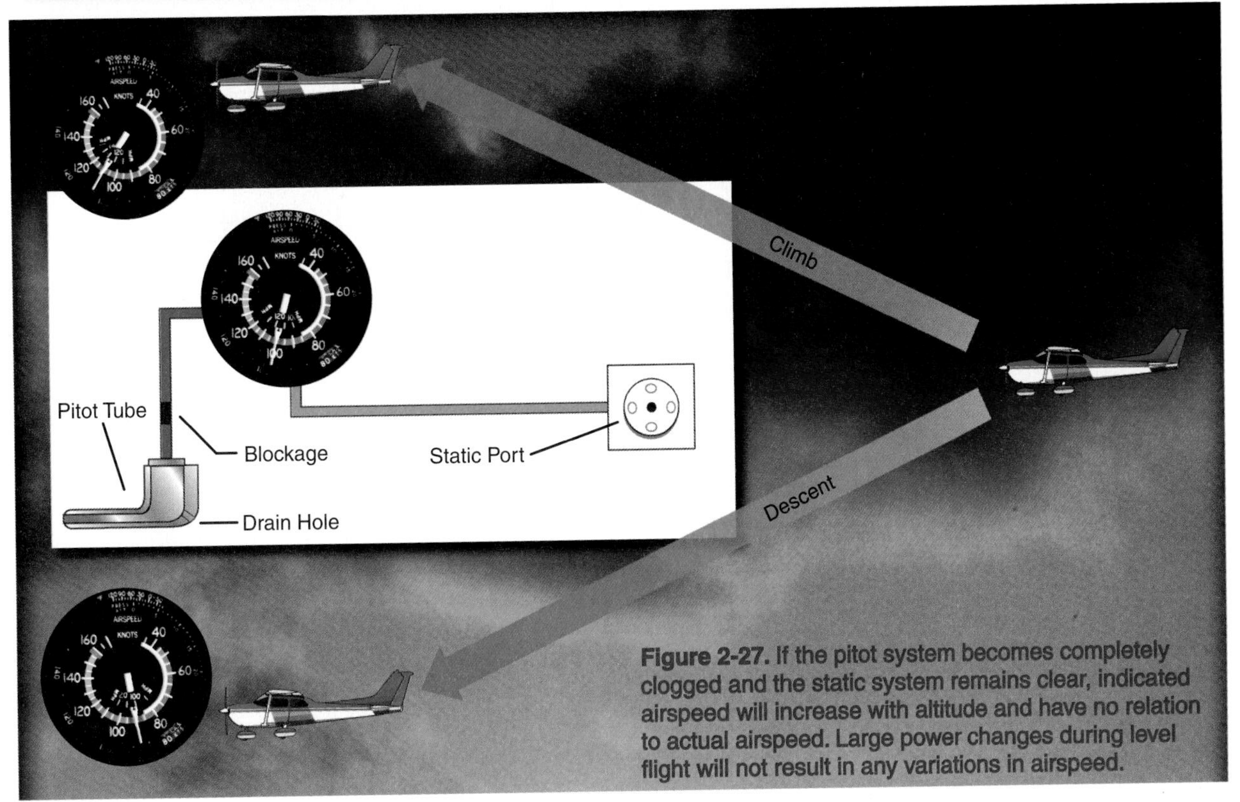

Figure 2-27. If the pitot system becomes completely clogged and the static system remains clear, indicated airspeed will increase with altitude and have no relation to actual airspeed. Large power changes during level flight will not result in any variations in airspeed.

STATIC BLOCKAGE

If the static system becomes clogged, the airspeed indicator will continue to react to changes in airspeed, since ram air pressure is still being supplied by the pitot tube, but the readings will not be correct. When you are operating above the altitude where the static port became clogged, the airspeed will read lower than it should. Conversely, when you operate at a lower altitude, a faster-than-actual airspeed will be displayed due to the relatively low static pressure trapped in the system. The amount of error is proportional to the distance from the altitude where the static system became clogged. The greater the difference, the greater the error. [Figure 2-28]

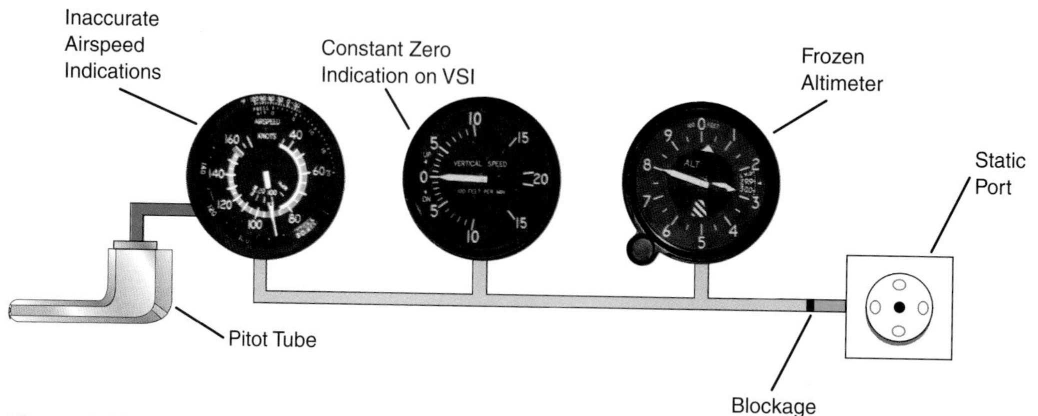

Figure 2-28. A blocked static system affects all pitot-static instruments.

Since the altimeter determines altitude by measuring ambient air pressure, any blockage of the static port will freeze the altimeter in place and make it unusable. The VSI freezes at zero, since its only source of pressure is from the static port. After verifying a blockage of the static system by cross-checking the other flight instruments, you should find an alternate source of static pressure.

If the static ports are iced over, the VSI pointer will remain at zero, regardless of the actual rate of descent or climb.

In many aircraft, an alternate static source is provided as a backup for the main static source. In nonpressurized aircraft, the alternate source usually is located inside the aircraft cabin. Due to the slipstream, the pressure inside the cabin is usually less than that of outside air. Normally, when you select the alternate static source, the altimeter will read a little higher and the airspeed a little faster than normal, while the vertical speed indicator will show a momentary climb. However, this is not always the case. Your airplane's POH may contain information regarding variations in airspeed and altimeter readings due to changes in airplane configuration and use of the alternate static source. In the case of a pressurized aircraft with a static line leak inside the pressurized compartment, the altimeter will read lower than the actual flight altitude, due to the increased static pressure. The airspeed may also read lower than it should, and the vertical speed indicator may indicate a momentary descent.

If, while in level flight, it becomes necessary to use an alternate source of static pressure vented inside the airplane, the altimeter may read a little higher and the airspeed a little faster than normal, while the vertical speed indicator may show a momentary climb.

If the aircraft is not equipped with an alternate static source, you can break the glass of the vertical speed indicator to allow ambient air pressure to enter the static system. Of course, this makes the VSI unusable and will not help in a pressurized airplane; but in an IFR emergency, depressurization is an option to consider.

SUMMARY CHECKLIST

✓ The instruments which provide information about the airplane's attitude, direction, altitude, and speed are collectively referred to as the flight instruments.

✓ The gyroscopic instruments in your aircraft are the attitude indicator, heading indicator, and turn coordinator. Gyroscopic instrument operation is based on rigidity in space and precession.

✓ The attitude indicator, or artificial horizon, is the only instrument that gives you an immediate and direct indication of the airplane's pitch and bank attitude.

✓ The heading indicator, when properly set, is your primary source of heading information. You must align it with the magnetic compass before flight and recheck it periodically during flight.

✓ Turn indicators allow you to establish and maintain standard-rate turns of three degrees per second, or in the case of certain high performance aircraft, half-standard-rate turns.

✓ Both turn coordinators and turn-and-slip indicators indicate rate of turn, but because of the improved design of the turn coordinator, this instrument also indicates rate of roll as you enter a turn.

✓ The inclinometer is the part of the turn indicator that indicates whether you are using the correct angle of bank for the rate of turn. Step on the ball to correct a slipping or skidding condition.

✓ The magnetic compass is the only direction-seeking instrument in most light airplanes, but it is susceptible to a number of errors.

✓ The pitot-static instruments are the airspeed indicator, altimeter, and vertical speed indicator. Blockages in both the pitot and static systems affect the airspeed indicator, while the remaining instruments are affected only by static system blockage.

✓ Calibrated airspeed (CAS) is indicated airspeed corrected for installation and instrument errors. Equivalent airspeed (EAS) is calibrated airspeed corrected for compressibility. True airspeed (TAS) is the actual speed your airplane moves through undisturbed air. Mach is the ratio of the aircraft's true airspeed to the speed of sound at the temperature and altitude in which the aircraft is flying.

✓ The most common altimeter error is failure to keep the current barometric pressure set. The altimeter indicates high when the actual pressure is lower than what is set in the window. The altimeter also indicates high when in colder-than-standard temperature conditions.

✓ Pressure altitude is displayed on the altimeter when it is set to the standard sea level pressure of 29.92 in. Hg. This also is the altimeter setting when operating at or above 18,000 feet MSL.

✓ Before an IFR flight, verify that the altimeter indicates within 75 feet of the actual field elevation when set to the current altimeter setting.

✓ The vertical speed indicator instantly alerts you of changes in vertical speed, and gives an accurate indication of the rate of climb or descent a few seconds after a change in vertical speed.

✓ Complete blockage of the pitot tube can cause the airspeed indicator to react opposite of normal, showing an increase in airspeed as you climb, and extremely low airspeed in a descent.

KEY TERMS

Rigidity in Space

Gimbals

Precession

Pendulous Vanes

Standard-Rate Turn

Turn-and-Slip Indicator

Turn Coordinator

Inclinometer

Slip

Skid

Magnetic Dip

Northerly Turning Error

Indicated Airspeed

Calibrated Airspeed

Equivalent Airspeed

True Airspeed

Mach

Sensitive Altimeter

Indicated Altitude

Pressure Altitude

Standard Datum Plane

True Altitude

International Standard Atmosphere (ISA)

Absolute Altitude

Rate Information

Trend Information

QUESTIONS

1. Which flight instrument is not legally required for flight under IFR?

   A. Slip-skid indicator
   B. Vertical speed indicator
   C. Gyroscopic heading indicator

2. Name the gyroscopic flight instruments. Upon which two principles do they depend?

3. Which flight instrument gives you an instantaneous display of both pitch and bank information?

4. True/False. The gyroscopic instruments are the only flight instruments that provide bank information.

5. Why is the turn coordinator a good backup for the attitude and heading indicators in most small airplanes?

6. What is the approximate bank angle required to maintain a standard-rate turn at 90 knots?

7. How long does it take to make a 360° standard-rate turn?

   A. One minute
   B. Two minutes
   C. Four minutes

8. Which of the above turn coordinators shows too much rudder pressure being used for the amount of bank?

9. Which of the above turn coordinators shows too little rudder pressure being used for the amount of bank?

10. What condition leads to unreliable operation of the heading indicator and attitude indicator?

    A. Low vacuum pressure.
    B. Short in the electrical system
    C. Pitot-static system leak

11. What should you use to correct for magnetic deviation?

    A. Compass correction card.
    B. Adjustment knob on the heading indicator.
    C. Isogonic lines on instrument charts.

12. During a compass turn, what heading should you use for roll-out if making a right turn to a heading of 360° at a latitude of 45°N? Assume you are using a 16° bank angle.

13. A blocked static source affects which instruments?

For questions 14 through 16, match the V-speed abbreviations with the appropriate definition.

14. V_A
15. V_{LO}
16. V_{S1}

A. Maximum landing gear operating speed
B. Design maneuvering speed
C. Maximum speed with landing gear extended
D. Stalling speed in a specified configuration
E. Stalling speed in landing configuration

17. How does colder-than-standard temperature affect the relationship between indicated altitude and true altitude?

SECTION B
ATTITUDE INSTRUMENT FLYING

Attitude instrument flying is controlling an aircraft by reference to flight instruments, rather than outside visual reference. You were introduced to attitude instrument flying during your private pilot course. In your instrument training, you will refine this skill to the point where you can maintain the precise control of an aircraft by instrument reference while carrying on the many additional duties of IFR flight.

FUNDAMENTAL SKILLS

Attitude instrument flying is one of the most important skills you will acquire as a pilot. Before allowing you to move on to other tasks, your instructor will insist you practice attitude flying until it becomes second nature. To achieve positive aircraft control and follow a desired flight path in IFR conditions you must master three fundamental skills — instrument cross-check, instrument interpretation, and aircraft control.

INSTRUMENT CROSS-CHECK

Instrument scan, or cross-check, is the first fundamental skill. A good scan requires logical and systematic observation of the instrument panel. It saves time and reduces the workload of instrument flying because you look at the pertinent instruments as you need information.

SCANNING TECHNIQUE

Regardless of your scanning technique, the attitude indicator is very important because it replaces the natural horizon in instrument conditions. [Figure 2-29] You normally cross check the attitude indicator with other instruments that provide information about pitch, bank, and power and verify that indicated attitude yields the desired results. With instruction and practice, you will learn what instruments you need to scan in order to maintain a particular flight attitude.

Figure 2-29. As a substitute for the natural horizon, the attitude indicator provides basic attitude reference. It is the only instrument that provides instant and direct aircraft attitude information.

"You Got a Bunch of Guys About to Turn Blue."

One of the most watched displays of crew coordination and attitude instrument flying occurred on July 20, 1969 when pilot, Buzz Aldrin, and commander, Neil Armstrong, safely guided the lunar module, *Eagle*, to a safe landing on the moon's Sea of Tranquility. [Figures A and B] During the landing approach, Aldrin kept his eyes focused on the spacecraft's computer display while Armstrong maneuvered the lunar module into position by following Aldrin's instructions and looking through a window scribed with a vertical scale. (A photo of the lunar module cockpit is shown in figure C.) Armstrong used the scale, which was graduated in degrees, to determine where the computer thought the lunar module would land. The following transcript begins 4 days, 6 hours, 44 minutes, 45 seconds after liftoff and about a minute before the historic landing.

04:06:44:45 — ALDRIN: *100 feet, 3 1/2 down, 9 forward. 5%.* [Fuel Remaining]

04:06:44:54 — ALDRIN: *Okay. 75 feet. Looking good. Down a half, 6 forward.*

04:06:45:02 — DUKE [Charlie Duke, the CapCom (Spacecraft Communicator) for the landing, located in Houston, TX]: *60 seconds.* [Fuel Remaining]

04:06:45:04 — ALDRIN: *Light's on.* [Fuel Quantity Light]

04:06:45:08 — ALDRIN: *Down 2 1/2. Forward . . . forward . . . good.*

A

04:06:45:17 — ALDRIN: *40 feet, down 2 1/2. Kicking up some dust.*

04:06:45:21 — ALDRIN: *30 feet, 2 1/2 down.*

04:06:45:25 — ALDRIN: *4 forward . . . 4 forward. Drifting to the right a little. Okay. Down a half.*

04:06:45:31 — DUKE: *30 seconds.*

04:06:45:32 — ARMSTRONG: *Forward drift?*

04:06:45:33 — ALDRIN: *Yes.*

04:06:45:34 — ALDRIN: *Okay.*

B

04:06:45:40 — ALDRIN: *Contact Light.* [The 10-foot long probes, which hang from 3 of the lunar module's footpads, have touched the moon's surface.]

04:06:45:43 — ALDRIN: *Okay. Engine stop.*

04:06:45:45 — ALDRIN: *ACA, out of detent.* [The ACA is the Attitude Control Assembly, or control stick.]

04:06:45:46 — ARMSTRONG: *Out of detent.*

04:06:45:47 — ALDRIN: *Mode control, both auto. Descent engine command override, off. Engine arm, off.*

04:06:45:52 — ALDRIN: *413 is in.* [413 is a code which tells the AGS (Abort Guidance System) that the lunar module has landed.]

04:06:45:57 — DUKE: *We copy you down,* Eagle.

04:06:45:59 — ARMSTRONG: *Houston, Tranquility Base here.*

04:06:46:04 — ARMSTRONG: *The* Eagle *has landed.*

04:06:46:06 — DUKE: *Roger, Tranquility. We copy you on the ground. You got a bunch of guys about to turn blue. We're breathing again. Thanks a lot.*

04:06:46:16 — ARMSTRONG: *Thank you.*

C

Six and a half hours later, an estimated 600 million people around the world watched as Neil Armstrong became the first person to set foot on the moon. While Jimmy Doolittle could hardly have envisioned space travel at the time of his first blind flight, the tremendous achievements of Apollo 11 would not have been possible without the application of the basic attitude instrument flight techniques he pioneered nearly 40 years earlier.

Emblem and photos courtesy of NASA

FREQUENT ERRORS

As your proficiency increases, you will scan primarily from habit, adjusting your scan rate and sequence to suit the demands of the situation. However, if you do not continue to maintain your proficiency through practice, you may find that your scan breaks down. This lapse in instrument cross-checking is usually the result of one or more of the three common scanning errors — fixation, omission, and emphasis. [Figure 2-30]

Fixation occurs when you stare at a single instrument. There may be a good reason, such as when you suspect an instrument malfunction, but fixation invariably yields poor results. The aircraft's attitude can rapidly deteriorate while you neglect other essential instruments, and errors in aircraft control can accumulate quickly.

Omission takes place any time you exclude one or more pertinent instruments from your scan. It often follows a change in pitch attitude or bank angle. If you fail to monitor the bank instruments during this time, you may deviate from your intended heading.

Emphasis on a single instrument instead of a combination of instruments is a common cross-check error during initial training. It is not as severe as fixation because you still maintain some scan, but control is degraded because of inappropriate reliance on one instrument. For example, you may be able to hold the airplane reasonably on altitude using the attitude indicator, but you cannot hold a precise altitude without including the altimeter in your scan.

Figure 2-30. Fixation, omission, and emphasis dramatically reduce the effectiveness of your scan.

INSTRUMENT INTERPRETATION

The second fundamental skill is **instrument interpretation**.This requires less effort if you have studied and observed how each instrument operates, and become aware of the instrument indications that represent the desired pitch and bank attitudes for your aircraft. Good interpretation skills also help your scan. If you know the limitations of an instrument, you will know what other instruments to cross check to get the complete picture. For example, a level pitch indication on the attitude indicator does not necessarily mean you are in a level flight attitude; you need to refer to the altimeter, vertical speed indicator (VSI), and airspeed indicator to confirm you are in level flight. To confirm bank attitude, refer to the heading indicator and turn coordinator in addition to the attitude indicator.

AIRCRAFT CONTROL

The third attitude instrument flying skill is aircraft control. This is the action you take as a result of cross checking and interpreting the flight instruments. Based on the information you receive from the instruments, adjust the pitch, bank, and power to achieve a desired flight path. The control inputs for attitude instrument flying are the same as for visual flying except that you cannot see the results using outside visual references.

During your training, you will develop a feel for how much control pressure achieves a desired change in pitch or bank attitude. To be precise, it is important that you maintain a light touch on the controls, [Figure 2-31] and keep the airplane properly trimmed. If you are constantly holding pressure on the controls, you will not be able to apply the precise pressures needed for controlled changes in attitude. An improperly trimmed airplane increases tension, interrupts your cross-check, and may result in abrupt or erratic control. On a properly trimmed airplane, you can easily adjust the attitude with gentle pressure on the controls.

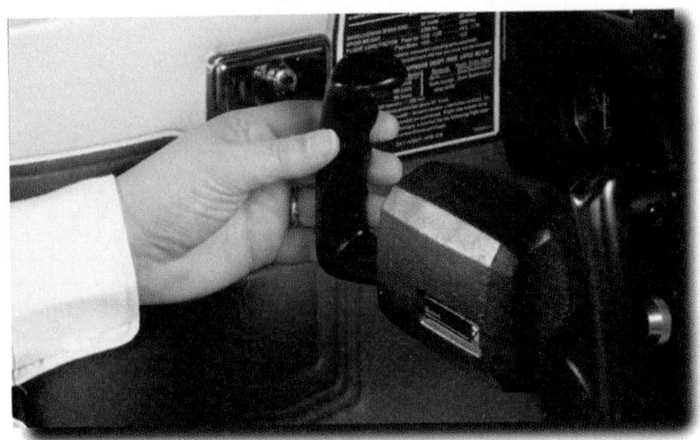

The correct sequence in which to apply the three skills used in instrument flying is cross-check, instrument interpretation, and aircraft control.

Figure 2-31. It is easier to control the airplane if you keep a relaxed grip on the control yoke.

ATTITUDE INSTRUMENT FLYING CONCEPTS

The following information will help you develop the thought processes for effective scanning and interpretation of flight instruments. You may wish to read this information again later in your training to make sure you are still on track. There are two generally accepted methods of teaching attitude instrument flying: the control and performance concept and the primary/support concept.

These two methods of scanning the same flight instruments generally yield the same control responses, but they differ in the degree of reliance on the attitude indicator. The control and performance concept, which designates certain flight instruments as control and others as performance instruments, relies heavily on the attitude indicator during most maneuvers. You will learn more about this method at the end of this section.

PRIMARY/SUPPORT CONCEPT

The primary/support concept divides the panel into pitch instruments, bank instruments, and power instruments. For a given maneuver, there are specific instruments you should use to control the airplane and obtain the desired performance. Those instruments which provide the most essential information during a given flight condition are called primary instruments. Instruments which help you maintain the desired indications on the primary instruments are called supporting instruments. Because the primary/support concept emphasizes monitoring primary instruments, this method prepares a pilot to maintain aircraft control if the gyroscopic attitude indicator fails. [Figure 2-32]

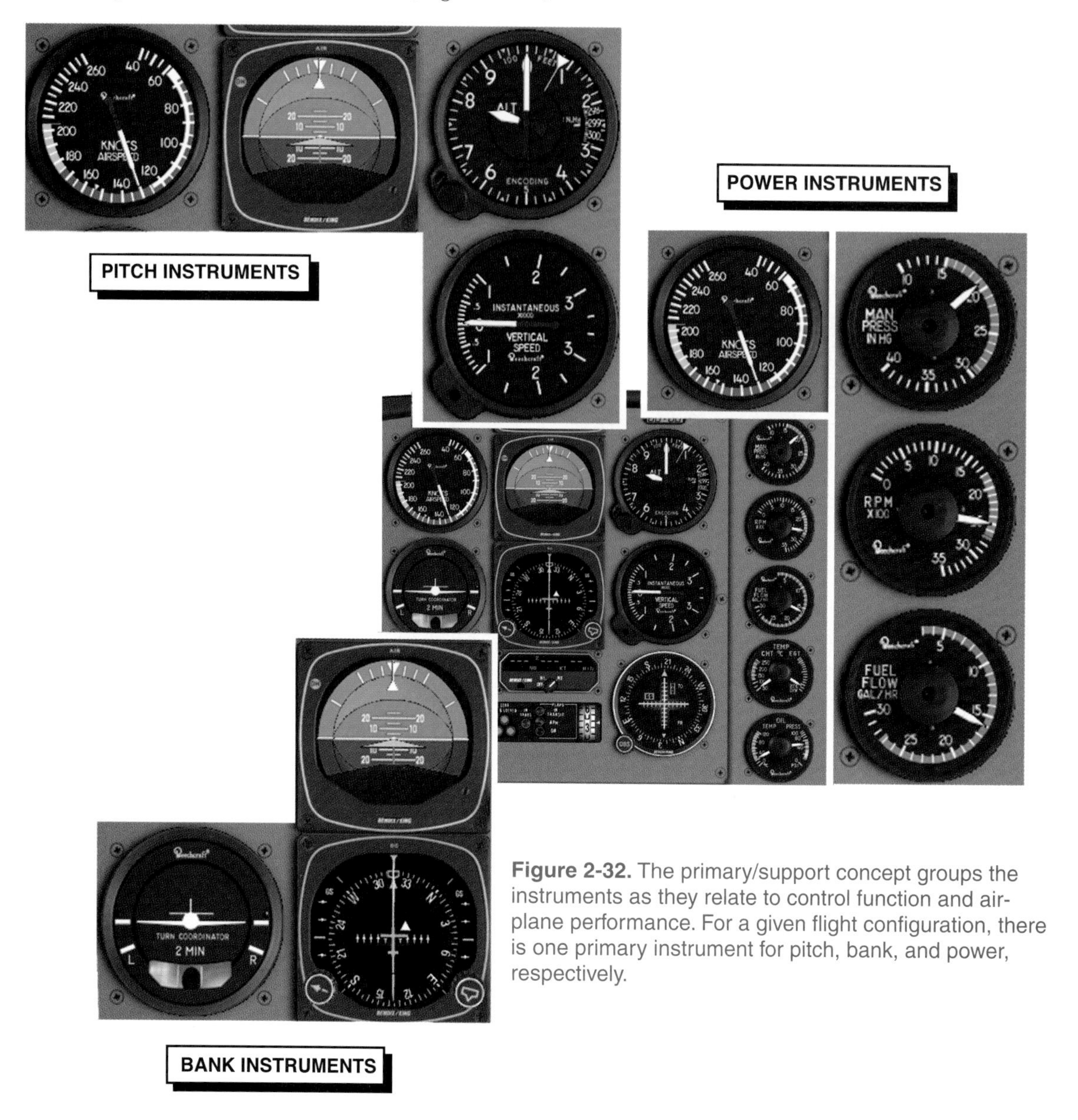

Figure 2-32. The primary/support concept groups the instruments as they relate to control function and airplane performance. For a given flight configuration, there is one primary instrument for pitch, bank, and power, respectively.

The attitude indicator, although often a supporting instrument, is essential and central to your scan. It is the only instrument that provides instant and direct aircraft attitude information, and it is the primary instrument during pitch or bank changes. After using the attitude indicator to establish a new attitude, other instruments become primary, and the attitude indicator becomes a supporting instrument. Nonetheless, loss of the attitude indicator in instrument conditions can result in a distress situation. If this happens, notify ATC immediately, since your ability to comply with clearances may be limited. Flying partial panel is covered later in this section.

The altimeter, airspeed indicator, and vertical speed indicator, in addition to the attitude indicator, are pitch instruments.

BASIC FLIGHT MANEUVERS

The following basic flight maneuvers are presented using the primary/support concept, with the primary instruments identified in each situation. The supporting instruments are the remaining instruments that provide information about pitch, bank, or power.

STRAIGHT-AND-LEVEL FLIGHT

Once established in straight-and-level flight, your primary objective is to maintain a specific altitude and heading at a specific airspeed. The instruments that can tell you unequivocally whether you have the correct pitch, bank, and power are the altimeter, heading indicator, and airspeed indicator, respectively. That is why these are primary instruments for pitch, bank, and power. While it is possible to maintain straight-and-level flight with only these instruments, it becomes more difficult without supporting instruments like the attitude indicator (pitch and bank), VSI (pitch), and turn coordinator (bank). [Figure 2-33]

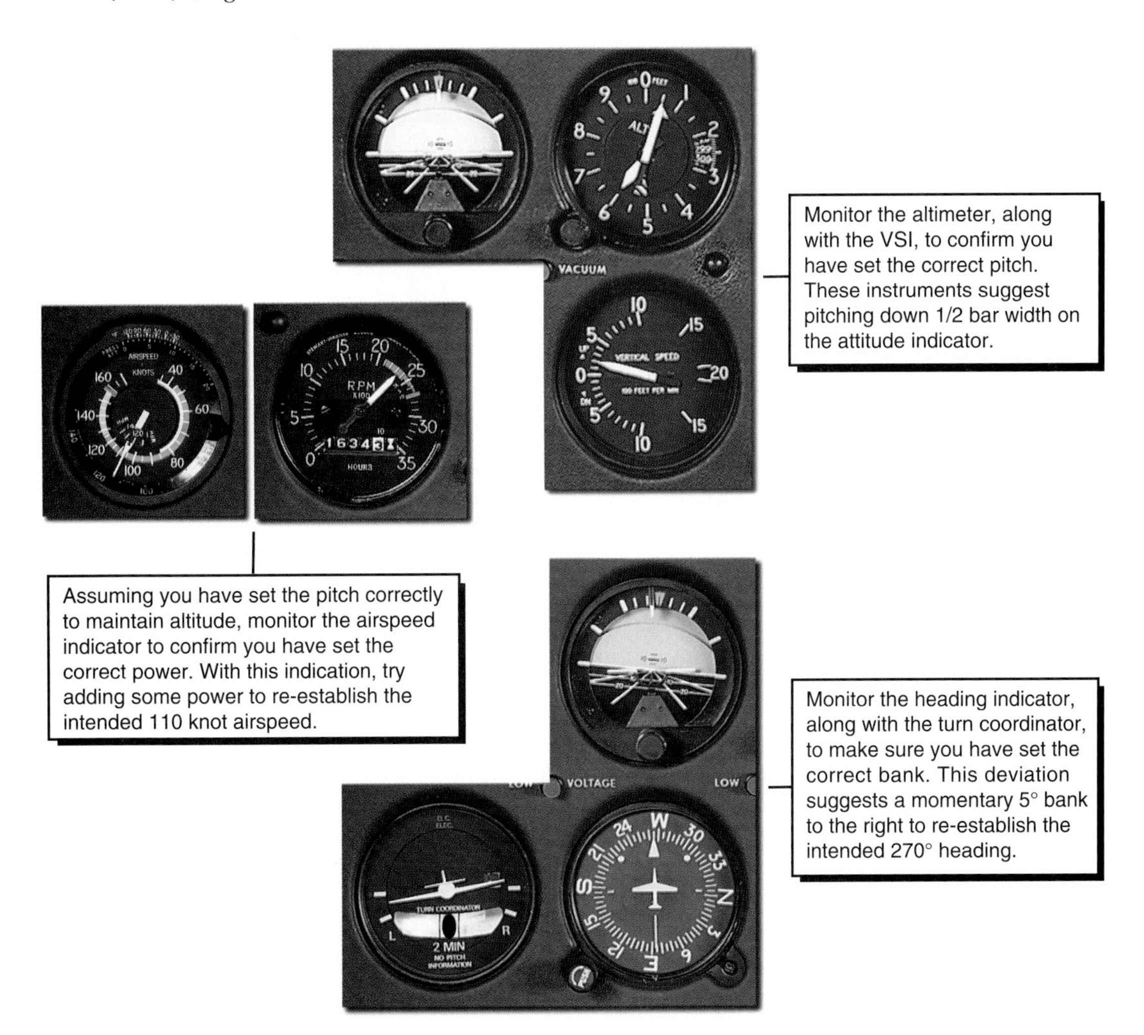

Figure 2-33. In straight-and-level flight, as in any flight maneuver, the primary instruments are those that supply the most pertinent information regarding pitch, bank, and power. The supporting instruments backup and supplement the primary instruments.

PITCH CONTROL

The attitude indicator gives you an instant and direct indication of relative pitch attitude, enabling you to precisely make pitch changes as small as one-half degree. Prior to takeoff, align the miniature aircraft on the attitude indicator with the horizon bar to indicate approximate level flight at normal cruise speed. Since the required level-flight pitch attitude varies, the miniature airplane may need periodic adjustment to the horizon bar. [Figure 2-34]

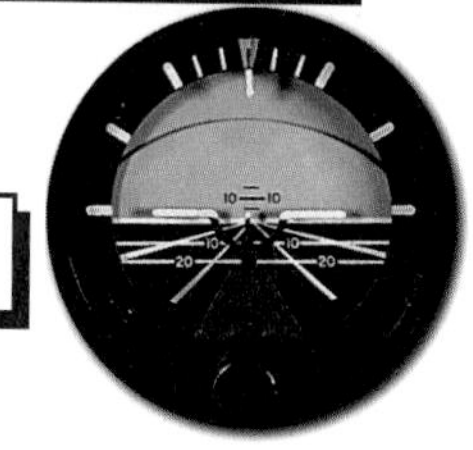

Figure 2-34. The pitch attitude required to maintain level flight is affected by airspeed, air density, and aircraft weight.

The three conditions which determine pitch attitude required to maintain level flight are airspeed, air density, and aircraft weight. For maintaining level flight at constant thrust, the attitude indicator would be the least appropriate pitch instrument for determining the need for a pitch change.

The vertical speed indicator promptly moves up or down with pitch, independently displaying even smaller deviations than the attitude indicator. Unlike the attitude indicator, the VSI measures the actual climb or descent performance resulting from a given pitch attitude, and normally does not need adjustment to provide reliable information.

To control pitch during straight-and-level flight, make sure the airplane is trimmed to maintain the desired pitch attitude without any control pressure. Monitor the attitude indicator for deviations from the desired pitch attitude, just as you would watch the horizon in VFR conditions. Scan the altimeter and VSI to verify that the pitch attitude you have chosen maintains the desired altitude.

When an altitude deviation occurs, use your judgment and experience in the aircraft to determine the rate of correction. Try a half-bar-width adjustment on the attitude indicator when 100 feet or less from your desired altitude, and correct back at twice the rate of your deviation. For example, if you are 100 feet below the desired altitude, climb back at 200 f.p.m. If you are descending when you discover the deviation, stop your descent first, then pitch up to climb back to the proper altitude. You may need to adjust the power if your altitude is off more than 200 feet.

BANK CONTROL

The heading indicator is primary for bank during straight flight, whether level, climbing, or descending, because this is the instrument that can tell you which direction you need to adjust your bank to maintain the correct heading. The attitude indicator and turn coordinator are supporting bank instruments. Deviations in heading are not as eye-catching as altitude deviations, and, for that reason, require more careful monitoring.

The altimeter provides the most pertinent information (primary) for pitch control in straight-and-level flight. As a rule of thumb, altitude corrections of less than 100 feet should be made using a half-bar width correction on the attitude indicator, and confirmed on the altimeter and VSI.

Maintain the desired heading in straight-and-level coordinated flight by establishing a zero bank on the attitude indicator and by monitoring the heading indicator for deviations from the desired heading. Cross check the turn coordinator to see if you are turning. When you see a heading deviation, use the attitude indicator to establish an angle of bank equal to the degrees deviation from heading. For example, if the aircraft has drifted 10° off the desired heading, establish a bank of 10°. For larger corrections, you should normally limit the bank angle so you do not exceed a standard-rate turn.

POWER CONTROL

During level flight you normally adjust pitch to maintain altitude and power to get the desired airspeed. The airspeed indicator is the only instrument that can tell you whether you are using the correct power to obtain the desired airspeed in straight-and-level flight. That makes it the primary power instrument. If you are not achieving the desired airspeed, estimate the power setting that is needed, and use your power instruments (manifold pressure gauge or tachometer) to make the adjustment. During a change in airspeed, the altimeter and heading indicator remain primary for pitch and bank, because your objective still is to maintain altitude and heading. The power instruments are primary for power during the adjustment, but be careful not to fixate on these instruments. Make a rough adjustment of the power controls and continue your scan. Look back to the power indicator and make final adjustments after the power instrument's indications have stabilized. If you have a fixed-pitch propeller, the sound of the engine helps you make rough adjustments without staring at the tachometer. When your power setting is established, check the airspeed indicator for the desired results. Continue monitoring your pitch instruments to verify you are maintaining altitude; otherwise, your airspeed indications will be affected. Watch your bank instruments closely because changes in power induce turning forces in most propeller airplanes that you must control. Remember to re-trim the airplane after the airspeed stabilizes following a power change.

The instrument which provides the most pertinent information (primary) for bank control in straight-and-level flight is the heading indicator.

To make suitable power adjustments, you need to know the approximate power required for a desired airspeed. Experiment with various flight configurations, and memorize the required power settings. Expect your instructor to help you with this early in your training. If you are making a large increase in airspeed, advance the power beyond the setting required for the new airspeed, just as you would do when accelerating your car to freeway speed. A similar procedure applies to large airspeed reductions. [Figure 2-35]

As power is reduced to change airspeed from high to low cruise in level flight, the instruments which are primary for pitch, bank, and power respectively, are the altimeter, heading indicator, and manifold pressure gauge or tachometer.

Figure 2-35. Underpowering can expedite a large airspeed reduction.

If your airspeed is too high and your altitude too low, or vice versa, you can trade one for the other by adjusting pitch. You may not need to adjust the power under these conditions. If both airspeed and altitude are high or low, you will need to make a larger power adjustment to correct the two conditions. [Figure 2-36]

Figure 2-36. Depending on the situation, a deviation in airspeed and altitude may or may not require you to adjust the power. Therefore, make sure you cross-check before making a power adjustment.

STANDARD-RATE TURNS

In small airplanes, you normally turn at a standard rate of three degrees per second. To enter a level turn, use the attitude indicator to establish an angle of bank you expect will result in a standard-rate turn. The attitude indicator is primary for bank while rolling into the turn. The altimeter is the primary instrument for pitch during roll-in and throughout the turn because there is no intended change in altitude.

BANK CONTROL

Whether conducting a level, climbing, or descending turn, the turn coordinator is the primary instrument for bank once established in the turn. This is because it is the only instrument that can tell you whether the angle of bank you have established is maintaining the desired rate of turn. The attitude indicator is a supporting bank instrument; use it to correct any deviations identified on the turn coordinator. [Figure 2-37]

The primary bank instrument while transitioning to any standard-rate turn is the attitude indicator. The primary pitch instrument during roll-in and throughout a constant altitude turn is the altimeter. The turn coordinator is a supporting bank instrument during roll-in, and becomes primary once the turn is established.

Rate of turn varies with true airspeed and the angle of bank. You can quickly estimate the approximate angle of bank required for a standard-rate turn by dividing the true airspeed in knots by 10 and adding 5 to the result. For example, according to this rule of thumb, the angle of bank required for a standard-rate turn at 110 knots is 11 plus 5, or 16°. The rate of turn at any given airspeed depends on the amount of sideward force causing the turn; that is, the horizontal component of lift. This varies directly in proportion to the bank in a coordinated turn, so the rate of turn at a given airspeed increases as the angle of bank increases, and the turn radius decreases. [Figure 2-38]

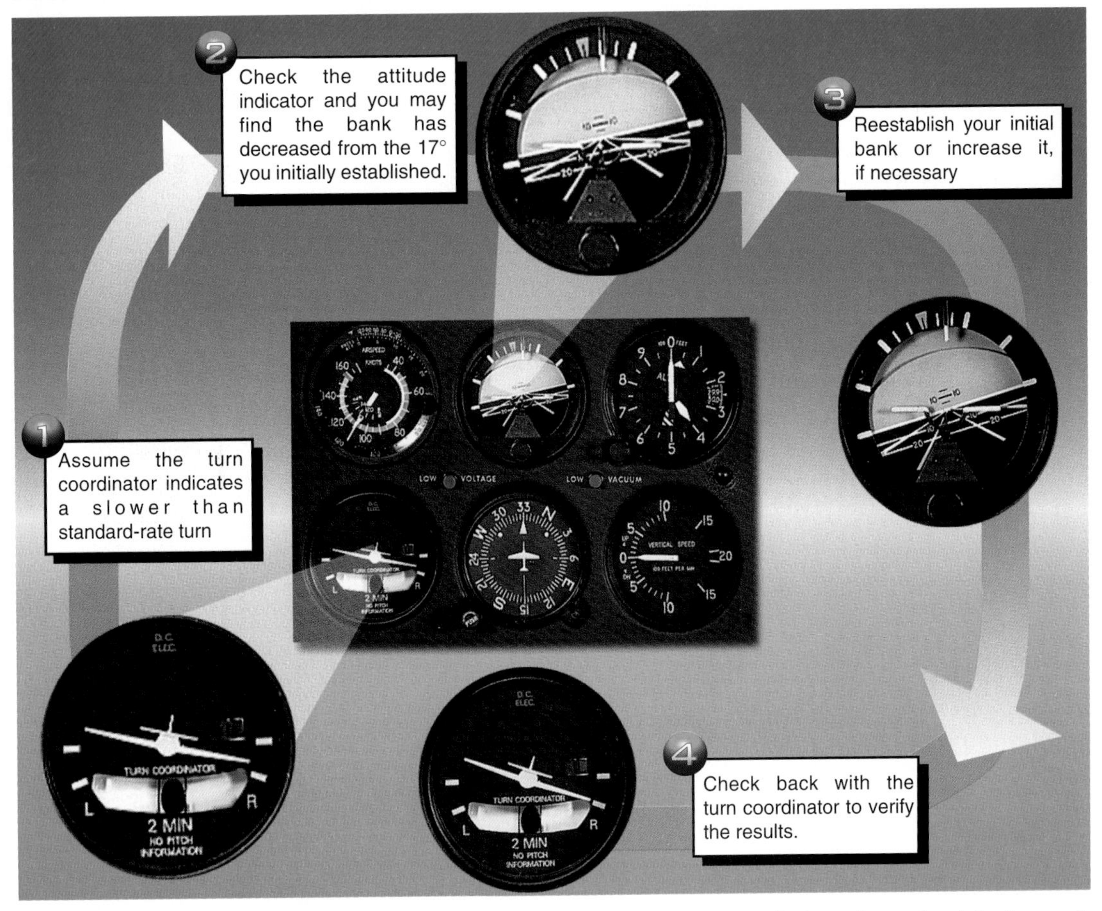

Figure 2-37. Once established in a standard-rate turn the turn coordinator is the primary bank instrument. This is because it supplies the most pertinent information about bank. In stabilized flight conditions, like this one, the primary instrument is one on which you try to maintain a constant indication.

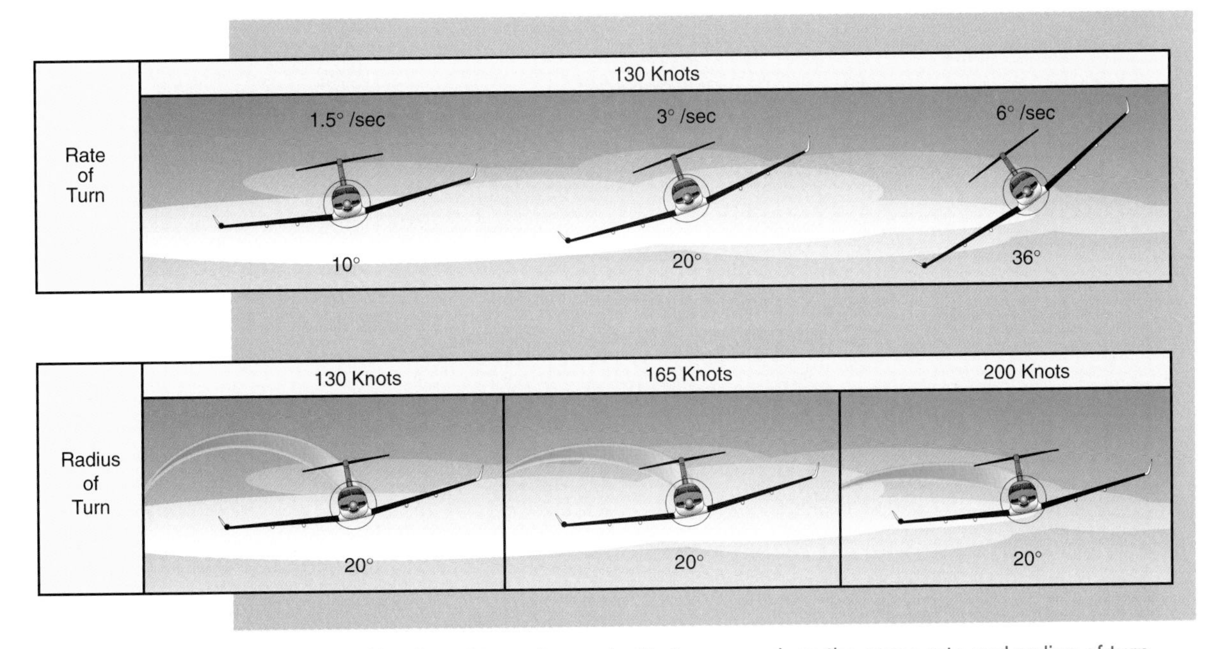

Figure 2-38. A specific angle of bank and true airspeed will always produce the same rate and radius of turn, regardless of aircraft type. If you increase only angle of bank, the rate of turn will increase. If you increase only the true airspeed, the radius of turn will increase. You can see that, as you increase airspeed during a level turn, the turn rate gets slower and the turn radius gets larger. On the other hand, a reduction in airspeed and/or an increase in bank angle will result in a faster rate of turn with a smaller radius.

To stop the turn on the desired heading, lead your roll-out by about one-half the angle of bank. For example, if in a 16° bank, begin your roll-out 8° before you reach the desired heading. With experience, you may choose a different leadpoint for your rate of roll-out, depending on your technique.

The rate of turn at any airspeed is dependent upon the horizontal component of lift. It can be increased and the radius of turn decreased by increasing the bank and/or decreasing airspeed. During a constant bank level turn, an increase in airspeed would cause the rate of turn to decrease, and the radius of turn to increase.

PITCH CONTROL

Remember to scan your pitch instruments during roll-in, roll-out, and throughout the turn. It is easy to get an unintended pitch change because of the loss of vertical lift in a turn. The nose tends to pitch down, but since most pilots automatically apply back pressure, overcompensation and a nose-up deviation can also occur. Careful monitoring of the attitude indicator helps you apply precise control adjustments because this instrument simultaneously displays pitch and bank information. Monitor the altimeter (the primary pitch instrument) to verify the correct pitch attitude. [Figure 2-39]

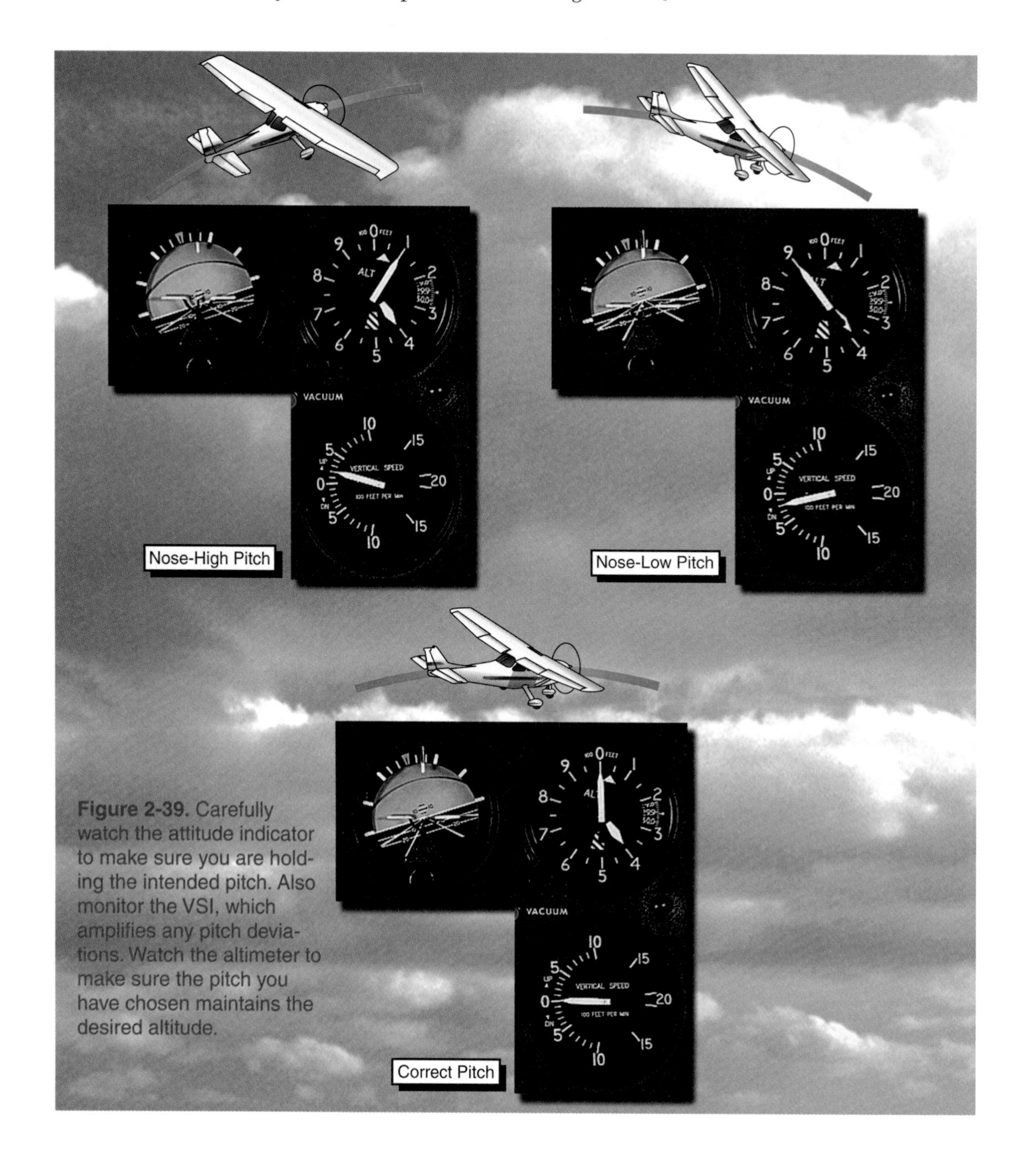

Figure 2-39. Carefully watch the attitude indicator to make sure you are holding the intended pitch. Also monitor the VSI, which amplifies any pitch deviations. Watch the altimeter to make sure the pitch you have chosen maintains the desired altitude.

If you reduce power to decrease airspeed in a level turn, you will lose some lift and must either increase the angle of attack and/or decrease the angle of bank to maintain altitude. Decreasing the angle of bank also is necessary to maintain the same rate of turn at a slower airspeed. Conversely, increasing the airspeed results in excess available lift, which requires a decrease in angle of attack and/or an increase in bank angle to avoid climbing. During airspeed changes, as with any maneuver where maintaining altitude is the objective, the altimeter is primary for pitch, supported by the attitude indicator and VSI. When rolling out of a turn, you must reduce the back pressure or trim that you used to maintain altitude during the turn. Use the attitude indicator with the VSI to adjust your pitch and monitor the results on the altimeter.

The primary reason the angle of attack must be increased to maintain a constant altitude during a coordinated turn is because the vertical component of lift has decreased as the result of the bank.

When airspeed is decreased in a turn, the angle of bank must be decreased and/or the angle of attack increased to maintain level flight. Conversely, when airspeed is increased during a level turn, additional vertical lift is generated. To avoid climbing, you must increase the angle of bank and/or decrease the angle of attack.

POWER CONTROL

An airplane tends to lose airspeed in a level turn because the increased angle of attack results in an increase in induced drag. To maintain speed, you need additional power. Although this effect is negligible at cruise speed, it can become significant as you slow to approach speed, where you should increase the power a certain amount while establishing a turn, rather than waiting for the airspeed to bleed off during the turn. [Figure 2-40]

When your objective is to maintain airspeed in a turn, your airspeed indicator is the primary instrument for power.

The power indicator is your primary power instrument for the brief period when you are adjusting the power in anticipation of a turn, or to change airspeed during a turn.

The attitude indicator and VSI are supporting instruments for pitch during a change of airspeed in a level turn. The airspeed indicator is considered primary for power as the airspeed reaches the desired value.

Figure 2-40. To control airspeed, use the power instrument(s) to make initial power adjustments, then use the airspeed indicator to confirm that the adjustment is correct.

STEEP TURNS

During instrument training, any turn greater than a standard rate is considered steep. Normally, you practice these turns with a 45° angle of bank. Although you normally would not fly this type of turn in IFR conditions, practicing it during your training helps you master controlling the airplane by instrument reference with greater-than-normal bank attitudes. This enables you to react smoothly, quickly, and confidently to unexpected abnormal flight attitudes in instrument flight conditions.

Steep turn techniques are similar to shallower turns, but the need for a higher angle of attack to compensate for the greatly reduced vertical component of lift magnifies any errors in pitch control. You need to speed up your cross-check, interpret the instruments accurately, and apply prompt, smooth control pressures. [Figure 2-41]

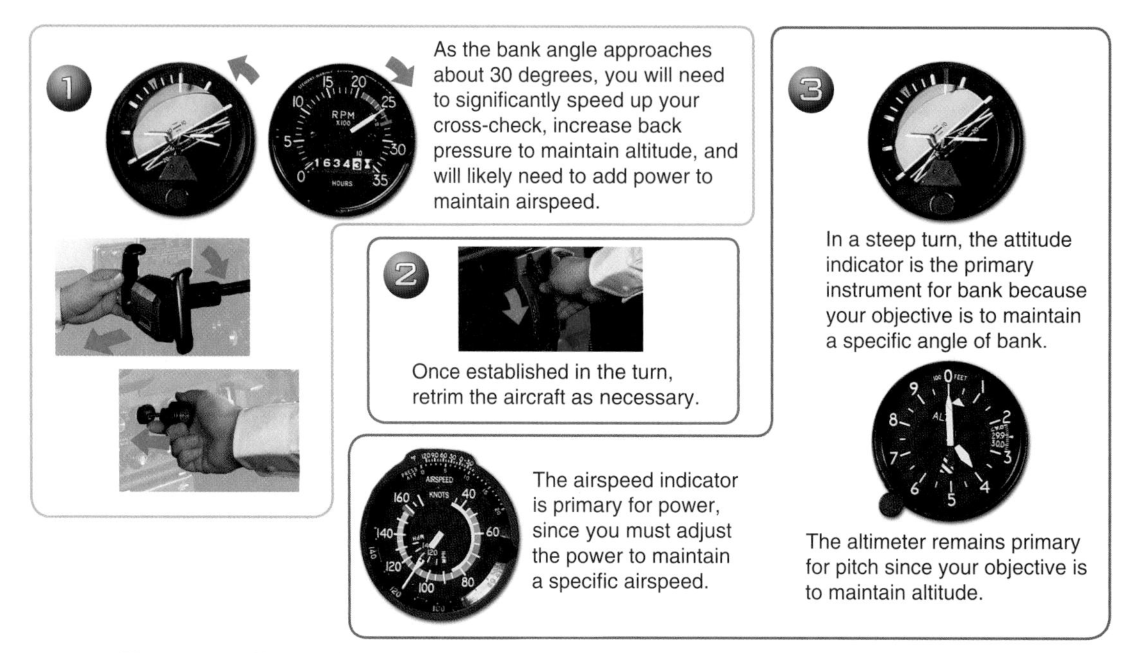

Figure 2-41. To establish a steep turn, roll in slowly, using the attitude indicator for pitch and bank reference.

Because the pitch indications of the attitude indicator can be more difficult to interpret in a steep-banked turn, the VSI becomes more important as a supporting pitch instrument. If it starts showing an undesirable trend, use the attitude indicator to make a specific pitch correction, then refer to the VSI and altimeter to verify you have arrested the altitude deviation.

The airplane usually will have a tendency to climb when you roll out of a steep turn because of the nose-up trim applied during the turn. Plan on pushing forward on the yoke during roll-out and maintaining an increased cross-check until the airplane is again trimmed for straight-and-level flight.

CLIMBS AND DESCENTS

There are two general categories — constant airspeed and constant rate. In practice, you normally pursue the best available rate of climb at a specific airspeed and power setting. When descending on an approach, you normally establish both a specific airspeed and rate of descent. To prepare for actual instrument flight, you will practice climbs and descents on constant headings and while turning.

CLIMBS

When entering a climb from straight-and-level flight at cruise airspeed, simultaneously increase the power to the climb setting and smoothly apply back pressure. Since the airplane tends to pitch up with additional power, a smooth, slow power application ensures you need only slight control pressure to make the desired pitch change.

CONSTANT AIRSPEED CLIMBS

Constant airspeed climbs include cruise climbs and climbs at best rate-of-climb (V_Y) and best angle-of-climb (V_X) speed. You add power, adjust the pitch attitude for the desired airspeed, and accept the resulting rate of climb. When established in a constant airspeed climb, your objective is to maintain a desired airspeed for a specific power setting. Whether climbing straight or turning, the airspeed indicator is the primary pitch instrument once you are stabilized, because its indications tell you whether pitch adjustments are necessary. [Figure 2-42]

During the pitch transition to a straight climb, the attitude indicator is the primary pitch instrument. Once you are established in a climb, the airspeed indicator becomes the primary pitch instrument with support from the attitude indicator. Throughout the climb, the attitude indicator acts as a supporting bank instrument.

Figure 2-42. To enter a constant airspeed climb, adjust the miniature airplane on the attitude indicator one to two bar widths above the horizon, depending on the aircraft and on the desired airspeed. The primary and supporting roles of the attitude indicator and airspeed indicator change from the transition to the steady-state climb.

CONSTANT RATE CLIMBS

In a **constant rate climb** you maintain a specific vertical velocity in addition to controlling airspeed. During IFR flight, you might use a constant rate climb within 1,000 feet of your assigned altitude, where ATC expects you to climb at 500 f.p.m. Because of limited horsepower, your training aircraft will likely be restricted to certain airspeed and climb rate combinations which you will learn with practice.

The proper way to transition from cruise flight to a climb at a specific speed is to increase back elevator pressure until the attitude indicator shows the approximate pitch attitude for that speed climb. The attitude indicator is primary for pitch during a change in pitch and primary for bank during a change in bank. Once stabilized in a straight or turning climb at cruise-climb airspeed, the primary pitch instrument is the airspeed indicator. See figure 2-42.

Since the VSI may lag six or more seconds, use the airspeed indicator to fine tune the pitch until the VSI stabilizes on the actual climb rate. If you intend to climb straight ahead, monitor the heading indicator (the primary bank instrument) throughout the transition and climb to make sure you are not turning. In addition, you should cross-check the supporting bank instruments, the attitude indicator and turn coordinator, for heading control.

Although the instrument indications for constant rate climbs are very similar to those for constant airspeed climbs, you emphasize different instruments in your scan. Once established in the climb, the VSI becomes the primary pitch instrument; if it indicates a lower-than-desired rate of climb, you need to pitch up further. The airspeed indicator is the primary power instrument; if it indicates low, you need to try a higher power setting, if available. [Figure 2-43] Be sure to consider the indications of both the VSI and the airspeed indicator before adjusting power. If airspeed is high and climb rate is low, pitching up, without adjusting the power, may correct the situation. If both airspeed and climb rate are low, you will need a larger power adjustment to restore the desired climb rate and maintain airspeed while pitching up.

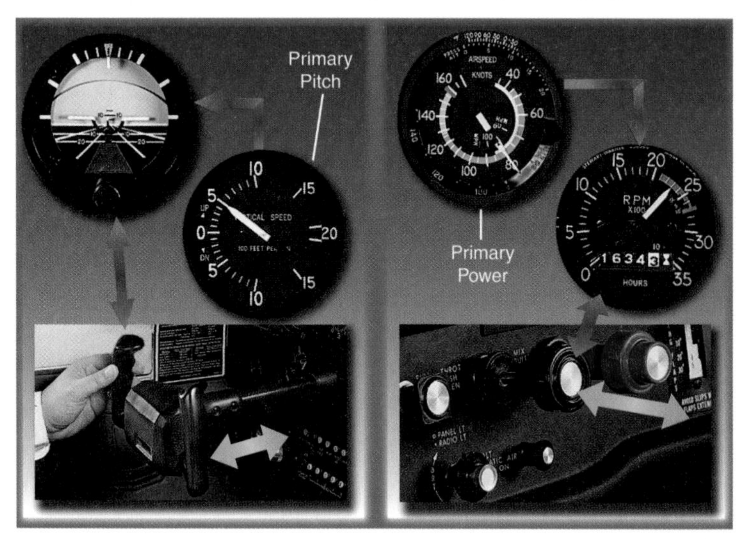

Figure 2-43. The primary pitch and power instruments for stabilized constant rate climbs differ from those for steady-state constant airspeed climbs. During a constant rate climb, monitor the VSI and make any necessary adjustments using the attitude indicator. For power, monitor the airspeed indicator and make any needed corrections using the manifold pressure gauge or tachometer.

DESCENTS

The attitude indicator and turn coordinator are supporting bank instruments during a straight, stabilized climb at a constant rate.

You can choose a constant airspeed descent when you are not concerned with achieving a specific rate of descent. Typically, you will use this procedure, sometimes called a cruise descent, when you descend from your assigned cruising altitude for arrival at your destination. During the approach phase, where precise control of the rate of descent is also important, you need to use a constant rate descent. [Figure 2-44]

Figure 2-44. The objective of a particular descent is reflected by the primary pitch instruments.

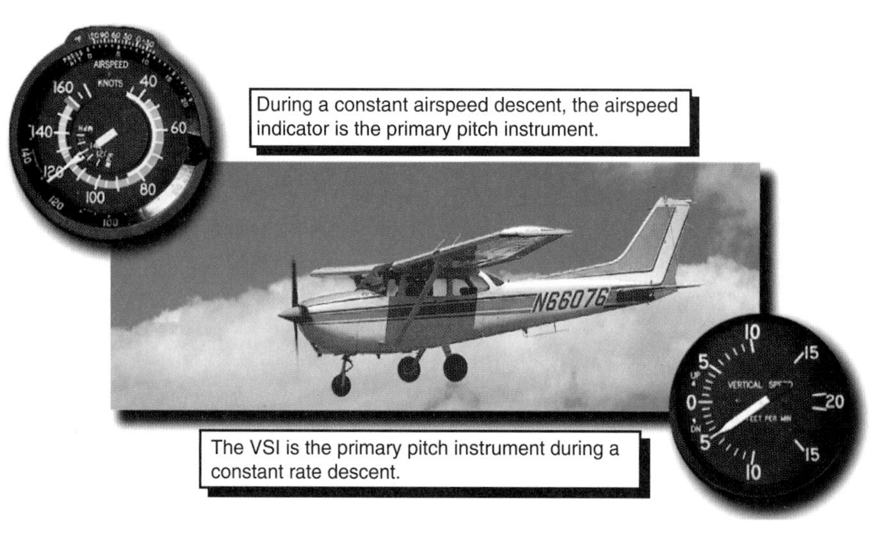

CONSTANT AIRSPEED DESCENTS

To enter a constant airspeed descent, reduce the power to the recommended value, simultaneously pitch down approximately one bar width to maintain airspeed, and accept the resulting rate of descent. When you reduce power, the aircraft may have a tendency to turn right, so expect to use slight left rudder pressure.

Depending on the airplane, if ATC wants you to maintain maximum forward speed, you may choose little or no power reduction when entering the descent. Even though you do not seek a specific vertical velocity on a constant airspeed descent, you should choose a pitch and power combination that results in a reasonable rate of descent. Use caution not to operate in the yellow arc, unless you are certain of smooth air throughout the descent. If you have a fixed-pitch propeller, make sure you reduce power enough to avoid excessive engine RPM. On high-performance piston engines, you will likely need to reduce power gradually, in accordance with the POH, to avoid damage to the engine from rapid cooling.

CONSTANT RATE DESCENTS

A constant rate descent differs from a constant airspeed descent in that you use pitch to control the rate of descent and power to control airspeed. This is true whether descending at cruise speed or approach speed. As with the constant airspeed descent, begin by simultaneously lowering the nose of the miniature airplane just below the attitude indicator's horizon bar and reducing the power to a predetermined setting.

To enter a constant airspeed descent from level cruising flight, and maintain cruising airspeed, you should simultaneously reduce power and adjust the pitch using the attitude indicator as a reference to maintain the cruising airspeed.

The VSI is the primary instrument for pitch once you are established in the descent, and its indication has stabilized. If, for example, it indicates too low a rate of descent, use the attitude indicator to pitch down further, then scan the VSI to verify the results. The airspeed indicator is primary for power. If the speed is too slow, add power. As with a constant rate climb, be sure to consider the indications of both the VSI and the airspeed indicator before adjusting power, since these factors affect each other.

LEVELOFF FROM CLIMBS AND DESCENTS

Leveloff procedures are similar whether climbing or descending. To avoid overshooting your desired altitude, it is important to lead your leveloff. The amount of lead depends on the aircraft, pilot technique, and the desired leveloff speed. As a guide, use 10% of the vertical velocity when you intend to maintain the descent speed during leveloff. For example, if your vertical velocity is 500 f.p.m., begin the leveloff 50 feet before you reach the desired altitude. In some situations, you may need to level off at an airspeed higher than your descent airspeed. If descending at 500 f.p.m., add power when you are 100 to 150 feet above the desired altitude.

The attitude indicator is the primary pitch instrument during the transition to level flight. The tachometer or manifold pressure gauge is primary for power as you move the throttle to the setting you estimate will give you the desired airspeed in level flight.

CLIMBING AND DESCENDING TURNS

Climbing and descending turns are combinations of the straight climb and descent procedures with turns. As always, expect changes in pitch forces when rolling into and out of turns, and use the attitude indicator to correct any deviations identified on the primary instruments.

COMMON ERRORS

Gaining proficiency in basic attitude instrument flight requires considerable practice. It is not unusual for beginning instrument students to encounter several areas of difficulty. You can go a long way to becoming proficient at attitude instrument flying if you can recognize a recurring problem and learn how to prevent it. Many solutions to common errors encountered by students are shown in figure 2-45.

To level off from a 500 f.p.m. descent while maintaining airspeed, lead the desired altitude by approximately 50 feet. Use approximately 10% of the vertical velocity to determine how far to lead the leveloff from a climb or descent. To level off at an airspeed higher than the descent speed, add power at approximately 100 to 150 feet above the desired altitude.

SOME COMMON ERRORS RELATED TO ATTITUDE INSTRUMENT FLYING

Error	Solution
Not correcting for pitch deviations during roll-out from a turn	Do not fixate on the heading indicator.
Large deviations from the desired attitude, heading, and airspeed	Do not tolerate small deviations. If you detect an error, initiate a correction immediately.
Consistent loss of altitude during turn entries	During turn entry, apply back pressure to compensate for the loss of vertical lift.
Consistent gain in altitude when rolling out from a turn	Make sure you relax back pressure during roll-out. If you added nose-up trim during the turn, use some forward pressure until you can trim for level flight.
Chasing the vertical speed indications	Use a proper cross-check of other pitch instruments and increase your understanding of the instrument characteristics.
Using excessive pitch corrections for the altimeter deviation	Do not rush the pitch correction. If you aggressively apply a large correction, you are more likely to aggravate the existing error.
Failure to maintain established pitch corrections	Continue to maintain your scan after making a correction. Also, make sure you trim off any control pressures.
Erratic control of airspeed and power	This can occur when you fixate on the airspeed indicator or manifold pressure/tachometer. Make sure your continue a good scan.
Overshooting the desired roll-out heading	Do not use too much bank angle for the amount of heading change. Also, continue to scan and begin your roll-out with the proper lead.
Unable to maintain heading consistently	Make sure you continue to cross-check the heading indicator and turn coordinator, particularly during changes in power and pitch attitude.
Failure to keep the heading indicator properly set	Periodically, include the magnetic compass in your scan.
Generally overcontrolling the aircraft	Maintain a light touch on the controls so you can feel the pressures.
Excessive trim control	Use the trim frequently and in small amounts.

Figure 2-45. Understanding common instrument flying errors and the ways to avoid them is beneficial to all instrument pilots.

COPING WITH INSTRUMENT FAILURE

The general aviation industry, the NTSB, and the FAA are concerned about the number of fatal aircraft accidents involving spatial disorientation of instrument-rated pilots who have attempted flight in the clouds with inoperative gyroscopic heading and attitude indicators. Light aircraft generally are not equipped with dual, independent, gyroscopic heading or attitude indicators and in many cases have only a single vacuum source. Therefore, the Instrument Rating Practical Test Standards stresses the need for instrument pilots to acquire and maintain adequate partial panel instrument skills and that they be cautioned to avoid over reliance on the gyroscopic instrument systems.

IDENTIFYING AN INSTRUMENT FAILURE

Although your training gives you some preparation in flying with a gyroscopic instrument failure, it is difficult to accurately simulate instrument failure the same way it might occur during actual IFR flight. While you know an instrument failure is being simulated when your instructor covers the attitude and heading indicators, actual failures are not nearly as obvious. An attitude indicator failure can be as subtle as continuing to indicate wings level while the airplane gradually drifts off into a diving spiral. It is important that you continually monitor all the instruments, to make sure the attitude indications correspond to other instrument indications. When there is a conflict, you must apply your knowledge and experience to determine which instrument(s) has failed.

If you suspect a problem with the attitude indicator, try a small control input, and see if the instrument responds. If it does not, then write it off immediately, and also suspect failure of the heading indicator, the other vacuum instrument. If it does respond as expected, it may be working, but you should still suspect an inaccurate indication due to precession. Ultimately, you must analyze the instrument indications from different instruments and systems and accept the instruments that correlate with each other. [Figure 2-46]

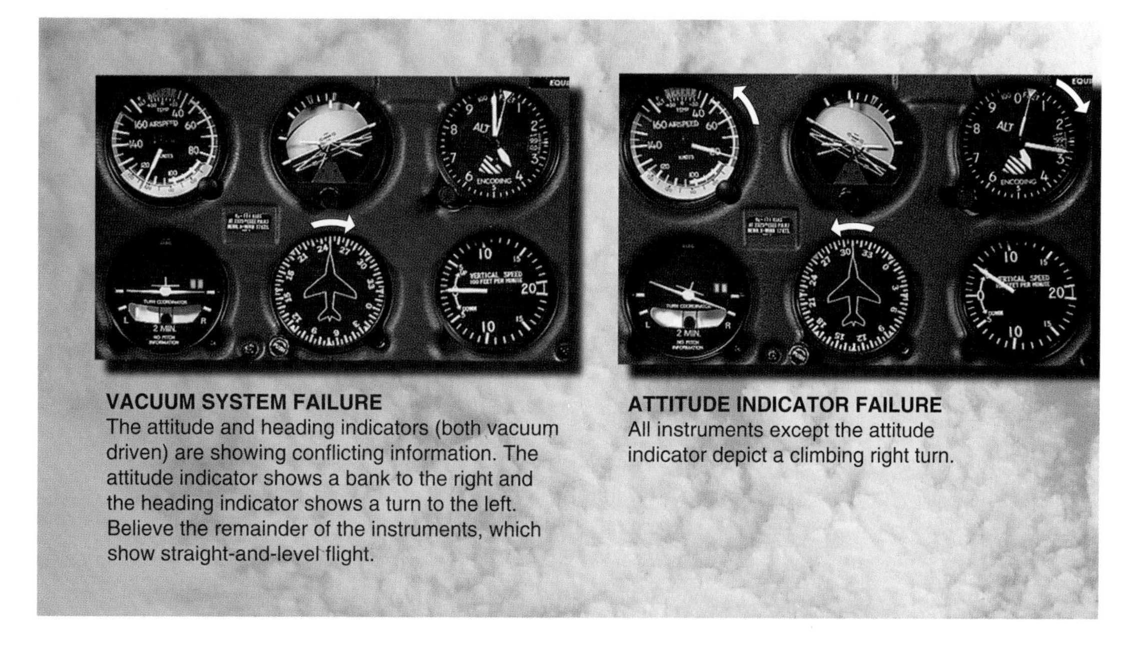

VACUUM SYSTEM FAILURE
The attitude and heading indicators (both vacuum driven) are showing conflicting information. The attitude indicator shows a bank to the right and the heading indicator shows a turn to the left. Believe the remainder of the instruments, which show straight-and-level flight.

ATTITUDE INDICATOR FAILURE
All instruments except the attitude indicator depict a climbing right turn.

Figure 2-46. When instruments give inconsistent indications, look for agreement between pitch instruments and between bank instruments.

ATTITUDE INDICATOR FAILURE

Jet and military aircraft usually have electrically powered attitude indicators and back-up systems. Most airliners are required by regulations to have three attitude indicators, and the third attitude indicator must have a separate source of power. Electric attitude indicators are equipped with an OFF flag which appears when the electrical power fails or the gyro is not operating at the proper speed. Some newer piston aircraft also have OFF flags on vacuum-powered attitude indicators. [Figure 2-47] If an electric attitude indicator fails, reset the circuit breakers in an attempt to restore the instrument. Do not let solving this problem distract you from aircraft control, which is a critical issue if flying in IFR conditions. Unless your aircraft is equipped with a manually activated backup vacuum pump, there is nothing you can do about the failure of a vacuum-powered attitude indicator. A recommended practice is to cover the failed instrument with a sticky note or other piece of paper to prevent it from distracting you with conflicting indications.

To identify an instrument failure, analyze the instrument indications from different instruments and systems and accept the indications of instruments that agree with each other. [Figure 2-46]

Figure 2-47. Cessna's new light aircraft have OFF flags on their vacuum-powered attitude indicators.

HEADING INDICATOR FAILURE

The heading indicator also tends to fail in a subtle way. You may believe you are doing an excellent job controlling your heading when, in fact, the heading indicator has simply failed. In many cases, you can avoid being caught by surprise if you check the heading indicator against the magnetic compass at 15 minute intervals.

While flying in IFR conditions, your first sign of a heading indicator failure may be when you experience difficulty tracking a VOR or GPS course. Suspect a failure of your heading indicator if your heading adjustments do not produce the expected results on your navigation instruments. If your VOR indicates an unusual deviation, you can easily make a small correction by using your turn coordinator to make a standard-rate turn for several seconds. Remember to make course corrections a little at a time and give the VOR needle some time to respond before making additional corrections. You will review VOR tracking procedures in Section C of this chapter.

PARTIAL PANEL FLYING

If you experience failure of your vacuum-powered gyroscopic instruments in IFR conditions, you will need to transition to flying partial panel, that is, you will be controlling the aircraft primarily with reference to the altimeter, airspeed indicator, turn coordinator, VSI, and magnetic compass. To help prepare you for this possibility, you will practice partial panel flying during your training. In addition, you will be required to demonstrate basic attitude instrument flying on partial panel and be required to fly a partial panel instrument approach during the practical test for your instrument rating.

If your gyroscopic attitude indicator and/or heading indicator fails in instrument conditions, you should immediately inform ATC even if control of the aircraft is not an immediate problem. Depending on the conditions, loss of the attitude indicator can be particularly serious, since it may be difficult to comply immediately and accurately with ATC instructions. In many cases, your best option is to divert to a nearby airport with more favorable weather conditions. If you must conduct a partial panel approach in IFR conditions, try to get a radar approach. These approaches, during which ATC provides

you horizontal, and, in some cases, vertical course guidance, are covered in detail in Chapter 10, Section A, IFR Emergencies.

STRAIGHT-AND-LEVEL FLIGHT

Since the altimeter normally is your primary pitch instrument, you still have all the information you need to maintain level flight when flying partial panel. However, if your attitude indicator fails, flying straight-and-level is more difficult because the remaining instruments do not provide you with an instantaneous pitch indication. Besides the altimeter, the VSI and the airspeed indicator are your next best instruments for pitch information. Yet, because of lag, these instruments also do not immediately indicate the exact result of a pitch adjustment. Therefore, when flying partial panel, it is even more important to apply gentle, precise control inputs, and patiently watch for the results on your primary instruments. If you inadvertently enter a climb or descent, your first objective is to reestablish level flight, then gently correct back to your intended altitude. If you pay attention to the speed at which the altimeter is moving up or down, it can help you determine the amount of deviation from level flight attitude and the approximate correction you need to make.

Without the heading indicator, you must use the magnetic compass and turn coordinator to maintain heading. While doing so, it is important to keep the miniature airplane as level as possible. Even barely visible deviations on the turn coordinator can easily result in heading errors of 30° or more within one minute, and the problem is even worse with the older, less precise, turn and slip indicators. Before making adjustments, give the compass time to stabilize in straight-and-level, unaccelerated flight, unless it is obvious that you are significantly off your desired heading. If the compass card bounces or swings, you can average the readings and correct back to your desired heading using timed, standard-rate turns. If you are 10° away from your desired heading, try a three second standard-rate turn.

TURNS

Use your turn coordinator to establish and maintain partial panel turns. Control of bank is easier than pitch, even with the loss of both the attitude indicator and heading indicator. The turn coordinator, which is your primary instrument for bank, responds quickly enough to adjustments in bank to enable good bank control without the attitude indicator.

If the gyroscopic heading indicator is inoperative, the primary bank instrument in unaccelerated straight-and-level flight is the magnetic compass.

Without the support of the attitude indicator, you must use the altimeter, VSI, and airspeed indicator to maintain pitch the same as in straight-and-level flight. But, without the attitude indicator, it is more challenging to accurately compensate for the pitch-change tendencies while maneuvering or during power/airspeed transitions. Gentle control technique, with some knowledge and experience of the required control pressures, helps avoid overcontrolling. Having the airplane properly trimmed before entering a turn is another key factor. Finally, you should increase your scan rate to compensate for the lack of the attitude indicator.

COMPASS TURNS

When you use the compass to turn to a northerly heading in the northern hemisphere, remember to roll out before the compass reaches the desired heading. When you turn to a southerly heading, wait until the compass passes the desired heading. As you learned in Section A of this chapter, compass turns should be at standard rate, and the amount of lead or lag approximately equals your latitude. Turning error is small when turning to an easterly or westerly heading; no special correction is needed. When you turn to north-

Creating a Renaissance in General Aviation

In the mid 1990s, the Advanced General Aviation Transport Experiments (AGATE) consortium was formed with the goal of developing a blueprint for a simpler, more viable medium range (150-700 nautical miles) aircraft than is currently provided in the general aviation market. This joint government/industry team comprises more than 70 members, including participants from NASA, the FAA, universities, non-profit organizations, and over 30 companies from the aviation industry.

The thrust of the AGATE initiative is to develop affordable new technology, as well as the industry standards and certification methods for airframe and cockpit design, flight training systems, and airspace infrastructure for the next generation single-pilot, 4-6 place, near all-weather light airplanes. The new aircraft [Figure A] is envisioned to revitalize a stagnant general aviation industry and provide transportation as safe, comfortable, cost-effective, and reliable as travel in an automobile or on an airliner.

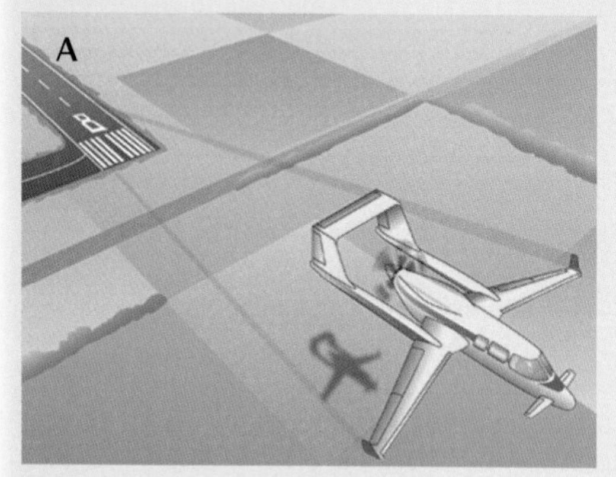

In response to the AGATE mandate, a futuristic cockpit layout has been developed. [Figure B] In the design, traditional light aircraft instrumentation has been replaced by a more intuitive multi-function display (MFD). Through the MFD, the pilot will be able to access up-linked weather data, GPS navigation information, data-linked ATC communications, terrain and traffic advisories, a PC-based flight planning package, and a multitude of other features. Critical attitude instrument flight information is shown on the primary flight display (PFD) and will also be available through a head-up display in front of the windscreen, facilitating the transition from an instrument scan to visual flight. The PFD is also called the "Highway in the Sky" since it provides a visual pathway for the pilot to follow.

One of the integral parts of the AGATE project is a training system which will use advanced technologies and techniques to allow a student to progress through pilot training more quickly and at a cost that competes with other transportation options. If all goes according to plan, learning to fly on instruments and maintaining your proficiency will be easier than ever. Nevertheless, the basics of instrument flight are unlikely to undergo significant changes in the near future. A thorough understanding of the airplane's instrument systems, as well as the ability to maintain a solid scan of pertinent data will still be necessary. In the end, AGATE may allow general aviation to really take off, yet a pilot lacking in fundamental instrument flight techniques will be stuck on the ground.

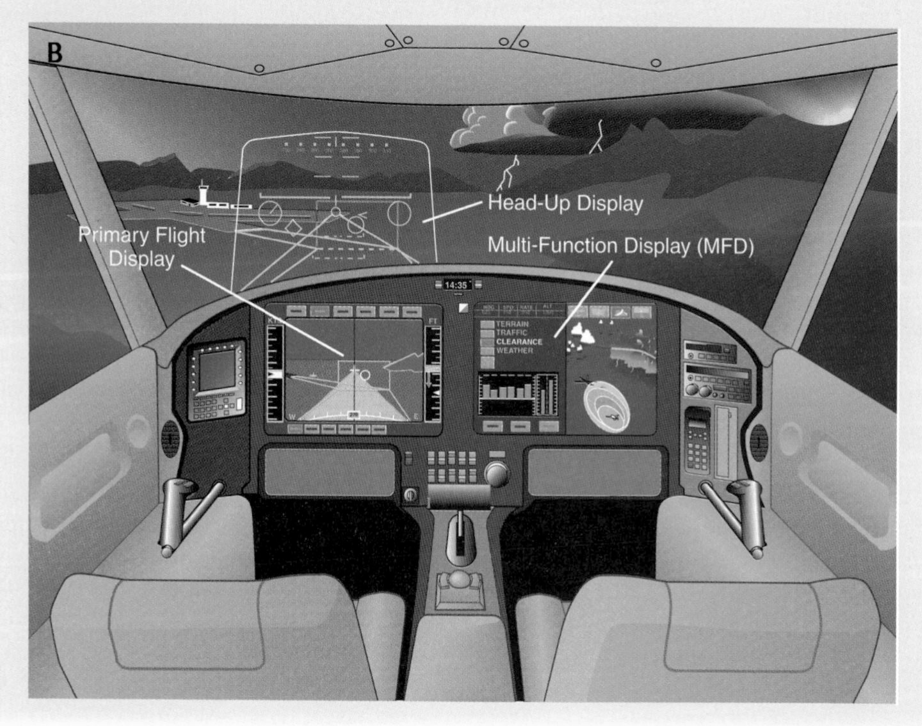

west, northeast, southwest and southeast headings, it may be necessary to use some lead or lag to account for turning error.

While executing compass turns, it is common for instrument students to fixate on the compass during rollout. Remember that until the airplane is stabilized in straight-and-level, unaccelerated flight, the indicated heading is not accurate. It is better to concentrate on other instruments to maintain straight-and-level flight before checking the accuracy of your turn.

TIMED TURNS

A **timed turn** is the most accurate way to turn to a specific heading without the heading indicator. In a timed turn, you use the clock instead of the compass card to determine when to roll out. For example, using a standard rate turn (3° per second), an airplane turns 45° in 15 seconds. You can still use the magnetic compass to back up the clock, when determining the time to roll out of a turn.

Prior to practicing timed turns, you need to determine the accuracy of the turn coordinator. Establish a standard-rate indication on the instrument for 30 seconds and determine whether the airplane turns 90°. If not, keep repeating the turn, adjusting the bank until you find the turn indication that corresponds to an actual standard-rate turn in each direction. Make a mental note of that indication and use it for standard-rate turns.

Divide the degrees of desired heading change by three degrees per second for a standard-rate turn, to get the number of seconds of turn. Start the roll-in when the clock second hand passes a cardinal point, hold the turn at the calibrated standard-rate indication, and begin the roll-out after the computed number of seconds. Do not count the time to roll in and out of the turn, and consider using half-standard-rate turns for small heading changes.

CLIMBS AND DESCENTS

Without the support of the attitude indicator, it is easier to control the transition to a climb if you first slow the airplane to a reasonable climb speed. Although you do not have the benefit of an instantaneous attitude indication, you still have instruments to set climb or descent power. Accurately setting power helps the airplane react predictably and makes it easier to control. However, avoid fixating on the power indicator, especially in partial panel conditions when you need an increased scan rate. Glance at the tachometer or manifold pressure gauge, move the throttle an estimated amount to correct any deviations, and move on with your scan. Check the results of the adjustment in a couple of seconds when your scan returns to the power instrument(s).

Use the altimeter, VSI, and airspeed indicator in place of the attitude indicator to make changes in pitch. Because of the lag of these instruments, it is very important to make smooth, gradual control inputs and allow a few moments for the change in pitch to be reflected on these instruments. The rate of movement of the altimeter also gives you indirect pitch information, but requires more interpretation than the VSI.

Partial panel bank control during climbing and descending turns is essentially the same as during level turns. Use the compass turn techniques described earlier to determine when to roll out of a turn. Although the most challenging part of partial panel turns is controlling pitch, it is essential you closely monitor your turn coordinator to prevent overbanking which could lead to a diving spiral. Carefully monitor the altimeter, VSI, and airspeed indicator to determine if rolling in or out of the turn has caused undesired deviations.

PITOT-STATIC INSTRUMENT FAILURES

Most of your partial panel training will focus on failure of gyroscopic flight instruments. It is almost impossible in flight to simulate the types of failures that affect the airspeed indicator, altimeter, and VSI discussed in Section A of this chapter. You may wish to consider practicing some of these instrument failures in a simulator, flight training device, or personal computer aviation training device (PCATD). If you do not experience these failures during your training, they will likely be unexpected and confusing when they occur.

The most insidious of these failures is total blockage of the pitot system, discussed in more detail in Section A of this chapter. Blockage of both the ram air inlet and the drain hole causes the airspeed indicator to react like an altimeter, moving in the opposite direction as you would expect the airspeed indicator to move in a climb or descent. [Figure 2-48] In-flight breakups of aircraft have occurred because the pilot dove the airplane at speeds exceeding red line while the airspeed indication dropped toward stall speed. It is important to use your experience in the aircraft when evaluating the indications of the airspeed indicator. If the airplane sounds like it is going fast, and the airspeed indicator decreases even more rapidly when pushing forward on the yoke, you should suspect your airspeed is not really dropping. If you do not hear the stall warning as the airspeed indicator drops to the bottom of the green arc, and the altimeter shows a descent, it is time to revise your thinking about what the airspeed indicator is telling you.

Blocked Pitot System
The airspeed indicator is increasing, implying a dive, which conflicts with the climb information shown by the attitude indicator, altimeter, and VSI. You can trust the attitude indicator, because it agrees with every instrument but the airspeed indicator.

Failing Airspeed Indicator
All instruments except the airspeed indicator suggest a level right turn. The pitot ram air input may be icing over.

Figure 2-48. A good instrument cross-check can help you identify instrument failures.

Although the altimeter is not immune to failure, it cannot fail in such a way as to show a climb when you are actually descending. When you discover the airspeed indicator is deceiving you and realize you are in a dangerous high-speed dive, immediately reduce power, verify wings level and very slowly adjust pitch to a level flight attitude.

UNUSUAL ATTITUDE RECOVERY

An unusual flight attitude can occur even with a properly trained instrument pilot, due to failure of the attitude indicator, disorientation, wake turbulence, lapse of attention, or abnormal trim. To become instrument-rated, you must demonstrate recovery from unusual attitudes with or without the attitude indicator.

To practice **unusual attitude recoveries**, your instructor will have you pass the controls and close your eyes, while he/she maneuvers the aircraft in such a way as to induce disorientation. The instructor will then have you open your eyes and take control of the airplane to make a proper recovery. In most cases, you should recover from unusual attitudes by executing the appropriate corrective control applications in sequence, but nearly simultaneously.

NOSE-HIGH ATTITUDE

The indications of a nose-high attitude are decreasing airspeed, rapid gain in altitude, a high rate of climb, and a nose-high attitude on the attitude indicator. To recover, you should add power, apply forward elevator pressure to prevent a stall, and correct the bank by applying coordinated aileron and rudder pressure to level the miniature aircraft and center the ball of the turn coordinator. [Figure 2-49]

If the airspeed is decreasing or below the desired airspeed, add power. Use the attitude indicator to lower the nose and level the wings.

If you suspect the attitude indicator is inoperative, apply just enough forward pressure to reverse the movement of the airspeed indicator and altimeter and to start the VSI moving toward zero.

If you suspect the attitude indicator is inoperative, use the turn coordinator to level the wings.

Figure 2-49. If your attitude indicator indicates a nose-high attitude confirmed by other instruments, you can quickly recover using the attitude indicator. If your first indication of an unusual attitude is indications from other instruments, and those instruments agree with each other, then disregard the attitude indicator and use the other instruments for recovery.

NOSE-LOW ATTITUDE

The indications of a nose-low attitude are increasing airspeed, rapid loss of altitude, a high rate of descent, and a nose-low attitude on the attitude indicator. The objective of a nose-low unusual attitude recovery is to avoid a critically high airspeed and load factor, as well as preventing the loss of altitude. To correct you should reduce power to prevent excessive airspeed and loss of altitude. Then, use coordinated aileron and rudder pressure to level the miniature

The correct sequence for recovery from the unusual attitude indicated in figure 2-49 is to add power, lower the nose, level the wings, and return to the original attitude and heading.

If the airspeed is increasing, or above the desired airspeed, reduce power to prevent excessive airspeed. Use the attitude indicator to first level the wings and then gently raise the nose to a level pitch attitude.

If you suspect the attitude indicator is inoperative, use the turn coordinator to level the wings.

Figure 2-50. If your attitude indicator indicates a nose-low attitude confirmed by other instruments, you can quickly recover using the attitude indicator. If your first indication of an unusual attitude is from other instruments, and those instruments agree with each other, then disregard the attitude indicator and use the other instruments for recovery.

If you suspect the attitude indicator is inoperative, use the altimeter and VSI to recover from the dive after leveling the wings. Apply just enough back pressure to reverse the movement of the airspeed indicator and altimeter and start the VSI moving toward zero.

SPIRALING OUT OF CONTROL

From the files of the NTSB. . .

Aircraft: *Cessna P210N*
Injuries: *5 Fatal*

Narrative: *The pilot called center while airborne and requested an IFR clearance. About 15 minutes after receiving the clearance, he was given a vector for traffic. The pilot said he needed help and asked if he was on course, then said he was in an unusual attitude and couldn't determine why. About 15 seconds later the pilot indicated he might be in a spin. The controller said he was showing the airplane as level and asked if the pilot was having problems, to which he responded, "yes our vacuum . . . our artificial gyro is showing us at a very unusual attitude." The airplane was observed coming out of the clouds with wings separating from the airplane. The airplane had a vacuum system problem which was recorded as repaired and checked as OK about 2 1/2 months before the accident. The pilot had about 4 hours of actual and about 56 hours simulated instrument time. The vacuum pump was destroyed in the impact.*

The NTSB listed the following probable causes of the accident:
1) A partial loss of the airplane's vacuum system which resulted in the total loss of the directional gyro and the attitude indicator, 2) the pilot not maintaining aircraft control due to spatial disorientation, and 3) the pilot not correcting for a spiral by using proper recovery techniques which resulted in the overload failure of the wings. A factor in the accident was the pilot's lack of total instrument flight experience.

At the first indication of an instrument failure, you should concentrate on maintaining control of the airplane using the properly functioning instruments. By flying the airplane first, then navigating, and finally communicating, you provide yourself with the best opportunity to maintain situational awareness and avoid instances where you become spatially disoriented.

If an unusual attitude situation develops despite your best efforts, it is essential that you remain calm and not suddenly pull back on the yoke when the airspeed is running away or the aircraft is in a turn. Reduce power to idle and level the wings before trying to pull out of an unusual diving attitude. You may even consider extending the landing gear to help regain control of the airspeed.

aircraft and center the ball in the turn coordinator. Finally, use smooth back pressure to raise the nose to a level flight attitude. [Figure 2-50]

PARTIAL PANEL UNUSUAL ATTITUDE RECOVERY

During your training, your instructor may require you to recover from unusual attitudes while flying partial panel. When recovering from an unusual attitude with partial panel, use the turn coordinator to stop a turn, and the pitot-static instruments to arrest an unintended climb or descent. You know you are passing through a level pitch attitude when you stop and reverse the direction of the altimeter and airspeed indicator, and start the VSI needle moving back toward zero. At this point, use the elevator pressure that maintains this pitch, give the instruments a moment to stabilize, then gently correct back to your desired altitude.

STALLS

You may practice stalls during instrument training to demonstrate that the recognition and recovery procedures under instrument conditions are the same as under visual conditions. Recover from a stall by immediately reducing the angle of attack and increasing power to the maximum allowable value. [Figure 2-51] Be careful to keep the ball centered on the turn coordinator throughout the stall because the wing will drop on the side opposite to the ball displacement in an uncoordinated stall and can easily result in an unusual attitude.

If an airplane is in an unusual flight attitude and the attitude indicator has exceeded its limits, the airspeed and altimeter should be relied on to determine pitch attitude before starting recovery. When recovering without the aid of the attitude indicator, approximate level pitch attitude is first attained when the airspeed and altimeter stop their movement and the VSI reverses its trend.

Figure 2-51. To enter a stall, establish and hold a pitch attitude that results in a stall. To recover, pitch down to a level or slightly nose-low flight attitude and add power.

CONTROL AND PERFORMANCE CONCEPT

Widely used by military pilots, the control and performance concept of attitude instrument flying focuses on controlling attitude and power as necessary to produce the desired performance. This concept is well suited to high-performance airplanes since these aircraft tend to have more precisely calibrated attitude indicators as well as backup attitude indicators, so heavy reliance on that instrument is more prudent. The control and performance concept divides the instruments into three categories: control, performance, and navigation. [Figure 2-52]

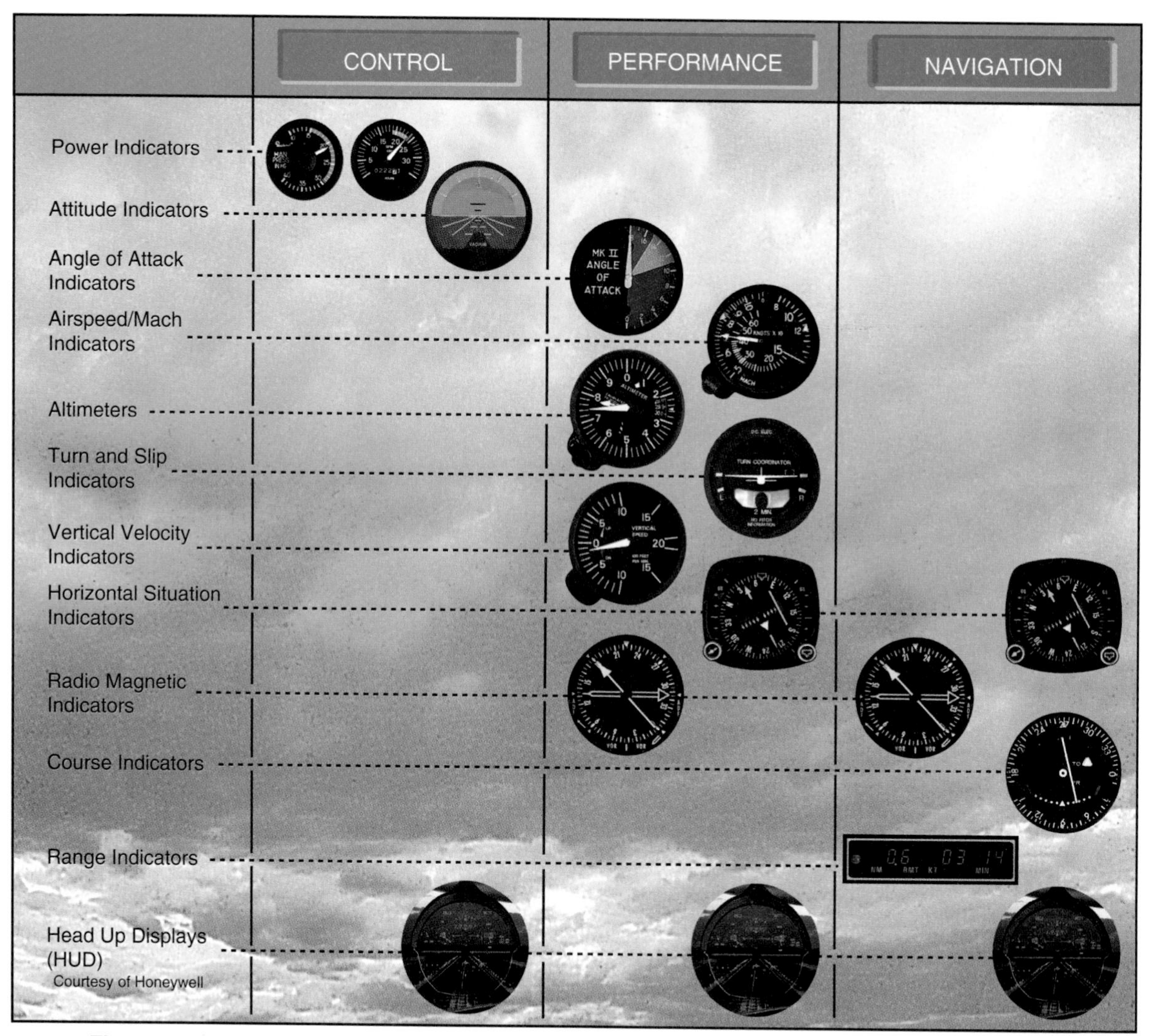

Figure 2-52. Control instruments directly indicate pitch and power. Performance instruments indicate how the aircraft responds to changes in pitch and power. Navigation instruments (discussed in Section C of this chapter) indicate the position of the aircraft relative to a facility or fix.

Essentially, a pilot using the control and performance method is controlling the angle of attack and thrust or drag relationship of the aircraft. Since the attitude indicator provides an immediate and direct indication of any change in aircraft pitch and bank attitude, proper control and performance technique requires heavy reliance on the attitude indicator. Much of your attention is devoted to precisely maintaining the desired indications on this instrument. The attitude indicator is the only instrument you should observe continuously for any appreciable length of time and the instrument you should check the greatest number of times. [Figure 2-53]

Figure 2-53. The recommended scan for the control and performance method is often compared to a wagon wheel where the attitude indicator is the hub and the spokes represent the performance instruments. While scanning, you glance from the attitude indicator to a performance instrument, back to the attitude indicator, then to another performance instrument, back to the attitude indicator, and so on.

Under the control and performance concept, you set the power and attitude using the power indicator and attitude indicator, the **control instruments**, and then trim to neutralize the control pressures. Next, check the **performance instruments** for the desired results. If a correction is necessary, use the control instruments to make any needed adjustments. [Figure 2-54]

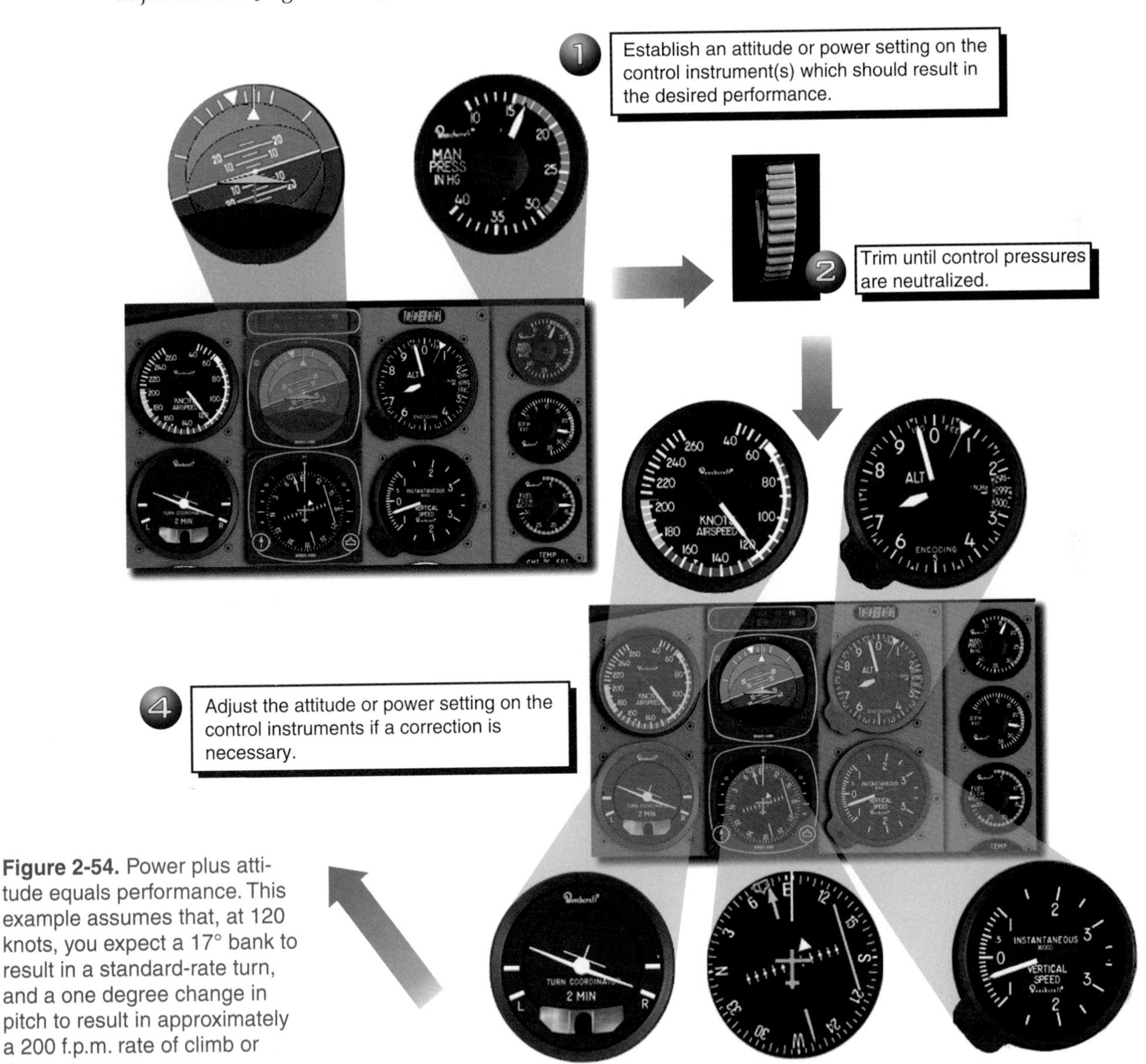

Figure 2-54. Power plus attitude equals performance. This example assumes that, at 120 knots, you expect a 17° bank to result in a standard-rate turn, and a one degree change in pitch to result in approximately a 200 f.p.m. rate of climb or descent.

The control and performance concept relies on the fact that power and pitch combinations yield certain performance results. Power control is necessary to maintain airspeed in conjunction with attitude changes. From experience, you learn how far to move the throttle to make a given change in power. Then, cross check the power indicators to establish a more precise setting. Good trim technique is essential for maintaining a constant attitude, and it allows you to devote more time to other cockpit duties. As you know, changes in attitude, power, or configuration usually require a trim adjustment. With the control and performance method, there is emphasis on monitoring the control instruments, and confidence that the desired results will follow from well-executed control inputs.

SUMMARY CHECKLIST

✓ Attitude instrument flying consists of three fundamental skills. These are instrument cross-check, instrument interpretation, and aircraft control.

✓ Instrument cross-check, or scan, requires logical and systematic observation of the instrument panel. The most common scanning errors are fixation, omission, and emphasis.

✓ Effective instrument interpretation requires a good working knowledge of how each instrument operates.

✓ Aircraft control is the result of instrument cross-check and interpretation. It requires that the airplane be kept properly trimmed so small flight control movements can achieve precise adjustments to pitch, bank, and power.

✓ Primary instruments provide the most pertinent pitch, bank, and power information for a given flight condition. Supporting instruments provide additional pitch, bank, and power information to help you maintain the desired indications on the primary instruments.

✓ Supporting instruments are no less important than primary instruments. The attitude indicator, although usually a supporting instrument, is essential and central to your scan.

✓ The attitude indicator is the primary pitch instrument during any change in pitch and the primary bank instrument during any change in bank.

✓ The altimeter is the primary pitch instrument any time your objective is to maintain altitude. The heading indicator is the primary bank instrument any time your objective is to maintain straight flight.

✓ The vertical speed indicator (VSI) is the primary pitch instrument any time your objective is to maintain a specific rate of climb or descent. The turn coordinator is the primary bank instrument any time your objective is to maintain a specific rate of turn.

✓ The airspeed indicator is the primary power instrument any time your objective is to maintain a constant airspeed during level flight. It is the primary pitch instrument during a constant airspeed climb or descent.

✓ In a constant airspeed climb, set climb power, pitch up to get a specific airspeed, and accept the resulting rate of climb. In a constant rate climb, maintain a specific vertical velocity in addition to controlling airspeed.

- ✓ To enter a constant airspeed descent, reduce the power, pitch down to maintain airspeed, and accept the resulting rate of descent. In a constant rate descent, control the rate of descent with pitch and control airspeed with power.

- ✓ Although you will become proficient in partial panel instrument flying, loss of the attitude indicator in IFR conditions is a potential distress situation under which you should advise ATC.

- ✓ Instrument failures can be subtle. If you suspect an instrument failure, look for corresponding indications among various instruments.

- ✓ A timed turn is the most accurate way to turn to a specific heading without the heading indicator.

- ✓ Use the VSI and airspeed indicator to make changes in pitch when flying with inoperative gyroscopic instruments. Use smooth, gradual control inputs and allow a few moments for the change in pitch to be reflected on these instruments.

- ✓ When recovering from a nose-high unusual attitude, your objective is to avert a stall. When recovering from a nose-low unusual attitude, your objective is to avoid overstressing the airplane structure, as well as an excessive loss of altitude. When recovering partial panel, use the turn coordinator to stop a turn, and the pitot-static instruments to arrest an unintended climb or descent.

- ✓ In the control and performance concept, you use the control instruments, such as the manifold pressure gauge and the attitude indicator, to set up power/attitude combinations for specific maneuvers. Then, you check the performance instruments for the desired effect.

KEY TERMS

Cross-Check

Fixation

Omission

Emphasis

Instrument Interpretation

Aircraft Control

Primary/Support Concept

Pitch Instruments

Bank Instruments

Power Instruments

Primary Instruments

Supporting Instruments

Constant Airspeed Climb

Constant Rate Climb

Constant Airspeed Descent

Constant Rate Descent

Partial Panel

Timed Turn

Unusual Attitude Recoveries

Control and Performance Concept

Control Instruments

Performance Instruments

QUESTIONS

Match the following concepts with the appropriate statements.

1. Interpretation	A. Looking at only one instrument
2. Cross-check	B. Instrument that offers the most pertinent information about pitch, bank, or power
3. Fixation	C. Requires good working knowledge of instruments
4. Control/Performance Concept	D. Control instrument
5. Primary Instrument	E. Looking at the right instrument at the right time
6. Attitude Indicator	F. Power plus attitude equals performance

7. What is the difference between the scanning errors of fixation and emphasis?

8. What are three things which can affect the pitch attitude required to maintain level flight?

Match the instruments in the photos with the descriptions in Questions 9-14. List all instruments that apply; you may use an instrument more than once.

9. Pitch instruments

10. Bank instruments

11. Power instruments

12. Control instruments

13. Performance instruments

14. Primary transition instrument

15. What is the primary pitch instrument during straight-and-level flight, and during level turns?

    A. Altimeter
    B. Attitude indicator
    C. Airspeed indicator

16. What is the primary bank instrument during straight-and-level flight and during straight climbs and descents?

    A. Turn coordinator
    B. Heading indicator
    C. Attitude indicator

17. What is the primary power instrument during straight-and-level flight?

    A. Tachometer
    B. Airspeed indicator
    C. Manifold pressure gauge

18. What is the primary pitch instrument during a transition from straight-and-level flight to a straight constant airspeed climb? What is the primary bank instrument?

19. What is the primary bank instrument while rolling into a level standard-rate turn? What is the primary pitch instrument?

20. Describe the proper sequence for recovering from a nose-low, turning, increasing airspeed, unusual flight attitude.

SECTION C
INSTRUMENT NAVIGATION

This section assumes you know how to use a VOR for VFR navigation. For IFR operations you must navigate more precisely than you would as a VFR pilot. You also need to refine your VOR interpretation skills to the point where you can easily visualize your position without any outside visual reference. In addition, you are more likely be flying complex or high-performance airplanes with more sophisticated navigation equipment and instrumentation.

VOR NAVIGATION

The cornerstone of aeronautical instrument navigation is the VOR. It is likely the first navigation system you learned, and is expected to be a part of the U.S. airspace system well into the twenty-first century. Even if VOR is eventually supplanted as the primary tool of instrument navigation, the instruments used to indicate your position probably will continue to look and work similar to the present VOR indicators. [Figure 2-55]

Figure 2-55. This VOR display is currently switched to a global positioning system (GPS) receiver. Today's VOR indicators are likely to appear in airplanes for many years to come.

There are various types of indicators for VOR navigation, including the basic VOR indicator, the horizontal situation indicator (HSI) and the radio magnetic indicator (RMI). Although these instruments are somewhat different from a functional standpoint, they all provide a means of orienting to a desired course, tracking that course, and showing

your direction of travel to or from a station. This section reviews VOR navigation with emphasis on an HSI presentation. You also will see how some of these indications appear on a conventional VOR indicator, so you can observe the similarities and differences between the instruments.

HORIZONTAL SITUATION INDICATOR

The name of the **horizontal situation indicator (HSI)** describes its major advantage. The design of this instrument solves nearly all reverse sensing and other visualization problems associated with a conventional VOR indicator. With an HSI, it is not necessary to mentally rotate the airplane to a heading that agrees with your selected course to get a clear picture of your situation; the HSI display combines the VOR navigation indicator with a heading indicator, so the display is automatically rotated to the correct position for you. [Figure 2-56] An HSI normally is mounted in the panel in place of the heading indicator.

On a conventional VOR indicator, left/right and to/from are ambiguous concepts because they must be interpreted in the context of the selected course. When an HSI is tuned to a VOR station, left and right always mean left and right, regardless of the course selected. The TO/FROM indicator is replaced with a simple arrow

As on a conventional VOR indicator, each dot on an HSI course deviation scale is a 2° deviation from course when navigating with a VOR. Each dot from center indicates a deviation of 200 feet per nautical mile. See figure 2-56.

The HSI contains a rotating **compass card** which indicates the aircraft's current magnetic heading. In situations where a standard VOR indicator gives you reverse sensing, the HSI compass card turns to provide normal sensing.

The **heading select bug** is used with an autopilot to automatically turn the aircraft to a newly selected heading.

The airplane's heading is displayed under the **heading index**, also called a lubber line.

The **course indicating arrow** visually shows the orientation of the selected course relative to your current heading. Because of this, left and right indications on the course deviation bar are always properly oriented.

The **course deviation bar** performs the same function as the CDI on a basic VOR indicator, depicting how far you are off course. When you are on course, the course deviation bar is aligned with the course arrow.

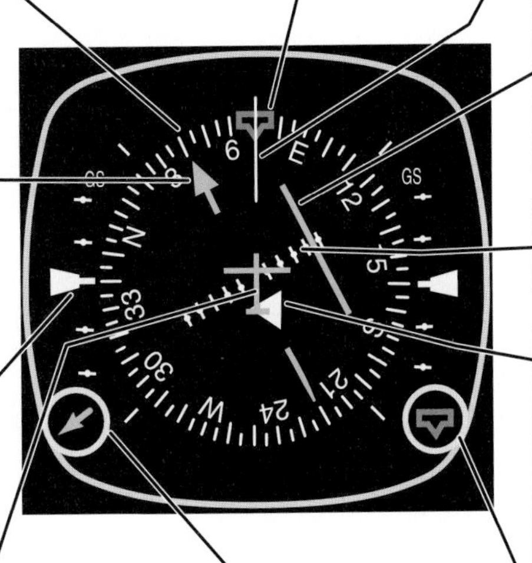

Each dot on the **course deviation scale** represents 2° for VOR navigation.

The **glide slope deviation pointer** indicates aircraft position relative to an instrument landing system (ILS) glide slope. When the pointer is below the center position, the aircraft is above the glide slope and vice versa.

The **TO/FROM indicator** on an HSI points to the head of the course arrow when the selected course is inbound to the navigation facility. When the selected course is outbound from the navaid, the TO/FROM indicator points away from the course arrowhead.

The **symbolic aircraft** shows your position in relation to the selected course as though you are above the aircraft looking down.

The **course set knob** controls the position of the course indicating arrow.

The **heading set knob** is used to position the heading select bug.

Figure 2-56. On an HSI, you always can see the course to be intercepted with respect to your aircraft, regardless of heading. This instrument shows the aircraft on a 070° heading, which is a 30° intercept angle for the 040° course from the station, which is currently 9° to the aircraft's right.

that always points toward the VOR station or other waypoint. If the arrow points ahead of you, it means TO, and if it points behind you, it means FROM. [Figure 2-57]

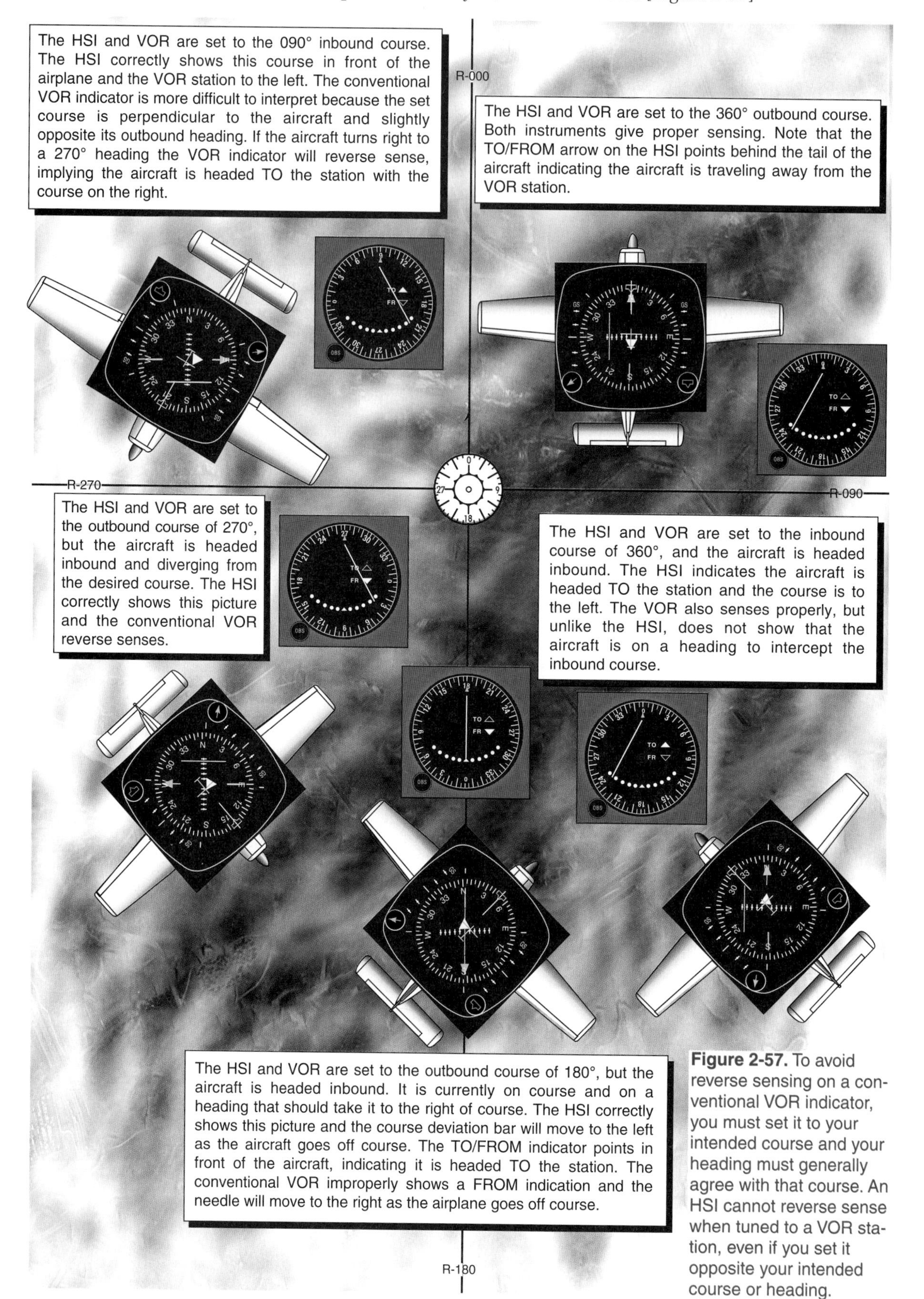

Figure 2-57. To avoid reverse sensing on a conventional VOR indicator, you must set it to your intended course and your heading must generally agree with that course. An HSI cannot reverse sense when tuned to a VOR station, even if you set it opposite your intended course or heading.

Flying a heading that is reciprocal to the bearing selected on the OBS would result in reverse sensing on a conventional VOR indicator.

There is one situation in which you need to watch out for reverse sensing on an HSI. As you will learn in Chapter 8, Section B, the course indications on a basic VOR indicator are not affected by the OBS setting when tuned to a localizer (a navigation aid for an approach to an airport runway). However, with an HSI tuned to a localizer, the course selector must be set to the inbound front course for both front and back course approaches or the display will be inverted, resulting in reverse sensing.

The HSI is a marvelous instrument, but because of its expense, is not standard equipment on many instrument training airplanes. Nonetheless, the FAA expects you to understand its operation, and improving technology should make this instrument more widely available. [Figure 2-58] You can already practice with an HSI on many flight simulators, training devices, and personal computer-based aviation training devices (PCATDs).

An HSI, unlike a conventional VOR indicator, gives information about your aircraft heading and its relationship to your intended course. See Figure 2-57.

Figure 2-58. Many GPS receivers, such as this VFR-only unit, incorporate an HSI presentation into the display.

The HSI: A Simplified Moving Map

One of the most exciting advances in general aviation navigation technology is the moving map. Several companies offer software that includes moving map capability for preflight and in-flight use. A moving map normally shows your aircraft at the center, with a chart or map moving under it to reflect the aircraft's current position. It is the ultimate situational awareness tool.

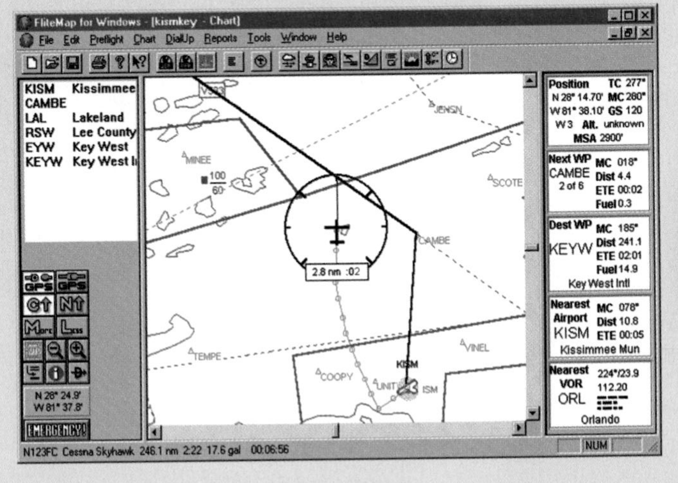

Most personal computer-based moving maps offer data about the aircraft's current position, as well as time, bearing and distance to the next fix and to your final destination. Normally, the moving map will display your planned course, and show your position relative to that course. If set to track-up mode, as shown here, you can easily see course intercepts. Notice the similarity between the course intercept shown on this moving map presentation and the intercept shown on the horizontal situation indicator.

Advanced aircraft with flight management systems (FMS) often have electronic flight information system (EFIS) displays that can be switched between a simple moving map and an HSI.

INTERCEPTING A RADIAL

A successful intercept starts with visualizing your present position and where you want to go. Then, you select a magnetic heading which will intercept the radial or course at a specific angle. During the intercept, scan your navigation instruments carefully and judge when to start your turn on course based on CDI deflection and rate of movement. [Figure 2-59]

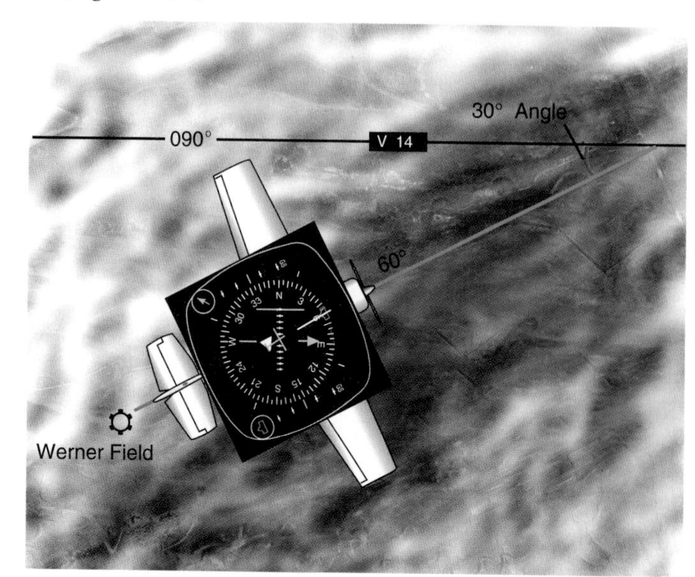

Figure 2-59. The intercept angle you use normally depends on a variety of factors such as your groundspeed, proximity to the navigation aid, and whether you will track to or from the facility.

After departing Werner Field, assume ATC clears you to intercept V-14 eastbound. Choosing a 60° heading for a 30° intercept promptly gets you on the airway. The CDI needle will show a full-scale deflection until you are within 10° of your course. Note how fast the needle is centering when determining when to start your turn on course.

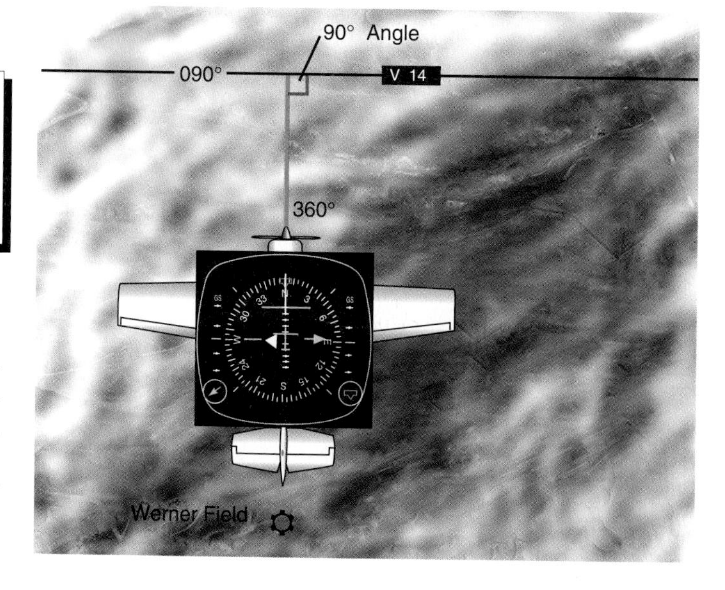

Assume ATC gives you a 90° intercept to separate you from traffic ahead. At this angle, be prepared to promptly begin your turn on course when the needle begins to center, or you will overshoot your course before completing the turn.

TRACKING

As you know, tracking a VOR radial is a trial and error process in which you establish a heading and watch whether it holds the desired course. Your preflight planning and your experience will help you estimate an initial heading. If you do not know the wind direction, simply try your intended course as your heading and watch the CDI needle. If it moves off course, turn 20° toward the needle and hold the heading correction until the needle centers. Reduce the drift correction to 10°, note whether this drift correction angle keeps the CDI centered, and make subsequent smaller corrections as needed. [Figure 2-60]

DETERMINING YOUR PROGRESS

In addition to navigating on course, you can use VORs to help you check your progress along your route. If you have two VOR receivers, you can determine your position simply by tuning your second VOR receiver to a station located to the side of your route. This can be especially helpful as you approach navigation fixes which are identified by off-route VORs. If you have only one VOR receiver, then carefully hold the heading that tracks your course from the first VOR while you tune to the second station whose radial intersects your course. [Figure 2-61]

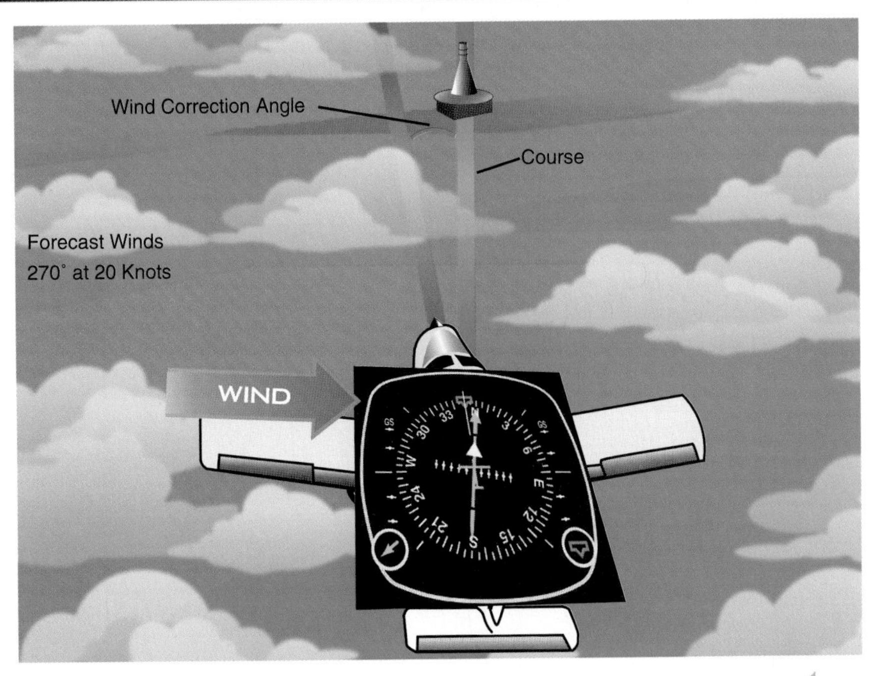

Figure 2-60. Visualizing the wind direction on the face of your heading indicator or HSI, can help you determine an initial heading to try for wind correction. A 20-knot wind perpendicular to this 120-knot airplane's course requires about 10° of correction.

The conventional VOR indicator, when set to the Anthony 060° radial with a FROM indication, points toward the VOR station (the airplane's left) while heading toward WENDL, and away from the VOR station (the airplane's right) when heading away from WENDL.

Although it is easier to identify an intersection when you have two VOR receivers, one VOR receiver is the minimum equipment needed. See figure 2-61.

The number 2 nav is tuned to Anthony VOR. As the aircraft passes WENDL intersection, the HSI shows the Anthony 060° radial moving from in front to behind the aircraft.

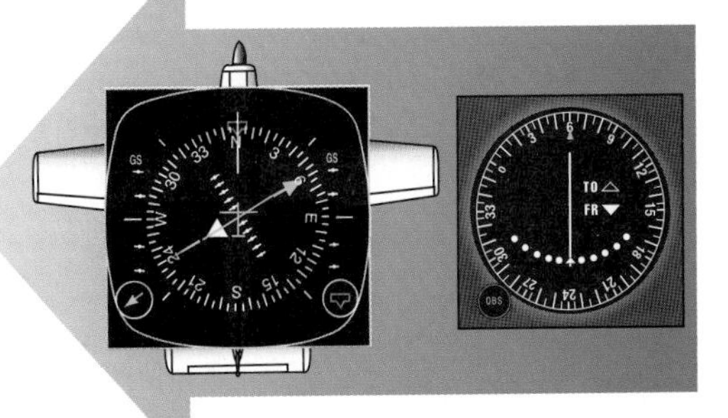

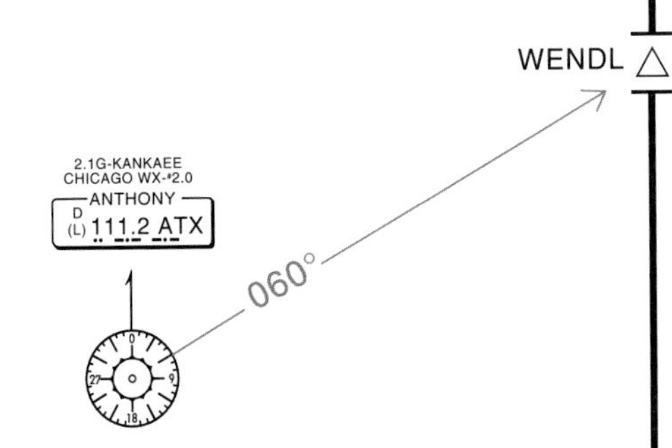

Figure 2-61. The HSI and VOR needles center as you reach an intersection defined by a radial from a VOR station to the side of the route.

TIME AND DISTANCE TO A STATION

While navigating using a VOR, you can calculate your time and/or distance to the station using a set of formulas or basic geometry. The formula method involves turning to place the station 90° from the aircraft heading and measuring the time it takes to travel to a new radial. The longer it takes to traverse a given number of radials, the farther you are from the station and the longer it will take to get there. Similarly, the smaller number of radials you cross in a given time, the farther you are from the station. Once you determine your time to reach a specific radial, you can calculate the time and distance to the station using the formulas shown in figure 2-62.

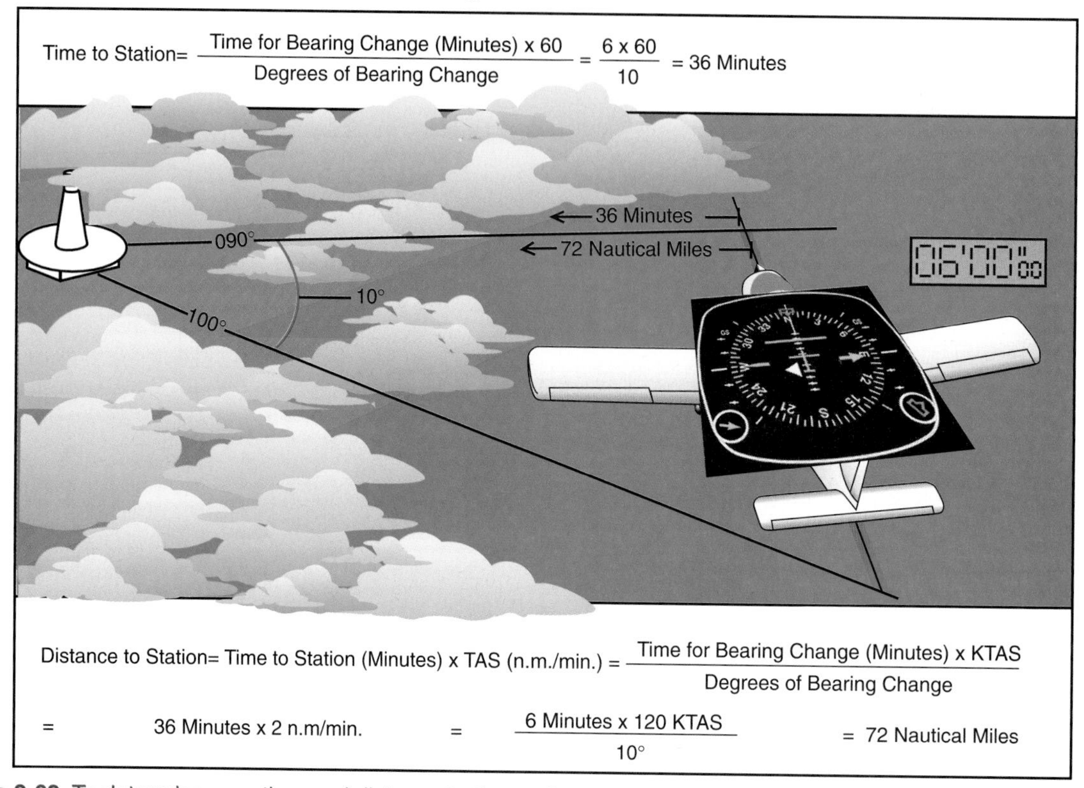

Figure 2-62. To determine your time and distance to the station, turn to place the station off the airplane's wingtip. Then, measure the time to move between 2 selected radials (a 10° change is easy to measure and calculate). In this example, the aircraft, traveling 120 knots (2 n.m./min.) traverses 10° in 6 minutes. The depicted procedure and accompanying formulas can be used to calculate time and distance to both a VOR and an NDB.

The isosceles triangle method uses a fundamental geometric principle to determine the time to a station. Procedurally, you turn 10° (or any angle) to the side of your course and twist your course selector the same amount in the opposite direction. Time to station is the same as the time it takes for your CDI to center (assuming no wind). [Figure 2-63]

You can calculate the time and distance to a station by turning perpendicular to the direct course to the station and measuring the time to move a specific number of degrees to a new radial. See figure 2-62.

Figure 2-63. While using the isosceles triangle method to determine the time to a station may take longer than employing the formula method, it does not consume as much time and fuel since you continue generally toward the station.

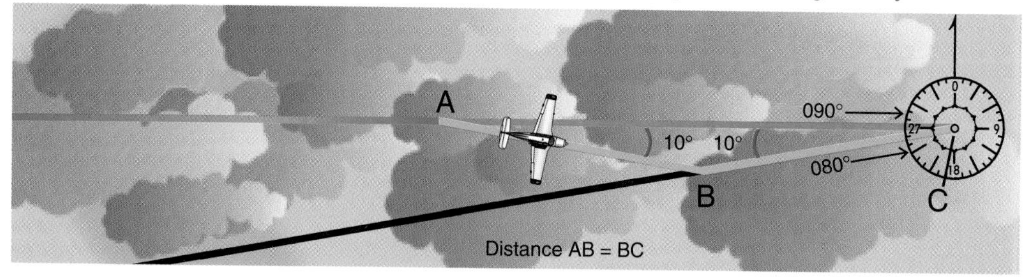

STATION PASSAGE

You can use the isosceles triangle method to determine time to a station. See figure 2-63.

VOR station passage is the first positive, complete reversal of the TO/FROM indicator.

As you get close to a station, in an area called the **cone of confusion**, the CDI and TO/FROM indicators fluctuate. Station passage is indicated by the first positive, complete reversal of the TO/FROM indicator. If you can determine you are getting close to a station where you plan a significant course change, you may wish to begin your turn early to avoid excessive course corrections after station passage. Since FARs require you operate along the centerline of an airway, this can be particularly important in higher performance aircraft. Depending on factors such as the amount of course change required, wind direction and velocity, and the pilot's technique, aircraft operating in excess of 290 KTAS can exceed the normal airway or route boundaries if a turn is initiated after passing a fix.

Ready, Set, Go!

You may be familiar with the story of the tortoise and the hare. This is the same type of tale, with a twist. A loggerhead turtle is thrown into the Atlantic Ocean. A pilot in Minneapolis is given a basic GPS receiver and a business jet that can scream through the sky at almost the speed of sound. Now, without a map, which one will get to the warm, sandy beaches of Florida first? There's no way the turtle can compete, right?

The answer to the question lies in the nature of the information each of the participants has available. Both contestants know their positions — the pilot gets a precise latitude/longitude readout; the sea turtle. . .well. . . it *just knows.* From the initial location, the best the pilot can do is merely point the airplane in a general direction and hope for the best. On the other hand, the turtle can easily navigate the warm waters of the Gulf Stream to the precise beach where it was born.

How does the sea turtle find its way? Two researchers, Ken and Catherine Lohmann, think they may have solved this mystery. Several years ago, the Lohmanns suggested that turtles could return to nesting beaches as much as 30 years after they left due to their innate ability to sense the earth's magnetic field. To validate their theories, the Lohmanns put several sea turtles in a large tank which was wrapped with coils of wire to control the magnetic field. When first put in the tank, the reptiles headed northeast but when the apparent magnetic field was reversed, the turtles turned around and headed the other direction. This discovery helped explain how the turtles could determine a general direction, but not how they could navigate to a specific location. Further tests were needed.

Working on the assumption that the loggerheads must be guided by some sort of internal mapping mechanism, the Lohmanns experimented with changing the angles of magnetic force. When the lines of magnetic force were changed to match those at Florida beaches, the turtles would swim east. If the turtles sensed (through a small crystal of magnetic material, called magnetitie, in their brains) that the magnetic force was altered to match a more northerly location, the loggerheads would turn south.

The Lohmanns experiments seem to verify not only that some sea turtles can sense changes in the earth's magnetic field, but that they also maintain a mental map of the ocean to help them find their way home. With this ability, the loggerhead turtle can do something the pilot in this story can't — confidently and accurately navigate to a destination. Simply knowing your position, no matter how accurately, sometimes just is not enough. Even a pilot in today's high-tech world needs precise mapping guidance to complement the information provided by modern positioning equipment.

ADF NAVIGATION

The automatic direction finder (ADF) is useful for supplemental navigation information, and as a backup when other aircraft or ground equipment is unavailable. Your ADF can be tuned to any low/medium frequency (L/MF) nondirectional radio beacon (NDB) or commercial broadcast stations of the amplitude modulation (AM) class. Although commercial broadcast stations are not approved for IFR operations, you may use them for VFR navigation. An NDB that is collocated with a marker beacon on a precision instrument approach is called a compass locator. These facilities alert you as you reach specific points on some instrument approaches.

Interpreting ADF indications can be more challenging than VOR indications because you must analyze both heading and bearing information to determine your position. As you know, a fixed-card indicator has zero at the top and directly indicates the angle between the nose of the aircraft and the station, or **relative bearing**. To determine your magnetic bearing (MB) to the station, you add magnetic heading (MH) and relative bearing (RB). [Figure 2-64]

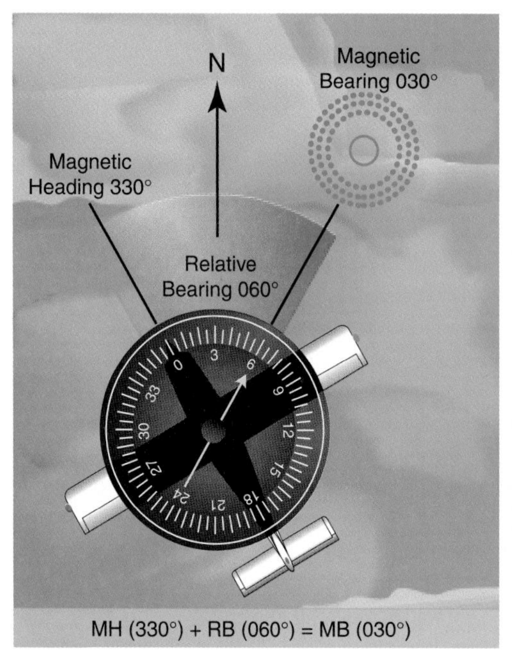

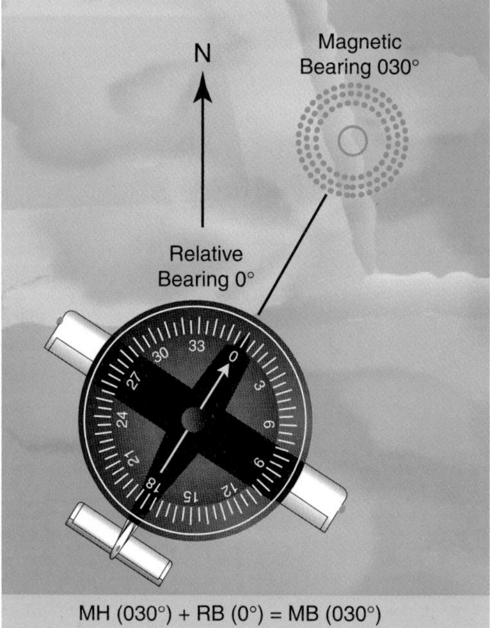

Figure 2-64. If you want to fly directly to the station, you should calculate your magnetic bearing to the station using the formula, MH + RB = MB. If the sum is greater than 360°, you will need to subtract 360° to determine the magnetic bearing to the station.

On a movable-card ADF, you can rotate the scale so your heading appears at the top. This allows you to directly read your magnetic heading to the station under the arrowhead and the bearing from the station under the tail of the arrow. However, where you are frequently changing heading, it may not be practical to keep the ADF card adjusted to your heading. One useful technique is to mentally superimpose the ADF needle over the heading indicator. [Figure 2-65]

Regardless of the type of ADF pointer, the relative bearing is the angle between the needle and the aircraft nose reference. Magnetic heading (MH) plus relative bearing (RB) equals magnetic bearing to the station (MB). On a properly set movable-card ADF, the arrowhead indicates MB to the station and the tail of the arrow indicates MB from the station.

RADIO MAGNETIC INDICATOR

A special instrument, called a **radio magnetic indicator (RMI)**, makes it easy to determine your position in relation to an NDB by combining a slaved compass card and bearing pointer in the same instrument. An RMI is like a movable-card ADF on which the card automatically rotates to reflect the aircraft's magnetic heading.

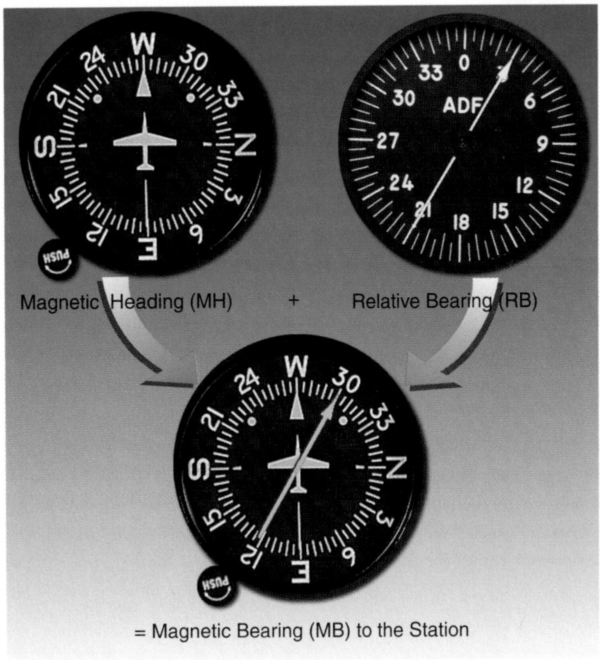

Magnetic Heading (MH) + Relative Bearing (RB)

= Magnetic Bearing (MB) to the Station

Figure 2-65. You can quickly visualize the magnetic bearing to a station by mentally superimposing the ADF needle over your heading indicator.

An RMI, like an HSI, contains a slaved compass card, which receives heading information from a magnetic flux valve mounted at a remote position on the aircraft. On older units, it was occasionally necessary to align the slaved compass card(s) by selecting free gyro mode and pressing buttons to rotate the card clockwise (for a left heading correction) or counterclockwise (to correct to a higher number heading). This is totally automatic on newer systems.

Unlike a standard ADF display, most RMIs have two bearing pointer needles, either one of which can be set to point to an NDB or VOR station. Because the aircraft heading appears at the top of the scale, the instrument always displays the bearing to a station at the head of the arrow, and the bearing from a station at the tail of the arrow. [Figure 2-66]

The tail of an RMI needle set to a VOR station indicates the radial you are on FROM the station. The head of an RMI needle set to an ADF shows the bearing TO the station and the tail of the needle indicates the bearing FROM the station.

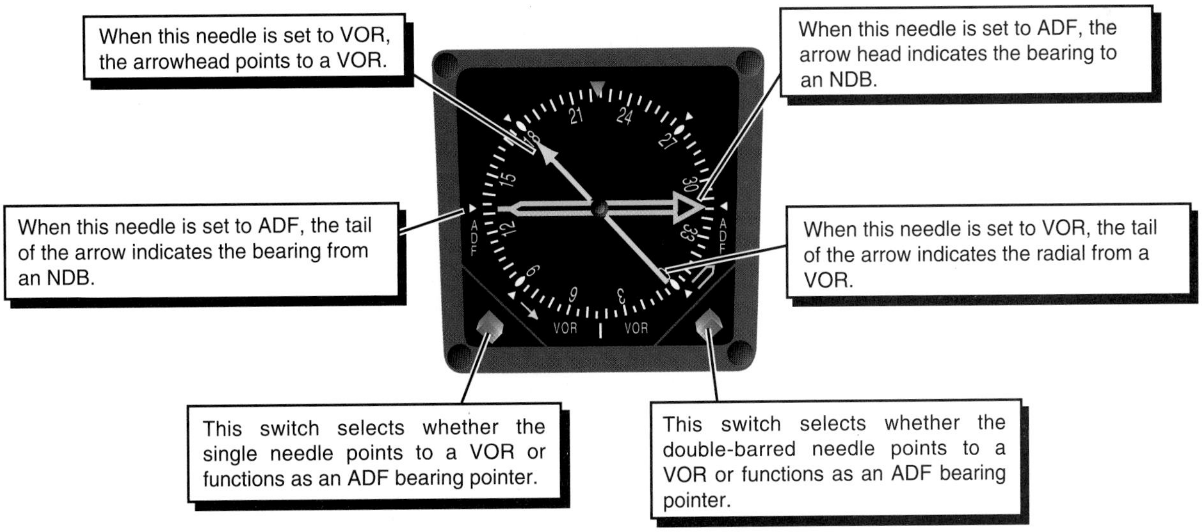

When this needle is set to VOR, the arrowhead points to a VOR.

When this needle is set to ADF, the arrow head indicates the bearing to an NDB.

When this needle is set to ADF, the tail of the arrow indicates the bearing from an NDB.

When this needle is set to VOR, the tail of the arrow indicates the radial from a VOR.

This switch selects whether the single needle points to a VOR or functions as an ADF bearing pointer.

This switch selects whether the double-barred needle points to a VOR or functions as an ADF bearing pointer.

Figure 2-66. An RMI incorporates two bearing pointer needles superimposed over a slaved compass card which indicate heading. One or both of the needles can be set to point to a VOR or an NDB.

INTERCEPTING A BEARING

Intercepting a bearing is easiest if you choose an angle, such as 45°, that is easy to read on the compass card. To establish a 45° intercept, turn so that the bearing to be intercepted appears over the heading indicator reference mark 45° to the left or right of the aircraft nose. Precisely maintain this heading and look for the ADF needle to also point 45° to the left or right of the aircraft's nose. [Figure 2-67]

Figure 2-67. When flying this 45° intercept, the aircraft intercepts the 100° bearing to the station when the ADF needle indicates a 45° relative bearing.

You can easily see when you have intercepted your bearing if using a 45° intercept or other easy-to-see value such as 30°. If using a 30° intercept, you will have intercepted your bearing to the station when the needle points 30° to the left or right of the aircraft's nose. Sometimes it can be challenging to determine your position when the needle does not exactly indicate the intercept angle. Just remember that the head of the needle is always moving toward the tail of the aircraft. If the needle is pointing in front of your intercept angle, you have not yet reached the bearing to be intercepted. If it points behind the intercept position, you have passed through your bearing.

TRACKING

For most instrument approaches, you must track a specific path to a station. The basic tracking procedure is to start with a heading you expect will keep you on course. In a no-wind, headwind, or unknown wind situation, this heading will be the same as your course. Precisely hold the heading and watch for the needle to drift to the left or right. The ADF needle's indications when tracking inbound tell which way you need to turn to capture your course. Double the ADF relative bearing when turning toward your course.

If your heading equals your course and the needle points 10° left, turn 20° left. When holding a heading 20° to the left of your desired course, watch for the ADF needle to move 20° right of the nose. You are on course when the relative bearing equals your course correction. Assuming a wind blew you right of course, you might try a heading 10° left of course in order to track inbound. This will result in a relative bearing of 10° right as long as you maintain course. [Figure 2-68]

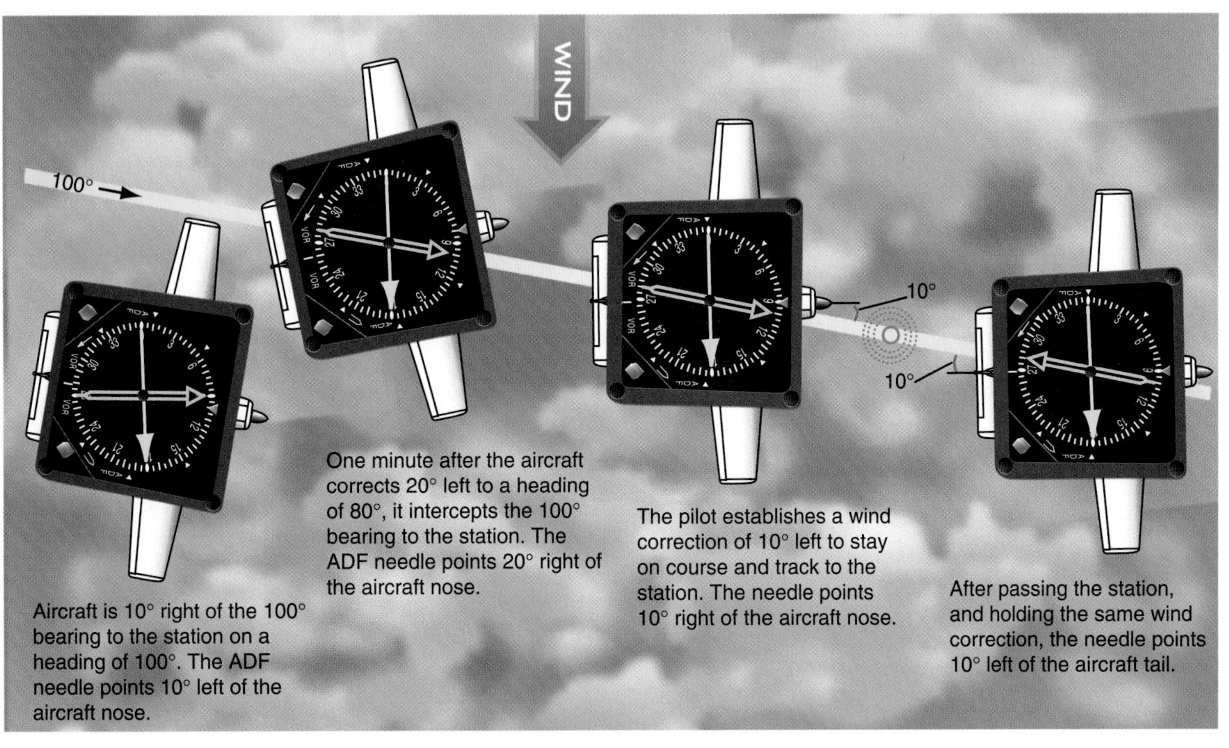

Figure 2-68. When paralleling your course, the ADF needle indicates whether the course is to your left or right. When on course, the relative bearing to the nose or tail of the aircraft equals the course correction.

If you attempt to correct for a crosswind by continually adjusting your magnetic heading to keep the ADF needled pointed at the airplane's nose (0° relative bearing), you will home to station instead of tracking to it. This will cause you to fly a curved path over the ground.

Unlike VOR navigation, ADF does not provide accurate position information independent of heading. Unless you are using an RMI with a compass card that automatically aligns itself with magnetic north, you could find yourself substantially off course if you do not check and set your heading indicator against your magnetic compass every 15 minutes.

When on the desired track outbound with the proper drift correction established, the ADF pointer will be deflected to the windward side of the tail position.

Homing to a station in a crosswind results in a curved path to the station.

TIME AND DISTANCE TO A STATION

You can calculate your time and distance to a station by turning perpendicular to the inbound course and measuring the time to make a 10° (or any angle) change in bearing. You calculate the time and distance to the station using the same formulas you learned in the VOR discussion earlier in this section. To determine the time to the station, you can also use the double-the-angle-on-the-bow method, which employs the same geometric principle as the isosceles triangle concept discussed earlier. If you hold a constant heading, the time to the station is simply equal to the amount of time it takes for the relative bearing to double. For example, if you hold a constant heading and the ADF needle moves from a relative bearing of 045° to 090° in 5 minutes, the time to the station is 5 minutes.

If you hold a constant heading, the time to the station is equal to the time it takes for the relative bearing to double.

STATION PASSAGE

As you approach the station, even small deviations from your desired track can result in large needle fluctuations. Therefore, it is important that you do not chase the needle, but use heading corrections no greater than about 5°. As the needle begins to rotate steadily toward a wingtip position, or shows erratic oscillations to the left or right, you should hold your last corrected heading constant. Station passage is considered to have occurred when the needle either points to a wingtip or settles at or near the 180° position. Depending on your altitude, it may take from a few seconds up to 3 minutes from the time that the needle begins to oscillate until you obtain a positive indication of station passage.

DISTANCE MEASURING EQUIPMENT

The easy way to measure your distance to or from a VOR is with **distance measuring equipment (DME)**. If your aircraft is equipped with a DME receiver, you can obtain distance information from several types of ground-based navaids including VOR/DME, VORTAC (which obtains distance information from the associated TACAN facility), instrument landing system (ILS)/DME, and localizer (LOC)/DME stations. These facilities provide course and distance information from collocated components under a frequency pairing plan. You tune a DME the same way as a VOR receiver. This is because DME receivers usually display the frequency of the paired VOR rather than the actual DME UHF frequency (962 MHz to 1,213 MHz). Some DME units can be remoted to a VOR receiver so they automatically tune in the DME or TACAN station that is paired with the selected VOR ground facility.

DME receivers work by transmitting paired pulses to a ground station. After measuring the time for the station's reply to return to the receiver, the receiver calculates the distance, and displays the result in nautical miles. Many DME receivers also provide a groundspeed readout which indicates the rate of change of the aircraft's distance from the station. DME groundspeed agrees with actual groundspeed only when you are headed directly toward or away from the station. [Figure 2-69]

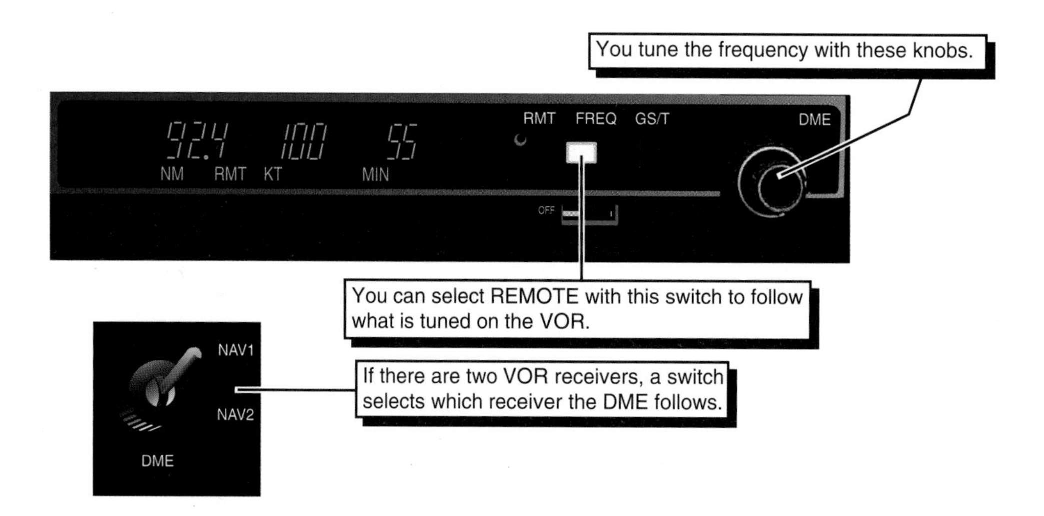

Figure 2-69. A typical DME's display includes distance, groundspeed, and time to station.

Since the DME signal travels in a straight line to and from the ground station, the DME receiver displays slant-range, not horizontal, distance. [Figure 2-70] Except for this error, DME is accurate to within 1/2 mile or 3%, whichever is greater, and can be received at line-of-sight distances up to 199 nautical miles.

DME indicates slant-range distance with greatest error at high altitudes close to a VORTAC. The indication should be 1 n.m. when you are directly over a VORTAC site at approximately 6,000 feet AGL. Generally, you can consider DME accurate when you are at least one horizontal mile from the station for each 1,000 feet above the site elevation.

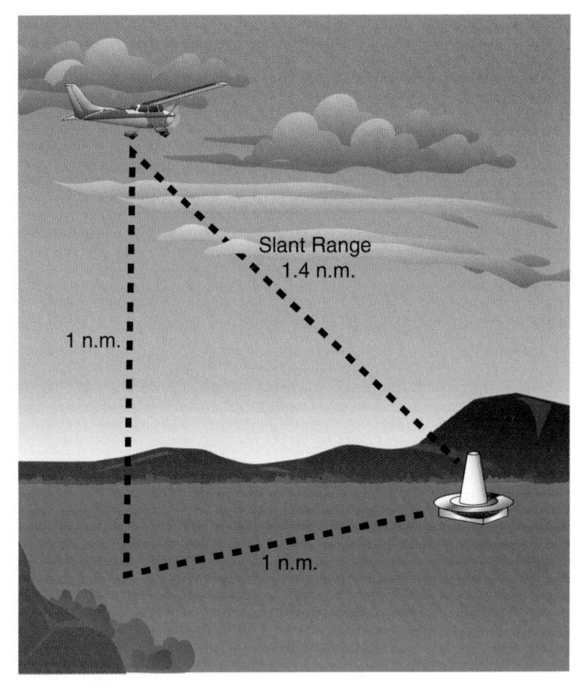

Figure 2-70. If you are flying at an altitude of 1 nautical mile at a horizontal distance of 1 nautical mile from the station, your DME will indicate a distance of 1.4 nautical mile. Slant range error is negligible if the aircraft is 1 mile or more from the ground facility for every 1,000 feet of altitude above the station.

DME ARCS

Many instrument approach procedures incorporate **DME arcs** for transition from the enroute phase of flight to the approach course. This is useful, for example, if your airway approaches an airport from the east, the VOR is on the airport, and you need to fly an approach from the north. A DME arc can position you for the approach without having to fly to the VOR, then fly outbound and reverse course.

It is easiest to fly a DME arc using an RMI. Generally, you intercept the arc while flying to or from a VOR/DME or VORTAC. To join the arc, you should turn approximately 90° from your inbound or outbound course, making sure you begin your turn early enough so you do not overshoot the arc. For groundspeeds of 150 knots or less, a lead of about 1/2 mile is usually sufficient.

In calm wind conditions, you could theoretically follow the arc by continuously adjusting your heading so that the RMI needle points at the wingtip. In practice, it is easiest to fly slightly inside the arc in a series of straight flight segments, forming a polygon, with course corrections every 10° to 20°. You begin each segment by turning slightly toward the VOR, putting the RMI pointer 5° to 10° ahead of the wingtip.

Then, you maintain heading until the bearing pointer moves 5° to 10° behind the wingtip at which point you repeat the process to fly another segment. [Figure 2-71] As you complete the arc, plan on leading the turn to your inbound or outbound course. The amount of lead you use should be fairly close to what you used to turn onto the arc.

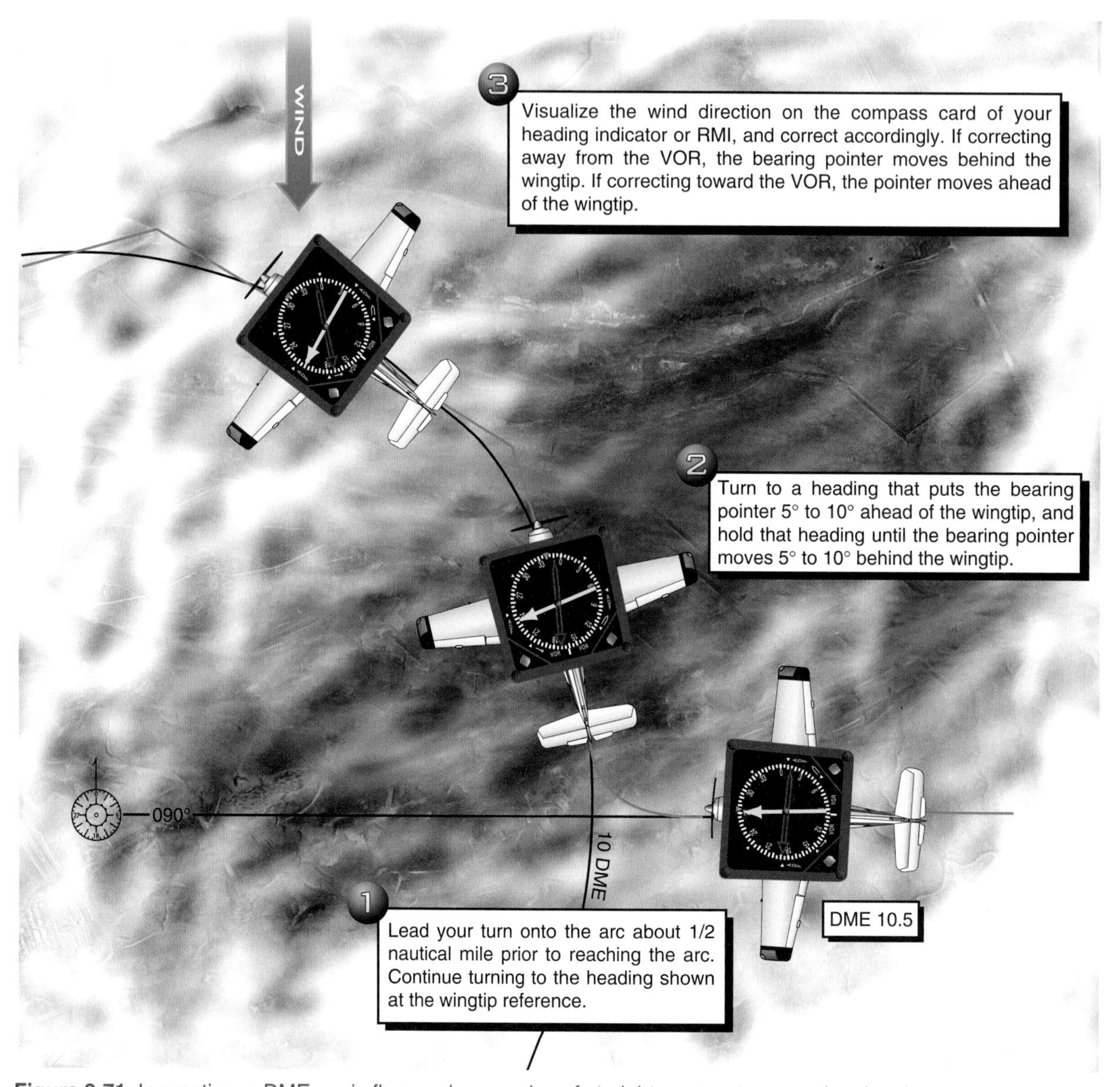

Figure 2-71. In practice, a DME arc is flown using a series of straight segments approximating the arc.

CORRECTING FOR A CROSSWIND

At times, you may find that a crosswind causes you to drift either away from or toward the station. If you are drifting away from the facility, turn to place the bearing pointer ahead of the wingtip; if you are drifting toward the station, turn to place the bearing pointer behind the wingtip. As a general rule, you should change the relative bearing 10° to 20° for each 1/2 mile you are away from your desired arc when abeam the station. For example, if you are 1/2 mile inside the arc and the RMI needle is pointing to the wingtip, turn 10° to 20° away from the facility to return to the arc.

As you turn toward the VOR to compensate for a crosswind, the bearing pointer moves ahead of the wingtip reference. Use 10° to 20° of correction if you drift 1/2 nautical mile outside the arc when abeam the station. When correcting away from the VOR, the bearing pointer moves behind the wingtip.

If you are using a conventional VOR indicator, the recommended procedure is to set the OBS to a radial 20° ahead of your present position. You should then turn and maintain a heading 100° from the radial you have just crossed. When your CDI centers or you reach the arc, you should repeat the process to fly another segment. As with the RMI, this technique will maintain a track slightly inside your desired arc in no-wind conditions. If there is a crosswind, you should compensate by adjusting your heading toward or away from the station as appropriate.

OPERATIONAL CONSIDERATIONS

When using VOR, DME, or ADF equipment for navigation, you need to know the strengths and limitations of each of these systems. When flying under IFR, it is particularly important that you positively identify ground stations, and that you verify proper operation of your navigation equipment and corresponding ground facilities.

GROUND FACILITIES

A VOR utilizes VHF frequencies like those used by FM radio and broadcast television. Although these signals are limited to line of sight, you can plan on receiving a reliable signal at altitudes published on instrument charts.

VOR facilities operate within the 108.0 to 117.95 MHz frequency band and are classified according to their usable range and altitude, or standard service volume (SSV). [Figure 2-72] Terminal VORs (TVOR) are short-range facilities which are normally placed in terminal areas to be used primarily for instrument approaches. High altitude VORs (HVORs) and low altitude VORs (LVORs) are used for navigation on most airways, and

Figure 2-72. High altitude VORs typically have power outputs of 200 watts and transmit usable signals up to 130 nautical miles. Since a distant station on the same frequency can interfere with the local signal, HVORs on the same frequency are adequately separated to avoid interference up to 45,000 feet, where usable range drops from 130 to 100 nautical miles. Terminal VORs have only 50 watts of power and a usable range of 25 nautical miles.

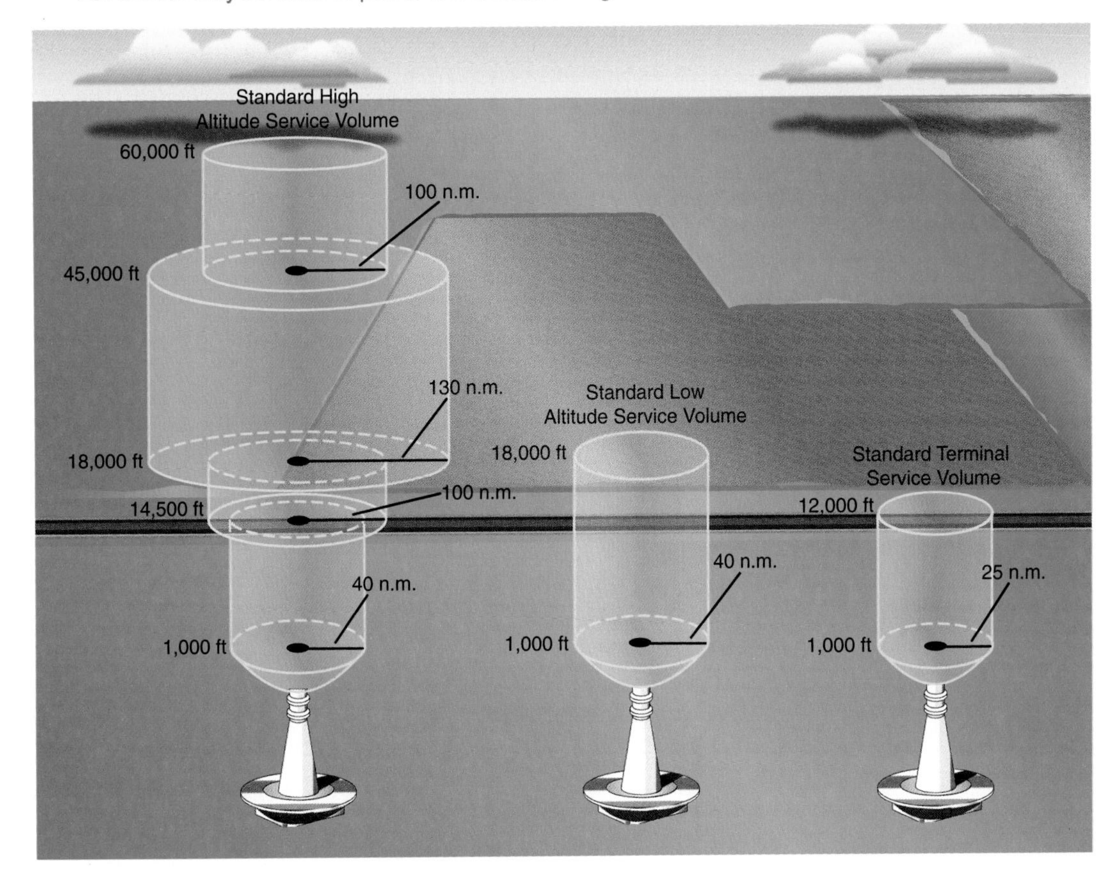

can also function as approach facilities when located on or near airports. You can find the SSV which applies to a particular VOR in the *Airport/Facility Directory.*

Your ADF uses nondirectional beacons which transmit in the low/medium frequency (L/MF) range between 190 and 535 kHz. The stations normally transmit a simple 400 or 1,020 Hz tone modulated with a Morse code identifier. [Figure 2-73] Radio beacons are subject to disturbances that may result in erroneous bearing information. Such disturbances result from such factors as lightning, precipitation static, etc. At night, radio beacons are vulnerable to interference from distant stations. Nearly all disturbances which affect the ADF bearing also affect the facility's identification.

At altitudes between 14,500 and 18,000 feet MSL, an (H) Class VORTAC has a usable signal range of 100 nautical miles. Therefore, for direct routes off established airways at these altitudes, the facilities should be no farther apart than 200 nautical miles. You can refer to the *Airport/Facility Directory* to determine what type of VOR facility is shown on a chart and whether the standard service volume applies for that facility.

FACILITY	POWER OUTPUT	USABLE RANGE
Compass Locator	25 Watts	15 Nautical Miles
MH Radio Beacon	Less Than 50 Watts	25 Nautical Miles
H Radio Beacon	50 to 1,999 Watts	Up to 50 Nautical Miles
HH Radio Beacon	2,000 Watts or More	75 Nautical Miles

Figure 2-73. Although NDBs do not suffer the line-of-sight limitations of VHF and UHF facilities, interference and L/MF wave propagation characteristics can limit their reception range.

VOR CHECKS

You could theoretically plan an IFR flight with no navigation equipment at all if your intended route would allow radar vectoring from takeoff through letdown. In practice, you need a VOR receiver, and to use the VOR for flight under IFR, it must have been tested for accuracy within the preceding 30 days. If it has been more than 30 days, you must perform this check before you can take off under IFR using VOR navigation.

For IFR flight, you are required to have working navigation equipment appropriate to the ground facilities to be used. It is your responsibility as pilot in command to make sure that the VOR check has been accomplished within the past 30 days.

VOR TEST FACILITIES

VOR test facilities (VOTs) enable you to make precise VOR accuracy checks from most locations on an airport. This is possible because VOTs broadcast a signal for only one radial — 360°. The airborne use of certain VOTs is also permitted; however, their use is strictly limited to those areas and altitudes specifically authorized in the *Airport/Facility Directory.* Tune your VOR receiver and listen for a series of dots or a continuous tone that identifies the facility as a VOT. Determine that the needle centers, ±4° (ground or airborne), when the course selector is set to 180° with a TO indication, or 360° with a FROM indication. If using an RMI, the bearing pointer should indicate 180°, ±4°.

You can find a VOT frequency for a particular airport in the *Airport/Facility Directory* or on the A/G Voice Communication Panel of the NOS Enroute Low Altitude Chart. When checking your VOR using a VOT, the CDI should be centered and the OBS should indicate that the aircraft is on the 360° radial, ±4°.

VOR CHECKPOINTS

You also can determine VOR accuracy using ground or airborne **VOR checkpoints**. On the ground, you can taxi your aircraft to a specific point on the airport designated in the

VOR Receiver Check section of the A/FD. After centering the CDI, compare your VOR course indication to the published radial for that checkpoint. The maximum permissible error is ±4°. Airborne checkpoints, also listed in the A/FD, usually are located over easily identifiable terrain or prominent features on the ground. With an airborne checkpoint, the maximum permissible course error is ± 6°. You also can perform an airborne check by selecting a VOR radial that defines the centerline of an airway. Then, using a sectional chart, locate a prominent terrain feature under the centerline of the airway, preferably 20 miles or more from the facility. Maneuver your aircraft directly over the point, twist the OBS to center the CDI needle and note what course is set on the VOR indicator. The permissible difference between the published radial and the indicated course is ± 6°.

To make a VOR receiver check when the aircraft is located on the designated checkpoint on the airport surface, set the OBS on the designated radial. The CDI must center within ±4° of that radial with a FROM indication. The allowable error when using an airborne checkpoint is ±6°.

DUAL SYSTEM CHECK

You also can conduct a VOR check by comparing the indications of two VOR systems which are independent of each other (except for the antenna). If your aircraft is equipped with two VOR radios, set both to the same VOR facility and note the indicated readings on each. When you check one against the other, the difference should not exceed 4°.

The allowable error when comparing the indications of two VOR receivers, whether in the air or on the ground, is 4° between receivers.

IDENTIFICATION

It is extremely important to positively identify a VOR, DME, or NDB facility before using it for navigation. When you are busy, it can be easy to listen to the identifying tone without making sure you have the correct Morse code identifier. It is essential that you always check the identifier against your chart to make sure you are tuned to the correct facility.

VOR AND DME

Since VOR and DME are separate components, each transmits its own identification signal on a time sharing basis. The VOR transmits a 1,020 Hz Morse code identifier and possibly a voice identifier several times for each DME identifier, which is a 1,350 Hz tone at approximately 30 second intervals. If one of these tones is missing, you should not use the associated portion of the facility for navigation.

VOR facilities undergoing maintenance may transmit the word, TEST (– • ••• –), or they may transmit no audio at all. The presence of an identifier signal verifies the proper operation of the facility. You should not use a ground station for navigation if it does not transmit an identifier, even if you appear to be receiving valid navigation indications.

If, when tuning to a VORTAC, you receive a single coded identification approximately once every 30 seconds, it means the DME component is operative and the VOR component is inoperative. The reverse is true if you hear the 1,020 Hz VOR signal several times and the 1,350 Hz DME tone is missing over a 30-second interval.

Even if you are receiving navigation indications from a VOR, DME, or NDB facility, the lack of a coded identification indicates the station is undergoing maintenance and is unreliable.

NDB

Just as you do with a VOR, you should positively identify an NDB before you use it for navigation by listening to the Morse code identifier. In contrast to VOR navigation, however, you should continuously monitor the NDB's identification since ADF receivers do not have a flag to warn you when erroneous bearing information is being displayed.

AREA NAVIGATION

Area navigation (RNAV) allows you to fly direct to your destination without the need to overfly VORs or other ground facilities. Some of the available direct navigation systems include long range navigation, inertial navigation, and VOR-based systems. The newest system, GPS, will be discussed separately. Area navigation courses are defined by **waypoints**, which are predetermined geographical positions used for route/instrument approach definition, or progress reporting purposes. They can be defined relative to VOR/DME or VORTAC stations or in terms of latitude/longitude coordinates.

WHEN YOUR IN-FLIGHT COMPUTER RESETS

The electronic flight information systems (EFIS) found in modern advanced aircraft cockpits offer pilots a tremendous amount of information on a colorful, easy-to-read display. Although these systems are generally very reliable, they are driven by computers that can have quirks.

Courtesy of Honeywell

The NTSB was concerned when it learned the primary flight displays on an Airbus A300 blinked off for 2-3 seconds during a critical moment in flight. The flight crew had reported severe turbulence and wind shear, and the EFIS symbol generator unit (SGU) software triggered a reset when the aircraft was suddenly tossed into a steep bank attitude. Apparently, the software assumed the aircraft would never experience a roll angle in excess of 40° per second, and the reset was designed to fix what the computer thought was inaccurate data. As it turns out, the pilots were without their primary flight displays at the moment they needed them to recover from an unusual flight attitude. Fortunately, the pilots were able to use their standby attitude indicators.

It is easy for a pilot to become complacent because overall, electronic navigation systems are extremely reliable. However, it is essential that pilots monitor their instruments carefully and know their options if the instruments malfunction.

VORTAC-BASED AREA NAVIGATION

RNAV based on VORTAC and VOR/DME facilities has been in use for some time. It is accomplished with a course-line computer that creates **phantom VORs** at convenient locations for your route. You navigate to and from these phantom VORs the same way as actual VORs. To create a phantom VOR in a course-line computer (CLC) you tune in a real VORTAC or VOR/DME, then specify a radial and distance for the location of the phantom VOR. The CLC requires DME to calculate the location of the phantom VOR. Since RNAV does not extend a VOR's standard service volume, if you are flying below 14,500 feet using low or high altitude VORs, you must still remain within 40 nautical miles of the physical facilities to receive a reliable navigation signal.

INERTIAL NAVIGATION SYSTEM

An inertial navigation system (INS) is a self-contained system which uses gyros, accelerometers, and a navigation computer to calculate position. By programming a series of waypoints, the system will navigate along a predetermined track. INS is extremely accurate when set to a known position upon departure. However, without recalibration, INS accuracy degrades 1 to 2 nautical miles per hour. To maintain precision, many INS systems automatically update their position by incorporating inputs from VOR, DME, and/or other navigation systems. INS systems can be approved as a sole means of navigation or used in combination with other systems.

LONG RANGE NAVIGATION

Long range navigation (LORAN), uses a network of land-based radio transmitters developed to provide all-weather navigation for marine users along the U.S. coasts and in the Great Lakes. To serve the needs of aviation, LORAN coastal facilities were augmented in 1991 to provide signal coverage over the entire continental U.S. When used for IFR navigation, the current system, LORAN-C, is accurate within 0.25 nautical mile. The *Aeronautical Information Manual* provides an extensive description of the operation of LORAN, and its use in the IFR environment. While LORAN is a reliable and fairly accurate navigation system, the advent of GPS has prompted the FAA to begin canceling all LORAN approaches.

GLOBAL POSITIONING SYSTEM

The area navigation system you are most likely to use is the global positioning system (GPS). GPS is a satellite-based radio navigational, positioning, and time transfer system operated by the U.S. Department of Defense (DOD). It offers an extremely precise positioning service (PPS) to authorized U.S. and allied military, federal government, and civil users who can satisfy specific U.S. requirements. For the rest of us, GPS provides standard positioning service (SPS) with a 95% probability of horizontal accuracy within 100 meters (328 feet), and 99.99% probability of accuracy within 300 meters (984 feet). On GPS's worst day, it is more accurate than any other area navigation system and many approach systems.

Although 100 meters is excellent horizontal accuracy, it is inadequate for vertical navigation. You could not safely fly an approach with this amount of error. And due to the geometry of the satellite constellation, the GPS vertical error is even larger than the horizontal error. With a 95% probability of vertical accuracy of approximately 150 meters (492 feet), GPS cannot replace your barometric altimeter as a source of vertical position information.

RECEIVER AUTONOMOUS INTEGRITY MONITORING

When flying IFR, you need the same assurance of a reliable GPS signal as you require from a VOR. One requirement for IFR approval is that a GPS receiver continuously verifies the integrity (usability) of the signals received from the GPS constellation through receiver autonomous integrity monitoring (RAIM). You cannot conduct a GPS approach unless your receiver meets stringent approach RAIM requirements, as indicated by an annunciator light in an approved IFR installation. Some receivers will not even go into approach mode unless RAIM is passed.

RAIM messages vary between receivers; generally there are two types. The first indicates that there are not enough satellites available to provide RAIM integrity monitoring and the other indicates that the RAIM integrity monitor has detected a potential error that exceeds the limit for the current phase of flight. To simply warn you that the system is unreliable requires one additional satellite to compare to the basic four that are required

Do You Have the Correct Time?

The GPS constellation consists of 24 satellites orbiting about 20,000 kilometers above the earth. Each satellite transmits a course/acquisition (CA) code, which contains information on the satellite's position, the GPS system time (from on-board atomic clocks), and the health and accuracy of the transmitted data. The GPS receiver calculates a pseudo-range to the satellite by comparing the reported time the signal left the satellite with the time it arrived at the receiver, based on the receiver's system clock. Since an error of one millionth of a second equates to a distance of 300 meters, the system must have a way of reconciling any differences between the atomic clocks on the satellite and the less accurate clocks in GPS receivers. To see how this is done, assume that when your GPS receiver thinks it's 10 a.m., the real time is 09:59.59. Using only 2 satellites, this inaccuracy would result in a false position estimate as shown in figure A.

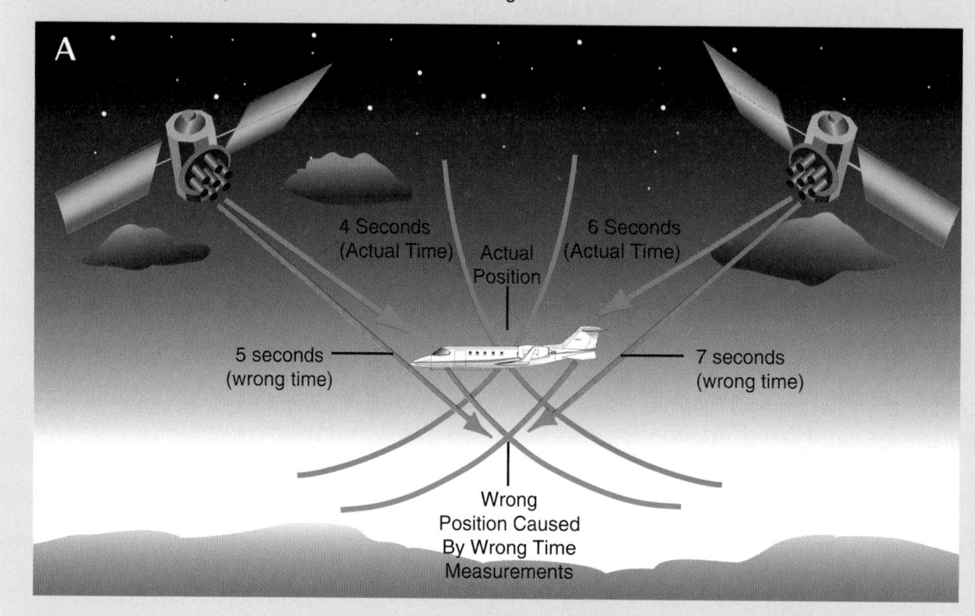

The calculated position shown in figure A would seem perfectly OK to us since we would have no way of knowing that our GPS receiver's clock was slightly out-of-sync. In reality, we could be miles away from where we thought we were. To solve this dilemma, the GPS receiver seeks out another opinion in the form of a 3rd satellite. Since the 3 arcs do not intersect at a common point, it may seem that adding another satellite makes matters worse. [Figure B] In fact, it provides the extra data the GPS receiver needs. You see, not only are GPS receivers programmed to sense that something's not right, but they also know how to correct for the problem.

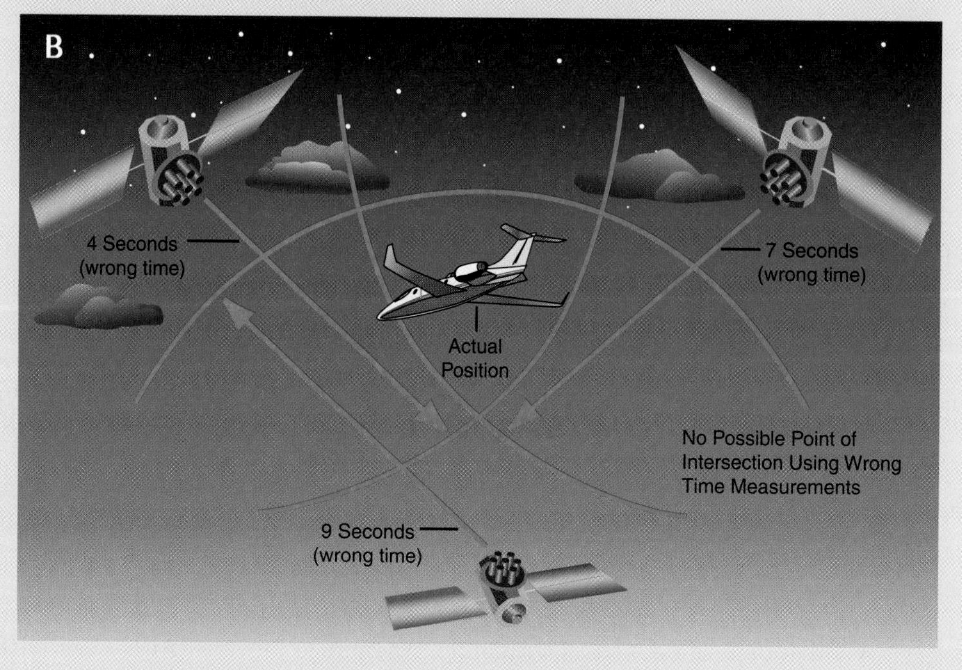

In simple terms, the computer in the receiver keeps subtracting time from all the measurements until it finds a point at which the 3 circles could intersect. This cancels out any consistent clock error present in the GPS receiver and defines the correct 2-dimensional position. For your receiver to calculate an accurate 3-dimensional position, a 4th satellite is added to the equation. In addition to improving GPS accuracy, this method also makes GPS affordable. Simply by applying a little trigonometry, the average GPS receiver can use a moderately accurate timepiece instead of a $100,000 atomic clock.

for three-dimensional navigation. Some receivers have the additional ability to isolate the corrupt satellite and continue reliable operation with the remaining satellites. This requires an additional satellite beyond what is required to identify a RAIM error, a total of six. Since the GPS constellation ensures only 5 satellites in view, you can count on this enhanced level of RAIM protection only if the GPS receiver has access to independent altitude information.

Baro-aiding is a method of augmenting the GPS integrity solution by using information from your Mode C altimeter. Many IFR-approved GPS installations require you to enter an altimeter setting when you turn them on. This altitude input replaces one of the satellites required for RAIM confirmation.

REGULATORY REQUIREMENTS FOR IFR GPS NAVIGATION

If an IFR-certified GPS receiver is installed with the appropriate technical standard order (TSO) C-129 authorization, it can be used for certain IFR operations. An approved system can replace the INS or other long-range navigation system required for short oceanic routes, and can replace one of the required dual INS systems required for longer transoceanic routes.

Aircraft using GPS navigation equipment under IFR must be equipped with an alternate means of navigation appropriate to the flight. You do not have to actively monitor the alternative navigation equipment if the GPS receiver uses RAIM for integrity monitoring. However, if RAIM capability of the GPS equipment is lost, active monitoring of an alternate means of navigation is required. Based on current satellite signals, most IFR-certified GPS receivers can predict whether RAIM capability will exist at the destination at the ETA. In the event loss of RAIM capability is predicted to occur, you must use other navigation equipment, or delay or cancel the flight.

DIFFERENCES IN RECEIVERS

One of the main challenges in using GPS for IFR is dealing with the lack of standardization between different makes and models of GPS receivers. Although this section discusses basic principles of GPS use, you must become thoroughly familiar with the particular GPS equipment installed in the aircraft you are using. Spend some time with the receiver operation manual, and the FAA-approved aircraft flight manual (AFM) supplement that covers this equipment installation. You should use the receiver's simulation mode to become familiar with its operation prior to flying with it. It is also a good idea to use the equipment in flight under VFR conditions prior to attempting IFR operation. [Figure 2-74]

Figure 2-74. A number of IFR-certified GPS receivers offer amazing capabilities. The challenge to pilots is dealing with the lack of standardization in the operation of these units.

BASIC GPS NAVIGATION

While the GPS receiver internally determines latitude, longitude, altitude, and groundspeed, it is difficult to navigate with only this information. That is why aviation GPS units contain databases and the ability to calculate course and distance from your present position to any point in the database. Although the navigation presentations vary between different GPS manufacturers and models, you will find information about ground track, course, course deviation, groundspeed and time to fix on almost all units. Many models enhance your situational awareness by also offering simple moving map presentations. [Figure 2-75]

WHERE YOU ARE NOW
The magnetic bearing from the present position to DBL VORTAC is 265°. When you are on course, BRG agrees with DTK. The distance to DBL is 31.1 n.m. Unlike DME, GPS displays horizontal, not slant-range, distance to the next waypoint.

ROUTE SEGMENT
Shows waypoints you are traveling between (Denver-Centennial Airport and Red Table VORTAC in this example). The distance for this leg is 96 n.m., and the desired track (DTK) is 262°. The DTK is equivalent to a course set on a VOR's OBS.

MESSAGE
The GPS will indicate if it has a message for you. These messages may request you to set a certain course on your HSI, enter an altimeter setting, or warn you of loss of RAIM. Press this button to read the message and normally press it again to dismiss the message.

CURSOR KEY
With certain data pages displayed, you press the cursor key to get a cursor that can be moved over various data fields on the page using the large data entry knob. You can change the data or option under the cursor using the small data entry knob. The cursor key is also called a SELECT (SEL) key on some units. Others use the ENTER (ENT) key to perform this function.

CLEAR KEY
This key typically blanks out a data entry field, or backs out of an operation.

ENTER KEY
When you have selected the data you want using the data entry knobs, press this key to complete the operation.

KAPA TO DBL 96.0 DTK 262°
BRG 265° DIST 31.1 NM
TRK 292° GS 154 KTS
FLY RIGHT 3.1 NM
ETE 13:59 ETA 14:18Z FUEL 3.1 GAL
JX-90 MSG CRSR CLR ENT PWR FPL WPT NAV NRST SELECT CHANGE

WHERE YOU ARE GOING
Your current magnetic track over the ground is 292°. This is a 30° intercept for the 262° course, or DTK to DBL. Groundspeed is 154 knots. Unlike DME, GPS groundspeed reflects actual groundspeed regardless of direction.

DATA ENTRY KNOBS
You can select waypoints from the database, as well as control other options, using these knobs. By convention, the large outer knob moves between pages or fields of data, and the small knob changes the data highlighted by the cursor.

YOUR COURSE
Most GPS receivers have a course deviation indicator similar to a VOR. In an IFR approach-certified installation, this indication is usually remoted to a VOR or HSI indicator in your panel. Since that makes the GPS CDI indication redundant, you may choose another setting which describes your deviation from course in another way, as shown below the CDI here.

TIME AND FUEL TO NEXT WAYPOINT
Given groundspeed and distance, the GPS computes the time to the waypoint, the time of arrival, and, if you provide fuel-flow information, expected fuel use.

DIRECT TO
After pressing this button, the receiver asks you to enter a destination waypoint. It then provides navigation indications to that waypoint. This is the easiest way to enter a simple route on almost every GPS.

WAYPOINT
Use this function to access information about waypoints in the database, such as frequencies for airport towers or VORs. On many units, the airport screens are where you access departures, arrivals, and approaches.

NEAREST
Almost all GPS receivers can quickly locate the nearest airports in the event of an emergency, and help you navigate there.

FLIGHT PLAN
This function is labeled ROUTE (RTE) on some receivers. For IFR, you need more detailed routing than simply DIRECT TO your destination. This function allows you to enter a list of waypoints that makes up a route.

NAVIGATION
Once you have programmed a route or selected DIRECT TO, this function brings up information that helps you navigate the route.

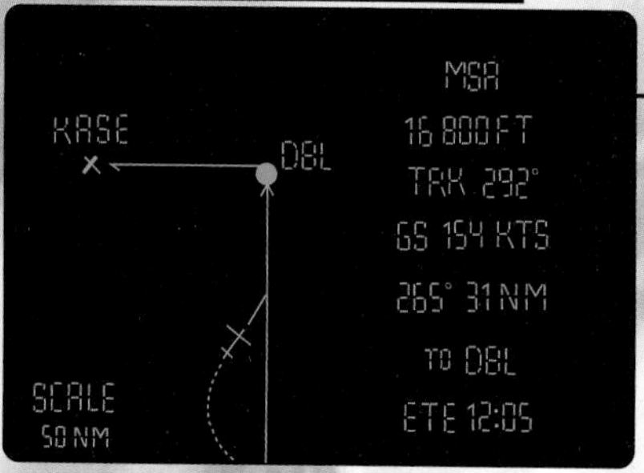

MOVING MAP SCREEN
Many GPS Receivers offer a moving map that shows the route visually with some essential navigation information. Some allow you to switch between north up, track up, and desired track (DTK) up (shown).

Figure 2-75. Almost all GPS receivers present the same basic navigation information, although normally it does not all fit on the data screen at the same time. You can cycle through various screens and customize individual screens to display the data you prefer. This sample receiver has a large screen that shows data that might appear on several different screens of a typical GPS receiver. This sample receiver contains buttons and knobs you might find on several different makes and models.

Generally, you access the various types of GPS navigation information by pressing a NAV button on the front of the unit. On some units, you select the main navigation pages using the large knob and use the small knob to obtain detailed navigation information. You also may use the knobs in conjunction with the cursor key to customize the information shown on individual screens. For example, if your GPS is connected to a VOR indicator, you might choose to display different information in place of the course deviation indicator on the GPS data screen.

PROGRAMMING AND FLYING ROUTES

Using a GPS receiver for navigation can be as simple as pressing the DIRECT TO button, dialing in your destination, following the GPS course-deviation indications and monitoring the distance and time to the destination. The problem with this method on long flights is that pilots may not know their position relative to nearby cities, airports, and terrain.

IFR-approved GPS receivers can be programmed with routes containing dozens of waypoints. [Figure 2-76] A simple direct route is just not sufficient for IFR navigation. You can enter a route based on existing VORs, or you can choose more direct routes based on your own waypoints. On direct routes below FL390, you should define your route with one route description waypoint for each ARTCC you will pass through. These waypoints must be located within 200 nautical miles of the preceding center's boundary.

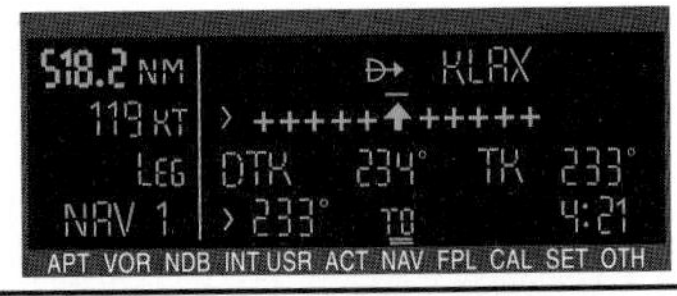

A direct route to a distant destination gives you little information about your present position.

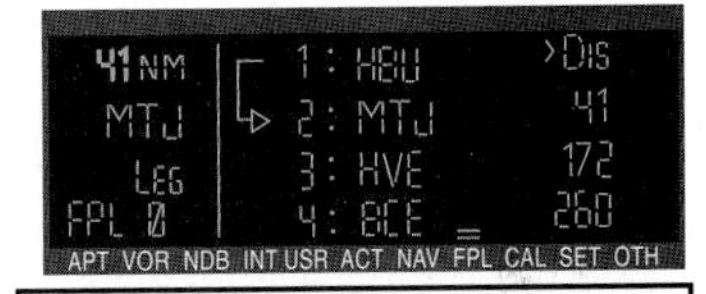

A route containing waypoints at reasonable intervals helps you monitor overall trip progress better.

Figure 2-76. If entering a simple route direct to your destination from your present position, you may not be able to maintain adequate situational awareness. When entering and following a route based on conventional navaids, you have more information about nearby cities, airports and terrain.

Normally, the most complex parts of your route are the departure, arrival, and approach. IFR flights are rarely a straight line to your destination; you must fly a specific flight path that properly sequences traffic and avoids obstacles. An approach-certified GPS contains waypoints and routes that define departure, arrival, and nonprecision approach procedures. You will learn more about GPS approaches in Chapter 8, Section C — GPS and RNAV Approaches.

GPS COURSE DEVIATION INDICATIONS

In an installation approved for IFR approaches, the GPS unit is usually connected to one of your conventional VOR indicators. Normally, a switch selects whether the indications are from the VOR or GPS. Unlike a VOR, which indicates your angle off course, RNAV systems like GPS tell your absolute deviation from course in nautical miles. To approximate the behavior of a VOR, which gets more sensitive as you get closer to the station, the GPS scale changes as you move from enroute, to terminal, to the approach phase of flight. When you are at least 30 nautical miles away from your destination, the GPS sensitivity is 5 nautical miles when the CDI is fully deflected to one side (1 nautical mile per dot), just like a VOR at 30 nautical miles. When the receiver is armed for the approach, and you are within 30 nautical miles of the destination, the sensitivity increases to 1 nautical mile, full scale. As you reach the final approach waypoint (FAWP), the sensitivity increases to 0.3 nautical mile full scale.

AUTOMATIC AND MANUAL COURSE SELECTION

One important difference between GPS and VOR navigation is that the GPS selects your courses for you, based on which waypoints you are flying between. There are situations where you need to manually select a course, as you do with the OBS on your VOR. ATC may clear you to a fix via a different course than what is published and contained in the GPS database. For this reason, IFR-certified GPS equipment allows manual selection of a course. When switching to this mode, you can typically select the course using the OBS on the VOR head you are using with the GPS.

DME ARCS

Most early GPS units were not capable of easily navigating DME arcs. However, many modern approach-certified GPS receivers provide superior navigation capability compared to conventional equipment without an RMI. The GPS may actually give you a CDI indication to fly the curved path, with recommended headings to stay on the arc as your route progresses. Since procedures may vary, you should check your GPS operating manual and/or the POH/AFM supplement for details.

AUTO SEQUENCING

When a departure, arrival, approach, or other route is programmed into your GPS, the receiver senses when you pass a waypoint and automatically cycles to the next waypoint. Unlike VOR navigation, there is no need to constantly tune and identify new facilities and change the OBS settings. **Auto sequencing** can work against you if you are not careful. It is essential you disable this feature if conducting maneuvers in which you might fly over or abeam a waypoint more than once and in different directions. If auto sequencing is on at these times, the receiver may cycle to a waypoint before you are ready to proceed to it, and it may be difficult to reprogram the unit back to the previous waypoint when you are busy flying an approach.

DIFFERENTIAL GPS

Differential GPS (DGPS) promises accuracy within one meter! The principle is simple. You put a stationary GPS receiver at a precisely surveyed location. This reference receiver compares its GPS satellite-derived position with its known location and transmits a list of timing errors for all satellites it sees. This correction signal could be coded into signals from existing transmitters, such as VOR or radar signals sent to your transponder. When the DGPS receiver receives these corrections, it applies them to its position solution to get an incredibly accurate result. A DGPS correction signal is accurate for distances of hundreds of kilometers from the reference receiver. This means that you can add precision approach capability to airports anywhere in the world, with the installation of relatively few, inexpensive reference receivers.

OTHER GPS PROMISES AND PRECAUTIONS

GPS offers a high level of cockpit automation to the general aviation pilot. When coupled to an autopilot, the GPS has the capability of flying your airplane from shortly after takeoff until just before touchdown. Once you are comfortable with the equipment's advanced features, and you invest the additional preflight effort programming the system, it can handle most of your navigation burden throughout the flight. Nevertheless, you should be careful not to rely on high technology equipment without paying enough attention to the details of the flight.

SUMMARY CHECKLIST

✓ The horizontal situation indicator (HSI) display combines a conventional VOR navigation indicator with a heading indicator.

✓ VOR station passage is indicated by the first positive, complete reversal of the TO/FROM indicator.

✓ The angle between the nose of the aircraft and an NDB is the relative bearing.

✓ Magnetic heading (MH) plus relative bearing (RB) equals magnetic bearing (MB) to the station.

✓ A radio magnetic indicator (RMI) always displays the bearing to a station at the head of the arrow, and the bearing from a station at the tail of the arrow.

✓ NDB station passage is considered to have occurred when the needle either points to a wingtip or settles at or near the 180° position.

✓ DME is accurate to within 1/2 mile or 3% (whichever is greater), as long as you are at least 1 mile or more from the DME facility for every 1,000 feet of altitude above the station.

✓ VOR facilities are classified according to their usable range and altitude, or standard service volume (SSV). You can find the SSV which applies to a particular VOR in the *Airport/Facility Directory.*

✓ When checking your VOR using a VOT, the CDI should be centered and the OBS should indicate that the aircraft is on the 360° radial, ±4°. When using a VOR checkpoint, the CDI must center within ±4°. The allowable error using an airborne checkpoint is ±6°. When you conduct a dual system check, the difference between VOR systems should not exceed 4°.

✓ Area Navigation (RNAV) allows you to fly direct to your destination without the need to overfly VORs or other ground facilities.

✓ The global positioning system (GPS) provides standard positioning service (SPS) with a 95% probability of horizontal accuracy within 100 meters (328 feet), and 99.99% probability of accuracy within 300 meters (984 feet).

✓ Receiver autonomous integrity monitoring (RAIM) continuously verifies the integrity (usability) of the signals received from the GPS constellation.

✓ You do not have to actively monitor alternative navigation equipment if your GPS receiver uses RAIM for integrity monitoring. However, if RAIM capability of the GPS equipment is lost, you must monitor an alternate means of navigation.

KEY TERMS

Horizontal Situation Indicator (HSI)

Cone of Confusion

Relative Bearing

Radio Magnetic Indicator (RMI)

Distance Measuring Equipment (DME)

DME Arcs

VOR Test Facilities (VOTs)

VOR Checkpoints

Area Navigation (RNAV)

Waypoints

Phantom VORs

Inertial Navigation System (INS)

Long Range Navigation (LORAN)

Global Positioning System (GPS)

Receiver Autonomous Integrity Monitoring (RAIM)

Auto Sequencing

Differential GPS (DGPS)

QUESTIONS

Refer to the following figure to answer questions 1-3.

1. What displacement from course is indicated by this instrument when used in conjunction with a VOR?

   A. 2-1/2°
   B. 5°
   C. 10°

2. What displacement from course is indicated by this instrument when connected to a GPS during enroute operations?

   A. 0.15 nautical mile
   B. 0.5 nautical mile
   C. 2.5 nautical miles

3. What displacement from course is indicated by this instrument when connected to a GPS and on final approach?

   A. 0.15 nautical mile
   B. 0.5 nautical mile
   C. 2.5 nautical miles

4. Why is it impossible for a horizontal situation indicator to provide reverse sensing when tuned to a VOR?

5. True/False. Station passage is indicated when the TO/FROM indicator first starts fluctuating.

6. If it takes 1 minute to fly from the 090° radial to the 080° radial on a perpendicular course, how long will it take to arrive at the VORTAC when you intercept the 070° radial and fly direct to the station at the present speed (assuming no wind)? If your speed is 100 KTAS, what is the distance to the station?

Use the following figure to answer question 7.

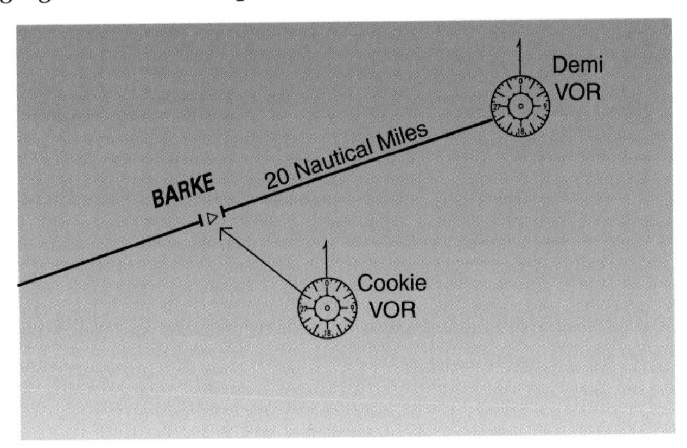

7. You are tracking an airway inbound to Demi VORTAC and tune your only VOR receiver to Cookie VORTAC to identify BARKE Intersection. After the CDI centers at BARKE, you tune back to Demi VORTAC and the needle indicates a one dot deviation to the left. Assuming you have set your CDI for proper sensing, what is your position (in nautical miles) relative to the airway?

8. Why is a 45° angle commonly used when intercepting a bearing with an ADF?

9. Select the true statement regarding an RMI?

   A. You must constantly adjust the RMI compass card to match your magnetic heading.
   B. The number under the head of the bearing pointer is the magnetic bearing from the station.
   C. Relative bearing is the angle between the head of the bearing pointer and the aircraft heading index.

10. When you are tracking outbound from an NDB with the proper drift correction established, which direction will the head of the ADF bearing pointer be deflected?

   A. To the windward side of the aircraft's tail
   B. To the downwind side of the aircraft's tail
   C. To the downwind side of the aircraft's nose

11. What ADF or RMI indication would you expect when tracking on a bearing toward an NDB with a 10° left wind correction?

    A. 0° relative bearing
    B. 10° left of the aircraft's nose
    C. 10° right of the aircraft's nose

12. True/False. Normally, when flying a DME arc, the bearing pointer reaches the wingtip position when the airplane is slightly outside the arc.

13. What is the maximum distance at which you can expect a reliable signal from a high altitude VOR. At what altitudes does this maximum range occur?

14. True/False. The maximum allowable error on a VOR check using a VOT is 4°.

15. What is the maximum allowable difference when checking two VORs against each other?

16. True/False. When tuning a VOR and hearing the correct 1,020 Hz Morse code identifier signal, you are assured that both the VOR and DME are properly tuned and usable for navigation.

17. Which RNAV system is completely self-contained, requiring no inputs from ground-based or space-based facilities?

18. True/False. Without periodic position updates, INS accuracy degrades 1 to 2 nautical miles per hour.

19. What horizontal position accuracy can you expect from GPS 99.99% of the time?

    A. Within 100 meters
    B. Within 200 meters
    C. Within 300 meters

20. True/False. If your airplane has an IFR-certified GPS (with RAIM capability), your airplane does not have to be equipped with an alternate means of navigation.

CHAPTER 3

THE FLIGHT ENVIRONMENT

Instrument/Commercial
Part I, Chapter 3 — The Flight Environment

SECTION A

AIRPORTS, AIRSPACE, AND FLIGHT INFORMATION

Much of the information in this section is a review of subject areas you have studied in previous training programs. If you need additional information, you may want to refer back to your initial training materials.

THE AIRPORT ENVIRONMENT

An important aspect of IFR flight operations is a sound knowledge of the airport environment, which includes runway markings, airport signs, runway lighting, approach lighting systems, and associated airport lighting. The types of markings, signs, and lighting systems installed may vary from airport to airport, depending on size, traffic volume, and the types of operations and approaches authorized.

RUNWAY MARKINGS

Markings vary between runways used solely for VFR operations and those used for IFR operations. A **visual runway** usually is marked only with the runway number and a centerline, but threshold markings may be included if the runway is used, or intended to be used, for international commercial operations, and aiming point markings may be included on runways 4,000 feet or longer used by jet aircraft. [Figure 3-1]

Figure 3-1. The common types of runway markings for visual, nonprecision, and precision instrument runways are shown here.

Runways used for instrument operations have additional markings. A **nonprecision instrument runway** is used with an instrument approach that does not have an electronic glide slope for approach glide path information. This type of runway has the visual runway markings, plus the threshold and aiming point markings.

As shown in figure 3-1, touchdown zone markings are located 500 feet from the beginning of the runway. Aiming point markings are 500 feet beyond the touchdown zone markings.

An Airport Project That Didn't Stay Afloat

In the 1920's, the economic range of a commercial airplane was approximately 500 to 600 miles. To fly farther than that nonstop, an airplane had to carry so much weight as fuel that it was difficult for companies to turn a profit. To make transatlantic air travel successful, Edward Armstrong, an engineer, had an idea to construct seadromes — floating airports strung across the ocean.

Armstrong designed a structure which consisted of a deck, approximately 1,500 feet long and 300 feet wide, placed on 28 steel columns, each 170 feet long. The seadrome was designed to float with the airfield deck 70 feet above the water's surface. A buoyancy chamber would be built into each column 100 feet below the deck, and 60 feet below that would be a ballast chamber filled with iron ore. The seadrome would be attached to a buoy connected by two 18,000-foot steel cables to a concrete mushroom anchor. This system reduced the great strain that would have been caused by connecting the anchor line directly to the structure, and ensured that the seadrome was free to swing about into the wind. Propellers would allow the seadrome to maneuver while anchored and to navigate in an emergency.

Beneath the deck of the seadrome, Armstrong envisioned hangars, which would use a large elevator to transport airplanes to and from the flight deck. In addition, there were plans for repair shops, a radio station, and a hotel. The 50-room hotel would feature a gymnasium, swimming pool, miniature golf course, billiard room, movie theater, bowling alley, and several tennis courts.

The seadrome project had enthusiastic approval from famous aviators such as Charles Lindbergh. However, it was discontinued by the U.S. government when concern arose about the difficulty of maintaining adequate protection of seadromes built over international waters.

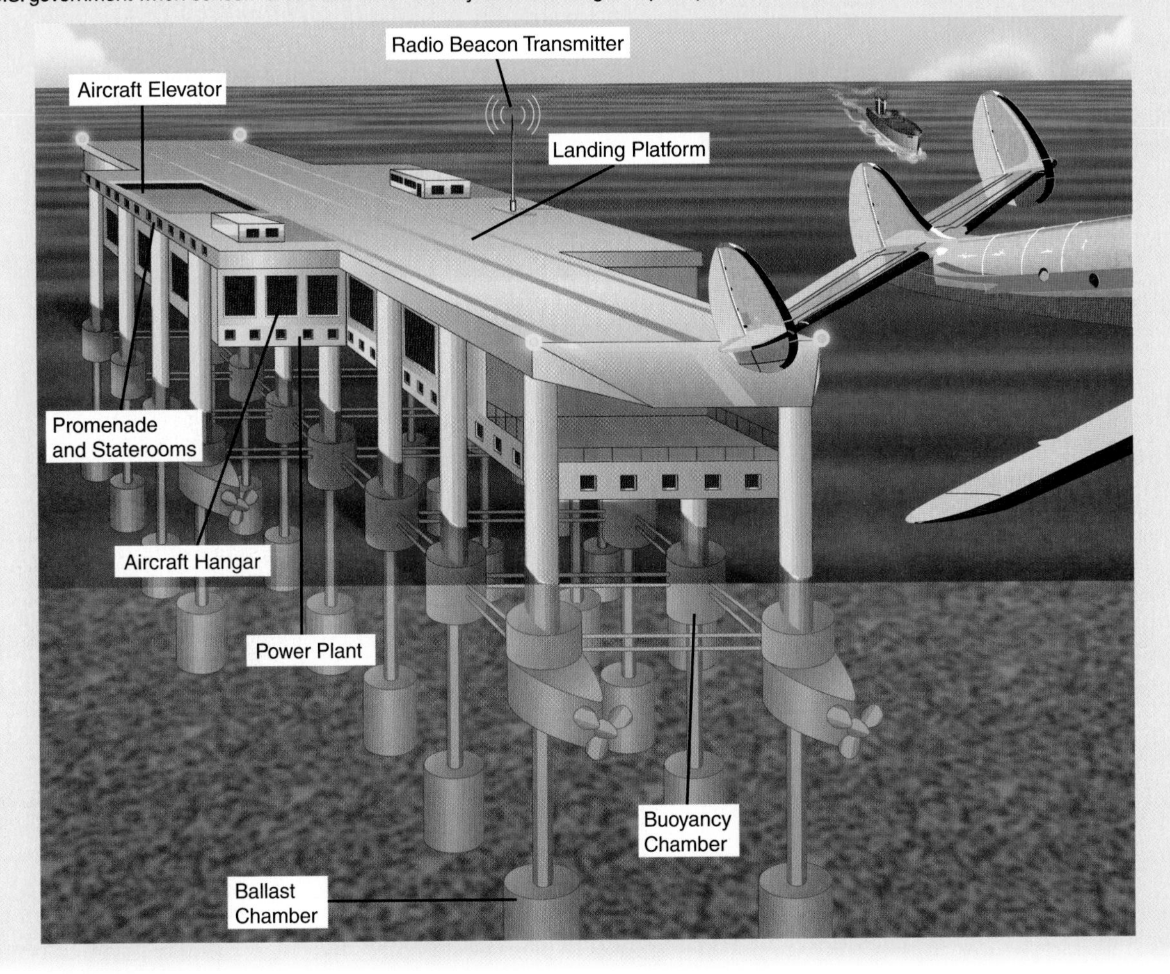

Precision instrument runways are served by nonvisual precision approach aids, such as the instrument landing system (ILS). The ILS uses an electronic glide slope to provide glide path information during the approach. The associated runways are marked so you can receive important visual cues, especially during periods of extremely low visibility. Besides the threshold markings, touchdown zone markings are coded to provide distance information in 500 foot increments. Aiming point markings are located approximately 1,000 feet from the landing threshold.

TAXIWAY MARKINGS

The links between the airport parking areas and the runways are the **taxiways**.They are easily identified by a continuous yellow centerline stripe. At some airports, taxiway edge markings are used to define the edge of the taxiway and are normally used to separate the taxiway from pavement that is not intended for aircraft use. Runway holding position markings, or **hold lines**, are used to keep aircraft clear of runways and, at controlled airports, serve as the point that separates the responsibilities of ground control from those of the tower. Hold lines are usually placed between 125 and 250 feet from the runway centerline. In addition, hold lines may be located at taxiway intersections. At some airports, hold signs may be used instead of, or in conjunction with, the hold lines painted on the taxiways.

At an uncontrolled airport, you should stop and check for traffic and cross the hold line only after ensuring that no one is on an approach to land. At a controlled airport, the controller may ask you to hold short of the runway for landing traffic. In this case, you should stop before the hold line and proceed only after you are cleared to do so by the controller, and you have checked for traffic. [Figure 3-2]

Hold line markings at the intersection of taxiways and runways consist of four yellow lines (two dashed lines and two solid lines). The dashed lines are nearest the runway.

At airports equipped with an instrument landing system, it is possible for aircraft near the runway to interfere with the ILS signal. If this is the case, the hold line may be placed farther from the runway to prevent any interference, or you may find two hold lines for a runway. The one closest to the runway is the normal hold line, while the one farthest away is the ILS hold line. At other locations, only an ILS hold line may be used. [Figure 3-3]

Figure 3-2. When you exit the runway after landing, be sure to cross the hold line before stopping to ensure that you are clear of the runway.

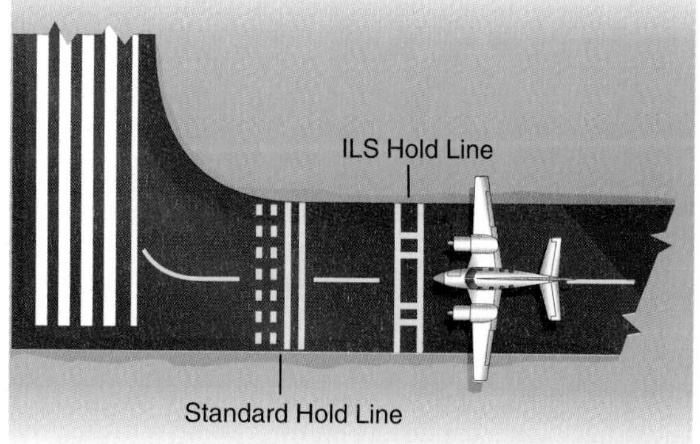

Figure 3-3. When ILS approaches are in progress, you may be asked by the controller to *". . . hold short of the ILS critical area."*

ADDITIONAL MARKINGS

In addition to the markings previously discussed, you may encounter other markings on runways and taxiways which place restrictions on their use. The following paragraphs provide a brief review of some of these markings. [Figure 3-4]

A **displaced threshold** is marked by a solid white line extending across the runway perpendicular to the centerline. It marks the point at which all normal takeoff and landing operations are permitted. The operations permitted prior to this point depend on the type of restriction imposed. Taxi, takeoff, and rollout areas are marked by white arrows leading to a displaced threshold. When landing, you must touch down beyond the displaced threshold. The paved area prior to the displaced threshold is not designed for landing. A taxi-only area is a designated portion of a runway to be used only for taxi operations. It is marked by a yellow taxi line leading to a displaced threshold. When departing such a runway, you may use the entire length. However, plan your landing beyond the displaced threshold marking. When departing from or arriving on the opposite end, you may not consider this area as usable for takeoff or landing, except as an overrun during an aborted takeoff.

On runways with a displaced threshold, the beginning portion of the landing zone is marked with a solid white line with white arrows leading up to it. Although the pavement leading up to a displaced threshold may not be used for landing, it may be available for taxiing, the landing rollout, and takeoffs.

Blast pad/stopway areas are marked by yellow chevrons and may not be used for taxiing, takeoffs, or landings. The blast pad area allows propeller or jet blast to dissipate without creating a hazard to others. In the event you must abort a takeoff, the stopway provides additional paved surface for you to decelerate and stop. On some runways with a displaced threshold, a **demarcation bar** separates the displaced threshold area from a blast pad, stopway, or taxiway that precedes the runway.

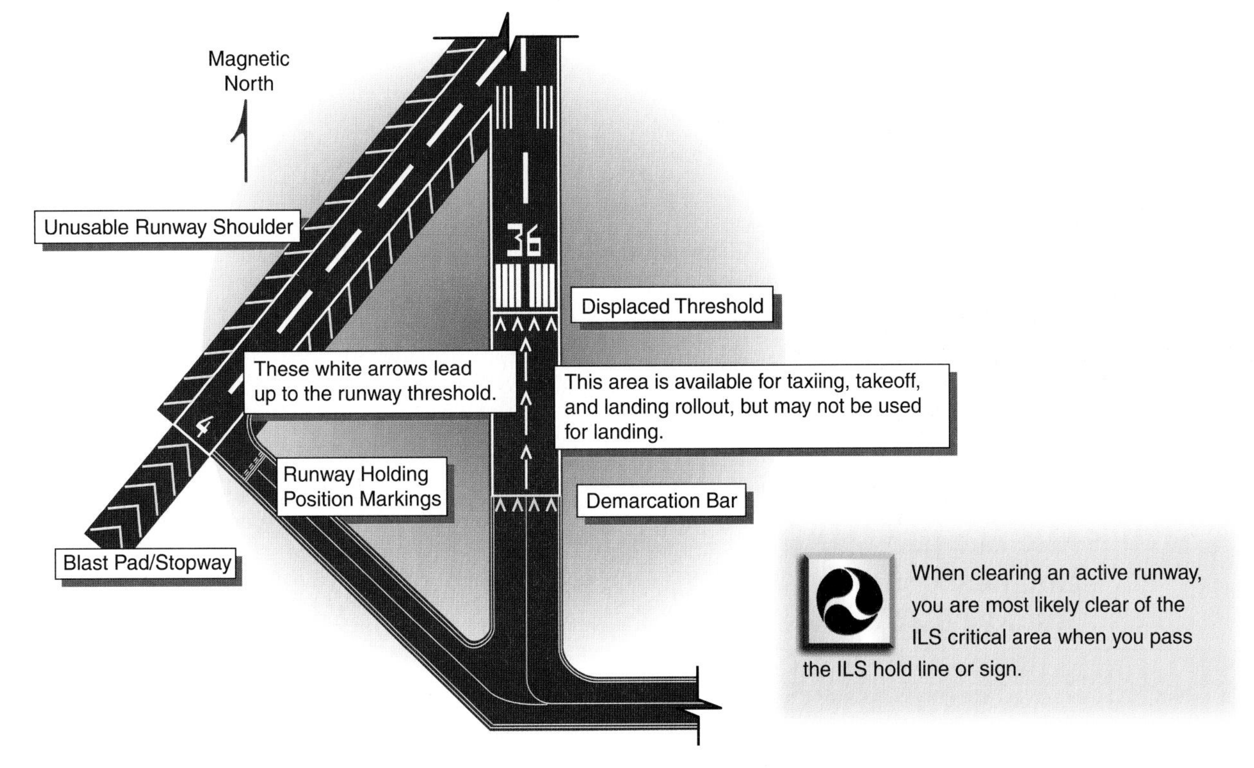

When clearing an active runway, you are most likely clear of the ILS critical area when you pass the ILS hold line or sign.

Figure 3-4. This illustration shows some samples of different types of runway restrictions. You may wish to refer to it as each element is discussed.

A closed runway is marked with a large yellow X at each end. Although the closed, or temporarily closed runway may not be used, other runways and taxiways which cross it are not affected unless specifically marked. A closed taxiway may be marked by Xs, or it may simply be blocked off.

AIRPORT SIGNS

Major airports usually have complex taxi routes, multiple runways, and widely dispersed parking areas. In addition, vehicular traffic in certain areas may be quite heavy. As shown in figure 3-5, most airfield signs are standardized to make it easy for you to identify taxi routes, mandatory holding positions, and boundaries for critical areas. Another benefit, if you fly outside the United States, is that the U.S. standards are practically the same as ICAO specifications. The International Civil Aviation Organization (ICAO) is a specialized agency of the United Nations whose objective is to develop standard principles and techniques of international air navigation and to promote development of civil aviation. Specifications for airport signs include size, height, location, and illumination requirements. [Figure 3-6]

As shown in figure 3-5, a mandatory instruction sign has white lettering on a red background.

Sometimes the installation of a sign is not practical so a surface-painted sign may be used. Surface painted signs may include directional guidance or location information. For example, the runway number may be painted on the taxiway pavement near the taxiway hold line.

9-27

Mandatory Instruction Signs denote an entrance to a runway, a critical area, or an area prohibited to aircraft. These signs are red with white letters or numbers. An example of a mandatory instruction sign is a runway holding position sign which is located at the holding position on taxiways that intersect a runway or on runways that intersect other runways.

Location Signs identify either the taxiway or runway where your aircraft is located. These signs are black with yellow inscriptions and a yellow border. Location signs also identify the runway boundary or ILS critical area for aircraft exiting the runway.

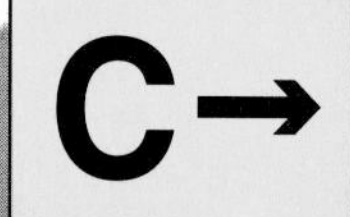

Direction Signs indicate directions of taxiways leading out of an intersection. They have black inscriptions on a yellow background and always contain arrows which show the approximate direction of turn.

←INTL

Destination Signs indicate the general direction to a location on the airport, such as civil aviation areas, military areas, international areas, or FBOs. They have black inscriptions on a yellow background and always contain an arrow.

Noise Sensitive Area Located Southeast of Runway 9/27

Information Signs advise you of such things as areas that cannot be seen from the control tower, applicable radio frequencies, and noise abatement procedures. These signs use yellow backgrounds with black inscriptions.

3

Runway Distance Remaining Signs are used to provide distance remaining information to pilots during takeoff and landing operations. The signs are located along the sides of the runway, and the inscription consists of a white numeral on a black background. The signs indicate the distance remaining in thousands of feet. Runway distance remaining signs are recommended for runways used by turbojet aircraft.

Figure 3-5. There are six basic types of airport signs — mandatory, location, direction, destination, information, and runway distance remaining.

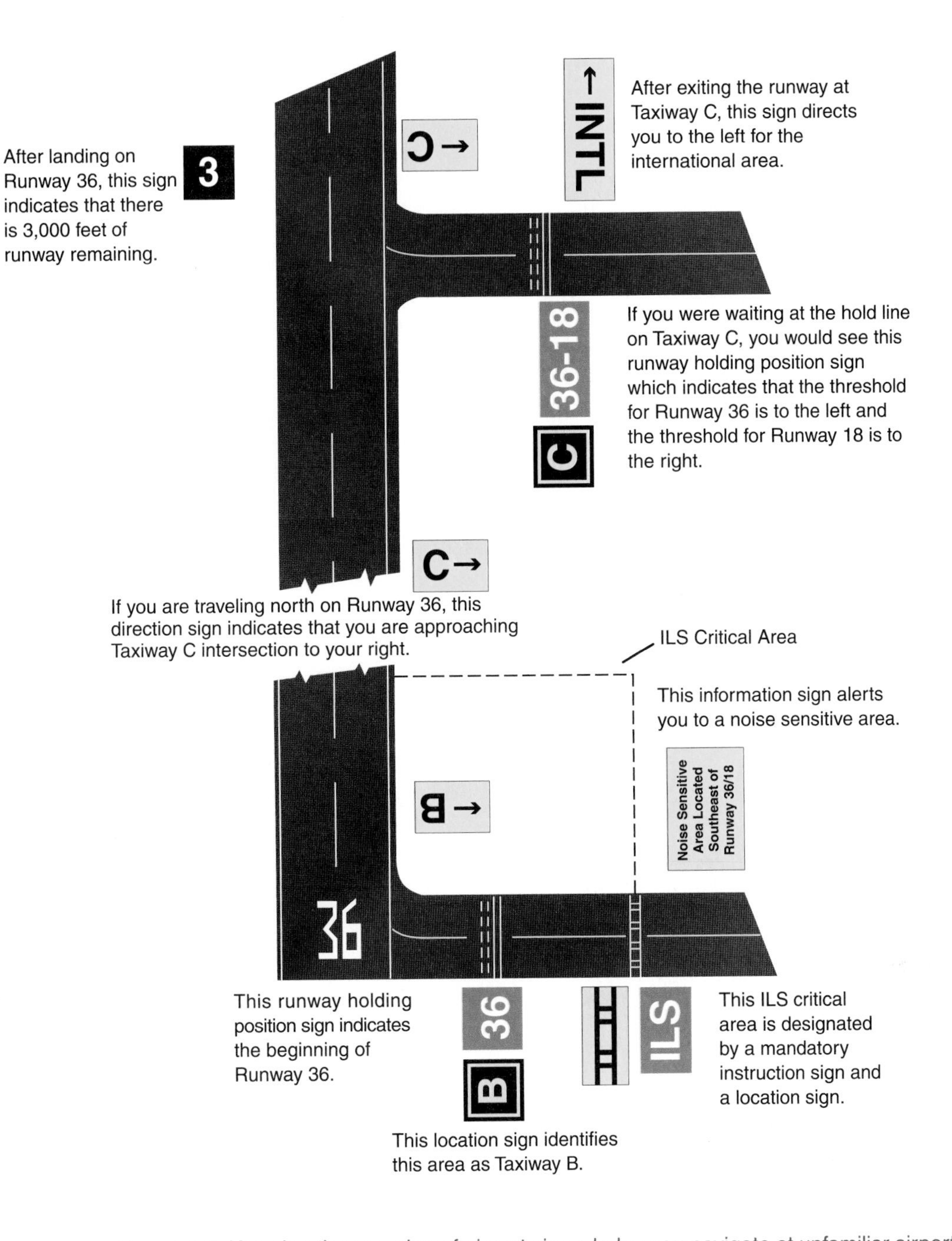

Figure 3-6. Knowing the meaning of airport signs, helps you navigate at unfamiliar airports.

RUNWAY INCURSION AVOIDANCE

The official definition of a runway incursion is "any occurrence at an airport involving an aircraft, vehicle, person, or object on the ground that creates a collision hazard or results in loss of separation with an aircraft taking off or intending to take off, landing, or intending to land." Runway incursions are primarily caused by errors associated with clearances, communication, airport surface movement, and positional awareness. There are several procedures that you can follow and precautions that you can take to avoid a runway incursion.

1. During your preflight planning, ensure that you have all the pertinent information regarding airport construction and lighting.
2. Complete as many checklist items as possible before taxi or while holding short.
3. Strive for clear and unambiguous pilot-controller communication. Read back (in full) all clearances involving active runway crossing, hold short, taxi into position, and hold instructions.
4. While taxiing, concentrate on your primary responsibilities. Don't become absorbed in other tasks, or conversation, while the aircraft is moving.
5. If unsure of your position on the airport, stop and ask for assistance. At a controlled airport, you can request progressive taxi instructions.
6. When possible, while in a run-up area or waiting for a clearance, position your aircraft so you can see landing aircraft.
7. Monitor the appropriate radio frequencies for information or other aircraft cleared onto your runway for takeoff or landing. Be alert for aircraft which may be on other frequencies or without radio communication.
8. After landing, stay on the tower frequency until instructed to change frequencies.
9. To help others see your aircraft during periods of reduced visibility or at night, use your exterior taxi/landing lights, when practical.
10. Report deteriorating or confusing airport markings, signs, and lighting to the airport operator or FAA officials. Also report confusing or erroneous airport diagrams and instructions.
11. Make sure you understand the required procedures if you fly into or out of an airport where LAHSO is in effect.

LAND AND HOLD SHORT OPERATIONS

When **land and hold short operations (LAHSO)** are in effect at controlled airports, ATC may clear you to land and hold short of an intersecting runway, an intersecting taxiway, or some other designated point on the runway. You can accept this clearance if you determine that you can safely land and stop within the available landing distance (ALD). ALD data is published in the special notices section of the applicable *Airport/Facility Directory* and in the *U.S. Terminal Procedures Publication.* Controllers also will provide ALD information upon request. A hold line/holding sign, as well as a row of lights at some airports, will indicate where you should hold short to remain clear of traffic. As part of your preflight planning, you should determine if your destination has LAHSO and decide if the procedure can be safely accomplished based on your airplane's required landing distance. You have the option to decline a LAHSO clearance if you feel that it compromises flight safety. [Figure 3-7]

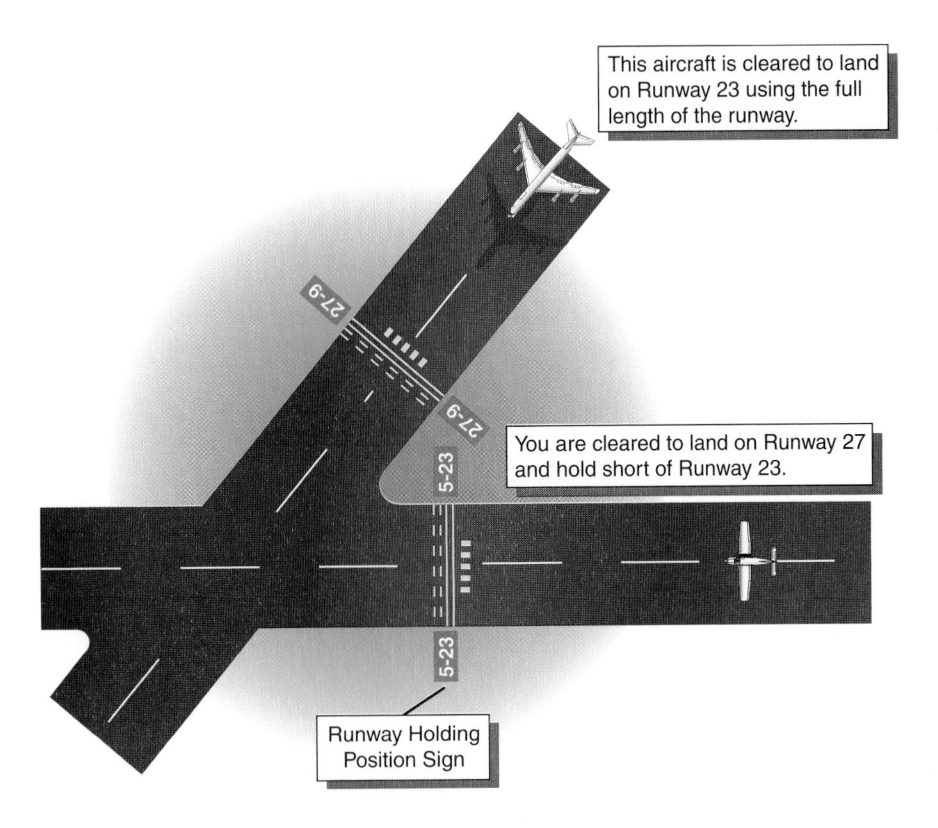

Figure 3-7. If you accept a LAHSO clearance, you should land and exit the runway at the first convenient taxiway before reaching the hold short line. If this is not possible, you must stop at the hold short point. It is crucial that you maintain situational awareness if you are participating in LAHSO.

NOTE: Refer to the NOTAMs section in the back of this manual for further details on LAHSO. In addition, you should consult the Aeronautical Information Manual (AIM) for the latest LAHSO policies and procedures.

LIGHTING SYSTEMS

Airport lighting systems range from the simple lighting needed for VFR night landings to sophisticated systems which guide you to the runway in IFR conditions. You should familiarize yourself with each type of lighting and its significance to VFR, as well as IFR operations.

APPROACH LIGHT SYSTEM

The approach light system (ALS) helps you transition from instrument to visual references during the approach to landing. It makes the runway environment more apparent in low visibility conditions and helps you maintain correct alignment with the runway. Approach light systems use a configuration of lights starting at the landing threshold and extending into the approach area. Normally, they extend outward to a distance of 2,400 to 3,000 feet from precision instrument runways and 1,400 to 1,500 feet from nonprecision instrument runways.

Some approach light systems include **sequenced flashing lights (SFL)** or **runway alignment indicator lights (RAIL)**. SFL and RAIL consist of a series of brilliant blue-white bursts of flashing light. From your viewpoint, these systems give the impression of a

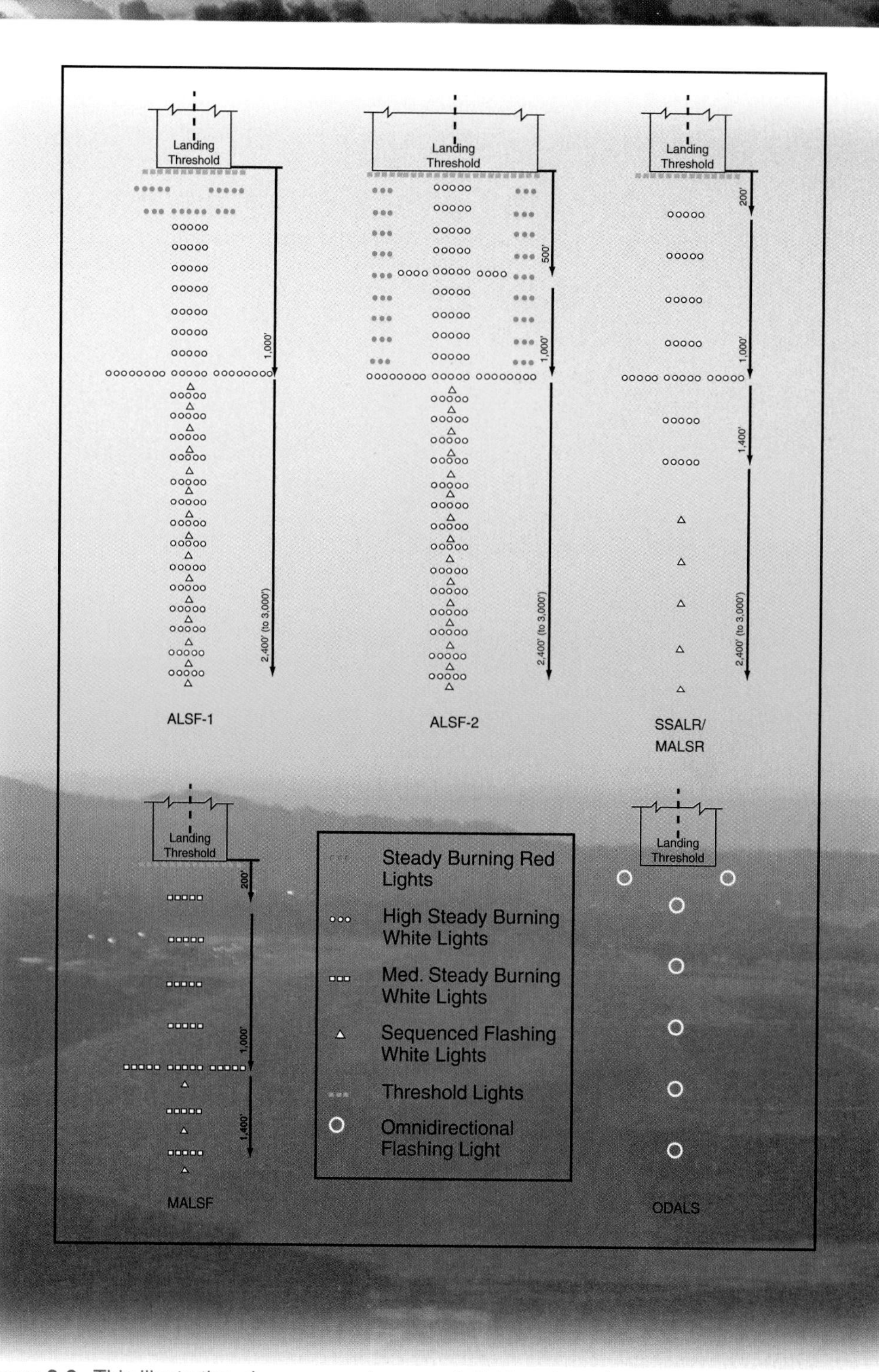

Figure 3-8. This illustration shows several approach light systems in various configurations. ALSF-1 and ALSF-2 configurations are typical for runways that have instrument landing systems, although SSALR/MALSR also may be used. The MALSF configuration and ODALS are used in conjunction with nonprecision instrument runways.

ball of light traveling at high speed toward the approach end of the runway. SFL and RAIL usually are incorporated into other approach light systems. [Figure 3-8]

At some locations, a high intensity white strobe light is placed on each side of the runway to mark the threshold. These lights, called **runway end identifier lights (REIL)**, can be used in conjunction with the green threshold lights. Their purpose is to provide you with a means of rapidly identifying the approach end of the runway during reduced vis-

ibility. They are normally aimed 10° up and 15° away from the runway centerline.

REIL refers to a pair of synchronized flashing lights which provide rapid identification of the approach end of the runway during low visibility conditions.

VISUAL GLIDE SLOPE INDICATORS

Once you have the runway environment in sight, visual glide slope indicators help you establish and maintain a safe descent path to the runway. Their purpose is to provide a clear visual means to determine if you are too high, too low, or on the correct glide path. These indicators are extremely useful during low visibility, or at night, when it may be difficult to judge the descent angle accurately due to a lack of runway contrast. Several different visual glide slope indicator systems are used, but one of the most common is the two-bar **visual approach slope indicator (VASI)**. The two-bar system provides one visual glide path, normally set to 3°. Staying on the VASI glide path assures you of safe obstruction clearance within ±10° of the extended runway centerline and out to 4 nautical miles from the threshold. When using a VASI, you should attempt to fly your approach so the far bars indicate red and the near bars show white. These are the proper light indications for maintaining the glide slope. VASI lights are visible from 3 to 5 miles during the day and up to 20 miles at night. [Figure 3-9]

If you remain on the proper glide path of a VASI, you are assured safe obstruction clearance in the approach area. Two-bar VASIs, which normally have an approach angle of three degrees, show red over white when you are on the glide path. See figure 3-9.

Some airports are equipped with three-bar VASI systems consisting of three sets of light sources forming near, middle, and far bars. These systems provide two visual glide paths to the same runway. The first uses the near and middle bars. This glide path is the same as that provided by a standard two-bar VASI installation. The second uses the middle and far bars. This upper glide path is intended for use only by pilots of

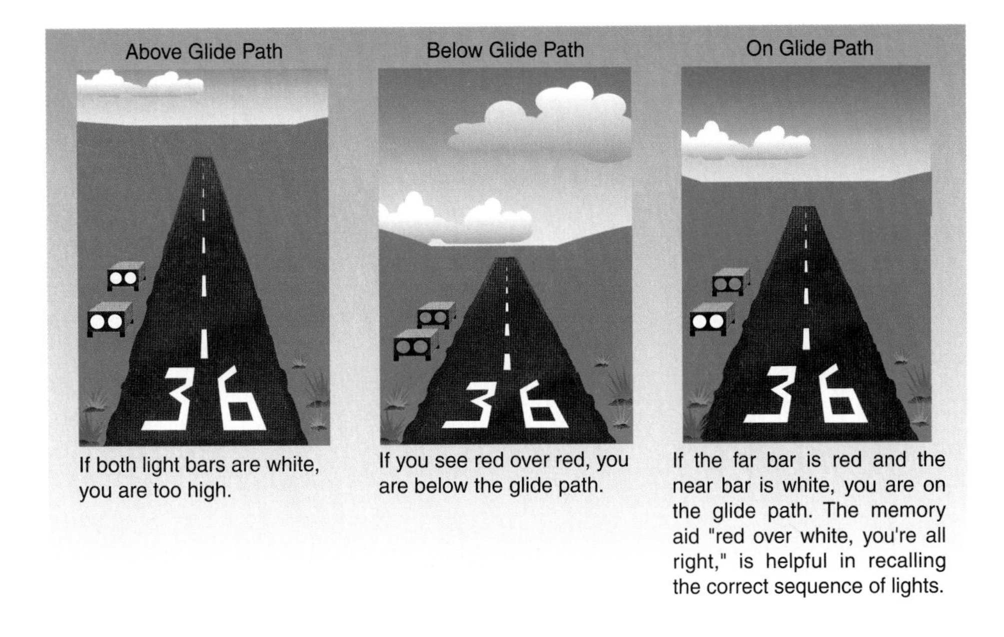

Figure 3-9. The two-bar VASI system consists of two light boxes on each side of the runway. A simplified, abbreviated version, called SAVASI, may use only a single set of boxes, usually installed on the left side. If you are approaching a runway and all of the VASI lights appear to be red, as shown in the center, you should level off momentarily to intercept the proper approach path.

The middle and far bars of a three-bar VASI can be used to descend on the upper glide path, which is usually .25° steeper than the lower glide path. See figure 3-10.

high-cockpit aircraft and is about one-quarter of a degree steeper than the first. The far bars are located approximately 700 feet beyond the middle bars. When on the upper glide path, the pilot sees red, white, and white. [Figure 3-10]

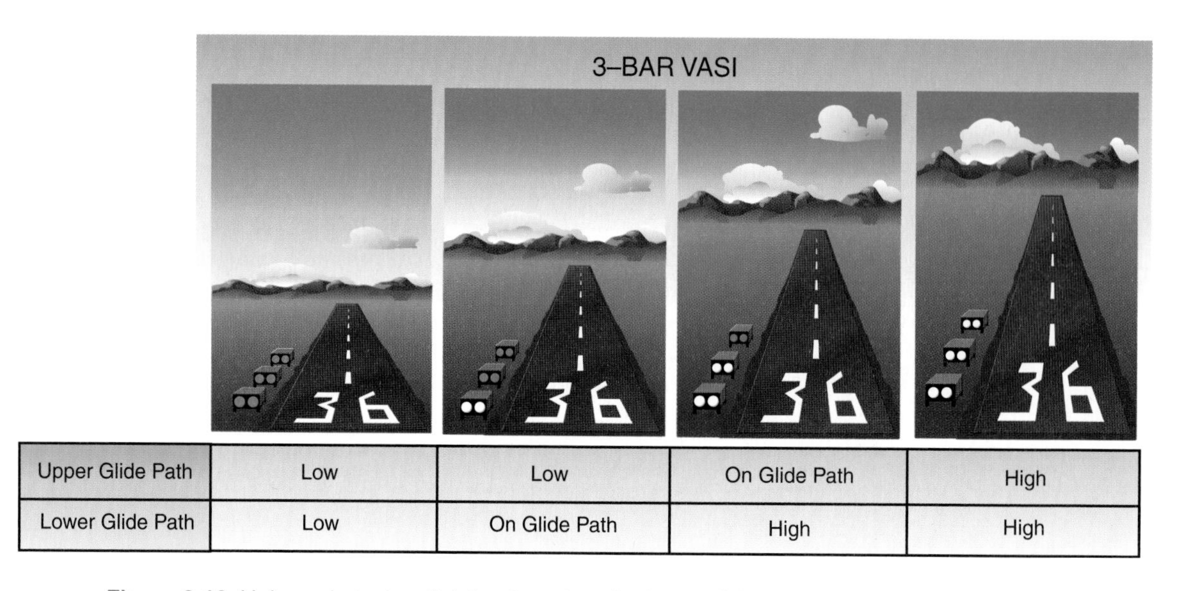

Upper Glide Path	Low	Low	On Glide Path	High
Lower Glide Path	Low	On Glide Path	High	High

Figure 3-10. Unless obstacles dictate otherwise, the lower glide path is usually set at 3°.

Some airports may have a **pulsating approach slope indicator (PLASI)** which projects a two-color visual approach path into the final approach area. A pulsating red light indicates you are below the glide path. A steady red light indicates you are slightly below the glide path, while an above glide path is indicated by a pulsating white light. The on-glide path indication is a steady white light. The useful range is about 4 miles during the day and up to 10 miles at night.

An on-glide path indication from a PAPI is two red lights and two white lights. See figure 3-11.

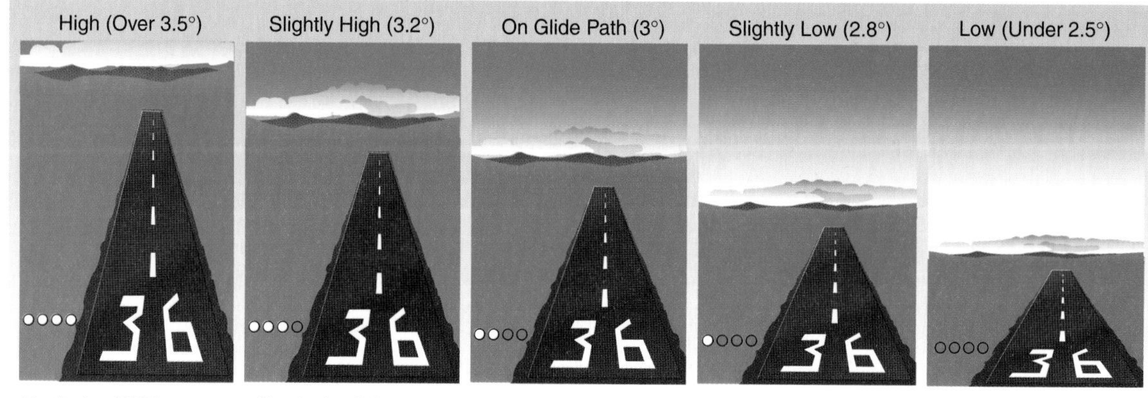

If all the PAPI system lights are white, you are too high.

If only the light on the far right is red and the other three are white, you are slightly high.

When you are on the glide path, the two lights on the left are white and the two lights on the right are red.

If you are slightly low, only the light on the far left is white.

If you are below the glide path, all four of the lights are red.

Figure 3-11. The PAPI is normally located on the left side of the runway and can be seen up to 5 miles during the day and 20 miles at night.

The precision approach path indicator (PAPI) uses lights similar to VASI, but the lights are installed in a single row of two- or four-light units. PAPI normally is located on the left side of the runway. [Figure 3-11]

TRI-COLOR VASI

Tri-color VASIs normally consist of one light projector with three colors — amber, green, and red. See figure 3-12.

Another system, which consists of a single light unit projecting a three-color visual glide path into the final approach area of the runway, is referred to as tri-color VASI. Depending on visibility conditions, this type of approach slope indicator has a useful range of approximately one-half to one mile during the day and up to five miles at night. [Figure 3-12]

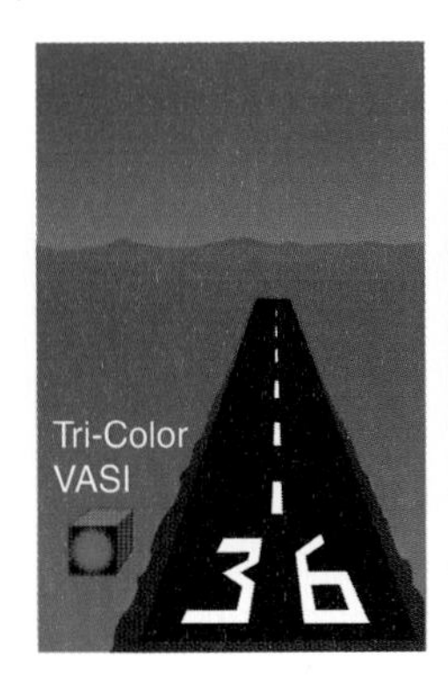

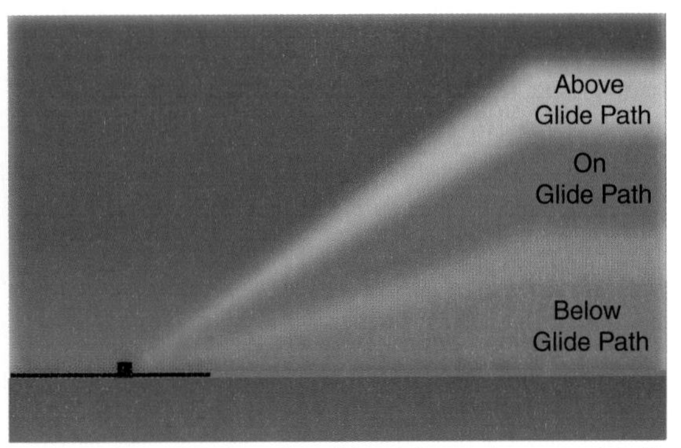

As you decend below the glide path, you may see dark amber during the transition from green light to red, so you should not be deceived into thinking you are too high.

Figure 3-12. The tri-color VASI uses amber, green, and red lights to show your position with respect to the glide path.

RUNWAY LIGHTING

Runway lights outline the landing area by clearly defining its boundaries. Some of these systems have bidirectional features which help you judge your position from the ends of the runway. A thorough understanding of runway lighting is important, particularly during low-visibility, IFR operations.

Runway edge lights are used to outline the runway during periods of darkness or restricted visibility. They are classified according to their brightness — high intensity runway lights (HIRL), medium intensity runway lights (MIRL), and low intensity runway lights (LIRL). The HIRL and MIRL systems have variable intensity controls which may be adjusted from the control tower or by the pilot using the CTAF or UNICOM frequency. The LIRL system normally has only one intensity setting. Runway edge lights are white, except on instrument runways where amber replaces white on the last 2,000 feet or half the runway length, whichever is less. The amber lights along the final 2,000 feet or one half of the instrument runway indicate a caution zone.

Bidirectional threshold lights mark the ends of each runway. As you approach for landing, the lights are green, indicating the beginning of the runway. As you take off, the lights are red, marking the departure end of the runway. These lights are inside the runway edge lights and perpendicular to the runway centerline.

Lights also help you identify a displaced threshold during low visibility conditions or at night. Displaced threshold lights, which appear green during an approach to a landing, are located on each side of the runway. They are outboard from the runway edges and perpendicular to the runway centerline. No landings are permitted short of the green displaced threshold lights. Additionally, the absence of runway edge lights in this area indicates that no operations are authorized short of the displaced threshold. If the displaced runway area is usable for specific operations (taxi, takeoff, and rollout), runway edge lights are installed to outline this area short of the displaced threshold.

You may use the area short of the displaced threshold lights for taxi, takeoff, or rollout purposes when the runway edge lights appear in one of the following combinations:

1. When looking down the runway from the displaced threshold portion, such as during a takeoff, the runway edge lights appear red, and the displaced threshold lights are visible.

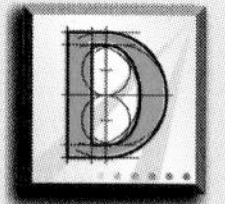

What's New at Newark?

In 1928, the city of Newark, New Jersey transformed an area of marshland into the Newark Metropolitan Airport which had the first hard-surfaced runway of any commercial airport in the United States. Designated as the eastern airmail terminal and official airport for the metropolitan area, Newark Metropolitan soon gained the reputation as the busiest airport in the world. Newark became a testing ground for airport control systems, and experiments conducted there aided the development of instrument landing approaches. Initially, Newark's traffic was controlled by an official who stood near the runway and waved aircraft in and out with flags.

At night, the airport was illuminated with floodlights mounted on a platform. Since it was too costly to keep these powerful lights on all night, an invention called Televox was tested in 1929. A pilot approaching the darkened field cranked the handle of a siren in the cockpit and the sound activated a device that switched on the floodlights. The Televox system was adopted at many airports. Newark also experimented with lights embedded in the center of the runway. In addition, wires were installed at right angles to the runway which emitted signals heard as clicks in the pilot's headset. The clicks indicated at what point the airplane was on the approach.

2. If looking from the runway to the displaced threshold area, such as during a landing and rollout, the runway edge lights appear white. In this case, you would not be able to see any displaced threshold lights.

Red runway edge lights signify a displaced threshold, where taxi, takeoff, and rollout operations are permitted.

Touchdown zone lighting (TDZL) helps you identify the touchdown zone when visibility is reduced. It consists of a series of white lights flush-mounted in the runway. They begin approximately 100 feet from the landing threshold and extend 3,000 feet down the runway or to the midpoint of the runway, whichever is less. These lights are visible only from the approach end of the runway.

Runway centerline lights (RCLS) are flush-mounted in the runway to help you maintain the centerline during takeoff and landing. They are spaced at intervals of 50 feet, beginning 75 feet from the landing threshold and extending to within 75 feet of the opposite end of the runway. As you approach the runway, the centerline lights first appear white. They change to alternating red and white lights when you have 3,000 feet of remaining runway, then they show all red for the last 1,000 feet of runway. These lights are bidirectional, so you see the correct lighting from either direction.

Land and hold short lights are a row of five flush-mounted flashing white lights installed at the hold short point, perpendicular to the centerline of the runway on which they are installed. When land and hold short operations are conducted continuously, the land and hold short lights will normally be on during that period. Therefore, departing pilots and pilots which are cleared to land using the full length of the runway, should ignore the lights.

Taxiway lead-off lights are similar to runway centerline lights. They generally are flush-mounted alternating green and yellow lights spaced at 50-foot intervals. They define the curved path of an aircraft from a point near the runway centerline to the center of the intersecting taxiway. When installed, taxiway centerline lights are green and taxiway edge lights are blue.

Pilot-controlled lighting is designed primarily to conserve energy and may be found at some airports which do not have a full-time tower or an FSS. Typically, you control the lights by keying the aircraft microphone a specified number of times in a given number of seconds. For example, you may key the microphone 7 times in 5 seconds to turn on the lights to maximum intensity. To reduce the lighting level, key the microphone the number of times specified. However, you should be aware that using the lower intensity on some installations may turn the runway end identifier lights completely off. The lights normally turn off automatically 15 minutes after they were last activated. You can find information on pilot-controlled lighting and the airports where they are installed in the *Airport/Facility Directory*, the *Jeppesen Airport Directory*, and on applicable instrument approach procedure charts.

AIRPORT BEACON AND OBSTRUCTION LIGHTS

Some of the other types of lights that are located at or near airports include the airport beacon and obstruction lighting. The beacon is designed to help you locate the airport at night and during conditions of reduced visibility. Operation of the beacon during daylight hours at an airport within controlled airspace (Class B, C, D, and E surface areas) may indicate that the ground visibility is less than 3 statute miles and/or the ceiling is less than 1,000 feet. However, since they are often turned on by photoelectric cells or time clocks, you must not rely on the airport beacon to indicate that the weather is below VFR minimums. Remember, an ATC clearance is required if you wish to take off or land when the weather is below VFR minimums.

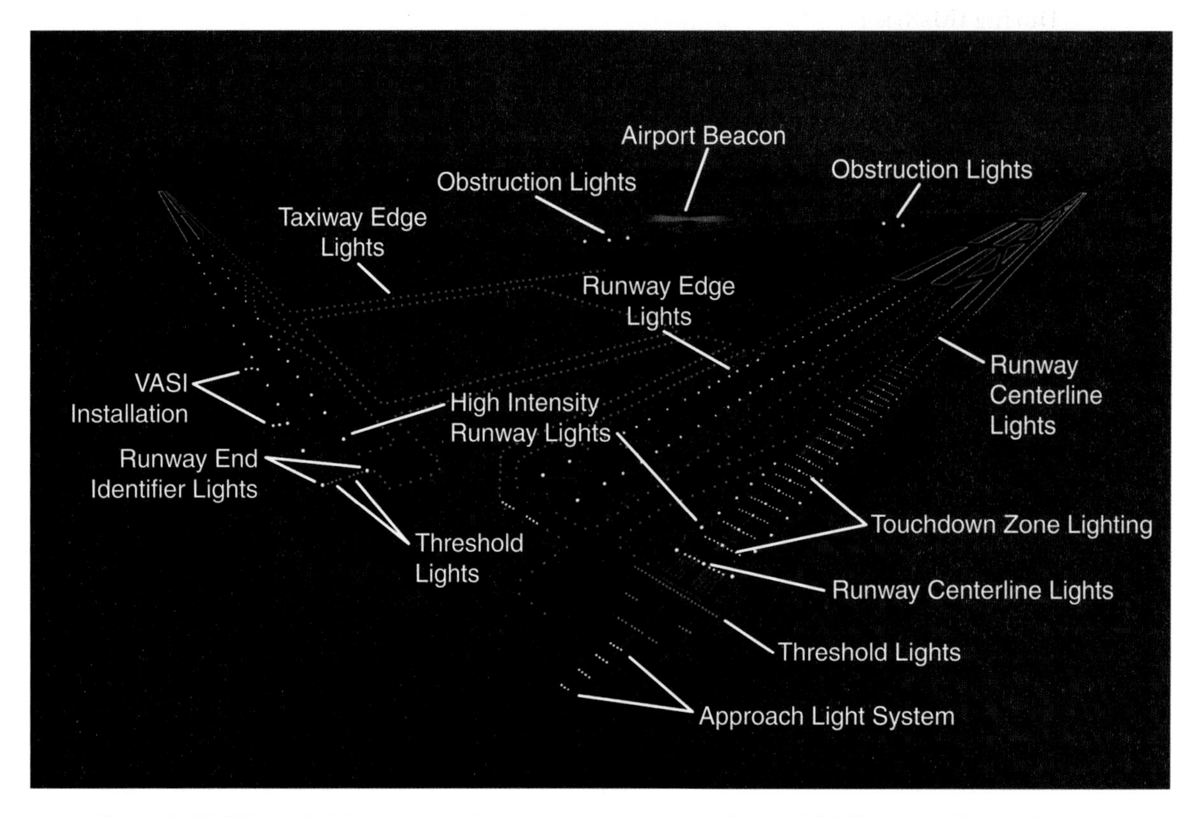

Figure 3-13. This pictorial summary shows the various types of airport lighting typically found at large, controlled airports.

Obstruction lights are installed on prominent structures such as towers, buildings and, occasionally, even powerlines. Bright red and high intensity white lights are typically used and may flash on and off to warn you of obstructions. [Figure 3-13]

AIRSPACE

Within the United States, airspace is classified as either controlled or uncontrolled. Special use airspace and other airspace areas are additional classifications which may include both controlled and uncontrolled segments. As pilot in command, you must know which flight restrictions or aircraft equipment requirements are applicable in these different types of airspace.

If the airport is located within controlled airspace, operation of the airport beacon during daylight hours may indicate that the ground visibility is less than 3 statute miles and/or the ceiling is less than 1,000 feet. An ATC clearance is required for takeoffs and landings if the weather conditions are less than VFR.

CONTROLLED AIRSPACE

Controlled airspace means an airspace of defined dimensions within which air traffic control service is provided to IFR flights and to VFR flights in accordance with the airspace classification. Controlled airspace is a generic term which covers Class A, Class B, Class C, Class D, and Class E airspace. As a routine measure, when you are operating under IFR, your flight must conform with ATC clearances from takeoff to touchdown, and your transponder must be on, including Mode C if installed.

As shown in figure 3-14, VFR requirements for flight visibility and distance from clouds change depending in which controlled airspace you are flying.

During this time, ATC provides separation between your aircraft and all other IFR flights. If workload permits, ATC also provides traffic advisories for VFR operations. It is very important to remember that controllers are not required to separate your aircraft from VFR flights and cannot provide separation from aircraft which do not appear on their radar display. Therefore, it is your responsibility to see and avoid other aircraft when you are operating under IFR in VFR weather conditions. [Figure 3-14]

VFR IN CONTROLLED AIRSPACE		
Altitude	**Flight Visibility**	**Distance From Clouds**
Class A	Not Applicable	Not Applicable
Class B	3 Statute Miles	Clear of Clouds
Class C and Class D	3 Statute Miles	500 Feet Below 1,000 Feet Above 2,000 Feet Horizontal
Class E: Less Than 10,000 Feet MSL	3 Statute Miles	500 Feet Below 1,000 Feet Above 2,000 Feet Horizontal
At or Above 10,000 Feet MSL	5 Statute Miles	1,000 Feet Below 1,000 Feet Above 1 Statute Mile Horizontal

Figure 3-14 The basic weather minimums which apply in Class A, B, C, D, and E controlled airspace are listed in FAR 91.155. Notice that VFR flight is not permitted in Class A airspace.

To fly in controlled airspace within the contiguous United States, your aircraft must meet certain equipment requirements. The FARs require that you have an operating transponder with Mode C capability in Class A airspace, Class B airspace, within 30 nautical miles of Class B primary airports, and in and above Class C airspace. In addition, you must have a Mode C transponder when flying at or above 10,000 feet MSL, excluding the airspace at and below 2,500 feet AGL. This requirement applies in all airspace (controlled or uncontrolled) within the 48 contiguous states and the District of Columbia. For flights under IFR, you must operate Mode C at all times unless ATC directs otherwise. [Figure 3-15]

An operable 4096-code transponder with Mode C capability is required while operating within Class A airspace, Class B airspace, within 30 nautical miles of Class B primary airports, and Class C airspace.

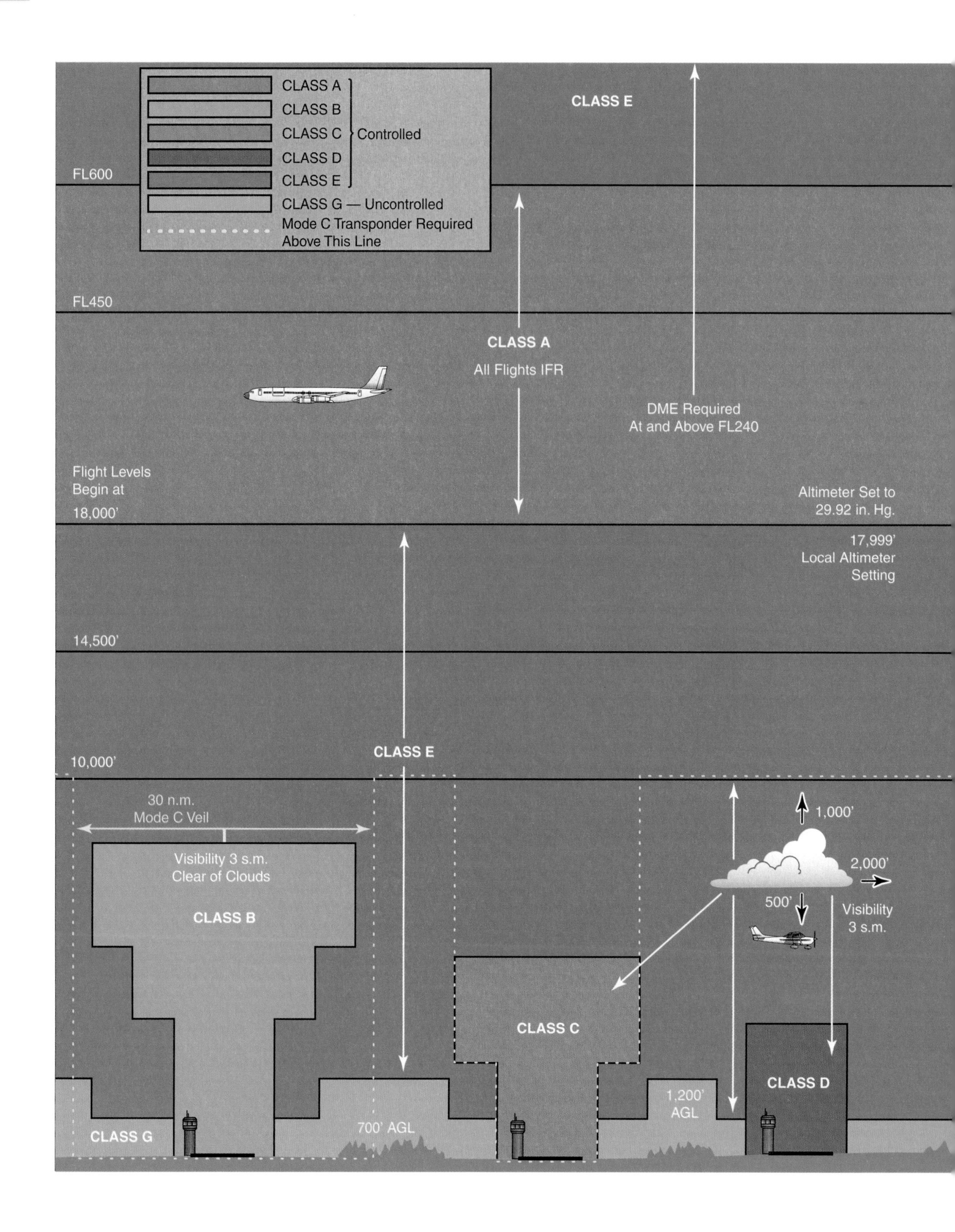
CLASS A
CLASS B
CLASS C
CLASS D
CLASS E
Controlled
CLASS G — Uncontrolled
Mode C Transponder Required Above This Line
CLASS E
FL600
FL450
CLASS A
All Flights IFR
DME Required At and Above FL240
Flight Levels Begin at 18,000'
Altimeter Set to 29.92 in. Hg.
17,999' Local Altimeter Setting
14,500'
10,000'
CLASS E
30 n.m. Mode C Veil
Visibility 3 s.m. Clear of Clouds
CLASS B
1,000'
2,000'
500'
Visibility 3 s.m.
CLASS C
CLASS D
1,200' AGL
700' AGL
CLASS G

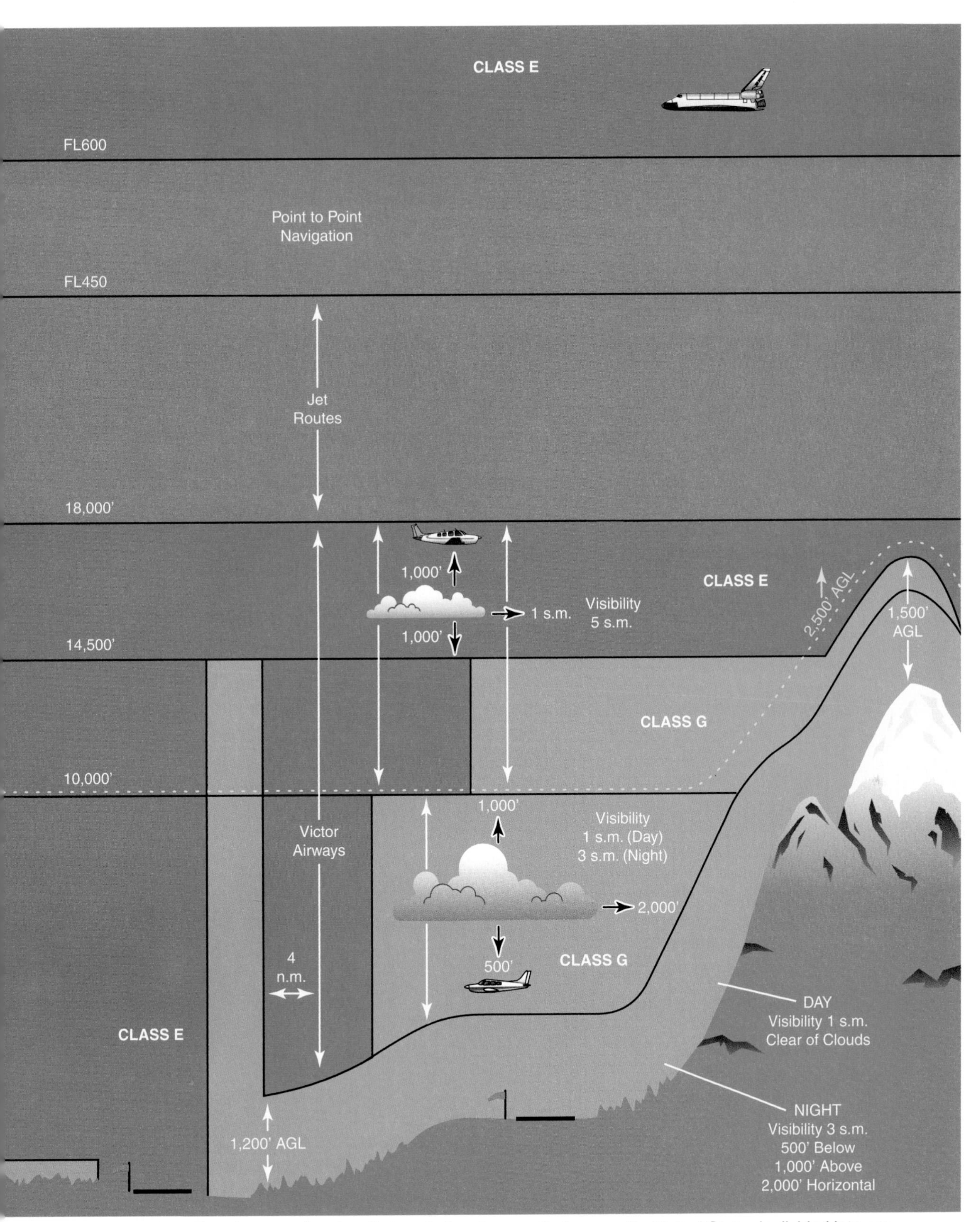

Figure 3-15. To enhance safety for all aircraft, the airspace that covers the United States is divided into controlled and uncontrolled airspace. Operational requirements depend on the class of airspace you are flying in, as well as your altitude. Along with these requirements, you also must be aware of the minimum day/night flight visibilities and cloud clearance requirements that apply at various altitudes in controlled and uncontrolled airspace.

CLASS A AIRSPACE

Within the contiguous United States, **Class A airspace** extends from 18,000 feet MSL up to and including FL600. Since VFR flight is not permitted in this area, your instrument training may provide your first opportunity to fly in Class A airspace. Instrument high altitude enroute charts must be used for flights in Class A airspace.

Aircraft operating in all airspace of the 48 contiguous States and the District of Columbia at and above 10,000 feet MSL must be equipped with an operable transponder with Mode C, except when operating at or below 2,500 feet AGL.

Because aircraft in Class A airspace operate at such high speeds, it would be impractical for pilots to reset their altimeters every 100 nautical miles. So, within Class A airspace, you are required to use a standard setting of 29.92 in. Hg. This means that all pilots are maintaining their assigned altitudes using the same altimeter reference. In addition, altitudes are prefaced by the letters FL, meaning flight level, with the last two zeros omitted. For example, 35,000 feet is referenced as FL350.

Over most of the United States , Class A airspace extends from 18,000 feet MSL to FL600.

To fly in Class A airspace, you must adhere to the following guidelines:

1. If acting as pilot in command, you must be rated and current for instrument flight.

2. You must operate under an IFR flight plan and in accordance with an ATC clearance at specified flight levels.

3. Your aircraft must be equipped with instruments and equipment required for IFR operations, including an encoding altimeter and transponder. You are also required to have a radio providing direct pilot/controller communication on the frequency specified by ATC for the area concerned. In addition, you must have navigation equipment appropriate to the ground facilities to be used.

To fly in Class A airspace, you must be instrument-rated, current, and on an IFR flight plan. The aircraft must be IFR equipped and, in most cases, DME is required at or above 24,000 feet MSL.

4. When VOR equipment is required for navigation, your aircraft must also be equipped with distance measuring equipment (DME) if the flight is conducted at or above 24,000 feet MSL. If the DME fails in flight, you must immediately notify ATC. Then, you may continue to operate at or above 24,000 feet MSL and proceed to the next airport of intended landing where repairs can be made.

When flying at 24,000 feet, if DME is required and it fails, you must notify ATC, but you can continue to the next airport of intended landing and have it repaired.

CLASS B AIRSPACE

At some of the country's busiest airports, **Class B airspace** has been established to separate all arriving and departing traffic. Generally, this airspace is from the surface to 10,000 feet MSL. The airspace consists of a surface area and two or more layers, which are unique for each Class B airspace since they are designed to facilitate traffic separation at a particular terminal. Pilot participation is mandatory, and an ATC clearance must be received before you enter a Class B area. Some of the Class B airspace areas have VFR corridors to allow pilots of VFR aircraft to pass through them without contacting ATC.

To operate under VFR in Class B airspace, your aircraft must have a two-way radio, and a 4096-code transponder with Mode C automatic altitude reporting. However, a Mode S

transponder can be used anywhere a Mode C transponder is required. For IFR operations in Class B airspace, you must also have a VOR receiver. In addition, to take off or land at certain large airports listed in Appendix D of FAR 91, you must hold at least a private pilot certificate. With certain exceptions, a transponder with altitude reporting capability is required anytime you are operating within 30 nautical miles of the primary airport from the surface upward to 10,000 feet MSL.

The upper limit for most Class B airspace areas is 10,000 feet MSL.

To operate in Class B airspace, you must be at least a private pilot, or a student pilot with the appropriate endorsement. However, at certain large airports, student pilot operations are not allowed.

Among other requirements for flight within Class B airspace, your aircraft must be equipped with either a Mode S or a 4096-code transponder that has Mode C altitude reporting equipment.

CLASS C AIRSPACE

Class C airspace areas are designated at certain airports where ATC is equipped to provide radar service for all aircraft. Normally, Class C airspace designations consist of 2 circular areas which extend outward from the primary airport and are referred to as the 5 nautical mile radius core area and the 10 nautical mile radius shelf area. In addition, an outer area extends 10 nautical miles beyond the outer circle. Although it is not required, pilot participation for flights within the outer area is strongly encouraged. Before operating within the core and shelf areas, you must establish two-way communication with the ATC facility having jurisdiction over the area and maintain radio contact at all times. If you depart a satellite airport located within Class C airspace, you must establish two-way communication with ATC as soon as practicable. [Figure 3-16]

As a minimum, you must have two-way radio communication equipment and a Mode C transponder for flight operations within Class C airspace.

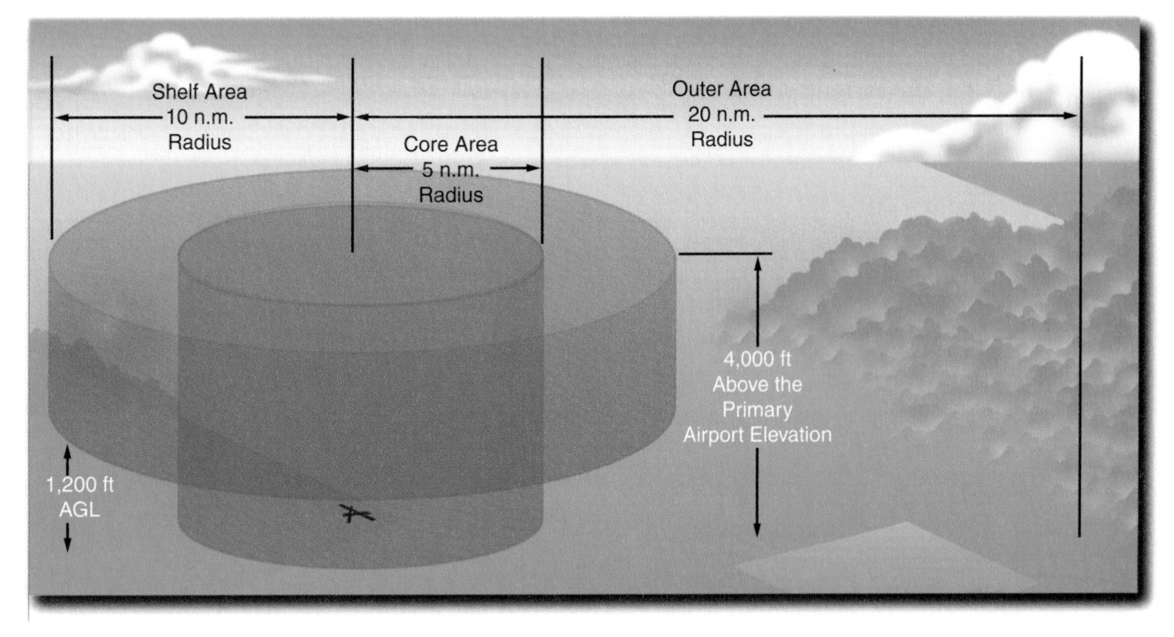

Figure 3-16. The core area typically begins at the surface and has a 5 nautical mile radius from the primary airport, while the shelf area normally starts at 1,200 feet AGL and has a 10 nautical mile radius. Both have upper limits of 4,000 feet above the primary airport. The outer area usually has a radius of 20 nautical miles and extends from the lower limits of radar/radio coverage up to the ceiling of approach control's delegated airspace, excluding the Class C airspace itself.

All aircraft operating in Class C airspace, and in all airspace above the ceiling and within the lateral boundaries extending upward to 10,000 feet MSL, must be equipped with an operable transponder with Mode C. Aircraft operating in the airspace beneath Class C airspace will not be required to have a transponder with Mode C.

CLASS D AIRSPACE

Class D airspace areas are designated at airports with operating control towers which are not associated with Class B or C airspace. Before you enter Class D airspace you must establish and maintain two-way radio communication with the control tower. When departing the primary airport within Class D airspace, you also must establish and maintain communication with the tower. It is important to note that airspace at an airport with a part-time control tower is classified as a Class D airspace only when the control tower is in operation.

You are required to establish communication with the tower before entering Class D airspace.

At some locations, there may be a satellite airport within the same Class D airspace designated for the primary airport. If the satellite airport also has a control tower, similar radio communication requirements with that tower prevail for arrivals and departures. If the satellite airport is a nontower field, arriving aircraft must establish contact with the primary airport's control tower. Departures from a nontower satellite must establish communication with the ATC facility (tower) having jurisdiction over the Class D airspace as soon as practicable after departing. To the maximum extent practical and consistent with safety, satellite airports have been excluded from Class D airspace. [Figure 3-17]

At part-time tower locations, Class D airspace normally becomes Class E airspace when the tower is closed. However, if weather observations and reporting are not available, the airspace becomes Class G.

Figure 3-17. At some locations, a satellite airport without an operating control tower might have airspace carved out of the Class D airspace, or it could be placed under a shelf of the Class D airspace.

The ceiling of a Class D airspace area is usually 2,500 feet above the surface of the airport converted to mean sea level, and rounded to the nearest 100-foot increment. If conditions of a particular airspace area warrant, the ceiling may be raised or lowered as appropriate. The ceiling of Class D airspace is shown in hundreds of feet MSL on sectional charts. Laterally, Class D airspace (depicted as blue dashed lines on sectional charts) normally consists of a circular area with a 4 nautical mile radius. However, because the airspace is based on the instrument procedures for which the controlled airspace is established, the lateral dimensions may be larger or smaller and can be irregular in shape.

Normally, the upper limit of Class D airspace is 2,500 feet AGL and the lateral limits are approximately 4 nautical miles.

CLASS E AIRSPACE

Much of the remaining controlled airspace is designated as **Class E airspace**, which includes several different segments. One portion of Class E consists of the airspace covering the 48 contiguous states, District of Columbia, and Alaska. Also included is the airspace out to 12 nautical miles from the coastlines. Unless designated at a lower altitude, Class E airspace begins at 14,500 feet MSL and extends up to, but not including, the base of the Class A airspace at 18,000 feet MSL. The only exceptions are the Alaska peninsula west of 160° west longitude and airspace within 1,500 feet of the surface.

During flight in Class E airspace between 14,500 and 18,000 feet MSL, you have no additional operating requirements beyond those mentioned previously. For example, you must operate the Mode C feature of your transponder when at or above 10,000 feet MSL, excluding the airspace at or below 2,500 feet AGL, and apply the appropriate cloud clearance and visibility requirements when flying under VFR. Remember that you cannot fly VFR above FL180, which is Class A airspace.

Another segment of Class E airspace is the low altitude airway system connecting one navaid to another. These routes are used by VFR, as well as IFR aircraft, and are called Federal airways, or Victor airways. These airways are based on VOR or VORTAC navigation aids and are identified by a V and the airway number. A few airways are based on L/MF (low/medium frequency) navigation aids, or NDBs. The only L/MF airways still in use are in Alaska and coastal North Carolina. Airways are usually 8 nautical miles wide, begin at 1,200 feet AGL, and extend up to, but not including 18,000 feet MSL. Some airway segments, such as those over mountainous terrain, may have a floor greater than 1,200 feet AGL, which is designated on sectional charts. VFR cloud clearance and visibility requirements on an airway depend on your cruising altitude. [Figure 3-18]

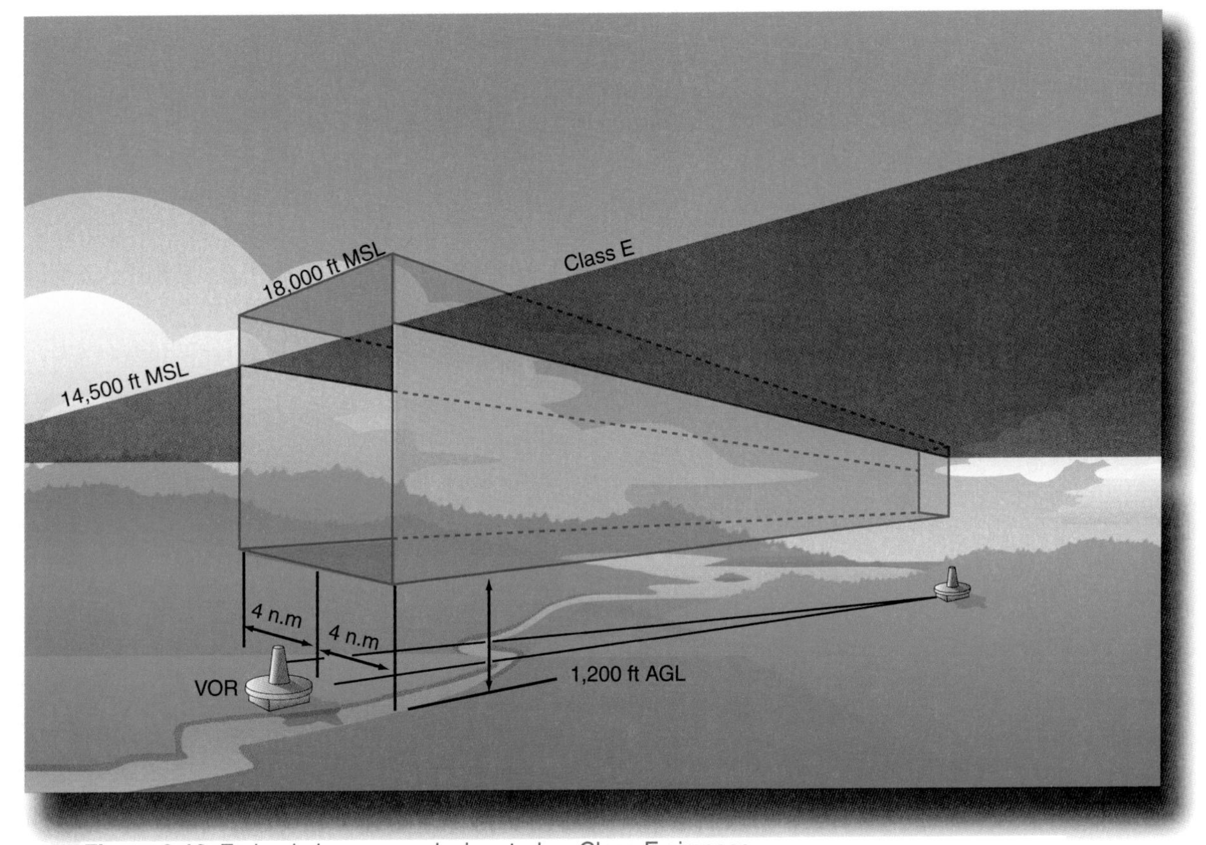

Figure 3-18. Federal airways are designated as Class E airspace.

Class E airspace transitional areas have also been established between airports and the airway route system to allow IFR traffic to remain in controlled airspace while transitioning between the enroute and airport environments. These segments of Class E airspace usually begin at 1,200 feet AGL if they are associated with an airway. Transitional areas are outlined on sectional charts only if they border uncontrolled airspace.

Federal airways normally begin at 1,200 feet above the surface, extend upward to 18,000 feet MSL, and include the airspace within 4 nautical miles each side of the airway centerline.

Some airports located in Class E or Class G airspace areas have operating control towers. You are required by FARs to establish radio communication prior to 4 nautical miles from the airport (up to and including 2,500 feet AGL) and to maintain radio contact with the tower while in the area.

At nontower airports which have an approved instrument approach procedure, Class E transition airspace often begins at 700 feet above the surface. At some nontower airports, Class E airspace extends upward from the surface, and usually encompasses airspace surrounding the airport, in addition to the extensions to accommodate arrivals and departures. Both of these types of Class E airspace are depicted on aeronautical charts. [Figure 3-19]

The floor of the Class E airspace which is used as a transition area for an airport with an approved instrument approach procedure is 700 feet AGL and extends to the overlying controlled airspace.

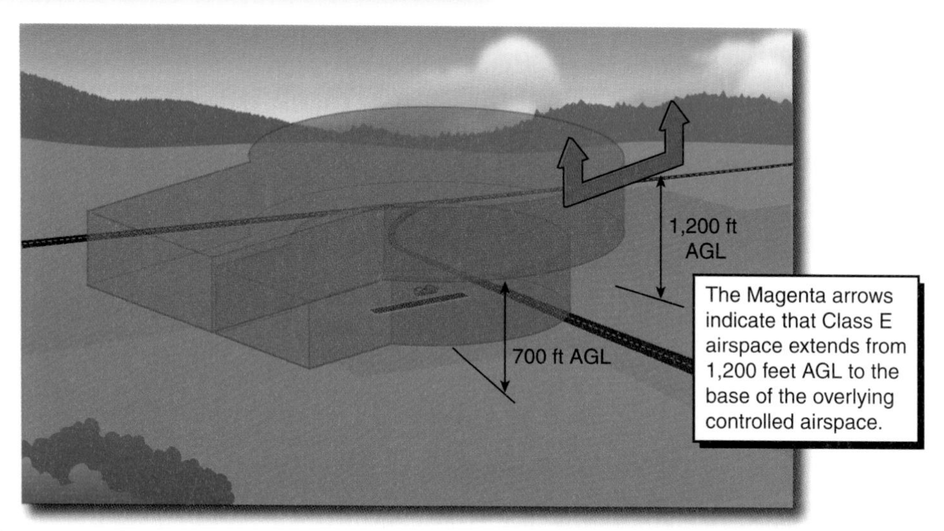

Figure 3-19. This nontower airport is surrounded by Class E airspace which extends from the surface and adjoins Class E airspace which begins at 700 feet AGL.

SPECIAL VFR

In addition to maintaining the VFR minimums specified in figure 3-14, you may only operate within the areas of Class B, C, D, or E airspace which extend to the surface around an airport, when the ground visibility is at least 3 statute miles and the cloud ceiling is at least 1,000 feet AGL. If ground visibility is not reported, you can use flight visibility. When the weather is below these VFR minimums, and there is no conflicting IFR traffic, a **special VFR clearance** may be obtained from the ATC facility having jurisdiction over the affected airspace. A special VFR clearance may allow you to enter, leave, or

Special VFR clearances require you to maintain a minimum ground visibility (or flight visibility, if ground visibility is not reported) of one mile and remain clear of clouds.

operate within most Class D and Class E surface areas and some Class B and Class C surface areas if the flight visibility is at least 1 statute mile and you can remain clear of clouds. At least one statute mile ground visibility is required for takeoff and landing; however, if ground visibility is not reported, you must have at least 1 statute mile flight visibility.

To operate an airplane under special VFR at night within Class D airspace, you must hold an instrument rating and the airplane must be equipped for instrument flight.

Special VFR is not permitted between sunset and sunrise unless you have a current instrument rating and the aircraft is equipped for instrument flight. In addition, special VFR clearances are not issued to fixed-wing aircraft (day or night) at the nation's busier airports which are listed in Section 3 of Appendix D of FAR 91.

CLASS G AIRSPACE (UNCONTROLLED)

Class G airspace is that area which has not been designated as Class A, B, C, D, or E airspace and is essentially uncontrolled by ATC. For example, the airspace below a Class E airspace area or below a Victor airway is normally uncontrolled. Most Class G airspace terminates at the base of Class E airspace at 700 or 1,200 feet AGL, or at 14,500 feet MSL. An exception to this rule occurs when 14,500 feet MSL is lower than 1,500 feet AGL. In this situation, Class G airspace continues up to 1,500 feet above the surface. The amount of uncontrolled airspace has steadily declined because of the expanding need to coordinate the movement of aircraft.

Unless below 1,500 feet AGL, the maximum altitude for Class G airspace is 14,500 feet MSL.

Except when associated with a temporary control tower, ATC does not have responsibility for or authority over aircraft in Class G airspace; however, most of the regulations affecting pilots and aircraft still apply. For example, although a flight plan is not required for IFR operations in Class G airspace, both pilot and aircraft must still be fully qualified for IFR flight. In addition, in several cases, the day weather minimums for VFR flight are reduced from those in controlled airspace. [Figure 3-20]

As shown in figures 3-14 and 3-20, the VFR minimums at or above 10,000 feet MSL (and more than 1,200 feet AGL) are the same for Class G and Class E.

Except for temporary control towers, ATC does not exercise control of air traffic in Class G airspace.

VFR CLASS G AIRSPACE (UNCONTROLLED)		
Altitude	**Flight Visibility**	**Distance From Clouds**
1,200 feet or less above the surface (regardless of MSL altitude)	Day: 1 s.m.	Clear of Clouds
	Night: 3 s.m.	500 Feet Below 1,000 Feet Above 2,000 Feet Horizontal
More than 1,200 feet above the surface, but less than 10,000 feet MSL	Day: 1 s.m. Night: 3 s.m.	500 Feet Below 1,000 Feet Above 2,000 Feet Horizontal
More than 1,200 feet above the surface and at or above 10,000 feet MSL	Day and Night: 5 s.m.	1,000 Feet Below 1,000 Feet Above 1 s.m. Horizontal

Figure 3-20. Weather minimums for VFR flight in Class G airspace are shown in this illustration.

AIRCRAFT SPEED LIMITS

As you move into faster aircraft, you must also be concerned with aircraft speed limitations. For example, unless otherwise authorized by air traffic control, you generally may not operate an aircraft below 10,000 feet MSL at a speed greater than 250 knots indicated airspeed (KIAS). Further, unless otherwise authorized or required by ATC, you may not operate an aircraft at or below 2,500 feet above the surface within 4 nautical miles of the primary airport of a Class C or Class D airspace area at a speed greater than 200 KIAS. The 200 KIAS limit also applies to the airspace underlying a Class B airspace area or in a VFR corridor designated through such airspace.

Normally, the maximum indicated airspeed permitted when at or below 2,500 feet AGL within 4 nautical miles of the primary airport of a Class C or Class D airspace is 200 knots.

SPECIAL USE AIRSPACE

Special use airspace is used to confine certain flight activities and to place limitations on aircraft operations which are not part of these activities. The various types of airspace may be designated as prohibited, restricted, warning, alert, military operations areas, controlled firing areas, and national security areas. With the exception of controlled firing areas, the dimensions of special use airspace are depicted on aeronautical charts. The information about the hours of operation and effective altitudes may be listed directly on the aeronautical chart, or indexed by area number on a chart panel.

Prohibited areas contain airspace within which the flight of aircraft is prohibited. Such areas are established for security or other reasons associated with the national welfare. **Restricted areas** often have invisible hazards to aircraft such as artillery firing, aerial gunnery, or flight of guided missiles. Permission to fly through a restricted area must be granted by the controlling agency. If ATC issues you an IFR clearance (including a clearance to maintain VFR-on-top) which takes you through restricted airspace, such a clearance constitutes authorization to penetrate the airspace. In this case, you need take no further action other than to comply with the clearance, as issued, and maintain normal vigilance.

A **warning area** is airspace of defined dimensions, extending from three nautical miles outward from the coast of the United States, that contains activities that may be hazardous to nonparticipating aircraft. The purpose of such warning areas is to warn nonparticipating pilots of the potential danger. A warning area may be located over domestic or international waters or both. **Alert areas** may contain a high volume of pilot training or an unusual type of aerial activity, such as parachute jumping or glider towing. Flight within alert areas is not restricted, but you are urged to exercise extreme caution. Pilots of participating aircraft, as well as pilots transiting the area, are equally responsible for collision avoidance and compliance with the FARs.

Military operations areas (MOAs) are established to separate certain military training activities from IFR traffic. When you are flying under IFR, you may be cleared through an active MOA if ATC can provide separation. Otherwise, ATC will reroute or restrict your flight operations. Before entering an active MOA under VFR, you should contact the controlling agency for traffic advisories. Information regarding route activity is available from any FSS within 100 nautical miles of the area.

MOAs are established to separate certain training activities from IFR traffic.

The distinguishing feature of a **controlled firing area**, compared to other special use airspace, is that its activities are discontinued immediately when a spotter aircraft, radar, or

ground lookout personnel determine an aircraft might be approaching the area. Since nonparticipating aircraft are not required to change their flight path, controlled firing areas are not depicted on aeronautical charts.

National security areas (NSAs) are established at locations where there is a requirement for increased security and safety of ground facilities. You are requested to voluntarily avoid flying through an NSA. At times, flight through an NSA may be prohibited to provide a greater level of security and safety. A NOTAM is issued to advise you of any changes in an NSA's status.

OTHER AIRSPACE AREAS

Other airspace areas include airport advisory areas, military training routes, and areas where temporary restrictions or limitations/prohibitions apply. Recommended procedures for operating in these areas are outlined in the *Aeronautical Information Manual.*

An **airport advisory area** encompasses the airspace within 10 statute miles of an airport where an FSS is located and there is no operating control tower. At these locations, the FSS provides local airport advisory (LAA) service. **Military training routes (MTRs)** are established below 10,000 feet MSL for both VFR and IFR operations at speeds in excess of 250 knots. However, some route segments may be at higher altitudes. **Parachute jump aircraft areas** are tabulated in the *Airport/Facility Directory.* Times of operation are local, and MSL altitudes are listed unless otherwise specified.

Temporary flight restrictions are imposed by the FAA to protect persons and property on the surface or in the air. For example, the FAA will normally issue a NOTAM to provide a safe environment for rescue/relief operations and to prevent unsafe congestion above an incident or event which may generate high public interest. The restricted airspace for rescue/relief operations is usually limited to within 2,000 feet above the surface within a three nautical mile radius. Incidents near controlled airports are handled through existing procedures and normally do not require issuance of a NOTAM. However, NOTAMs are issued to restrict flight in the vicinity of space flight operations and in the proximity of the President, Vice President, and other public figures.

Terminal radar service areas (TRSAs) do not fit into any of the U.S. airspace classes. Originally part of the terminal radar program at selected airports, TRSAs have never been established as controlled airspace and, therefore, FAR Part 91 does not contain any rules for TRSA operations. However, the airspace surrounding the primary airport within a TRSA becomes Class D airspace, while the remaining portion of the TRSA overlies other controlled airspace which is normally Class E airspace beginning at 700 or 1,200 feet AGL and established to transition to and from the enroute/terminal environment. Therefore, you must abide by the rules established for these areas. By contacting approach control, you can receive radar services within a TRSA, but participation is not mandatory.

ADIZ

Aircraft entering U.S. domestic airspace from points outside must provide identification prior to entry. **Air defense identification zones (ADIZs)** have been established to facilitate this early identification. You must file a flight plan, with an appropriate facility such as an FSS, to penetrate or operate within a coastal or domestic ADIZ. If flying VFR, you file a **defense VFR** (DVFR) flight plan. It contains information similar to local flight plans, but helps to identify your aircraft as you enter the country. Unless otherwise authorized by ATC, a transponder with Mode C (or Mode S) capability is required, and the transponder must be turned on and operable. You are also required to have a two-way radio and periodically give ATC reports of your location while inbound toward the ADIZ.

The following table (figure 3-21), summarizes the features and requirements of the different classes of airspace.

Airspace Features	Class A	Class B	Class C	Class D	Class E	Class G
Operations Permitted	IFR	IFR and VFR	IFR and VFR	IFR and VFR	IFR and VFR	IFR and VFR
VFR Entry and Equipment Requirements	IFR Flight Plan and IFR Clearance Required	ATC Clearance Transponder with Mode C	Establish Radio Communication Transponder with Mode C	Establish Radio Communication	None	None
Minimum Pilot Qualifications	Instrument Rating	Private Pilot Certificate Student Pilot Certificate Endorsement	Student Pilot Certificate	Student Pilot Certificate	Student Pilot Certificate	Student Pilot Certificate
Two-way Radio Communications	Yes	Yes	Yes	Yes	Yes for IFR Operations	No
VFR Min. Vis. and Distance from Clouds 1,200 ft AGL or less (Regardless of MSL Altitude)	N/A	N/A	N/A	N/A	N/A	Day 1 s.m. Clear of Clouds Night 3 s.m. 500 ft Below 1,000 ft Above 2,000 ft Horizontal
VFR Minimum Visibility	N/A	3 Statute Miles	3 Statute Miles	3 Statute Miles	**Below 10,000 ft MSL** – 3 s.m. **At or Above 10,000 MSL** – 5 s.m.	**Below 10,000 ft MSL** – Day 1 s.m. Night 3 s.m. **At or Above 10,000 MSL** – 5 s.m. (above 1,200 ft AGL)
VFR Minimum Distance from Clouds	N/A	Clear of Clouds	500 ft Below 1,000 ft Above 2,000 ft Horizontal	500 ft Below 1,000 ft Above 2,000 ft Horizontal	**Below 10,000 ft MSL** – 500 ft Below 1,000 ft Above 2,000 ft Horizontal **At or Above 10,000 ft MSL** – 1,000 ft Below 1,000 ft Above 1 s.m. Horizontal	**Below 10,000 ft MSL** – 500 ft Below 1,000 ft Above 2,000 ft Horizontal (above 1,200 ft AGL) **At or Above 10,000 ft MSL** – 1,000 ft Below 1,000 ft Above 1 s.m. Horizontal (above 1,200 ft AGL)
ATC Services	All Aircraft Separation	All Aircraft Separation	IFR/IFR Separation IFR/VFR Separation VFR Traffic Advisories (workload permitting)	IFR/IFR Separation VFR Traffic Advisories (workload permitting)	IFR/IFR Separation VFR Traffic Advisories on Request (workload permitting)	VFR Traffic Advisories on Request (workload permitting)

Figure 3-21. You can use this table as a quick reference to the weather minimums and operating requirements of each class of airspace.

FLIGHT INFORMATION

Regulations require you to familiarize yourself with all available information concerning each flight. The following review of flight information publications is designed to help you fulfill this requirement. The publications in this section include the *Airport/Facility Directory, Aeronautical Information Manual, Notices to Airmen,* the *International Flight Information Manual,* Advisory Circulars, and the Jeppesen Information Services. [Figure 3-22]

Figure 3-22. Use these publications to help you stay current with the latest changes in the aviation industry.

AIRPORT/FACILITY DIRECTORY

The **Airport/Facility Directory (A/FD)** is a series of regional books which includes a tabulation of all data on record with the FAA for public-use civil airports, associated terminal control facilities, air route traffic control centers, and radio aids to navigation. A comprehensive legend sample is printed in the first few pages of each book. The legend provides you with a breakdown of all the information in the A/FD. [Figure 3-23]

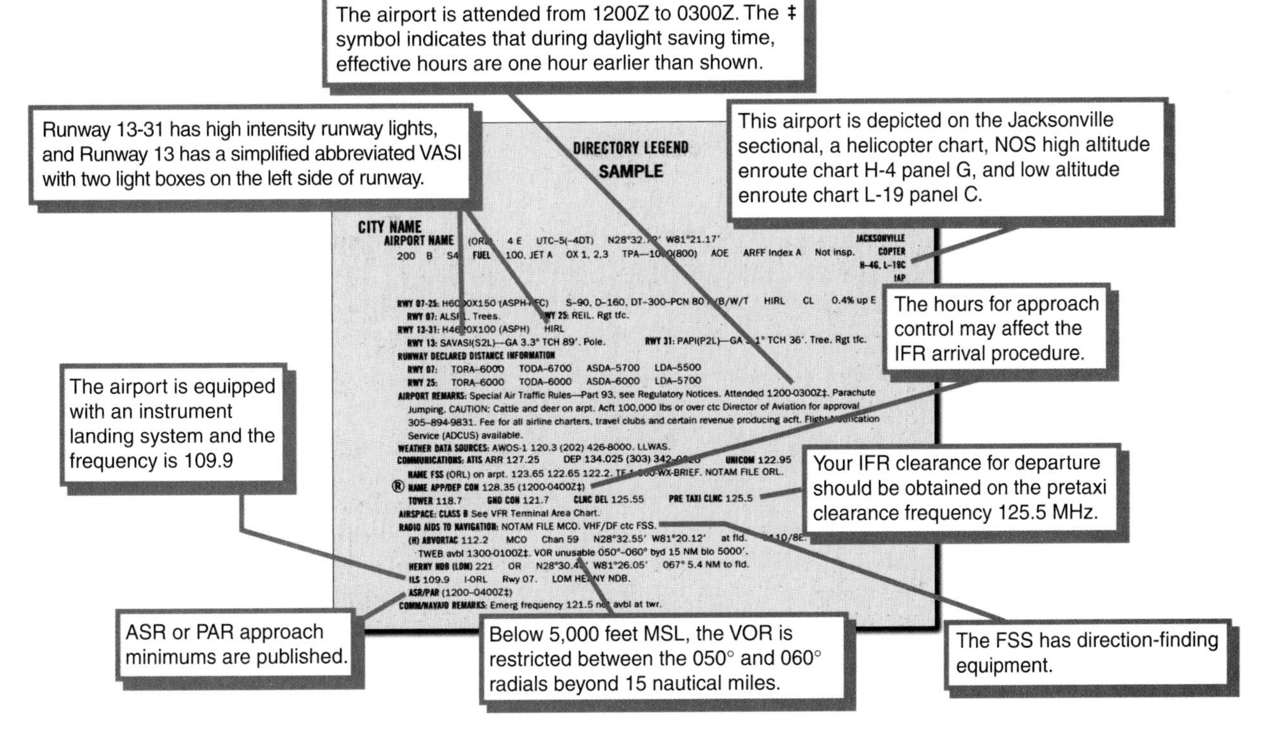

Figure 3-23. The A/FD provides information in the facility listing which is of special interest to you as an instrument-rated pilot.

Although airport and facility data make up the bulk of the directory, there are several other sections that contain essential information. Many of these areas pertain to IFR flight operations. The special notices section furnishes information regarding subjects such as the civil use of military fields, newly certified airports, continuous power facilities, and special flight procedures. Preferred IFR routes, special notices, LAHSO data, and telephone numbers for the FSS and NWS outlets, as well as numbers for PATWAS and TWEB are provided in the A/FD. You also can find a listing of VOR receiver checkpoints and VOT facilities for each region. If you have not used the A/FD recently, you should get a copy and thoroughly familiarize yourself with it. [Figure 3-24]

As shown in figure 3-23, the airport is attended from 1200Z - 0300Z.

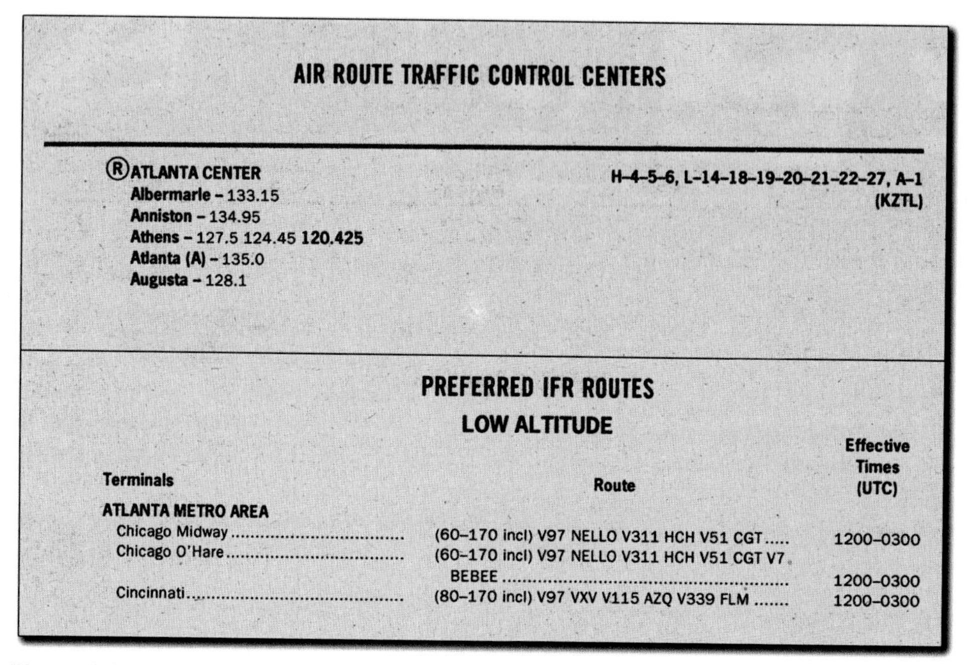

AIR ROUTE TRAFFIC CONTROL CENTERS

®ATLANTA CENTER H-4-5-6, L-14-18-19-20-21-22-27, A-1 (KZTL)
Albermarle - 133.15
Anniston - 134.95
Athens - 127.5 124.45 **120.425**
Atlanta (A) - 135.0
Augusta - 128.1

PREFERRED IFR ROUTES

LOW ALTITUDE

Terminals	Route	Effective Times (UTC)
ATLANTA METRO AREA		
Chicago Midway..................................	(60–170 incl) V97 NELLO V311 HCH V51 CGT.....	1200–0300
Chicago O'Hare..................................	(60–170 incl) V97 NELLO V311 HCH V51 CGT V7 BEBEE	1200–0300
Cincinnati..	(80–170 incl) V97 VXV V115 AZQ V339 FLM	1200–0300

Figure 3-24. Areas of the A/FD you may not have used before include the ARTCC sector frequencies and the low altitude preferred IFR routes.

AERONAUTICAL INFORMATION MANUAL

The *Aeronautical Information Manual* (AIM) contains fundamental information required for both VFR and IFR flight operations within the National Airspace System. It is revised several times each year and is an excellent source of operational information which you should study and review periodically. For example, the AIM describes the capabilities, components, and procedures required for each type of air navigation aid and includes a discussion of radar services, capabilities, and limitations. You can also find a comprehensive description of current airport lighting and runway markings, airspace, and ATC services. Additional coverage addresses altimetry, wake turbulence, and potential flight hazards, as well as safety reporting programs, medical facts for pilots, and aeronautical charts. The Pilot/Controller Glossary promotes a common understanding of the terms used in the ATC system. International terms that differ from the FAA definitions are listed after their U.S. equivalents.

NOTICES TO AIRMEN

Notices to Airmen or NOTAMs contain time-critical, aeronautical information that could affect your decision to make a flight. The information is either temporary in nature or unknown in time for publication on aeronautical charts and/or in other documents.

Airport or primary runway closures, changes in the status of navigational aids, radar service availability, and other data essential to enroute, terminal, or landing operations are examples of information that may be included in NOTAMs. NOTAMs are divided into three categories, NOTAM(D) (distant), NOTAM(L) (local), and Flight Data Center (FDC) NOTAMs.

NOTAM(D) information is disseminated for all navigational facilities that are part of the National Airspace System, all public use airports, seaplane bases, and heliports listed in the *Airport/Facility Directory*. The complete file of all NOTAM(D) information is maintained in a computer data base at the Weather Message Switching Center located in Atlanta, Georgia. Most air traffic facilities, primarily FSSs, have access to the entire database of NOTAM(D)s which remain available for the duration of their validity or until published.

The latest status of airport conditions can be determined from the *Airport/Facility Directory*, as well as NOTAM(D) and NOTAM(L) information.

NOTAM(L) information is distributed locally and includes items such as taxiway closures, construction activities near runways, snow conditions, and changes in the status of airport lighting, such as VASI, that do not affect instrument approach criteria. A separate file of local NOTAMs is maintained at each FSS for facilities in their area. You must specifically request NOTAM(L) information directly from other FSS areas that have responsibility for the airport concerned.

FDC NOTAMs are used to disseminate information that is regulatory in nature. Examples are amendments to aeronautical charts, changes to instrument approach procedures, and temporary flight restrictions. FSSs are responsible for maintaining unpublished FDC NOTAMs concerning conditions within 400 nautical miles of their facilities. FDC NOTAM information that affects conditions more than 400 miles from the FSS, or that is already published, is provided to a pilot only upon request.

FDC NOTAMs are issued to advise pilots of changes in flight data which affect instrument approach procedures, aeronautical charts, and flight restrictions prior to normal publication.

The *Notices to Airmen* publication is issued every 28 days and includes NOTAM(D)s that are expected to remain in effect for an extended period and current FDC NOTAMs. Once published in the *Notices to Airmen*, this information is taken off the distribution circuits and will not be provided in a pilot weather briefing unless you specifically request it. Data of a permanent nature is sometimes printed in *Notices to Airmen* as an interim step prior to publication in the appropriate aeronautical chart or *Airport/Facility Directory*.

You may obtain FDC NOTAMs from an FSS when they are issued too late for inclusion in the *Notices to Airmen* publication.

INTERNATIONAL FLIGHT INFORMATION MANUAL

The ***International Flight Information Manual*** contains the requirements and instructions for flying outside the United States. It is intended to be used as a preflight planning guide by nonscheduled operators. Airport of entry, passport/visa, and customs procedures for each country are detailed. In addition, routes, such as the established North American Routes (NARs), minimum navigation equipment, long-range navigation information, and other planning data are listed.

ADVISORY CIRCULARS

To provide current aviation information on a recurring basis, the Department of Transportation publishes and distributes **Advisory Circulars**. These circulars provide information and procedures which are necessary for good operating practice, but which are not binding to the public unless they are incorporated into a regulation. For ease of reference, Advisory Circulars use a coded numbering system that corresponds to the subject areas of the FARs. An *Advisory Circular Checklist*, which is issued periodically as AC 00-2, contains the subjects covered and the availability of each circular. It also provides you with pricing information. While many advisory circulars are free of charge, you must purchase others.

JEPPESEN INFORMATION SERVICES

Many commercial publishers offer information for pilots which is comparable to that found in government sources. For example, Jeppesen publishes both a printed and CD ROM version of the FAR/AIM, each of which is revised annually. In addition, **Jeppesen Information Services** provide revision subscriptions for several flight information publications including the *Jeppesen AIM*, *Jeppesen FARs for Pilots*, the *Jeppesen Airport Directory*, *JeppGuide,* and the *GPS/LORAN Coordinate Directory.* The AIM, FARs, airport directory, and *JeppGuide* are printed in loose-leaf format for ease of revision.

JEPPESEN FARS FOR PILOTS

By subscribing to the FARs service, you receive updates to the regulations as they are amended. The FARs service includes the Parts most commonly used by pilots, including the Hazardous Materials Regulations (HMR), National Transportation Safety Board (NTSB) Part 830, and current Special FARs. [Figure 3-25]

"Pilotage" means navigation by visual reference to landmarks.

"Pilot in command" means the person who:
(1) Has final authority and responsibility for the operation and safety of the flight;
(2) Has been designated as pilot in command before or during the flight; and
(3) Holds the appropriate category, class, and type rating, if appropriate, for the conduct of the flight.

"Pitch setting" means the propeller blade setting as determined by the blade angle measured in a manner, and at a radius, specified by the instruction manual for the propeller.

"Positive control" means control of all air traffic, within designated airspace, by air traffic control.

Figure 3-25. The revised paragraphs, which are actually amendments to the regulations, are clearly indicated by arrows or brackets in the margin of the affected pages.

JEPPESEN AIM

The *Jeppesen AIM* service which provides revisions twice a year contains the information published in the government edition, including the Pilot/Controller Glossary, as well as Tables and Codes, and Entry Requirements. Because of the comprehensive coverage of the material and the frequency of revision, FAR Part 135.81 authorizes air taxi and commercial operators to use the *Jeppesen AIM* as a substitute for the government edition.

JEPPESEN AIRPORT DIRECTORY AND JEPPGUIDE

The *Jeppesen Airport Directory* service includes a choice of regional coverages or full U.S. coverage updated twice a year. To avoid duplication of diagrams on IFR approach charts, only VFR airport diagrams are published in these directories. However, pertinent airport data are included for both IFR and VFR airports. When appropriate, general operational information, airport remarks, fixed base operator (FBO) services, car rental services, and lodging and restaurant availability are noted. Airports within each state are listed by city name, followed by the airport name, when different. The *JeppGuide* depicts one airport per page and provides diagrams for both VFR and IFR airports. [Figure 3-26]

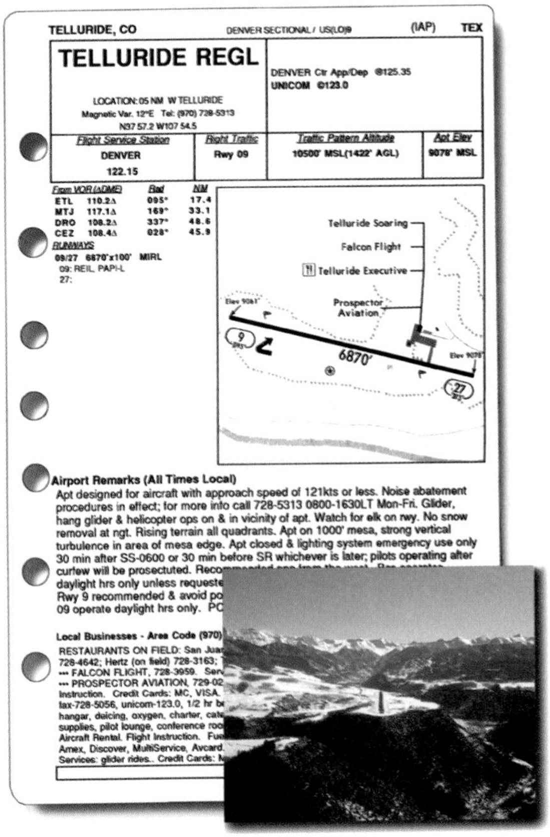

Figure 3-26. In addition to airport data, *JeppGuide* provides extensive, detailed listings of FBO services, including availability of restaurants at the airport and nearby, fuel services, repair facilities, credit card acceptance, lodging, and rental cars.

GPS/LORAN COORDINATE DIRECTORY

An important reference document for users of GPS and LORAN equipment, this directory contains coordinates for all public-use airports, navaids, and waypoints within the contiguous 48 states. Airports are arranged by state, city, and airport name. An FAA identifier is included, as well as the airport reference point (ARP) coordinates. The ARP is usually located in the center of the airport. Waypoints are listed by name, and airports and navaids are listed by name and three-letter identifier. Each listing is indicated by an edge-marked tab. [Figure 3-27]

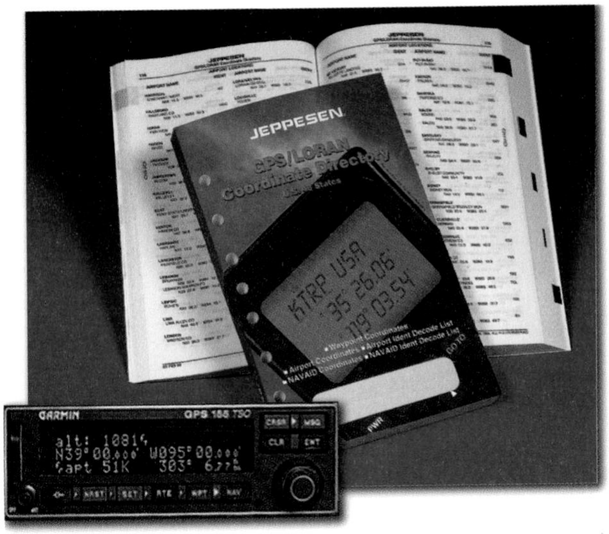

Figure 3-27. Unlike the other Jeppesen Information Services, the *GPS/LORAN Coordinate Directory* is published in a bound book (5-1/2″ x 8-1/2″). It is updated and reissued on the FAA's 56-day cycle.

ELECTRONIC FLIGHT PUBLICATIONS

You can access a variety of flight information by using a modem-equipped personal computer and the appropriate software or Internet service. The government uses electronic bulletin boards and world wide web sites to make flight publications and information updates more readily accessible to the general aviation community. For example, the FAA Home Page contains information on various aviation subjects and links to other home pages, such as FedWorld Information Network. Through FedWorld, you can access and order government publications, such as advisory circulars, FARs, practical test standards, and the *Federal Register*. [Figure 3-28]

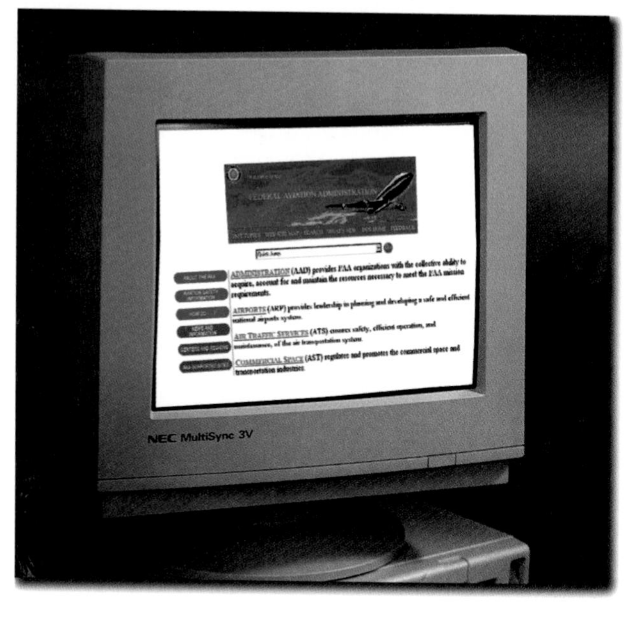

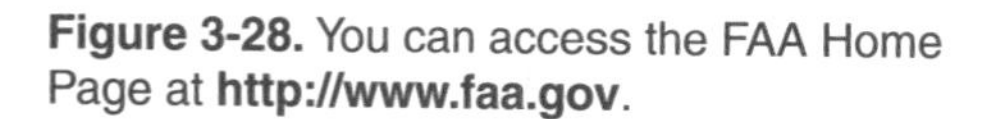
Figure 3-28. You can access the FAA Home Page at **http://www.faa.gov**.

SUMMARY CHECKLIST

✓ A visual runway normally is marked only with the runway number and a dashed white centerline. It may include additional markings for specific operations. A runway used for instrument approaches has additional markings, such as threshold markings, touchdown zone markings, aiming point markings, and side stripes.

✓ Additional markings for displaced thresholds include taxi, takeoff, and rollout areas, as well as blastpad/stopway areas. Closed runways and taxiways may be marked by Xs at some airports.

✓ Hold lines keep aircraft clear of the runways, and at controlled airports, serve as the point that separates the responsibilities of ground control from those of the tower.

✓ There are six basic types of airport signs — mandatory, location, direction, destination, information, and runway distance remaining.

✓ A variety of lighting systems, including approach light systems, sequenced flashing lights, runway alignment indicator lights, and runway end identifier lights are used at airports to aid the pilot in identifying the airport environment, particularly at night and in low visibility conditions.

✓ Various visual glide slope indicators, such as the visual approach slope indicator (VASI), precision approach slope indicator (PAPI), and tri-color VASI help pilots establish and maintain a safe descent path to the runway.

✓ Runway and taxiway lighting are installed at some airports to assist you in landing and taxing at night or during low visibility conditions. This lighting can consist of runway edge lights, threshold lights, displaced threshold lights, touchdown zone

lights, runway centerline lights, taxiway lead-off lights, taxiway centerline lights, and taxiway edge lights.

✓ Pilot-controlled lighting is the term used to describe systems that you can activate by keying the aircraft's microphone on a specified radio frequency.

✓ While operating in controlled airspace (Class A, Class B, Class C, Class D, and Class E), you are subject to certain operating rules, as well as pilot qualification and aircraft equipment requirements. In addition, specific VFR weather minimums apply to each class of airspace.

✓ The FARs require that you have an operating transponder with Mode C capability when flying at or above 10,000 feet MSL, excluding the airspace at and below 2,500 feet AGL.

✓ To operate within Class A airspace, you must be instrument-rated, your aircraft transponder-equipped, be operating under an IFR flight plan, and controlled by ATC.

✓ Flight levels instead of MSL altitudes are used in Class A airspace.

✓ To operate in Class B airspace, you are required to obtain a clearance from ATC.

✓ Prior to entering Class C airspace, you must establish two-way radio communication with the ATC facility having jurisdiction and maintain it while you are operating within the airspace.

✓ You must establish two-way radio communication with the tower prior to entering Class D airspace and maintain radio contact during all operations to, from, or on that airport.

✓ Federal airways are normally 8 nautical miles wide, begin at 1,200 feet AGL and extend up to but not including 18,000 feet MSL.

✓ A special VFR clearance must be obtained from ATC to operate within the surface areas of Class B, C, D, or E airspace when the ground visibility is less than 3 statute miles and the cloud ceiling is less than 1,000 feet AGL.

✓ Class G airspace typically extends from the surface to 700 or 1,200 feet AGL. In some areas, Class G extends from the surface to 14,500 feet MSL. ATC normally does not exercise control of air traffic in uncontrolled, or Class G airspace.

✓ Since the airspace at lower altitudes, and especially in the vicinity of airports, tends to be congested, the FAA has established aircraft speed limits.

✓ Prohibited areas are established for security or other reasons associated with the national welfare and contain airspace within which the flight of aircraft is prohibited.

✓ Restricted areas often have invisible hazards to aircraft, such as artillery firing, aerial gunnery, or flight of guided missiles. You must obtain permission from the controlling agency to fly through a restricted area. An IFR clearance which takes you through a restricted area constitutes authorization to penetrate the airspace.

✓ Warning areas extend from three nautical miles outward from the coast of the United States and contain activity which may be hazardous to nonparticipating aircraft.

✓ Alert areas may contain a high volume of pilot training or an unusual type of aerial activity.

✓ A military operations area (MOA) is a block of airspace in which military training and other military maneuvers are conducted.

✓ Activities within a controlled firing area are discontinued immediately when a spotter aircraft, radar, or ground lookout personnel determine an aircraft might be approaching the area.

✓ Airport advisory areas extend 10 statute miles from airports where there is an FSS located on the field and no operating control tower.

✓ Generally, military training routes (MTRs) are established below 10,000 feet MSL for operations at speeds in excess of 250 knots.

✓ Temporary flight restrictions are imposed by the FAA to protect persons or property on the surface or in the air from a specific hazard or situation.

✓ Air defense identification zones (ADIZs) are established to facilitate early identification of aircraft in the vicinity of U.S. international airspace boundaries.

✓ The *Airport/Facility Directory* contains a descriptive listing of all airports, heliports, and seaplane bases which are open to the public.

✓ The *Aeronautical Information Manual* (AIM) contains basic flight information, a detailed description of the National Airspace System, ATC procedures, and other items of special interest to pilots, such as medical facts and flight safety information.

✓ NOTAM(D)s are disseminated for all navigational facilities which are part of the National Airspace System, all public use airports, seaplane bases, and heliports listed in the A/FD.

✓ NOTAM(L)s, which are locally distributed, contain information such as taxiway closures, personnel and equipment near or crossing runways, and airport rotating beacon and lighting aid outages.

✓ FDC NOTAMs contain regulatory information such as temporary flight restrictions or amendments to standard instrument approach procedures and aeronautical chart revisions.

✓ Advisory circulars (ACs) provide nonregulatory guidance and information in a variety of subject areas. ACs also explain methods for complying with FARs.

✓ The *International Flight Information Manual* contains the requirements and instructions for flying outside of the United States.

✓ Jeppesen Information Services provide revisions for the *Jeppesen AIM*, *Jeppesen FARs for Pilots*, the *Jeppesen Airport Directory*, *JeppGuide*, and the *GPS/LORAN Coordinate Directory*.

KEY TERMS

Visual Runway

Nonprecision Instrument Runway

Precision Instrument Runway

Taxiways

Hold Lines

Displaced Threshold

Blast Pad/Stopway Area

Demarcation Bar

Land and Hold Short Operations (LAHSO)

Sequenced Flashing Lights (SFL)

Runway Alignment Indicator Lights (RAIL)

Runway End Identifier Lights (REIL)

Visual Approach Slope Indicator (VASI)

Pulsating Approach Slope Indicator (PLASI)

Precision Approach Path Indicator (PAPI)

Tri-Color VASI

Runway Edge Lights

Threshold Lights

Displaced Threshold Lights

Touchdown Zone Lighting (TDZL)

Runway Centerline Lights (RCLS)

Land and Hold Short Lights

Taxiway Lead-Off Lights

Taxiway Centerline Lights

Taxiway Edge Lights

Pilot-Controlled Lighting

Class A Airspace

Class B Airspace

Class C Airspace

Class D Airspace

Class E Airspace

Special VFR Clearance

Class G Airspace

Prohibited Area

Restricted Area

Warning Area

Alert Area

Military Operations Area (MOA)

Controlled Firing Area

National Security Area

Airport Advisory Area

Military Training Route (MTR)

Parachute Jump Aircraft Area

Temporary Flight Restrictions

Terminal Radar Service Area (TRSA)

Air Defense Identification Zone (ADIZ)

Defense VFR

Airport/Facility Directory (A/FD)

Aeronautical Information Manual (AIM)

Notices to Airmen

NOTAM(D)

NOTAM(L)

FDC NOTAM

International Flight Information Manual

Advisory Circulars (ACs)

Jeppesen Information Services

QUESTIONS

For questions 1 through 5, match the lettered callouts in the accompanying illustration to identify the appropriate runway markings.

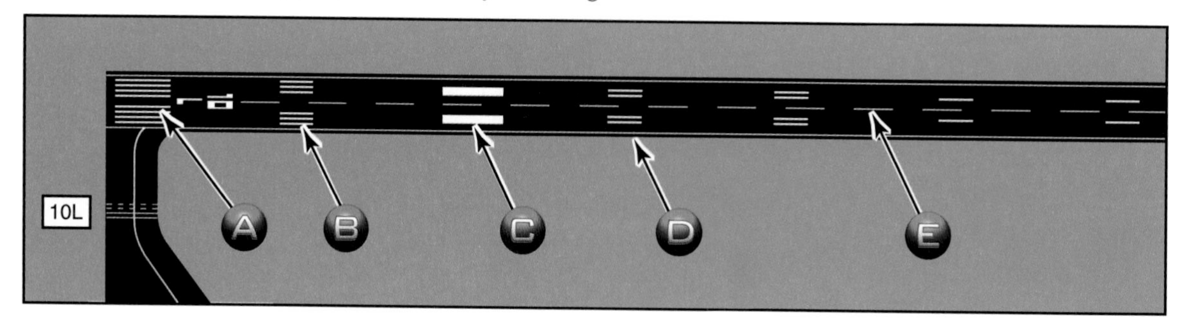

1. Side stripe

2. Threshold markings

3. Runway centerline

4. Aiming point marking

5. Touchdown zone marking

6. True/False. Runway holding position signs are black with yellow inscriptions and yellow borders.

7. Name the high intensity white strobe lights located laterally, one on each side of the runway threshold.

   A. Sequenced flashing lights
   B. Runway alignment indicator lights
   C. Runway end identifier lights

For questions 8 through 11, match each illustration to the correct glide slope description.

8. VASI, on glide path

9. PAPI, slightly low

10. PAPI, on glide path

11. VASI, high

12. As you approach for landing, what color are the runway threshold lights? What color are they when departing the runway?

13. How many feet of runway remain when the centerline lights change from white to alternating red and white lights?

14. What is indicated when the airport beacon is illuminated during daylight hours?

15. Excluding the airspace at and below 2,500 feet AGL, transponders with altitude encoding capability are required in all airspace (controlled or uncontrolled) of the contiguous 48 states and the District of Columbia at and above what altitude?

    A. 10,000 feet AGL
    B. 10,000 feet MSL
    C. 14,500 feet MSL

16. What are the dimensions of Class A airspace?

For questions 17 through 23, match the type of airspace with the appropriate description.

A. Federal Airways B. Class B C. Class C D. Class D E. Class E F. Class G

17. That portion of the airspace within which ATC does not control air traffic

18. Airspace at and above 14,500 feet MSL over the 48 contiguous states, District of Columbia, and Alaska east of 160° west longitude, but not including the airspace within 1,500 feet of the surface

19. An airspace segment that is normally 8 nautical miles wide and extends from 1,200 feet AGL (or in some cases higher) up to but not including 18,000 feet MSL

20. Circular airspace segments around airport, plus any extensions necessary to include instrument approach and departure paths, that extend upward from the surface to a vertical limit, charted in hundreds of feet MSL

21. Airspace within which equipment requirements include an operable VOR (for IFR operations), two-way radio capable of communicating with ATC, and a 4096-code transponder with Mode C automatic altitude reporting

22. Airspace generally consisting of circular areas extending to 4,000 feet above the primary airport where the shelf area has a radius of 10 nautical miles, the core area has a radius of 5 nautical miles, and two-way radio communication is required

23. Airspace outside Class B or Class C airspace which normally extends from the surface up to 2,500 feet above the elevation of the airport and having charted, but possibly irregular, lateral dimensions at which a control tower is operating

24. What is the maximum indicated aircraft speed limit (in knots) below 2,500 feet when within 4 nautical miles of the primary airport of a Class C airspace area?

25. True/False. Military training route flights are generally limited to operations under VFR.

Refer to the *Airport/Facility Directory* excerpt to answer questions 26 through 30.

NORTH PLATTE REGIONAL AIRPORT LEE BIRD FLD (LBF) 3 E UTC–6(–5DT) **OMAHA**
N41°07.56′ W100°41.23′ **H–10, L–11A**
2778 B S4 **FUEL** 100LL, JET A OX 4 ARFF Index Ltd. **IAP**
RWY 12L–30R: H8000X150 (CONC–GRVD) S–75, D–110, DT–190 HIRL
RWY 12L: VASI(V4L)—GA 3.0° TCH 55′. **RWY 30R:** MALSR. Rgt tfc.
RWY 12R–30L: H4925X100 (ASPH) S–42, D–58, DT–106 MIRL
RWY 12R: Rgt tfc. **RWY 30L:** Tree.
RWY 17–35: H4436X100 (ASPH) S–28, D–48, DT–86 MIRL
RWY 17: Road. **RWY 35:** REIL. VASI(V4L)—GA 3.0° TCH 41′. Thld dsplcd 234 . Berm.
AIRPORT REMARKS: Attended 1200–0500Z‡. 5 foot dike 100′ from approach end Rwy 35. Waterfowl and deer on and in the vicinity of the arpt. PPR 24 hours for unscheduled air carrier ops with more than 30 passenger seats call arpt manager 308–532–1900. ACTIVATE HIRL Rwy 12L–30R, MIRL Rwy 17–35, VASI Rwy 12L and Rwy 35, MALSR Rwy 30R and REIL Rwy 35—CTAF. For MIRL Rwy 12R–30L ctc arpt manager 308–532–1900.
WEATHER DATA SOURCES: ASOS 118.425 (308) 534–1617.
COMMUNICATIONS: CTAF/UNICOM 123.0
COLUMBUS FSS (OLU) TF 1–800–WX–BRIEF. NOTAM FILE LBF.
LEE BIRD RCO 122.5 (COLUMBUS FSS)
Ⓡ **DENVER CENTER APP/DEP CON** 132.7 **CLNC DEL** 132.7
RADIO AIDS TO NAVIGATION: NOTAM FILE LBF.
(L) VORTACW 117.4 LBF Chan 121 N41°02.92′ W100°44.83′ 019° 5 4 NM to fld. 3050/11E.
PANBE NDB (LOM) 416 LB N41°04.10′ W100°34.35′ 295° 6.3 NM to fld. Unmonitored.
ILS 111.5 I–LBF Rwy 30R LOM PANBE NDB. LOM and MM unmonitored.

26. In December, what are the hours (in local time) when the airport is attended?

27. What type of approach lighting system is installed on Runway 30R?

28. How far is the threshold displaced on Runway 35?

29. What is the name of the approach/departure control facility that serves North Platte Regional Airport Lee Bird Field?

30. What is the frequency and identification for the ILS approach?

31. Select the information which you can obtain by referencing the *Aeronautical Information Manual* (AIM).

    A. The official text of regulations issued by the agencies of the Federal government
    B. ATC procedures, a description of the airspace system, and flight safety information
    C. Information regarding specific airports, including runway lengths, communication frequencies, and airport services

32. Explain the differences between NOTAM(L)s, NOTAM(D)s, and FDC NOTAMs.

33. What is the primary source document for identifying and ordering advisory circulars?

SECTION B

AIR TRAFFIC CONTROL SYSTEM

To operate efficiently within the IFR environment, it is important that you understand the **air traffic control (ATC)** system. In a broad sense, the ATC system may be thought of as a network of radar and nonradar facilities that provides nationwide traffic separation during all phases of IFR flight. These facilities include the air route traffic control centers, approach and departure controls, control towers, ground controls, and clearance deliveries. An air route traffic control center is the main ATC facility for enroute operations while the others operate in the terminal area.

AIR ROUTE TRAFFIC CONTROL CENTER

The facilities which provide air traffic control service to aircraft operating on IFR flight plans in controlled airspace are the **air route traffic control centers (ARTCCs)**. They are also the central authority for issuing IFR clearances, and they provide nationwide monitoring of each IFR flight, primarily during the enroute phase. Each ARTCC (center), due to its size, is divided into sectors. Each sector is manned by one or more controllers, who maintain lateral and vertical separation of aircraft within its airspace boundaries. In addition, they coordinate traffic arriving and departing their assigned areas. Frequently, sectors are further stratified by altitude. For example, there may be low altitude sectors that extend from the floor of controlled airspace to altitudes of 18,000 to 24,000 feet MSL. Above these levels, one or more high altitude sectors may be established. [Figure 3-29]

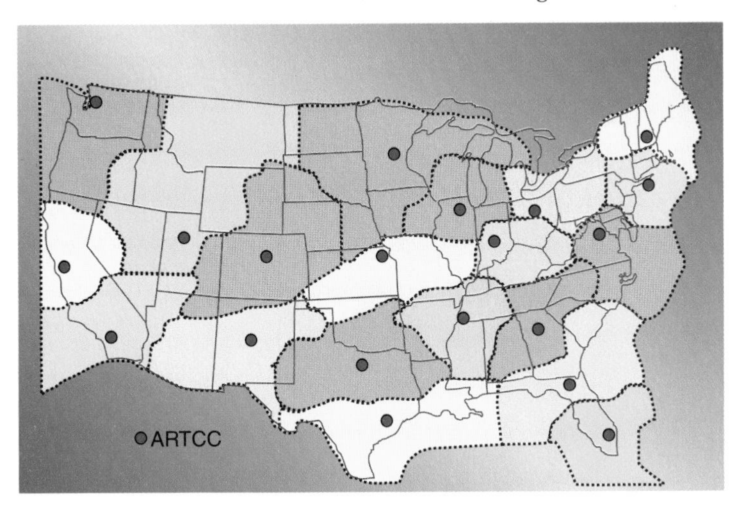

Figure 3-29. The dashed lines portray the various ARTCC boundaries within the United States, while the dots indicate the physical location of each center. Appropriate radar and communication sites are connected to the centers by microwave links and telephone lines.

ARTCC TRAFFIC SEPARATION

Although the ATC system provides numerous services, among the highest priorities is the safe separation of known IFR traffic and the issuance of safety alerts. Separation of aircraft operating under VFR is conducted on a workload-permitting basis. There are many methods of traffic separation, but the elimination of traffic conflicts actually begins when you file your IFR flight plan.

PROCESSING THE IFR FLIGHT PLAN

Prior to departure from within or prior to entering controlled airspace, you must submit a complete flight plan and receive an ATC clearance if weather conditions are below VFR minimums. Instrument flight plans can be submitted to the nearest flight service station (FSS) or air traffic control tower either in person, by telephone, or by radio if no other means are available. You can also file domestic VFR and IFR flight plans through a DUATS provider. Once your flight plan is filed, it is processed by the center in which the flight originates. This processing begins when the flight plan is entered into the center's computer, where it is scanned for preferred routes. The route is then analyzed by ATC controllers for any restrictions that may be in effect, such as traffic conflicts, inoperable navaids, special use airspace penetration, known or projected delays, and flow control restrictions. Due to the time it takes to check these variables, you should plan to file your IFR flight plan at least 30 minutes prior to your intended departure time.

Once your flight plan has been processed, a clearance will be available when the appropriate facility at your departure airport requests it. In addition, after the route is finalized, it is sent to the various centers covering your flight via the nationwide ARTCC computer system. Your flight data is then printed out approximately 30 to 45 minutes prior to your entry into that center's airspace. This provides the controller with time to

Flow Control Restrictions

Flow control was originally implemented to help regulate traffic at busy airports during the 1981 air traffic controller's strike. It is now an integral part of the National Airspace System, and is managed by the Air Traffic Control System Command Center (ATCSCC).

Each airport has an arrival rate which is determined by ATC personnel located on the airport. Depending on many factors such as the number of runways, instrument approaches, weather conditions, etc., the rate may be over a hundred aircraft per hour. However, this number could drop to just a few during bad weather conditions. ATCSCC takes a look at the anticipated number of aircraft expected to land during a specific time frame. (This number includes the aircraft enroute and those on the ground awaiting departure.) If the number of arriving aircraft is less than the anticipated number, there is no problem. However, if there are more aircraft inbound than the arrival rate allows, the aircraft either have to hold or divert to an alternate. This is not a good thing. Instead, ATCSCC will issue EDCTs (expect departure clearance times). This means that aircraft still on the ground will be held on the ground until they can be worked into the system. So next time you are being held on the ground at an airport where the sun is shining, it is probably due to the fact that your destination is anticipating more arrivals than it can handle.

Courtesy of Chicago Dept. of Aviation

analyze your flight for any of the above mentioned restrictions and amend your clearance, if necessary.

When you are departing an airport with an operating control tower, you normally request your clearance through ground control or, if appropriate, clearance delivery. If you are departing an airport served only by a flight service station, the FSS will request your clearance and forward it to you before your departure. In the event you are departing an airport without an ATC or FSS facility on the field, you may request your clearance by contacting the local FSS by telephone prior to departure. If necessary, you can obtain your clearance after departing VFR, if weather conditions permit. In this situation, you could contact the ARTCC sector where your flight originated, or you could request your clearance from the local FSS.

To prevent computer saturation, most centers will delete an IFR flight plan a minimum of one hour after the proposed departure time. To ensure your flight plan remains active, advise ATC of your revised departure time if you will be delayed one hour or more.

Free Flight

Most IFR flights are currently confined to airways established between navaids. This usually means that your route becomes a zigzag course instead of a straight line. Free flight is a concept that would allow you the same type of freedom you have during a VFR flight. Instead of a National Airspace System that is rigid in design, pilots would be allowed to choose their own routes, or even change routes and altitudes at will to avoid icing, turbulence, or even to take advantage of winds aloft. Complicated clearances would become unnecessary. However, flight plans would still be required for traffic planning purposes and as a fall-back in the event of lost communication.

Free flight will be possible with the use of advanced avionics, such as GPS navigation and datalinks between your aircraft, other aircraft, and controllers. Separation would be maintained by establishing two airspace zones around each aircraft. The protected zone, which is the one closest to the aircraft, can never meet the protected zone of another aircraft. The alert zone extends well beyond the protected zone, and aircraft can maneuver freely until alert zones touch. If alert zones do touch, a controller may provide the pilots with course suggestions, or onboard traffic displays may be used to resolve the conflict. The size of the zones will be based on the aircraft's speed, performance, and equipment.

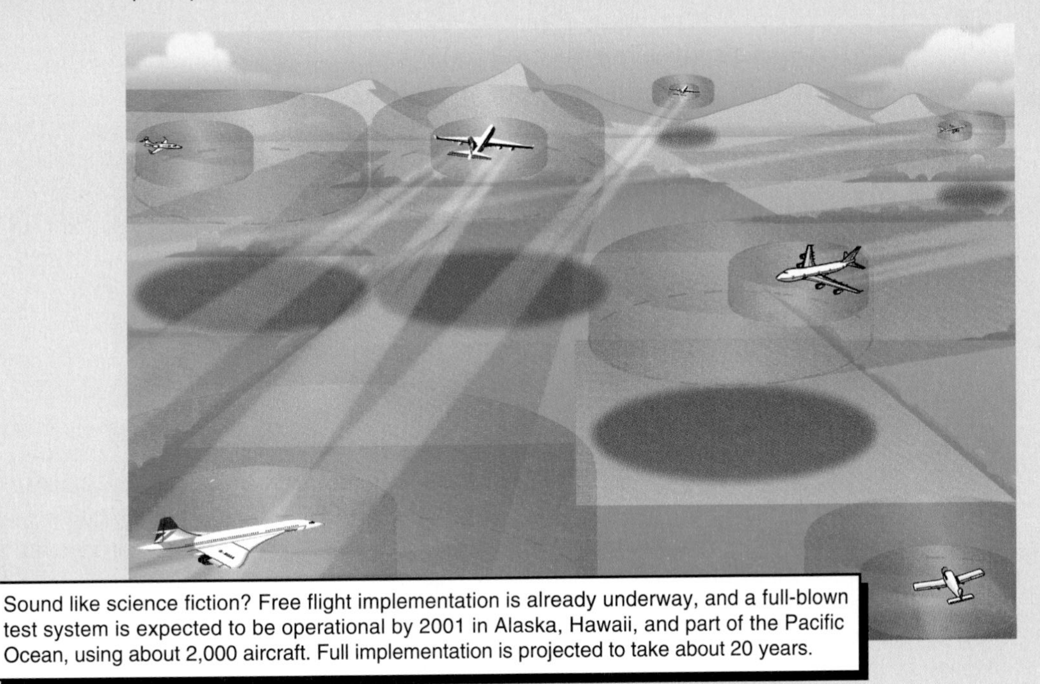

Sound like science fiction? Free flight implementation is already underway, and a full-blown test system is expected to be operational by 2001 in Alaska, Hawaii, and part of the Pacific Ocean, using about 2,000 aircraft. Full implementation is projected to take about 20 years.

Enroute IFR Traffic Separation

During the enroute phase of the IFR flight, certain situations may require the controller to amend your clearance so adequate separation is maintained between aircraft. Among these are deviations due to weather, unplanned pilot requests, flow control restrictions, and aircraft emergencies. If required, the controller will employ various techniques to ensure adequate separation standards. Some of the most common of these include route changes, radar vectoring, altitude crossing restrictions for navaids and intersections, altitude changes, and speed adjustments.

Speed adjustments are most commonly used during the arrival phase of a flight. For example, ATC may advise you to "*reduce speed to 100*" as you near your destination. This means ATC has requested you to decrease your indicated airspeed to 100 knots and to maintain that speed within 10 knots.

An ATC request for a speed reduction means you should maintain the new indicated airspeed within 10 knots.

Pilot Responsibilities

Although ATC has strict requirements with regard to the separation of IFR aircraft, keep in mind there are certain pilot responsibilities as well. For example, you must know the requirements for IFR flight, and you must know when an IFR clearance is required. Regulations state that you may not act as pilot in command of a flight conducted under IFR unless you hold an instrument rating and meet the recency of experience requirements for instrument flight as specified under FAR Part 61. In addition, your aircraft must meet the equipment and inspection requirements of FAR Part 91. If weather conditions are below VFR minimums, you must file an IFR flight plan and obtain an IFR clearance before departing from within, or prior to entering controlled airspace.

Additional ARTCC Services

In addition to separation of all IFR traffic, ARTCCs provide other safety-related services. These services include separating IFR aircraft from other traffic known to the controller, weather information, safety alerts, and emergency assistance.

Separation From Other Traffic

ATC's first priority is the separation of all IFR aircraft from one another. However, if workload permits, the controller may advise you of VFR aircraft which might affect your flight. It is important for you to realize the controller is not obligated to provide traffic advisory service. In addition, some aircraft in your area may not appear on the controller's radar display. For this reason, FARs require every pilot to see and avoid other aircraft whenever possible, even when they are operated under positive radar control, as in Class B airspace. When you are operating under IFR in VFR conditions, you must continually search for all other aircraft, regardless of the radar service being provided.

Weather Information

Although providing weather information is of lower priority than the separation of IFR traffic, center controllers will make every effort to update you on current conditions along your route of flight. ARTCCs are phasing in computer-generated digital radar displays. This system provides the controller with two distinct levels of weather intensity by assigning radar display symbols for specific precipitation densities. To the extent possible, controllers will issue on request, pertinent information on significant weather areas and assist you in avoiding such areas. Keep in mind that frequency congestion and limitations of ATC radar may affect the controller's ability to provide this type of service.

ATC radar limitations and frequency congestion may limit a controller's capability to provide in-flight weather avoidance assistance.

To ensure that timely weather information is available, each center has a meteorologist who monitors the weather within the airspace of that center. One of the tools the center meteorologist uses for providing real-time weather is a network of radar sites for detecting coverage, intensity, and movement of precipitation. The network is supplemented by FAA and Department of Defense radar sites in the western sections of the country. Local warning radar sites augment the network by operating on an as needed basis to support warning and forecast programs. When appropriate, weather information is distributed to each controller and to other facilities. If the center meteorologist finds it necessary, a **center weather advisory (CWA)** is issued. This report is used to alert pilots to existing or forecast adverse weather conditions that may affect terminal or enroute operations. These reports may later be issued as part of an in-flight advisory. A CWA is passed to the sector controllers who broadcast the information to all affected aircraft. The center weather advisory is an important subject and will be discussed in more detail in Chapter 9.

SAFETY ALERTS

A safety alert will be issued by a center controller when it becomes apparent that your flight is in unsafe proximity to terrain, obstructions, or other aircraft. A **terrain or obstruction alert** is issued when your Mode C altitude readout indicates your flight is below the published minimum safe altitude for that area. In general, you will be requested to check your altitude immediately. The controller will then provide the minimum altitude required in your area of flight. [Figure 3-30]

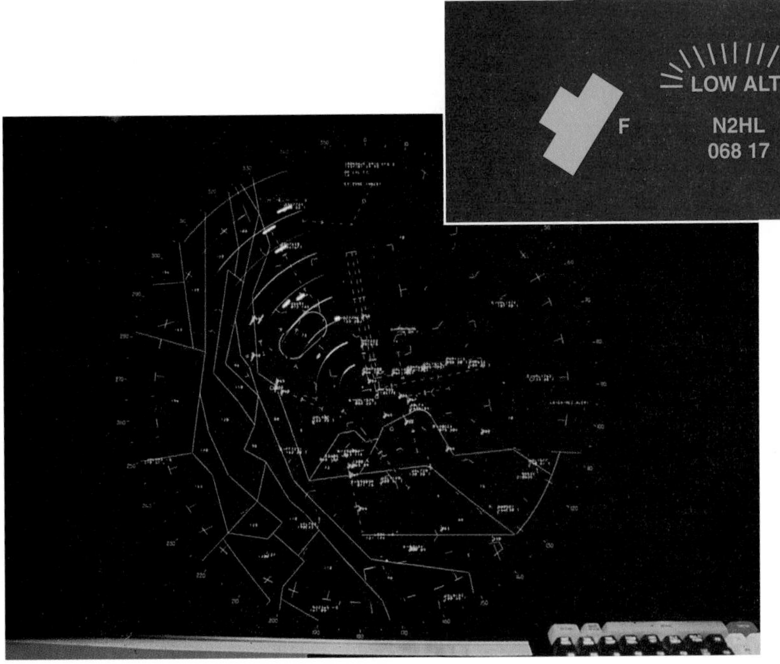

Figure 3-30. Based on the Mode C information provided by this aircraft's transponder, the computer has determined the aircraft is below the minimum safe altitude for this area. This is called a minimum safe altitude warning (MSAW). The flashing LOW ALT alerts the controller to this potential danger; the controller, in turn, alerts the pilot.

The second type of safety alert is called an **aircraft conflict alert**. This service is provided when the controller determines that the minimum separation between an aircraft being controlled and another aircraft could be compromised. If a conflict alert is issued to you, the controller will advise you of the position of the other aircraft and a possible alternate course of action. Keep in mind that for either of these alert services to be available, your aircraft must be under radar control and your Mode C transponder must be fully operational. Both types of safety alerts may also be issued by terminal radar facilities.

EMERGENCY ASSISTANCE

One advantage of IFR flight is the continual radio contact with ATC. In addition, throughout most of the United States, your flight is continuously in radar contact. If a problem arises, ATC is immediately available to render a wide variety of services, from simply clearing conflicting traffic to providing radar vectors and, if required, a radar approach to the nearest suitable airport. If a serious situation requiring an immediate landing develops, ATC will alert search and rescue (SAR) agencies in the area. [Figure 3-31]

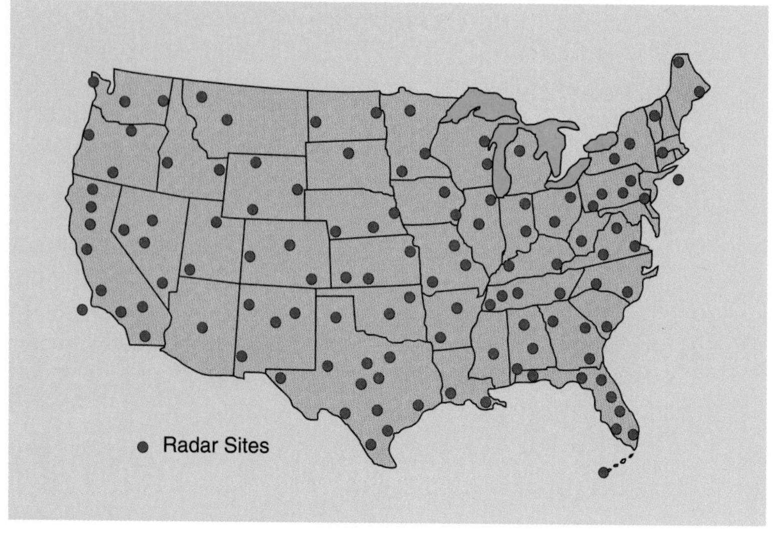

Figure 3-31. This map represents the air route surveillance radar (ARSR) sites used by ARTCC to monitor enroute operations. Except for areas where radar signals are blocked by terrain, radar coverage extends throughout the contiguous United States. Since new sites are being commissioned and old sites decommissioned, this map is constantly changing.

TERMINAL FACILITIES

Within the air traffic control system, each terminal facility is closely linked with the associated ARTCC to integrate the flow of IFR departures and arrivals. Terminal services include ATIS, ground control, clearance delivery, control tower, approach and departure control, and FSS.

ATIS

At busy airports, airport advisory information is provided by automatic terminal information service (ATIS). This continuous, recorded broadcast of noncontrol information helps to improve controller effectiveness and to reduce frequency congestion. At larger airports, there may be one ATIS frequency for departing aircraft and another one for arriving aircraft. ATIS is updated whenever any official weather is received, regardless of content change. It is also updated whenever airport conditions change. When a new ATIS is broadcast, it is changed to the next letter of the phonetic alphabet, such as Information Bravo or Information Charlie and so on. [Figure 3-32]

As indicated in figure 3-32, the absence of the sky condition and visibility on an ATIS broadcast specifically implies the ceiling is more than 5,000 feet AGL and the visibility is more than 5 statute miles.

CLEARANCE DELIVERY

In order to relieve congestion on the ground control frequency at busier airports, a discrete clearance delivery frequency may be provided. This facility allows you to receive an IFR clearance prior to contacting ground control for taxi. Additionally, this service may be

Regardless of content change, ATIS broadcasts are updated upon receipt of any official weather.

Following the airport name and ATIS phonetic letter identifier, the broadcast will state the time of the current weather report,	*Centennial Airport Information Tango, 1655 Zulu weather,*
magnetic wind direction and velocity,	*wind 070 at 12,*
visibility, obstructions to visibility, and ceiling/sky condition,	*visibility 2 light snow and mist, ceiling 1,200 broken, 2,000 overcast,*
temperature and dewpoint (if available),	*temperature 0, dewpoint 0,*
and altimeter setting.	*altimeter 29.74.*
Next, the instrument approach and runways in use are indicated.	*ILS Runway 35 Right is in use landing and departing Runway 35 Right.*
The ATIS broadcast also contains any other pertinent remarks relating to operations on or near the airport, such as closed runways or temporary obstructions.	*Pilot weather reports; bases 7,200 during approach and departure, light rime icing encountered between 7,000 and 9,000 feet by a King Air and a Learjet. A Boeing 727, 8 miles west of Denver, encountered light rime icing below 9,000 feet. All services, including taxi and IFR clearance available on the tower frequency 118.9.*
The phonetic letter identifier is restated at the end of the broadcast.	*Advise on initial contact you have Information Tango.*

Figure 3-32. If the cloud ceiling is above 5,000 feet AGL and the visibility is more than 5 statute miles, inclusion of the ceiling/sky condition, visibility, and obstructions to vision in an ATIS message is optional.

used by VFR pilots to receive an ATC clearance when departing an airport within Class B or Class C airspace. Clearance delivery can also be used to receive a departure control frequency and transponder code when departing an airport with a radar departure control.

CONTROL TOWER

Towers are responsible for the safe, orderly, and expeditious flow of all traffic which is landing, taking off, operating on and in the vicinity of an airport, and when the responsibility has been delegated, towers can also provide for the separation of IFR aircraft in terminal areas. When a control tower is operational, you are required to obtain a clearance prior to operating in a movement area. This clearance can be from ground control or the tower operator. **Movement areas** are defined as runways, taxiways, and other areas which are used for taxiing, takeoffs, and landings, exclusive of loading ramps and parking areas. Ground control, when available, usually issues clearances for areas other than the active runway. When communicating with ground control or the tower, make sure they are aware of your position on the airport. One such position report which you are required to make is when your are ready for takeoff from a runway intersection.

When departing from a runway intersection, you should always state your position when calling the tower for takeoff.

Aircraft that are departing IFR are integrated into the departure sequence by the tower. Prior to takeoff, the tower controller coordinates with departure control to assure adequate aircraft spacing. After takeoff, you are required to remain on the tower frequency until you are instructed to contact departure control. Within Class D airspace, pilots of VFR aircraft are required to remain on the tower's frequency unless directed otherwise.

During a takeoff in IFR conditions, contact departure control only after you are advised to do so by the tower controller.

When arriving IFR at a controlled airport, you are sequenced by approach control for spacing and then advised to contact the tower for landing clearance. The tower controller will issue your landing clearance which may include wind direction, wind velocity, current visibility and, if appropriate, special instructions. This information may be omitted at an airport served by ATIS. If you are arriving VFR, you should contact the tower approximately 15 miles from the airport.

Approach and Departure Control

Approach and departure control coordinate very closely with the ARTCC to integrate arrival traffic from the enroute stage to the terminal area, and transition departure traffic to the enroute phase. In addition to coordinating IFR traffic, radar-equipped terminal areas also provide optional radar service to VFR aircraft. These services are designated as either basic or TRSA (terminal radar service area) service. In addition, radar service, called Class B or Class C service, is provided to all aircraft operating within Class B or C airspace areas.

IT CAN HAPPEN TO ANYONE

Next time you think that you are invincible and that ATC will never allow your aircraft to get close to another, just remember that it can happen to anyone, even Air Force One. Flying west of Washington National Airport, Air Force One was heading west and climbing through 7,500 feet MSL. An MD-88 commercial flight was also west inbound to Washington National on an easterly heading. A 737 commercial flight was southwest of the airport descending from 8,500 feet MSL and circling away from the airport for spacing. Controllers turned Air Force One to the southwest to ensure separation from the MD-88. However, the turn placed Air Force one in the vicinity of the 737. It was determined that Air Force One and the 737 were separated at their closest point by 900 feet and a little over 2 miles.

While this incident was resolved quickly, the established separation of 3 nautical miles and 1,000 feet in a terminal area was penetrated. Therefore, when flying under IFR conditions, you should always listen to what is happening with other traffic and visualize where you are in relation to them. If you ever have a concern that you are being vectored into an area near another aircraft, express your concern to the controller.

Radar Service for VFR Aircraft

Basic radar service provides safety alerts, traffic advisories, limited vectoring, and sequencing at certain locations. When you are approaching an airport for landing, you should contact approach control and state your position, altitude, aircraft call sign, type of aircraft, radar beacon code, and destination. Approach control will issue wind and runway information, except when you state "*have numbers,*" or indicate you have the current ATIS information. For sequencing service, contact approach control when you are approximately 25 miles from the airport. At airports within a terminal radar service area (TRSA), air traffic controllers provide **TRSA service** which includes radar vectoring, sequencing, and separation for all IFR and participating VFR aircraft.

Use of basic and TRSA radar service is not mandatory for VFR operations, but pilot participation is strongly urged. Participation does not relieve you of your responsibilities regarding terrain/obstruction clearance, vortex exposure, and to see and avoid other aircraft when operating under VFR. In addition, an ATC instruction to follow a preceding aircraft does not authorize you to comply with any ATC clearance or instruction issued to the preceding aircraft. A clearance issued by the radar controller which will cause you to violate a rule, such as entering a cloud under VFR, must be declined. In this situation,

you should advise the controller of your inability to comply with the issued clearance and request a revised clearance or instruction.

CLASS C SERVICE AREAS

Class C airspace areas are established by regulation at locations where traffic conditions warrant. Participation in Class C radar service is mandatory. You are not permitted to operate within Class C airspace unless you have established two-way radio communication with the ATC facility having jurisdiction over the area. In addition, you must maintain radio contact while operating within its limits. You can expect ATC to provide sequencing of all arriving aircraft to the primary airport. ATC also provides separation between IFR and VFR aircraft, traffic advisories, and safety alerts, if appropriate.

Departure control provides separation of all aircraft within a Class C airspace area.

CLASS B SERVICE AREAS

Class B airspace areas are established to accommodate arrivals and departures at the nation's busiest terminal areas. Since you may not operate within Class B airspace unless you obtain a specific ATC clearance, participation in Class B radar service is mandatory. Besides the basic radar services, Class B service provides separation of aircraft based on whether the flight is IFR or VFR, and/or on aircraft weight criteria. More space is required behind heavy aircraft. Sequencing of all arriving aircraft is also provided with Class B service.

TRAFFIC ADVISORIES

No matter what type of ATC radar facility you work with, controllers follow certain conventions when calling traffic to your attention. Normally, you are told the position (azimuth) of the traffic relative to your aircraft, its distance in nautical miles, its direction of movement, the type of aircraft, and its altitude, if known. When calling out traffic, controllers describe the position of the traffic in terms of the 12-hour clock. For example, "*traffic at 3 o'clock*" indicates the aircraft lies off your right wing. Keep in mind that the issuance of traffic information is based on the observation of your ground track and the position of the traffic. [Figure 3-33]

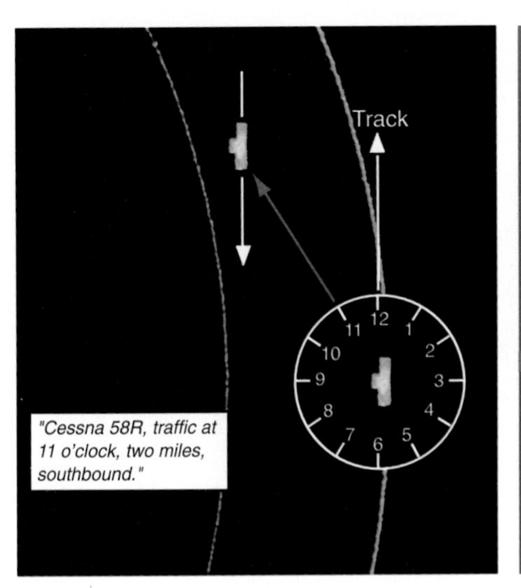

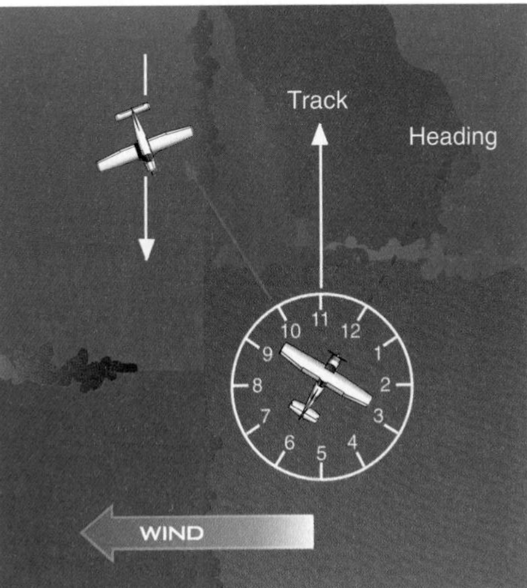

Figure 3-33. A radarscope cannot account for the amount of wind correction you may be using to maintain your track over the ground. Since the controller is unable to determine your actual heading, you must adjust the traffic callout for any wind correction angle you may be using. In this example, the controller advises you about traffic at your 11 o'clock position, but your wind correction angle places it closer to your 10 o'clock position.

As shown in figure 3-33, you should look for traffic based on your ground track, not your heading.

FLIGHT SERVICE STATIONS

Flight service stations (FSSs) at selected locations also provide a number of essential functions for both IFR and VFR aircraft. In addition to conducting weather briefings and handling flight plans, they also provide local airport advisory (LAA). This service is provided by an FSS that is located on an airport which does not have a control tower or where the tower operates on a part-time basis. At these locations, the FSS provides official weather information, and also relays clearances from ATC. When inbound under VFR, you should report when you are approximately 10 miles from the airport and provide your altitude and aircraft type. Also, state your location relative to the airport, whether landing or overflying, and request a local airport advisory. Departing aircraft should state the aircraft type, full identification number, type of flight (VFR or IFR), and the planned destination or direction of flight.

Local airport advisories are provided by flight service at FSS airports not served by an operating control tower.

While LAA is useful, you should be aware that airport advisory areas and the associated services have been reduced as automated flight service stations have replaced nonautomated facilities. Information on the availability of LAA is published on sectional charts and in the *Airport/Facility Directory*.

Air Traffic Control System

While the numbers are constantly changing, it takes a concerted effort of many people to handle the task of managing millions of flight operations. The FAA employs approximately 17,000 controllers and 7,800 field maintenance personnel to operate 20 contiguous U.S. air route traffic control centers, 200 terminal radar approach control facilities (TRACONs), 500 airport traffic control towers, and 100 flight service stations.

SUMMARY CHECKLIST

- ✓ The air traffic control (ATC) system consists of enroute and terminal facilities. The main enroute facility is air route traffic control center (ARTCC), while approach and departure control, the control tower, ground control, and clearance delivery are terminal facilities.
- ✓ Two of ATC's priorities are the separation of known IFR traffic and the issuance of safety alerts.
- ✓ IFR flight plans should be filed at least 30 minutes before departure.
- ✓ Flight plans are processed by the ARTCC in which the flight originates.
- ✓ IFR flight plans are usually deleted from the ARTCC computer if they are not activated within one hour of the proposed departure time.
- ✓ Due to weather, unplanned pilot requests, flow control restrictions, etc., controllers may alter your clearance to maintain proper aircraft separation.
- ✓ Regardless of whether operating under VFR or IFR, it is the pilot's responsibility to see and avoid other aircraft whenever weather conditions permit.
- ✓ If adverse weather exists or is forecast, an on-site meteorologist at the ARTCC may issue a center weather advisory (CWA).
- ✓ A safety alert is issued when in the controller's judgment, an aircraft is in unsafe proximity to terrain, an obstruction, or another aircraft.
- ✓ To improve controller effectiveness and to reduce frequency congestion, automatic terminal information service (ATIS) is available in selected high activity terminal areas.
- ✓ To relieve congestion on ground control frequencies, clearance delivery is used for ATC clearances at busier airports.
- ✓ Control towers are responsible for the safe, orderly, and expeditious flow of all traffic operating on and in the vicinity of an airport.
- ✓ At airports with an operating control tower, you are required to obtain a clearance before operating in a movement area, which is defined as an area on the airport, other than a parking area and loading ramp, used for taxiing, takeoff, and landing.
- ✓ During a takeoff in IFR conditions, contact departure control only after you are advised to do so by the tower controller.
- ✓ Terminal radar service for VFR aircraft includes basic radar service, terminal radar service area (TRSA) service, Class C service, and Class B service.
- ✓ Basic radar service for VFR aircraft includes safety alerts, traffic advisories, and limited radar vectoring. Sequencing also is available at certain terminal locations.
- ✓ Traffic advisories from ATC are based on your aircraft's actual ground track, not on your aircraft's heading.
- ✓ A local airport advisory (LAA) is provided by flight service at FSS airports not served by an operating control tower, or when the tower is closed.

KEY TERMS

Air Traffic Control (ATC)

Air Route Traffic Control Center (ARTCC)

Center Weather Advisory (CWA)

Terrain or Obstruction Alert

Aircraft Conflict Alert

Movement Areas

Basic Radar Service

TRSA Service

Local Airport Advisory (LAA)

QUESTIONS

1. Air traffic control centers (ARTCCs) were established to provide air traffic control service to aircraft operating on what type of flight plan?

   A. IFR
   B. VFR
   C. both IFR and VFR

2. What are the two highest priorities of the air traffic control system?

3. Within a given area, what facility is the central authority for processing an IFR flight plan?

   A. Flight service station
   B. Air traffic control tower
   C. Air route traffic control center

4. At least how many minutes prior to your planned departure should you file your IFR flight plan?

5. True/False. As a general rule, you cannot obtain an IFR clearance through a flight service station.

6. IFR flight plans filed in the ARTCC computer are usually deleted if they are not activated within what time period?

   A. 30 minutes
   B. 1 hour
   C. 2 hours

7. What facility issues center weather advisories?

   A. The nearest flight service station
   B. A national weather service office
   C. An ARTCC

8. When does ATC issue a safety alert?

9. What is the name of the facility that allows you to receive an IFR clearance before you contact ground control for taxi?

For questions 10 through 16, match the ATC facility or service with its description.

10. Relays IFR clearances to pilots departing uncontrolled airports
11. Controls traffic on and in the vicinity of a controlled airport
12. Transitions aircraft between the control tower and ARTCC
13. Controls the surface movement on the airport other than on the active runways
14. Broadcasts a prerecorded message of airport information
15. Issues IFR clearances at large airports
16. Controls all enroute IFR air traffic

A. ATIS
B. ARTCC
C. Control Tower
D. Ground Control
E. Clearance Delivery
F. Flight Service Station
G. Approach/Departure Control

17. Which services are included in basic radar service for VFR aircraft?

    A. Separation, limited radar vectoring, and traffic advisories
    B. Safety alerts, traffic advisories, sequencing, and separation
    C. Safety alerts, traffic advisories, and limited radar vectoring

18. ATC gives you the following traffic advisory: "*traffic 2 o'clock, 3 miles northbound.*" Refer to the accompanying illustration. Which traffic position corresponds to the advisory?

19. True/False. An ATC advisory, such as "*traffic 2 o'clock, 3 miles northbound*" is based on the position of that traffic relative to your aircraft's true heading.

20. If a control tower and a flight service station are on the same airport, what type of advisory service is provided by selected flight service stations when the tower is closed?

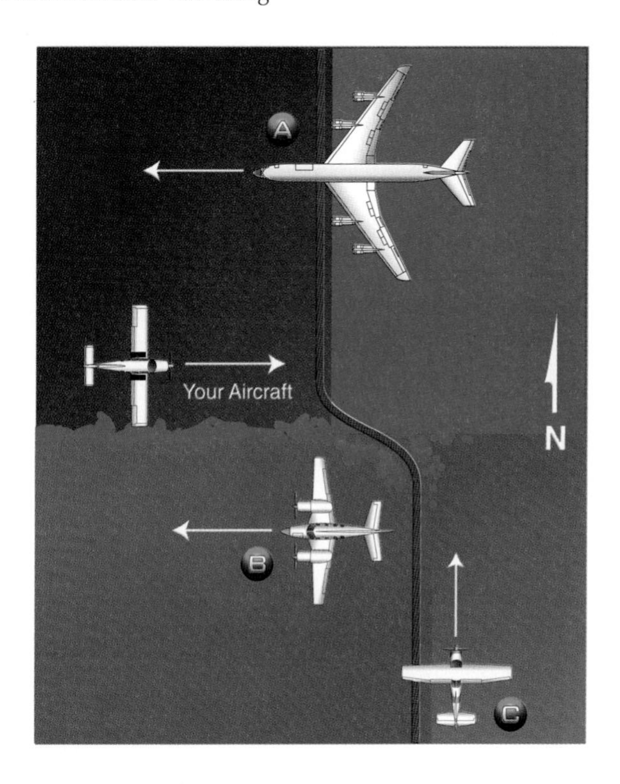

SECTION C
ATC CLEARANCES

An ATC clearance constitutes an authorization for you to proceed under a specified set of conditions within controlled airspace. It is the means by which ATC exercises its responsibility to provide separation between aircraft. It is not, however, an authorization for you to deviate from any regulation or minimum altitude, nor to conduct unsafe operations.

PILOT RESPONSIBILITIES

When ATC issues a clearance, regulations specify that you are not to deviate from it, except in an emergency, unless an amended clearance is received or unless complying with that clearance will cause you to violate a rule or regulation. Therefore, before you accept a clearance, you must determine if you can safely comply with that clearance. Points to consider include whether compliance will cause you to break the rules, such as being vectored into a cloud when you are operating under VFR, or to exceed the performance capabilities of yourself or your aircraft. If, in your opinion, a clearance is unsafe or not appropriate, it is your responsibility to promptly request an amended clearance.

While operating under VFR, if ATC assigns an altitude or heading that will cause you to enter clouds, you should avoid the clouds and inform ATC that the altitude or heading will not permit VFR.

If you find it necessary to deviate from a clearance due to an emergency, or compliance with the clearance would place your aircraft in jeopardy, you must notify ATC as soon as possible. This also includes a deviation in response to a traffic alert and collision avoidance system (TCAS) resolution advisory (RA). However, if you have to deviate, you should not disrupt existing traffic flow. In addition, if ATC gives your aircraft priority because of an emergency, you may be requested to submit a written report within 48 hours to the manager of that ATC facility. [Figure 3-34]

You may not deviate from a clearance unless you experience an emergency or the clearance will cause you to violate an FAR. If you deviate, you must notify ATC as soon as possible.

Figure 3-34. If you deviate from an ATC clearance, you must notify ATC as soon as possible.

. . . Climb! Climb!

FAR Part 91.123 authorizes deviations from an ATC clearance when responding to a traffic alert and collision avoidance system resolution advisory (TCAS RA). It also requires pilots to notify ATC as soon as possible if they deviate from a clearance in response to a TCAS RA.

TCAS airborne equipment interrogates transponders of other aircraft which have entered monitored airspace. The size of the airspace area being monitored is established by the flight crew using various ranges on the TCAS equipment. By computer analysis of the replies, TCAS equipment determines which transponder-equipped aircraft are potential collision hazards and provides appropriate visual and oral advisory information to the flight crew.

There are two types of TCAS systems. TCAS I is used primarily by corporate and commuter aircraft and TCAS II is used by commercial airliners. TCAS I only provides traffic advisories (TAs), to assist in the visual acquisition of intruder aircraft. There are no recommended avoidance maneuvers provided. TCAS II, on the other hand, provides traffic advisories (TAs) and resolution advisories (RAs). Resolution advisories provide recommended maneuvers in a vertical direction (climbs or descents only) to avoid conflicting traffic. When an RA occurs, the pilot should maneuver as indicated on the RA displays unless doing so would jeopardize the safe operation of the flight or unless the flight crew has definitive visual acquisition of the aircraft causing the RA.

A collision alert device, known as TCAD, has been developed for the general aviation environment. It is a passive system that receives all transponder replies in the local area and then presents the altitude and distance of targets within the designated monitoring range. With a TCAD system, there is no bearing information.

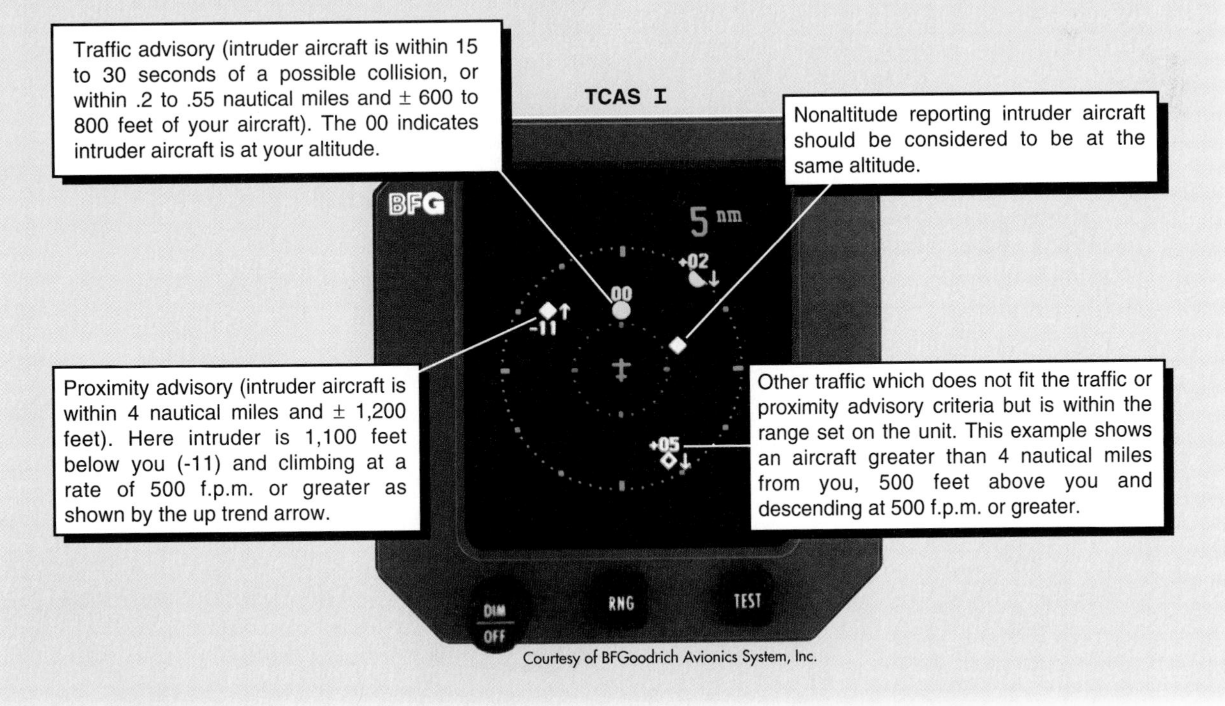

Courtesy of BFGoodrich Avionics System, Inc.

SEE AND AVOID

When meteorological conditions permit, regardless of the type of flight plan and whether or not you are under the control of an ATC facility, you are responsible to see and avoid other aircraft, terrain, or obstacles. This rule has been established solely in the interest of safety. ATC radar is not capable of detecting every aircraft which may pose a hazard to your flight. Also, one of ATC's primary responsibilities is the separation of IFR traffic. Traffic advisories are provided for pilots operating under VFR conditions on a workload-permitting basis only.

In VFR conditions, you are required to see and avoid all aircraft, regardless of the type of flight plan you have filed.

IFR CLIMB CONSIDERATIONS

ATC expects you to maintain a continuous rate of climb of at least 500 f.p.m. to your assigned cruising altitude. If you are unable to maintain this climb rate, you should notify ATC of your reduced rate of climb. Unless ATC advises "*At pilot's discretion,*" you are expected to climb at an optimum rate consistent with your airplane's performance to within 1,000 feet of your assigned altitude. Then attempt to climb at a rate of between 500 and 1,500 f.p.m. for the last 1,000 feet of climb.

You should climb at an optimum rate to within 1,000 feet of your assigned altitude. Then, you should climb at a rate of between 500 and 1,500 feet per minute.

When established on an airway, FAR Part 91 specifies that you must fly the centerline of that airway during climb, cruise, and descent. However, the regulation further provides that you are not prohibited from maneuvering the aircraft to pass well clear of other aircraft in VFR conditions. In addition, the FAA recommends that, while climbing in VFR conditions, you make gentle turns in each direction so you can continuously scan the area around you. As previously discussed, whenever you are operating on an IFR flight plan in VFR conditions, you are responsible for collision avoidance.

While climbing on an airway, you are required by regulation to maintain the centerline except when maneuvering in VFR conditions to detect and/or avoid other air traffic.

IFR FLIGHT PLAN AND ATC CLEARANCE

Prior to flying in controlled airspace when the weather is below VFR minimums and in Class A airspace regardless of the weather, you are required to file an IFR flight plan. Keep in mind that you may not file an IFR flight plan unless you hold an instrument rating for the category of aircraft you are flying and you are instrument current as specified in FAR Part 61. In addition, the aircraft to be used must be approved for IFR flight and must have the navigation equipment appropriate to the navigation aids to be used. You may cancel an IFR flight plan anytime you are operating under VFR conditions outside of Class A airspace. However, once you cancel IFR, the flight must be conducted strictly in VFR conditions from that point on. Should you again encounter IFR weather, you must remain in VFR conditions while you file a new flight plan and obtain an IFR clearance.

You may cancel an IFR flight plan anytime you are operating in VFR conditions outside of Class A airspace.

An ATC clearance is required before entering Class A and Class B airspace regardless of the weather conditions. When the weather is below VFR minimums, an ATC clearance is also required in Class C, D, and E airspace.

Prior to flying in controlled airspace when weather conditions are below VFR minimums, and in Class A airspace regardless of the weather, you must have filed an IFR flight plan and received an ATC clearance.

ELEMENTS OF AN IFR CLEARANCE

An IFR clearance is made up of one or more instructions. Knowing the order in which ATC issues these instructions makes it easier to understand a clearance. [Figure 3-35]

Figure 3-35. The following items, when appropriate, are contained in an initial IFR clearance in the order shown.

CLEARANCE LIMIT

The clearance issued prior to departure normally authorizes you to fly to your airport of intended landing. However, because of delays at your destination, you may be cleared to a fix short of your destination. If this happens, you will be given an **expect further clearance (EFC)** time. At some locations, you may be given a **short-range clearance**, whereby a clearance is issued to a fix within or just outside of the departure terminal area. A short-range clearance contains the frequency of the air route traffic control center which will issue your long-range clearance. A short-range clearance is often used in a nonradar environment to get you to a location where you can be identified by radar.

DEPARTURE PROCEDURE

You may be issued specific headings to fly and altitude restrictions to separate your aircraft from other traffic in the terminal area. Where the volume of traffic warrants, instrument departure procedures (DPs) have been developed. DPs and standard terminal arrival routes (STARs) are essentially charted procedures that help simplify the issuance of a clearance. ATC assumes you have all applicable DPs and STARs and will issue them, as appropriate, without request. If you do not possess DPs and STARs, or do not wish to use them, you should include the phrase "No DP/STAR" in the remarks sections of your IFR flight plan. DPs and STARs will be discussed in more detail in Chapters 4 and 6.

ROUTE OF FLIGHT

Clearances are normally issued for the altitude or flight level and route filed by the pilot. However, due to traffic conditions, it is sometimes necessary for ATC to specify an altitude/flight level or route different from that requested. In addition, flow patterns have been established in certain congested areas, or between congested areas, whereby capacity is increased by routing all traffic on preferred routes. Information on these flow patters is available in offices where preflight briefings are furnished or where flight plans are accepted.

When required, clearances include data to assist in identifying reporting points. It is your responsibility to notify ATC immediately if your navigation equipment cannot receive the type of signals needed to comply with the clearance.

ALTITUDE DATA

The altitude or flight level instructions in an ATC clearance normally require that you "*maintain*" the altitude or flight level at which the flight will operate when in controlled airspace. Altitude or flight level changes while enroute should be requested prior to the time the change is desired.

When possible, if the altitude assigned is different from the altitude requested, ATC will inform you when to expect a climb or descent clearance or when to request an altitude change from another facility. If a new altitude assignment has not been received prior to leaving the area, and you still desire a different altitude, you should reinitiate the request with the next facility.

A **cruise clearance** may be issued by ATC in situations where the route segment is relatively short and traffic congestion is not a consideration. In this type of clearance, the controller uses the word "*cruise*" instead of the word "*maintain*" when issuing an altitude assignment. [Figure 3-36]

Figure 3-36. This is an example of a cruise clearance.

The significance of a cruise clearance is that you may operate at any altitude, from the minimum IFR altitude up to and including, but not above, the altitude specified in the clearance. You may climb, level off, descend, and cruise at an intermediate altitude at any time. Each change in altitude does not require a report to ATC. However, once you begin a descent and report leaving an altitude, you may not climb back to that altitude without obtaining an ATC clearance. Another important aspect of a cruise clearance is that it also authorizes you to proceed to and execute an approach at the destination airport. In other words, you do not need to request, and ATC will not issue, a separate approach clearance at the destination airport when you have been issued a cruise clearance. When you are operating in uncontrolled airspace on a cruise clearance, you are responsible for determining the minimum IFR altitude. In addition, your descent and landing at an airport in uncontrolled airspace is governed by the applicable FARs for flight under VFR.

A cruise clearance is an authorization to conduct flight at any altitude from the minimum IFR altitude up to and including the assigned altitude without a further clearance. In addition, you may vacate an altitude/flight level within the cruise clearance block of airspace without notifying ATC.

HOLDING INSTRUCTIONS

If you have been cleared to a fix other than the destination airport and a delay is expected, it is the responsibility of ATC to issue complete holding instructions, unless the pattern is charted on the enroute chart or approach procedure. In addition, the controller should issue an EFC time, and a best estimate of any additional enroute or terminal delay. If the holding pattern is charted and the controller does not issue complete holding instructions, you are expected to hold as depicted on the appropriate chart. [Figure 3-37]

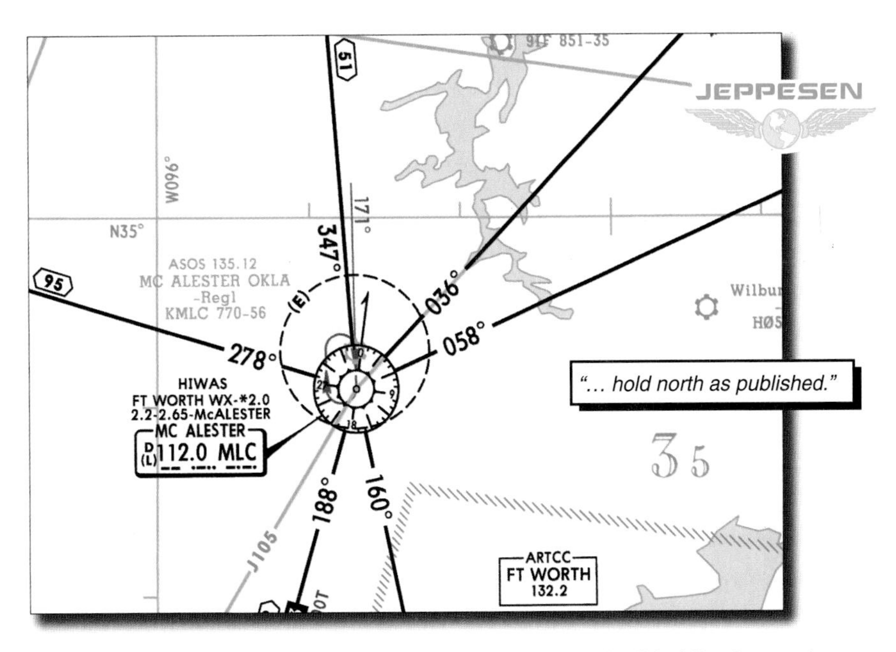

Figure 3-37. When the pattern is charted, the controller may omit all holding instructions except the charted holding direction and the statement "*as published*." However, controllers will always issue complete holding instructions if you request them.

ABBREVIATED IFR DEPARTURE CLEARANCE

In order to decrease radio congestion and controller workload, ATC issues an **abbreviated IFR departure clearance** whenever possible. This type of clearance uses the phrase "*cleared as filed*" to indicate you have been cleared to fly the route as contained in your IFR flight plan. This technique is particularly useful when numerous navigation fixes and Victor airways are contained in the flight plan, and ATC can accommodate the routing as you filed it with little or no change. However, if ATC finds it necessary to change your requested routing, a full route clearance is issued and the abbreviated clearance procedure is not used.

In the event you have filed a DP in your flight plan or a DP is in use at the departure airport, the DP or DP transition to be flown is included in the abbreviated clearance. Although DPs are included in the abbreviated clearance, STARs are considered part of the routing and normally are not stated in the body of the clearance. If, for example, you filed a flight plan which included a STAR, and ATC did not amend it in your clearance, you should plan to fly the entire route, including the STAR, when you are cleared as filed.

An abbreviated clearance only applies to the route segment of the clearance. Besides the statement "*cleared as filed*," it always contains the name of the destination airport or a clearance limit; any applicable DP name, number, and transition; your assigned enroute

Figure 3-38. Shown are examples of abbreviated IFR departure clearances.

altitude, and any additional instructions such as the departure control frequency or transponder code assignment. [Figure 3-38]

An abbreviated clearance contains the name of your destination airport or clearance limit; the assigned enroute altitude; DP information, if appropriate; and it may include a departure frequency or transponder code assignment.

VFR On Top

In some situations, it may be to your advantage to request VFR on top during an IFR flight. This type of clearance does not cancel your IFR flight plan. Instead, it allows you more flexibility with regard to altitude assignments. Basically, a **VFR-on-top clearance** allows you to fly in VFR conditions and at appropriate VFR cruising altitudes of your choice. You may only request this clearance if you are in VFR conditions and below Class A airspace, which begins at 18,000 feet. ATC can not initiate a VFR-on-top clearance. Once ATC approves your request, you must maintain VFR flight conditions at all times. Altitude selection must comply with the VFR cruising altitude rules, which are based on the magnetic course of the aircraft. You may not, however, select an altitude which is less than the minimum enroute altitude prescribed for the route segment. [Figure 3-39]

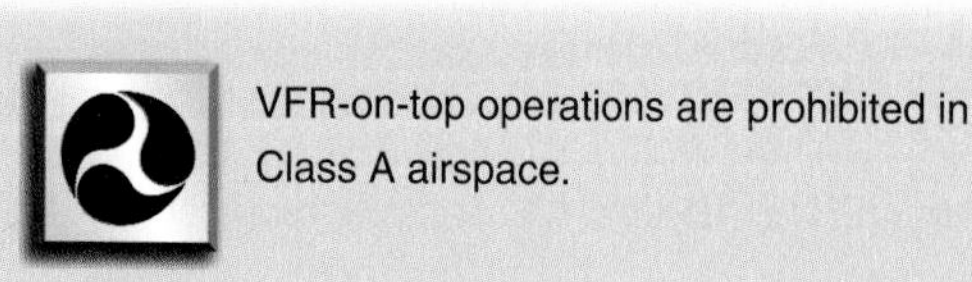

VFR-on-top operations are prohibited in Class A airspace.

A VFR-on-top clearance can only be assigned by ATC if it has been requested by the pilot and conditions are indicated to be suitable for that type of flight.

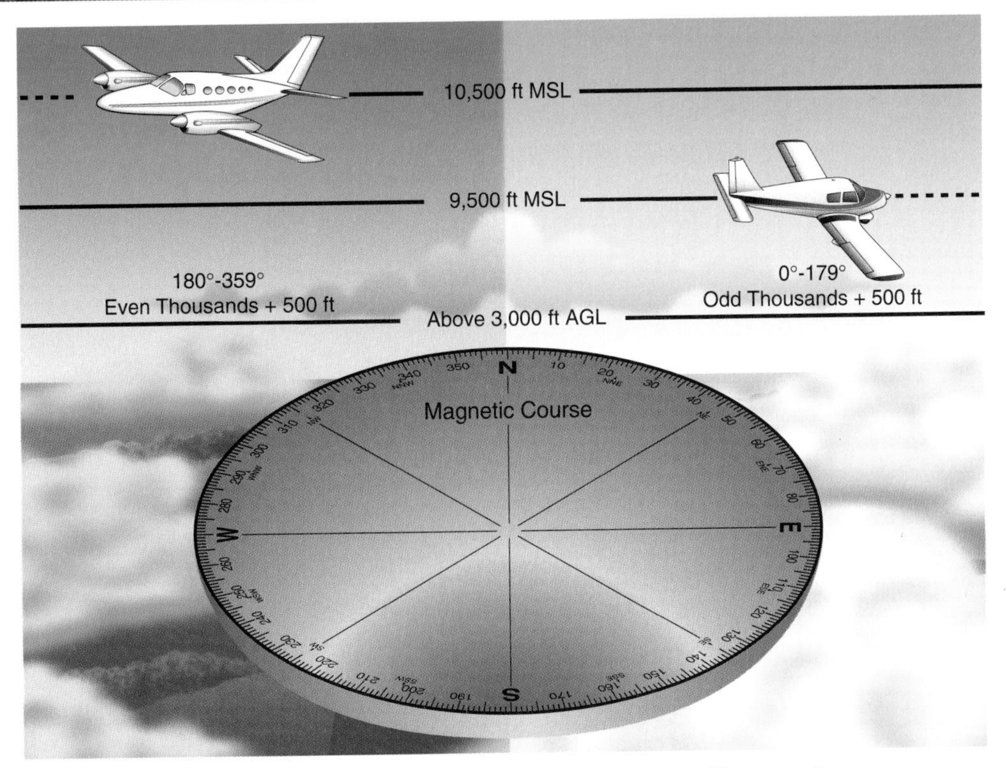

Figure 3-39. The VFR cruising altitude rule does not guarantee traffic separation.

An ATC authorization to maintain VFR on top does not literally restrict you to on-top operations. You may operate VFR on an IFR flight plan when you are above, below, or between layers, or in the clear. This type of clearance simply allows you to change altitude in VFR conditions after advising ATC of the intended altitude changes. Keep in mind, however, that all the rules applicable to instrument flight, such as minimum IFR altitudes, position reporting, radio communication, and adherence to ATC clearances, must still be followed. If at any time VFR conditions cannot be maintained, you must inform ATC and receive a new clearance before you enter IFR conditions.

Both VFR and IFR rules apply to a VFR-on-top flight. However, when flying VFR on top, you are required to comply with the VFR cruising altitude rules and the basic VFR weather minimums. See figure 3-39.

CLIMB TO VFR ON TOP

A variation of the VFR-on-top clearance is a request to **climb to VFR on top**. You would request this type of clearance when departing an airport where you wanted to climb through an area of restricted visibility, such as a haze or fog layer, and then proceed enroute under VFR. If you are departing a controlled airport, it may not be necessary to file a complete IFR flight plan with a request to VFR on top. You can generally make your request directly with ground control or clearance delivery. However, if an ATC facility is not available, your request should be made with a filed flight plan.

When ATC issues a clearance to climb to VFR on top, you should expect a clearance limit and an ATC request to report reaching VFR on top. Also, if you elect to file an IFR flight plan, the term "*VFR on top*" is used instead of an altitude assignment when you receive your clearance. If necessary, ATC may restrict your climb to maintain traffic separation. [Figure 3-40]

A climb-to-VFR-on-top clearance should be requested in order to climb through a cloud layer or an area of reduced visibility and then continue the flight VFR.

Figure 3-40. This represents a typical climb-to-VFR-on-top clearance.

APPROACH CLEARANCES

You should be aware of a few peculiarities with the issuance of an instrument approach clearance. First, if only one approach procedure exists or if you are authorized by ATC to execute the approach procedure of your choice, you are issued a clearance, such as ". . . *cleared for approach.*" If more than one approach procedure is available at the destination airport or if ATC restricts you to a specific approach, the controller specifies, ". . . *cleared for ILS Runway 35 Right approach.*" If you are established on a route or approach segment that has a published minimum altitude, your approach clearance generally will not specify an altitude to maintain. [Figure 3-41]

When the landing will be made on a runway that is not aligned with the approach being flown, the controller may issue a circling approach clearance. In this case, the controller will specify, ". . . *cleared for VOR Runway 17 approach, circle to land Runway 23.*"

A contact or visual approach may be used in lieu of conducting a standard instrument approach procedure.

If conditions permit, you can request a contact approach, which is then authorized by the controller. A contact approach cannot be initiated by ATC. This procedure may be used instead of the published procedure to expedite your arrival, as long as the airport has a standard or special instrument approach procedure, the reported ground visibility is at least one statute mile, and you are able to remain clear of clouds with at least one mile flight visibility throughout the approach. Some advantages of a contact approach are that it usually requires less time than the published instrument procedure, it allows you to retain your IFR clearance, and provides separation for IFR and special VFR traffic. On the other hand, obstruction clearances and VFR traffic avoidance becomes your responsibility.

". . . turn right heading 320°, maintain 2,000 until established on the localizer. Cleared for ILS Runway 36 approach."

Figure 3-41. If you are being radar vectored to the final approach course, your approach clearance should always include an altitude to maintain.

When it is operationally beneficial, ATC may authorize you to conduct a **visual approach** to the airport in lieu of the published approach procedure. A visual approach can be initiated by you or the controller. Before issuing a visual approach clearance, the controller must verify that you have the airport, or a preceding aircraft which you are to follow, in sight. In the event you have the airport in sight but do not see the aircraft you are to follow, ATC may issue the visual approach clearance but will maintain responsibility for aircraft separation. Once you report the aircraft in sight, you assume the responsibilities for your own separation and wake turbulence avoidance.

Keep in mind that the visual approach clearance is issued to expedite the flow of traffic to an airport. It is authorized when the ceiling is reported or expected to be at least 1,000 feet AGL and the visibility at least 3 statute miles. You must remain clear of the clouds at all times while conducting a visual approach. At a controlled airport, you may be cleared to fly a visual approach to one runway while others are conducting VFR or IFR approaches to another parallel, intersecting, or converging runway. Also, when radar service is provided, it is automatically terminated when the controller advises you to change to the tower or advisory frequency.

During a visual approach, radar service is terminated automatically when ATC instructs the pilot to contact the tower.

The main differences between a visual approach and a contact approach are: a pilot must request a contact approach, while a visual approach may be assigned by ATC or requested by the pilot; and a contact approach may be approved with 1 mile visibility if the flight can remain clear of clouds, while a visual approach requires the pilot to have the airport in sight, or a preceding aircraft to be followed, and the ceiling must be at least 1,000 feet AGL with at least 3 statute miles visibility.

Without prior pilot request, ATC may issue DPs, STARs, and visual approach clearances.

VFR RESTRICTIONS TO AN IFR CLEARANCE

During the issuance of an ATC clearance, the controller may direct you to "***maintain VFR conditions***." However, this restriction is only issued when you request it. For example, a VFR restriction is issued when you request a VFR climb or descent. This is the case when you are departing or arriving in VFR conditions and you wish to avoid a complicated, time-consuming departure or arrival procedure. In the case of a VFR departure, for example, you may request a VFR climb, which would allow you to avoid the departure procedure and climb on course. However, you should fully understand that while operating on an IFR flight plan with a VFR restriction, you must remain in VFR conditions and maintain your own traffic separation during the VFR portion of your clearance.

VFR restrictions to an IFR flight can only be initiated by the pilot.

COMPOSITE FLIGHT PLAN

A **composite flight plan** is a request to operate IFR on one portion of a flight and VFR for another portion. When you file a composite flight plan, the route should include all normal IFR route segments, as well as the clearance limit fix where you anticipate the IFR portion of the flight will terminate. You may file a composite flight plan any time a portion of the flight will be flown under VFR. For example, if the first portion of your flight is VFR, you should activate your VFR flight plan with the nearest FSS after departure. As you near the point where you planned to activate your IFR flight plan, you must contact the nearest

Figure 3-42. When you file a composite flight plan, check both the VFR and IFR boxes in block 1 of the flight plan.

FSS, close your VFR flight plan, and request your IFR clearance. Keep in mind that you must remain in VFR conditions until you receive your IFR clearance. In the event your flight dictates IFR for the first portion and VFR for the latter portion, you will normally be cleared IFR to the point where the change is proposed. As you near this point and are operating in VFR conditions, you should cancel your IFR flight plan and contact the nearest FSS to activate your VFR flight plan. If you want to continue past your clearance limit on an IFR flight plan, you must contact ATC five minutes before you reach the limit to request a further clearance. Should you reach your clearance limit without receiving a clearance to continue, you are expected to enter a holding pattern and wait for the clearance. [Figure 3-42]

For a composite flight plan, check both the VFR and IFR boxes under type of flight. See figure 3-42.

You may file a composite flight plan anytime a portion of your flight will be VFR. If the VFR portion is first, contact the nearest FSS prior to transitioning to the IFR portion, close the VFR portion, and request an ATC clearance.

When filing a composite flight plan where the first portion of the flight is IFR, include all normal IFR route segments and the clearance limit fix where you anticipate the IFR portion will end.

Tower Enroute Control Clearance

Tower enroute control (TEC) is an alternative IFR procedure which permits you to fly short, low altitude routes between terminal areas. TEC routes are published for certain portions of the United States in the *Airport/Facility Directory* and in the Enroute Section of the *Jeppesen Airway Manual Services.* Essentially, a flight is transferred from departure control at one airport to successive approach control facilities. In most cases, TEC routes are generally intended for nonturbojet aircraft operating below 10,000 feet MSL with a flight duration of normally less than two hours. If you want to fly a TEC route, include the acronym TEC in the remarks section of the flight plan.

Departure Restrictions

In order to separate IFR departure traffic from other traffic in the area, or to restrict or regulate the departure flow of traffic, ATC may place time restrictions on your clearance.

Some of these restrictions include a release time, a hold for release time, and a clearance void time. A **release time** specifies the earliest time you may depart. This type of restriction is generally a result of traffic saturation, weather, or ATC departure management procedures. Occasionally, you may be advised to "***hold for release***." When ATC issues a hold for release, you may not depart until you receive a release time or you are given additional instructions. Generally, the additional instructions will include the expected release time and the length of the departure delay. Once the local conditions and traffic permit, you will be released for departure.

If you are operating at an airport not served by an operating control tower, ATC may find it necessary to issue a **clearance void time** in conjunction with your IFR departure clearance. The wording, "*clearance void if not off by . . . ,*" indicates that ATC expects you to be airborne by a certain time. A common situation for the issuance of a void time is when inbound traffic is expected to arrive at approximately the same time as your departure. In this case, the required traffic separation cannot be achieved without restricting your departure. In the event you do not depart by the void time, you must advise ATC of your intentions as soon as possible, but no later than 30 minutes after the void time. Your failure to contact ATC within the allotted time after the clearance void time could result in your aircraft being considered overdue, and search and rescue procedures initiated.

When departing from an airport not served by a control tower, the issuance of a clearance containing a void time indicates that you must advise ATC as soon as possible, but no later than 30 minutes, of your intentions if not off by the void time.

CLEARANCE READBACK

Although there is no requirement that an ATC clearance be read back, you are expected to read back those parts of any clearance which contain altitude assignments, radar vectors, or any instructions requiring verification. Additionally, controllers may request that you read back a clearance when the complexity of the clearance or any other factors indicate a need.

You should read back that portion of a clearance containing altitude assignments, radar vectors, or any instruction requiring clarification.

As the pilot in command, you should read back the clearance if you feel the need for confirmation. Even though it is not specifically stated, it is generally expected that you will read back the initial enroute clearance you receive from clearance delivery, ground control, or a flight service station. When you receive your IFR clearance, the phraseology may be slightly different, depending upon the facility issuing the clearance. The terms "*ATC clears,*" "*ATC advises,*" or "*ATC requests*" are used only when a facility other than air traffic control is used to relay information originated by ATC.

CLEARANCE SHORTHAND

To operate efficiently in the IFR environment, you must be able to copy and thoroughly understand clearances. Copying IFR clearances becomes easy with practice. Although numerous changes have been made since the acceptance of the first shorthand used by early instrument pilots, many of the original symbols have been retained. The shorthand symbols in this chapter are considered by the FAA and experienced instrument pilots to be the best. [Figure 3-43]

SHORTHAND SYMBOLS

Words and Phrases	Shorthand
Above	ABV
Above (*"Above six thousand"*)	$\underline{60}$
Advise	ADV
After (passing)	< or AFT
Airport	A
(Alternate Instructions)	( )
Altitude 6,000 — 17,000	60 — 170
And	&
Approach	AP
Final	F
Instrument Landing System	ILS
Localizer Back Course	LBC
Localizer Only	LOC
Nondirectional Beacon	NDB
Precision Approach Radar	PAR
Surveillance Radar	ASR
VOR	VOR or ⊙
Approach Control	APC
As Filed	AF
As Published	APUB
At	@
(ATC) Advises	CA
(ATC) Clears or Cleared	C
(ATC) Requests	CR
Bearing	BRG or BR
Before (reaching, passing)	>
Below	BLO
Below (*"Below six thousand"*)	$\overline{60}$
Center	CTR
Cleared or (ATC) Clears	C
Cleared As Filed	CAF
Cleared to Land	CL
Climb (to)	↑
Contact	CT
Contact Approach	CAP
Course	CRS
Cross (crossing)	X
Cruise	→
Depart (departure)	DP
Departure Control	DPC
Descend (to)	↓
Direct	DR or D→
DME Fix (15 DME mile fix)	Ⓓ or D15
Each	EA
Eastbound	EB
Established	ESTB
Expect	EX

Words and Phrases	Shorthand
Expect Further Clearance (time or location)	EFC
Flight Level	FL
Flight Planned Route	FPR
For Further Clearance	FFC
From	FR or FRM
Heading	HDG
Hold (direction) (*"Hold west"*)	H-W
Holding Pattern	⬭
Inbound	IB
Intercept	△̸ or INT
Intersection	△ or XN
Landing	LDG
Maintain (or magnetic)	M
Middle Marker	MM
Compass Locator at Middle Marker	LMM
No (or not) Later Than	NLT
On Course	OC
Outbound	OB
Outer Marker	OM
Compass Locator at Outer Marker	LOM
Over (ident over the line)	$\underline{OKC}$
Procedure Turn	PT
Radar Contact	RCT
Radar Vector	RV
Radial (092 radial)	092R
Report	R
Report Leaving	RL
Report On Course	R-CRS
Report Over	RO
Report Passing	RP
Report Reaching	RR
Report Starting Procedure Turn	RSPT
Reverse Course	RC
Runway (number)	RWY 26
Squawk	SQ
Standby	STBY
Takeoff	TO
Tower	Z OR TWR
Track	TR
Turn Left, or Turn Left After Departure	↰ or LT
Turn Right, or Turn Right After Departure	↱ or RT
Until	til or U
Until Advised (by)	UA
Until Further Advised	UFA
VFR On Top	$\underline{VFR}$
Victor (airway number)	$\overline{V294}$
VOR	⊙
VORTAC	Ⓣ

Figure 3-43. The purpose of these shorthand symbols is to allow you to copy IFR clearances as fast as they are read. The more clearances you copy, the easier the task becomes.

CAUSES OF COMMUNICATION BREAKDOWN

Controllers are asking, "*what is going on up there*?" as they complain about pilot errors in clearance readbacks, and pilots are asking, "*what is going on down there*?" as they are being informed that they busted the altitudes which they had dutifully read back in their clearance.

Based on studies conducted by the Aviation Safety Reporting System (ASRS), there seems to be four major problem areas causing readback errors.

1. **Similar aircraft call signs** — There could be many aircraft with similar call signs operating on the same frequency, at the same time and in the same airspace

2. **Only one pilot listening on ATC frequency** — In two-pilot operations, both pilots should be listening to clearances. If one pilot is getting the ATIS report or talking to another facility, backup monitoring is lost.

3. **Slips of mind and tongue** — The typical human errors in this category include being advised of traffic at another flight level or altitude and accepting the information as clearance to that altitude; confusing "*one zero*" and "*one one thousand*;" confusing left and right in parallel runways; and the interpretation of "*maintain two five zero*" as an altitude instead of an airspeed limitation.

4. **Mind-set, preprogrammed for . . ., and expectancy factors** — The airmen who request "*higher*" or "*lower*" tend to be spring-loaded to hear what they want to hear upon receipt of a blurred call sign transmission.

But why didn't the controller catch the pilot error in the readback?

The main problem here seems to be overload or working too many aircraft. At busy airports, controllers can have a rush of departures/arrivals at the same time they are working land-lines and phones coordinating hand-offs of traffic. This is further complicated by stepped on transmissions where only partial clearances or readbacks are heard.

So what can be done? When pilots read back a clearance, they are asking a question: "*Did we get it right*?" Unfortunately, ASRS reports reveal that ATC is not always listening. Contrary to many pilots' assumptions, controller silence is not confirmation of a readback's correctness, especially during peak traffic periods.

As a pilot, you can take several precautions to reduce the likelihood of readback/hearback failures:

- Ask for verification of any ATC instruction about which there is doubt. Don't read back a best guess at a clearance, expecting ATC to catch any mistakes.

- Be aware that being off ATC frequency while picking up the ATIS or while talking to another facility is a potential communication trap for a two-pilot crew.

- Use standard communication procedures in reading back clearances. "*Okay*," "*Roger*" and microphone clicks are poor substitutes for readbacks.

- Avoid interpreting altitudes mentioned for purposes other than a clearance as an instruction to proceed to that altitude.

- Be aware of similar sounding call signs, and make sure the clearance is for you.

The preceding discussion was compiled from an ASRS *Directline* article by Bill Monan.

Proficiency in copying clearances is the result of practice and knowing clearance terminology. Figure 3-44 shows samples of clearances as they might be issued. Each is followed by its appropriate clearance shorthand.

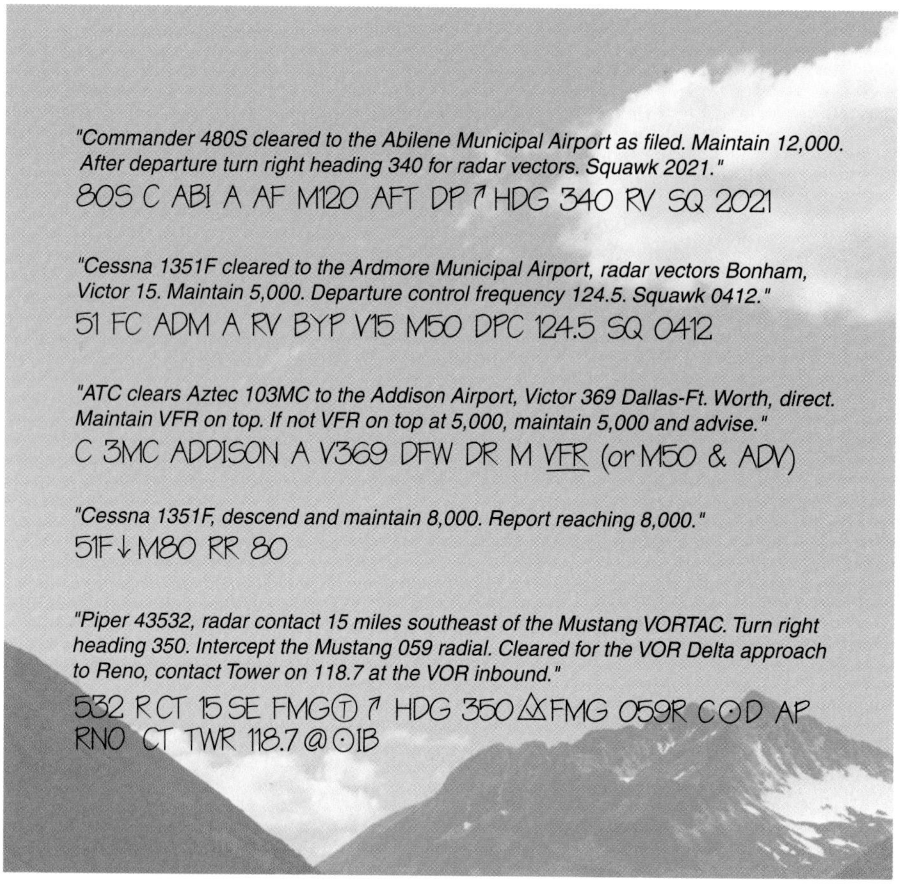

"Commander 480S cleared to the Abilene Municipal Airport as filed. Maintain 12,000. After departure turn right heading 340 for radar vectors. Squawk 2021."

80S C ABI A AF M120 AFT DP ↱ HDG 340 RV SQ 2021

"Cessna 1351F cleared to the Ardmore Municipal Airport, radar vectors Bonham, Victor 15. Maintain 5,000. Departure control frequency 124.5. Squawk 0412."

51 FC ADM A RV BYP V15 M50 DPC 124.5 SQ 0412

"ATC clears Aztec 103MC to the Addison Airport, Victor 369 Dallas-Ft. Worth, direct. Maintain VFR on top. If not VFR on top at 5,000, maintain 5,000 and advise."

C 3MC ADDISON A V369 DFW DR M VFR (or M50 & ADV)

"Cessna 1351F, descend and maintain 8,000. Report reaching 8,000."

51F ↓ M80 RR 80

"Piper 43532, radar contact 15 miles southeast of the Mustang VORTAC. Turn right heading 350. Intercept the Mustang 059 radial. Cleared for the VOR Delta approach to Reno, contact Tower on 118.7 at the VOR inbound."

532 RCT 15 SE FMGⓉ ↱ HDG 350 △ FMG 059R C⊙D AP RNO CT TWR 118.7 @ ⊙IB

Figure 3-44. When using clearance shorthand, the most important consideration is not what symbols you use, but that you can still interpret them after a period of time.

SUMMARY CHECKLIST

✓ An ATC clearance is an authorization for you to proceed under a specified set of conditions within controlled airspace.

✓ You may not deviate from an ATC clearance unless you experience an emergency or the clearance will cause you to violate a rule or regulation.

✓ If you deviate from an ATC clearance, you must notify ATC as soon as possible. If you are given priority over other aircraft you may be requested to submit a written report to the manager of the ATC facility within 48 hours.

✓ Anytime you are in VFR conditions, it is your responsibility to see and avoid all other traffic, regardless of the type of flight plan you are on.

✓ An IFR flight plan is required before flying into Class A airspace or any other controlled airspace when the weather is below VFR minimums.

✓ You must receive an ATC clearance before entering Class A or B airspace regardless of the weather and in Class C, D, and E airspace when the weather is below VFR minimums.

✓ The elements of an ATC clearance are: aircraft identification, clearance limit, departure procedure, route of flight, altitudes/flight levels in the order to be flown, holding instructions, any special information, and frequency and transponder code information.

✓ A cruise clearance authorizes you to operate at any altitude from the minimum IFR altitude up to and including the altitude specified in the clearance without reporting the change in altitude to ATC.

✓ An abbreviated clearance can be issued when your route of flight has not changed substantially from that filed in your flight plan. An abbreviated clearance always contains the words "*cleared as filed*" as well as the name of the destination airport or clearance limit; any applicable DP name, number and transition; the assigned enroute altitude; and any additional instructions such as departure control frequency or transponder code assignment.

✓ A VFR-on-top clearance allows you to fly in VFR conditions and at the appropriate VFR cruising altitudes of your choice. VFR on top is prohibited in Class A airspace.

✓ A contact approach must be initiated by the pilot, it cannot be initiated by ATC.

✓ In order to fly a contact approach, the reported ground visibility must be at least one statute mile, and you must be able to remain clear of clouds with at least one statute mile flight visibility.

✓ A visual approach may be initiated by the controller or the pilot when the ceiling is at least 1,000 feet and the visibility is at least 3 statute miles and the pilot has the airport or the aircraft to follow in sight.

✓ VFR restrictions to an IFR clearance can only be initiated by the pilot.

✓ A composite flight plan should be filed when you wish to operate IFR on one portion of a flight and VFR on another portion.

✓ A tower enroute control clearance (TEC) is intended to be used by nonturbojet aircraft at altitudes less than 10,000 feet MSL where the duration of the flight is less than 2 hours.

✓ Departure restrictions, such as a release time, hold for release time, and a clearance void time may be imposed to separate IFR departure traffic from other traffic in the area or to regulate the flow of IFR traffic.

✓ You should read back those parts of a clearance which contain altitude assignments, radar vectors, or any instructions requiring verification.

✓ Shorthand should be used to quickly copy IFR clearances. The type of shorthand you use is not as important as whether you can read the clearance at a later time.

KEY TERMS

Expect Further Clearance (EFC)

Short-Range Clearance

Cruise Clearance

Abbreviated IFR Departure Clearance

Cleared As Filed

VFR-On-Top Clearance

Climb To VFR On Top

Circling Approach Clearance

Contact Approach

Visual Approach

Maintain VFR Conditions

Composite Flight Plan

Tower Enroute Control (TEC)

Release Time

Hold For Release

Clearance Void Time

QUESTIONS

1. An ATC clearance is an authorization to proceed under specified conditions within what type of airspace?

2. In the event you deviate from a clearance due to an emergency, when must you notify ATC?

   A. As soon as possible
   B. Immediately after landing
   C. Within 24 hours after landing

3. If ATC provides you with priority service because of an emergency, you may be required to submit a written report to the manager of that ATC facility within what time frame?

   A. Immediately after landing
   B. Within 24 hours after landing
   C. Within 48 hours after landing

4. True/False. You are always required to see and avoid other aircraft while operating in VFR conditions, even on an IFR flight plan.

5. True/False. To operate under IFR within controlled airspace, you must file an IFR flight plan and obtain an ATC clearance.

6. If you depart from an airport in Class G airspace in less than VFR conditions, when must you file an IFR flight plan and receive a clearance?

   A. Before takeoff
   B. Before entering IFR conditions
   C. Before entering controlled airspace

7. Regardless of weather conditions, you are required to file an IFR flight plan before you can legally fly within what class of airspace?

   A. Class A
   B. Class B
   C. all controlled airspace

For questions 8 through 15, use the associated letter of the IFR departure clearance items to arrange them in the correct sequence.

8. __________
9. __________
10. __________
11. __________
12. __________
13. __________
14. __________
15. __________

A. Aircraft identification
B. Altitudes, in the order to be flown
C. Any special instructions
D. Clearance limit
E. Departure procedure
F. Frequency and transponder code information
G. Holding instructions
H. Route of flight

16. True/False. ATC will not issue an instrument departure procedure (DP) unless you request it.

17. True/False. When ATC issues a clearance with significant changes to your requested routing, you can expect an abbreviated clearance.

18. What is the significance of the following clearance: "*...cruise 5,000*"?

   A. You can fly at any altitude from 5,000 feet MSL up to the base of Class A airspace.
   B. You can conduct your flight at any altitude from the minimum IFR altitude up to and including 5,000 feet MSL.
   C. You can fly at any altitude from the minimum IFR altitude up to and including 5,000 feet MSL, but you must report leaving each altitude to ATC.

19. Before a contact approach is approved, the reported ground visibility must be at least what?

    A. One statute mile
    B. One nautical mile
    C. Three statute miles

20. Select the true statement when cleared to fly VFR on top on an IFR flight.

    A. Your IFR flight plan is automatically cancelled.
    B. You are required to comply with VFR cruising altitude rules.
    C. You may fly below the minimum enroute altitude prescribed for the route segment you are flying as long as you remain VFR.

21. When you are flying on a composite flight plan, who should you contact to cancel the VFR portion and request a clearance to proceed under IFR?

22. When ATC issues a clearance void time in conjunction with a departure clearance, in what time frame must you depart?

    A. No later than the clearance void time
    B. No later than 30 minutes after the clearance void time
    C. 30 minutes or later after the clearance void time

23. True/False. You should read back all portions of an ATC clearance that contain specific instructions, such as altitude assignments or radar vectors.

24. Decipher the following clearance. (CYS is Cheyenne Airport)

    C CYS A AF ↑M 80 DPC 120.9 SQ5417

25. Rewrite the following clearance using shorthand symbols.

    "Gulfstream 37R, cleared to the Dallas Love Airport (DAL) direct Bonham (BYP) VORTAC. Descend and maintain 12,000, report passing 15,000. Depart Bonham VORTAC heading 210 for vectors to Runway 31 Right ILS final approach course. Landing Runway 31 Right."

PART II
Instrument Charts and Procedures

I didn't develop the charts to get famous. I did it to stay alive.

— Elrey B. Jeppesen

PART II

When Elrey B. Jeppesen began recording aeronautical information in the early 1930s, he did it simply as a means of survival. While it may not have been his original intention, Captain Jepp also managed to revolutionize air travel by making it a more reliable and safer form of transportation. As an instrument pilot, you will use charts inspired by Captain Jepp's sketches to routinely fly in conditions which ordinarily would keep you on the ground. Of course, reading and understanding the charts is only half of the story. To be a competent instrument pilot, you must also know how to fly the procedures depicted on the charts. To help you interpret the charts and translate the instructions into action, Part II breaks instrument flight into several broad phases. The departure phase, which takes you from the airport to the enroute structure is covered in Chapter 4. After leaving the departure phase, you begin the enroute portion of your flight, which is discussed in Chapter 5. As you near your destination, you enter the arrival phase which may begin with a published procedure similar to those shown in Chapter 6. In most cases, your IFR flights will end with an instrument approach to the runway. To prepare you for these operations, Chapter 7 covers approach charts and general approach procedures while Chapter 8 discusses specific types of approaches and the associated procedures.

CHAPTER 4

DEPARTURE

Instrument/Commercial
Part II, Chapter 4 — Departure

SECTION A
DEPARTURE CHARTS

Departure charts are published to help simplify complex clearance delivery procedures, reduce frequency congestion, ensure obstacle clearance, and control the flow of traffic around an airport. In some cases, they help reduce fuel consumption, and may include noise abatement procedures. This section analyzes instrument departure procedure charts to help you become better acquainted with the symbols and information they contain. The practical application of flying these procedures is covered in the next section.

Prior to any discussion of chart format and published instrument procedures, you should understand that the chart examples depicted in this manual are representative of the current published formats, and that they may not be the most current editions and should not be used for navigation. Prior to any flight under IFR, you must obtain the most current charts for your area of operation.

OBTAINING CHARTS

Instrument charts are published by Jeppesen Sanderson, Inc. and the National Ocean Service (NOS) which is part of the National Oceanic and Atmospheric Administration (a division of the United States Department of Commerce). While the NOS primarily publishes instrument charts for the United States, its territories, and possessions, Jeppesen has a broader scope, producing instrument charts on a worldwide basis. [Figure 4-1]

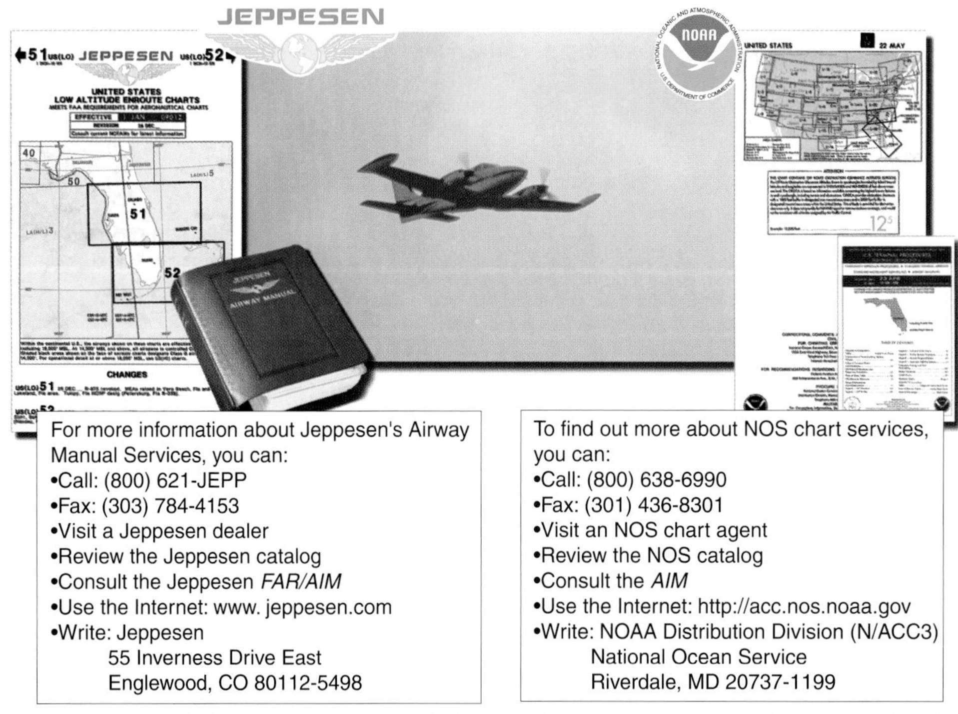

Figure 4-1. Many pilots find a chart subscription service a convenient way to ensure that they always have the most up-to-date charts. You can procure charts from Jeppesen or the NOS either by subscription or as a one time purchase.

DEPARTURE STANDARDS

IFR departures are designed according to the criteria established in the **U.S. Standard for Terminal Instrument Procedures (TERPs)**. In part, TERPs sets standards for a specific clearance from obstacles at a given distance from the runway based on an aircraft climbing at least 200 feet per nautical mile. [Figure 4-2] If obstacles penetrate a slope of 152 feet per nautical mile, beginning no higher than 35 feet above the departure end of the runway, a minimum ceiling and/or climb gradient may be required. In some cases, the aircraft may have to be maneuvered to avoid obstacles. Some departures may require a combination of these restrictions to ensure a safe departure.

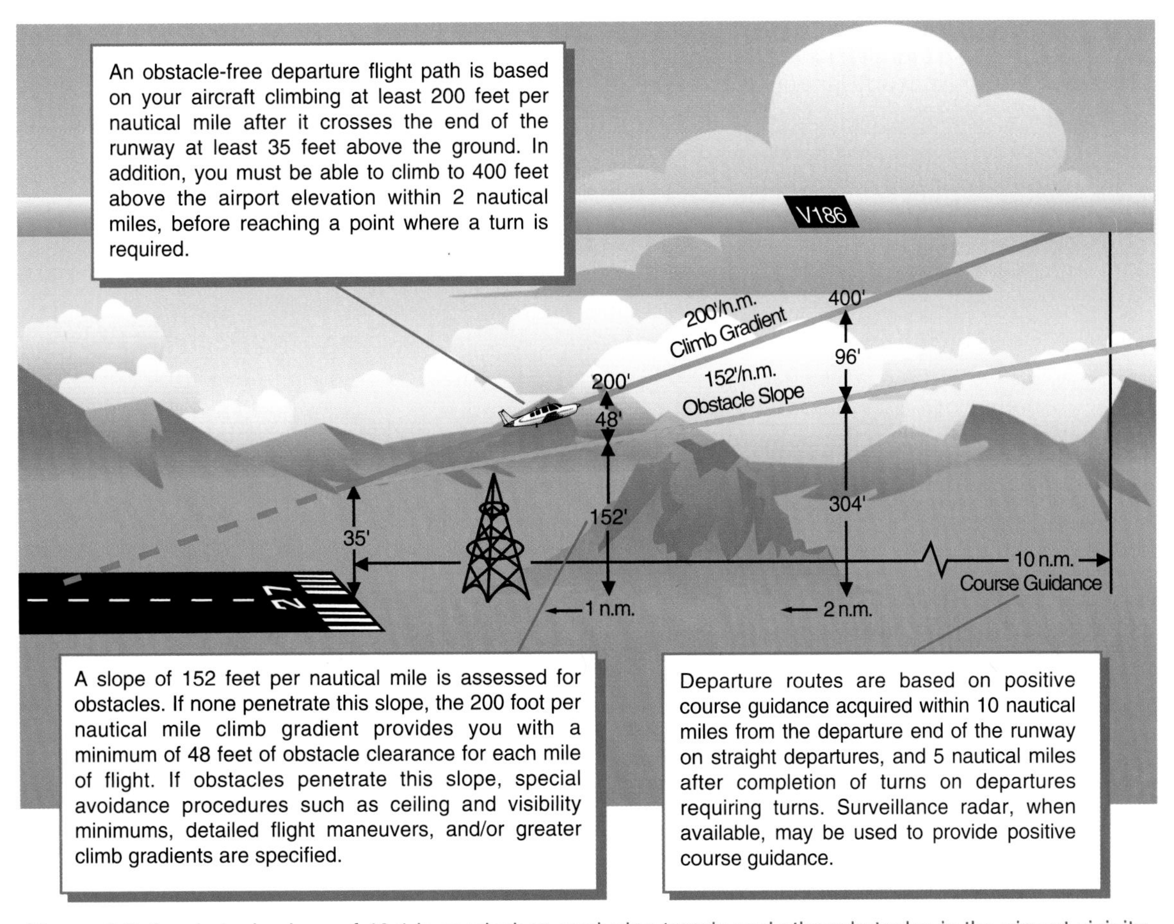

Figure 4-2. An obstacle slope of 40:1 is used when analyzing terrain and other obstacles in the airport vicinity.

INSTRUMENT DEPARTURE PROCEDURES

The **instrument departure procedure (DP)** is used after takeoff to provide a transition between the airport and the enroute structure. You can find DP charts grouped with the associated airport approach charts in both Jeppesen and NOS coverages.

To illustrate how DPs can be used to simplify a complex clearance and reduce frequency congestion, consider the following departure clearance issued to a pilot who was about to depart from Terps, California. *"Cessna 1732G, cleared to the Oakville Airport via the Terps 170° radial Terps, Victor 195 Thor, Thor 285° radial and Barnsdall 103° radial to Barnsdall, then as filed. Maintain 8,000 feet. After departure, maintain runway heading*

until reaching 3,000 feet, then turn right to 120° and intercept the Terps 170° radial. Cross Terps at or above 5,000 feet."

Now consider how this same clearance is issued when a DP exists for this departure. *"Cessna 1732G, cleared to Oakville Airport, Thor One Departure, Barnsdall Transition, then as filed. Maintain 8,000 feet."* This brief transmission conveys the same information as the longer example.

DPs require specific aircraft performance to guarantee obstacle clearance. When you are issued a clearance that contains a DP, you are obligated to comply with the provisions listed for the DP and must ensure your aircraft is capable of achieving the performance requirements. For example, when necessary for obstruction clearance, DPs specify a climb gradient in feet per nautical mile. To convert this figure to rate of climb in feet per minute, you can use the table provided on the Jeppesen DP or the chart provided in the front of the NOS *Terminal Procedures Publication.* [Figure 4-3] Alternatively, you can determine your climb rate mathmatically by dividing the groundspeed by 60 and multiplying by the climb gradient. For example, if a DP requires an altitude gain of 200 feet per nautical mile and your planned groundspeed is 150 knots, the required rate of climb is 500 feet per minute ((150 ÷ 60) 200 = 500).

You also should recognize that some DPs require you to maintain a climb gradient to altitudes in excess of 10,000 feet. Therefore, the calculated continuous climb performance must be valid to the altitude required in the DP. For example, if a DP ends along a high altitude, or jet route, most light aircraft will have difficulty complying with it, since jet routes have minimum enroute altitudes (MEAs) of at least 18,000 feet. This could be a significant factor if you encounter adverse weather conditions, such as high density altitude, turbulence, and/or icing during the climb to cruising altitude. Another consideration is the average winds aloft during your departure. Since climb gradient is based on groundspeed, a tailwind requires an even greater rate of climb. Your responsibility as pilot in

Minimum climb gradients are given in feet per nautical mile and must be converted to feet per minute for use during departure.

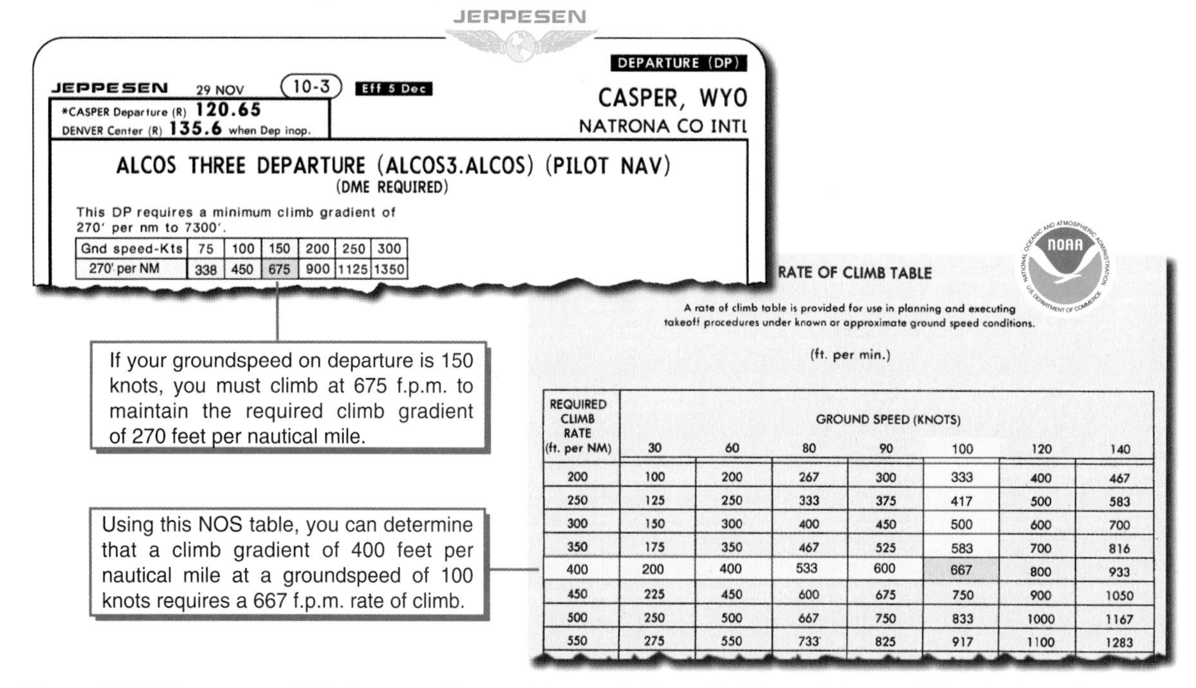

JEPPESEN

JEPPESEN 29 NOV (10-3) Eff 5 Dec

DEPARTURE (DP)

*CASPER Departure (R) **120.65**
DENVER Center (R) **135.6** when Dep inop.

CASPER, WYO
NATRONA CO INTL

ALCOS THREE DEPARTURE (ALCOS3.ALCOS) (PILOT NAV)
(DME REQUIRED)

This DP requires a minimum climb gradient of 270' per nm to 7300'.

Gnd speed-Kts	75	100	150	200	250	300
270' per NM	338	450	675	900	1125	1350

RATE OF CLIMB TABLE

A rate of climb table is provided for use in planning and executing takeoff procedures under known or approximate ground speed conditions.

(ft. per min.)

REQUIRED CLIMB RATE (ft. per NM)	GROUND SPEED (KNOTS) 30	60	80	90	100	120	140
200	100	200	267	300	333	400	467
250	125	250	333	375	417	500	583
300	150	300	400	450	500	600	700
350	175	350	467	525	583	700	816
400	200	400	533	600	667	800	933
450	225	450	600	675	750	900	1050
500	250	500	667	750	833	1000	1167
550	275	550	733	825	917	1100	1283

Figure 4-3. Tables are available for converting a minimum climb gradient to a minimum climb rate, based on your groundspeed.

command is to review each DP, make sure your aircraft can comply with the procedures, and refuse any DP that is beyond the limits of your aircraft.

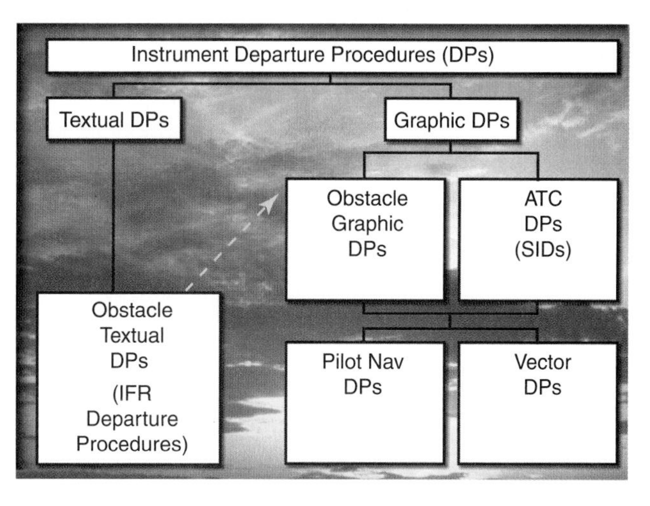

DPs can be either textual, or graphic. Textual procedures, referred to as IFR departure procedures outside the U.S., exist only for obstacle clearance and can normally be found at the bottom of Jeppesen airport charts. If ATC does not assign a graphic DP or other instructions, you can use the textual DP for obstacle clearance to depart the airport, and in some cases, to transition to the enroute structure.

The FAA is converting some complex textual departure procedures into graphic procedures to improve situational awareness. Graphic procedures are published as separate charts, and can be filed as part of a flight plan. When a graphic departure procedure is created only for obstacle clearance, it is labeled obstacle, and may contain the statement, "If not assigned a departure procedure by ATC, this procedure may be flown to provide obstacle clearance."

ATC departure procedures, referred to as standard instrument departures, or SIDs, outside the U.S., do not have the obstacle label. These procedures also provide obstacle clearance, but are established primarily to simplify pilot/controller communications, expedite the flow of traffic, and to help with noise abatement. Due to the dynamic nature of instrument procedures and charts, you may see changes to DPs in the future.

Clearances for instrument departure procedures are issued at the option of ATC. If you do not have the graphic or textual information for the DP with you, you should not accept the clearance. If you do not want to use a DP, you should indicate "NO DP" in the remarks section of your flight plan.

Obstacle graphic and ATC DPs are one of two types, either pilot nav, or vector. A pilot nav DP is designed to allow you to navigate along a specified route with minimal ATC communication. A vector DP is established where ATC provides radar navigation guidance.

PILOT NAV DP

A **pilot nav DP** usually contains an initial set of instructions that apply to all aircraft, followed by one or more **transition** routes that require you to navigate to the appropriate fix within the enroute structure. Many pilot nav DPs include a radar vector segment that helps you join the DP. The symbols used on both Jeppesen and NOS DPs are very similar to the symbols found on the respective enroute, area, and approach charts. Both chart publishers include a textual description of the initial takeoff and transition procedures, and a graphic, or plan view, of the routing. In some cases, a textual description may not be provided for a simple transition. [Figure 4-4]

If you intend to use only the basic portion of the DP shown in figure 4-4 that ends at the DAWNN Intersection, you should list the computer identification code, DAWNN1.DAWNN, as the first part of your routing on your IFR flight plan. This code lets ATC know that you intend to use the DP to get from the airport to DAWNN Intersection. Following the DP code, you should list the remainder of the route from DAWNN. Your ATC clearance might sound like this: *"Cessna 1732G, cleared to Athens Airport, Dawnn One Departure, then as filed. Maintain 9,000."*

If your planned route allows you to follow one of the transitions, it is usually to your benefit to file for the transition. For example, suppose your route of flight takes you over the Louisville VORTAC. In this case, you can include the Louisville Transition in your flight plan. When you file for the Louisville transition, the first part of your routing should list the computer identification code, DAWNN1.IIU. This tells ATC you plan to fly the DAWNN ONE DEPARTURE (DAWNN1) and the Louisville Transition (.IIU). Listing the code exactly as it appears on the DP helps ATC enter it into the computer and reduces the time required to process your flight plan. The remainder of your route, starting at Louisville, is entered following this code. Your clearance from ATC might sound like this: *"Cessna 1732G, cleared to Athens Airport, Dawnn One Departure, Louisville Transition, then as filed. Maintain 9,000."*

Using the computer identification code for a transition in your flight plan informs ATC you intend to fly both the DP and the appropriate transition.

When you file for a DP, be sure to use the appropriate computer identification code in your flight plan.

At times, you may be instructed to maintain runway heading during a DP. In these cases, you are expected to maintain the heading that corresponds with the extended runway centerline, and not apply a drift correction. For example, if the actual magnetic value of Runway 4 centerline is 044°, you should maintain a heading of 044°.

VECTOR DP

During a **vector DP**, ATC provides radar vectors that start just after takeoff and continue until you reach your filed or assigned route or one of the fixes shown on the chart. The vector DP chart is similar to the pilot nav DP, except for the absence of departure routes and transitions. The chart usually contains an initial set of instructions, such as a heading to fly and an altitude for initial climb. When ATC establishes radar contact, they

When special lost communication procedures are necessary for a DP, they are included on the chart.

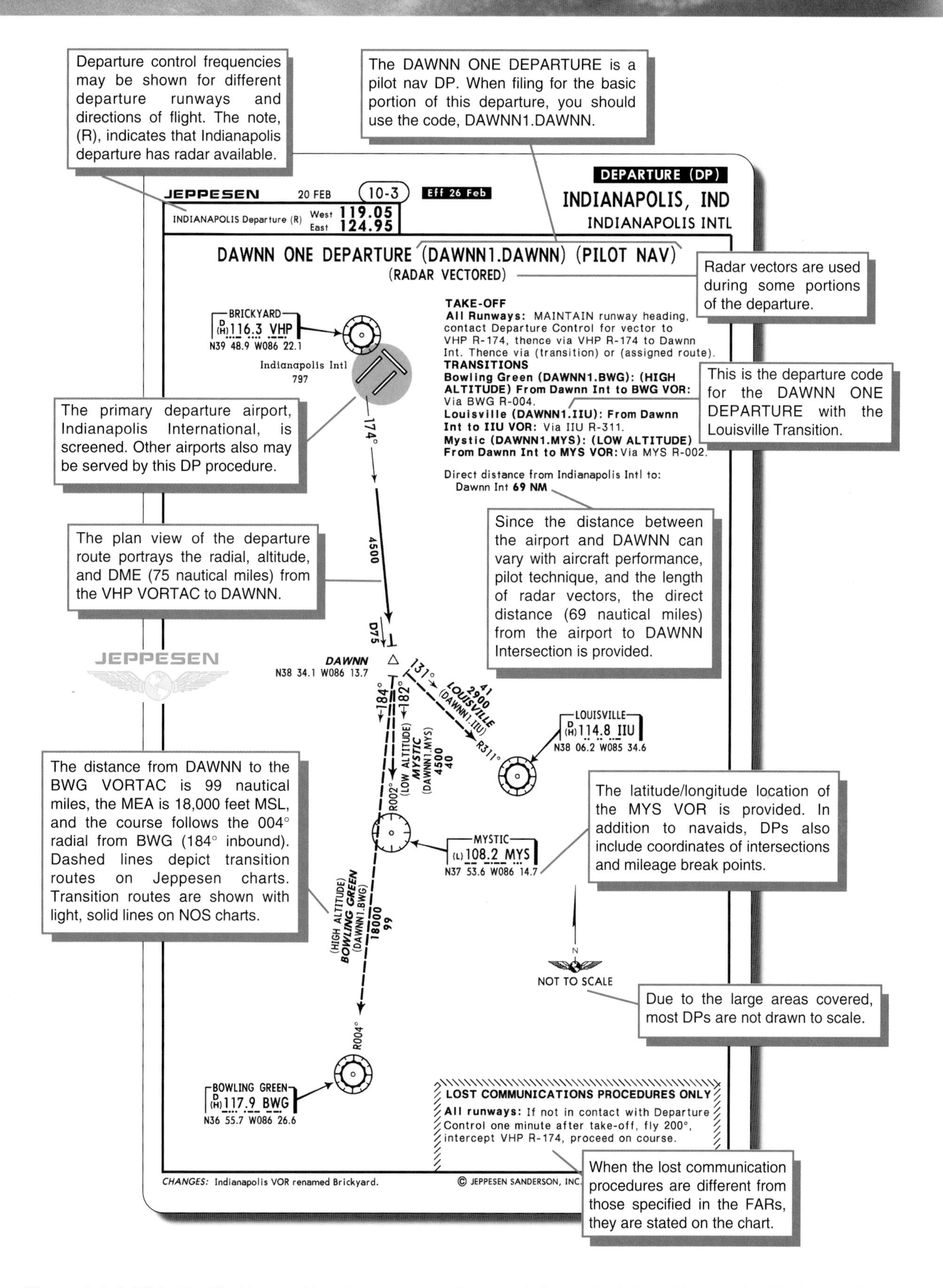

Figure 4-4. A DP is identified by an abbreviated name and numeral, then a dot followed by the identifier/name of the exit or transition fix. When a significant change in the procedure occurs, the number increases by one. After 9, the DP number reverts to 1.

provide vectors to help you reach one of several fixes portrayed on the chart. The textual description provides initial procedures to follow after takeoff as well as nonstandard lost communication procedures. [Figure 4-5]

On NOS DPs, the applicable NOS enroute charts are listed below the navaid information boxes.

As with pilot nav DPs, when obstacle clearance cannot be guaranteed by the standard required climb gradient of 200 feet per nautical mile on a vector DP, a minimum climb gradient may be specified. Since climb gradients are given in feet per nautical mile, they must be converted to feet per minute to be usable during departure. Jeppesen provides a conversion table on each DP chart that has a minimum climb gradient. NOS includes a conversion table that is located in the front portion of the *Terminal Procedures Publication.*

The Wylyy One

As aircraft and avionics technology improves, DP charts are enhanced to accommodate these advancements, as shown on this FMS RNAV DP chart excerpt for Boston, Massachusetts, Logan International Airport. This flight management system, or FMS, procedure requires /E or /F area navigation system aircraft certification. The /E aircraft equipment suffix listed on an IFR flight plan refers, in part, to dual FMS with barometric vertical navigation (VNAV), oceanic, enroute, terminal, and approach capability. Equipment requirements include, in part, a flight director and autopilot control system, dual inertial reference units, or IRUs, a database, and an electronic map. A /F refers to a single FMS. Note that the DP chart is restricted to turbojets, and requires a navigation system latitude/longitude coordinate update prior to departure.

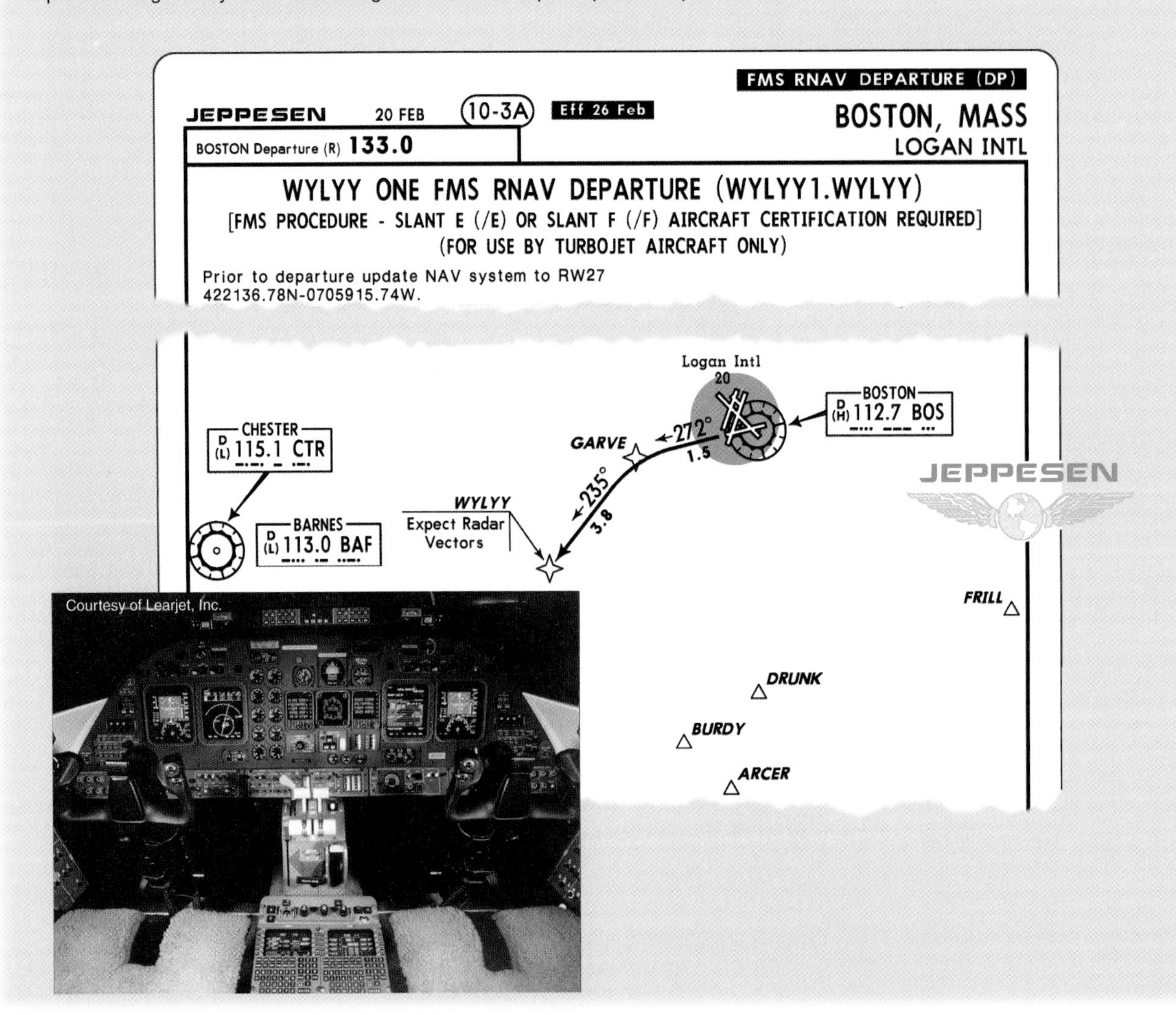

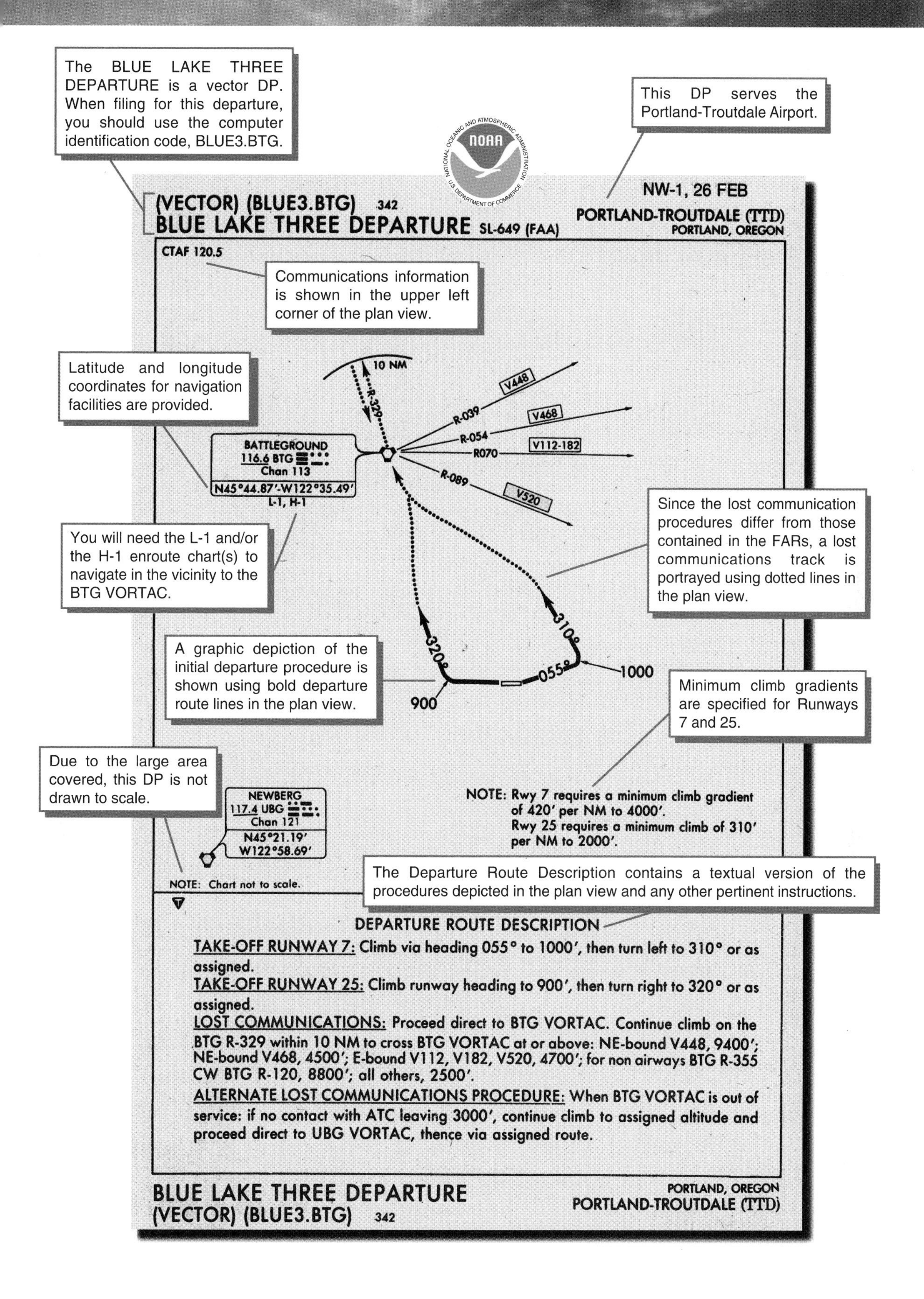

Figure 4-5. A vector DP is normally established at a radar facility except when terrain, ATC coordination, or other safety related factors require the use of a pilot nav DP.

SUMMARY CHECKLIST

✓ Charted departure procedures help simplify clearances, reduce frequency congestion, ensure obstacle clearance, and control traffic flow around an airport. They also help reduce fuel consumption, and may include noise abatement procedures.

✓ Instrument departure procedures (DPs) are used after takeoff to provide a transition between the airport and enroute structure.

✓ When you are issued a DP, you must ensure your aircraft is capable of achieving the DP performance requirements.

✓ DPs require minimum climb gradients of at least 200 feet per nautical mile, to ensure you can clear departure path obstacles.

✓ DPs may specify a minimum ceiling and visibility to allow you to see and avoid obstacles, a climb gradient greater than 200 feet per mile, detailed flight maneuvers, or a combination of these procedures.

✓ When you accept a DP in a clearance, or file one in your flight plan, you must possess the DP chart or the textual description.

✓ To avoid being issued DPs, enter the phrase "NO DP" in the remarks section of your flight plan.

✓ Pilot nav DPs allow you to navigate along a route with minimal ATC communications. They usually contain instructions to all aircraft, followed by transition routes to navigate to an enroute fix, and may include radar vectors that help you join the DP.

✓ Jeppesen and NOS list the airport served by the procedure, the name, and the type of DP at the top of the chart.

✓ If you are instructed to maintain runway heading, it means you should maintain the magnetic heading of the runway centerline.

✓ DP initial takeoff procedures may apply to all runways, or apply only to the specific runway identified.

✓ Since the actual mileage between a given runway and the first fix varies with aircraft performance, pilot technique, and the length of the radar vector, Jeppesen charts include the direct distance from the airport to the first fix.

✓ DP transition routes are shown with dashed lines on Jeppesen charts and with light, solid lines on NOS charts.

✓ The computer identification code for a transition in your flight plan informs ATC you intend to fly both the DP and the appropriate transition.

✓ Because of the large area covered, most DPs are usually not drawn to scale.

✓ Vector DPs exist where ATC provides radar navigation guidance. They usually contain a heading to fly, and an altitude for initial climb. When ATC establishes radar contact, they provide vectors to help you reach fixes portrayed on the chart.

✓ Minimum climb gradients are given in feet per nautical mile and must be converted to feet per minute for use during departure.

KEY TERMS

U.S. Standard for Terminal Instrument Procedures (TERPs)

Instrument Departure Procedure (DP)

Pilot Nav DP

Transition

Vector DP

QUESTIONS

You will be departing San Diego International — Lindbergh Airport (KSAN) using the BORDER FOUR DEPARTURE. Refer to the accompanying departure procedure chart to answer the following questions.

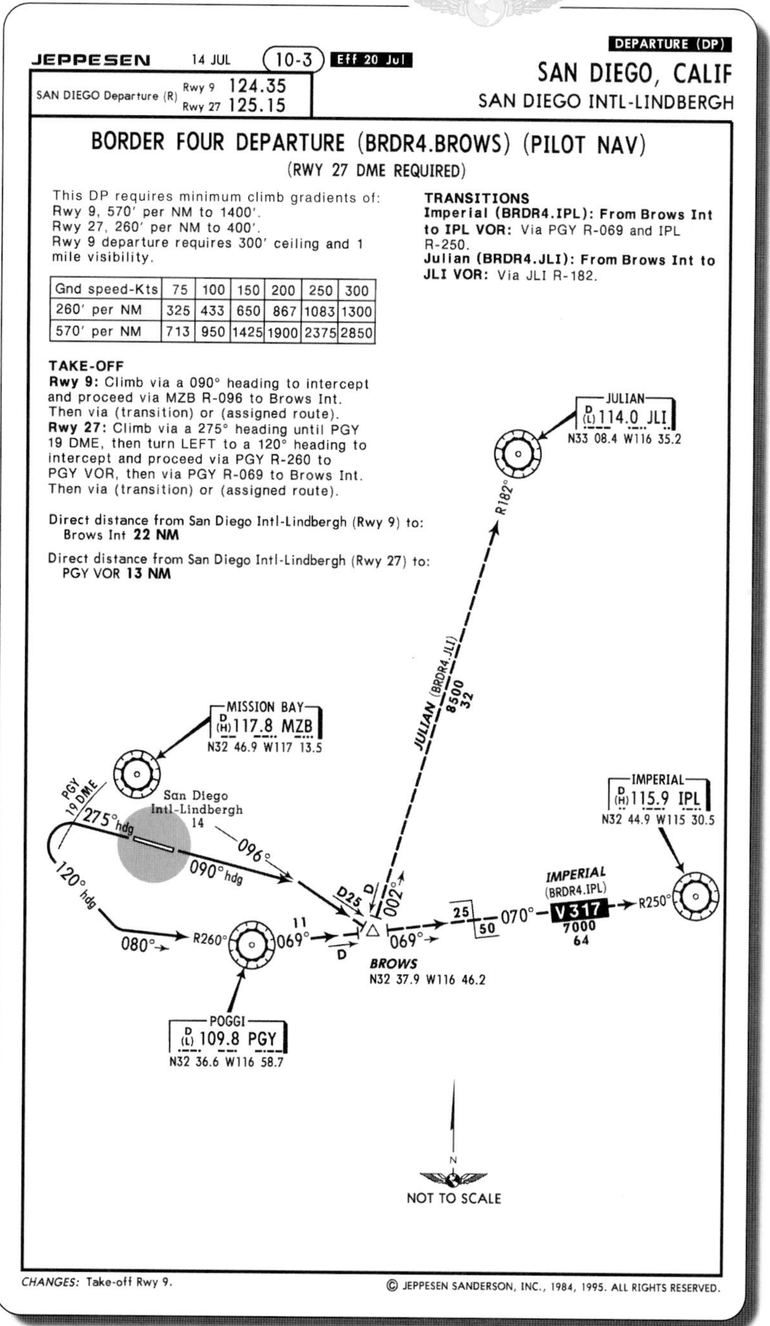

1. Assume that the reported weather conditions at the time of departure are 500 feet overcast, 2 miles visibility, and the wind is calm. If you determine that the climb performance for your airplane is 300 feet per nautical mile, which runway should you choose for departure?

2. When filing your flight plan, what code should you enter on the flight plan for the BORDER FOUR DEPARTURE with Julian Transition?

   A. BRDR4.IPL
   B. BRDR4.JLI
   C. BFDR4.BROWS.JLI

3. If you depart Runway 27, what heading should you use until you reach PGY 19 DME?

4. Where does the basic portion of the BORDER FOUR DEPARTURE procedure end?

   A. At the Poggi VORTAC
   B. At BROWS Intersection
   C. At the Imperial VORTAC

5. What is the departure control frequency for Runway 9 at KSAN?

6. Where does the Imperial Transition begin?

   A. At the Poggi VORTAC
   B. At BROWS Intersection
   C. AT the 19-mile DME fix from the Poggi VORTAC

7. What is the minimum enroute altitude for the Julian Transition?

8. What is the length of the Imperial Transition?

   A. 25 nautical miles
   B. 64 nautical miles
   C. 75 nautical miles

9. If you choose to fly the departure procedure to the BROWS Intersection, what is the appropriate code to enter in your IFR flight plan?

   A. BRDR4.
   B. BRDR4.PGG11
   C. BRDR4.BROWS

10. If you depart Runway 27 on the BORDER FOUR DEPARTURE, Imperial Transition, what initial altitude can your expect?

    A. 7,000 feet at the Poggi VORTAC
    B. 7,000 feet at BROWS Intersection
    C. Your initial altitude will be assigned by the controller

SECTION B
DEPARTURE PROCEDURES

An IFR departure in a radar environment may be as simple as holding the headings assigned by the departure controller while monitoring your position from local navaids. In other cases, you may be adhering to a detailed instrument departure procedure (DP). At remote locations, you may fly the entire departure without the benefit of radar vectors or a graphic DP. This section covers important considerations for IFR departures, beginning with takeoff minimums.

TAKEOFF MINIMUMS

FAR Part 97 prescribes the standard instrument approaches for airports in the United States. Whenever an approach is established, the FAA establishes rules which provide takeoff minimums for large aircraft and all commercially operated aircraft. In cases where takeoff minimums for a particular airport are not published but an instrument approach procedure is prescribed, standard minimums apply — one statute mile visibility for single- and twin-engine airplanes and one-half statute mile visibility for aircraft with more than two engines.

Visibility, which is the typical basis for takeoff minimums, may be expressed in a variety of ways, including prevailing visibility, runway visibility value, and runway visual range. **Prevailing visibility** is the greatest distance a weather observer or tower personnel can see throughout one-half the horizon. This visibility, which need not be continuous, is reported in statute miles or fractions of miles and recorded on the aviation routine weather report (METAR).

Runway visibility value (RVV) is the visibility determined for a particular runway by a device, called a transmissometer, located near the runway. RVV, which is reported in statute miles or fractions of miles, is used in lieu of prevailing visibility in determining minimums for a particular runway.

Runway Visual Range (RVR), in contrast to prevailing or runway visibility, is based on what a pilot in a moving aircraft should see when looking down the runway from the approach end. It is based on the measurement of a transmissometer near the instrument runway and is reported in hundreds of feet. RVR is used in lieu of RVV and/or prevailing visibility in determining minimums for a particular runway. The primary instrument runways at major airports may have as many as three transmis-

RVR represents the horizontal distance a pilot will see when looking down the runway from the approach end.

Definition	Conversion		
Touchdown RVR is the RVR visibility readout values obtained from RVR equipment serving the runway touchdown zone.	**RVR (ft)**		**Visibility (s.m.)**
Mid-RVR is the RVR readout values obtained from RVR equipment located midfield of the runway.	1,600	....	1/4
Roll-out RVR is the RVR readout values obtained from RVR equipment located nearest the roll-out end of the runway.	2,400	....	1/2
	3,200	....	5/8
	4,000	....	3/4
	4,500	....	7/8
	5,000	....	1
	6,000	....	1 1/4

Figure 4-6. At some airports, RVR may be measured at three points along a runway. If RVR is not reported, you can convert the published values to statute miles or fractions thereof.

someters providing RVR readings, which include touchdown RVR, mid-RVR, and roll-out RVR. [Figure 4-6]

It is important to remember that the prevailing visibility or RVR shown in the aviation routine weather report should normally be used only for informational purposes. Since RVR is updated approximately once every minute, controllers can provide you with more accurate visibility information for determining compliance with takeoff minimums.

In some cases, the required visibility for takeoff may be greater than standard due to terrain, obstructions, or departure procedures. In addition, some airports may require both a minimum ceiling and visibility before you can take off. Climb performance also may be a factor. When a ceiling is required, it is generally due to obstructions that you must avoid during the departure. However, the ceiling and visibility requirements may be waived in some cases where a minimum climb gradient is published and you can comply with it. The climb gradient specified ensures obstruction clearance, so standard visibility minimums are adequate. If you note nonstandard visibility or ceiling requirements for takeoff minimums, it should alert you to the fact that some type of operational limitation exists and you will need additional information before departure. [Figure 4-7]

A ∇ in the minimums section of an NOS approach chart indicates takeoff minimums are not standard and/or IFR departure-procedures are published.

As an FAR Part 91 operator, you are not required to comply with published IFR takeoff minimums. However, you should realize that these minimums have been established for the safe operation of commercial aircraft. These aircraft are piloted by highly trained and skilled individuals, and are usually operated by two pilots. Although legal, it is unwise to initiate a takeoff in weather conditions that would keep a commercial flight grounded.

DEPARTURE OPTIONS

In general, you have four alternatives when departing an airport on an IFR flight. Your options include a graphic instrument departure procedure (DP), a textual DP, a radar departure, or a VFR departure.

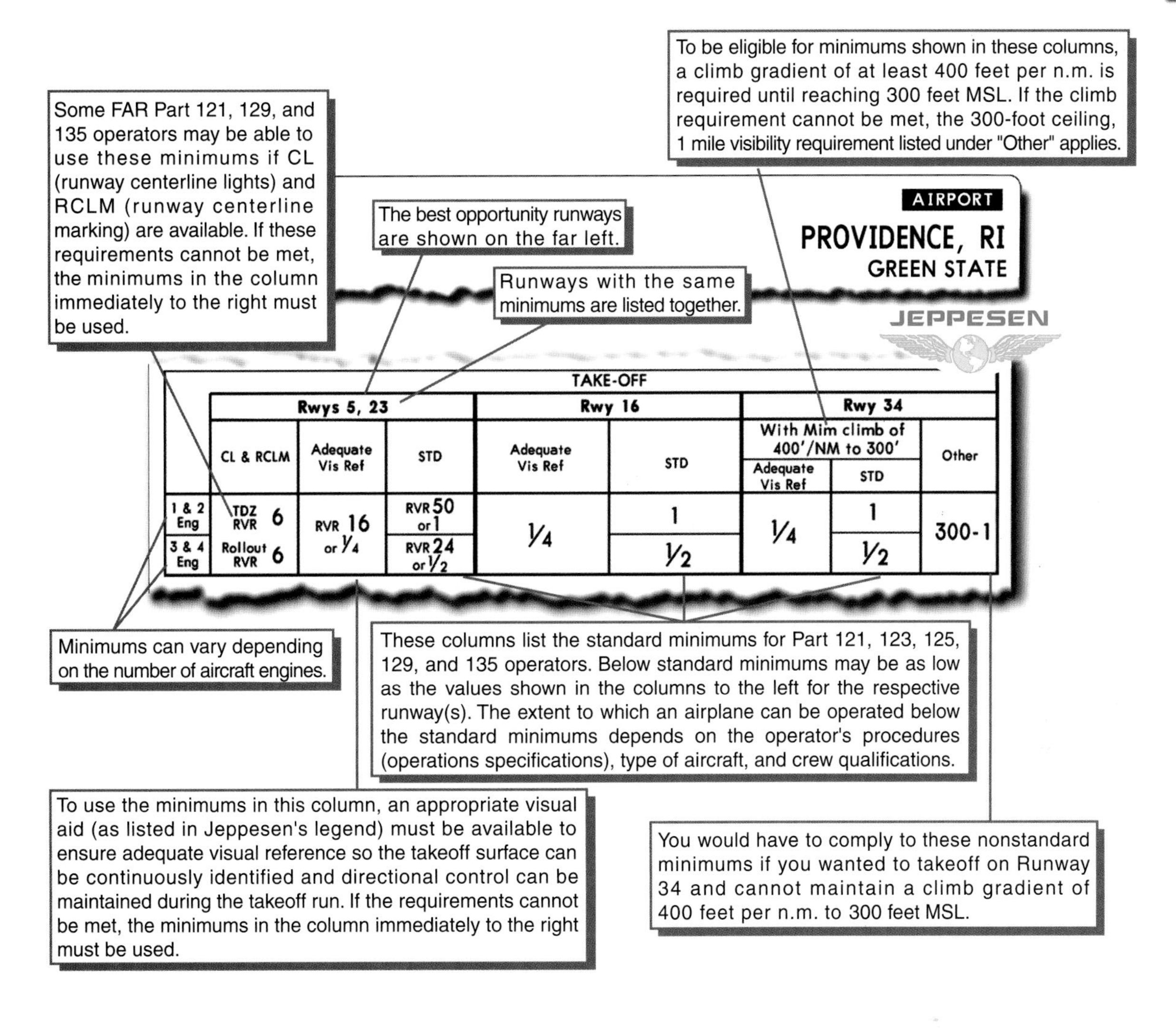

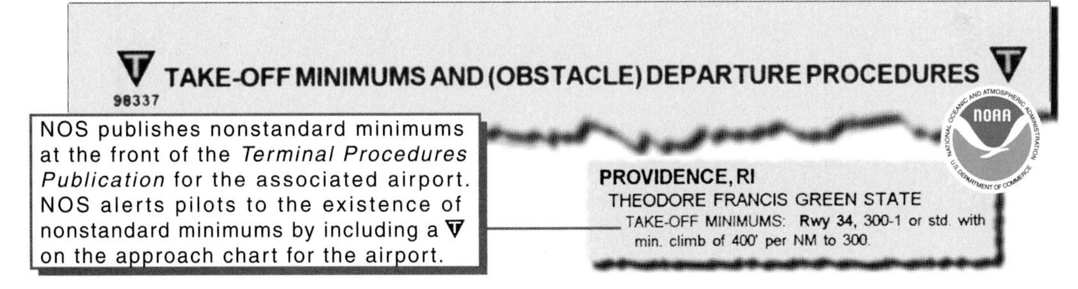

Figure 4-7. In addition to the nonstandard takeoff minimums NOS publishes, Jeppesen also includes non-standard minimums applicable to some commercial operators. Although these minimums do not specifically apply to private aircraft operating IFR under FAR Part 91, good judgment dictates that you use the standard (or higher) minimums as your guide.

GRAPHIC DEPARTURE PROCEDURES

Graphic departure procedures, commonly referred to simply as DPs, are coded departure routes established to expedite departures at airports with a large volume of traffic. In general, they are intended to simplify clearance delivery and departure procedures for both you and air traffic control, however, they may be designed solely for obstacle clearance. If you are operating from an airport where DPs are published, you can expect to have a DP routing included in your ATC clearance. Your acceptance of a DP is not mandatory, but you must advise ATC if you do not wish to use one. The recommended procedure is to file

"NO DP" in the remarks section of your flight plan. The least desirable method is to advise ATC verbally when the DP is assigned. If you wish to fly a DP, you must have either the charted procedure or at least the textual description of the appropriate DP in your possession. [Figure 4-8]

To fly a DP, you must have the charted procedure or at least the textual description in your possession; otherwise, you should file "NO DP" in your flight plan.

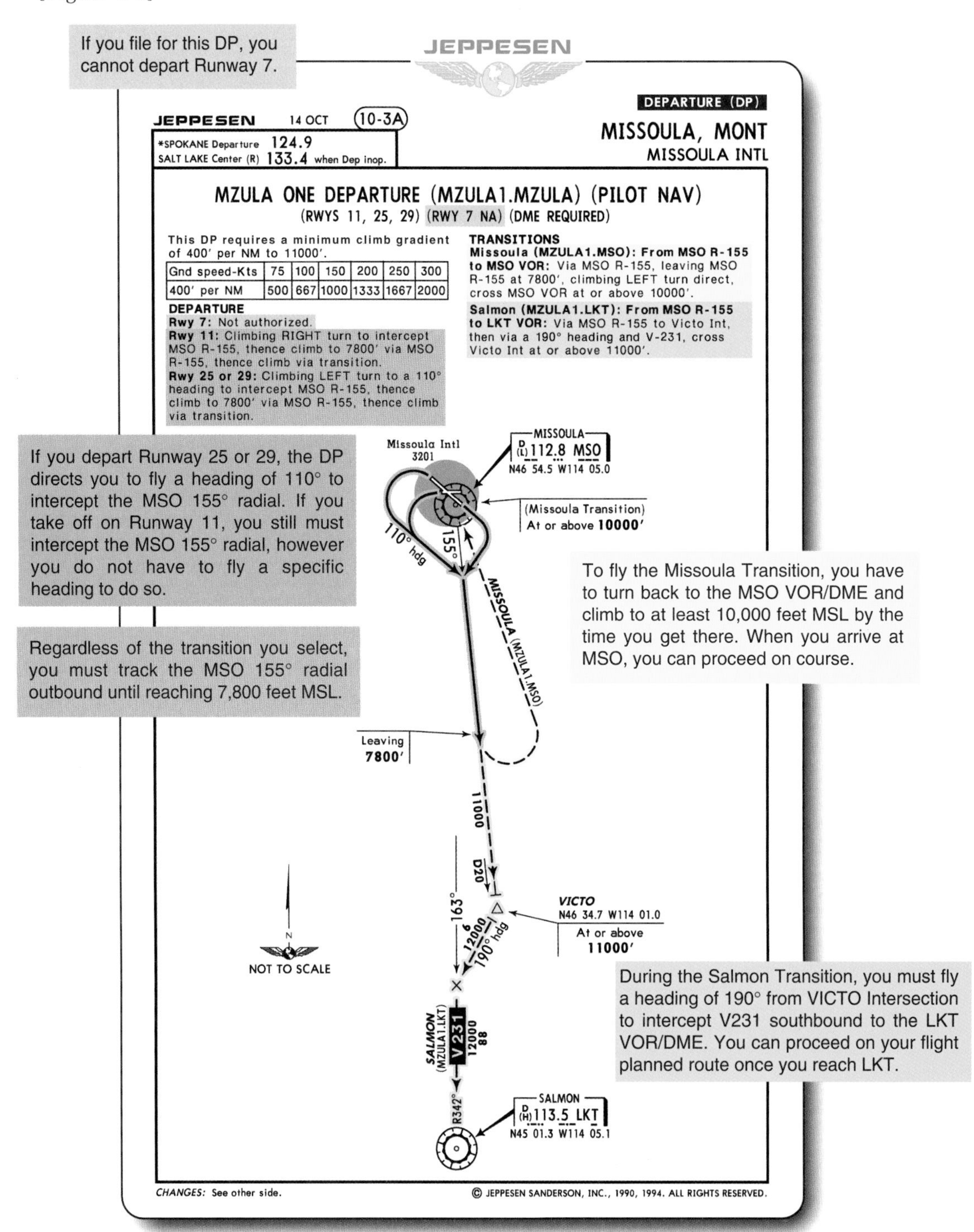

Figure 4-8. To fly this DP, your aircraft must be able to achieve a minimum climb gradient of 400 feet per nautical mile to 11,000 feet MSL.

TEXTUAL DEPARTURE PROCEDURES

In addition to graphic DPs, **textual departure procedures** also are established when necessary for airports that have published instrument approaches. When you are using NOS charts, the textual departure procedures are tabulated in the front of each *Terminal Procedures Publication.* [Figure 4-9] If you are using Jeppesen approach charts, the procedure is printed on the airport chart which is usually located on the reverse side of the first approach chart for each airport. At larger terminals, the airport chart may be printed on a separate sheet.

A textual DP, unlike a graphic DP, is not assigned as a portion of your IFR clearance unless it is required for separation purposes. In general, it is your responsibility to

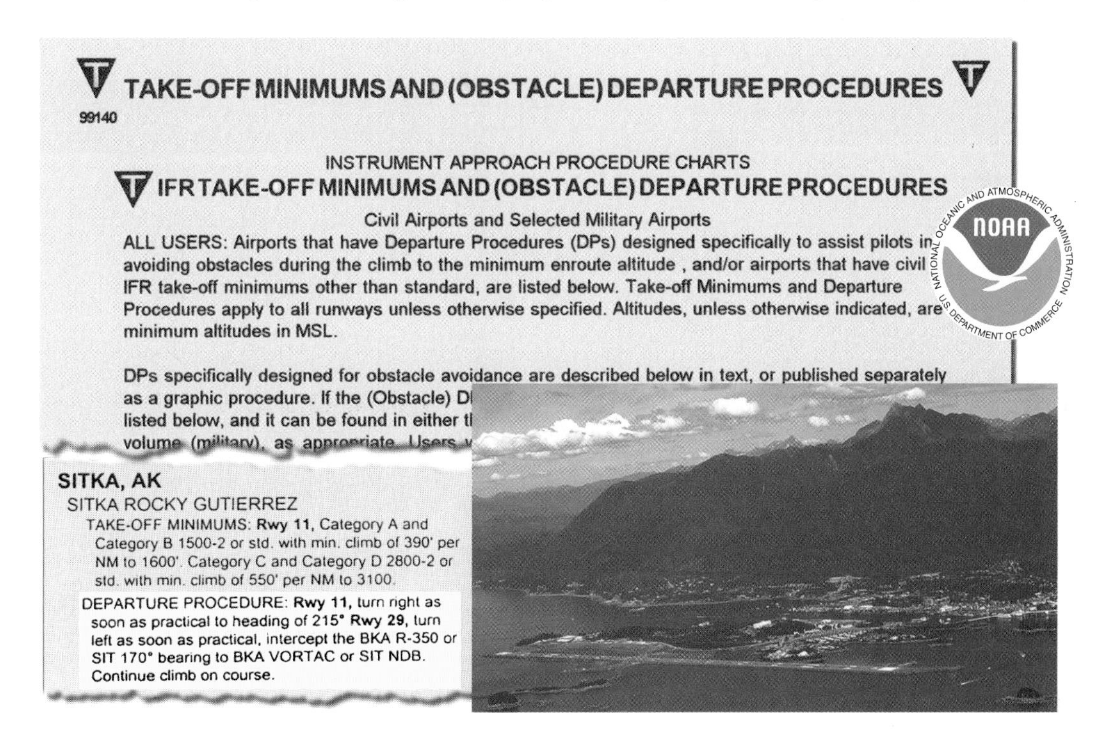

TAKE-OFF MINIMUMS AND (OBSTACLE) DEPARTURE PROCEDURES

99140

INSTRUMENT APPROACH PROCEDURE CHARTS

IFR TAKE-OFF MINIMUMS AND (OBSTACLE) DEPARTURE PROCEDURES

Civil Airports and Selected Military Airports

ALL USERS: Airports that have Departure Procedures (DPs) designed specifically to assist pilots in avoiding obstacles during the climb to the minimum enroute altitude , and/or airports that have civil IFR take-off minimums other than standard, are listed below. Take-off Minimums and Departure Procedures apply to all runways unless otherwise specified. Altitudes, unless otherwise indicated, are minimum altitudes in MSL.

DPs specifically designed for obstacle avoidance are described below in text, or published separately as a graphic procedure. If the (Obstacle) D listed below, and it can be found in either t volume (military), as appropriate. Users

SITKA, AK

SITKA ROCKY GUTIERREZ

TAKE-OFF MINIMUMS: **Rwy 11,** Category A and Category B 1500-2 or std. with min. climb of 390' per NM to 1600'. Category C and Category D 2800-2 or std. with min. climb of 550' per NM to 3100.

DEPARTURE PROCEDURE: **Rwy 11,** turn right as soon as practical to heading of 215° **Rwy 29,** turn left as soon as practical, intercept the BKA R-350 or SIT 170° bearing to BKA VORTAC or SIT NDB. Continue climb on course.

Figure 4-9. While these textual DPs are not depicted graphically, the intent is the same as a graphic DP — to ensure obstacle clearance and a safe transition from takeoff to the IFR enroute structure.

determine if one has been established, then comply with it. In IFR conditions, the departure procedure is the only method of ensuring terrain and obstacle clearance. However, if the weather is VFR, you can expedite your departure by requesting a clearance to climb in VFR conditions from the appropriate ATC facility. This relieves you from flying the entire textual DP, although you are then responsible for your own terrain clearance and traffic separation during the climb. When a clearance is issued to climb in VFR conditions, the controller may abbreviate the term by saying, *". . . climb VFR, . . ."* and if required, *". . . climb VFR between 4,000 and 10,000."* If there is reason to believe that flight in VFR conditions may become impractical, your clearance also may provide for that possibility.

RADAR DEPARTURES

A **radar departure** is often assigned at radar-equipped approach control facilities and requires close coordination with the tower. If your flight is to be radar vectored immediately after takeoff, the tower will advise you of the heading to be flown, but not necessarily the purpose of the heading. This type of advisory will be issued to you either in your initial IFR routing clearance or by the tower just before takeoff. Once you have received your takeoff clearance, you should understand that coordination of your flight is the responsibility of the tower controller. After you are airborne, you can expect a handoff to the departure controller. [Figure 4-10]

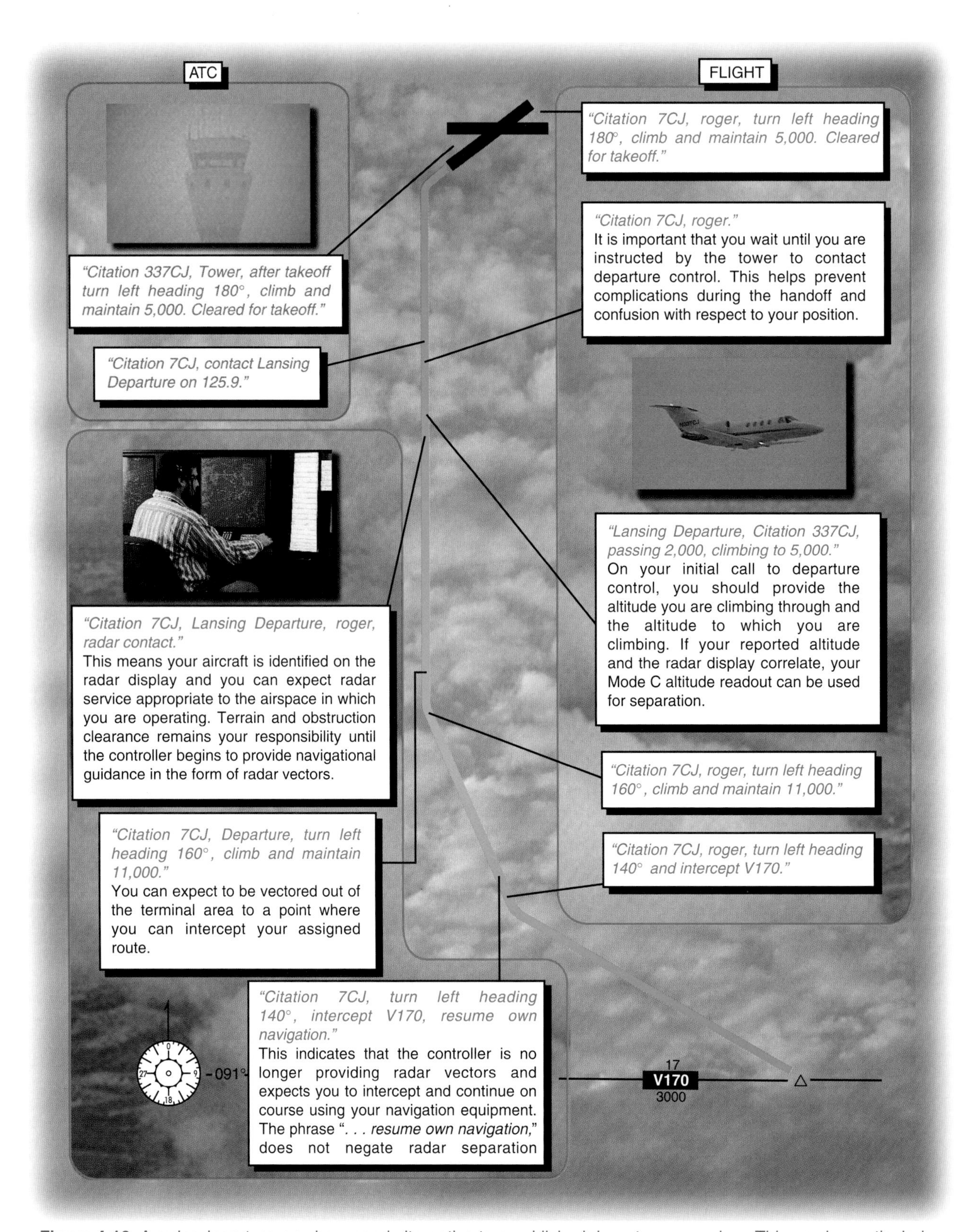

Figure 4-10. A radar departure may be a good alternative to a published departure procedure. This can be particularly true when none of the available departure procedures are convenient for your planned route of flight.

In certain circumstances, you may be issued a vector which takes your flight off a previously assigned route. When this occurs, you will be advised briefly what the vector is to achieve. Generally, it is issued for weather avoidance, terrain clearance, or traffic separation. Radar vectoring will be provided until your aircraft has been reestablished on course and you have been advised of your position. In some cases, a handoff may be made to another radar controller with continuing radar surveillance capabilities. If you feel that any vector is given in error, you are encouraged to question the controller and verify the purpose of the vector. In addition, you should maintain orientation to your present position at all times so you are prepared to resume your own navigation, if necessary.

During the IFR departure, you should not contact departure control until advised to do so by the tower.

The term "*radar contact*" is used by ATC to advise you that your aircraft has been identified and radar flight following will be provided until radar identification has been terminated.

If radar contact is lost for some reason, you can expect the controller to request additional reports from you in order to monitor your flight progress. These requested reports may include crossing a particular navigation fix, reaching an altitude, or intercepting and proceeding on course.

"Resume own navigation" is a phrase used by ATC to advise you to assume responsibility for your own navigation.

VFR DEPARTURES

When conditions permit, you may be able to depart an airport under VFR and obtain your clearance from ARTCC after takeoff. However, you must remain aware of your position relative to terrain and obstructions. In addition, you should always maintain VFR conditions until you have obtained your IFR clearance and have ATC approval to proceed on course in accordance with the clearance. If you accept a clearance while below the minimum altitude for IFR operations in the area, you are responsible for your own terrain/obstruction clearance until you reach that altitude.

SELECTING A DEPARTURE METHOD

As you prepare to depart an airport on an IFR flight, you should assess the situation and determine which type of departure is best suited for your circumstances. After analyzing the type of terrain and other obstacles on or in the vicinity of your departure airport, you should determine whether a textual DP and/or graphic DP is available. If one exists, does it allow you to proceed expeditiously on your route? If not, does departure control/center have the ability to provide you with a radar guided departure? Does the weather and terrain allow you to initially depart VFR? Once you have answered these basic questions, you will be ready to select a course of action, familiarize yourself with the associated procedures, obtain your IFR clearance, and execute your departure.

Like Ships Passing on the Potomac

Development of radio detection and ranging, commonly referred to as radar, began as early as 1922. Researchers at the Naval Aircraft Laboratory in Washington, D.C. observed radio signals reflecting from ships passing in the Potomac River. [Figure A] Reports on radio echo signals from moving objects led to British involvement with radar in 1935. In 1940, MIT Radiation Laboratory, in conjunction with the British Tizard Mission, mounted a crash program to make microwave radar sets for British night fighter airplanes. These efforts led to further developments such as the military Identification of Friend and Foe (IFF) system, and a talk-down blind landing system for aircraft called Ground Controlled Approach (GCA).

The prototype GCA used microwave radar which provided airplane coordinates to a small analog computer called a director. The director compared the coordinates to those of an ideal glide path and developed error signals on meters monitored by the controller. The controller would give the pilot right-left steering instructions and adjustments to rate of descent until the pilot could see the runway and land.

One of the most spectacular successes for GCA occurred when the Russians blockaded Berlin during the rainy fall and winter of 1948. With all road, rail, and canal links to West Germany severed, a military GCA set operated around the clock, bringing in a steady stream of planes carrying thousands of tons of food and fuel. Figure B shows one such aircraft, a Douglas C-54, flying a relief mission during the Berlin Air Lift. In the end, the GCA was credited for making the airlift successful and breaking the blockade.

Courtesy of the United Technologies Archive

VERTICAL SITUATIONAL AWARENESS

During a recent 26-year period, more than 50% of the business aircraft accidents which resulted in fatalities occurred when the pilots unknowingly flew their airplanes into the ground. Over two thirds of these controlled flight into terrain (CFIT) accidents resulted from a lack of vertical situational awareness on the part of the crew. As you make a decision about what type of departure to use, you should not only consider the procedures, but also why the procedures exist.

Suppose you plan on departing Reno/Tahoe International Airport's Runway 16R on an IFR flight to the south. If you were to use the textual DP, you would need to fly to the VOR east of the airport and climb in a holding pattern to a safe altitude before proceeding on course. [Figure A] A more viable alternative would be to file for the WAGGE ONE DEPARTURE which allows you to continue southbound as long as you can maintain a minimum climb gradient. [Figure B] Using this departure would not only simplify your clearance, but it would also save you time and fuel. Neither option, however, would provide you with a picture of the surrounding topography. To develop your vertical situational awareness prior to departure, you may consult a variety of sources, including the Jeppesen area chart [Figure C], and the associated sectional chart [Figure D].

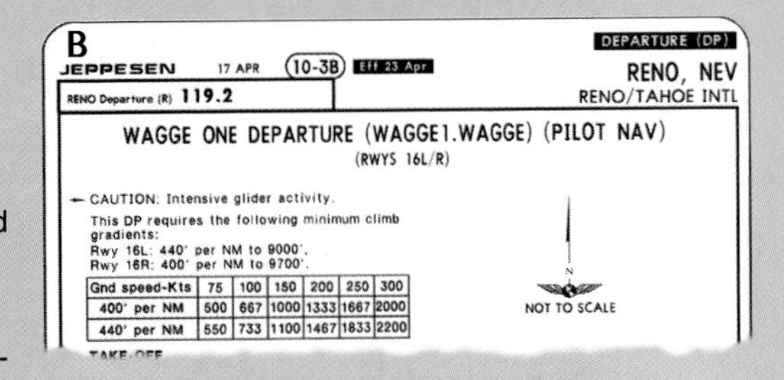

B

JEPPESEN 17 APR (10-3B) Eff 23 Apr DEPARTURE (DP)

RENO Departure (R) 119.2 RENO, NEV RENO/TAHOE INTL

WAGGE ONE DEPARTURE (WAGGE1.WAGGE) (PILOT NAV)

(RWYS 16L/R)

CAUTION: Intensive glider activity.

This DP requires the following minimum climb gradients:
Rwy 16L: 440' per NM to 9000'.
Rwy 16R: 400' per NM to 9700'.

Gnd speed-Kts	75	100	150	200	250	300
400' per NM	500	667	1000	1333	1667	2000
440' per NM	550	733	1100	1467	1833	2200

NOT TO SCALE

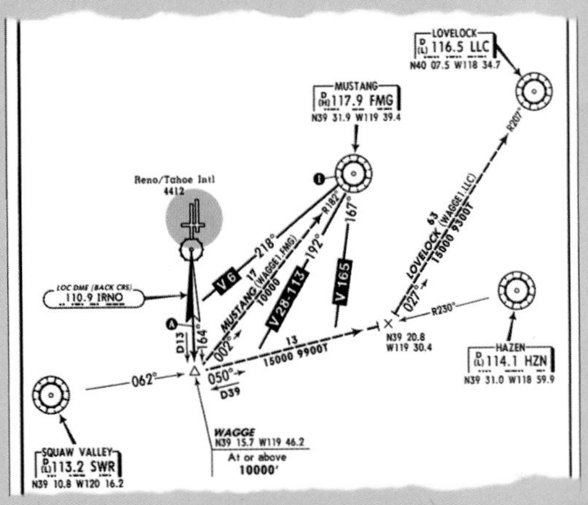

A

OBSTACLE DEPARTURE PROCEDURE

Rwy 7: NA. Rwy 16 L/R: Climb via IRNO LOC south course to 5500', then climbing left turn direct FMG VOR. Rwy 25: Turn right, climb direct FMG VOR. Rwy 34L/R: Climb via IRNO LOC north course to 7500', then climbing right turn direct FMG VOR. All aircraft cross FMG VOR at or above 8000'. All aircraft climb in FMG VOR holding pattern (hold northeast, left turns, 221° inbound) to depart FMG VOR: R-260 clockwise R-170 at or above 10000', R-171 clockwise R-195 at or above 10500', R-196 clockwise R-259 at or above 12000'.

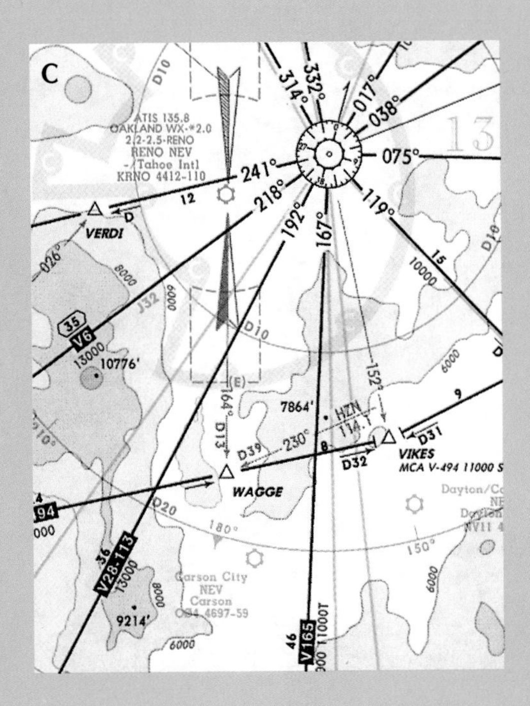

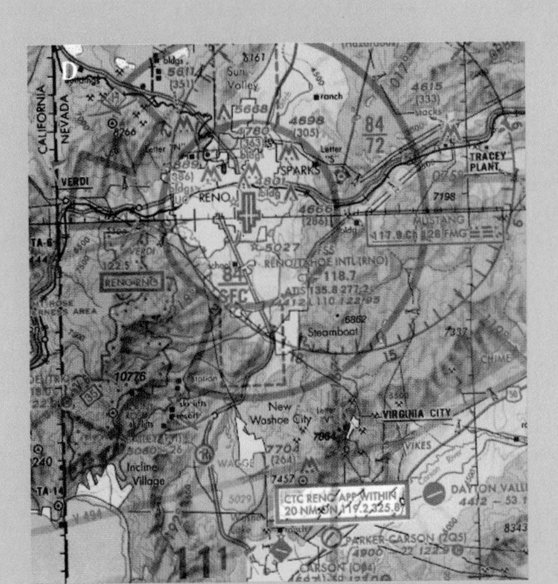

SUMMARY CHECKLIST

✓ Runway visibility value (RVV) is reported in statute miles or fractions of miles.

✓ Runway visual range (RVR) represents the distance you can expect to see down the runway from a moving aircraft.

✓ When RVR is out of service, convert published RVR values to visibility in statute miles.

✓ Prevailing visibility or RVR in the aviation routine weather report should normally be used only for informational purposes. The current visibility at the time of departure is the value you should use for determining compliance with takeoff minimums.

✓ IFR takeoff minimums do not apply to private aircraft operating under IFR and Part 91, but good judgment should dictate compliance.

✓ Standard takeoff weather minimums are usually based on visibility. Greater than standard takeoff minimums may be due to terrain, obstructions, or departure procedures.

✓ If you wish to fly a graphic DP, you must possess the charted DP procedure or at least the textual description .

✓ Textual DPs are not assigned as a portion of your IFR clearance unless required for separation purposes.

✓ During the IFR departure, you should not contact departure control until advised to do so by the tower.

✓ Radar departures are often assigned at radar-equipped approach control facilities and require close coordination with the tower.

✓ The term *"radar contact"* means your aircraft has been identified and radar flight following will be provided until radar identification has been terminated.

✓ During departure, terrain and obstruction clearance remains your responsibility until the controller begins to provide navigational guidance in the form of radar vectors.

✓ "*Resume own navigation"* is a phrase used by ATC to advise you to assume responsibility for your own navigation.

KEY TERMS

Prevailing Visibility

Runway Visibility Value (RVV)

Runway Visual Range (RVR)

Graphic Departure Procedure

Textual Departure Procedure

Radar Departure

QUESTIONS

1. True/False. Runway visibility value (RVV) is normally reported in hundreds of feet.

2. What is the statute mile equivalent of 1,600 feet RVR?

3. True/False. IFR takeoff minimums do not apply to private aircraft operating under FAR Part 91, but good judgment dictates compliance.

4. What is the recommended procedure if you do not wish to use a DP?

   A. Advise departure control upon initial contact.
   B. Enter "NO DP" in the remarks section of the IFR flight plan.
   C. No action is necessary, since ATC will not assign a DP unless you specifically request it.

Refer to the POAKE ONE DEPARTURE to answer questions 5 through 9.

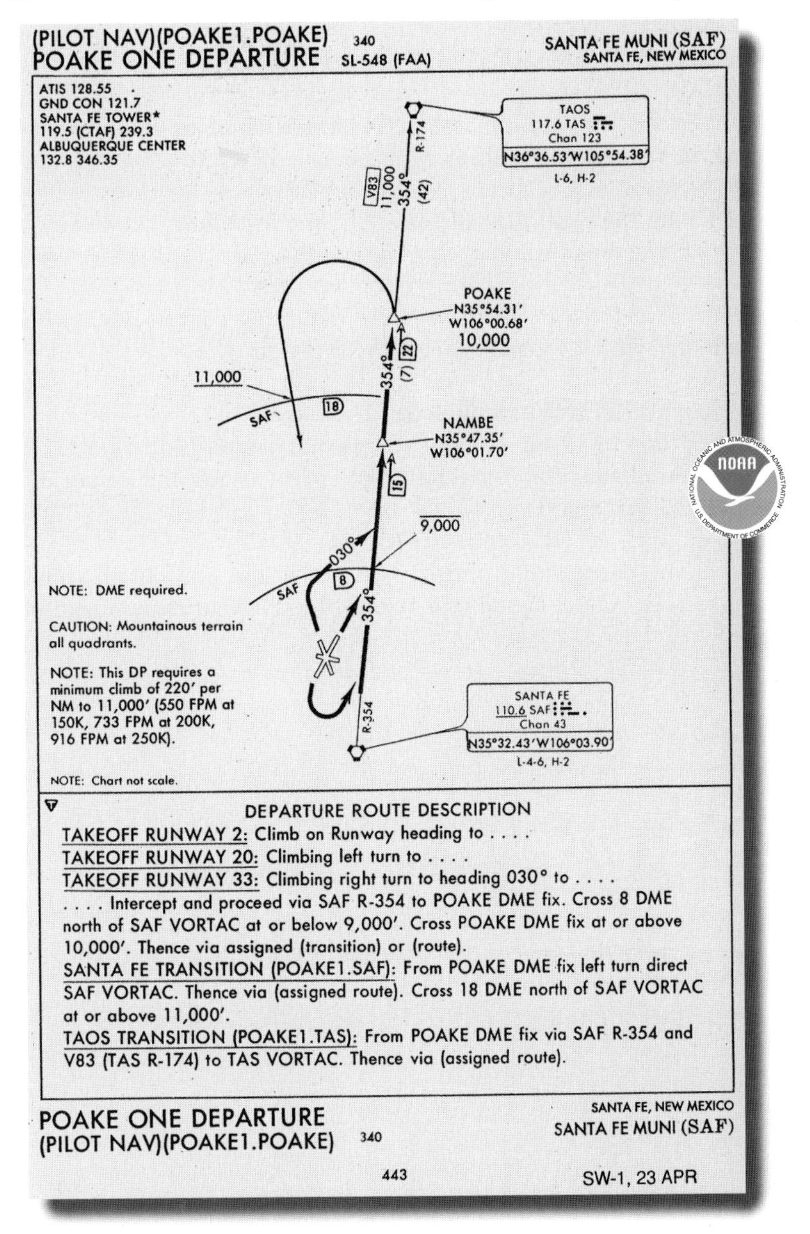

5. If your groundspeed is 150 knots, what rate of climb (in feet per minute) must you maintain to 11,000 feet MSL?

6. If you takeoff on Runway 33, what heading should you fly to intercept the SAF 354° radial?

7. What is the required altitude when crossing the SAF 8 DME arc northbound?

8. When flying the Santa Fe Transition, when must you be at or above 11,000 feet MSL?

   A. At the Santa Fe VORTAC
   B. Crossing the SAF 18 DME arc northbound
   C. Crossing the SAF 18 DME arc southbound

9. When flying the Taos Transition, where do you intercept V83?

10. True/False. Textual DPs are published for all airports where terrain and/or obstacles could compromise a safe departure.

11. Where can you find textual DPs published by NOS?

12. True/False. Textual DPs are not assigned as a portion of an IFR clearance unless it is required for separation purposes.

13. True/False. After the controller advises, *". . . radar contact,"* you can assume that terrain and obstruction clearance will be provided.

14. You have been vectored to an airway, and departure control advises you to *". . . resume own navigation."* What is meant by this term?

    A. Radar service has been terminated.
    B. You are still in radar contact, but you must make position reports.
    C. You should intercept and maintain the airway centerline by use of your own navigation equipment.

15. Assume that you depart an airport in VFR conditions and obtain your IFR clearance after takeoff. How long are you responsible for your own terrain/obstruction clearance?

CHAPTER 5

ENROUTE

Instrument/Commercial
Part II, Segment 1, Chapter 5 — Enroute

SECTION A
ENROUTE AND AREA CHARTS

The increase in the number of navaids and the complexity of the airway and airspace system has made specialized enroute charts a necessity for IFR flight. [Figure 5-1] In addition to helping you keep track of your position, enroute charts provide the information you need to maintain a safe altitude and ensure navigation signal reception. Area charts show major terminal areas in more detail, and are primarily used during the transition to or from the enroute structure. You will see examples of both Jeppesen and NOS charts in this section. Each chart system has its own set of symbols, but you will find them simple to learn and interpret. While this coverage presents a general description of the symbols, be sure to familiarize yourself thoroughly with the legend for the charts you use, since chart enhancements and symbology improvements are ongoing.

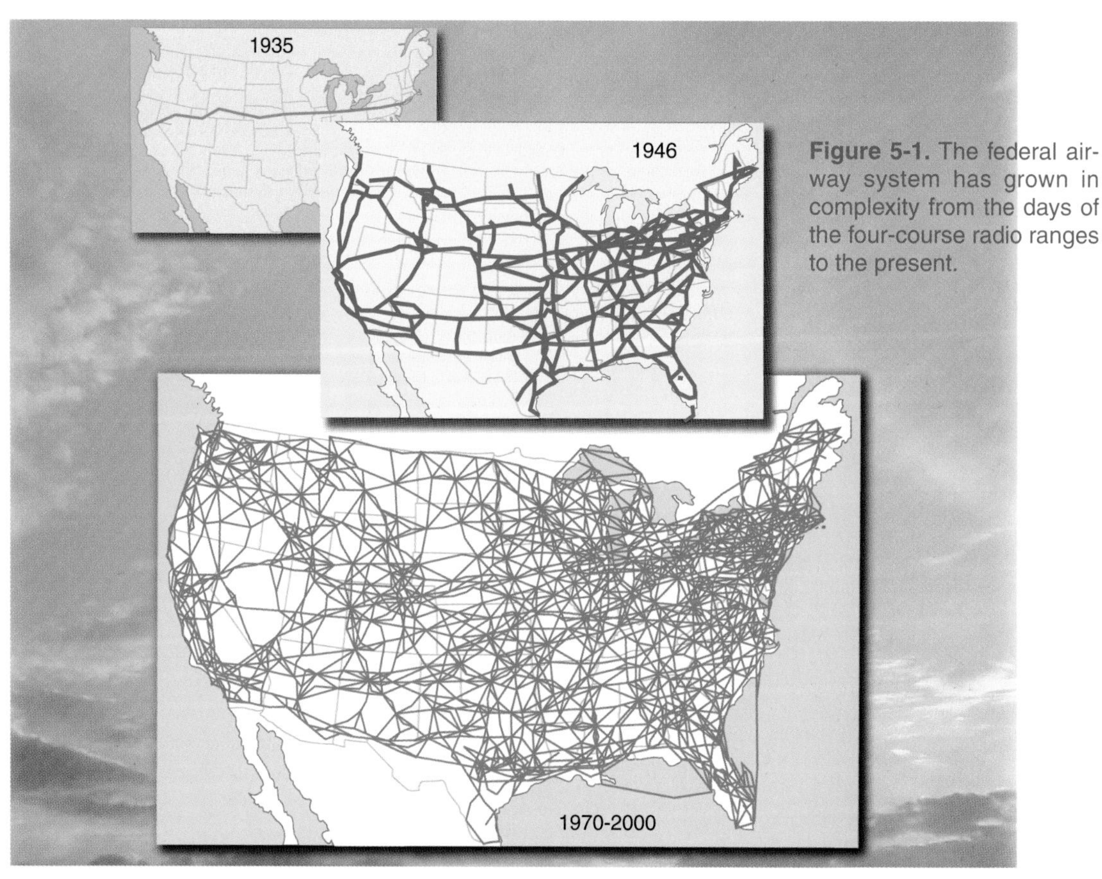

Figure 5-1. The federal airway system has grown in complexity from the days of the four-course radio ranges to the present.

ENROUTE CHARTS

In the United States, 18,000 feet MSL is the lower boundary of Class A airspace, so it is a convenient place to establish the division between the low and high altitude airway structures. Airways below 18,000 feet MSL are depicted on **low altitude enroute charts** and are called **Victor airways**. Those at and above 18,000 feet MSL and up to FL450 are shown on **high altitude enroute charts** and are called **jet routes**

This section concentrates on low altitude enroute charts, since most of your initial instrument flying will take place below 18,000 feet MSL. High altitude charts use similar symbols, but show only the jet routes. Because aircraft using the jet route system are usually operating at higher speeds, these charts cover larger areas at a smaller scale. [Figure 5-2]

Since Class A airspace begins at 18,000 feet MSL, it is not shown on low altitude enroute charts. All airways at 18,000 feet MSL and above are jet routes.

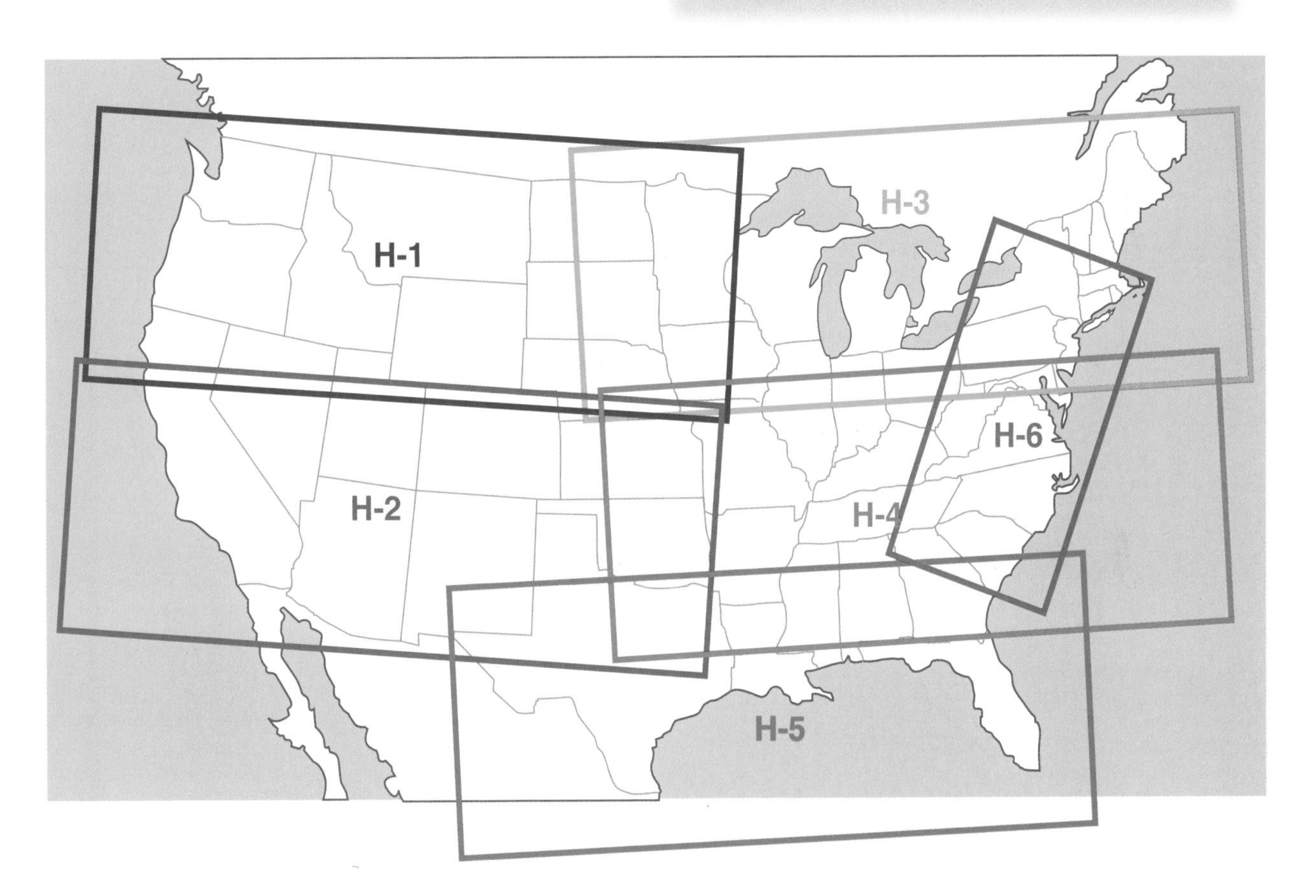

Figure 5-2. Because of the smaller scale of high altitude charts, the entire continental U.S. is covered by six charts printed on three pieces of paper.

Compared to WAC or sectional charts, the enroute chart is greatly simplified. Some of the symbols are similar, but most of the information depicted on visual navigation charts is missing from enroute charts. The topography features, contour lines, obstruction heights, roads, cities, and towns depicted on visual charts are not included on IFR enroute charts. In fact, the only surface features shown on enroute charts are major bodies of water and airports. Since you will usually follow airways between navaids, and terrain and obstruction clearance is guaranteed by flying minimum IFR altitudes, much of the detail can be omitted to provide space for other information necessary for IFR navigation. [Figure 5-3]

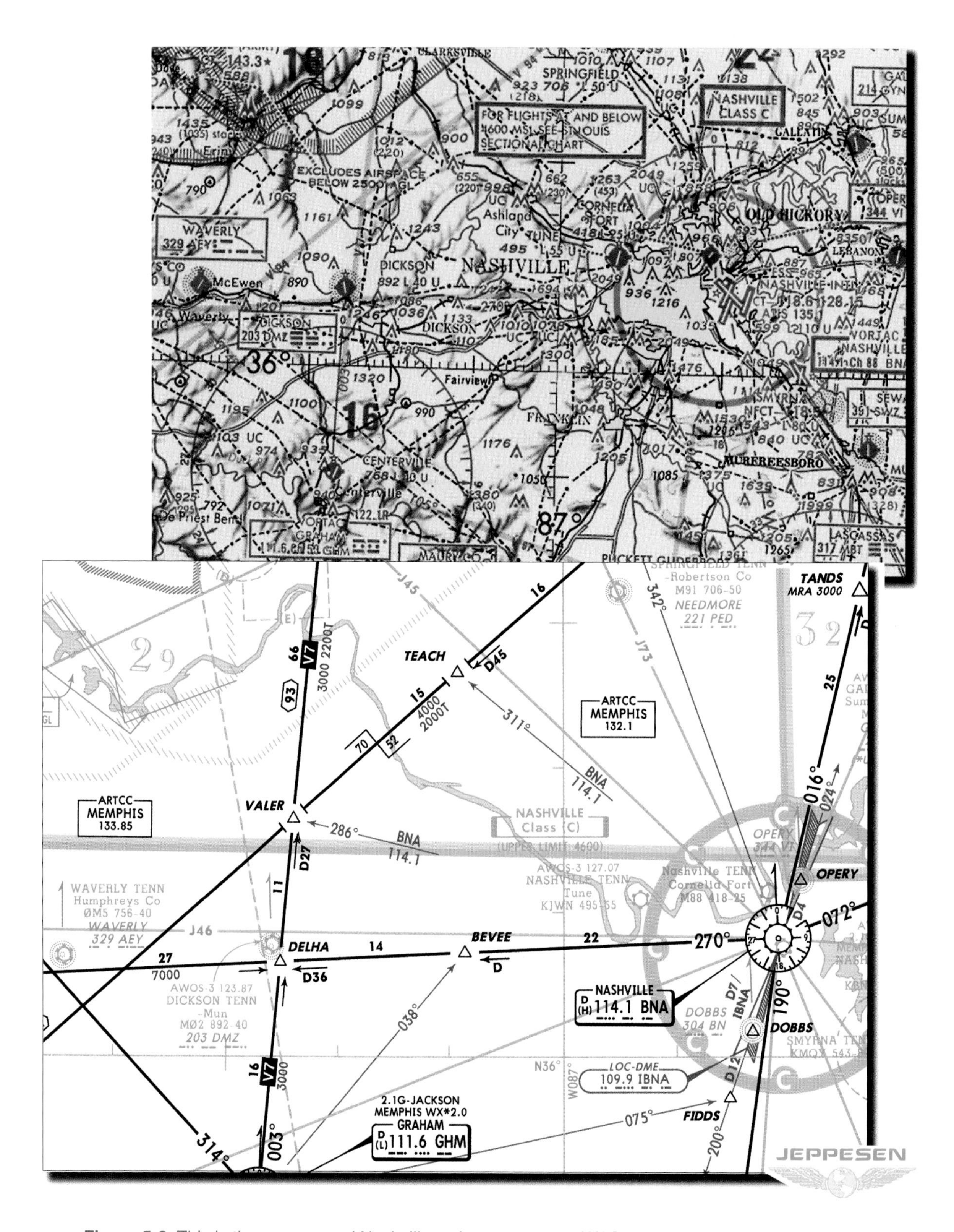

Figure 5-3. This is the area around Nashville as it appears on a WAC chart and on a Jeppesen Low Altitude Enroute Chart. The scale of enroute charts may vary from one chart to the next, so be careful to use the correct scale if you measure distances with a plotter.

FRONT PANEL

The front panels of both Jeppesen and NOS show the area covered by that chart. Although both chart systems present similar information, the area covered by individual charts varies. [Figure 5-4]

Although they do not appear on the chart itself, state boundaries and/or major cities are shown on the coverage diagram to help you find the appropriate chart more easily. Jeppesen uses gray shading on the front panel to show where area charts are available. NOS shows cities that have area charts in black type. Time zone boundaries are included on both charts. NOS uses a series of dots, while Jeppesen uses a series of Ts.

Both NOS and Jeppesen furnish legends to help you interpret the symbols on enroute charts. Jeppesen provides a comprehensive legend in a separate introduction section. Most NOS charts have the legend right on the chart. You should maintain a working knowledge of chart symbology and review the appropriate legend periodically for updates and improvements.

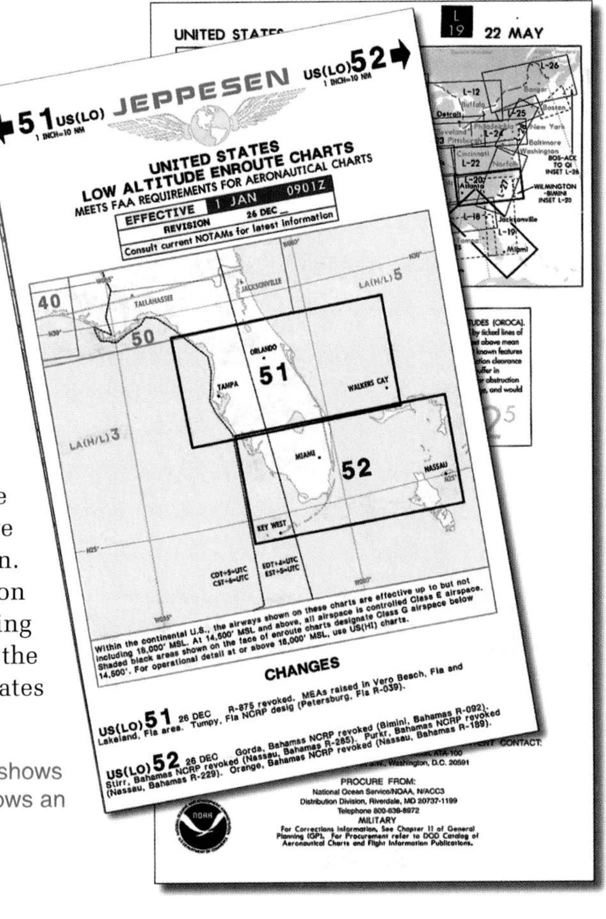

Figure 5-4. The front panel of this NOS chart shows the whole lower 48 states, while Jeppesen shows an area somewhat larger than the chart.

NAVIGATION AIDS

Since all airways are defined by electronic navigation aids, you will want to become familiar with the corresponding chart symbols. Note the similarities and differences in how the navaids are depicted on NOS and Jeppesen charts.

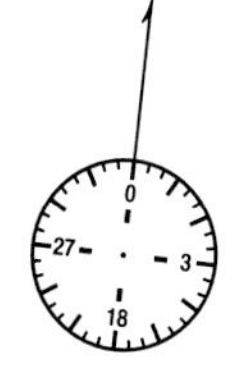

VOR — The VOR symbol on enroute charts is a small compass rose. The pointer on the VOR symbol indicates magnetic north, making it easier to measure bearings with a plotter. The center of the NOS symbol is similar to the VOR symbol used on WAC and sectional charts.

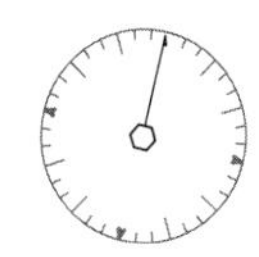

TACAN — Most TACAN stations without a collocated VOR can be used by civilian DME units. Jeppesen's symbol, a serrated circle, represents both TACAN and DME facilities.

VORTAC and VOR/DME — Since these facilities are functionally identical for civilian users, Jeppesen uses a single symbol for both by simply combining the VOR and TACAN/DME symbols. NOS shows VORTACs and VOR/DMEs by adding the familiar symbols from the WAC and sectional charts to the center of a compass rose.

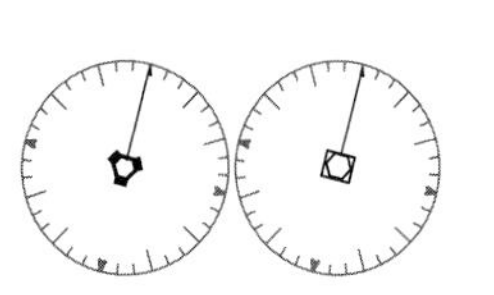

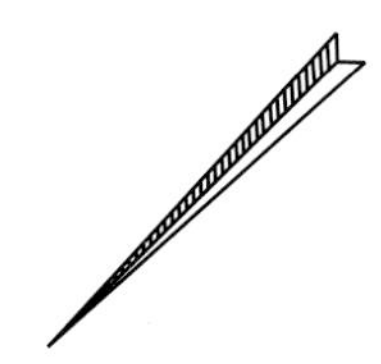

NDB — Nondirectional beacons are also shown with a magnetic north arrow to help you measure magnetic bearings with your plotter. A smaller version of this symbol indicates a compass locator beacon. On Jeppesen charts, compass locators are shown only when the facility provides an enroute function or TWEB information.

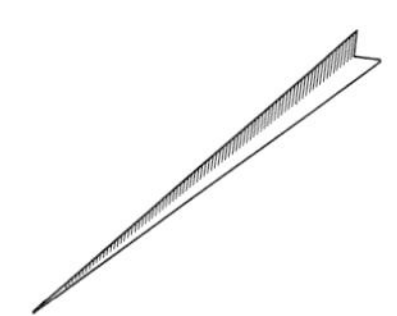

ILS Localizer — Localizer symbols are used to show ILS, MLS, LDA, and SDF facilities. NOS uses them only when they serve an enroute ATC function, but Jeppesen shows their availability to assist pilots with flight planning. On Jeppesen charts, when the facility serves an enroute function, the symbol has the frequency and identifier nearby in a round-ended box. The localizer back course is sometimes used to establish a fix, and is labeled as such on the chart.

On NOS enroute charts, localizers and back courses are shown only when they serve an enroute ATC function, such as establishing a fix or intersection.

Remote Communication Outlet (RCO) — This symbol is used for FSS remote communication outlets (RCOs) when they are not located adjacent to navaids or airports.

Facility Information — This provides you with the name, frequency, and identifier of navigation aids. Morse code for the identifier is also provided to make recognition easier. On Jeppesen charts, a box indicates that the facility is part of an airway, while off-airway navaid information is not boxed. On NOS charts, information on each navaid is boxed. A variety of other data may be shown by letters or symbols near the navaid symbol; a few minutes with the chart legend should acquaint you with the abbreviations.

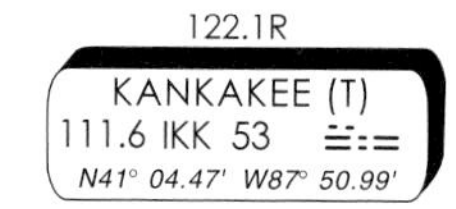

VICTOR AIRWAYS

The V in Victor airways stands for VHF, because these airways connect VOR, VORTAC, and VOR/DME stations. When the first VORs were commissioned, airways between them were given the V designation to distinguish them from the established network of low frequency airways. The number of the airway indicates its general direction. Even-numbered airways usually run more or less east and west, while odd-numbered airways are generally oriented north and south, much like the Interstate highway system. When more than one airway shares a common route segment, all of the airway numbers are shown. Now, of course, the Victor airways dominate the domestic airway route structure, but low frequency airways defined by NDBs are still used in Alaska and along the Atlantic coast.

The width of an airway is normally 8 nautical miles, 4 on each side of its centerline. When an airway segment is more than 102 nautical miles long, additional airspace is allocated. [Figure 5-5]

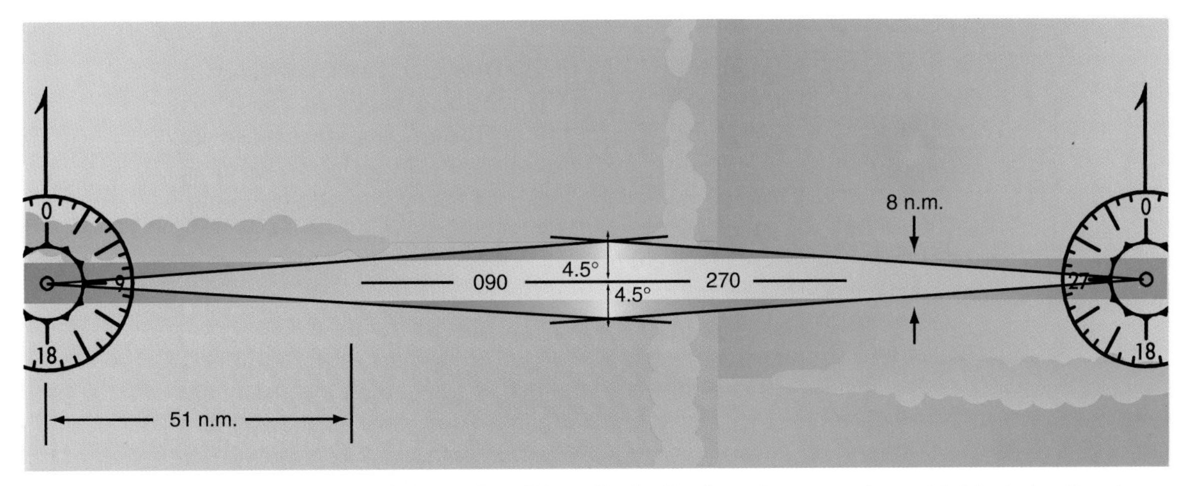

Figure 5-5. When the airway extends more than 51 nautical miles from the nearest navaid, it includes the airspace between lines diverging at angles of 4.5° from the center of each navaid. These lines extend until they intersect the diverging lines from the navaid at the other end of the segment.

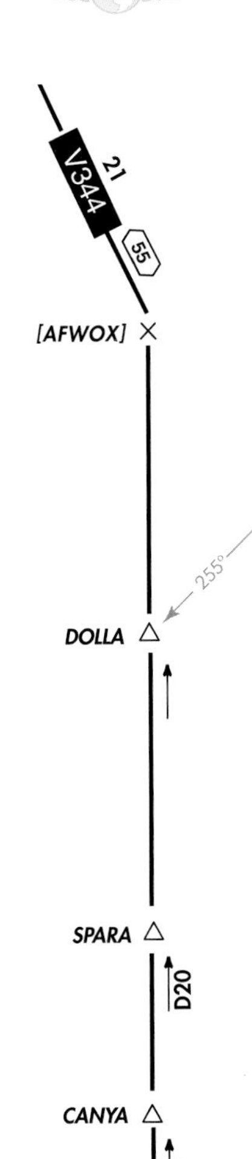

All distances on enroute charts are in nautical miles. On Jeppesen charts, a number in an outlined hexagonal box indicates total mileage between navaids. NOS uses the outlined box to show total mileage between navaids and/or compulsory reporting points. A number without an outlined box indicates the mileage between any combination of intersections, navaids, or mileage break points.

A **mileage break point** is shown on a chart by a small x on the airway. Generally, this symbol indicates a point on the airway where the course changes direction and where no intersection is designated. It may also designate a computer navigation fix with no ATC function.

Intersections are checkpoints along an airway that provide a means for you and ATC to check the progress of your flight. They are often located at points where the airway turns or where you need a positive means of establishing your position. Intersections are given five-letter names, and the actual location of an intersection may be based on two VOR radials, DME, or other navaids, such as a localizer or a bearing to an NDB. Arrows are placed next to the intersection pointing from the navaids that form the intersection.

On Jeppesen charts, an intersection that can be defined by DME is indicated by an arrow with the letter D below it, while NOS uses an open arrow. If it is the first intersection from the navaid, the mileage is found along the airway as a standard mileage number.

When it is not obvious, DME mileage to an intersection follows the letter D on Jeppesen charts. On NOS charts, it is enclosed within a D-shaped outline. This number represents the total distance from the navaid to the fix, as you would see it on your DME.

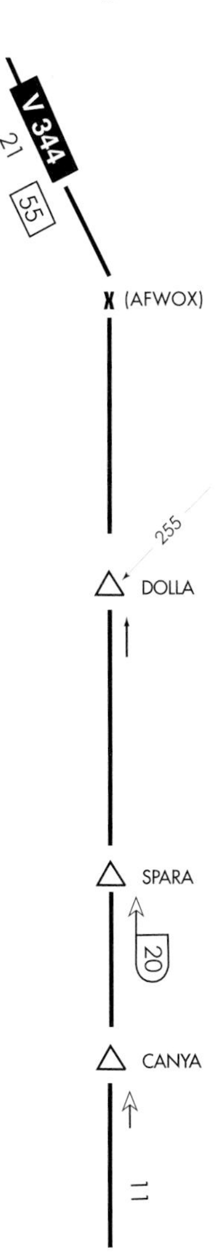

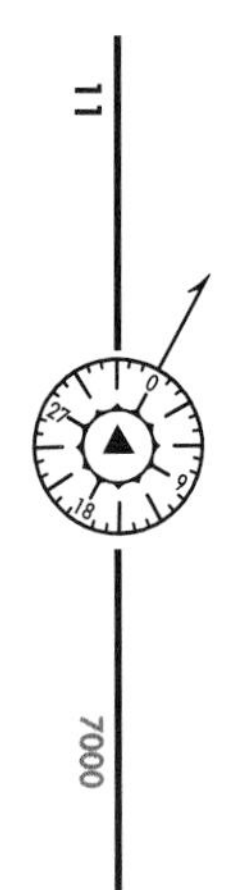

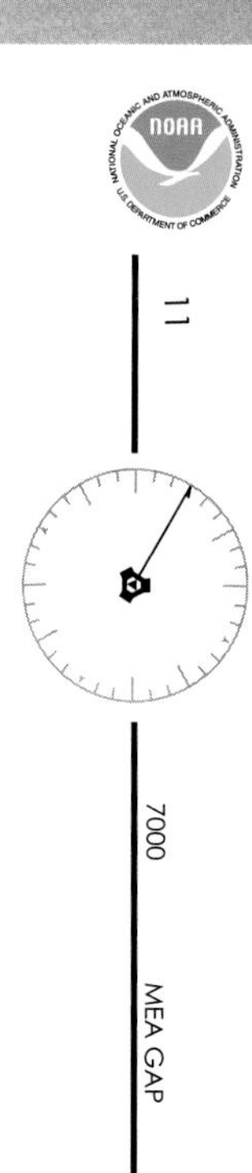

Intersections and navaids are designated as either compulsory or noncompulsory reporting points. Noncompulsory reporting points are identified by open triangles, and position reports are not required, unless requested by ATC. A compulsory reporting point is identified by a solid triangle. In a nonradar environment, you are required to make a position report when you pass over this point. When a navaid is a compulsory reporting point, a black triangle is placed in the center of the navaid symbol.

The minimum enroute altitude (MEA) is ordinarily the lowest published altitude between radio fixes that guarantees adequate navigation signal reception and obstruction clearance (2,000 feet in mountainous areas and 1,000 feet elsewhere). It is normally the lowest altitude you would use during an IFR flight on airways. You can generally expect adequate communication at the MEA, although it is not guaranteed. Although the MEA is defined as providing acceptable navigational signal coverage, under certain circumstances the MEA may have a gap in signal coverage of up to 65 miles. These gaps are noted on Jeppesen charts with a symbol, and with the words MEA GAP on NOS charts.

An MEA generally guarantees both obstruction clearance and navigation signal coverage, and is normally the lowest altitude you use on an airway.

A minimum obstruction clearance altitude (MOCA) is shown for some route segments. On Jeppesen charts, it is identified by the letter T following the altitude. On NOS charts, an asterisk precedes the altitude. The major difference between an MEA and a MOCA is that the MOCA ensures a reliable navigation signal only within 22 nautical miles of the facility; conversely, the MEA ordinarily provides reliable navigation signals throughout the entire segment. Since you may not be able to receive the facility or navigate along the airway beyond 22 nautical miles, ATC will only issue the MOCA as an assigned altitude when you are close enough to the navaid.

ATC may assign the MOCA when certain special conditions exist, and when within 22 nautical miles of a VOR. A MOCA does not guarantee you will receive a reliable navigation signal if you are more than 22 nautical miles from the facility. MOCAs are preceded by an asterisk on NOS charts.

When you are not following an airway, such as during a direct segment of an IFR flight, you are responsible for determining your own minimum altitude in accordance with FAR Part 91. Basically, you must remain at least 1,000 feet above the highest obstacle within a horizontal distance of 4 nautical miles from your intended course. In designated mountainous areas, the minimum altitude is increased to 2,000 feet and the distance from the

. . . RADAR CONTACT LOST . . .

There are dozens of instances every year when ATC computers go down, power fails, radar scopes go dark, or radio communications fail, leaving pilots without ATC guidance. In other cases, there may be gaps in radar coverage along your route of flight. Accepting vectors from controllers does not relieve you of responsibility for the safety of your flight. Sometimes controllers accidentally vector airplanes toward other traffic or terrain. You must maintain a safe altitude and keep track of your position, and it is your obligation to question controllers, request an amended clearance, or, in an emergency, deviate from their instructions if you believe that the safety of your flight is in doubt.

Of course, keeping track of your altitude and position are basic elements of situational awareness, but sometimes you may feel that ATC has assumed some of your responsibility, particularly when things are busy or stressful. Until you fly with the airlines or corporate flight departments, you probably will not have equipment such as a ground proximity warning system (GPWS) and traffic alert and collision avoidance system (TCAS) on board to help you detect and correct unsafe altitudes and traffic conflicts. Because of this, be sure to keep careful track of your position and altitude, and pay attention to the locations of traffic around you.

course remains the same. Remember to consider the range limitations of the navigation facilities and your communication requirements when you establish your minimum altitude. Both kinds of charts provide a minimum off-route altitude for each quadrangle of the latitude-longitude grid on the chart. NOS uses the abbreviation OROCA, for off-route obstruction clearance altitude, and Jeppesen uses MORA, for minimum off-route altitude. The concept is similar to the maximum elevation figures (MEF) shown on sectional charts, except that MEFs show the approximate height of the highest terrain or obstruction in the quadrangle, while MORAs/OROCAs provide 1,000 feet of clearance above the highest terrain or man-made structure within the quadrangle. As with MEAs, the clearance in mountainous areas increases to 2,000 feet.

In mountainous areas where no other minimum altitude is prescribed, IFR operations must remain 2,000 feet above the highest obstacle within a horizontal distance of 4 nautical miles from the intended course.

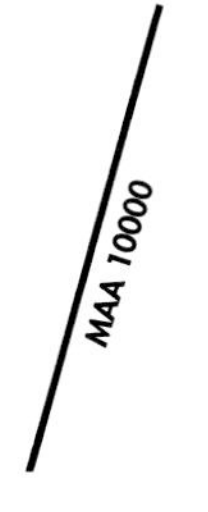

Occasionally, it is necessary to establish a **maximum authorized altitude (MAA)** for a route segment. At higher altitudes, you might be able to receive two or more VOR stations simultaneously on the same frequency, making the signals unreliable for navigation. MAA is the highest altitude you can fly based on the line-of-sight transmitting distance of VOR or VORTAC stations using the same frequency. It guarantees that you will only receive one signal at a

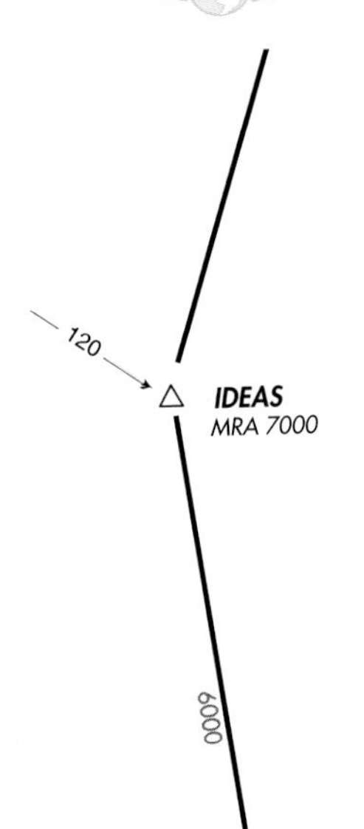

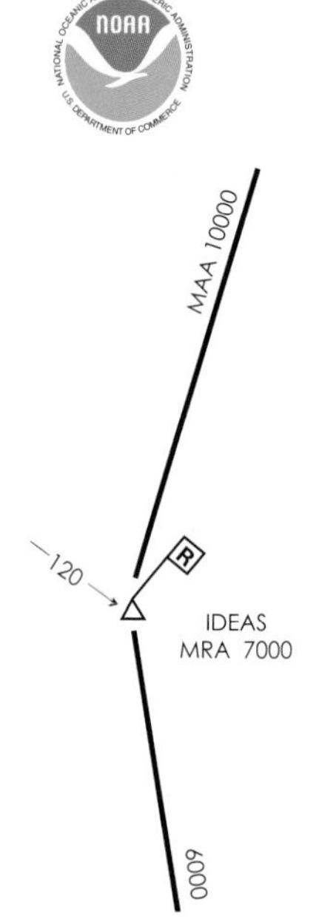

time on a given frequency. A maximum authorized altitude is shown on the chart with the letters MAA, followed by the altitude, either in feet or as a flight level.

On the other hand, the **minimum reception altitude (MRA)** is the lowest altitude that ensures adequate reception of the navigation signals forming an intersection or other fix. The MEA provides reception for continuous course guidance, but a higher altitude may be necessary to receive signals from the navaids off the airway being flown to enable you to identify a specific position. Operating below an MRA does not mean you will be unable to maintain the airway centerline, only that you may not be able to identify a particular fix or intersection. On both Jeppesen and NOS charts, the letters MRA precede the minimum reception altitude. NOS also alerts you to an MRA by enclosing the letter R in a flag.

An MRA is designated where a minimum altitude is needed to receive a navaid away from the airway being flown in order to identify an intersection.

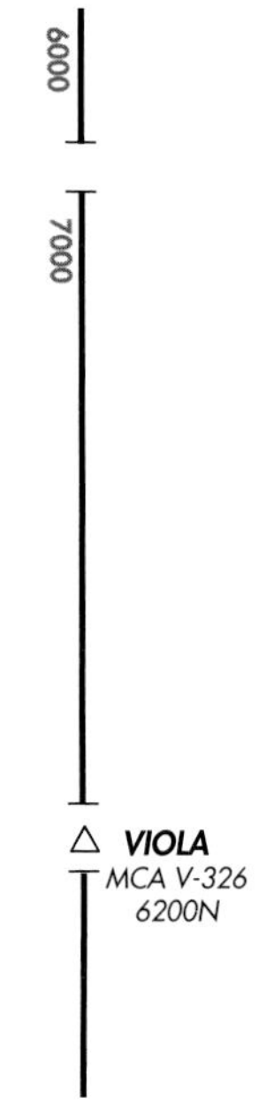

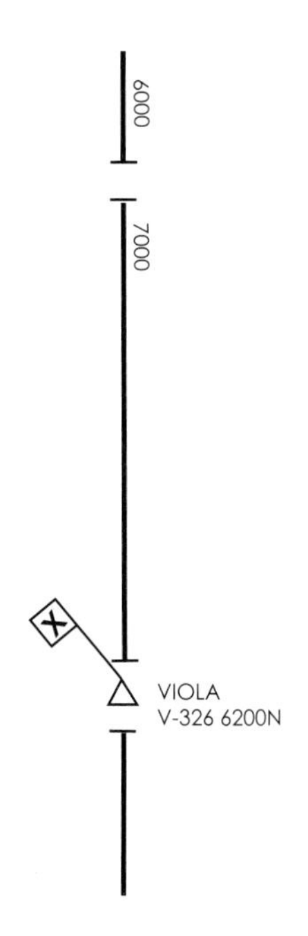

A bar symbol crossing an airway at an intersection indicates a change in MEA. This symbol can also be used to indicate a change in MAA or to show a change in the MOCA when an MEA is not published for the route. When you see this symbol, be sure to compare MEAs and MOCAs along the entire route and look for an MAA to determine the basis for the change.

When an MEA changes to a higher altitude, you normally begin your climb upon reaching the fix where the change occurs. If you are able to maintain a climb of at least 120 feet/n.m., you should have adequate obstruction clearance. In some cases, rising terrain, obstacles, or navigation signal reception may dictate a **minimum crossing altitude (MCA)** at the fix. You must begin climbing prior to reaching the fix in order to arrive over the fix at the MCA. [Figure 5-6] NOS uses an X enclosed in a flag to alert you to an MCA restriction. Do not confuse this with the flag symbol used to show a minimum reception altitude (MRA).

Jeppesen usually shows an MCA next to the intersection, along with the airway number, altitude, and direction. To reduce clutter around navaids, a reference number in a black circle on the facility box is used to indicate MCAs. The MCAs are listed in a box nearby. [Figure 5-7]

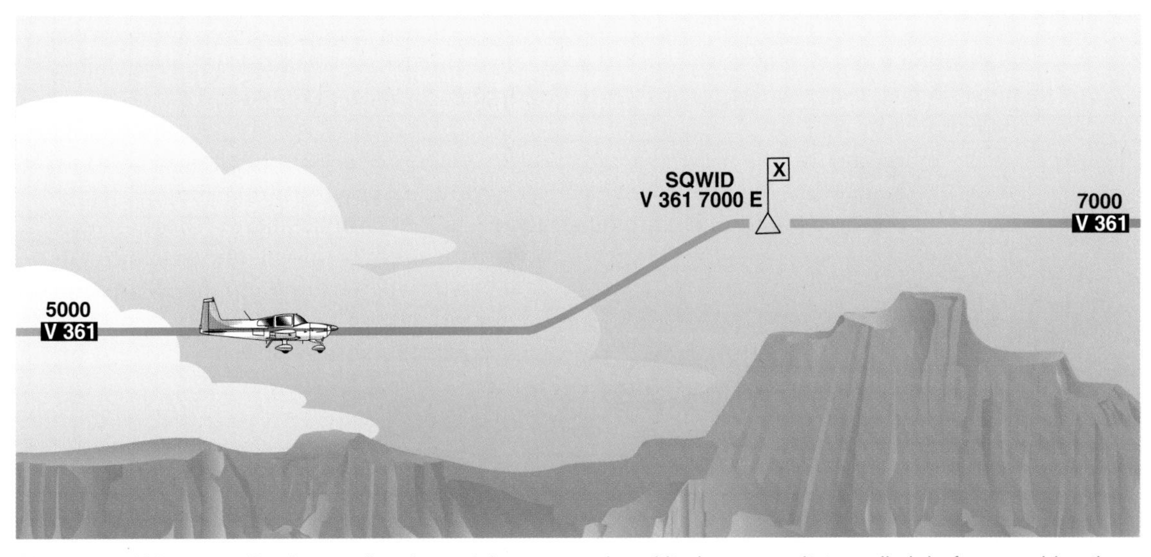

Figure 5-6. When traveling in one direction, minimum crossing altitudes necessitate a climb before reaching the fix. In this example, eastbound flights at the lower MEA must climb before reaching the fix (NOS symbols shown).

A flag with an X signifies the MCA on NOS charts. The altitude and applicable flight direction appear near the symbol. Plan your climb so that you will reach the MCA before crossing the fix. See figure 5-6.

The MAA, MCA, MRA, MOCA, and MEA all guarantee 1,000 feet of obstacle clearance in non-mountainous areas. In designated mountainous areas the clearance is 2,000 feet.

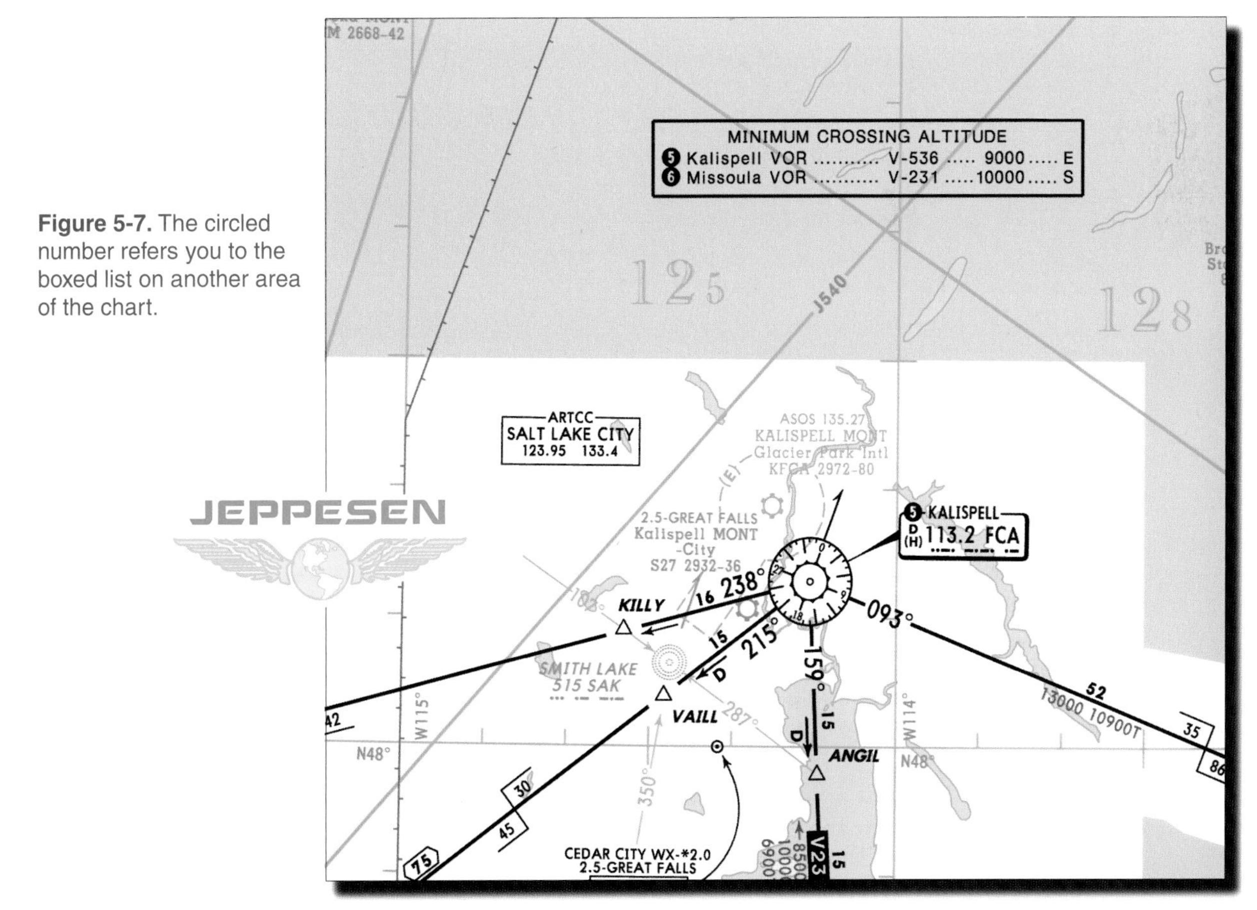

Figure 5-7. The circled number refers you to the boxed list on another area of the chart.

When you are following an airway, you normally change frequencies midway between navigation aids. However, there are times when this is not practical. When a change must be made somewhere other than the midpoint, a **changeover point (COP)** is established. [Figure 5-8]

To help you make the transition between low altitude and high altitude airways, Jeppesen includes the high altitude enroute structure on low altitude charts. These airways are printed in green and include the appropriate jet route number. The MEA for all jet routes is 18,000 feet MSL, unless otherwise specified.

You normally change frequencies midway between navaids, unless a changeover point (COP) is designated. The COP symbols are illustrated in figure 5-8.

The MEA along jet routes is 18,000 feet MSL, unless otherwise specified.

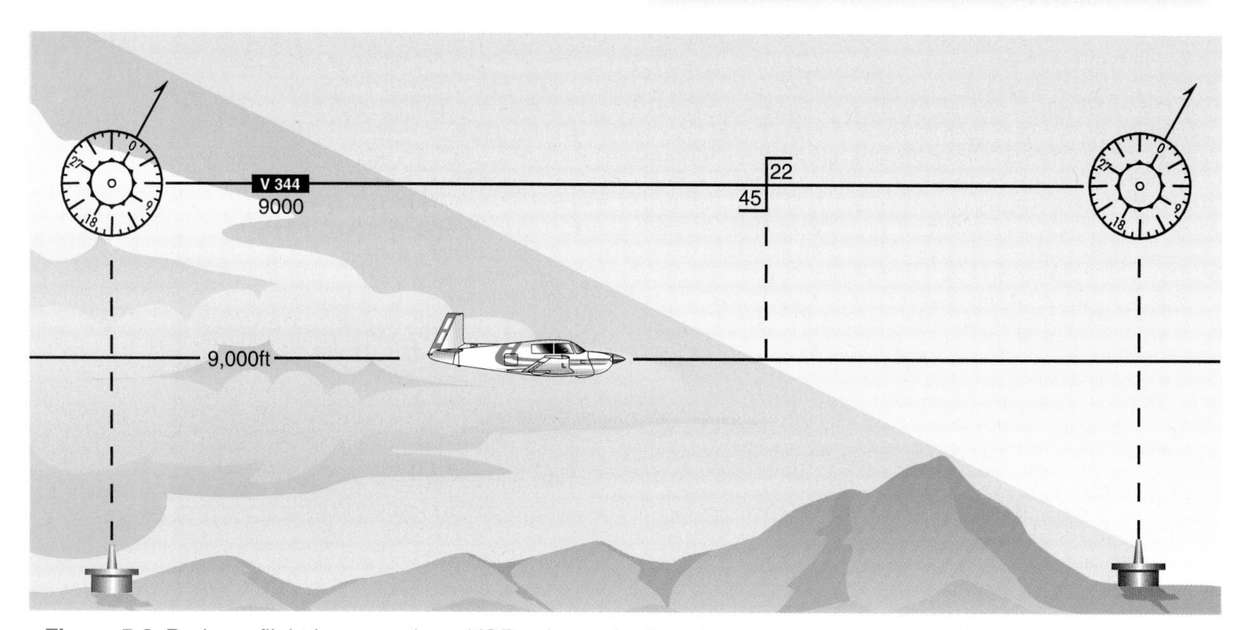

Figure 5-8. During a flight between these VORs, the navigation signals cannot be received from the second VOR at the midpoint of the route, so a changeover point is depicted. COPs indicate the point where a frequency change is necessary and show the distance in nautical miles to each navaid.

Communication

On Jeppesen charts, FSS communication frequencies are found above navaid facility boxes. Since these frequencies are always in the 120 MHz range, Jeppesen displays only the last two or three digits, thus 122.2 is shown as 2.2. NOS charts show the complete frequency. NOS uses a shadow box to indicate an FSS. The standard FSS frequencies are normally available with other frequencies shown above the box, while Jeppesen includes 122.2 with the others. The emergency frequency, 121.5, is not shown on either Jeppesen or NOS charts, since it is normally available at all FSSs.

An additional feature of Jeppesen charts is the identifier and frequency of the FSS offering enroute flight advisory service (EFAS) in the area. If HIWAS or TWEB is available at a

HIWAS is indicated by a small circled H in the upper right-hand corner of the navaid box on NOS charts, while Jeppesen places the acronym itself above the box.

GPS: Sole Navaid in the 21st Century?

The amazing accuracy, flexibility, and accessibility of the GPS system has prompted the FAA to plan the phaseout of existing ground-based navigation aids. Omega, a very low frequency long range navigation system, has already been turned off, and plans call for LORAN-C, VOR, NDB, and DME stations to be shut down eventually. The good news for the future is that pilots would only have to learn to use one kind of black box for all their navigation and approach needs.

But what about the reliability of the satellite constellation and its signals? Of the 38 GPS satellites launched by the U.S. through 1997, 12 have failed. Nonetheless, the spacecraft have proven themselves reliable and rugged, and there are usually three operational spares in orbit when a satellite is lost. The signal itself is subject to interference, both natural and man-made. Solar activity and ionospheric disturbances, accidental or intentional jamming, and even interference from other radios inside your airplane can render the weak satellite signal unusable. GPS augmentation systems and newer dual frequency satellites should help solve some of these problems. Many users advocate maintaining elements of existing navaid systems as an adjunct and backup to GPS.

Since airways link navaids, does that mean that future IFR enroute charts will be blank except for airports with their latitude and longitude coordinates? Hardly. Standardized GPS waypoints will likely replace many navaids and intersections, especially around busier airports. ATC communication frequencies and sector boundaries will still be important, as well as boundries between airspace classes. Since MEAs would disappear with the airways, you could expect more terrain and obstruction information to enable you to find a safe altitude for your chosen course.

particular facility, it is indicated above the box on Jeppesen charts. NOS places a small circled H in the upper right-hand corner of the facility box to indicate HIWAS is available, and a circled T in the upper right-hand corner signifies a TWEB at the NAVAID.

Most FSSs are able to use 122.2, as well as the emergency frequency, 121.5. Additional frequencies are shown above navaid boxes.

The boundaries between air route traffic control centers (ARTCCs) are designated by distinctive lines on both Jeppesen and NOS charts, with the names of the controlling centers on each side. **Remote communication outlets (RCOs)** have been set up to provide adequate communication coverage throughout the area served by the center. An RCO is shown with a box that lists the name of the center and the frequencies used. NOS charts include the UHF frequencies. RCOs are also used by Flight Service Stations to extend their area of coverage. An RCO not associated with a navaid is shown with a thin-lined box. The name of the FSS and the appropriate frequency also are listed. [Figure 5-9]

Both charts use distinctive lines to mark the edges of adjacent ARTCCs. Look for ARTCC discrete frequencies in boxes with the name of the controlling center. An RCO for an FSS will have the name of the FSS and the frequency in a communication box. See figure 5-9.

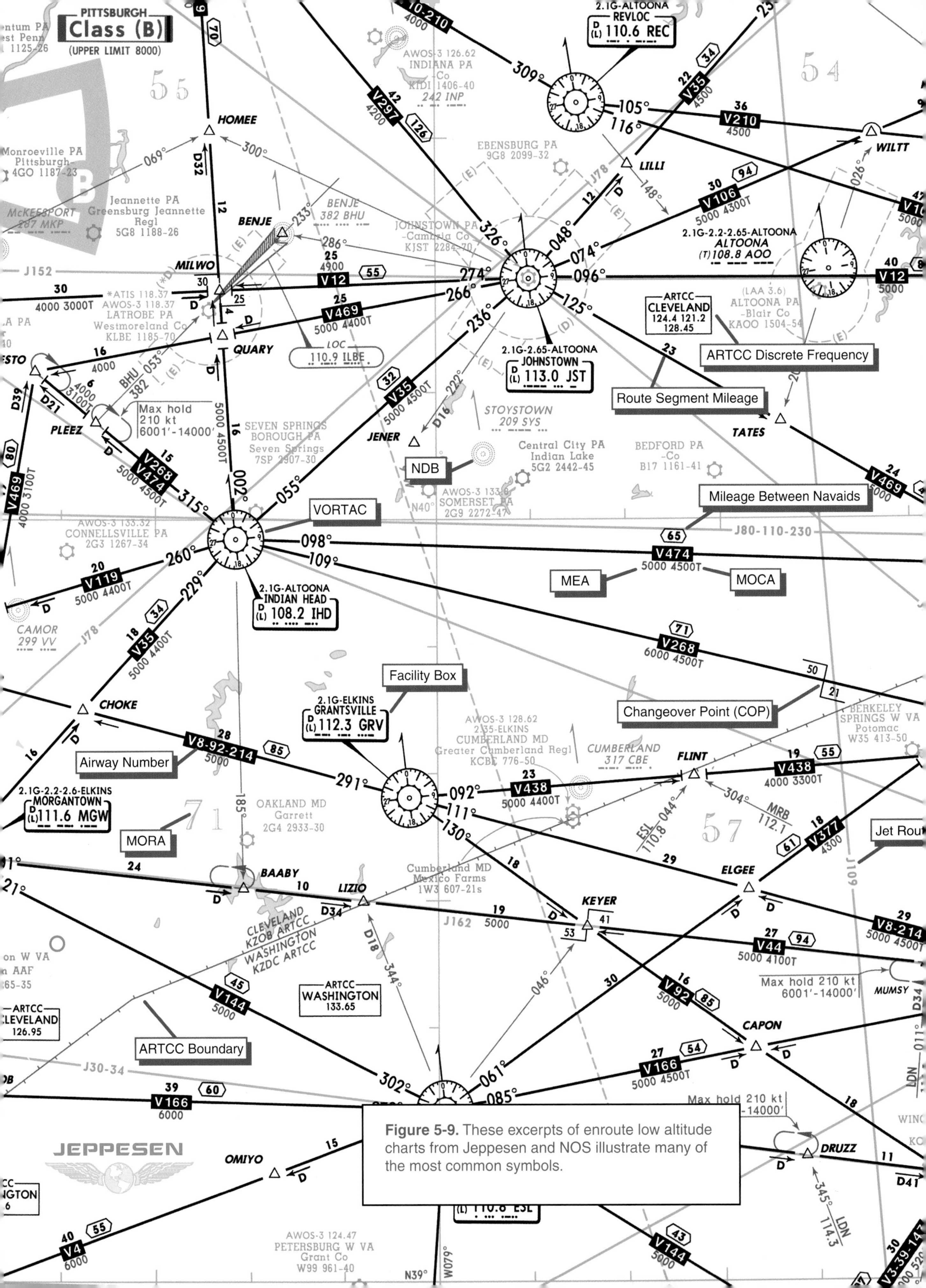

Figure 5-9. These excerpts of enroute low altitude charts from Jeppesen and NOS illustrate many of the most common symbols.

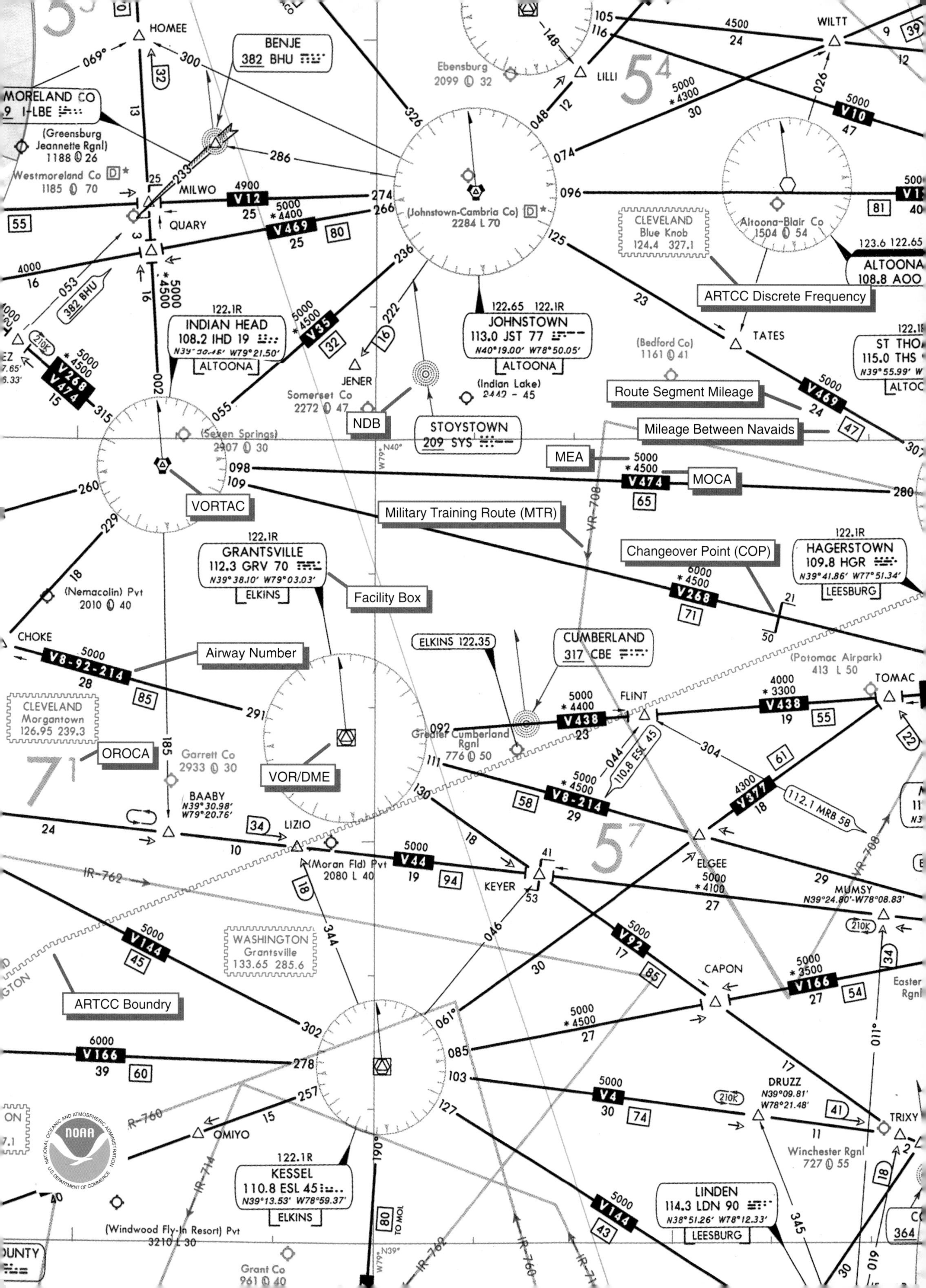

ARTCC Discrete Frequency
Route Segment Mileage
Mileage Between Navaids
MEA
MOCA
NDB
VORTAC
Military Training Route (MTR)
Changeover Point (COP)
Facility Box
Airway Number
OROCA
VOR/DME
ARTCC Boundry
BENJE 382 BHU
HOMEE
MILWO
QUARY
JOHNSTOWN 113.0 JST 77
N40°19.00′ W78°50.05′
ALTOONA
(Johnstown-Cambria Co) 2284 L 70
CLEVELAND Blue Knob 124.4 327.1
Altoona-Blair Co 1504 54
ALTOONA 108.8 AOO
INDIAN HEAD 108.2 IHD 19
STOYSTOWN 209 SYS
JENER
TATES
WILTT
LILLI
Ebensburg 2099 32
GRANTSVILLE 112.3 GRV 70
N39°38.10′ W79°03.03′
ELKINS
HAGERSTOWN 109.8 HGR
N39°41.86′ W77°51.34′
LEESBURG
CUMBERLAND 317 CBE
ELKINS 122.35
FLINT
TOMAC
CHOKE
CLEVELAND Morgantown 126.95 239.3
Garrett Co 2933 30
BAABY
N39°30.98′
W79°20.76′
LIZIO
KEYER
ELGEE
MUMSY
N39°24.80′-W78°08.83′
CAPON
WASHINGTON Grantsville 133.65 285.6
DRUZZ
N39°09.81′
W78°21.48′
TRIXY
Winchester Rgnl 727 55
KESSEL 110.8 ESL 45
N39°13.53′ W78°59.37′
LINDEN 114.3 LDN 90
N38°51.26′ W78°12.33′
OMIYO
(Windwood Fly-In Resort) Pvt 3210 L 30
Grant Co 961 40
V12
V469
V35
V268
V474
V8-92-214
V438
V8-214
V377
V44
V92
V144
V166
V4
V10

AIRPORTS

Although there are many ways to classify airports, they are divided into two categories on instrument charts — those with a published instrument approach procedure and those without. If an airport has an instrument approach, Jeppesen prints the airport symbol and related information in blue. In addition, the city and state name are in capital letters. Airports without an approach are printed in green, using upper and lower case letters. NOS also uses color to distinguish between the two types. Airports with an instrument approach are printed in blue or green; airports without an instrument approach are printed in brown. [Figure 5-10]

Basic information about each airport is portrayed on enroute charts using symbols from the chart legend. Additional information about airports with instrument approaches is found on the end panels of the chart.

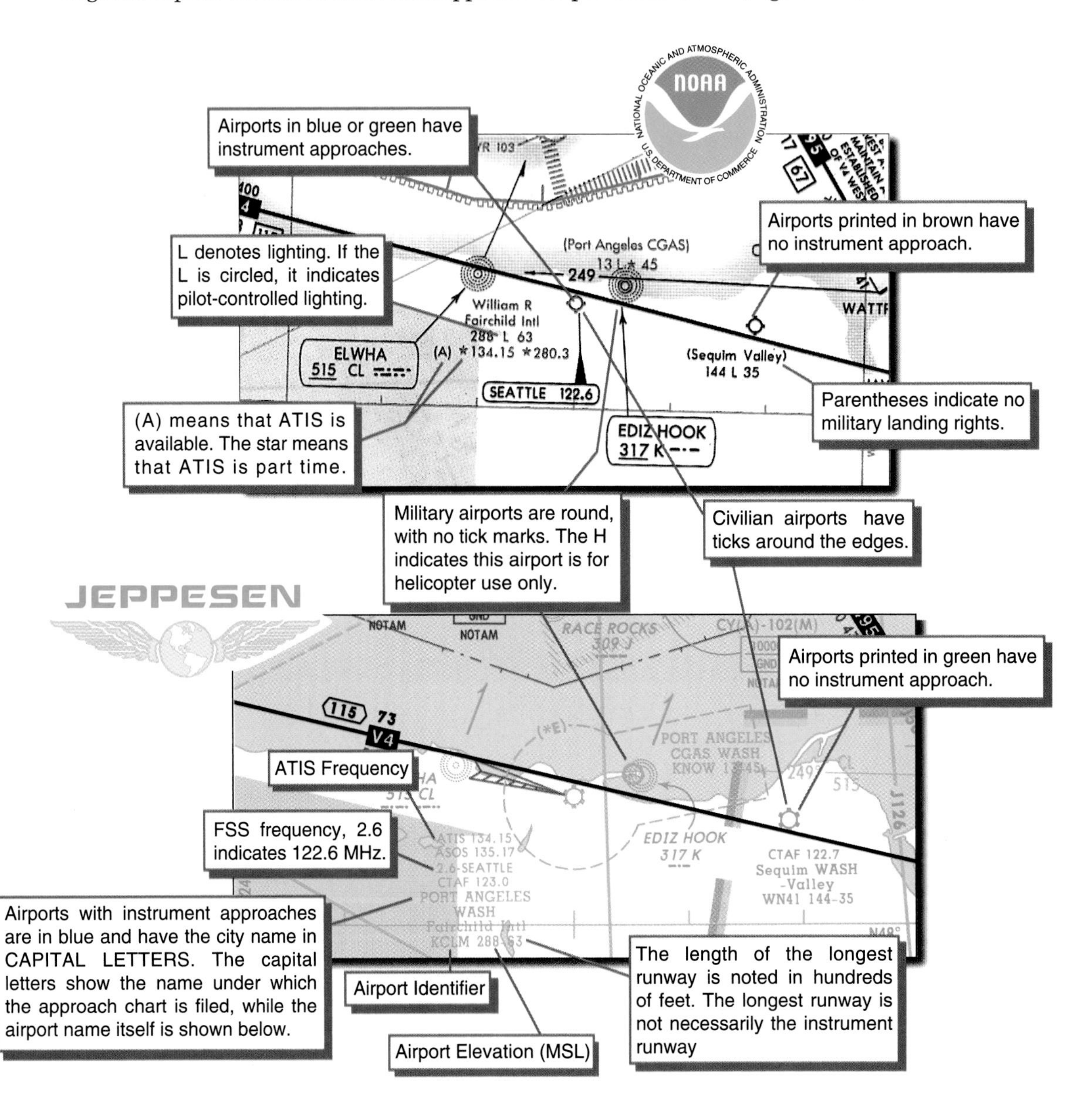

Figure 5-10. Jeppesen and NOS provide basic information about each airport, such as field elevation and length of the longest runway. Note the differences in how some information is portrayed; for example, Jeppesen shows airports without instrument approaches in green, while NOS charts them in brown. Additional information about airports with instrument approaches can be found on the end panels of both Jeppesen and NOS charts.

AIRSPACE

Within the contiguous United States, all airspace at and above 14,500 feet MSL, excluding the airspace within 1,500 feet of the ground, is controlled airspace. Below this altitude, the airspace may be either controlled or uncontrolled. Both Jeppesen and NOS use color to indicate the different types of airspace. Areas in white show controlled airspace, which includes Class B, C, D, and E airspace. Uncontrolled airspace, Class G, is shaded gray on Jeppesen charts and brown on NOS charts.

Class B airspace is outlined on Jeppesen charts by a maroon shaded band with the letter B repeated at intervals around the inside. The altitude limits for the various sectors also are shown. Class C is shown with a blue shaded outline with the letter C. NOS uses solid blue lines filled with light blue shading to identify Class B, and blue shading with a broken blue outline to denote Class C. In addition, Mode C areas are shown on NOS charts with blue and white stripes.

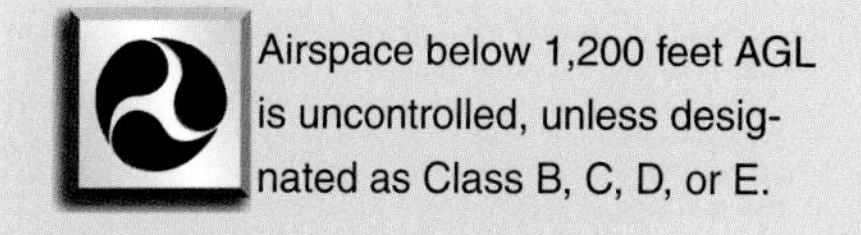

The letter C or D in a box following the airport name indicates Class C or D airspace on NOS charts. Jeppesen charts show the outlines of Class D and Class E airspace with a dashed blue line and the appropriate letter. An asterisk indicates that the airspace classification is part time, and a tabulation elsewhere on the chart shows the effective hours. Areas where fixed-wing, special VFR clearances are not available are outlined with small squares in either maroon or blue on Jeppesen charts. NOS follows the convention of sectional charts, placing the notation NO SVFR above the airport name.

On Jeppesen charts, prohibited and restricted areas have maroon hatched outlines, while warning, alert, or military operations areas have green hatching. NOS uses blue hatching around the edges of all special use airspace except military operations areas, which are shown with brown hatched edges. Military training routes (MTRs) also are depicted on NOS low altitude enroute charts. [Figure 5-11] Information concerning each area is listed in or near the airspace, or it can be found in tabular form on a chart panel.

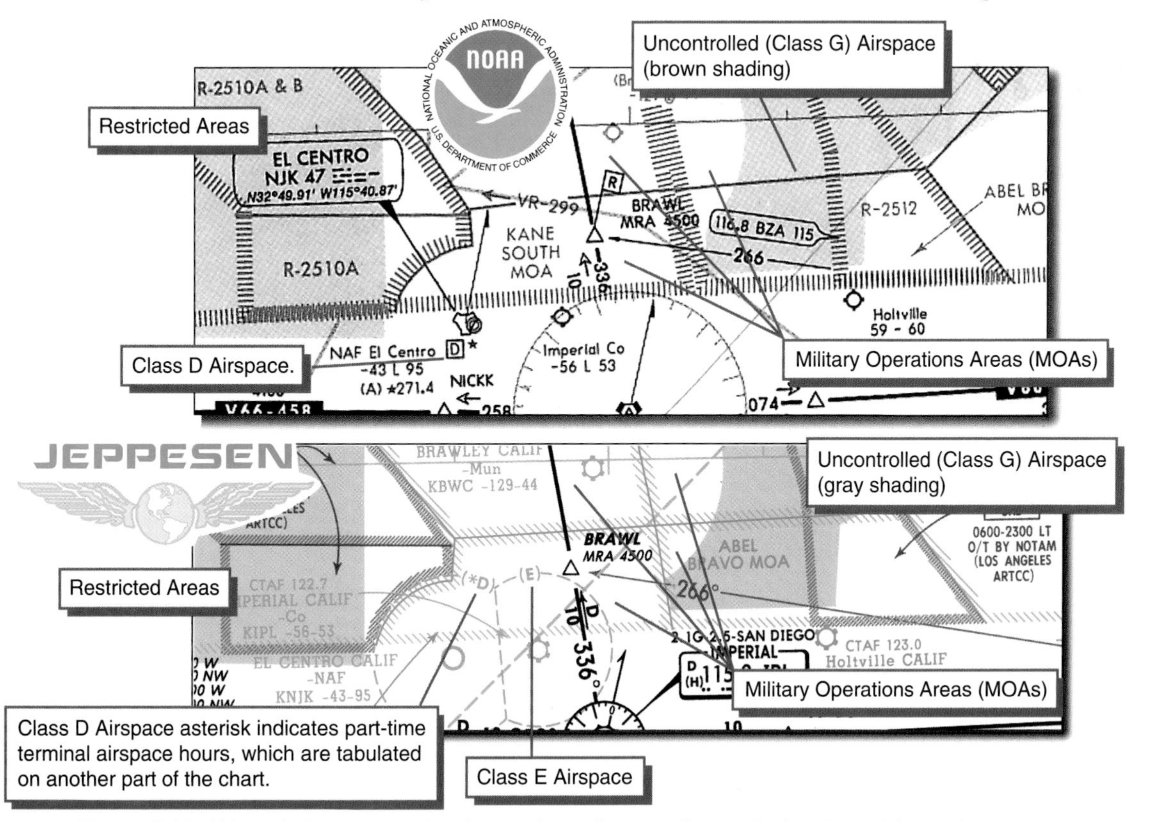

Figure 5-11. Although these examples do not show all types of controlled and special use airspace, they give you an idea of the differences between the NOS and Jeppesen portrayals.

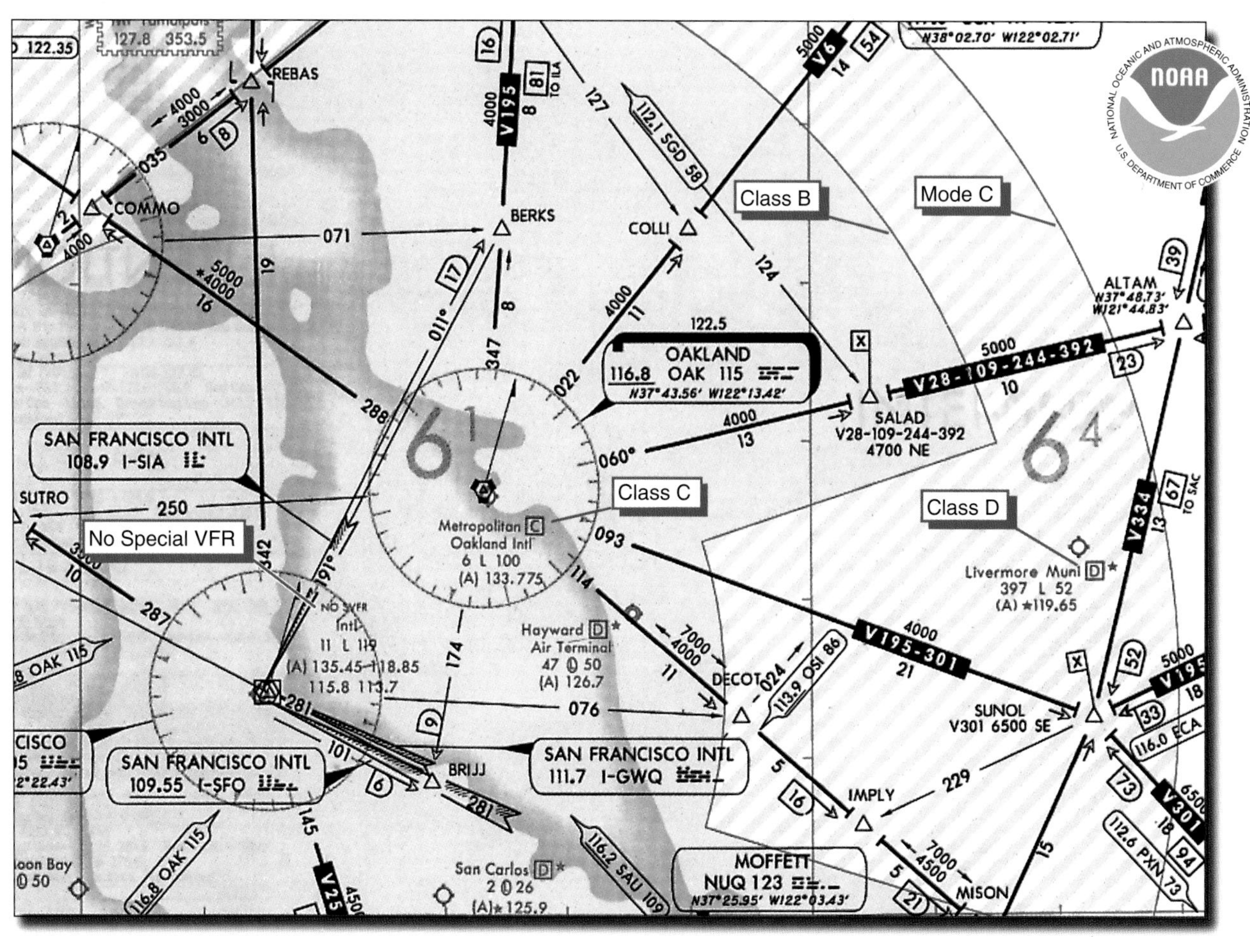

JEPPESEN

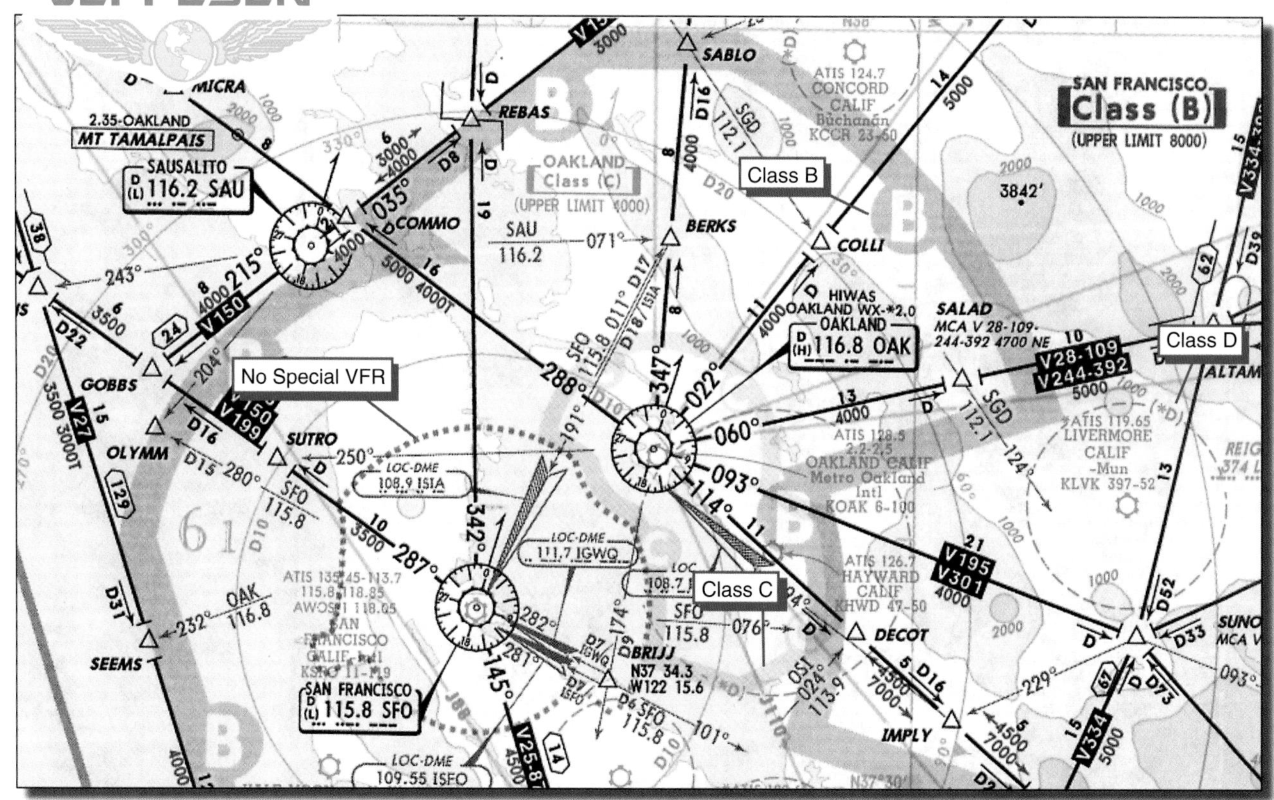

Figure 5-12. Compare these excerpts from the NOS and Jeppesen area charts for San Francisco.

AREA CHARTS

Near several major air traffic hubs, the density of information on the enroute chart can make it difficult to read and interpret. Area charts are created to portray these locations in a larger scale, to improve readability and provide more detail. You might think of these charts as the IFR equivalent of VFR Terminal Area Charts. Area charts do not provide approach or departure information, but can help with the transition from departure to the enroute structure, and from enroute to approach. Even when flying enroute, you should always refer to area charts to navigate through their coverage areas, since they provide important information that may not be shown on the enroute chart. [Figure 5-12]

Area chart coverage is shown on both the front panel and the face of enroute charts. On the front panel, Jeppesen uses a gray screen tint to show the actual area covered by area charts, while NOS prints the names of cities with area charts in black. On the faces of both Jeppesen and NOS enroute charts, a screened gray dashed line shows the area covered by a separate area chart. On the enroute chart, the information within the outlined area may be limited. If your departure or destination airport is within the boundary, you should refer to the area chart for information on arrival and departure routes, speed limit points, and other handy information.

Most of the symbology on area charts is the same as on enroute charts. Jeppesen includes several additional features not found on NOS area charts, for example, select area charts with terrain in excess of 4,000 feet above the main airport elevation may contain generalized contour information. Gradient tints in brown are used to indicate the elevation change between contour intervals, with lighter tints depicting lower elevations. Some spot elevations for high terrain may also be included. Keep in mind that this contour information does not ensure clearance around either terrain or man-made obstructions. Terrain contours are intended to help you orient yourself and visualize the layout of the terrain in the area. There may be higher uncharted terrain or man-made structures within the same vicinity. You must still comply with all minimum IFR altitudes dictated by the airway and route structure. [Figure 5-13]

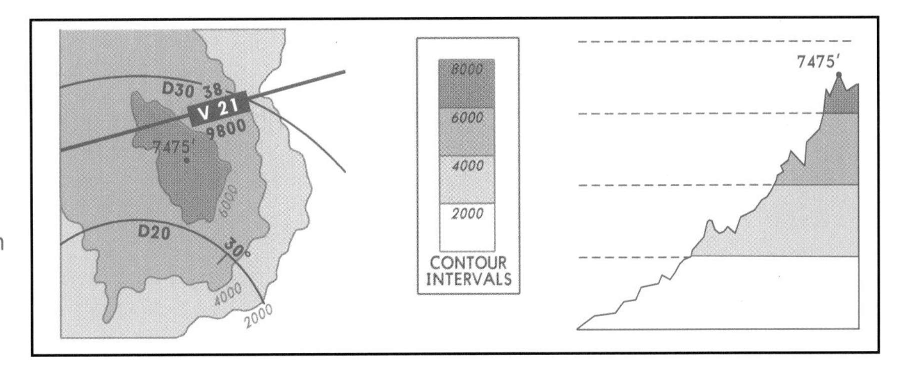

Figure 5-13. Within each contour interval, terrain may exist up to but not exceeding the elevation of the next higher contour interval. These contours are provided only as an aid in orientation, and do not provide detailed terrain information. In certain cases, a high terrain reference point is portrayed, along with its MSL elevation, when the highest terrain point within the closed contour exceeds the closed contour elevation by more than 50% of the chart's contour interval. DME arcs are included for situational awareness.

SUMMARY CHECKLIST

✓ Airways below 18,000 feet MSL are Victor airways. Airways at and above 18,000 feet MSL are jet routes.

✓ Airways are 8 nautical miles wide within 51 nautical miles of a navaid. At distances greater than 51 miles, the airway widens, and is defined by lines diverging at 4.5° from the center of each navaid.

✓ Intersections are defined by two navaids, or by a navaid and a DME distance. All intersections can be used as reporting points. Compulsory reporting points are charted as filled triangles.

✓ The minimum enroute altitude (MEA) generally guarantees both obstruction clearance and navigation signal coverage for the length of the airway segment.

✓ The minimum obstruction clearance altitude (MOCA) has the same terrain and obstruction clearance specifications as MEAs, and OROCAs/MORAs, but only promises reliable navigation signal coverage within 22 nautical miles of the facility.

✓ To provide obstruction clearance when flying outside of established airways, NOS and Jeppesen provide off-route obstruction clearance altitudes on enroute low altitude charts. NOS uses the term off-route obstruction clearance altitudes (OROCAs), and Jeppesen calls them minimum off-route altitudes (MORAs).

✓ The maximum authorized altitude (MAA) keeps you from receiving more than one VOR station at a time.

✓ The minimum reception altitude (MRA) ensures reception of both of the navaids that establish a fix. Below the MRA and above the MEA you still have course guidance, but may not be able to receive the off-course navaid that establishes the intersection fix.

✓ The minimum crossing altitude (MCA) reminds you to climb to a higher altitude prior to crossing a fix when rising terrain does not permit a safe climb after passing the fix.

✓ A changeover point (COP) is established where the navigation signal coverage from a navaid is not usable to the midpoint of an airway segment. Instead of changing frequencies at the midpoint of the route segment, you should tune to the next navaid at the COP.

✓ ARTCC boundaries are shown with distinctive lines on both Jeppesen and NOS charts.

✓ Colors are used to differentiate between airports with approach procedures and airports without instrument approaches.

✓ Class G airspace is uncontrolled and shown with gray shading on Jeppesen charts and brown shading on NOS charts.

✓ Area charts are usually larger-scale depictions of major terminal areas. They should be referred to whenever you are in their coverage area, since they may show details that have been omitted from enroute charts.

✓ High terrain is sometimes shown with gradient-tinted contours on select Jeppesen area charts.

KEY TERMS

Low Altitude Enroute Charts

Victor Airways

High Altitude Enroute Charts

Jet Routes

Mileage Break Point

Intersections

Noncompulsory Reporting Point

Compulsory Reporting Point

Minimum Enroute Altitude (MEA)

Minimum Obstruction Clearance Altitude (MOCA)

Maximum Authorized Altitude (MAA)

Minimum Reception Altitude (MRA)

Minimum Crossing Altitude (MCA)

Changeover Point (COP)

Remote Communications Outlet (RCO)

Area Charts

QUESTIONS

1. True/False. Low altitude enroute charts generally depict localizers that have only approach functions.

Match the following navaid symbols with the appropriate facility names.

2. NDB
3. TACAN
4. Localizer
5. VORTAC
6. VOR

A.

B.

C.

D.

E.

7. What is the symbol for a noncompulsory reporting point on both NOS and Jeppesen charts?

   A. A filled triangle
   B. An open triangle
   C. The letter R in a flag

8. True/False. A small circled H in the upper right-hand corner of a VOR facility box on an NOS chart indicates that the station broadcasts hazardous inflight weather advisory service (HIWAS) information.

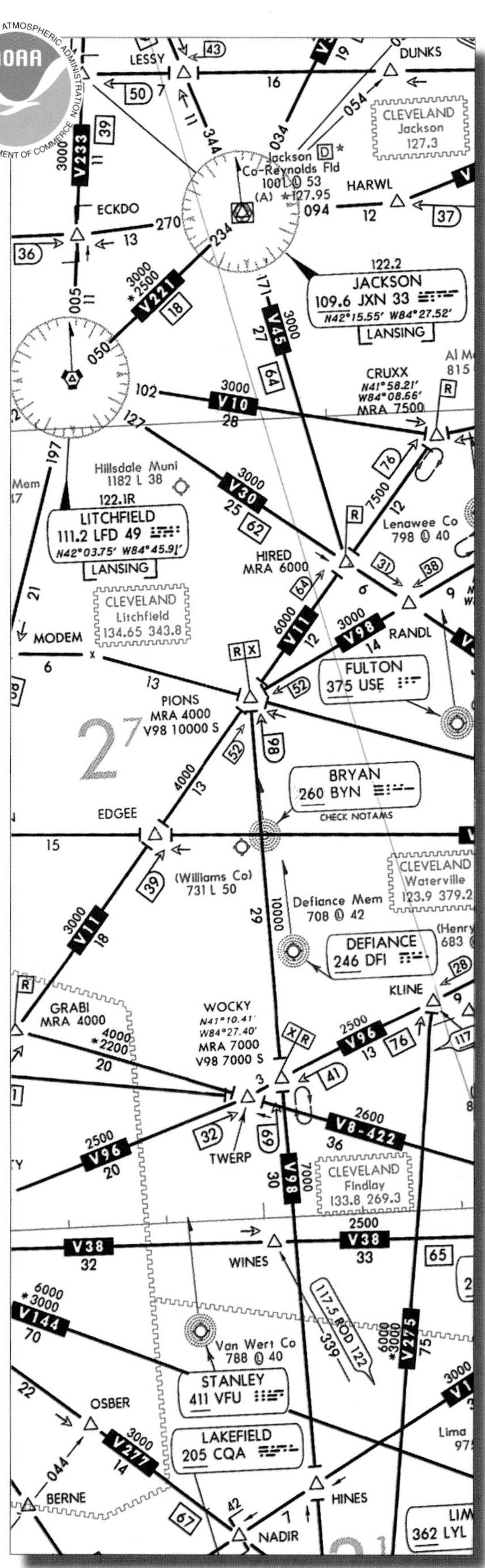

Refer to the accompanying low altitude enroute chart excerpt to answer questions 9 through 13.

9. What is the minimum enroute altitude (MEA) for V98 from PIONS Intersection to WOCKY Intersection?

   A. 4,000 feet
   B. 7,000 feet
   C. 10,000 feet

10. What is the minimum crossing altitude (MCA) at PIONS Intersection when flying southbound on V98?

    A. 4,000 feet
    B. 7,000 feet
    C. 10,000 feet

11. True/False. When flying northwestbound on V30 at HIRED Intersection, you must be at 6,000 feet MSL or more to receive LFD.

12. True/False. Williams County airport has no approved instrument approach procedure.

13. What is the MOCA on V221 between Litchfield VORTAC and Jackson VOR/DME?

    A. 1,800 feet
    B. 2,500 feet
    C. 3,000 feet

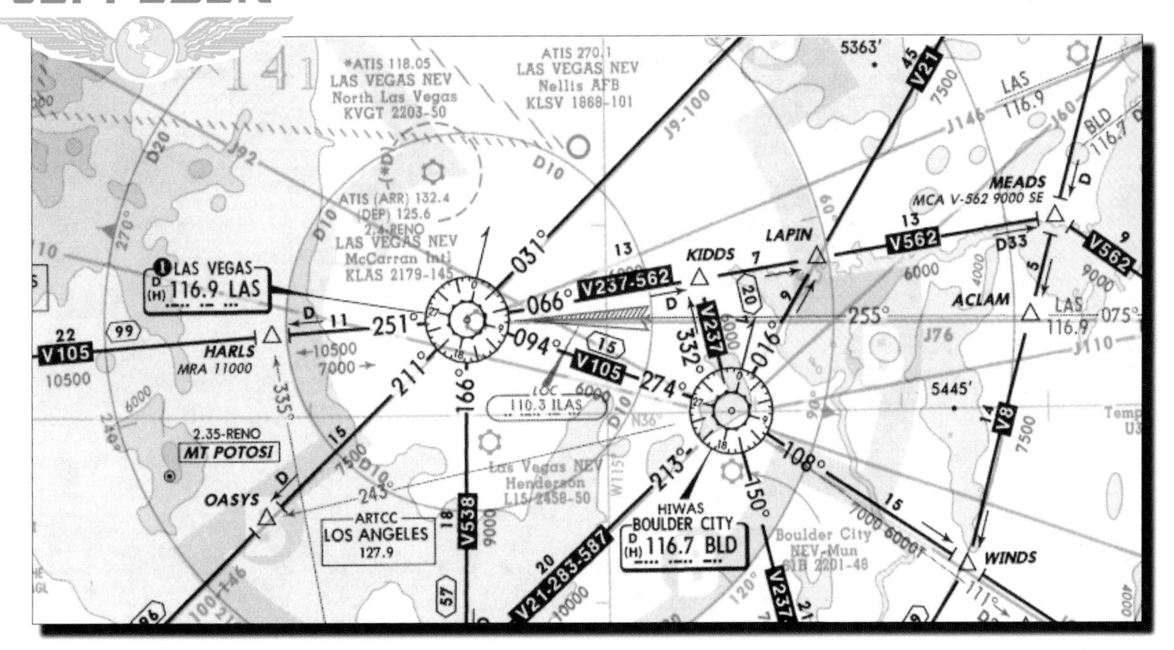

Refer to the accompanying area chart excerpt to answer questions 14 through 16.

14. When flying east from Las Vegas on V562, what is the minimum crossing altitude at MEADS Intersection?

    A. 6,000 feet
    B. 7,500 feet
    C. 9,000 feet

15. True/False. The minimum altitude at which you can expect to receive usable navigation signals from LAS when crossing HARLS Intersection is 11,000 feet.

16. What is the airspace class at the surface on North Las Vegas Airport?

    A. Class B
    B. Class D
    C. Class D (part time)

SECTION B

ENROUTE PROCEDURES

Before the days of nationwide radar coverage, enroute aircraft were separated from each other primarily by specific altitude assignments and position reporting procedures. Much of the pilot's time was devoted to inflight calculations, revising ETAs, and relaying position reports within the airway structure. Today, pilots have far more information and better tools to make inflight computations easier, and, with the expanded use of ATC, position reports are only necessary as a backup in case of radar failure. Still, the enroute phase of an IFR flight involves more than simply flying the airplane from one point to another. Staying in communication with ATC, making necessary reports, and responding to clearances take up a good portion of your time. The rest should be devoted to monitoring your position and staying abreast of any changes to the airplane's equipment status or weather. [Figure 5-14]

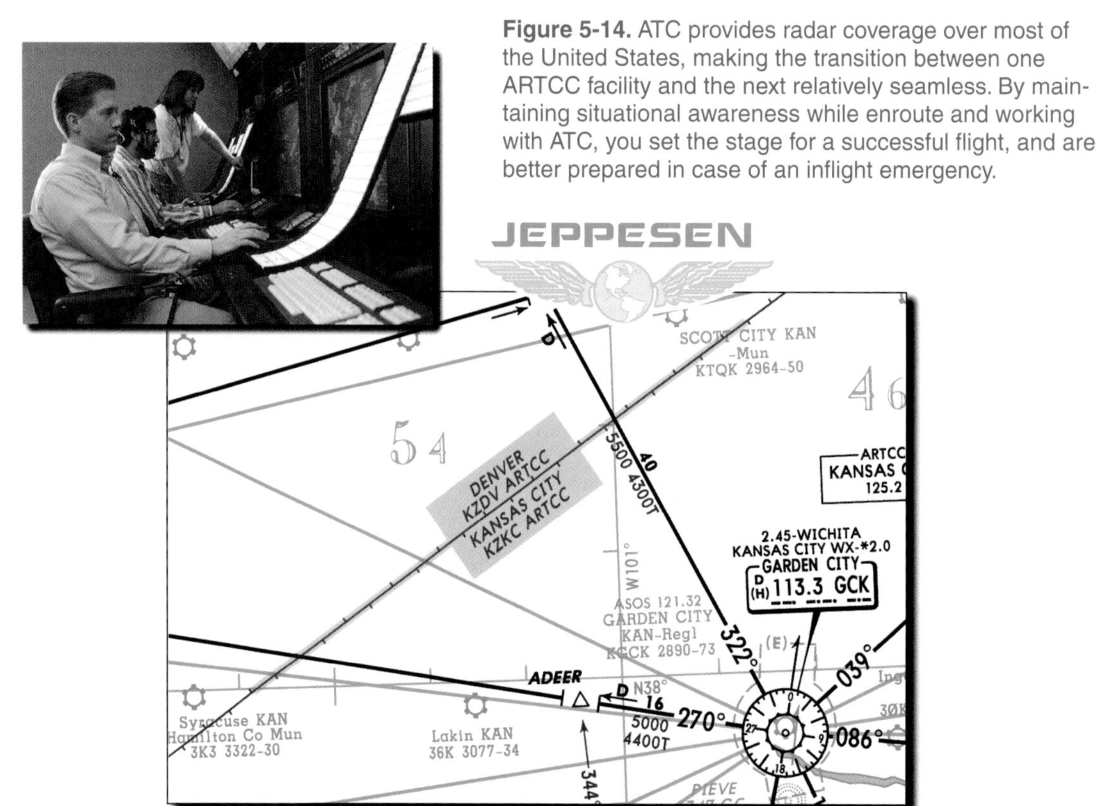

Figure 5-14. ATC provides radar coverage over most of the United States, making the transition between one ARTCC facility and the next relatively seamless. By maintaining situational awareness while enroute and working with ATC, you set the stage for a successful flight, and are better prepared in case of an inflight emergency.

TO FILE OR NOT TO FILE

"(I) called for a weather briefing. Both San Francisco and Oakland (reported) 1400 feet broken and greater than 6 miles visibility . . . pretty strong winds with a not-so-bad storm blowing through. I didn't file an IFR flight plan when asked by the briefer as the weather sounded OK. The forecast was for improvement, although it had been slow in coming as of that time . . . during preflight (I) noticed mist forming. It was also getting humid. I didn't think it could get foggy or misty with all of that wind. It seemed to be drizzling too. (I) departed Runway 30, and at about 300 feet, (I) started going through wisps of mist. . . (I) looked back and could see some lights along the bay. . .but nothing ahead. This was the last chance I had to stay anything near VFR.

The pilot then flew out over San Francisco Bay, and tried to pick up a clearance.

"They asked me to climb to no greater than 3500 feet in VFR conditions. I told them I was unable, but I could climb to 2000 feet (even though I was nowhere near VFR). . .I didn't tell them the truth of course — I was so worried about getting in trouble. Within about 5 minutes or so I had the clearance. . .got a vector of 080° for VOR Runway 9R (at) Oakland. . .fumbling with my charts the whole time. What a night! Pilots: don't do this to yourself, just file the IFR flight plan — you can cancel if you don't need it. And thank you, Bay Approach."

The pilot in this ASRS incident report may have felt that he could save time and avoid hassles by going VFR. However, filing IFR when weather conditions are marginal VFR normally is the wisest and safest decision. NTSB reports often show continued flight into marginal VFR conditions as a causal factor in weather-related accidents. The irony is that the pilots involved in these accidents are instrument-rated more times than not.

ENROUTE RADAR PROCEDURES

When you operate as pilot in command under IFR in controlled airspace, the FARs require you to continuously monitor an appropriate center or control frequency. After takeoff, your IFR flight is either in contact with a radar-equipped local departure control or, in some areas, an ARTCC facility. As your flight transitions to the enroute phase, you typically can expect a handoff from departure control to a center frequency if you are not already in contact with center. If you are using the tower enroute control procedure (TEC), you will receive a handoff from departure control to approach control at the next facility. The handoff between two radar facilities normally is a quick and simple procedure, but using the correct technique makes the process easier.

COMMUNICATION

As your flight leaves the departure controller's airspace, either by radar vector or your own navigation, you will be instructed to contact the center. The instructions issued by the controller include the name of the facility, the appropriate frequency, and any pertinent remarks. When you receive instructions to change frequencies, you should acknowledge the change by repeating it back to the controller. This readback also verifies that you understand the instruction and have received the correct frequency. [Figure 5-15]

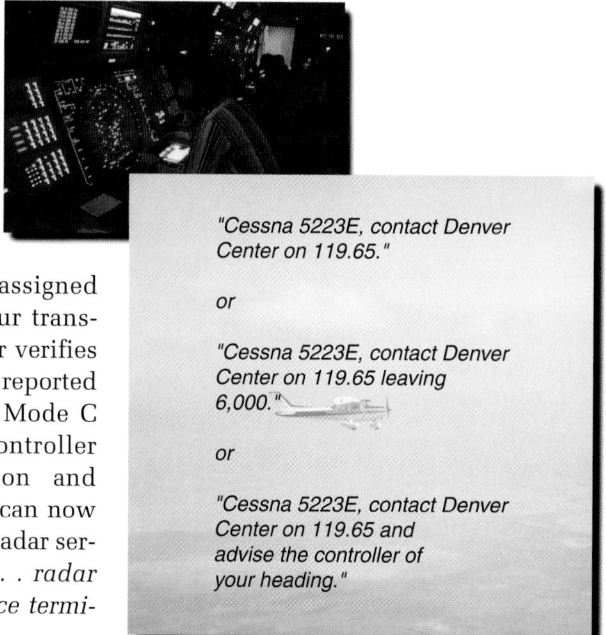

Figure 5-15. In addition to stating the facility and frequency, the controller may issue specific instructions during the handoff.

Your initial callup to the center should include the facility identification, your aircraft identification, altitude, and assigned altitude. [Figure 5-16] When your transmission is received, the controller verifies your position and compares your reported altitude to that shown by your Mode C equipment. Once verified, the controller acknowledges your transmission and states, " . . . *radar contact.*" You can now expect radar flight following and radar services until you are advised, " . . . *radar contact lost*" or " . . . *radar service terminated.*"

Figure 5-16. Your callup procedure to the center is basically the same as the one used when contacting departure control.

If you make a required frequency change and the center does not acknowledge your callup, return to the previously assigned frequency. ARTCC facilities are subject to transmitter/receiver failures, though this is rare, and radio contact may be momentarily lost. Each ARTCC frequency has at least one backup transmitter and receiver which can be put into service quickly, with little or no disruption of service. Technical problems of this type may cause a delay, but the switch-over process rarely takes more than one minute. Therefore, you should wait at least that long before deciding that the center's radio has failed. If you cannot establish contact using the newly assigned frequency, return to the one previously assigned and request an alternate frequency. If you are still unable to establish radio contact, try again on any ARTCC frequency. Failing that, contact the nearest FSS in the area for further instructions.

REPORTING PROCEDURES

In addition to acknowledging a handoff to another controller, there are reports that you should make without a specific request from ATC. Certain reports should be made at all times regardless of whether you are in radar contact with ATC, while others are necessary only if radar contact has been lost or terminated.

RADAR/NONRADAR REPORTS

Whether or not you are in radar contact, when you are cleared from an altitude to one newly assigned, you should report leaving the previous altitude. However, when you reach the newly assigned altitude, you are not required to report unless ATC requests you to do so. If you are operating on a VFR-on-top clearance, you should advise ATC of an altitude change. You should report your time and altitude upon reaching a holding fix or clearance limit and report when you are leaving a holding fix or clearance limit. If you cannot continue your instrument approach to a landing, you must report missed approach and request a clearance for a specific action. You may request to proceed to an alternate airport or try another approach. Changes in your aircraft's performance also warrant a report. For example, you should notify ATC if you are unable to climb or descend at a rate of at least 500 feet per minute, or if your average true airspeed at cruising altitude varies by 5% or 10 knots (whichever is greater) from that filed in your flight plan.

If you experience any loss of VOR, TACAN, ADF, low frequency navigation receiver capability, complete or partial loss of ILS receiver capability or impairment of air/ground communication capability, you are required by the FARs to advise ATC. This report should include your aircraft identification, the equipment affected, the degree to which your capability to operate under IFR in the ATC system is impaired, and the nature and extent of assistance you desire. ATC uses these reports to help regulate traffic and to avoid any conflicts which might develop. For example, if you lose your DME equipment, the controller will know that you cannot accept a hold or approach that requires DME at your destination. You also are required by regulation to advise ATC if you encounter weather conditions which have not been forecast or hazardous weather, as well as any information relating to the safety of flight.

You should advise ATC when your airspeed changes by 5% or 10 knots, whichever is greater. You should also inform ATC if your DME fails.

NONRADAR REPORTS

If radar contact has been lost or radar service terminated, the FARs require you to provide ATC with position reports over designated VORs and intersections along your route of flight. These compulsory reporting points are depicted on IFR enroute charts by solid triangles. Position reports over fixes indicated by open triangles are only necessary when requested by ATC. If you are on a direct course that is not on an established airway, report over the fixes used in your flight plan that define the route. Compulsory reporting points also apply when you conduct your IFR flight in accordance with a VFR-on-top clearance. [Figure 5-17]

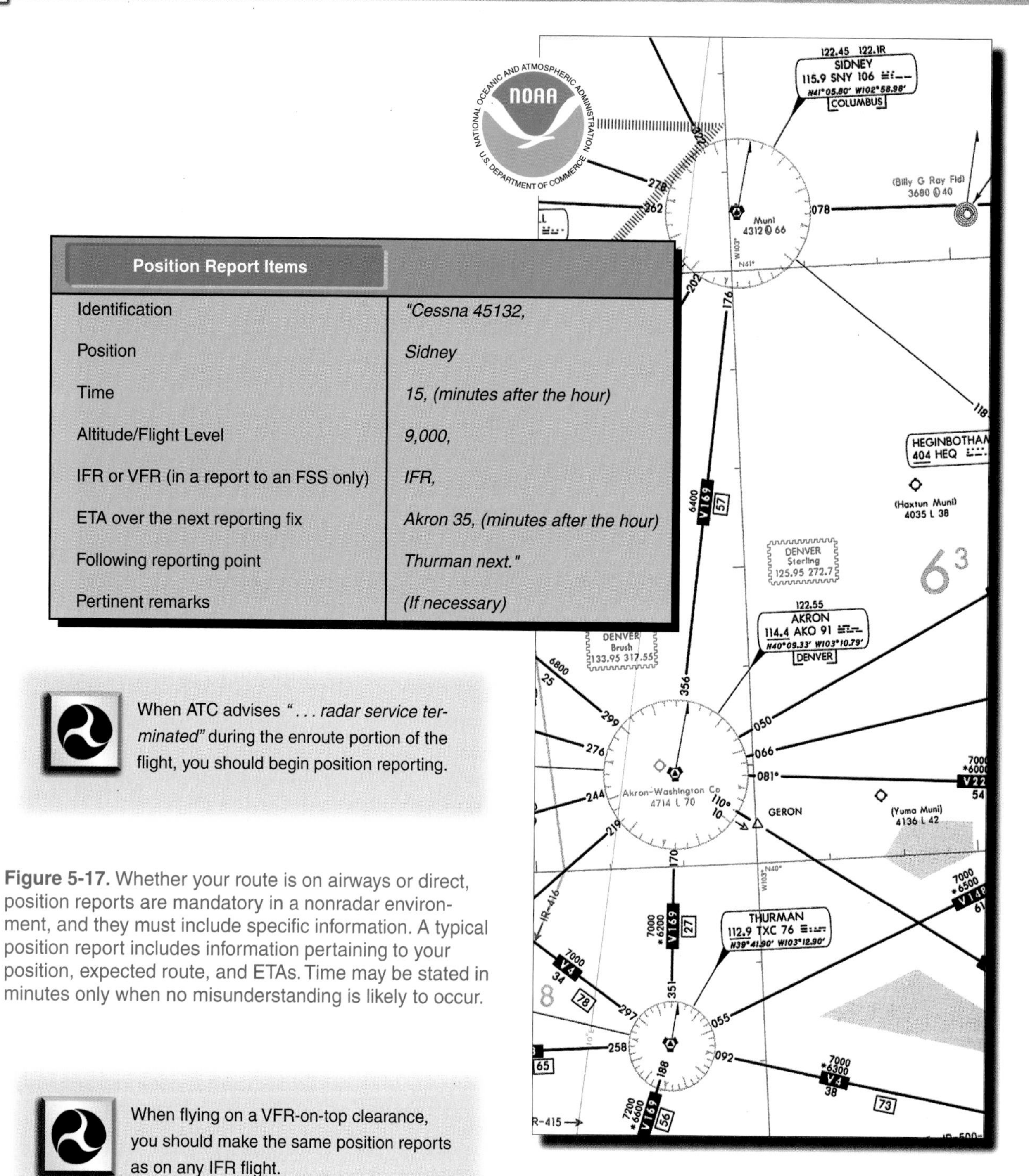

Position Report Items	
Identification	*"Cessna 45132,*
Position	*Sidney*
Time	*15, (minutes after the hour)*
Altitude/Flight Level	*9,000,*
IFR or VFR (in a report to an FSS only)	*IFR,*
ETA over the next reporting fix	*Akron 35, (minutes after the hour)*
Following reporting point	*Thurman next."*
Pertinent remarks	*(If necessary)*

When ATC advises *". . . radar service terminated"* during the enroute portion of the flight, you should begin position reporting.

Figure 5-17. Whether your route is on airways or direct, position reports are mandatory in a nonradar environment, and they must include specific information. A typical position report includes information pertaining to your position, expected route, and ETAs. Time may be stated in minutes only when no misunderstanding is likely to occur.

When flying on a VFR-on-top clearance, you should make the same position reports as on any IFR flight.

There are several additional reports that you should make if you are not in radar contact with ATC. When it becomes apparent that an estimated time that you previously submitted to ATC will be in error in excess of 3 minutes, you should notify the controller. In addition, you should report the final approach fix (FAF) inbound on a nonprecision approach, and when

While flying on a direct route, the fixes that define the route become compulsory reporting points.

you leave the outer marker (OM), or fix used in lieu of it, on a precision approach. FAFs and other approach terms are covered in more detail in Chapter 7. [Figure 5-18]

RADAR/NONRADAR REPORTS

These reports should be made at all times without a specific ATC request.

Leaving one assigned flight altitude or flight level for another	*"Cessna 45132, leaving 8,000, climb to 10,000."*
VFR-on-top change in altitude	*"Cessna 45132, VFR-on-top, climbing to 10,500."*
Leaving any assigned holding fix or point	*"Cessna 45132, leaving FARGO Intersection."*
Missed approach	*"Cessna 45132, missed approach, request clearance to Chicago."*
Unable to climb or descend at least 500 feet per minute	*"Cessna 45132, maximum climb rate 400 feet per minute."*
TAS variation from filed speed of 5% or 10 knots, whichever is greater	*"Cessna 45132, advises TAS decrease to140 knots."*
Time and altitude or flight level upon reaching a holding fix or clearance limit	*"Cessna 45132, FARGO Intersection at 05, 10,000, holding east."*
Loss of nav/comm capability (required by FAR 91.187)	*"Cessna 45132, ILS receiver inoperative."*
Unforecast weather conditions or other information relating to the safety of flight (required by FAR 91.183)	*"Cessna 45132, experiencing moderate turbulence at 10,000."*

NONRADAR REPORTS

When you are not in radar contact, these reports should be made without a specific request from ATC.

Leaving FAF or OM inbound on final approach	*"Cessna 45132, outer marker inbound, leaving 2,000."*
Revised ETA of more than three minutes	*"Cessna 45132, revising SCURRY estimate to 55."*
Position reporting at compulsory reporting points (required by FAR 91.183)	*See figure 5-17 for position report items.*

Figure 5-18. You are assisting ATC in maintaining aircraft separation by reporting changes in altitude, aircraft performance, and navigation equipment status, as well as by making position reports in the nonradar environment.

When not in radar contact on a nonprecision approach, report to ATC any time you leave a final approach fix inbound on final approach.

ENROUTE NAVIGATION USING GPS

Over the next several years, you will find global positioning system (GPS) receivers in more and more of the aircraft you fly. You may already possess a handheld version. In fact, the FAA plans to eventually phase out certain ground-based navigation systems, like VORTACs and NDBs in favor of the GPS system. Though only panel-mounted GPS units may be approved for IFR enroute flight, both the IFR and the VFR-only models are useful sources of additional information during instrument flight. A VFR-only GPS unit may be used as a reference, but never as the primary source of navigation under IFR.

If you have a panel-mounted, IFR enroute-approved GPS, you may use it as your primary means of point-to-point navigation. However, your aircraft must be equipped with an alternate means of navigation, such as VOR-based equipment, appropriate to the flight. Active monitoring of the alternate navigation equipment is not required if the GPS receiver uses receiver autonomous integrity monitoring (RAIM). Due to the traffic saturation in crowded airspace, such as in the northeast United States, you most likely will use the GPS to fly on published airways. In less congested areas, you may opt to file a direct route between published waypoints or those you program yourself. [Figure 5-19]

GPS IFR ENROUTE AND TERMINAL REQUIREMENTS

Operation	GPS Requirements
Domestic GPS IFR Enroute and Terminal Area	• GPS[1] • Traditional navigation equipment (VOR, DME, TACAN, and/or NDB) appropriate to the route of flight must be installed and operational, but it does not need to be on and monitored unless RAIM fails. • All the compatible underlying navaids along the route of flight must be operational.
Oceanic GPS IFR Route Type: • Routes that require dual long range navigation systems, such as dual INS or dual OMEGA	• GPS[1] • One other long range navigation system compatible with the route of flight
• Routes that require only one long range navigation system	• GPS[1] • One other means of navigation compatible with the route of flight

[1]GPS units that are approved, in accordance with FAA TSO C129, for IFR enroute, terminal, and approach procedures

Figure 5-19. When using GPS for enroute, terminal, and some approach operations, you need additional navigation equipment on your aircraft. The type of equipment depends on the operation. You also may need to ensure that ground-based navaids are available.

If your route takes you from one center facility to another, you should file the waypoint latitude/longitude coordinates on your flight plan along with the departure point and destination. Center facilities possess maps of the airspace they control, but only portions of the surrounding airspace, making location of a point in another center's area difficult unless coordinates are provided. When flying direct, you should program an airway route into your GPS unit as a backup, since ATC can request a switch to traditional airways at any time. [Figure 5-20]

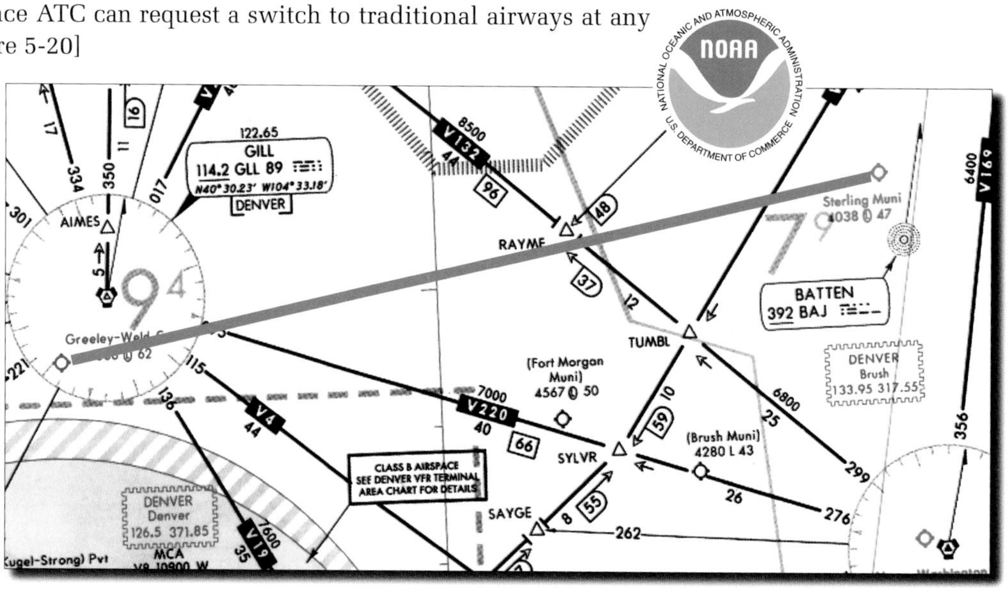

Figure 5-20. A flight between Greeley and Sterling, Colorado would be awkward if flown on VOR radials. Area navigation, such as GPS, makes a direct route feasible.

GPS operation must be conducted in accordance with the FAA-approved aircraft flight manual (AFM) or flight manual supplement. Prior to using a GPS for IFR enroute flights, you must be thoroughly familiar with the particular GPS equipment installed in your aircraft, the receiver operation manual, and the AFM. Unlike traditional navigation equipment, such as the VOR receiver, basic operation, receiver presentation, and capabilities of GPS units can vary greatly. Most receivers have a simulator mode which allows you to become familiar with the equipment's operation prior to using it in the aircraft. In addition, using the equipment in flight under VFR conditions prior to attempting IFR operation is advised. Reviewing appropriate GPS NOTAMs prior to a flight should alert you to any satellite outages. LORAN-C receivers, if approved for IFR use, may also be used in the same manner as GPS, though their future is uncertain.

Across the Pond

Since Charles Lindbergh made the first successful solo Atlantic crossing in 1927, advances in technology have made the overwater trip much less of an epic. Commercial airlines make the now routine trip daily, and the skies are busy enough that jets must follow specific routes across the Atlantic to keep them separated from traffic in the opposite direction. Since radar coverage does not extend over entire oceanic routes, pilots must resume position reporting in order to ensure the separation necessary for safety.

A long flight out of range from land still requires careful preparation and consideration. For light aircraft, the trip is usually either completed for business reasons, to deliver an airplane to an overseas customer, or to fulfill a personal dream. Several contingencies need to be addressed. First, pilots need to carry an auxiliary fuel tank, as most production singles and light twins do not hold enough fuel for a safe reserve, even on the shortest leg. Then, pilots need to consider navigation. VOR and NDB stations are widely separated by featureless ocean. Pilots used to make the transatlantic journey using only dead reckoning and celestial navigation. Now, precise navigation is much easier through the use of GPS, LORAN, or self-contained navigation equipment such as inertial navigation systems.

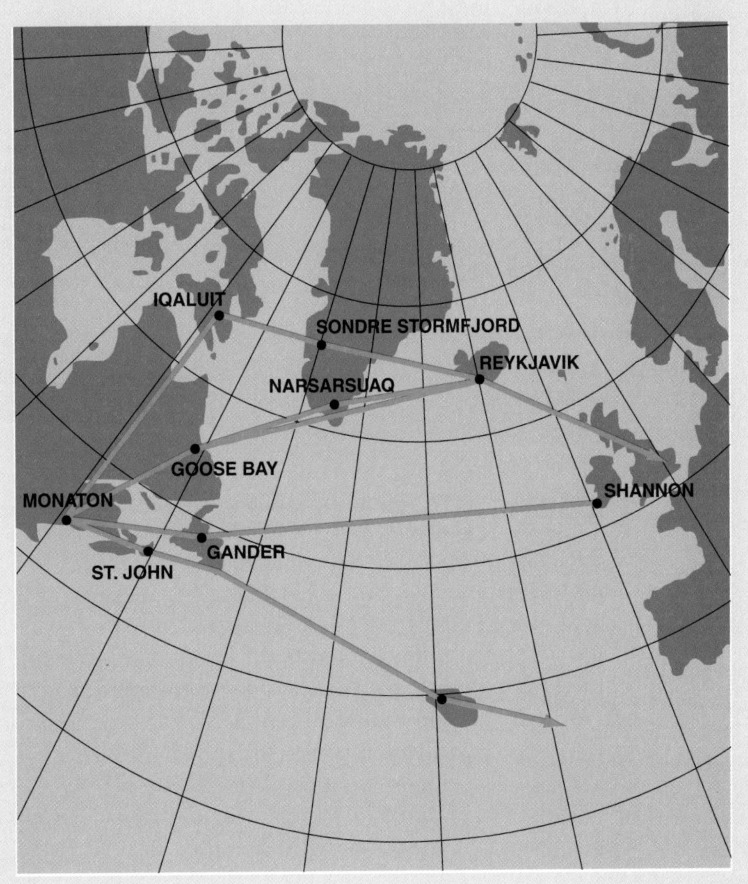

Light aircraft, however, rarely take a direct route. Pilots fly from Canada to Greenland, Iceland, and England on their way to Europe, depending on winds and weather. Some commonly flown routes below FL195 are shown. Transport Canada, which regulates all transatlantic flights departing from Canada, requires that aircraft carry immersion suits, a life raft, and a survival kit which provides a means for shelter, purifying water, and visual distress signal devices so that those who must ditch have a chance of survival in the Atlantic's frigid water. The trip requires a lot from a pilot — long hours and the chance of bad weather somewhere along the way. Before you undertake a solo transoceanic journey, flying the first trip with a pilot who has experience in long-distance, overwater operations is the best approach.

SPECIAL USE AIRSPACE

With the exception of controlled firing areas, special use airspace areas are depicted on IFR enroute charts and should be considered during preflight planning. If your filed route of flight takes you through a restricted area that is active, ATC normally issues you a revised routing clearance which ensures that your flight avoids the affected airspace. ATC normally will not issue an IFR route clearance that crosses an active restricted area so if you can file a course that circumvents an active area, you may avoid rerouting. Inactive areas are often released for the appropriate FAA controlling agency to use, so if ATC issues you an IFR clearance (including a VFR-on-top clearance) which takes you through restricted airspace, such a clearance constitutes permission to penetrate the airspace. [Figure 5-21]

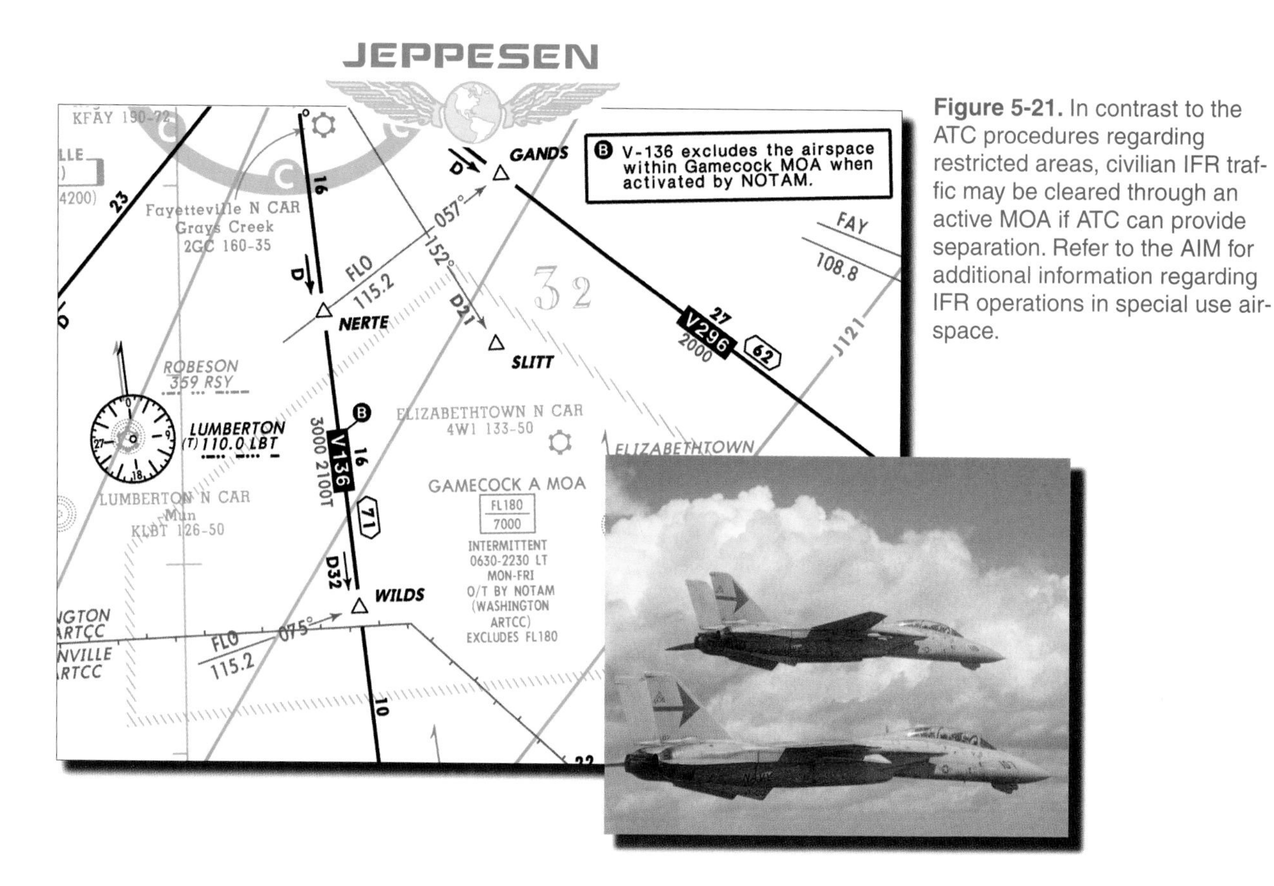

Figure 5-21. In contrast to the ATC procedures regarding restricted areas, civilian IFR traffic may be cleared through an active MOA if ATC can provide separation. Refer to the AIM for additional information regarding IFR operations in special use airspace.

IFR CRUISING ALTITUDES

FAR 91.179 provides information on IFR cruising altitudes. When operating under IFR in controlled airspace in level cruising flight, you must fly at the altitude or flight level assigned by ATC. For IFR flight in uncontrolled airspace, FAR 91.179 specifies altitudes which you must maintain based on your magnetic course. This hemispheric rule states that when operating below 18,000 feet MSL on a magnetic course of zero through 179°, you must fly an odd thousand foot MSL altitude, and for courses of 180° through 359°, an even thousand foot altitude must be maintained. Out of common practice, pilots normally file, and ATC usually assigns, IFR altitudes which agree with the hemispheric rule. However, in controlled airspace, you may request and receive altitude assignments that do not comply with east/west rules. [Figure 5-22]

You normally will file, and receive a clearance for, a cruising altitude which agrees with the hemispheric rule. See figure 5-22.

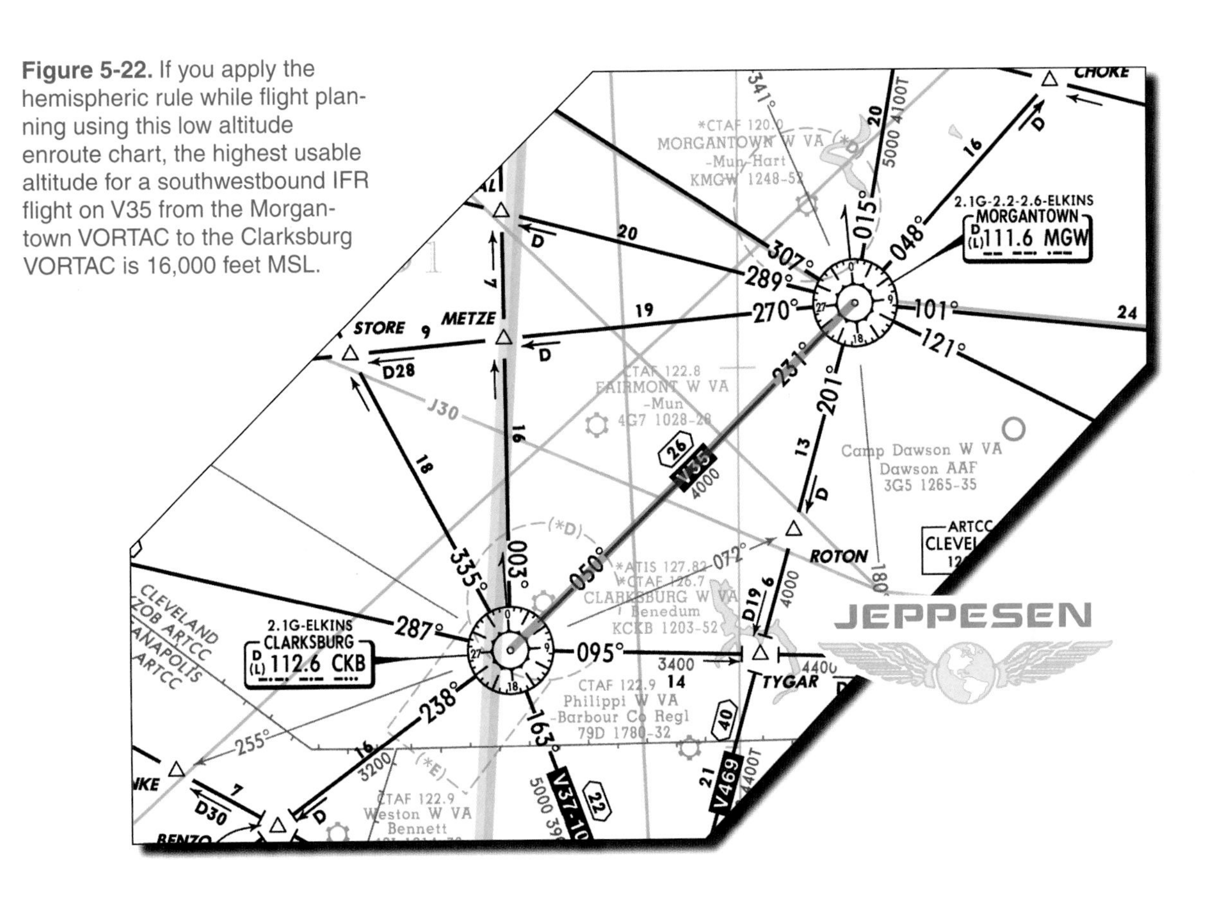

Figure 5-22. If you apply the hemispheric rule while flight planning using this low altitude enroute chart, the highest usable altitude for a southwestbound IFR flight on V35 from the Morgantown VORTAC to the Clarksburg VORTAC is 16,000 feet MSL.

If you are on an IFR flight plan below 18,000 feet MSL, and operating on a VFR-on-top clearance, you may select any VFR cruising altitude appropriate to your direction of flight between the MEA and 18,000 feet MSL, that allows you to remain in VFR conditions. Of course, you must still report any change in altitude to ATC and comply with all other IFR reporting procedures. VFR-on-top is not authorized in Class A airspace.

Although you are on an IFR flight plan when assigned a VFR-on-top clearance, you should fly at an appropriate VFR cruising altitude.

When you cruise below 18,000 feet MSL, keep your altimeter adjusted to the current setting, as reported by a station within 100 nautical miles of your position. In areas where weather reporting stations are more than 100 nautical miles from your route, you may use the altimeter setting of a station that is closest to you. During IFR flight, ATC advises you periodically of the current altimeter setting, but it remains your responsibility to update your altimeter in a timely manner.

At or above 18,000 feet MSL, you will be operating at **flight levels**, and the FARs require you to set your altimeter to 29.92. A flight level is defined as a level of constant atmospheric pressure related to a reference datum of 29.92 in. Hg. Each flight level is stated in three digits which represent hundreds of feet. For example, FL250 represents an altimeter indication of 25,000 feet. Conflicts with traffic operating below 18,000 feet MSL may arise when actual altimeter settings along the route of flight are lower than 29.92. Therefore, FAR 91.121 specifies the **lowest usable flight levels** for a given altimeter setting range. [Figure 5-23]

Table of Usable Altitudes	
Current Altimeter Setting	**Lowest Usable Flight Level**
29.92 (or higher)	180
29.91 — 29.42	185
29.41 — 28.92	190
28.91 — 28.42	195
28.41 — 27.92	200
27.91 — 27.42	205
27.41 — 26.92	210

Figure 5-23. As local altimeter settings fall below 29.92, a pilot operating in Class A airspace must cruise at progressively higher indicated altitudes to ensure separation from aircraft operating in the low altitude structure.

DESCENDING FROM THE ENROUTE SEGMENT

As you near your destination, ATC will issue a **descent clearance** so that you arrive in the approach control's airspace at an appropriate altitude. There are two basic descent clearances that ATC issues. ATC may ask you to descend to and maintain a specific altitude. Generally, this clearance is for enroute traffic separation purposes, and you need to respond to it promptly. Descend at the optimum rate for your aircraft, until you are 1,000 feet above the assigned altitude. You should report vacating any previous altitude for a newly assigned altitude, and your last 1,000 feet of descent should be made at a rate of 500 to 1,500 f.p.m. The second type of descent clearance allows you to descend *". . . at pilot's discretion."* When ATC issues you a clearance **at pilot's discretion**, you may begin the descent whenever you choose. You also are authorized to level off, temporarily, at any intermediate altitude during the descent. However, once you leave an altitude, you may not return to it.

A descent clearance also may include a segment where the descent is at your discretion, such as *". . . cross the Joliet VOR at or above 12,000, descend and maintain 5,000."* This clearance authorizes you to descend from the assigned altitude whenever you choose, so long as you cross the Joliet VOR at or above 12,000 feet MSL. After that, you should descend at a normal rate until you reach the assigned altitude of 5,000 feet MSL. [Figure 5-24]

Clearances to descend at pilot's discretion are not just an option for ATC. You may also request this type of clearance so that you can operate more efficiently. If you are enroute above an overcast layer, you may ask for a descent at your discretion, which would allow you to remain above the clouds for as long as possible. You may find this particularly

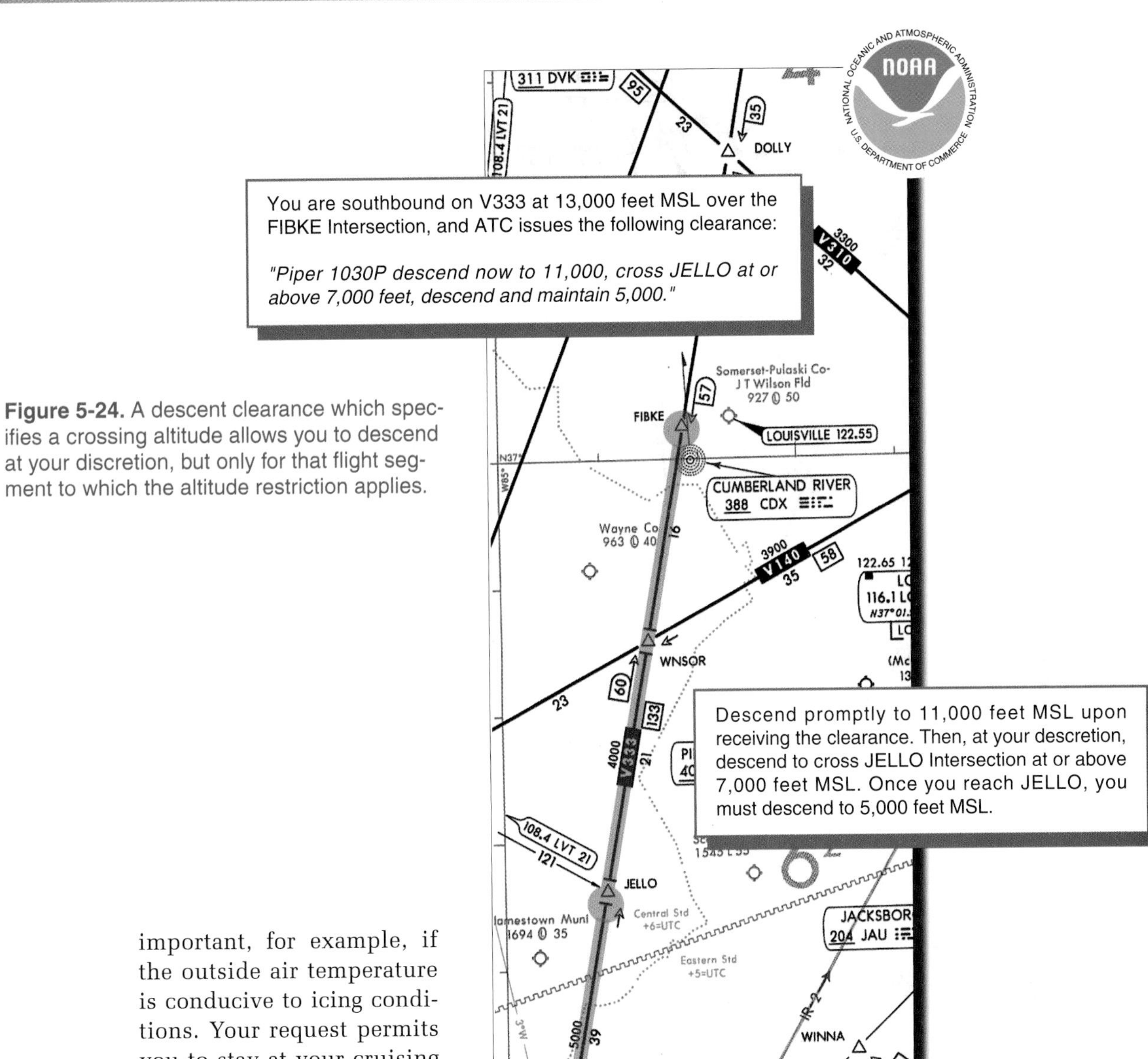

Figure 5-24. A descent clearance which specifies a crossing altitude allows you to descend at your discretion, but only for that flight segment to which the altitude restriction applies.

important, for example, if the outside air temperature is conducive to icing conditions. Your request permits you to stay at your cruising altitude longer, in order to conserve fuel and avoid prolonged periods of IFR flight in icing conditions. This type of descent also minimizes your exposure to turbulence, by allowing you to level off at an altitude where the air is smoother.

SUMMARY CHECKLIST

✓ During a radar handoff, the controller may advise you to give the next controller certain information, such as a heading or altitude.

✓ If you cannot establish contact using a newly assigned frequency, return to the one previously assigned and request an alternate frequency.

✓ You should make the following reports to ATC at all times: leaving an altitude, an altitude change if VFR-on-top, time and altitude upon reaching a holding fix or clearance limit, leaving a holding fix or clearance limit, missed approach, inability to climb or descend at a rate of at least 500 feet per minute, and change in true airspeed by 5% or 10 knots (whichever is greater).

- ✓ You are required by regulation to report a loss of airplane navigational capability, unforecast or hazardous weather conditions, and any other information relating to the safety of flight.
- ✓ If radar contact has been lost or radar service terminated, the FARs require you to provide ATC with position reports over compulsory reporting points.
- ✓ The compulsory reporting points on a direct route include those fixes that define the route.
- ✓ The standard position report includes your identification, current position, time, altitude, ETA over the next reporting fix, the following reporting point, and any pertinent remarks.
- ✓ In a nonradar environment, you should report when you reach the final approach fix inbound on a nonprecision approach, and when you leave the outer marker inbound on a precision approach. In addition, a report is necessary when it becomes apparent that an estimated time that you previously submitted to ATC will be in error in excess of 3 minutes.
- ✓ To use panel-mounted, IFR enroute-approved GPS as your primary means of point-to-point navigation, your aircraft must be equipped with an alternate means of navigation, such as VOR-based equipment, appropriate to the flight.
- ✓ Active monitoring of alternate navigation equipment is not required if the GPS receiver uses receiver autonomous integrity monitoring (RAIM).
- ✓ ATC usually does not issue an IFR route clearance that crosses an active restricted area, but inactive areas are often released for use.
- ✓ Though you may request and be assigned any altitude in controlled airspace, most pilots file flight plan altitudes that correspond to the hemispheric rule.
- ✓ Lowest usable altitudes are specified for use above 18,000 feet MSL when the barometric pressure is below certain values.
- ✓ When you are given a descent clearance *". . . at pilot's discretion,"* you are authorized to begin the descent whenever you choose, and level off temporarily during the descent, but you cannot return to an altitude once you vacate it.

KEY TERMS

Position Report

Compulsory Reporting Point

IFR Cruising Altitude

Flight Level

Lowest Usable Flight Level

Descent Clearance

At Pilot's Discretion

QUESTIONS

1. What does it mean when ATC advises you that your aircraft is in *". . . radar contact?"*

   A. You can expect radar flight following.
   B. You no longer need to watch for other traffic, even if you are in VFR conditions.
   C. The controller no longer needs you to advise of any change in airspeed or performance.

2. If you attempt to contact the center after a handoff from departure control, and you get no response, what action should you take?

3. You filed a true airspeed of 130 knots. While enroute, you calculate your true airspeed to be 135 knots. Do you need to advise ATC?

4. True/False. You do not need to report a malfunction of your ADF receiver to ATC if you are in radar contact.

5. True/False. When you are operating VFR-on-top, you should report to ATC when you are leaving an altitude.

6. What reports should you make only if you are in a nonradar environment?

7. You are operating in a nonradar environment on a direct course that is not an established airway. Select the true statement regarding position reports in this situation.

   A. You should report over the fixes used to define your route.
   B. You should report your position to ATC every 30 minutes until radar contact has been reestablished.
   C. You do not need to report your position until you are established on a Victor airway over a compulsory reporting point.

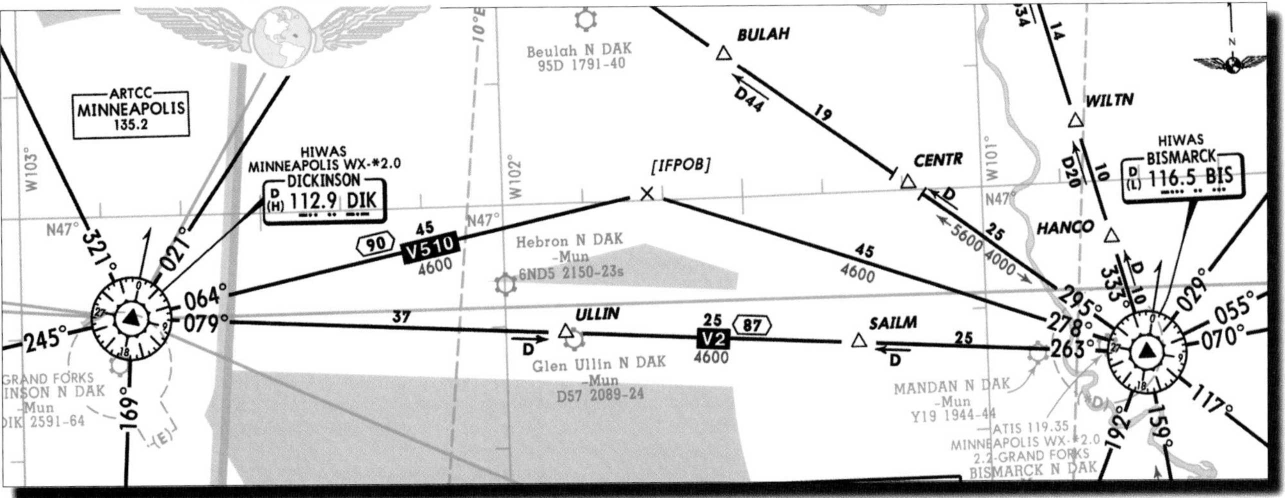

8. You are on an IFR flight in Aztec 3490R, heading east on V2. Prior to reaching Dickinson, ATC advises you that radar contact has been lost. You are instructed to report ULLIN Intersection. Assuming the time is 1330, your groundspeed is 150 knots, and you are cruising level at 9,000 feet MSL, give a position report to Minneapolis Center over Dickinson VOR.

9. If you apply the hemispheric rule, what is the highest usable altitude on an eastbound IFR flight on V2 between Dickinson VOR and Bismarck VOR?

10. Select the true statement regarding the use of GPS for IFR enroute navigation.

    A. You can use a handheld GPS unit as your primary means of navigation for an IFR flight.
    B. Your aircraft must be equipped with an alternative mode of navigation, such as VOR-based equipment, to use GPS as your primary method of enroute navigation.
    C. It is not necessary to learn about the features of every different GPS unit you use since the operation, receiver presentation, and capabilities of every GPS model are basically the same.

11. True/False. If the route filed on your IFR flight plan requires you to penetrate a restricted area, you must request a special clearance from the restricted area's controlling agency.

SECTION C

HOLDING PROCEDURES

Holding patterns are a method of delaying airborne aircraft to help maintain separation and provide a smooth flow of traffic. When the volume of traffic becomes overwhelming, or in the event of a radar failure, the need for holding patterns increases. There are a few areas that do not have ATC radar coverage, and holding patterns routinely help controllers manage the traffic in those areas. Holding patterns also are used when you reach a clearance limit, or may be required following a missed approach. There may be times when you will want to request a hold, for instance, when you need to climb to reach an assigned altitude, or when weather at your destination is improving and a few more minutes might make the difference between diverting to an alternate and a successful landing at your destination. Thus, it is important that you know what is expected of you and how to properly execute this classic instrument maneuver.

CAN HOLDING PATTERNS MAKE YOU DIZZY?

Disorientation can occur at any time, but is more likely in low visibility or instrument conditions. In some cases, disorientation can be so severe that the pilot cannot control the aircraft. Instrument-rated pilots are just as susceptible as noninstrument-rated pilots. The frequent turns and continuous workload of flying a holding pattern in turbulent, windy conditions can create a prime environment for both vestibular and spatial disorientation.

To minimize the risk in such situations, avoid rapid head movements, which can precipitate an attack of vestibular disorientation. This is where good cockpit management really helps. Arrange to have necessary charts and pilot supplies easily available. Think before you reach for the next chart, lean over to switch fuel tanks, or look for a dropped pencil.

THE STANDARD HOLDING PATTERN

Holding patterns are shaped like the oval racetracks you see on sectionals. In a standard holding pattern, the turns are to the right, while a nonstandard holding pattern uses left turns. Below 14,000 feet MSL, a holding pattern is usually two standard-rate 180° turns separated by 1 minute straight segments. Thus, with no wind, each circuit takes 4 minutes. Above 14,000 feet MSL, the straight legs are flown for 1 and 1/2 minutes, so each trip around takes 5 minutes with no wind. For simplicity in this section, assume holding patterns below 14,000 feet. The physical size of the holding pattern varies with your speed. Discounting the effects of wind, each circuit takes the same amount of time whether you fly fast or slow. [Figure 5-25]

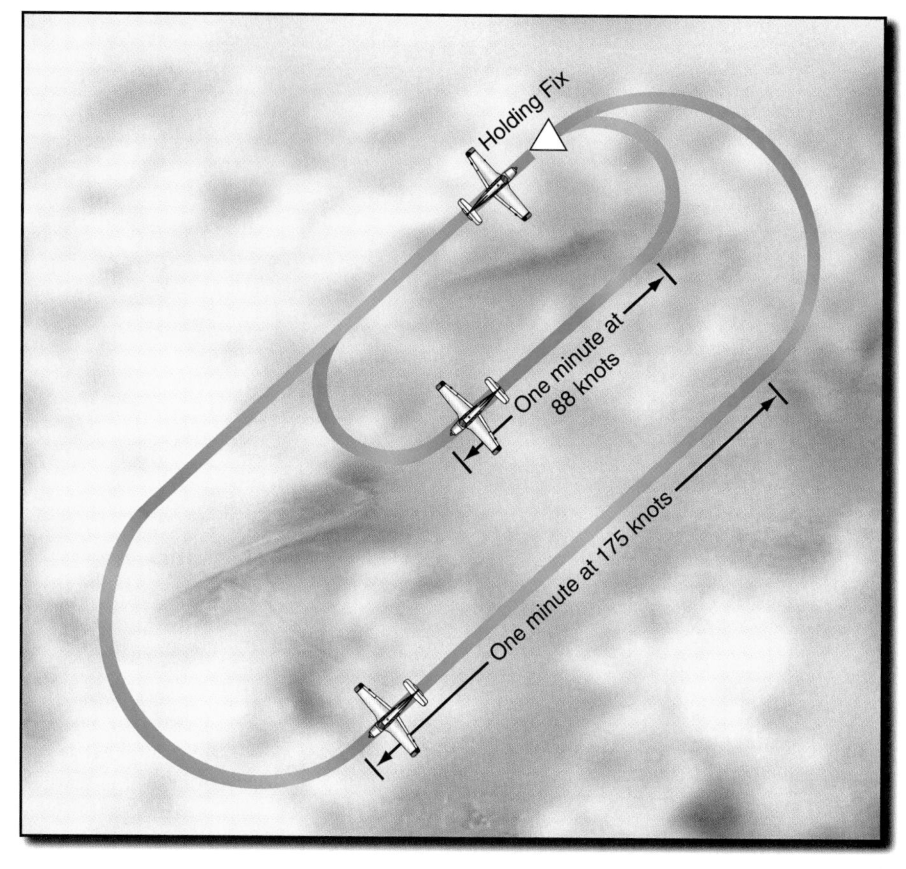

Figure 5-25. At 175 knots, the straight legs of the holding pattern are almost 3 nautical miles long with no wind. At 88 knots, the legs are about 1.5 miles long. It's important to remember that the holding pattern is a time delay, so there is no advantage to flying any faster than necessary. Slowing down saves fuel and uses less airspace.

Each holding pattern has a fix, a direction from the fix, and a line of position (NDB bearing or VOR radial) on which to fly one leg of the pattern. These elements, along with the direction of the turns, define the holding pattern. Each circuit of the holding pattern begins and ends at the holding fix. The holding fix may be an intersection, a navaid, or a certain DME distance from a navaid. The inbound leg of the pattern is flown toward the fix on the holding course. As you would expect, the side of the holding course where the pattern is flown is called the holding side. [Figure 5-26]

Turns are made to the right in a standard holding pattern, and to the left in a nonstandard holding pattern.

OUTBOUND AND INBOUND TIMING

The challenge during holding is to make your inbound legs one minute long. Since it is difficult to predict the effects of wind, use the first trip around to find some approximate correction factors for subsequent circuits. Begin timing the outbound leg when you are abeam the holding fix. If you cannot identify the abeam position, you should start timing when you complete the turn outbound. After turning inbound, time your inbound leg to

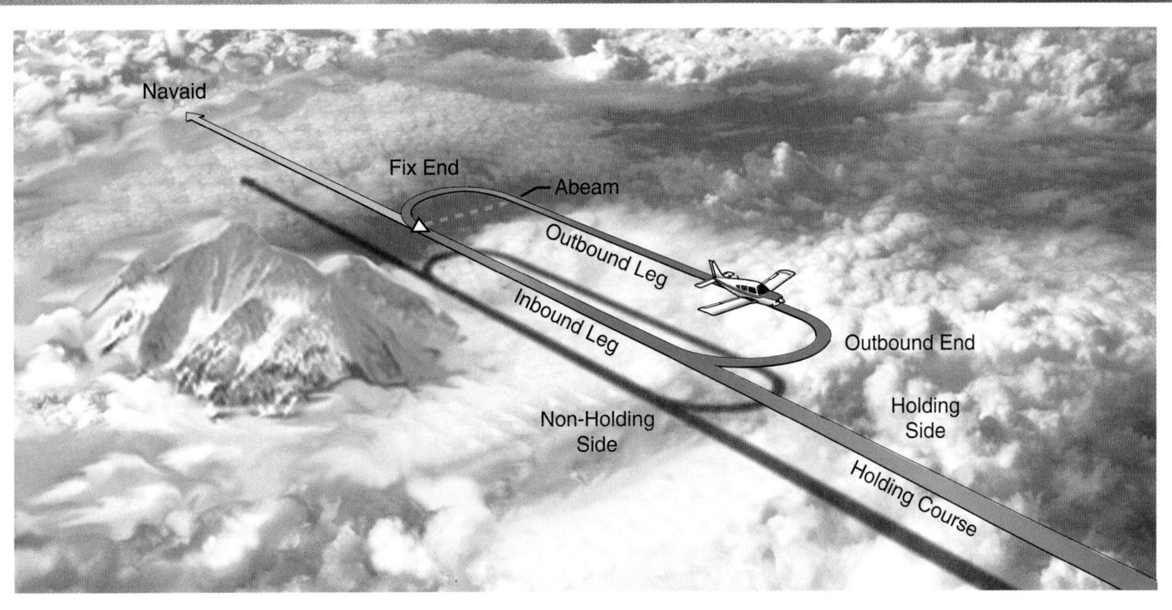

Figure 5-26. The airspace is protected on the holding side. If the holding pattern is flown correctly, you are guaranteed obstacle clearance and separation from other air traffic.

gauge the effect of the wind, and adjust the timing for subsequent outbound legs to achieve an inbound time of one minute. A longer inbound leg indicates that you should shorten the outbound leg, and vice versa. [Figure 5-27]

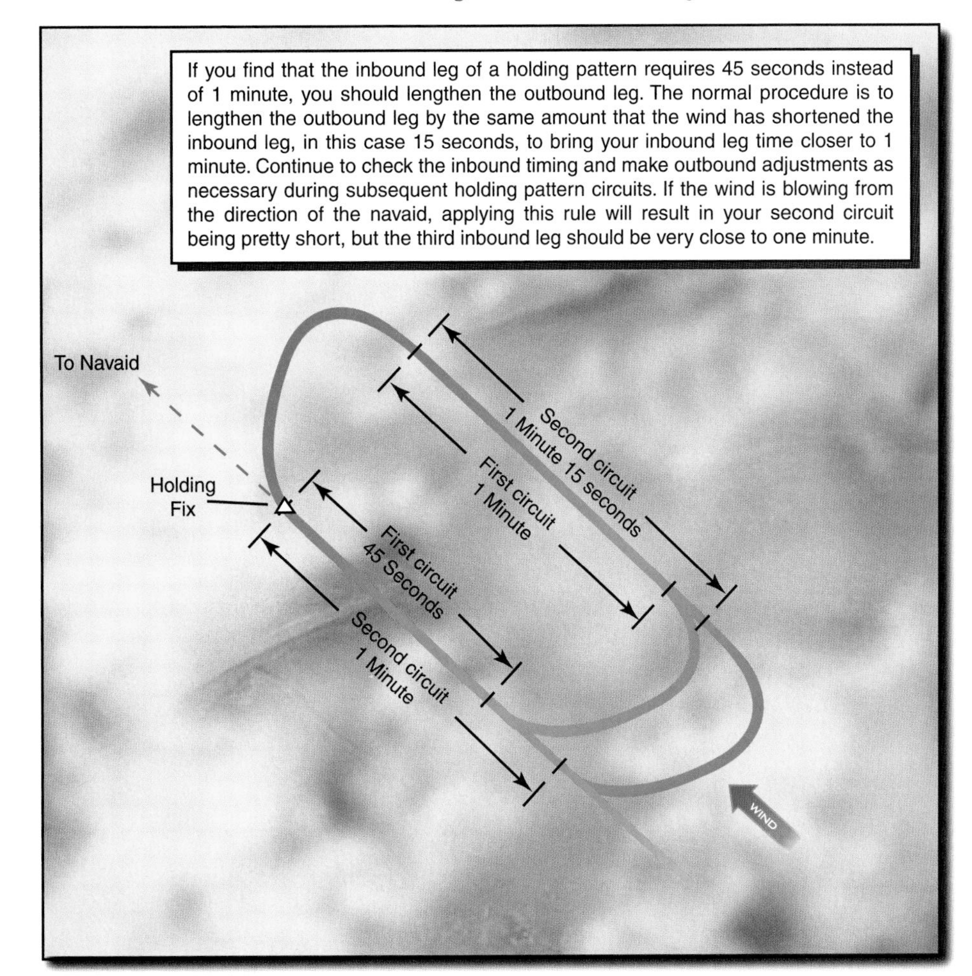

Figure 5-27. If you have a tailwind blowing you directly toward the navaid, you will need a longer outbound leg to make your inbound leg come out to one minute.

So Many Planes, So Little Time

As air commerce expanded in the early 1930s, the rapid increase in the number of commercial, military, and private flights began to create traffic conflicts and congestion problems, so in July of 1936, the federal government took over control of air traffic. Just as traffic lights delay some ground traffic in order to allow other traffic to move, pilots were often instructed to circle over a designated spot so that another airplane could complete its approach. By the early 1940s, part of the standard holding procedure called for pilots to fly along the right edge of the on-course signal to a radio range station so that departing traffic could fly out on the other side of the same on-course signal. The holding pattern itself was flown between the holding fix and a point four minutes flying time away.Although different navaids have replaced the low frequency radio ranges and now radar covers most of the country, holding patterns are still used as backup procedures today.

Timing for the outbound leg of either a standard or nonstandard holding pattern should begin abeam the holding fix. If the abeam position cannot be identified, start timing the outbound leg at the completion of the turn outbound.

Sometimes DME distances are used instead of timing the inbound and outbound legs. When DME is used, the same holding procedures apply, but the turns are initiated at specified DME distances from the station. The holding fix is on a radial at a designated distance. If you are asked to hold at a DME fix and the holding pattern lies between the DME fix and the navaid, remember that you are holding at the fix, not the navaid. Your inbound leg to the holding fix will be outbound from the navaid. [Figure 5-28]

CROSSWIND CORRECTION

If you fly your holding pattern without correcting for crosswind drift, you could inadvertently stray from the protected airspace area or have difficulty coming back to the inbound course before passing the fix. To avoid these problems, use your normal bracketing and drift correction techniques to determine the amount of drift correction necessary during the inbound leg. Once you determine the wind correction angle (WCA) required to maintain the inbound course, triple the correction for the outbound leg. [Figure 5-29]

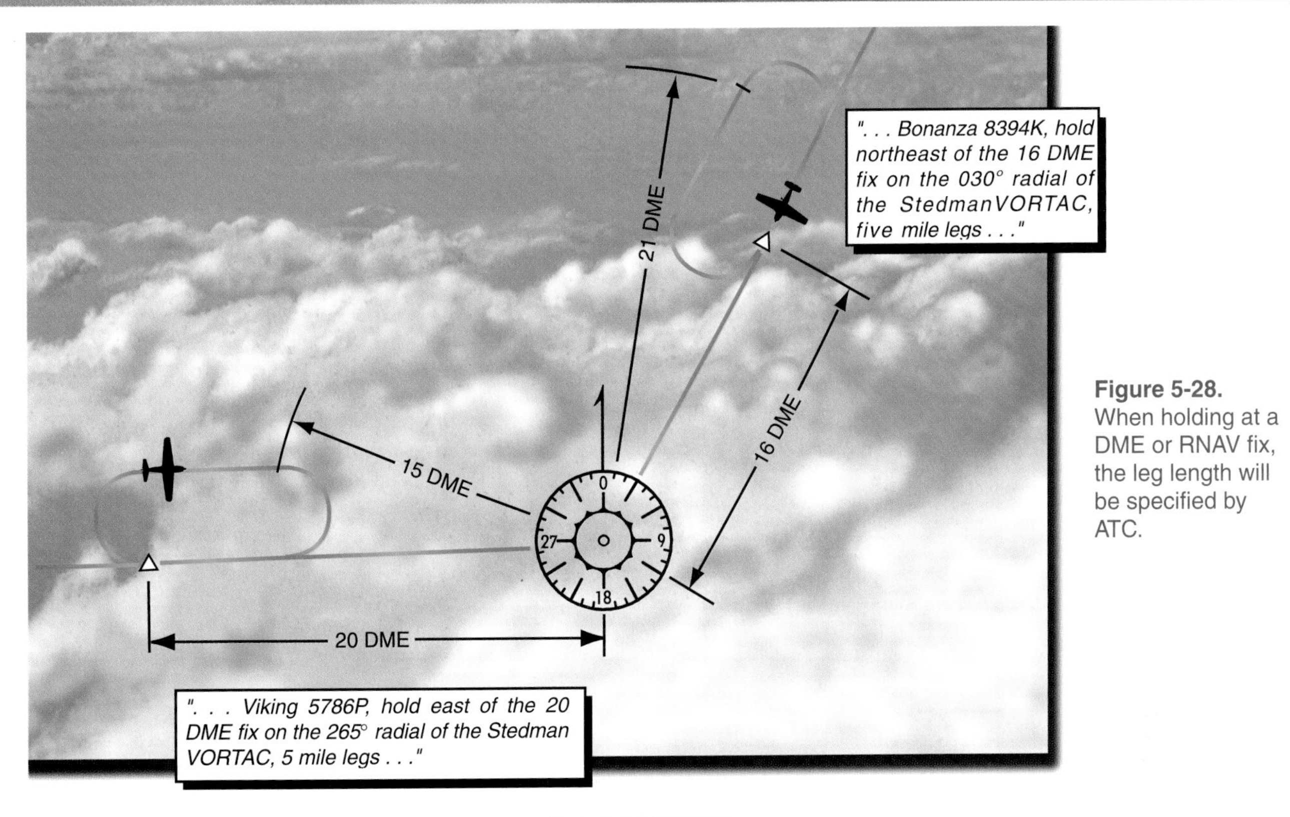

Figure 5-28. When holding at a DME or RNAV fix, the leg length will be specified by ATC.

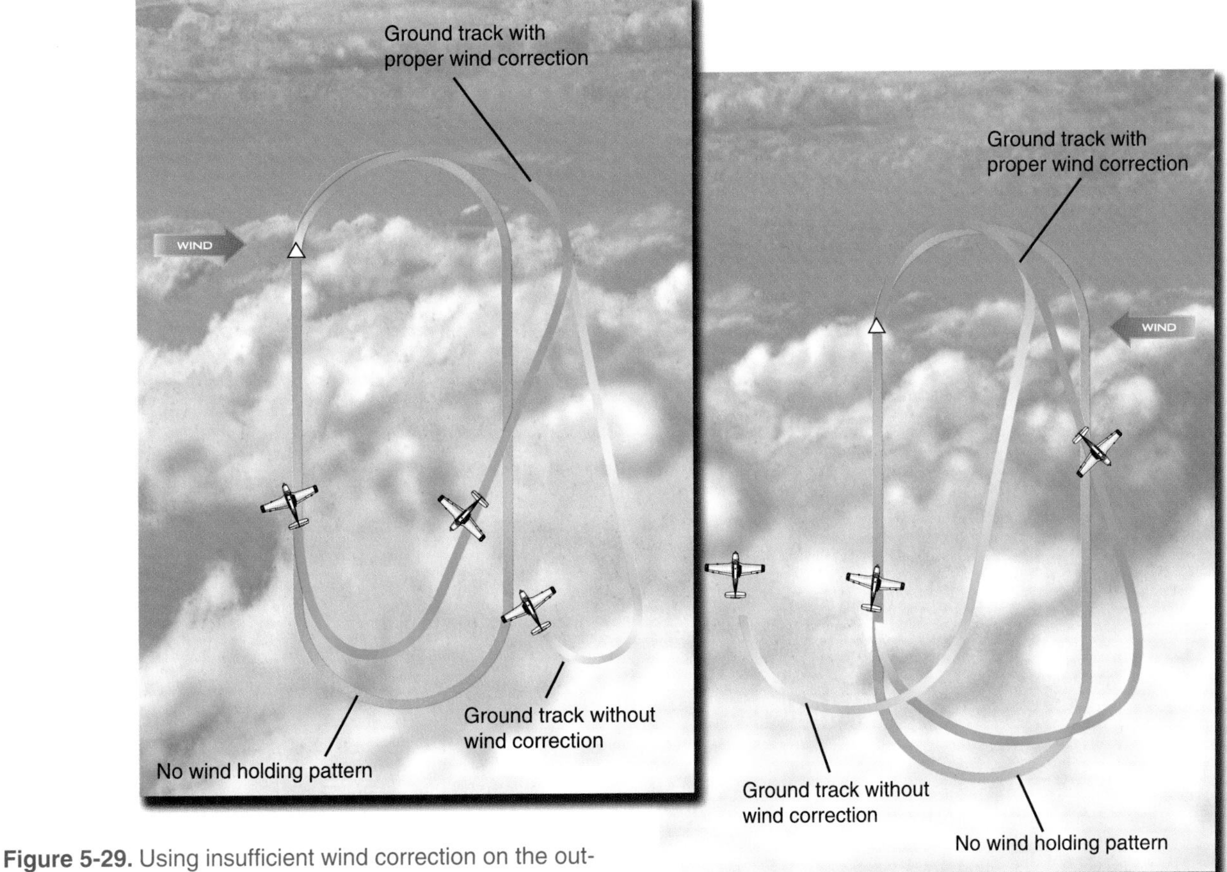

Figure 5-29. Using insufficient wind correction on the outbound leg causes the aircraft to over- or undershoot the course during the turn inbound. Tripling the inbound correction on the outbound heading should give you room to intercept the inbound course again while remaining on the holding side of the course.

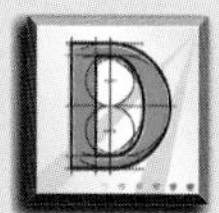

Things to Think About While Holding

The first radio stations for aviation use were operated by the Post Office, and went into service in the early 1920s. DeHaviland mail planes could home on these beacons with primitive battery-powered direction finding receivers.

In 1923, the lighted airway beacon system was begun, using electric or acetylene lamps. By 1941 there were 2,274 airway light beacons over 32,679 airway miles.

The first four-course radio ranges began transmitting in 1929. These provided a specific on-course signal about 3° wide to define the airways. Terms like *"radio beams"* and *"flying on the beam"* entered the common vocabulary. Low powered marker beacons along the routes gave position fixes. That is why the white marker beacon indicator light in many older airplanes is labeled AIR-WAY.

The first air traffic control facility was formed by four airlines in 1935 to coordinate their traffic around Newark, New Jersey. American, Eastern, TWA, and United worked together to provide separation for instrument traffic. When the government took over air traffic control a few months later, there were two additional centers, in Chicago, Illinois and Cleveland, Ohio.

Adcock 4 course range station

Runway numbering was not standardized until the late 1930s. Before that, each airport would number its runways as Number 1, Number 2, etc., usually according to the order in which they were built. Many pilots felt that the new system based on magnetic headings was confusing, since the same runway now had two numbers, depending on which direction was being used.

The first VOR airway was created in 1950, and within a few years there were more than 500 VOR stations. DME stations multiplied at a slower rate. As the new Victor airways were created, the radio ranges and lighted airways were dismantled.

MAXIMUM HOLDING SPEED

As you have seen, the size of the holding pattern is directly proportional to the speed of the airplane. In order to limit the amount of airspace that must be protected by ATC, **maximum holding speeds** have been designated for specific altitude ranges. [Figure 5-30] Even so, some holding patterns may have additional speed restrictions to keep faster airplanes from flying out of the protected area. If a holding pattern has a nonstandard speed restriction, it will be depicted by an icon with the limiting airspeed. If the holding speed limit is less than you feel is necessary, you should advise ATC of your revised holding speed. Also, if your indicated airspeed exceeds the applicable maximum holding speed, ATC expects you to slow to the speed limit within three minutes of your ETA at the holding fix. Often pilots can avoid flying a holding pattern, or reduce the length of time spent in the holding pattern, by slowing down on the way to the holding fix.

The maximum holding airspeeds for civil aircraft are: 200 KIAS at 6,000 feet MSL and below; 230 KIAS from 6,001 feet MSL through 14,000 feet MSL; and 265 KIAS above 14,001 feet MSL.

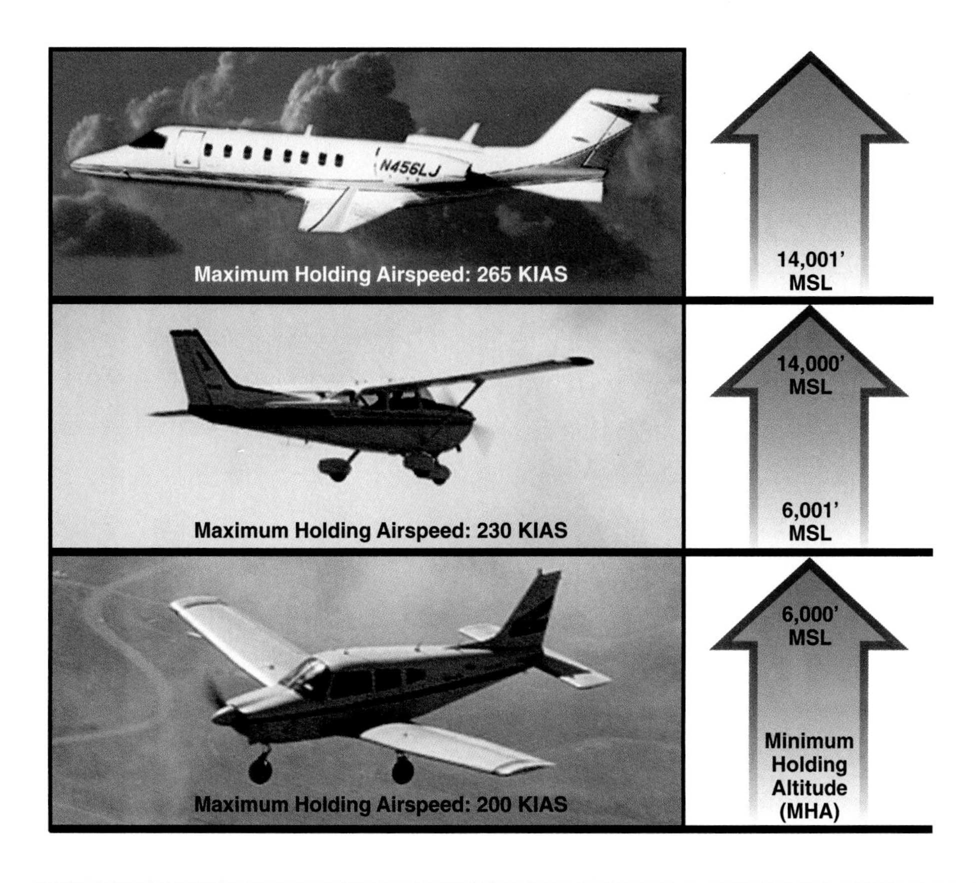

Figure 5-30. This diagram shows the maximum holding airspeeds in knots indicated airspeed (KIAS) for each altitude range.

High Performance Holding

The maximum holding airspeeds in figure 5-30 may be above the red line airspeed for the airplane you use for instrument training, but in time your instrument flying may involve higher performance aircraft. Certain limitations come into play when you operate at higher speeds; for instance, aircraft do not make standard rate turns in holding patterns if their bank angle will exceed 30°. If the aircraft is using a flight director system, the bank angle is limited to 25°. Since any aircraft must be traveling at over 210 knots TAS for the bank angle in a standard rate turn to exceed 30°, this limit applies to relatively fast airplanes. An aircraft using a flight director would have to be holding at more than 170 knots TAS to come up against the 25° limit. These true airspeeds correspond to indicated airspeeds of about 183 and 156 knots, respectively, at 6,000 feet in a standard atmosphere.

Since some military aircraft need to hold at higher speeds than the civilian limits, the maximum at military airfields is higher. For example, the maximum holding airspeed at USAF airfields is 310 KIAS.

HOLDING PATTERN ENTRIES

Three holding pattern entry procedures have been developed to get you headed in the right direction on the holding course without excessive maneuvering. The entry you use depends on your magnetic heading relative to the holding course when you arrive at the fix. [Figure 5-31] The entries for standard holding patterns are shown. Entries to non-standard patterns are mirror images of those illustrated.

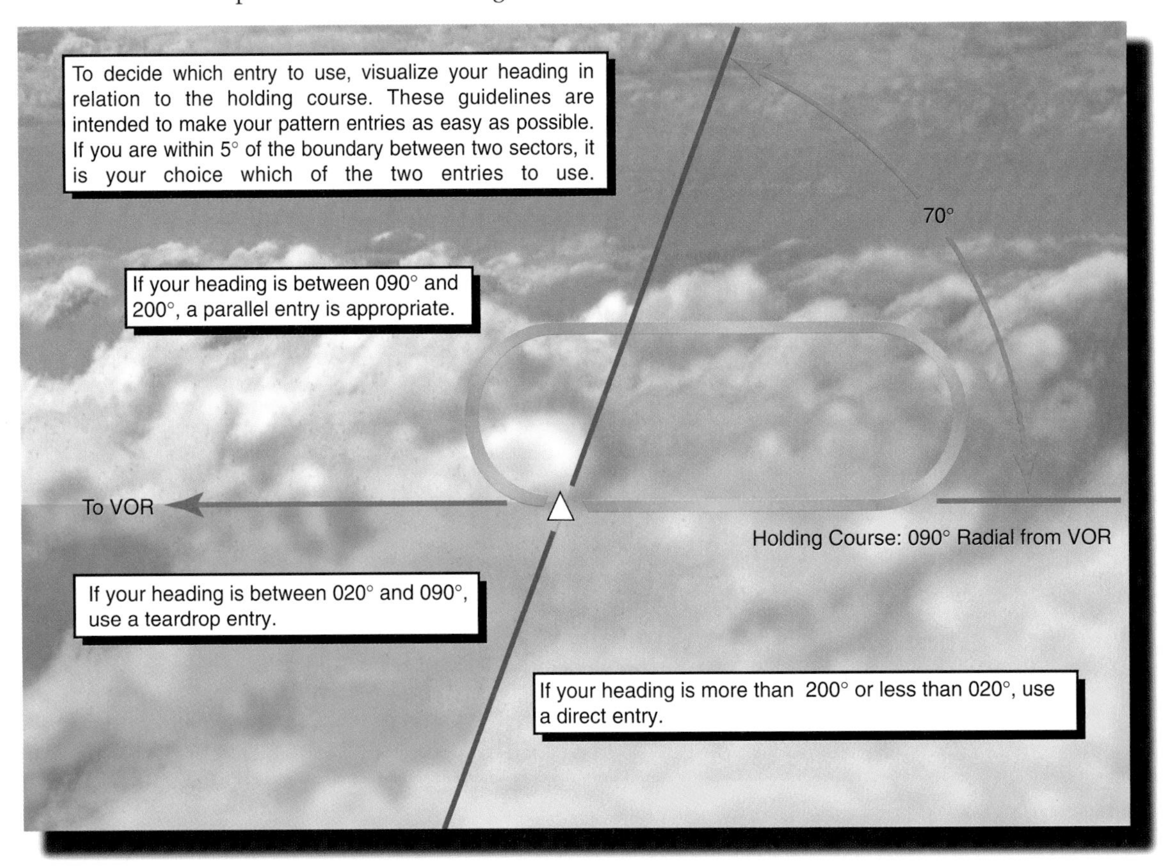

Figure 5-31. Entry sectors are established by imagining a line at 70° across the holding course. For example, if the holding course is the 090° radial from a VOR, the entry sectors are defined by a line through the fix coinciding with headings of 020° and 200°.

DIRECT ENTRY

The **direct entry** procedure is the most often used, because it can be applied throughout 180°. When you use the direct entry, you simply fly across the fix, turn right to the outbound heading, and fly the pattern. [Figure 5-32]

Figure 5-32. After you have passed the fix, simply turn right to a heading of 090° and time for 1 minute, then turn right again and intercept the 090° radial inbound.

TEARDROP ENTRY

The **teardrop entry**, is similar to the direct entry, with the exception that after crossing the fix, you turn to a heading which is approximately 30° away from the holding course outbound on the holding side of the pattern. Once you are established on this heading, fly outbound for approximately 1 minute; then, turn right to intercept the holding course inbound, and return to the fix. This procedure is necessary to give you some maneuvering room for your turn inbound onto the holding course [Figure 5-33]

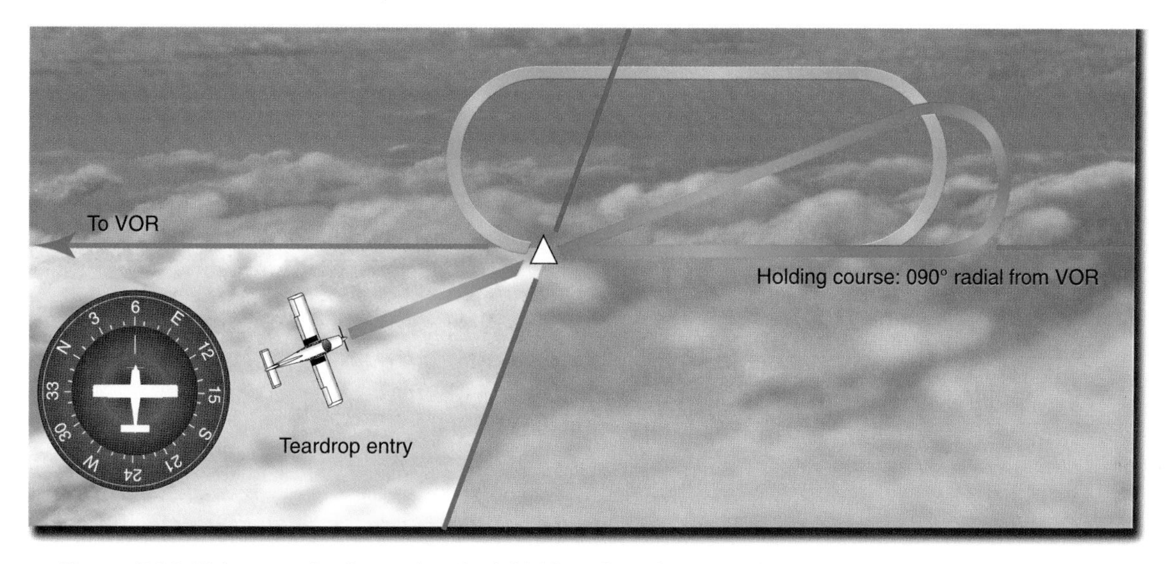

Figure 5-33. This example shows that the initial heading of 060° positions the airplane away from the holding course on the holding side, to allow sufficient room for the inbound turn.

PARALLEL ENTRY

The **parallel entry** involves paralleling the holding course outbound on the nonholding side. After crossing the holding fix, turn the airplane to a heading that parallels the holding course and begin timing for one minute. Then, begin a left turn and return to the fix or reintercept the course from the holding side and proceed to the holding fix. [Figure 5-34]

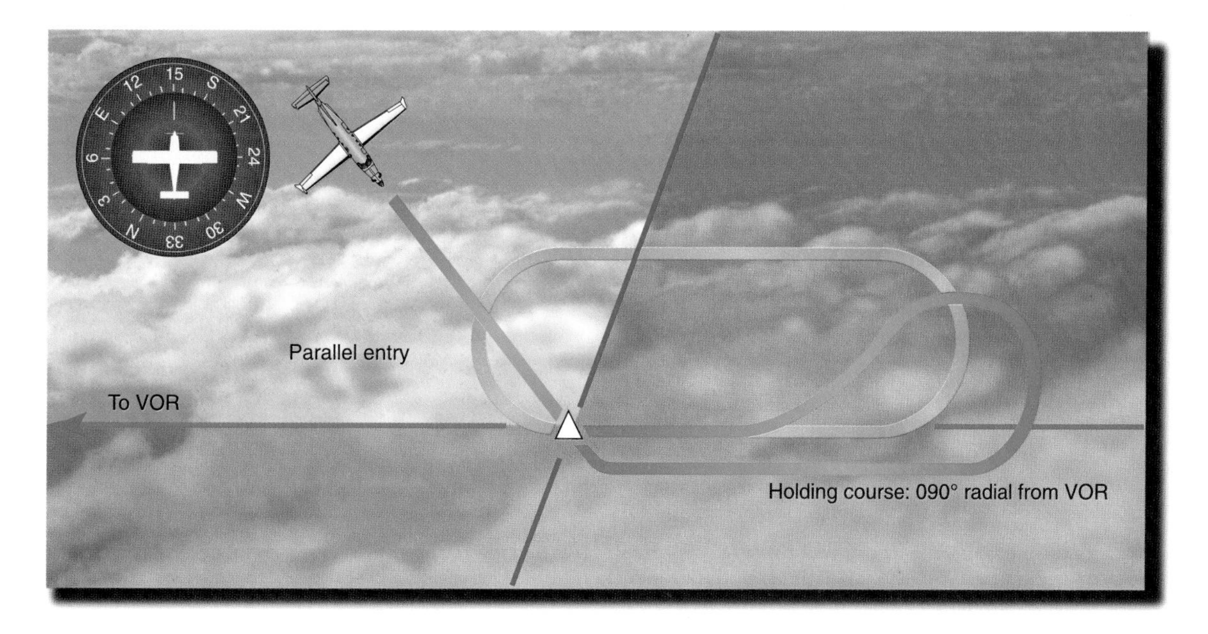

Figure 5-34. In this example, make the first turn parallel to the outbound course using a heading of 090°. After one minute, make a left turn and return to the holding fix or intercept the holding course inbound.

VISUALIZING ENTRY PROCEDURES

The key to easy holding pattern entries is to accurately visualize your position relative to the holding pattern before you arrive over the fix. Many methods have been invented to make this easier, including sketching the holding pattern on your aeronautical chart, using the wind side of a flight computer, employing the aircraft heading indicator, referring to a holding pattern inscription on a plotter, or using specially designed pattern entry computers. Different pilots prefer different methods, but the value of any technique depends on how well it helps you visualize the holding pattern and the appropriate entry.

Here is a general rule of thumb which does not require any separate paraphernalia. Visualize your arrival over the holding fix. If the holding course is behind the aircraft when you arrive at the fix, a direct entry is usually appropriate. When the holding course is ahead and to the right of the aircraft, use a teardrop entry. If the holding course is ahead and to the left, make a parallel entry. You can apply this method to nonstandard holding patterns by exchanging the parallel and teardrop entry sectors. For nonstandard patterns, the parallel sector is ahead and to the right, and the teardrop sector is ahead and to the left. The direct entry sector is still behind the aircraft. [Figure 5-35]

The entry procedure for a holding pattern depends on your heading relative to the holding course. The recommended entries are shown in figure 5-31.

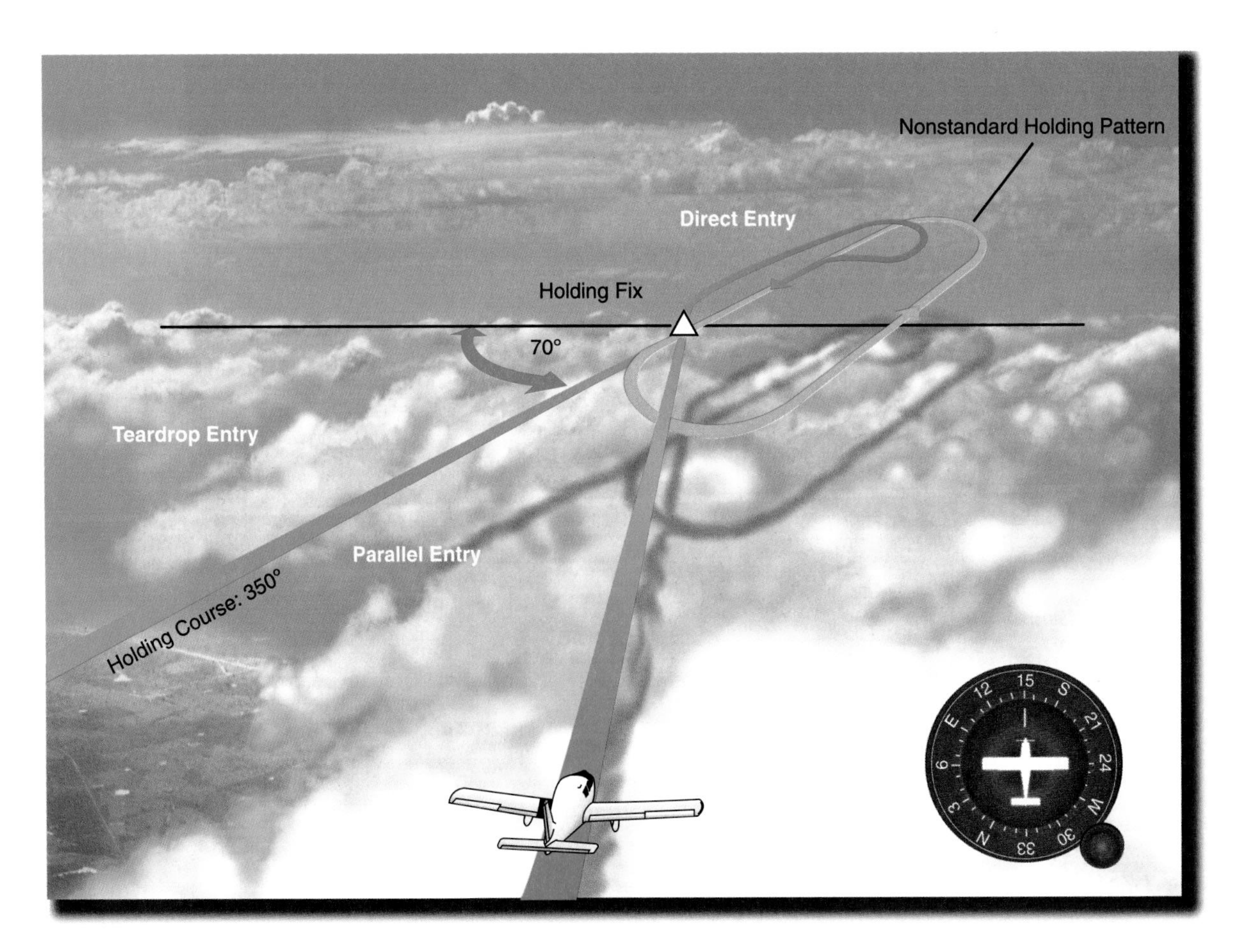

Figure 5-35. Visualizing your heading in relation to the holding course clarifies which entry to use. In this example, the holding pattern is nonstandard, and the holding course is ahead and to the right, so a parallel entry is correct.

ATC HOLDING INSTRUCTIONS

When controllers anticipate a delay at a clearance limit or fix, you will usually be issued a holding clearance at least five minutes before your ETA at the clearance limit or fix. If the holding pattern assigned by ATC is depicted on the appropriate aeronautical chart, you are expected to hold as published, unless advised otherwise by ATC. In this situation, the controller will issue a holding clearance which includes the name of the fix, directs you to hold as published, and includes an expect further clearance (EFC) time. An example of such a clearance is: *"Cessna 1124R, hold east of MIKEY Intersection as published, expect further clearance at 1521."* When ATC issues a clearance requiring you to hold at a fix where a holding pattern is not charted, you will be issued complete holding instructions. This information includes the direction from the fix, name of the fix, course, leg length, if appropriate, direction of turns (if left turns are required), and the EFC time. You are required to maintain your last assigned altitude unless a new altitude is specifically included in the holding clearance, and you should fly right turns unless left turns are assigned. Note that all holding instructions should include an expect further clearance time. If you lose two-way radio communication, the EFC allows you to depart the holding fix at a definite time. Plan the last lap of your holding pattern to leave the fix as close as possible to the exact time. [Figure 5-36]

There are at least three items in a clearance for a charted holding pattern:

- Direction to hold from the holding fix — *"...Hold southeast*
- Holding fix — *of PINNE Intersection as published.*
- Expect further clearance time — *Expect further clearance at 1645."*

Figure 5-36. Some elements of the holding clearance depend on whether the holding pattern is published on the chart.

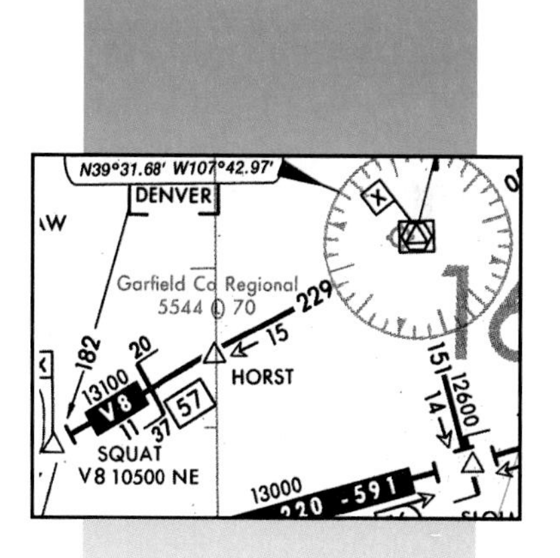

A clearance for an uncharted holding pattern contains additional information:

- Direction to hold from holding fix — *"...Hold west*
- Holding fix — *of Horst Intersection*
- The holding course (a specified radial, magnetic bearing, airway or route number) — *on Victor 8*
- The outbound leg length in minutes or nautical miles when DME is used — *5 mile legs*
- Nonstandard pattern, if used — *left turns*
- Expect further clearance time — *expect further clearance at 1430."*

If you are approaching your clearance limit and have not received holding instructions from ATC, you are expected to follow certain procedures. First, call ATC and request further clearance before you reach the fix. If you cannot obtain further clearance, you are expected to hold at the fix in compliance with the published holding

pattern. If a holding pattern is not charted at the fix, you are expected to hold on the inbound course using right turns. This procedure ensures that ATC will provide adequate separation. [Figure 5-37]

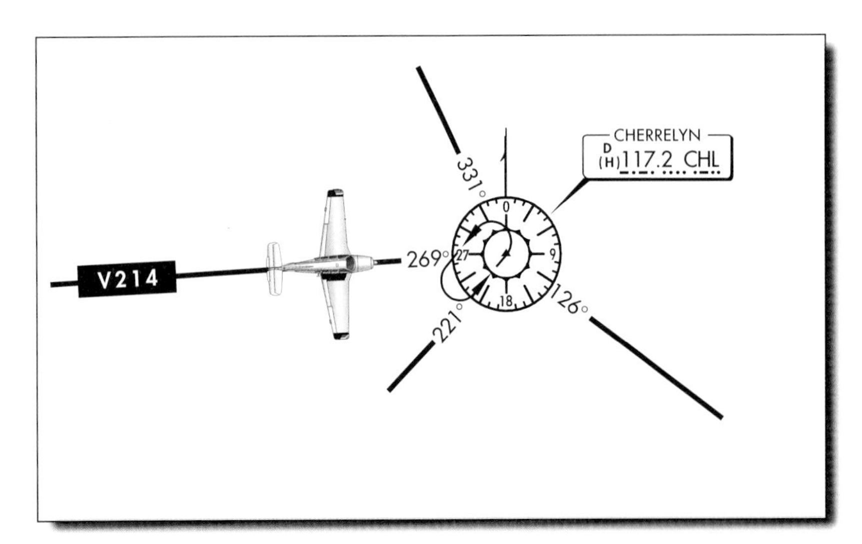

Figure 5-37. Assume you are eastbound on V214 and the Cherrelyn VORTAC is your clearance limit. If you have not been able to obtain further clearance and have not received holding instructions, you should plan to hold southwest on the 221° radial using left-hand turns, as depicted. If this holding pattern was not charted, you would hold west of the VOR on V214 using right-hand turns.

SUMMARY CHECKLIST

✓ A holding pattern is a time delay used by ATC to help maintain separation and smooth out the traffic flow.

✓ You may request a hold, for example, to wait for weather conditions to improve.

✓ Holding pattern size is directly proportional to aircraft speed; doubling your speed doubles the size of your holding pattern.

✓ Turns are to the right in standard holding patterns, and to the left in nonstandard holding patterns.

✓ Each circuit of the holding pattern begins and ends at the holding fix.

✓ Adjust the timing of your outbound leg to make your inbound leg one minute long.

✓ To correct for crosswind drift in the holding pattern, triple your inbound wind correction angle on the outbound leg.

✓ To keep the volume of the protected airspace for a holding pattern within reasonable limits, maximum holding airspeeds are designated according to altitude.

✓ The entry procedure for a holding pattern depends on your heading relative to the holding course. The three recommended procedures are direct, teardrop, and parallel.

✓ A holding clearance should always contain the holding direction, the holding fix, and an expect further clearance (EFC) time. If the holding pattern is not published, the clearance also contains the holding course. For nonstandard patterns, left turns are specified. For patterns using DME, the clearance gives the outbound leg length in nautical miles.

KEY TERMS

Standard Holding Pattern

Nonstandard Holding Pattern

Holding Fix

Inbound Leg

Holding Course

Holding Side

Maximum Holding Speeds

Direct Entry

Teardrop Entry

Parallel Entry

Holding Clearance

Expect Further Clearance (EFC) Time

QUESTIONS

1. True/False. Turns are made to the left in nonstandard holding patterns.

2. Above what altitude are the straight legs for a standard holding pattern one and a half minutes?

   A. 10,000 feet MSL
   B. 12,000 feet MSL
   C. 14,000 feet MSL

3. What is the speed limit in holding patterns below 6,000 feet MSL?

4. The first complete circuit of your standard holding pattern at 10,000 feet MSL takes a total of 3 minutes and 52 seconds. To achieve an inbound leg time of 1 minute on your next circuit, assuming there is no crosswind, approximately how many seconds should you fly outbound before turning?

   A. 52 seconds
   B. 60 seconds
   C. 68 seconds

5. Assume you are flying toward the Kelly VOR on a magnetic heading of 015°. What kind of holding pattern entry would be appropriate for the following clearance? ". . . *cleared to the Kelly VOR. Hold southwest on the 215 radial . . .*"

   A. Direct
   B. Parallel
   C. Teardrop

6. If you are flying west on V345 toward the ENNUI Intersection, and are instructed to "*. . . hold west of ENNUI on Victor 345, left turns,*" which holding pattern entry is recommended?

   A. Direct only
   B. Parallel or Teardrop
   C. Teardrop or Direct

7. Inbound to the PJG VORTAC on a magnetic heading of 060°, you receive the following clearance: "*. . . cleared to the PJG VORTAC, hold south of the VOR on the 160 radial, left turns . . .*" Which holding pattern entry should you use?

   A. Direct
   B. Parallel
   C. Teardrop

8. If you were cleared to hold northwest of the 15 DME fix on the 323° radial from Belleview VORTAC, which illustration in the accompanying figure correctly depicts your holding pattern?

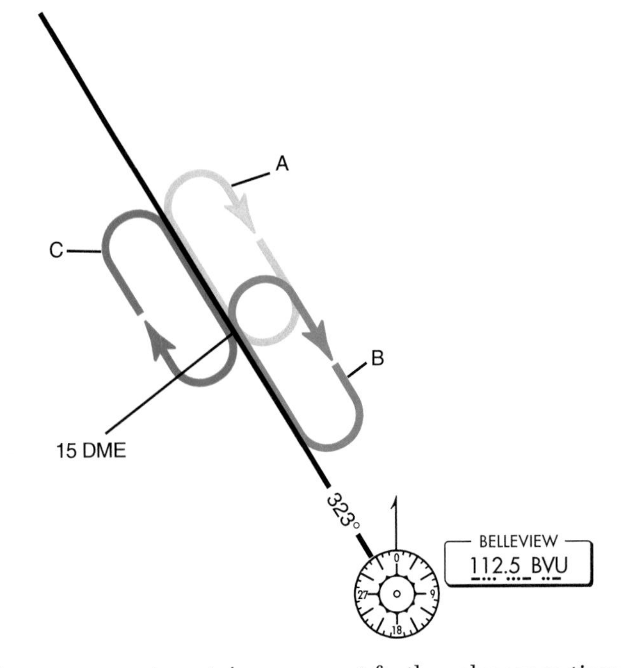

   A. A
   B. B
   C. C

9. True/False. All holding clearances must contain an expect further clearance time.

CHAPTER 6
ARRIVAL

Instrument/Commercial
Part II, Segment 1, Chapter 6 — Arrival

SECTION A
ARRIVAL CHARTS

The ability to file IFR gives you easier access to large airports that you might not have ventured into under visual flight rules. As traffic into an airport increases, procedures for sequencing aircraft to an instrument approach must be established. Arrival charts provide you with a smooth transition between the enroute structure and busy terminal areas, simplifying complex clearances and supplying an expected plan of action to both pilots and controllers. Knowing how to integrate these charts into your flight plan will enable you to fly IFR into the most congested airspace with confidence. [Figure 6-1]

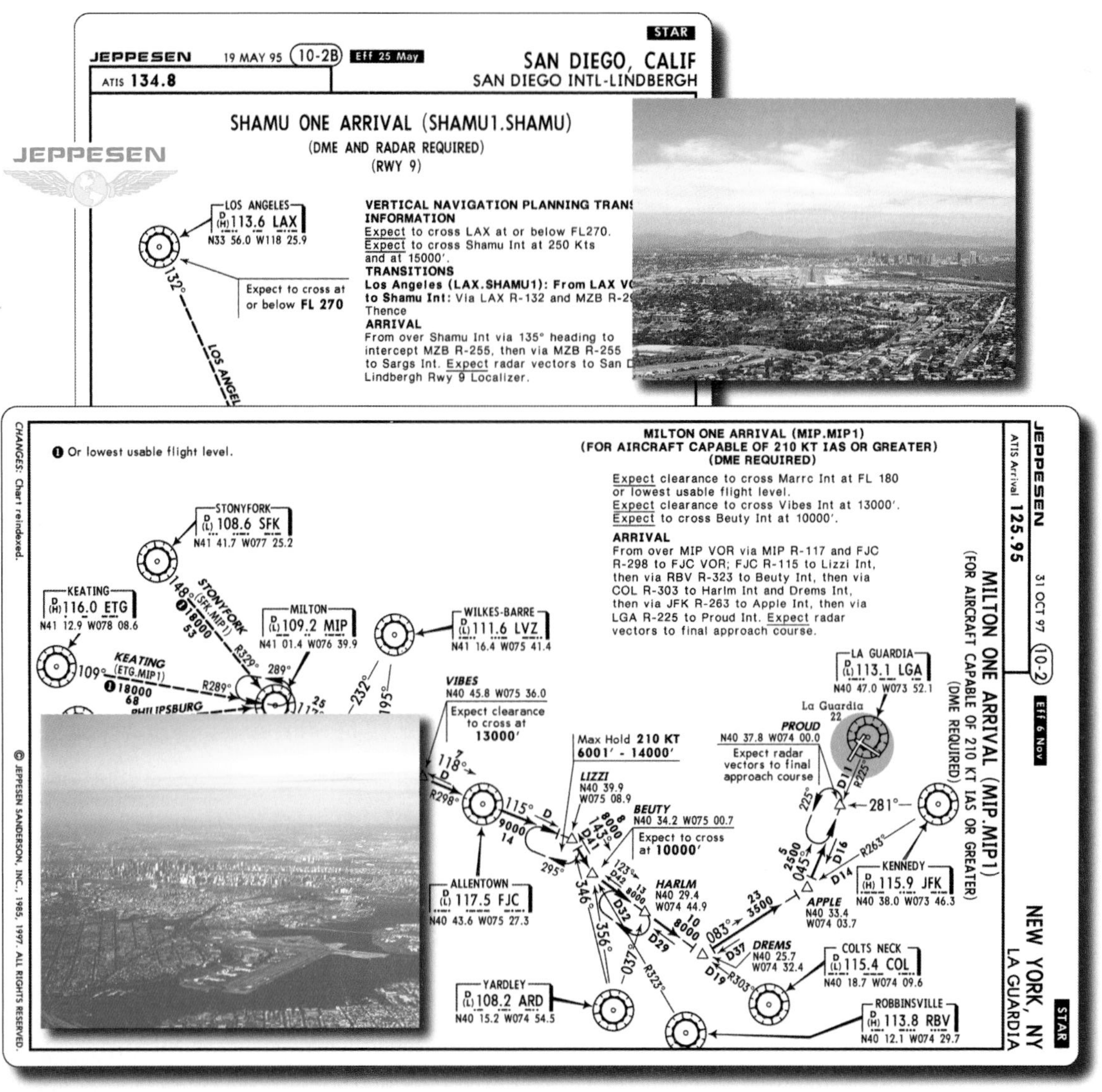

Figure 6-1. Arrival charts give you preferred routes into high volume terminal areas.

STANDARD TERMINAL ARRIVAL ROUTE

The **standard terminal arrival route (STAR)** provides a common method for departing the enroute structure and navigating to your destination. STARs usually terminate with an instrument or visual approach procedure. Both Jeppesen and the National Ocean Service (NOS) publish STARs for airports with procedures authorized by the FAA. Jeppesen incorporates the applicable arrival charts into the basic terminal chart subscription, and these charts are filed with the airport's approach charts. NOS includes STARs at the front of each *Terminal Procedures Publication* regional booklet.

STARs are established to simplify clearance delivery procedures.

To illustrate how STARs can be used to simplify a complex clearance and reduce frequency congestion, consider the following arrival clearance issued to a pilot flying to Seattle, Washington. *"Cessna 32G, cleared to the Seattle/Tacoma International Airport as filed. Maintain 12,000. At the Ephrata VOR intercept the 221° radial to CHINS Intersection. Intercept the 284° of the Yakima VOR to SNOMY Intersection. Cross SNOMY at 10,000. Continue via the Yakima 281° radial to AUBRN Intersection. Expect radar vectors to the final approach course."*

REDUCING PILOT/CONTROLLER WORKLOAD

Safety is enhanced when both pilots and controllers know what to expect. Being able to rehearse a proposed route or clearance in advance means that the pilot can pay more attention to situational awareness during the corresponding portion of a flight. Effective communication increases with the reduction of repetitive clearances, freeing up congested control frequencies.

To accomplish this, STARs are developed according to the following criteria:

- STARs must be simple, easily understood and, if possible, limited to one page.
- A STAR transition should be able to accommodate as many different types of aircraft as possible. That way, both jets and single-engine piston aircraft may be able to use the same chart for arrival.
- VORTACs are used wherever possible, so that military and civilian aircraft can use the same arrival.
- DME arcs within a STAR should be avoided as not all aircraft in the IFR environment are so equipped.
- Altitude crossing and airspeed restrictions are included when they are assigned by ATC a majority of the time.

Now consider how this same clearance is issued when a STAR exists for this terminal area. *"Cessna 32G, cleared to Seattle/Tacoma International Airport as filed, then CHINS TWO ARRIVAL, Ephrata Transition. Maintain 10,000 feet."* A shorter transmission conveys the same information. A transition is one of several routes that bring traffic from different directions into one STAR.

INTERPRETING THE STAR

STARs use much of the same symbology as departure and approach charts. In fact, a STAR may at first appear identical to a similar graphic DP, except that the direction of flight is reversed, and the procedure ends at an approach fix. The STAR officially begins at the common navaid or intersection where all the various transitions to the arrival come together. This way, arrivals from several directions can be accommodated on the

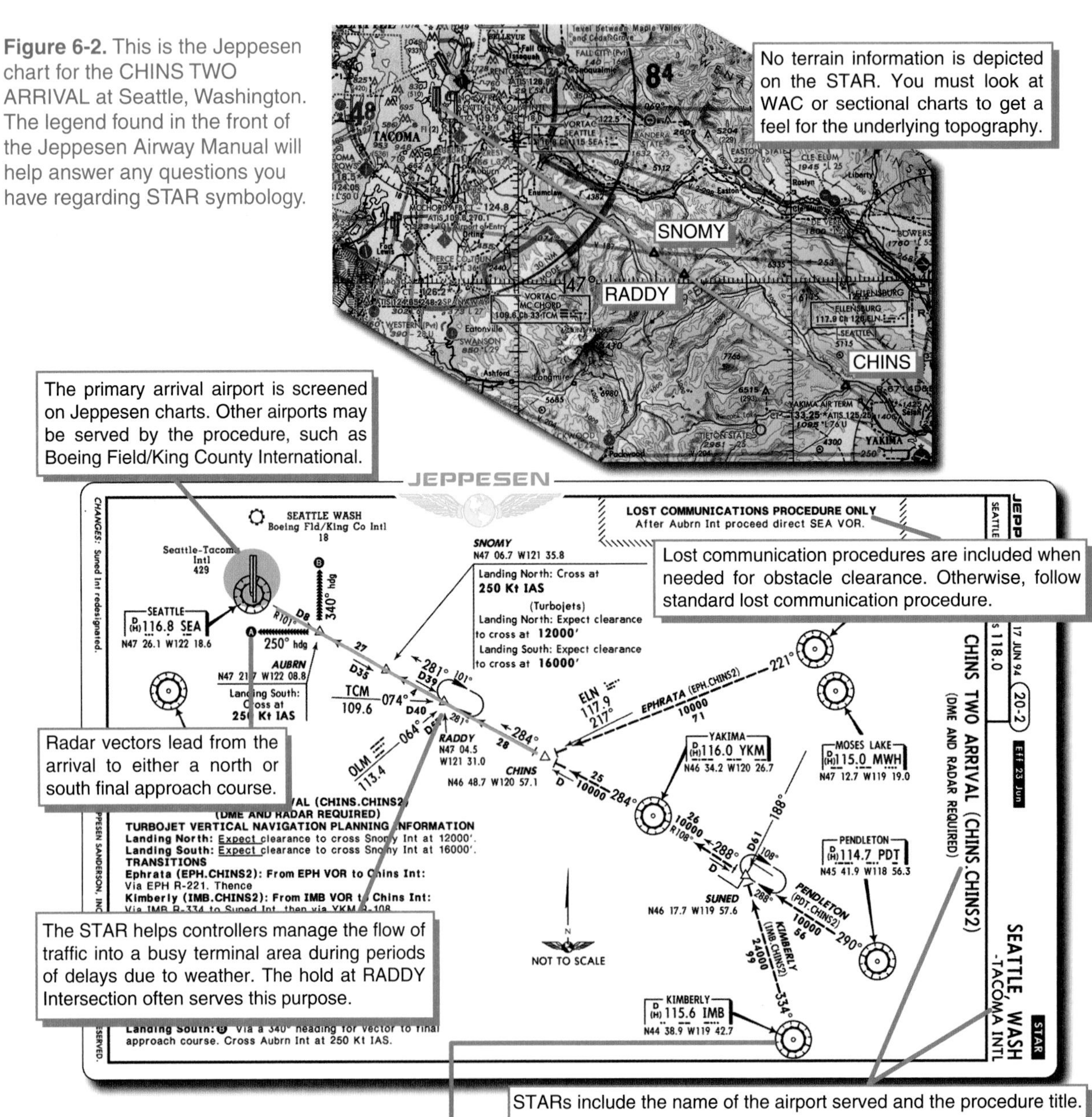

Figure 6-2. This is the Jeppesen chart for the CHINS TWO ARRIVAL at Seattle, Washington. The legend found in the front of the Jeppesen Airway Manual will help answer any questions you have regarding STAR symbology.

same chart, and traffic flow is routed appropriately within the congested airspace. [Figure 6-2]

STARs are usually named according to the point at which the procedure begins. A STAR that commences at the CHINS Intersection becomes the CHINS ONE ARRIVAL. When a significant portion of the arrival is revised, such as an altitude, a route, or data concerning the navaid, the number of the arrival changes. For example, the CHINS ONE ARRIVAL is now the CHINS TWO ARRIVAL due to modifications in the procedure.

A STAR begins at a navaid or intersection where all arrival transitions join.

Studying the STARs for an airport can clue you into the specific topography of the area. Note the initial fixes, and where they correspond to fixes on the enroute chart. Arrivals may incorporate stepdown fixes when necessary to keep aircraft within airspace boundaries, or for obstacle clearance. Routes between fixes contain courses, distances, and minimum altitudes, alerting you to possible obstructions or terrain under your arrival path. Air-

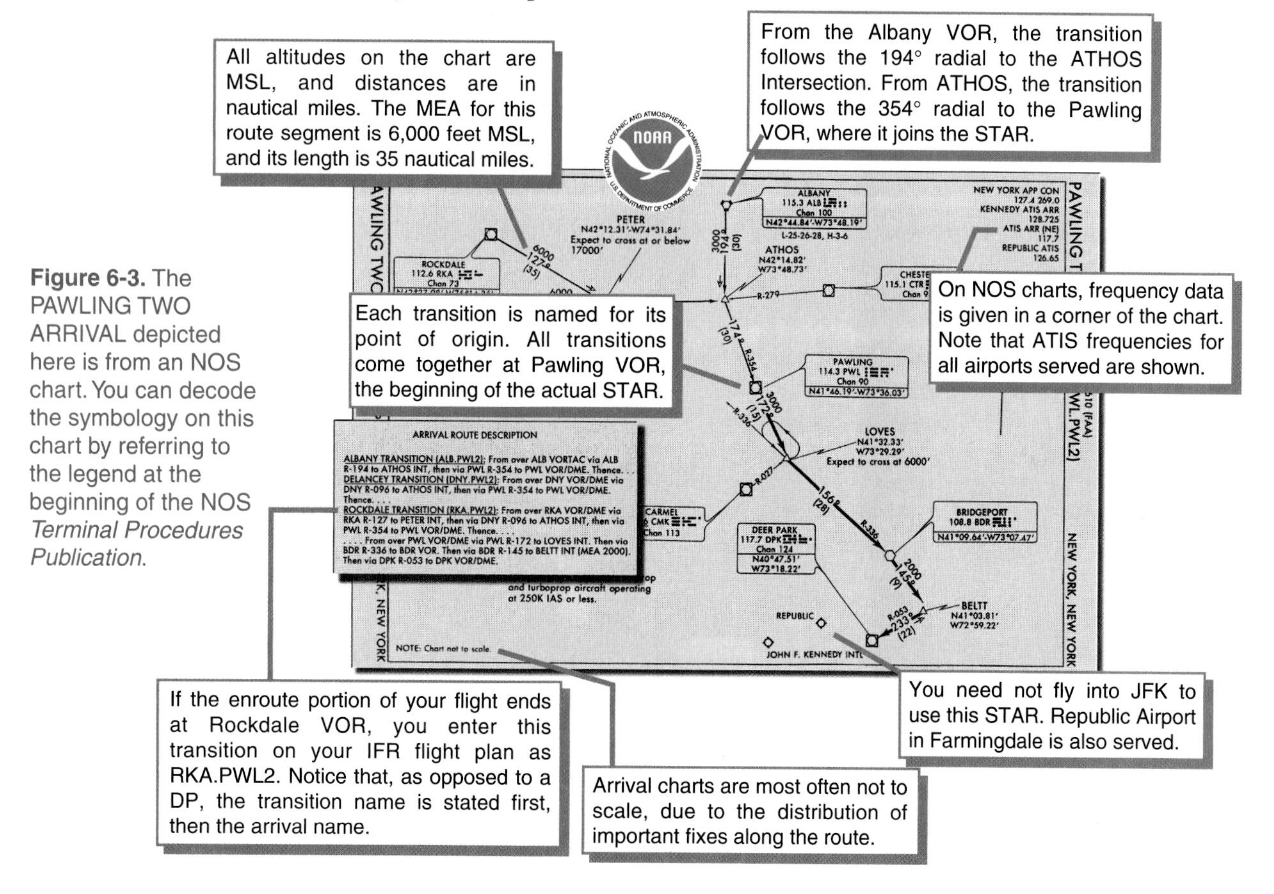

Figure 6-3. The PAWLING TWO ARRIVAL depicted here is from an NOS chart. You can decode the symbology on this chart by referring to the legend at the beginning of the NOS *Terminal Procedures Publication.*

Arrival route headings on an NOS STAR are depicted by large numerals and a heavyweight line, while Jeppesen STARs use the abbreviation **hdg** next to the direction arrow. See figures 6-2 and 6-3.

Frequencies on which to contact the proper approach controller are found in a corner of the NOS chart. See figure 6-3.

speed restrictions also appear where they aid in managing the traffic flow. In addition, some STARs require that you use DME and/or ATC radar. [Figure 6-3]

VERTICAL NAVIGATION PLANNING

Included within certain STARs is information on **vertical navigation planning**. This information is provided to reduce the amount of low altitude flying time for high-performance aircraft, like jets and turboprops. An expected altitude is given for a key fix along the route. By knowing an intermediate altitude in advance, the pilot of a high-performance aircraft can plan the power settings and aircraft configurations that will result in the most efficient descent, in terms of time, fuel requirements, and engine wear. [Figure 6-4]

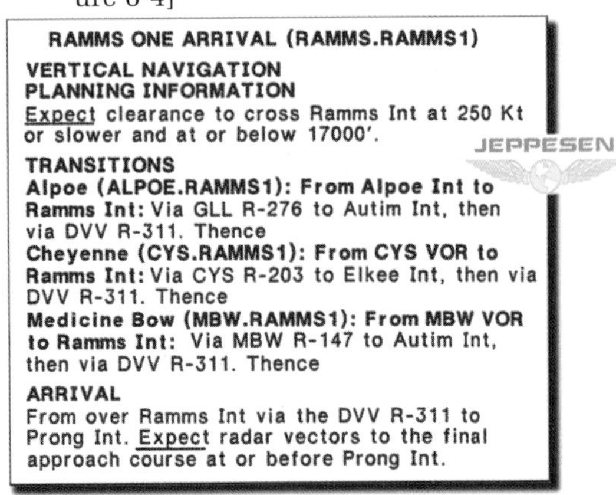

RAMMS ONE ARRIVAL (RAMMS.RAMMS1)

VERTICAL NAVIGATION PLANNING INFORMATION
Expect clearance to cross Ramms Int at 250 Kt or slower and at or below 17000'.

TRANSITIONS
Alpoe (ALPOE.RAMMS1): From Alpoe Int to Ramms Int: Via GLL R-276 to Autim Int, then via DVV R-311. Thence
Cheyenne (CYS.RAMMS1): From CYS VOR to Ramms Int: Via CYS R-203 to Elkee Int, then via DVV R-311. Thence
Medicine Bow (MBW.RAMMS1): From MBW VOR to Ramms Int: Via MBW R-147 to Autim Int, then via DVV R-311. Thence

ARRIVAL
From over Ramms Int via the DVV R-311 to Prong Int. Expect radar vectors to the final approach course at or before Prong Int.

JEPPESEN

Figure 6-4. The vertical navigation planning information from the RAMMS ONE ARRIVAL at Denver, CO, is used by pilots of larger aircraft to plan their descents.

Computer Navigation Fixes

Beginning in 1998, the FAA began assigning five-letter Computer Navigation Fix (CNF) names to previously unnamed airspace fixes and mileage break points on DPs, enroute, area, and standard terminal arrival charts. These named fixes are printed in italic type within brackets near an X along airways and routes, and include latitude and longitude coordinates. [Figure A] NOS DP charts do not include the CNF name. [Figure B] CNFs are for reference to navigation systems using electronic databases only, and provide these systems with the information necessary to identify required positions using only this equipment. [Figure C]

Computer navigation fixes have no air traffic control functions. This means that ATC will not request reports or assign holding in conjunction with a CNF. Further, you should not use CNFs to define routing in IFR flight plans or in direct requests to ATC for route changes while in flight.

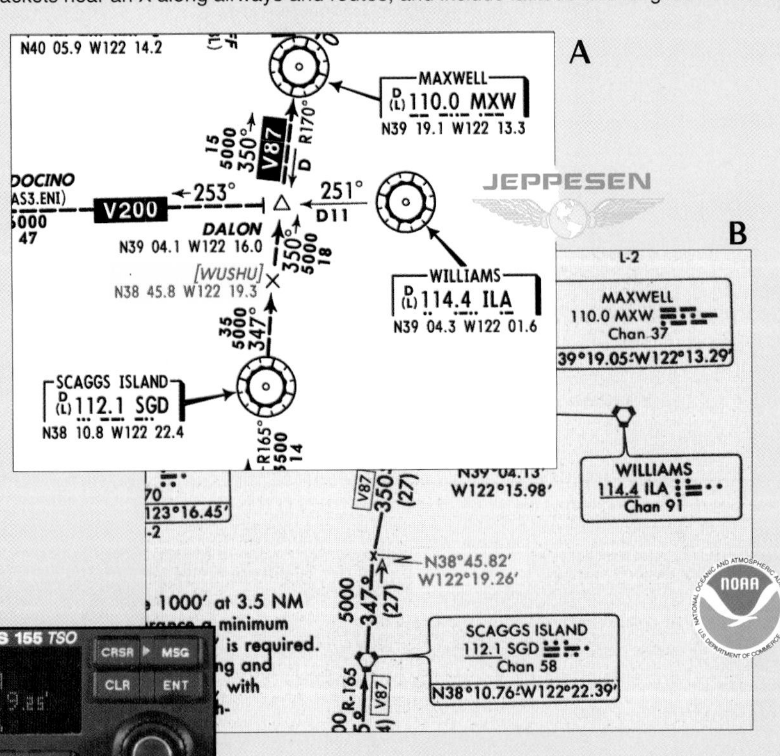

SUMMARY CHECKLIST

✓ Standard terminal arrival routes (STARs) provide a standard method for leaving the enroute structure and entering a busy terminal area.

✓ STARs are grouped along with other airport charts in a Jeppesen subscription, and appear in the front of NOS booklets. Legends are found in the front of the corresponding book.

✓ If you accept a STAR, you must have at least a textual description of the procedure in your possession. A graphic description is preferable.

✓ Writing "No STAR" in the remarks section of your flight plan will alert ATC that you do not wish to use these procedures during your flight. You also may refuse a clearance containing a STAR, but avoid this practice if possible.

✓ STARs use symbology that is similar to that on graphic DPs. Altitudes are given in reference to mean sea level, and distances are in nautical miles.

✓ A STAR begins at a navaid or intersection where all arrival transitions join.

✓ STARs are named according to the point where a procedure begins. They are revised in numerical sequence.

✓ Arrival route headings on an NOS STAR are depicted by large numerals within a heavyweight line, while those on Jeppesen STARs are depicted with the abbreviation **hdg** next to the heading in degrees.

✓ Frequencies on which to contact the proper approach controller are found in the corner of an NOS chart.

✓ Vertical navigation planning information is given for pilots of turboprop and jet traffic, to aid them in making efficient descents from the enroute structure to approach fixes.

KEY TERMS

Standard Terminal Arrival Route (STAR)

Transition

Vertical Navigation Planning

QUESTIONS

Refer to the reprint of the WISKE ONE ARRIVAL to answer questions 1 through 5.

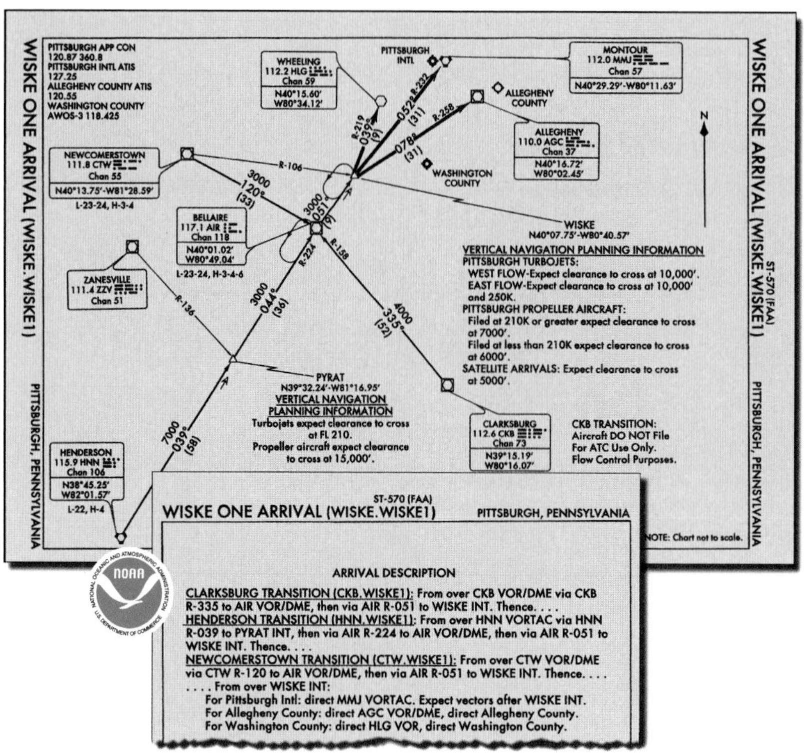

1. At what point does the WISKE ONE ARRIVAL begin?

2. Which airports are served by this STAR?

   A. Only Pittsburgh International
   B. Pittsburgh International and Wheeling
   C. Pittsburgh International, Allegheny County and Washington County

3. What is the MEA for the STAR segment from Henderson VORTAC to PYRAT Intersection within the Henderson Transition?

4. How would you indicate that you wanted to use the WISKE ONE ARRIVAL, Clarksburg Transition on an IFR flight plan?

5. If you are flying the WISKE ONE ARRIVAL in a propeller-driven aircraft, at what altitude should you expect to cross the WISKE Intersection, if you filed for a TAS of less than 210 knots?

6. True/False. ATC can include a STAR in your clearance even if you have not requested one.

7. Select the appropriate phrase to enter in the Remarks section of your flight plan if you do not wish to use a STAR during your flight.

   A. "Use no STARs"
   B. "NO STAR"
   C. "No STARs approved"

8. Where can you find arrival charts in an NOS format? Where can you find arrival charts in a Jeppesen format?

SECTION B
ARRIVAL PROCEDURES

Preparation for a well-executed arrival and approach begins long before you descend from the enroute phase of the flight. Planning your approach in advance, while there are fewer demands on your attention, leaves you free to concentrate on precise control of the airplane during the approach. You also will be better equipped to deal with any problems that might arise during the last segment of the flight.

PREPARING FOR THE ARRIVAL

You may accept a STAR within a clearance, or you may file for one in your flight plan. As you near your destination airport, ATC may add a STAR procedure to your original clearance. Keep in mind that ATC can assign a STAR even if you have not requested one. If you accept the clearance, you must have at least a textual description of the procedure in your possession. If you do not want to use a STAR, you must specify "No STAR" in the remarks section of your flight plan. You may also refuse the STAR when it is given to you verbally by ATC, but the system works better if you advise ATC ahead of time.

ATC will issue a STAR when they deem one appropriate, unless you request "No STAR."

As mentioned before, STARs include navigation fixes that are used to provide transition and arrival routes from the enroute structure to the final approach course. They may also lead to a fix where radar vectors will be provided to intercept the final approach course. **Minimum crossing altitudes** and airspeed restrictions appear on some STARs when they are used 75 percent or more of the time. These expected altitudes and airspeeds are not part of your clearance until ATC includes them verbally. A STAR is simply a published routing, and it does not have the force of a clearance until issued specifically by ATC. For example, MEAs printed on STARs are not valid unless stated within an ATC clearance or in cases of lost communication.

As early as practical, listen to ATIS to obtain current weather information, runways in use, and NOTAMs involving the destination airport. You also may ask to monitor AWOS or ASOS, or to contact the UNICOM or FSS on the field, for updated weather and airport information when no ATIS is available. If you are landing at an airport with approach control services which has two or more published instrument approach procedures, you will receive advance notice of the instrument approaches in use. This information will be broadcast by either ATIS or a controller, however, it may not be provided when the visibility is three miles or better and the ceiling is at or above the highest initial approach altitude established for any instrument approach procedure for the airport. [Figure 6-5]

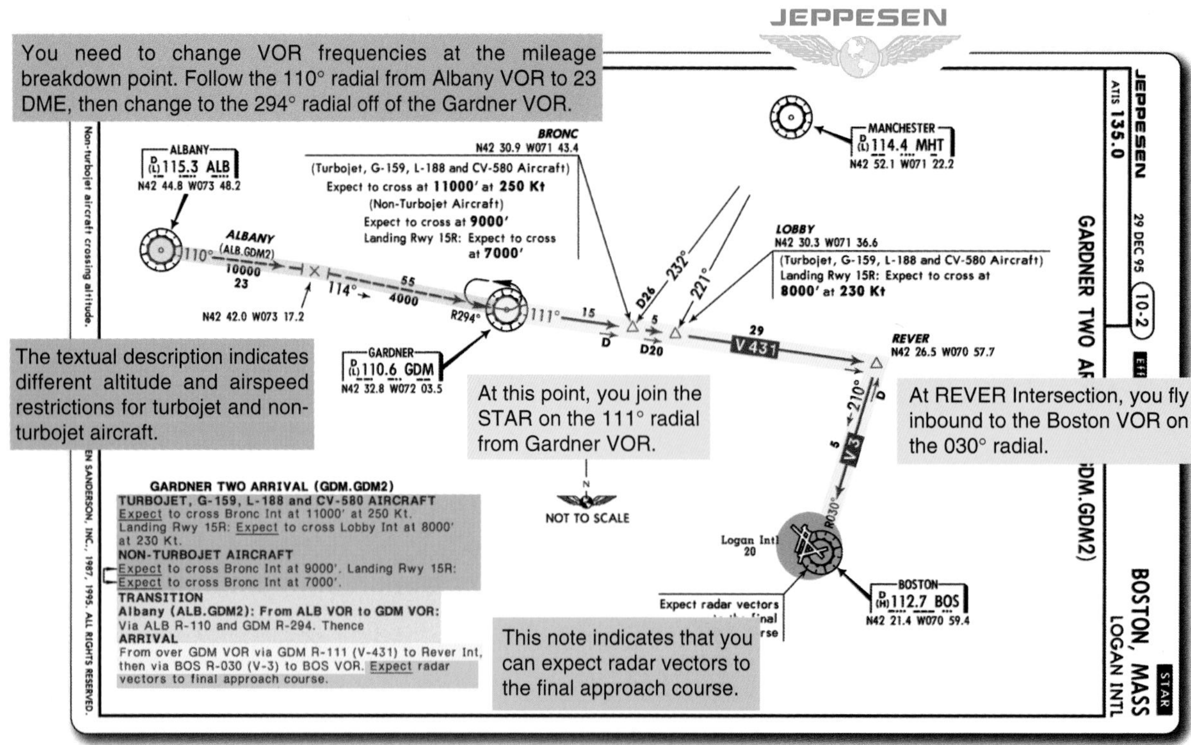

Figure 6-5. After receiving your arrival clearance, you should review the assigned STAR procedure.

Arriving on the Space Shuttle Orbiter

Arrival procedures for the space shuttle begin on the opposite side of the planet. The deorbit burn, which will bring the orbiter back to Kennedy Space Center, occurs about one hour before landing. Thirty minutes before touchdown, the orbiter begins to enter the atmosphere. At approximately 45,000 feet, the orbiter starts maneuvers to intercept the landing approach corridor at the desired altitude and velocity.

Normally, re-entry will follow one of two general patterns, depending on the mission and the shuttle's orbital parameters. One pattern begins with a deorbit burn over the Indian Ocean off the western coast of Australia. Then, the flight path normally continues across the Pacific Ocean to the Baja Peninsula, across Mexico and southern Texas, over the Gulf of Mexico, on to the west coast of Florida, and across the central part of the state.

The second major pattern varies, depending on where re-entry occurs. The ground track takes the orbiter across Canada and the eastern United States or above the Southern Pacific and across South America. The orbiter will either parallel the northeast Florida coast after cutting across Georgia or will fly over the Florida Everglades and up the southeast coast. The sonic booms that are generated when the orbiter enters the atmosphere can be heard across the width of Florida.

REVIEWING THE APPROACH

Once you have determined which approach to expect, review the corresponding approach chart thoroughly before you enter the terminal area. Your review should include radio frequencies, the inbound course, descent minimums, the missed approach point, and the missed approach procedure. Rehearse the approach in your mind, with careful attention paid to the order in which you will perform the approach. Establish your personal minimums, and take into account the weather conditions and your fatigue level when determining your plan of action.

When you have completed the approach chart review, consider moving on to the descent and approach checklists, as appropriate to your aircraft. Check your fuel level, and make sure a prolonged hold or increased headwinds have not cut into your reserve. There is always a chance that you will have to make a missed approach, or go to an alternate. By completing important prelanding items early in the arrival stage, you free yourself to concentrate on maneuvering and navigating while on the approach.

ALTITUDE

During your arrival in the terminal area ATC will either clear you to a specific altitude, or they will give you a descend via clearance, which instructs you to follow the altitudes published on the STAR. [Figure 6-6] If ATC amends the altitude or route to one that is different from the published procedure, the rest of the charted descent procedure is canceled. ATC will assign you any further route, altitude, or airspeed clearances, as necessary. [Figure 6-7]

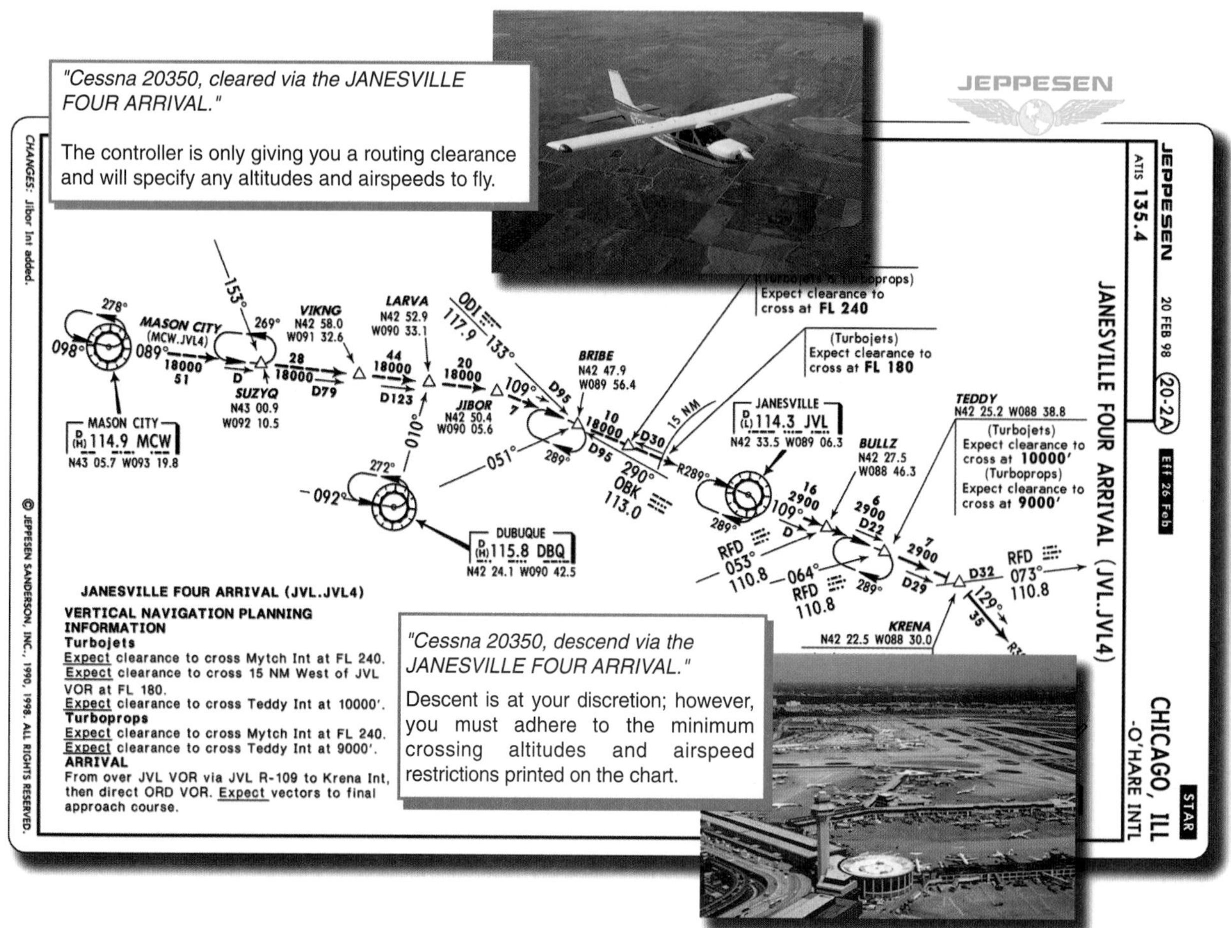

Figure 6-6. You are not authorized to leave your last assigned altitude unless specifically cleared to do so.

AIRSPEED

During the arrival, expect to make adjustments in indicated airspeed at the controller's request. When you fly a high-performance airplane on an IFR flight plan, ATC may ask you to adjust your airspeed to achieve proper traffic sequencing and separation and to reduce the amount of vectoring required in the terminal area. When you operate a reciprocating-engine or turboprop within 20 miles of your destination airport, 150 knots is usually the lowest speed you will be assigned. If your aircraft cannot maintain the assigned airspeed, you must advise ATC. In this case, the controller may ask you to

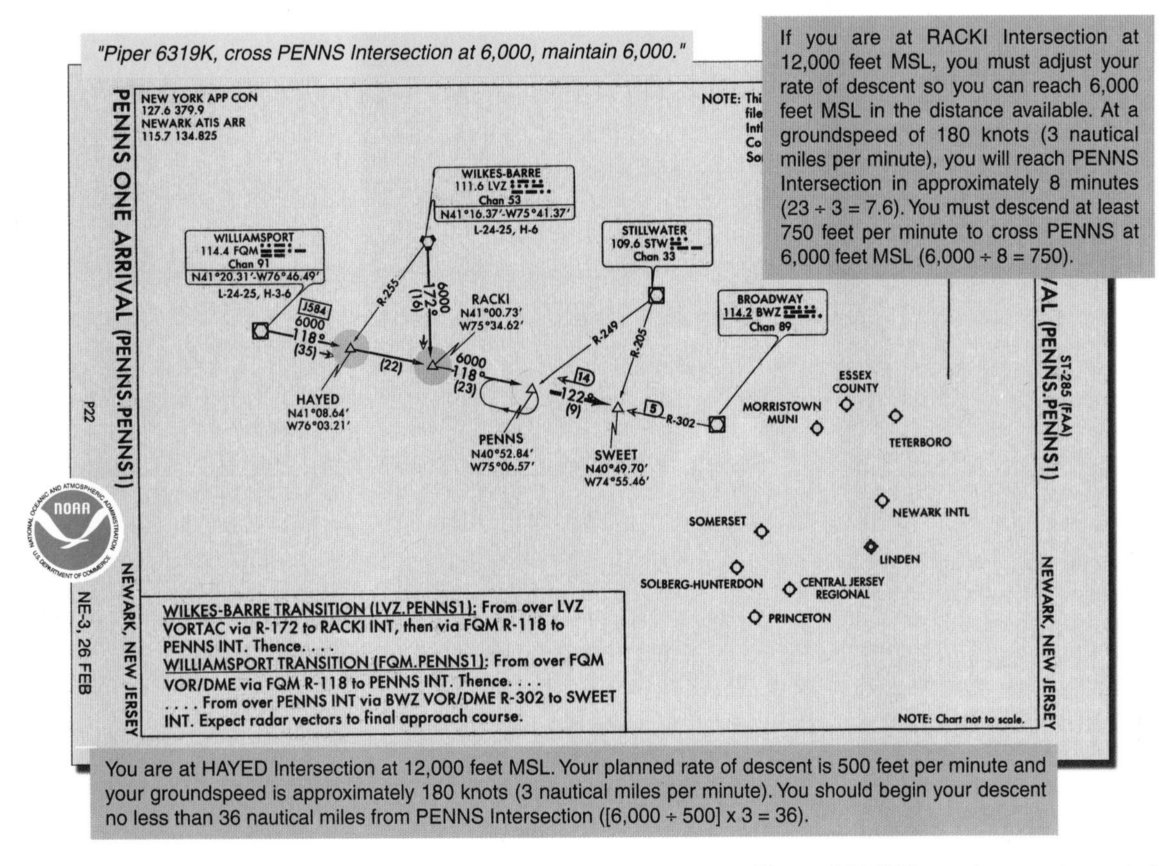

Figure 6-7. ATC may issue a descent clearance which includes a crossing altitude restriction. In this example, you are authorized to descend at your discretion, as long as you cross the PENNS Intersection at 6,000 feet MSL.

maintain the same airspeed as those aircraft ahead of you or behind you on the approach. ATC expects you to maintain the specified airspeed within ±10 knots. At other times, ATC may ask you to increase or decrease your speed by 10 knots, or multiples thereof. When ATC no longer requires the speed adjustment, they will advise you to "*...resume normal speed.*"

Keep in mind that the maximum speeds specified in FAR 91.117 still apply during speed adjustments. It is your responsibility, as pilot in command, to advise ATC if an assigned speed adjustment would cause you to exceed these limits. However, for operations in Class C or D airspace at or below 2,500 feet AGL, within 4 nautical miles of the primary airport, ATC has the authority to request or approve a higher speed than those prescribed in FAR 91.117.

CONFLICTING SIGNALS

On April 10, 1989, a Fairchild FH-227 crashed into a fog-shrouded mountainside in southeastern France. The twin-engine turboprop was on a published arrival route into Valence, flying in light rain and a stratified cloud layer. The crew had both an NDB and a VOR tuned in, but only the ADF receiver was tuned to the proper frequency. The VOR receiver was tuned to a VOR located 13 miles to the northeast, putting the FH-227 off of the intended course by 10 miles. Ninety seconds before impact, the captain noted on the cockpit voice recorder that there was a 30° difference between the ADF and VOR indications. His fatal assumption? That "the ADF (was) no good." The crew and 19 passengers perished in the crash.

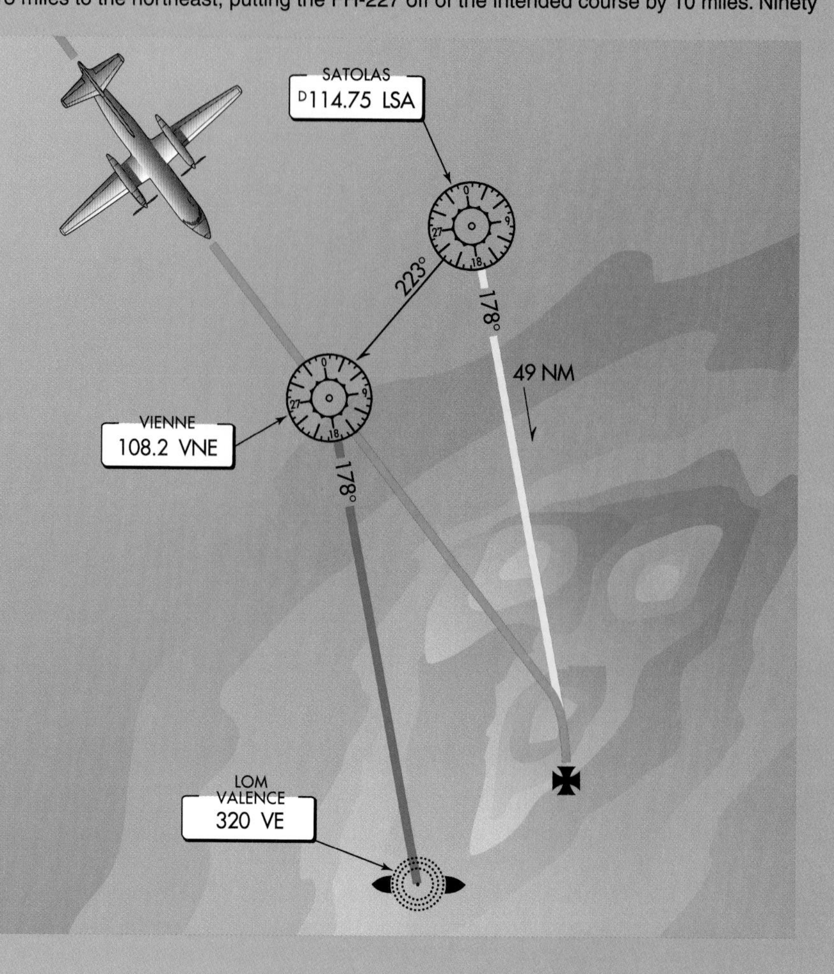

One of the most critical safety concerns on board any instrument flight is instrument cross-check. If two navigation receivers give you conflicting messages, one of them may be incorrect. However, do not assume that you know which one is incorrectly tuned, or inoperative, until you double check all frequencies, volumes, and warning lights. A thorough look at the chart also may reveal the true problem. Use all your resources to maintain situational awareness and avoid the trap of a mind-set regarding your position. You may be in a completely different position than you think.

SUMMARY CHECKLIST

✓ ATC may assign a STAR at any time, and it is your responsibility to accept or refuse the procedure.

✓ Altitudes and airspeeds published on the STAR are not considered restrictions until verbally given by ATC as part of a clearance.

✓ After receiving the arrival clearance, certain tasks can be completed before starting your approach, including gathering weather information and accomplishing the descent and approach checklists.

✓ After you determine the approach in use, review the appropriate chart and create a plan of action.

✓ A descend via clearance instructs you to follow the altitudes published on the STAR, with descent at your discretion.

✓ ATC may issue a descent clearance which includes a crossing altitude. Comply by estimating the distance and rate of descent required.

✓ Expect to make airspeed adjustments as required by ATC. Responsibility for complying with FAR 91.117 is up to the pilot in command.

KEY TERMS

Minimum Crossing Altitudes

Descend Via

QUESTIONS

1. When should you plan to obtain weather information for your approach during an IFR flight?

2. What does a descend via clearance authorize you to do?

3. When does a minimum crossing altitude become a restriction when a STAR is used on an IFR flight plan?

   A. When issued verbally as part of a clearance
   B. When the STAR is filed in an IFR flight plan
   C. Only if the altitude is lower than the minimum altitude at the initial approach fix

4. True/False. ATC is responsible for ensuring you are not cleared for an airspeed that violates FAR 91.117.

5. Assume you are cruising at 12,000 feet MSL. You are cleared to cross the FREDY Intersection at 8,000 feet MSL, which is 30 nautical miles ahead of you. If your current groundspeed is 180 knots and you expect a descent rate of 500 feet per minute, where should you begin your descent?

CHAPTER 7

APPROACH

Instrument/Commercial
Part II, Segment 2, Chapter 7 — Approach

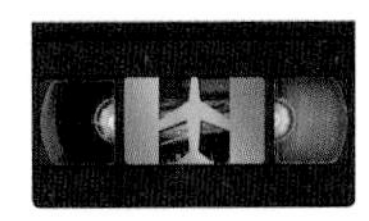

SECTION A
APPROACH CHARTS

The standard **instrument approach procedure (IAP)** allows you to descend safely by reference to instruments from the enroute altitude to a point near the runway at your destination from which a landing may be made visually. A **precision approach** procedure provides vertical guidance through means of an electronic glide slope, as well as horizontal course guidance. The instrument landing system (ILS) and the precision approach radar (PAR) procedure are examples of precision approaches. A **nonprecision approach**, such as a VOR or NDB approach, provides horizontal course guidance with no glide slope information.

Although there are many different types of approaches in use, most incorporate common procedures and chart symbology. Therefore, your ability to read one approach chart generally means you will be able to read others. However, due to the dynamic nature of charting, you must familiarize yourself with the chart legend and review chart symbology on a continuing basis to stay abreast of any changes. This section is designed to help you become proficient at interpreting chart information. The most common approaches you will fly are the ILS, localizer, VOR, NDB, and GPS. Other approaches you may encounter include LDA, SDF, VOR/DME, RNAV, and MLS procedures. Specific approach types and procedures are examined in more detail in Section B of this chapter and Chapter 8 — Instrument Approaches.

APPROACH SEGMENTS

Before looking at approach chart symbology, it is helpful to have a basic understanding of approach procedures. An instrument approach may be divided into as many as four approach segments: initial, intermediate, final, and missed approach. **Feeder routes**, also referred to as approach transitions or terminal routes, technically are not considered approach segments but are an integral part of many instrument approach procedures. They provide a link between the enroute and approach structures. Although an

approach procedure may have several feeder routes, you generally use the one closest to your enroute arrival point. When a feeder route is shown, the chart provides the course or bearing to be flown, the distance, and the minimum altitude. [Figure 7-1]

Figure 7-1. Approach segments often begin and end at designated fixes. However, they also can begin and end at the completion of a particular maneuver, such as a course reversal.

MISSED APPROACH SEGMENT
Upon reaching the MAP, if you are unable to continue the approach to a landing, you follow the missed approach segment back to the enroute structure if you are continuing to an alternate airport.

ENROUTE FIX

INITIAL APPROACH FIX (IAF)

FINAL APPROACH FIX (FAF)

INTERMEDIATE APPROACH FIX (IF)

AIRPORT

MISSED APPROACH POINT (MAP)

FINAL APPROACH SEGMENT
The final approach segment ends at the runway, airport, or missed approach point (MAP).

INTERMEDIATE APPROACH SEGMENT
From the IF, you follow the intermediate approach segment to the final approach fix (FAF).

INITIAL APPROACH SEGMENT
Next, you follow an initial approach segment to an intermediate fix (IF).

FEEDER ROUTE
A feeder route may be used to take you from the enroute structure to an initial approach fix (IAF).

INITIAL APPROACH SEGMENT

The purpose of the initial approach segment is to provide a method for aligning your aircraft with the approach course. This is accomplished by using an arc procedure, a course reversal such as a procedure turn or holding pattern, or by following a route which intersects the final approach course. The initial approach segment begins at an **initial approach fix (IAF)** and usually ends where it joins the intermediate approach segment. The letters IAF on an approach chart indicate the location of an initial approach fix and course, distance, and minimum altitudes are provided for initial approach segments. A given procedure may have several initial approach segments. Where more than one exist, each will join a common intermediate segment, although not necessarily at the same location.

The letters IAF indicate the location of an initial approach fix.

Charting Evolution

The nature of instrument flight cartography involves continuous chart enhancement. Captain Elrey B. Jeppesen's original little black book from the 1930s depicts slopes, drainage patterns, terrain, and airport layouts, as well as provides information on field lengths, lights, and obstacles. [Figure A]

As instrument approach procedures improved, a standard format was developed which emerged into an identifiable instrument approach chart look. Shown here is the Buffalo Municipal Airport instrument approach chart, revised August 2, 1945. Note that the approach chart depicts the old range airway system, utilizing the Morse code letters A and N. [Figure B]

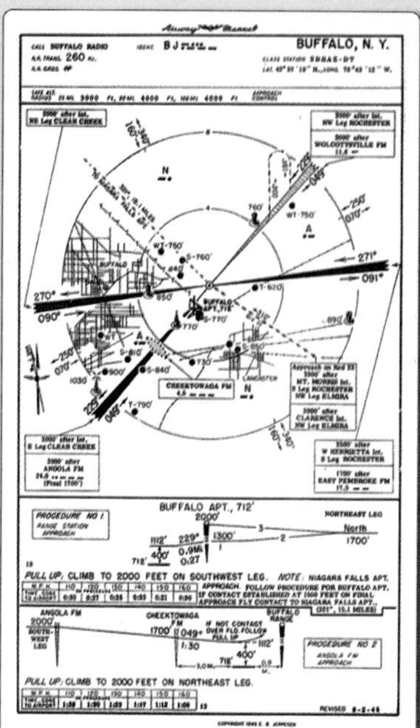

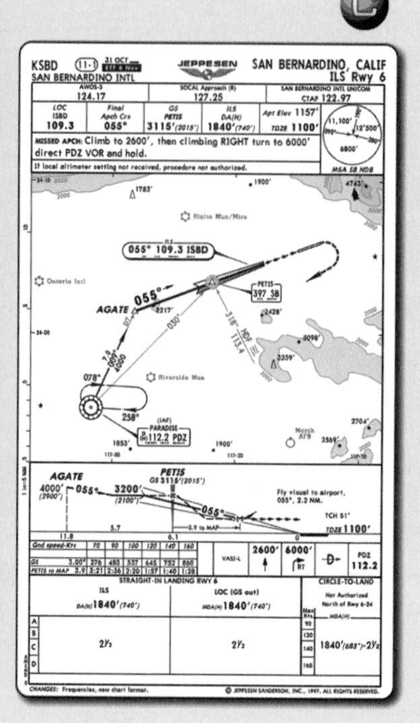

Enhancements in today's highly refined standard instrument approach procedures (SIAPs) and instrument chart formats are driven by complex changes in airspace, air traffic control, advanced technology, human factors, and many other issues. Jeppesen's briefing strip charts represent many instrument approach chart advancements. [Figure C]

Occasionally, a chart may depict an IAF, although there is no initial approach segment for the procedure. This usually occurs where the intermediate segment begins at a point located within the enroute structure. In this situation, the IAF signals the beginning of the intermediate segment.

INTERMEDIATE APPROACH SEGMENT

The intermediate segment is designed primarily to position your aircraft for the final descent to the airport. On this segment, you typically reduce your airspeed to or near the approach airspeed, complete the before landing checklist, and make a final review of the approach procedure and applicable minimums. Like the feeder route and initial approach segment, the chart depiction of the intermediate segment provides you with course, distance, and minimum altitude information.

The intermediate segment, normally aligned within 30° of the final approach course, begins at the **intermediate fix (IF)**, or point, and ends at the beginning of the final approach segment. In some cases, an intermediate fix is not shown on an approach chart. In this situation, the intermediate segment begins at a point where you are proceeding

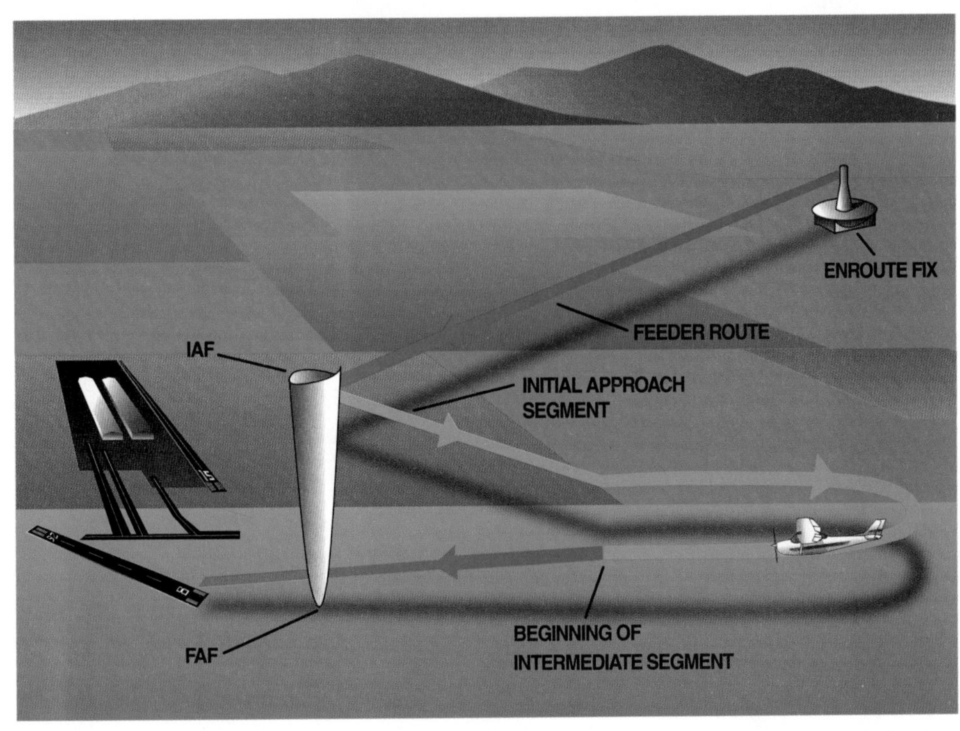

Figure 7-2. An instrument approach that incorporates a procedure turn is the most common example of an approach which may not have a charted intermediate approach fix. The intermediate segment, in this example, begins when you intercept the inbound course after completing the procedure turn.

inbound to the final approach fix, are properly aligned with the final approach course, and are located within the prescribed distance from the final approach fix. [Figure 7-2]

FINAL APPROACH SEGMENT

The purpose of the final approach segment is to allow you to navigate safely to a point at which, if the required visual references are available, you can continue the approach to a landing. If you cannot see the required cues at the missed approach point, you must execute the missed approach procedure. The final approach segment for a precision approach begins where the glide slope is intercepted at the minimum glide slope intercept altitude shown on the approach chart. If ATC authorizes a lower intercept altitude, the final approach segment begins upon glide slope interception at that altitude. For a nonprecision approach, the final approach segment begins either at a designated **final approach fix (FAF)** or at the point where you are established on the final approach course. When an FAF is not designated, such as on an approach which incorporates an on-airport VOR or NDB, this point is typically where the procedure turn intersects the final approach course inbound. This point is referred to as the **final approach point (FAP)**. The final approach segment ends either at the designated missed approach point or when you land.

Although the charted final approach segment provides you with course and distance information, many factors influence the minimum altitude to which you can descend. These include the type of aircraft being flown, the aircraft's equipment and approach speed, the operational status of navaids, the airport lighting, the type of approach being flown, and local terrain features.

MISSED APPROACH SEGMENT

The purpose of the missed approach segment is to allow you to safely navigate from the missed approach point to a point where you can attempt another approach or continue to another airport. Every instrument approach has a missed approach segment with appropriate heading, course, and altitude information. During a radar approach, ATC gives you missed approach instructions.

The missed approach segment begins at the **missed approach point (MAP)** and ends at a designated point, such as an initial approach or enroute fix. The actual location of the

missed approach point depends upon the type of approach you are flying. For example, during a precision approach, the MAP occurs when you reach a designated altitude on the glide slope called the decision height (DH). For nonprecision approaches, the missed approach point occurs either at a fix defined by a navaid or after a specified period of time has elapsed since you crossed the final approach fix.

CHART LAYOUT

Both Jeppesen and NOS use charts to portray the instrument approaches which are available at a given airport. Jeppesen charts are filed in a loose-leaf format by state, then by city within each state. NOS charts are published in regional volumes referred to as *Terminal Procedures Publications*, with each airport filed alphabetically by the name of the associated city. Generally, both chart formats present the same information; however, the symbology and chart layout vary. The chart used throughout the following discussion is the ILS Runway 1 approach to Key Field at Meridian, Mississippi. [Figure 7-3]

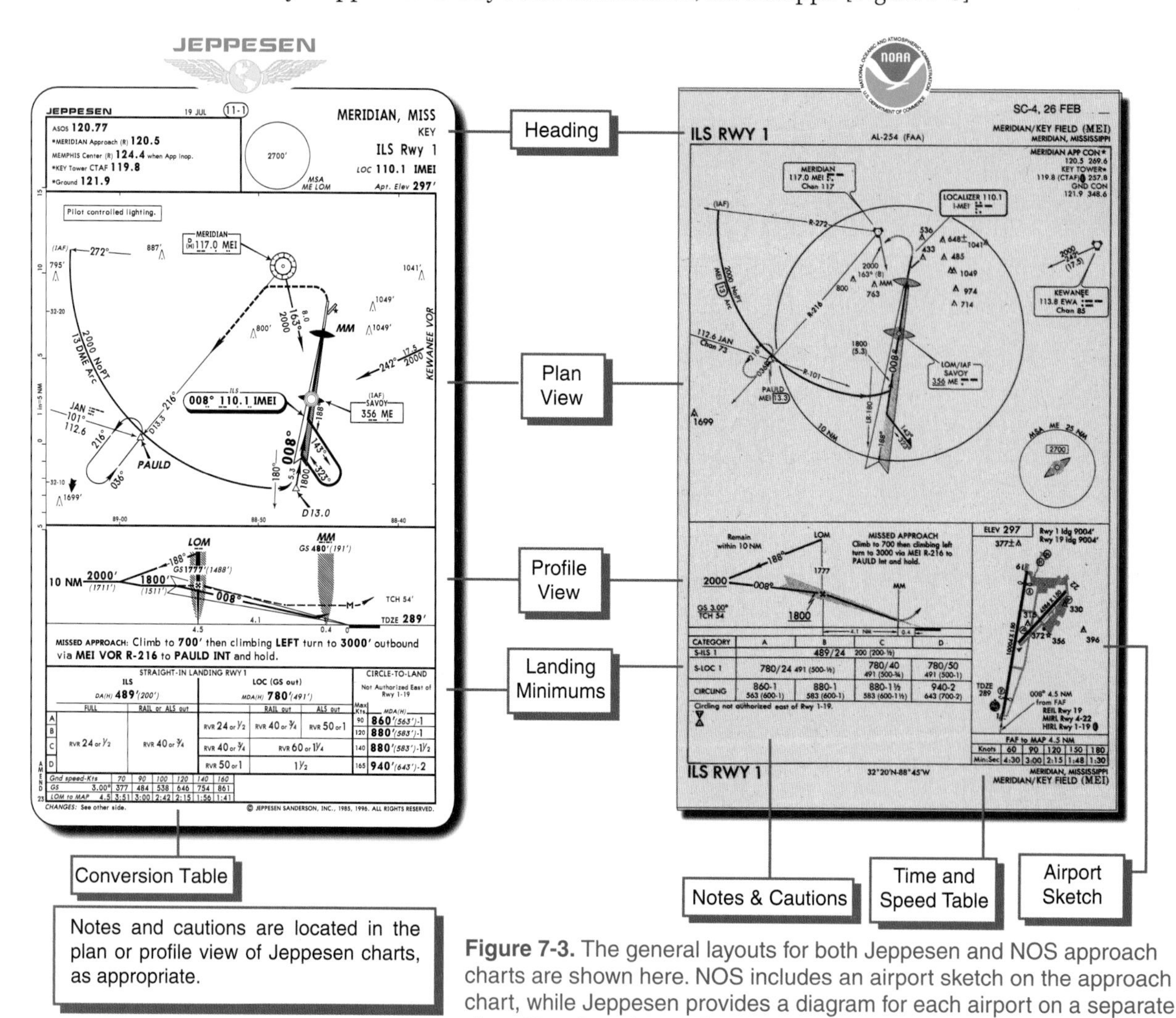

Figure 7-3. The general layouts for both Jeppesen and NOS approach charts are shown here. NOS includes an airport sketch on the approach chart, while Jeppesen provides a diagram for each airport on a separate chart. Knowing where information is located is just as important as interpreting the chart symbology.

The following discussion employs a standard format to make it easy to understand the features of approach charts. Each section of the chart is introduced followed by general information and definitions of terms which apply to that section. Then examples of the chart section are shown in Jeppesen and NOS formats and specific features and symbology are described through the use of text boxes. You may wish to refer to the chart

excerpts as you are reading the general information and definitions. NOS charts and Jeppesen's original chart format are presented initially followed by an examination of Jeppesen's new briefing strip charts at the conclusion of this chapter.

HEADING SECTION

On Jeppesen charts, the heading section identifies the city and airport, the instrument approach procedure title, the primary approach facility, communication frequencies, and if available, minimum safe altitude information. On NOS charts, some of this information is located in the heading section, with the rest located on the plan view or airport sketch.

ATC uses the procedure title when clearing you for the approach. The procedure title indicates the type of approach system used and the equipment required to fly the approach. For example, a procedure title such as ILS DME Rwy 1 requires you to have a DME receiver in addition to ILS equipment to fly the approach. When a letter appears in the procedure title, such as VOR-A, it means the procedure does not meet the criteria for a straight-in landing. In this case, a turning maneuver, called a circling approach, may be required to complete the landing.

The procedure title indicates the type of approach system used and the equipment required to fly the approach.

Both Jeppesen and NOS use chart index numbers to identify the chart. The index number on Jeppesen charts helps you file a chart and identify certain features. For example, the first digit represents an airport number. Normally, this is the number 1. However, where more than one airport shares the same city and state name, the first airport is given the number 1, the second the number 2, and so on. The second digit identifies the type of chart. [Figure 7-4] The third digit is used to index charts with the same approach types. For example, the first ILS approach at the second airport is given an index number of 21-1. The second ILS at the same airport receives the number 21-2. The first VOR approach at this airport receives the number 23-1.

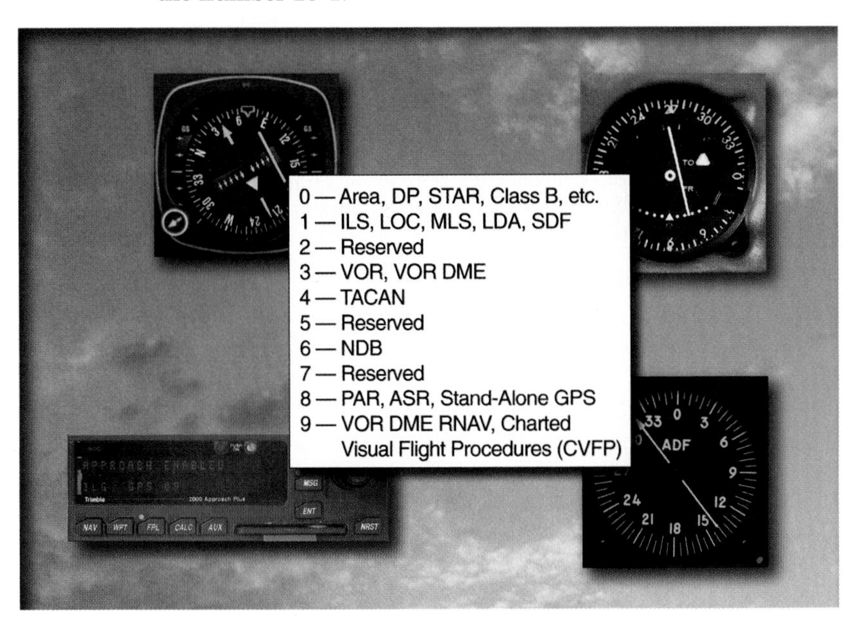

Figure 7-4. The second digit of the Jeppesen chart index number is used to identify the type of chart.

The **minimum safe altitude (MSA)**, shown on approach charts, provides 1,000 feet of obstruction clearance within 25 nautical miles of the indicated facility, unless some other distance is specified. The 1,000-foot criteria applies over both mountainous and nonmountainous terrain. The MSA may be divided into several sectors which are defined by radials, or bearings to the navaid. Each MSA is applicable only to the approach on which it is displayed, and it may not be used for any other approach.

There are several important features of the MSA. First, it only provides obstruction clearance within the sector. Neither navigation nor communication coverage is guaranteed. Second, the MSA is designed only for use in an emergency or during VFR flight, such as during a VFR approach at night. And third, an MSA is not listed for every approach. Its omission may be due to the lack of an easily identifiable facility upon which to orient the MSA circle. [Figures 7-5 and 7-6]

The MSA provides 1,000 feet of obstruction clearance within a specified distance, usually 25 nautical miles, from the facility. However, neither navigation nor communication coverage is guaranteed within this distance. See figures 7-5 and 7-6.

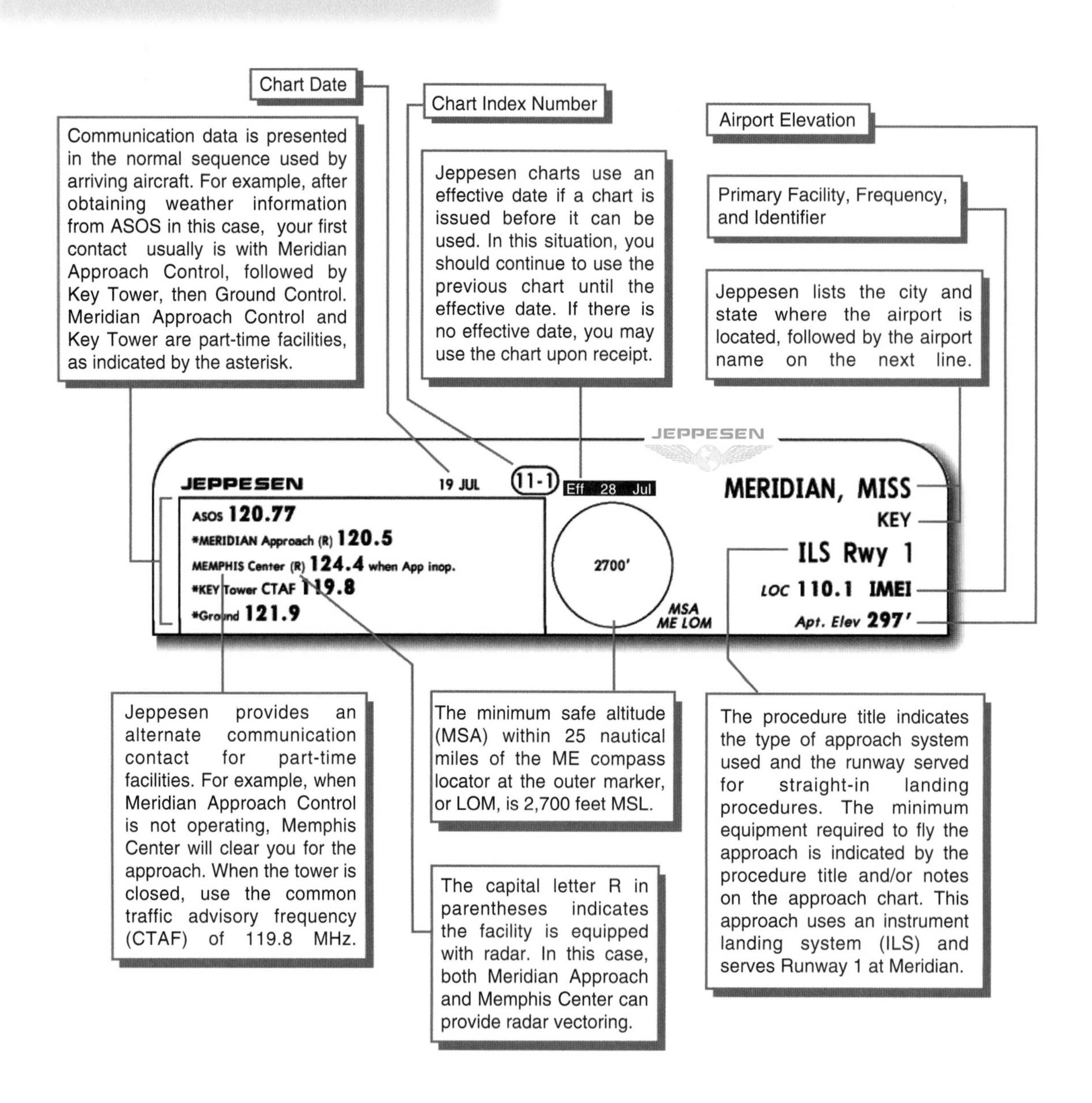

Figure 7-5. Jeppesen Heading Section — Big and bold type is selectively used in the heading section of Jeppesen charts to highlight specific information.

Communication frequencies are listed on approach charts in the normal sequence used by arriving aircraft. See figures 7-5 and 7-6.

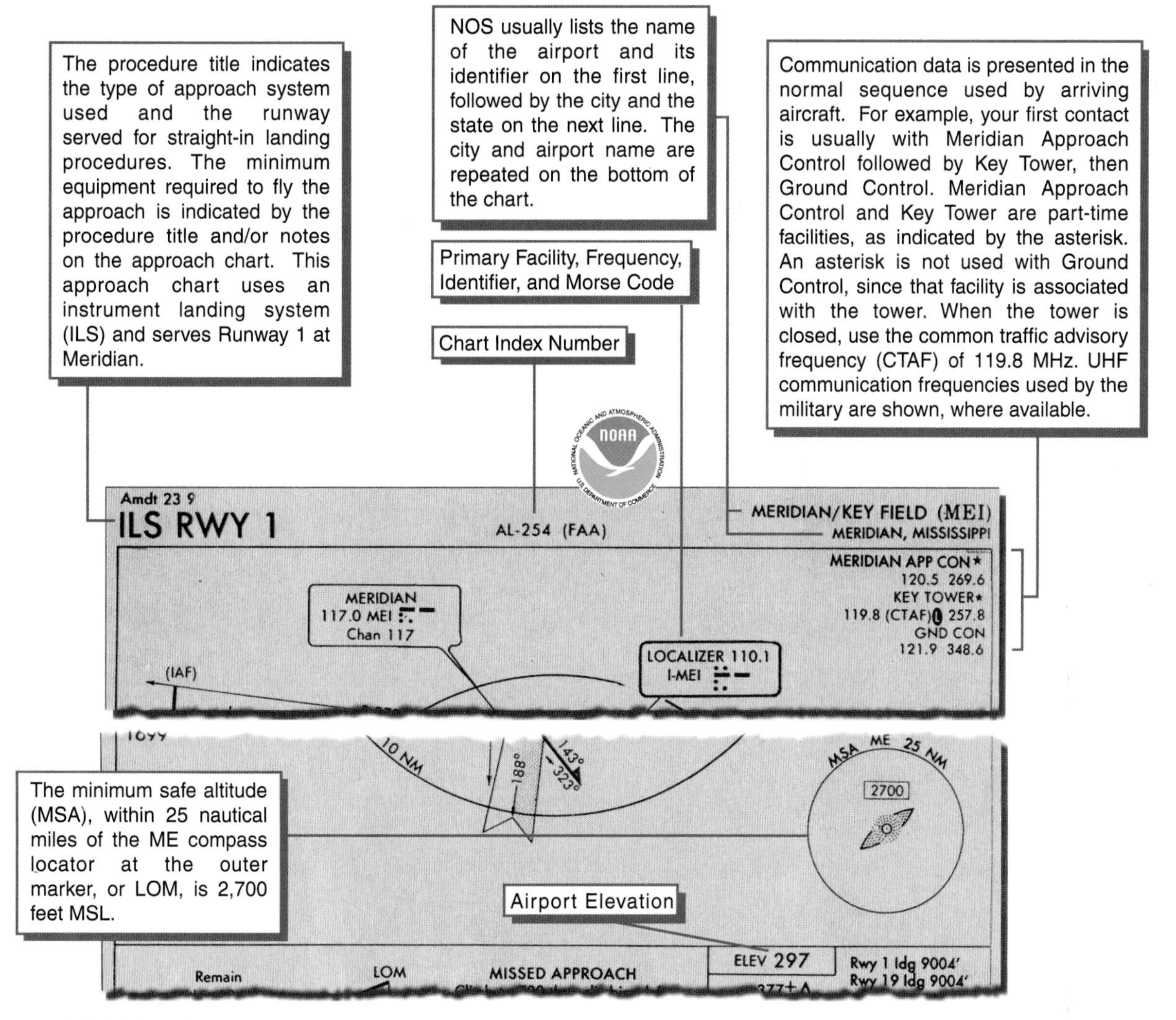

Figure 7-6. NOS Heading Section — The communication frequencies, primary approach facility, and minimum safe altitude information are shown on the plan view of NOS charts, while the airport elevation is indicated on the airport sketch.

PLAN VIEW

The plan view is an overhead presentation of the entire approach procedure. Since approach charts are intended for use during instrument weather conditions, they show only limited terrain and obstruction information. However, obstruction clearance is provided throughout the approach when the procedure is flown as depicted.

A procedure turn is a standard method of reversing your course. When the procedure turn is depicted on the plan view, it means you may reverse course any way you desire as long as the turn is made on the same side of the approach course as the symbol, the turn is completed within the distance specified in the profile view, and you remain within protected airspace. If a holding or teardrop pattern is shown instead of a procedure turn, it is the only approved method of course reversal. If a procedure turn, holding or teardrop pattern is not shown, a course reversal is not authorized.

Absence of a procedure turn or holding pattern indicates a course reversal is not authorized. See figures 7-7 and 7-8.

On both NOS and Jeppesen charts, some, but not all, reference points/terrain high points, such as natural or man-made objects, are depicted with their elevations. These reference points cannot be relied on for terrain or obstruction avoidance since there may be higher uncharted terrain or obstructions within the same vicinity. Generally, on Jeppesen charts, reference points less than 400 feet above the airport elevation are not depicted. Minimum altitudes on the approach procedures provide clearance of terrain and obstructions along the depicted flight tracks. [Figures 7-7 and 7-8]

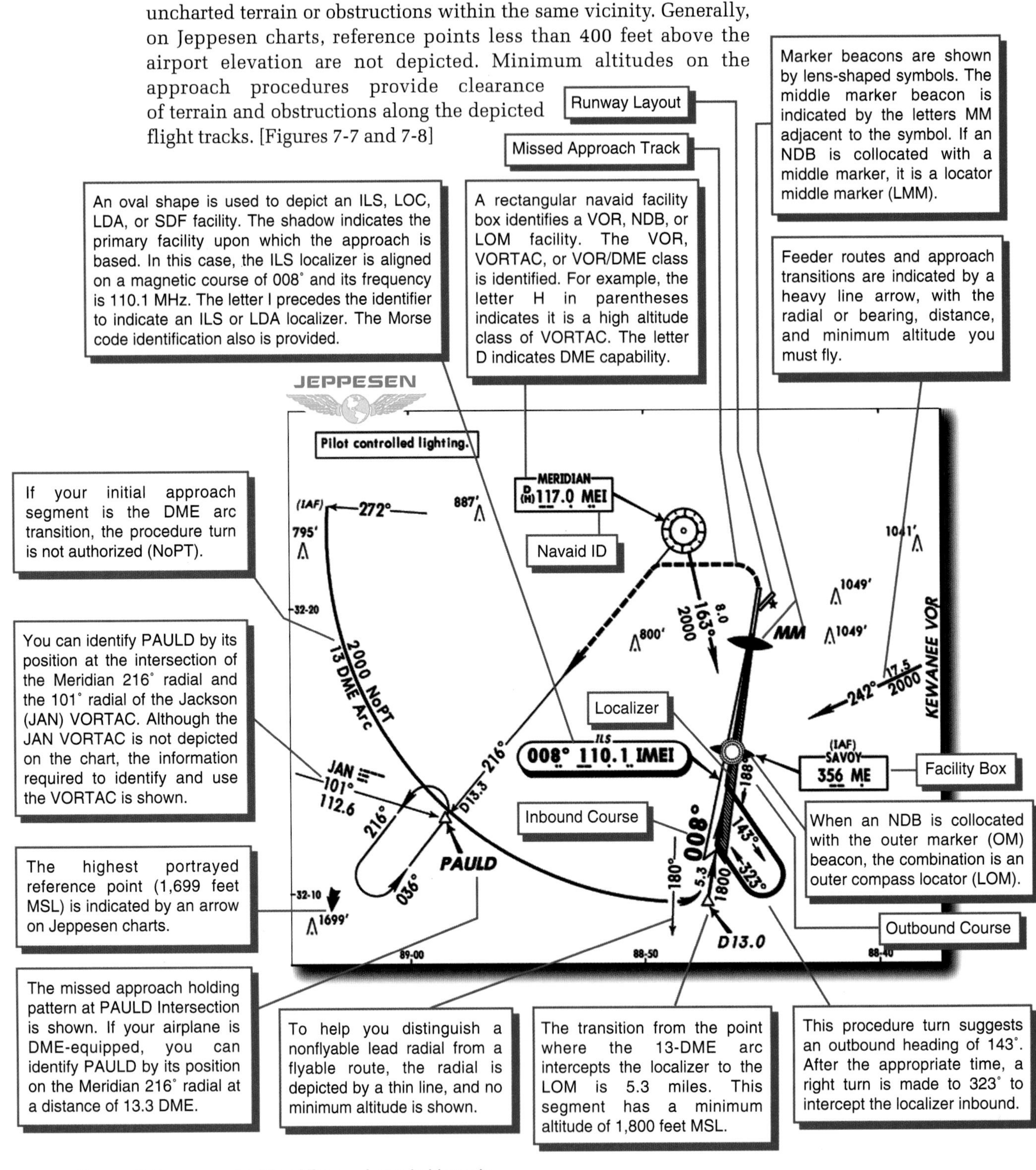

Figure 7-7. Jeppesen Plan View — Large bold type is used on the plan view to highlight the primary navaid frequency and identifier, as well as the inbound approach course shown in the ILS oval box. The inbound approach course also is shown in large type along the final approach path and names of fixes which are part of the procedure are enhanced. Secondary airports are subdued in a lighter color to differentiate them from the primary airport and reduce visual clutter.

Adherence to the minimum altitudes depicted on approach charts provides terrain and obstacle clearance.

Approach chart symbology provides information about navaids, such as the availability of DME or voice capability. See figures 7-7 and 7-8.

Figure 7-8. NOS Plan View

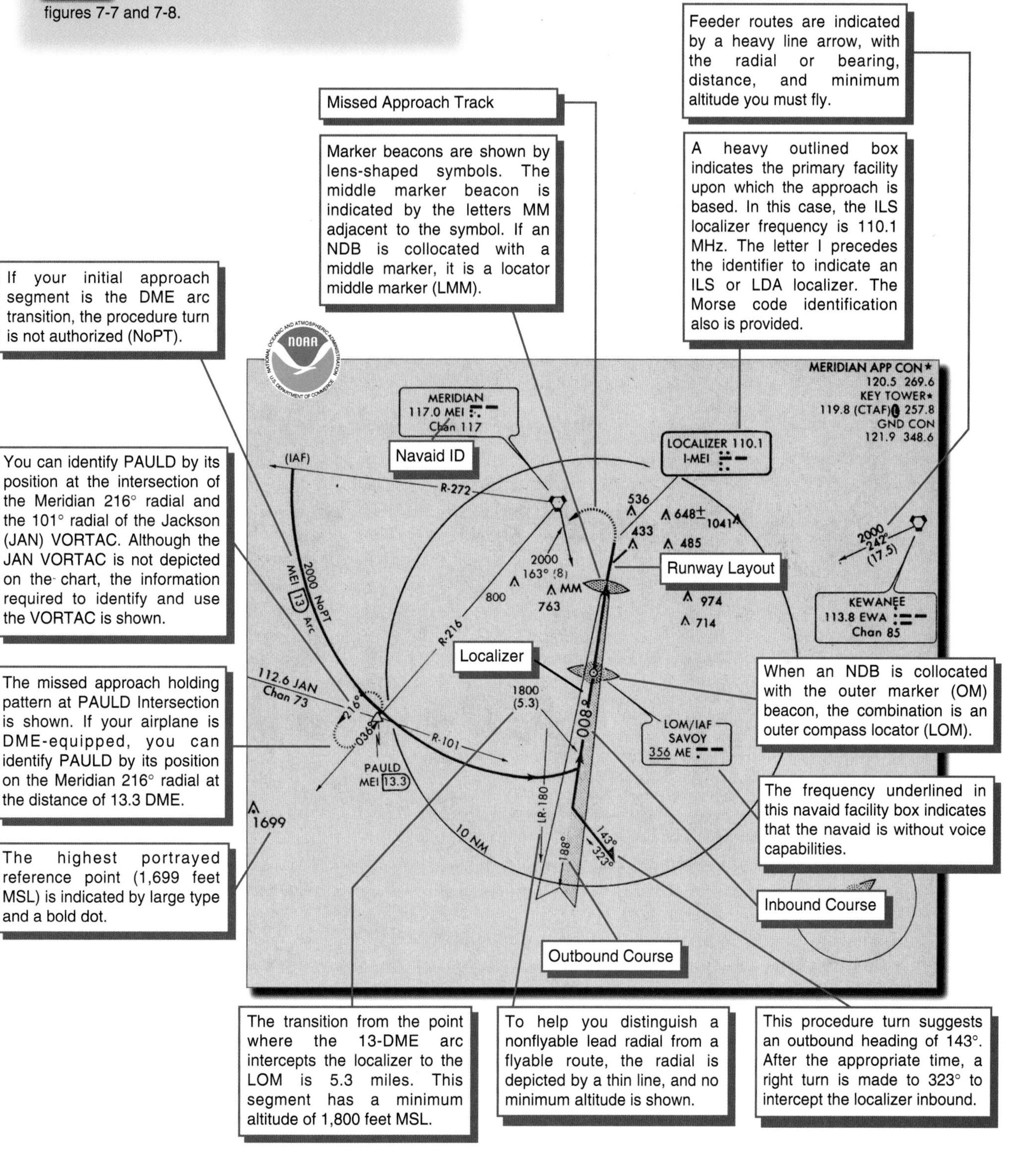

On selected Jeppesen approach chart plan views, generalized terrain contour lines, values, and gradient tints may be depicted in brown. This may occur when terrain within the approach chart plan view exceeds 4,000 feet above the airport elevation, or when terrain within 6 nautical miles of the airport reference point rises to at least 2,000 feet above the airport elevation. Gradient tints indicate the elevation change between

contour intervals. This information does not ensure clearance above or around the terrain and must not be relied on for descent below the minimum altitudes dictated by the approach procedure. Furthermore, the absence of terrain contour information does not ensure the absence of terrain or structures. [Figure 7-9]

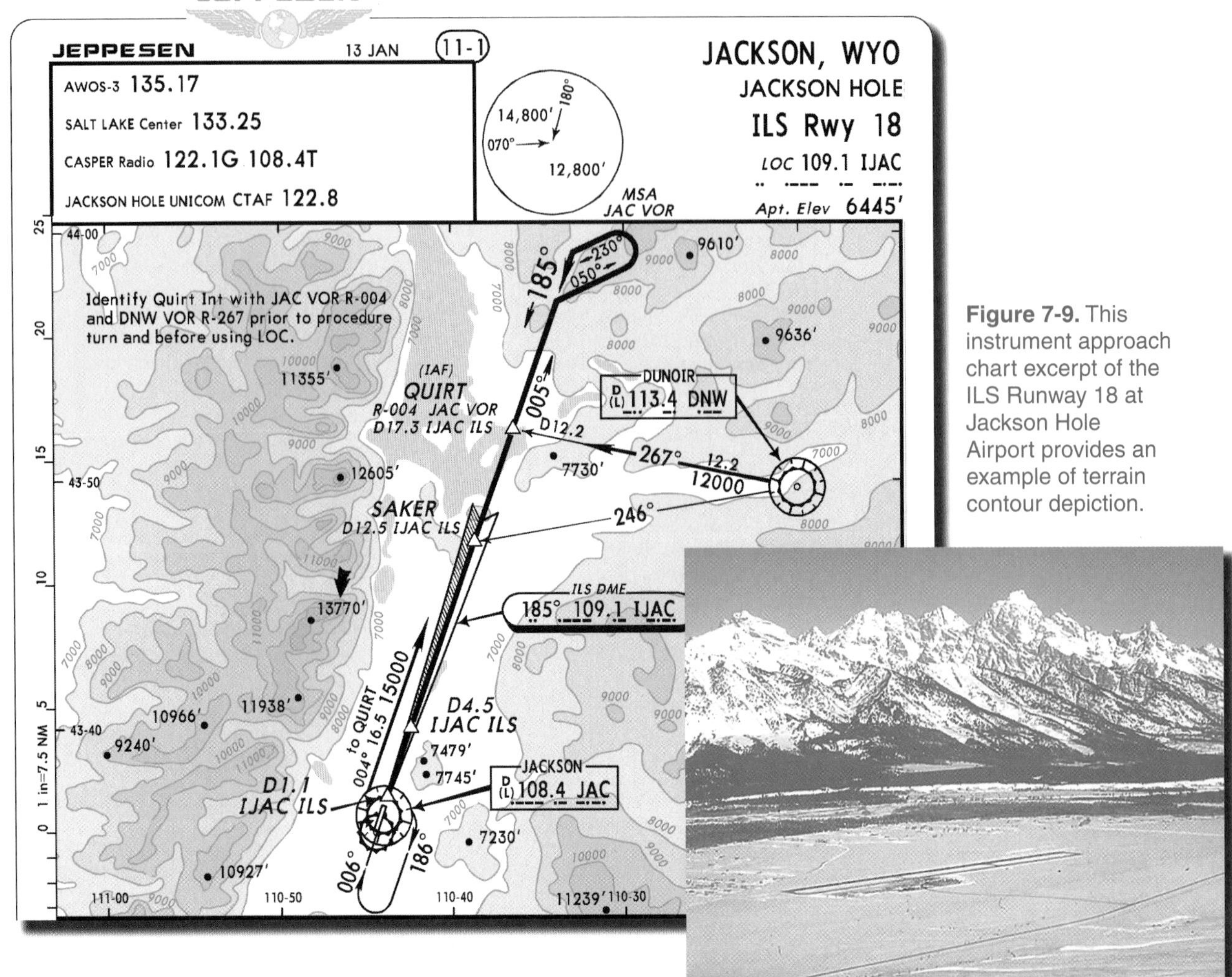

Figure 7-9. This instrument approach chart excerpt of the ILS Runway 18 at Jackson Hole Airport provides an example of terrain contour depiction.

PROFILE VIEW

The profile view shows the approach from the side and displays the flight path and facilities, as well as minimum altitudes in feet MSL. There are several terms used to describe the altitudes commonly shown on the profile view. The touchdown zone elevation (TDZE) is the highest elevation in the first 3,000 feet of the landing surface. The height above touchdown (HAT) is measured from the touchdown zone elevation or the threshold elevation of the runway served by the approach. The height above airport (HAA) is measured above the official airport elevation which is the highest point of an airport's usable runways. The threshold crossing height (TCH) is the altitude at which you cross the runway threshold when established on the glide slope centerline. [Figures 7-10 and 7-11]

The touchdown zone elevation (TDZE) is the highest elevation in the first 3,000 feet of the landing surface. The TDZE is depicted on Jeppesen approach charts in the profile view and on NOS charts in the airport sketch. See figures 7-10 and 7-11.

The threshold crossing height (TCH) is the altitude at which you cross the runway threshold when established on the glide slope centerline. See figures 7-10 and 7-11.

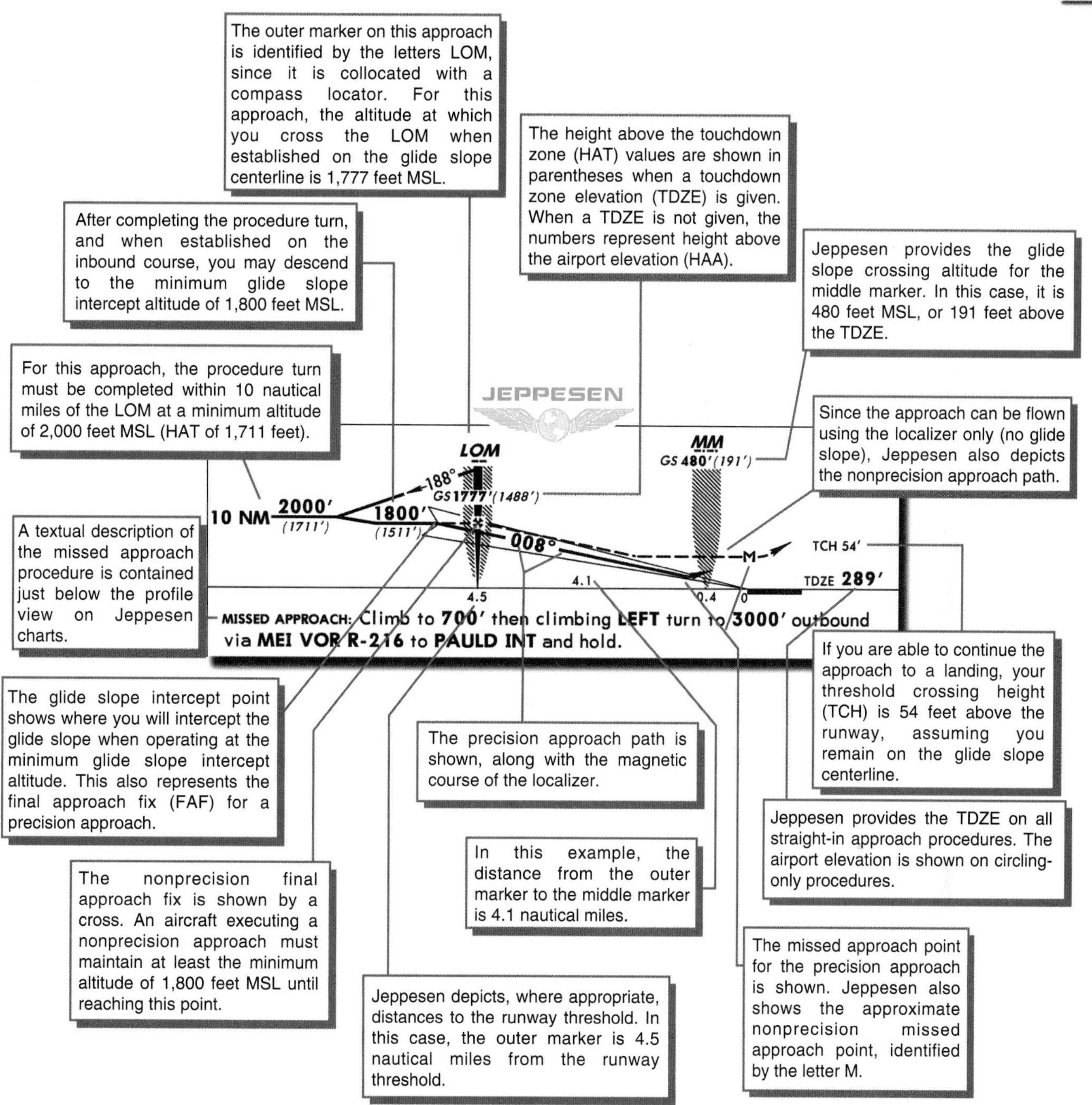

Figure 7-10. Jeppesen Profile View — Large bold type is used for navaids and fixes, TDZEs when charted, final approach courses, minimum altitudes, glide slope interception altitudes, and the procedure turn limits. The missed approach text contains big bold type for altitudes, headings or courses, and navaid or fix identifiers. The airport elevation is omitted when the TDZE or runway threshold is charted. All the altitudes given on Jeppesen approach charts represent minimum altitudes, unless noted as max, maximum, mandatory, or recommended.

The procedure turn must be completed within the prescribed distance from the facility. See figures 7-10 and 7-11.

Distances between fixes along the approach path and the runway threshold are shown on the profile view. See figures 7-10 and 7-11.

The precision approach FAF is located at the minimum glide slope intercept point. See figures 7-10 and 7-11.

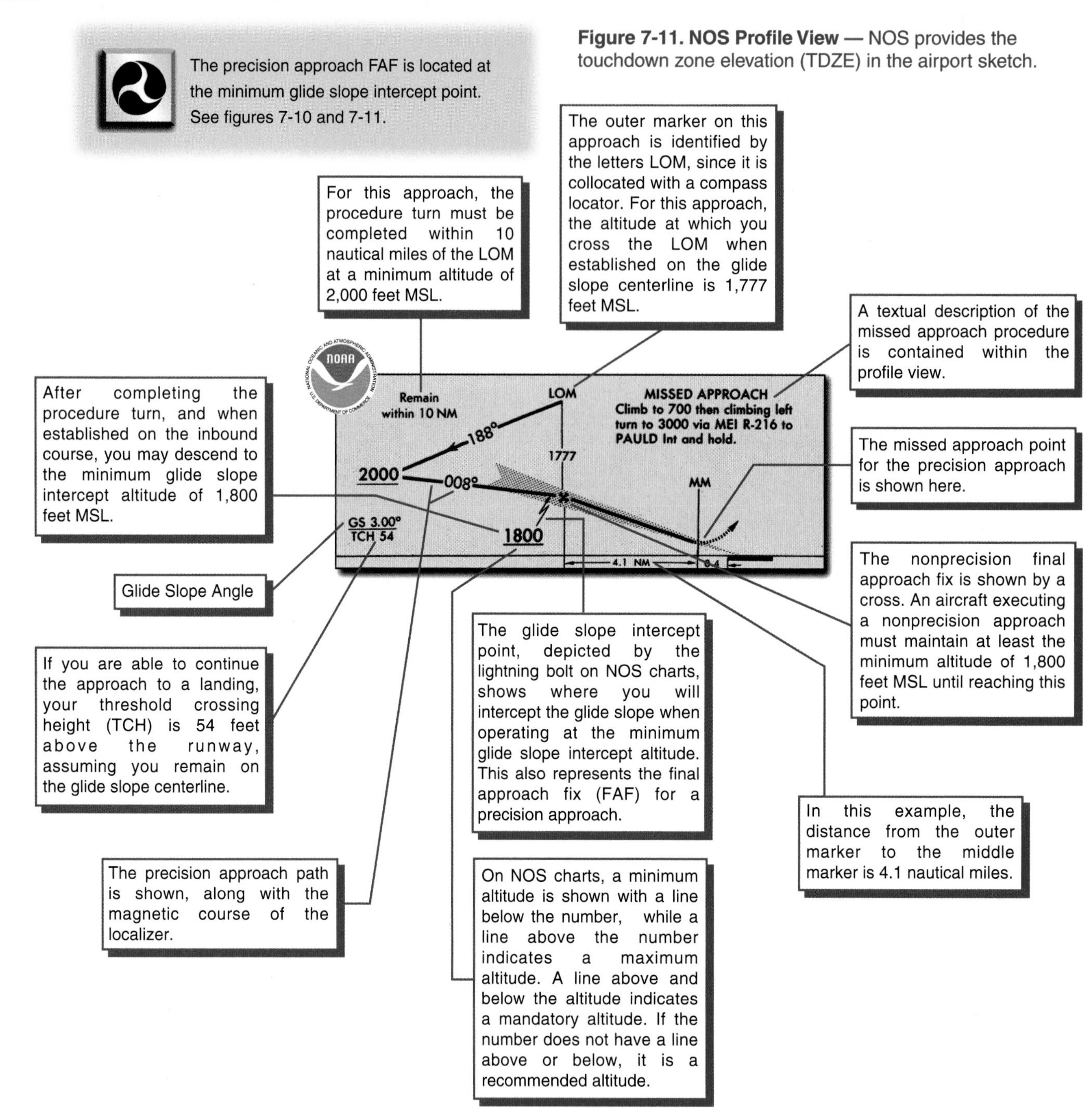

Figure 7-11. NOS Profile View — NOS provides the touchdown zone elevation (TDZE) in the airport sketch.

STEPDOWN FIX AND VISUAL DESCENT POINT

Many approaches incorporate one or more **stepdown fixes**. They are commonly used along approach segments to allow you to descend to a lower altitude as you overfly various obstacles, or in conjunction with the design of local air traffic control procedures. Only one stepdown fix normally is permitted between the final approach fix and the missed approach point. A **visual descent point (VDP)** also may be depicted in the profile view. [Figure 7-12]

Your ability to identify selected stepdown fixes may permit lower landing minimums in some cases. When you cannot identify a stepdown fix, you must use the minimum altitude given just prior to the fix. See figure 7-12.

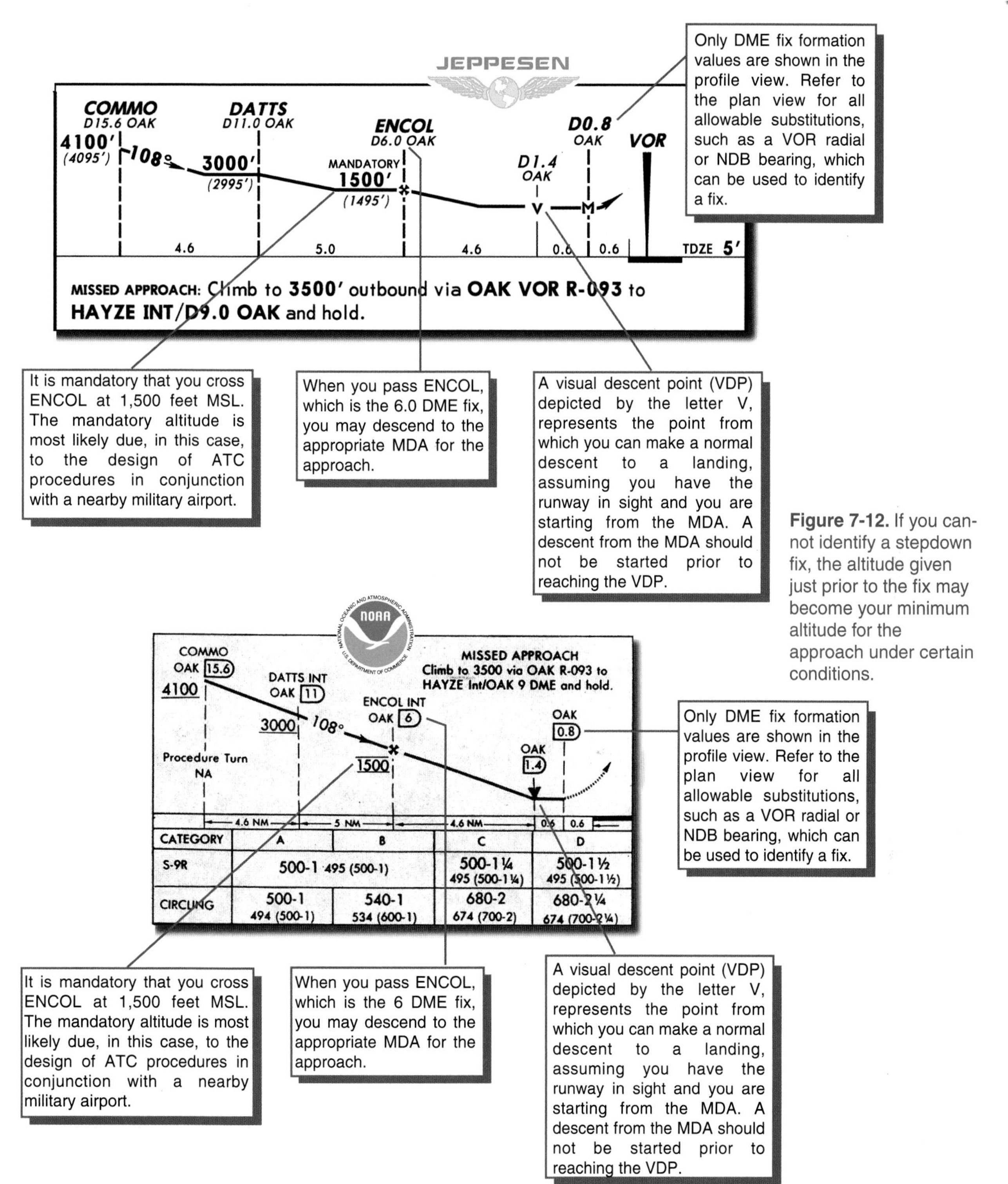

Figure 7-12. If you cannot identify a stepdown fix, the altitude given just prior to the fix may become your minimum altitude for the approach under certain conditions.

LANDING MINIMUMS

Landing minimums, which contain both minimum visibility and minimum altitude requirements, have been established for each approach at a given airport. Factors which affect these minimums include the type of approach equipment installed, type of approach lights installed, and obstructions in the approach or missed approach paths. Landing minimums also are affected by the equipment on board your aircraft, your approach speed, and whether you are executing a straight-in landing or flying a circling approach. A **circling approach** is a procedure which involves executing an approach to

Landing minimums published on instrument approach charts consist of both minimum visibility and minimum altitude requirements. See figures 7-15 and 7-16.

one runway and then landing on another. Separate circle-to-land minimums are published in the landing minimums section if this procedure is authorized.

Another procedure listed on some charts is a variation of the circling maneuver. During a **sidestep maneuver**, you are cleared for an approach to one runway with a clearance to land on a parallel runway. The minimums for this procedure usually are higher than the minimums for the straight-in landing runway, but lower than the circling minimums. When executing a sidestep maneuver, you are expected to begin the procedure as soon as possible after sighting the sidestep runway environment.

AIRCRAFT APPROACH CATEGORIES

Each aircraft is placed into an **aircraft approach category** based on its computed **approach speed**. This speed equals 130% of the aircraft's power-off stall speed in the landing configuration at the maximum certificated landing weight ($1.3V_{S0}$). For example, if the stalling speed of your aircraft is 65 knots, its computed approach speed is 85 knots. [Figure 7-13]

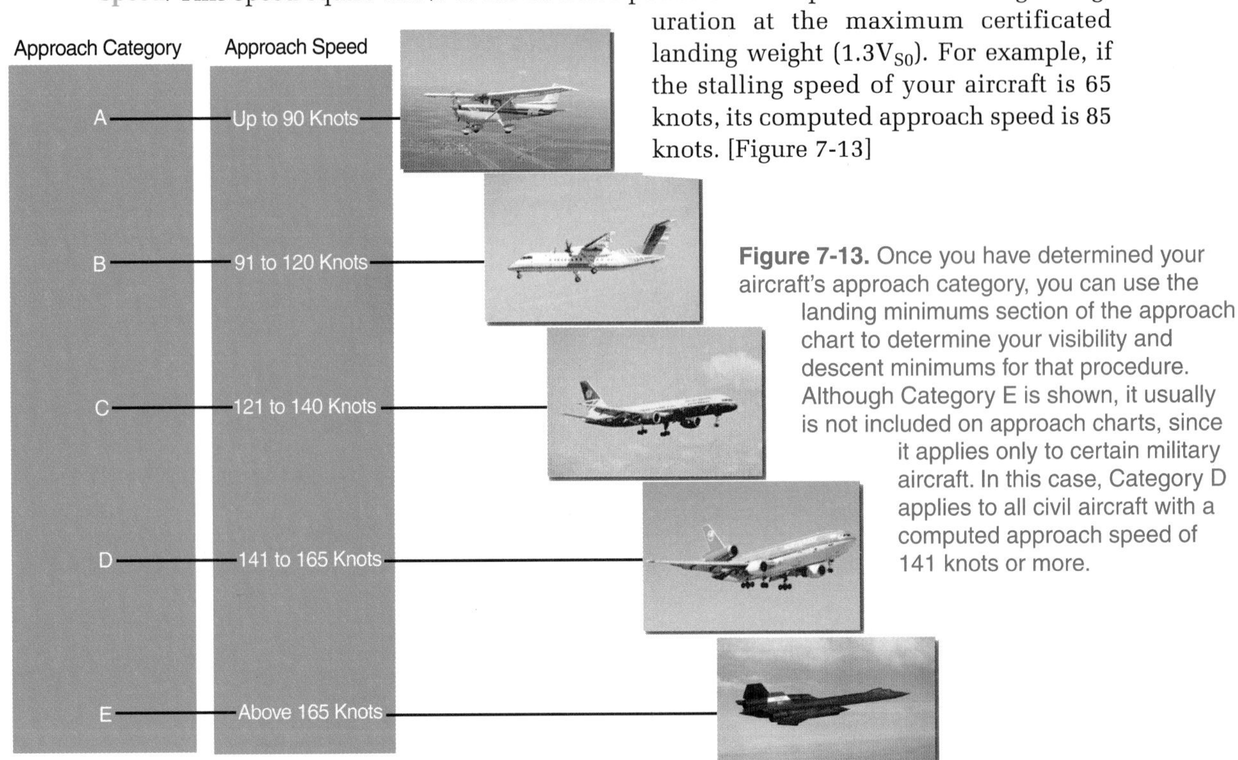

Figure 7-13. Once you have determined your aircraft's approach category, you can use the landing minimums section of the approach chart to determine your visibility and descent minimums for that procedure. Although Category E is shown, it usually is not included on approach charts, since it applies only to certain military aircraft. In this case, Category D applies to all civil aircraft with a computed approach speed of 141 knots or more.

MINIMUM DESCENT REQUIREMENTS

The minimum altitude to which you are permitted to descend while executing the approach procedure is depicted in the landing minimums section. The terms used to describe this altitude depend on the type of approach — precision or nonprecision. During a precision approach, the height where you must make the decision to continue the approach or execute a missed approach is referred to in the FARs as the **decision height (DH)**. NOS charts show the decision height as an MSL altitude with the height above touchdown (HAT) listed after the visibility requirement. The International Civil Aviation Organization (ICAO) uses the terms decision altitude (DA), which is referenced to mean sea level, and decision height (DH) which is referenced to the threshold elevation. Jeppesen charts reflect the ICAO terminology by showing a **decision altitude (height) — DA(H)** as an MSL altitude followed by the height above touchdown (HAT) in parentheses.

Aircraft approach categories used to determine landing minimums are based on $1.3V_{S0}$. See figures 7-13, 7-15, and 7-16.

When flying a nonprecision approach, upon crossing the final approach fix or stepdown fix, as appropriate, you descend to the minimum altitude shown on the approach chart and remain at this altitude until you identify the runway environment or execute a missed approach. On Jeppesen charts, the term **minimum descent altitude (height) — MDA(H)** is used to indicate this altitude. The MDA is stated first in feet MSL with the HAT (in the case of a straight-in landing) or HAA (in the case of a circling approach) shown in parentheses following it. NOS uses the FAA term minimum descent altitude (MDA) and includes the HAT after the visibility requirement. Throughout this manual, the FAA terms decision height (DH) and minimum descent altitude (MDA) generally are used to describe minimum descent requirements unless the text is specifically referencing Jeppesen charts.

VISIBILITY REQUIREMENTS

Under Part 91, you can descend below the approach minimums only if the flight visibility is not less than the visibility prescribed in the approach procedure. Visibility is listed on approach charts in either statute miles or hundreds of feet. When it is expressed in miles or fractions of miles, it is usually a prevailing visibility that is reported by an accredited observer such as tower or weather personnel. When the visibility is expressed in hundreds of feet, it is determined through the use of runway visual range (RVR) equipment. If RVR minimums for landing are prescribed for an instrument approach procedure, but RVR is inoperative and cannot be reported for the intended runway at the time, RVR minimums should be converted and applied as ground visibility.

If RVR minimums for landing are prescribed for an instrument approach procedure, but RVR is inoperative and cannot be reported for the intended runway at the time, RVR minimums should be converted and applied as ground visibility. For example, RVR 24 translates to 1/2 statute mile visibility. See figures 7-15 and 7-16.

INOPERATIVE COMPONENTS

Landing minimums usually increase when a required component or visual aid becomes inoperative. If more than one component is inoperative, each minimum is raised to the highest minimum required by any single component that is inoperative. ILS glide slope inoperative minimums are published on instrument approach charts as localizer minimums. While Jeppesen depicts inoperative component minimums with the normal minimums, NOS provides a separate inoperative components table in each chart volume. [Figure 7-14]

(1) ILS, MLS, and PAR

Inoperative Component or Aid	Approach Category	Increase Visibility
ALSF 1 & 2, MALSR, & SSALR	ABCD	¼ mile

(2) ILS with visibility minimum of 1,800 RVR.

ALSF 1 & 2, MALSR, & SSALR	ABCD	To 4000 RVR
TDZI RCLS	ABCD	To 2400 RVR
RVR	ABCD	To ½ mile

If both approach lights and the touchdown zone lights are inoperative, the visibility must be increased to 4000 RVR.

(3) VOR, VOR/DME, VORTAC, VOR (TAC), VOR/DME (TAC), LOC, LOC/DME, LDA, LDA/DME, SDF, SDF/DME, GPS, RNAV, and ASR

Inoperative Visual Aid	Approach Category	Increase Visibility
ALSF 1 & 2, MALSR, & SSALR	ABCD	½ mile
SSALS, MALS, & ODALS	ABC	¼ mile

(4) NDB

ALSF 1 & 2, MALSR & SSALR	C	½ mile
	ABD	¼ mile
MALS, SSALS, ODALS	ABC	¼ mile

Figure 7-14. To determine the appropriate corrections to the landing minimums when an approach component is inoperative, you should consult the Inoperative Components or Visual Aids Table in the NOS *Terminal Procedures Publication.*

Regulations permit you to make substitutions for certain components when the component is inoperative or is not utilized during an approach. For example, on an ILS approach, a compass locator or precision radar may be substituted for the outer marker. This allows you to fly the approach without increasing the landing minimums. When DME, VOR, or NDB fixes are authorized in the approach procedure or, where surveillance radar is available, they also may be substituted for the outer marker. [Figures 7-15 and 7-16]

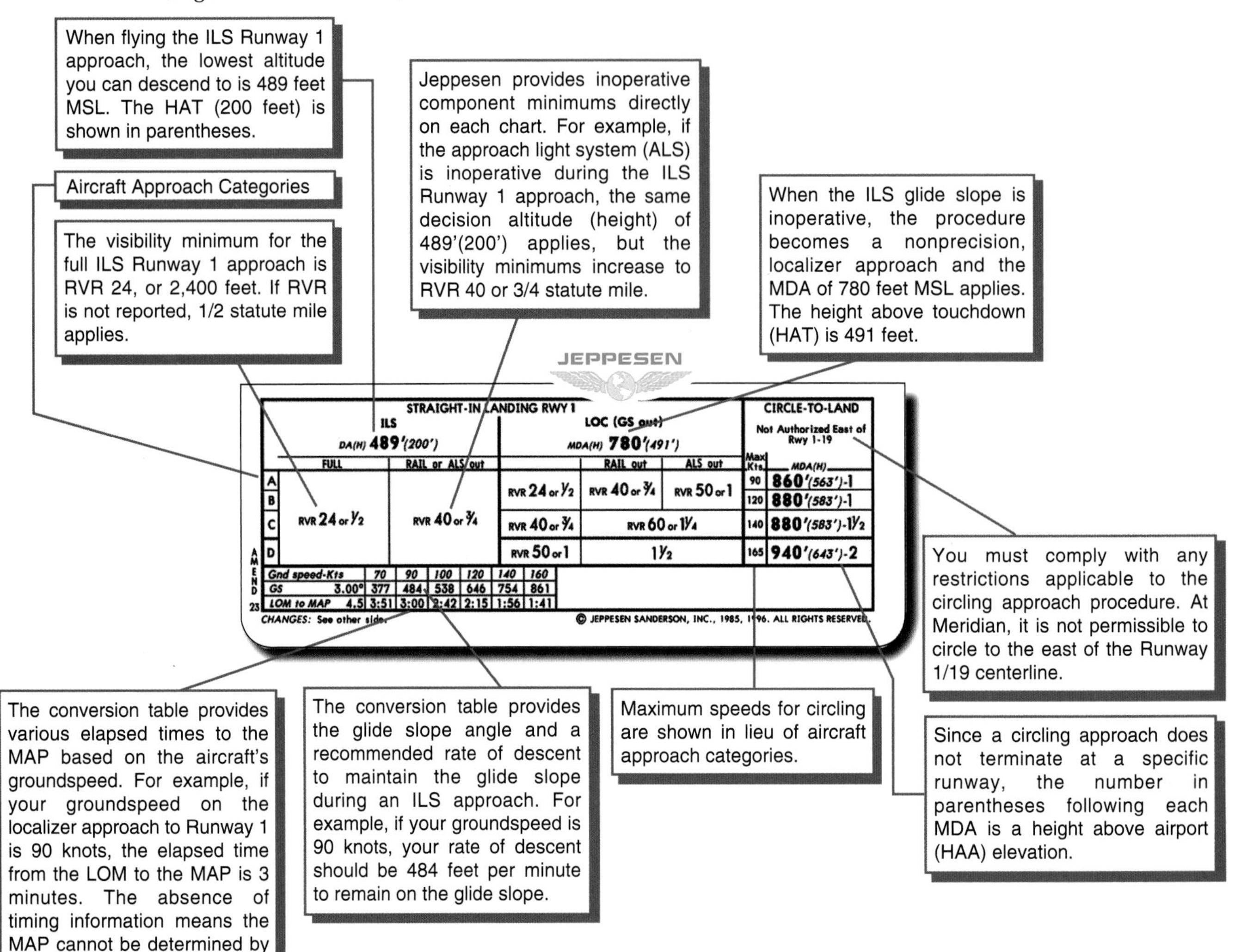

STRAIGHT-IN LANDING RWY 1	ILS DA(H) 489'(200')		LOC (GS out) MDA(H) 780'(491')			CIRCLE-TO-LAND Not Authorized East of Rwy 1-19	
	FULL	RAIL or ALS out		RAIL out	ALS out	Max Kts	MDA(H)
A	RVR 24 or 1/2	RVR 40 or 3/4	RVR 24 or 1/2	RVR 40 or 3/4	RVR 50 or 1	90	860'(563')-1
B						120	880'(583')-1
C			RVR 40 or 3/4	RVR 60 or 1 1/4		140	880'(583')-1 1/2
D			RVR 50 or 1	1 1/2		165	940'(643')-2

Gnd speed-Kts	70	90	100	120	140	160
GS 3.00°	377	484	538	646	754	861
LOM to MAP 4.5	3:51	3:00	2:42	2:15	1:56	1:41

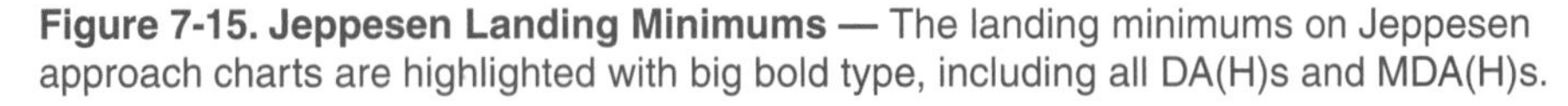
Figure 7-15. Jeppesen Landing Minimums — The landing minimums on Jeppesen approach charts are highlighted with big bold type, including all DA(H)s and MDA(H)s.

Substitutions for certain ILS components, when the component is inoperative or is not utilized during an approach, are permitted. See figures 7-15 and 7-16.

Restrictions may exist applicable to the circle-to-land procedure. For example, a circle-to-land procedure might not be authorized in a specific area. See figures 7-15 and 7-16.

If the glide slope becomes inoperative during an ILS procedure, localizer minimums are used. See figures 7-15 and 7-16.

Figure 7-16. NOS Landing Minimums

The visibility minimum for the full ILS Runway 1 approach is RVR 24, or 2,400 feet. If RVR is not reported, 1/2 statute mile applies. NOS provides a separate table in the front of each *Terminal Procedures Publication* volume to convert RVR to statute miles.

When flying the ILS Runway 1 approach, the lowest altitude you can descend to is 489 feet MSL.

On NOS charts, the HAT (200 feet) is shown after the visibility.

ircraft Approach Categories

When the ILS glide slope is inoperative, the procedure becomes a nonprecision, localizer approach and the MDA of 780 feet MSL applies. The height above touchdown (HAT) is 491 feet.

ince a circling approach does ot terminate at a specific unway, the number following ach MDA is a height above irport (HAA) elevation.

CATEGORY	A	B	C	D
S-ILS 1	489/24 200 (200-½)			
S-LOC 1	780/24 491 (500-½)		780/40 491 (500-¾)	780/50 491 (500-1)
CIRCLING	860-1 563 (600-1)	880-1 583 (600-1)	880-1½ 583 (600-1½)	940-2 643 (700-2)

Circling not authorized east of Rwy 1-19.

TDZE 289

Nonstandard takeoff and alternate minimums apply.

ou must comply with any estrictions applicable to the ircling approach procedure. At leridian, it is not permissible o circle to the east of the unway 1/19 centerline.

FAF to MAP 4.5 NM					
Knots	60	90	120	150	180
Min:Sec	4:30	3:00	2:15	1:48	1:30

ILS RWY 1 32°20'N-88°45'W MERIDIAN, MISSISSIPPI MERIDIAN/KEY FIELD (MEI)

The ceiling and visibility figures in parentheses on NOS charts are used for planning purposes by the military.

Latitude and Longitude

The time and speed table provides various elapsed times to the MAP based on the aircraft's groundspeed. For example, if your groundspeed on the localizer approach to Runway 1 is 90 knots, the elapsed time from the LOM to the MAP is 3 minutes. The absence of timing information means the MAP cannot be determined by timing and a timed approach is not authorized.

(ft. per min.)

ANGLE OF DESCENT (degrees and tenths)	GROUND SPEED (knots)						
	30	45	60	75	90	105	120
2.0	105	160	210	265	320	370	425
2.5	130	200	265	330	395	465	530
3.0	160	240	320	395	480	555	635
3.5	185	280	370	465	555	650	740

In a separate descent table contained in each *Terminal Procedures Publication*, NOS provides the glide slope angle and a recommended rate of descent to maintain the glide slope during an ILS approach. In the descent table excerpt shown, if your groundspeed is 90 knots with a 3.0° glide slope, your rate of descent should be approximately 480 feet per minute to remain on the glide slope.

By using an average groundspeed on the final approach segment and the glide slope angle, a corresponding rate of descent can be determined to initially establish the airplane on the glidepath for an ILS approach procedure. See figures 7-15 and 7-16.

When on the glide slope during a precision approach, the missed approach point is the decision height. See figures 7-15 and 7-16.

AIRPORT CHART

Both Jeppesen and NOS include a diagram of each airport for which they publish an instrument approach procedure. NOS places an airport sketch in the lower right-hand corner of each approach chart, and also provides a full-page airport diagram for selected airports to assist the movement of ground traffic where complex runway and taxiway configurations exist. Jeppesen uses an entire page, called the airport chart, for each airport. This chart is usually located on the reverse side of the first approach chart for a given airport. At larger airports, Jeppesen provides even more detail than the standard airport chart. [Figure 7-17]

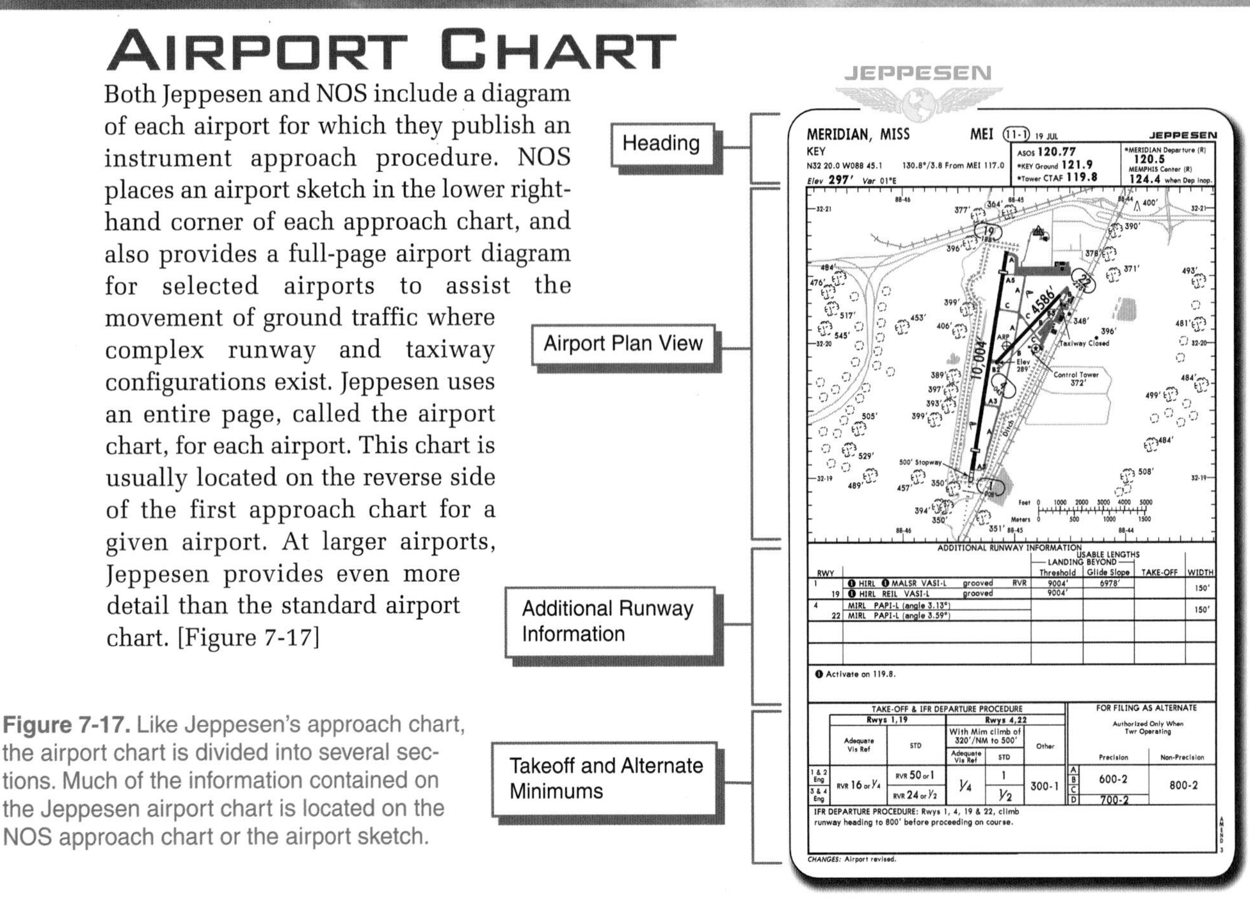

Figure 7-17. Like Jeppesen's approach chart, the airport chart is divided into several sections. Much of the information contained on the Jeppesen airport chart is located on the NOS approach chart or the airport sketch.

HEADING SECTION

The heading section of the Jeppesen airport chart identifies the airport, its location, elevation, magnetic variation, and outbound communication frequencies. There are many similarities between this section and the heading section of a Jeppesen approach chart. [Figure 7-18]

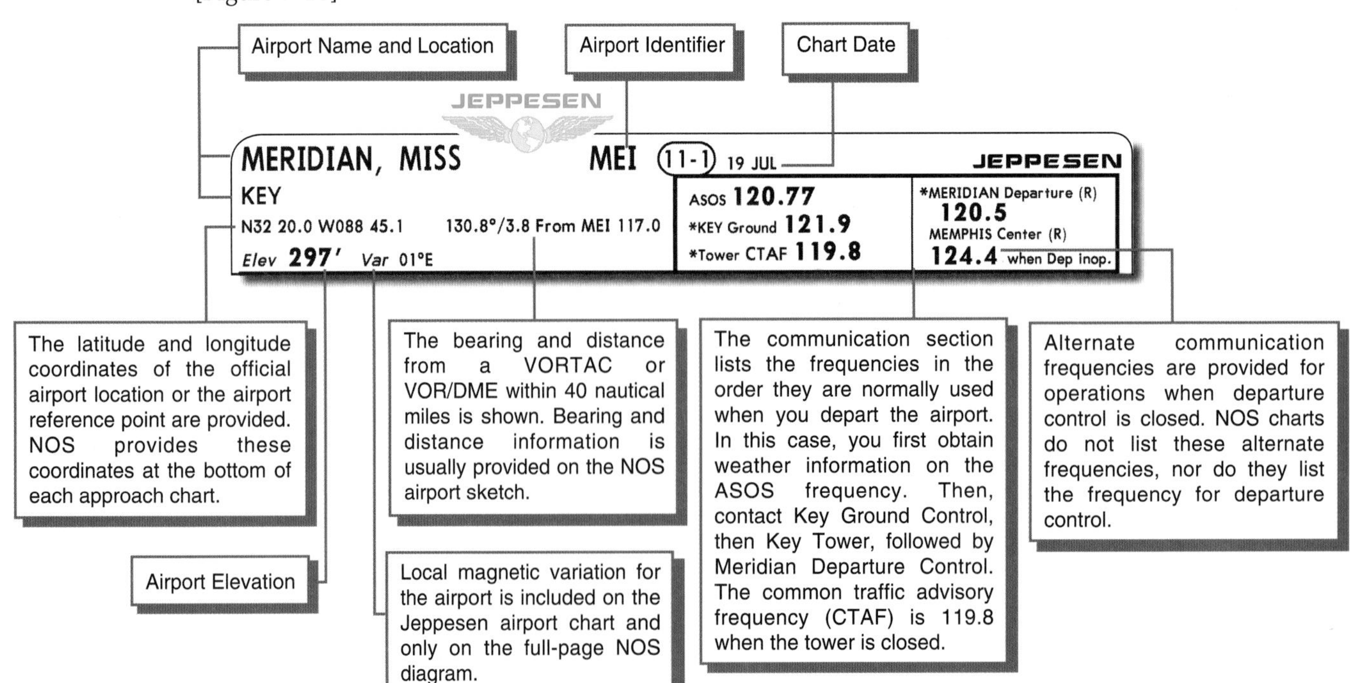

Figure 7-18. Here is an example of the Jeppesen heading section of the airport chart for Key Field at Meridian, Mississippi. Except as noted, most of the information contained on this chart is listed on an NOS approach chart.

PLAN VIEW AND ADDITIONAL RUNWAY INFORMATION

The plan view portrays an overhead view of the airport. Its purpose is to provide you with information about the airport, such as its runways and lighting systems. The **airport reference point (ARP)**, shown on Jeppesen charts, is the approximate geometric center of all usable runway surfaces, and is where the official latitude and longitude coordinates are derived. [Figures 7-19 and 7-20]

Figure 7-19. Jeppesen lists some airport information below the airport diagram.

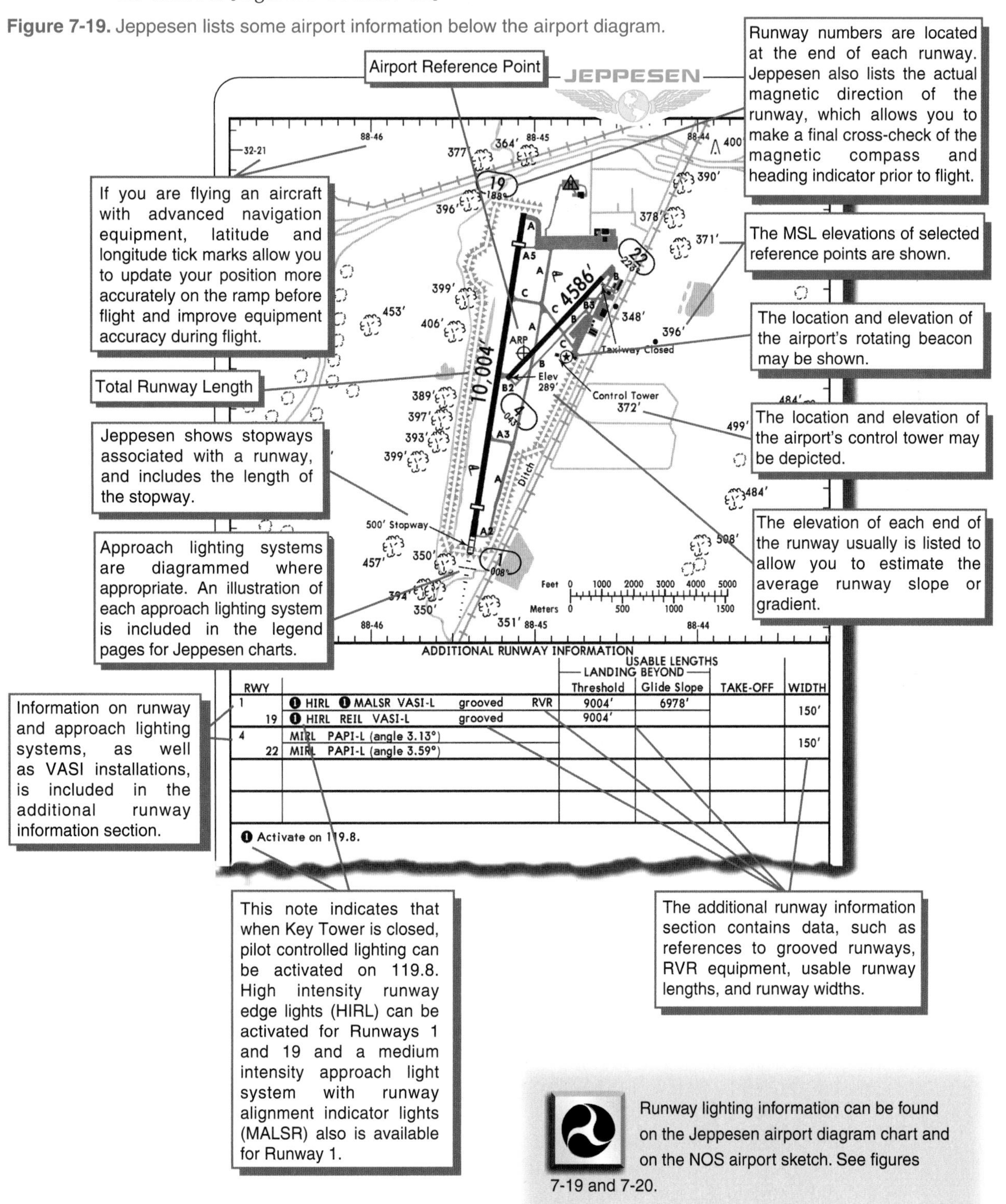

RWY		USABLE LENGTHS LANDING BEYOND Threshold	Glide Slope	TAKE-OFF	WIDTH
1	❶ HIRL ❶ MALSR VASI-L grooved RVR	9004'	6978'		150'
19	❶ HIRL REIL VASI-L grooved	9004'			
4	MIRL PAPI-L (angle 3.13°)				150'
22	MIRL PAPI-L (angle 3.59°)				

❶ Activate on 119.8.

Runway lighting information can be found on the Jeppesen airport diagram chart and on the NOS airport sketch. See figures 7-19 and 7-20.

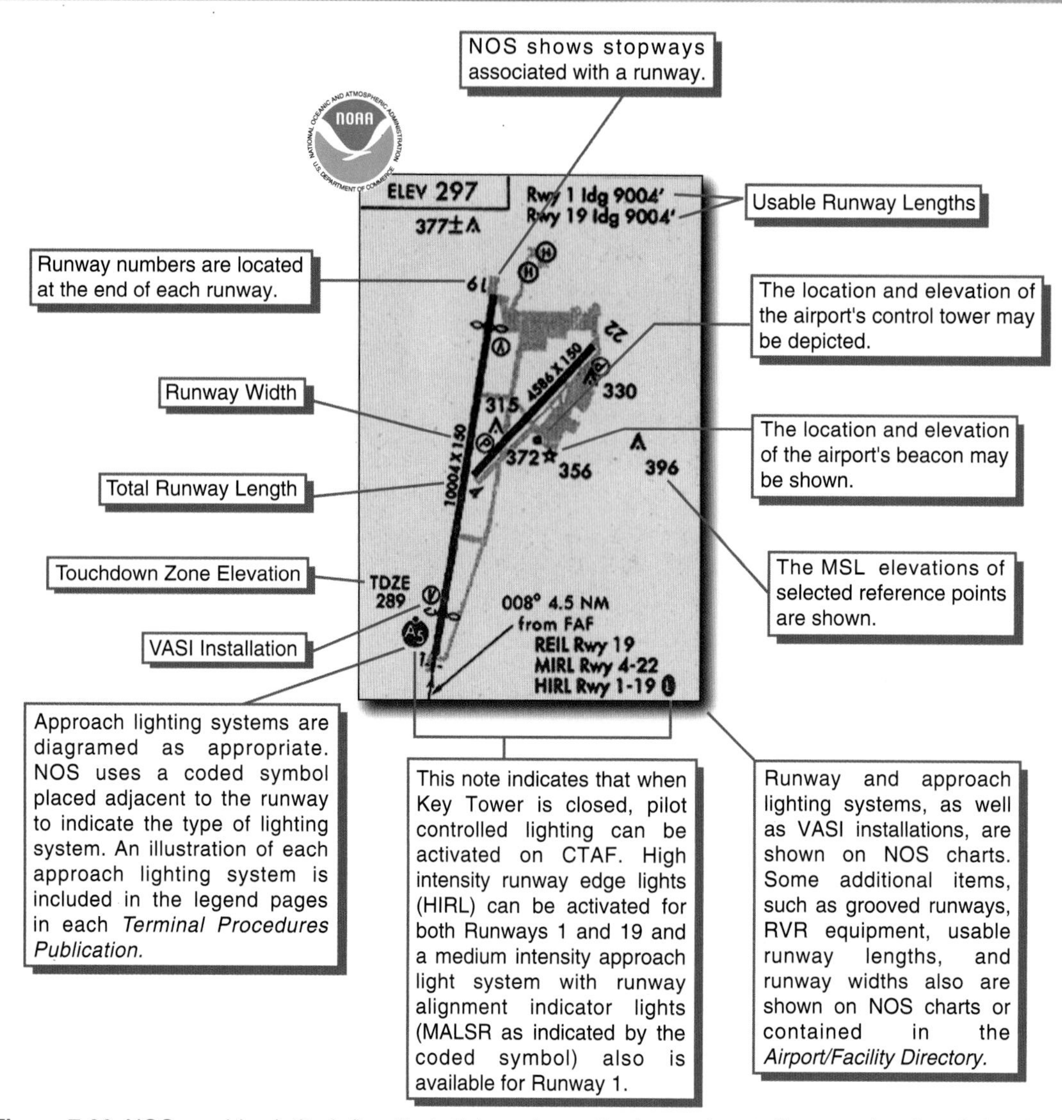

Figure 7-20. NOS provides latitude/longitude tick marks on the large airport diagrams for aircraft that have advanced navigation equipment. This allows you to update your position more accurately on the ramp before flight and improves equipment accuracy during flight. Runway gradient information appears on the NOS airport diagram/sketch when it exceeds 0.3%.

TAKEOFF AND ALTERNATE MINIMUMS

Both Jeppesen and NOS charts provide information regarding nonstandard takeoff and alternate minimums. The FAA has established FAR Part 91 rules for selecting an alternate airport for an IFR flight, including standard alternate minimums and nonstandard alternate minimums. If the forecast weather at your estimated time of arrival, plus or minus 1 hour, indicates a ceiling of less than 2,000 feet or a visibility of less than 3 statute miles, you must list an alternate airport on your IFR flight plan. To qualify as an alternate, the airport you select and the forecast weather for your arrival time, must meet certain conditions. The standard alternate minimums for precision approaches are a ceiling of 600 feet and a visibility of 2 statute miles. For nonprecision approaches, an 800-foot ceiling and 2 statute miles visibility apply.

There are numerous chart differences between Jeppesen and NOS regarding the portrayal of takeoff and alternate minimums. Jeppesen lists both standard and nonstandard takeoff and alternate minimums and departure procedures on the airport chart. If you are using NOS charts, you must refer to separate sections of the *Terminal Procedures Publication* to find the nonstandard takeoff and nonstandard alternate minimums. A

symbol on an NOS approach chart indicates when nonstandard takeoff minimums apply and/or IFR departure procedures are published, and a symbol alerts you to nonstandard alternate minimums. [Figure 7-21]

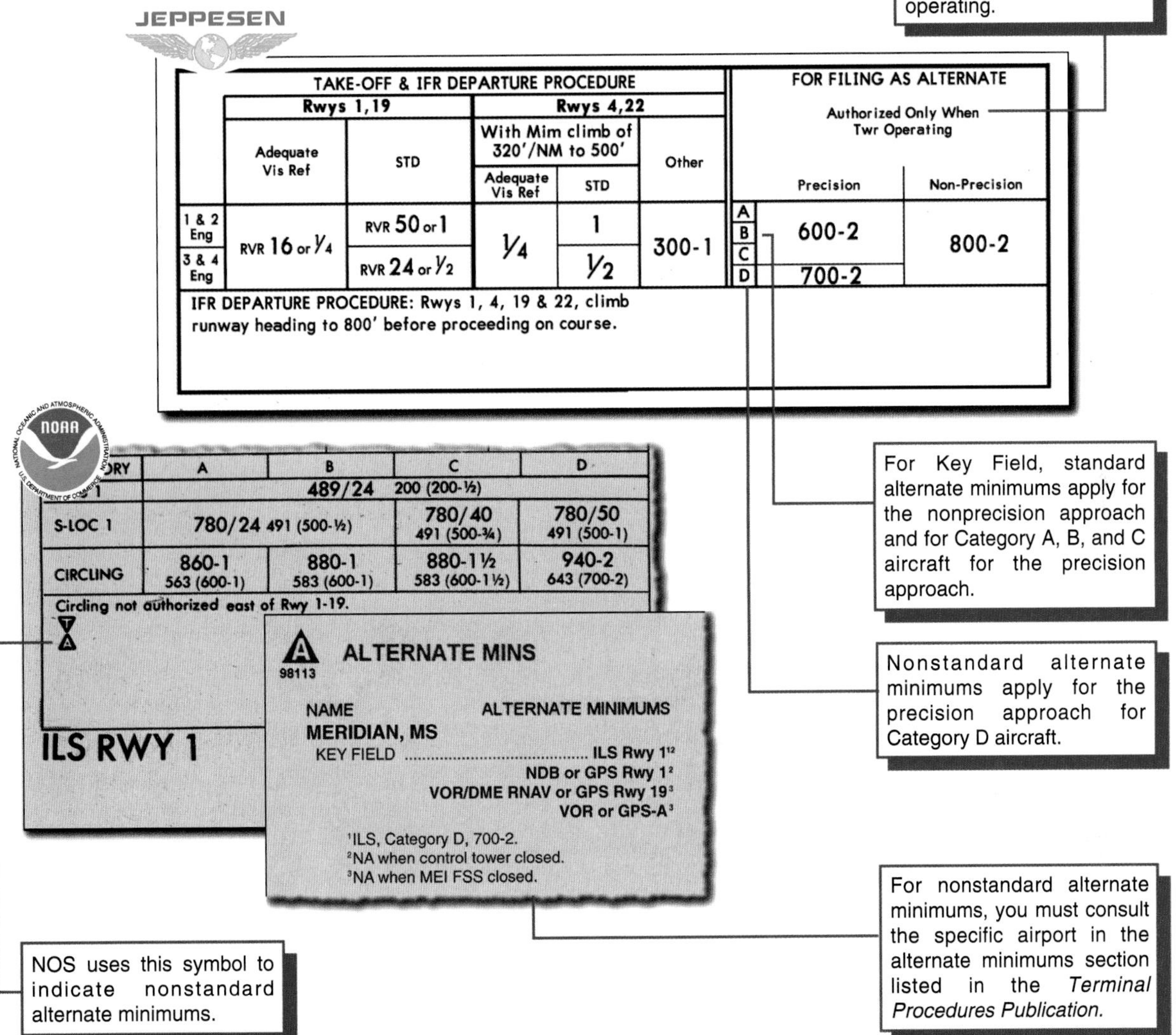

Figure 7-21. This portion of the Jeppesen airport chart can be thought of as three separate sections: takeoff minimums, departure procedures, and alternate minimums. Takeoff minimums and departure procedures are discussed in Chapter 4, Section B — Departure Procedures.

Standard alternate minimums for a precision approach are a 600-foot ceiling and 2 statute miles visibility. For a nonprecision approach, the minimums are an 800-foot ceiling and 2 statute miles visibility. See figure 7-21.

NEW APPROACH CHART FORMAT

Prior to flying an instrument approach, a thorough review of the procedure must be completed. The FAA recommends an approach briefing as a standard flight deck procedure, especially for airline, corporate, and other commercial flight crews. Also, in recent years, crew resource management (CRM) concepts have become more widely accepted and used. Flight safety is enhanced when all flight crew members are involved and familiar with the approach procedure. Jeppesen's new approach chart format, referred to as the Briefing Strip™ approach chart, incorporates human factors research, a standard approach briefing sequence of information, crew resource management (CRM) techniques, and an emphasis on usability and legibility. [Figure 7-22]

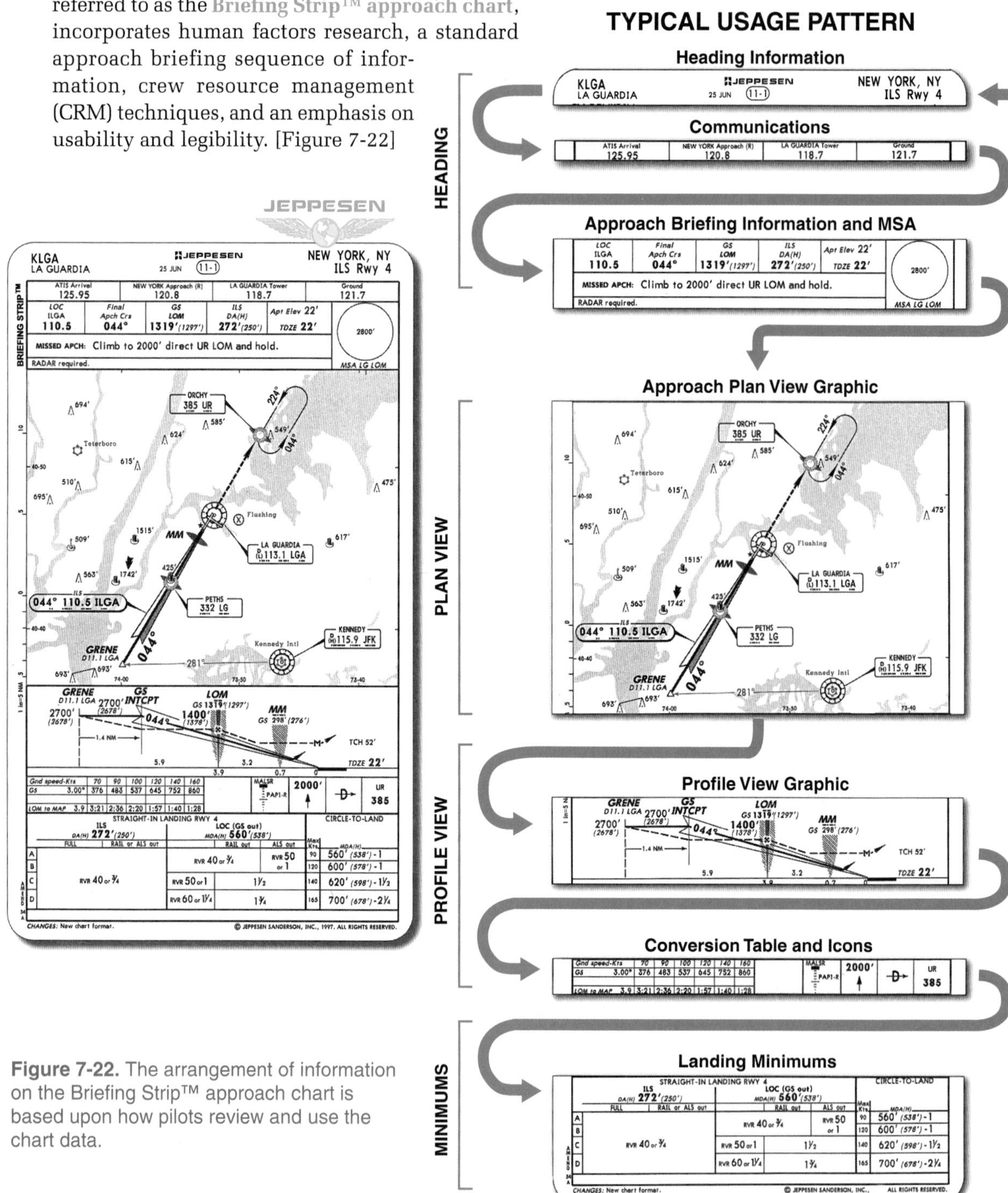

Figure 7-22. The arrangement of information on the Briefing Strip™ approach chart is based upon how pilots review and use the chart data.

HEADING

The main feature of the chart arrangement is placement of basic information in a common location in the heading section for more convenient use during the approach briefing. However, you should always review the entire chart for complete information. [Figure 7-23]

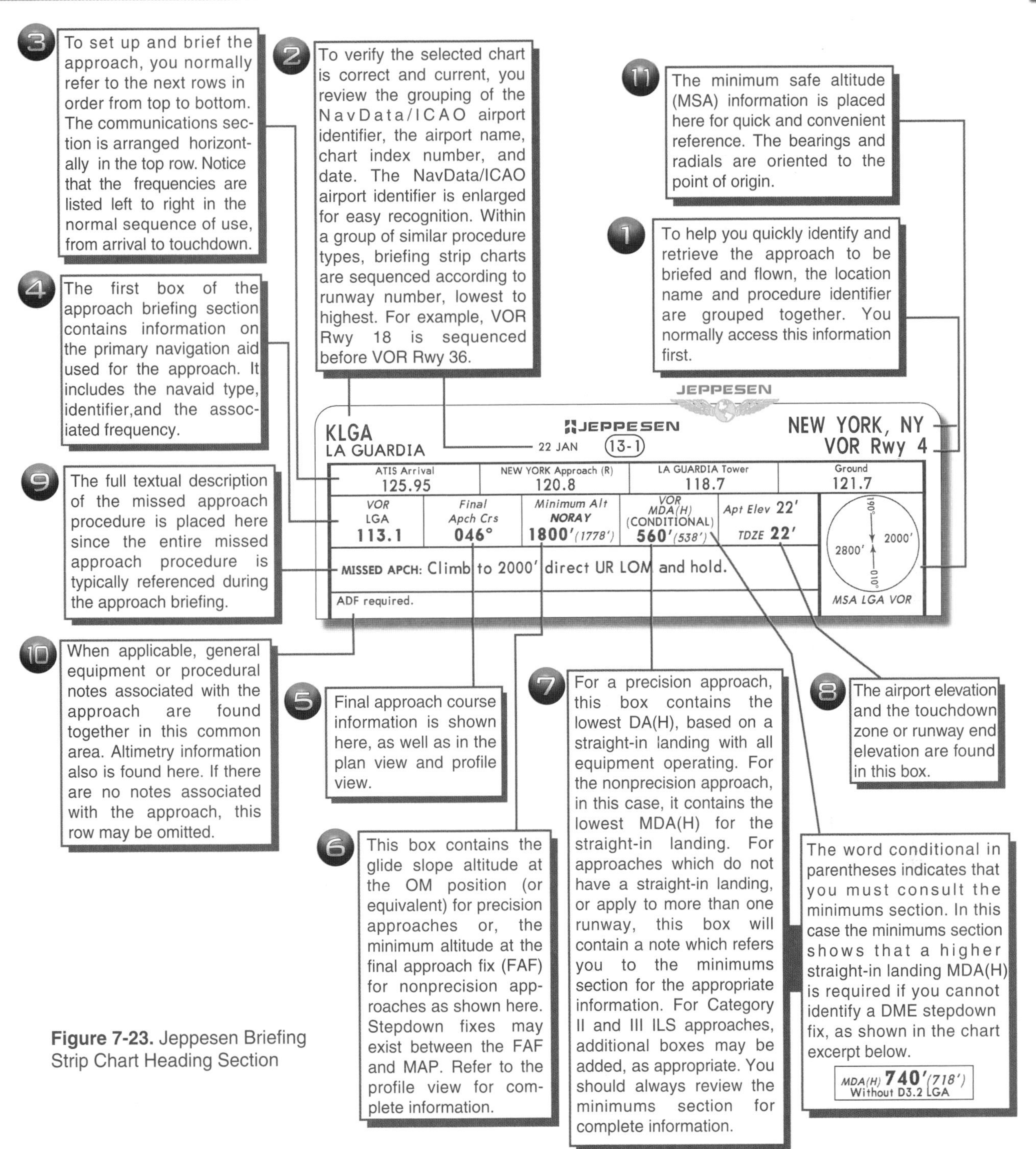

Figure 7-23. Jeppesen Briefing Strip Chart Heading Section

PLAN VIEW

The plan view section of the new format is essentially the same as the original chart format, with the exception of a few subtle changes in appearance. For example, leader lines are thinner than the original chart format and contain no arrowhead.

PROFILE VIEW

The profile view includes style changes with shading, along with some noticeable changes, such as the relocation of the conversion table, a graphic representation of the

Figure 7-24. Jeppesen Briefing Strip Chart Profile View

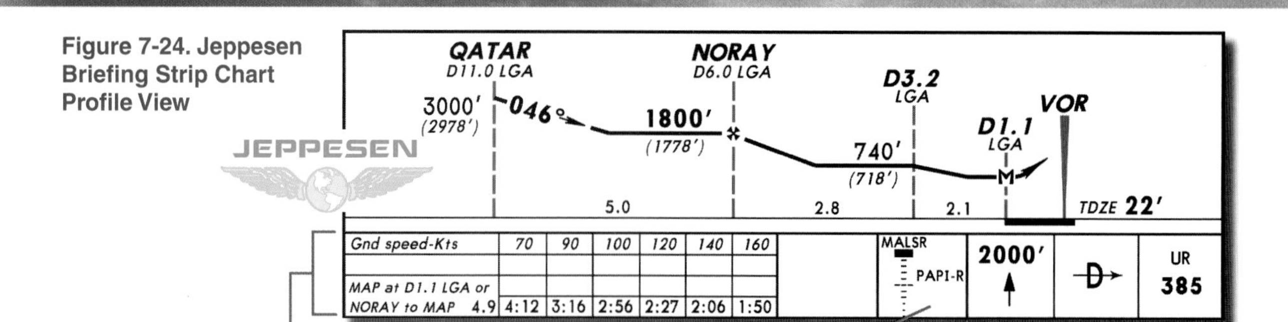

The conversion table is positioned here to reinforce the relationship between the profile view graphic and the conversion table information.

This box graphically depicts the applicable approach light system (ALS) and/or visual descent lighting aid for the straight-in landing runway. When REIL, VASI, or PAPI systems are available, they are shown on the appropriate side of the runway. In this example, Runway 4 is equipped with a medium intensity approach light system with runway alignment indicator lights (MALSR), and PAPI on the right side of the runway.

Missed approach icons represent initial pilot actions in the event of a missed approach. They provide symbolic information about the initial up and out maneuvers only, and improve the connection between the profile view graphic and the initiation of a missed approach procedure. Always refer to the missed approach instructions in the heading section and the plan view graphic for complete information about the missed approach procedure.

approach lighting system, and the symbolic depiction of the initial up and out maneuvers for the missed approach procedure. [Figures 7-24 and 7-25]

LIGHTING BOX

The lighting box displays the approach lights (ALS), visual glideslope indicators (VASI or PAPI), and runway end identifier lights (REIL) for the straight-in landing runway. The lighting box is omitted when ALS, VASI, PAPI or REIL are not installed.

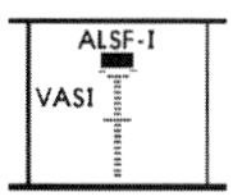

Approach Lights and VASI—VASI and PAPI are depicted in their relative position; left, right or both sides of centerline.

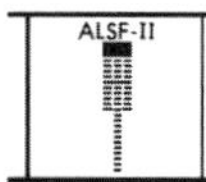

Approach Lights

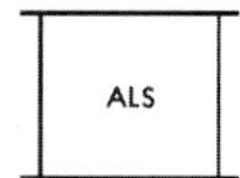

Approach Lights (Configuration Unknown)

REIL and VASI

MISSED APPROACH ICONS

Missed approach icons include a wide variety of initial action instructions. A representative sample of icons are shown below;

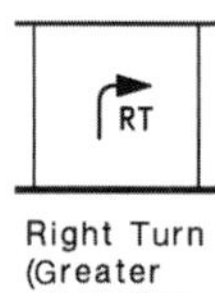

Right Turn (Greater Than 45°)

LT

Left Turn (Greater Than 45°)

LT

Left Turn (Less Than 45°)

Climb

6000'

Climb to Altitude

Direct

270° hdg

Fly Heading

ANY 117.9 R-270

Track Radial

PODUK

To Specified Fix

ANY 117.9

To Specified Navaid

090° RT

Turn to Specified Course

7000' LT

Turn to Specified Altitude

285 kts max

Airspeed Limit

RT within ANY 9.0 DME

Right Turn with Limit

Figure 7-25. Jeppesen Briefing Strip Chart Legend — This excerpt from the new approach chart format legend depicts profile view examples of the lighting box and missed approach icons.

Consider the Source

How will your approach be affected if the part-time tower is closed and you must use an altimeter setting reported for an airport nearby? A higher MDA or DH above that required for obstacle clearance may be required. The *U.S. Standard for Terminal Instrument Procedures* (TERPs) prescribes standardized methods for use in designing instrument flight procedures, including takeoff and landing minimums.

Generally, when the altimeter setting is obtained from a source more than 5 nautical miles from the airport reference point (ARP) for an airport, landing minimums may be affected. In the case of a remote altimeter setting source for the VOR or GPS-A approach at San Diego, California, Brown Municipal Airport, a circle-to-land minimums adjustment of 120 feet is applied as shown in the chart excerpts. Where intervening terrain does not adversely influence atmospheric pressure patterns, the formula used to compute the basic adjustment in feet is as follows:

Adjustment = $2.30d_R + 0.14e$,

where d_R is the horizontal distance in nautical miles from the altimeter source to the airport reference point, and e is the differential in feet between the elevation of the remote altimeter setting source and the elevation of the airport.

Distance from Brown Municipal to San Diego International - Lindbergh Field = 15 nautical miles

Lindbergh Field airport elevation = 15 feet MSL

Brown Municipal airport elevation = 524 feet MSL

$2.3 \times 15 + .14 \times 509 = 105'$ adjustment

MDA(H) 1220′(696′) + 105′ = 1325′(801′)

Round to next higher 20-foot increment (add 15 feet to 105 feet = 120 feet adjustment) = 1340′(816′)

CIRCLE-TO-LAND
Not Authorized South of Runway 8R-26L

	Max Kts.	When Twr Operating MDA(H)	With Lindbergh Field Altimeter Setting MDA(H)
A	90		
B	120		
C	140	1220′(696′)-2¾	1340′(816′)-2¾
D	165		

MINIMUMS

The presentation of landing minimums is unchanged in the new approach chart format. The notes applicable to landing minimums are commonly located below the landing minimums section.

AIRPORT CHART

Minimal changes have been made to airport charts. The headings have been slightly rearranged to conform with the new approach chart format. Communication frequencies are arranged horizontally in the normal sequence of use. The airport plan view graphic includes a new magnetic variation symbol. Additionally, the airport bearing and distance from a nearby VORTAC is depicted.

SUMMARY

Starting with the 19 September 97 revision, the new Jeppesen approach chart format began to appear in Airway Manual revisions. Within a few years, all Jeppesen approach and airport charts worldwide will be converted to the new briefing strip format. As the revision process develops, this textbook will be revised in accordance with the implementation schedule. In the meantime, it is important to supplement your

knowledge from the regular approach chart legend by referring to the approach chart legend for the new format in the Jeppesen Airway Manual. This legend includes only the items unique to the new format.

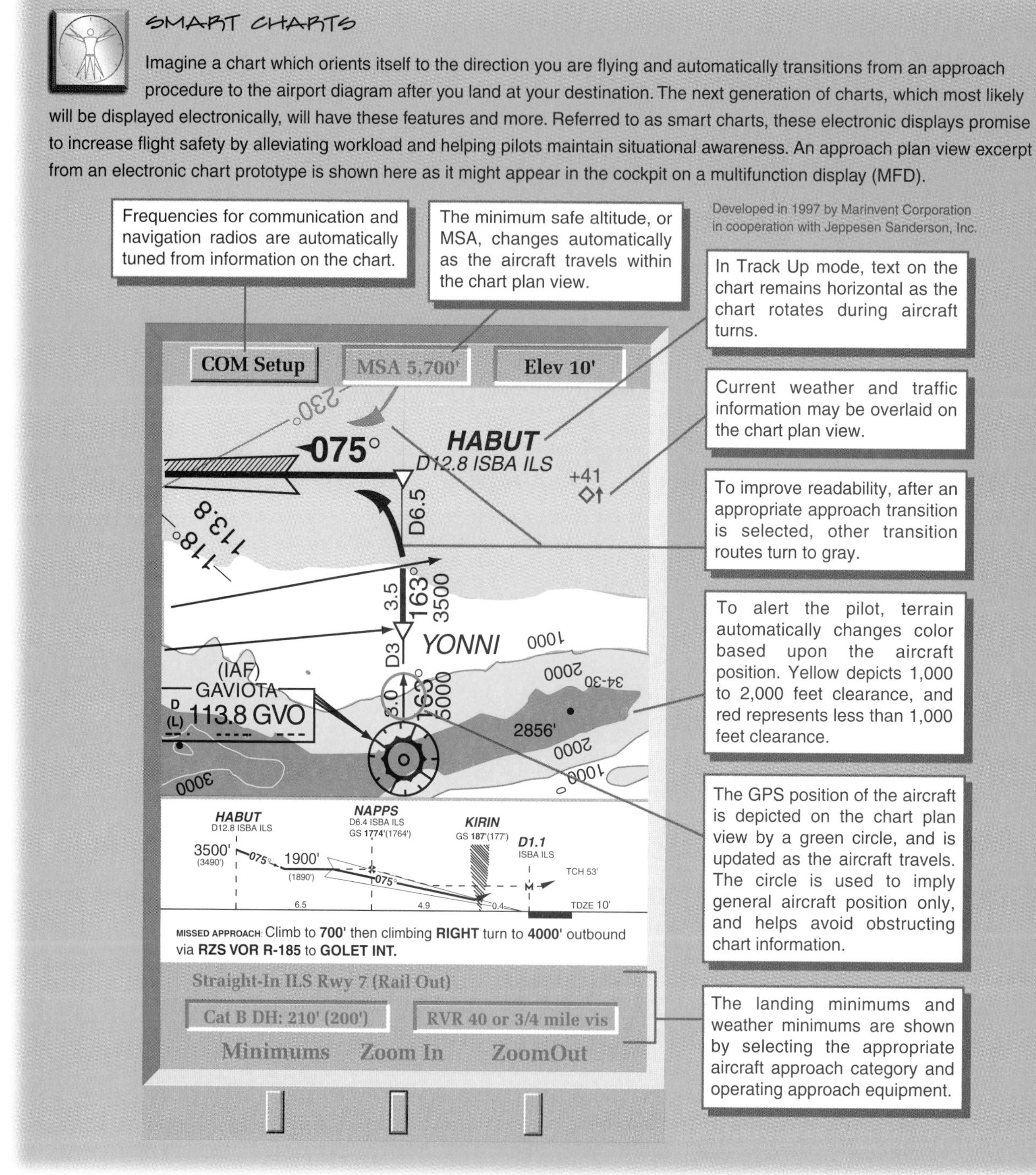

SUMMARY CHECKLIST

✓ The standard instrument approach procedure (IAP) allows you to descend safely by reference to instruments from the enroute altitude to a point near the runway at your destination from which a landing may be made visually. An IAP may be divided into as many as four segments: initial, intermediate, final, and missed approach.

- ✓ A precision approach, such as an ILS or PAR procedure provides vertical guidance through means of an electronic glide slope, as well as horizontal course guidance. A nonprecision approach, such as a VOR or NDB approach, provides horizontal course guidance with no glide slope information.

- ✓ Feeder routes provide a link between the enroute and approach structures.

- ✓ The purpose of the initial approach segment is to provide a method for aligning your aircraft with the approach course. The intermediate segment primarily is designed to position your aircraft for the final descent to the airport.

- ✓ The final approach segment allows you to navigate safely to a point at which, if the required visual references are available, you can continue the approach to a landing.

- ✓ The final approach segment for a precision approach begins where the glide slope is intercepted at the minimum glide slope intercept altitude shown on the approach chart. For a nonprecision approach, the final approach segment begins either at a designated final approach fix (FAF) or at the point where you are aligned with the final approach course, the final approach point (FAP).

- ✓ The missed approach segment begins at the missed approach point (MAP) and ends at a designated point, such as an initial approach or enroute fix.

- ✓ Jeppesen approach charts are filed in a loose-leaf format by state, then by city within each state. NOS charts are published in regional volumes referred to as *Terminal Procedures Publications*, with each airport filed alphabetically by the name of the associated city. Generally, both charts present the same information; however, the symbology and chart layout vary.

- ✓ Jeppesen and NOS present communication data in the normal sequence of use, and both use index numbers for chart identification.

- ✓ The minimum safe altitude, or MSA, provides 1,000 feet of obstruction clearance within 25 nautical miles of the indicated facility, unless some other distance is specified.

- ✓ The plan view is an overhead presentation of the entire approach procedure, including navaid facility boxes and reference points, such as natural or man-made objects.

- ✓ A procedure turn is a standard method of reversing your course. When a holding or teardrop pattern is shown instead of a procedure turn, it is the only approved method of course reversal. If a procedure turn, holding or teardrop pattern is not shown, a course reversal is not authorized.

- ✓ Minimum altitudes on approach procedures provide clearance of terrain and obstructions along the depicted flight tracks.

- ✓ Feeder routes are indicated with a heavy line arrow on both Jeppesen and NOS charts. Each flyable route lists the radial or bearing, the distance, and the minimum altitude you must fly to reach an IAF.

- ✓ A lead radial is a common example of a nonflyable radial. A thin line is used to depict the nonflyable lead radial, and it does not have a minimum altitude.

✓ The touchdown zone elevation (TDZE) is the highest centerline altitude for the first 3,000 feet of the landing runway.

✓ The profile view shows the approach from the side and displays flight path, facilities, and minimum altitudes. Height above touchdown (HAT) is measured from the touchdown zone elevation of the runway. Height above airport (HAA) is measured above the official airport elevation, which is the highest point of an airport's usable runways.

✓ The threshold crossing height (TCH) is the altitude at which you cross the runway threshold when established on the glide slope centerline.

✓ The procedure turn, as depicted on the profile view, must be completed within the prescribed distance from the facility.

✓ As shown on the profile view, the precision approach FAF is the point at which you will intercept the glide slope while operating at the minimum glide slope intercept altitude.

✓ The missed approach point on a precision approach is shown on both Jeppesen and NOS charts. Jeppesen also shows the approximate missed approach point on a nonprecision approach.

✓ Many approaches incorporate one or more stepdown fixes, used along approach segments to allow you to descend to a lower altitude as you overfly various obstacles.

✓ A visual descent point (VDP) represents the point from which you can make a normal descent to a landing, assuming you have the runway in sight and you are starting from the minimum descent altitude.

✓ An approach procedure to one runway with a landing on another is a circling approach. Seperate circle-to-land minimums are published in the landing minimums section if this procedure is authorized.

✓ During a sidestep maneuver, you are cleared for an approach to one runway with a clearance to land on a parallel runway.

✓ Aircraft are placed into an approach category based on computed approach speed. This speed equals 130% of the aircraft's power-off stall speed in the landing configuration at the maximum certificated landing weight ($1.3V_{S0}$).

✓ During a precision approach, the height where you must make the decision to continue the approach or execute a missed approach is referred to in the FARs as the decision height (DH). NOS charts show the decision height as an MSL altitude with the height above touchdown (HAT) listed after the visibility requirement. Jeppesen charts reflect ICAO terminology by showing a decision altitude (height) — DA(H) as an MSL altitude followed by the height above touchdown (HAT) in parentheses.

✓ The minimum descent requirement for a nonprecision approach is shown as a minimum descent altitude (MDA) on NOS charts and as a minimum descent altitude (height) — MDA(H) on Jeppesen charts.

✓ Visibility is listed on approach charts in statute miles, usually as a prevailing visibility reported by an accredited observer such as tower or weather personnel, or in hundreds of feet determined through the use of runway visual range (RVR) equipment.

✓ Both Jeppesen and NOS use tables to provide various elapsed times to the MAP based on the aircraft's groundspeed.

✓ Landing minimums usually increase when a required component or visual aid becomes inoperative. Regulations permit you to make substitutions for certain components when the component is inoperative or is not utilized during an approach.

✓ The airport diagram plan view portrays an overhead view of the airport, including runways and lighting systems. The official latitude and longitude coordinates are derived from the airport reference point, or ARP.

✓ If the forecast weather at your estimated time of arrival, plus or minus 1 hour, indicates a ceiling of less than 2,000 feet or a visibility of less than 3 miles, you must list an alternate airport on your IFR flight plan.

✓ Standard alternate minimums for precision approaches are a 600-foot ceiling and 2 statute miles visibility. For nonprecision approaches, an 800-foot ceiling and 2 statute miles apply.

✓ Jeppesen's new briefing strip charts incorporate human factors research, a standard approach briefing sequence of information, crew resource management (CRM) techniques, and an emphasis on usability and legibility.

✓ The heading section of briefing strip charts includes communication frequencies arranged horizontally, approach information, and the MSA circle.

✓ Briefing strip charts depict missed approach icons which represent initial pilot actions for up and out maneuvers only.

KEY TERMS

Instrument Approach Procedure (IAP)

Precision Approach

Nonprecision Approach

Feeder Route

Initial Approach Fix (IAF)

Intermediate Fix (IF)

Final Approach Fix (FAF)

Final Approach Point (FAP)

Missed Approach Point (MAP)

Minimum Safe Altitude (MSA)

Touchdown Zone Elevation (TDZE)

Height Above Touchdown (HAT)

Height Above Airport (HAA)

Airport Elevation

Threshold Crossing Height (TCH)

Stepdown Fix

Visual Descent Point (VDP)

Circling Approach

Sidestep Maneuver

Aircraft Approach Category

Approach Speed

Decision Height (DH)

Decision Altitude (Height) — DA(H)

Minimum Descent Altitude (Height) — MDA(H)

Airport Reference Point (ARP)

Briefing Strip™ Approach Chart

QUESTIONS

1. Name the four segments which are applicable to instrument approach procedures.
2. True/False. The route which provides a transition from the enroute structure to the approach structure is called a feeder route.
3. What is the missed approach point (MAP) for a precision approach?

   A. Final approach fix
   B. Minimum descent altitude
   C. Decision height while on the glide slope

Refer to the NOS chart for Dallas Love Airport to answer questions 4 through 18.

4. What is the appropriate frequency to receive advanced information regarding instrument approach procedures in progress at Dallas Love Airport?
5. What is the elevation of the Dallas Love Airport?

   A. 485 feet MSL
   B. 487 feet MSL
   C. 685 feet MSL

6. What is the ILS localizer magnetic course alignment and frequency?
7. True/False. During the transition from WADES Intersection to the NITER OM, you must use caution due to the proximity of JIFFY OM to the localizer course. The 5.6 DME fix from the IDAL localizer can help you identify NITER OM.
8. What is the minimum glide slope intercept altitude when you are inbound on the intermediate approach segment?

   A. 1,856 feet MSL
   B. 1,900 feet MSL
   C. 2,500 feet MSL

9. At what altitude will you cross the outer marker when you are established inbound on the glide slope?

   A. 685 feet MSL
   B. 1,856 feet MSL
   C. 1,900 feet MSL

10. True/False. If the glide slope becomes inoperative, resulting in a localizer approach, the final approach fix is now located at the OM, which is 4.5 nautical miles from the end of the runway.
11. At what altitude will you cross the runway threshold if you remain on the glide slope centerline and continue the approach to a landing?

    A. 50 feet
    B. 52 feet
    C. 200 feet

12. What is the difference in feet between the touchdown zone elevation and airport elevation?

    A. 2 feet
    B. 5 feet
    C. 10 feet

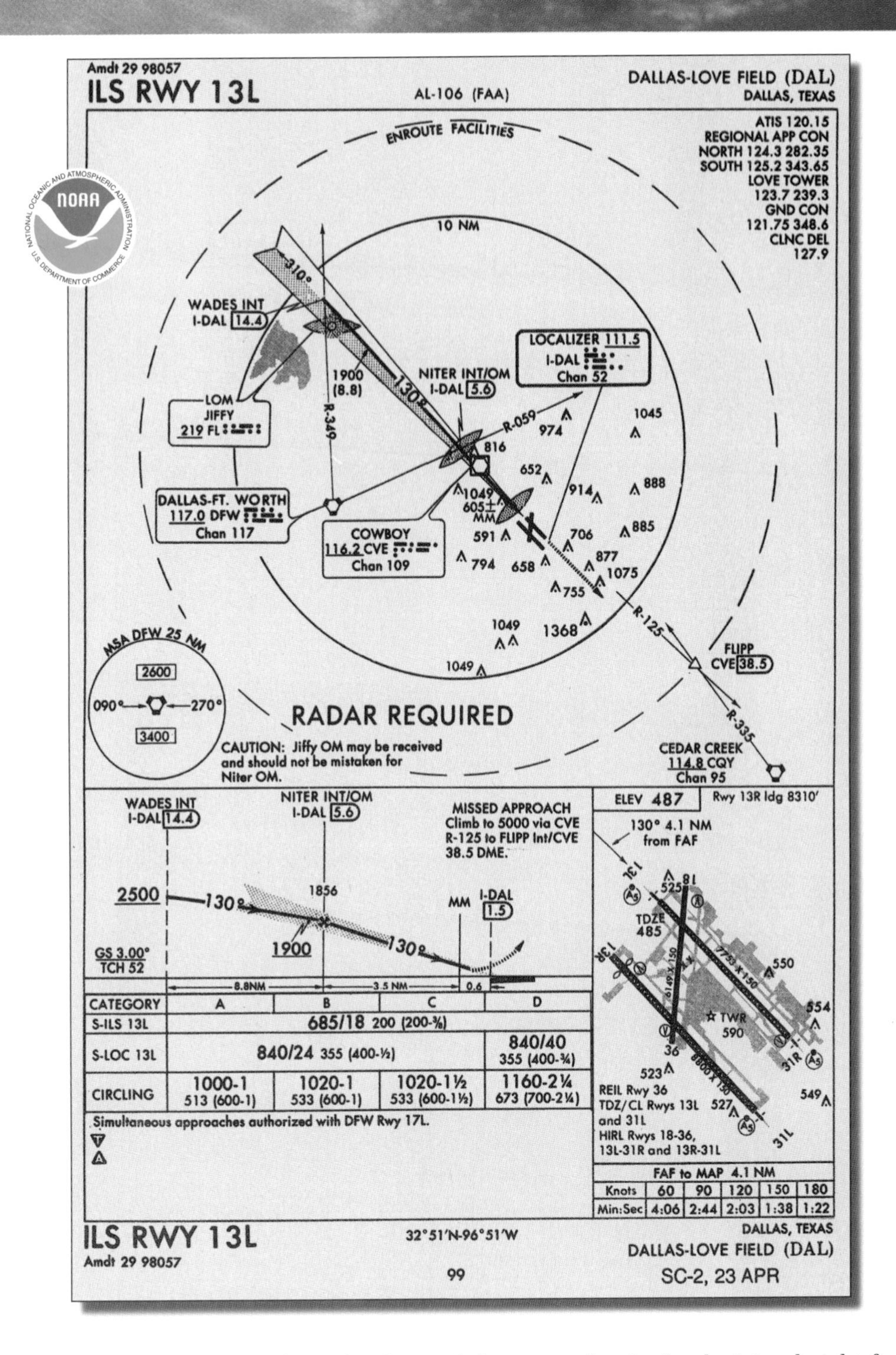

CATEGORY	A	B	C	D
S-ILS 13L	685/18 200 (200-½)			
S-LOC 13L	840/24 355 (400-½)			840/40 355 (400-¾)
CIRCLING	1000-1 513 (600-1)	1020-1 533 (600-1)	1020-1½ 533 (600-1½)	1160-2¼ 673 (700-2¼)

FAF to MAP 4.1 NM					
Knots	60	90	120	150	180
Min:Sec	4:06	2:44	2:03	1:38	1:22

13. According to straight-in landing minimums, what is the decision height for a Category A aircraft?

    A. 685 feet MSL, or 200 feet above the TDZE
    B. 840 feet MSL, or 355 feet above the TDZE
    C. 1020 feet MSL, or 533 feet above the TDZE

14. If the glide slope is not operational, what is the minimum descent altitude?

15. If you maintain a groundspeed of 90 knots while flying a localizer approach, what is the elapsed time from NITER OM to the MAP?

16. What is the MDA for a Category C aircraft performing a circling approach?

17. Which runways are equipped with VASI installations?

    A. 18/36 only
    B. 31R, 13R only
    C. 18/36, 31R, 13R

18. True/False. When you are filing Dallas Love as an alternate airport, standard alternate minimums apply for all aircraft categories and all runways.

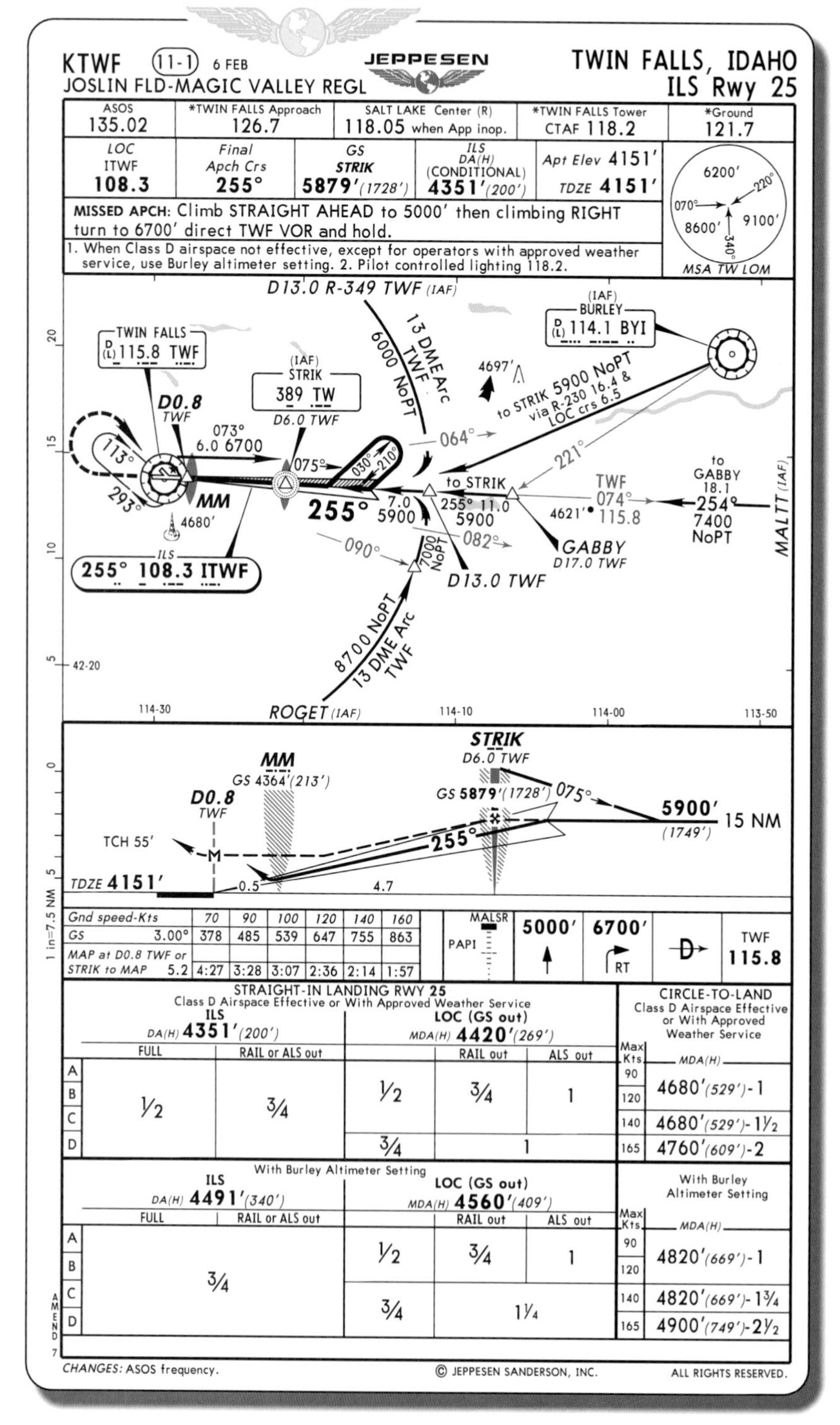

Refer to the Jeppesen Twin Falls, Idaho ILS Runway 25 briefing strip approach chart to answer questions 19 through 22.

19. On the LOC (GS out) approach, how can the MAP be identified?

    A. When crossing the middle marker
    B. At 0.8 DME from the Twin Falls VORTAC only
    C. At 0.8 DME from the Twin Falls VORTAC, or timing from STRIK

20. True/False. PAPI is available on the left side of Runway 25.

21. Describe the purpose of the missed approach icons in the profile view.

22. What does the bold letter D with an arrow mean in the missed approach icons?

    A. DME
    B. Direct
    C. Distance

SECTION B

APPROACH PROCEDURES

Under VFR, you fly the traffic pattern to prepare for landing and align your aircraft with the final approach course. In IFR conditions, you will need to fly an instrument approach procedure (IAP) to accomplish the same goal. While a standard rectangular traffic pattern is used at most airports, the direction and placement of the pattern, the altitude at which it is flown, and the procedures for entering and exiting the pattern may vary. A similar situation exists for instrument approaches. While IAPs have many basic features in common, each approach procedure is unique. [Figure 7-26]

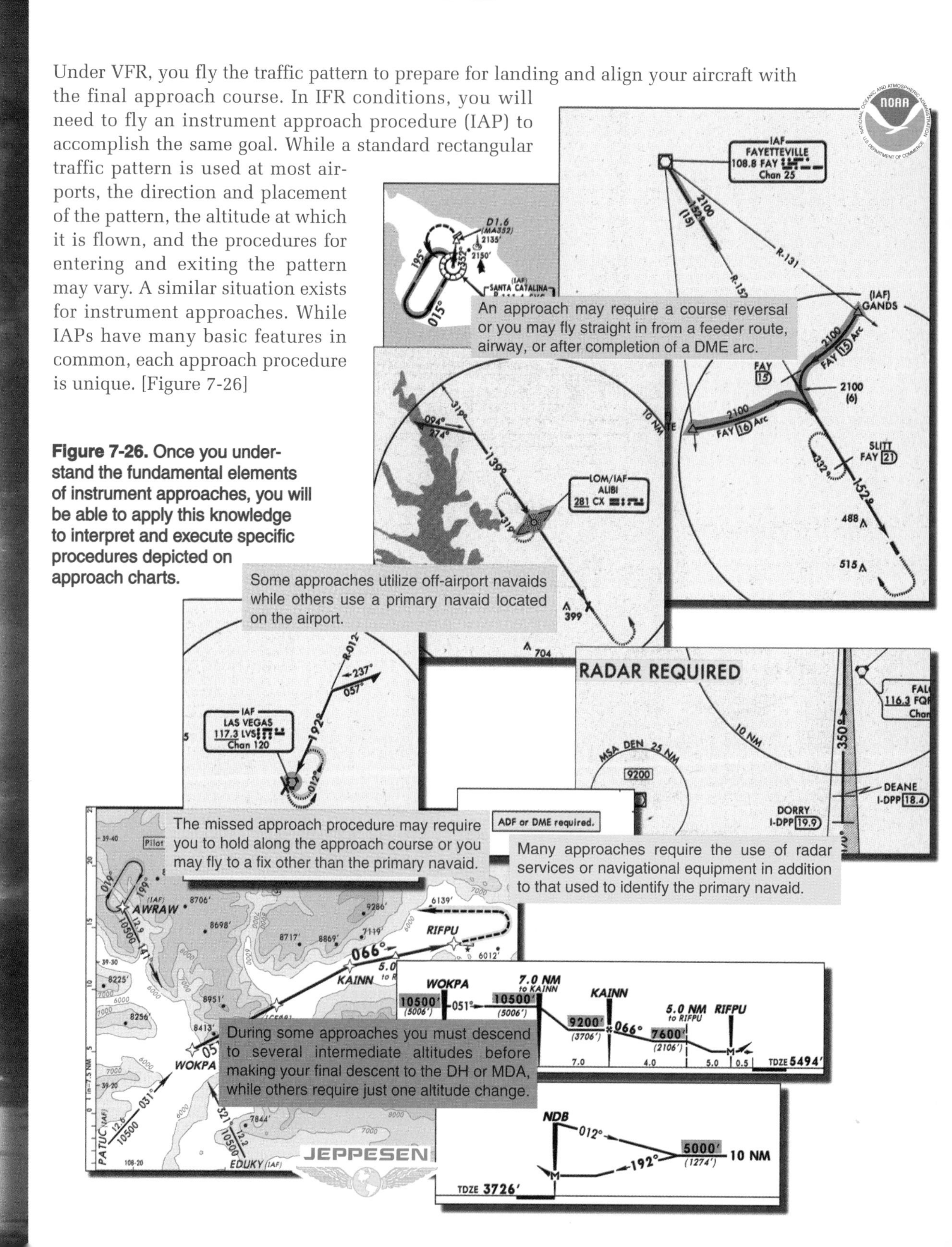

Figure 7-26. Once you understand the fundamental elements of instrument approaches, you will be able to apply this knowledge to interpret and execute specific procedures depicted on approach charts.

PREPARING FOR THE APPROACH

During your IFR flight planning, you should examine the approach charts for your destination to review procedures and symbology. As you approach your destination, the type of approach to expect will be provided by a controller or broadcast on ATIS. The purpose of this information is to help you plan your arrival actions; however, it is not an ATC clearance or commitment and is subject to change. For example, a shift in wind direction, fluctuations in weather conditions, and a blocked runway are conditions which may result in changes to approach information previously received.

APPROACH CHART REVIEW

After you have been advised as to which approach to expect, you should conduct a thorough **approach chart review** to familiarize yourself with the specific approach procedure. In addition to reviewing the primary elements of the charted procedure, you should tune and identify the appropriate communication and navigation frequencies well before you initiate the approach. [Figure 7-27]

Procedure Title

Communication Frequencies

Primary Navaid Frequency

Inbound Course

Altitude at Which You Will Cross the FAF or FAP (Nonprecision Approach) or Glide Slope Intercept Altitude (Precision Approach)

DA(H) — Precision Approach or MDA(H) — Nonprecision Approach

Airport Elevation

Touchdown Zone Elevation

Missed Approach Instructions

Special Notes/Procedures

Figure 7-27. An approach chart review should always be conducted with emphasis on the key information shown here, as well as additional features which are unique to the specific procedure. Briefing strip charts provide an easy way to quickly identify the primary items during an approach chart review.

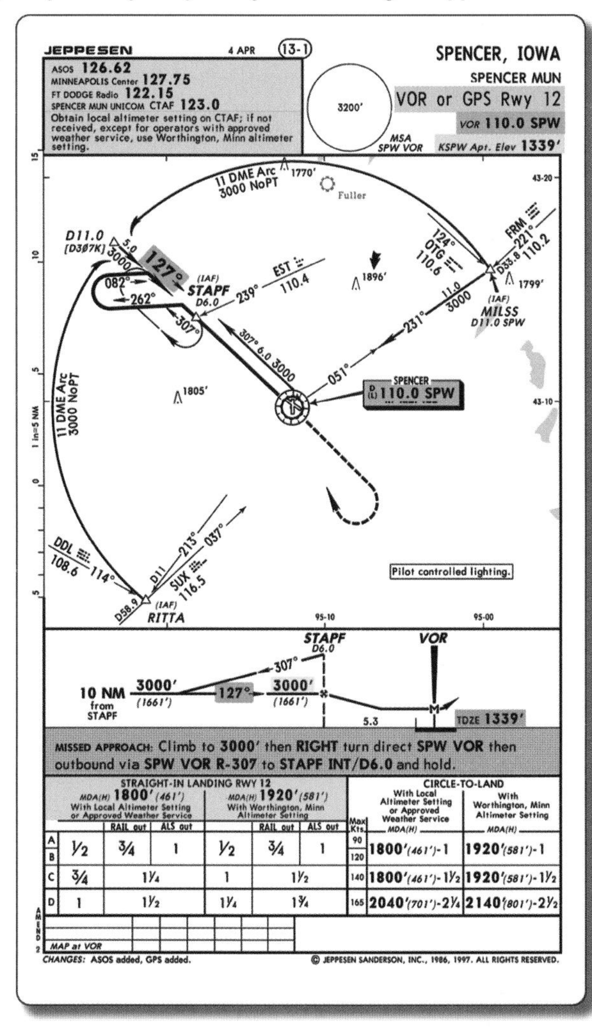

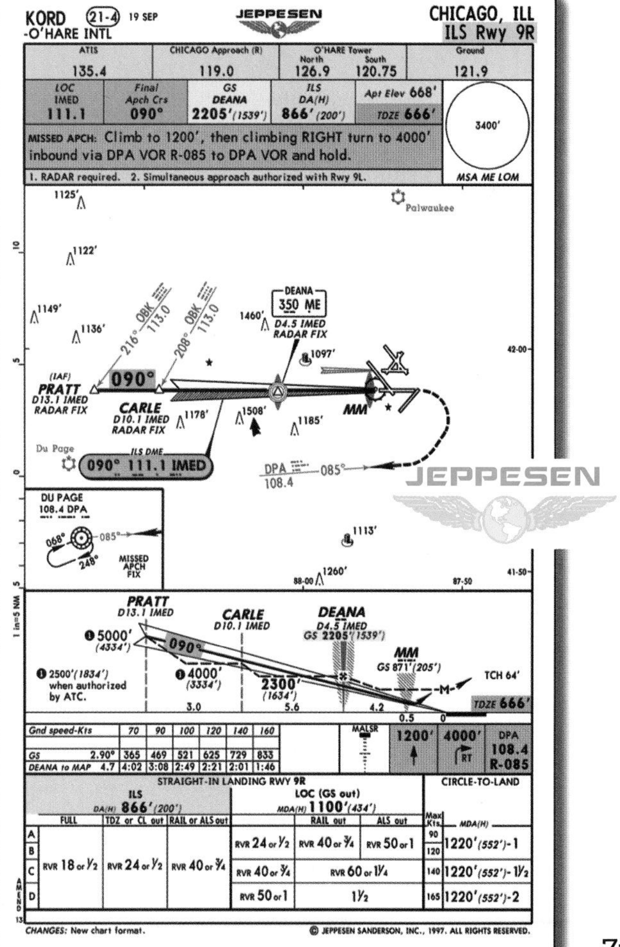

APPROACH CLEARANCE

Since several instrument approach procedures, using a variety of navaids, may be authorized for an airport, ATC may clear you for a particular approach procedure, primarily to expedite traffic. You should notify the controller immediately if you desire a different approach; however, it may be necessary for ATC to withhold clearance for the approach until such time as traffic conditions permit. If ATC does not specify a particular approach but states, *"cleared for approach,"* you may execute any one of the authorized IAPs for that airport. This clearance does not permit you to fly a contact or visual approach.

Except when you are being radar vectored to the final approach course, you must execute the entire procedure commencing at an IAF or an associated feeder route unless an appropriate new or revised ATC clearance is received, or the IFR flight plan is canceled. **Feeder routes** provide a transition from the enroute structure to the IAF or to a facility from which a course reversal is initiated. [Figure 7-28]

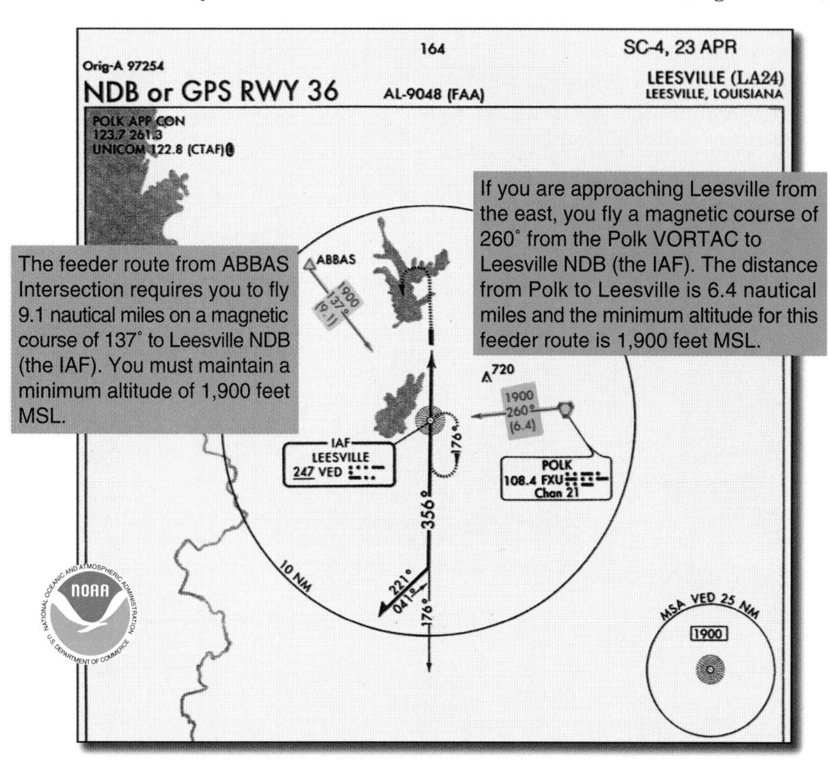

Figure 7-28. There may be several feeder routes designated for an instrument approach procedure to cater to airplanes arriving from different directions. However, many procedures do not require a feeder route. For example, a feeder route is not necessary if the enroute structure ends at an initial approach fix.

EXECUTING THE APPROACH

As you transition from the enroute to the terminal environment, your workload increases. To execute an instrument approach, you must perform an extensive sequence of tasks and normally more precise flying and navigation skills are required than during enroute operations. Some general procedures for executing approaches are introduced here, while more detailed descriptions of specific nonprecision and precision approach procedures are presented in Chapter 8.

STRAIGHT-IN APPROACHES

The terms straight-in approach and straight-in landing are often used interchangeably although they do have specific definitions when used in ATC clearances or in reference to landing minimums. **Straight-in landing** minimums normally are used when the final approach course is positioned within 30° of the runway and a minimum of maneuvering is required to align the airplane with the runway. If the final approach course is not

STATE YOUR QUALIFICATIONS

The airport is located on a 1,000-foot mesa. Strong vertical turbulence and variable crosswinds may exist in the area of the mesa's edge.

The high landing minimums and special missed approach procedures are designed to allow for clearance of high terrain near the airport.

These advisories are included on an airport qualification chart for Telluride Regional Airport. Airport qualification charts were developed to help pilots meet the requirements of FAR 121.445, Pilot in Command Airport Qualification: Special Areas and Airports. This regulation requires that pilots in command of Part 121 carriers meet special qualifications to operate at selected airports due to surrounding terrain, obstructions, or complex approach or departure procedures.

One way to meet these qualifications is for the pilot in command or second in command to have performed a takeoff and landing at the applicable airport as a flight crewmember within the preceding 12 calendar months. Another way is through the use of acceptable pictorial means and that is where airport qualification charts come in.

Airport qualification charts are valuable tools to help pilots maintain situational awareness during approaches to challenging airports. These charts provide photos of the airport, graphic and textual descriptions of the surrounding terrain and obstacles, as well as typical weather conditions and other unique airport information of concern to pilots. While they were originally developed for Part 121 airlines, airport qualification charts are an effective resource used by pilots in a wide variety of operating environments, including corporate and general aviation flights.

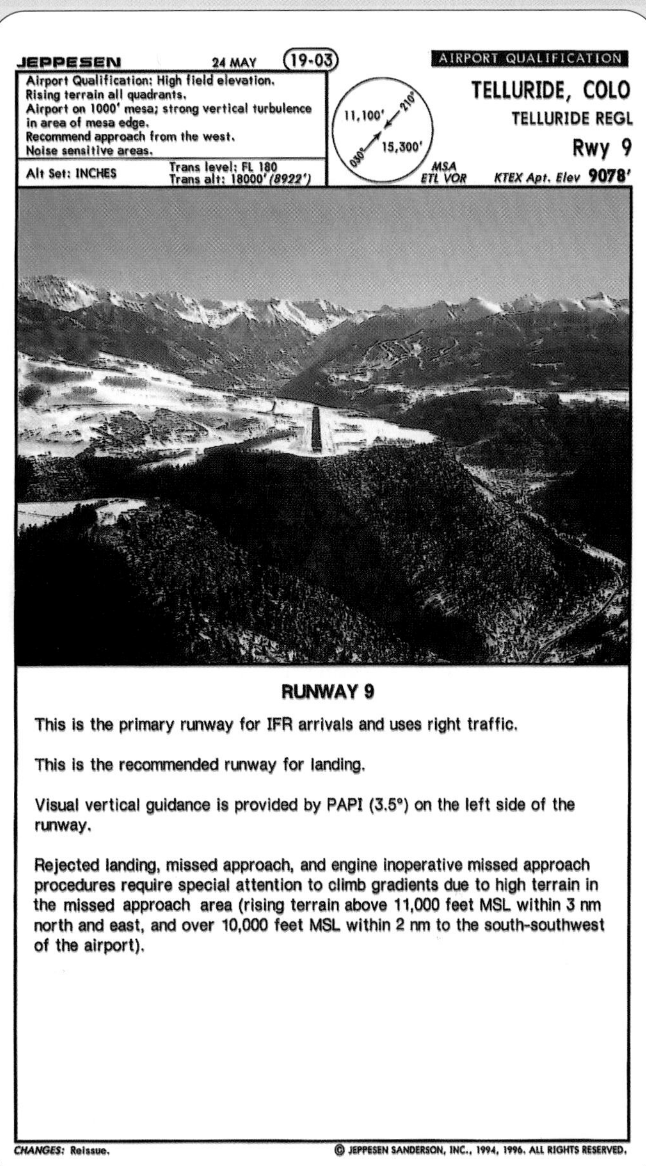

JEPPESEN 24 MAY (19-03) AIRPORT QUALIFICATION

TELLURIDE, COLO
TELLURIDE REGL
Rwy 9

Airport Qualification: High field elevation.
Rising terrain all quadrants.
Airport on 1000' mesa; strong vertical turbulence in area of mesa edge.
Recommend approach from the west.
Noise sensitive areas.

Alt Set: INCHES | Trans level: FL 180 | Trans alt: 18000' (8922')

MSA ETL VOR: 11,100' 210° – 030° 15,300'

KTEX Apt. Elev 9078'

RUNWAY 9

This is the primary runway for IFR arrivals and uses right traffic.

This is the recommended runway for landing.

Visual vertical guidance is provided by PAPI (3.5°) on the left side of the runway.

Rejected landing, missed approach, and engine inoperative missed approach procedures require special attention to climb gradients due to high terrain in the missed approach area (rising terrain above 11,000 feet MSL within 3 nm north and east, and over 10,000 feet MSL within 2 nm to the south-southwest of the airport).

CHANGES: Reissue.

properly aligned, or if it is desirable to land on a different runway, a circling approach may be executed and circle-to-land minimums apply. While most approach procedures provide landing minimums for both straight-in and circling maneuvers, some may be limited to circle-to-land minimums only. For example, an approach from a VOR facility not closely aligned with any of the runways on the airport would authorize only circling maneuvers.

In contrast to a straight-in landing, the controller terminology, *"cleared for straight-in approach . . ."* means that you should not perform any published procedure to reverse your course, but does not reference landing minimums. For example, you could be *"cleared for straight-in ILS Runway 25 approach, circle to land Runway 34."* A

straight-in approach may be initiated from a fix closely aligned with the final approach course, may commence from the completion of a DME arc, or you may receive vectors to the final approach course. [Figure 7-29]

A straight-in approach does not require nor authorize a procedure turn or course reversal.

Figure 7-29. Straight-in landings and straight-in approaches should not be confused. If you are not being radar vectored, generally you begin a straight-in approach at an outlying IAF, then fly the initial and intermediate segments, which places you on the final approach segment.

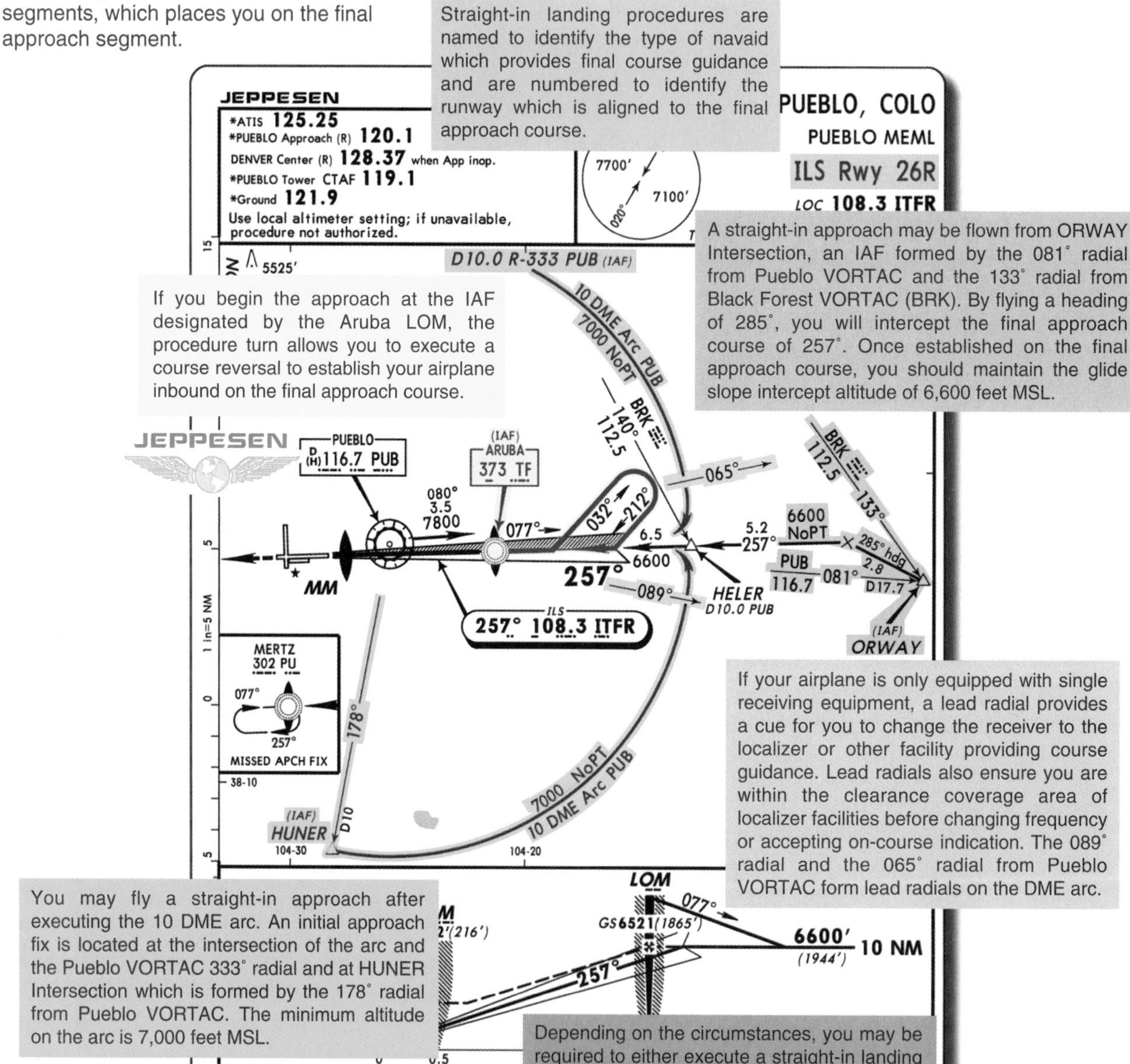

Figure 7-29. Continued

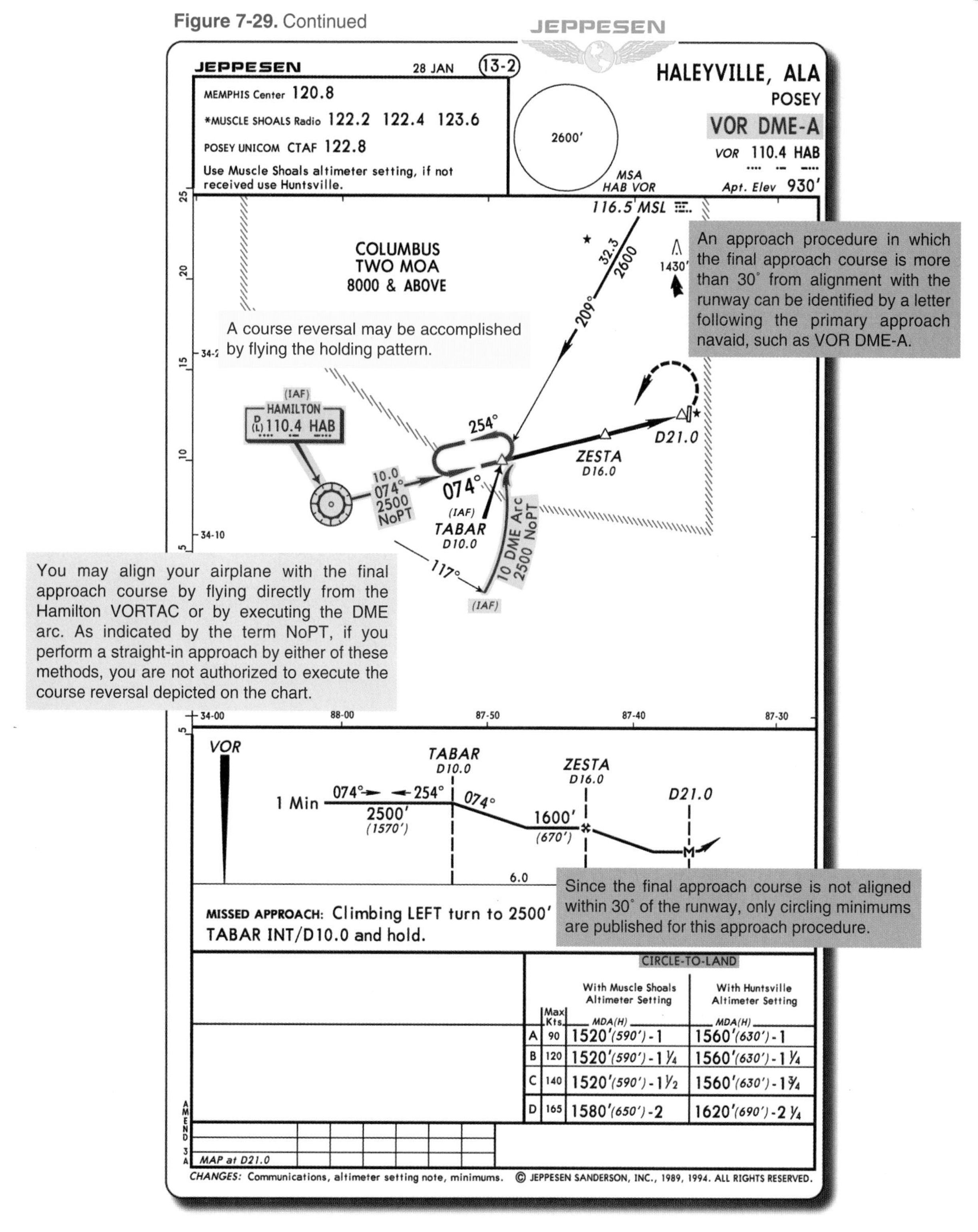

A straight-in approach also is required if an airway provides a means of navigating directly to the final approach course. When this occurs, the approach chart may show a NoPT (no procedure turn) arrival sector formed by airways leading to an enroute navaid which also serves as an initial approach fix. [Figure 7-30]

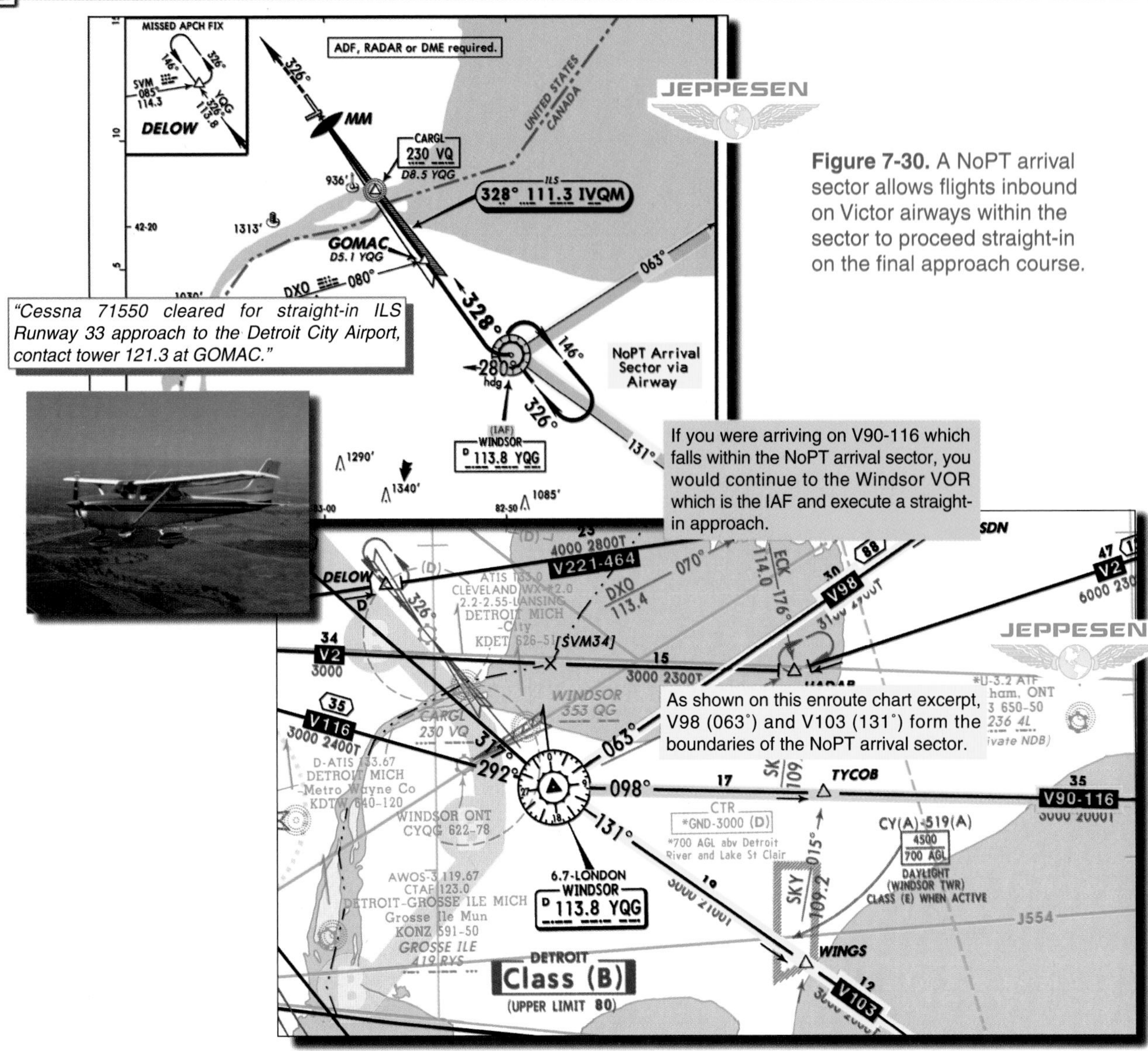

Figure 7-30. A NoPT arrival sector allows flights inbound on Victor airways within the sector to proceed straight-in on the final approach course.

Use of ATC Radar for Approaches

In locations where radar is approved for approach control service, it also may be used in conjunction with published instrument approach procedures to provide guidance to the final approach course or to the traffic pattern for a visual approach. In addition, approach control radar is used for airport surveillance radar (ASR) and precision approach radar (PAR) approaches and to monitor nonradar approaches. ATC provides azimuth guidance and range information during an ASR approach and navigational guidance in azimuth, range, and elevation during a PAR approach. These approach procedures may be used in an emergency situation, such as during a loss of gyroscopic instruments. ASR and PAR approaches are discussed in greater detail in Chapter 10, Section A — IFR Emergencies.

ATC radar approved for approach control service is used for course guidance to the final approach course, ASR and PAR approaches, and the monitoring of nonradar approaches.

Radar vectors to the final approach course provide a method of intercepting and proceeding inbound on the published instrument approach procedure. During an arrival, you normally are cleared to the airport or an outer fix that is appropriate to your arrival route. Once you have been handed off to approach control, you continue inbound

to the airport or to the fix in accordance with your last route clearance. ATC will advise you to expect radar vectors to the final approach course for a specific approach procedure unless that information is included in ATIS. [Figure 7-31]

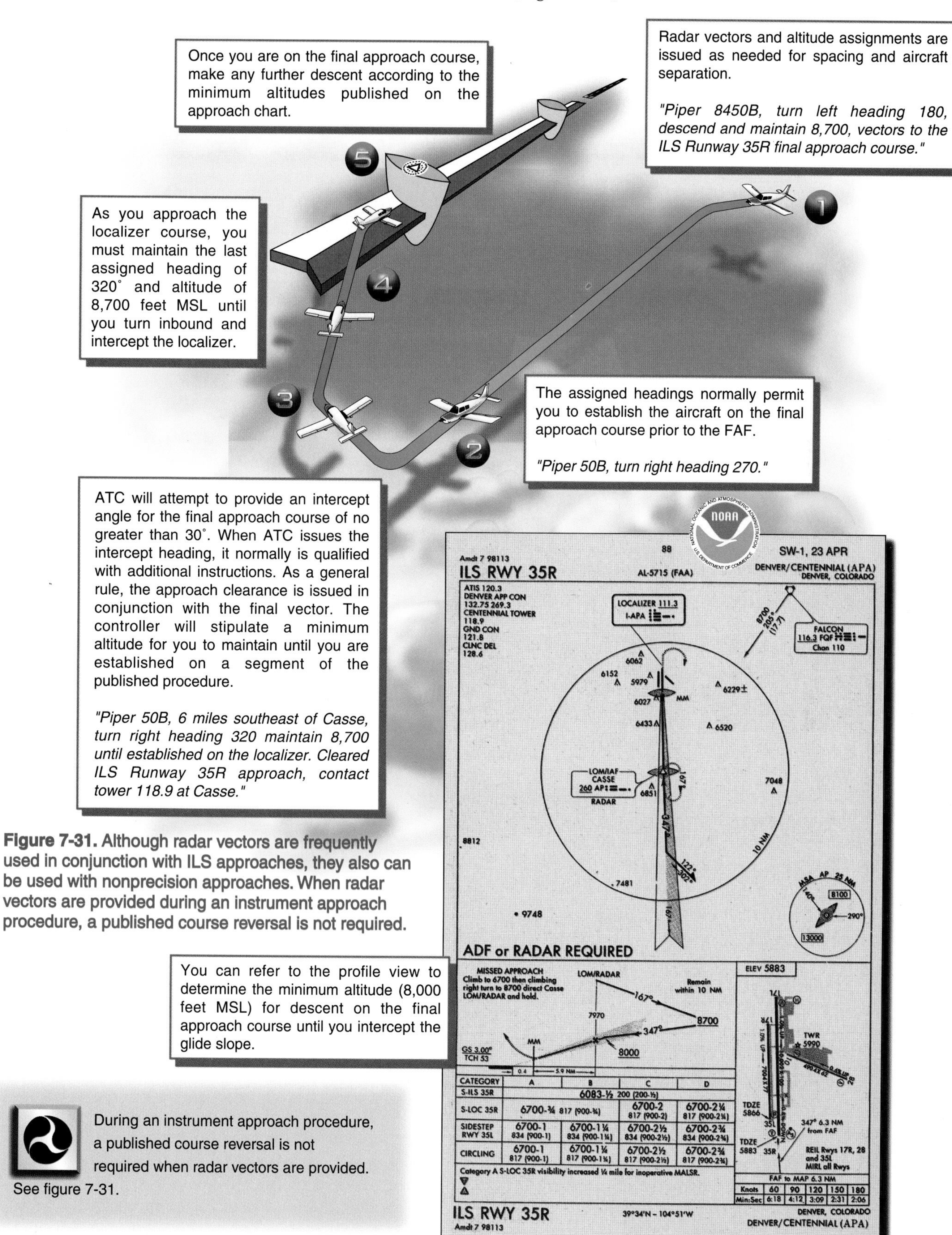

Figure 7-31. Although radar vectors are frequently used in conjunction with ILS approaches, they also can be used with nonprecision approaches. When radar vectors are provided during an instrument approach procedure, a published course reversal is not required.

During an instrument approach procedure, a published course reversal is not required when radar vectors are provided.

See figure 7-31.

When you are cleared for an approach while being radar vectored, you must maintain your last assigned altitude until established on a segment of the published approach. See figure 7-31.

If it becomes apparent an assigned heading will cause you to pass through the final approach course, you should maintain that heading and question the controller.

While being radar vectored, you should not turn inbound on the final approach course until you are cleared for the approach. If it appears imminent that an assigned heading will cause you to cross through the final approach course and you have not been advised that you will be vectored through it, question the controller. The controller may intentionally vector you through the final approach course to achieve traffic separation. If this is the case, the controller should advise you and state the reason such as, "*. . . expect vector across final approach course for spacing.*"

During the process of radar vectoring, the controller is responsible for assigning altitudes that are at or above the **minimum vectoring altitude (MVA)**. These altitudes are established in terminal areas to provide terrain and obstruction clearance. The MVA in a given sector may be lower than the nonradar MEA, MOCA, or other minimum altitude shown on instrument charts. [Figure 7-32]

Figure 7-32. The altitudes you are assigned by ATC when being radar vectored should provide terrain and obstruction clearance, as well as maintain your aircraft within controlled airspace.

Approaches Which Require Course Reversal

Some approach procedures do not provide for straight-in approaches unless you are being radar vectored. In these situations, you are required to complete a **course reversal**, generally within 10 nautical miles of the primary navaid or fix designated on the approach chart, to establish your aircraft inbound on the intermediate or final approach segments. [Figure 7-33]

The maximum speed in a procedure turn is 200 knots IAS.

If you are above the altitude designated for the course reversal, you may begin descent as soon as you cross the IAF. See figure 7-33.

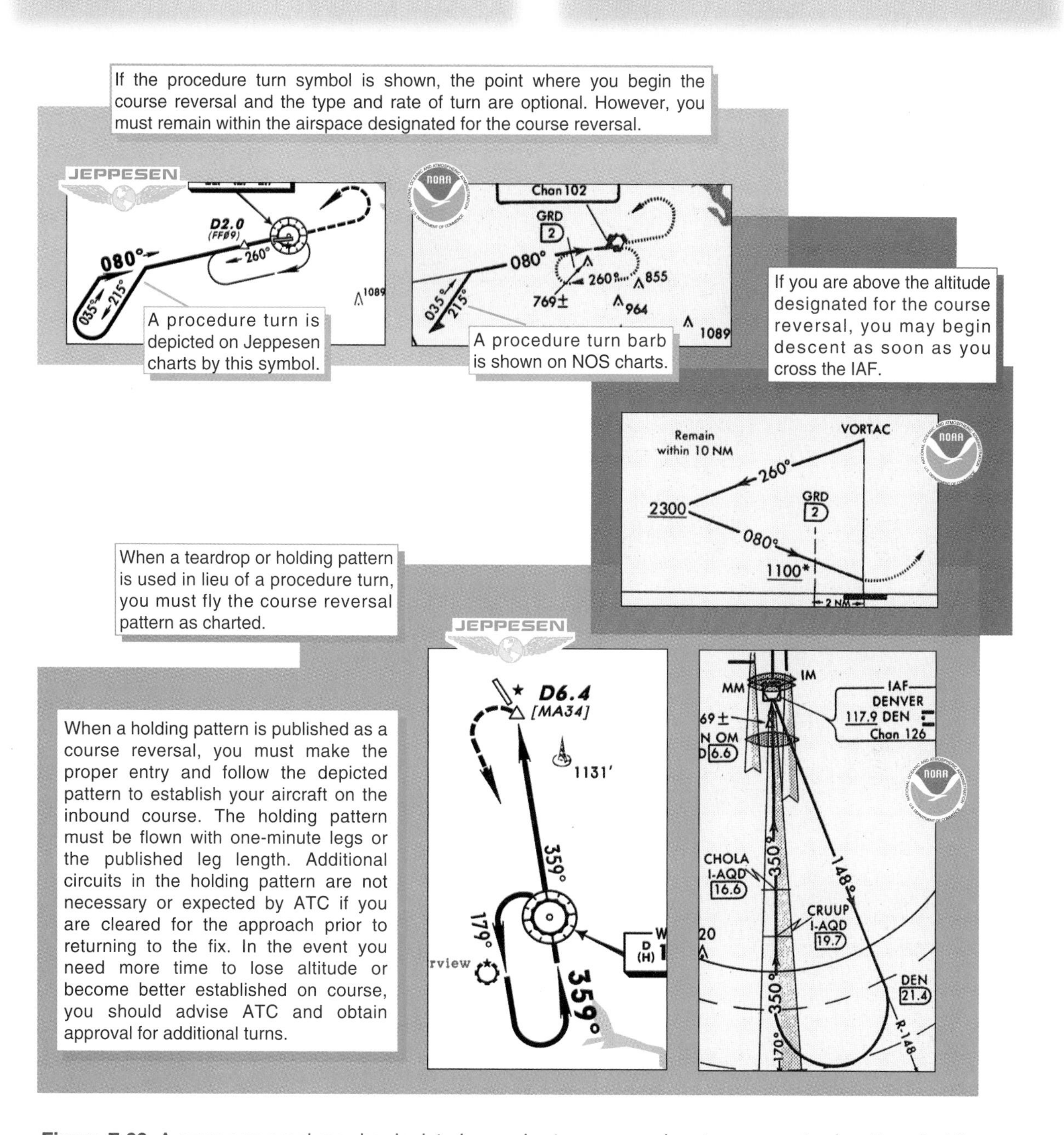

Figure 7-33. A course reversal may be depicted on a chart as a procedure turn, a racetrack pattern (holding pattern), or a teardrop procedure. The maximum speed in a course reversal is 200 knots IAS.

Course reversals must be completed within the distance specified on the chart which is typically 10 nautical miles from the primary navaid or fix indicated on the approach chart. See figure 7-33.

When more than one circuit of the holding pattern is needed to lose altitude or become better established on course, the additional circuits can be made only if you advise ATC and ATC approves. See figure 7-33.

Often the terms procedure turn and course reversal are used interchangeably. For example, if you are flying an approach segment shown on the chart with the notation, NoPT (no procedure turn) you may not execute the course reversal depicted whether it is indicated by the procedure turn symbol, holding pattern course reversal, or teardrop pattern.

If a teardrop or holding pattern is shown on an approach chart, you must execute the course reversal as depicted. See figure 7-33.

TIMED APPROACHES FROM A HOLDING FIX

Timed approaches from a holding fix are generally conducted at airports where the radar system for traffic sequencing is out of service or is not available and numerous aircraft are waiting for approach clearance. [Figure 7-34] ATC may not specifically state that timed approaches are in progress. However, if you are issued a time to depart the holding fix inbound, it means that timed approaches are being used. The holding fix may be the FAF on a nonprecision approach, while a precision approach may utilize the outer marker or a fix used in lieu of the outer marker for holding. [Figure 7-35]

Timed approaches from a holding fix may be conducted if the following conditions are met.

- A control tower is in operation at the airport where the approaches are conducted.
- Direct communication is maintained between you and the center or approach controller until you are instructed to contact the tower.
- If more than one missed approach procedure is available, none require course reversal.
- If only one missed approach procedure is available, course reversal is not required, and reported ceiling and visibility are equal to or greater than the highest prescribed circling minimums for the IAP.
- When cleared for the approach, you shall not execute a procedure turn.

10:29:00

Figure 7-34. Timed approaches from a holding fix are not performed at all airports. ATC initiates timed approaches only if certain conditions are met.

If more than one missed approach procedure is available, a timed approach from a holding fix may be conducted if none require a course reversal. See figure 7-34.

Timed approaches from a holding fix are only conducted at airports which have operating control towers. See figure 7-34.

If only one missed approach procedure is available, a timed approach from a holding fix may be conducted if the reported ceiling and visibility minimums are equal to or greater than the highest prescribed circling minimums for the IAP. See figure 7-34.

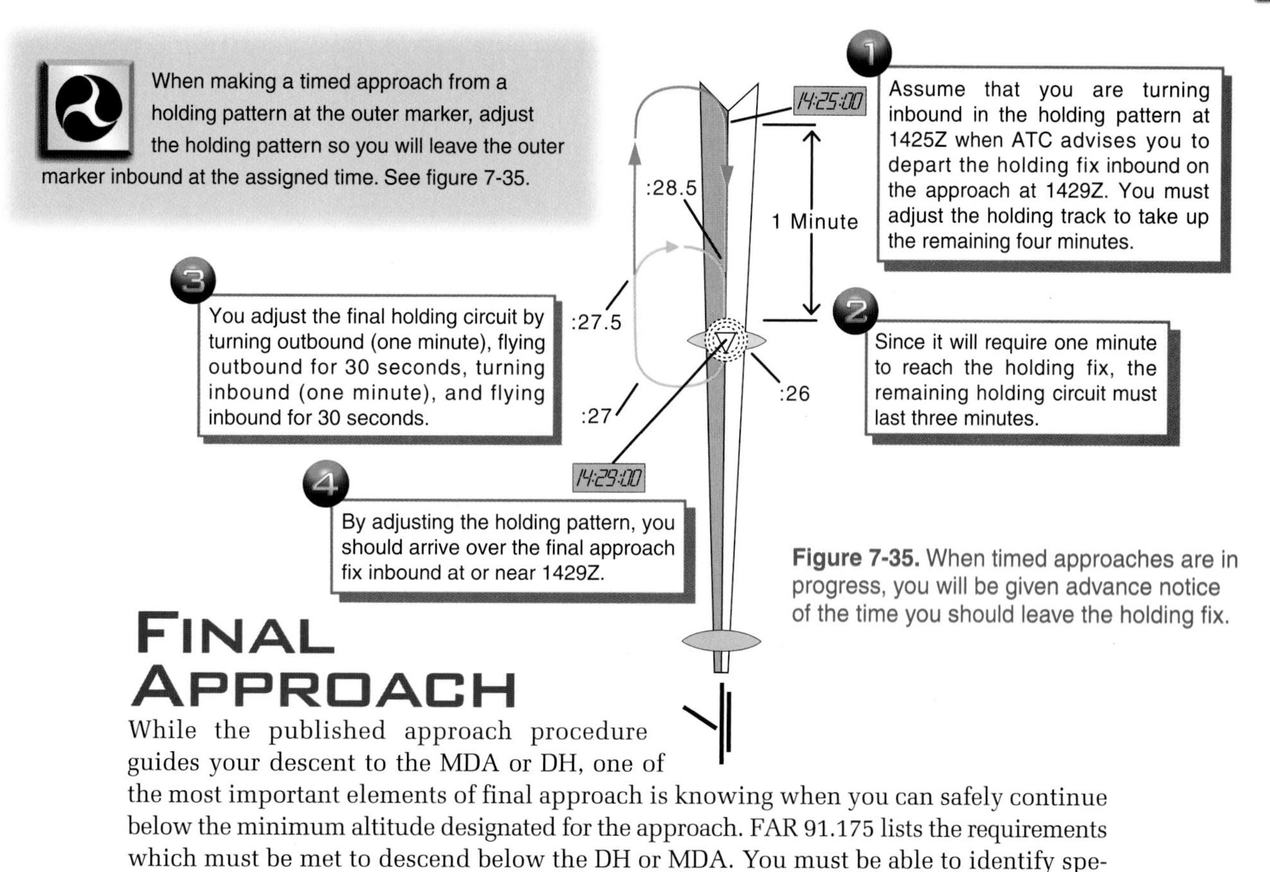

Figure 7-35. When timed approaches are in progress, you will be given advance notice of the time you should leave the holding fix.

FINAL APPROACH

While the published approach procedure guides your descent to the MDA or DH, one of the most important elements of final approach is knowing when you can safely continue below the minimum altitude designated for the approach. FAR 91.175 lists the requirements which must be met to descend below the DH or MDA. You must be able to identify specific visual references of the runway environment, as well as comply with visibility and operating requirements. [Figure 7-36]

Figure 7-36. If you are using the approach lights for reference, you may not descend lower than 100 feet above the touchdown zone elevation unless the red terminating bars or the red side row bars are also distinctly visible and identifiable. Red terminating bars and side row bars are used in ALSF-1 and ALSF-2 approach lighting systems. The red terminating bars located near the threshold pertain to ALSF-1 systems, and the red side row bars are included in the ALSF-2 system.

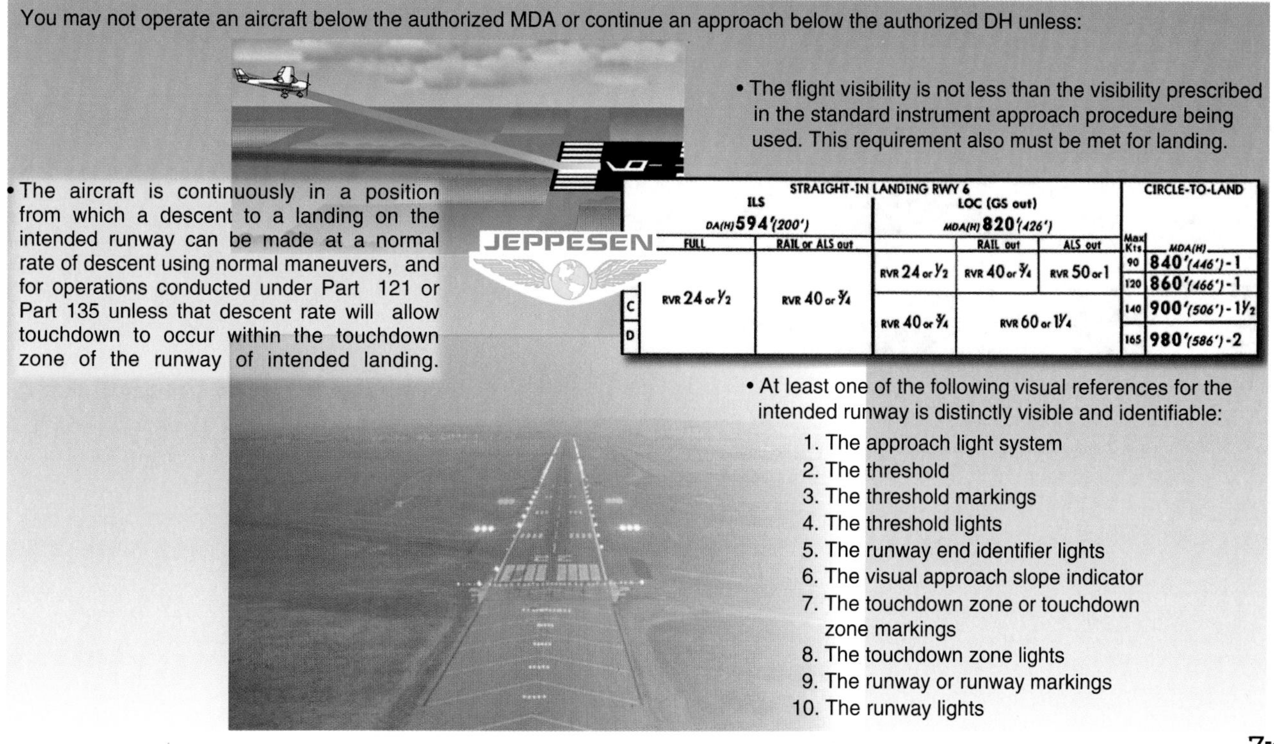

	STRAIGHT-IN LANDING RWY 6					CIRCLE-TO-LAND	
	ILS DA(H) 594'(200')		LOC (GS out) MDA(H) 820'(426')				
	FULL	RAIL or ALS out		RAIL out	ALS out	Max Kts	MDA(H)
	RVR 24 or ½	RVR 40 or ¾	RVR 24 or ½	RVR 40 or ¾	RVR 50 or 1	90	840'(446')-1
						120	860'(466')-1
C			RVR 40 or ¾	RVR 60 or 1¼		140	900'(506')-1½
D						165	980'(586')-2

The advantage of an ILS approach is that you can simply descend on the electronic glide slope to the decision height, and if you meet the requirements for descent below the DH, you are aligned with the runway at an appropriate altitude and position for executing a landing. If you are flying a nonprecision approach, you must plan your descent from the final approach fix to the MDA so the aircraft is in a position to land. [Figure 7-37]

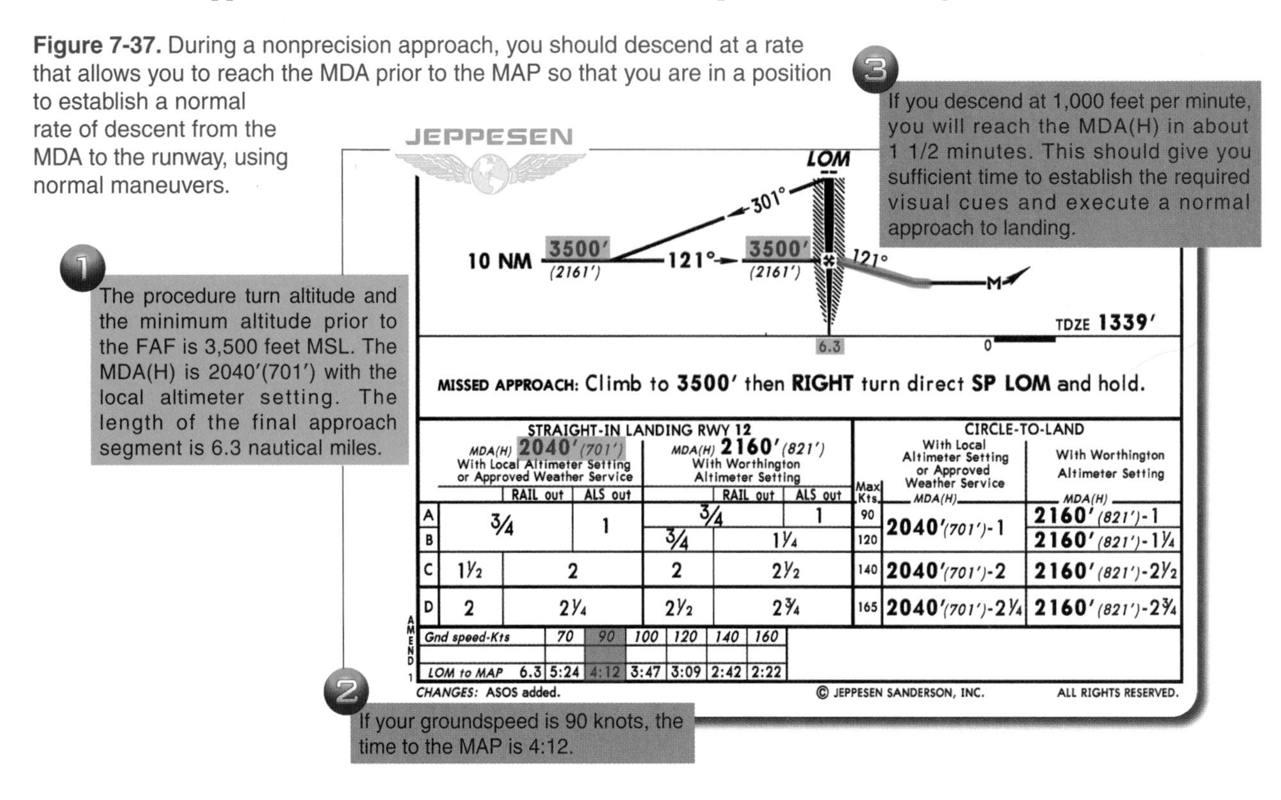

Figure 7-37. During a nonprecision approach, you should descend at a rate that allows you to reach the MDA prior to the MAP so that you are in a position to establish a normal rate of descent from the MDA to the runway, using normal maneuvers.

VASI lights can help you maintain the proper descent angle to the runway once you have established visual contact with the runway environment. If a glide slope malfunction occurs when you are in IFR conditions during an ILS approach, you must apply the localizer-only minimums (MDA) and report the malfunction to ATC. Once you have established visual references, you may continue the descent at or above the VASI glide path for the remainder of the approach. [Figure 7-38]

You are executing an ILS approach and are past the OM to a runway which has VASI. If the glide slope malfunction occurs, and you have the VASI in sight, you may continue the approach using the VASI glide slope in place of the electronic glideslope.

Figure 7-38. At airports with operating control towers, you must maintain a glide path at or above the VASI while approaching a runway served by a VASI installation.

Once you have reached visual conditions on the approach, you should be aware that certain **landing illusions** may be apparent as you approach the runway. These visual illusions are the product of various runway conditions, terrain features, and atmospheric

A Perfect Approach Plus a Bounced Landing Equals Success

At 4:35 a.m. local solar time (11:34 a.m. Pacific time) on July 4, 1997, the Mars Pathfinder lander sent its first radio transmission from the surface of the red planet. The celebrations that followed were well deserved for the Pathfinder had just completed the most challenging part of its mission — the approach and landing. During the five minutes in which the spacecraft transitioned from the enroute portion of its flight to the rigors of atmospheric entry, descent, and landing, more than 50 critical events had to occur at precisely the right times. Pathfinder embarked on its journey from Earth seven months earlier — its mission not only to study Mars but to open a new era of exploration by demonstrating the feasibility of developing and launching a low-cost spacecraft and mobile vehicle in a relatively short period of time (three and a half years).

Courtesy of NASA

Pathfinder's final approach began at an altitude of 130 kilometers and a speed of 27,000 kilometers per hour as it entered the atmosphere behind a protective aeroshell (including a heat shield). After a parachute was deployed to slow the lander, the heat shield separated and the lander was lowered beneath its backshell on a 20-meter-long bridle, or tether. To further decrease its speed as it neared the surface, Pathfinder's radar altimeter triggered the firing of three solid rocket motors contained in the backshell. Next, giant air bags inflated around each face of the lander and the bridle was cut. After a perfect approach, a smooth landing was not in Pathfinder's future. As planned, the ballooned-covered lander bounced at least 15 times before coming to rest on the Martian terrain. Surface operations began after the air bags were deflated and the petals of the tetrahedron-shaped lander were opened.

Cruise-Stage Separation
Entry
Parachute Deployment
Lander Separation/ Bridle Deployment
Heat-Shield Separation
Radar Ground Acquisition
Air-Bag Inflation
Rocket Ignition
Bridle Cut
Landing
Air-Bag Deflation
Opening of Petals

In addition to demonstrating an innovative approach and landing sequence, Pathfinder's mission exceeded expectations. Over 2 billion bits of new data was received from Mars, including 16,500 lander and 550 rover images and over 8 million temperature, pressure, and wind measurements. The rover Sojourner explored over 200 square meters and obtained measurements of rock and soil chemistry, as well as completed numerous soil-mechanics and technology experiments. While its scientific and technological accomplishments were impressive, perhaps Pathfinder's greatest achievement was to fuel the public's sense of adventure and ignite our passion for knowledge and exploration. The largest Internet event in history, the mission garnered 566 million hits in one month.

Courtesy of NASA

phenomena which can create the sensation of incorrect height above the runway or incorrect distance from the runway threshold. [Figure 7-39]

Situation	Illusion	Result
Upsloping Runway or Terrain	Greater Height	Lower Approaches
Narrower-Than-Usual Runway	Greater Height	Lower Approaches
Featureless Terrain	Greater Height	Lower Approaches
Rain on Windscreen	Greater Height	Lower Approaches
Haze	Greater Height	Lower Approaches
Downsloping Runway or Terrain	Less Height	Higher Approaches
Wider-Than-Usual Runway	Less Height	Higher Approaches
Bright Runway and Approach Lights	Less Distance	Higher Approaches
Penetration of Fog	Pitching Up	Steeper Approaches

Figure 7-39. You should pay particular attention to illusions which lead to a lower-than-normal final approach profile.

Due to a visual illusion, when landing on a narrower-than-usual runway, the aircraft will appear to be higher than actual, leading to a lower-than-normal approach. See figure 7-39.

An upsloping runway creates the same effect as a narrower-than-usual runway. See figure 7-39.

CIRCLING APPROACHES

Several situations may require you to execute a **circling approach**. For example, a circling approach is necessary if the instrument approach course is not aligned within 30° of the runway. At times, you may find that unfavorable wind or a runway closure makes a straight-in landing impractical. Circling minimums appropriate to each aircraft approach category are established in accordance with TERPs criteria. Each circling approach is confined to a protected area which is defined by TERPs and also published in the AIM. The size of this area varies with aircraft approach category. Only if you remain within the protected area are you assured obstacle clearance at the MDA during circling maneuvers. In addition, you must remain at or above the circling MDA unless the aircraft is continuously in a position from which a descent to a landing on the intended runway can be made, using a normal rate of descent and normal maneuvering. [Figure 7-40]

As you know, approach categories are based on 1.3 times the stall speed in the landing configuration. When executing a circling approach, just as you do during a straight-in landing, you must use the appropriate approach category minimums for your aircraft. However, if you operate at a higher speed than is designated for your aircraft approach category, you should use the minimums of the next higher category. [Figure 7-41]

The circling approach protected area is established by the connection of arcs drawn from each runway end. The radii (r) which define the size of the areas vary with the approach category. As the approach speeds increase, the turn radii increase which often results in higher circling MDAs.

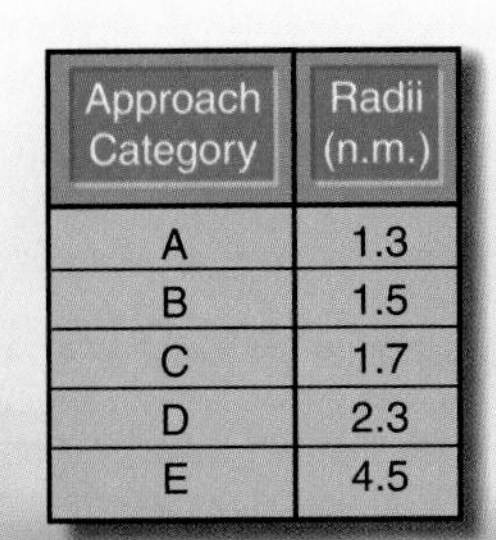

Approach Category	Radii (n.m.)
A	1.3
B	1.5
C	1.7
D	2.3
E	4.5

JEPPESEN

CIRCLE-TO-LAND	
Max Kts	MDA(H)
90	6260′(501′)-1
120	6280′(521′)-1
140	6700′(941′)-2¾
165	6740′(981′)-3

If obstacles are present within the protected area, a procedural note may be added which prohibits circling within a portion of that area.

CIRCLE-TO-LAND
Not Authorized North of Rwy 10-28

"Cessna 52241, cleared VOR Runway 1 approach, circle to land Runway 28."

If you are circling to land in a Category A aircraft at the appropriate MDA, you must remain within 1.3 nautical miles from the ends of the runways, even if the visibility minimum for the approach is greater than this distance. For example, the Category A aircraft circle-to-land visibility minimum for this approach is 2 3/4 statute miles, or approximately 2.4 nautical miles. Higher visibility minimums do not constitute authorization to leave the protected area while circling to land.

CATEGORY	A	B	C	D
CIRCLING	9180-2¾ 1507 (1600-2¾)		9180-3 1507 (1600-3)	NA

Figure 7-40. During a circling approach, you are provided obstacle clearance at the MDA as long as you maneuver within the protected area.

$1.3\ V_{SO}$ places this airplane in Category B.

1.3 x 72 knots = 94 knots

Approach Speed	Approach Category
A	Up to 90 knots
B	91 to 120 knots
C	121 to 140 knots
D	141 to 165 knots
E	Above 165 knots

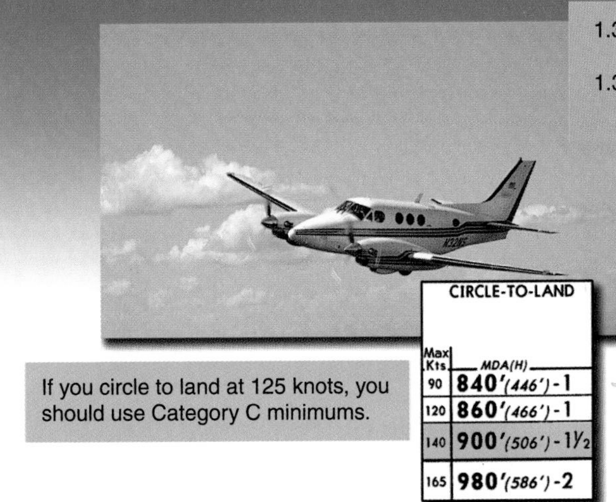

CIRCLE-TO-LAND	
Max Kts	MDA(H)
90	840′(446′)-1
120	860′(466′)-1
140	900′(506′)-1½
165	980′(586′)-2

JEPPESEN

If you circle to land at 125 knots, you should use Category C minimums.

Figure 7-41. If you are circling in a Category B aircraft, but you are operating at a speed above the Category B speed limit, you should use the MDA and visibility requirement appropriate to Category C.

In simple terms, the circling approach procedure involves flying the approach, establishing visual contact with the airport, and positioning the aircraft on final approach to the runway of intended landing. However, the circling approach is not a simple maneuver, since you are required to fly at a low altitude at a fairly slow airspeed while remaining within a specifically defined area. Remember that you fly the circling approach at or above the MDA, and you cannot descend from the MDA until the airplane is properly positioned to make a normal descent to the landing runway. [Figure 7-42]

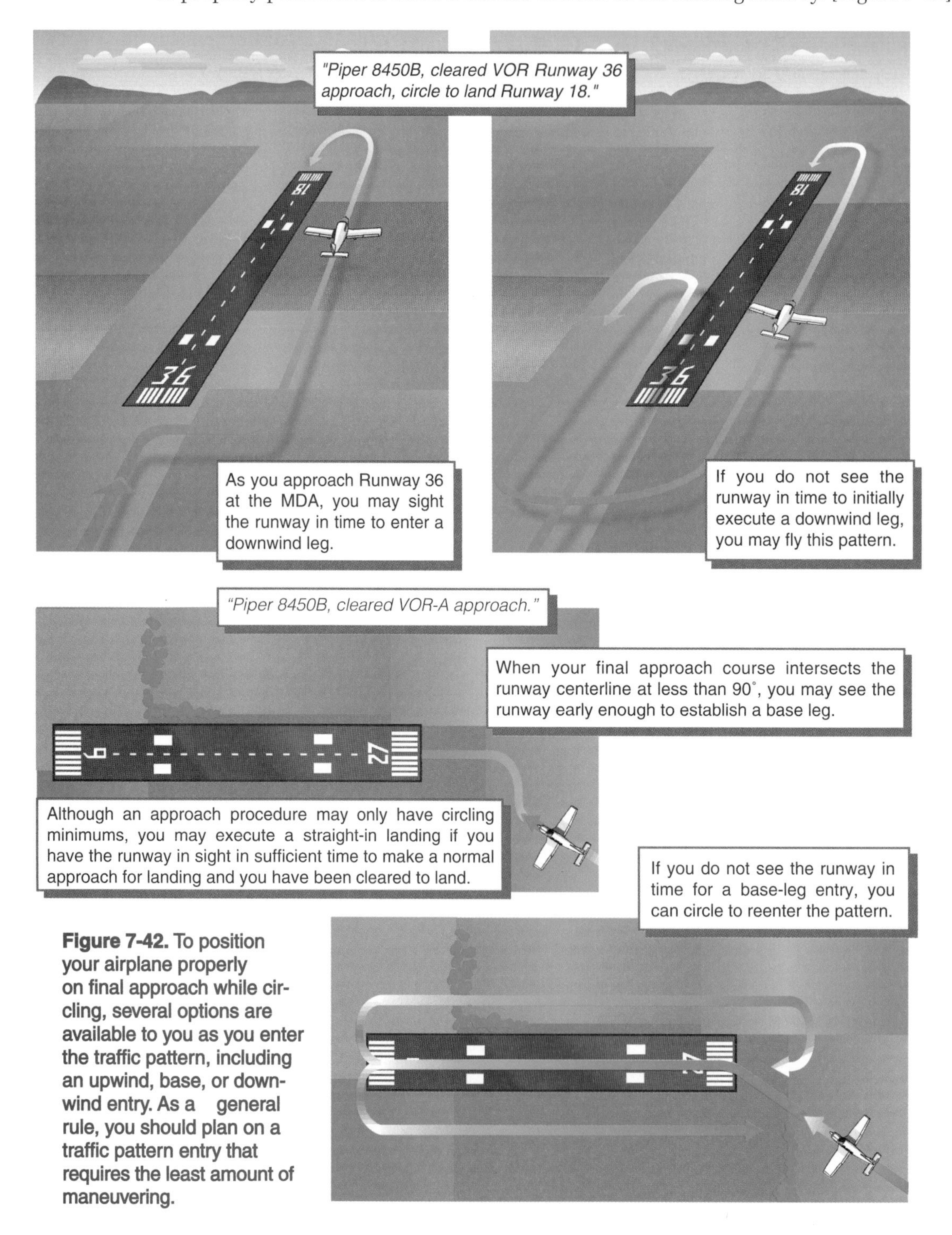

Figure 7-42. To position your airplane properly on final approach while circling, several options are available to you as you enter the traffic pattern, including an upwind, base, or downwind entry. As a general rule, you should plan on a traffic pattern entry that requires the least amount of maneuvering.

You should be very cautious when making circling approaches since they can be extremely hazardous when combined with such factors as low visibility, hilly or

mountainous terrain, and/or night operations. You must remain within the protected area for your aircraft approach category while circling, and once you descend below the MDA, obstacle clearance is your own responsibility. If conditions prevent you from seeing well enough to guarantee obstruction clearance during the final descent, you should not continue the approach. Many accidents have occurred during circling approaches because the pilot could see lights in the vicinity but could not see hilly or mountainous terrain between the aircraft and the airport.

You may execute a straight-in landing if the IAP has only circling minimums if you have the runway in sight in sufficient time to make a normal approach for landing and you have been cleared to land. See figure 7-42.

SIDESTEP MANEUVER

Under certain conditions, your approach clearance may include a **sidestep maneuver**. At some airports where there are two parallel runways that are 1,200 feet or less apart, you may be cleared to execute an approach to one runway followed by a straight-in landing on the adjacent runway. [Figure 7-43]

When cleared to execute a published sidestep maneuver for a landing on a parallel runway, you should commence the maneuver as soon as possible after the runway or runway environment is in sight.

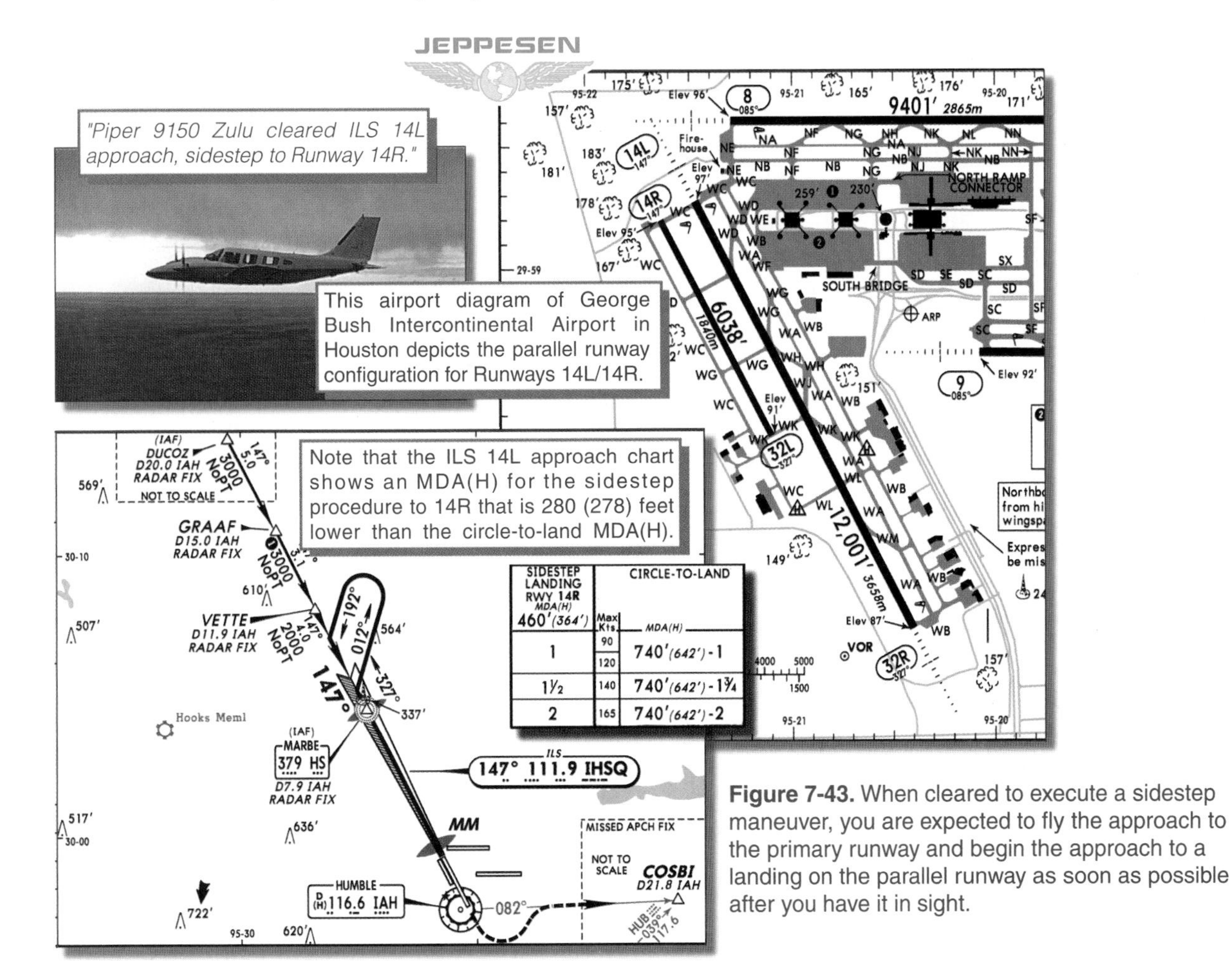

Figure 7-43. When cleared to execute a sidestep maneuver, you are expected to fly the approach to the primary runway and begin the approach to a landing on the parallel runway as soon as possible after you have it in sight.

MISSED APPROACH PROCEDURES

The most common reason for a missed approach is low visibility conditions that do not permit you to establish required visual cues. [Figure 7-44] A published missed approach procedure is carefully designed and flight tested so you will have adequate obstacle clearance throughout the missed approach segment. Each procedure is unique to the airport and to the particular approach. Depending on obstacles and surrounding terrain, a missed approach segment may designate a straight climb, a climbing turn, or a climb to a specified altitude, followed by a turn to a specified heading, navaid, or navigation fix. [Figure 7-45]

Figure 7-44. Several situations may require you to execute a missed approach.

Regardless of the reason for a missed approach, it is very important that you can maneuver the aircraft safely throughout the missed approach segment. This is generally not difficult when you begin the missed approach at the missed approach point. In this situation, you simply fly the procedure as described and depicted on the approach chart. On occasion, however, you will be required to initiate a missed approach from a position that is not at the missed approach point and may not be on the missed approach segment. For example, assume you decide to make a missed approach before reaching the MAP. In this situation, remember that no consideration for an abnormally early turn is given in the procedure design. If the procedure you are flying requires a turn, you must fly the approach to the MAP before turning. There is no prohibition against climbing early, but you must delay the turn until you are over the MAP.

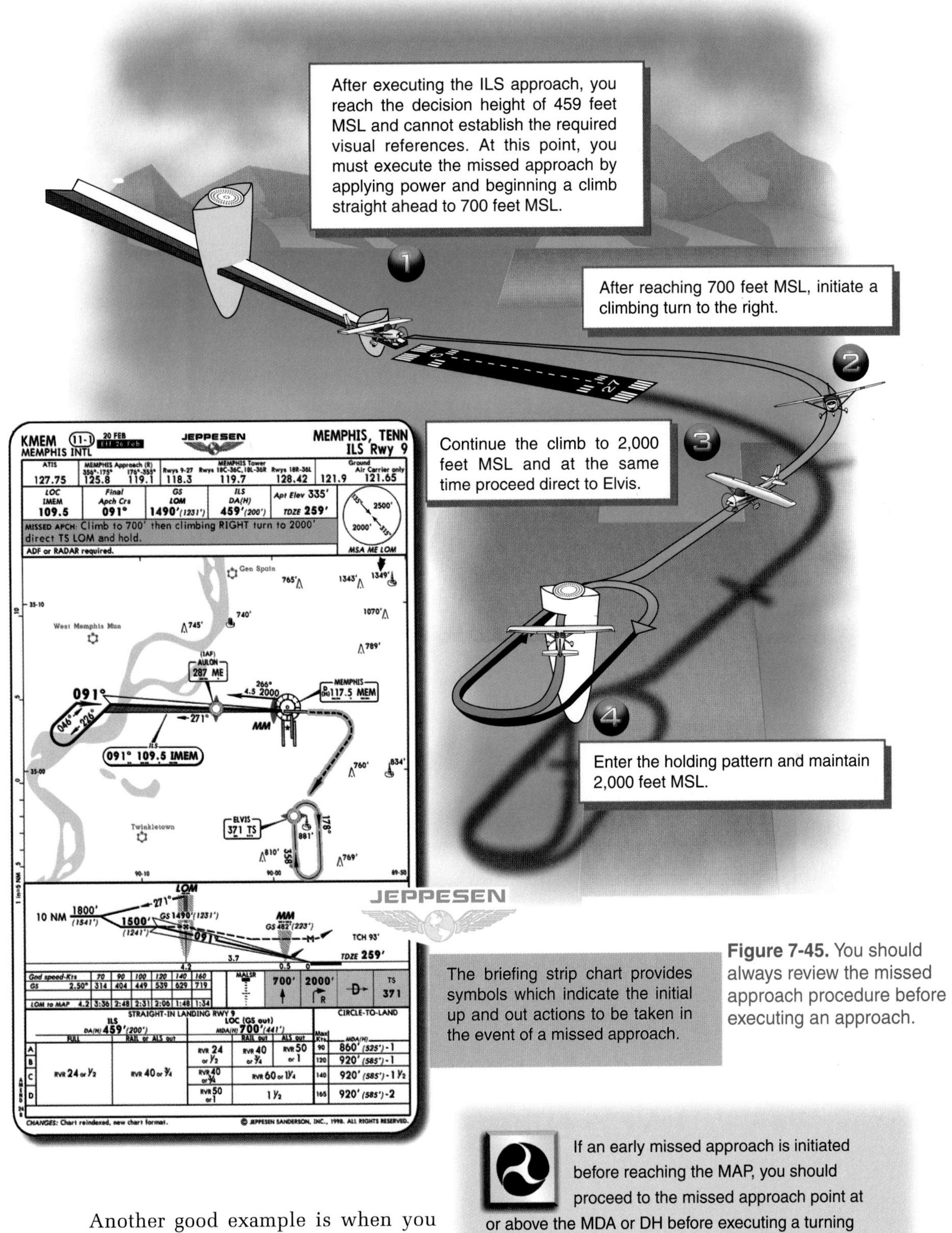

Figure 7-45. You should always review the missed approach procedure before executing an approach.

If an early missed approach is initiated before reaching the MAP, you should proceed to the missed approach point at or above the MDA or DH before executing a turning maneuver.

Another good example is when you are executing a circling maneuver and suddenly lose sight of the runway. According to the AIM, you should make an initial climbing turn toward the landing runway to become established on the missed approach course. The airspace over the airport affords you the greatest obstacle clearance protection. Since the missed approach point for a nonprecision approach usually is the runway threshold, a turn toward the runway will keep you over the airport and may position you very close to the actual missed approach point. [Figure 7-46]

If you lose visual reference while circling to land from an instrument approach and ATC radar service is not available, you should initiate a missed approach by making a climbing turn toward the landing runway and continue the turn until established on the missed approach course.

Figure 7-46. Since a circling maneuver may be accomplished in more than one direction, different patterns are required to become established on the missed approach course. The one to use depends on the position of the aircraft at the time visual reference is lost. These patterns are shown in the AIM and are intended to ensure that an aircraft will remain within the circling and missed approach obstruction clearance areas.

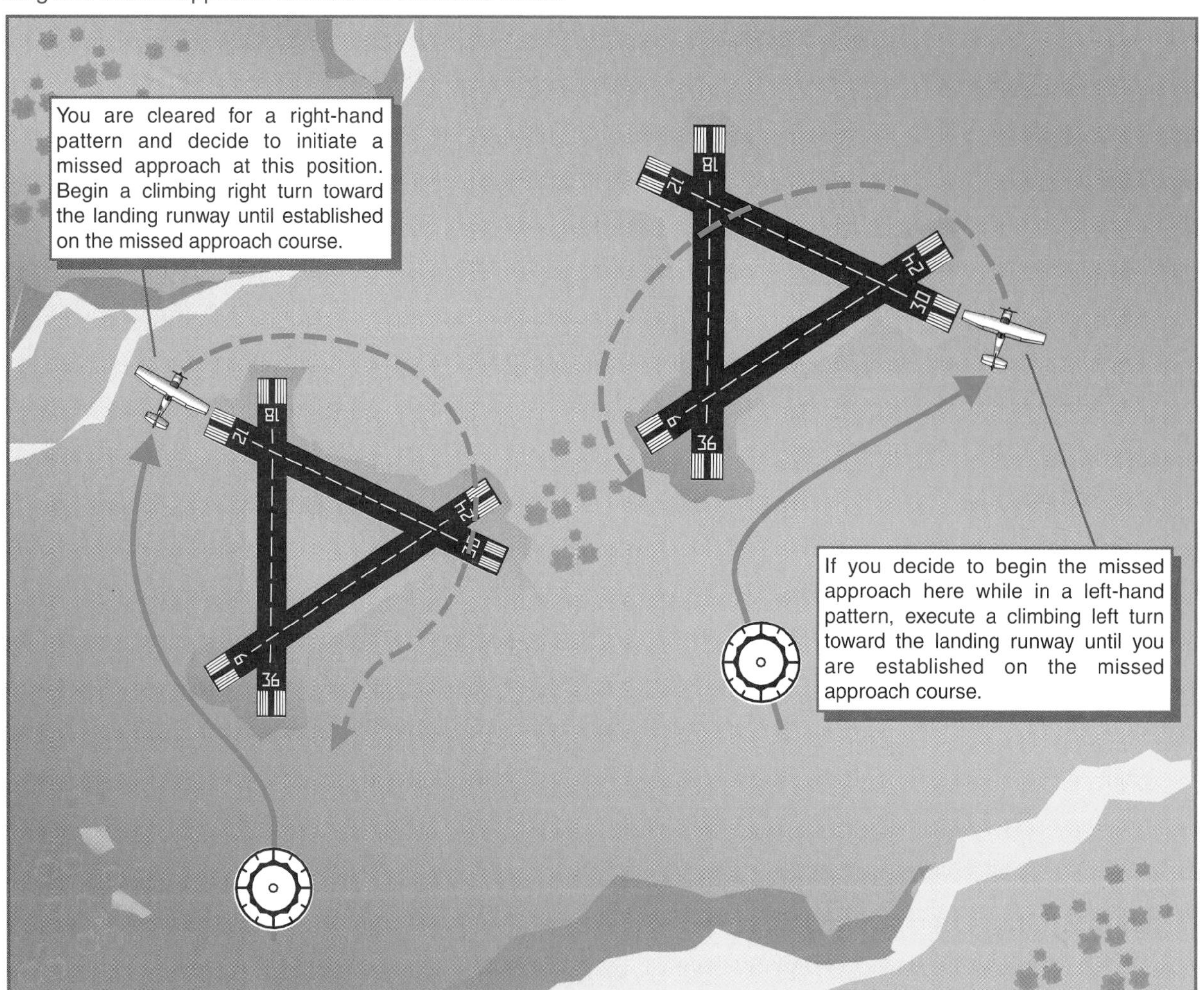

VISUAL AND CONTACT APPROACHES

To expedite traffic, ATC may clear you for a **visual approach** in lieu of the published approach procedure if flight conditions permit. Requesting a **contact approach** may be to your advantage since it requires less time than the published instrument procedure, allows you to retain your IFR clearance, and provides separation from IFR and special VFR traffic. Visual and contact approaches were introduced in Chapter 3, Section C — ATC Clearances. [Figure 7-47]

Charted visual flight procedures (CVFPs) may be established at some controlled airports for environmental or noise considerations, as well as when necessary for the safety and

efficiency of air traffic operations. Designed primarily for turbojet aircraft, CVFPs depict prominent landmarks, courses, and recommended altitudes to specific runways.

Visual Approach

- A visual approach can be initiated by ATC or you may request the approach.
- The reported ceiling must be at least 1,000 feet AGL and visibility must be at least 3 statute miles.
- At airports without an operating control tower and no weather reporting facility, ATC will authorize a visual approach only if there is a reasonable assurance that the ceiling is at least 1,000 feet AGL and the visibility is 3 statute miles.
- You must have the airport or preceding aircraft in sight.
- If you report the preceding aircraft in sight, you are responsible for maintaining separation from that aircraft and avoiding the associated wake turbulence.
- In the event you have the airport in sight, but do not see the aircraft you are following on the approach, ATC is responsible for maintaining aircraft and wake turbulence separation.
- You must remain clear of the clouds at all times while conducting a visual approach.

Figure 7-47. Pilot and controller responsibilities differ significantly between visual and contact approaches.

Contact Approach

- ATC cannot initiate a contact approach.
- ATC can issue a clearance for a contact approach upon your request when the reported ground visibility at the airport is 1 statute mile or greater.
- Contact approaches are issued only for airports with published approach procedures (public or private).
- During a contact approach you are responsible for your own obstruction clearance, but ATC provides separation from other IFR or special VFR traffic. Separation from normal VFR traffic is not provided.
- You must maintain 1 statute mile flight visibility, and remain clear of clouds while flying the approach.

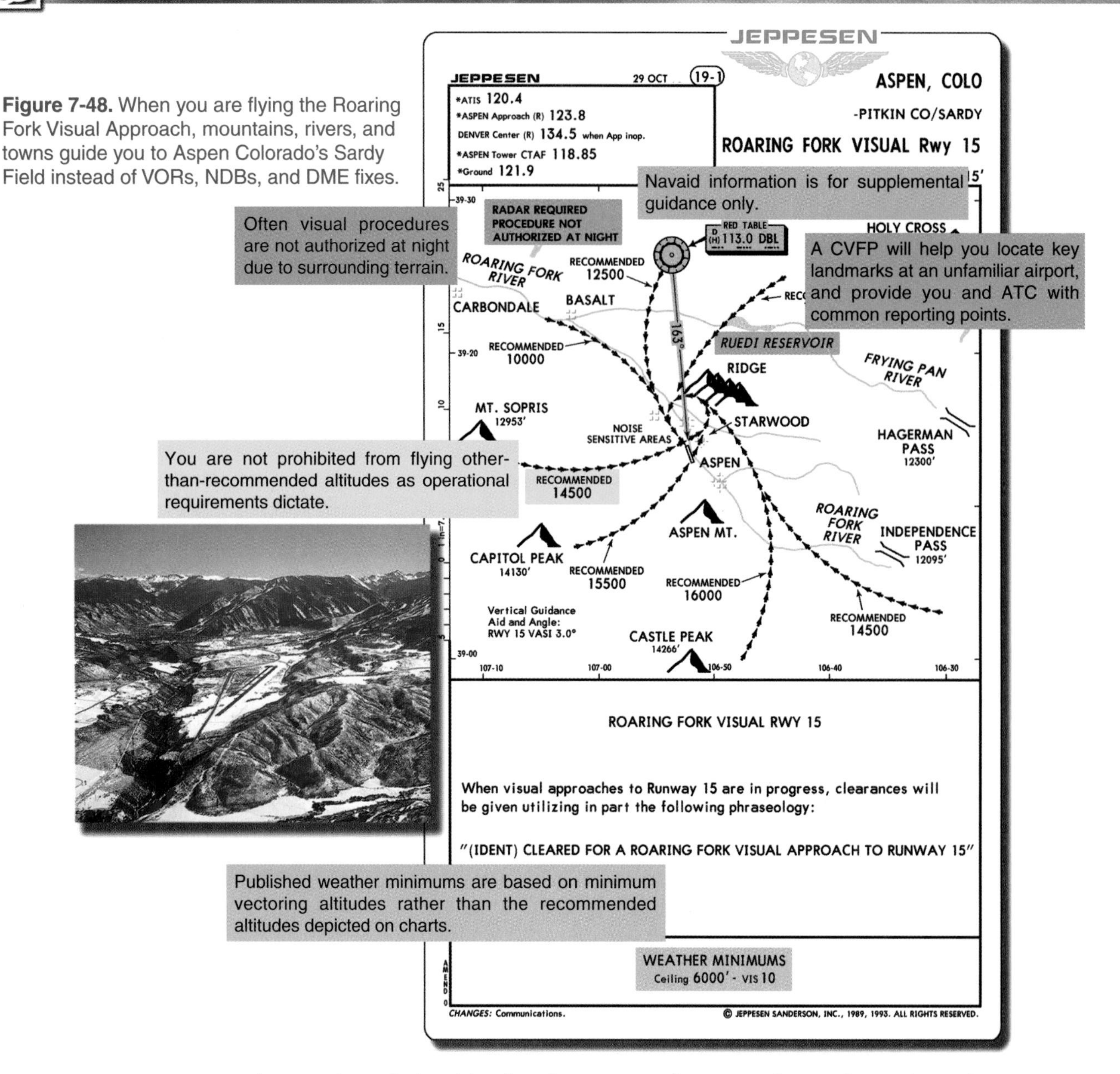

Figure 7-48. When you are flying the Roaring Fork Visual Approach, mountains, rivers, and towns guide you to Aspen Colorado's Sardy Field instead of VORs, NDBs, and DME fixes.

You must have a charted visual landmark or a preceding aircraft in sight, and weather must be at or above the published minimums before ATC will clear you for a CVFP. When instructed to follow a preceding aircraft, you are responsible for maintaining a safe approach interval and wake turbulence separation. You should advise ATC if at any point you are unable to continue a charted visual approach or lose sight of a preceding aircraft. [Figure 7-48]

SUMMARY CHECKLIST

- ✓ After you have been advised as to which approach to expect, you should conduct a thorough approach chart review to familiarize yourself with the specific approach procedure.
- ✓ If ATC does not specify a particular approach but states, *"cleared for approach,"* you may execute any one of the authorized IAPs for that airport.
- ✓ Feeder routes provide a transition from the enroute structure to the IAF or to a facility from which a course reversal is initiated.

- ✓ The terms straight-in approach and straight-in landing have specific definitions when used in ATC clearances or in reference to landing minimums.
- ✓ Although most approach procedures provide landing minimums for both straight-in and circling maneuvers, some may be limited to circling only.
- ✓ A straight-in approach may be initiated from a fix closely aligned with the final approach course, may commence from the completion of a DME arc, or you may receive vectors to the final approach course.
- ✓ A straight-in approach does not require nor authorize a procedure turn or course reversal.
- ✓ A NoPT arrival sector allows flights inbound on Victor airways within the sector to proceed straight in on the final approach course.
- ✓ ATC radar approved for approach control service is used for course guidance to the final approach course, ASR and PAR approaches, and the monitoring of nonradar approaches.
- ✓ Radar vectors to the final approach course provide a method of intercepting and proceeding inbound on the published instrument approach procedure.
- ✓ During an instrument approach procedure, a published course reversal is not required when radar vectors are provided.
- ✓ If it becomes apparent the heading assigned by ATC will cause you to pass through the final approach course, you should maintain that heading and question the controller.
- ✓ A course reversal may be depicted on a chart as a procedure turn, a racetrack pattern (holding pattern), or a teardrop procedure. The maximum speed in a course reversal is 200 knots IAS.
- ✓ When more than one circuit of a holding pattern is needed to lose altitude or become better established on course, the additional circuits can be made only if you advise ATC and ATC approves.
- ✓ Timed approaches from a holding fix are generally conducted at airports where the radar system for traffic sequencing is out of service or is not available and numerous aircraft are waiting for approach clearance.
- ✓ When timed approaches are in progress, you will be given advance notice of the time you should leave the holding fix.
- ✓ To descend below the DH or MDA, you must be able to identify specific visual references, as well as comply with visibility and operating requirements which are listed in FAR 91.175
- ✓ You should usually descend at a rate that allows you to reach the MDA prior to the MAP so that you are in a position to establish a normal rate of descent from the MDA to the runway, using normal maneuvers.
- ✓ VASI lights can help you maintain the proper descent angle to the runway once you have established visual contact with the runway environment.
- ✓ Visual illusions are the product of various runway conditions, terrain features, and atmospheric phenomena which can create the sensation of incorrect height above the runway or incorrect distance from the runway threshold.

- ✓ A circling approach is necessary if the instrument approach course is not aligned within 30° of the runway. In addition, you may find that an unfavorable wind or a runway closure makes a circling approach necessary.
- ✓ Each circling approach is confined to a protected area which varies with aircraft approach category.
- ✓ When executing a circling approach, if you operate at a higher speed than is designated for your aircraft approach category, you should use the minimums of the next higher category.
- ✓ When cleared to execute a sidestep maneuver, you are expected to fly the approach to the primary runway and begin the approach to a landing on the parallel runway as soon as possible after you have it in sight.
- ✓ The most common reason for a missed approach is low visibility conditions that do not permit you to establish required visual cues.
- ✓ If an early missed approach is initiated before reaching the MAP, you should proceed to the missed approach point at or above the MDA or DH before executing a turning maneuver.
- ✓ If you lose visual reference while circling to land from an instrument approach and ATC radar service is not available, you should initiate a missed approach by making a climbing turn toward the landing runway and continue the turn until established on the missed approach course.
- ✓ If the ceiling is at least 1,000 feet AGL and visibility is at least 3 statute miles, ATC may clear you for a visual approach in lieu of the published approach procedure.
- ✓ ATC can issue a clearance for a contact approach upon your request when the reported ground visibility at the airport is 1 statute mile or greater. ATC cannot initiate a contact approach.
- ✓ Charted Visual Flight Procedures (CVFPs) may be established at some controlled airports for environmental or noise considerations, as well as when necessary for the safety and efficiency of air traffic operations.

KEY TERMS

Approach Chart Review

Feeder Route

Straight-In Landing

Straight-In Approach

Radar Vectors

Minimum Vectoring Altitude (MVA)

Course Reversal

Timed Approaches From a Holding Fix

Landing Illusions

Circling Approach

Sidestep Maneuver

Missed Approach

Visual Approach

Contact Approach

Charted Visual Flight Procedure (CVFP)

QUESTIONS

1. As a minimum, what items should be included in the approach chart review?
2. True/False. A feeder route provides a transition from the enroute structure to the IAF or to a facility from which a course reversal is initiated.
3. Explain the difference between the terms straight-in approach and straight-in landing.

Refer to the ILS Runway 34 instrument approach chart for Easterwood Airport to answer questions 4 through 9.

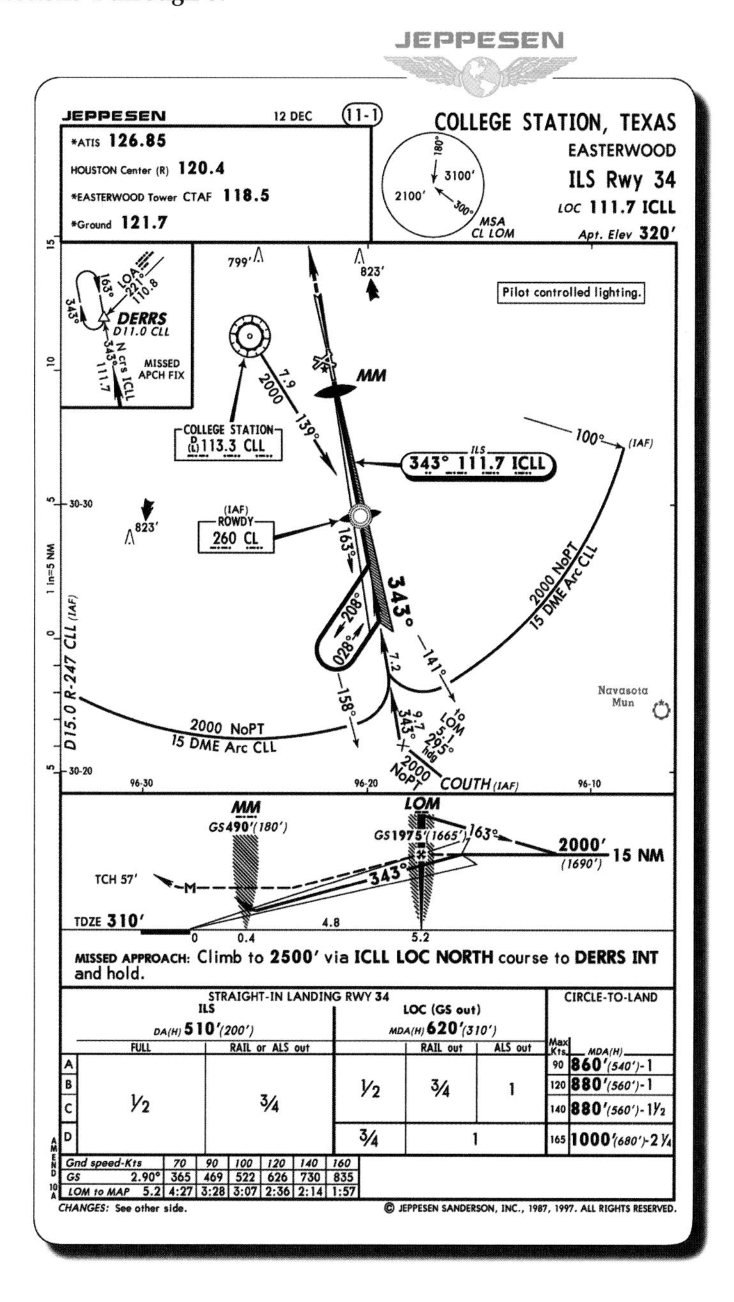

	ILS FULL	ILS RAIL or ALS out	LOC	LOC RAIL out	LOC ALS out	Max Kts	MDA(H)
A	1/2	3/4	1/2	3/4	1	90	860'(540')-1
B						120	880'(560')-1
C						140	880'(560')-1 1/2
D			3/4	1		165	1000'(680')-2 1/4

Gnd speed-Kts	70	90	100	120	140	160
GS 2.90°	365	469	522	626	730	835
LOM to MAP 5.2	4:27	3:28	3:07	2:36	2:14	1:57

4. What magnetic course and minimum altitude applies to the feeder route depicted on this approach chart?

5. A straight-in approach can be executed from which initial approach fix?

   A. Rowdy LOM
   B. COUTH Intersection
   C. The intersection of the 15 DME arc and the 141° radial from College Station VORTAC

6. What are the functions of the 158° radial and the 141° radial from College Station VORTAC?

7. True/False. You must execute the procedure turn exactly as it is depicted on the chart.

8. To remain in protected airspace during the procedure turn, you must not fly beyond how many nautical miles from Rowdy LOM?

   A. 7.2 nautical miles
   B. 10 nautical miles
   C. 15 nautical miles

9. If you are circling to land at 100 knots in a Category A aircraft, what is your MDA(H) and visibility requirement?

10. What is the maximum airspeed you can use when executing a course reversal?

11. What conditions must be met for timed approaches from a holding fix to be conducted at an airport?

    A. A control tower must be in operation.
    B. Course reversal is required for the missed approach procedure.
    C. The ceiling must be at least 1,000 feet AGL and visibility must be at least 3 statute miles.

12. The flight visibility is greater than the visibility prescribed in the standard instrument approach procedure being used; however, at the MDA you cannot identify any of the runway references described in FAR 91.175. Can you descend below the MDA for landing?

13. True/False. When landing on a narrower-than-usual runway, the tendency is to fly an approach which is too high.

14. If you made a decision to execute a missed approach prior to the MAP and the procedure specifies a climbing right turn to 5,000 feet, what action should you take?

    A. Immediately initiate a climbing right turn.
    B. Climb straight ahead to 5,000 feet, then initiate a climbing right turn.
    C. Proceed to the MAP at or above the MDA or DH, then initiate a climbing right turn to 5,000 feet.

15. Explain the difference between a visual approach and a contact approach.

CHAPTER 8

INSTRUMENT APPROACHES

Instrument/Commercial
Part II, Segment 2, Chapter 8 — VOR and NDB Approaches

SECTION A

VOR AND NDB APPROACHES

In Chapter 7, you learned about instrument approach charts and procedures. Now you will become oriented to different types of approach procedures beginning with nonprecision approaches. VOR approaches typically are easier to fly, but are less standardized than precision approach procedures. NDB approaches also are basic procedures, once you are comfortable interpreting the ADF or RMI indications.

There are two basic types of VOR or NDB approaches — those that use a navaid located beyond the airport boundaries and those with the navaid located on the airport. You can easily determine whether the procedure uses an off-airport facility or on-airport facility by looking at the approach chart profile. [Figure 8-1]

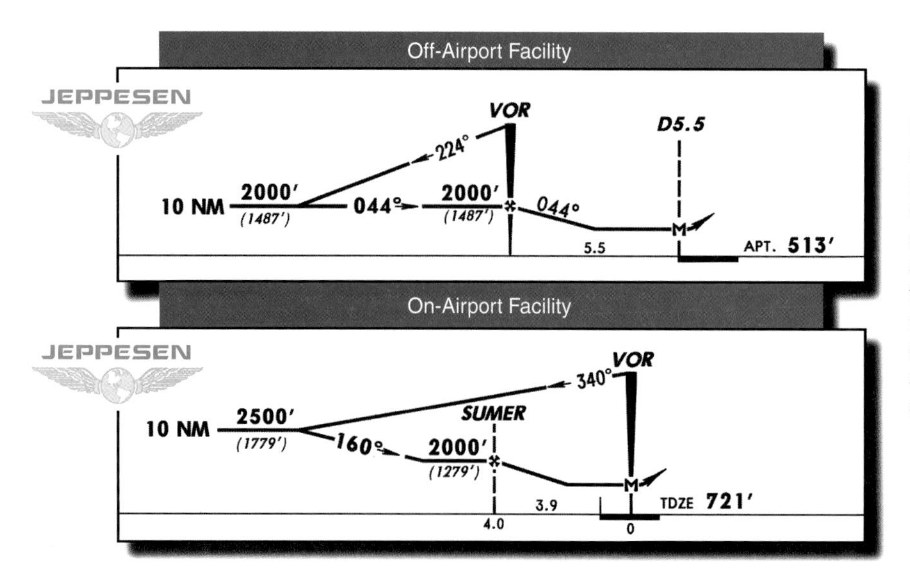

Figure 8-1. These approach chart profiles show the basic differences in approaches with off-airport and on-airport facilities. When the navaid is not located on the airport, it often serves as both the IAF and FAF.

VOR APPROACHES

VOR approaches provide for final descents as low as 250 feet above the runway. However, the MDAs for VOR approach procedures typically range from 500 to 1,000 feet above touchdown zone elevation. This section presents examples of approaches that use off-airport and on-airport VORs. DME is not required on the first two examples, since DME does not appear in either procedure title. However, many approaches utilize fixes which can be more easily identified with DME, even though it is not required to fly the procedure.

OFF-AIRPORT FACILITY

Kalispell, Montana's Glacier Park International Airport (KFCA) is in a mountainous area, and requires carefully designed and executed approach procedures. Assume you are approaching from the southwest on V536. [Figure 8-2]

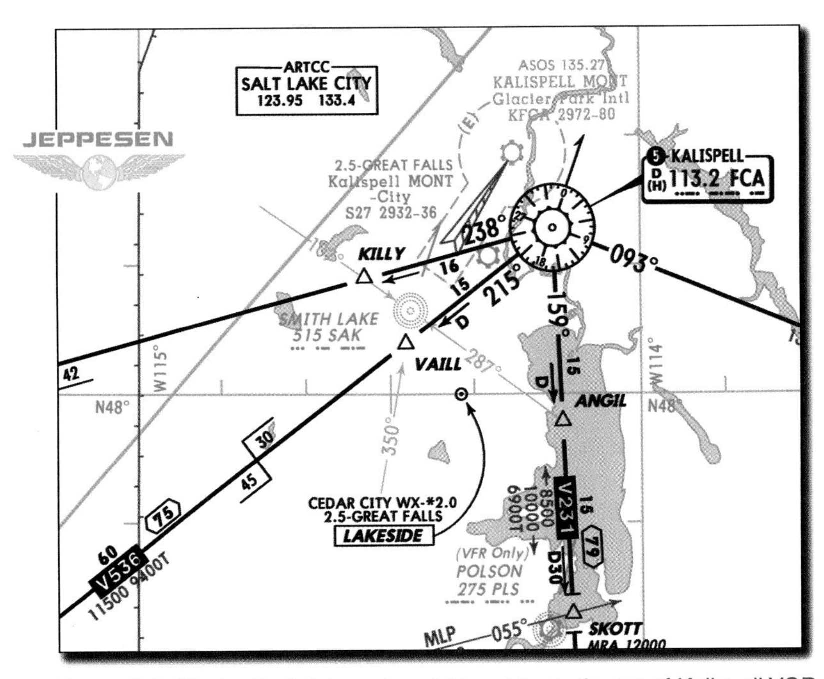

Figure 8-2. Glacier Park International Airport is northwest of Kalispell VOR.

YOUR CLEARANCE

Salt Lake Center has given you information from the last METAR. The surface winds were reported from the northwest at 20 gusting to 30 knots. The ceiling was last reported at 600 broken with 2 miles visibility in rain showers. ATC gives you the altimeter setting, and clears you as follows: *"Bonanza 1523S, cleared for approach, Glacier Park International. Change to advisory frequency approved."* The winds clearly favor landing on Runway 30, so you choose the VOR or GPS Rwy 30 approach and begin a review of the approach procedure.

PREPARING FOR THE APPROACH

Prior to conducting an approach, you need to conduct a chart review to familiarize yourself with certain procedural information. During your approach chart review, you should check for any unusual procedures. [Figure 8-3]

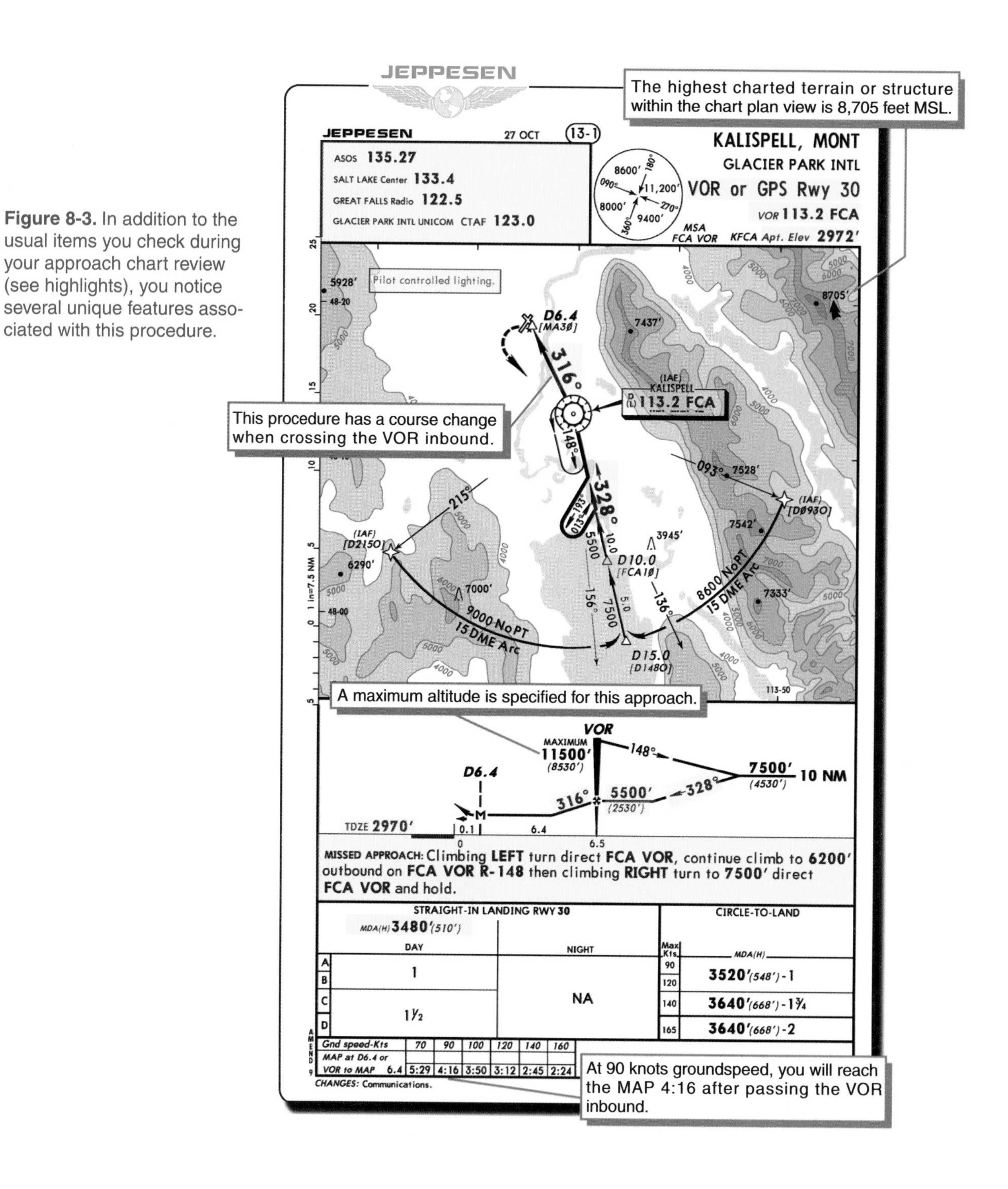

Figure 8-3. In addition to the usual items you check during your approach chart review (see highlights), you notice several unique features associated with this procedure.

It is also a good practice to check the associated airport diagram for each procedure. [Figure 8-4] Runway 30 is 3,500 feet long, with 80 to 90-foot trees near the approach end. There is a PAPI on the left side of the runway that can be used for a stabilized descent once visual contact is established.

Upon monitoring ASOS, you learn the surface weather conditions have not changed since the last hourly report. UNICOM confirms that traffic has been landing on Runway 30. You plan a straight-in approach and a challenging landing with gusty winds.

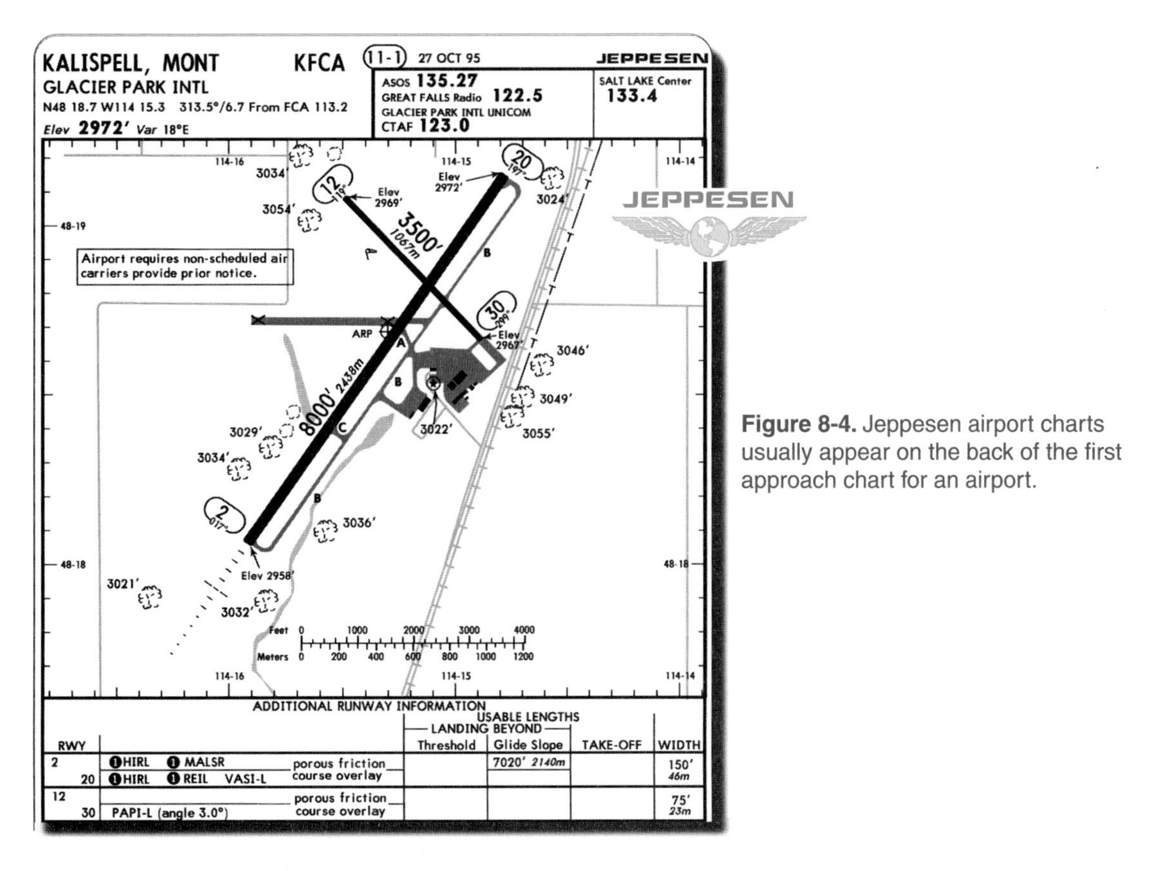

Figure 8-4. Jeppesen airport charts usually appear on the back of the first approach chart for an airport.

In spite of the strong surface winds, the forecast and observed winds aloft are relatively light, so no special adjustment of approach speed or rate of descent is needed. With a 600-foot ceiling, you expect to descend to minimums just after breaking out of the clouds. Because you have some concern about whether you can complete the approach successfully, make sure you understand the missed approach procedure and are prepared to fly it.

Since you are tracking toward Kalispell VOR on the airway, your VOR receiver is already set to the correct frequency and the station has previously been identified. You may wish to listen to the identifier again to confirm the VOR is set correctly and the ground facility still is operating properly.

DESCENDING PRIOR TO THE IAF

Your clearance allows you to begin your descent as appropriate to conduct your approach. The MEA on V536 approaching Kalispell is 11,500 feet MSL, which also is the maximum altitude for initiating the approach procedure. While you could approach the VOR at 11,500 feet MSL, a descent rate of nearly 1,000 feet per minute would be required to reach 7,500 feet MSL during the 4 to 5 minutes it normally takes to fly outbound and complete the procedure turn. You might prefer to start the approach from a lower altitude if one is authorized.

Notice there is a MOCA of 9,400 feet MSL published for this airway route segment, which guarantees terrain clearance. However, you are not guaranteed a reliable navigation signal below the 11,500-foot MEA until you are within 22 nautical miles of Kalispell VOR, and you may not begin a descent to the MOCA until that point. VAILL Intersection is 15 nautical miles from Kalispell VOR and can be positively identified if you have an ADF receiver. Beginning your descent here will easily get you down to 9,400 feet MSL prior to crossing the VOR.

When cleared for the VOR Rwy 30 approach to Kalispell [Figure 8-3] with no altitude restrictions, you can descend to the MOCA within 22 nautical miles of the VOR, and to the minimum altitude shown for each route segment on the approach procedure.

Outbound on the Procedure Turn

As you cross Kalispell VOR, promptly turn right to intercept the 148° radial outbound, and note the time. Keeping track of your time will help you remain within 10 nautical miles of the VOR, as required by the note in the profile. Throttle back and trim as needed to begin a 500- to 750-foot-per-minute descent at approach speed to 7,500 feet MSL. Due to the large difference in magnetic course between V536 and the outbound approach radial, a 45° intercept angle is appropriate. Turn right to 193° and twist the OBS to 148°. As the CDI begins to center, start a left turn to 148°.

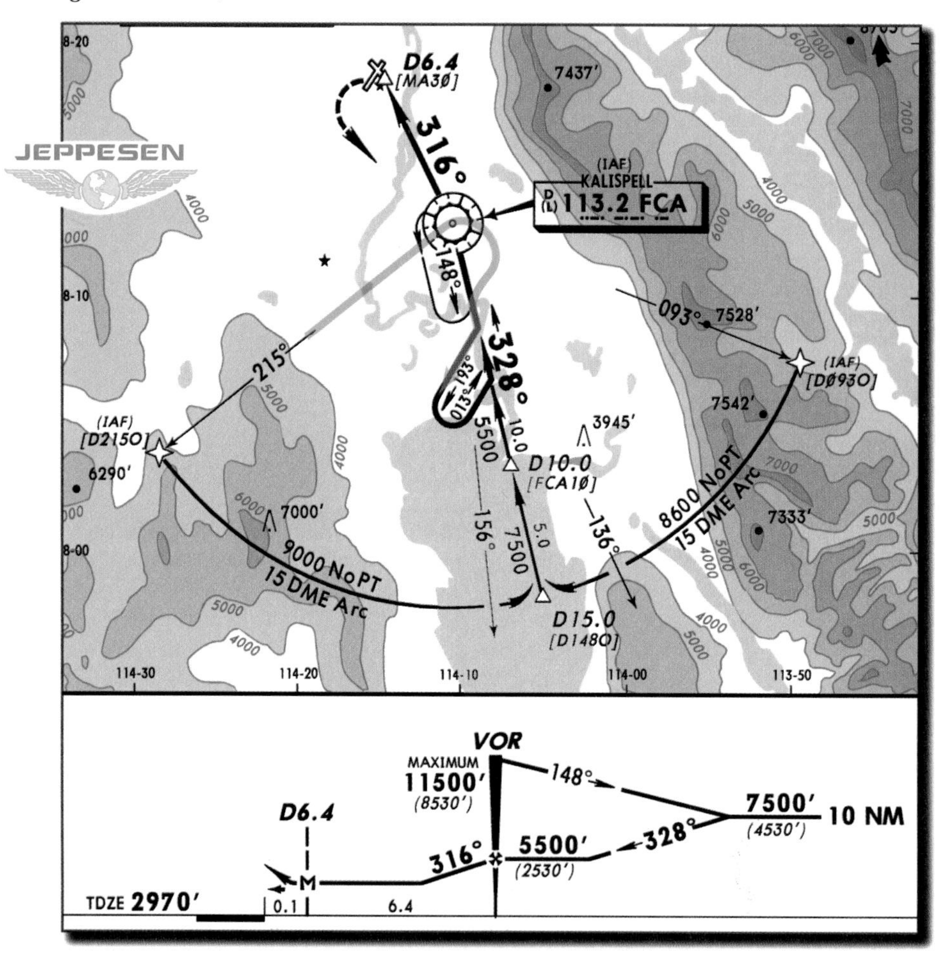

Enter the procedure turn after flying outbound from the VOR for 2 minutes. When you fly at 90 to 120 knots, beginning the course reversal at this time easily keeps you within 10 nautical miles of Kalispell VOR. Turn 45° right, twist the OBS to the inbound course of 328°, and fly for one minute on a heading of 193°. Level off as you reach 7,500 feet MSL. After flying outbound on the procedure turn for one minute, begin a 180° left turn to 013°.

Inbound to the Final Approach Fix

Complete your before landing checklist, with the possible exception of the landing gear. Depending on the aircraft, you may have already extended the gear to help slow to approach speed. If that is the case, verify at this time that it is down and locked. On other aircraft, it may be preferable to extend the landing gear when beginning the final descent.

Often, the additional drag causes the airplane to enter an appropriate rate of descent with only minor adjustment in power and trim. In any case, make sure you extend the landing gear and verify it is locked upon reaching the final approach fix.

Maintaining a heading of 013°, watch for the VOR needle to center on the 328° course. As the needle centers, turn left to track the inbound course. When established on the inbound course to the VOR, it is safe to begin your descent to 5,500 feet MSL. Since you are about 3 minutes from the VOR, a 750-foot-per-minute rate of descent is required to reach 5,500 feet MSL before you reach the VOR. If you reach 5,500 feet MSL before crossing the VOR, level off until you pass over the station.

When the TO/FROM indicator changes, reset your stopwatch, turn left to track the 316° course away from the VOR, and begin your final descent. On most approaches, it is not necessary to turn when crossing the VOR, but it is very important to check for a course change and comply with it on procedures that incorporate one. Twist the OBS to 316° and track outbound as you descend. Report your position on the CTAF: *"Glacier Park Traffic, Bonanza 1523S, Kalispell VOR, inbound for landing on the VOR Runway 30 approach, Glacier Park."*

Final Approach Segment

If you fly this approach without DME, it is important that you carefully control your groundspeed. You will establish the missed approach point based on your time from the VOR, using the conversion table at the bottom of the chart. You should be aware of the winds at the surface and at your altitude, and factor those into your groundspeed. If you are using a speed that lies between values in the table, interpolate, or use the time for the higher speed.

At 90 knots groundspeed you should descend promptly from 5,500 feet MSL in order to reach the MDA before 4:16 elapses. Keep in mind the missed approach point (MAP) is only 0.1 nautical mile from the threshold. If you descend at only 500 feet per minute,you will likely reach the MAP at the same time as the MDA. It will be impossible to land straight ahead if you break out of the clouds at the MAP, and, if you are below circling minimums, you will have no choice but to conduct a missed approach. At 90 knots, a 750-foot-per-minute descent will get you down to the MDA in less than 3 minutes. This allows a confortable margin after breaking out of the clouds to look for the airport and set up a normal landing approach. If you use a higher approach speed, plan on using an even higher rate of descent.

Missed Approach

It is essential that you do not descend below 3,480 feet MSL until you are in a position to make a normal landing approach to Runway 30 with adequate visual reference. In this case, a good time to descend for landing is when you intercept the glide path on the PAPI with the runway continuously in sight. If your approach speed is 120 knots or less, you need a visibility of 1 statute mile to land.

MISSED APPROACH: Climbing **LEFT** turn direct **FCA VOR**, continue climb to **6200'** outbound on **FCA VOR R-148** then climbing **RIGHT** turn to **7500'** direct **FCA VOR** and hold.

	STRAIGHT-IN LANDING RWY 30 MDA(H) 3480'(510') DAY	NIGHT	Max Kts	CIRCLE-TO-LAND MDA(H)
A	1	NA	90	3520'(548')-1
B			120	
C	1½		140	3640'(668')-1¾
D			165	3640'(668')-2

JEPPESEN

Gnd speed-Kts	70	90	100	120	140	160
MAP at D6.4 or VOR to MAP 6.4	5:29	4:16	3:50	3:12	2:45	2:24

It is possible for the ASOS to measure a 600-foot ceiling over the airport, when there are lower cloud layers in the vicinity. On this attempt, imagine you are flying in and out of the clouds as you near the missed approach point, but, as your stopwatch reaches 4:16, you do not have the runway environment in sight. You must immediately execute the missed approach procedure.

You should have memorized the first step of the missed approach procedure and be prepared to carry it out now. You need to make a climbing turn to the left and proceed direct to the Kalispell VOR. Concentrate on aircraft control as you add climb power. Make sure you retract the landing gear as soon as you have a positive rate of climb. Scan your flight instruments carefully and concentrate on executing an approximate 180° turn at the best rate-of-climb airspeed. Once you have the aircraft established in the climbing turn, twist the OBS to center the needle with a TO indication, and track back to the VOR while climbing. Tune the radio back to Salt Lake Center and report executing the missed approach. At this point you should request a clearance for another approach or a clearance to your alternate airport.

This missed approach procedure specifies an exact procedure for entering the holding pattern upon reaching the VOR. It is similar to the parallel entry you would normally use, but instead of flying outbound on a 148° heading for one minute, you must track outbound on the 148° radial until reaching 6,200 feet MSL before turning back toward the VOR. Continue to climb to 7,500 feet MSL as you return to the VOR and make a direct entry into the nonstandard holding pattern.

TRYING AGAIN

A decision whether to attempt this approach again depends on a number of factors. Perhaps the rain showers were dissipating at the time of your first attempt and you are confident your second attempt will be successful. Among other factors, you also need to consider your ability, distance to and weather at the alternate, available fuel, and your level of fatigue. For a more detailed discussion of these issues, see Chapter 10, Section B — IFR Decision Making.

If you are cleared to conduct the approach to Kalispell again, simply continue in the holding pattern until you are outbound on the 148° heading, then turn left to intercept the approach course from the VOR. Maintain 7,500 feet MSL until intercepting the 328° course inbound, and fly the rest of the approach as before.

FLYING THE PROCEDURE WITH DME

Having DME on board allows you to use the DME arc procedure, which lines you up on the final approach course more directly than when you perform the procedure turn. The notation, NoPT, along the DME arc segments reminds you that a procedure turn is neither expected nor authorized when flying the arc. Use of DME also enables you to accurately determine your position throughout the approach without timing.

Assume you are inbound on V536 as in the previous scenario. You will turn right onto the 15 DME arc 1/2 to 1 nautical mile prior to reaching VAILL Intersection. On the approach chart, the location that corresponds to VAILL Intersection is the IAF labeled [D215O]. You should descend to 9,000 feet MSL once established on the arc. A lead radial of 156° assists you in turning to intercept the 328° course to the VOR.

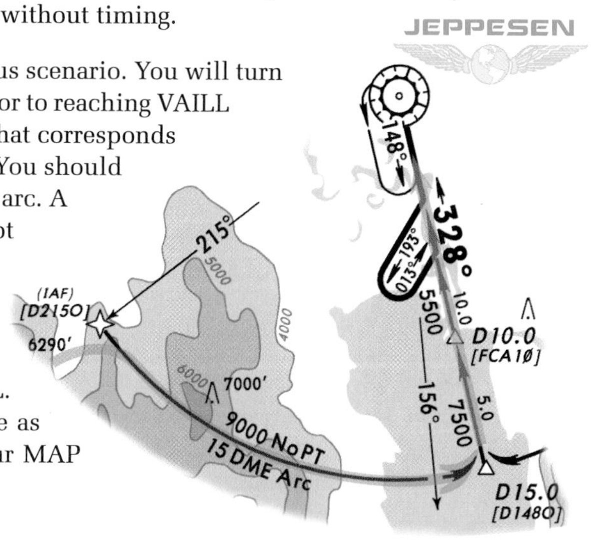

When established on the 328° inbound course, descend to and maintain 7,500 feet MSL until reaching 10 DME, then descend to 5,500 feet MSL. From this point, the approach is exactly the same as without DME, except that you can determine your MAP

JEPPESEN

12 MILES, CLEARED FOR THE APPROACH?

A light aircraft was just passing 12 DME inbound to Runway 26, when the approach controller cleared the flight crew for a 12 DME arc approach to Runway 26 instead of the straight-in approach that the crew had planned. Because the aircraft was already inside the 12 DME arc, and because of unclear phraseology from the controller, the crew proceeded with a straight-in approach. Fortunately, the discrepancy was discovered without incident; the crew obtained an amended clearance and completed the approach. The following excerpt was taken from the ASRS report which was filed subsequent to the incident.

"While tuned to approach control we were just passing 12 DME on the inbound radial upon being told "12 miles, cleared VOR Runway 26 approach, descend to 3,000 feet." The first officer read back "cleared VOR Runway 26, descend to 3,000 feet," to which approach rogered. I then briefed the VOR Runway 26 approach. At approximately 6 DME the controller inquired as to our DME, which we replied. Controller said that we were cleared for the 12 mile arc VOR Runway 26 approach. We informed him that we never heard arc or DME approach and would execute the VOR Runway 26 approach due to our proximity. The approach was then uneventful. It has been my experience that when you are to fly an arc approach you are given clearance at least 2 miles prior to the arc to commence your turn."

"Communication is the culprit here. Precise terminology should be used and the approach should be renamed (or referred to by the controller as) *12 DME Arc Runway 26 approach."*

The ASRS contains frequent reports of controllers using nonstandard terminology in clearances. In some cases, the controller made modifications to published departure procedures which confused the flight crew. Unfortunately, pilot requests for clarification were not always satisfactorily handled, at least according to pilots filing these reports.

It is not only your right as a pilot to get clarification for a confusing clearance, it is your duty. There are times when you may need to be assertive. Refuse a clearance if necessary, and ensure the controller provides a clearance that makes sense to you.

simply by looking for 6.4 DME when outbound from the VOR on the 316° radial. Even when you use DME, it is still good operating practice to reset your stopwatch when crossing the VOR, in case DME fails before you reach the MAP.

ON-AIRPORT FACILITY

When approach procedures utilize on-airport facilities, no final approach fix (FAF) is designated, unless DME or another means is available for identifying such a fix. However, a final approach point (FAP) is designated and identifies the beginning of the final approach segment. This point is where the aircraft is established inbound after completing the procedure turn. The FAP serves as the FAF.

Without DME, special procedures are required to ensure that you are in an obstruction-free area as you descend to the airport. Generally, this is accomplished by overflying the VOR at a safe altitude, proceeding outbound, and turning around and tracking back toward the airport/VOR as you descend to the MDA. Hopefully, you will break out of the clouds prior to reaching the facility and land, otherwise you will cross over the airport and circle to land or, if you don't break out, conduct your missed approach.

In this example, you are on your way to John F. Kennedy Memorial Airport in Ashland, Wisconsin. You are approaching GRASS Intersection from the southwest on V345. [Figure 8-5]

Figure 8-5. Ashland (ASX) VOR is located at John F. Kennedy Memorial Airport. Notice that V345 terminates at the VOR.

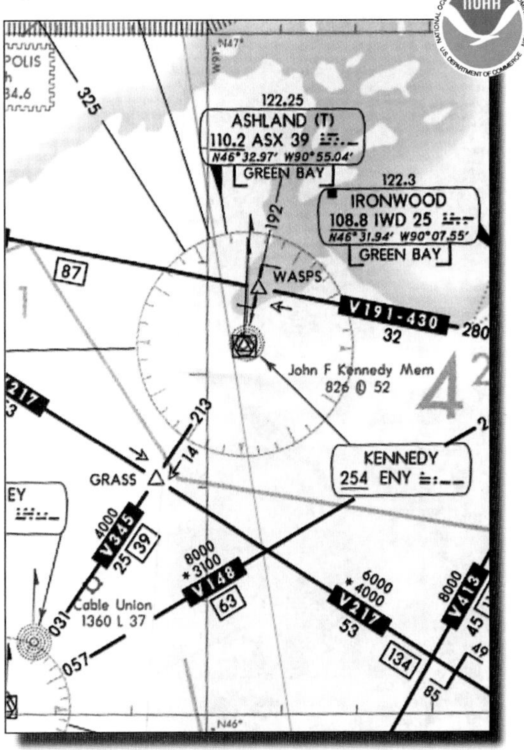

PREPARING FOR THE APPROACH

As always, take a few minutes before you begin the approach to familiarize yourself with essential procedural information. One important piece of information you will gather is that your ability to obtain an altimeter setting from UNICOM will significantly affect the minimums for this procedure. [Figure 8-6]

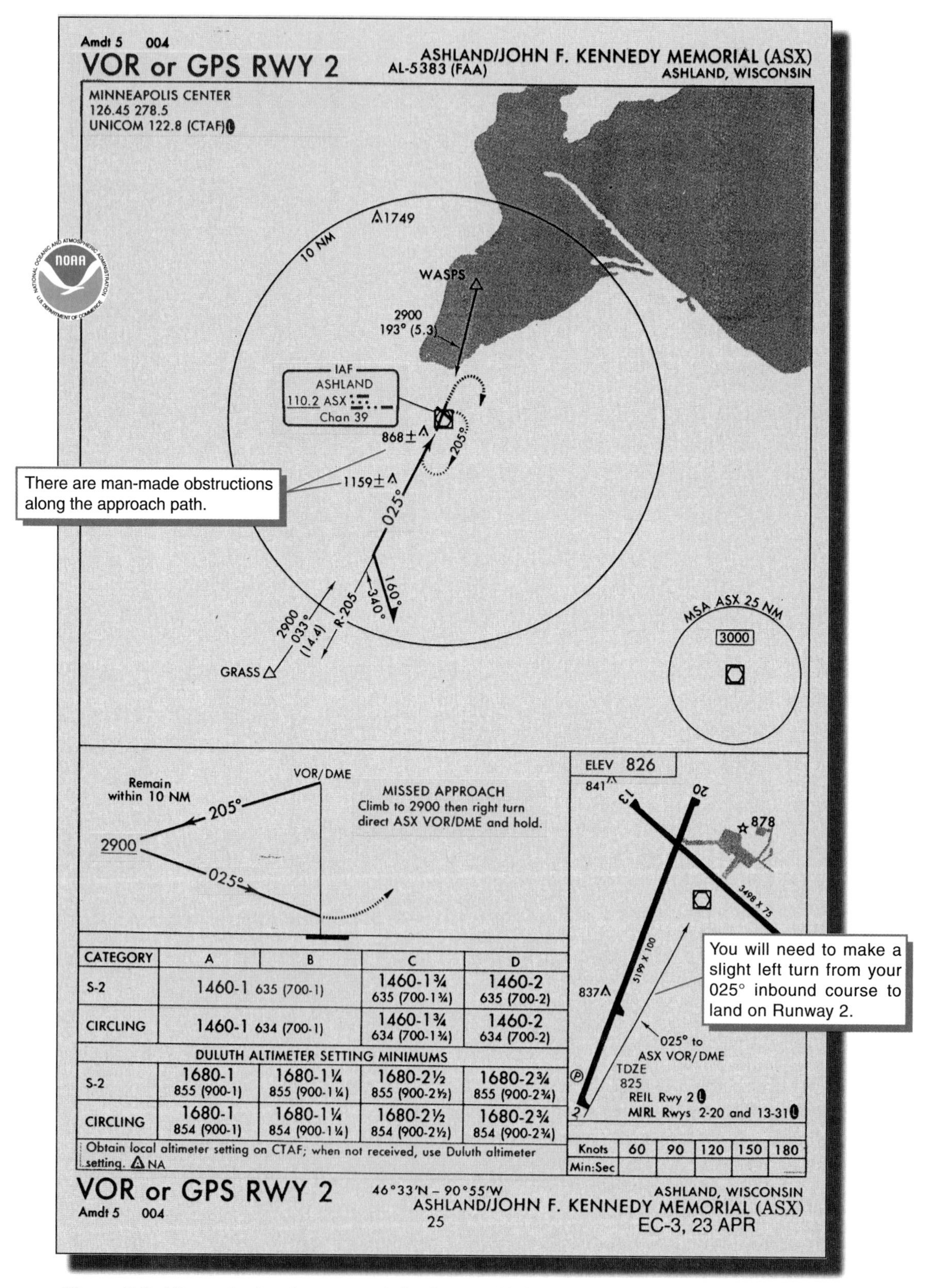

Figure 8-6. After reviewing the approach for the usual information (see highlights), you look for other items of interest.

Typically, if this were an approach based on an off-airport VOR, you would not have to execute a procedure turn when approaching from GRASS. However, with the VOR on the field, you need to fly to the VOR and proceed outbound to establish your position prior to descending to the airport.

Since you are tracking toward Ashland VOR on the airway, your VOR receiver is already set to the correct frequency and the station has previously been identified. You may wish to listen to the identifier again to confirm the VOR is set correctly and the ground facility is still operating properly.

YOUR CLEARANCE

Since there is only one IAF for this approach, you can expect your clearance while on V345. As you approach Grass, Minneapolis Center clears you as follows: *"Cessna 1773R, cleared for the VOR Runway 2 approach to Ashland, maintain 3,500 until outbound from Ashland VOR."*

DESCENDING PRIOR TO THE IAF

The MEA on V345 approaching GRASS Intersection is 4,000 feet MSL. On this eastbound course you are likely to be assigned an odd altitude of 5,000 feet MSL or higher. Upon passing GRASS and receiving your clearance for the approach, promptly begin your descent to 3,500 feet MSL, the altitude assigned by ATC, and continue tracking the 033° course to Ashland VOR.

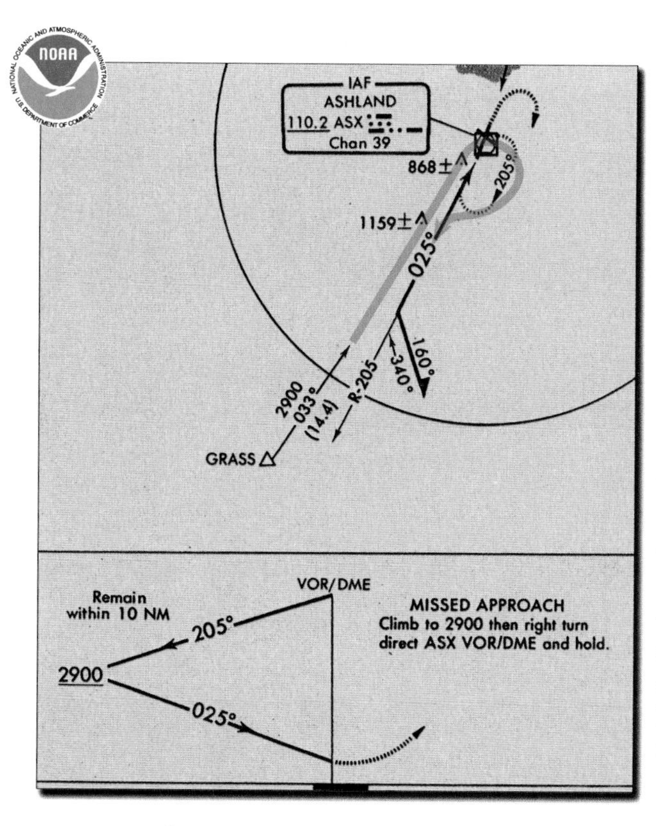

OUTBOUND ON THE PROCEDURE TURN

Upon crossing the Ashland VOR, begin a right turn to intercept the 205° radial outbound and throttle back to begin a descent to 2,900 feet MSL. Twist your OBS to a 205° course, watch for the TO/FROM indicator to change from TO to FROM as you pass abeam the VOR southbound, and note the time. A heading of 250° provides a 45° intercept which should center the needle promptly. Level off at 2,900 feet MSL, and track outbound for 2 minutes after passing the VOR before initiating the procedure turn.

Turn left approximately 45° to a heading of 160°. Fly this heading for one minute and then begin a 180° right turn to 340°. Complete your before landing checklist, although you may postpone extending the landing gear until beginning your descent for the airport.

INBOUND TO THE AIRPORT

You are now on a 45° intercept to the 025° final approach course to the VOR. When the needle centers, begin your descent to 1,460 feet MSL. You need a descent rate of about 750 feet per minute to be sure of making it down to this altitude before arriving at the airport. Your missed approach point is where the TO/FROM indicator changes from TO to FROM, directly over the airport. Since the circling minimums are the same as the

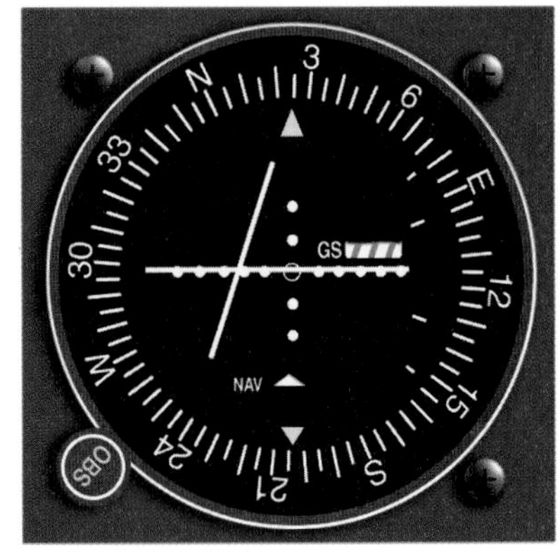

straight-in minimums, you can circle as needed to land if you break out of the clouds right at the missed approach point. Do not descend below the MDA unless you are in position to land without excessive maneuvering and have the required visibility and visual references. When circling, positioning yourself on a normal base or final with a clear view of the approach end of the runway generally is sufficient to descend below the MDA and land.

MISSED APPROACH

In the event you do not see the touchdown area of the runway when you reach the MAP, or that you lose sight of it while circling to land, you should execute the missed approach procedure immediately. When flying straight in to Runway 2, climb on the runway heading to 2,900 feet MSL, then turn right and proceed to the VOR. If you are circling and lose sight of the runway, you must turn toward the landing runway and interpret the missed approach instructions with reference to your final approach course. In this case, continue turning to an approximate heading of 025° which you used during final approach, and climb to 2,900 feet MSL on that heading.

HOLDING PATTERN ENTRY

When approaching the VOR from the northeast, you would use a parallel entry to this standard holding pattern. When crossing the VOR, turn left to a heading of 205° and fly outbound one minute. Then, make a left turn direct to the VOR. When arriving at the VOR turn right to approximately 205°, depending on the wind. This puts you outbound in the holding pattern.

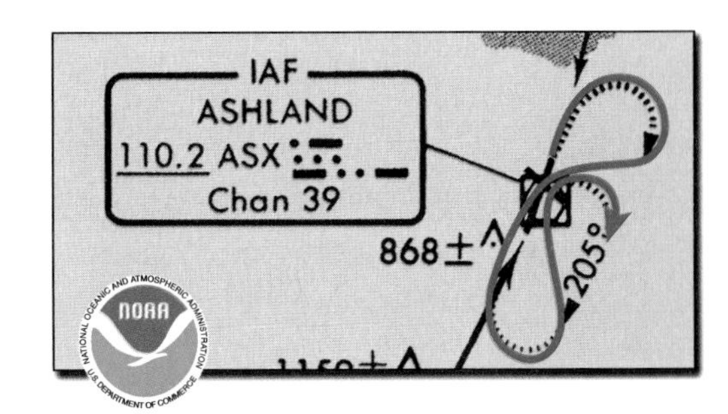

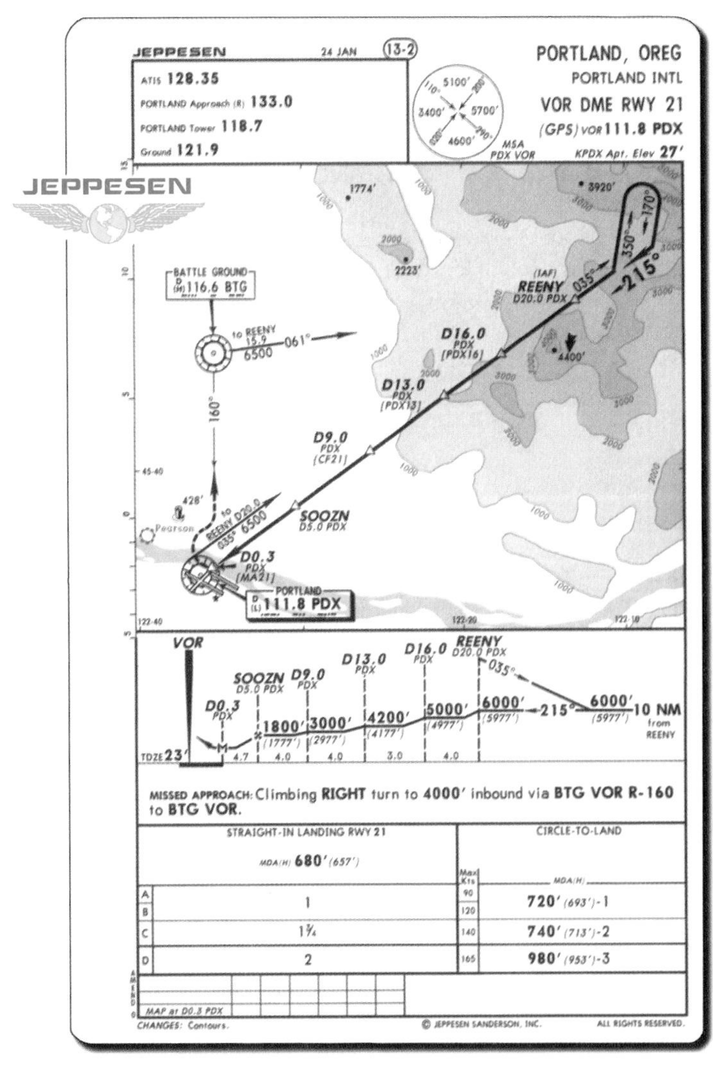

VOR/DME PROCEDURES

When the procedure title includes the words, VOR DME, it means the use of DME equipment is mandatory for the approach. One procedure that requires DME is the VOR DME Rwy 21 approach to Portland International Airport (KPDX) in Oregon. [Figure 8-7]

Figure 8-7. Stepdown fixes on this approach can only be identified with DME.

According to the profile view, you can descend to 5,000 feet MSL after passing REENY inbound. This intersection is defined as 20 DME from the Portland (PDX) VOR on the 035° radial. Progressive stepdown fixes also are defined using DME. There is no other way of identifying these fixes, unless you are conducting the procedure using an approach-certified GPS. The FAF is at SOOZN (5.0 DME), and the MAP is at 0.3 DME.

The minimums section of the Portland VOR DME Rwy 21 approach contains only one straight-in MDA(H). On VOR approaches that do not require DME but which still use supplemental DME fixes, two straight-in MDAs may be authorized. The lowest MDA applies only if you can identify the DME stepdown fix. In addition, you may encounter a stepdown fix and a visual descent point (VDP) on the same approach. [Figure 8-8]

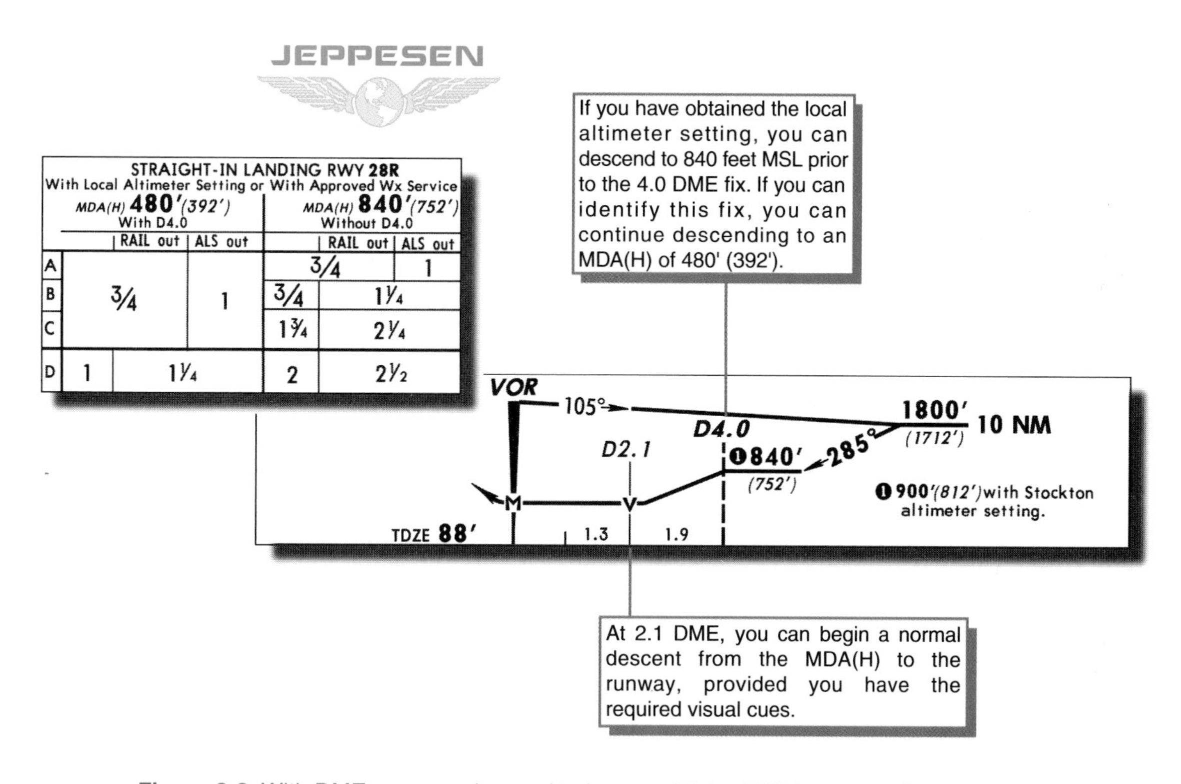

Figure 8-8. With DME, you can descend to lower published MDAs on certain approaches.

NDB APPROACHES

Procedurally, an NDB approach is similar to a VOR approach without DME. You navigate to an initial approach fix, usually perform a course reversal, and track to the airport while descending. Although the accuracy of an NDB approach depends on your skill using ADF navigation, NDB approaches offer some advantages over those using a VOR. For example, it is not necessary to readjust a course selector as you may during a VOR approach. In addition, position orientation is easier because the bearing pointer needle always points to the station.

The NDB or GPS Rwy 35R approach to Denver's Centennial Airport (KAPA) is an example of an NDB approach with an off-airport navaid. [Figure 8-9] The procedure title indicates that either an ADF or an approach-certified GPS is required for this approach.

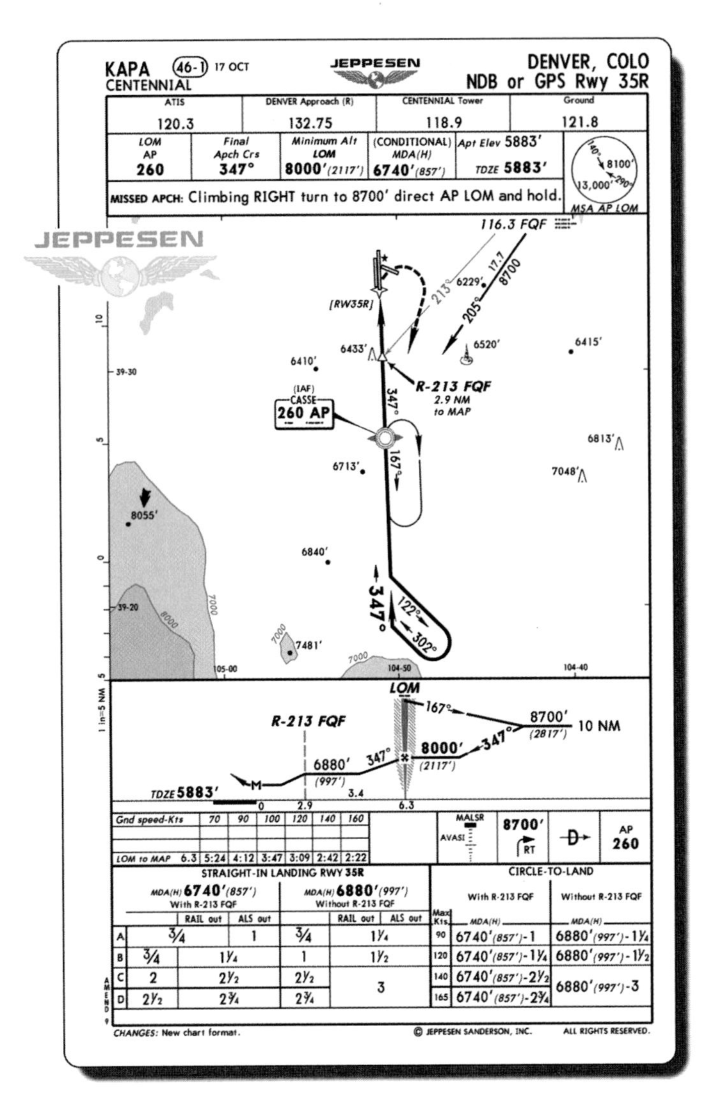

Figure 8-9. On this approach to Denver's Centennial Airport, the NDB is both the IAF and FAF.

Radar Vectors to the Approach

If you lost communication with ATC, you might fly this approach by proceeding to Falcon (FQF) VOR, which is part of the enroute structure, flying a 205° course to Casse LOM while descending to 8,700 feet MSL, and then flying the procedure turn and approach. Like many approaches in a radar environment, you rarely fly the charted procedure turn. It is more expedient for you and better for overall traffic flow if you are vectored onto the final approach course. As you arrive from the east, Denver Approach instructs you to maintain 8,700 feet MSL and a heading of 270°, and tells you these are vectors to the final approach course.

Preparing for the Approach

This chart uses Jeppesen's briefing strip format. The first line shows the communication frequencies in the order you will use them. Essential components of the procedure appear on the next line. Your primary navaid is the Casse (AP) LOM at 260 kHz. The final approach

ATIS	DENVER Approach (R)	CENTENNIAL Tower	Ground
120.3	132.75	118.9	121.8

LOM AP	Final Apch Crs	Minimum Alt LOM	(CONDITIONAL) MDA(H)	Apt Elev 5883'	MSA AP LOM
260	347°	8000'(2117')	6740'(857')	TDZE 5883'	140° 8100' 290° 13,000'

MISSED APCH: Climbing RIGHT turn to 8700' direct AP LOM and hold.

Approach to a 1,000-Foot Wide Runway?

Yes, they do publish approaches to seaplane bases, although there are less than half a dozen such procedures. Here is the NDB approach to Rangeley Lake, Maine.

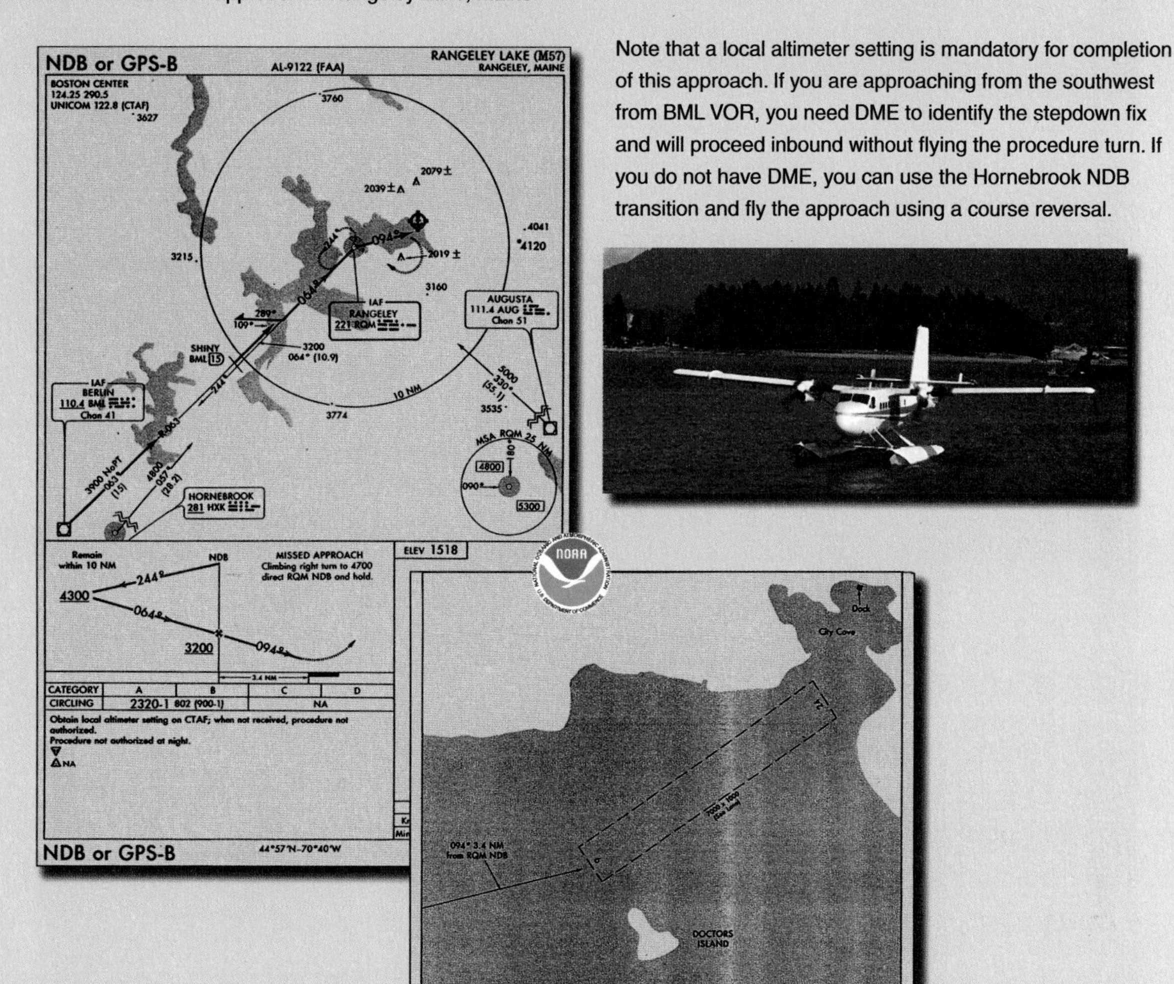

Note that a local altimeter setting is mandatory for completion of this approach. If you are approaching from the southwest from BML VOR, you need DME to identify the stepdown fix and will proceed inbound without flying the procedure turn. If you do not have DME, you can use the Hornebrook NDB transition and fly the approach using a course reversal.

course is 347° and the minimum altitude inbound to the LOM is 8,000 feet MSL. The MDA(H) is 6,740′ (857′). However, this MDA(H) is conditional; checking the minimums section reveals you must be able to identify the Falcon 213° radial to descend to this MDA. The touchdown zone elevation in this case is the same as the airport elevation, 5,883 feet MSL. The minimum safe altitude within 25 miles of Casse is 8,100 feet MSL if approaching from the north through southeast, otherwise it is 13,000 feet MSL.

JEPPESEN

Gnd speed-Kts	70	90	100	120	140	160		MALSR AVASI	8700′ RT	-D→	AP 260
LOM to MAP 6.3	5:24	4:12	3:47	3:09	2:42	2:22					

The textual description of the missed approach procedure appears next. It is a climbing right turn back to the NDB. Icons showing this procedure appear in the profile view. Runway 35R has a medium intensity approach light system with runway alignment indicator lights (MALSR) and an abbreviated VASI (AVASI) on the left side of the runway.

Set your ADF to 260 and positively identify Casse (AP). Leave the volume turned up so that you can continue to listen to the identifier. Unlike a VOR, an ADF has no OFF flag, so you need to monitor the identifier throughout the approach. For this example, assume you have a VOR receiver. Set it to Falcon (116.3 MHz) and identify the facility. Twist the OBS to 213° so that you can identify the stepdown fix 2.9 nautical miles prior to the MAP. During the approach, check your heading indicator against the magnetic compass often. Unlike a VOR approach, the accuracy of your NDB approach directly depends on your heading indicator being set precisely.

As you continue inbound, obtain ATIS information on 120.3 MHz. Either monitor ATIS on a second radio or make sure you get approval to leave Denver's frequency. Analyze the ATIS information and consider whether you are likely to successfully complete the approach, given the ceiling and other weather conditions. Set your altimeter and make a judgment as to the effect the winds will have on your groundspeed during the approach. Determine whether you will be able to land straight ahead on Runway 35R or if you have to circle to another runway.

YOUR CLEARANCE

Denver Approach instructs you to turn right to a heading of 320° and clears you as follows: *"Cessna 86BJ, you are five miles from Casse, maintain 8,700 until established on the final approach course, cleared for the Centennial NDB Runway 35R approach. Contact the tower 118.9 at Casse."* Slow the airplane to approach speed. Denver Approach has given you a heading which is a 27° intercept to the final approach course. As the relative bearing approaches 27°, turn on course and begin your descent to 8,000 feet MSL prior to the NDB. Apply any corrections needed to track the course with current wind conditions. [Figure 8-10]

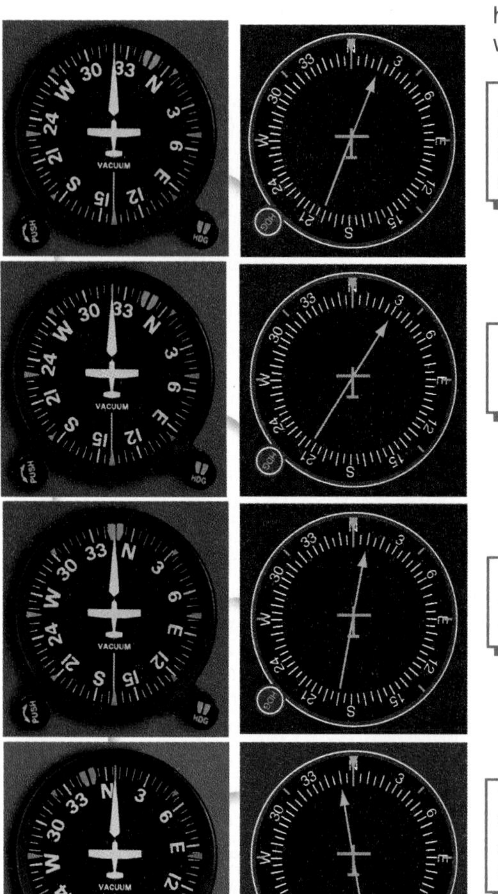

Figure 8-10. ATC typically vectors you in such a way as to facilitate interception of the final approach course, which is 347°. Setting the heading bug to 347° makes it easy to see your course intercept and wind correction angles.

ATC has assigned a heading of 320°, a 27° intercept course. The ADF needle, which always moves toward the bottom, indicates course interception will occur shortly.

As the ADF needle approaches a 27° relative bearing, turn right to a heading of 347° to parallel the final approach course.

Check the results on the ADF. This indicator suggests the turn was late or that there is a wind out of the east that has blown you left of the desired course.

Turn right to correct back to the final approach course. With this heading 20° right of course, you will be on course when the ADF needle moves to 20° left of the nose.

INBOUND TO THE FAF

Make sure you complete your before landing checklist early in the approach, prior to the FAF. If you have retractable landing gear, extend it where appropriate to slow the airplane down or begin your descent, and confirm that it is down and locked upon reaching the FAF.

Level off at 8,000 feet MSL, adding power to precisely maintain approach speed. When the ADF needle swings around to point behind you, begin your descent to 6,880 feet MSL and set your stopwatch. Contact the tower on 118.9, and report Casse inbound. Although you are not required to report the FAF when you are in radar contact, it is an appropriate report when establishing contact with the tower.

FINAL APPROACH SEGMENT

Assume an average headwind of 10 knots during final approach and an approach airspeed of 100 knots. Since this gives you a groundspeed of 90 knots, use the time from the FAF to the MAP of 4:12.

Level off at 6,880 feet MSL and watch for the VOR needle to center, indicating the stepdown fix. [Figure 8-11] Without this fix, 6,880′(997′) would be your MDA(H). Upon identifying the fix and being cleared to land straight ahead on Runway 35R, you can descend another 140 feet to 6,740 feet MSL.

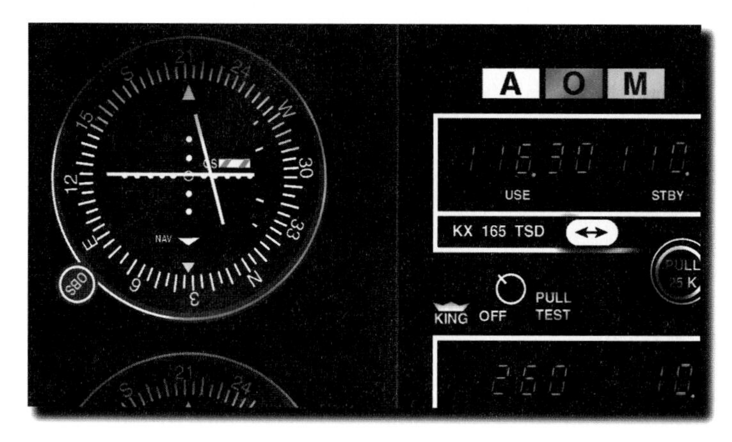

Figure 8-11. As you approach and pass the stepdown fix, the VOR needle moves from the right to left.

Upon leveling off at the MDA, check for the 4:12 time to the missed approach point. If you reach the MAP without the runway environment in sight, you should conduct a missed approach immediately. If you do see the runway, you may proceed below the MDA only when you are in position to descend for a safe landing. The VASI glidepath is a good visual cue for initiating final descent to the runway.

If cleared for the NDB Rwy 35R approach, Denver/Centennial, [Figure 8-9] and over FQF VOR, you would begin by flying the published transition to Casse, followed by the procedure turn. Descend to the altitudes shown for each route segment and conclude with the minimums appropriate for your approach and landing.

MISSED APPROACH

If it is necessary to perform a missed approach, immediately begin a right turn as you climb to 8,700 feet MSL. Turn to place the ADF needle off the aircraft's nose and proceed direct to the station. Report the missed approach to Centennial Tower. Perform a parallel entry to the depicted holding pattern, and await a further clearance from Denver Approach.

SUMMARY CHECKLIST

✓ VOR and NDB approaches primarily fall into two categories — those that use an on-airport facility and those with an off-airport facility. On approaches with on-airport navaids, the FAP often serves as the FAF.

✓ Preparation to fly an approach should begin well before flying the procedure. Determine which approaches are in use or likely to be in use at the destination airport, and review the approach procedures as early as possible. Obtain weather information, if possible, for the destination airport and analyze whether a successful approach is likely.

✓ ATC may clear you to fly the approach of your choice, but they will more likely clear you for a specific approach.

✓ A published procedure turn or similar course reversal is mandatory unless you are vectored to the final approach course by ATC, or unless your particular approach transition indicates NoPT. Typically, you accomplish a course reversal by flying outbound for two minutes, turning to a charted heading 45° left or right of your outbound course and flying for one minute, then making a 180° opposite direction turn back to re-intercept the inbound course.

✓ When cleared for an approach, you generally should descend promptly to the minimum altitude published for your current route segment or approach transition, or other altitude assigned by ATC.

✓ Complete your before landing checklist prior to the FAF, or if there is no FAF, before intercepting the final approach course.

✓ Make sure you know what rate of descent is required to reach stepdown altitudes or the MDA by the appropriate time.

✓ If you do not have the runway environment in sight when reaching the MAP, or if you lose sight of it at any time while circling, it is imperative that you immediately execute the missed approach procedure.

✓ When executing a missed approach, notify ATC, and, depending on your circumstances, request a clearance to fly the approach again, or request a clearance to your alternate.

✓ DME is required on certain approaches that indicate DME in the procedure title. Even on those approaches that do not require DME, using DME to identify stepdown fixes may allow lower minimums.

✓ NDB approach procedures are similar to VOR approaches. However, the precision with which you complete the approach is dependent on your skill in ADF tracking and on the accuracy of your heading indicator.

KEY TERMS

Off-Airport Facility

On-Airport Facility

Stepdown Fix

QUESTIONS

1. If you are flying a circling maneuver to the favored runway and the visibility is at or above the required minimums, when can you begin your descent below the MDA?

   A. After being cleared to land
   B. After you have the airport environment in sight
   C. When you are in a position from which you can make a normal descent to the runway with adequate visual reference

2. True/False. When using a DME arc transition, you can descend to an authorized lower altitude as soon as you begin the turn inbound.

3. What is true of an approach that contains the name, VOR/DME, in the procedure title?

   A. Use of DME is required to complete the procedure.
   B. Stepdown fixes in this procedure are identified with both VOR cross radials and DME values.
   C. Use of DME allows you to identify optional stepdown fixes and to descend to lower minimums.

4. As soon as practical after you begin a missed approach procedure, you should notify ATC and request what type of clearance?

Refer to the Chicago/Aurora VOR or GPS-A approach chart on page 8-20 to answer questions 5-16.

5. What minimum navigation equipment is required to fly this approach?

   A. One VOR receiver
   B. Two VOR receivers
   C. One VOR receiver plus DME

6. What is the primary navaid for this approach?

   A. JOT VOR, 112.3 MHz
   B. DPA VOR, 108.4 MHz
   C. DPA TACAN, Channel 21

7. What minimum altitude guarantees 1,000 feet of obstacle clearance anywhere within 25 nautical miles of Du Page VOR?

8. True/False. DME is required to identify HOLTA.

9. What approach transitions are provided for this procedure?

   A. The 343° course from JOT VOR, then direct to the initial approach fix at DPA VOR
   B. No approach transition is needed because the initial approach fix at DPA VOR is part of the enroute structure.
   C. This procedure has no approach transitions or initial approach fixes. Radar vectoring to the final approach course is required.

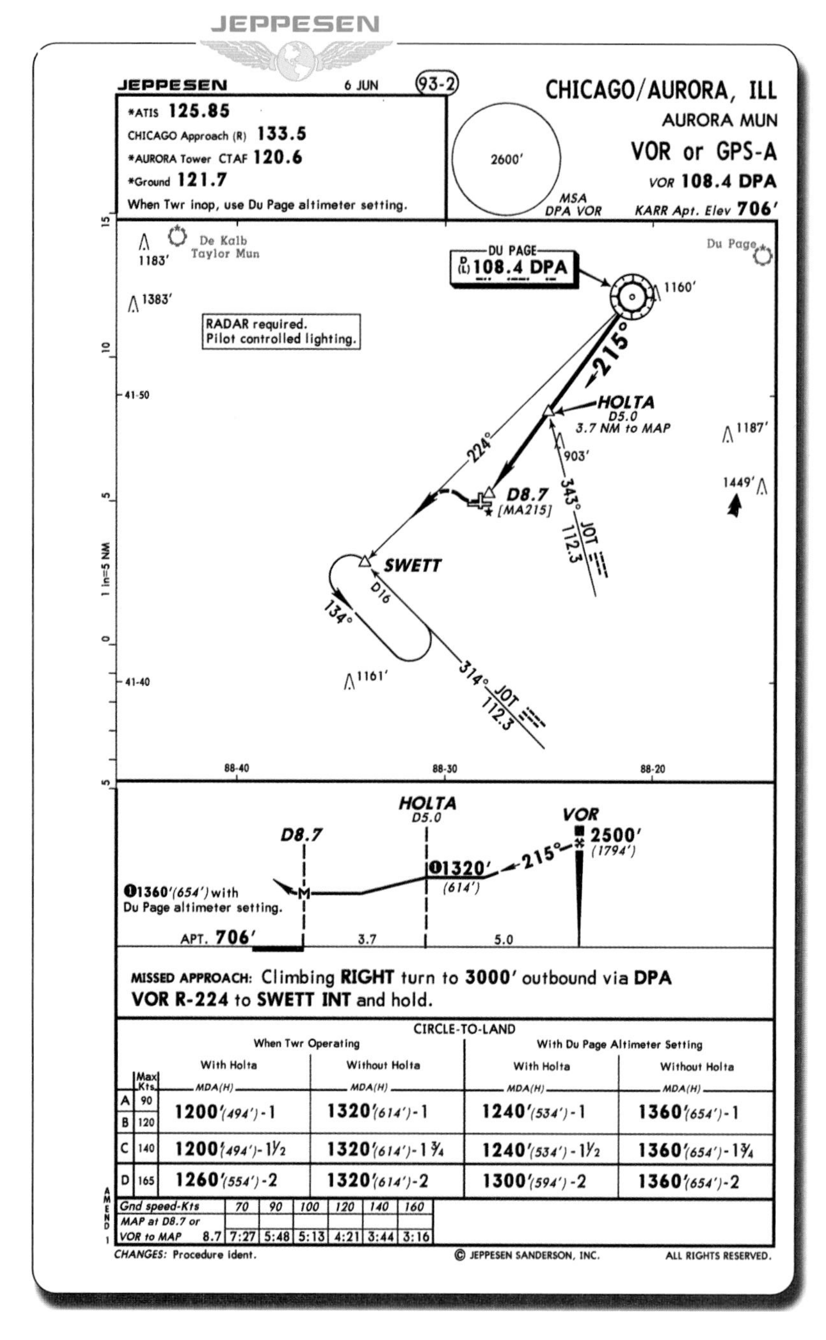

CIRCLE-TO-LAND	Max Kts	When Twr Operating: With Holta MDA(H)	When Twr Operating: Without Holta MDA(H)	With Du Page Altimeter Setting: With Holta MDA(H)	With Du Page Altimeter Setting: Without Holta MDA(H)
A	90	1200'(494')-1	1320'(614')-1	1240'(534')-1	1360'(654')-1
B	120				
C	140	1200'(494')-1½	1320'(614')-1¾	1240'(534')-1½	1360'(654')-1¾
D	165	1260'(554')-2	1320'(614')-2	1300'(594')-2	1360'(654')-2

Gnd speed-Kts	70	90	100	120	140	160
MAP at D8.7 or VOR to MAP 8.7	7:27	5:48	5:13	4:21	3:44	3:16

10. Which of the following is the reason no runway number is listed in the procedure title?

    A. The approach serves more than one runway.
    B. The final approach course is not within 30° of any runway heading.
    C. The GPS procedure that overlays this approach is not runway specific.

11. How do you obtain an altimeter setting when Aurora Tower is not operating?

    A. Monitor after-hours ATIS on 125.85.
    B. Monitor ASOS on the ATIS frequency.
    C. Obtain the Du Page altimeter setting from Chicago Approach and increase the MDA by 40 feet.

12. What is the MDA if you are flying an aircraft at its normal approach speed of 90 knots, with Aurora ATIS information, and you have identified the HOLTA step-down fix?

    A. 1,200 feet AGL
    B. 1,200 feet MSL
    C. 1,240 feet MSL

13. Where is the MAP if your TAS is 120 knots and your groundspeed is 100 knots?

    A. 8.7 DME from DPA VOR
    B. 4:21 after passing DPA VOR
    C. The 343° radial from JOT VOR

14. What heading should you initially turn to when conducting the missed approach procedure?

    A. 135°
    B. 224°
    C. Approximately 270°

15. What is the appropriate entry for the missed approach holding pattern?

    A. Direct
    B. Parallel
    C. Teardrop

16. True/False. Obstacle clearance is guaranteed if, while in the depicted plan view area, you fly high enough to clear the 1,449-foot MSL man-made structure shown near the right edge of the chart.

SECTION B
ILS APPROACHES

The instrument landing system (ILS) is a precision approach navigational aid which provides highly accurate course, glide slope, and distance guidance to a given runway. The ILS can be the best approach alternative in poor weather conditions for several reasons. First, the ILS is a more accurate approach aid than any other widely available system. Secondly, the increased accuracy generally allows for lower approach minimums. Third, the lower minimums can make it possible to execute an ILS approach and land at an airport when it otherwise would not have been possible using a nonprecision approach.

ILS CATEGORIES AND MINIMUMS

There are three general classifications of ILS approaches — Category I, Category II, and Category III. The basic ILS approach is a Category (CAT) I approach and requires only that you be instrument rated and current and that your aircraft be equipped appropriately. CAT II and CAT III ILS approaches typically have lower minimums and require special certification for operators, pilots, aircraft, and air/ground equipment. [Figure 8-12]

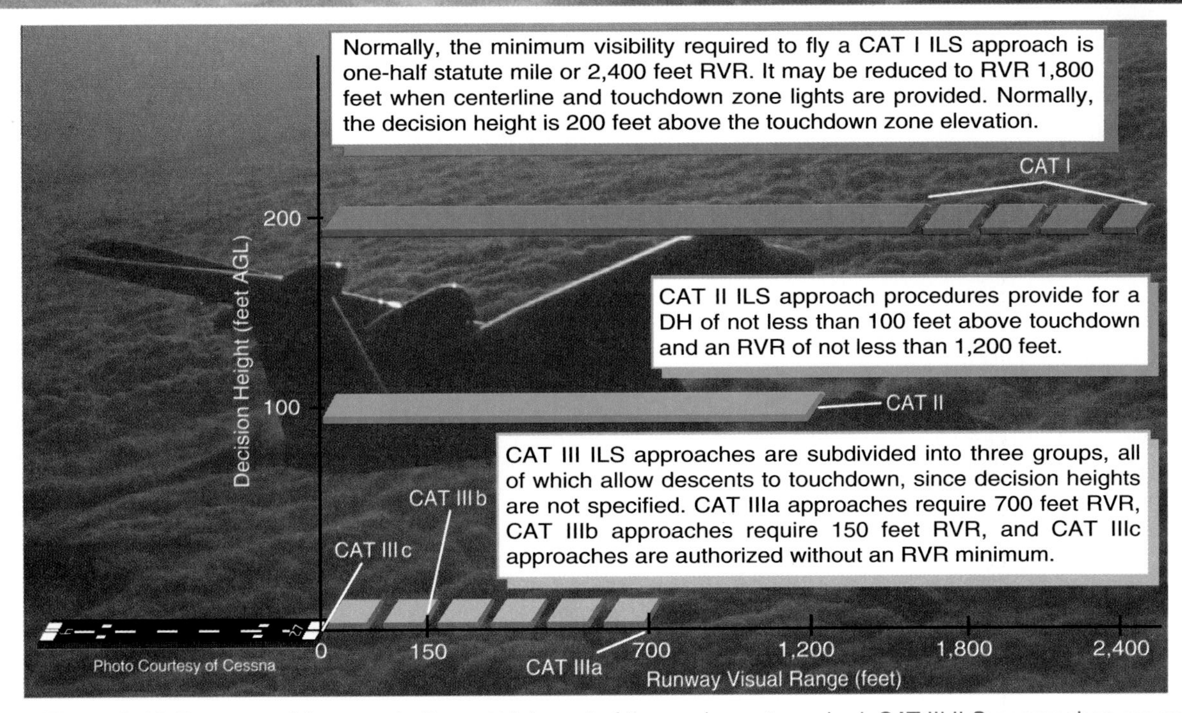

Figure 8-12. Because of the complexity and high cost of the equipment required, CAT III ILS approaches are used primarily in air carrier and military operations.

ILS COMPONENTS

The typical ILS installation is made up of several components. You receive guidance information from ground-based localizer and glide slope transmitters. To help you determine your distance from the runway, the ILS installation may provide DME fixes or marker beacons located along the ILS approach path. To facilitate the transition from instrument to visual flight as you approach the airport, runway and approach lighting systems are installed. [Figure 8-13]

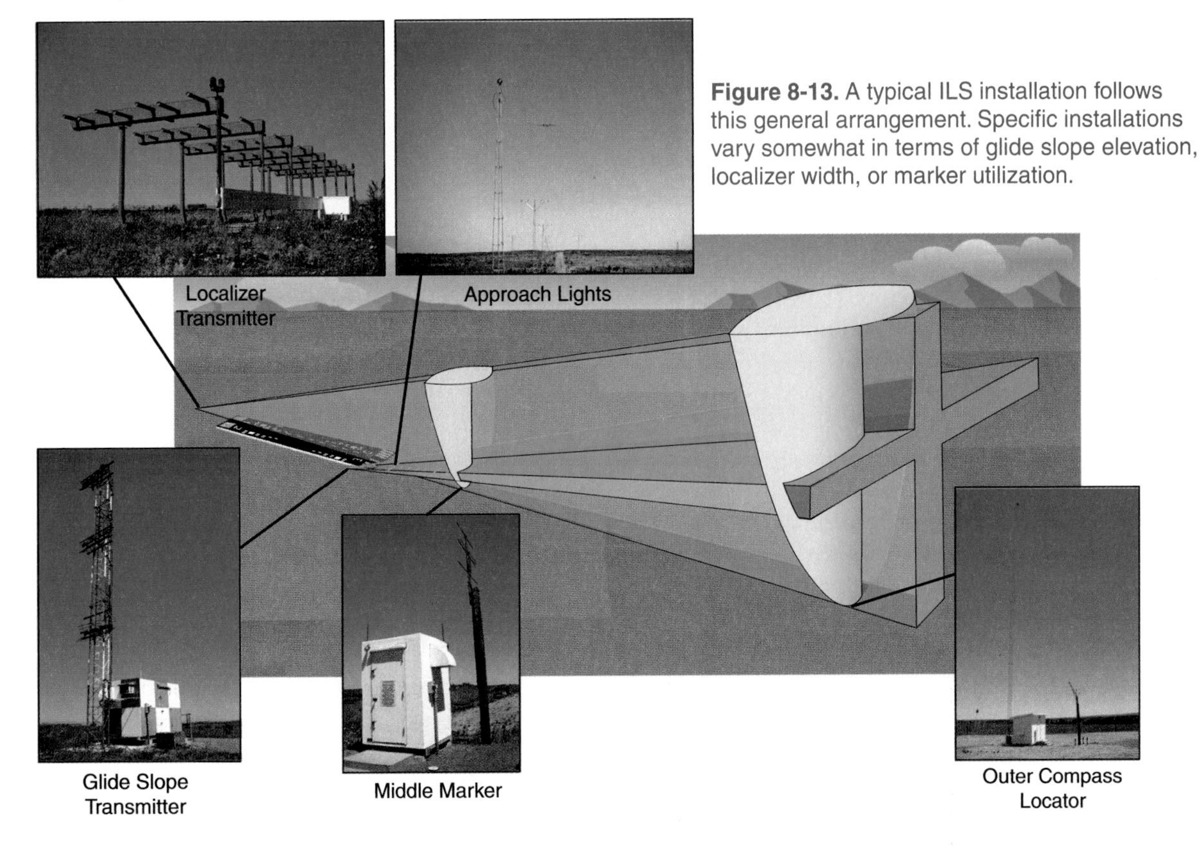

Figure 8-13. A typical ILS installation follows this general arrangement. Specific installations vary somewhat in terms of glide slope elevation, localizer width, or marker utilization.

LOCALIZER

The ILS uses a **localizer** transmitter to provide information regarding your alignment with the runway centerline. The localizer transmitter emits a navigational array from the far end of the runway, opposite the approach end. When these signals are used, the approach is said to be a front course approach. This transmitter can also provide a signal in the direction opposite the front course; however, you should not use these back course signals for navigation unless a back course approach is established and ATC has authorized you to execute the procedure. [Figure 8-14]

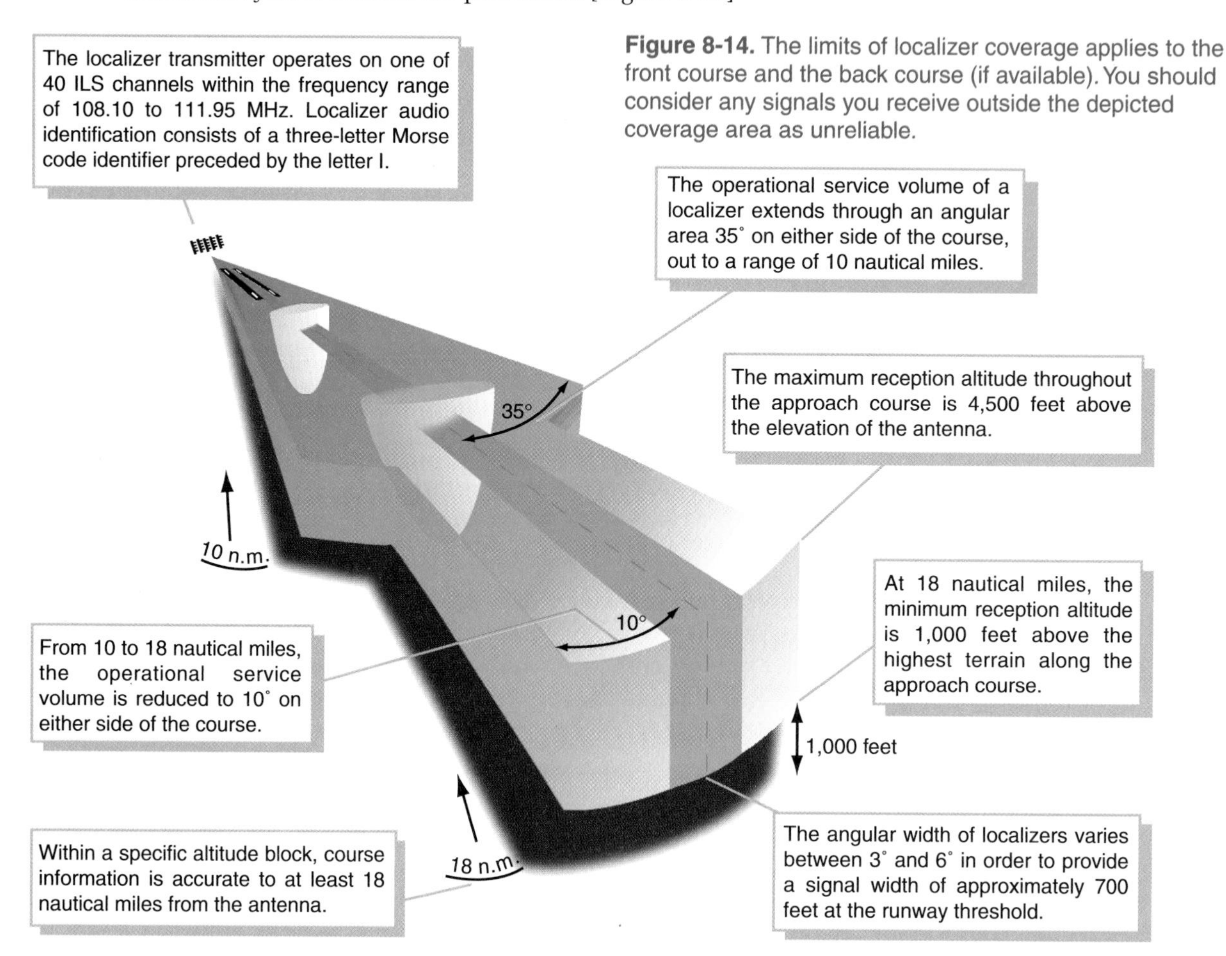

Figure 8-14. The limits of localizer coverage applies to the front course and the back course (if available). You should consider any signals you receive outside the depicted coverage area as unreliable.

The signal from the localizer transmitter represents only one magnetic course to the runway. Therefore, the course selected using the OBS of a basic VOR indicator does not affect course tracking. However, you may find it helpful to set the published course of the ILS on the navigation indicator as a reminder of the inbound course during tracking and heading corrections. Regardless of what course you select, the CDI senses off-course position only with respect to the localizer course.

If you are using an HSI for the approach, you normally must set it to the ILS front course for guidance. When you are tracking the front course of an ILS toward the runway using a basic VOR indicator, CDI sensing is normal; that is, right heading corrections are applied to right deflections of the CDI. Reverse CDI sensing occurs whenever the aircraft travels on the reciprocal heading of the localizer course. [Figure 8-15]

When using a basic VOR indicator, normal sensing occurs inbound on the front course and outbound on the back course. Reverse sensing occurs inbound on the back course and outbound on the front course. With an HSI, you can avoid reverse sensing by setting the published front course under the course index. This applies regardless of your direction of travel, whether inbound or outbound on either the front or back course. See figure 8-15.

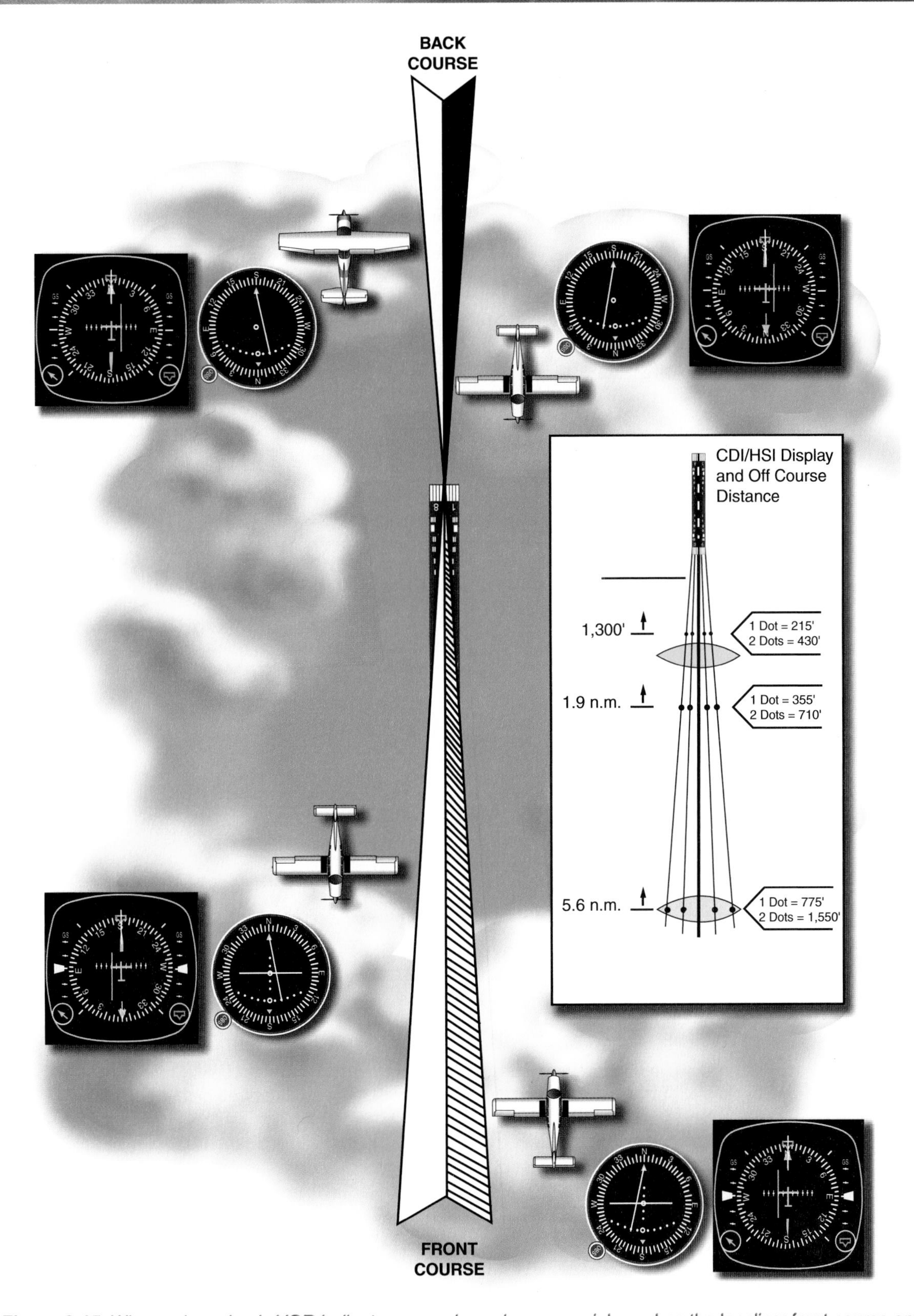

Figure 8-15. When using a basic VOR indicator, normal sensing occurs inbound on the localizer front course and outbound on the back course. Reverse sensing occurs inbound on the back course and outbound on the front course. With an HSI, you can avoid reverse sensing by setting the published front course under the course index. This applies regardless of your direction of travel, whether inbound or outbound on either the front or back course. Each dot of displacement on your CDI/HSI equates to a specific distance from the localizer centerline, depending on your distance from the runway (see inset).

As you fly on a localizer course you will notice that the CDI is more sensitive than during VOR navigation. For instance, when navigating using a VOR, full-scale deflection of the CDI represents a 20° radial span, or 10° to each side of course. However, when using localizer signals, the total span of the CDI is 5°, or 2.5° each side of the course. The advantage to greater CDI sensitivity is that you can track a localizer course with a theoretical accuracy four times that of a VOR radial. Because of this, however, CDI movement is more rapid during localizer tracking necessitating timely corrections for off-course indications. The corrections you make should be relatively small to help prevent overshooting your desired course.

GLIDE SLOPE

The horizontal needle of a VOR indicator or HSI provides the vertical guidance you need to maintain the glide slope. The glide slope transmitter is offset and somewhat elevated from the runway centerline and normally only directs signals to the front course approach. [Figure 8-16]

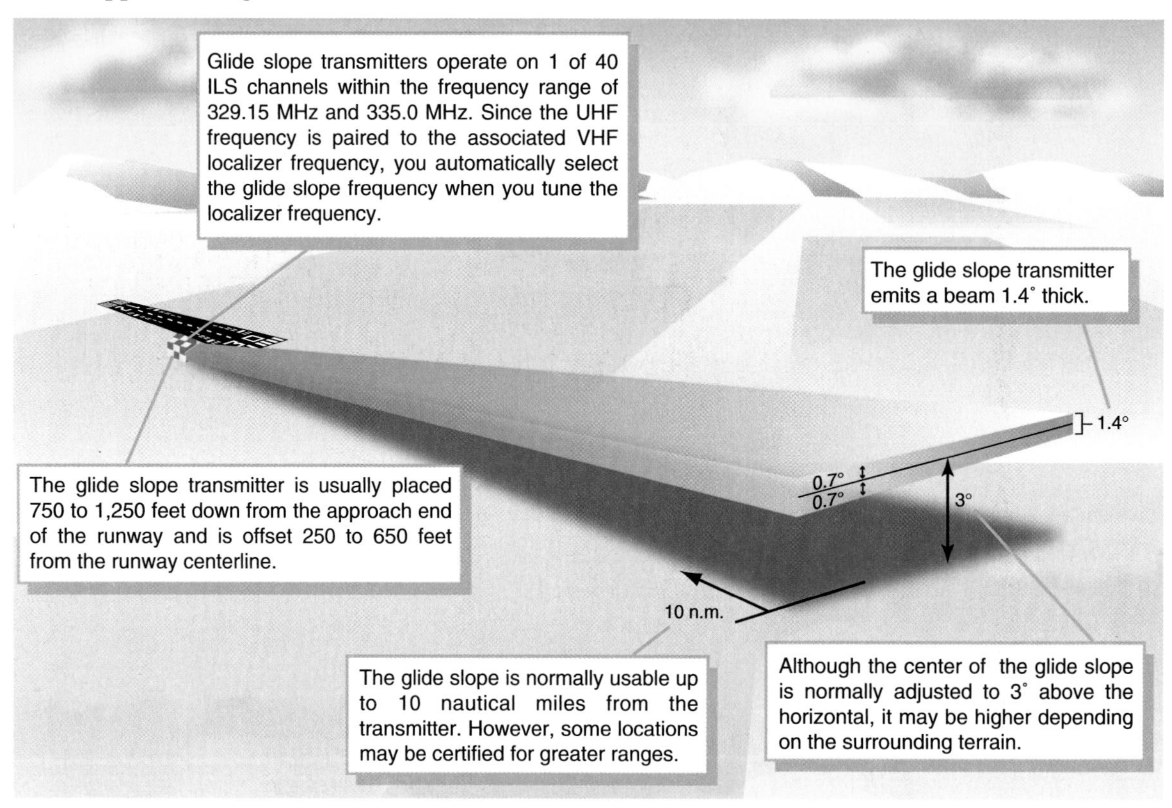

Figure 8-16. The glide slope transmitter is offset from the approach end of the runway to avoid being an obstacle to landing aircraft and to help prevent signal interference from aircraft on the ground.

The glide slope signal provides vertical navigation information for descent to the lowest authorized decision height for the associated approach procedure. If you receive glide slope guidance below decision height you should consider it unreliable. You may also receive other erroneous navigation information, such as false signals and reverse sensing, at high angles above the normal 3° glide slope projection. To avoid navigation errors, you should only rely on glide slope indications from the time you approach the glide slope intercept altitude shown on the approach chart until you reach decision height.

Since full-scale deviation of the glide slope needle is 0.7° above or below the center of the glide slope beam, a position only slightly off the glide slope centerline will produce large needle deflections on the navigation indicator. To fly a smooth approach, you must respond immediately to needle movement with heading, pitch, and/or power changes. [Figure 8-17]

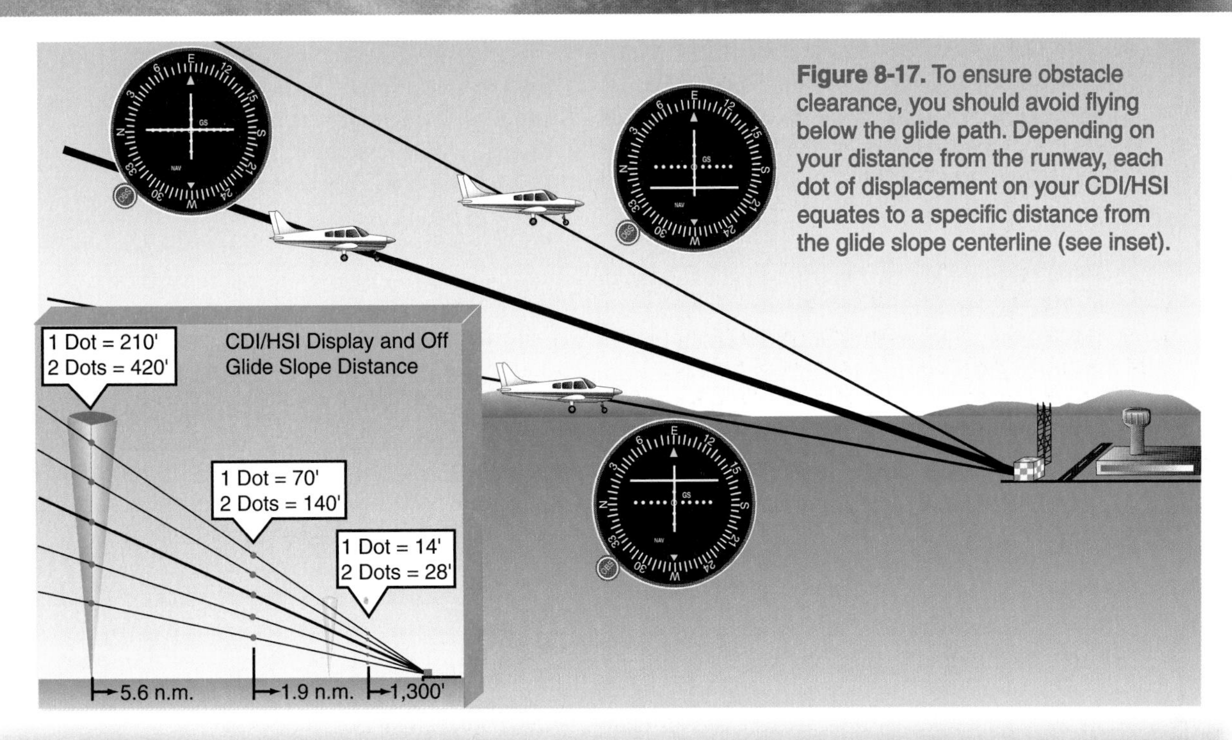

Figure 8-17. To ensure obstacle clearance, you should avoid flying below the glide path. Depending on your distance from the runway, each dot of displacement on your CDI/HSI equates to a specific distance from the glide slope centerline (see inset).

THE SAFER APPROACH

Is it safer to fly a precision approach than a nonprecision procedure? There is little question that if both types of approaches are flown correctly by a well trained pilot to a successful landing, the approaches can be considered equally safe. However, in some cases, specifically when analyzing accident statistics, the safety of an operation can be defined as the relative risk over a period of time. When taken in this context, it seems that precision approaches are indeed the safer approach.

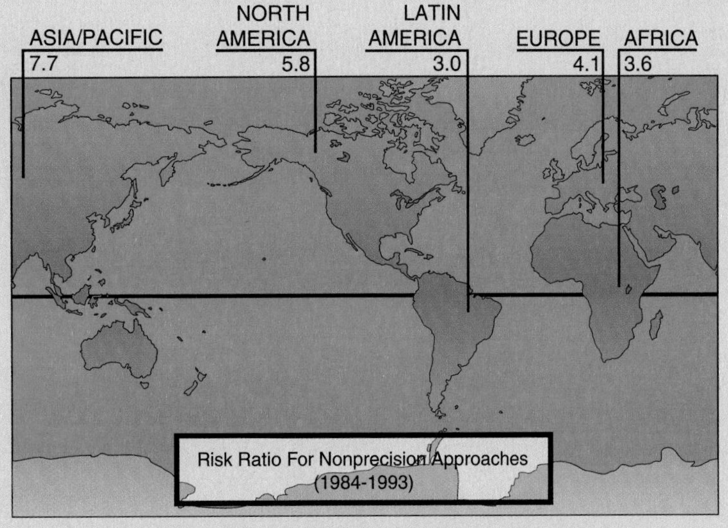

A recent study completed under the auspices of the Flight Safety Foundation concluded that an increase in approach and landing safety can be obtained through the use of precision approach and landing guidance facilities. In the course of the study, a risk ratio was calculated for commercial aircraft flying a nonprecision approach for several regions of the world between 1984 and 1993 (see the accompanying graph). When an aggregate was calculated for all regions, it was concluded that the accident risk was over five times greater when flying a nonprecision approach versus a precision procedure.

Operations experts believe that the lowest-risk approach results when the pilot establishes a stabilized approach. This can be accomplished with the addition of glide slope guidance to help pilots fly a steady descent to landing. Without the aid of a glide slope, the flight crew must more actively navigate vertically which can increase workload and translate to a degradation of situational awareness.

One might ask the logical follow-up question, “If precision approaches can increase safety, why don’t all approach procedures incorporate some sort of vertical guidance system?” The answer centers on two primary issues — terrain and economics. While the terrain surrounding an airport can limit the locations where precision approaches can be used, another, possibly more significant issue lies in the higher cost of precision guidance systems. This economic constraint limits precision approach placement to only the busiest airports, particularly in developing countries. Fortunately, new technologies, such as GPS, may soon provide a solution to these problems. In any case, regardless of the sophistication of the approach system, one fact will remain — no one thing can make an approach safer than a well trained pilot.

MARKER BEACONS

ILS marker beacons provide range information with respect to the runway during the approach. All ILS marker beacons project an elliptical array upward from the antenna site. At about 1,000 feet above the antenna, the array is 2,400 feet thick and 4,200 feet wide. At 1,000 feet AGL, with a groundspeed of 120 knots, you will receive the signal for about 12 seconds.

Usually, there are two marker beacons associated with an ILS — the outer marker (OM) and middle marker (MM). The placement of the OM varies from four to seven miles from the runway, depending on the installation. It usually is placed inside the point where an aircraft flying the ILS intercepts the glide slope. The MM is usually located 3,500 feet from the landing threshold, with its signal array intercepting a 3° glide slope at approximately 200 feet above the touchdown zone.

At some locations where Category II and III ILS operations have been certified, an inner marker (IM) is installed. The inner marker indicates the decision height on the CAT II glide slope and indicates progress on a CAT III approach. Occasionally, marker beacons may be used on the localizer back course to indicate the final approach fix.

The glide slope centerline normally intersects the middle marker at an altitude of approximately 200 feet above the touchdown zone.

In most general aviation aircraft, the marker beacon receiver is incorporated into the audio control console for the avionics. Marker beacon receivers have three separate lights to correspond to the types of marker beacons. The audio identification of each marker beacon can be heard over the speaker during station passage. Usually, a control is provided for high or low receiver sensitivity. [Figure 8-18]

As you pass over a marker beacon, the appropriate visual display and audio identification is activated. See figure 8-18.

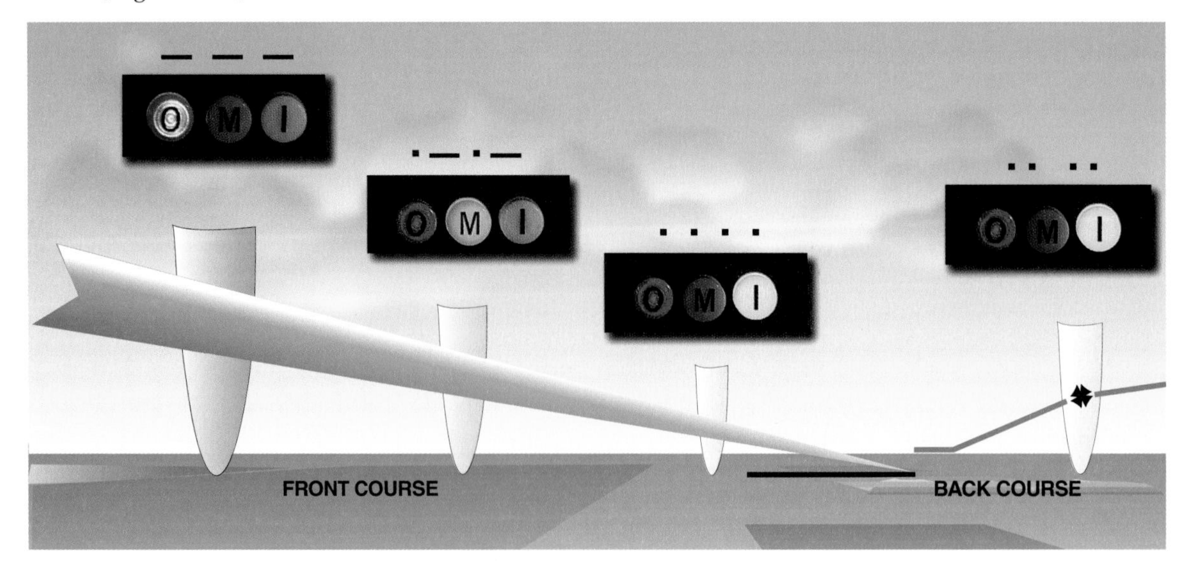

Figure 8-18. When the aircraft passes through the signal array of the respective marker beacon, the associated light flashes and the Morse code identification sounds.

COMPASS LOCATORS

Many ILS systems use a low power, low or medium frequency (L/MF) radio beacon, called a compass locator collocated with the outer or middle marker. When the compass locator is installed in conjunction with the outer marker, it is called an outer compass locator (LOM). Similarly, when the compass locator is associated with the middle

marker, it is referred to as a middle compass locator (LMM). Compass locators usually have a power output of less than 25 watts, resulting in a reception range of at least 15 miles. At some locations, high powered NDBs of up to 400 watts may be utilized as outer compass locators.

The first two letters of the localizer identification group designate an LOM, and the last two letters are used for an LMM.

The frequency range for compass locators is 190 to 535 kHz. Compass locators transmit a two-letter Morse code identifier taken from the last three letters of the localizer identifier group. For example, if the localizer identifier is IXYZ, the LOM would transmit the first two letters of the localizer identifier group (XY). If an LMM is installed, it would transmit the second two letters of the localizer identifier group (YZ).

DME

On many ILS procedures, a DME transmitter is placed at or near the localizer or glide slope transmitter to provide runway distance information. When installed on an ILS approach, you may use DME in lieu of the outer marker or to identify other published fixes on a localizer front or back course. In some cases, you also may use DME information from a separate facility, such as a VORTAC. On Jeppesen charts you can determine if DME is associated with the localizer frequency by looking for the ILS/DME notation on top of the facility box. NOS charts include the DME/TACAN channel in the localizer frequency box.

When DME is available through the localizer frequency, Jeppesen charts publish the notation, ILS/DME, on the top of the facility box. On NOS charts, a DME/TACAN channel is shown in the facility box.

VISUAL INFORMATION

Since you commonly fly ILS approaches in periods of poor visibility, visual information in the form of approach and runway lighting systems is provided to help you acquire the runway environment and transition to visual flight. Since some combinations of visual navigation aids can make the runway environment more apparent than others, certain lighting configurations can qualify precision approaches for lower minimums. [Figure 8-19] Complete information about lighting systems is contained in Chapter 3, Section A — Airports, Airspace, and Flight Information.

PRECISION APPROACH LIGHTING AND MINIMUMS

Approach Light Configurations	Height Above Touchdown	Categories A,B,C, and D	
		Visibility	RVR
No Lights	200 feet	3/4 mile	4,000 feet
MALSR, SSALR, or ALSF-1	200 feet	1/2 mile	2,400 feet
MALSR, SSALR, or ALSF-1 **and** TDZ/CL	200 feet	—	1,800 feet

Figure 8-19. Whenever the minimum landing visibility for an ILS approach is specified as 1,800 feet RVR, touchdown zone and runway centerline lights are required. Anytime RVR is used, HIRL are required. In addition, runway edge lights are required for night operations.

INOPERATIVE COMPONENTS

The lowest landing minimums on an approach are authorized when all components and visual aids are operating. If some components are inoperative, higher landing minimums may be required. If more than one component is inoperative, you apply only the greatest increase in altitude and/or visibility required by the failure of a single component.

According to FAR 91.175, certain substitutions for equipment outages are authorized. A compass locator or precision radar may be substituted for the outer marker or middle marker. DME, VOR, or NDB fixes authorized in the standard instrument approach procedure or surveillance radar may be substituted for the outer marker. However, when the glide slope becomes inoperative (GS out), localizer-only minimums must be used. When the localizer is out or when a substitution for the outer marker cannot be made, an ILS approach is not authorized.

If more than one component of an ILS is unusable, use the highest minimum required by any single component that is unusable.

A compass locator, precision radar, surveillance radar, or published DME, VOR, or NDB fixes may be substituted for the outer marker.

If the glide slope is inoperative or fails during your approach, the localizer (GS out) minimums apply and you may continue the approach to the applicable MDA.

When any basic ILS ground component (except the middle marker or the localizer) or required visual aid is inoperative, unusable, or not utilized, the standard straight-in landing minimums prescribed by the approach are raised. Jeppesen charts show the increase in minimums directly on the approach chart. If you use NOS charts, you compute the higher minimums by consulting the Inoperative Components or Visual Aids Table located in the front of the *Terminal Procedures Publication*. [Figure 8-20]

If an ILS visual aid is inoperative, the visibility requirements are raised on approaches where 1,800 RVR is authorized.

INOP COMPONENTS

INOPERATIVE COMPONENTS OR VISUAL AIDS TABLE

Landing minimums published on instrument approach procedure charts are based upon full operation of all components and visual aids associated with the particular instrument approach chart being used. Higher minimums are required with inoperative components or visual aids as indicated below. If more than one component is inoperative, each minimum is raised to the highest minimum required by any single component that is inoperative. ILS glide slope inoperative minimums are published on instrument approach charts as localizer minimums. This table may be amended by notes on the approach chart. Such notes apply only to the particular approach category(ies) as stated. See legend page for description of components indicated below.

(1) ILS, MLS, and PAR

Inoperative Component or Aid	Approach Category	Increase Visibility
ALSF 1 & 2, MALSR, & SSALR	ABCD	¼ mile

(2) ILS with visibility minimum of 1,800 RVR.

ALSF 1 & 2, MALSR, & SSALR	ABCD	To 4000 RVR
TDZI RCLS	ABCD	To 2400 RVR
RVR	ABCD	To ½ mile

Figure 8-20. This excerpt from the NOS Inoperative Components or Visual Aids Table shows the minimums increase for the failure of certain components associated with a precision approach. For example, suppose you plan on flying an ILS with a visibility minimum of 1,800 RVR and subsequently are informed that the ALSF 1 is out of service. In this case, you should raise the visibility minimum to 4000 RVR.

FLYING THE ILS

ILS approach training introduces vertical guidance by radio navigation. While tracking the localizer, your CDI senses horizontal movement of the aircraft away from the course. Simultaneously, a precise descent path to the runway is provided through glide slope information displayed on the same instrument. [Figure 8-21]

Figure 8-21. When flying an ILS, you track the line formed by the intersection of the glide slope and localizer courses.

Prior to intercepting the ILS glide slope, you should concentrate on stabilizing airspeed and altitude while establishing a magnetic heading which will maintain the aircraft on the localizer centerline. Flying at a constant airspeed is not only desirable, but essential for smooth, accurate descents to decision height. At glide slope interception, you need to initiate a descent to stay on glide slope. In fixed-gear aircraft, you will usually need to reduce power, although the amount of reduction depends on whether you lower the flaps. In retractable gear aircraft, the amount you adjust pitch and power depends on when you lower the landing gear and flaps. [Figure 8-22]

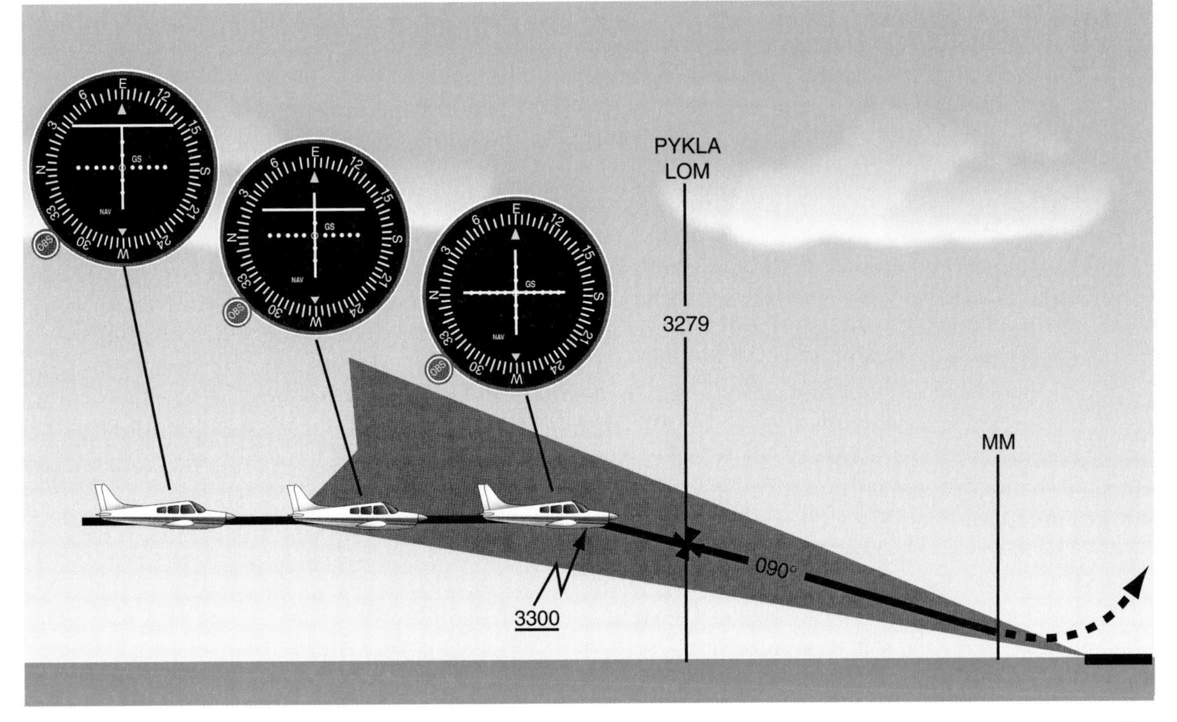

Figure 8-22. Since your aircraft is usually below the glide slope during the intermediate approach segment, the glide slope indicator will display a full-up needle deflection. You should observe the initial downward movement of the indicator and lead the descent to intercept the glide slope centerline accordingly.

Intercepting the glide slope at the proper speed will make the descent more stable. However, if your airspeed is too high after glide slope interception, you will need to make a further power reduction. As the airspeed decreases, you should make a pitch adjustment to keep the aircraft from going below the glide slope. After the descent rate stabilizes, use power as necessary to maintain a constant approach speed. You can use small pitch changes to maintain the glide slope. However, if the glide slope indicator approaches full scale deflection, you should respond immediately with pitch and power adjustments to re-intercept the glide slope.

If the glide slope and localizer are centered but your airspeed is too fast, your initial adjustment should be to reduce power.

The rate of descent you should maintain primarily depends on your groundspeed. For the same glide slope angle, you will need a lower rate of descent as your groundspeed decreases, and vice versa. For example, in no-wind conditions, a 555-foot per minute rate of descent will keep you on a 3° glide slope if you maintain 105 KIAS. However, with a 15-knot headwind (90-knot groundspeed), you will have to reduce your rate of descent to 480 feet per minute to stay on a 3° glide slope.

If your groundspeed decreases, the rate of descent required to stay on glide slope must also decrease, and vice versa.

While inbound on the localizer, establish required drift corrections before you reach the outer marker. As the course narrows, make small drift corrections and reduce them proportionately. By the time you reach the outer marker, your drift correction should be established accurately enough to permit you to complete the approach with heading corrections no greater than 2° in calm wind conditions.

Localizer and glide slope indications become more sensitive as you get closer to the runway. On a well executed ILS in calm wind conditions, you should not need heading corrections greater than 2° after you have passed the outer marker.

If you maintain the glide slope for an approach, you should reach the decision height at approximately the middle marker (if installed). Although you may reach the decision height at or near the middle marker, the charted MAP for an ILS approach is the point where the glide slope intercepts the decision height. This point may, or may not, be at the middle marker.

During actual IFR weather conditions, it is usually apparent when you can continue the approach visually. However, prior to DH or MDA (GS out), you should continue your instrument cross-check with only brief glances outside until you are sure that you have established positive visual contact with the runway environment. It is not unusual on an ILS approach to establish visual contact at 500 to 600 feet AGL and then lose outside visual references as the descent continues. This may be due to a very low fog layer that allows visual contact from above but causes loss of visual cues on a horizontal plane within the layer. For this reason, you should avoid descents below the glide slope before you reach DH, even though you have visual contact with the runway.

The conditions under which you can descend below DH or MDA on an approach are specified in FAR 91.175 and discussed in Chapter 7, Section B — Approach Procedures. If the required conditions are not met at the MAP, you must execute a missed approach. A missed approach also is required if you cannot maintain these requirements all the way to touchdown. During a missed approach, you must comply with the published procedure unless ATC specifies otherwise.

The DH on the glide slope is the MAP on a precision approach. If you have not established the required visual references at the DH on an ILS approach, you must execute the missed approach.

STRAIGHT-IN (NOPT) ILS APPROACH

As with the other types of approaches, each ILS approach is designed for the individual airport. Many ILS procedures allow you to fly the approach without executing the depicted course reversal. These straight-in (NoPT) initial approach segments can be a convenient and efficient way of positioning the aircraft on the final approach course. The following discussion assumes you are inbound from the north for landing at Lebanon Municipal Airport, Lebanon, New Hampshire.

PREPARING FOR THE APPROACH

Your route takes you over the Montpelier VOR/DME, an IAF for Lebanon's only ILS approach. The current ATIS reveals that the weather is 500 broken, 2 miles visibility in rain showers, wind 140° at 7 knots, and the altimeter setting is 29.94. The ATIS also advises that ILS Runway 18 approaches are in progress. With this information in mind, conduct an approach chart review so you are familiar with the procedure. [Figure 8-23]

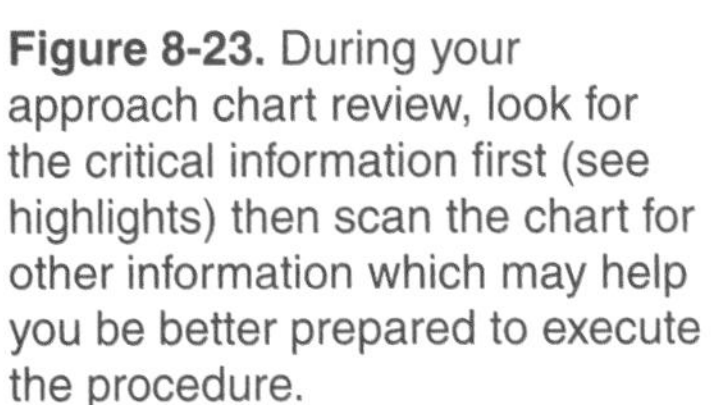

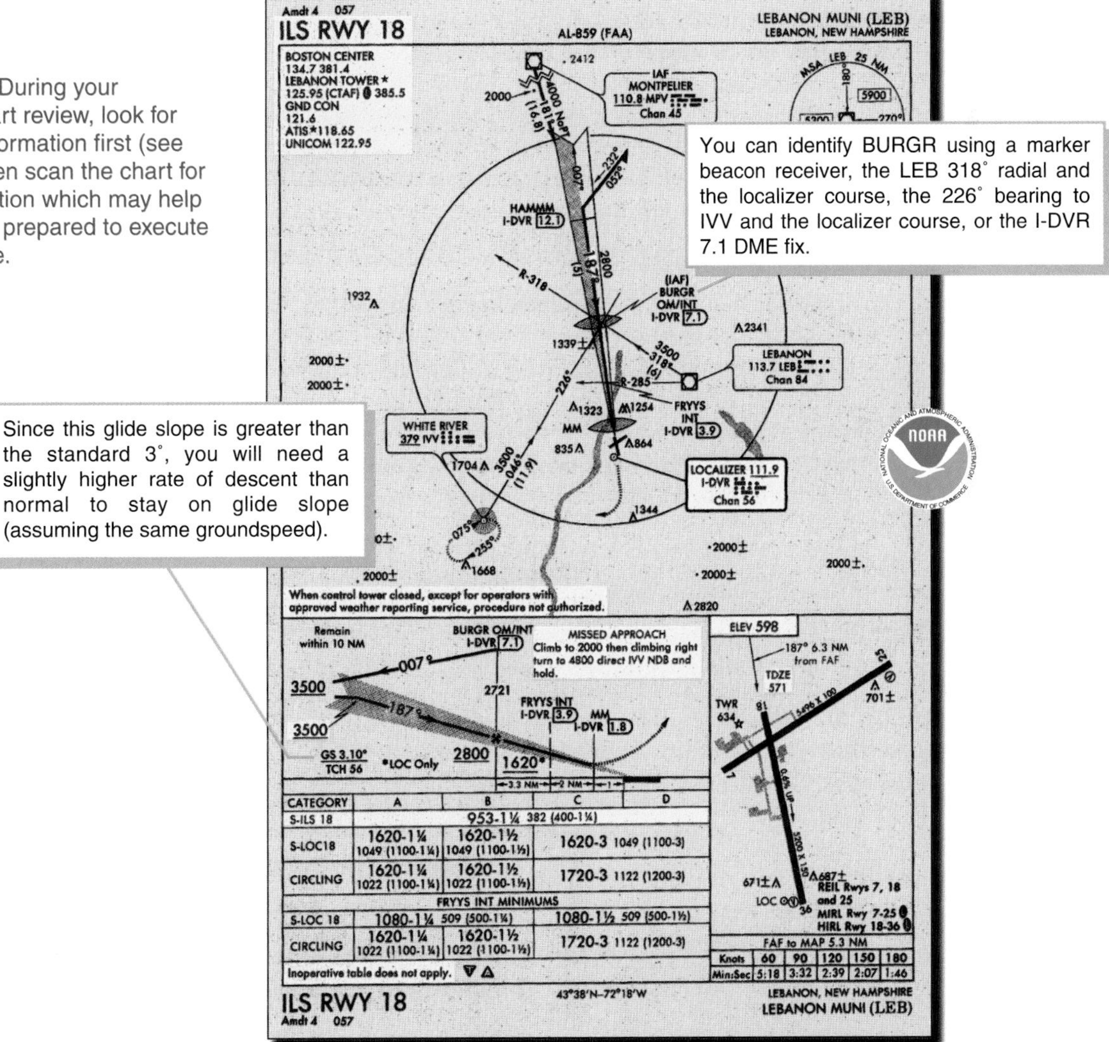

Figure 8-23. During your approach chart review, look for the critical information first (see highlights) then scan the chart for other information which may help you be better prepared to execute the procedure.

To prepare for the approach, reset your altimeter and tune your navaids for the approach. Set the ADF to 379 kHz for the IVV NDB and the VOR/LOC receiver to 111.9 MHz for the localizer. Tune the number two communications radio to the tower frequency of 125.95 MHz and leave the other VOR receiver tuned to the MPV VOR/DME. Verify all navaid frequencies, including the DME, by listening to the identification feature, then turn on the marker beacon receiver and test it for proper operation.

YOUR CLEARANCE

Prior to reaching MPV, Boston Center advises, *"Arrow 659MR, you are cleared for the ILS Runway 18 approach to Lebanon. Maintain 6,000 until established on the approach. Contact Lebanon Tower at BURGR."* With this clearance, you can begin the approach upon arrival at the IAF.

DESCENT PRIOR TO THE FINAL APPROACH COURSE

Since the MPV VOR/DME is an IAF, you can begin a descent to 4,000 feet MSL once you are established on a magnetic course of 181°. Using a 500-foot per minute rate of descent at 120 knots, you should reach 4,000 feet in 4 minutes and level off about 12 miles prior to reaching HAMMM. This is a good time to complete your before landing checklist, with the possible exception of the landing gear and/or flaps. As you approach HAMMM, scan the VOR indicator (or HSI) for localizer indications. As the localizer needle begins to center, you time your turn to roll out on the final approach course.

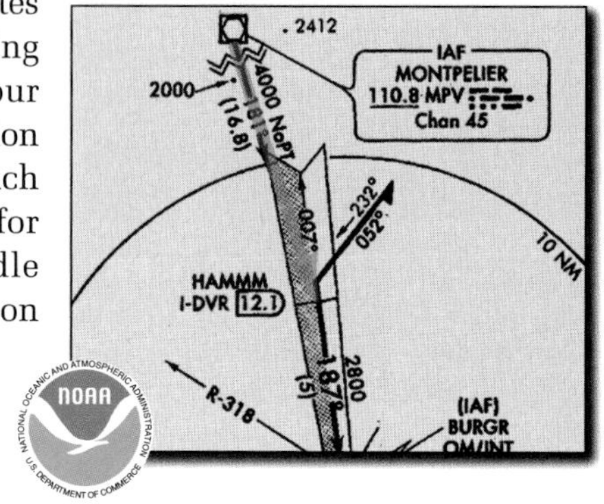

INBOUND TO THE OUTER MARKER

Once you are established on the localizer, begin a descent to 3,500 feet MSL, as shown in the chart profile view. Continue to scan your VOR indicator (or HSI) since you may have to use a wind correction to stay on course. As you level off (about 3 miles from BURGR), slow to approach speed and maintain altitude until you intercept the glide slope from below. Begin a stabilized descent to remain on glide slope while simultaneously tracking the localizer. At a groundspeed of 90 knots, a descent rate of 480 feet per minute will keep you on the standard 3° glide slope. If you have not already lowered the landing gear and/or flaps, do so prior to reaching the outer marker, BURGR. You can monitor your progress toward BURGR using DME as well as scanning for the outer marker on-glide-slope altitude of 2,721 feet MSL.

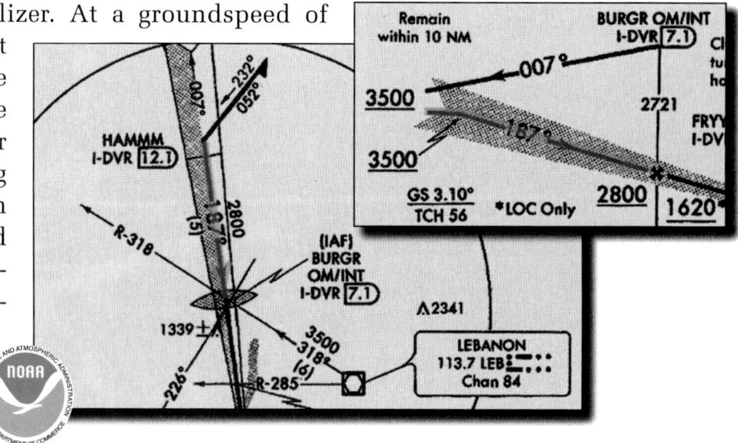

FINAL APPROACH SEGMENT

You know you have arrived at the OM when your marker beacon receiver flashes a blue light and emits a series of dash tones. As directed earlier, contact Lebanon tower: *"Lebanon Tower, Arrow 659MR at BURGR inbound on the ILS Runway 18 approach."* Depending on traffic, you may receive landing clearance at this time. At 3.9 DME, you reach FRYYS, which is only significant if you are flying the approach without the aid of a glide slope. As you continue the approach, make corrections as necessary to stay on glide slope and localizer. [Figure 8-24]

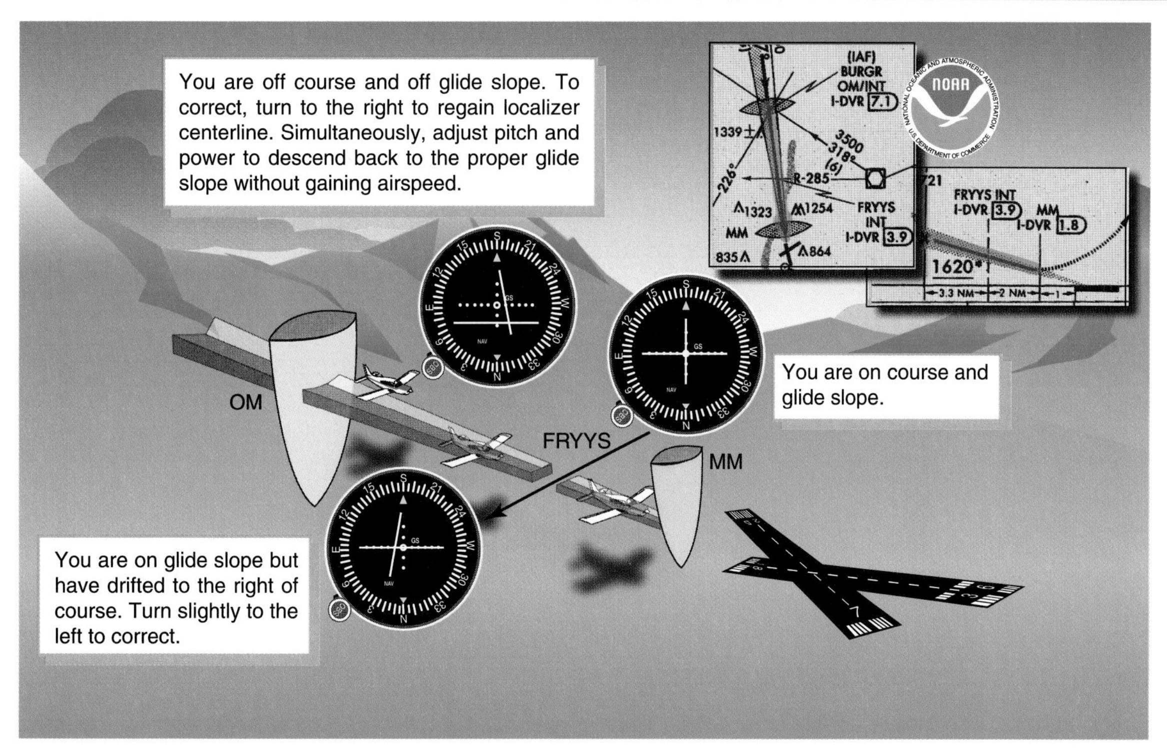

Figure 8-24. Once you are established on ILS final approach course you should make small corrections to remain on course and glide slope. Remember to use pitch attitude to control your position relative to the glide slope and adjust the throttle to maintain airspeed.

While descending toward the runway, systematically monitor the altimeter so you know precisely when you reach the DH of 953 feet MSL. If you establish the required visual contact with the runway environment prior to, or at, the DH, you can continue for a straight-in landing.

MISSED APPROACH

If you reach decision height and determine that you must execute a missed approach, add power, increase pitch attitude, and raise the gear and/or flaps, as appropriate. When you report the missed approach to tower you will probably be told to contact Boston Center to request clearance for another approach or routing to your alternate. To continue the missed approach procedure, climb straight ahead to 2,000 feet MSL before beginning a right turn direct to the IVV NDB. Keep climbing until you reach 4,800 feet MSL. Depending on winds and your climb rate, you will probably use a parallel entry into the published holding pattern at IVV.

ILS APPROACH WITH A COURSE REVERSAL

Some ILS procedures, particularly those serving airports without radar coverage, necessitate the use of a course reversal. You might encounter a good example of this if you were flying into West Yellowstone, Montana.

PREPARING FOR THE APPROACH

In this example, your planned route takes you south over WAIDE Intersection at 16,000 feet inbound to West Yellowstone. After obtaining the local altimeter setting and other airport information from UNICOM, you conduct a chart review for the approach in use. [Figure 8-25]

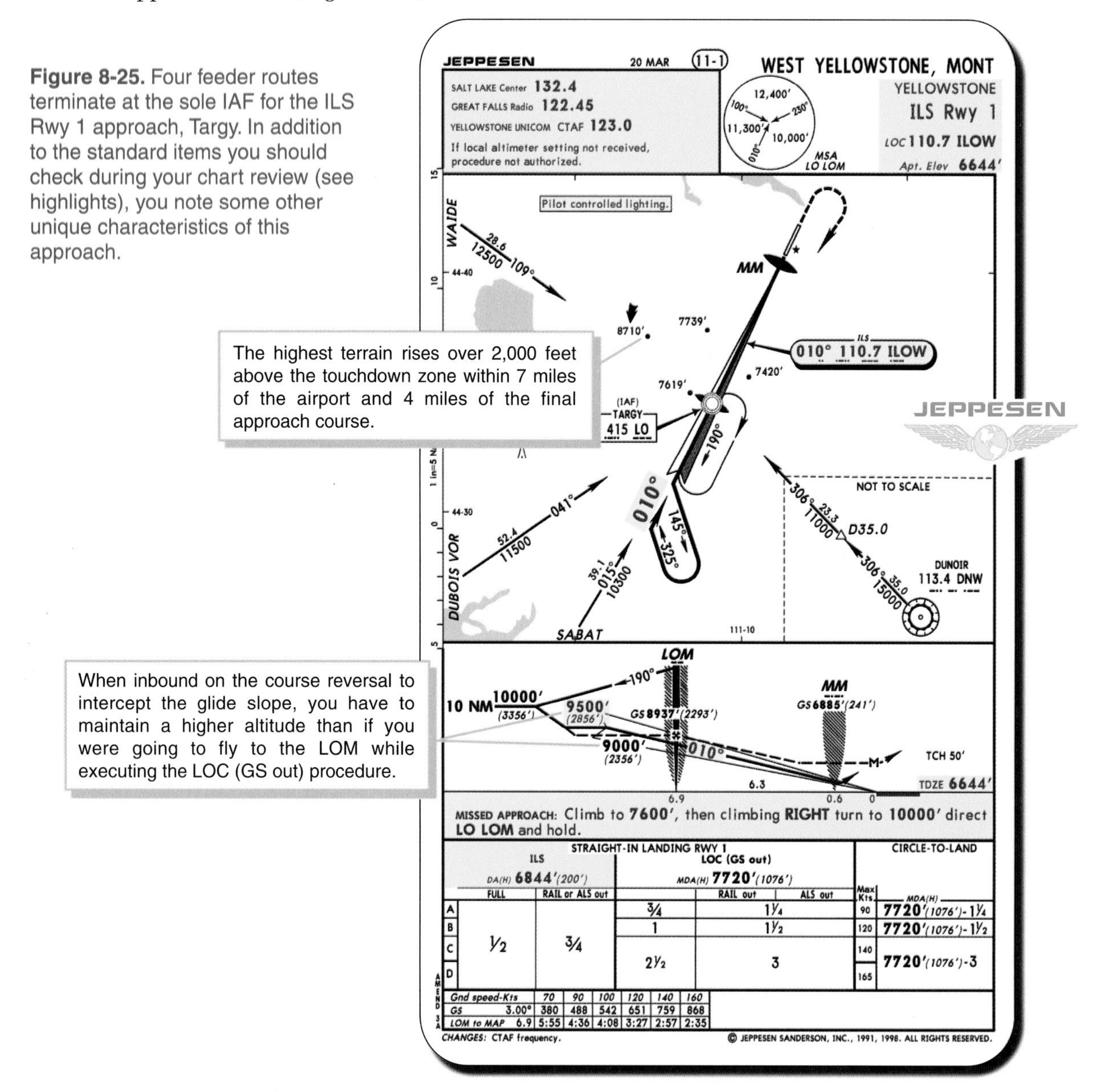

Figure 8-25. Four feeder routes terminate at the sole IAF for the ILS Rwy 1 approach, Targy. In addition to the standard items you should check during your chart review (see highlights), you note some other unique characteristics of this approach.

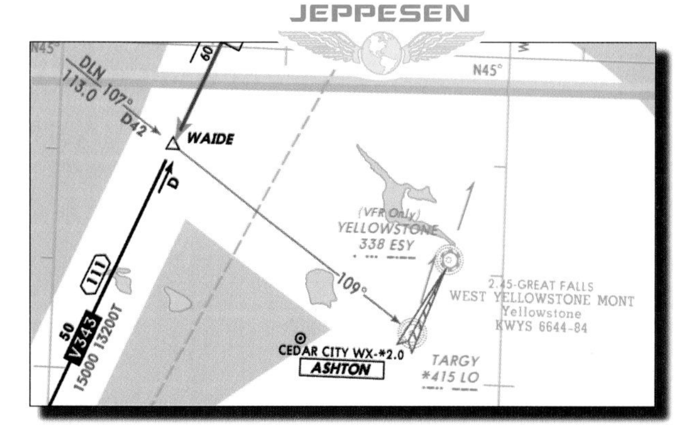

As you prepare to transition from the enroute portion of your flight, tune your ADF receiver to 415 kHz for the Targy LOM, identify the station, and check the bearing indicator for proper operation. With two VOR receivers, you can tune one of them to the localizer, ILOW. At this distance from the airport, however, it is unlikely you will be able to receive the signal. Therefore, you make a mental note to identify the station as you get closer.

YOUR CLEARANCE

While headed south on V343, ATC asks you to report WAIDE Intersection. Upon arrival at WAIDE, Salt Lake Center issues you an approach clearance: *"Arrow 659MR, proceed direct to Targy. Descend and maintain 13,000. Cleared ILS Runway 1 approach to Yellowstone. Report Targy outbound."*

DESCENT PRIOR TO THE IAF

Since WAIDE Intersection is not an IAF, you may only descend to your last assigned altitude until you reach the IAF, Targy. From WAIDE, you should fly a magnetic course of 109° to the LOM. To help you maintain your course, you can check the ADF bearing indicator for any apparent wind drift as you fly toward the LOM and, if necessary, apply a wind correction. Prior to reaching the LOM, make sure you identify the localizer you previously tuned.

OUTBOUND ON THE PROCEDURE TURN

After you arrive at Targy, turn to intercept the localizer outbound. Since you are crossing over the outer marker, the ADF bearing indicator will begin to move away from the nose position and the blue marker beacon light will flash. As you turn outbound on the localizer course, reverse CDI sensing will be apparent. As previously directed, report outbound: *"Salt Lake Center, Arrow 659MR Targy outbound."* At this point Center will typically instruct you to contact the CTAF. When the ATC controller advises *"Change to advisory frequency approved,"* you should switch to the CTAF frequency and broadcast your intentions, including the type of approach being executed, your position when over the FAF inbound (nonprecision approach) or when over the outer marker or fix used in lieu of the outer marker (precision approach) inbound.

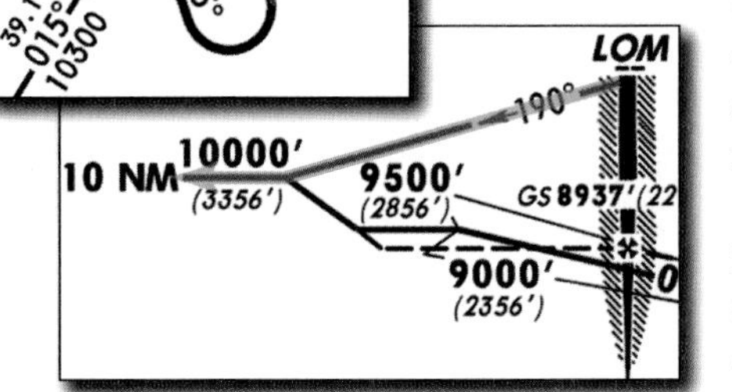

During this segment of the approach, begin a descent to 10,000 feet MSL. Within two minutes of tracking outbound on the localizer, turn left to a heading of 145°. You may want to complete the before landing checklist (except for landing gear and/or flaps) during this portion of the approach.

INBOUND TO THE LOM

After flying for approximately one minute on the procedure turn outbound, reverse course by completing a 180° standard rate right turn to a heading of 325°. It is a good idea to set the course selector to 010° as a reminder of the inbound course. As the ADF bearing indicator moves toward a 045° relative bearing, watch the CDI (which now displays proper

sensing) for the localizer needle to begin to center. As the needle centers, time your right turn to 010° so that you roll out on course. Once you are established on the localizer inbound begin a descent to 9,500 feet MSL. As you continue the approach, apply any necessary wind correction to keep the localizer needle centered. Keep scanning for glide slope interception, which you can expect to occur before you reach the LOM.

FINAL APPROACH SEGMENT

Upon crossing the LOM, start your timer so that you can identify the localizer-only missed approach point in the event that the glide slope fails. As you fly the final approach segment from Targy, apply normal ILS procedures to stay on course and glide slope. Continue to scan the altimeter for your DA(H) of 6,844′(200′). [Figure 8-26]

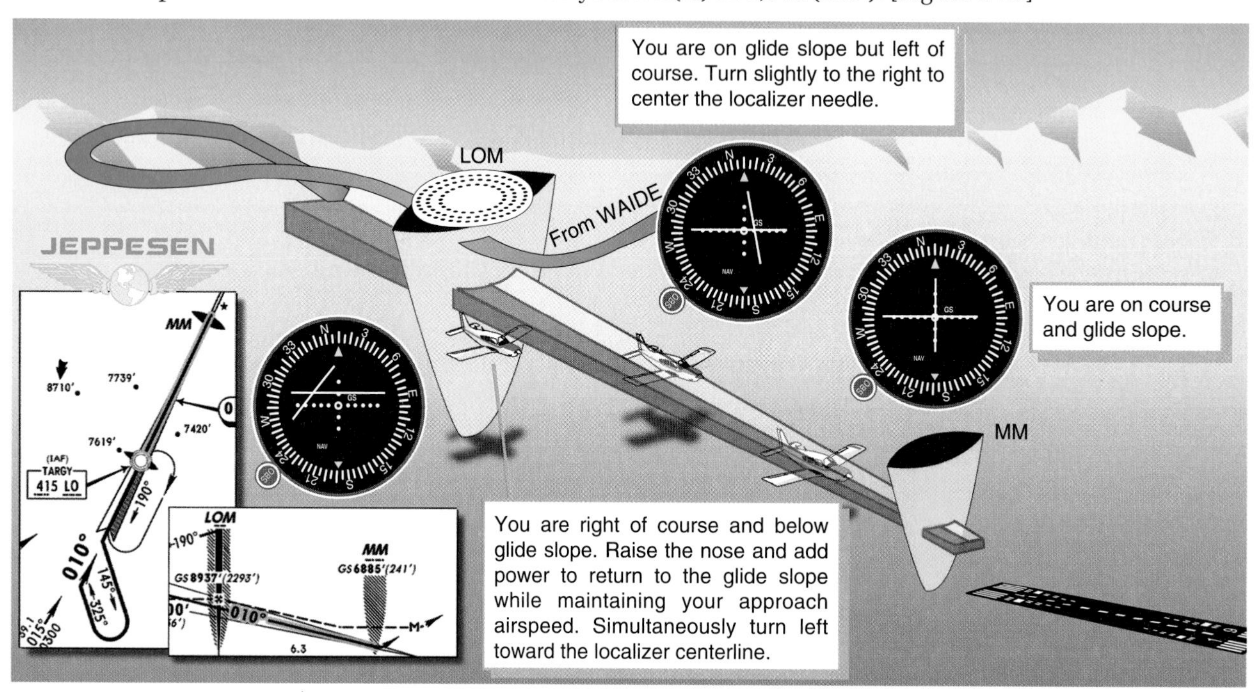

Figure 8-26. You should always correct immediately for any below glide slope indications.

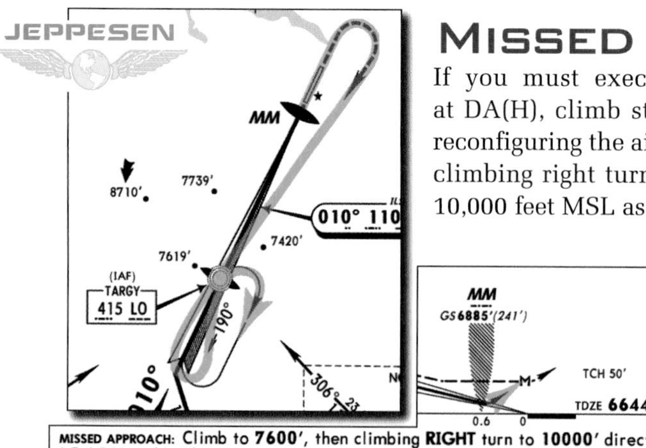

MISSED APPROACH

If you must execute a missed approach upon arrival at DA(H), climb straight ahead to 7,600 feet MSL while reconfiguring the aircraft. Upon reaching 7,600 feet, begin a climbing right turn and track direct to Targy. Level off at 10,000 feet MSL as you fly toward the LOM. In most cases, you will have to make a parallel entry into the depicted holding pattern. Make sure you report the missed approach to Salt Lake Center and request another approach or clearance to your alternate.

ILS/DME APPROACH

Although separate ILS approach procedures share many similarities, there also are many differences due to facilities, equipment, terrain, and airport location. Since every instrument approach is unique, you should always study each approach carefully. The procedures you use to fly an ILS/DME approach are essentially the same as any other ILS approach except for the requirement to identify approach and/or missed approach fixes

using DME. As an example, assume you have been cleared for the ILS DME Rwy 32 approach from the Crazy Woman VOR (IAF) to the Sheridan County Airport, Sheridan, Wyoming. [Figure 8-27]

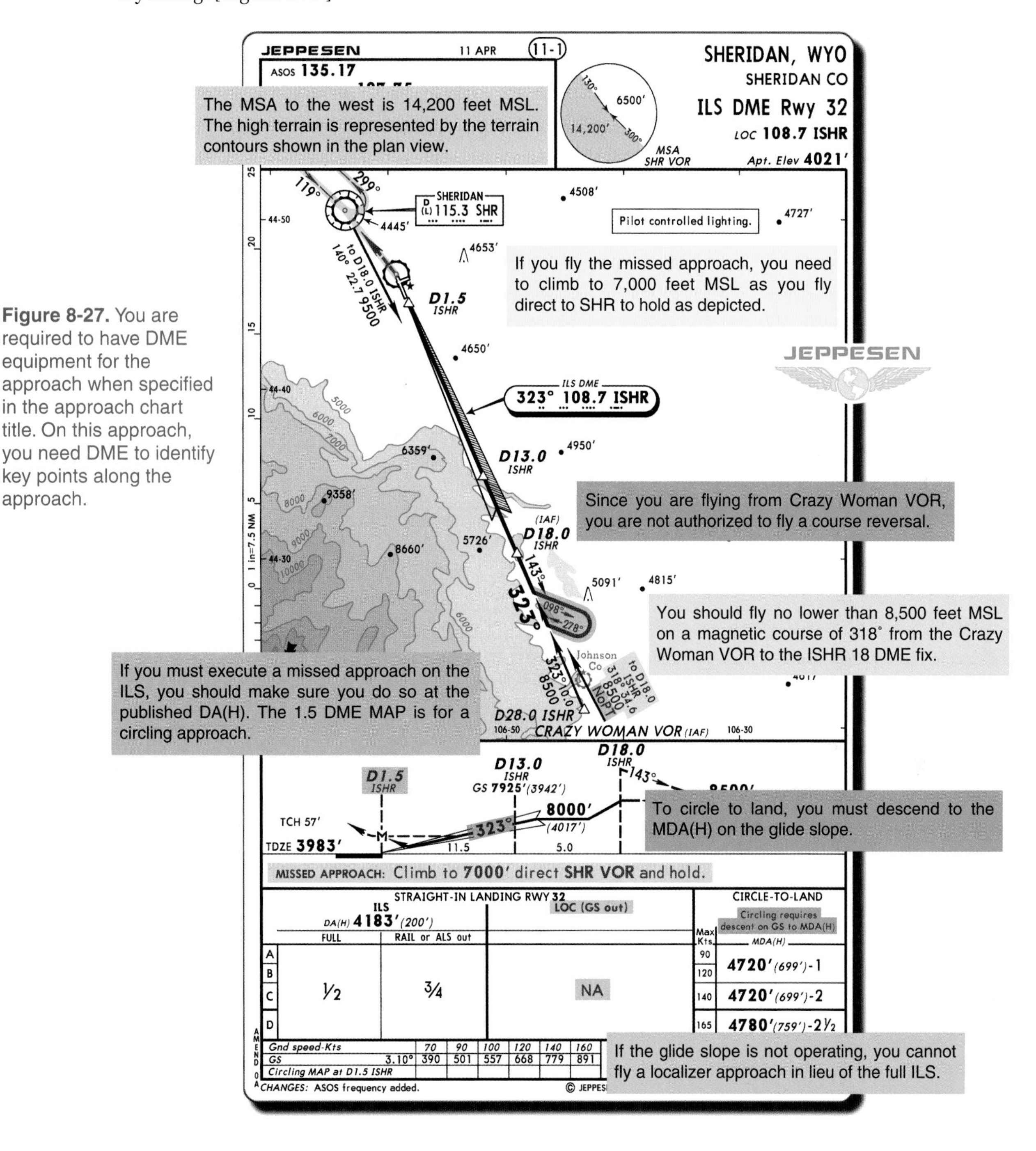

Figure 8-27. You are required to have DME equipment for the approach when specified in the approach chart title. On this approach, you need DME to identify key points along the approach.

RADAR VECTORS TO ILS FINAL

When radar is approved for approach control service, it is also used to provide vectors to published instrument approach procedures, such as the ILS. Radar vectors can be provided for course guidance and for expediting traffic to the final approach course of any established instrument approach procedure. Whenever ATC is providing radar vectors to the ILS approach course, you normally will be advised by the controller or

through ATIS. [Figure 8-28] The procedures related to flying radar vectors to the final approach course are covered in detail in Chapter 7, Section B — Approach Procedures.

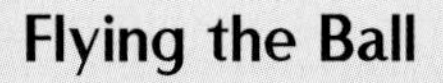

Figure 8-28. Although you can typically be radar vectored to the final approach course from any direction, it would be particularly efficient when approaching Mobile Regional Airport from the east. Once you are established on the final approach course, you fly the ILS just as you would without radar available.

Flying the Ball

You have just completed your assigned mission. You are hundreds of miles from land but only have enough fuel to fly for about another 45 minutes. Your job is to make an approach and land on a 400-foot runway which is moving away from you and pitching up and down as much as 30 feet. If you are a carrier-based Naval Aviator, this is probably just another day at the office. [Figure A]

Courtesy of U.S. Navy

Day or night, Navy pilots rely heavily on an alignment device, called a fresnel lens optical landing system (FLOLS), to provide approach guidance. The stabilized system consists of a yellow light, commonly referred to as the meatball (or simply the ball), which appears to move vertically, depending on the aircraft's position relative to the ideal glide slope. When on the proper approach angle, the ball is centered between two rows of fixed green lights. The FLOLS, which is the Navy's primary landing guidance system, is located on the port (left) side of the deck edge. [Figure B]

B

Courtesy of U.S. Navy

The glide slope coverage of the FLOLS is 1.7° high by 40° wide. If flown correctly, the aircraft will cross over the ramp of the carrier through a window 5 feet high and 20 feet wide. By keeping the ball in the center, the pilot flies the aircraft to a point which results in the aircraft's tailhook catching the 3rd of 4 arresting wires. To accomplish this feat, every pilot on approach must scan 3 critical items — the aircraft's angle of attack, lineup on the flight deck marking and/or lighting, and the ball. The pilot receives additional guidance from a Landing Signal Officer (LSO) stationed on the flight deck. [Figure C] While you may not have the same tools available to you as a carrier pilot, you also don't have to land on a pitching, rolling, heaving runway.

C

Courtesy of U.S. Navy

ILS APPROACHES TO PARALLEL RUNWAYS

At airports which have two or three parallel runway configurations ILS approach operations may be authorized on each runway. ILS approaches to parallel runways are divided into three classes depending on runway centerline separation as well as ATC procedures and capabilities. [Figure 8-29]

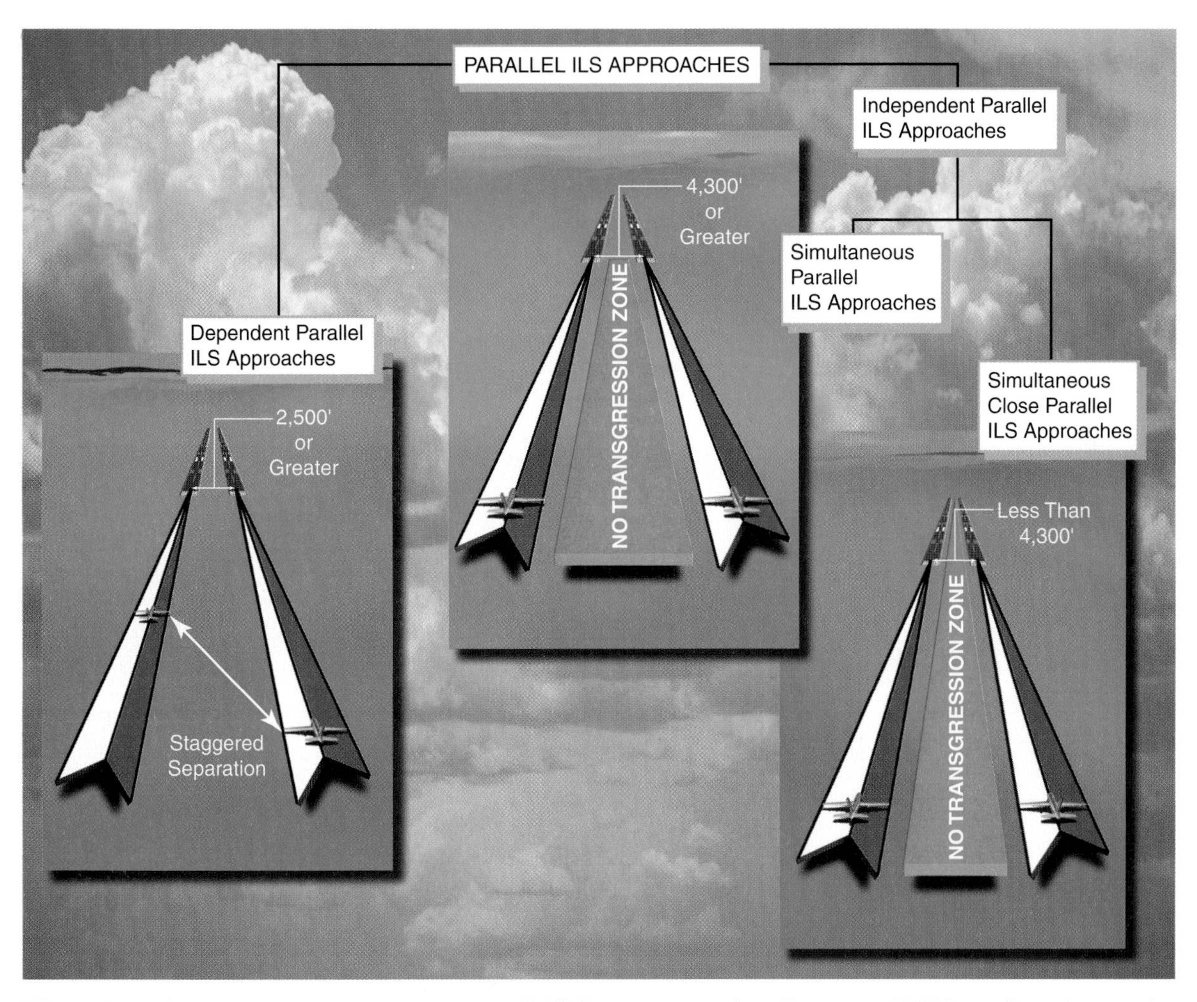

Figure 8-29. At some airports, one or more parallel ILS courses may be offset up to 3°. This configuration results in an increase in decision height of 50 feet and a loss of CAT II capabilities (if applicable).

Due to the close proximity of aircraft during some parallel procedures, it is imperative that you maintain situational awareness and strict radio discipline. During the approach, you should avoid lengthy and/or unnecessary radio transmissions to allow final approach controllers time to issue critical instructions as needed. In addition, to help eliminate any confusion among aircraft, you should always use your full call sign when responding to ATC.

PARALLEL (DEPENDENT) ILS APPROACH

Parallel (dependent) ILS approach operations may be conducted to parallel runways with centerlines at least 2,500 feet apart. However, due to the close proximity of the final approach courses, aircraft are separated by a minimum of 1.5 miles diagonally. If the runway centerlines are more than 4,300 feet but no more than 9,000 feet apart, ATC must maintain a diagonal separation between aircraft of at least 2 miles. The procedures you use to fly the approach are the same as a standard (nonparallel) ILS approach. [Figure 8-30]

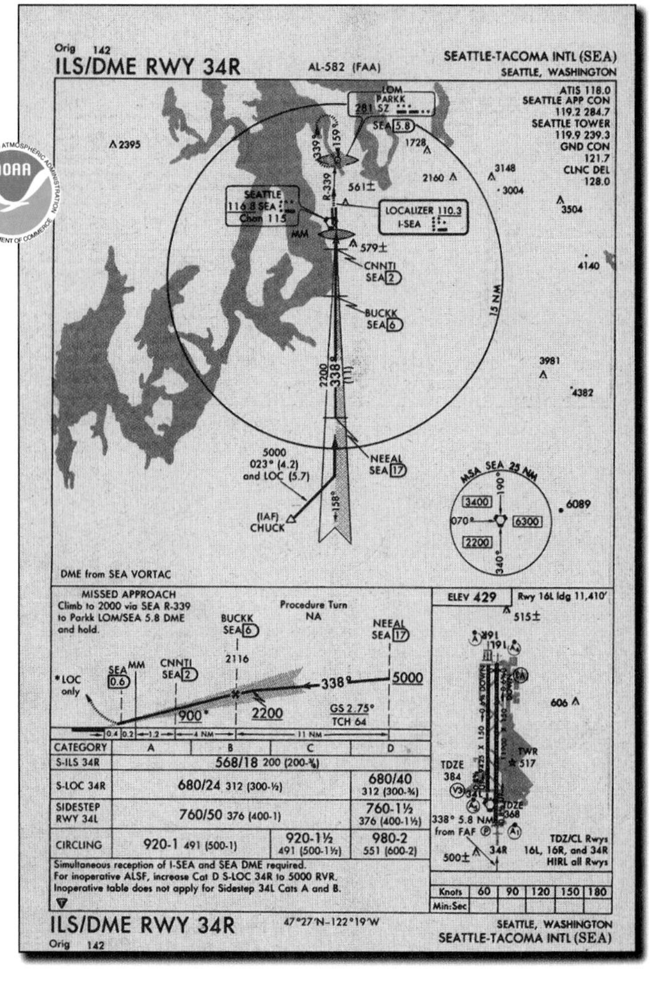

Figure 8-30. While the approach chart is not required to indicate that parallel approaches may be conducted, you will be informed by ATC or through the ATIS broadcast if parallel approaches are in progress.

A parallel ILS approach provides aircraft with a minimum of two miles separation between successive aircraft on the adjacent localizer course.

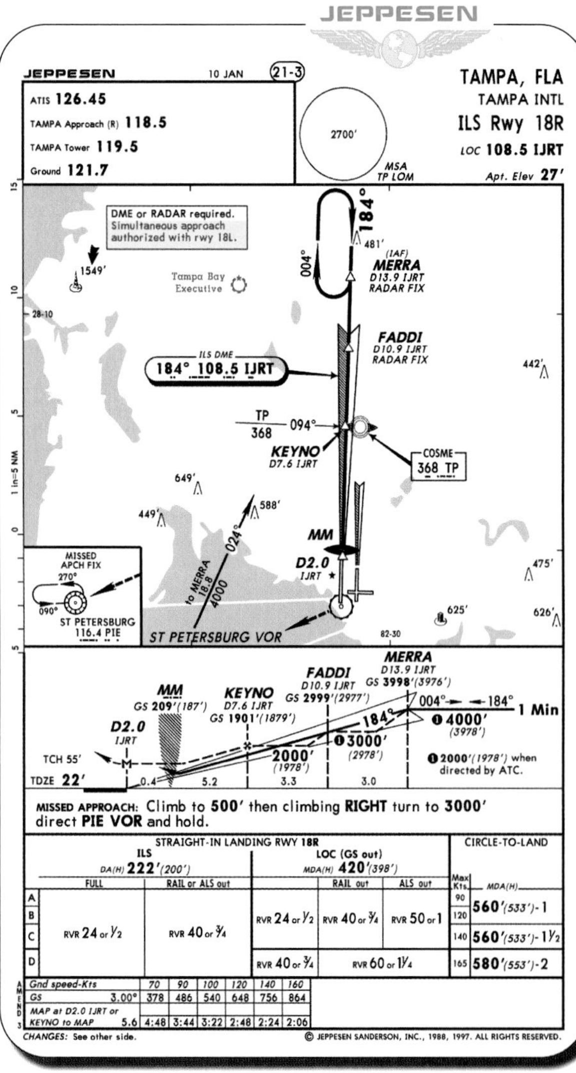

SIMULTANEOUS (INDEPENDENT) PARALLEL ILS APPROACH

A **simultaneous (independent) parallel ILS approach** differs from a parallel (dependent) approach in that the runway centerlines are separated by 4,300 to 9,000 feet and the approaches, which do not require staggered separation, are monitored by dedicated final controllers. The final monitor controllers track aircraft position and issue instructions to pilots of aircraft observed deviating from the localizer course. [Figure 8-31]

Figure 8-31. A note is included on the approach chart of an ILS approach which is certified for simultaneous operations. In addition, ATC or the ATIS broadcast will advise you if simultaneous parallel approaches are in progress. If you do not want to fly this type of approach, you should advise ATC immediately.

During the approach, ATC is not required to maintain staggered aircraft separation. You will be instructed to monitor tower frequency for advisories while on final approach. The final monitor controller has the ability to override the tower controller if needed to issue instructions to aircraft deviating from the localizer course. If an aircraft does not respond correctly or enters the no transgression zone (NTZ) between runways, the final monitor controller may issue breakout or missed approach instructions. If necessary, the controller may also instruct the aircraft on the adjacent approach to alter course.

When simultaneous approaches are in progress, each pilot may receive radar advisories on tower frequency.

You fly a simultaneous parallel ILS approach as you would any other ILS. You will be radar monitored during the approach; however, once you report the runway environment in sight, or you are one mile or less from the threshold, or when visual separation is applied, radar monitoring will be terminated. In any case, you should not expect that the final monitor controller will advise you when radar monitoring is terminated.

SIMULTANEOUS CLOSE PARALLEL ILS APPROACH

Typically, a single-runway approach procedure can support 29 arrivals per hour. Two simultaneous independent flows of approach traffic can double an airport's capacity on two parallel runways. Historically, parallel runways had to be at least 4,300 feet apart for ATC to authorize simultaneous parallel ILS approaches; however, this is no longer the case with the advent of a system called the **precision runway monitor (PRM)**.

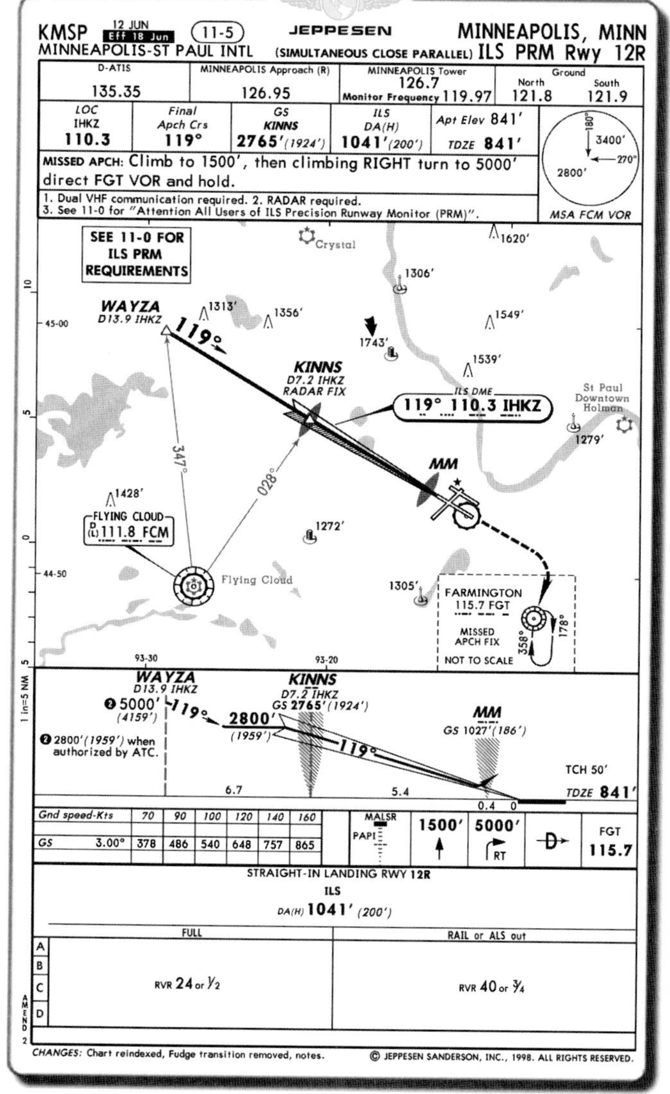

The PRM uses a radar offering one second (or faster) updates on targets, a high-resolution color ATC display, audio and visual alert systems for controllers, and software for projecting aircraft track vectors. This system, which can display turns as they occur, does not require any extra equipment on the aircraft or sensors on the airport other than the PRM electronically guided antenna. [Figure 8-32]

Figure 8-32. Although many requirements must be met before you can execute a simultaneous close parallel ILS approach, you fly it using normal ILS procedures. The photo shows a PRM mounted on a tower.

Aircraft on **simultaneous close parallel ILS approaches** are monitored by a radar controller to ensure that neither aircraft enters, or blunders, into the NTZ which lies between the runways. If the 10-second projected track indicates that an aircraft is going to enter the NTZ, the system alerts the controller. If the aircraft's displayed position enters the NTZ, the controller immediately issues breakout instructions to the pilots. Breakout instructions involve discontinuing the approach, turning outward, climbing to minimum safe altitude, and executing a missed approach.

The PRM has been able to provide a solution to increased air traffic around busy airports with closely spaced parallel runways. The system is not only safe and efficient, but it is also a much more cost effective and practical alternative to building new runways.

SIMULTANEOUS CONVERGING INSTRUMENT APPROACH

Airports which have runways situated at an angle of 15° to 100° to each other may have approval for conducting simultaneous instrument approaches to the converging runways. The development criteria for **simultaneous converging instrument approaches** requires that the approaches have missed approach points at least 3 miles apart and missed approach procedures which do not overlap. Other requirements vary somewhat depending on whether or not the runways intersect. While you should fly the converging approach as you would any other similar approach, you will notice that converging approaches all terminate with a straight-in landing. [Figure 8-33]

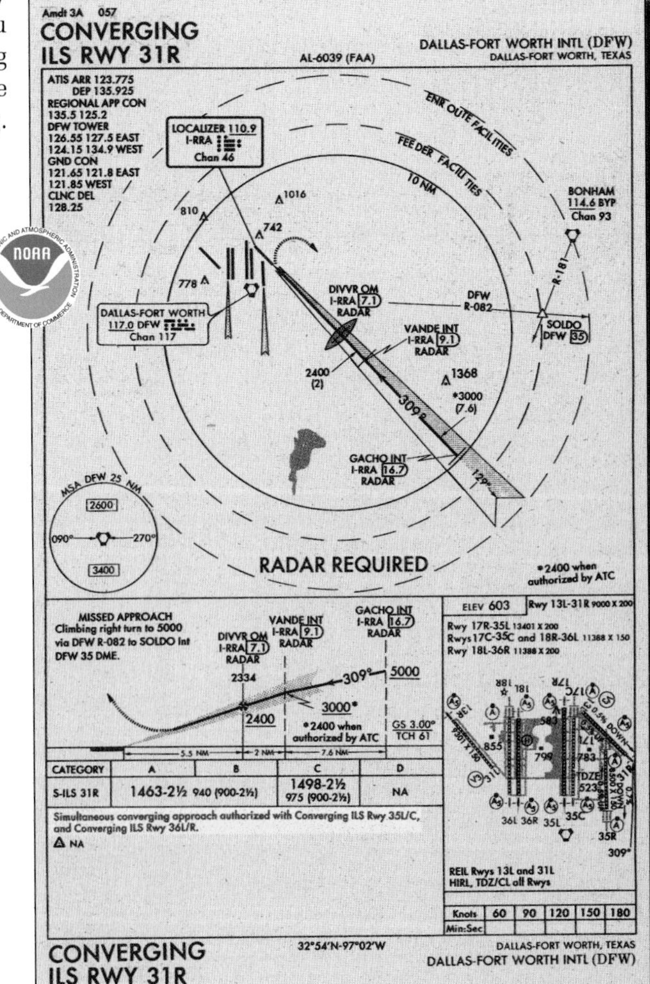

Figure 8-33. As with other simultaneous approaches, you will be informed of converging simultaneous operations on initial contact with the controller or via ATIS. In addition, authorization for converging simultaneous approaches will be noted on the associated approach charts.

LOCALIZER APPROACH

There are two circumstances where you may execute a localizer approach. First, if it is authorized, you can fly the localizer portion of an ILS approach if you cannot, or choose not, to use glide slope guidance. The second instance occurs on approaches designed specifically as nonprecision procedures using a localizer transmitter. [Figure 8-34] Since the characteristics of the localizer are the same as those associated with an ILS approach, you fly a localizer approach using the same procedures to maintain course as you would for an ILS.

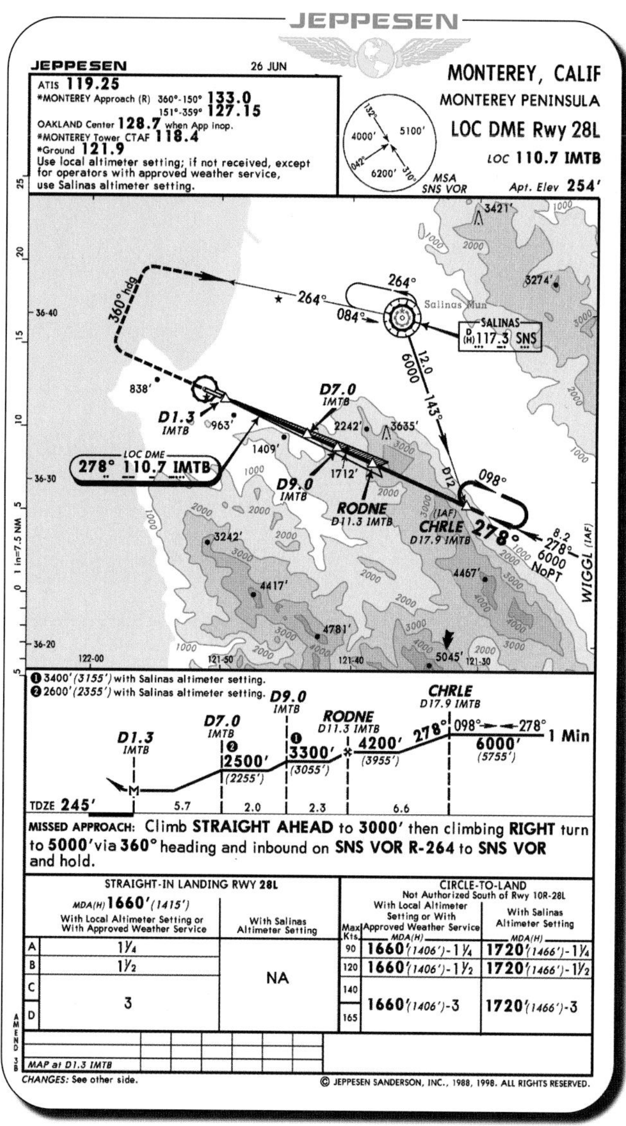

Figure 8-34. The main difference between a localizer and an ILS approach, other than the lack of a glide slope, is that since the localizer is a nonprecision approach, you descend to a minimum descent altitude instead of a decision height.

LOCALIZER BACK COURSE APPROACH

At airports where ILS equipment is installed, it is not unusual to find localizer back course approaches. Although a localizer back course approach does not have an associated glide slope, you may find that you receive false glide slope signals from the front course ILS equipment. It is important that you disregard any glide slope information while executing a back course approach. Another important consideration is the type of equipment you have onboard your aircraft. Remember, reverse sensing will occur on a back course localizer inbound when you are using a basic VOR indicator. With an HSI, you can avoid reverse sensing if you set the front course on the course selector. Except for these considerations, flying a back course localizer is very much like executing any other localizer approach. [Figure 8-35]

Figure 8-35. Assume you are inbound to Boise Air Terminal and are cleared for the LOC (BACK CRS) Rwy 28L approach, sidestep to Runway 28R. After conducting a chart review and setting your navaids and communications equipment in preparation for the approach, you arrive at the IBOI 20 DME fix at 6,800 feet MSL.

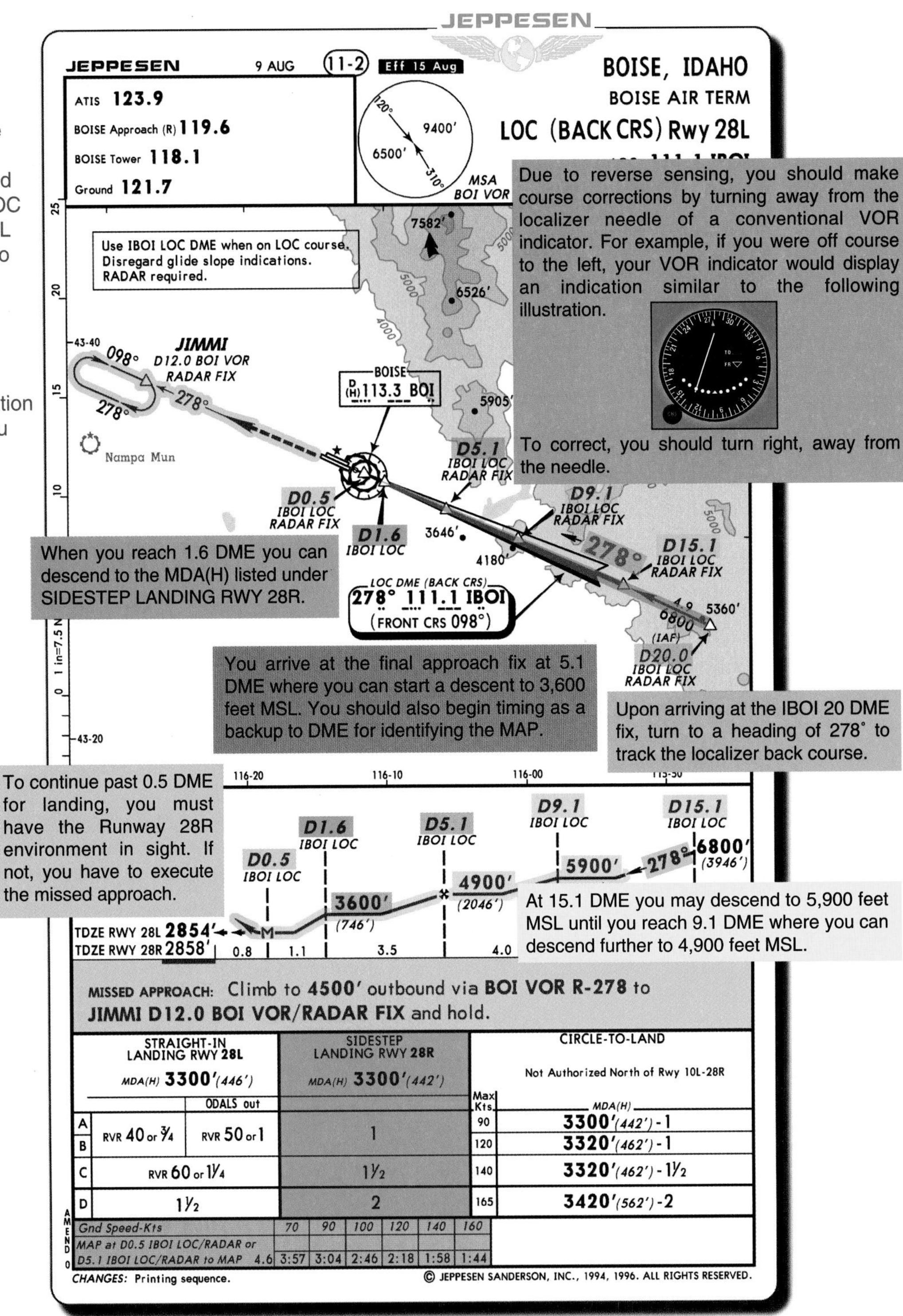

	STRAIGHT-IN LANDING RWY 28L		SIDESTEP LANDING RWY 28R	Max Kts	CIRCLE-TO-LAND MDA(H)
		ODALS out			
A	RVR 40 or ¾	RVR 50 or 1	1	90	3300'(442')-1
B				120	3320'(462')-1
C	RVR 60 or 1¼		1½	140	3320'(462')-1½
D	1½		2	165	3420'(562')-2

Gnd Speed-Kts	70	90	100	120	140	160
MAP at D0.5 IBOI LOC/RADAR or D5.1 IBOI LOC/RADAR to MAP 4.6	3:57	3:04	2:46	2:18	1:58	1:44

LDA, SDF, AND MLS APPROACHES

Approaches which are similar to the localizer or ILS include the nonprecision localizer-type directional aid (LDA) and simplified directional facility (SDF) approaches as well as the precision microwave landing system (MLS) approach. While you can fly an LDA or SDF approach if your aircraft is equipped to track a localizer, you need a special receiver to execute an MLS procedure.

LDA APPROACH

A localizer-type directional aid (LDA) can be thought of as a localizer approach system that is not aligned with the runway centerline. In fact, the identifier for an LDA begins with an I followed by a 3-letter group, just like a localizer. The LDA course width is the same as a localizer associated with an ILS — between 3° and 6° wide. Some LDA approaches have an electronic glide slope, although this is not a requirement for the system. If the final approach course is aligned to within 30° of the runway centerline, straight-in landing minimums also may be available. [Figure 8-36]

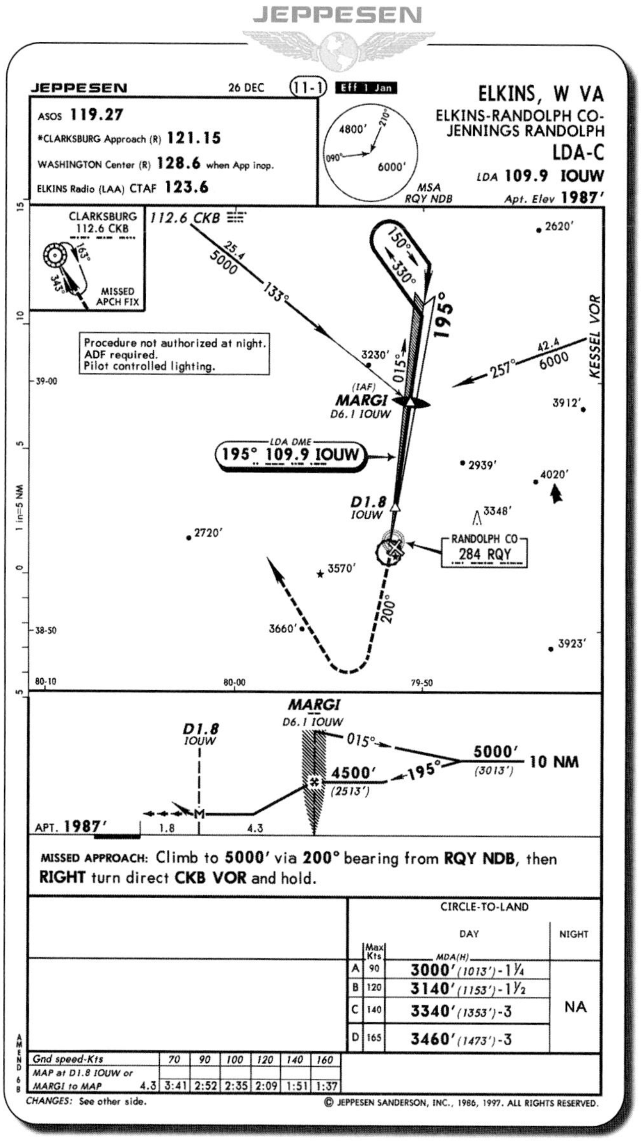

JEPPESEN

JEPPESEN 26 DEC (11-1) Eff 1 Jan

ELKINS, W VA
ELKINS-RANDOLPH CO-
JENNINGS RANDOLPH
LDA-C
LDA 109.9 IOUW
Apt. Elev 1987'

ASOS 119.27
*CLARKSBURG Approach (R) 121.15
WASHINGTON Center (R) 128.6 when App inop.
ELKINS Radio (LAA) CTAF 123.6

MSA RQY NDB 4800' 6000'

CLARKSBURG 112.6 CKB
MISSED APCH FIX
112.6 CKB
25.4 5000 133°
Procedure not authorized at night.
ADF required.
Pilot controlled lighting.
150° 330° 195° 015°
3230'
2620'
42.4 6000 257°
KESSEL VOR
(IAF) MARGI D6.1 IOUW
3912'
LDA DME 195° 109.9 IOUW
2939' 4020'
D1.8 IOUW
3348'
2720'
RANDOLPH CO 284 RQY
3570'
200°
3660'
3923'

MARGI D6.1 IOUW
D1.8 IOUW
015° 195°
5000' (3013') 10 NM
4500' (2513')
APT. 1987' 1.8 4.3

MISSED APPROACH: Climb to 5000' via 200° bearing from RQY NDB, then RIGHT turn direct CKB VOR and hold.

CIRCLE-TO-LAND

	Max Kts	DAY MDA(H)	NIGHT
A	90	3000' (1013')-1¼	NA
B	120	3140' (1153')-1½	
C	140	3340' (1353')-3	
D	165	3460' (1473')-3	

Gnd speed-Kts	70	90	100	120	140	160
MAP at D1.8 IOUW or MARGI to MAP 4.3	3:41	2:52	2:35	2:09	1:51	1:37

CHANGES: See other side.

An LDA approach is comparable to a localizer, but it is not aligned with the runway. The LDA course width is between 3° and 6°.

Figure 8-36. Since this LDA course is not aligned within 30° of the runway bearing, only circling minimums are published. You fly an LDA approach the same as you would a localizer approach.

SDF APPROACH

The simplified directional facility (SDF) approach system does not incorporate an electronic glide slope and may offer less accuracy than the LDA. While the typical ILS or LDA localizer is between 3° and 6° wide, an SDF localizer course is fixed at either 6° or 12° wide. The lateral limits of SDF course guidance are 35° either side of centerline; you should disregard any SDF navigation information you receive beyond 35°. As with any other navaid, you should tune and identify the facility prior to using it for navigation. While similar to a localizer or LDA, an SDF 3-letter identifier is not preceded by an I.

Once you are on SDF final approach course, you should use essentially the same procedures you use when flying a localizer or LDA approach. However, it is important to remember that you may experience reduced CDI sensitivity during the approach. [Figure 8-37]

An SDF course (which is either 6° or 12° wide) may be offset from the runway centerline.

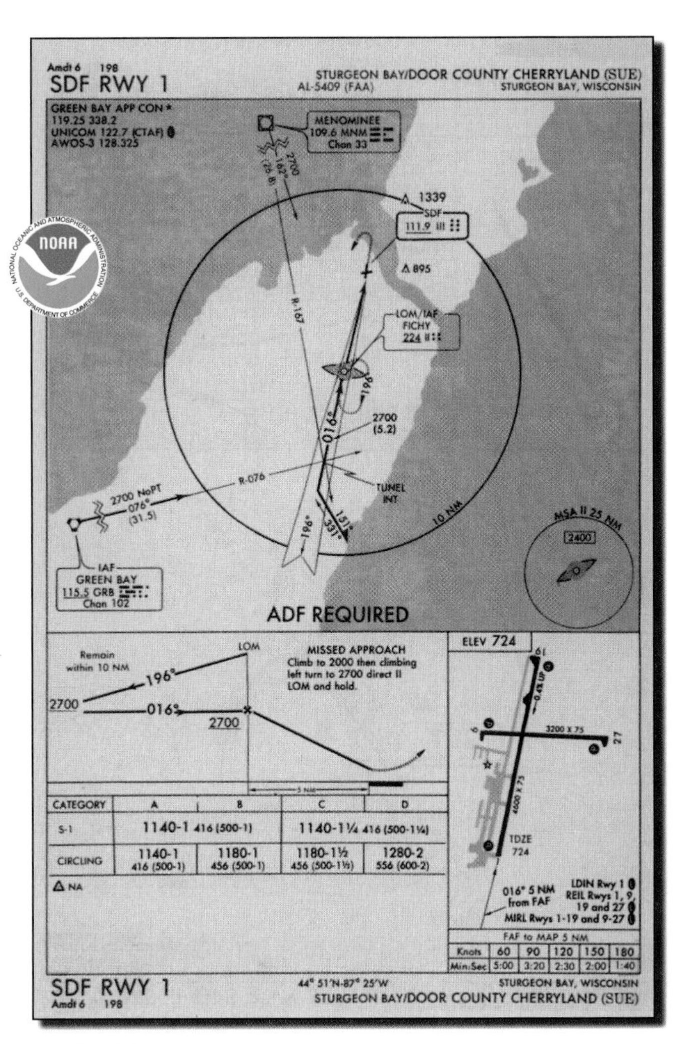

Figure 8-37. Since most SDF courses are aligned within 3° of the runway bearing, you will typically find that SDF approaches are published with straight-in minimums.

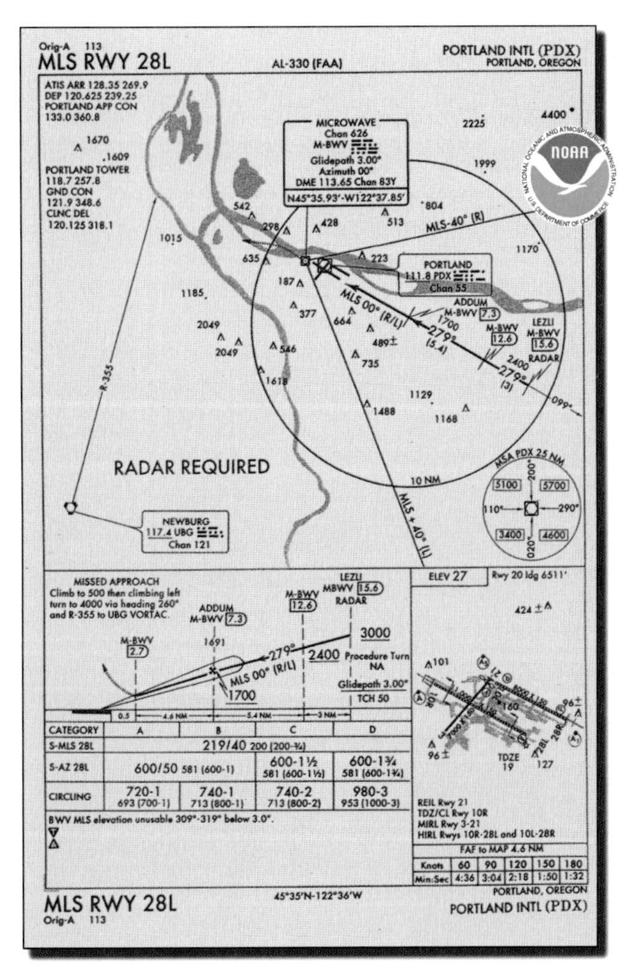

MLS APPROACH

The **microwave landing system (MLS)** approach is a precision approach procedure which provides azimuth, elevation, and distance information. A back azimuth may be included for guidance during the missed approach and departure. Due to the unique characteristics of the MLS, procedures can be designed with multiple final approach paths, including curved paths, and multiple glide slope angles. [Figure 8-38]

Figure 8-38. You must have special equipment to receive MLS signals (which can be identified by the Morse code for the letter M, preceding the three-letter Morse code identifier). MLS navigation guidance can be displayed on a conventional VOR indicator and DME display or incorporated into a multi-purpose cockpit presentation.

An MLS is identified by the Morse code for the letter M followed by the three-letter identifier for the facility.

The microwave landing system is less susceptible to interference from obstacles and high power FM stations. This makes MLS well suited to areas in which the installation of ILS is difficult or impossible. While originally thought to be a replacement for ILS, the future of MLS has been put in doubt as the promise of precision GPS approaches has emerged. Because of this uncertainty, no new installations of MLS are planned for the U.S. Internationally, however, the use of MLS may continue to increase until an acceptable alternative becomes available.

SUMMARY CHECKLIST

✓ ILS approaches are classified as Category I, Category II, or Category III.

✓ The ILS localizer transmitter emits a navigational array from the far end of the runway to provide you with information regarding your alignment with the runway centerline.

✓ When using a basic VOR indicator, normal sensing occurs inbound on the front course and outbound on the back course. Reverse sensing occurs inbound on the back course and outbound on the front course.

✓ You can avoid reverse sensing when using an HSI by setting the published front course under the course index. This applies regardless of your direction of travel, whether inbound or outbound on either the front or back course.

✓ The glide slope signal provides vertical navigation information for descent to the lowest authorized decision height for the associated approach procedure. The glide slope may not be reliable below decision height.

✓ Full-scale deviation of the glide slope needle is 0.7° above or below the center of the glide slope beam.

✓ Usually, an ILS includes an outer marker (OM) and middle marker (MM). An inner marker (IM) is installed at locations where Category II and III ILS operations have been certified.

✓ When a compass locator is installed in conjunction with the outer marker, it is called an outer compass locator (LOM). When a compass locator is collocated with the middle marker, it is referred to as a middle compass locator (LMM).

✓ The glide slope centerline normally intersects the middle marker approximately 200 feet above the touchdown zone.

✓ When the aircraft passes through the signal array of a marker beacon, a colored light flashes on the marker beacon receiver and a Morse code identification sounds.

✓ Certain approach and runway lighting configurations can qualify precision approaches for lower minimums.

✓ Higher landing minimums may be required if some components of an ILS are inoperative. If more than one component is not available for use, you should adjust the minimums by applying only the greatest increase in altitude and/or visibility required by the failure of a single component.

✓ If the glide slope is inoperative or fails during your approach, the localizer (GS out) minimums apply. In this case, you may continue the approach to the applicable MDA.

- ✓ Prior to intercepting the ILS glide slope, you should concentrate on stabilizing airspeed and altitude while establishing a magnetic heading which will maintain the aircraft on the localizer centerline. Once your descent rate stabilizes, use power as needed to maintain a constant approach speed.
- ✓ The rate of descent you must maintain to stay on glide slope must decrease if your groundspeed decreases, and vice versa.
- ✓ Since localizer and glide slope indications become more sensitive as you get closer to the runway, you should strive to fly an ILS approach so that you do not need heading corrections greater than 2° after you have passed the outer marker.
- ✓ On an ILS approach, you must execute a missed approach if you have not established the required visual references at the DH.
- ✓ When advised to change to advisory frequency, you should broadcast your intentions on CTAF, including the type of approach being executed, your position when over the FAF inbound (nonprecision approach) or when over the outer marker or fix used in lieu of the outer marker (precision approach) inbound.
- ✓ The procedures you use to fly an ILS/DME approach are essentially the same as any other ILS approach except for the requirement to identify approach fixes using DME.
- ✓ Parallel (dependent) ILS approach operations may be conducted on parallel runways with centerlines at least 2,500 feet apart. Simultaneous (independent) parallel ILS approaches may be conducted to airports with parallel runway centerlines separated by 4,300 to 9,000 feet. When certain requirements are met, including the installation of a precision runway monitor, simultaneous close parallel ILS approach procedures may be established at airports with parallel runway centerlines less than 4,300 feet apart.
- ✓ You will be informed by ATC or through the ATIS broadcast if parallel approaches are in progress.
- ✓ Reverse sensing will occur on a back course localizer inbound when you are using a basic VOR indicator.
- ✓ A localizer-type directional aid (LDA) is an approach system which uses a localizer course that is not aligned with the runway centerline. If the final approach course is aligned to within 30° of the runway centerline, straight-in landing minimums may be available.
- ✓ A simplified directional facility (SDF) course is fixed at either 6° or 12° wide. Since most SDF courses are aligned within 3° of the runway bearing, SDF approaches are typically published with straight-in minimums.
- ✓ The microwave landing system (MLS) offers a precision approach alternative to airports where interference from obstacles and/or high power FM stations makes the installation of ILS difficult or impossible.

KEY TERMS

Instrument Landing System (ILS)

Localizer

Glide Slope

ILS Marker Beacon

Outer Marker (OM)

Middle Marker (MM)

Inner Marker (IM)

Marker Beacon Receiver

Compass Locator

Parallel (Dependent) ILS Approach

Simultaneous (Independent) Parallel ILS Approach

Precision Runway Monitor (PRM)

Simultaneous Close Parallel ILS Approach

Simultaneous Converging Instrument Approach

Localizer Approach

Localizer Back Course Approach

Localizer-Type Directional Aid (LDA)

Simplified Directional Facility (SDF)

Microwave Landing System (MLS)

QUESTIONS

1. What are the normal minimums for a CAT I ILS approach with all components operative?

   A. RVR 2400 and a DH of 200 feet MSL
   B. RVR 2400 and a DH of 400 feet AGL
   C. RVR 2400 and a DH of 200 feet AGL

2. What is the full-scale deflection of a CDI when tuned to a localizer?

   A. 5° either side of course
   B. 10° either side of course
   C. 2.5° either side of course

3. Refer to the accompanying illustration to match each navigation indicator with the appropriate aircraft positions. More than one aircraft position may apply to each CDI.

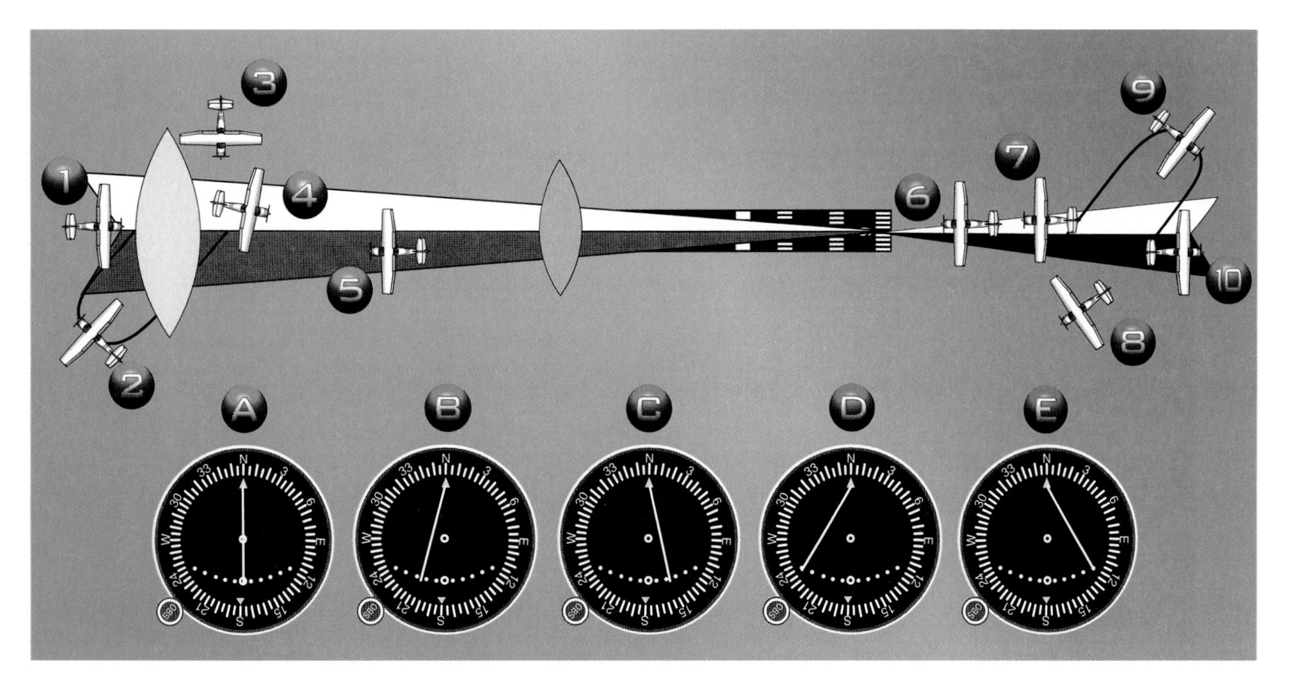

4. Refer to the accompanying illustration to match each navigation indicator with the appropriate aircraft positions. More than one aircraft position may apply to each HSI.

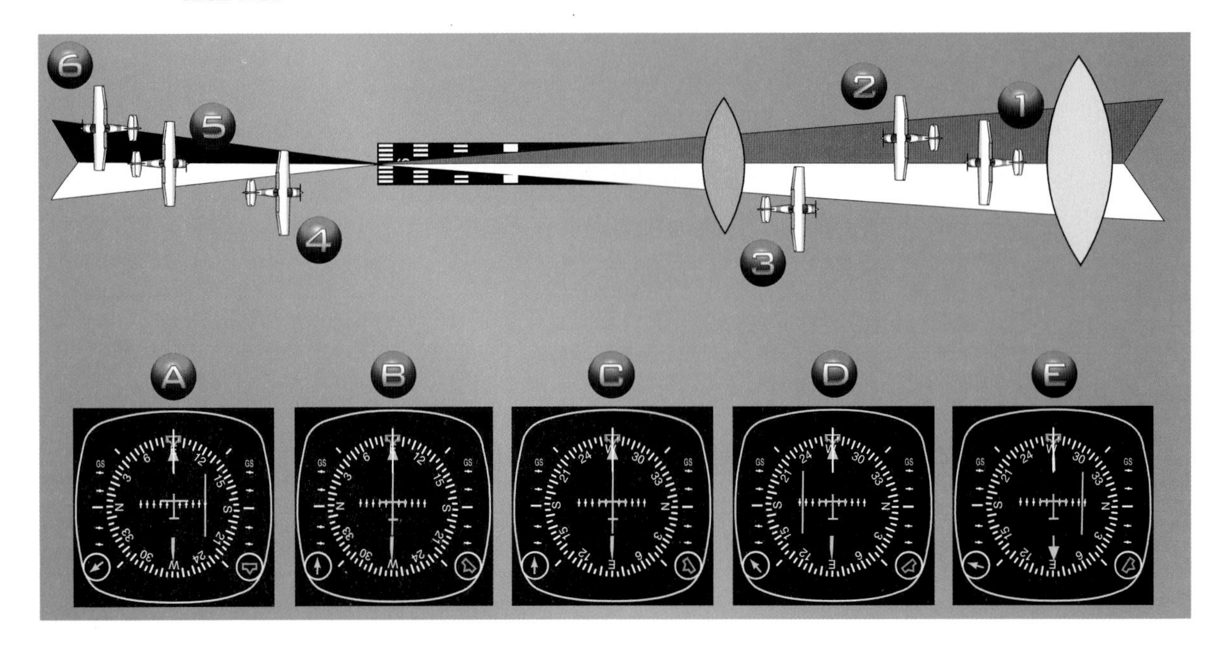

Refer to the following illustration to answer question 5.

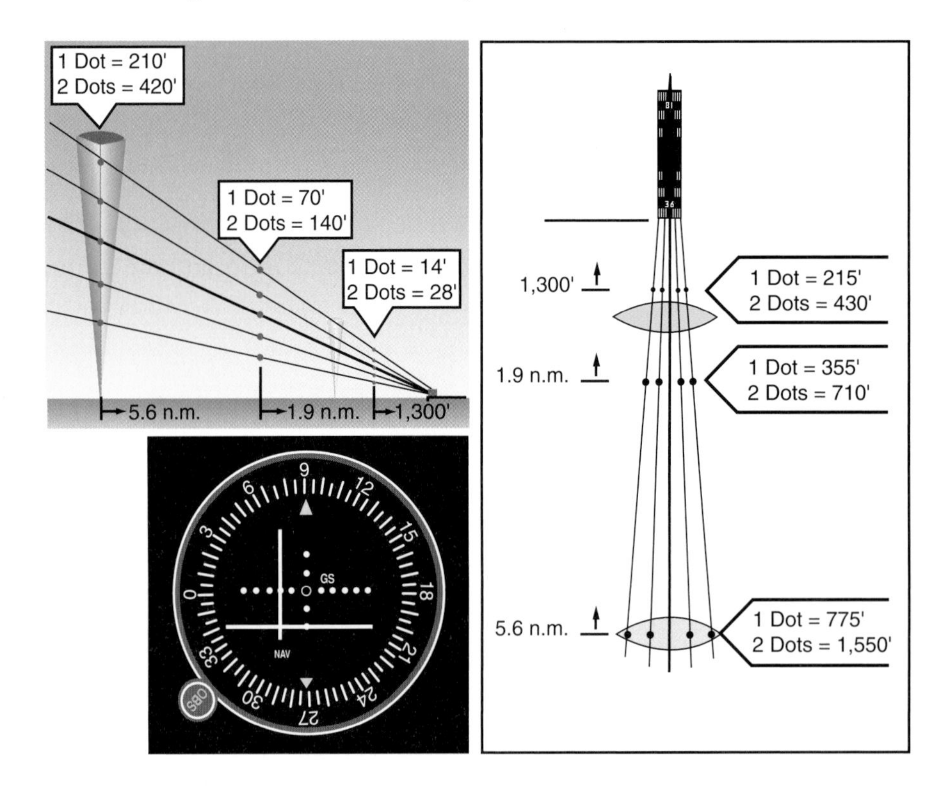

5. If you are 1.9 nautical miles from the runway and your VOR indicator matches the display shown, how many feet are you displaced from the localizer centerline and the glide slope?

6. If your airspeed is too high but you are on course and on glide slope, should you initially adjust pitch or power?

Refer to the ILS Rwy 17R approach to Will Rogers World Airport, Oklahoma City, Oklahoma to answer questions 7 through 10.

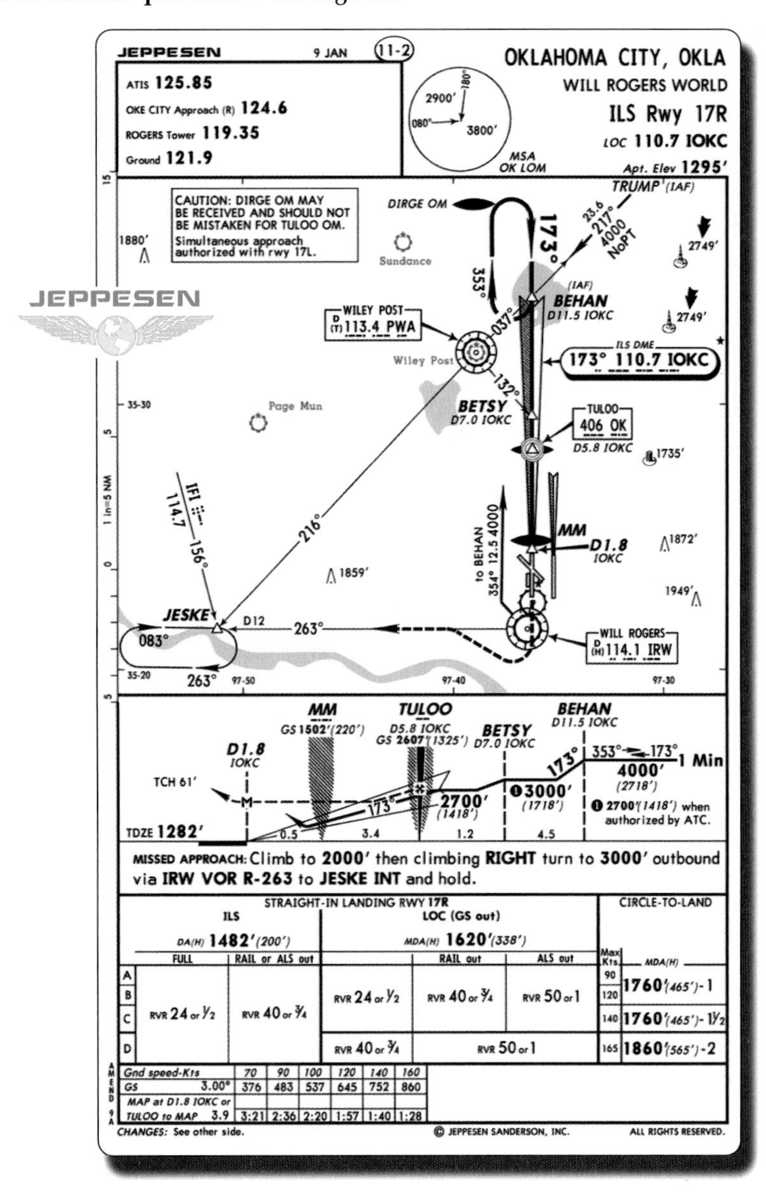

7. When established on the localizer, IOKC, how can you identify BEHAN Intersection?

8. If you are on glide slope at the LOM, what altitude should your altimeter display?

   A. 2,607 feet
   B. 2,700 feet
   C. 4,000 feet

9. If the glide slope flag appears in your CDI after passing Tuloo, what is the minimum altitude to which you can descend? How can you identify the MAP?

10. What are the three ways you can identify JESKE? While executing the missed approach, what type of entry into the published holding pattern can you expect?

Refer to the LOC BC RWY 35L approach to Will Rogers World Airport, Oklahoma City, Oklahoma to answer questions 11 through 14.

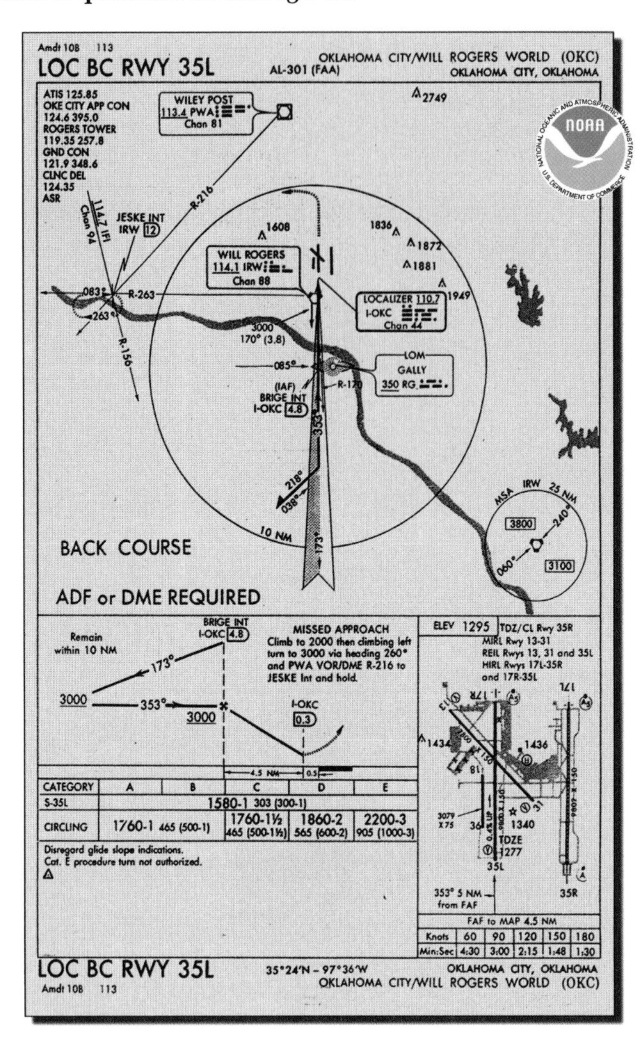

11. What magnetic course should you fly from IRW VORTAC to the IAF?

    A. 085°
    B. 170°
    C. 173°

12. If you do not have DME, how can you identify the FAF when on the localizer?

13. If you are inbound on the back course ILS approach and have the HSI indication shown in the accompanying illustration, which way should you turn to return to localizer centerline?

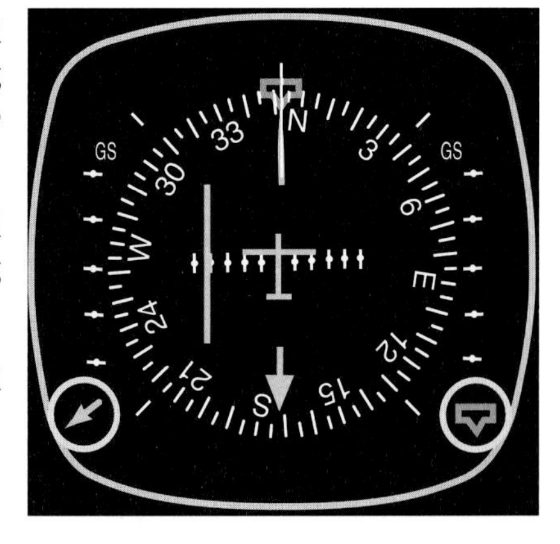

14. True/False. Immediately after executing a missed approach from the MDA, you should initiate a climbing left turn to 3,000 feet MSL.

15. What are the course widths associated with an LDA and SDF, respectively?

SECTION C

GPS AND RNAV APPROACHES

Instrument approaches based on area navigation (RNAV) systems are becoming increasingly popular for many reasons, including the ability to improve IFR accessibility to airports and offer better situational awareness for pilots. One type of area navigation system uses VOR/DME, VORTAC, or TACAN facilities to establish RNAV approach procedures. These types of procedures, which are collectively identified on approach charts as VOR/DME RNAV approaches, offer greater design flexibility over non-RNAV approaches since the routes do not have to pass directly over ground-based navaids. Another type of approach procedure, based on the satellite-based global positioning system (GPS), provides even greater versatility. Through the use of GPS, nonprecision approach capability can be added to smaller airports without the need to invest in ground-based navigation equipment.

The fundamental procedures involved in flying GPS or VOR/DME RNAV approaches are similar to those you use for other nonprecision approaches. However, due to the complexity of the associated navigation equipment, you should not attempt a GPS or VOR/DME RNAV approach unless you having a thorough working knowledge of the system(s) installed in your aircraft. Most of the following information is procedurally oriented. Prior to reading this section, you may want to review the introductory discussion regarding GPS and VOR/DME RNAV navigation contained in Chapter 2, Section C — Instrument Navigation.

APPROACH DESIGN

Transitions from the enroute to the terminal environment have traditionally been restricted to specific ground tracks based on the limitations of ground-based navaids. The proliferation of area navigation systems has led to the development of a standardized approach segment configuration known as the Basic T. [Figure 8-39]

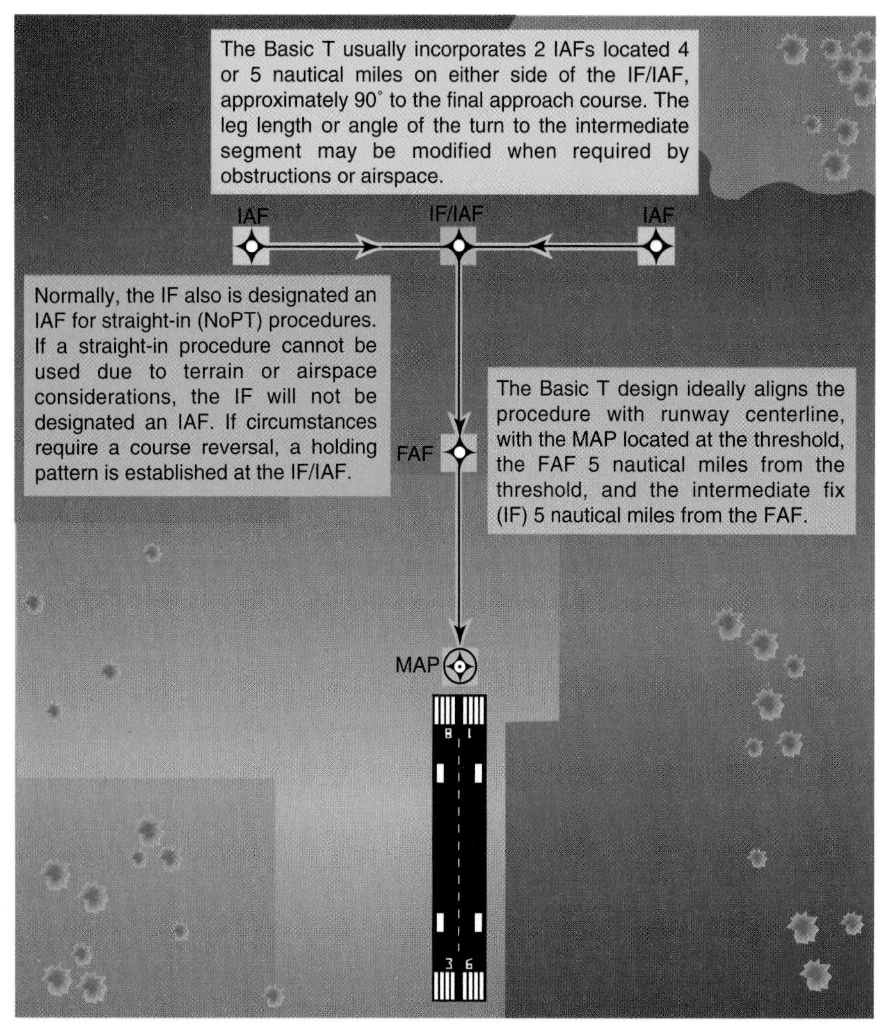

Figure 8-39. The Basic T approach design provides an efficient method of routing aircraft to a destination. Although the FAA is making an effort to standardize area navigation procedures using the Basic T design, slight modification of the configuration is sometimes required due to nearby obstacles or special use airspace.

GPS APPROACHES

There are two major categories of GPS approaches — overlay and stand alone. Properly installed and certified GPS equipment can be used to fly many nonprecision approaches based on conventional navaids, with the exception of those using localizers (LOC), localizer-type directional aids (LDA), and simplified directional facilities (SDF).

NONPRECISION APPROACHES

The utilization of GPS to fly existing procedures is made possible by the **GPS overlay program.** The program was implemented by the FAA to accelerate the adoption of GPS in approach applications. The GPS overlay program made it possible to offer a large number of GPS approaches without the time consuming and expensive process of creating, test flying, and charting a library of new approaches. When the FAA declared GPS operational for civil operations in February 1994, Phase I of the GPS overlay program ended and Phase II began. Phase II of the overlay program uses existing approach charts,

and requires the underlying ground navaids and associated aircraft navigation equipment to be operational, but not monitored during the approach as long as the GPS meets RAIM accuracy requirements. Phase III of the GPS overlay program, which now runs concurrently with Phase II, eliminates the requirement for conventional navigation equipment to be operational during the approach, although that equipment is required for other portions of IFR flight. Phase III approach charts incorporate additional information for GPS navigation and can usually can be identified by the words "or GPS" in the approach title. [Figure 8-40]

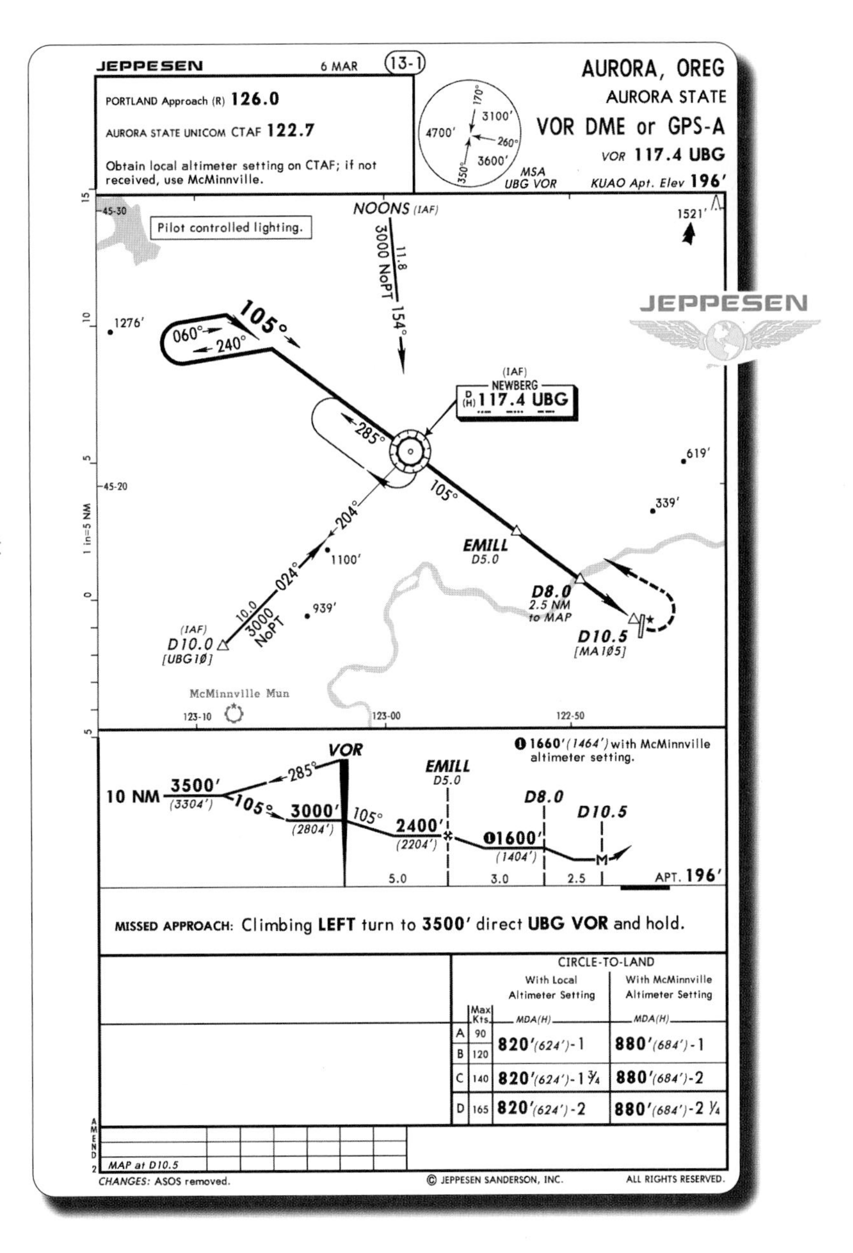

	Max Kts	With Local Altimeter Setting MDA(H)	With McMinnville Altimeter Setting MDA(H)
A	90	820'(624')-1	880'(684')-1
B	120		
C	140	820'(624')-1 3/4	880'(684')-2
D	165	820'(624')-2	880'(684')-2 1/4

Figure 8-40. You can recognize that this is a Phase III GPS overlay approach by the words "or GPS" in the title. With a properly certified GPS installation, you could conduct this approach with Newburg VOR out of service.

Some Phase III approach charts published by Jeppesen may not include "or GPS" in the title until the next major revision. If a procedure appears in the GPS electronic database and the printed Jeppesen chart does not include the words "or GPS" in the title, you can determine if the approach falls under Phase II or Phase III requirements by referring to the Phase III listing beginning on Terminal page US-1 in the Jeppesen coverage. If the approach title does not appear in the listing, you can assume it is a Phase II approach and that the associated equipment requirements apply.

A **GPS stand alone approach** is designed solely for GPS and offers more efficient routing than is possible with some ground-based navaids. To capitalize on the unique capabilities of GPS, stand alone procedures use the Basic T structure where possible. [Figure 8-41]

You are not required to monitor or have conventional navigation equipment for stand alone GPS approaches. However, you still must have conventional navigation equipment aboard your aircraft as a backup for enroute navigation, and to fly to an alternate airport if it becomes necessary. While you can conduct an approach to an alternate airport using GPS, you must have the capability of conducting the approach using conventional equipment in the event GPS reliability becomes an issue.

When using GPS for navigation and instrument approaches, any required alternate airport must have an approved operational instrument approach procedure other than GPS.

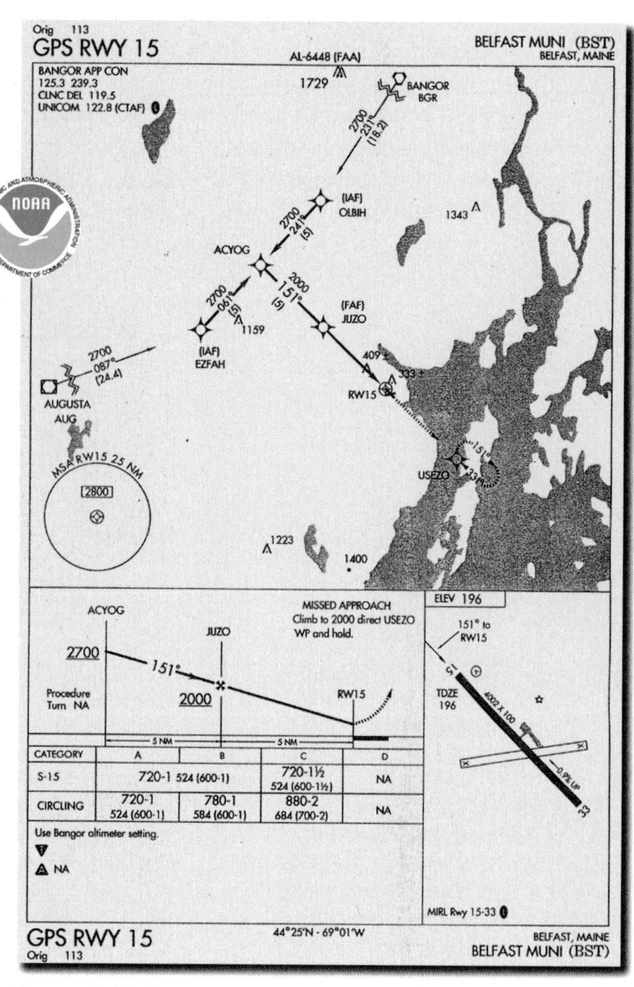

Figure 8-41. You will see very few, if any, references to traditional navaids on a GPS stand alone approach.

PRECISION APPROACHES

Without additional ground equipment, standard GPS does not provide the vertical accuracy necessary for precision approaches. The FAA is investigating a wide area augmentation system (WAAS) that could provide precision GPS navigation over a large area with relatively few ground facilities. Such a system could bring precision approach capability to any airport where terrain and other operational considerations permit such procedures, without the need to equip these airports with an ILS or other costly equipment.

GPS EQUIPMENT REQUIREMENTS

Before you can use a GPS receiver as a sole source of navigation for a nonprecision GPS approach, you must make sure that both the receiver and the way it is installed in the aircraft meet FAA requirements. The GPS must be approved in accordance with Technical Standard Order (TSO) C-129 and must be installed in accordance with AC 20-138, which provides guidance for airworthiness approval of GPS equipment. You can determine the allowable uses for a GPS installation by refer-

You can determine if a GPS is approved for IFR enroute and approach operations by referring to the Airplane Flight Manual (AFM) supplement.

ring to the supplements section of the Airplane Flight Manual (AFM). You cannot use handheld GPS receivers for IFR flight.

RECEIVER AUTONOMOUS INTEGRITY MONITORING (RAIM)

A GPS receiver continuously monitors the reliability of the GPS signal using a system known as receiver autonomous integrity monitoring (RAIM). Although basic GPS positioning is almost always available worldwide, the satellite measurement redundancy necessary to ensure integrity of the GPS position is neither worldwide nor continuous. This means you may only find four satellites the required angle above the horizon (the mask angle) at certain times and locations. While this is enough to triangulate your position, it does not provide the extra satellite you need for integrity monitoring.

In addition to real-time integrity monitoring, IFR-certified GPS receivers can also predict RAIM availability at the intended location and time of the approach. The receivers make this prediction based on the satellites that are functioning at the time you request the prediction. GPS NOTAMs, available from AFSSs or DUATS, also contain information on scheduled satellite outages. If a scheduled outage will make one of the currently functioning satellites unavailable at your ETA, you should program your receiver to exclude that satellite from the RAIM prediction. In practice, the easiest and most reliable way to ensure you will have RAIM for the approach at your ETA is to specifically request GPS RAIM availability information from your FSS weather briefer. If flying a published GPS departure, you should also request a RAIM prediction for the departure airport.

I See Your Satellite System and Raise You One

Did you know that GPS is not the only satellite-based positioning system? It's true. The Russian Federation has a navigation satellite system similar to the U.S. GPS system. The Russian system is called GLONASS which stands for Global Navigation Satellite System. GPS and GLONASS are alike in many ways, from the number of spacecraft to the radio frequency bands used. Although both systems operate on the same principles, there are significant differences in some of the details. For example, the GLONASS system provides somewhat higher autonomous positioning accuracy than GPS (60 meters for GLONASS vs. 100 meters for GPS), primarily due to the intentional degradation of the GPS signal.

In an effort to get the best of both worlds, some avionics manufacturers offer integrated GPS+GLONASS receivers that can use signals from both systems. Although positional accuracy can be somewhat improved when you receive more satellites, airborne units do not benefit much from using both systems, since either constellation usually allows plenty of satellites to be in view at a given time. The integrated systems are most valuable in areas where satellite signals are blocked from large parts of the sky, such as in mountainous regions.

Although dual-use avionics can help, both positioning systems are still under independent military command and control. To avoid unexpected service interruptions, the ideal may be to create a truly global system under the control of one agency. In pursuit of this end, the European Commission is actively hammering out the details of an integrated system using GPS, GLONASS, and the ground-based augmentation and monitoring elements of both systems. This future system, known as GNSS (Global Navigation Satellite System) will be under international civilian control and will be dedicated to providing position information for aviation, maritime, highway, railroad, and surveying applications.

External Annunciators and Course Deviation Indicators

An IFR approved GPS installation requires that essential annunciators and a CDI display be located in the pilot's primary field of view. Generally, this means the system must be able to display course deviation indications on a conventional VOR or HSI display separate from the receiver's built-in display. Most installations have the ability to switch the primary navigation display between a VOR and GPS receiver. An annunciator is required to be installed on a switchable display to indicate when the GPS is selected. It is essential that you verify the VOR indicator or HSI is switched to the proper navigation system before beginning your approach. External annunciator lights are also used to advise of GPS receiver messages, waypoint passage, approach mode enabled, and waypoint hold. [Figure 8-42]

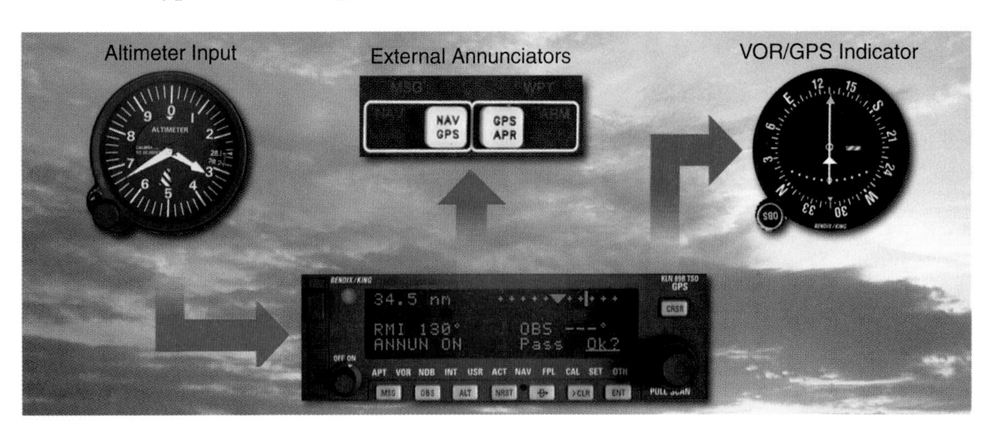

Figure 8-42. Before a GPS installation can be approved for IFR approaches, it normally must be connected to the altimeter, to external annunciator lights, and to an external VOR or HSI display.

The Navigation Database

A GPS receiver that is approved for IFR operations must have a **navigation database** that can be updated. You are not allowed to edit this database, although a receiver can allow you to add user-defined waypoints. The waypoints, fixes, and station identifiers used for GPS navigation are loaded into the receiver, usually by means of a magnetic card containing the information. The database is updated on a regular basis, and the GPS receiver is required to have current data before it is used for IFR navigation.

If an IAF or other fix has a pronounceable five-letter name, that name is used on the chart and in the database. DME fixes are typically given a five-character identifier which consists of the letter D, a radial, and an alphabet letter which corresponds to the DME from the station. For example, the IAF waypoint D271H is a DME fix on the 271° radial at 8 DME, since H is the eighth letter of the alphabet. Some DME fixes at a greater distance from the station may be coded with two letters from the VOR identifier followed by the radial, such as AX141.

When a waypoint is at a navaid, it is coded using the navaid identifier. When a MAP is based on timing or a DME distance, a **missed approach waypoint (MAWP)** is coded and added to the database. The identifier usually consists of either RW followed by the runway number, or, if the MAP is not at the runway threshold, MA followed by the final approach course.

Another type of coded waypoint which appears on some Jeppesen GPS approach charts is the **sensor final approach waypoint**. The GPS unit needs this waypoint in its database so that it can switch to ±0.3 nautical mile CDI sensitivity at the proper point in the approach. The sensor waypoint is included on approaches that do not have a final approach fix defined and usable as the **final approach waypoint (FAWP)**. An example is

a VOR approach where the MAP is at a VOR located on an airport. If a stepdown fix exists and is greater than 2 nautical miles to the MAP, the stepdown fix is coded in the database as the FAWP. Otherwise, a sensor FAWP is added at least 4 nautical miles from the MAP. Although sensor FAWPs are coded into the database, they do not appear on Phase II or Phase III NOS charts. They are shown on Jeppesen charts which have been updated for GPS. [Figure 8-43]

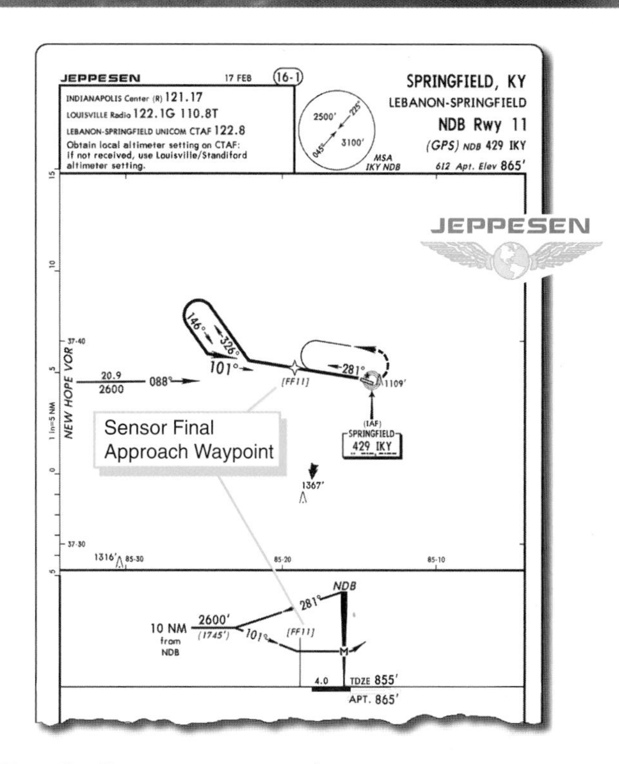

Figure 8-43. This approach procedure does not have a final approach fix. The sensor FAWP (FF11) has been added 4 nautical miles from the missed approach point.

SPECIAL GPS NAVIGATION CONSIDERATIONS

Although using GPS can be an efficient and easy alternative for IFR flight, there are many unique characteristics of GPS navigation with which you should be familiar. In most cases, a complete understanding of the procedures outlined in your GPS operating manual should allow you to maintain situational awareness throughout your IFR flights.

DISTANCE INFORMATION

When flying an overlay approach with GPS, you frequently must interpret distance information differently than when using conventional navigation equipment to identify certain fixes. Since GPS normally displays distance TO the next waypoint, GPS distances may be different numbers than DME distances. [Figure 8-44]

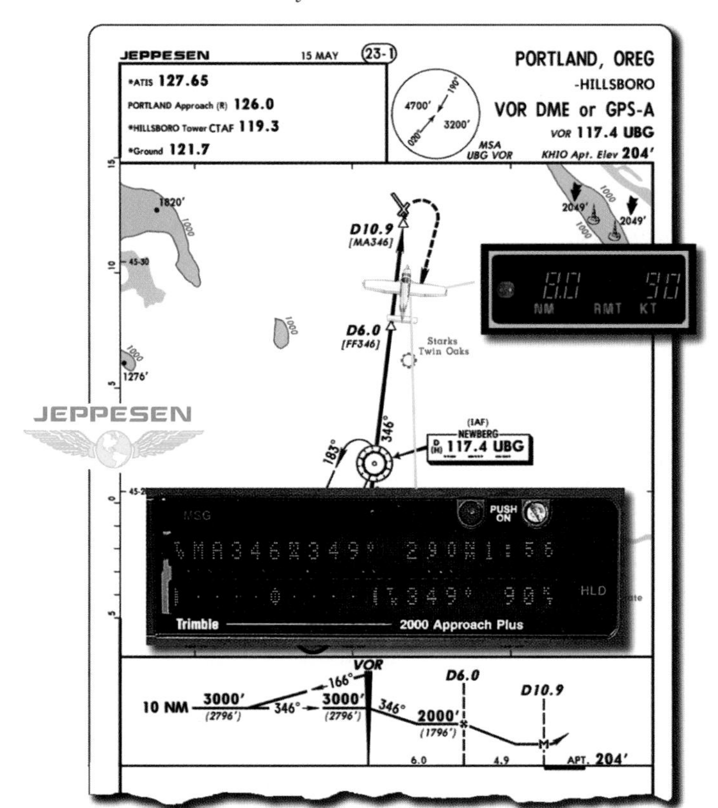

Figure 8-44. On this approach, the FAF is at 6.0 DME from Newburg VORTAC, and the MAP is at 10.9 DME. In this example, the GPS unit displays distance and time TO the next GPS waypoint, the MAWP, as 2.9 nautical miles. At the same position on the approach, a DME unit would indicate a distance of 8.0 nautical miles from the Newberg VORTAC.

At times, you may notice a disparity between the distance displayed on your GPS receiver versus the distance published on the accompanying procedure. This occurs because GPS uses a straight line, or **along track distance (ATD)**, between waypoints while the DME data published on instrument charts is based on slant range to the respective station. The difference between ATD and DME will vary depending on your altitude as well as your proximity to the navaid.

HEADING INFORMATION

When flying a GPS overlay procedure or along an airway, you may notice a small variance between the heading information published on the instrument chart and your GPS display. While the charted magnetic tracks defined by a VOR radial are determined by the application of magnetic variation at the VOR, GPS computers usually use an algorithm to apply magnetic variation at the current position. Although this process can produce small differences between GPS displayed data and charted information, both operations should produce the same ground track.

WAYPOINT SEQUENCING

Normal operation of the GPS unit is referred to as TO-TO navigation. The navigation indications are similar to flying a localizer in that you receive guidance TO the next point in your flight plan. There are certain occasions when you need to navigate a specific course away from a waypoint, such as when flying a procedure turn. To allow for this situation, GPS receivers incorporate a hold mode. When passing a waypoint in hold mode, the external VOR indicator changes from TO to FROM, and the GPS does not autosequence to the next waypoint. Some built-in digital displays will continue to show the bearing TO the waypoint. This bearing will change 180° as you pass the waypoint.

Errors in operating the GPS receiver are especially critical during approaches. In some cases, an incorrect entry can cause the receiver to leave the approach mode. If this occurs after the FAWP, you will have to execute a missed approach. For the most part, you should not touch your GPS unit between the final approach fix and the missed approach point.

BANK ANGLE/TURN RATE

To ensure smooth tracking of flight routes, it is important that you use the proper turn method whenever possible. The bank angle/turn rate you should use depends on how your GPS receiver calculates turn anticipation. To avoid greatly overshooting or undershooting approach legs, you should use the turn technique recommended in your GPS operating manual.

RAIM FAILURE

If your GPS receiver detects an integrity problem, you will generally be provided with one of two messages. One indicates that not enough satellites are available to provide RAIM integrity monitoring. The other type of message indicates that the RAIM integrity monitor has detected a potential error which exceeds tolerances for the current phase of flight. If RAIM is not available when you set up a GPS approach, you should use another type of navigation and approach system.

If RAIM is not available when you set up a GPS approach, you should select another type of navigation and approach system.

If you receive a RAIM failure indication prior to the FAWP, you should not descend to MDA, but should proceed to the MAWP via the FAWP, perform a missed approach, and contact ATC as soon as practical. If a RAIM failure occurs after the FAWP, the GPS receiver is allowed to continue operating without an annunciation for up to 5 minutes to allow you to complete the approach. Once a RAIM failure message appears, however, you should immediately execute a missed approach.

MISSED APPROACH ROUTING

You should also be aware that missed approach routings in which the first track is via a specific course rather than direct to the missed approach holding waypoint require additional actions to set the proper course. This is very important because if the missed approach procedure is not a direct course, flying from the MAP directly to the **missed approach holding waypoint (MAHWP)** may not provide sufficient obstacle clearance. You should always fly the full missed approach procedure as published on the approach chart.

GPS OVERLAY APPROACH

Since GPS overlay approaches are, by definition, based on preexisting approach procedures, the methods you use to fly a GPS overlay approach highly depends on the design of the underlying procedure. The limited amount of GPS related information published on the chart demands that you are thoroughly familiar with the operation of your GPS receiver and that you strive to maintain the highest degree of situational awareness. The following example assumes that you are approaching Escanaba VORTAC (ESC) at 5,000 feet MSL enroute to your destination, Delta County Airport, Escanaba, Michigan. For the purpose of this discussion, the scenario ends with a missed approach. The procedures on this and the other GPS approach example in this section are shown using a Trimble 2000 Approach Plus GPS Navigator. Keep in mind that the specific actions required to select and fly an approach can vary between different makes and models of GPS receivers.

PREPARING FOR THE APPROACH

As you fly toward ESC, you tune the AWOS-3 and find that the latest weather is reported to be 800 broken, 2 miles visibility with the wind from 100° at 6 knots. Since your aircraft is equipped with an approach certified GPS receiver, you plan on requesting the GPS RWY 9 approach. [Figure 8-45]

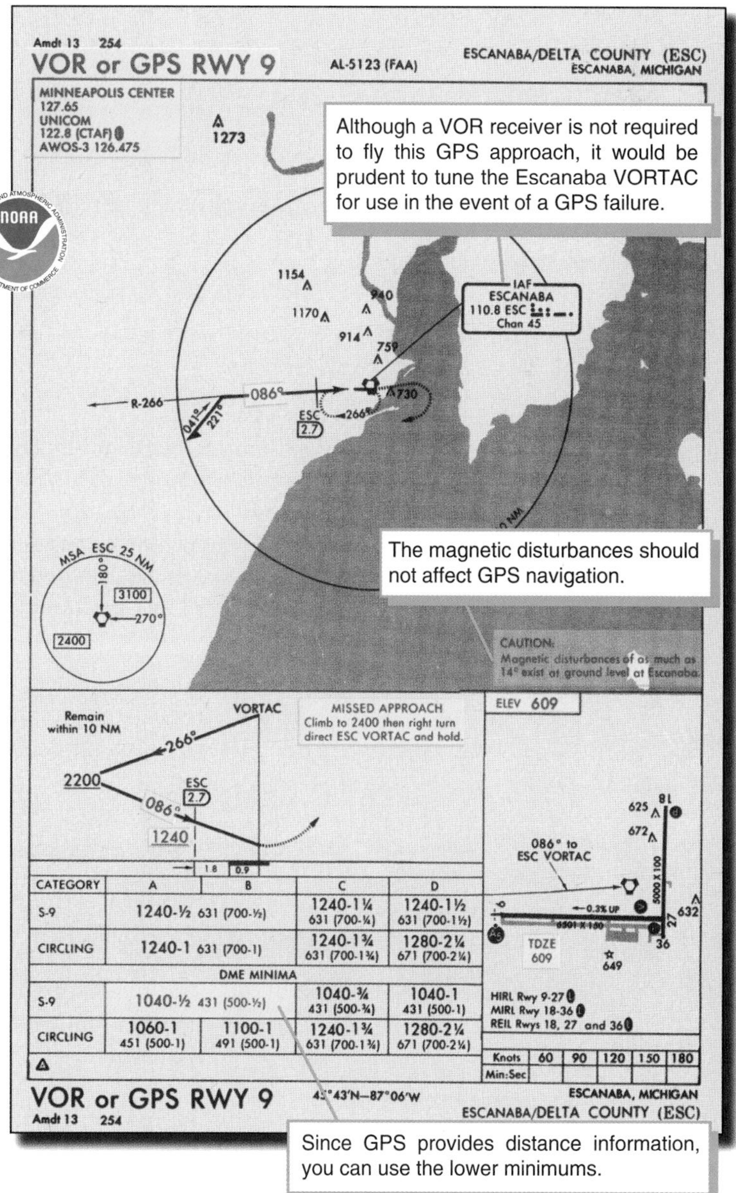

Figure 8-45. During your chart review (see highlights) you look for anything which might be a factor while using GPS to fly the approach.

Your next step is to set your GPS receiver so you are ready to begin the approach when you arrive at the IAF. As you select the approach, make sure to choose the IAF at which you expect to begin the approach. Since there is only one IAF in this case, selecting the approach allows you to view the only routing available. [Figure 8-46]

Figure 8-46. Cycling through the waypoints shows you the sequence of waypoints from the IAF through the MAHWP.

YOUR CLEARANCE

About 15 miles from your destination, Minneapolis Center clears you for the approach: *"Beech 653MC, descend and maintain 3,000. Cleared GPS Runway 9 approach to Delta County. Maintain 3,000 until established on the approach. Change to advisory frequency approved."* After receiving this clearance and announcing your intentions on CTAF, you should ensure the approach mode is activated. (This is sometimes referred to as arming the approach mode.) Since you are within 30 nautical miles of the airport reference point, the CDI sensitivity gradually begins to increase from ±5.0 nautical miles to ±1.0 nautical mile. This change in sensitivity occurs automatically once the approach mode is activated.

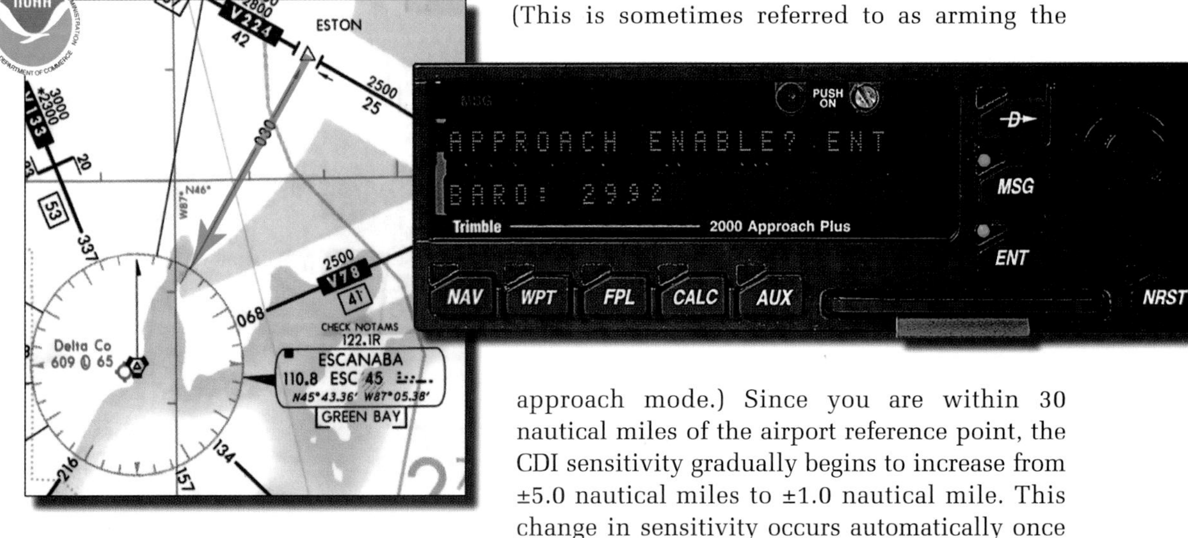

IAF OUTBOUND

As you cross the IAF, the GPS receiver prompts you to turn to intercept the outbound course of 266°. At this point, acknowledge the message, start your turn, and begin a descent to the procedure turn minimum altitude of 2,200 feet MSL. With only 800 feet to lose, you should level off prior to beginning the procedure turn.

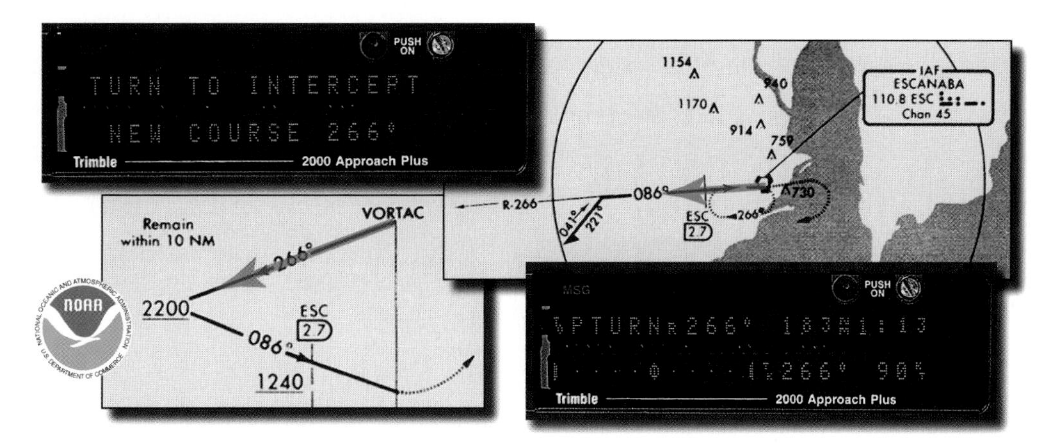

PROCEDURE TURN

While outbound from the IAF, the GPS receiver prompts you to begin the procedure turn and indicates the outbound heading of 221°. Fly the procedure turn as you would on a standard VOR approach, referring to the GPS for guidance relative to the outbound course. At this time, you can complete the before landing checklist with the possible exception of the landing gear and/or flaps.

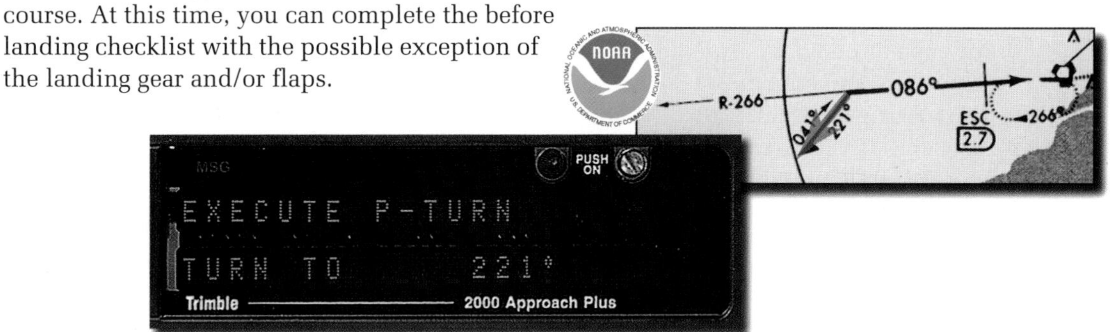

FINAL APPROACH COURSE

Once inbound on the final approach course, the GPS receiver provides guidance to the 2.7 DME stepdown fix, which acts as the FAWP on this approach. Once you are established on the inbound course, start a descent to 1,240 feet MSL and complete any remaining items on the before landing checklist.

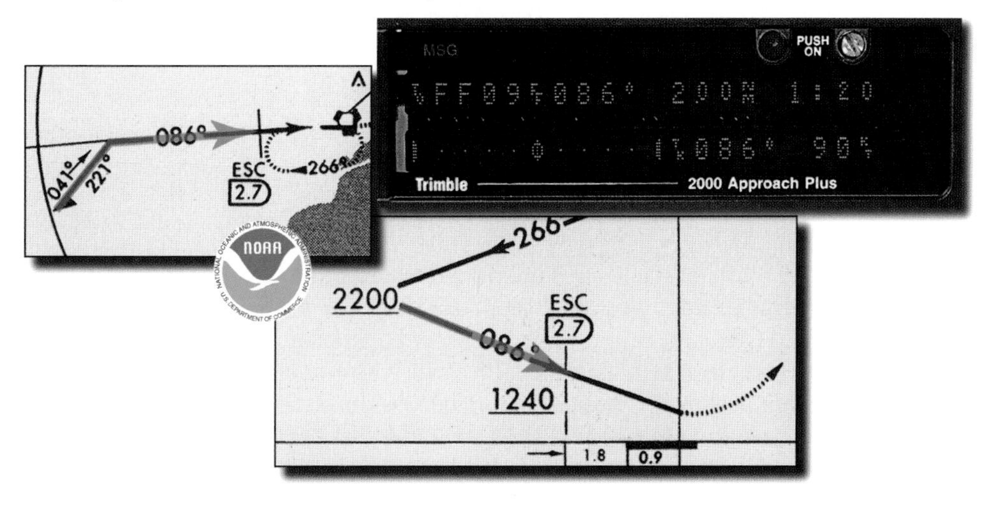

At 2 nautical miles prior the FAWP, the GPS receiver performs a RAIM check. Since RAIM is available, the receiver is allowed to enter the approach mode and CDI sensitivity begins to smoothly increase to ±0.3 nautical mile by the time you reach the FAWP. Since the GPS display provides an along track distance TO the waypoint, FF09, you expect to cross the FAWP when the distance on your GPS display reads 0.0, not 2.7.

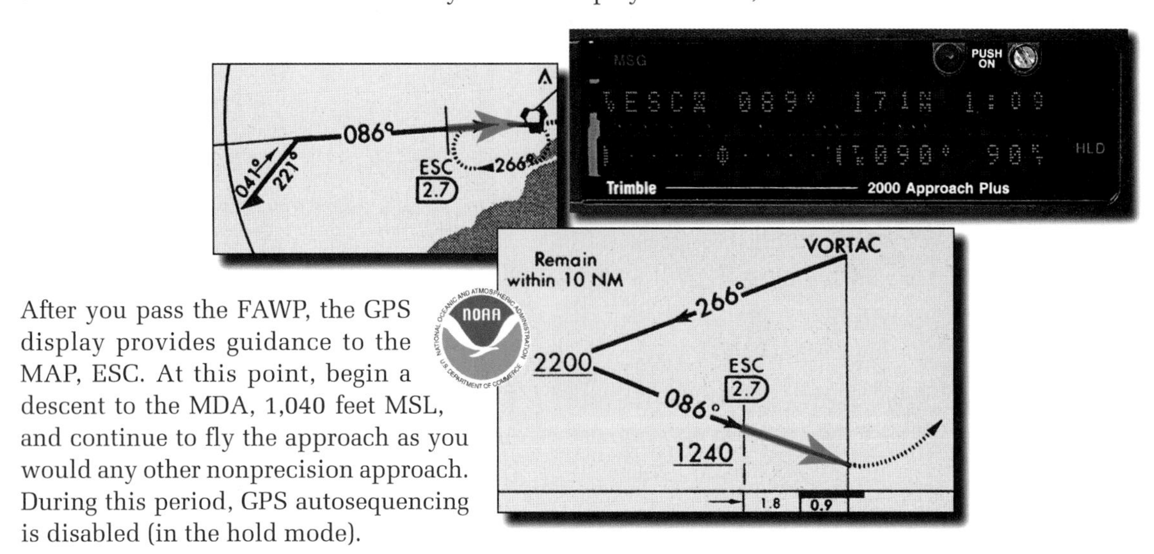

After you pass the FAWP, the GPS display provides guidance to the MAP, ESC. At this point, begin a descent to the MDA, 1,040 feet MSL, and continue to fly the approach as you would any other nonprecision approach. During this period, GPS autosequencing is disabled (in the hold mode).

Missed Approach

As you reach the MAWP, you determine that while you can see the runway, you are not in a position to safely land on Runway 9 and therefore must execute the missed approach. You announce your intentions on CTAF and begin to fly the published missed approach procedure. Since the GPS receiver is in the hold mode, you must manually sequence to the first waypoint in the missed approach procedure. When you sequence to the MAHWP, the GPS receiver remains in the hold mode and provides course guidance to ESC. As you pass over

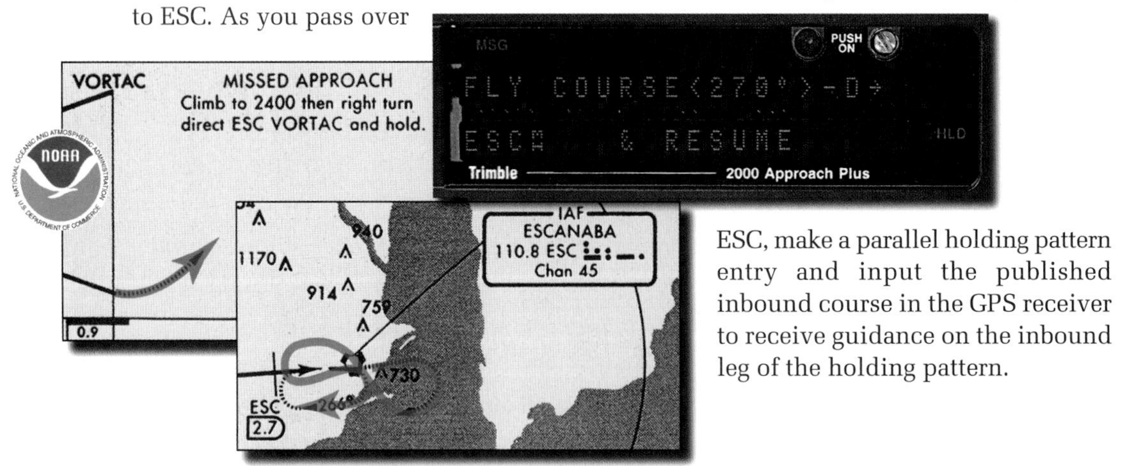

ESC, make a parallel holding pattern entry and input the published inbound course in the GPS receiver to receive guidance on the inbound leg of the holding pattern.

GPS Stand Alone Approach

GPS stand alone approaches are designed specifically to take advantage of the strengths of area navigation. While obstacles and airspace considerations may preclude its use in some instances, the Basic T approach configuration, or a modification thereof, is commonly used for GPS stand alone approaches. A good example of a GPS stand alone approach which uses the Basic T is the GPS Rwy 9 approach to New Castle County Airport, Wilmington, Delaware. The following discussion assumes you are approaching New Castle from over the Modena VORTAC at 3,000 feet MSL. Again, for illustrative purposes, the example ends with a missed approach.

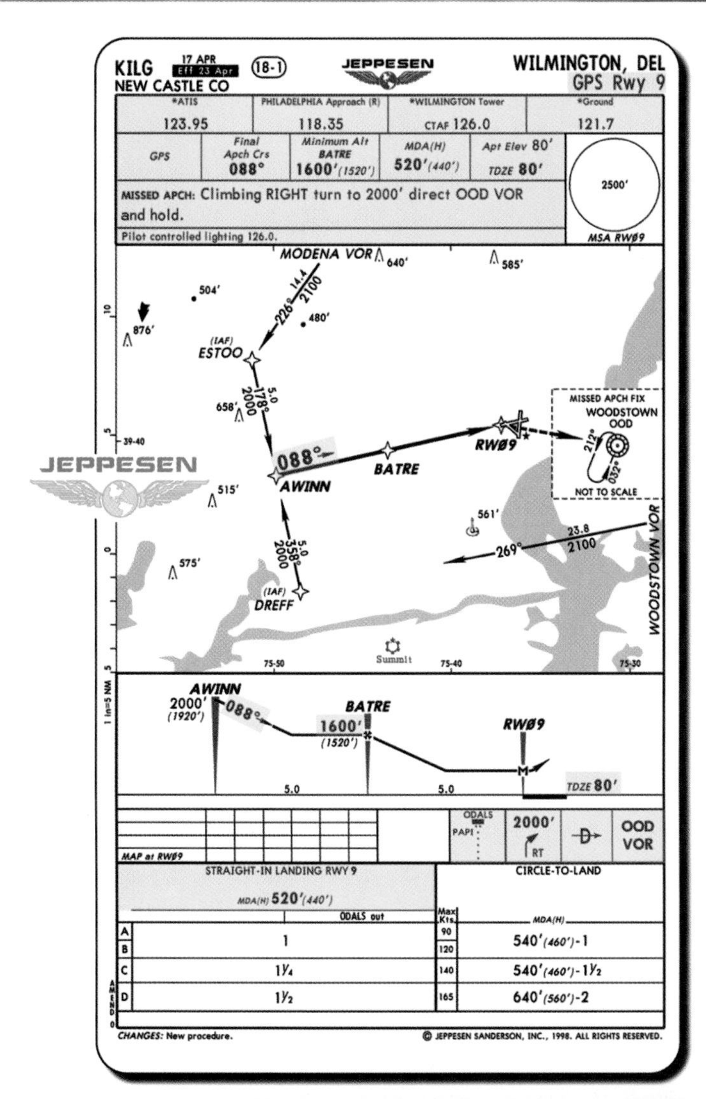

PREPARING FOR THE APPROACH

As you approach your destination, you tune the ATIS and find that the latest reported weather is 600 broken, 1,000 overcast, and 2 miles visibility in rain showers. The wind was reported to be 090° at 6 knots. Since approaches to Runway 9 are in progress, you plan on requesting the GPS Rwy 9 approach from ATC. [Figure 8-47]

Figure 8-47. In preparation for your arrival, you conduct a chart review.

After examining the approach chart, you program your GPS receiver to provide guidance on the approach from ESTOO, the IAF which you anticipate using. In addition to routing from the Modena VORTAC, your GPS unit displays the waypoints you will need on the approach, including the MAHWP. [Figure 8-48]

Figure 8-48. Before beginning a GPS approach, you should ensure that your GPS receiver displays the waypoints you expect to use during the approach.

YOUR CLEARANCE

As you arrive over the Modena VORTAC, you receive clearance from Philadelphia Approach: *"Mooney 66RM, descend and maintain 2,100 and proceed direct ESTOO. Cleared GPS Runway 9 approach to New Castle County. Contact Wilmington Tower on 126.0 at BATRE."* With clearance for the approach, make sure the GPS receiver's approach mode is activated (armed).

DESCENT PRIOR TO THE IAF

While flying from the Modena VORTAC to ESTOO, descend to 2,100 feet MSL. As you level off, complete the before landing checklist with the possible exception of the landing gear and/or flaps.

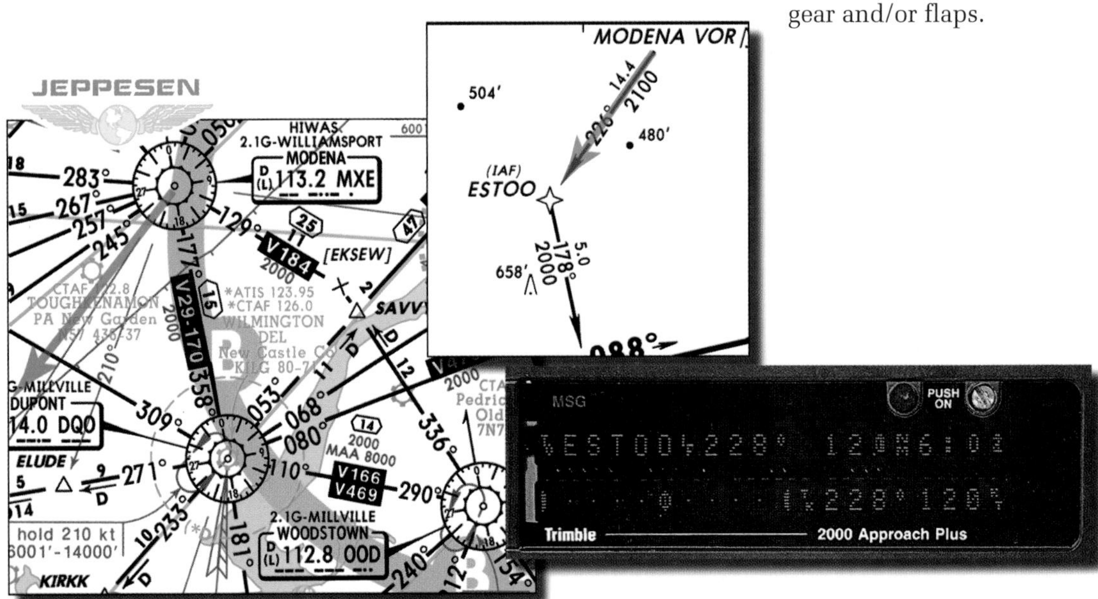

INBOUND TO THE FAF

As you approach ESTOO, the GPS display prompts you to turn and fly to AWINN. Enroute to AWINN descend 100 feet to 2,000 feet MSL, the minimum altitude for this leg of the approach. Just prior to reaching AWINN, you are prompted again to begin a turn. As you fly toward the FAWP, BATRE, begin a descent to 1,600 feet MSL. After you level off, complete any remaining items on your before landing checklist.

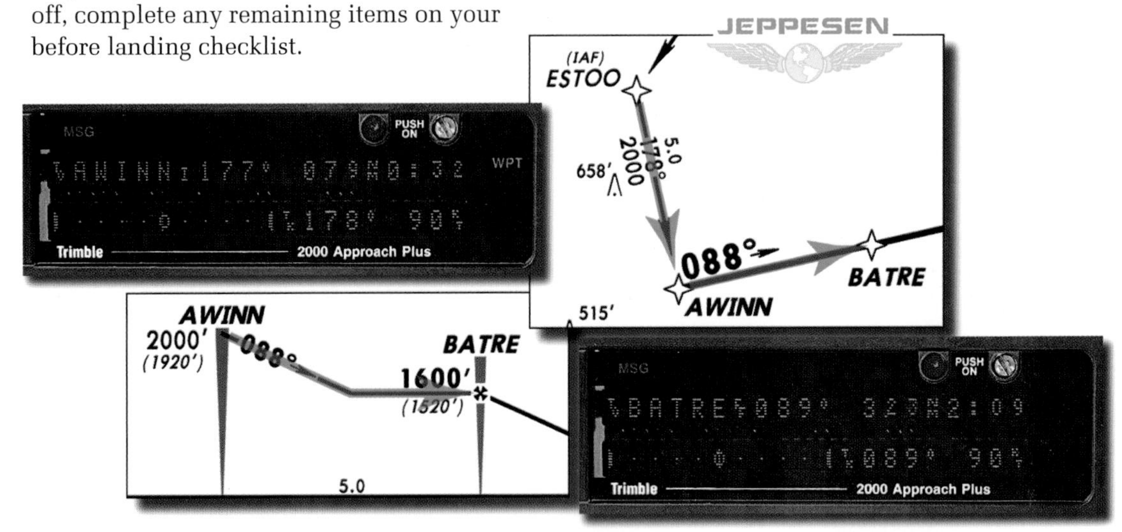

FINAL APPROACH COURSE

As you cross BATRE, start to descend to the MDA(H) and contact Wilmington Tower: *"Wilmington Tower, Mooney 66RM, BATRE inbound on the GPS Runway 9 approach."* As expected, the tower clears you for landing on Runway 9, confirming that you should continue your descent to the straight-in MDA(H) of 520′(440′). Continue inbound and periodically scan ahead for the runway environment to come into view.

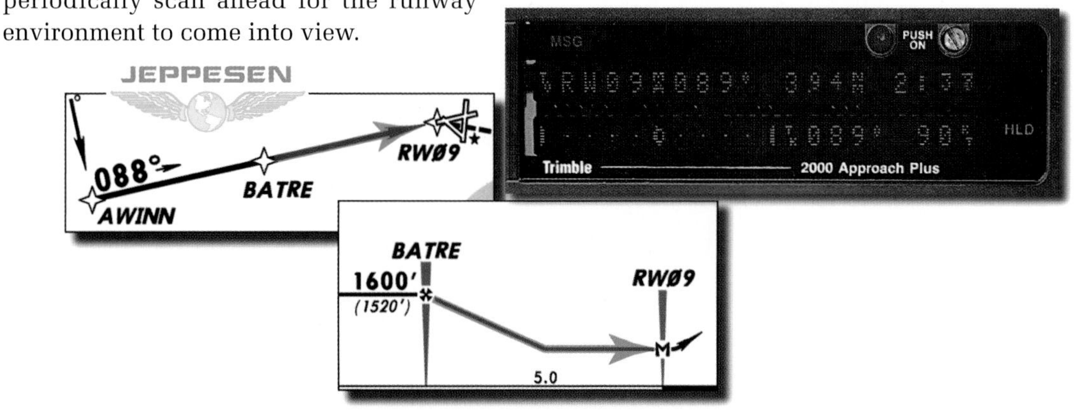

MISSED APPROACH

Since you do not have the runway environment in sight at the MAWP, RW09, you execute a missed approach by starting a climbing right turn to 2,000 feet MSL. Since your GPS receiver automatically switched to the hold mode after BATRE, you must manually sequence to the MAHWP, OOD. Once you do so, the GPS display provides guidance to the MAHWP. Use a parallel entry into the published holding pattern and set the inbound holding course, 032°, in the GPS unit to receive navigation guidance inbound to the MAHWP.

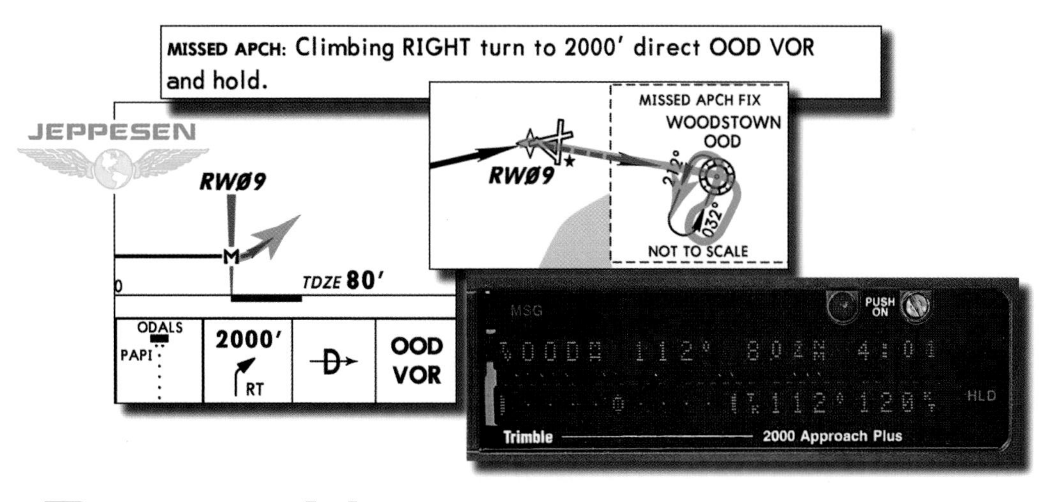

RADAR VECTORS TO A GPS APPROACH

When you receive radar vectors to final, you generally will have to manually sequence ahead and select the leg to which you are being vectored. In accordance with TSO C-129, you can usually choose any leg of an approach to intercept except for the FAWP to MAWP leg. GPS receivers are also programmed to reject any requests to proceed direct to the MAWP.

If you receive vectors to the final approach course, some GPS manufacturers recommend that you place the receiver in the nonsequencing mode on the FAWP and manually set the course to intercept. Since this technique provides an extended final approach course, you should ensure that you are on a published portion of the approach before beginning any descents, unless otherwise authorized by ATC. To avoid confusion, it is also advisable to avoid accepting or requesting radar vectors which will cause you to intercept the final approach course within 2 nautical miles of the FAWP. If you attempt to intercept the final approach course while the CDI sensitivity is increasing to ±0.3 nautical mile, the CDI may be moving further away from center even though you are getting closer to the final approach course.

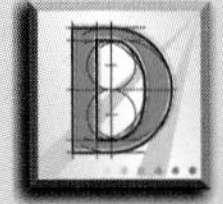

GPS Satellite Anatomy 101

Have you ever wondered about the GPS satellites themselves? What do they look like? How big are they? How do they work? Well, there have been several different types of GPS spacecraft since the original NAVSTAR (Navigation

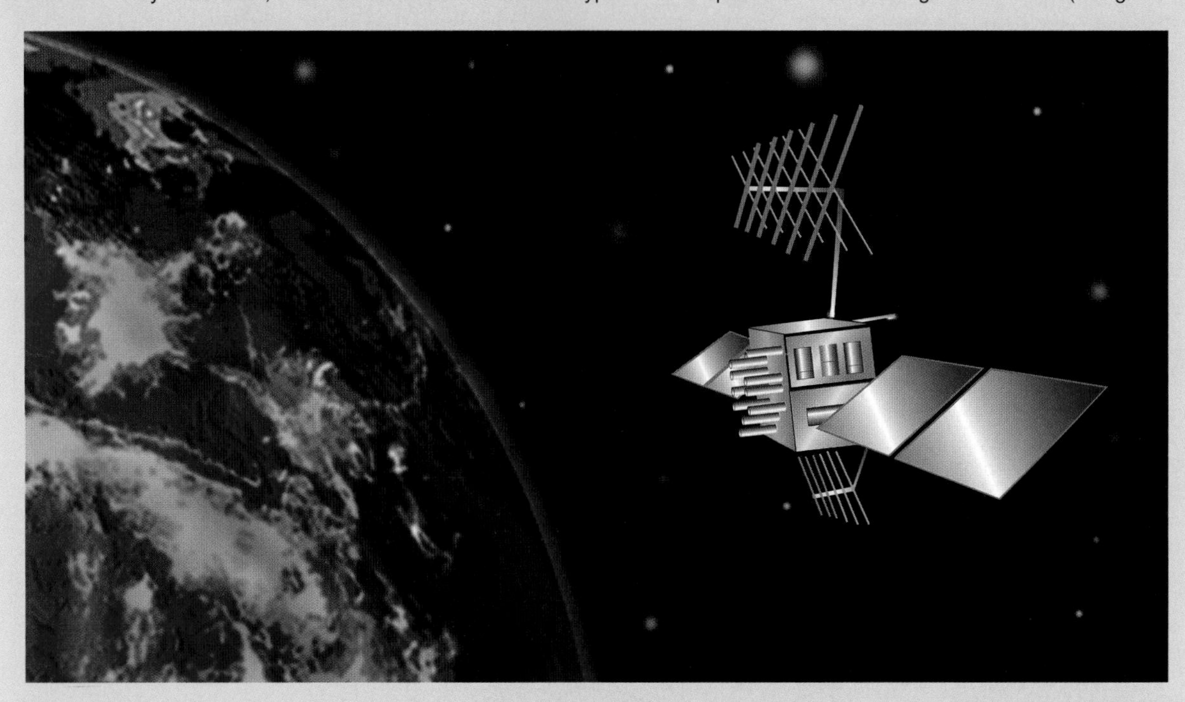

System with Timing and Ranging) launch in 1978. The first 10 satellites in orbit were the Block 1 spacecraft, built by Rockwell Space Systems and weighing 945 pounds each. All of them have since worn out, but they averaged almost eight years each in service. The majority of the current GPS constellation is made up of Rockwell's newer Block 2 and Block 2A spacecraft [see photo] which were launched from 1989 through 1996. The next series of satellites is the Block 2R model produced by Lockheed Martin and scheduled to continue entering service through approximately 2002. The follow-on generation of GPS satellites, Block 2F spacecraft, will be constructed by Boeing with an inaugural launch scheduled for 2001.

Each spacecraft is basically a box-shaped central structure with large solar panels on each side to provide electrical power. On Block 2R spacecraft, the solar panels span more than 30 feet and provide about 1,136 watts of electrical power. The satellite is stabilized in all 3 axes so it always points straight down. Attitude is maintained by electrically actuated reaction wheels. Driving the reaction wheel in one direction causes the spacecraft to rotate in the opposite direction. By arranging three reaction wheels to correspond to the 3 axes of rotation, the spacecraft can be kept in any desired attitude. A system of hydrazine propulsion thrusters is also used to make changes in orbital position of the spacecraft.

As you know, timing is at the heart of the GPS concept, so each satellite has four atomic clocks on board, two rubidium clocks and two cesium clocks. The accurate time signals are of no use if they are not broadcast to your receiver, so each spacecraft has an array of 12 L-band antennas for the downlink transmitters. The spacecraft use S-band radios to communicate with ground controllers, and UHF to communicate with each other. The satellites also carry nuclear detonation detectors, and are protected from laser and nuclear radiation. They are military spacecraft, after all.

VOR/DME RNAV

RNAV based on VOR-DME is a proven system that has provided IFR approach capability for many years. Although GPS may replace RNAV in new equipment installations, there is a good probability that an aircraft you will fly will be equipped with a VOR/DME RNAV system.

OPERATING PRINCIPLES

The heart of a VOR/DME RNAV system is a computer that can determine position based on nearby VORTACs. The computer handles the geometric calculations necessary for accurate navigation on straight line courses not directly to or from ground facilities.

Many jet aircraft utilize flight management systems (FMS) based on VOR/DME RNAV. These systems contain navigation databases and can be programmed for point-to-point enroute and approach operations similar to GPS receivers. The RNAV system you are likely to see in light general aviation airplanes, however, consists of a comparatively simple course line computer (CLC) in which you can program a small number of waypoints based on azimuth and distance from VORTACs. These waypoints are called phantom VORs, which you navigate to and from like you would using actual VORs. [Figure 8-49].

One difference between RNAV CDI indications and conventional VOR navigation indications is that an RNAV CDI indicates absolute deviation in nautical miles rather than angular deviation from course. VOR/DME RNAV sensitivity is not as well-defined as it is for GPS; each dot on the horizontal scale may have any value, such as 0.5, 1, 2, or 10 nautical miles.

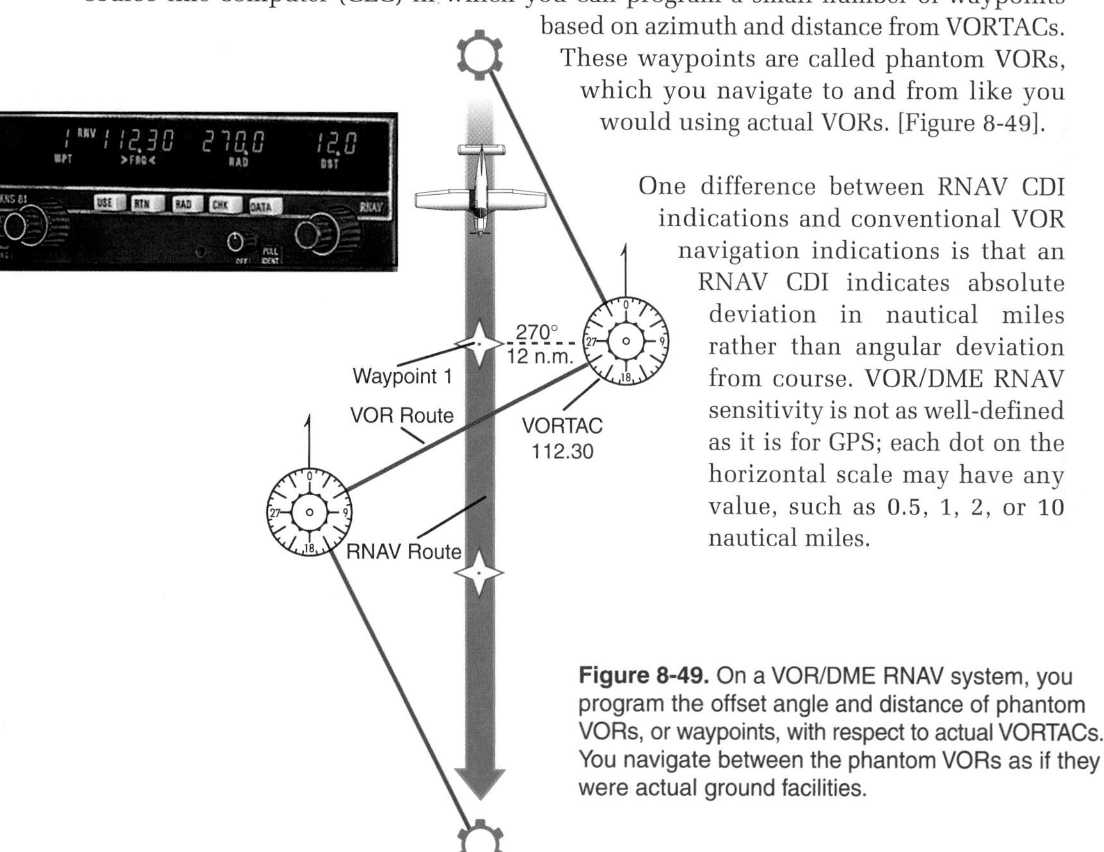

Figure 8-49. On a VOR/DME RNAV system, you program the offset angle and distance of phantom VORs, or waypoints, with respect to actual VORTACs. You navigate between the phantom VORs as if they were actual ground facilities.

VOR/DME RNAV APPROACHES

Many VOR/DME RNAV approaches have been overlaid with GPS data. In other cases, stand alone GPS approaches have been published in addition to the VOR/DME RNAV procedure. Since GPS is another type of RNAV system, there is great similarity between the procedures. Older charts referred to VOR/DME RNAV procedures simply as RNAV approaches. To differentiate this type of RNAV system from GPS and others, updated charts include the words "VOR/DME RNAV". [Figure 8-50]

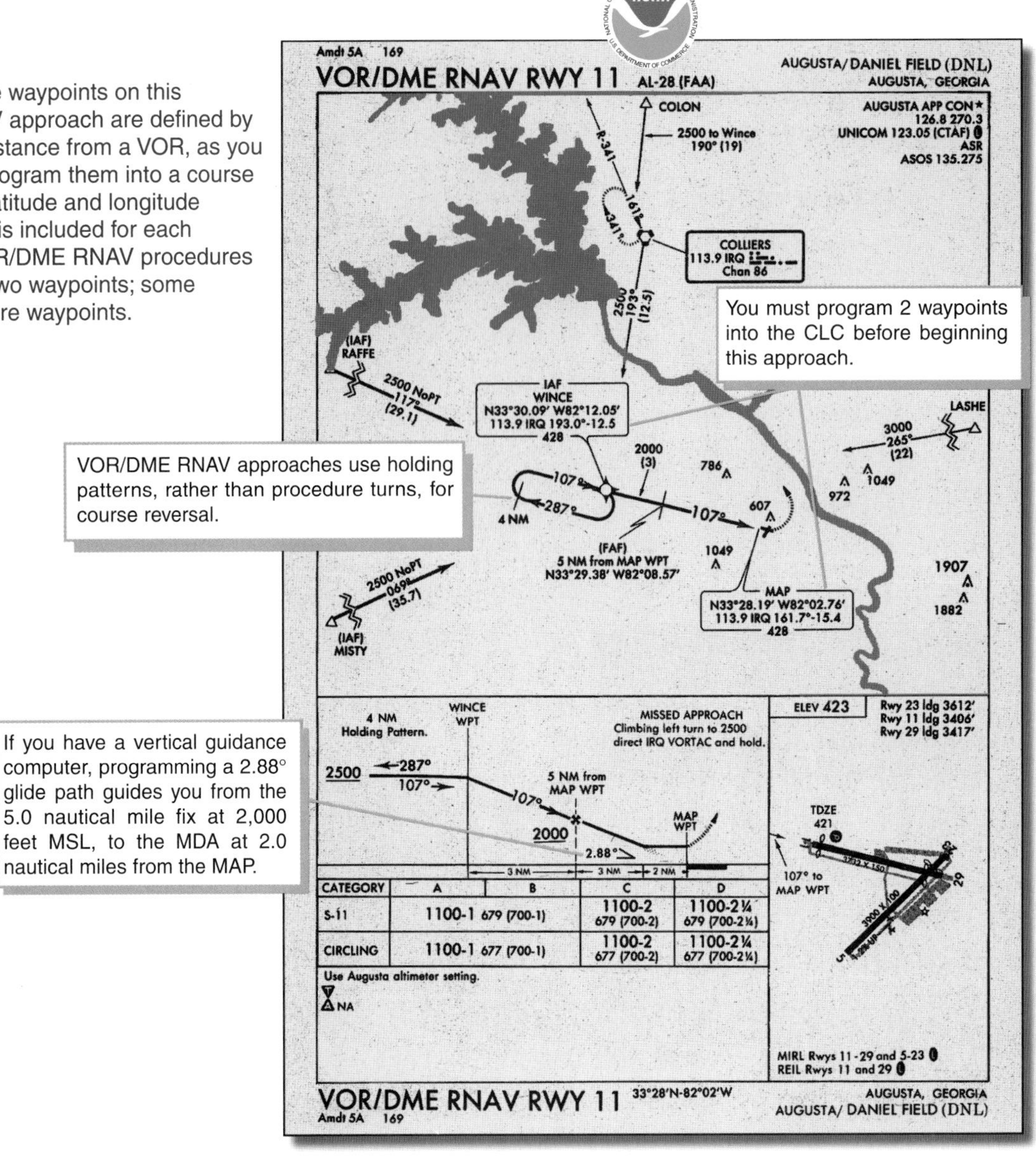

Figure 8-50. The waypoints on this VOR/DME RNAV approach are defined by a bearing and distance from a VOR, as you would need to program them into a course line computer. Latitude and longitude information also is included for each waypoint. All VOR/DME RNAV procedures require at least two waypoints; some require six or more waypoints.

Waypoints are predetermined geographical positions used for RNAV routes or approaches. They are marked with special symbols on approach charts. One example is WINCE, in figure 8-50, which is 193° and 12.5 nautical miles from the reference facility, Colliers VOR. The number of waypoints required to fly a VOR/DME RNAV approach varies with the procedure, but is always at least two.

Once the CLC is programmed and you are cleared for the approach shown in figure 8-50, you would proceed to WINCE, reverse course and begin a descent, if necessary, then intercept and track the 107° final approach course. When passing WINCE inbound, you can descend to 2,000 feet MSL. At 5.0 nautical miles prior to the MAP you can descend to the MDA, 1,100 feet MSL, and continue tracking the 107° course to the MAP. When you reach the MAP, your VOR TO/FROM indicator changes. If you do not have the runway in sight at this point, you should commence the missed approach procedure to Colliers VORTAC.

Some RNAV systems incorporate a vertical guidance feature in addition to normal azimuth/distance guidance. With vertical guidance equipment, you can select a waypoint not only at a designated surface location, but also at a desired altitude. The

An approved RNAV receiver is the minimum navigation equipment needed to fly the Augusta VOR/DME RNAV approach. [Figure 8-50] You may begin your descent to the MDA 5.0 nautical miles prior to the MAP waypoint. You know you have reached the missed approach point when the TO/FROM indicator changes.

On an NOS approach chart, the glidepath angle for vertical guidance systems appears in the profile view between the FAF and the MAP.

equipment provides horizontal and vertical guidance to navigate to that point in space. VOR/DME RNAV approach charts provide the final approach angle for equipment with vertical navigation capability. Vertical guidance capability is not a requirement for standard VOR/DME RNAV approaches.

SUMMARY CHECKLIST

✓ Phase II of the overlay program uses existing approach charts, and requires the underlying ground navaids and associated aircraft navigation equipment to be operational, but not monitored during the approach as long as the GPS meets RAIM accuracy requirements. Phase III of the GPS overlay program eliminates the requirement for conventional navigation equipment to be operational during the approach to your destination airport.

✓ You are not required to monitor or have conventional navigation equipment for stand alone GPS approaches to your destination airport.

✓ You must have conventional navigation equipment aboard your aircraft as a backup for enroute navigation, and to fly to an alternate airport if it becomes necessary. While you can conduct an approach to an alternate airport using GPS, you must have the capability of conducting the approach using conventional equipment.

✓ You can determine the allowable uses for the GPS installation by referring to the supplements section of the Airplane Flight Manual (AFM).

✓ The GPS continuously monitors the reliability of the GPS signal using a system known as receiver autonomous integrity monitoring (RAIM).

✓ Your GPS receiver is required to have current data before it is used for IFR navigation.

✓ A sensor waypoint is included on approaches that do not have a final approach fix defined and usable as the FAWP.

✓ You may need to compute the along track distance (ATD) to stepdown fixes and other points due to the receiver showing ATD to the waypoint rather than DME from the VOR or other ground station.

✓ There may be a variance between the distance displayed on your GPS receiver and the distance published on the accompanying procedure. This occurs because GPS uses a straight line ATD between waypoints while the DME data published on instrument charts is based on slant range to the respective station. The difference between ATD and DME will vary depending on your altitude as well as your proximity to the navaid.

- ✓ While the charted magnetic tracks defined by a VOR radial are determined by the application of magnetic variation at the VOR, GPS computers usually use an algorithm to apply magnetic variation at the current position. Although this process can produce small differences between GPS displayed data and charted information, both operations should produce the same ground track.
- ✓ Normal operation of the GPS unit is referred to as TO-TO navigation. However, when passing a waypoint in hold mode, the external VOR indicator changes from TO to FROM, and the GPS does not autosequence to the next waypoint.
- ✓ Flying from the MAP directly to the MAHWP may not provide sufficient obstacle clearance. You should always fly the full missed approach procedure as published on the approach chart.
- ✓ When you receive radar vectors to final you generally will have to manually sequence ahead and select the leg to which you are being vectored. You should avoid accepting or requesting radar vectors which will cause you to intercept the final approach course within 2 nautical miles of the FAWP.
- ✓ On a VOR/DME RNAV system, you program the offset angle and distance of phantom VORs, or waypoints, with respect to actual VORTACs.
- ✓ An RNAV CDI indicates absolute deviation in nautical miles, rather than angular deviation from course.
- ✓ All VOR/DME RNAV procedures require at least two waypoints; some require six or more waypoints.
- ✓ VOR/DME RNAV approach charts provide the final approach angle for equipment with vertical navigation capability.

KEY TERMS

GPS Overlay Program

GPS Stand Alone Approaches

Navigation Database

Missed Approach Waypoint (MAWP)

Sensor Final Approach Waypoint

Final Approach Waypoint (FAWP)

Along Track Distance (ATD)

Missed Approach Holding Waypoint (MAHWP)

QUESTIONS

1. Which operation requires conventional navigation equipment (VOR, localizer, or NDB) be installed in the aircraft and that the appropriate ground facilities operational?

   A. Stand alone GPS approach
   B. Phase III overlay approach
   C. Stand alone GPS approach to an alternate airport

2. True/False. Although conventional navigation equipment is not required as a backup on certain approach procedures, it is required during other phases of flight.

3. True/False. FAA regulations require monitoring of conventional navigation equipment during GPS operations in approach mode on the final segment of a Phase II approach.

4. How can you determine if a GPS receiver is approved for IFR enroute and approach operations?

5. True/False. If GPS navigation data is out of date, you can still conduct a GPS approach in IFR conditions if you manually enter the waypoints from a current approach chart.

6. What is along track distance (ATD)?

   A. The slant range distance to a waypoint
   B. The straight line distance to a waypoint
   C. The difference between GPS and DME distance measurements

7. What is the normal purpose of the hold mode on a GPS receiver?

   A. To provide TO-TO navigation
   B. To provide TO-FROM navigation
   C. To save frequently used flight plans

Refer to the VOR/DME RNAV or GPS RWY 23 approach to Buffalo Niagara International Airport, Buffalo, New York to answer questions 8-13.

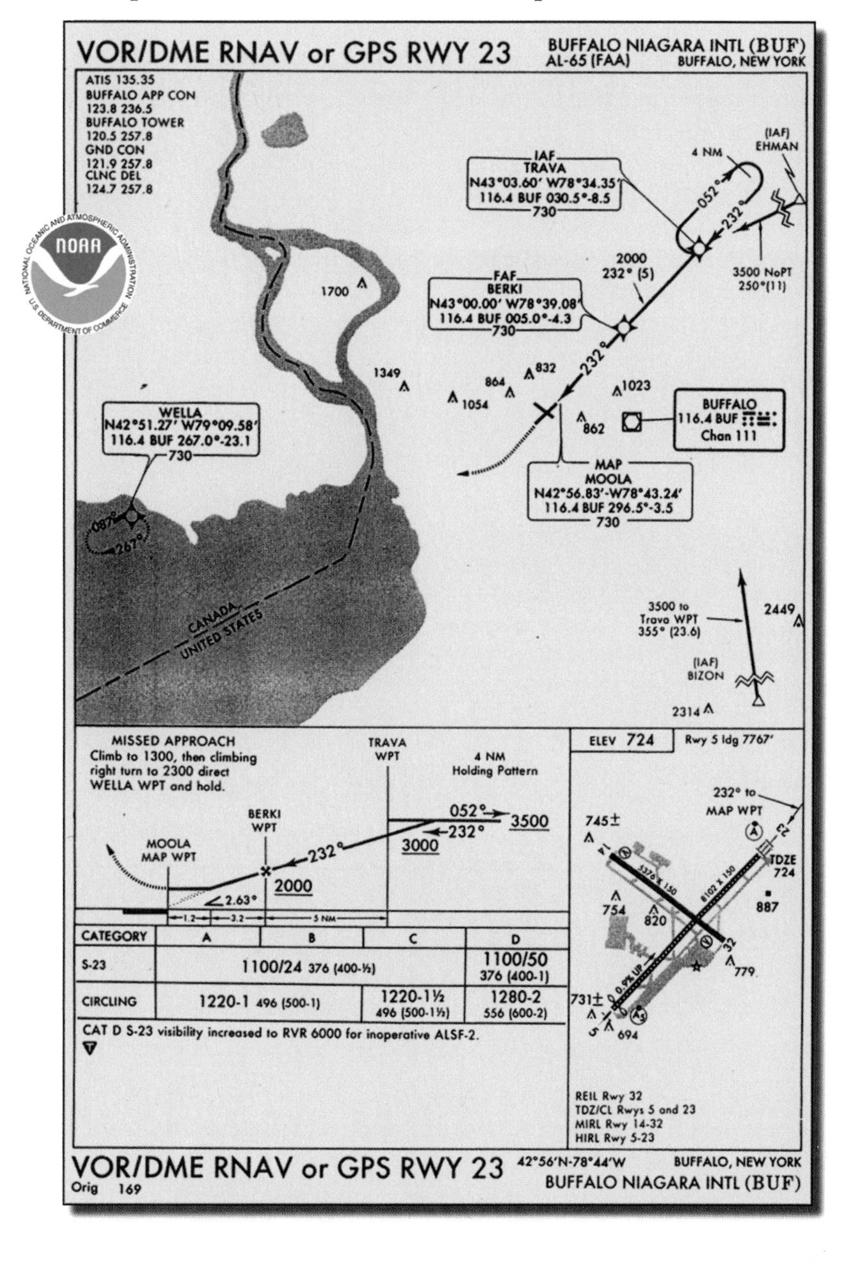

CATEGORY	A	B	C	D
S-23	1100/24 376 (400-½)			1100/50 376 (400-1)
CIRCLING	1220-1 496 (500-1)		1220-1½ 496 (500-1½)	1280-2 556 (600-2)

CAT D S-23 visibility increased to RVR 6000 for inoperative ALSF-2.

8. True/False. If you fly the approach using GPS, you must program the locations of TRAVA, BERKI, MOOLA, and WELLA in your GPS receiver.

9. What is the minimum altitude for the course reversal?

   A. 2,000 feet MSL
   B. 3,000 feet MSL
   C. 3,500 feet MSL

10. Select the true statement regarding the approach segment beginning at BIZON.

    A. BIZON is included in the GPS database.
    B. Since BIZON is an IAF, you can proceed direct to TRAVA or BERKI.
    C. You do not have to execute the course reversal upon arrival at TRAVA.

11. When using GPS for the approach, when can you expect CDI sensitivity to begin to increase to ±0.3 nautical mile?

12. What is the final approach angle you should use for VOR/DME RNAV equipment capable of vertical navigation?

13. True/False. When executing the missed approach from the MDA, you should fly the direct route from MOOLA to WELLA as shown on your GPS receiver.

14. What is the difference between the CDI sensitivity of GPS and conventional non-RNAV VOR indicators?

15. What is the minimum number of waypoints that can be used in the design of a VOR/DME RNAV approach?

    A. 2
    B. 4
    C 6

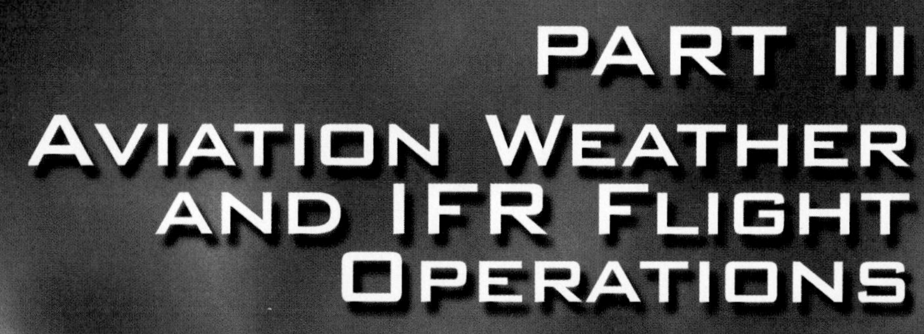

PART III
Aviation Weather and IFR Flight Operations

Weather bothers a pilot only in a few basic ways. It prevents him from seeing; it bounces him around to the extent that it may be difficult to keep the airplane under control and in one piece; and by ice, wind, or large temperature variations, it may reduce the airplane's performance to a serious degree.

— Robert N. Buck

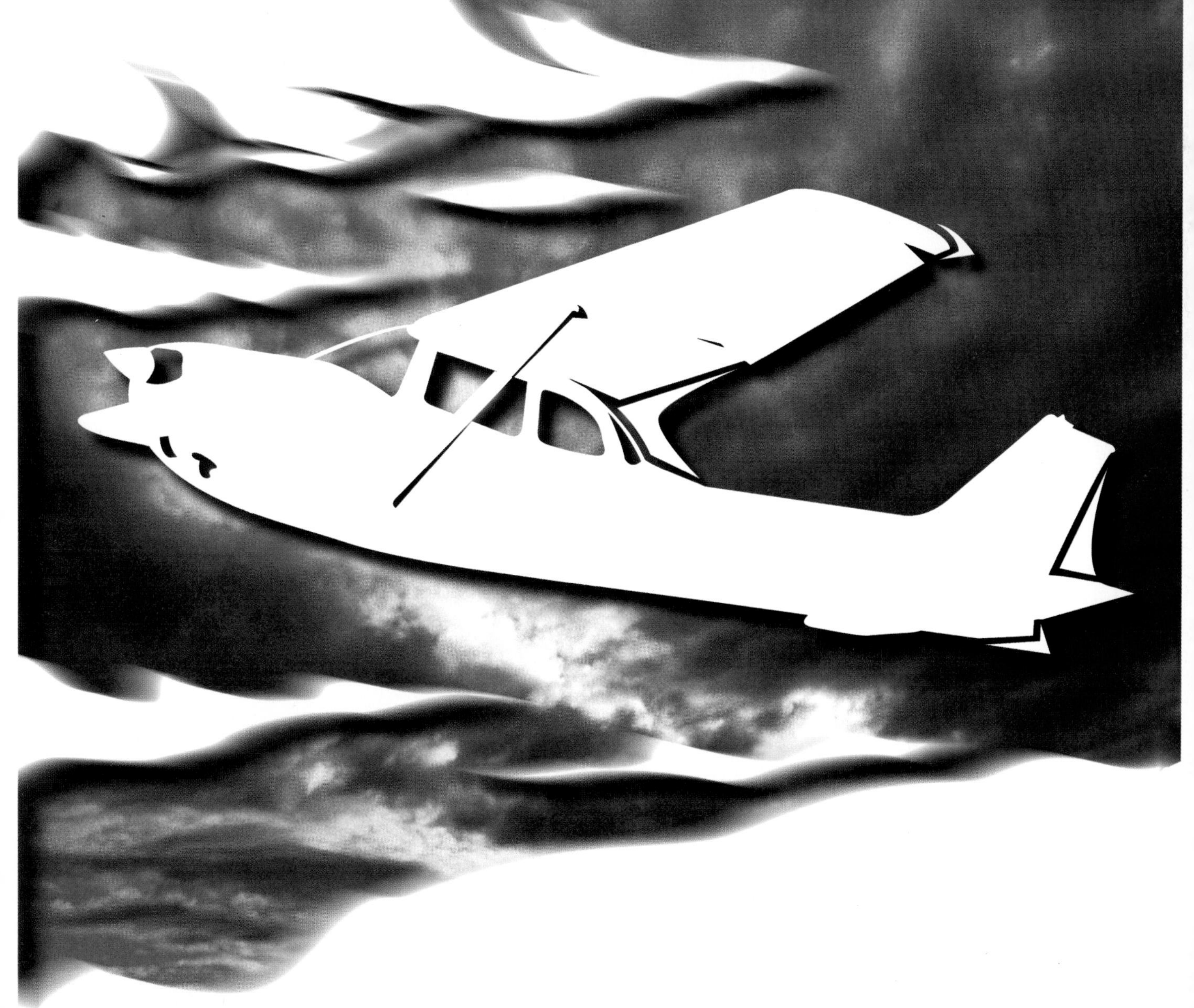

PART III

Decision making for VFR pilots is relatively easy; if there is any chance of getting caught in IFR weather conditions, you cancel your flight. Although you have more options as an instrument-rated pilot, your ability to fly IFR does not mean you can operate in any weather conditions. You must exercise better judgment because you have the opportunity to experience weather hazards that do not affect VFR pilots. Chapter 9 reviews basic weather theory, discusses hazards that affect IFR operations, and reviews weather reports, forecasts, graphic weather products, and sources of weather information. Chapter 10 looks at emergencies unique to instrument flight, and gives you the tools you need for effective planning and decision making in the IFR environment.

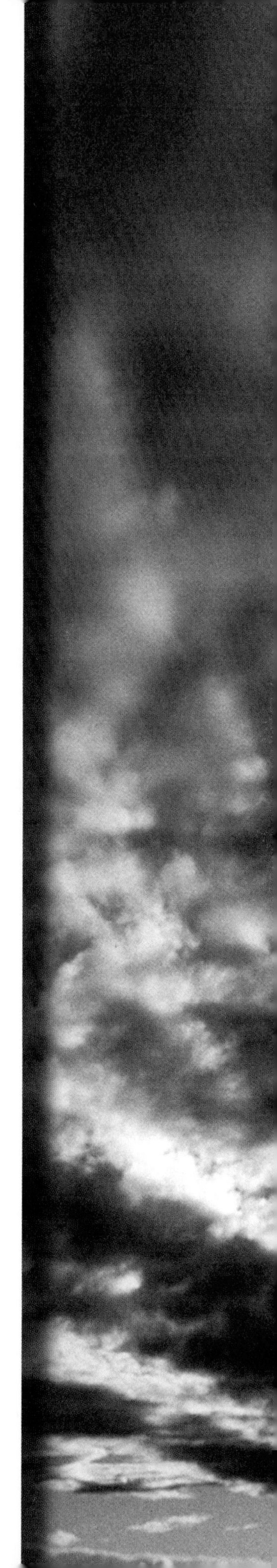

CHAPTER 9

METEOROLOGY

Instrument/Commercial
Part III, Chapter 9 — Meteorology

SECTION A
WEATHER FACTORS

The reports and forecasts you obtain in a weather briefing do not always give you the complete picture of the weather conditions along your entire route of flight. For example, the conditions between reporting points may be difficult to determine, especially in areas with dramatically changing topography. In situations like these, you need to have a solid understanding of basic weather theory, because the determination of whether the flight can be made will most likely be based on your own observations.

THE ATMOSPHERE

The atmosphere is a remarkable mixture of life-giving gases surrounding our planet. Without the atmosphere there would be no protection from X rays, ultraviolet rays, and other harmful radiation from the sun. Though this protective blanket is essential to life on earth, it is extraordinarily thin — almost all of the earth's atmospheric mass is within 30 miles (50 km) of the surface. [Figure 9-1] As a comparison, the thickness of the atmosphere is roughly equivalent to a piece of paper wrapped around a beach ball. The atmosphere does not have a clearly defined upper limit, but simply fades away with increasing altitude.

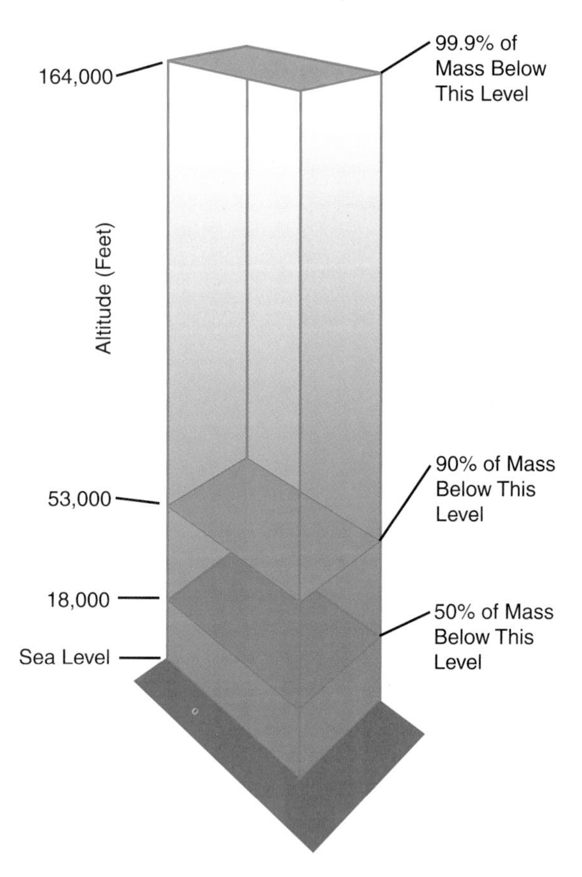

Figure 9-1. Almost all of the earth's atmosphere exists within 50 km (164,000 feet) of the surface. 90% of the atmospheric mass exists below 16 km (53,000 feet).

The most common way of classifying the atmosphere is according to its thermal characteristics. The **troposphere** is the layer from the surface to an altitude which varies between 24,000 and 50,000 feet. It is characterized by a decrease in temperature with altitude. The top of the troposphere is called the **tropopause**. The abrupt change in temperature lapse rate at the tropopause acts as a lid which confines most water vapor, and the associated weather, to the troposphere. Severe thunderstorms are one of the few phenomena that extend into the next layer, the **stratosphere**. The uppermost layers, which contain almost no atmospheric gases, are the mesosphere and thermosphere. [Figure 9-2]

In the troposphere, temperatures decrease with altitude up to the tropopause, where an abrupt change in the temperature lapse rate occurs. The average height of the troposphere in the middle latitudes is 36,000 to 37,000 feet. As shown in figure 9-2, the temperature in the lower part of the stratosphere (up to approximately 66,000 feet) experiences relatively small changes in temperature with an increase in altitude.

Additional layer designations help identify the vertical structure of the atmosphere. The ozone layer is characterized by a high concentration of O_3, with maximum concentration at about 80,000 feet MSL. This special type of oxygen molecule absorbs harmful solar radiation and accounts for the increase in temperature with altitude in that part of the atmosphere.

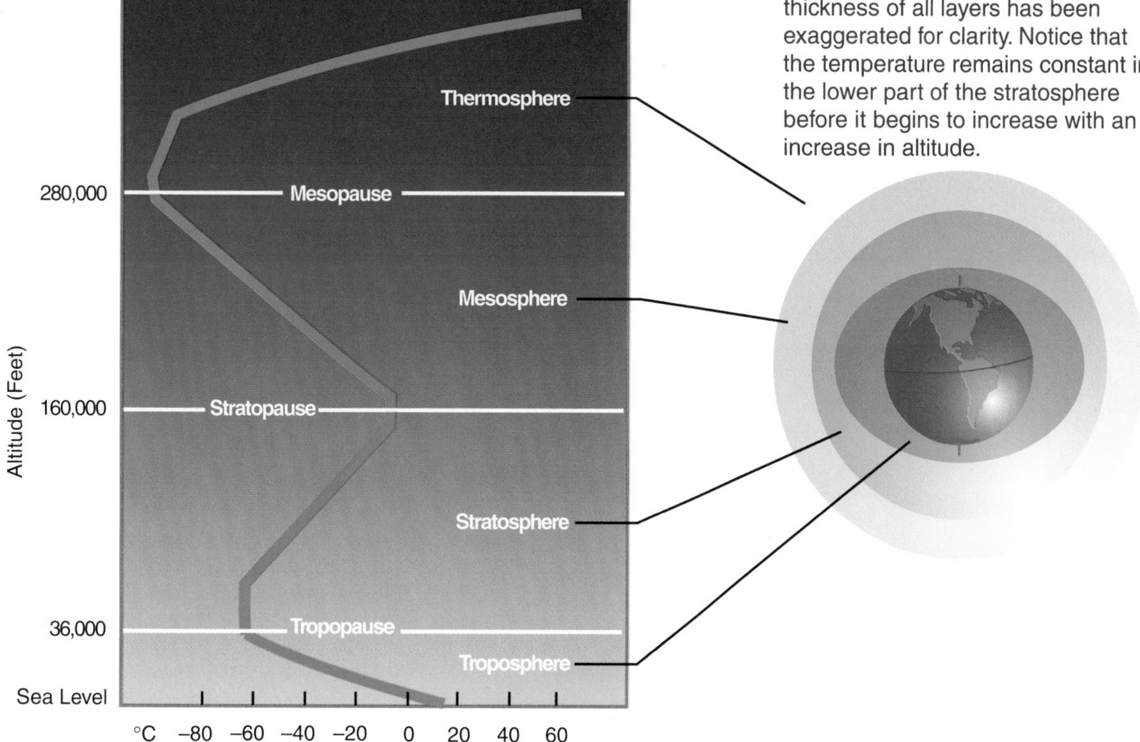

Figure 9-2. The height of the tropopause varies with the season and location over the globe. It tends to be higher where it is warmer. The thickness of all layers has been exaggerated for clarity. Notice that the temperature remains constant in the lower part of the stratosphere before it begins to increase with an increase in altitude.

The ionosphere is a deep layer of charged particles beginning about 30 miles above the surface. The electrical characteristics of the ionosphere can affect radio communications around sunrise and sunset, and during periods of increased solar activity.

ATMOSPHERIC CIRCULATION

Solar radiation strikes the earth at different angles at different locations, resulting in uneven heating of the earth's surface. This uneven heating is the driving force behind all weather. Because of the tilt of the earth's axis, the northern and southern hemispheres

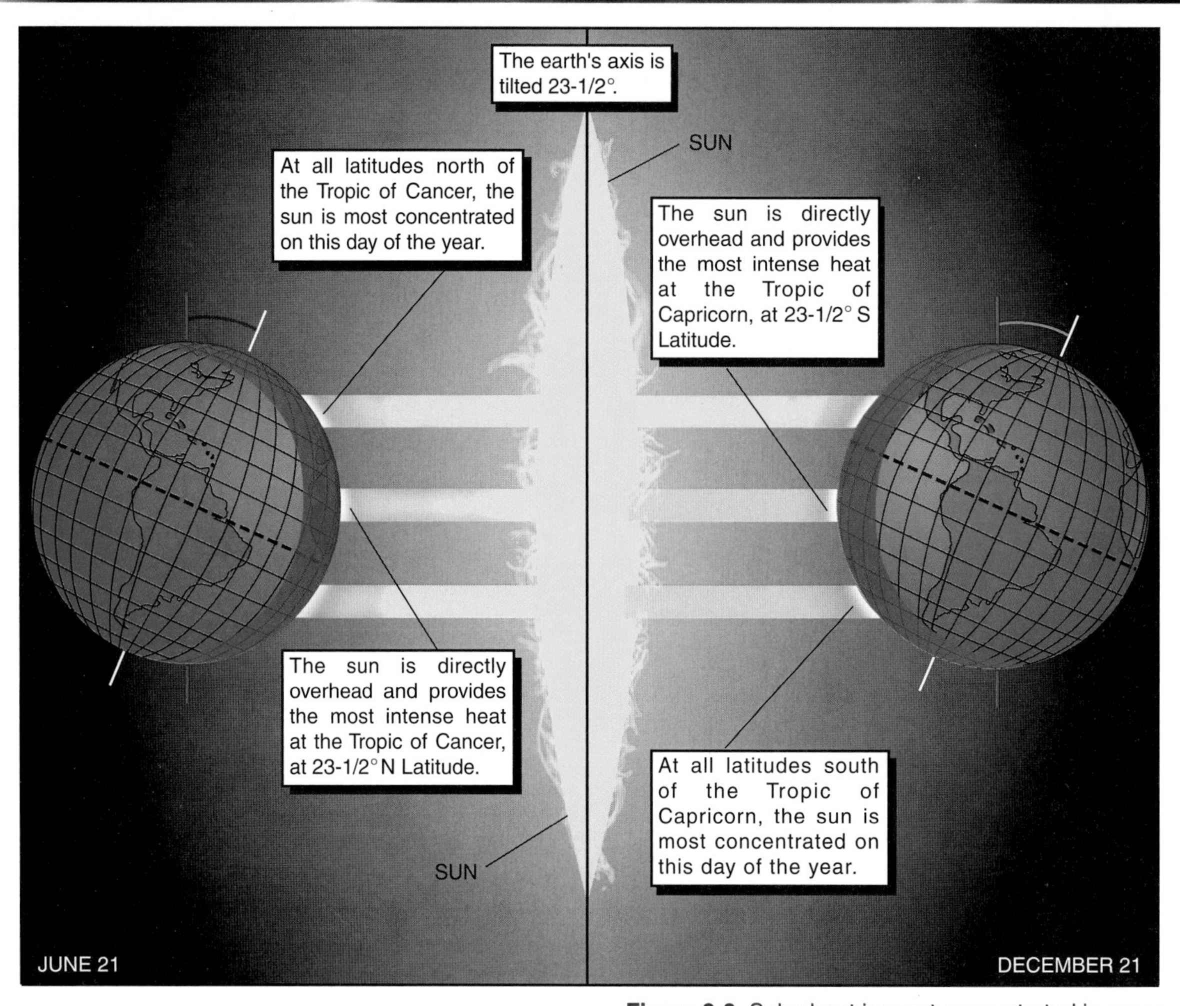

Figure 9-3. Solar heat is most concentrated in areas where the sun's rays strike the earth most nearly perpendicular to the surface.

Every physical process of weather is accompanied by or is the result of a heat exchange. The primary cause of all changes in the earth's weather is the variation of solar energy received by the earth's regions.

receive disproportionate amounts of solar radiation throughout the year. In general, the most direct rays of the sun strike the earth in the vicinity of the equator while the poles receive the least direct light and energy from the sun. [Figure 9-3]

PRESSURE AND WIND PATTERNS

The unequal heating of the surface modifies air density which results in differences in pressure. Meteorologists plot pressure readings from weather reporting stations on charts and connect points of equal pressure with lines called **isobars**. These lines are normally labeled in millibars. The resulting pattern reveals the **pressure gradient**, or change in pressure over distance. When isobars are spread widely apart, the gradient is considered to be weak, while closely spaced isobars indicate a strong gradient.

Isobars help identify pressure systems, which are classified as highs, lows, ridges, troughs, and cols. A **high** is a center of high pressure surrounded on all sides by lower pressure. A **low** is an area of low pressure surrounded by higher pressure. A **ridge** is an elongated area of high pressure and a **trough** is an elongated area of low pressure. A **col** can designate either a neutral area between two highs or two lows, or the intersection of a ridge and a trough. [Figure 9-4] Low pressure areas are areas of rising air, which can encourage bad weather, while high pressure areas consist of descending air which encourages good weather. [Figure 9-5]

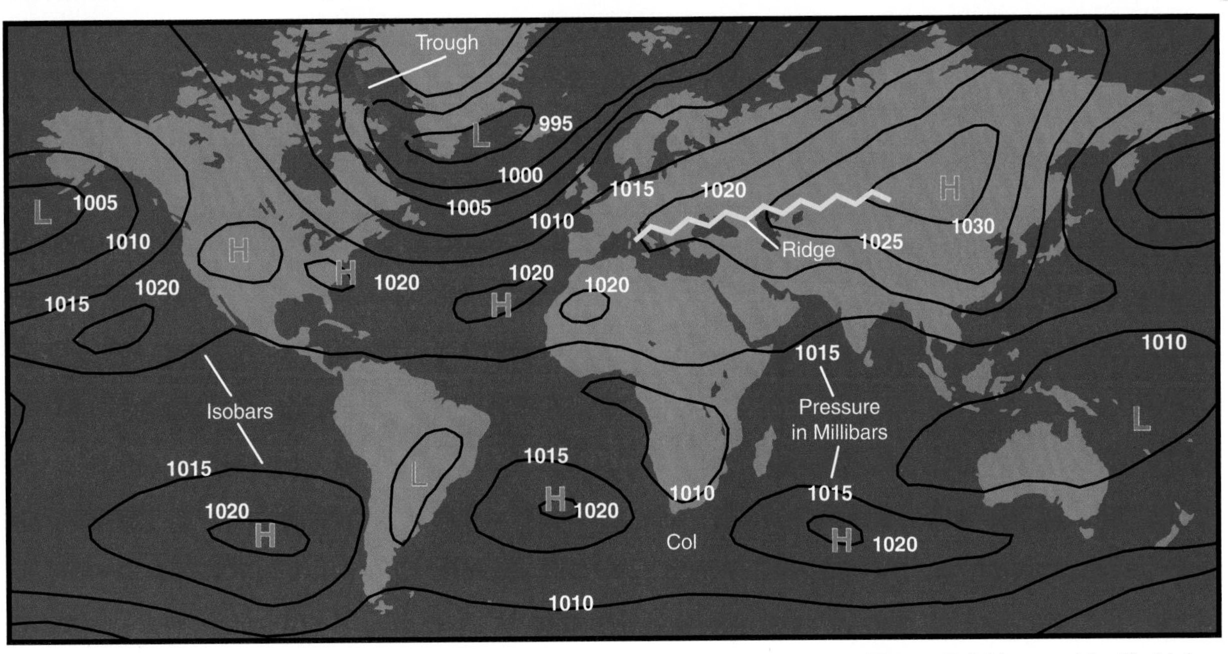

Figure 9-4. You can identify highs, lows, ridges, troughs, and cols when isobars are plotted on a weather chart.

Figure 9-5. The weather associated with high and low pressure areas is affected by descending and rising air.

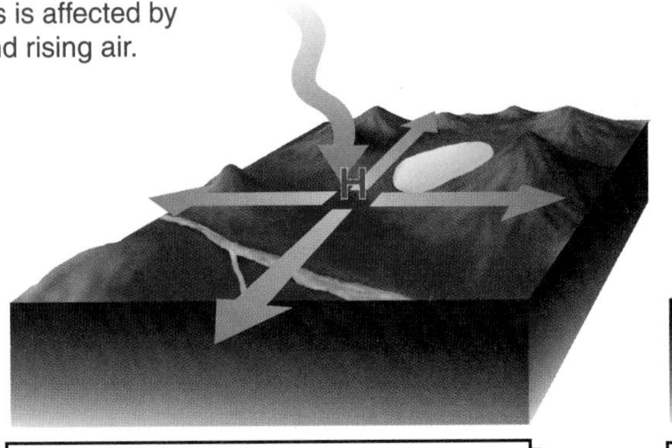

In high pressure systems, air moves outward, and must be replenished from aloft. As a result, a high or ridge is characterized by descending air which favors dissipation of cloudiness. This is why a high or a ridge is generally associated with good visibility, calm or light winds, and few clouds.

When air converges into a low, it cannot go anywhere but up. This is why a low or trough is an area of rising air. Rising air is conducive to cloudiness, precipitation, poor visibility, gusty winds, and turbulence. Weather may be very violent in the area of a trough.

Wind is caused by airflow from cool, dense high pressure areas into warm, less dense, low pressure areas. The speed of this wind depends on the pressure gradient force. The stronger the gradient force, the stronger the wind. As the earth rotates beneath this airflow, **Coriolis force** counterbalances the pressure gradient force and deflects airflow to the right as it flows out of a high pressure area in the northern hemisphere. This results in clockwise circulation leaving a high and counterclockwise, or **cyclonic**, circulation entering a low.

A low pressure area or trough is an area of rising air, while a high pressure area or ridge is characterized by descending air. See figure 9-5.

Coriolis force affects air that flows independent of the earth's surface. While this force deflects winds aloft parallel to the isobars, airflow near the surface is influenced by friction with the surface, which weakens the effects of Coriolis force. As a result, pressure gradient force causes surface winds to cross the isobars at an angle. For this reason, wind direction tends to shift when you descend to within 2,000 feet of the surface. [Figure 9-6]

Wind is caused by pressure differences as air flows outward from a high pressure area to a low pressure area. However, the wind does not flow directly from a high to a low because of Coriolis force, which deflects air to the right in the northern hemisphere. The result is a wind that flows in a clockwise direction leaving a high and in a counterclockwise, or cyclonic, direction when entering a low.

Winds aloft parallel the isobars because Coriolis force tends to counterbalance the pressure gradient force. Surface winds, however, cross isobars at an angle and are weaker because of surface friction.

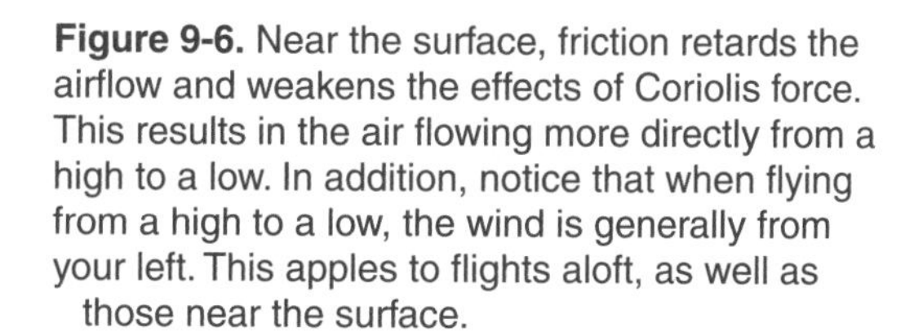

Figure 9-6. Near the surface, friction retards the airflow and weakens the effects of Coriolis force. This results in the air flowing more directly from a high to a low. In addition, notice that when flying from a high to a low, the wind is generally from your left. This apples to flights aloft, as well as those near the surface.

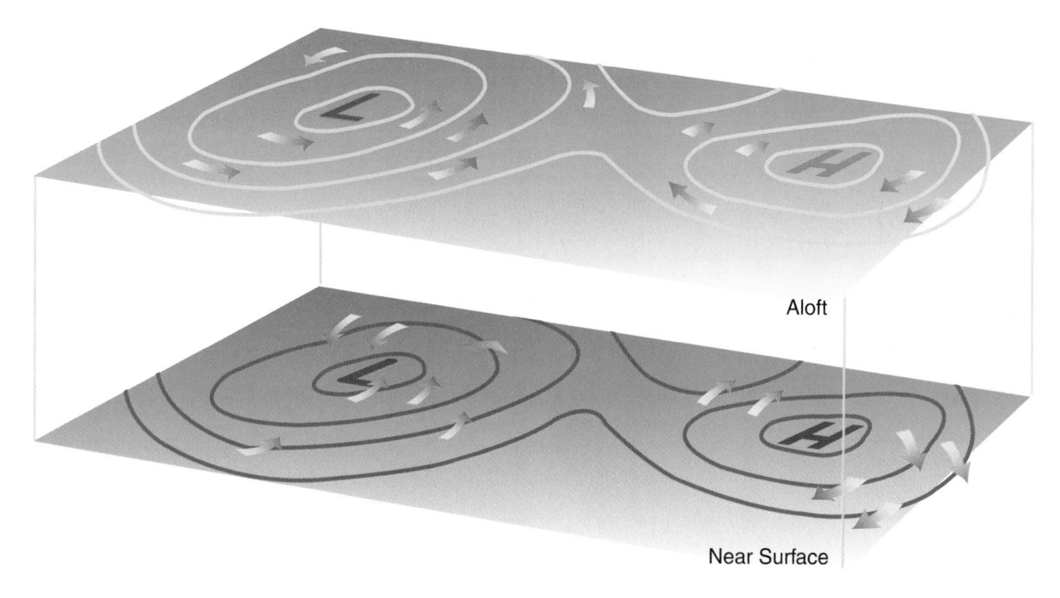

Local Convective Circulation

As shown in figure 9-6, when flying to an area of low pressure, which is generally an area of unfavorable weather conditions, you will most likely experience a crosswind from the left. As pressure gradient gets stronger toward the center of the low, the winds increase.

Winds near bodies of water are caused by differences in temperature between the land and water surfaces. Since land surfaces warm or cool more rapidly than water surfaces, land usually is warmer than water during the day. This creates a sea breeze, which is a wind that blows from cool water to warmer land. As afternoon heating increases, the sea breeze can reach speeds of 10 to 20 knots. At night, land cools faster than water, and a land breeze blows from the cooler land to the warmer water. Since the temperature contrasts are smaller at night, the land breeze is generally weaker than the sea breeze. [Figure 9-7]

Convective circulation patterns associated with sea breezes are caused by land absorbing and radiating heat faster than water. Cool air must sink to force warm air upward. See figure 9-7.

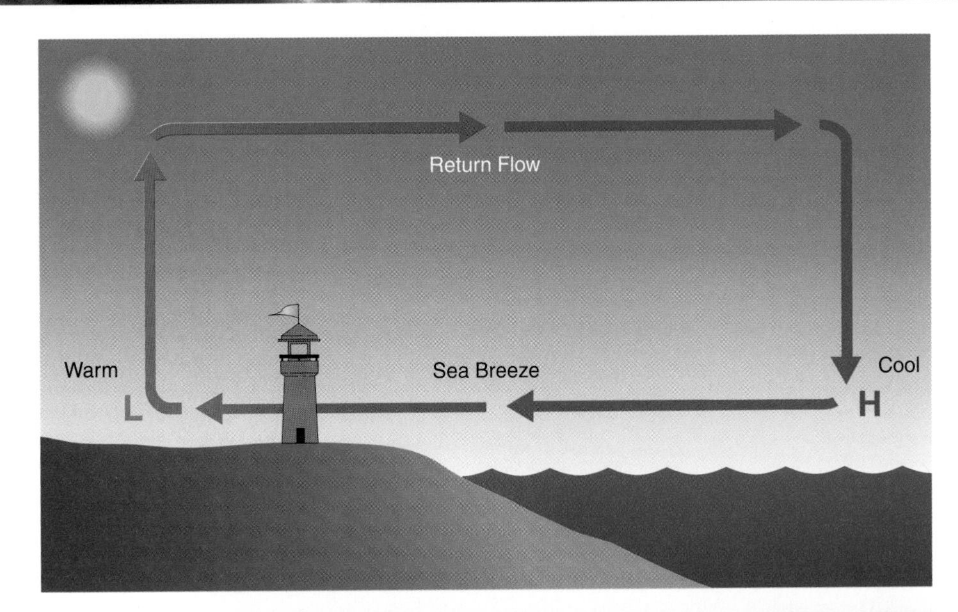

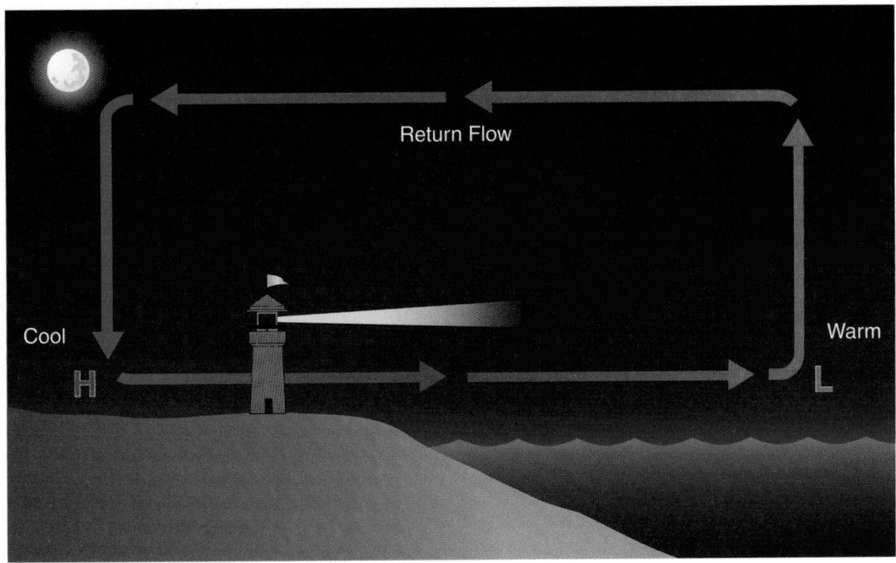

Figure 9-7. During the warm part of the day, air rises over the relatively warm land mass and sinks over the cooler water. This circulation pattern results in an onshore flow. At night, when the land cools more than the water, the circulation pattern reverses.

MOISTURE, PRECIPITATION, AND STABILITY

While the sun provides the energy that drives the earth's weather, water with its special thermal properties, stores and releases this energy in ways that dramatically affect the weather. Water can exist in a solid (ice), liquid, or gaseous (vapor) state. Water vapor is added to the atmosphere through evaporation and sublimation. **Evaporation** occurs when heat is added to liquid water, changing it to a gas. **Sublimation** is the changing of ice directly to water vapor, bypassing the liquid state. Water vapor is removed from the atmosphere by condensation and deposition. **Condensation** occurs when the air becomes saturated, and water vapor in the air becomes liquid. **Deposition** is when water vapor freezes directly to ice. Of course, liquid water can also freeze, and ice can melt into liquid water.

Moisture is added to a parcel of air by evaporation and sublimation.

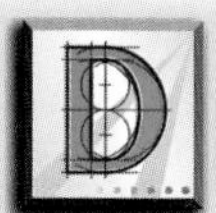

The Ultimate Instrument Airplane

The Research Aviation Facility (RAF) of the National Center for Atmospheric Research (NCAR) operates two National Science Foundation (NSF) aircraft used for weather research.

The Lockheed EC-130Q Hercules and the L-188 Electra are four-engine turboprops that can carry up to 16 researchers in addition to the 3 flight crewmembers. The Hercules offers nearly 90 kilowatts (kw) of electric power for research equipment and can stay aloft for 10 hours. The Electra powers up to 50 kw of equipment and can stay aloft 8-1/2 hours.

NASA's Dryden Flight Research Center is currently using the L-188 in a study on detecting and forecasting clear air turbulence (CAT), discussed in the next section. Other typical research applications include oceanographic investigations, air-sea interaction studies, cloud physics studies, tropospheric profiling, atmospheric chemistry, and aerosol studies. Qualified researchers interested in utilizing these aircraft can submit a formal application to NCAR and the National Science Foundation.

Courtesy of NCAR/ Research Aviation Facility

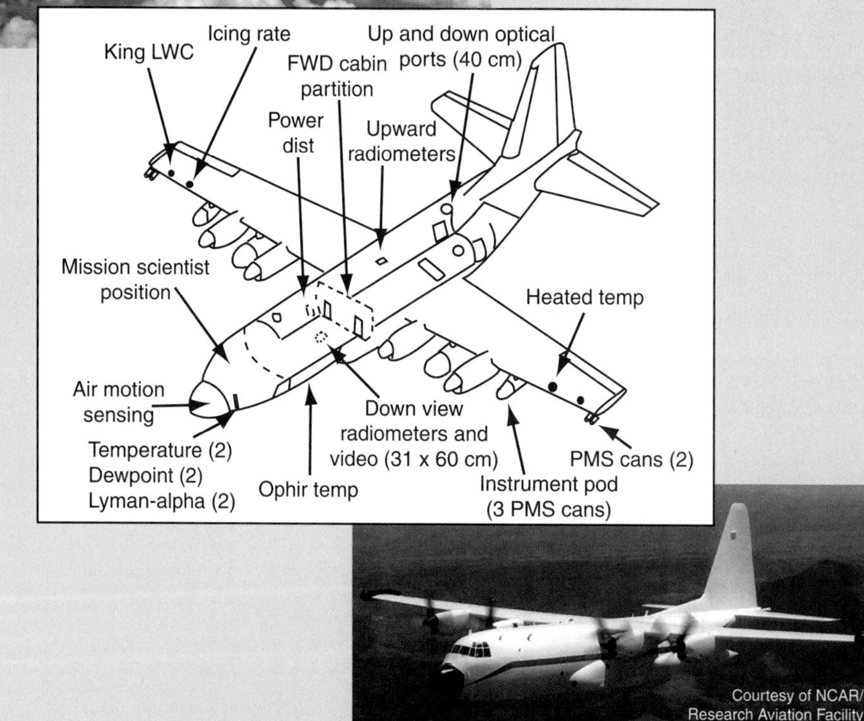

Courtesy of NCAR/ Research Aviation Facility

DEWPOINT

The amount of water vapor the air can hold decreases with the air's temperature. When the air cools to the **dewpoint**, it contains all the moisture it can hold at that temperature, and is said to be saturated. Relative humidity increases as the temperature/dewpoint spread decreases. When the air is saturated, the relative humidity is 100%. On cool, still nights, surface features and objects may cool to a temperature below the dewpoint of the surrounding air. When this happens, dew condenses on the cold surfaces. Frost forms when water vapor changes directly to ice on a surface that is below freezing.

Clouds are composed of very small droplets of water or ice crystals. When they form near the surface, clouds are referred to as fog. You can anticipate the formation of fog or very low clouds when the temperature/dewpoint spread is decreasing below 2°C, or about 4°F.

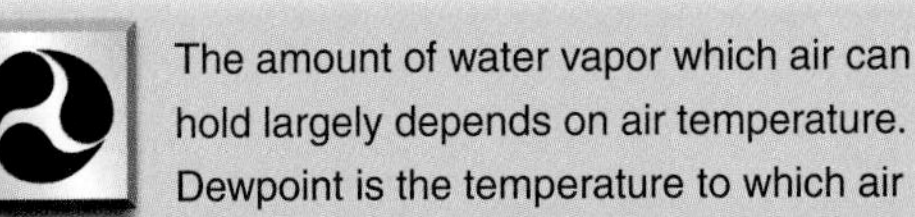

The amount of water vapor which air can hold largely depends on air temperature. Dewpoint is the temperature to which air must be cooled to become saturated. The temperature and dewpoint spread decreases as the relative humidity increases. At 100% humidity, water vapor condenses, forming clouds, fog or dew. Frost forms when the temperature of the collecting surface is below the dewpoint and the dewpoint is below freezing.

PRECIPITATION

When condensed water droplets grow to a size where the atmosphere can no longer support their weight, they fall as precipitation. Water droplets that remain liquid fall as drizzle or rain. With low relative humidity, rain may evaporate before it reaches the surface. When this occurs, it is called virga. [Figure 9-8]

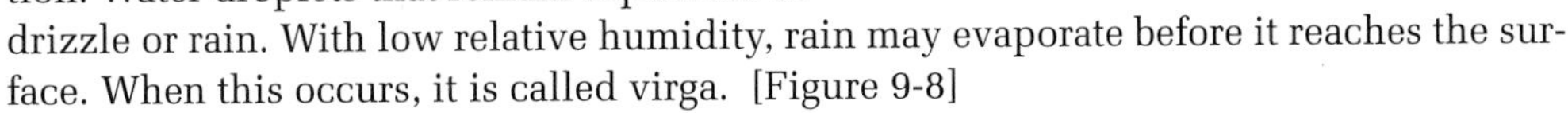

Figure 9-8. Virga appears as streamers of precipitation trailing from clouds.

Sometimes water droplets can remain in liquid form even though they are cooled below freezing. When this **supercooled water** strikes an object, such as an airplane in flight or the earth's surface, it immediately turns to ice, or freezing rain. Ice pellets, by contrast, freeze as they fall through cold air and are more likely to bounce off your aircraft rather than freeze to it. The presence of ice pellets generally indicates the existence of freezing rain and warmer air at higher altitudes.

Virga is best described as streamers of precipitation trailing beneath clouds which evaporate before reaching the ground.

The presence of ice pellets normally indicates freezing rain at higher altitudes.

Unlike ice pellets, which fall directly to the ground, hail forms in clouds with strong vertical currents. The freezing water droplets are carried up and down increasing in size as they collide and freeze with other water droplets. When they become too large for the air currents to support, they finally fall as hail. [Figure 9-9] If the air currents are particularly strong, hailstones can grow as large as 5 inches in diameter and weigh up to 1-1/2 pounds. Obviously, large hailstones are very dangerous and can cause tremendous damage.

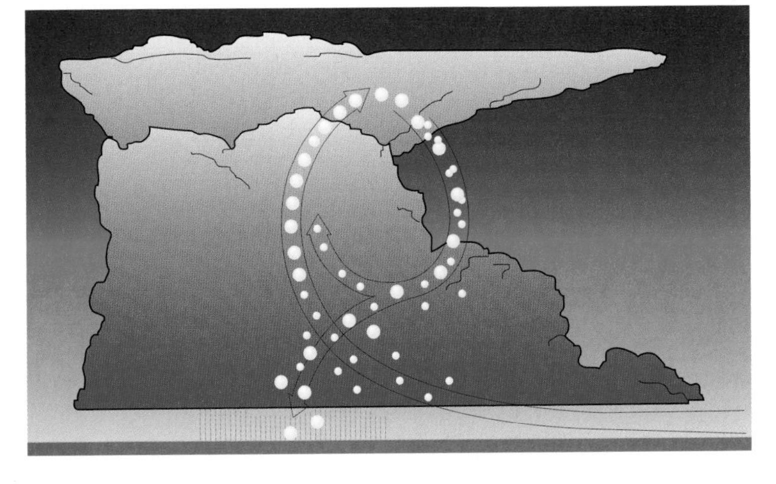

Figure 9-9. One way hail can grow large is by being recirculated through a storm.

Upward currents enhance the growth rate of precipitation.

Snow forms through the process of deposition, rather than condensation. It differs from ice pellets and hail in that it does not start as liquid water and then freeze. Snow grains are the solid equivalent of drizzle. They are very small, white, opaque particles of ice different from ice pellets in that they are flatter and they neither shatter nor bounce when they strike the ground. If the temperature of the air remains below freezing, the snow falls to the ground as snow; otherwise, it melts and turns to rain.

The presence of wet snow indicates the temperature is above freezing at your flight altitude.

LATENT HEAT OF WATER

The difference between the **latent heat** of water vapor and liquid water results in significant temperature differences which dramatically influence weather. It takes 540 calories (2,260 joules) of heat to vaporize one gram of water. This is referred to as the latent heat of evaporation. That is why on hot days it is cooler near lakes and rivers, and why you feel cold when getting out of the shower. In more familiar terms, about 5/8 kilowatt hour (kwh) is needed to vaporize one liter (1 kg) of water. Conversely, when one liter of water condenses, it gives back 5/8 kwh of heat. This latent heat of condensation is the reason such violent energy is released when thousands of tons of moisture condense into a thunderstorm cloud.

The latent heat of water also changes between freezing and liquid states. However, the heat exchange between melting and freezing is small, only 80 calories/gram. While this is enough to keep a soft drink cold, it has relatively little effect on weather.

STABILITY

Stability is the atmosphere's resistance to vertical motion. The stability of a parcel of air determines whether it rises or sinks in relation to the air around it. Stable air resists vertical movement, while unstable air has a tendency to rise. The combined effects of temperature and moisture determine the stability of the air and, to a large extent, the type of weather produced. The greatest instability occurs when the air is both warm and moist. Tropical weather, with its almost daily thunderstorm activity, is a perfect example of weather that occurs in unstable air. Air that is both cool and dry resists vertical movement and is very stable. A good example of this can be found in the polar regions during winter.

Air that is lifted expands due to lower atmospheric pressure. Lifting can occur orographically, as when air is forced up a mountain slope, or it can be caused by frontal activity. Lifting is also caused by convective currents that are generated by uneven heating of the earth's surface. As the air expands, it cools through a process known as adiabatic cooling. Conversely, air compresses and its temperature increases as it sinks. [Figure 9-10]

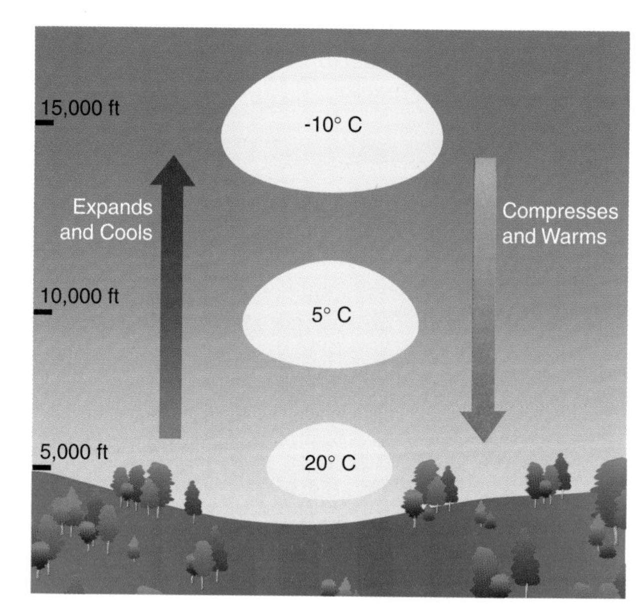

Figure 9-10. When air, or any gas, expands, the temperature decreases because of decreased molecular density.

The **dry adiabatic lapse rate (DALR)** is 3°C (5.4°F) per 1,000 feet that a parcel of unsaturated air is lifted. To find out whether a parcel of air is unstable, you must determine whether it will be warmer than the surrounding air after it is lifted. Because of the latent heat contained in water vapor, stability is strongly related to the moisture of the lifted air.

When unsaturated air is forced to ascend a mountain slope, it cools at the rate of approximately 3°C per 1,000 feet.

When condensation occurs in a parcel of rising air, adiabatic cooling is partially offset by warming due to the release of latent heat. Keep in mind that latent heat never completely offsets adiabatic cooling. A saturated parcel continues to cool as it rises, but at a slower rate than if it were dry. The rate of cooling of a rising, saturated parcel is called the moist, or **saturated adiabatic lapse rate (SALR)**. Although DALR is a constant 3°C per 1,000 feet, SALR is variable. It is about the same as DALR at extremely cold temperatures (–40°F), but is only a third of the DALR value at very hot temperatures (100°F). This is because saturated air holds much more water vapor at high temperatures, so there is much more latent heat to release when condensation occurs. [Figure 9-11]

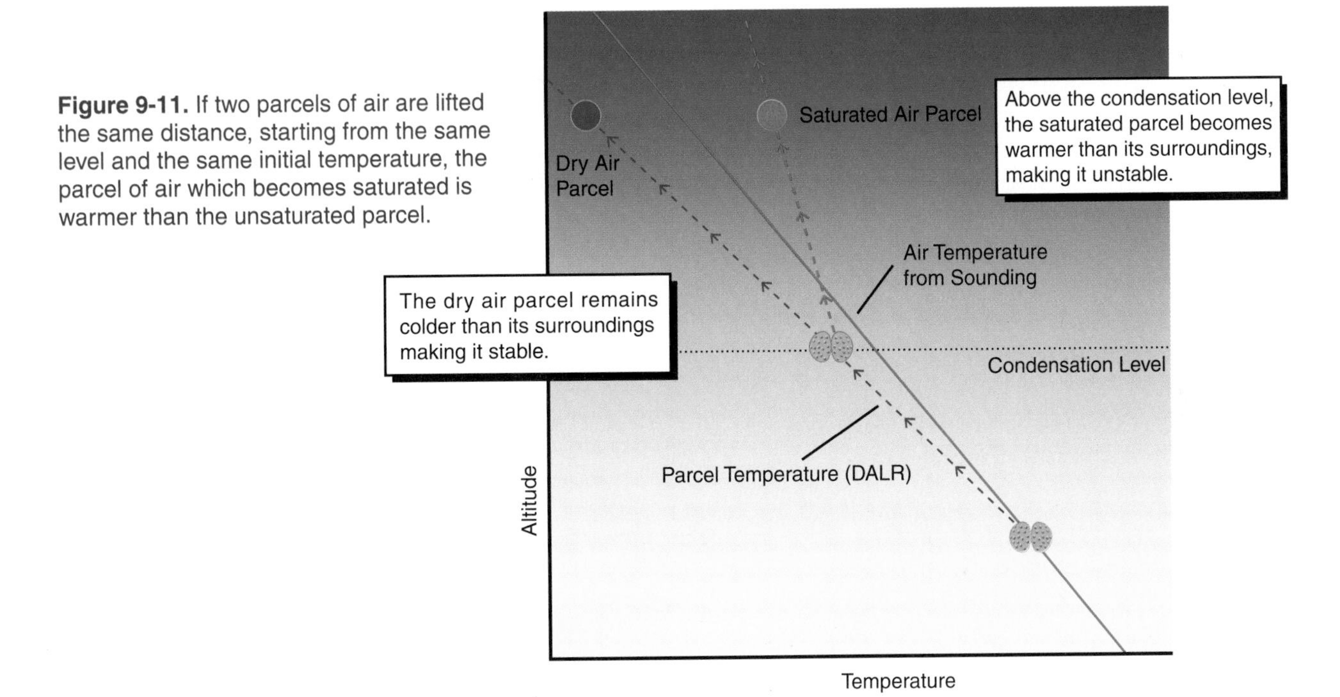

Figure 9-11. If two parcels of air are lifted the same distance, starting from the same level and the same initial temperature, the parcel of air which becomes saturated is warmer than the unsaturated parcel.

Air is unstable when its adiabatic lapse rate is less than the ambient air lapse rate. The standard temperature at sea level is 15°C and decreases at an average rate of 2°C (3.5°F) per 1,000 feet, but this can vary. When the lapse rate causes the ambient, or surrounding, air to be colder than a lifted parcel of air, lifted air tends to rise and be unstable. When there is an inversion in the lapse rate, causing the air above to be warm, it discourages a parcel of air from rising, contributing to stability.

Since the standard temperature at sea level is 15°C, and it decreases at an average rate of 2°C per 1,000 feet, the standard temperature at 10,000 feet is –5°C (15°C – 20°C = –5°C).

Measuring the Lapse Rate

Radiosondes are instruments, often encased in Styrofoam, that measure pressure, temperature and humidity when launched into the upper atmosphere on a weather balloon. Wind speed and direction are measured by monitoring the balloon's progress from ground level to altitudes that often exceed 20 miles. The observed data is transmitted to ground equipment that processes the data into weather information. Less than half of the radiosondes launched by the National Weather Service are recovered and reused.

Courtesy of Vaisala Corporation

Courtesy of Vaisala Corporation

The ambient lapse rate allows you to determine atmospheric stability.

Inversions usually are confined to fairly shallow layers and may occur near the surface or at higher altitudes. They act as a lid for weather and pollutants. When the humidity is high, visibility often is restricted by fog, haze, smoke, and low clouds. Temperature inversions normally occur in stable air with little or no wind and turbulence. One of the most familiar types of ground- or surface-based inversions forms from radiation cooling just above the ground on clear, cool nights. A frontal inversion occurs when cool air is forced under warm air, or when warm air spreads over cold air.

A common type of ground- or surface-based temperature inversion is that which is produced by ground radiation on clear, cool nights with calm or light winds. An inversion normally forms only in stable air. When humidity is high, you can expect poor visibility due to fog, haze, or low clouds.

The **condensation level** is the level at which the temperature and dewpoint converge, and a cloud forms in rising air. Below the condensation level, the rising air parcel cools at the DALR; above that level, it cools at the SALR. With dewpoint decreasing at 1°F per 1,000 feet, and a DALR of 5.4°F per 1,000 feet, the temperature and dewpoint converge at about 4.4°F (2.5°C) per 1,000 feet.

To estimate cloud bases, divide the surface temperature/dewpoint spread by the rate that the temperature approaches the dewpoint. For example, if the surface temperature is 25°C and the surface dewpoint is 0°C, you would divide the 25°C spread by

2.5°C to get the approximate height of the cloud base in thousands of feet. In this case, 10,000 feet AGL.

To estimate the bases of cumulus clouds, in thousands of feet, divide the temperature/dewpoint spread at the surface by 2.5°C (4.4°F). If using the quick estimate method, divide the temperature/dewpoint spread by 4°F (2.2°C).

When air is forced to ascend, the cloud structure, whether stratiform or cumuliform, is dependent upon the stability of the air being lifted.

If you know the stability of an airmass, you can predict its characteristics. Stable air is associated with stratus clouds, poorer visibility and lack of turbulence. Unstable air supports cumulus clouds, good visibility outside the clouds, and generally more extreme weather such as icing, heavy rain, hail, and turbulence. [Figure 9-12]

	Stable Air	Unstable Air
Clouds	Wide areas of layered or stratiform clouds or fog; gray at low altitude, thin white at high altitude	Cumuliform with extensive vertical development; bright white to black;
Precipitation	Small droplets in fog and low-level clouds; large droplets in thick stratified clouds; widespread and lengthy periods of rain or snow	Large drops in heavy rain showers; showers usually brief; hail possible
Visibility	Restricted for long periods	Poor in showers or thundershowers, good otherwise
Turbulence	Usually light or nonexistent	Moderate to severe
Icing	Moderate in mid-altitudes; freezing rain, rime, or clear ice	Moderate to severe clear ice
Other	Frost, dew, temperature inversions	High or gusty surface winds, lightning, tornadoes

Figure 9-12. You can determine much about the clouds and weather if you know the stability of an airmass.

CLOUDS

As air cools to its saturation point, condensation and sublimation change invisible water vapor to a visible state. Most commonly, this visible moisture takes the form of clouds or fog. Clouds are composed of very small droplets of water or, if the temperature is low enough, ice crystals. Condensation and sublimation are facilitated by **condensation nuclei**, which can be dust, salt from evaporating sea spray, or products of combustion.

As shown in figure 9-12, when moist, stable air is forced upwards through convection, frontal activity, or by orographic lifting, the result is stratiform clouds, with continuous precipitation and little or no turbulence. Stability contributes to smoke, dust, and haze concentrated at the lower levels with resulting poor visibility. Lifting of unstable, moist air results in cumuliform type clouds (those with extensive vertical development), good visibility outside the cloud, showery precipitation, turbulence, and possible clear icing in clouds.

The four families of clouds are high, middle, low, and those with extensive vertical development.

TYPES OF CLOUDS

Clouds are divided into four basic groups, or families, depending upon their characteristics and the altitudes where they occur. These are low, middle, high, and clouds with vertical development.

The suffix nimbus, used in naming clouds, means a rain cloud.

Cloud names are based on the terms, cumulus (heap), stratus (layer), nimbus (rain), and cirrus (ringlet). The prefixes alto and cirro denote cumulus and stratus clouds from the middle and high families, respectively. The prefix nimbo and the suffix nimbus denote clouds that produce rain.

Fair weather cumulus clouds indicate turbulence at and below the cloud level.

Cumulus clouds form when moist air is lifted and condenses. They usually have flat bottoms and dome-shaped tops. Widely spaced cumulus clouds that form in otherwise clear skies are called fair weather cumulus. These clouds generally form in unstable air, but are capped at the top by stable air. At and below the cloud level, you can expect turbulence, but little icing or precipitation. If you can fly above a fair weather cumulus cloud, you can expect smooth air at that altitude.

Stratus clouds are associated with stable air. They frequently produce low ceilings and visibilities, but usually have little turbulence. Icing conditions are possible if temperatures are at or near freezing. Stratus clouds may form when air is cooled from below, or when stable air is lifted up sloping terrain. They also may form along with fog when rain falls through cooler air and raises the humidity to the saturation point. Nimbostratus clouds are stratus clouds that produce rain. They can be several thousand feet thick and contain large quantities of moisture. If temperatures are near or below freezing, they may create heavy icing.

LOW CLOUDS

Low clouds extend from near the surface to about 6,500 feet AGL. Low clouds usually consist almost entirely of water but sometimes may contain supercooled water which can create an icing hazard for aircraft. Types of low clouds include stratus, stratocumulus, and nimbostratus. [Figure 9-13]

Fog is a low cloud which has its base within 50 feet of the ground. [Figure 9-14] If the fog is less than 20 feet deep, it is called ground fog. Having an instrument rating does

LOW CLOUDS

Stratus Clouds
Stratus clouds are layered clouds that form in stable air near the surface due to cooling from below. They have a gray, uniform appearance and generally cover a wide area.

Nimbostratus Clouds
Nimbostratus clouds can be several thousand feet thick and contain large quantities of moisture. These clouds vary from gray to black, depending on thickness and moisture content.

Stratocumulus Clouds
Stratocumulus clouds are white, puffy clouds that form as stable air is lifted. They often form as a stratus layer breaks up or as cumulus clouds spread out.

Figure 9-13. Low clouds are found at altitudes extending from the surface to about 6,500 feet AGL.

not eliminate fog as a flight hazard. Aside from its ability to reduce visibility even below safe IFR minimums, fog can contain icing hazards. Fog is covered in detail in the next section.

Figure 9-14. Fog is a cloud that forms next to the ground.

MIDDLE CLOUDS

Middle clouds have bases that range from about 6,500 to 20,000 feet AGL. They are composed of water, ice crystals, or supercooled water, and may contain moderate turbulence and potentially severe icing. Altostratus and altocumulus are classified as middle clouds. [Figure 9-15]

MIDDLE CLOUDS

Figure 9-15. Middle clouds are found at altitudes extending from 6,500 feet to 20,000 feet AGL.

Altostratus Clouds
Altostratus clouds are flat, dense clouds that cover a wide area. They are a uniform gray or gray-white in color. Although they produce minimal turbulence, they can contain moderate icing.

Altocumulus Clouds
Altocumulus clouds are gray or white, patchy clouds of uniform appearance that often form when altostratus clouds start to break up. They may produce light turbulence and icing.

HIGH CLOUDS

High clouds have bases beginning above 20,000 feet AGL. They are generally white to light gray in color and form in stable air. They are composed mainly of ice crystals and seldom pose a serious turbulence or icing hazard. The three basic types of high clouds are called cirrus, cirrostratus, and cirrocumulus. [Figure 9-16]

Cirrus clouds are thin, wispy clouds composed mostly of ice crystals that usually form above 30,000 feet. White or light gray in color, they often exist in patches or narrow bands

HIGH CLOUDS

Figure 9-16. High clouds are found at altitudes above 20,000 feet AGL.

Cirrus Clouds
Cirrus clouds are composed mostly of ice crystals that usually form above 30,000 feet.

Cirrostratus Clouds
Cirrostratus clouds also are thin, white clouds that often form in long bands or sheets against a deep blue background. Although they may be several thousands of feet thick, moisture content is low and they pose no icing hazard.

Cirrocumulus Clouds
Cirrocumulus clouds are white patchy clouds that look like cotton. They form as a result of shallow convective currents at high altitude and may produce light turbulence.

A high cloud is composed mostly of ice crystals.

across the sky. Since they are sometimes blown from the tops of thunderstorms they can be an advance warning of approaching bad weather.

CLOUDS WITH VERTICAL DEVELOPMENT

When lifting and instability are present, cumulus clouds may build vertically into towering cumulus or cumulonimbus clouds. The bases are typically at 1,000 to 10,000 feet MSL, and their tops sometimes exceed 60,000 feet MSL. [Figure 9-17]

Towering cumulus clouds indicate convective turbulence.

Towering cumulus clouds indicate a fairly deep layer of unstable air. They contain moderate to heavy convective turbulence with icing and often develop into thunderstorms.

Cumulonimbus clouds, or thunderstorms, are large, vertically developed clouds that form in moist, unstable air. They are gray-white to black in color and contain large amounts of moisture, turbulence, icing, and lightning. You can think of them as severe versions of towering cumulus. Thunderstorms are discussed in more detail in the next section.

CLOUDS WITH VERTICAL DEVELOPMENT

Figure 9-17. Clouds with vertical development, or cumuliform clouds, indicate instability.

Cumulus Clouds
These puffy white clouds usually have flat bottoms and dome-shaped tops.

Towering Cumulus
Towering cumulus clouds are similar to cumulus clouds, except they have more vertical development.

Cumulonimbus Clouds
Cumulonimbus clouds, or thunderstorms, are large, vertically developed clouds ranging from gray-white to black in color. They contain large amounts of moisture, turbulence, icing, and lightning.

AIRMASS

An airmass is a body of air that covers an extensive area and has fairly uniform properties of temperature and moisture.

An airmass is a large body of air with fairly uniform temperature and moisture content. It usually forms where air remains stationary or nearly stationary for at least several days. During this time, the airmass takes on the temperature and moisture properties of the underlying surface. The area where an airmass acquires the properties of temperature and moisture that determine its stability is called its source region. [Figure 9-18]

As an airmass moves over a warmer surface, its lower layers are heated, and vertical movement of the air develops. Depending on temperature and moisture levels, this can result in extreme instability, characterized by cumuliform clouds, turbulence, and good visibility outside the clouds. When an airmass flows over a cooler surface, its lower lay-

Safe From Lightning?

Some bolts of lightning are powerful enough to light a small city. With this much electricity hitting one spot, it is easy to see why lightning is so dangerous.

One common myth is that the rubber tires of a car act as an insulator to lightning. Although rubber is considered an electrical insulator, it is ineffective against a 300,000 volt per foot electrical charge that can jump up to two miles through air. In fact, when lightning strikes a car, it usually destroys the tires as it punches through them on the way to the ground.

Since it is nearly impossible to insulate against an electrical charge as powerful as lightning, the key to effective lightning protection is to conduct the electricity along a path in which it can do no harm. An enclosed metal vehicle can provide effective protection, because the vehicle frame conducts the electricity around, rather than letting it go through, the occupants. Motorcycles and convertibles provide no such protection.

Metal aircraft also protect their occupants from lightning. As you will learn in the next section, the primary aviation hazards from thunderstorms are not from lightning, but from icing and turbulence.

ers are cooled and vertical movement is inhibited. As a result, the stability of the air is increased. If the air is cooled to its dewpoint, low clouds or fog may form. This cooling from below creates a temperature inversion and may result in low ceilings and restricted visibility for long periods of time.

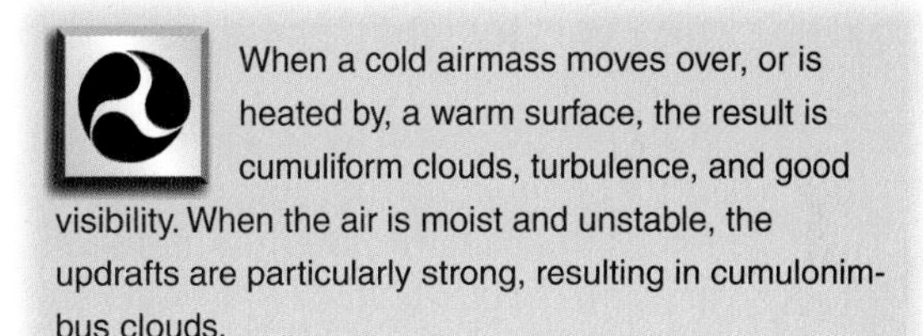

When a cold airmass moves over, or is heated by, a warm surface, the result is cumuliform clouds, turbulence, and good visibility. When the air is moist and unstable, the updrafts are particularly strong, resulting in cumulonimbus clouds.

Cooling from below increases the stability of an airmass and warming from below decreases it.

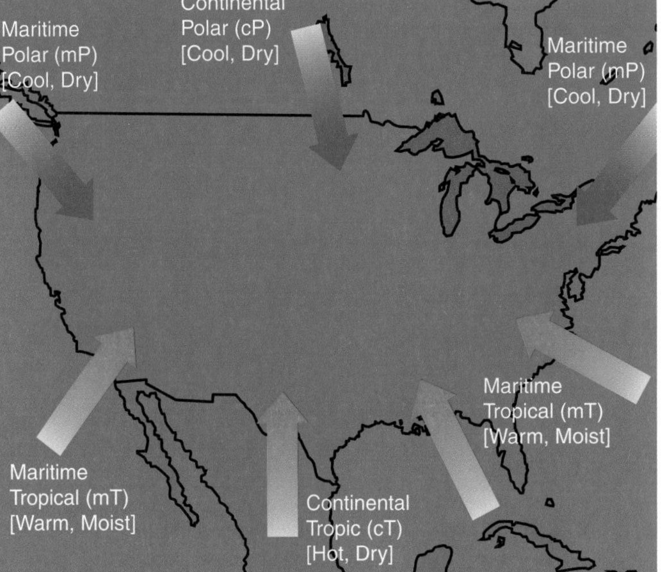

Figure 9-18. Airmass source regions surround North America. As airmasses move out of those regions, they often converge to form the continent's major weather systems.

FRONTS

When an airmass moves out of its source region, it comes in contact with other airmasses that have different moisture and temperature characteristics. The boundary between airmasses is called a front and often contains hazardous weather.

When you cross a front, you move from one airmass into another with different properties. The changes between the two may be very abrupt, indicating a narrow frontal zone. On the other hand, changes may occur gradually, indicating a wide and, perhaps, diffused frontal zone. These changes can give you important cues to the location and intensity of the front.

A change in the wind is always associated with the passage of a frontal system.

A change in the temperature is one of the easiest ways to recognize the passage of a front. At the surface, the temperature change usually is very noticeable and may be quite abrupt in a fast-moving front. With a slow-moving front, it is less pronounced. When you are flying through a front, you can observe the temperature change on the outside air temperature gauge. However, the change may be less abrupt at middle and high altitudes than it is at the surface.

The most reliable indications that you are crossing a front are a change in wind direction and, less frequently, wind speed. Although the exact new direction of the wind is difficult to predict, the wind always shifts to the right in the northern hemisphere as the front passes.

As a front approaches, atmospheric pressure usually decreases, with the area of lowest pressure lying directly over the front. Pressure changes on the warm side of the front generally occur more slowly than on the cold side. The important thing to remember is that you should promptly update your altimeter setting after crossing a front.

The type and intensity of frontal weather depend on several factors, including the availability of moisture, the stability of the air being lifted, and the speed of the frontal movement. Other factors include the slope of the front and the moisture and temperature variations between the two fronts. Although some frontal weather can be severe and hazardous, other fronts produce relatively calm weather.

CAN ARTHRITIS SUFFERERS PREDICT THE WEATHER?

Many of those with arthritis are convinced that their pain and stiffness increase with deteriorating weather. In a 1960 study, Dr. Joseph Hollander at the University of Pennsylvania found that 11 of his 12 subjects reported worsened symptoms 73 percent of the time they were simultaneously exposed to high humidity and falling barometric pressure. More recent studies have showed no such correlation.

According to the Arthritis Society, living in a cold, damp climate may make you feel your arthritis more than living in a hot, dry one. They go on to state that a rise in humidity and a fall in barometric pressure may also make the joints feel worse temporarily, but not everyone with arthritis can predict weather change.

It seems there is no consensus on this issue. Many researchers feel arthritis victims notice their pain more when the weather is bad because people feel worse anyway when it is dreary outside. On the other hand, if the joints ache, yet the weather stays nice, a person may feel better and forget that their joints predicted a bad day.

COLD FRONTS

A **cold front** separates an advancing mass of cold, dense, and stable air from an area of warm, lighter, and unstable air. Because of its greater density, the cold air moves along

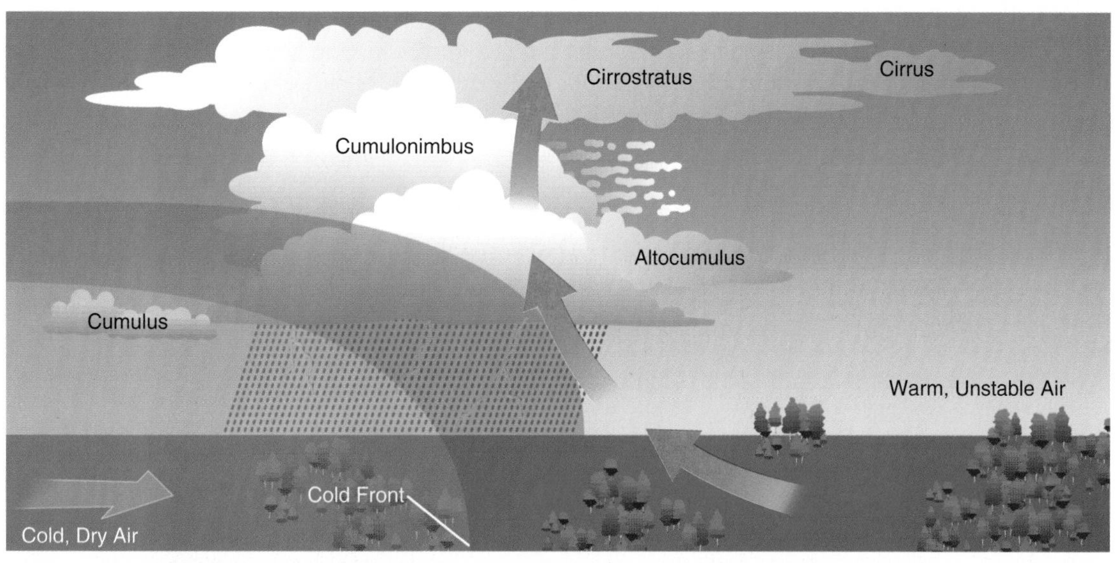

TYPICAL COLD FRONT WEATHER	Prior to Passage	During Passage	After Passage
Clouds	• Cirriform • Towering cumulus and/or cumulonimbus	• Towering cumulus and/or cumulonimbus	• Cumulus
Precipitation	• Showers	• Heavy showers • Possible hail, lightning, and thunder	• Slowly decreasing showers
Visibility	• Fair in haze	• Poor	• Good
Wind	• SSW	• Variable and gusty	• WNW
Temperature	• Warm	• Suddenly cooler	• Continued cooler
Dewpoint	• High	• Rapidly dropping	• Continued drop
Pressure	• Falling	• Bottoms out, then rises rapidly	• Rising

Figure 9-19. Cumuliform clouds and showers are common in the vicinity of cold fronts.

the surface and forces the less dense, warm air upward. In the northern hemisphere, cold fronts are usually oriented in a northeast to southwest line and may be several hundred miles long. Movement is usually in an easterly direction. A depiction of the typical cold front and a summary of its associated weather is shown in figure 9-19.

FAST-MOVING COLD FRONTS

Fast-moving cold fronts are pushed along by intense high pressure systems located well behind the front. Friction slows the movement at the surface, causing the front's leading edge to bulge out and steepen its slope. Because of the steep slope and wide differences in moisture and temperature between the two airmasses, fast-moving cold fronts are particularly hazardous.

Fast-moving cold fronts rapidly force warmer air to rise, which can cause widespread vertical cloud development along a narrow frontal zone. If sufficient moisture is present, an area of severe weather forms well ahead of the front, and usually clears quickly as the front passes. You often notice reduced cloud cover, improved visibility, lower temperatures, and gusty surface winds following the passage of a fast-moving cold front.

SLOW-MOVING COLD FRONTS

The leading edge of a slow-moving cold front is much shallower than that of a fast-moving front. This produces clouds extending far behind the surface front. A slow-moving cold front meeting stable air usually causes a broad area of stratus clouds to form behind the front. A slow-moving cold front meeting unstable air usually causes large numbers of vertical clouds to form at and just behind the front, creating hazards from icing and turbulence. Fair weather cumulus clouds can extend well behind the surface front.

WARM FRONTS

Warm fronts occur when warm air moves over the top of cooler air at the surface. They usually move at much slower speeds than cold fronts. The slope of a warm front is very gradual, and the warm air may extend over the cool air for several hundred miles ahead of the front. A depiction of the typical warm front and a summary of its associated weather is shown in figure 9-20.

The stability and moisture content of the air in a warm front determines what type of clouds will form. If the air is warm, moist, and stable, stratus clouds and steady precipitation can develop. If the air is warm, moist, and unstable, cumulus clouds and showery precipitation can develop.

Steady precipitation, in contrast to showers, preceding a front is an indication of stratiform clouds with little or no turbulence.

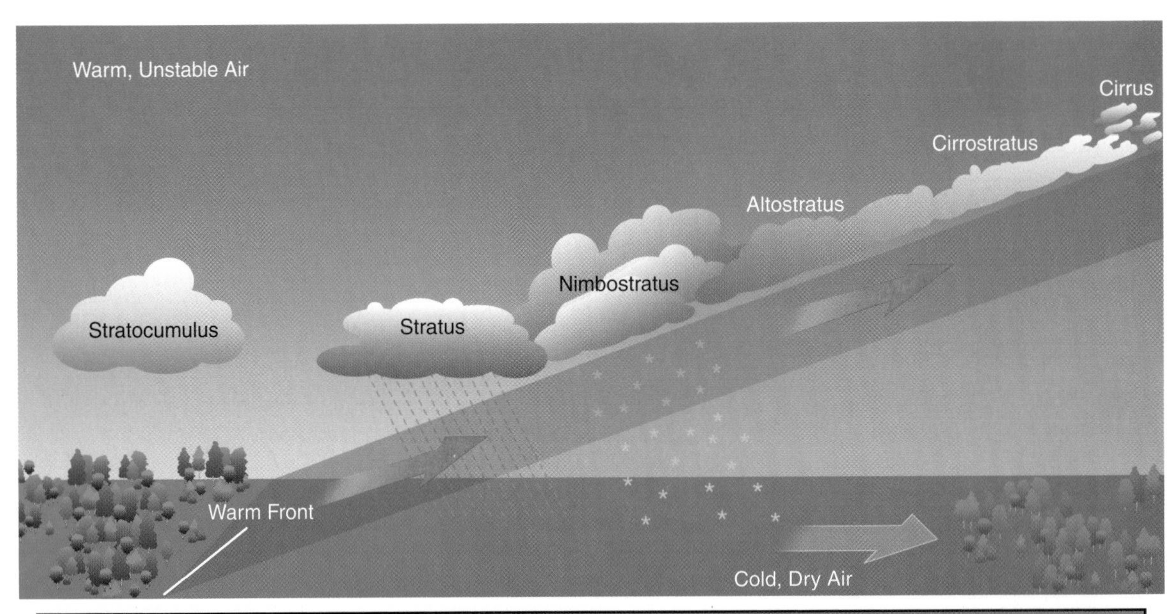

TYPICAL WARM FRONT WEATHER

	Prior to Passage	During Passage	After Passage
Clouds	• Cirriform • Stratiform • Fog • Possible cumulonimbus in the summer	• Stratiform	• Stratocumulus • Possible cumulonimbus in the summer
Precipitation	• Light-to-moderate rain, drizzle, sleet, or snow	• Drizzle, if any	• Rain or showers, if any
Visibility	• Poor	• Poor, but improving	• Fair in haze
Wind	• SSE	• Variable	• SSW
Temperature	• Cold to cool	• Rising steadily	• Warming, then steady
Dewpoint	• Rising steadily	• Steady	• Rising, then steady
Pressure	• Falling	• Becoming steady	• Slight rise, then falling

Figure 9-20. Although stratus clouds usually extend out ahead of a slow-moving warm front, cumulus clouds sometimes develop along and ahead of the surface front if the air is unstable.

STATIONARY FRONTS

When the opposing forces of two airmasses are balanced, the front that separates them may remain stationary and influence local flying conditions for several days. The weather in a stationary front usually is a mixture of that found in both warm and cold fronts.

OCCLUDED FRONTS

In a cold front occlusion, the air ahead of the warm front is warmer than the air behind the overtaking cold front.

A frontal occlusion occurs when a fast-moving cold front catches up to a slow-moving warm front. The difference in temperature within each frontal system strongly influences which type of front and weather are created. A cold front occlusion develops when the fast-moving cold front is colder than the air ahead of the slow-moving warm front. In this case, the cold air replaces the cool air at the surface and forces the warm front aloft. A warm front occlusion occurs when the fast-moving cold front is warmer than the air ahead of the slow-moving warm front. In this case, the cold front rides up over the warm front, forcing the cold front aloft. A depiction of the typical cold and warm front occlusions and a summary of their associated weather is shown in figure 9-21.

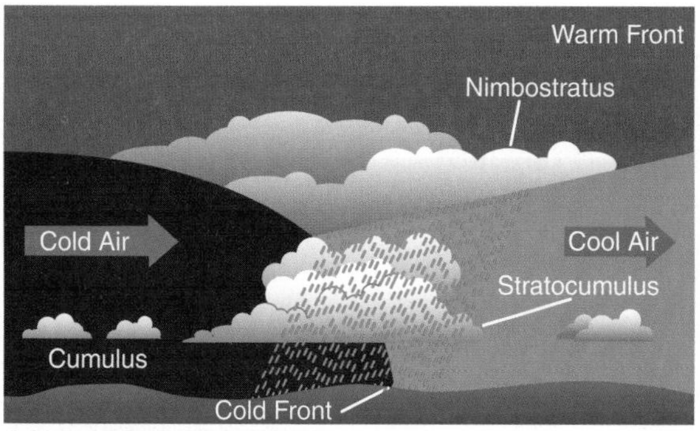

COLD FRONT OCCLUSION

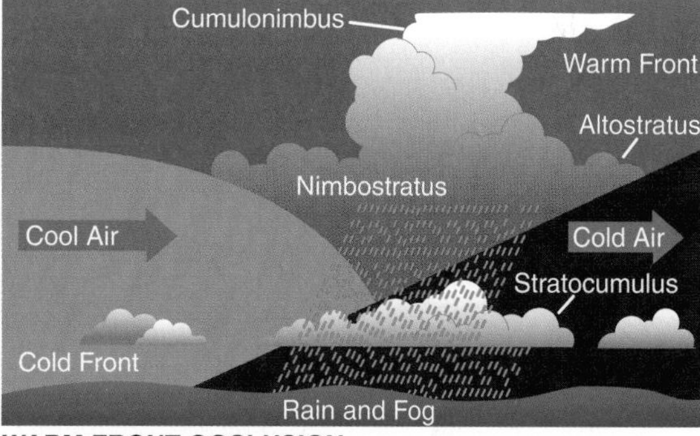

WARM FRONT OCCLUSION

TYPICAL OCCLUDED FRONT WEATHER	Prior to Passage	During Passage	After Passage
Clouds	• Cirriform • Stratiform	• Nimbostratus • Possible towering cumulus and/or cumulonimbus	• Nimbostratus • Altostratus • Possible cumulus
Precipitation	• Light-to-heavy precipitation	• Light-to-heavy precipitation	• Light-to-moderate precipitation, then clearing
Visibility	• Poor	• Poor	• Improving
Wind	• SE to S	• Variable	• W to NW
Temperature	• Cold Occlusion: Cold to Cool • Warm Occlusion: Cold	• Cold Occlusion: Falling • Warm Occlusion: Rising	• Cold Occlusion: Colder • Warm Occlusion: Milder
Dewpoint	• Steady	• Slight drop	• Rising, then steady
Pressure	• Falling	• Becoming steady	• Slight drop; however, may rise after passage of warm occlusion

Figure 9-21. When the air being lifted by a cold front occlusion is moist and stable, the weather will be a mixture of that found in both a warm and a cold front. When the air being lifted by a warm front occlusion is moist and unstable, the weather will be more severe than that found in a cold front occlusion.

THE FRONTAL CYCLONE

One important process by which fronts are set in motion is the frontal cyclone, sometimes referred to as an extratropical cyclone or a frontal low. As you may recall, cyclonic circulation is counterclockwise in the northern hemisphere. Because there is an excess of solar energy received at the equator and a deficit at the poles, a temperature gradient occurs and is concentrated in an area called the polar front. If that temperature gradient becomes excessive at some point along the polar front, a disturbance occurs to equalize the pressure. The circulation patterns that result from this activity act to transport the warm air toward the pole and cold air toward the equator and reduce the temperature gradient.

In addition to the polar fronts, other areas are conducive to the formation of frontal cyclones. For example, in winter, locally strong temperature gradients are found along some coastlines where cold continents are next to very warm oceans. This is the case for the U.S. just off the Gulf of Mexico and along the East Coast. When fronts move into these areas, the development of a low pressure area around which these fronts can circulate is common. [Figure 9-22]

Figure 9-22. Frontal cyclones develop in areas with strong temperature gradients. Regions with the highest frequency of cyclone development are shown in red.

Orographic lifting from large mountain chains and latent heat from condensation of moist air also can enhance cyclone development (cyclogenesis). These two processes frequently work together to produce lows on the east slopes of the Rocky Mountains.

STRUCTURE AND DEVELOPMENT

The surface development of a frontal cyclone in the northern hemisphere follows a distinctive life cycle. Before the **frontal wave** development begins, a stationary or slow-moving cold front is present in the area. The frontal zone is characterized by a change in wind speed and/or direction (wind shear) from the warm side to the cold side. As the cyclone development begins, pressure falls at some point along the original front, and counterclockwise circulation is generated. At this point, the cyclone is in the incipient, or wave cyclone, stage because the original front

Frontal waves normally form on slow-moving cold fronts or stationary fronts.

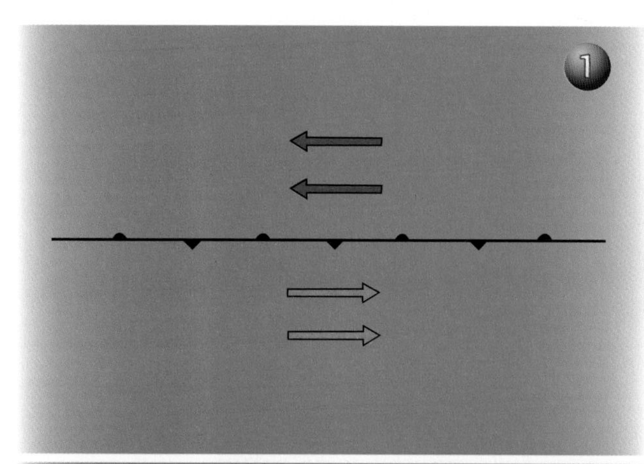

Pre-Development Stage
This stage is characterized by a stationary or slow-moving cold front with winds blowing in opposite directions on each side of the front.

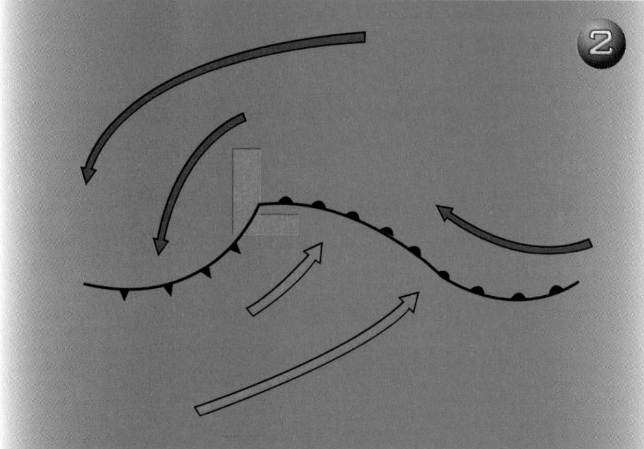

Incipient (Wave Cyclone) Stage
The counterclockwise circulation around the low generates a frontal wave. By definition, the cold air moving toward the warm air is called a cold front, and the warm air moving toward the cold air is a warm front.

Figure 9-23. The development of a frontal low is shown in chronological order. The incipient stage of frontal cyclone development typically takes 12 hours.

has been distorted into a wave shape in response to the developing circulation. [Figure 9-23]

Frontal cyclones do not necessarily develop beyond the incipient stage. These stable waves can simply move rapidly along the polar front, and finally dissipate. However, a cyclone that continues to develop moves northeastward at 15 to 25 knots, pushing warm air northward ahead of it, and bringing cold air to the south into the wake of the cyclone. As it progresses eastward, the central pressure continues to fall; the cyclone deepens, and the winds around it increase in response to the greater pressure gradient. About 12 hours after the initial appearance of the frontal low, the cold airmass trailing the cyclone

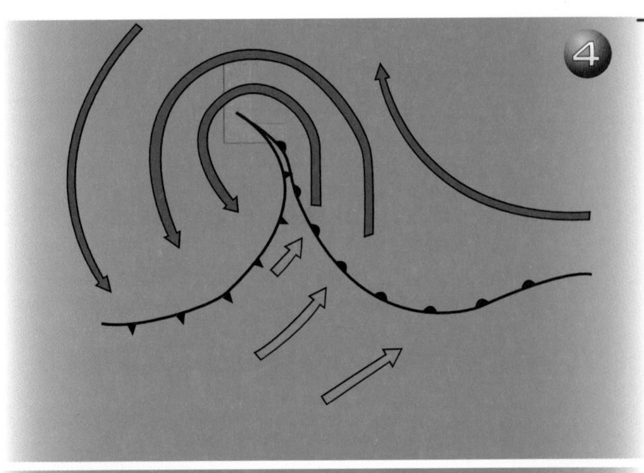

Occluded Stage
The winds around the cyclone increase in response to the greater pressure gradient. The cold airmass overtakes the retreating cold air ahead of the cyclone, resulting in an occlusion.

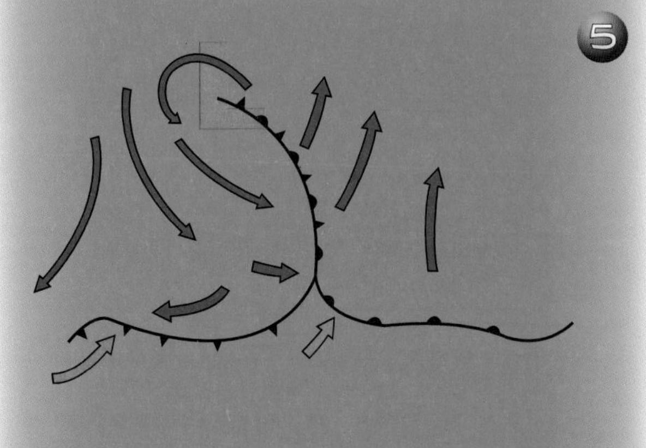

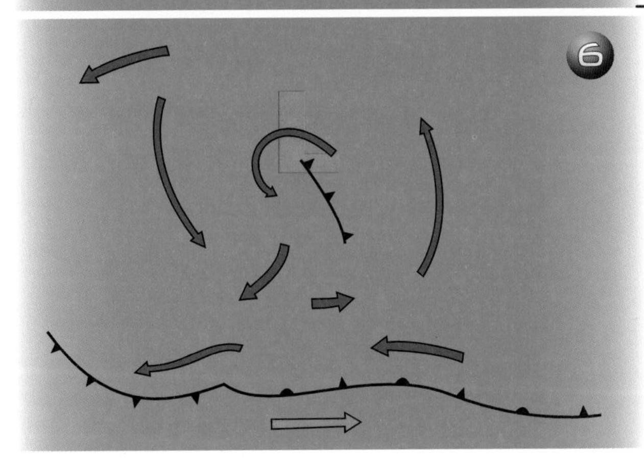

Dissipating Stage
In the occlusion process, the temperature gradient is destroyed, so the cyclone, now located entirely in the cold air, dies from the lack of an energy source.

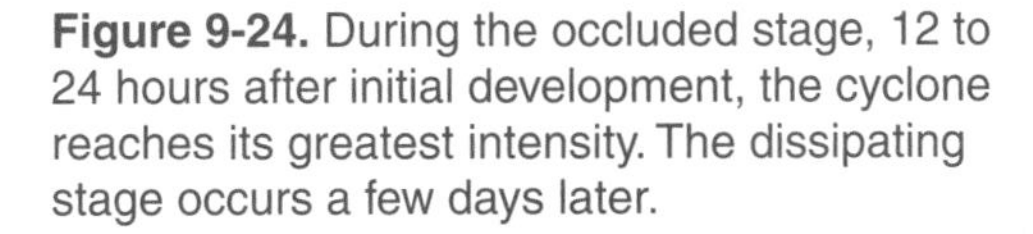
Figure 9-24. During the occluded stage, 12 to 24 hours after initial development, the cyclone reaches its greatest intensity. The dissipating stage occurs a few days later.

is swept around the low and overtakes the retreating cold air ahead of the cyclone. This process pushes the warm sector air aloft and the cyclone enters the occluded stage. [Figure 9-24]

As the cyclone enters the dissipating stage of its life cycle, the central pressure begins to rise. The weakening of the cyclone begins 24 to 36 hours after the initial formation of the disturbance and lasts for another few days. The weakening occurs because the temperature gradient, from which the cyclone draws its energy, has been diminished by the mixing of warm and cold air.

HIGH ALTITUDE WEATHER

If you transition into jet or turboprop aircraft, it will become important to understand weather patterns at and above the tropopause. The height of the tropopause varies

between 24,000 feet MSL near the poles and 50,000 feet MSL near the equator. The tropopause, which is the boundary between the troposphere and the stratosphere, is characterized by an abrupt change in the temperature lapse rate. In International Standard Atmospheric (ISA) conditions, the height of the tropopause is approximately 36,000 feet. From this altitude up to 66,000 feet in the standard atmosphere, the temperature remains constant at –57°C. The tropopause acts like a lid, since it resists the exchange of air between the troposphere and the stratosphere above. However, in the northern hemisphere, there are generally two breaks in the tropopause. One is between the polar and subtropical airmass and the other is between the subtropical and tropical airmass.

Jet streams often are embedded in the zone of strong westerlies at these breaks in the tropopause. [Figure 9-25] A **jet stream** is a narrow band of high speed winds that reaches its greatest speed near the tropopause. Typical jet stream speeds range between 60 knots and about 240 knots. Jet streams normally are several thousand miles long, several hundred miles wide, and a few miles thick.

A jetstream is defined as wind of 50 knots or greater.

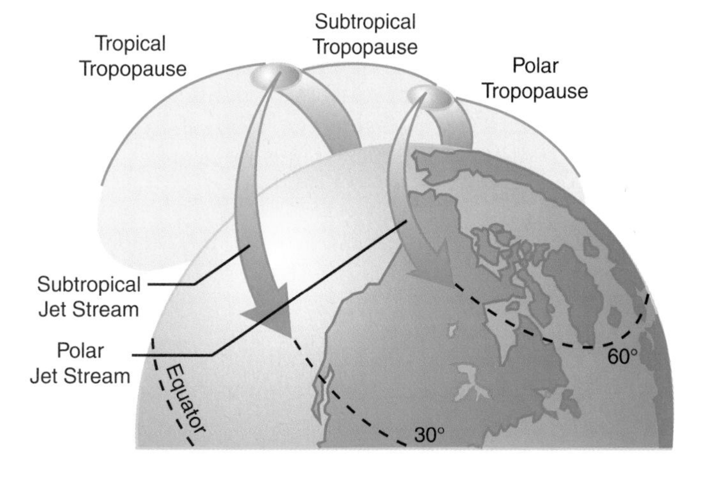

Figure 9-25. The jet stream occurs at breaks in the tropopause, which vary with the seasonal migration of airmass boundaries.

The strength and location of the jetstream is normally weaker and farther north in the summer. During the winter months in the middle latitudes, the jet stream shifts toward the south and speed increases.

The polar front jet stream occurs at about 30°N to 60°N latitude. Although it exists year round, the polar front tends to be higher, weaker, and farther north in the summer. The subtropical jet stream is found near 25°N latitude. It reaches its greatest strength in the wintertime and is nonexistent in the summer.

The tropopause slopes upward from polar to tropical regions. If you could stand at the breaks in the tropopause with the wind at your back, a distinctly higher tropopause would occur on the right side of each jet stream and a separate, lower tropopause on the left. This structure is reversed in the Southern Hemisphere.

SUMMARY CHECKLIST

✓ The atmosphere is commonly divided into a number of layers according to its thermal characteristics. The lowest layer, the troposphere, is where most weather occurs.

- ✓ Uneven heating of the earth's surface is the driving force behind all weather. The special characteristics of water also affect the release of heat into the atmosphere, and dramatically affect the weather.
- ✓ Atmospheric circulation patterns are caused by differences in pressure. Above the friction layer, Coriolis force diverts wind to the right as it flows out of a high pressure area in the northern hemisphere. Near the surface, wind flows more directly from a high to low pressure area, or across the isobars.
- ✓ The amount of water vapor that air can hold decreases with temperature. At the dewpoint, the air is saturated.
- ✓ Precipitation occurs when water vapor condenses out of the air and becomes heavy enough to fall to earth. The type of precipitation is influenced by the temperature and other conditions under which condensation occurs.
- ✓ Stability is the atmosphere's resistance to vertical motion. Air is stable when a lifted parcel of air is cooler than the ambient air. Dry air tends to cool more when lifted and tends to be more stable. Ambient air with a low or inverted lapse rate also contributes to stability.
- ✓ Clouds occur when water vapor condenses. They are divided into four basic families: low, middle, high, and clouds with vertical development.
- ✓ Cumulus clouds are formed when unstable air is lifted. Stratiform clouds are formed when stable air is lifted. The lifting of moist, unstable air results in good visibility outside the cloud, showery precipitation, and turbulence. However, the lifting of moist, stable air results in continuous precipitation, little or no turbulence, and poor visibility.
- ✓ An airmass is a large body of air with fairly uniform temperature and moisture content. A front is a discontinuity between two airmasses. A cold front occurs when cold air displaces warmer air. A warm front occurs when warm air overruns colder air.
- ✓ A frontal cyclone starts as a slow-moving cold front or stationary front and can end as a cold front occlusion with potentially severe weather.
- ✓ Jet streams are bands of strong westerly winds that occur at breaks in the tropopause in the northern hemisphere. While they can provide beneficial winds when flying west to east, they also can be associated with strong turbulence.

KEY TERMS

Troposphere

Tropopause

Stratosphere

Isobars

Pressure Gradient

High

Low

Ridge

Trough

Col

Coriolis Force

Cyclonic

Evaporation

Sublimation

Condensation

Deposition

Dewpoint

Supercooled Water

Latent Heat

Stability

Dry Adiabatic Lapse Rate (DALR)

Saturated Adiabatic Lapse Rate (SALR)

Condensation Level

Condensation Nuclei

Low Clouds

Middle Clouds

High Clouds

Towering Cumulus

Cumulonimbus Clouds

Cold Front

Warm Front

Stationary Front

Frontal Occlusion

Frontal Cyclone

Frontal Wave

Jet Stream

QUESTIONS

1. In which level of the atmosphere does most of the earth's weather occur? Why?

2. What is the major driving force behind the weather?
   A. Variations in moisture content
   B. Uneven heating of the earth's surface
   C. Rotation of the earth and its effect on the movement of high and low pressure areas

3. What is indicated by close spacing of isobars on a weather map?
   A. Weak pressure gradient and weak winds
   B. Weak pressure gradient and strong winds
   C. Strong pressure gradient and strong winds

4. Select the true statement regarding Coriolis force.
   A. Coriolis force is strongest within 2,000 feet of the surface.
   B. Coriolis force causes cyclonic circulation around high pressure areas.
   C. In the northern hemisphere, Coriolis force deflects wind to the right as it flows out of a high pressure area.

Match the following items with the associated weather characteristics.

5. Instability
6. Stability
7. Stratus
8. SALR
9. Supercooled water
10. Nimbus
11. Towering cumulus
12. DALR

A. Resistance to vertical motion
B. Layered clouds with little turbulence
C. Clouds with vertical development
D. Rain clouds
E. The result of a high ambient lapse rate combined with a low adiabatic lapse rate
F. 2°C per 1,000 feet
G. Liquid water that is colder than 0°C
H. 3°C per 1,000 feet
I. As low as 1°C per 1,000 feet

13. Select the characteristic(s) associated with the cloud shown in the accompanying photo. More than one characteristic may apply.

    A. Hail
    B. Drizzle
    C. Lightning
    D. Stable air
    E. Turbulence
    F. Restricted visibility for long periods

14. True/False. When an airmass is warmed from below, it becomes more stable.

15. True/False. Passage of a fast-moving cold front creates a narrow frontal zone with less severe weather than the passage of a slow-moving cold front.

16. Steady precipitation with little turbulence precedes what type of front?

    A. Cold front
    B. Warm front
    C. Occluded front

17. What is the most reliable indication that you have flown through a front?

    A. Change in pressure
    B. Change in temperature
    C. Change in wind direction

18. What conditions favor the formation of a frontal wave?

    A. A fast-moving warm front overtaking a cold front
    B. A deep low pressure area located northeast of a ridge
    C. A stationary front or slow moving cold front with a strong temperature gradient

19. Jet stream winds occur at which location?

    A. South of highs
    B. Parallel to troughs
    C. Breaks in the tropopause

SECTION B
WEATHER HAZARDS

This section covers various weather hazards that you need to be aware of to safely fly under both VFR and IFR. Instrument operations, in particular, require a sound knowledge of hazardous weather phenomena. When flying in instrument meteorological conditions, you are often unable to observe weather hazards directly. Knowledge of the conditions that produce hazardous weather increases your ability to recognize and avoid dangerous conditions.

THUNDERSTORMS

Thunderstorms produce some of the most dangerous weather elements in aviation and should be avoided. Remember, there are three conditions necessary to create a thunderstorm — air that has a tendency toward instability, some type of lifting action, and relatively high moisture content. [Figure 9-26]

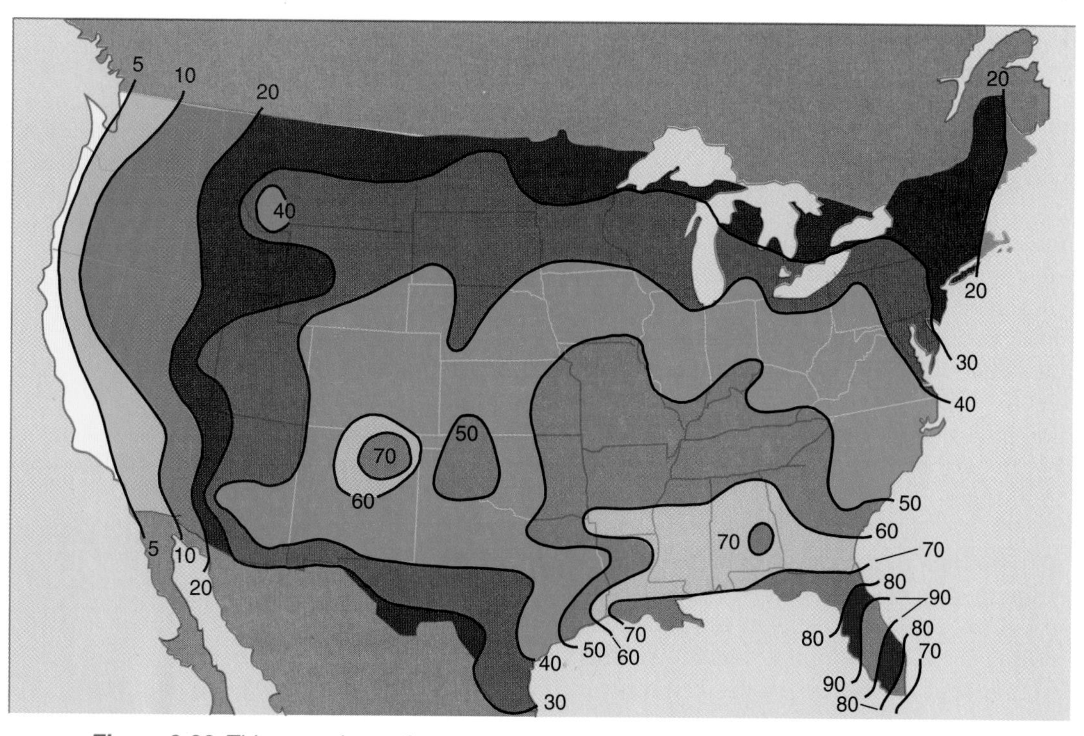

Figure 9-26. This map shows the average number of thunderstorms that occur each year in different parts of the United States.

Thunderstorm formation requires an unstable lapse rate, a lifting force, and a relatively high moisture level.

The lifting action may be provided by several factors, such as rising terrain (orographic lifting), fronts, or the heating of the earth's surface (convection). Thunderstorms progress through three definite stages — cumulus, mature, and dissipating. [Figure 9-27]. You can anticipate the development of thunderstorms and the associated hazards by becoming familiar with the characteristics of each stage. Be aware that other weather phenomena may prevent you from seeing their characteristic shapes. For example, a cumulonimbus cloud may be embedded, or contained within, other cloud layers making it impossible to see. Night operations also make it more difficult to see hazardous cloud formations.

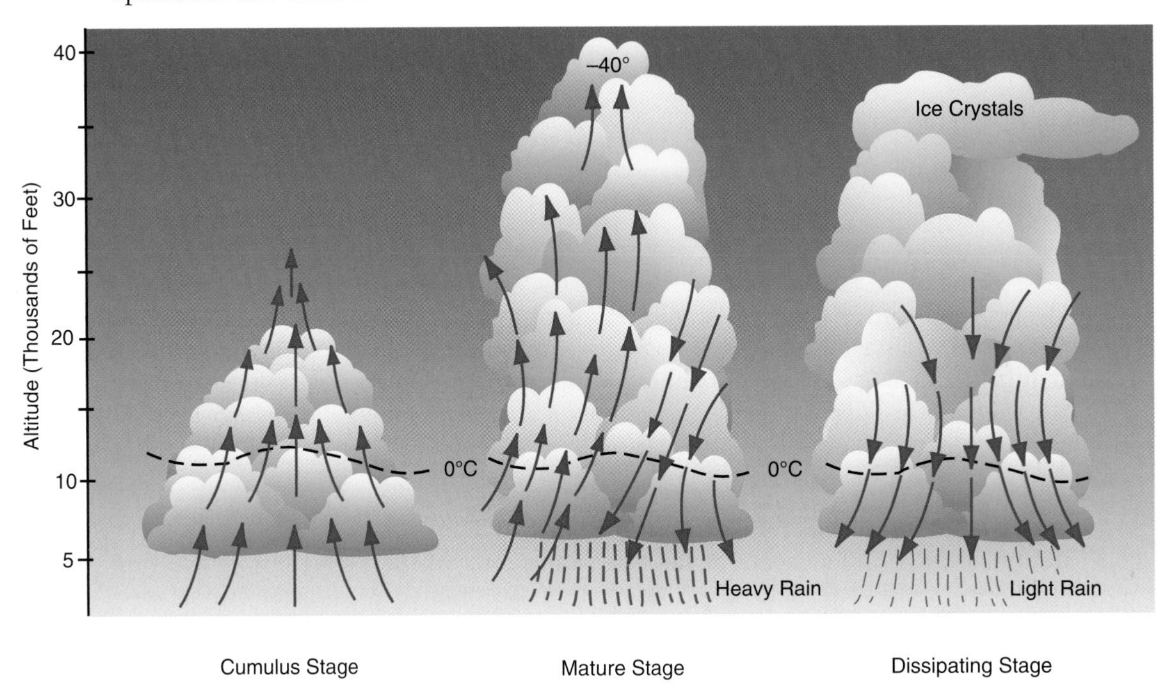

Figure 9-27. A typical airmass thunderstorm consists of three states — cumulus, mature, and dissipating.

An embedded thunderstorm is one which is obscured by massive cloud layers and cannot be seen.

In the **cumulus stage**, a lifting action initiates the vertical movement of air. As the air rises and cools to its dewpoint, water vapor condenses into small water droplets or ice crystals. If sufficient moisture is present, heat released by the condensing vapor provides energy for the continued vertical growth of the cloud. Because of strong updrafts, precipitation usually does not fall. Instead, the water drops or ice crystals rise and fall within the cloud, growing larger with each cycle. Updrafts as great as 3,000 f.p.m. may begin near the surface and extend well above the cloud top. During the cumulus stage, the convective circulation grows rapidly into a towering cumulus (TCU) cloud that typically grows to 20,000 feet in height and 3 to 5 miles in diameter. The cloud reaches the mature stage in about 15 minutes.

The cumulus stage is characterized by continuous updrafts.

As the drops in the cloud grow too large to be supported by the updrafts, precipitation begins to fall to the surface. This creates a downward motion in the surrounding air and signals the beginning of the **mature stage**. The resulting downdraft may reach a velocity of 2,500 f.p.m. The down-rushing air spreads outward at the surface, producing a sharp drop in temperature, a rise in pressure, strong gusty surface winds, and turbulent conditions.

Turbulence develops when air currents change direction or velocity rapidly over a short distance. The magnitude of the turbulence depends on the differences between the two air currents. Within the thunderstorm cloud, the strongest turbulence occurs in the shear between the updrafts and downdrafts. Near the surface, there is an area of low-level turbulence which develops as the downdrafts spread out at the surface. A **shear zone** is created between the surrounding air and the cooler air of the downdraft. This shear zone with its gusty winds and turbulence is not necessarily confined to the thunderstorm itself, but can extend outward for many miles from the center of the storm. The leading edge of the downdraft is referred to as a **gust front**. As the thunderstorm advances, a rolling, turbulent, circular-shaped cloud may form at the lower leading edge of the cloud. This is called the **roll cloud**. [Figure 9-28]

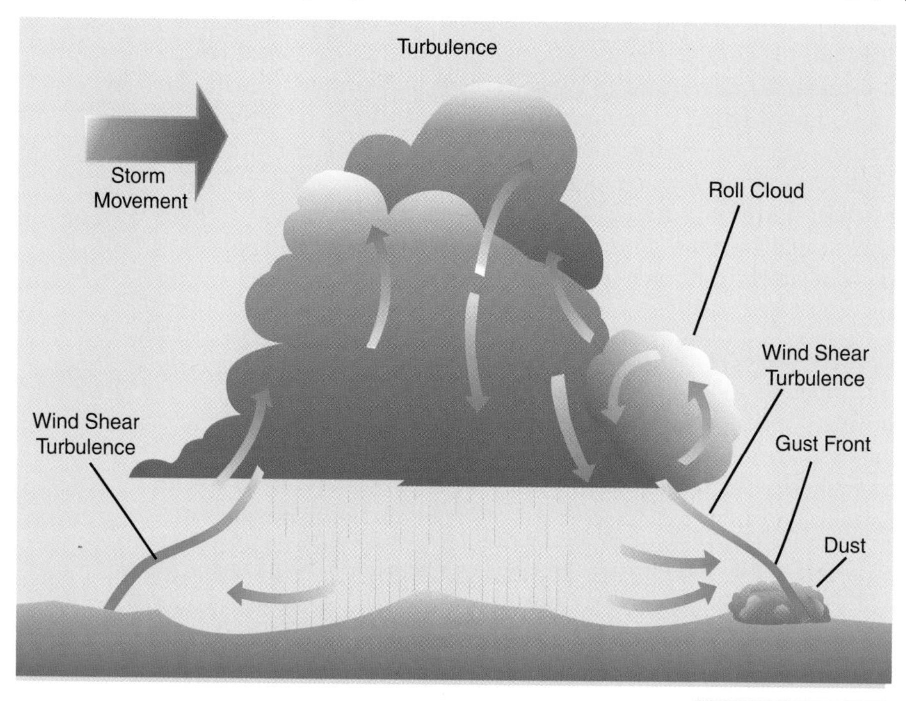

Figure 9-28. Early in the mature stage, the updrafts continue to increase up to speeds of 6,000 f.p.m. The adjacent updrafts and downdrafts cause severe turbulence. The most violent weather occurs during this phase of the life cycle.

Wind shear areas can be found on all sides of a thunderstorm, as well as directly under it.

Thunderstorms reach the greatest intensity during the mature stage, which is signaled by the beginning of precipitation at the surface.

As the mature stage progresses, more and more air aloft is disturbed by the falling drops. Eventually, the downdrafts begin to spread out within the cell, taking the place of the weakening updrafts. Because upward movement is necessary for condensation and the release of the latent energy, the entire thunderstorm begins to weaken. When the cell becomes an area of predominant downdrafts, it is considered to be in the **dissipating stage**. During this stage, the upper level winds often blow the top of the cloud downwind, creating the familiar anvil shape. [Figure 9-29]

Figure 9-29. An anvil shape usually forms at the top of a cumulonimbus cloud during the dissipating stage of a thunderstorm. However, severe weather can still occur well after its appearance.

A dissipating thunderstorm is characterized by predominant downdrafts.

Thunderstorms usually have similar physical features, but their intensity, degree of development, and associated weather do differ. Thunderstorms are generally classified as airmass or severe storms. **Airmass thunderstorms** generally form in a warm, moist airmass and are isolated or scattered over a large area. They are usually caused by solar heating of the land, which results in convection currents that lift unstable air. These thunderstorms are most common during hot summer afternoons when winds are light. They are also common along coastal areas at night. Airmass storms can also be caused by orographic lifting. Although they are usually scattered along individual mountain peaks, they may cover large areas. They also may be embedded in other clouds, making them difficult to identify when approached from the windward side of a mountain. Nocturnal thunderstorms can occur in late spring and summer during the late night or early morning hours when relatively moist air exists aloft. Usually found from the Mississippi Valley westward, nocturnal storms cover many square miles, and their effects may continue for hours at a given location. **Severe thunderstorms** are usually associated with weather patterns, such as fronts, converging winds, and troughs aloft and are more intense than an airmass thunderstorm with wind gusts of 50 knots or more, hail 3/4 of an inch or more in diameter, and/or strong tornadoes.

Convective currents are most active on warm summer afternoons when the winds are light.

Thunderstorms may exist as a **single cell** airmass storm lasting less than one hour, or as a **supercell** severe thunderstorm lasting approximately two hours. A **multicell** storm is usually a cluster of airmass thunderstorms in various stages of development. Because of the interaction of the various stages, the duration of a multicell storm is much longer than a single cell storm. [Figure 9-30]

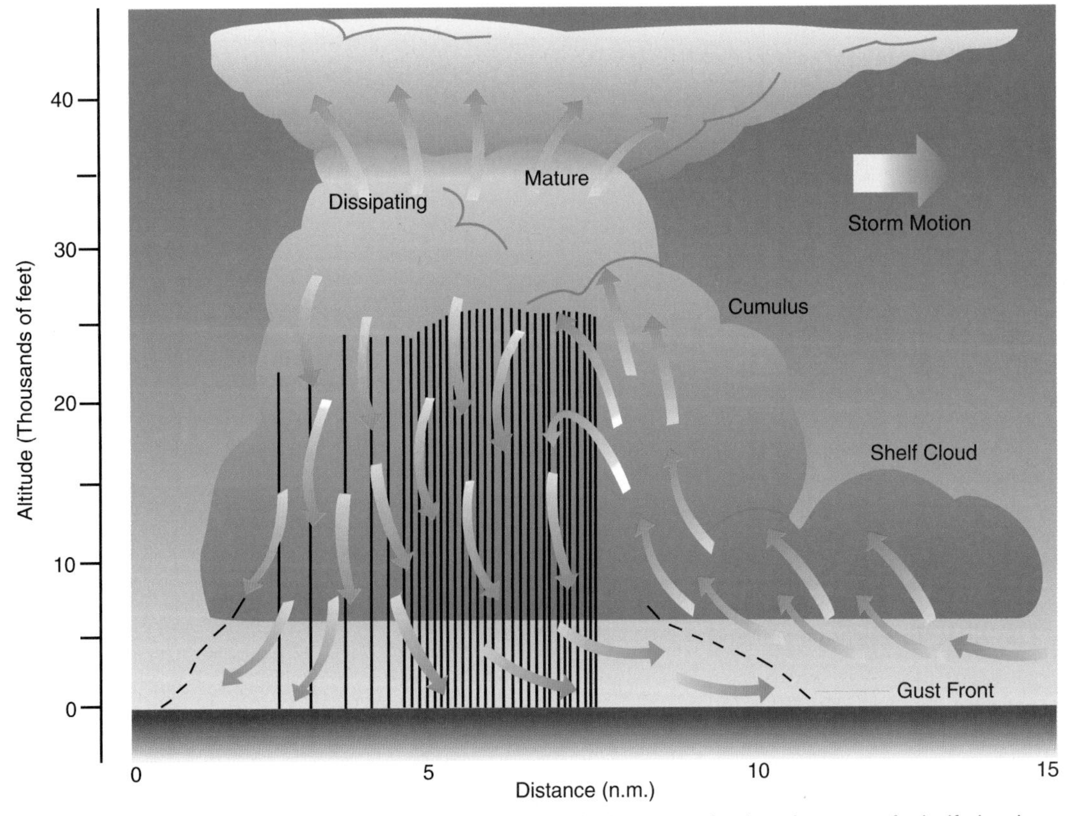

Figure 9-30. A multicell thunderstorm may consist of all stages of a thunderstorm. A shelf cloud often indicates the rising air over the gust front. The vertical lines represent precipitation intensity.

The term frontal thunderstorm is sometimes used to refer to storms which are associated with frontal activity. Those storms that occur with a warm front are often obscured by stratiform clouds. You should expect thunderstorms when there is showery precipitation near a warm front. In a cold front, the cumulonimbus clouds are often visible in a continuous line parallel to the frontal surface. Occlusions can also spawn storms. A squall line is a narrow band of active thunderstorms which normally contains very severe weather. While it often forms 50 to 200 miles ahead of a fast-moving cold front, the existence of a front is not necessary for a squall line to form.

A squall line is a band of thunderstorms that often forms several miles in front of a fast-moving cold front and contains some of the most severe types of weather-related hazards.

Thunderstorms typically contain many severe weather hazards, such as lightning, hail, turbulence, gusty surface winds, and even tornadoes. These hazards are not confined to the cloud itself. For example, you can encounter turbulence in VFR conditions as far as 20 miles from the storm. It may help to think of a cumulonimbus cloud as the visible part of a widespread system of turbulence and other weather hazards. In fact, the cumulonimbus cloud is the most turbulent of all clouds. Indications of severe turbulence within the storm system include the cumulonimbus cloud itself, very frequent lightning, and roll clouds.

Cumulonimbus clouds by themselves indicate severe turbulence. Other indications of turbulence are very frequent lightning and roll clouds. Turbulence can be encountered as far as 20 miles from the cumulonimbus cloud.

Lightning is one of the hazards which is always associated with thunderstorms and is found throughout the cloud. While it rarely causes personal injury or substantial damage to the aircraft structure in flight, it can cause temporary loss of vision, puncture the aircraft skin, or damage electronic navigation and communications equipment. Your aircraft can also be struck by lightning when you are clear of the thunderstorm, but still in the vicinity.

Lightning is always associated with thunderstorms.

Hail is another thunderstorm hazard. You can encounter it in flight, even when no hail is reaching the surface. In addition, large hailstones have been encountered in clear air several miles from a thunderstorm. Hail can cause extensive damage to your aircraft in a very short period of time.

Funnel clouds are violent, spinning columns of air which descend from the base of a cloud. Wind speeds within them may exceed 200 knots. If a funnel cloud reaches the earth's surface, it is referred to as a tornado. If it touches down over water, it is called a waterspout. [Figure 9-31]

Hail is most likely associated with a cumulonimbus cloud but it can be encountered several miles from the cloud.

Figure 9-31. The diameter of most tornadoes is between 300 and 2,000 feet, although there have been cases of tornadoes reaching a mile in diameter. Wind speeds within a tornado can exceed 250 knots.

How Lightning Forms

As a towering cumulus develops, a large electrical charge separation builds up within the cloud. Lightning results when this electrical charge becomes strong enough to jump from the cloud to the ground, to another cloud, or to an opposite electrical charge within the same cloud.

Although the process that creates lightning is not fully understood, it is enhanced substantially when the cloud grows above the freezing level. As the outer boundaries of water droplets start to freeze, positive ions, or particles with a positive electrical charge, flow to the area of ice formation. This creates an outer shell which is positively charged and a center which is negatively charged.

When the interior freezes and expands, it shatters the outer shell. The droplets can also be broken up by colliding with other particles. In either case, the lighter pieces are then carried in the updrafts to the top of the cloud. The heavier, negatively charged particles fall to the bottom of the cloud. This makes the top of the cloud have a net positive charge while the lower part has a net negative charge.

As the negative charge builds at the bottom of the cloud, it repels the negative charge on the earth's surface. This leaves the area below the cloud with a positive charge. This positive charged area acts like a shadow that follows the cloud as it moves.

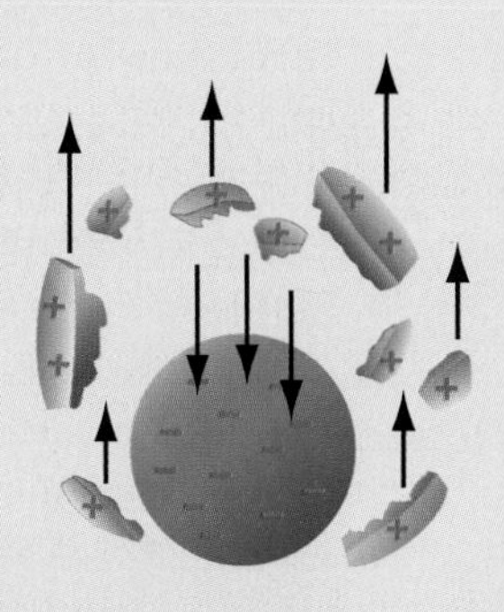

When the cloud has a very intense negative charge at its base, it seeks to neutralize itself by discharging to a positive area. This discharge is what we see as lightning. When it discharges it goes to the most accessible opposite charge. This is usually within the cloud itself, or between clouds. At times, however, it goes from the cloud to the ground.

Lightning begins when a negatively charged pilot leader descends from the cloud. This leader forms a conductive path approximately eight inches in diameter and from thirty-five to one hundred fifty feet long. At this point, the electrons in the cloud begin to descend down the path. This recharges the path and causes additional leaders to extend earthward. These leaders are called stepped leaders because they seek the most conductive path to the ground and may try several branches before the best is located. As the path is extended the electrons from the cloud extend further downward.

The final stepped leader takes place a few feet above the earth where it is met by a rising positive flow from the surface. With the path completed, the positive ions on the earth can now flow to the cloud and neutralize the lower portion of the cloud . This upward flow is referred to as the return stroke and because it energizes the air molecules and illuminates the path, you see it as lightning.

Before the path dissipates, the first stroke is followed by additional strokes which further neutralize the negative charges in the cloud. What appears to be a single lightning flash can actually be three or four strokes. The actual stroke of lightning takes less than a half second to occur and that includes the leaders and three or four return strokes.

The peak current in the channel can reach ten thousand amps and the air in the channel can be heated to a temperature hotter than the surface of the sun. This causes the air in the path to expand violently, producing the sound waves we hear as thunder.

THUNDERSTORM AVOIDANCE

When conditions permit, you should circumnavigate thunderstorms. Remember that hail and severe turbulence may exist well outside the storm cloud. During night operations, lightning may provide a clue as to a storm's location; however, lightning may not yet have developed in younger storms and may have ceased in older ones. Even without visible lightning, storms may still contain destructive turbulence or hail. The best approach during night operations or during flight under instrument conditions is to avoid areas where thunderstorms exist or are likely to develop.

If the aircraft you are flying is equipped with a weather avoidance system, such as weather radar, you can use it to avoid thunderstorms. With radar, you should avoid intense thunderstorm echoes by at least 20 miles. [Figure 9-32]

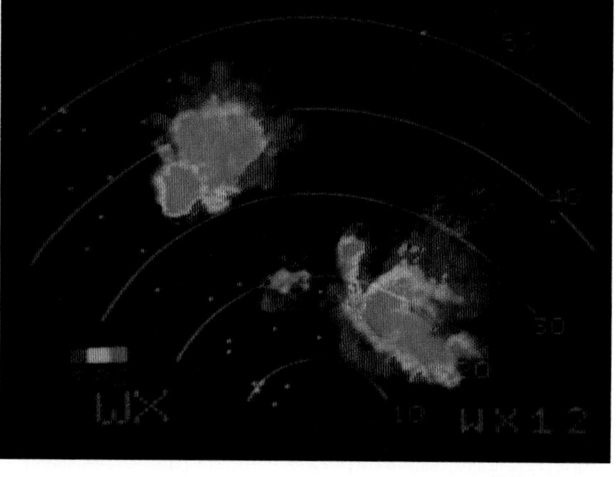

Figure 9-32. A general recommendation is that you should not fly between intense radar echoes unless they are at least 40 miles apart.

Avoid intense radar echoes by at least 20 miles and do not fly between them if they are less than 40 miles apart.

Airborne weather radar is designed for avoiding severe weather, not for penetrating it. Weather radar detects drops of precipitation; it does not detect minute cloud droplets. Therefore, it should not be relied on to avoid instrument weather associated with clouds and fog. Also, be sure you are familiar with the operation of the systems and the manufacturer's recommendations appropriate to your system.

Airborne weather radar provides no assurance of avoiding IFR weather conditions.

TURBULENCE

In addition to turbulence in and near thunderstorms, three other categories of turbulence affect aviation operations: low-level turbulence, clear air turbulence, and mountain wave turbulence. The effects of turbulence can vary from occasional light bumps to severe jolts which can cause personal injury to occupants and/or structural damage to the airplane. If you enter turbulence unexpectedly, unintentionally enter a thunderstorm, or expect that you may encounter it during flight, slow the airplane to maneuvering speed (V_A) or less, or the recommended rough air penetration speed. Then, attempt to maintain a level flight attitude and accept variations in airspeed and altitude. This helps avoid the high structural loads that may be imposed on the aircraft if you try to maintain a precise airspeed and altitude. If you encounter turbulent or gusty conditions during an approach to a landing, you should consider flying a power-on approach and landing at an airspeed slightly above the normal approach speed. This helps to stabilize the aircraft, which gives you more control.

If you encounter turbulence during flight, establish maneuvering or penetration speed, maintain a level flight attitude, and accept variations in airspeed and altitude.

When turbulence is encountered during the approach to a landing, it is recommended that you increase the airspeed slightly above normal approach speed to attain more positive control.

LOW-LEVEL TURBULENCE

While **low-level turbulence (LLT)** is often defined as turbulence below 15,000 feet MSL, most low-level turbulence originates due to surface heating or friction within a few thousand feet of the ground. LLT includes mechanical turbulence, convective turbulence, frontal turbulence, and wake turbulence.

MECHANICAL TURBULENCE

When obstacles such as buildings or rough terrain interfere with the normal wind flow, turbulence develops. This phenomenon, referred to as **mechanical turbulence**, is often experienced in the traffic pattern when the wind forms eddies as it blows around hangars, stands of trees, or other obstructions. As the winds grow stronger, mechanical turbulence extends to greater heights. For example, when surface winds are 50 knots or greater, significant turbulence due to surface effects can reach altitudes in excess of 3,000 feet AGL. [Figure 9-33]

Figure 9-33. Mechanical turbulence is produced downwind of obstructions such as a line of trees, buildings, and hills.

Mechanical turbulence also occurs when strong winds flow nearly perpendicular to steep hills or mountain ridges. In comparison with turbulence over flat ground, the relatively larger size of the hills produce greater turbulence. In addition, steep hillsides generally produce stronger turbulence because the sharp slope encourages the wind flow to separate from the surface. Steep slopes on either side of a valley can produce particularly dangerous turbulence for aircraft operations.

CONVECTIVE TURBULENCE

Convective turbulence, which is also referred to as thermal turbulence, is typically a daytime phenomena which occurs over land in fair weather. It is caused by currents, or thermals, which develop in air heated by contact with the warm surface below. This heating can occur when cold air is moved horizontally over a warmer surface or when the ground is heated by the sun. When the air is moist, the currents may be marked by build-ups of cumulus cloud formations. In some cases, you can find relief from this turbulence by climbing into the **capping stable layer** which begins at the top of the convective layer. This can sometimes be identified by a layer of cumulus clouds, haze, or dust. [Figure 9-34]

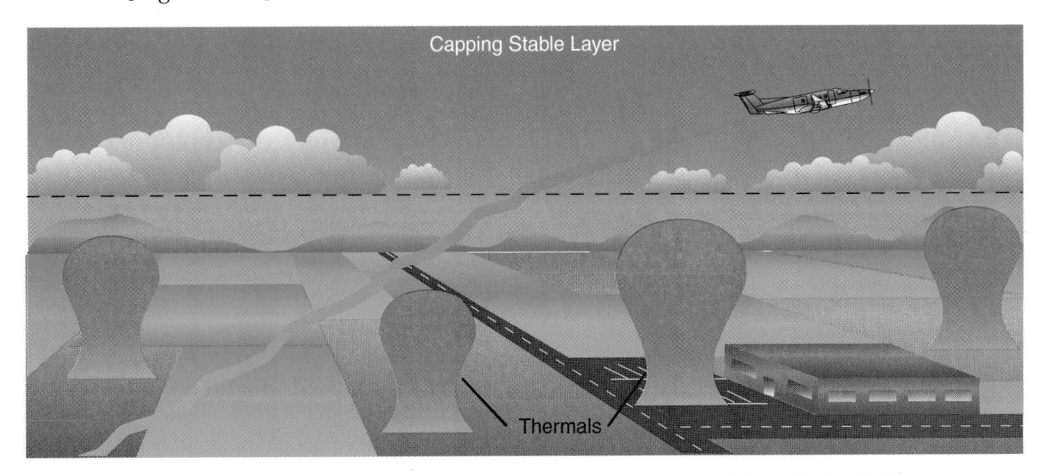

Figure 9-34. By climbing into the capping stable layer, you may be able to find relief from convective turbulence. The height of the capping layer is typically a few thousand feet above the ground, although it can exceed 10,000 feet AGL over the desert in the summer.

FRONTAL TURBULENCE

Frontal turbulence occurs in the narrow zone just ahead of a fast-moving cold front where updrafts can reach 1,000 f.p.m. When combined with convection and strong winds across the front, these updrafts can produce significant turbulence. Over flat ground, any front moving at a speed of 30 knots or more generates at least a moderate amount of turbulence. A front moving over rough terrain produces moderate or greater turbulence, regardless of its speed.

WAKE TURBULENCE

Whenever an airplane generates lift, air spills over the wingtips from the high pressure areas below the wings to the low pressure areas above them. This flow causes rapidly rotating whirlpools of air called **wingtip vortices**, or **wake turbulence**. The intensity of the turbulence depends on aircraft weight, speed, and configuration. [Figure 9-35]

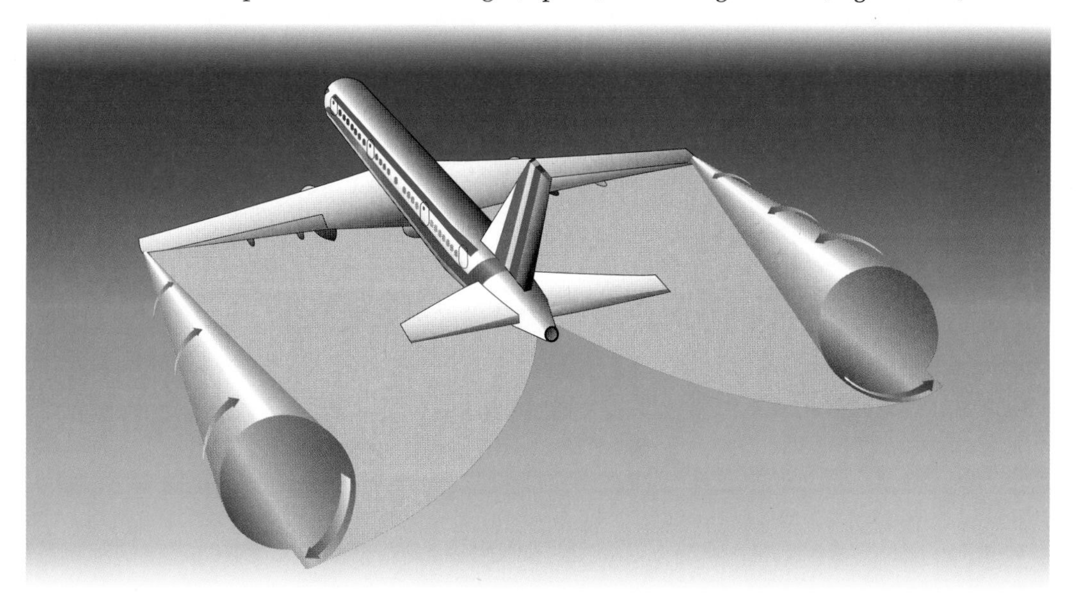

Figure 9-35. Wake vortices are created when lift is generated by the wing of an aircraft.

Wingtip vortices can exceed the roll rate of an aircraft, especially when flying in the same direction as the generating aircraft.

The greatest wake turbulence danger is produced by large, heavy aircraft operating at low speeds, high angles of attack, and in a clean configuration. Since these conditions are most closely duplicated on takeoff and landing, you should be alert for wake turbulence near airports used by large aircraft. In fact, wingtip vortices from large commercial jets can induce uncontrollable roll rates in smaller aircraft. Although wake turbulence settles, it persists in the air for several minutes, depending on wind conditions. In light winds of three to seven knots, the vortices may stay in the touchdown area, sink into your takeoff or landing path, or drift over a parallel runway. The most dangerous condition for landing is a light, quartering tailwind. It can move the upwind vortex of a landing aircraft over the runway and forward into the touchdown zone.

The greatest vortex strength occurs when the generating aircraft is heavy, slow, in a clean configuration, and operating at a high angle of attack.

If you are in a small aircraft approaching to land behind a large aircraft, controllers must ensure adequate separation. However, if you accept a clearance to follow an aircraft you have in sight, the responsibility for wake turbulence avoidance

Wingtip vortices tend to sink below the flight path of the aircraft which generated them. They are most hazardous during light, quartering tailwind conditions.

is transferred from the controller to you. On takeoff, controllers provide a two-minute interval behind departing heavy jets (three minutes for intersection takeoffs or takeoffs in the opposite direction on the same runway). You may waive these time intervals if you wish, but this is not a wise procedure. [Figure 9-36]

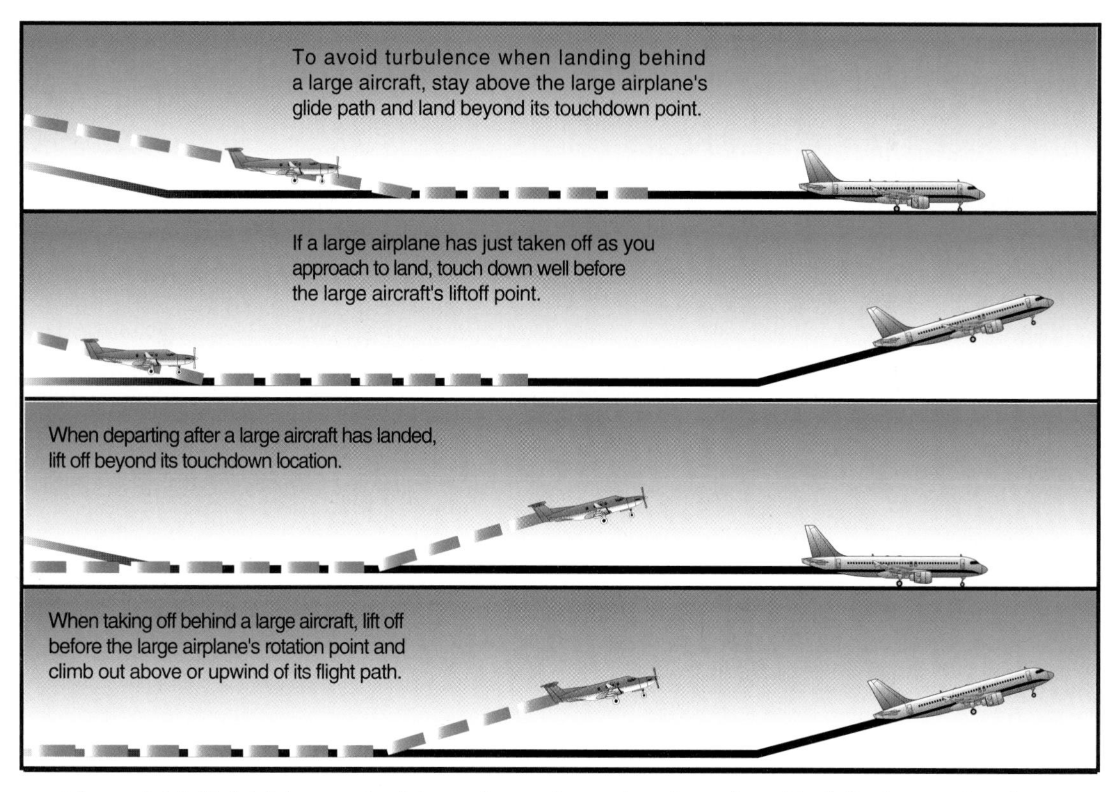

Figure 9-36. Maintaining a safe distance from a large aircraft can be critical. If a heavy aircraft is at your altitude and crosses your course, you should remain slightly above the path of the jet.

Jet engine blast is a related hazard. It can damage or even overturn a small airplane if it is encountered at close range. To avoid excessive jet blast, you must stay several hundred feet behind a jet with its engines operating, even when it is at idle thrust.

As shown in figure 9-36, you should avoid the area below and behind an aircraft generating wake turbulence, especially at low altitude where even a momentary wake encounter could be hazardous.

In a slow hover-taxi or stationary hover near the surface, helicopter main rotor(s) generate downwash producing high velocity outwash vortices to a distance approximately three times the diameter of the rotor. When rotor downwash hits the surface, the resulting outwash vortices have behavioral characteristics similar to wingtip vortices produced by fixed wing aircraft. However, the vortex circulation is outward, upward, around, and away from the main rotor(s) in all directions. If piloting a small aircraft, you should avoid operating within three rotor diameters of any helicopter in a slow hover-taxi or stationary hover. In forward flight, departing or landing helicopters produce a pair of strong, high-speed trailing vortices similar to the wingtip vortices of large fixed wing aircraft. As with large airplanes, use caution when operating behind or crossing behind landing or departing helicopters. [Figure 9-37]

Clear Air Turbulence

Clear air turbulence (CAT) is commonly thought of as a high altitude phenomenon. It usually is encountered above 15,000 feet, however, it can take place at any altitude and is often present with no visual warning. While its name suggests that it cannot occur except in clear skies, CAT can also be present in nonconvective clouds. Clear air

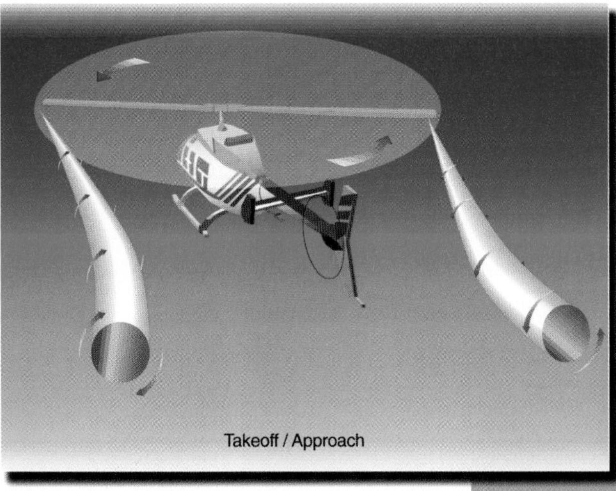

Figure 9-37. Avoid hovering or slow moving helicopters by at least a distance equal to three rotor diameters. Landing and departing helicopters produce the same vortex circulation as a large airplane.

Vortex circulation generated by helicopters in forward flight trail behind in a manner similar to wingtip vortices generated by airplanes.

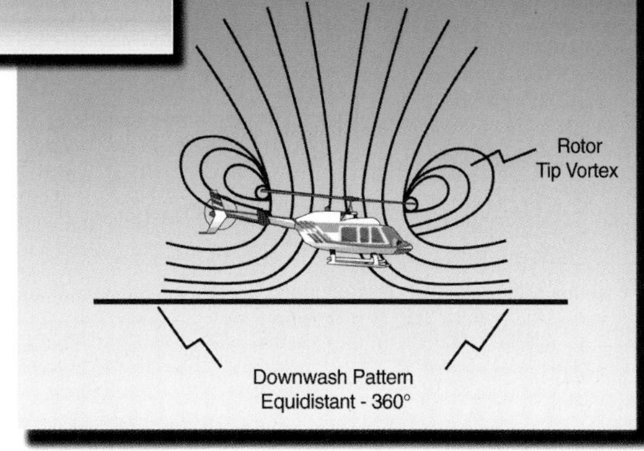

turbulence may be caused by the interaction of layers of air with differing wind speeds, convective currents, or obstructions to normal wind flow. It often develops in or near the **jet stream**, which is a narrow band of high altitude winds near the tropopause. CAT tends to be found in thin layers, typically less than 2,000 feet deep, a few tens of miles wide and more than 50 miles long. CAT often occurs in sudden bursts as aircraft intersect thin, sloping turbulent layers. [Figure 9-38]

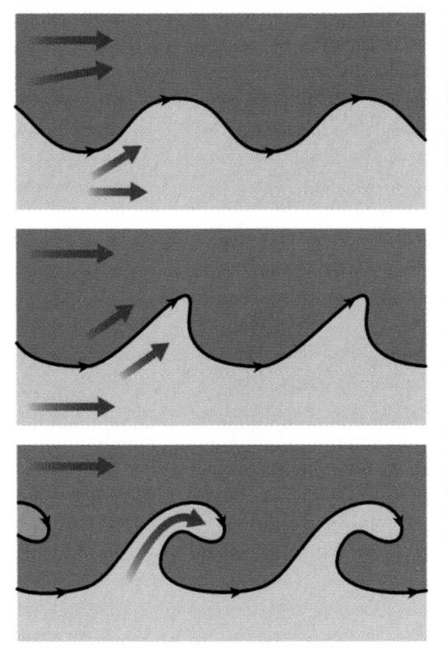

Figure 9-38. Clear air turbulence can form when a layer of air slides over the top of another, relatively slower moving layer. Eventually, the difference in speed may cause waves and, in some cases, distinctive clouds to form.

Turbulence that is encountered above 15,000 feet AGL that is not associated with cumuliform cloudiness, including thunderstorms, are reported as clear air turbulence.

Jet streams can sometimes be identified by long streams of cirrus cloud formations or high, windswept looking cirrus clouds. The turbulence associated with a jet stream can be very strong, and because it often occurs in clear air, it is difficult to forecast accurately. As a rule of thumb, clear air turbulence can be expected

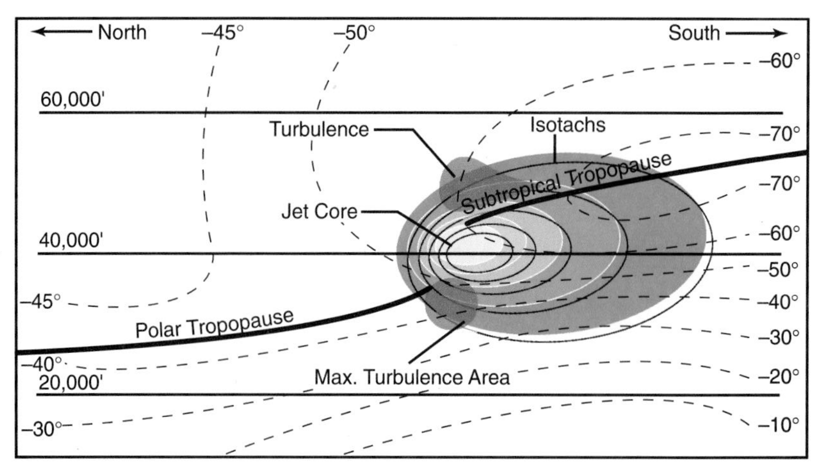

Figure 9-39. This is a cross section of a polar jet stream core. Note the wind speed gradient, shown by the spacing of the isotachs, or lines of equal wind velocity, is much stronger on the polar side of the jet. For this reason, wind shear or CAT is usually greater in an upper trough on the polar side. Precise analysis of the jet stream core is not possible, so you should anticipate CAT whenever you are near a jet stream.

when a curving jet is found north of a deep low pressure system. It can be particularly violent on the low pressure side of the jet stream when the wind speed at the core is 110 knots or greater. [Figure 9-39]

Where's the CAT?

On flights conducted under FAR Part 121, passengers are only required to have their seat belts fastened during takeoff and landing, and when the pilot in command instructs them to buckle up. However, there are compelling reasons to keep your seat belt fastened throughout any flight. One important reason is the potential for sudden, severe CAT, and, in the interest of further increasing airline safety, researchers are looking for ways to detect and predict the presence of CAT far enough ahead so that pilots can avoid the worst areas.

NCAR, in association with NOAA, NASA and other industry leaders, has developed a system that can detect the wind patterns that are associated with CAT before the airplane on which it is installed reaches the turbulent area. Airborne Coherent LiDAR (for Light Detection and Ranging) uses a form of laser technology to measure changes in the velocity of particles in the air up to 10 miles ahead, giving pilots up to 45 seconds to plan for the encounter, or to steer around it, if possible. [Figure A] At the present time, pilots are able to access CAT forecasts on the Internet through NOAA's website (http://orbit-net.nesdis.noaa.gov). These forecasts are issued for 12 and 24 hours and are valid for altitudes from FL 300 to 350. [Figure B]

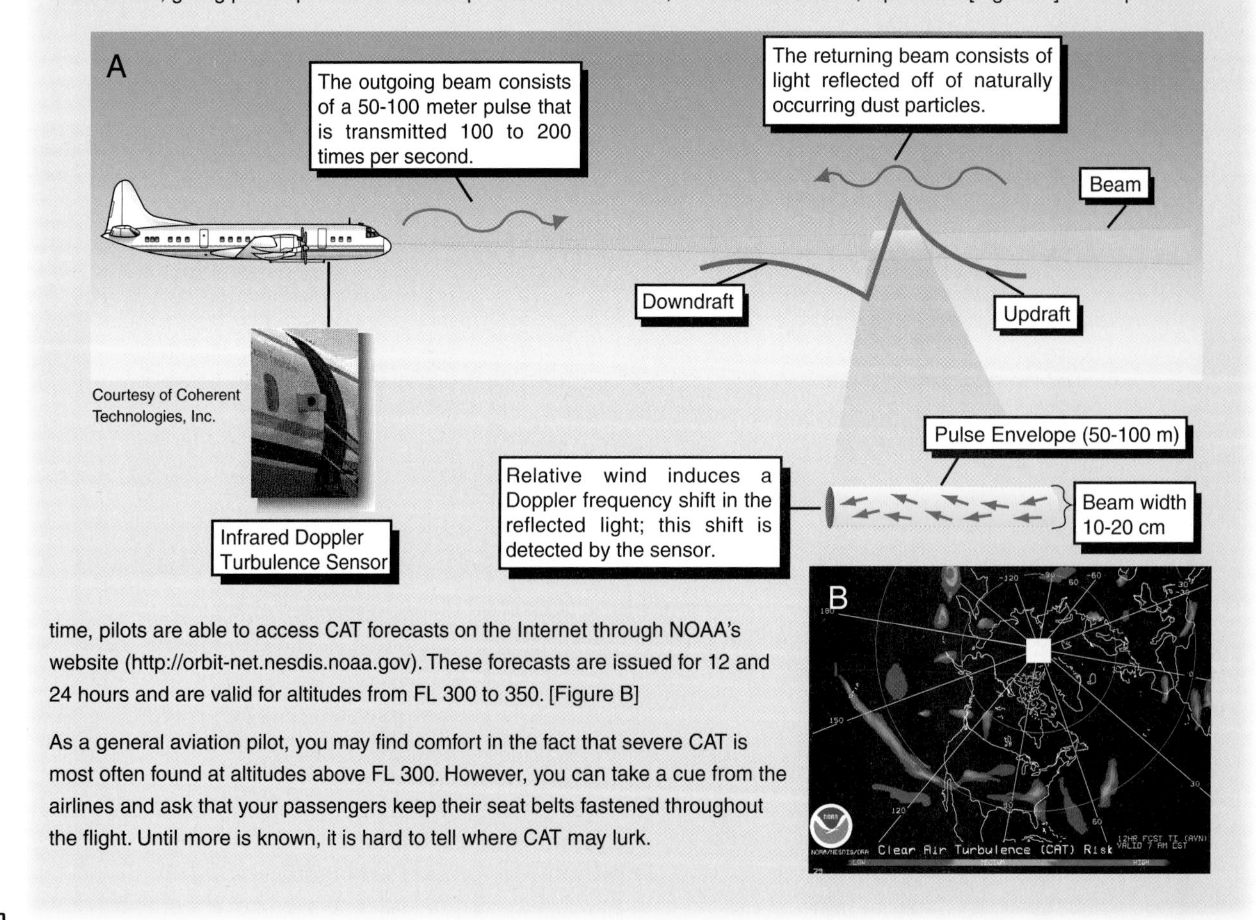

As a general aviation pilot, you may find comfort in the fact that severe CAT is most often found at altitudes above FL 300. However, you can take a cue from the airlines and ask that your passengers keep their seat belts fastened throughout the flight. Until more is known, it is hard to tell where CAT may lurk.

The jet stream and associated clear air turbulence can sometimes be visually identified in flight by long streaks of cirrus clouds.

Clear air turbulence has become a very serious operational factor to flight operations at all levels and especially to aircraft flying in excess of 15,000 feet. The best available information on this phenomena comes from pilots via the PIREP reporting procedures. All pilots encountering CAT conditions are urgently requested to report the time, location and intensity of the CAT to the FAA facility with which they are maintaining radio contact.

A curving jet stream associated with a deep low-pressure trough can be expected to cause great turbulence. In addition, a strong wind shear can be expected on the low-pressure side of a jet stream core when the speed at the core is stronger than 110 knots.

As shown in figure 9-39, a common location of clear air turbulence is in an upper trough on the polar side of a jet stream.

MOUNTAIN WAVE TURBULENCE

When stable air crosses a mountain barrier, the airflow is smooth on the windward side. Wind flow across the barrier is laminar — that is, it tends to flow in layers. The barrier may set up waves, called **mountain waves**. In order for mountain waves to form, the wind speed at the summit must be at least 20 knots. In addition, the wind direction should be roughly perpendicular to the range. Lift diminishes as the winds more nearly parallel the range. Wind speeds in excess of 40 knots can create very strong turbulence. The wave pattern may extend 100 miles or more downwind, and the crests may extend well above the highest peaks. Below the crest of each wave is an area of rotary circulation, or a rotor, which forms below the mountain peaks. Both the rotor and the waves can create violent turbulence.

The greatest turbulence normally occurs as you approach the lee side of mountain ranges, ridges, or hilly terrain in strong headwinds.

Mountain wave formation can be anticipated when the winds across a ridge are 20 knots or more, and the air is stable. Winds in excess of 40 knots can create strong turbulence.

If sufficient moisture is present, characteristic clouds warn you of the mountain wave. A rotor cloud (sometimes called a roll cloud) may form in the rotors. The crests of the waves may be marked by lens-shaped, or lenticular, clouds. Although the winds within the clouds may be 50 knots or greater, the clouds may appear stationary because they form in updrafts and dissipate in downdrafts. Because of this, they are sometimes referred to as standing lenticulars. Another cloud, which may signal the presence of mountain wave turbulence, is called a cap cloud. In some instances, cap clouds may obscure the mountain peaks. [Figure 9-40]

The presence of standing lenticular clouds and rotor clouds indicates the possibility of strong turbulence.

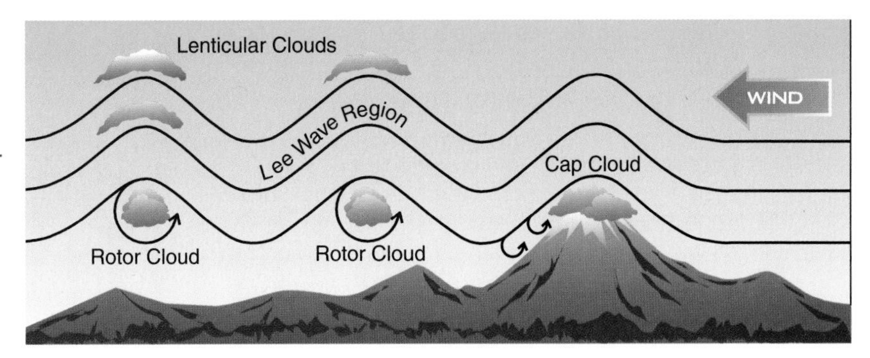

Figure 9-40. Mountain waves can create significant turbulence particularly along the lee slopes of the mountain, below the first lee wave trough, and in the rotor.

When conditions indicate a possible mountain wave, it is recommended that you fly at least 3,000 to 5,000 feet above the peaks. You should also climb to the selected altitude while approximately 100 miles from the range, depending on wind and aircraft performance. Another consideration is that you should approach the ridge from a 45° angle to permit a safer retreat if turbulence becomes too severe. If winds at the planned flight altitude exceed 30 knots, the FAA recommends against flight over mountainous areas in small aircraft. Since local conditions and aircraft performance vary widely, you should consider scheduling a thorough checkout by a qualified flight instructor if you plan on flying in mountainous terrain.

REPORTING TURBULENCE

You are encouraged to report encounters with turbulence, including the frequency and intensity. Turbulence is considered to be **occasional** when it occurs less than one-third of a given time span, **intermediate** when it covers one-third to two-thirds of the time, and **continuous** when it occurs more than two-thirds of the time. You can classify the intensity using the following guidelines:

Light — Slight erratic changes in altitude or attitude; slight strain against seat belts. **Light chop** is slight, rapid bumpiness without appreciable changes in altitude or attitude.

Moderate — Changes in altitude or attitude, but the aircraft remains in positive control at all times; usually causes variations in indicated airspeed; occupants feel definite strains against seat belts. **Moderate chop** is rapid bumps or jolts without appreciable changes in altitude or attitude.

Severe — Large abrupt changes in altitude or attitude; usually causes large variations in indicated airspeed; aircraft may be momentarily out of control; occupants forced violently against seat belts.

Extreme — Aircraft practically impossible to control; may cause structural damage.

Light turbulence momentarily causes slight, erratic changes in altitude and/or attitude.

Moderate turbulence causes changes in altitude and/or attitude, but aircraft control remains positive.

WIND SHEAR

Wind shear is a sudden, drastic shift in wind speed and/or direction that occurs over a short distance at any altitude in a vertical or horizontal plane. It can subject your aircraft to sudden updrafts, downdrafts, or extreme horizontal wind components, causing loss of lift or violent changes in vertical speeds or altitudes. Wind shear can be associated with convective precipitation, a jet stream, or a frontal zone. Wind shear also can materialize during a low-level temperature inversion when cold, still surface air is covered by warmer air which contains winds of 25 knots or more at 2,000 to 4,000 feet above the surface.

Wind shear is a sudden, drastic change in wind speed and/or direction and can exist at any altitude and may occur in all directions.

Generally, wind shear is most often associated with convective precipitation. While not all precipitation-induced downdrafts are associated with critical wind shears, one such downdraft, known as a **microburst**, is one of the most dangerous sources of wind shear. Microbursts are small-scale intense downdrafts which, on reaching the

Wind shear is an atmospheric condition that may be associated with a strong low-level temperature inversion with strong winds above the inversion, a jet stream, a thunderstorm, or a frontal zone.

surface, spread outward in all directions from the downdraft center. This causes the presence of both vertical and horizontal wind shear that can be extremely hazardous to all types and categories of aircraft, especially when within 1,000 feet of the ground.

A microburst downdraft is typically less than 1 mile in diameter as it descends from the cloud base to about 1,000 to 3,000 feet above the ground. In the transition zone near the ground, the downdraft changes to a horizontal outflow that can extend to approximately 2-1/2 miles in diameter. The downdrafts can be as strong as 6,000 feet per minute. Horizontal winds near the surface can be as strong as 45 knots resulting in a 90 knot shear as the wind changes to or from a headwind across the microburst. These strong horizontal winds occur within a few hundred feet of the ground. An individual microburst seldom lasts longer than 15 minutes from the time it strikes the ground until dissipation.

An important consideration for pilots, is that a microburst intensifies for about 5 minutes after it first strikes the ground, with the maximum intensity winds lasting approximately 2 to 4 minutes. Sometimes microbursts are concentrated into a line structure and, under these conditions, activity may continue for as long as an hour. Once microburst activity starts, multiple microbursts in the same general area are not uncommon and should be expected. [Figure 9-41]

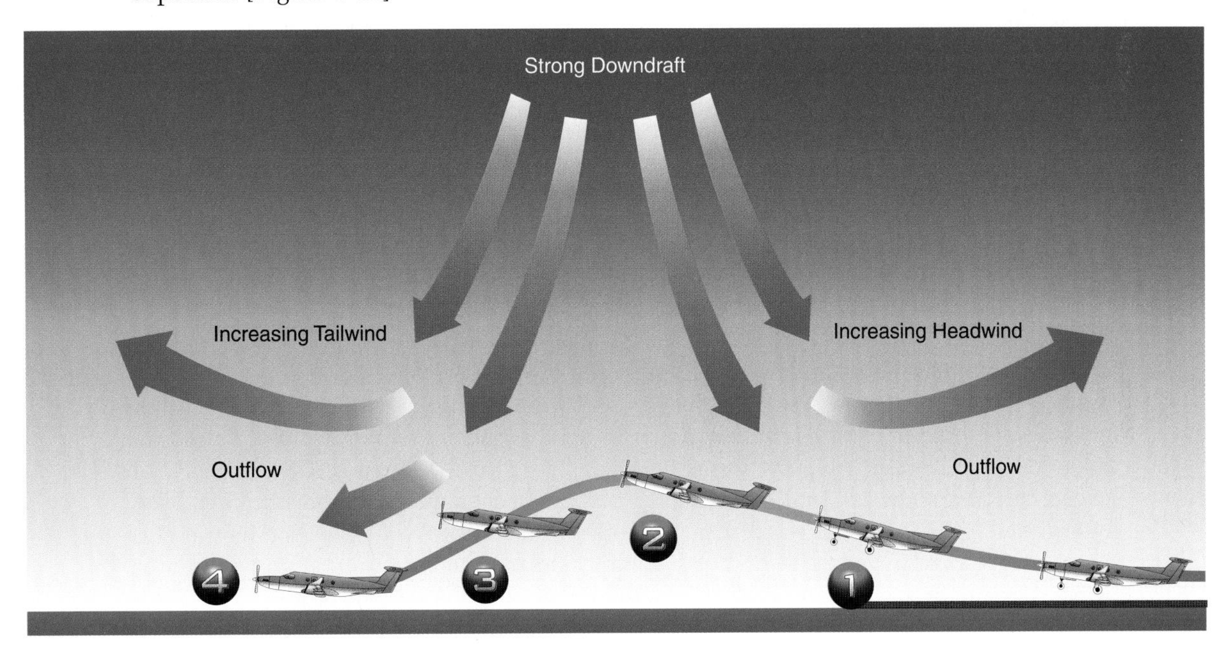

Figure 9-41. During a takeoff into a microburst, an aircraft first experiences a headwind which increases performance without a change in pitch and power (position 1). This is followed by a decreasing headwind and performance, and a strong downdraft (position 2). Performance continues to deteriorate as the wind shears to a tailwind in the downdraft (position 3). The most severe downdraft will be encountered between positions 2 and 3, which may result in an uncontrollable descent and impact with the ground (position 4).

Microbursts are intense, localized downdrafts seldom lasting longer than 15 minutes from the time the burst first strikes the ground until dissipation. The maximum downdrafts encountered in a microburst may be as strong as 6,000 feet per minute.

An aircraft that encounters a headwind of 45 knots within a microburst may expect a total shear across the microburst of 90 knots.

The performance of an aircraft changes drastically as it flies through a microburst. See figure 9-41.

On an approach, you should monitor the power and vertical velocity required to maintain the glide path. If you encounter an unexpected wind shear during an approach, it may be difficult to stay on the glide path at normal power and descent rates. If there is ever any doubt that you can regain a reasonable rate of descent and land without abnormal maneuvers, you should apply full power and make a go-around or missed approach. [Figure 9-42]

During an approach, monitoring the power and vertical velocity required to remain on the proper glideslope is the most important and most easily recognized means of being alerted to possible wind shear.

	WIND SHEAR			
	From Headwind	To Calm or Tailwind	From Tailwind	To Calm or Headwind
Indications				
Indicated Airspeed	Decreases		Increases	
Pitch Attitude	Decreases		Increases	
Aircraft	Tends to Sink		Balloons	
Groundspeed	Increases		Decreases	
Actions				
Power	Increase		Decrease	
Fly	Up to Glideslope		Down to Glideslope	
Be Prepared to	Reduce Power		Increase Power	
To Stay on Glide Path	Increase Rate of Descent (Due to faster groundspeed)		Decrease Rate of Descent (Due to slower groundspeed)	

Figure 9-42. This table reflects the recommendations of the Federal Aviation Administration should you encounter wind shear during a stabilized landing approach. Approaches should never be attempted into known wind shear conditions.

If you encounter a wind shear on an approach or departure, you are urged to promptly report it to the controller. An advanced warning of this information can assist other pilots in avoiding or coping with a wind shear on approach or departure. When describing conditions, avoid using the terms "*negative*" or "*positive*" wind shear. PIREPS of "*negative wind shear on final*" intended to describe loss of airspeed and lift, have been interpreted to mean that no wind shear was encountered. The recommended method for wind shear reporting is to state the loss or gain of airspeed and the altitudes at which it was encountered. "*Tulsa Tower, American 721 encountered wind shear on final, gained 25 knots between 600 and 400 feet, followed by a loss of 40 knots between 400 feet and surface.*" If you are unable to report wind shear in these specific terms, you are encouraged to make reports in terms of the effect upon your aircraft. "*Miami Tower, Gulfstream 403 Charlie encountered an abrupt wind shear at 800 feet on final, max thrust required.*"

See figure 9-42 for recommended actions should you encounter a wind shear during a stabilized approach.

With frontal activity, the most critical period for wind shear is either just before or just after frontal passage. With a cold front, wind shear occurs just after the front passes and for a short time afterward. Studies indicate the amount of wind shear in a warm front is generally much greater than in a cold front. The wind shear in a warm front may last for approximately six hours. The most critical period is before the front passes; the problem ceases to exist after it passes.

Look for wind shear before a warm front passes and after a cold front passes.

LOW-LEVEL WIND SHEAR ALERT SYSTEM

To help detect hazardous wind shear associated with microbursts, **low-level wind shear alert systems (LLWAS)** have been installed at many airports. The LLWAS uses a system of anemometers placed at strategic locations around the airport to detect variances in the wind readings. Many systems operate by sending individual anemometer readings every 10 seconds to a central computer which evaluates the wind differences across the airport. A wind shear alert is usually issued if one reading differs from the average by at least 15 knots. If you are arriving or departing from an airport equipped with LLWAS, you will be advised by air traffic controllers if an alert is posted. You also will be provided with wind velocities at two or more of the sensors. Consult the *Airport/Facility Directory* listings to determine if an airport has LLWAS.

TERMINAL DOPPLER WEATHER RADAR

In addition to LLWAS, **terminal doppler weather radar (TDWR)** systems are being installed at airports with high wind shear potential. These radar systems use a more powerful and narrower radar beam than conventional radar. The TDWR can provide a clearer, more detailed picture of thunderstorms which allows for better probability of predicting the presence of wind shear.

WEATHER SYSTEMS PROCESSOR

The **weather systems processor (WSP)** provides the controller, and ultimately the pilot, with the same products as the terminal doppler weather radar at a fraction of the cost of a TDWR. This is accomplished by utilizing new technologies to access the weather channel capabilities of an existing ASR-9 radar located on or near the airport, thus eliminating the requirements for a separate radar location.

IN-FLIGHT VISUAL INDICATIONS

In areas not covered by LLWAS or TDWR, you may only be able to predict the presence of wind shear using visual indications. In humid climates where the bases of convective clouds tend to be low, wet microbursts are associated with a visible rainshaft. In the drier climates of the deserts and mountains of the western United States, the higher thunderstorm cloud bases result in the evaporation of the rainshaft producing a dry microburst. The only visible indications under these conditions may be virga at the cloud base and a dust ring on the ground. It's important to note that since downdrafts can spread horizontally across the ground, low-level wind shear may be found beyond the boundaries of the visible rainshaft. You should avoid any area which you suspect could contain a wind shear hazard. [Figure 9-43]

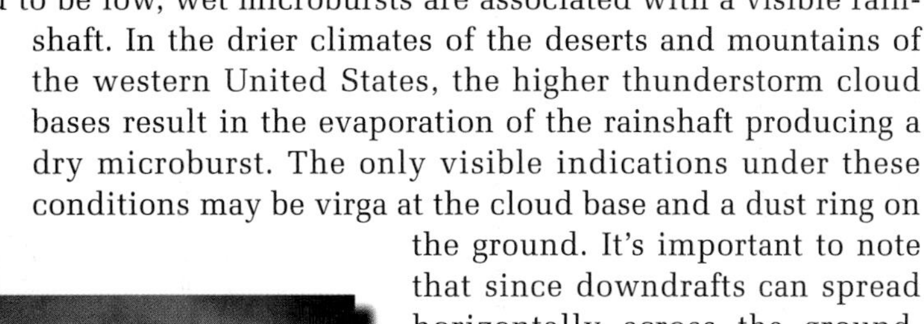

Figure 9-43. A visible rainshaft or virga indicates the possibility of a microburst.

LOW VISIBILITY

One of the most common aviation weather hazards is low visibility. As a pilot, you usually are concerned with two types of visibility — prevailing visibility and flight visibility. As you already know, prevailing visibility represents the greatest horizontal surface distance an observer can see and identify objects through at least half of the horizon. This is the visibility which is found in aviation routine weather reports (METARs). Flight visibility is defined as the average forward horizontal distance, from the cockpit of an aircraft in flight, at which prominent unlighted objects may be seen and identified by day and prominent lighted objects may be seen and identified by night. However, your most practical concern in flight operations often is slant-range visibility. The slant-range visibility may be greater or less than the surface horizontal visibility, depending on the depth of the surface condition. Slant-range visibility is important in the approach zone when you are landing from an instrument approach. Horizontal visibility at the surface is most important during IFR takeoff operations. Poor visibility creates the greatest hazard when it exists together with a low cloud ceiling.

RESTRICTIONS TO VISIBILITY

Particles which can absorb, scatter, and reflect light are always present in the atmosphere. The fact that the amount of particles in the air varies considerably explains why visibility is better on some days than on others. Restrictions to visibility can include fog, haze, smoke, smog, and dust. Fog requires both sufficient moisture and condensation nuclei on which the water vapor can condense. It is more prevalent in industrial areas, due to an abundance of condensation nuclei. One of the most hazardous characteristics of fog is its ability to form rapidly. It can completely obscure a runway in a matter of minutes.

Industrial areas typically produce more fog since they have more condensation nuclei.

Fog is classified according to the way it forms. **Radiation fog**, also known as **ground fog**, is very common. It forms over fairly level land areas on clear, calm, humid nights. As the surface cools by radiation, the adjacent air is also cooled to its dewpoint. Radiation fog usually occurs in stable air associated with a high pressure system. As early morning temperatures increase, the fog begins to lift and usually burns off by mid-morning. If higher cloud layers form over the fog, visibility will improve more slowly. [Figure 9-44]

Radiation fog forms over fairly flat land on clear, calm nights when the air is moist and there is a small temperature/dewpoint spread.

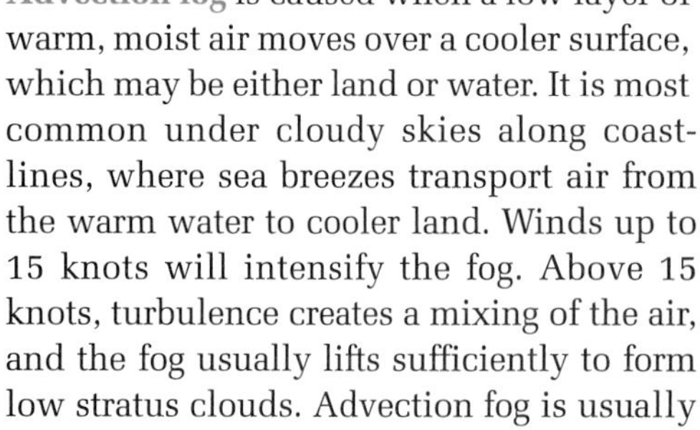

Advection fog is caused when a low layer of warm, moist air moves over a cooler surface, which may be either land or water. It is most common under cloudy skies along coastlines, where sea breezes transport air from the warm water to cooler land. Winds up to 15 knots will intensify the fog. Above 15 knots, turbulence creates a mixing of the air, and the fog usually lifts sufficiently to form low stratus clouds. Advection fog is usually

Figure 9-44. Radiation fog usually burns off by mid-morning.

Advection fog can appear suddenly during the day or night and is more persistent than radiation fog.

Advection fog is most likely to form in coastal areas when moist air moves over colder ground or water.

Surface winds stronger than 15 knots tend to dissipate or lift advection fog into low stratus clouds.

more persistent and extensive than radiation fog and can form rapidly during day or night. It commonly forms in winter when airmasses move inland from the coasts. **Upslope fog** forms when moist, stable air is forced up a sloping land mass. Like advection fog, upslope fog can form in moderate to strong winds and under cloudy skies.

Advection fog and upslope fog are both dependent upon wind for their formation.

Precipitation-induced fog may form when warm rain or drizzle falls through a layer of cooler air near the surface. Evaporation from the falling precipitation saturates the cool air, causing fog to form. This fog can be very dense, and usually does not clear until the rain moves out of the area. This type of fog may extend over large areas, completely suspending most air operations. It is most commonly associated with warm fronts, but can occur with slow moving cold fronts and stationary fronts.

Precipitation-induced fog is most commonly associated with warm fronts and is a result of saturation due to evaporation of precipitation.

Steam fog occurs as cool air moves over warmer water. It rises upward from the water's surface and resembles rising smoke. [Figure 9-45] **Ice fog** occurs in cold weather when the temperature is much below freezing and water vapor sublimates directly as ice crystals. Conditions favorable for its formation are the same as for radiation fog except for cold temperatures, usually -32°C (-25°F) or colder. It occurs mostly in the Arctic regions, but can occur in the middle latitudes during the cold season. Ice fog can be quite blinding to pilots flying into the sun.

Fogs differ as to the location where they form. For example, radiation fog is restricted to land areas, advection fog is most common along coastal areas, and steam fog forms over a water surface.

Haze, smoke, smog, and blowing dust or snow can also restrict your visibility. **Haze** is caused by a concentration of very fine dry particles. Individually, they are invisible to the naked eye, but in sufficient numbers, can restrict your visibility. Haze particles may be composed of a variety of substances, such as salt or dust particles. It occurs in stable atmospheric conditions with relatively light winds. Haze is usually no more than a few thousand feet thick, but it may occasionally extend to 15,000 feet. Visibility above the haze layer is usually good; however, visibility through the haze can be very poor. Dark objects tend to be bluish, while bright objects, like the sun or distant lights, have a dirty yellow or reddish hue. Haze can also create the illusion of being at a greater distance than actual from a runway, causing you to fly a lower approach.

Figure 9-45. As warm water evaporates and saturates a thin layer of cold air above the water, steam fog can form.

Smoke is the suspension of combustion particles in the air. The impact of smoke on visibility is determined by the amount of smoke produced, wind velocity, diffusion by turbulence, and distance from the source. You can often identify smoke by a reddish sky as the sun rises or sets, and an orange-colored sky when the sun is well above the horizon. When smoke travels distances of 25 miles or more, large particles fall out and the smoke tends to become more evenly distributed, giving the sky a grayish or bluish appearance, similar to haze.

Smog, which is a combination of fog and smoke, can create very poor visibility over a large area. In some geographical areas, topographical barriers, such as mountains, may combine with stable air to trap pollutants. This results in a build-up of smog, further reducing visibility.

Dust refers to fine particles of soil suspended in the air. When the soil is loose, the winds are strong, and the atmosphere is unstable, dust may be blown for hundreds of miles. Dust gives a tan or gray tinge to distant objects while the sun may appear as colorless, or with a yellow hue. Blowing dust is common in areas where dry land farming is extensive, such as the Texas panhandle.

Restrictions to visibility, such as haze, create the illusion of being at a greater distance above the runway, causing pilots to fly a lower approach.

VOLCANIC ASH

While lava from volcanoes generally threatens areas only in the immediate vicinity of the volcano, the ash cloud can affect a much more widespread area. Volcanic ash, which consists of gases, dust, and ash from a volcanic eruption, can spread around the world and remain in the stratosphere for months or longer. Due to its highly abrasive characteristics, volcanic ash can pit the aircraft windscreens and landing lights to the point they are rendered useless. Under severe conditions, the ash can clog pitot-static and ventilation systems as well as damage aircraft control surfaces. Piston aircraft are less likely than jet aircraft to lose power due to ingestion of volcanic ash, but severe damage is possible, especially if the volcanic cloud is only a few hours old. If you suspect that you are in the vicinity of an ash cloud you should attempt to stay upwind. If you inadvertently enter a volcanic ash cloud you should not attempt to fly straight through or climb out of the cloud since an ash cloud may be hundreds of miles wide and extend to great heights. You should reduce power to a minimum, altitude permitting, and reverse course to escape the cloud. [Figure 9-46]

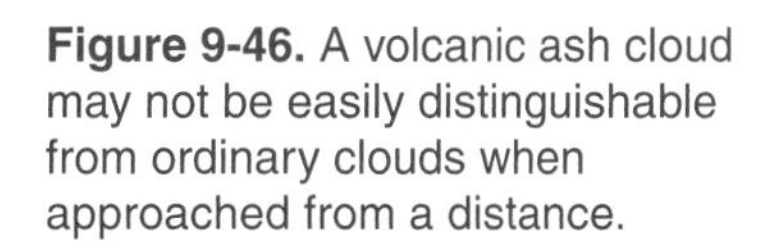

Figure 9-46. A volcanic ash cloud may not be easily distinguishable from ordinary clouds when approached from a distance.

ICING

There are two general types of icing with which you must be familiar — induction and structural. **Induction icing** includes carburetor icing as well as air intake icing. Carburetor icing forms in the carburetor venturi. It is most likely to occur when the outside air temperature is between approximately -7°C (20°F) and 21°C (70°F) and relative humidity is above 80%. Some aircraft may have a carburetor air temperature gauge to help detect potential carburetor icing conditions. A typical gauge is marked with a yellow arc between -15° and +5° Celsius. The yellow arc indicates the carburetor temperature range where carburetor icing can occur. However, it is dangerous for the indicator to be in the yellow arc only if the moisture content of the air, as well as the air temperature, is conducive to ice formation. Carburetor heat should be used as recommended in the pilot's operating handbook.

Air intake icing, like airframe icing, usually requires the aircraft surface temperature to be 0°C (32°F) or colder with visible moisture present. However, it also can form in clear air when relative humidity is high and temperatures are 10°C (50°F) or colder.

Structural icing builds up on any exposed surface of an aircraft, causing a loss of lift, an increase in weight, and control problems. There are two general types — rime and clear. Mixed icing is a combination of the two. **Rime ice** normally is encountered in stratus clouds and results from instantaneous freezing of tiny water droplets striking the aircraft surface. It has an opaque appearance caused by air being trapped in the water droplets as they freeze. The major hazard of rime ice is its ability to change the shape of an airfoil and destroy lift. Since rime ice freezes instantly, it builds up on the leading edge of airfoils, but it does not flow back over the wing and tail surfaces. **Clear ice** may develop in areas of large water droplets which are found in cumulus clouds or in freezing rain beneath a warm front inversion. Freezing rain means there is warmer air at higher altitudes. When the droplets flow over the aircraft structure and slowly freeze, they can glaze the aircraft's surfaces. Clear ice is the most serious of the various forms of ice because it has the fastest rate of accumulation, adheres tenaciously to the aircraft, and is more difficult to remove than rime ice.

If you encounter rain that freezes on impact, the temperatures are above freezing at some higher altitude.

The effects of ice buildup on aircraft are cumulative. Ice increases drag and weight, and decreases lift and thrust. Tests have shown that ice, snow, or frost with a thickness and roughness similar to medium or coarse sandpaper on the leading edge and upper surface of a wing can reduce lift by as much as 30% and increase drag by 40%. These changes in lift and drag significantly increase the stalling speed and reduce the angle of attack at which the airplane stalls. If you encounter icing, you must quickly alter your course or altitude to maintain safe flight. In extreme cases, two to three inches of ice can form on the leading edge of an aircraft in less than five minutes.

Freezing rain is most likely to have the highest rate of accumulation of structural icing.

Ice, snow, or frost having the thickness and roughness of sandpaper can reduce lift by 30% and increase drag by 40%.

Two conditions are necessary for a substantial accumulation of ice on an aircraft. First, the aircraft must be flying through visible moisture, such as rain or clouds. Second, the temperature of the water or of the aircraft must be 0°C (32°F) or lower. Keep in mind that aerodynamic cooling can lower the temperature of an airfoil to 0°C, even though the ambient temperature is a few degrees warmer.

When water droplets are cooled below the freezing temperature, they are in a **supercooled state**. They turn to ice quickly when disturbed by an aircraft passing through them. Clear icing is most heavily concentrated in cumuliform clouds in the range of temperature from

0°C to -10°C, (32°F to 14°F) usually from altitudes near the freezing level to 5,000 feet above the freezing level. However, you can encounter clear icing in cumulonimbus clouds with temperatures as low as -25°C (-13°F). In addition, supercooled water and icing have been encountered in thunderstorms as high as 40,000 feet, with temperatures of -40°C (-40°F). Small supercooled water droplets also can cause rime icing in stratiform clouds, although rime does not usually accumulate as fast as clear ice. Continuous icing can be expected in stratiform clouds in the temperature range from 0°C to -20°C (32°F to -4°F). You are least likely to encounter icing in the high cloud family, since these clouds are composed mainly of ice crystals. Icing may also occur as a mixture of both rime and clear. This commonly is the case with frontal systems which can produce a wide variety of icing conditions. [Figure 9-47]

High clouds are least likely to contribute to aircraft structural icing.

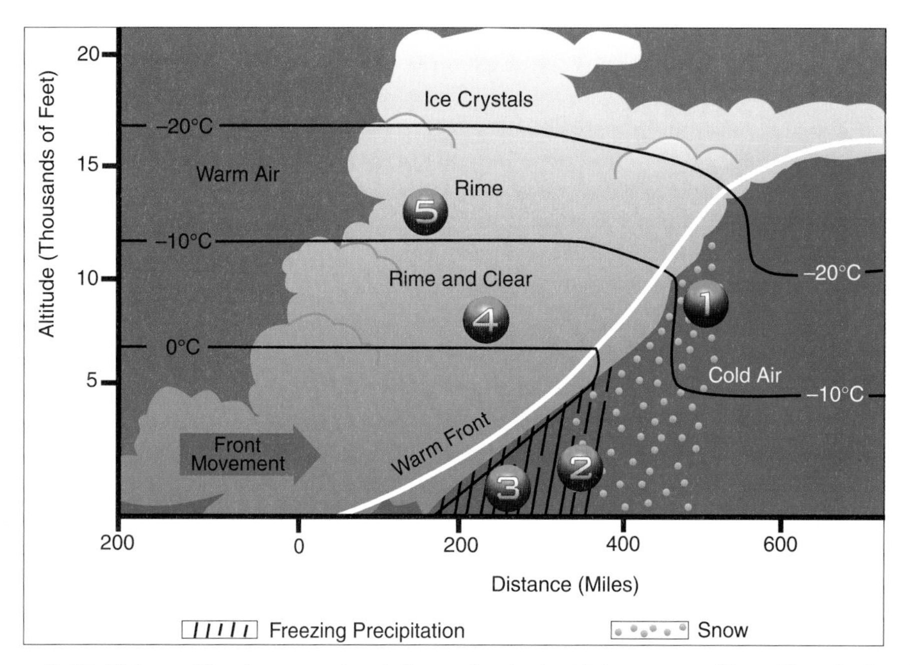

Figure 9-47. This profile of a warm front shows the typical icing areas. At position 1, rime ice is possible when you are flying in snow falling from stratus clouds. Wet snow also could be encountered below the freezing level. As you proceed under the warm front at position 2, you may encounter ice pellets formed by raindrops which have frozen while falling through colder air. Ice pellets normally do not adhere to an aircraft; however, they always indicate there is freezing rain at a higher altitude. The level of the freezing rain will be lower as you fly under the frontal zone. You should anticipate icing from freezing rain or drizzle at position 3. The area of freezing rain may be quite narrow, or it may extend for several miles. You should expect a mixture of rime and clear ice at position 4. If you are flying in the upper level of clouds associated with this warm front where the temperatures are far below freezing, rime ice is the predominant type of icing you will encounter at position 5.

ESTIMATING FREEZING LEVEL

Knowing the location of the freezing level is important when you select a cruising altitude for an IFR flight. You can estimate the freezing level by using temperature lapse rates. The standard, or average, temperature lapse rate is approximately 2°C (3.5°F) per 1,000 feet. If the surface temperature is 60°F, subtract 32°F and divide by 3.5°F to determine the freezing level (60 - 32 = 28 ÷ 3.5 = 8). This means the freezing level is approximately 8,000 feet AGL, assuming the actual lapse rate is close to 3.5°F.

The presence of ice pellets indicates a warm front is about to pass and that freezing rain probably exists at a higher altitude. See figure 9-47.

AVOIDING ICE ENCOUNTERS

Assuming a standard lapse rate, if the temperature is +8°C at an elevation of 1,350 feet, the approximate freezing level is 5,350 feet. (8 ÷ 2 = 4 or 4,000 + 1,350 = 5,350 feet).

Obviously, you should avoid areas where icing is forecast or expected. If you unintentionally get into an icing situation without ice removal equipment, there usually are only limited courses of action available. Also, remember that you must obtain approval from ATC if you need to make a diversion while operating on an IFR flight plan. When you encounter icing while flying in stratiform clouds, you can climb or descend out of the visible moisture or to an altitude where the temperature is above freezing or lower than -10°C (14°F).

If you encounter cumuliform clouds and suspect icing conditions, a simple change of course may be the best action. If this is impractical, descend to a lower altitude to avoid the ice. Overflying developed cumuliform clouds may be beyond the performance capability of your aircraft. Your remaining option is a 180° turn, which in some cases is the best choice.

If you encounter freezing rain, you must take immediate action. One option is to climb into the warmer air above. If you choose to do this, start the climb at the earliest possible moment. Otherwise, you may not have enough power to reach a higher altitude, because ice buildup quickly degrades your aircraft's performance. Also, if the frontal surface slopes up in the direction of your flight, you may not be able to climb fast enough to reach the warm air. Other options include making a 180° turn and reversing course or descending to a lower altitude. You may choose to make a descent if the free air temperature at lower altitudes is above freezing or cold enough to change the precipitation into ice pellets (sleet). A 180° turn or a descent is probably the best course of action if the altitude of warmer air aloft is unknown and your aircraft's performance will not permit a rapid and extended climb. If enough ice accumulates to make level flight impossible, descent is inevitable, so trade altitude for airspeed to maintain control.

In weather forecasts or pilot reports, aircraft structural icing is normally classified as trace, light, moderate, or severe depending on the accumulation rate. A **trace** means ice is perceptible, but accumulation is nearly balanced by its rate of sublimation. De-icing equipment is unnecessary, unless icing is encountered for an extended period of time. **Light ice** accumulation can be a problem during prolonged exposure (over one hour) if you do not have adequate de-icing/anti-icing equipment. In **moderate icing** conditions, even short encounters become potentially hazardous unless you use de-icing/anti-icing equipment. **Severe icing** produces a rate of accumulation greater than the reduction or control capabilities of the de-icing/anti-icing equipment.

Most small general aviation aircraft are not approved for flight into icing conditions. Those that are have special de-ice and anti-ice equipment to protect aircraft surfaces, as well as the induction system, and have been flight tested to demonstrate their ability to fly in icing conditions. However, even these aircraft cannot fly in severe icing, and prolonged flight in moderate icing can become hazardous due to ice accumulation on unprotected surfaces. Since ice protection systems vary widely in both operation and effectiveness, be sure to consult the POH for detailed information on system operation. A general discussion of de-ice and anti-ice systems is included in Chapter 11, Section B.

If frost is not removed from the wings before flight, it may cause an early airflow separation which decreases lift and increases drag. This causes the airplane to stall at a lower-than-normal angle of attack.

Frost is a related element which poses a serious hazard during takeoff. It interferes with smooth airflow over the wings and can cause early airflow separation, resulting in a loss of lift. This means the wing stalls at a lower-than-normal angle of attack. Although frost increases drag, the aircraft should be able to reach takeoff speed. The danger is that it may stall shortly after liftoff. Always remove all frost from the aircraft surfaces before flight.

HYDROPLANING

Hydroplaning is caused by a thin layer of standing water that separates the tires from the runway. It causes a substantial reduction in friction between the aircraft tires and the runway surface and results in poor or nil braking action at high speeds. If severe enough, it may result in your aircraft skidding off the side or end of the runway. Hydroplaning is most likely at high speeds on wet, slushy, or snow-covered runways which have smooth textures.

High aircraft speed, standing water, slush, and a smooth runway texture are factors conducive to hydroplaning.

The best remedy for hydroplaning is to prevent its occurrence. You should study hydroplaning speeds and recommended procedures for your aircraft. When selecting a runway, pick one which is longer than required and allows you to use only light braking pressures to stop the aircraft safely. Since it may take several seconds for the wheels to reach their rotational speed after landing, do not apply the brakes too quickly. More detailed information on hydroplaning is available in Chapter 12, Section B.

COLD WEATHER OPERATIONS

Prior to a flight in cold weather, there are additional precautions which you must observe. As mentioned above, you must remove any frost, snow, or ice on the aircraft. You should also inspect all control surfaces and their associated control rods and cables for snow or ice which may interfere with their operation. Be sure to check the crankcase breather lines, since vapor from the engine can condense and freeze, preventing the release of air from the crankcase. [Figure 9-48]

Figure 9-48. If ice or snow is removed by melting, be sure the water does not run into the control surface hinges and refreeze.

Another important consideration is whether or not to preheat the aircraft prior to flight. If temperatures are so low that you will experience difficulty starting the engine, you should preheat it. You should also consider preheating the cabin, as well as the engine compartment. This is recommended for proper operation of instruments, many of which are adversely affected by cold temperatures.

During preflight in cold weather, crankcase breather lines should receive special attention because they are susceptible to being clogged by ice from crankcase vapors that have condensed and subsequently frozen.

It is recommended that during cold weather operations, you should preheat the cabin, as well as the engine.

A final consideration before you try to start the engine is priming. Here, you should go by the manufacturer's recommended procedures in the POH. Overpriming may result in poor compression and difficulty in starting the engine. Another cold starting problem is caused by icing on sparkplug electrodes. This can occur when the engine fires for only a few revolutions and then quits. Heating the engine is the only remedy for frosted plugs, short of removing them from the engine and heating them individually.

SUMMARY CHECKLIST

✓ The life of a thunderstorm passes through three distinct stages. The cumulus stage is characterized by continuous updrafts. When precipitation begins to fall, the thunderstorm has reached the mature stage. As the storm dies during the dissipating stage, updrafts weaken and downdrafts become predominant.

✓ Airmass thunderstorms are relatively short-lived storms and are usually isolated or scattered over a large area. Severe thunderstorms contain wind gusts of 50 knots or more, hail 3/4 inch in diameter or larger, and/or tornadoes.

✓ A squall line is a narrow band of active thunderstorms that often forms 50 to 200 miles ahead of a fast moving cold front. It often contains very severe weather.

✓ Some weather hazards associated with thunderstorms, such as lightning, hail, and turbulence are not confined to the cloud itself.

✓ If you are using airborne weather radar, you should avoid intense radar echoes by at least 20 miles.

✓ If you encounter turbulence during flight, you should establish maneuvering or penetration speed and try to maintain a level flight attitude.

✓ Mechanical turbulence is often experienced in the traffic pattern when wind forms eddies as it blows over hangars, stands of trees, or other obstructions.

✓ When sufficient moisture is present, cumulus cloud build-ups indicate the presence of convective turbulence.

✓ Any front traveling at a speed of 30 knots or more produces at least a moderate amount of turbulence.

✓ Wake turbulence is created when an aircraft generates lift. The greatest vortex strength occurs when the generating aircraft is heavy, slow, in a clean configuration, and at a high angle of attack.

✓ A helicopter can produce vortices similar to wingtip vortices of a large fixed-wing airplane.

✓ Clear air turbulence often develops in or near the jet stream, which is a narrow band of high altitude winds near the tropopause.

✓ Strong mountain wave turbulence can be anticipated when the winds across a ridge are 40 knots or more, and the air is stable. The crests of mountain waves may be marked by lens-shaped, or lenticular, clouds.

✓ Wind shear can exist at any altitude and may occur in a vertical or horizontal direction. A microburst is one of the most dangerous sources of wind shear.

✓ Restrictions to visibility can include fog, haze, smoke, smog, and dust.

✓ Volcanic ash clouds are highly abrasive to aircraft and engines, and they also restrict visibility.

✓ The three types of structural ice are rime, clear, and mixed.

✓ The accumulation of ice on an aircraft increases drag and weight and decreases lift and thrust.

✓ Ice pellets usually indicate the presence of freezing rain at a higher altitude.

KEY TERMS

Cumulus Stage

Mature Stage

Shear Zone

Gust Front

Roll Cloud

Dissipating Stage

Airmass Thunderstorm

Severe Thunderstorm

Single Cell Thunderstorm

Supercell Thunderstorm

Multicell Thunderstorm

Frontal Thunderstorm

Squall Line

Lightning

Hail

Funnel Clouds

Tornado

Waterspout

Low-Level Turbulence (LLT)

Mechanical Turbulence

Convective Turbulence

Capping Stable Layer

Frontal Turbulence

Wingtip Vortices

Wake Turbulence

Jet Engine Blast

Clear Air Turbulence (CAT)

Jet Stream

Mountain Waves

Occasional Turbulence

Intermediate Turbulence

Continuous Turbulence

Light Turbulence

Light Chop

Moderate Turbulence

Moderate Chop

Severe Turbulence

Extreme Turbulence

Wind Shear

Microburst

Low-Level Wind Shear Alert System (LLWAS)

Terminal Doppler Weather Radar (TDWR)

Weather Systems Processor (WSP)

Radiation Fog

Ground Fog

Advection Fog

Upslope Fog

Precipitation-Induced Fog

Steam Fog

Ice Fog

Haze

Smoke

Smog

Dust

Volcanic Ash

Induction Icing

Structural Icing

Rime Ice

Clear Ice

Supercooled State

Trace Ice

Light Icing

Moderate Icing

Severe Icing

Frost

Hydroplaning

QUESTIONS

1. What are the three basic ingredients needed for the formation of a thunderstorm?

2. Continuous updrafts occur in a thunderstorm during what stage?

   A. Cumulus stage
   B. Mature stage
   C. Dissipating stage

3. Thunderstorms reach their greatest intensity during which stage?

   A. Cumulus stage
   B. Mature stage
   C. Dissipating stage

4. What is the term used to describe a narrow band of thunderstorms that normally contains the most severe types of weather-related hazards?

   A. Airmass thunderstorms
   B. Frontal boundary thunderstorms
   C. Squall line thunderstorms

5. What type of cloud is associated with the most severe turbulence?

   A. Cumulonimbus
   B. Cumulus
   C. Standing lenticular altocumulus

6. What hazard is always associated with a thunderstorm?

   A. Heavy rain
   B. Lightning
   C. Severe icing

7. True/False. You may encounter hail in clear air several miles from a thunderstorm.

8. When using on-board weather radar to avoid thunderstorms, you should avoid intense echoes by at least how many miles?

   A. 10 miles
   B. 20 miles
   C. 30 miles

9. What should you do if you unexpectedly encounter turbulence?

10. Wake turbulence is greatest from a large, heavy aircraft which is operating in what configuration?

    A. Low speeds and low angles of attack
    B. Low speeds and high angles of attack
    C. High speeds and high angles of attack

11. Should you plan to land before or after the touchdown point of a large, heavy aircraft?

12. True/False. There is no correlation between clear air turbulence and the jet stream.

13. What type of turbulence is indicated by the presence of rotor, cap and lenticular clouds?

    A. Convective
    B. Wake
    C. Mountain wave

14. Turbulence that causes changes in altitude or attitude resulting in definite strains against seatbelts is classified as what?

    A. Light
    B. Moderate
    C. Severe

15. During a stabilized landing approach, what happens if the wind unexpectedly shifts from a headwind to a tailwind?

    A. Pitch attitude decreases, IAS decreases, and the airplane tends to sink below the glide path
    B. Pitch attitude decreases, IAS increases, and the airplane tends to sink below the glide path
    C. Pitch attitude increases, IAS increases, and the airplane tends to rise above the glide path

16. Discuss the in-flight visual indications of possible wind shear.

17. What is the name of the fog that typically forms over fairly level land on clear, calm, humid nights?

    A. Steam fog
    B. Radiation fog
    C. Advection fog

18. Haze can create the illusion that you are

    A. closer to the runway than you actually are.
    B. at a greater distance from the runway than you actually are.
    C. lower on the approach than you actually are.

19. What is the recommended course of action if you inadvertently enter a volcanic ash cloud?

    A. Reverse course
    B. Attempt to climb up and out of the cloud
    C. Continue straight ahead to exit on the opposite side

20. Carburetor icing is most likely when the relative humidity is above 80% and the outside air temperature is between

    A. -18°C (0°F) and 1°C (34°F).
    B. -15°C (5°F) and 5°C (41°F).
    C. -7°C (20°F) and 21°C (70°F).

21. True/False. Airframe icing increases drag and weight and decreases lift and thrust.

22. True/False. Airframe icing cannot occur when the outside air temperature is above 0°C.

23. When the surface temperature is 53°F at sea level, the estimated freezing level is

    A. 4,700 feet AGL.
    B. 6,000 feet AGL.
    C. 10,500 feet AGL.

24. To avoid hydroplaning after landing on a wet runway, you should

    A. apply the brakes immediately.
    B. delay the application of brakes.
    C. carry a higher-than-normal power setting after touchdown.

SECTION C

PRINTED REPORTS AND FORECASTS

Weather information is available through a vast network of government agencies and commercial sources. While the proliferation of electronic media has generally increased the availability of weather data, you may find that access to weather briefers and NWS meteorologists to interpret the data might not always be readily available. Consequently, your ability to gather and decipher the multitude of printed reports and forecasts is more important than ever.

PRINTED WEATHER REPORTS

In general, a weather report is a record of observed atmospheric conditions at a particular location at a specific point in time. A variety of reports are used to disseminate information gathered by trained observers, radar systems, and other pilots. The types of reports you are likely to encounter include aviation routine weather reports, radar weather reports, and pilot weather reports.

AVIATION ROUTINE WEATHER REPORT

An **aviation routine weather report (METAR)** is an observation of surface weather which is reported in a standard international format. While the METAR code has been adopted worldwide, each country was allowed to make modifications or exceptions to the code. Usually these differences were minor but necessary to accommodate local procedures or particular units of measure. The following discussion covers the elements as reported in a METAR originating from an observation in the United States. [Figure 9-49]

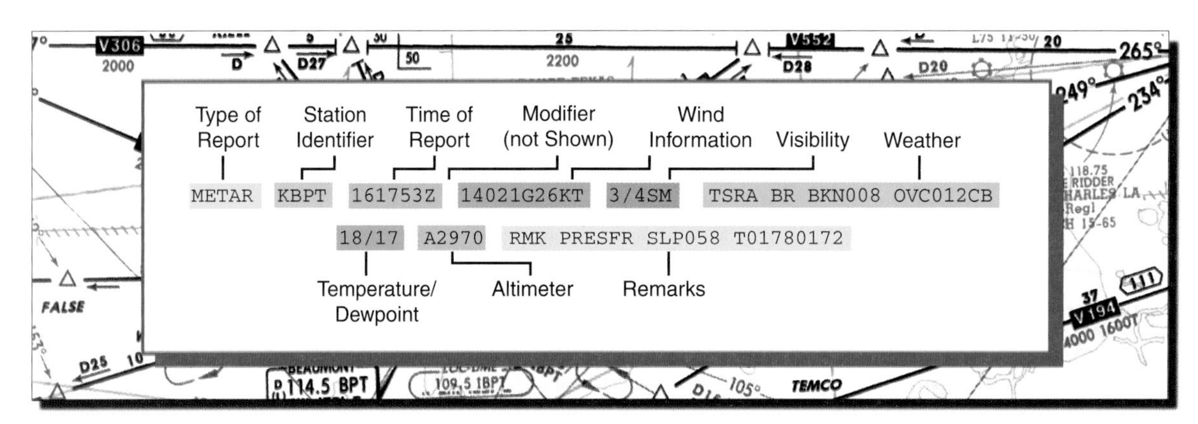

Figure 9-49. Although the content may vary somewhat, a typical METAR contains several distinct elements. If an element cannot be observed at the time of the report, it will be omitted.

TYPE OF REPORT

METAR KBPT...

The two types of reports are the METAR, which is taken every hour, and the **non-routine (special) aviation weather report (SPECI)**. The SPECI weather observation is an unscheduled report indicating a significant change in one or more of the reported elements.

STATION IDENTIFIER

METAR **KBPT** 161753Z...

Each reporting station is listed by its four-letter International Civil Aviation Organization (ICAO) identifier. In the contiguous 48 states, the letter K precedes the three-letter domestic airport identifier. For example, the domestic identifier for Beaumont/Port Arthur, Texas is BPT, and the ICAO identifier is KBPT. In other areas of the world, the first two letters indicate the region, country, or state. Alaska identifiers begin with PA, for Hawaii, they begin with PH, for Canada, the prefixes are CU, CW, CY, and CZ, and for Mexico, the preface is NM. A list of station identifiers is usually available at an FSS or National Weather Service (NWS) office.

DATE/TIME OF REPORT

...KBPT **161753Z** 14021G26KT...

The date (day of the month) and time of the observation follows the station identifier. The time is given in UTC or Zulu, as indicated by a Z following the time. The report in the example was issued on the 16th of the month at 1753Z.

MODIFIER

When a METAR is created by a totally automated weather observation station, the modifier AUTO follows the date/time element (e.g., ...251955Z AUTO 30008KT...). These stations are classified by the type of sensor equipment used to make the observations, and AO1 or AO2 will be noted in the remarks section of the report. An automated weather reporting station without a precipitation discriminator (which can determine the difference between liquid and frozen/freezing precipitation) is indicated by an AO1 in the remarks section of the METAR. An AO2 indicates that automated observing equipment with a precipitation discriminator was used to take the observation.

The modifier COR is used to indicate a corrected METAR and replaces a previously disseminated report. When the abbreviation COR is used in a report from an automated facility, the station type designator, AO1 or AO2, is removed from the remarks section. If a modifier is not included in the METAR (as in the example), it signifies that the observation was taken at a manual station or that manual input occurred at an automated station.

WIND

...161753Z **14021G26KT** 3/4SM...

The wind direction and speed are reported in a five digit group, or six digits if the speed is over 99 knots. The first three digits represent the direction from which the wind is blowing, in reference to true north. The letters VRB are used if the direction is variable.

The next two digits show the speed in knots (KT). Calm winds are reported as 00000KT. Gusty winds are reported with a G, followed by the highest gust. In the example, wind was reported to be from 140° true at 21 knots with gusts to 26 knots.

If the wind direction varies 60 degrees or more and the speed is above 6 knots, a variable group follows the wind group. The extremes of wind direction are shown, separated by a V. For example, if the wind is blowing from 020°, varying to 090°, it is reported as 020V090. [Figure 9-50]

CODED METAR DATA	EXPLANATIONS
00000KT	Wind calm
20014KT	Wind from 200° at 14 knots
15010G25KT	Wind from 150° at 10 knots, gusts to 25 knots
VRB04KT	Wind variable in direction at 4 knots
210103G130KT	Wind from 210° at 103 knots with gusts to 130 knots

Figure 9-50. This figure shows several examples of wind information as it appears on the METAR. The decoded wind direction and speed are shown to the right.

VISIBILITY

...14021G26KT **3/4SM** TSRA BR...

Prevailing visibility is the greatest distance an observer can see and identify objects through at least half of the horizon. When the prevailing visibility varies from one area of the sky to another, the visibility in the majority of the sky is reported. If visibility differs significantly between sectors, the observer can report individual sector visibility in the remarks section of the METAR. [Figure 9-51]

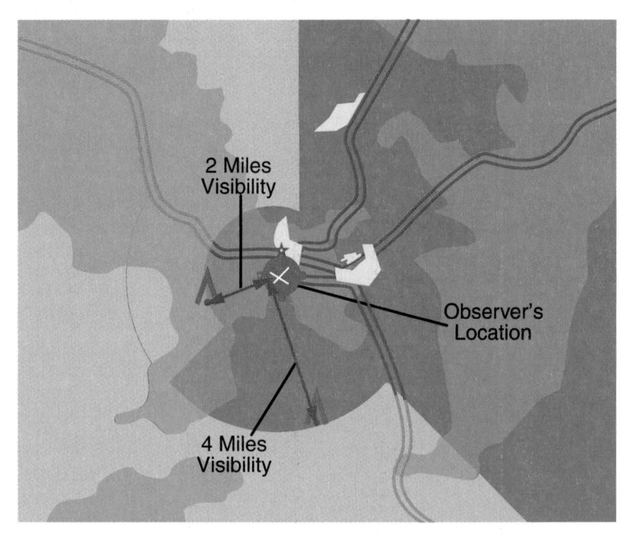

Figure 9-51. Prevailing visibility is determined by identifying distinctive objects, such as a tower or smokestack, which are at a known distance. At night, observers use lighted objects to determine visibility.

Visibility is reported in statute miles (SM). For example, 1/2SM indicates one-half statute mile and 4SM would be used to report four statute miles. At times, runway visual range (RVR) may be reported following prevailing visibility. RVR, in contrast to prevailing visibility, is based on what a pilot in a moving aircraft should see when looking down the runway. RVR is designated with an R, followed by the runway number, a slant (/), and the visual range in feet (FT). In the example, R36L/2400FT means Runway 36 Left visual range is 2,400 feet. Variable RVR is shown as the lowest and highest visual range values separated by a V. Outside the United States, RVR is normally reported in meters.

The runway visual range (RVR) value represents the horizontal distance a pilot should see down the runway from the approach end of a runway.

PRESENT WEATHER

The weather phenomenon, along with its qualifiers, that is present at the time of the observation is reported immediately after the visibility. The qualifiers can either be the intensity or proximity of the weather phenomena, and/or a descriptor. Weather phenomena falls into three categories: precipitation, obstruction to visibility, or any other phenomena.

Taking the Guesswork Out of Visibility

Without a frame of reference, determining how far you can see down a runway is just a guess. To determine ground visibility at some airports, weather observers have several selected objects at known distances from a specific location. When visibility is restricted, the farthest visible object determines the reported visibility..

A much better way of determining visibility is with an optical instrument called a transmissometer. A transmissometer measures runway visual range (RVR) down a specific runway. It is made up of two components. The first is a projector which sends a beam of light, at a known intensity, down the length of the runway. The second component, the receiver, contains a photoelectric cell which measures the amount of light penetrating the obscuring phenomena. A computer then converts the amount of light received into runway visual range.

QUALIFIERS

...3/4SM **TS**RA BR BKN008 OVC012CB...

Intensity of precipitation is shown immediately prior to the descriptor and weather phenomena code. The indicated intensity applies only to the first type of precipitation reported. Intensity levels are shown as light (–), moderate (no sign), or heavy (+). In this example there is no qualifier so the intensity of the rain (RA) is reported as moderate.

Weather obscurations or other phenomena occurring in the vicinity of the airport (between 5 and 10 statute miles) are shown by the letters VC preceding the code. If precipitation is not occurring at the point of the observation, but is within 10 statute miles of the observation point, it is reported as being in the vicinity. VC will not appear if an intensity qualifier is reported.

A description of the precipitation or obscurations is reported as a two-letter code. Only one descriptor is used for each type of precipitation or obscuration. In the example, a thunderstorm (TS) was reported.

DESCRIPTOR CODES	
TS – Thunderstorm	DR – Low Drifting
SH – Shower(s)	MI – Shallow
FZ – Freezing	BC – Patches
BL – Blowing	PR – Partial

WEATHER PHENOMENA CODES	
Precipitation	
RA — Rain	GR — Hail (> 1/4")
DZ — Drizzle	GS — Small Hail/Snow Pellets
SN — Snow	SG — Snow Grains
IC — Ice Crystals	PL — Ice Pellets
Obstructions to Visibility	
FG — Fog	PY — Spray
BR — Mist	SA — Sand
FU — Smoke	DU — Dust
HZ — Haze	VA — Volcanic Ash
Other Weather Phenomena	
SQ — Squall	SS — Sandstorm
DS — Duststorm	PO — Dust/Sand Whirls
FC — Funnel Cloud/Tornado/Waterspout	

WEATHER PHENOMENA

...3/4SM TS**RA BR** BKN008 OVC012CB...

Weather phenomena covers eight types of precipitation, eight kinds of obscurations, as well as five, rather uncommon, weather events. Up to three types of precipitation can be coded in a single grouping of present weather conditions. When more than one is reported, they are shown in order of decreasing dominance. In the example, moderate rain (RA) has been observed in connection with the thunderstorm.

Obscurations, which are anything other than precipitation that reduce horizontal visibility, are shown after any reported precipitation. Fog (FG) is listed when the visibility is less than 5/8 mile. However, if the visibility were to increase to between 5/8 and 6 miles, the code for mist (BR) would be used. In the example, mist (BR) is the obstruction to visibility. Shallow fog (MIFG), patches of fog (BCFG), or partial fog (PRFG) may be coded with prevailing visibility of 7 statute miles or greater. Following the obscurations, other weather phenomena, such as tornadoes, duststorms, and squalls may be listed when they occur. A squall is a sudden increase in the wind speed of at least 15 knots to a sustained speed of 20 knots or more for at least one minute.

SKY CONDITION

The sky condition groups describe the amount of clouds, if any, their heights, and, in some cases, their type. In addition, a vertical visibility may be reported if the height of the clouds cannot be determined due to an obscuration.

When a squall (SQ) is reported, you can expect a sudden increase in wind speed of at least 15 knots to a sustained wind speed of 20 knots or more for at least 1 minute.

AMOUNT

...TSRA BR **BKN**008 **OVC**012CB 18/17...

The amount of clouds covering the sky is reported in eighths, or octas, of sky cover. Each layer of clouds is described using a code which corresponds to the octas of sky coverage. If the sky is clear, it is designated by SKC in a manual report and CLR in an automated report. FEW is used when cloud coverage is 1/8 to 2/8 of the sky. However, any amount less than 1/8 can also be reported as FEW. Scattered clouds, which cover 3/8 to 4/8 of the sky, are shown by SCT. Broken clouds, covering 5/8 to 7/8 of the sky, are designated by BKN, while an overcast sky is reported as OVC. In the example, there was a layer of broken clouds (BKN) and an overcast layer (OVC).

The sky cover condition for a cloud layer represents total sky coverage, which includes any lower layers of clouds. In other words, when considering a layer of clouds, the observer must add the amount of sky covered by that layer to the amount of sky covered by all lower layers to determine the reported sky coverage for the layer being considered. [Figure 9-52]

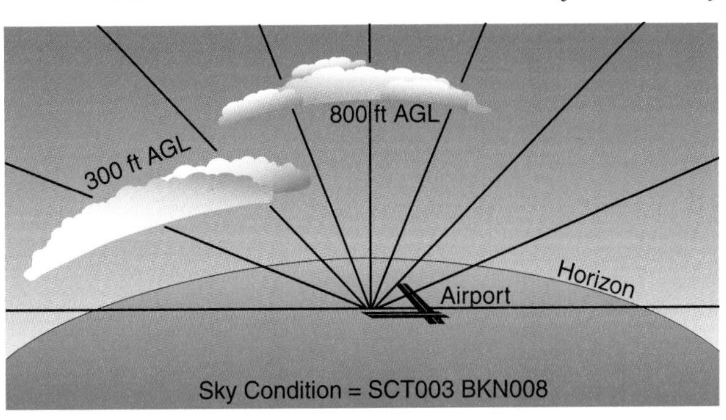

Figure 9-52. Note that the upper layer is reported as broken, even though that layer by itself covers less than 5/8 of the horizon. If the total coverage did not exceed 4/8, the upper layer would be reported as scattered.

HEIGHT, TYPE, AND VERTICAL VISIBILITY

...TSRA BR BKN**008** OVC**012CB** 18/17...

The height of cloud bases or the vertical visibility into obscuring phenomena is reported with three digits in hundreds of feet above ground level (AGL). To determine the cloud bases, add two zeros to the number given in the report. When more than one layer is present, the layers are reported in ascending order. Automated stations can only report a maximum of three layers at 12,000 feet AGL and below. Human observers can report up to six layers of clouds at any altitude. In this example, a broken layer is at 800 feet AGL, and an overcast layer is at 1,200 feet AGL.

If the MSL height of the top of a cloud layer is known, you can easily determine the layer's thickness by adding the airport elevation (MSL) to the height of the cloud base (AGL), and then subtracting this number from the height of the cloud tops.

METAR KMDW 121856Z AUTO 32005KT
1 1/2SM +RA BR OVC007 17/16 A2980

For example, in this METAR, the overcast begins at 700 feet AGL. If the field elevation is 620 feet MSL, and the tops of the overcast layer are reported at 6,500 feet MSL, then the cloud layer is 5,180 feet thick.

620 feet (MSL field elevation)
+700 feet (AGL height of cloud base)
1,320 feet (MSL height of cloud base)

6,500 feet (MSL top of overcast)
-1,320 feet (MSL height of cloud base)
5,180 feet (thickness of cloud layer)

In a manual report, a cloud type may be included if towering cumulus clouds (TCU) or cumulonimbus clouds (CB) are present. The code follows the height of their reported base. In the example, the base of the reported cumulonimbus clouds was at 1,200 feet AGL.

When more than half of the sky is covered by clouds, a ceiling exists. By definition, a **ceiling** is the AGL height of the lowest layer of clouds that is reported as broken or overcast, or the vertical visibility into an obscuration, such as fog or haze. Human observers may rely simply on their experience and knowledge of cloud formations to determine ceiling heights, or they may combine their experience with the help of reports from pilots, balloons, or other instruments. You can use the ceiling, combined with the visibility, to determine if IFR flight conditions exist. For example, if you are planning an IFR flight with a VFR departure from a controlled airport (without a special VFR clearance), you must ensure that the ceiling is at least 1,000 feet and there is at least 3 miles visibility.

A ceiling is defined as the height of the lowest layer of clouds or obscuring phenomena aloft that is reported as broken or overcast.

When conditions are below VFR, you need to check the published IFR takeoff, landing, and alternate minimums. If, for example, the takeoff minimums are 600-2, the ceiling must be at least 600 feet AGL and the visibility must be at least two statute miles. A reported sky condition of SCT002 BKN006 with a visibility of 3 statute miles is above minimums, since the lowest ceiling is 600 feet and the visibility is greater than two miles. The lower scattered layer at 200 feet does not constitute a ceiling and is, therefore, not restrictive to an IFR departure.

Unlike clouds, an obscuration does not have a definite base. An obscuration can be caused by phenomena such as fog, haze, or smoke which extend from the surface to an indeterminable height. In these instances, a total obscuration is shown with a VV followed by three digits indicating the vertical visibility in hundreds of feet. For example, VV006 describes an indefinite ceiling at 600 feet AGL. Obscurations which do not cover the entire sky may be reported in the remarks section of the METAR.

In a METAR observation VV008 indicates that the sky is obscured with a vertical visibility of 800 feet.

TEMPERATURE AND DEWPOINT

...BKN008 OVC012CB **18/17** A2970 RMK PRESFR SLP058 T01780172...

The observed air temperature and dewpoint (in degrees Celsius) are listed immediately following the sky condition. In the example, the temperature and dewpoint are reported as 18°C and 17°C, respectively. Temperatures below 0° Celsius are prefixed with an M to indicate minus. For instance 10° below zero would be shown as M10. Temperature and dewpoint readings to the nearest 1/10°C may be included in the remarks section of the METAR.

ALTIMETER

...BKN008 OVC012CB 18/17 **A2970** RMK PRESFR SLP058 T01780172...

The altimeter setting is reported in inches of mercury in a four digit group prefaced by an A. In this example, the altimeter setting was reported as 29.70 in. Hg.

REMARKS

...**RMK PRESFR SLP058 T01780172**

Certain remarks are included to report weather considered significant to aircraft operations. The types of information that may be included are wind data, variable visibility, beginning and ending times of a particular weather phenomenon, pressure information, and precise temperature/dewpoint readings. The start of the remarks section is identified by the code, RMK. The beginning of an event is shown by a B, followed by the time in minutes after the hour. If the event ended before the observation, the ending time in minutes past the hour is noted by an E. In this example, PRESFR indicates the barometric pressure is falling rapidly. Additionally, the sea level pressure (SLP) was 1005.8 millibars, or hectoPascals (hPa). A hectopascal is the metric equivalent of a millibar (1 mb = 1 hPa). Finally, the precise temperature/dewpoint (T) reading is 17.8°C and 17.2°C respectively. Examples of other coded remarks are shown in figure 9-53.

CODED DATA	EXPLANATIONS
A02	Automated station with precipitation discriminator
PK WND 20032/25	Peak wind from 200° at 32 knots, 25 minutes past the hour
VIS 3/4V1 1/2	Prevailing visibility variable 3/4 to 1 and 1/2 miles
FRQ LTG NE	Frequent lightning to the northeast
FZDZB45	Freezing drizzle began at 45 minutes past the hour
RAE42SNB42	Rain ended and snow began at 42 minutes past the hour
PRESFR	Pressure falling rapidly
SLP045	Sea level pressure in millibars (hPa), 1004.5 mb (hPa)
WSHFT 30 FROPA	Wind shift occurred at 30 minutes past the hour due to frontal passage
T00081016	Temperature/dewpoint in tenths °C, .8 °C/–1.6 °C (Since the first digit after the T is a 0, it indicates that the temperature is positive; the dewpoint in this example is negative since the fifth digit is a 1.)

Figure 9-53. Examples of coded remarks are shown in the left column, with the corresponding explanations on the right.

The beginning of the remarks section is indicated by the code RMK. The remarks section reports weather considered significant to aircraft operations, which are not covered in the previous sections of the METAR. See figure 9-53.

As shown in figure 9-54, the abbreviation MT is used to denote maximum tops of the precipitation in the clouds. Heights are reported in hundreds of feet MSL followed by the radial and distance in nautical miles from the reporting loca-

RADAR WEATHER REPORTS

General areas of precipitation, especially thunderstorms, are observed by radar on a routine basis. Most radar stations issue routine **radar weather reports (SDs)** thirty-five minutes past each hour, with intervening special reports as required.

A radar weather report not only defines the areas of precipitation, but also provides information on the type, intensity, and intensity trend. In addition, these reports normally include movement (direction and speed) of the precipitation areas, as well as maximum height of the precipitation. If the base is considered significant, it may also be reported. All heights in an SD are reported in hundreds of feet MSL. [Figure 9-54]

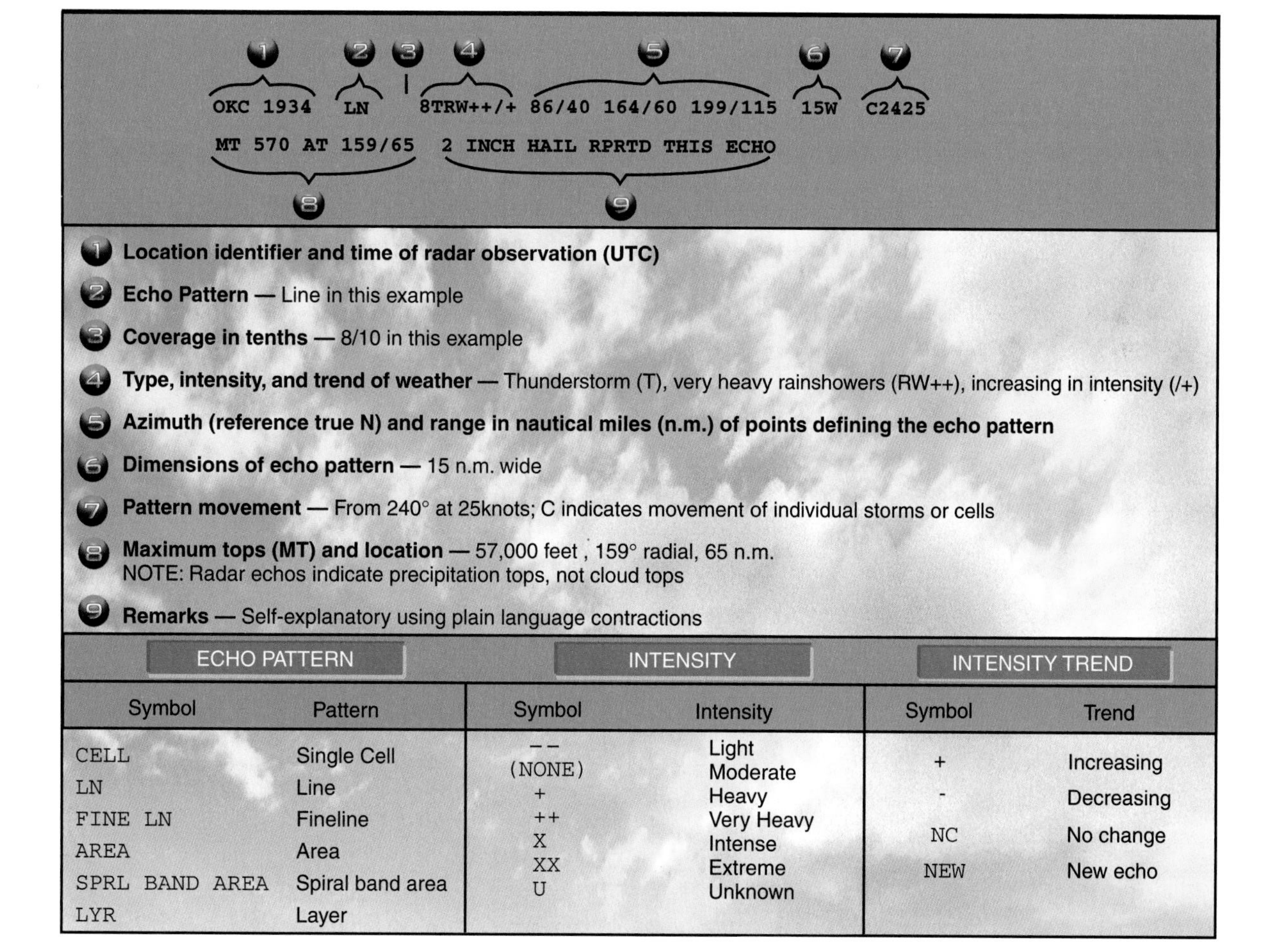

ECHO PATTERN		INTENSITY		INTENSITY TREND	
Symbol	Pattern	Symbol	Intensity	Symbol	Trend
CELL	Single Cell	--	Light	+	Increasing
LN	Line	(NONE)	Moderate	-	Decreasing
FINE LN	Fineline	+	Heavy	NC	No change
AREA	Area	++	Very Heavy	NEW	New echo
SPRL BAND AREA	Spiral band area	X	Intense		
LYR	Layer	XX	Extreme		
		U	Unknown		

Figure 9-54. Radar weather reports are useful in determining areas of severe weather. When the clouds extend to high levels, thunderstorms and the associated hazards are likely.

PILOT WEATHER REPORTS

Pilot weather reports (PIREPs) are often your best source to confirm such information as the bases and tops of cloud layers, in-flight visibility, icing conditions, wind shear, and turbulence. When significant conditions are reported or forecast, ATC facilities are required to solicit PIREPs. However, anytime you encounter unexpected weather conditions, you are encouraged to make a pilot report. If you make a PIREP, the ATC facility or the FSS can add your report to the distribution system so it can be used to brief other pilots or provide in-flight advisories. [Figure 9-55]

PIREPs are the best source for current weather, such as icing and turbulence, between reporting stations.

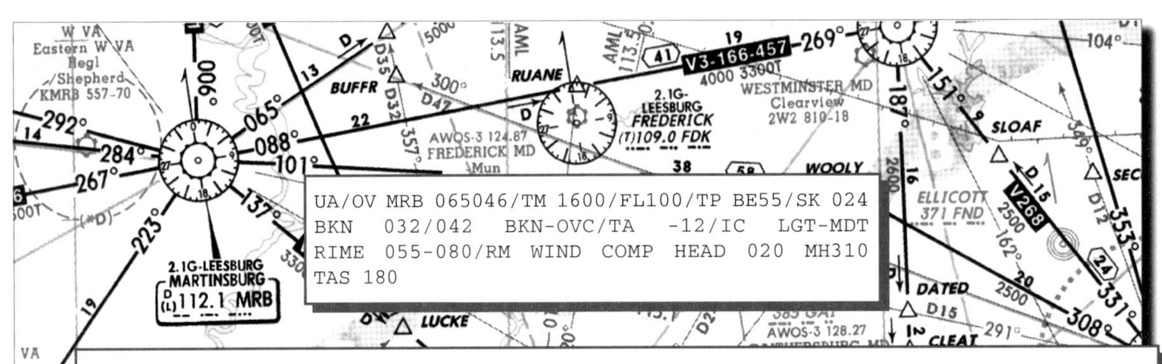

This PIREP decodes as follows: Pilot report, 46 n.m. on the 065° radial from Martinsburg at 1600UTC at 10,000 feet. Type of aircraft is a Beechcraft Baron. First cloud layer has a base at 2,400 feet broken with tops at 3,200 feet. The second cloud layer has a base at 4,200 broken occasionally overcast with no tops reported. Outside air temperature is -12 degrees Celsius. Light to moderate rime icing is reported between 5,500 and 8,000 feet. The headwind component is 20 knots. Magnetic heading is 310 degrees and the true airspeed is 180 knots.

PIREP FORM		
Pilot Weather Report		
3-Letter SA Identifier — — —	1. UA Routine Report	UUA Urgent Report
2. /OV	Location: In relation to a NAVAID	
3. /TM	Time: Coordinated Universal Time	
4. /FL	Altitude/Flight Level: Essential for turbulence and icing reports	
5. /TP	Aircraft Type: Essential for turbulence and icing reports	
Items 1 through 5 are mandatory for all PIREPs		
6. /SK	Sky Cover: Cloud height and coverage (scattered, broken, or overcast)	
7. /WX	Flight Visibility and Weather: Flight visibility, precipitation, restrictions to visibility, etc.	
8. /TA	Temperature (Celsius): Essential for icing reports	
9. /WV	Wind: Direction in degrees and speed in knots	
10. /TB	Turbulence: Turbulence intensity, whether the turbulence occurred in or near clouds, and duration of turbulence	
11. /IC	Icing: Intensity and Type	
12. /RM	Remarks: For reporting elements not included or to clarify previously reported items	

Figure 9-55. PIREPs are made up of several elements. Notice that MSL altitudes are used. Although PIREPs should be complete and concise, you should not be overly concerned with strict format or terminology. The important thing is to forward the report so other pilots can benefit from your observation. Further information relating to the coding and interpretation of PIREPs is contained in the *Aeronautical Information Manual.*

PIREPs use a standard format, as shown in figure 9-55. Note that altitudes are given in hundreds of feet above mean sea level (MSL).

Another type of PIREP is called an AIREP or air report. Disseminated electronically, these reports are used almost exclusively by commercial airlines. However, when accessing the internet for current weather you may come upon the abbreviation ARP followed by a weather report which was captured from an Aeronautical Radio Incorporated (ARINC) communications addressing and reporting system (ACARS) transmission.

Workload Management Through Cockpit Automation

As early as the 1970s there has been an effort by the world airlines to reduce the workload imposed on flight deck crews. ACARS is one of the answers. This technology provides bi-directional, non-verbal communications from the cockpit of an airborne aircraft to a dispatcher on the ground. Not only can the crew send and receive discrete company messages, but the system also provides many automated functions timed to support the crew in the different phases of flight. The automatic functions include weather updates, weight and balance information, runway data, and enroute engine performance. ACARS also sends block times, wheels-off, and wheels-on times, as well as present position automatically so the airline dispatcher knows the location of the aircraft at all times.

A manual input feature allows the crew to request weather reports and landing data. When there is a problem, maintenance information can be transmitted directly to the cockpit. When using the ACARS print capability, the crew can save time and reduce the possibility of copy errors from voice communications. By reducing the communications workload, pilots can more fully concentrate on their jobs.

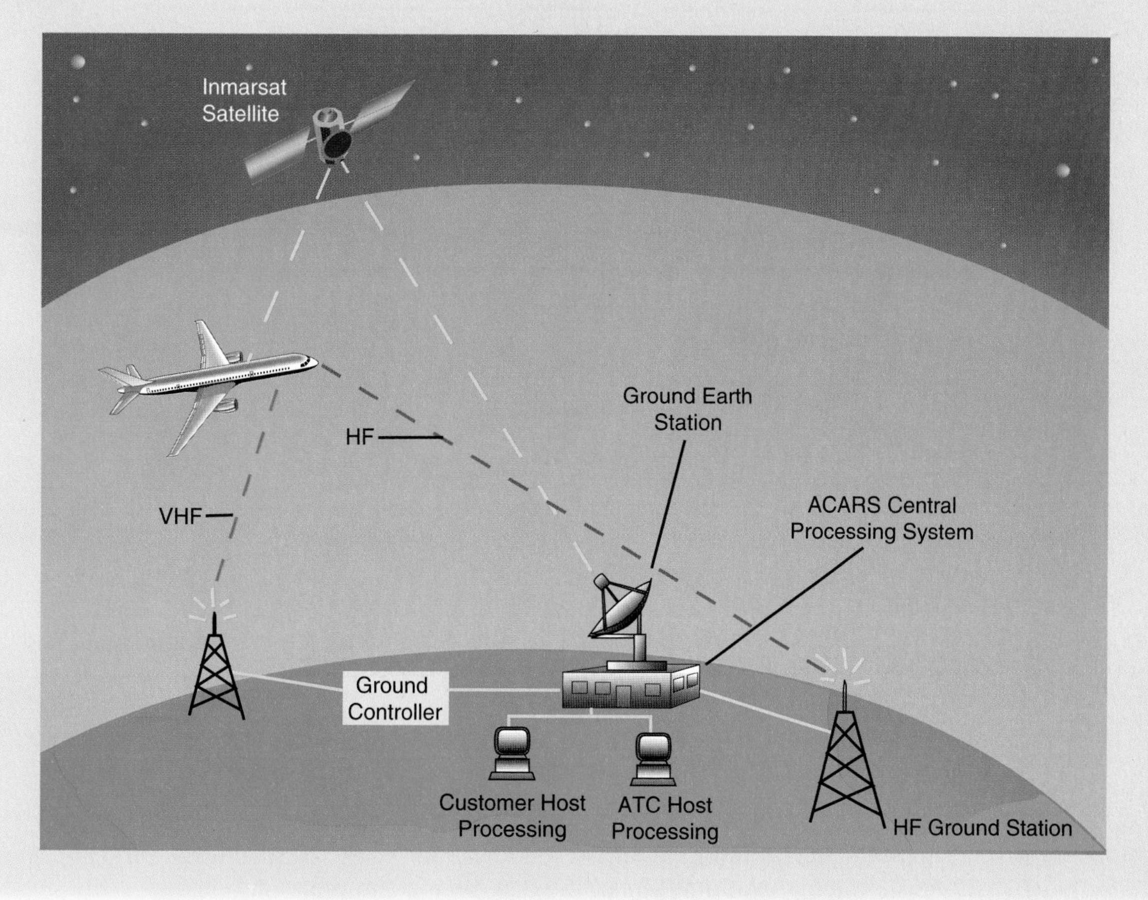

PRINTED WEATHER FORECASTS

Many of the reports of observed weather conditions are used to develop forecasts of future conditions. Every day, Weather Forecast Offices (WFOs) prepare over 2,000 forecasts for specific airports, over 900 route forecasts, and a variety of other forecasts for flight planning purposes. The printed forecasts with which you need to become familiar include the terminal aerodrome forecast, aviation area forecast, and the winds and temperatures aloft forecast.

TERMINAL AERODROME FORECAST

One of your best sources for an estimate of what the weather will be in the future at a specific airport is the terminal aerodrome forecast (TAF). TAFs normally are valid for a 24-hour period and scheduled four times a day at 0000Z, 0600Z, 1200Z, and 1800Z. Each TAF contains these elements: type, ICAO station identifier, issuance date and time, valid period, and the forecast. [Figure 9-56]

A terminal aerodrome forecast (TAF) is valid for a 24 hour period and is issued four times a day. A TAF should be your primary source of weather information for your destination.

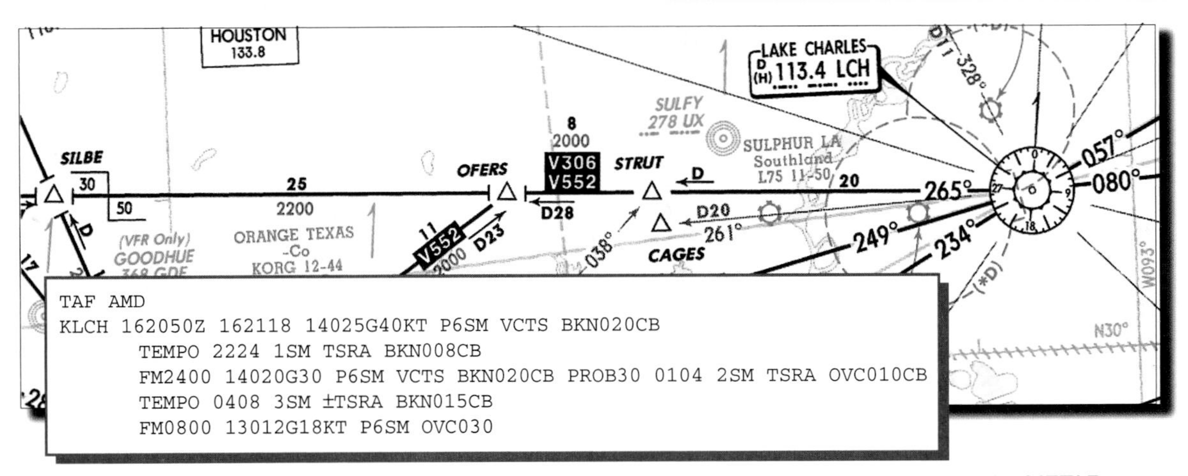


```
TAF AMD
KLCH 162050Z 162118 14025G40KT P6SM VCTS BKN020CB
        TEMPO 2224 1SM TSRA BKN008CB
        FM2400 14020G30 P6SM VCTS BKN020CB PROB30 0104 2SM TSRA OVC010CB
        TEMPO 0408 3SM ±TSRA BKN015CB
        FM0800 13012G18KT P6SM OVC030
```

Figure 9-56. With a few exceptions, the codes used in the TAF are similar to those used in the METAR.

TYPE OF FORECAST

TAF AMD

KLCH...

The TAF normally is a routine forecast. However, an amended TAF (TAF AMD) may be issued when the current TAF no longer represents the expected weather. TAF or TAF AMD appears in a header line prior to the text of the forecast. The abbreviations COR and RTD indicate a corrected and delayed TAF, respectively. The example depicts an amended TAF.

STATION IDENTIFIER AND ISSUANCE DATE/TIME

TAF AMD

KLCH 162050Z 162118...

The four letter ICAO location identifier code is the same as that used for the METAR/SPECI. The first two numbers of the date/time group represent the day of the month, and the next four digits are the Zulu time that the forecast was issued. The example TAF is a forecast for Lake Charles Regional Airport (KLCH) which was issued on the 16th day of the month at 2050Z.

VALID PERIOD

...162050Z **162118** 14025G40KT...

Normally, the forecast is valid for 24 hours. The first two digits represent the valid date. Next is the beginning hour of the valid time in Zulu, and the last two digits are the ending hour. The forecast for Lake Charles Regional Airport is valid from 2100Z on the 16th of the month to 1800Z the next day. Since the TAF in this example was amended, the valid period is less than 24 hours. At an airport that is open part time, the TAFs issued for that location will have the abbreviated statement NIL AMD NOT SKED AFT (closing time) added to the end of the forecast text. For TAFs issued when these airports are closed, the word NIL appears in place of the forecast text.

FORECAST

...162050Z 162118 **14025G40KT P6SM VCTS BKN020CB...**

The body of the TAF contains codes for forecast wind, visibility, weather, and sky condition. Weather, including obstructions to visibility, is added to the forecast when it is significant to aviation.

The forecast wind is depicted by a five digit group with the first three digits representing the wind direction and the last two digits the wind speed. If the wind speed is 100 knots or more, there will be three digits. The contraction KT follows to denote that the wind speed is in knots. Wind gusts are noted by the letter G appended to the wind speed followed by the highest expected gust. A variable wind direction is noted by VRB where the three digit direction usually appears. A calm wind (3 knots or less) is forecast as 00000KT. In this example, the winds are predicted to be from 140° at 25 knots with gusts to 40 knots.

In a TAF, the contraction VRB indicates that the wind direction is variable. A calm wind (3 knots or less) is indicated by 00000KT.

The next area is the expected prevailing visibility in miles, including fractions of miles. The SM indicates the measurement is in statute miles. Expected visibilities greater than 6 miles are forecast as a P6SM, which means plus 6 statute miles.

P6SM in terminal aerodrome forecast implies that the prevailing visibility is expected to be greater than 6 statute miles.

The weather phenomena and sky conditions are given next using the same format, qualifiers, and contractions as the METAR reports. If no significant weather is expected to occur during a specific time period in the forecast, the weather group is omitted for that time period. CLR is never used in the TAF. In this example, thunderstorms are forecast to be in the vicinity (VC) and a broken cloud layer, made up of cumulonimbus clouds, is forecast at 2,000 feet AGL.

The letters SKC are used in a terminal aerodrome forecast to indicate a clear sky.

Low-level wind shear which is not associated with convective activity may be included using the code WS followed by a three digit height (up to and including 2,000 feet AGL), a forward slash (/) and the winds at the height indicated. For example, WS010/18040KT indicates low-level wind shear at 1,000 feet AGL, wind 180° at 40 knots.

The letters WS indicates that low-level wind shear which is not associated with convective activity may be present during the valid time of the forecast. WS005/27050KT indicates that the wind at 500 feet AGL is 270° at 50 knots.

FORECAST CHANGE GROUPS

When a significant permanent change to the weather conditions is expected during the valid time, a change group is used. The changes can be temporary, rapid, or gradual. Each change indicator marks a time group within the TAF.

TEMPORARY FORECAST

...TEMPO 2224 1SM TSRA BKN008CB

FM2400 14020G30 P6SM VCTS BKN020CB PROB30 0104 2SM TSRA OVC010CB

TEMPO 0408 3SM -TSRA BKN015CB

FM0800 13012G18KT P6SM OVC030

Wind, visibility, weather, or sky conditions that are expected to last less than an hour are described in a temporary (TEMPO) group, followed by beginning and ending times. The first temporary group in the example predicts that between 2200Z and 2400Z, visibility is expected to be reduced to 1 statute mile in moderate rain associated with a thunderstorm. A broken layer of cumulonimbus clouds with bases at 800 feet is also predicted to occur. It's important to remember that these conditions only modify the previous forecast and are expected to last for periods of less than one hour. The second temporary group predicts that between 0400Z and 0800Z, the visibility is likely to be 3 statute miles in light rain associated with a thunderstorm. The ceiling is expected to be a broken layer of cumulonimbus clouds with the bases at 1,500 feet.

FROM OR BECOMING FORECAST

...TEMPO 2224 1SM TSRA BKN008CB

FM2400 14020G30 P6SM VCTS BKN020CB PROB30 0104 2SM TSRA OVC010CB

TEMPO 0408 3SM -TSRA BKN015CB

FM0800 13012G18KT P6SM OVC030

If a rapid change, usually within one hour, is expected, the code for from (FM) is used with the time of change. The conditions listed following the FM will continue until the next change group or the end of the valid time of the TAF. In the example, the first forecast change group indicates from 2400Z, the wind will be from 140° at 20 knots with gusts to 30 knots, the visibility will be greater than 6 statute miles, with thunderstorms in the vicinity and a cloud layer, made up of cumulonimbus clouds, is expected to be broken at 2,000 feet AGL. From 0800Z, the wind is forecast to be from 130° at 12 knots gusting to 18 knots, visibility will improve to greater than 6 statute miles and the sky will be overcast at 3,000 feet AGL.

A more gradual change in the weather, taking about two hours, is coded as BECMG, followed by beginning and ending times of the change period. The gradual change is expected to occur at an unspecified time within this time period. All items, except for the changing conditions shown in the BECMG group, are carried over from the previous time group. For instance, BECMG 2310 24007KT P6SM NSW indicates a gradual change in conditions is expected to occur between 2300Z and 1000Z. Sometime during this period, the wind will be from 240° at 7 knots, visibility is predicted to improve to greater than 6 miles with no significant weather (NSW). The NSW code is used only after a time period in which significant weather was forecast. NSW only appears in BECMG or TEMPO groups.

PROBABILITY FORECAST

...FM2400 14020G30 P6SM VCTS BKN020CB **PROB30 0104 2SM TSRA OVC010CB** TEMPO 0408 3SM -TSRA BKN0015CB...

TAFs may include the probability of thunderstorms or precipitation events with the associated wind, visibility, and sky conditions. A PROB group is used when the probability of occurrence is between 30 and 49%. The percentage is followed by the beginning and ending time of the period during which the thunderstorm or precipitation is expected. In the example, there is a 30% chance of a thunderstorm with moderate rain and 2 statute miles visibility between 0100Z and 0400Z. In addition, a 30% probability of an overcast layer of cumulonimbus clouds with bases at 1,000 feet AGL also is expected to occur during the four hour time period.

The term PROB40 2102 +TSRA in a terminal aerodrome forecast indicates that there is approximately a 40% probability of thunderstorms with heavy rain between 2100Z and 0200Z.

AVIATION AREA FORECAST

An aviation area forecast (FA) covers general weather conditions over several states or a known geographical area and is a good source of information for enroute weather. It also helps you determine the conditions at airports which do not have Terminal Aerodrome Forecasts. FAs are issued three times a day in the 48 contiguous states, and amended as required. NWS offices issue FAs for Hawaii and Alaska; however, the Alaska FA uses a different format. An additional specialized FA may be issued for the Gulf of Mexico by the National Hurricane Center in Miami, Florida. [Figure 9-57]

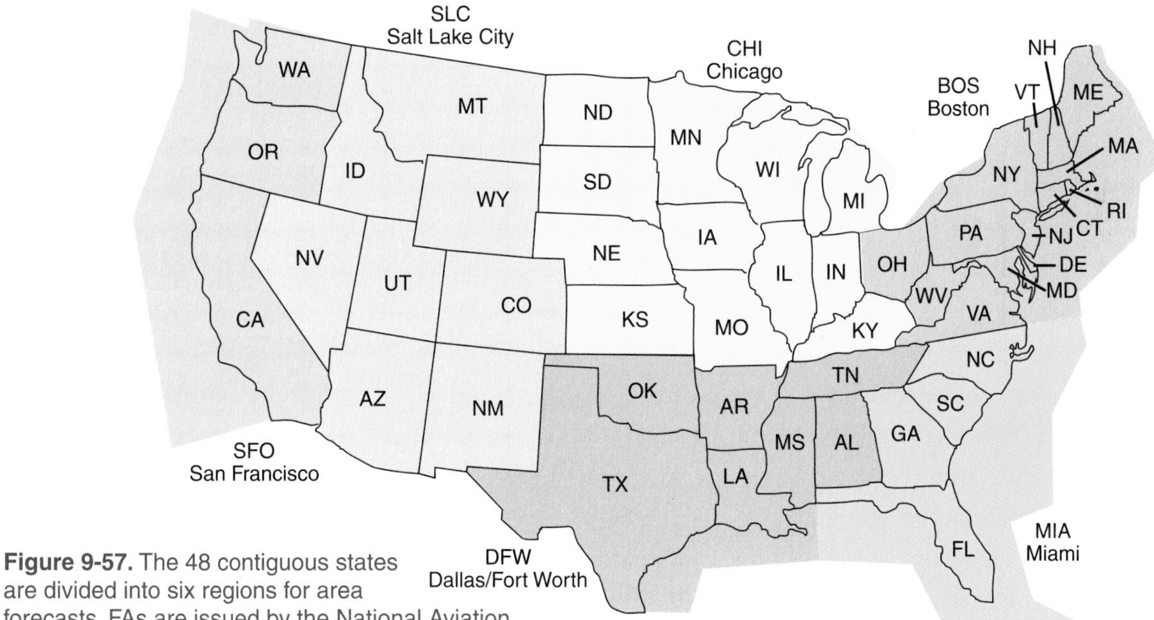

Figure 9-57. The 48 contiguous states are divided into six regions for area forecasts. FAs are issued by the National Aviation Weather Advisory Unit in Kansas City, Missouri. A specialized Gulf of Mexico FA is issued by the National Hurricane Center in Miami, Florida.

The FA consists of several sections: a communications and product header section, a precautionary statement section, and two weather sections (synopsis and VFR clouds and weather). Each area forecast covers an 18-hour period.

Aviation area forecasts are issued three times each day and generally include a total forecast period of 18 hours. They cover a geographical group of states or well known areas.

COMMUNICATIONS AND PRODUCT HEADERS

In the heading SLCC FA 141045, the SLC identifies the Salt Lake City forecast area, C indicates the product contains a clouds and weather forecast, FA means area forecast, and 141045 tells you this forecast was issued on the 14th day of the month at 1045Z. Since these forecasts are rounded to the nearest full hour, the valid time for the report begins at 1100Z. The synopsis is valid until 18 hours later, which is shown as the 15th at 0500Z. The clouds and weather section forecast is valid for a 12-hour period, until 2300Z on the 14th. The outlook portion is valid for six hours following the forecast, from 2300Z on the 14th to 0500 on the 15th. The last line of the header lists the states that are included in the Salt Lake City forecast area. [Figure 9-58]

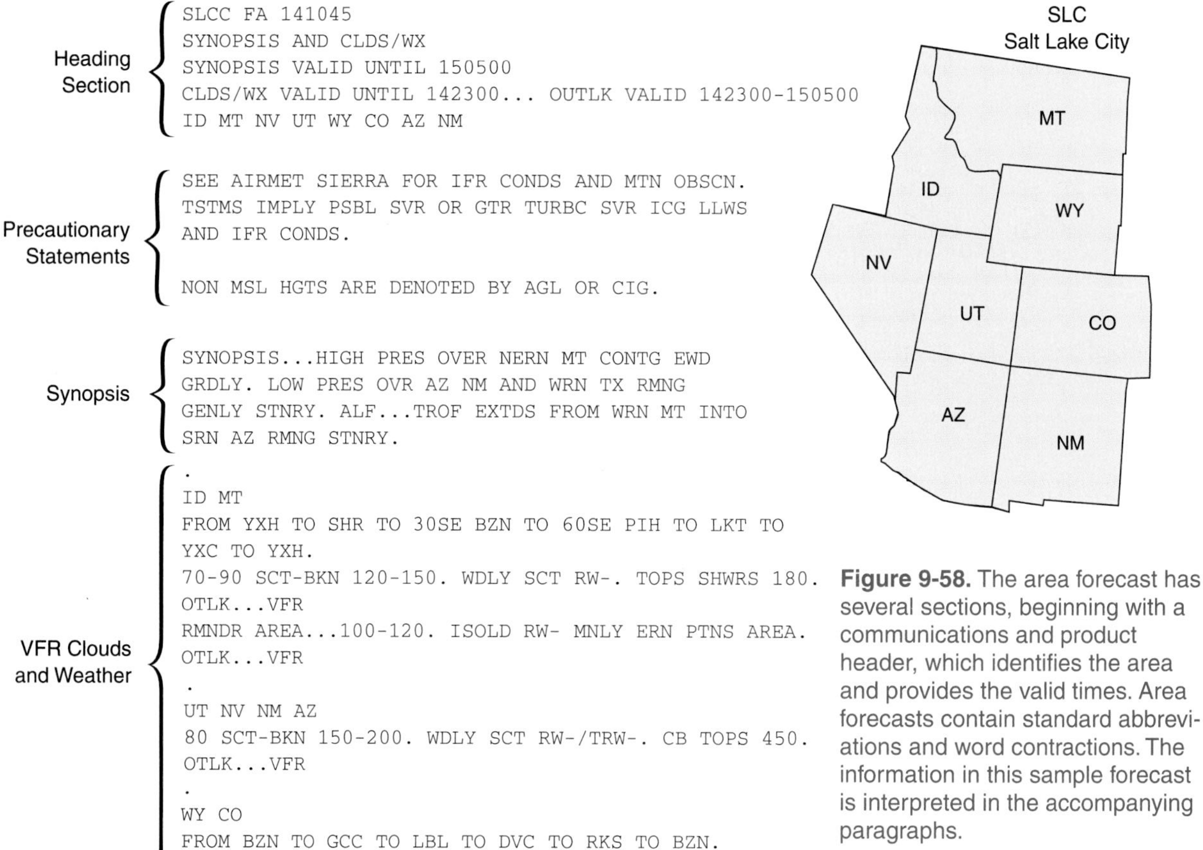

Figure 9-58. The area forecast has several sections, beginning with a communications and product header, which identifies the area and provides the valid times. Area forecasts contain standard abbreviations and word contractions. The information in this sample forecast is interpreted in the accompanying paragraphs.

Amendments to FAs are issued whenever the weather significantly improves or deteriorates based on the judgment of the forecaster. An amended FA is identified by the contraction AMD in the header along with the time of the amended forecast. When an FA is corrected, the contraction COR appears in the heading, along with the time of the correction.

PRECAUTIONARY STATEMENTS

Following the headers are three precautionary statements which are part of all FAs. The first statement alerts you to check the latest AIRMET Sierra, which describes areas of mountain obscuration which may be forecast for the area. The next statement is a reminder that thunderstorms imply possible severe or greater turbulence, severe icing, low-level wind shear, and instrument conditions. Therefore, when thunderstorms are forecast, these hazards are not included in the body of the FA. The third statement points out that heights which are not MSL are noted by the letters AGL (above ground level) or CIG (ceiling). All heights are expressed in hundreds of feet.

SYNOPSIS

The synopsis is a brief description of the location and movement of fronts, pressure systems, and circulation patterns in the FA area over an 18-hour period. When appropriate, forecasters may use terms describing ceilings and visibility, strong winds, or other phenomena. In the example, high pressure over northeastern Montana will continue moving gradually eastward. A low pressure system over Arizona, New Mexico, and western Texas will remain generally stationary. Aloft (ALF), a trough of low pressure extending from western Montana into southern Arizona is expected to remain stationary.

VFR CLOUDS AND WEATHER

The VFR clouds and weather portion is usually several paragraphs long, and broken down by states or geographical regions. It describes clouds and weather which could affect VFR operations over an area of 3,000 square miles or more. The forecast is valid for 12 hours, and is followed by a 6-hour categorical outlook (18 hours in Alaska).

The VFR clouds and weather section of an aviation area forecast summarizes sky conditions, cloud heights, visibility, obstructions to vision, precipitation, and sustained surface winds of 20 knots or greater.

When the surface visibility is expected to be six statute miles or less, the visibility and obstructions to vision are included in the forecast. When precipitation, thunderstorms, and sustained winds of 20 knots or greater are forecast, they will be included in this section. The term OCNL (occasional) is used when there is a 50% or greater probability, but for less than 1/2 of the forecast period, of cloud or visibility conditions which could affect VFR flight. The area covered by showers or thunderstorms is indicated by the terms ISOLD (isolated, meaning single cells), WDLY SCT (widely scattered, less than 25% of the area), SCT or AREAS (25% to 54% of the area), and NMRS or WDSPRD (numerous or widespread, 55% or more of the area). In addition, the term ISOLD is sometimes used to describe areas of ceilings or visibility which are less than 3,000 square miles.

The outlook follows the main body of the forecast, and gives a general description of the expected weather, using the terms VFR, IFR, or MVFR (marginal VFR). A ceiling less than 1,000 feet and/or visibility less than 3 miles is considered IFR. Marginal VFR areas are those with ceilings from 1,000 to 3,000 feet and/or visibility between 3 and 5 miles. Abbreviations are used to describe causes of IFR or MVFR weather.

In the example shown in figure 9-58, the area of coverage in the specific forecast for Wyoming and Colorado is identified using three-letter designators. This area extends from Bozeman, Montana, to Gillette, Wyoming, to Liberal, Kansas, to Dove Creek, Wyoming, to Rock Springs, Wyoming, and back to Bozeman. As mentioned previously under the header, the valid time begins on the 14th day of the month at 1100Z for a 12-hour period. A broken to overcast cloud layer begins between 7,000 to 9,000 feet MSL, with tops extending to 20,000 feet. Since visibility and wind information is omitted, the visibility is expected to be greater than six statute miles, and the wind less than 20 knots. However, the visibility (VSBY) is forecast to be occasionally 3 miles in light rain and fog (3R-F). After 2000Z, widely scattered thunderstorms with light rain showers are expected, with cumulonimbus (CB) cloud tops to 45,000 feet. The 6-hour categorical outlook covers the period from 2300Z on the 14th to 0500 on the 15th. The forecast is for marginal VFR weather due to ceilings (CIG) and rain showers (RW).

When the wind is forecast to be 20 knots or greater the categorical outlook in the aviation area forecast includes the contraction WND.

WINDS AND TEMPERATURES ALOFT FORECAST

A **winds and temperatures aloft forecast (FD)** provides an estimate of wind direction in relation to true north, wind speed in knots, and the temperature in degrees Celsius for selected altitudes. Depending on the station elevation, winds and temperatures are usually forecast for nine levels between 3,000 and 39,000 feet. Information for two additional levels (45,000 foot and 53,000 foot) may be requested from a FSS briefer or NWS meteorologist, but are not included on an FD. [Figure 9-59]

In a winds and temperatures aloft forecast, winds are given in true direction and speed is shown in knots.

Heading Information
The heading includes the type of forecast, the day of the month, and the time of transmission.

Time
The second line tells you the forecast is based on observations at 1200Z and is valid at 1800Z on the 15th. It is intended for use between 1700Z and 2100Z on the same day.

```
FD KWBC 151640

BASED ON 151200Z DATA

VALID 151800Z FOR USE 1700-2100Z TEMPS NEG ABV 24000
```

FD	3000	6000	9000	12000	18000	24000	30000
ALA			2420	2635-08	2535-18	2444-30	245945
AMA		2714	2725+00	2625-04	2531-15	2542-27	265842
DEN			2321-04	2532-08	2434-19	2441-31	235347
HLC		1707-01	2113-03	2219-07	2330-17	2435-30	244145

Winds and Temperatures
Since temperatures above 24,000 feet are always negative, a note indicates that the minus sign is omitted for 30,000 feet and above. The column on the left lists the FD location identifiers. The columns to the right show forecast information for each level appropriate to that location.

Figure 9-59. This excerpt from an FD shows winds only to the 30,000-foot level. As stated on the report, all temperatures above 24,000 feet are negative.

It is important to note that temperatures are not forecast for the 3,000-foot level or for any level within 2,500 feet of the station elevation. Likewise wind groups are omitted when the level is within 1,500 feet of the station elevation. At Denver (DEN), for example, the forecast for the lower two levels is omitted, since the station elevation at Denver is over 5,000 feet.

A winds and temperatures aloft forecast (FD) does not include winds within 1,500 feet of the station elevation. Likewise temperatures for the 3,000-foot level or for a level within 2,500 feet of the station elevation are omitted.

The presentation of wind information in the body of the FD is similar to other reports and forecasts. The first two numbers indicate the true direction from which the wind is blowing. For example, 2635-08 indicates the wind is from 260° at 35 knots and the temperature is –8°C. Quite often you must interpolate between two levels. For instance, if you plan to fly near Hill City (HLC) at 7,500 feet, you must interpolate. Refer to figure 9-59, since your planned flight altitude is midway between 6,000 and 9,000 feet, a good estimate of the wind at 7,500 feet is 190° at 10 knots with a temperature of –2°C.

Wind direction and speed information on an FD are shown by a four-digit code. The first two digits are the wind direction in tens of degrees. Wind speed is shown by the second two digits. The last two digits indicate the temperature in degrees Celsius. All temperatures above 24,000 feet are negative and the minus sign is omitted.

Wind speeds between 100 and 199 knots are encoded so direction and speed can be represented by four digits. This is done by adding 50 to the two-digit wind direction and subtracting 100 from the velocity. For example, a wind of 270° at 101 knots is encoded as 7701 (27 + 50 = 77 for wind direction, and 101 – 100 = 01 for wind speed). A code of 9900 indicates light and variable winds (less than five knots). However, wind speeds of 200 knots or more are encoded as 199.

To decode a forecast of winds between 100 and 199 knots, subtract 50 from the two-digit direction code and multiply by ten. Then add 100 to the two-digit wind speed code. The code 9900 indicates the winds are light and variable.

SEVERE WEATHER REPORTS AND FORECASTS

While much weather gathering activity is concerned with routine reports and forecasts, considerable effort is also devoted to monitoring and reporting severe weather conditions. The National Hurricane Center in Miami, Florida issues hurricane advisories, while the National Severe Storms Forecast Center in Kansas City, Missouri issues special reports and forecasts for other severe weather conditions. These include convective outlooks, and severe weather watch bulletins, AIRMETs, SIGMETs, and convective SIGMETs. AIRMETs, SIGMETs, and convective SIGMETs are discussed in detail in Section E.

HURRICANE ADVISORY

When a hurricane is located at least 300 nautical miles offshore, but threatens a coast line, a **hurricane advisory (WH)** is issued. The WH gives the location of the storm center, its expected movement, and the maximum winds in and near the storm center. It does not contain specific ceilings, visibility, and weather hazards. As needed, those details will be reported in Area Forecasts, Terminal Aerodrome Forecasts, and in-flight advisories. [Figure 9-60]

Figure 9-60. Hurricane Bonnie, shown in this Space Shuttle photograph, never reached landfall. Even as a category 1 hurricane, Bonnie likely would have been reported in a WH and its effects would have been evident on the eastern seaboard.

CONVECTIVE OUTLOOK

The **convective outlook (AC)** forecasts general thunderstorm activity for the next 24-hour period. ACs describe areas in which there is a risk of severe thunderstorms. Severe thunderstorm criteria include winds equal to or greater than 50 knots at the surface or hail equal to or greater than 3/4 inch in diameter, or tornadoes. Convective outlooks are useful for planning flights within the forecast period.

A convective outlook describes prospects for an area coverage of both severe and general thunderstorms during the following 24-hour period.

SEVERE WEATHER WATCH BULLETIN

A severe weather watch bulletin (WW) defines areas of possible severe thunderstorms or tornadoes. WWs are issued on an unscheduled basis and are updated as required. Since severe weather forecasts and reports may affect the general public, as well as pilots, they are widely disseminated through all available media. When it becomes evident that no severe weather will develop or that storms have subsided, cancellation bulletins are issued.

Severe weather watch bulletins are issued only when required.

In order to alert forecasters and weather briefers that a severe weather watch bulletin is being issued, a preliminary message, called an alert severe weather watch (AWW) is sent. Each AWW is numbered sequentially beginning with the first of January each year. [Figure 9-61]

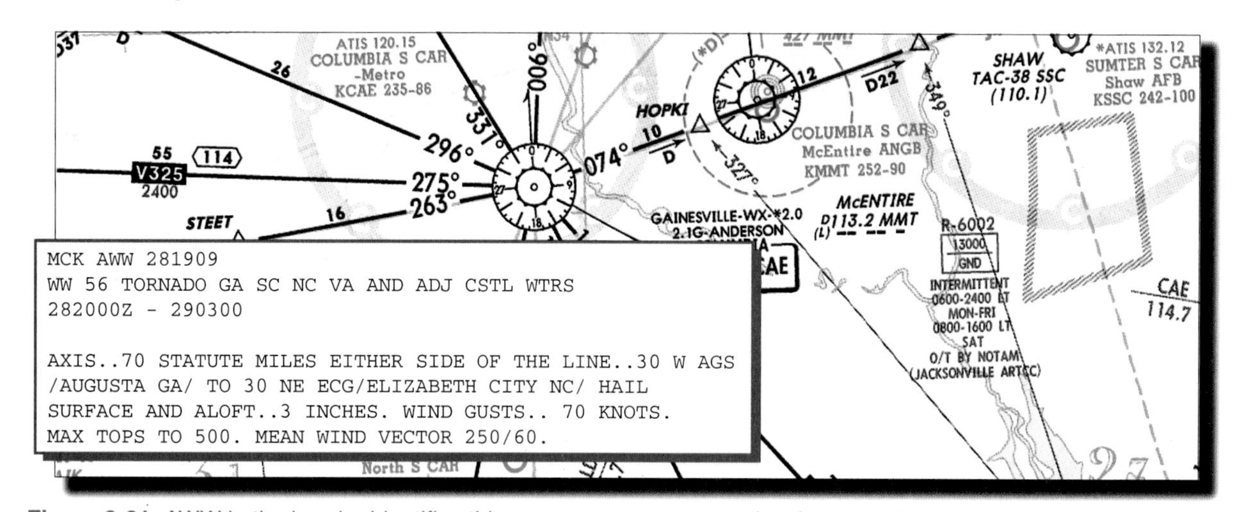

```
MCK AWW 281909
WW 56 TORNADO GA SC NC VA AND ADJ CSTL WTRS
282000Z - 290300

AXIS..70 STATUTE MILES EITHER SIDE OF THE LINE..30 W AGS
/AUGUSTA GA/ TO 30 NE ECG/ELIZABETH CITY NC/ HAIL
SURFACE AND ALOFT..3 INCHES. WIND GUSTS.. 70 KNOTS.
MAX TOPS TO 500. MEAN WIND VECTOR 250/60.
```


Figure 9-61. AWW in the header identifies this report as a severe weather forecast alert. This message warns of possible tornado activity, large hail, and winds in excess of 60 miles per hour. The watch area is defined by a line from a point 30 miles west of Augusta, Georgia to a point 30 miles northeast of Elizabeth City, North Carolina. A detailed severe weather watch bulletin (WW) immediately follows the alert message.

Tornado Myths

Many myths about tornadoes have been promoted as fact. One such misconception is that you should open the windows in your house to equalize the pressure difference as the storm approaches. In fact, your windows will probably get broken anyway because the winds in the walls of the vortex can reach 100 to 200 miles per hour. In addition, inside the tornado is a barrage of boards, stones, tree limbs, and anything else that the winds can move through the air.

Given the aerodynamics of a house, it is unlikely that open windows would allow the strong winds of a tornado to pass through the house and leave it undamaged. In reality the open windows allow the air to exert pressures on the structure from the inside, blowing it up like a balloon. The best thing for you to do is to leave the windows alone and move to the basement or other shelter as quickly as possible.

SUMMARY CHECKLIST

✓ An aviation routine weather report (METAR) is an observation of surface weather written in a standard format which typically contains 10 or more separate elements.

✓ A non-routine aviation weather report (SPECI) is issued when a significant change in one or more of the elements of a METAR has occurred.

✓ Prevailing visibility is the greatest distance an observer can see and identify objects through at least half of the horizon.

✓ Runway visual range (RVR) is based on what a pilot in a moving aircraft should see when looking down the runway. If included in a METAR, RVR is reported following prevailing visibility.

✓ A ceiling is the height above ground level of the lowest layer of clouds aloft which is reported as broken (BKN) or overcast (OVC), or the vertical visibility (VV) into an obscuration.

✓ Radar weather reports (SDs) define general areas of precipitation, particularly thunderstorms.

✓ The bases and tops of cloud layers, in-flight visibility, icing conditions, wind shear, and turbulence may be included in a pilot weather report (PIREP).

✓ Terminal aerodrome forecasts (TAFs) predict the weather at a specific airport for a 24-hour period of time.

✓ An aviation area forecast (FA) is a good source of information for weather at airports which do not have terminal aerodrome forecasts, as well as for enroute weather.

✓ An estimate of wind direction in relation to true north, wind speed in knots, and the temperature in degrees Celsius for selected altitudes can be found in the winds and temperatures aloft forecast (FD).

✓ A convective outlook (AC) forecasts general thunderstorm activity for the next 24-hour period.

✓ Areas of possible severe thunderstorms or tornadoes are defined by a severe weather watch bulletin (WW).

KEY TERMS

Aviation Routine Weather Report (METAR)

Non-Routine (Special) Aviation Weather Report (SPECI)

Prevailing Visibility

Runway Visual Range (RVR)

Ceiling

Radar Weather Report (SD)

Pilot Weather Report (PIREP)

Terminal Aerodrome Forecast (TAF)

Aviation Area Forecast (FA)

Winds And Temperatures Aloft Forecast (FD)

Hurricane Advisory (WH)

Convective Outlook (AC)

Severe Weather Watch Bulletin (WW)

Alert Severe Weather Watch (AWW)

QUESTIONS

Use the following METAR for Ponca City Municipal Airport (KPNC), to answer the questions 1 through 5.

METAR KPNC 161954Z 03015G27KT 2SM -RA BR BKN007 BKN017 OVC030 07/06 A2978 RMK PK WND 06030/51 SLP086 T00670061

1. What are the reported winds?

   A. 03° at 015 knots with gusts from 220° at 7 knots
   B. 300° at 15 knots with gusts at 27 knots
   C. 030° at 15 knots with gusts to 27 knots

2. What is the reported intensity of the rain in Ponca City, Oklahoma?

   A. Moderate
   B. Light
   C. Heavy

3. What is the height of the lowest ceiling at Ponca City Municipal?

   A. 700 feet
   B. 1,700 feet
   C. 3,000 feet

4. What is the actual temperature/dewpoint at Ponca City Municipal Airport?

5. What is the sea level pressure in millibars?

6. True/False. Altitudes given in PIREPs are in hundreds of feet MSL.

7. In a TAF what does the code TEMPO indicate?

Use the following TAF for Dallas Fort Worth International Airport (KDFW), to answer questions 8 through 11.

TAF
KDFW AMD 161849Z161918 14015G20KT P6SM VCTS BKN015CB OVC040
FM2100 17015G20KT P6SM VCTS BKN020CB OVC050
FM0200 20008KT P6SM BKN012 OVC030
TEMPO 0206 2SM -RA OVC008
FM0600 24012KT P6SM BKN012 OVC030 PROB30 0612 -RA
FM1400 24010KT P6SM SCT020 BKN060

8. What is the valid period for the KDFW TAF?

   A. 1900Z to 1800Z the following day
   B. 1600Z to 1900Z the following day
   C. 1600Z to 1800Z the following day

9. What weather conditions are forecast to exist between 0200Z and 0600Z?

10. What does the statement **PROB30 0612 -RA** indicate?

    A. There is a probability of light rain over 30% of the vicinity
    B. There is a 30% probability of light rain during the forecast
    C. There is a 30% probability of light rain occurring between 0600Z and 1200Z

11. True/False. After 1400Z the visibility at KDFW will be less than 6 statute miles.

Use the following area forecast excerpt to answer questions 12 through 15.

```
. . .
SLCC FA 191145
SYNOPSIS AND VFR CLDS/WX
SYNOPSIS VALID UNTIL 200600
CLDS/WX VALID UNTIL 200000 . . . OTLK VALID 200000-200600
ID      MT     NV     WY     CO     AZ     NM.
.
SEE AIRMET SIERRA FOR IFR CONDITIONS AND MTN OBSCN.
TSTMS IMPLY PSBL SVRE OR GTR TURBC SVR ICG LLWS
AND IFR CONDS.
NON MSL HTS ARE DENOTED BY AGL OR CIG.
.
ID      WY     MT
ERN WY 30 BKNV OVC. SCT SW-. OTLK . . . VFR.
RMDR AREA . . . 60-80 BKN. OCNL SW IN MTNS.
.
CO
WRN PTN 80-120 SCT. 18Z 60-100 BKN. VSBY 4-6 SW-.
OTLK . . . MVFR CIG SW. E OF DVD . . . 15-25 OVC WITH OCNL CIGS
BLO 20 OVC VSBYS BLO 5 R-S-. OTLK . . . MVFR CIG BCMG VFR BY 02Z.
```

12. What is the outlook for Wyoming?

    A. VFR
    B. IFR
    C. MVFR

13. True/False. In Colorado, the sky condition after 1800Z is forecast to be 6,000 feet to 10,000 feet broken.

14. What are the forecast ceilings east of the divide?

15. What is the outlook for Colorado by 0200Z?

    A. VFR
    B. IFR
    C. MVFR

16. In the winds and temperature aloft forecast at 30,000 feet, how is **751015** decoded?

SECTION D
GRAPHIC WEATHER PRODUCTS

Through the use of graphic weather products you can quickly assess large-scale weather patterns, identify trends, and predict the weather that might occur over time. The weather charts can help you identify areas of IFR weather at a glance and also see where IFR conditions are forecast within the next 12 to 24 hours. These graphic weather products use data gathered from ground observations, weather radar, satellites, and other sources, and provide valuable flight planning information for any flight whether VFR or IFR.

GRAPHIC REPORTS

Reports of observed weather are graphically depicted in a number of products. These include the surface analysis chart, weather depiction chart, radar summary chart, satellite weather pictures, and observed winds aloft chart.

SURFACE ANALYSIS CHART

The **surface analysis chart**, also referred to as a surface weather analysis chart, is a computer prepared graphic which covers the contiguous 48 states and adjacent areas. It shows weather conditions as of the valid time shown on the chart. [Figure 9-62] By reviewing this chart, you can get a picture of atmospheric pressure patterns at the earth's surface. These pressure patterns are depicted by solid lines called isobars. If you see isobars that are close together on the chart, the pressure gradient is higher than the surrounding area and the wind is usually stronger. Additionally, you also can see the locations of high and low pressure systems and associated fronts, as well as temperature and dewpoint, wind direction and speed, local weather, and obstruction to vision. This chart is transmitted every three hours.

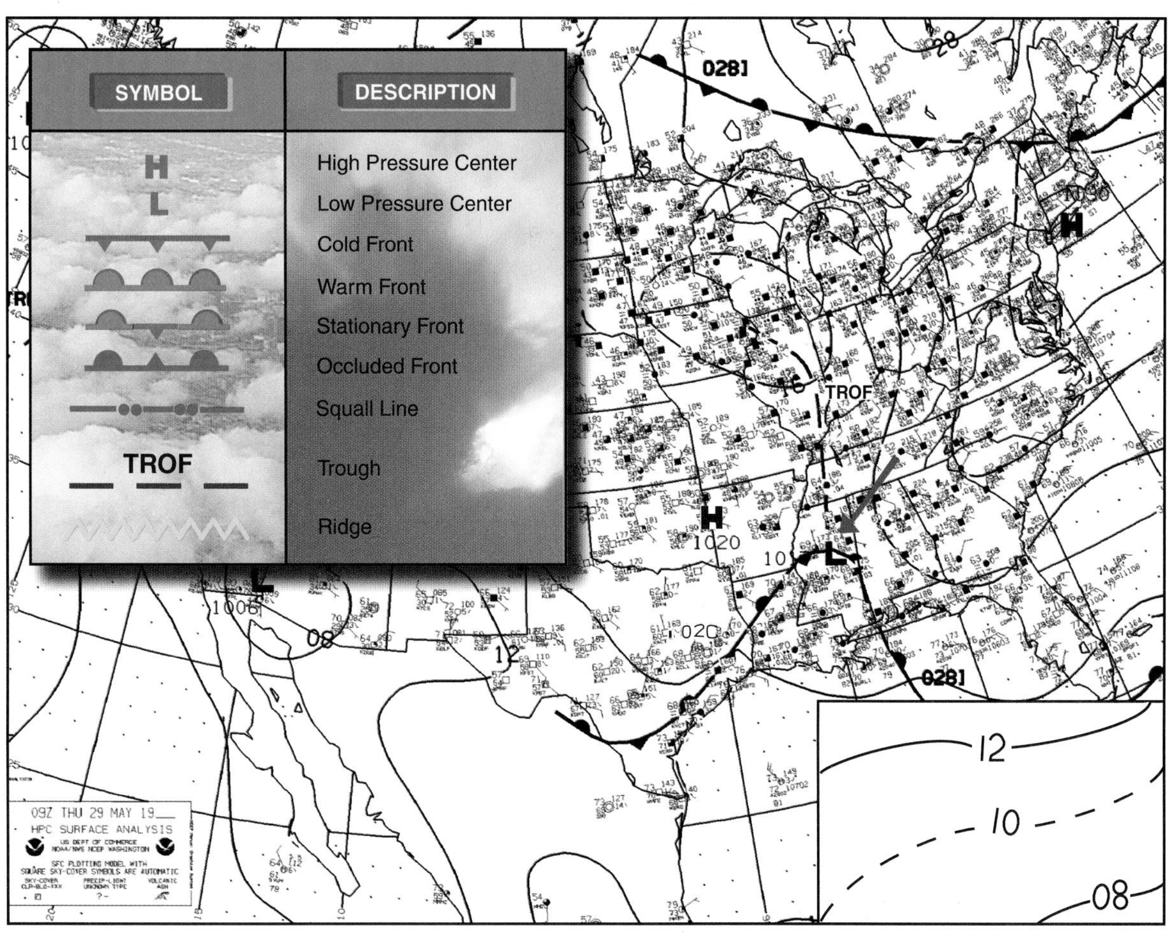

Figure 9-62. This excerpt shows a prominent low pressure center in the southeast portion of the United States. Isobars connecting points of equal pressure are usually drawn at four-millibar intervals. The inset on the right shows how they are numbered; the number 12 means the line connects points where the pressure is 1012 millibars. When the pressure gradient is weak, dashed isobars are sometimes inserted at two-millibar intervals to easily define the pressure patterns. Some common chart symbols are shown in the left inset. Weather service meteorologists sometimes add standard colors to make the chart easier to read.

When solid lines, called isobars, are close together on the chart, the pressure gradient is greater and wind velocities are stronger.

As shown in figure 9-62, a dashed line on a surface analysis chart indicates a weak pressure gradient.

A surface analysis chart is a good source for general weather information over a wide area, depicting the actual positions of fronts, pressure patterns, temperatures, dewpoint, wind, weather, and obstructions to vision at the valid time of the chart.

The surface analysis chart also provides surface weather observations for a large number of reporting points throughout the United States. Each of these points is illustrated by a **station model**. Round station symbols depict stations where observations are taken by human observers. Square station symbols indicate automated sites. Other models, which appear over water, display information gathered by ships, buoys, and offshore oil platforms. [Figure 9-63]

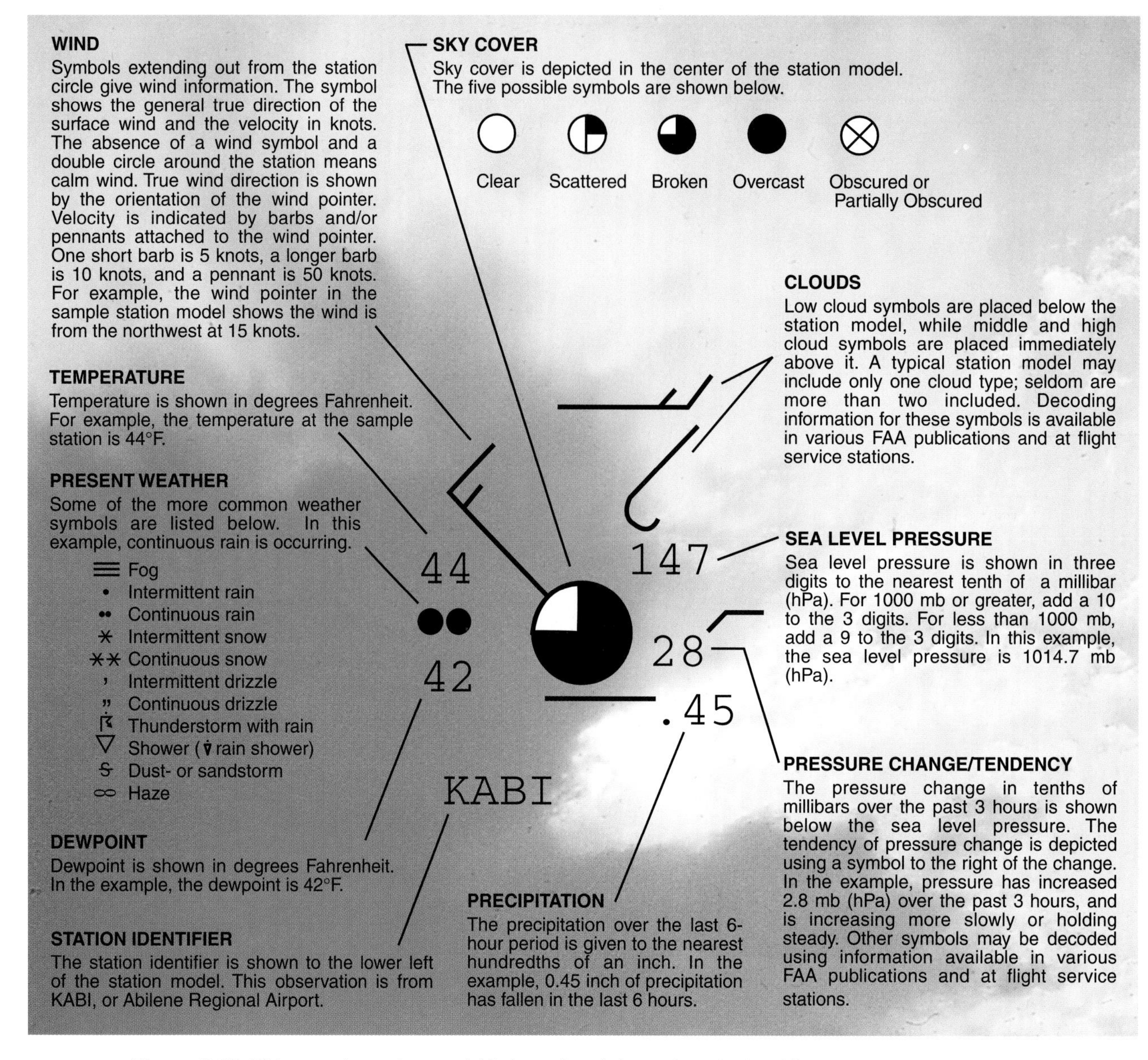

Figure 9-63. This sample station model is based on information obtained from a manual observation.

WEATHER DEPICTION CHART

The **weather depiction chart** provides an overview of favorable and adverse weather conditions for the chart time and is an excellent resource to help you determine general weather conditions during flight planning. Information plotted on this chart is derived from aviation routine weather reports (METARs). Like the surface chart, the weather depiction chart is prepared and transmitted by computer every three hours, and is valid at the time of the plotted data. Unlike the surface chart, pressure patterns and wind information are not provided; however, simplified station models are included. These station models are depicted as circles; a bracket symbol ( ]) to the right of the circle indicates an observation made by an automated station. [Figure 9-64]

A ( ]) plotted to the right of a station circle on the weather depiction chart means the station is an automated observation location.

As shown in figure 9-64, when total sky cover is FEW or scattered, the height shown on the weather depiction chart is the base of the lowest layer.

The weather depiction chart provides a graphic display of VFR and IFR weather, as well as the type of precipitation. See figure 9-64.

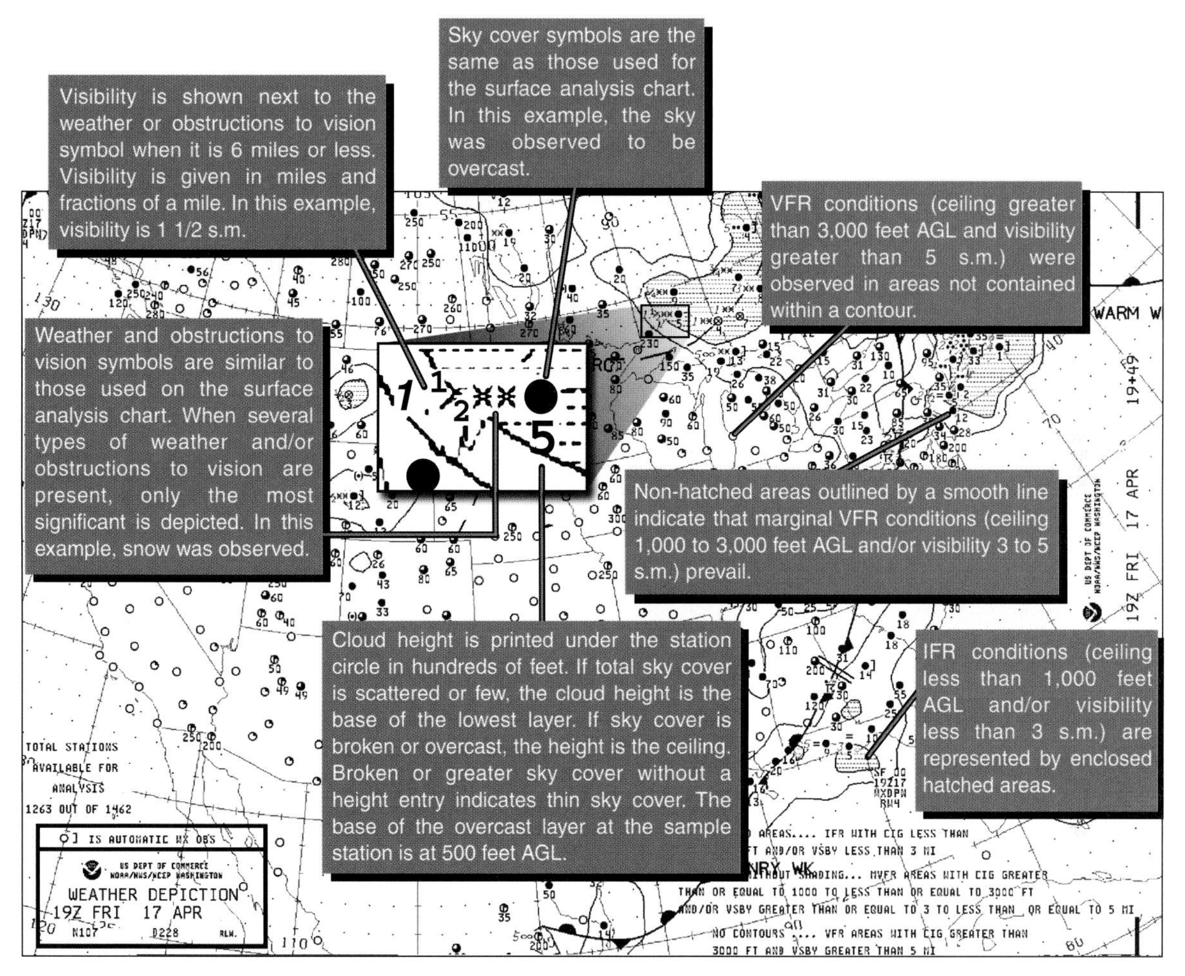

Figure 9-64. The weather depiction chart can give you a birds-eye view of areas of both VFR and IFR weather. However, the weather depiction chart may not represent the total enroute conditions because of variations in terrain and weather occurring between reporting stations.

RADAR SUMMARY CHART

The computer-generated radar summary chart graphically depicts a collection of radar weather reports (SDs). The information for both products is gathered by special weather radar systems. These systems transmit a pulse of radar energy from a rotating antenna in a circle around the radar site. When the signals encounter precipitation, they are reflected back to the antenna. The reflected signals, also called echoes, are then presented on a radar display which shows the strength (levels 1 through 6), and location of the radar return. The radar summary chart uses this data to depict the location, size, shape, and intensity of returns, as well as the intensity trend and direction of movement. On the chart, echo intensity is shown by contours. These contours identify three levels of intensity. The first contour represents levels one and two, weak to moderate returns (light to moderate precipitation); the second shows levels three and four, or strong to very strong returns (heavy to very heavy precipitation); the third contour outlines levels five and six, which represents intense and extreme returns (intense to extreme precipitation). In addition, the chart shows lines and cells of hazardous thunderstorms as well as echo heights of the tops and bases of precipitation areas. [Figure 9-65]

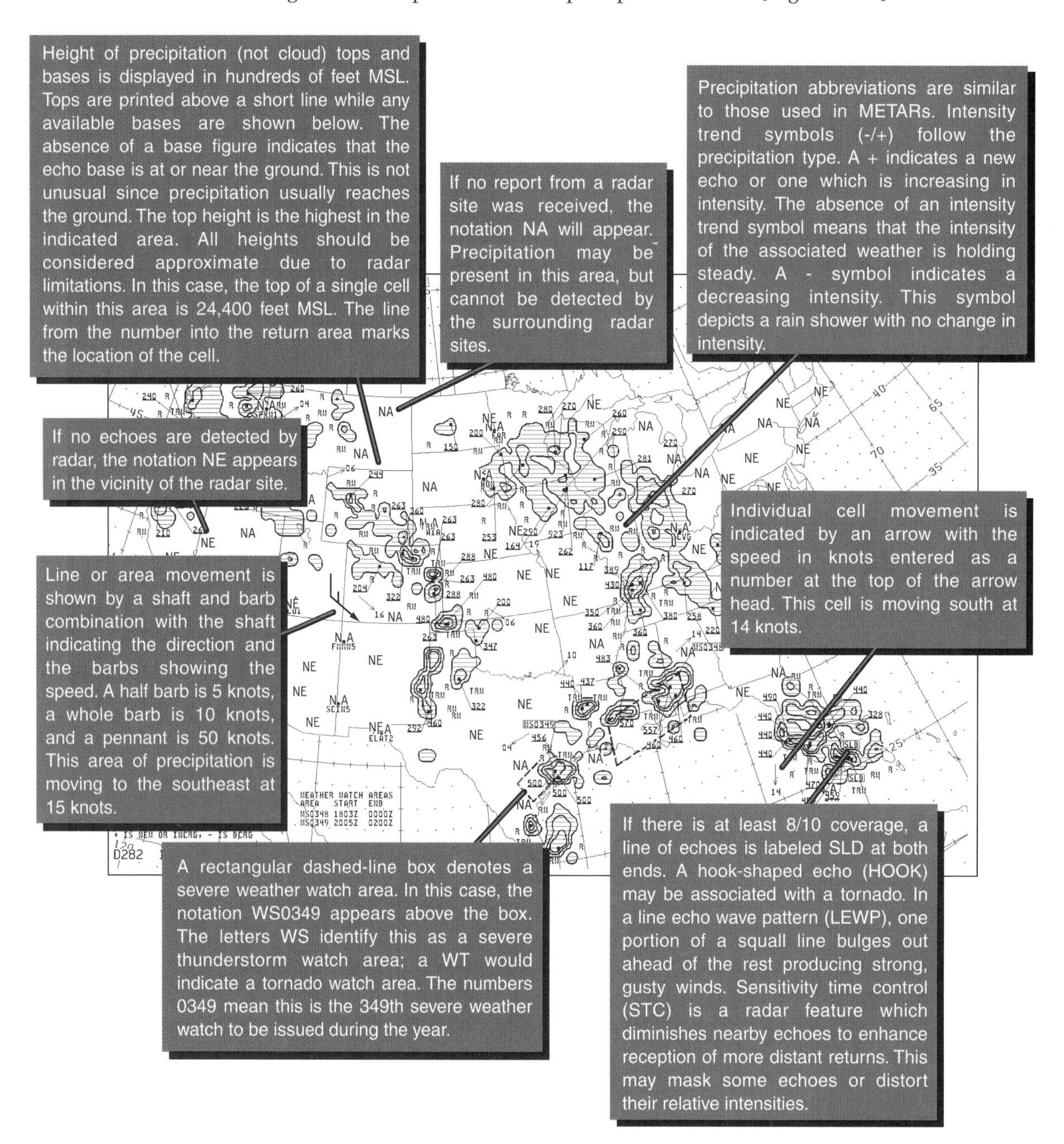

Figure 9-65. The absence of echoes on a radar summary chart does not guarantee clear conditions. To be most effective, radar summary charts should be used in combination with other reports and forecasts.

Radar summary charts are the only weather charts which show lines and cells of thunderstorms. You can also determine the tops and bases of the echoes, the intensity of the precipitation, and the echo movement. See figure 9-65.

While the radar summary chart is a valuable preflight planning tool, it has certain limitations. Since radar only detects precipitation, either in frozen or liquid form, it does not detect all cloud formations. For instance, fog is not displayed, and actual cloud tops may be higher or lower than the precipitation returns indicate. Also, you should keep in mind that it is an observation of conditions that existed at

the valid time. Since thunderstorms develop rapidly, you should examine other weather sources for current and forecast conditions that may affect your flight. [Figure 9-66]

A radar summary chart is most effective when used in combination with other charts, reports, and forecasts.

Figure 9-66. Thunderstorms, such as this single cell storm, can develop quickly and, therefore, may not be depicted on the radar summary chart.

Radar data is also used to produce computer-generated composite images which can provide you with valuable real time weather information. A single-site radar image has limited viewing distance and can contain extraneous information such as ground clutter. To resolve these problems, radar images are combined from several locations to produce a **composite radar image**, which effectively increases the viewing distance and gives you a more complete picture of the weather.

SATELLITE WEATHER PICTURES

Some of the most recognizable weather products come from satellites. Specialized weather satellites not only generate photos, but also record temperatures, humidities, wind speeds, and water vapor locations. Two types of images are available from weather satellites — visible and infrared (IR). Visible pictures are used primarily to determine the presence of clouds as well as the cloud shape and texture. IR photos depict the heat radiation emitted by the various cloud tops and the earth's surface. The difference in temperature between clouds can be used to determine cloud height. For example, since cold temperatures show up as light gray or white, the brightest white areas on an IR satellite photo depict the highest clouds. Both types of photos are transmitted every 30 minutes except for nighttime when visible photos are not available. [Figure 9-67]

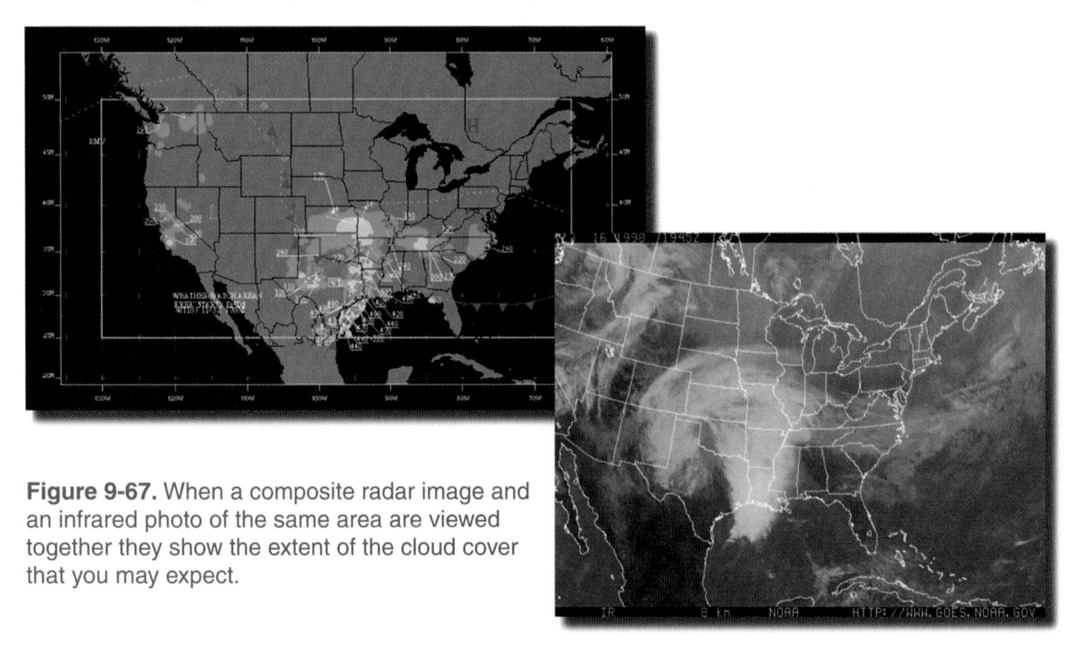

Figure 9-67. When a composite radar image and an infrared photo of the same area are viewed together they show the extent of the cloud cover that you may expect.

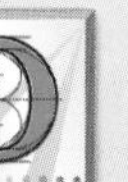

What's going on out there?

Satellite imaging is changing the way we look at the world. Weather satellite images which, for a time, were rarely seen by the public are now commonplace. Every evening, for example, the meteorologist on your local news runs a loop of the day's satellite pictures. But, have you ever wondered what it would be like if you could zoom in to your neighborhood or get a birds-eye view of your local airport?

Well, now you can. Earth Watch Incorporated is providing low-cost, high-resolution satellite images which can be delivered directly to your desktop personal computer. Digital images can have a resolution as high as 0.82 meters which means that each pixel of the image represents about 32 inches on the surface of the earth. This is near aerial photograph quality. Figure A is an example of a 0.82 meter satellite image of Concourse D at San Francisco International Airport.

Though this type of surface imaging will not replace an aerial photo, in some situations it can be of great value for planning. As a city planner or airport manager, this type of data would be invaluable if you need a digital image for planning the expansion of an runway, a terminal, or ensuring regulatory compliance and land ownership.

Even though satellite imaging is not a new technology, the advancements in current technology have improved quality and availability of images. Figure B is an infrared image of Canberra, Australia.

COMPOSITE MOISTURE STABILITY CHART

The **composite moisture stability chart** is an analysis chart derived from observed upper air data. Composed of four panels, this chart is available twice a day, with valid times of 0000Z and 1200Z. Each of the four panels present different information which can be used to determine the characteristics of a particular weather system in terms of stability, moisture, and possible aviation hazards, such as thunderstorms and icing.

STABILITY PANEL

The upper left corner of the chart is the stability panel which outlines the areas of stable and unstable air. [Figure 9-68] On this panel you will see sets of numbers which resemble fractions, the top number is the lifted index (LI), and the bottom number is the K index. The lifted index is the difference between the temperature of a parcel of air being lifted from the surface to the 500-millibar level (approximately 18,000 feet MSL) and the actual temperature at the 500-millibar level. If the number is positive, the air is considered to be stable. For example, a lifted index of +8 would be considered very stable and the potential for a severe thunderstorm would be weak. On the other hand, if the number is negative, the air is considered unstable. An index of -6 or less would be very unstable and have a strong potential for a severe thunderstorm. A zero index is neutrally stable.

The difference found by subtracting the temperature of a parcel of air theoretically lifted from the surface to 500 millibars and the existing temperature at 500 millibars is called the lifted index.

The K index is used primarily by meteorologists. This is the number which indicates whether the conditions are favorable for airmass thunderstorms. Unlike the lifted index, the K index is based on temperature, low-level moisture, and saturation. The

probability of airmass thunderstorms increases with high K values. A K index of 15 or less would be forecast as a 0% probability of airmass thunderstorms and an index of 40 or more is forecast as 100% percent probability.

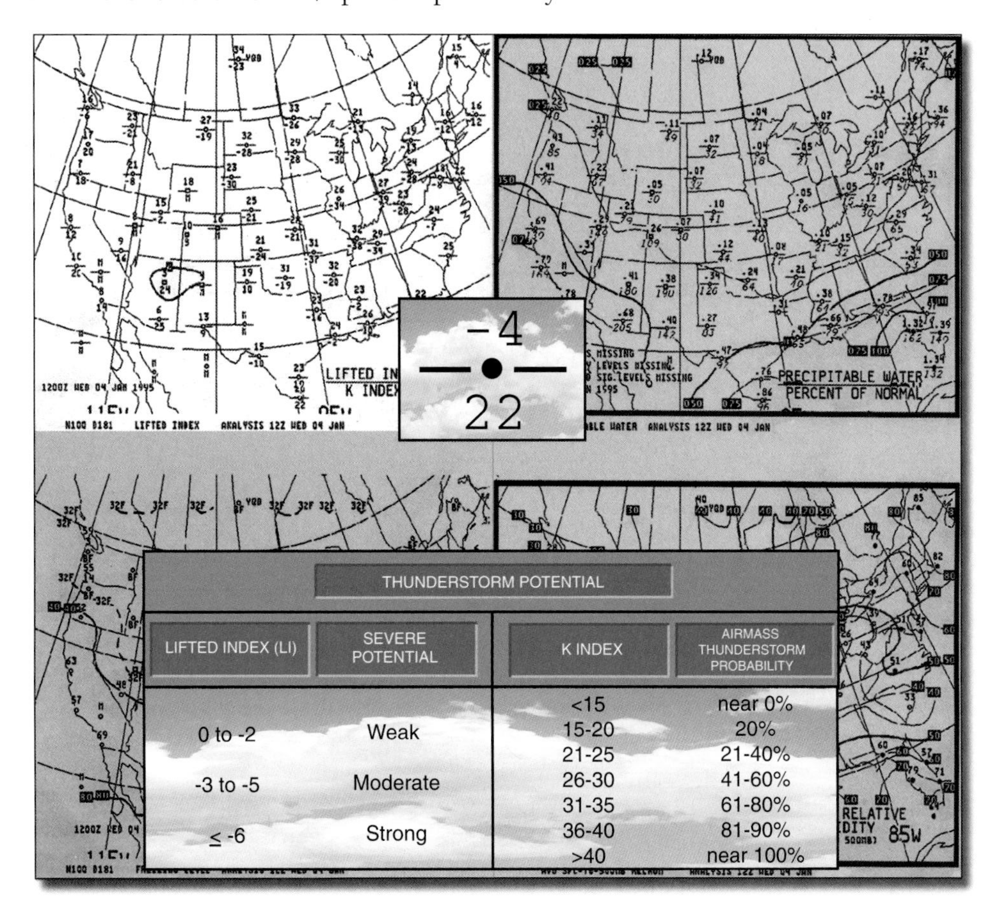

THUNDERSTORM POTENTIAL

LIFTED INDEX (LI)	SEVERE POTENTIAL
0 to -2	Weak
-3 to -5	Moderate
≤ -6	Strong

K INDEX	AIRMASS THUNDERSTORM PROBABILITY
<15	near 0%
15-20	20%
21-25	21-40%
26-30	41-60%
31-35	61-80%
36-40	81-90%
>40	near 100%

Figure 9-68. The station model for Miami, Florida shows that the upper atmosphere is very unstable with a moderate potential for severe thunderstorms. The K index of 22 indicates a 21% to 40% probability of airmass thunderstorms.

For quick reference, the chart is also marked to show areas of relative instability. This stability analysis is depicted in two ways. First the station circle is darkened when the lifted index is zero or less. Second, solid lines are used to enclose areas which have an index of +4 or less at intervals of 4 (+4, 0, -4, -8). The stability panel is an important tool for preflight planning because the stability of the airmass will give you an indication of the type of clouds and precipitation to expect. For example, if the airmass is stable, you should expect fairly smooth air and steady precipitation. On the hand, if the airmass is unstable, you should expect convective turbulence and showery precipitation.

FREEZING LEVEL PANEL

The lower left of the chart is the freezing level panel, which is a depiction of the observed freezing level data gathered from upper air observations. The freezing level (0° Celsius isotherm) at the surface is shown as a dashed contour line. The abbreviation BF is used to indicate stations with temperatures below freezing at the surface. As the freezing level increases in height it is shown by solid contour lines at 4,000-foot intervals which are labeled in hundreds of feet MSL. Since the

The freezing level panel of the composite moisture stability chart is an analysis of the observed freezing level data from upper air observations.

freezing level chart gives you an overall view of the isotherms, you can easily identify the altitudes where icing is probable. [Figure 9-69]

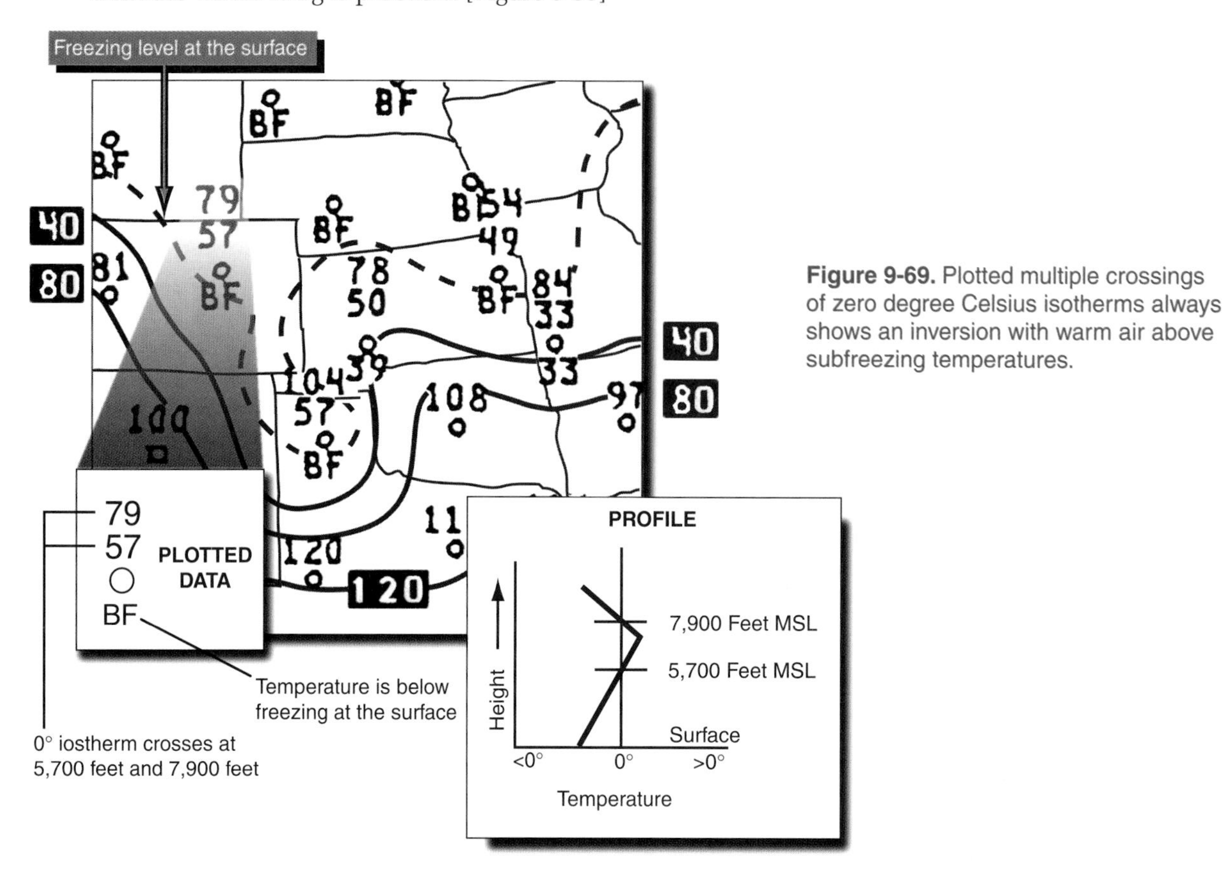

Figure 9-69. Plotted multiple crossings of zero degree Celsius isotherms always shows an inversion with warm air above subfreezing temperatures.

Precipitable Water Panel

The precipitable water panel is the upper right panel of the composite moisture stability chart. This panel graphically depicts the atmospheric water vapor available for condensation, from the surface to the 500-millibar level. Each station model is made up of two numbers. The number on the top is the amount of precipitable water in hundredths of an inch. Precipitable water is the amount of liquid precipitation that would result if all water vapor were condensed. The lower number is the percent of normal value for the month. For example, if the value is .44/90, there is 44 hundredths of an inch of precipitable water, which is 90% of the normal (below average) for any day during the month. [Figure 9-70]

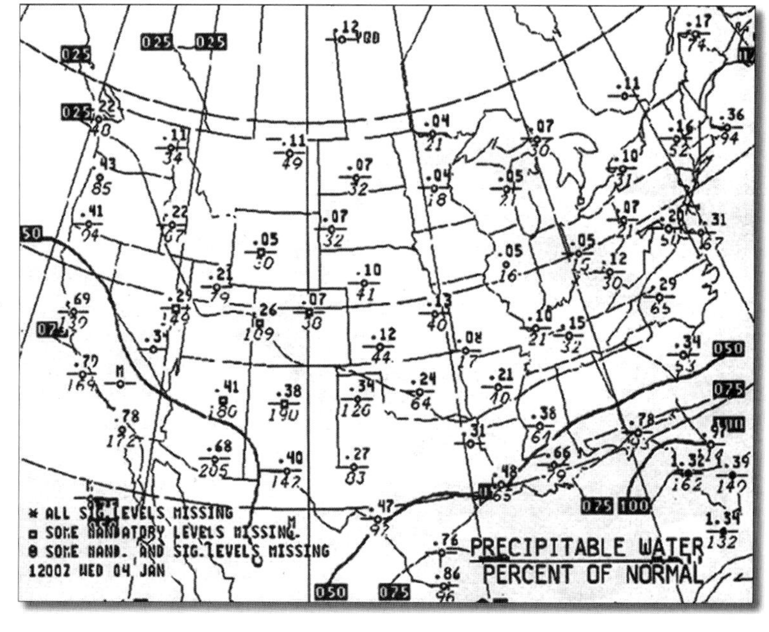

Figure 9-70. The precipitable water panel is used primarily by meteorologists who are concerned with predicting localized flooding.

For quick reference, the station symbol is filled in when the precipitable water value at a station is 1 inch or more. Isopleths (lines of equal value) are also drawn on the panel and labeled for every 1/4 inch. To differentiate the isopleths a heavier line is used to indicate 1/2 inch isopleths.

AVERAGE RELATIVE HUMIDITY PANEL

The lower right panel of the chart is an analysis of the average relative humidity from the surface to 500-millibars. The values are plotted as a percentage for each of the reporting stations. An M is used when the station report is missing. When a station is reporting a relative humidity higher than 50%, the station symbol is darkened. In addition, isopleths of relative humidity, called isohumes, are drawn and labeled every 10%, with heavier isohumes drawn for 10%, 50%, and 90%. [Figure 9-71]

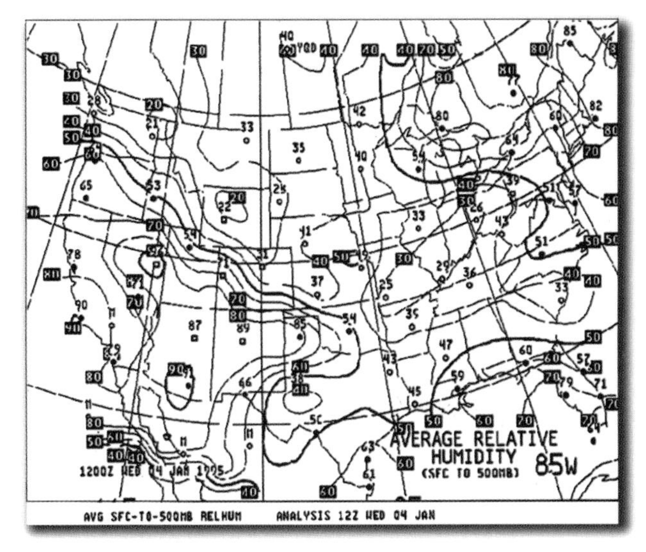

Figure 9-71. Average relative humidities of 50% or higher are quite frequently associated with areas of clouds and possible precipitation.

For preflight planning purposes, this panel is useful for determining the average air saturation from the surface to 500-millibars. However, an area with high relative humidity may or may not have high water vapor content (precipitable water). For example, Las Vegas could have the same average relative humidity as New Orleans, but if the precipitable water value is .43 inches in Las Vegas while New Orleans has 1.34 inches, you can expect greater precipitation in New Orleans.

CONSTANT PRESSURE ANALYSIS CHART

The constant pressure analysis chart is an upper air weather map on which all information is referenced to a specified pressure level. It is issued twice daily, with a valid time of 1200Z and 0000Z for each of five pressure altitude levels from 850 millibars (5,000 feet) to 200 millibars (39,000 feet). The metric equivalent of a millibar is hectoPascal (1 mb = 1 hPa). The abbreviation mb/hPa is commonly used in reference to constant pressure levels.

The observed data for each reporting location (at the specified altitude) is plotted on the chart. The information includes the observed temperature and temperature/dewpoint spread (°C), wind direction (true north) and wind speed (knots), height of the pressure surface (meters), and changes in height over the previous 12 hours. Although the station model allows you to determine specific

For flight planning purposes, constant pressure analysis charts provide a very good resource for observed winds, temperatures, and temperature/dewpoint spread along a route of flight.

information at various points along your route, the constant pressure chart is most useful for quickly determining winds and temperatures aloft for your flight. [Figure 9-72]

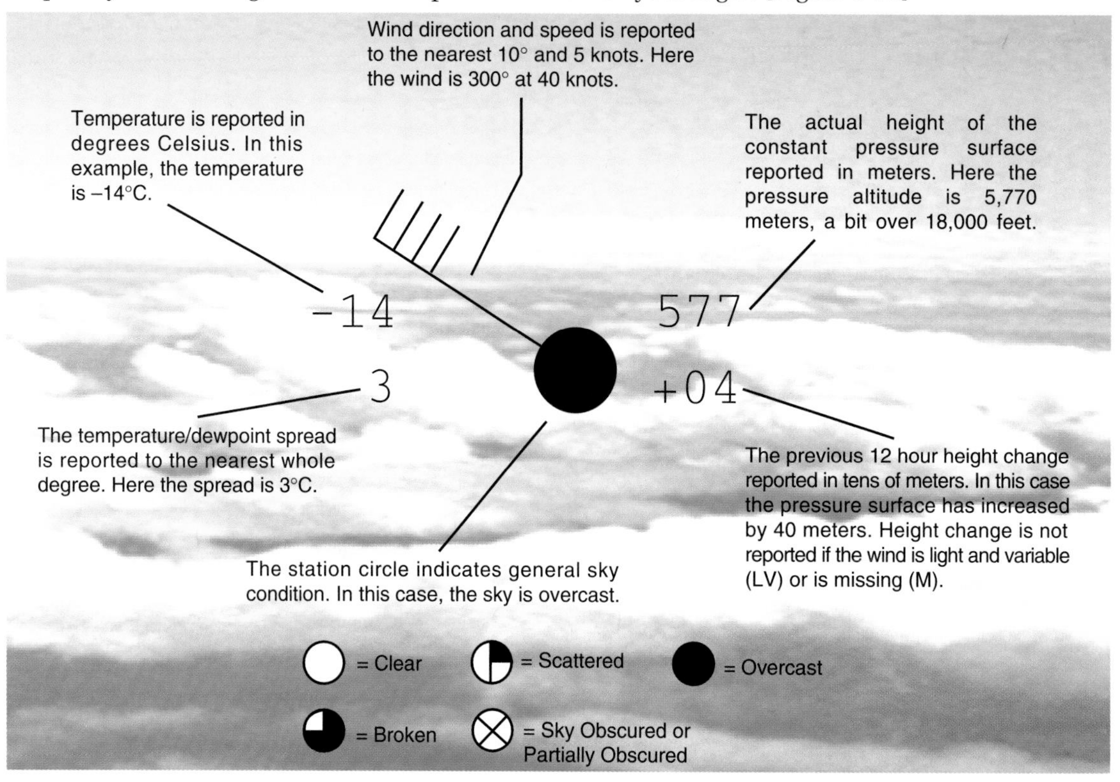

Figure 9-72. This is a sample station model with information from a radiosonde as it would be plotted on a constant pressure analysis chart.

Constant pressure charts also depict highs, lows, troughs, and ridges aloft with height contour patterns that resemble isobars on a surface map. Cold upper lows are producers of bad weather, since the injection of cold air from aloft creates instability. Thus, the area underneath cold upper lows will generally have poor flying conditions. Frontal positions are shown in the same way as on surface charts, provided the fronts extend up to the pressure level of the chart. Fronts are observed most frequently on the 850-millibar and 700-millibar charts which correspond to approximately 5,000 and 10,000 feet, respectively. **Isotherms** or lines of equal temperature show the horizontal temperature variations at that altitude.

The 850-millibar constant pressure chart portrays conditions for low-level flight at approximately 5,000 feet MSL. This is the general level of clouds associated with bad weather. This chart is valuable for forecasting thunderstorms, rain, snow, overcast, and heavy cloudiness.

The 700-millibar chart shows weather data in the vicinity of the 10,000-foot level. It shows wind conditions associated with heavy clouds and rain, but only well-developed fronts appear on this type of chart. The symbols used are the same as those used for the 850-millibar chart.

The 500-millibar chart portrays weather at about the 18,000-foot level and represents average troposphere conditions. These charts are useful in determining average wind and temperature conditions for long-range flights at or near FL180.

The 300- and 200-millibar constant pressure charts show conditions in the upper atmosphere that are significant to high altitude flight. The 300-millibar chart reflects weather at about the 30,000-foot level; the 200-millibar chart applies at the 39,000-foot level. Height of the pressure level, wind, and temperature can easily be determined from a constant pressure chart. However, to decode a constant pressure chart, you need to know the level for which the chart was issued. [Figure 9-73]

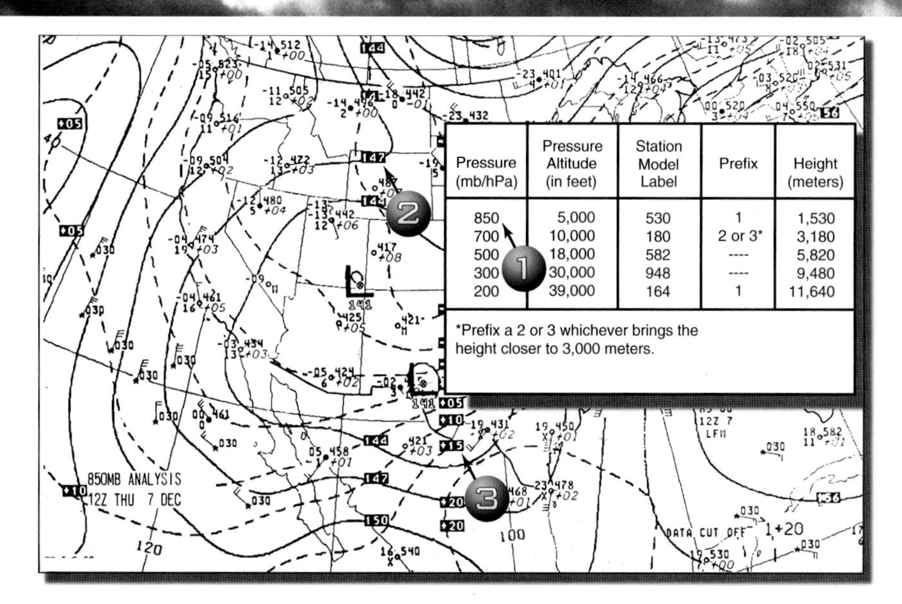

Pressure (mb/hPa)	Pressure Altitude (in feet)	Station Model Label	Prefix	Height (meters)
850	5,000	530	1	1,530
700	10,000	180	2 or 3*	3,180
500	18,000	582	----	5,820
300	30,000	948	----	9,480
200	39,000	164	1	11,640

*Prefix a 2 or 3 whichever brings the height closer to 3,000 meters.

Figure 9-73. The 850-millibar/hectoPascal chart represents a pressure altitude of 5,000 feet as shown in the table on the right (item 1). The solid line with the notation 147 (item 2) means the height of the 850-millibar level at all points along this line is 1,470 meters (4,823 feet). An isotherm labeled +15 (item 3) indicates a line of equal temperature at +15°C.

Isotachs, or lines of equal wind velocity, are drawn only for the 300- and 200-millibar charts. Cross-hatching is used on these charts to denote wind speeds of 70 to 110 knots. A clear area within a hatched area indicates winds of 111 to 150 knots. Another hatched area would denote an area of 151 to 190 knots.

Hatching on a constant pressure analysis chart indicates windspeeds between 70 knots and 110 knots.

The Profiler

What is a wind profiler? A wind profiler is a ground-based, remote sensing Doppler radar that is used to observe winds aloft. The system is designed to operate in nearly all weather conditions and without human intervention, providing nearly continuous measurements of vertical wind structure up to 53,000 feet in the atmosphere.

Starting in the early 1930s upper air measurements were taken with remote telemetry equipment which was carried aloft by a helium balloon. Radiosondes are still in use today. Even though wind profilers are being used on a limited basis at select locations, they represent the first major advance in quantitative upper air measurements since the radiosonde was introduced.

Unlike the twice per day labor intensive radio soundings, wind profilers provide continuous measurements automatically. The information gathered by the profiler aids in the forecast of severe storms, tornadoes, downbursts, and flash floods. For the weather researcher, the data is increasing the understanding of the origin and evolution of regional weather events. With the increase in weather knowledge and computer modeling of the atmosphere aloft, forecasters are able to make accurate weather forecasts at the surface. When used with other existing observational systems, wind profilers help provide the essential information needed to improve weather services.

OBSERVED WINDS AND TEMPERATURES ALOFT CHART

The **observed winds and temperatures aloft chart**, graphically depicts the winds at eight selected levels on two four panel charts. One four-panel chart covers 6,000, 9,000, 12,000, and 18,000 feet MSL, and a second chart covers 24,000, 30,000, 34,000, and 39,000 feet MSL. These observations are made twice a day at 1200Z and 0000Z. For flight planning, these charts help you choose a cruising altitude with the most favorable winds for your flight and to determine the temperature at altitude. However, since the winds and temperatures are based on observations that are reported only twice, they may not be valid at your time of departure. [Figure 9-74]

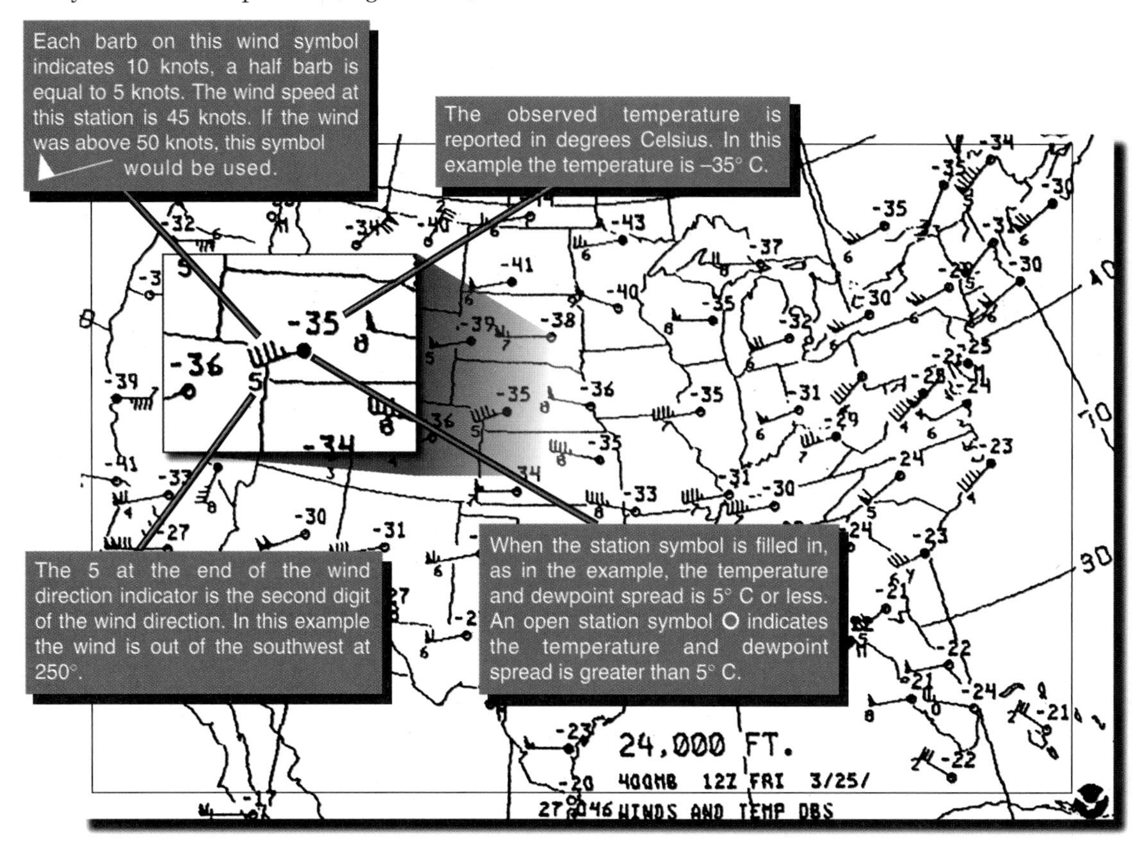

Figure 9-74. Wind direction and speed at each observing station are shown by arrows. When the winds are calm, the arrow is omitted, and the letters LV are positioned near the station. If the wind information is missing, an M is used.

GRAPHIC FORECASTS

Graphically formatted forecasts for a variety of weather elements are available for severeal time periods in the future. These products include U.S. low-level significant weather prognostic charts, U.S. high-level significant prognostic charts, severe weather outlook charts, forecast winds and temperatures aloft charts, tropopause data charts, and volcanic ash forecast charts.

The station model as depicted in figure 9-74 provides wind direction and speed, temperature, and dewpoint information in a graphic format.

LOW-LEVEL SIGNIFICANT WEATHER PROG CHART

The **U.S. low-level significant weather prog (prognostic) chart** is valid from the surface up to the 400-millibar pressure level (24,000 feet). It is designed to help you plan your flights to avoid areas of low visibilities and ceilings, as well as regions where turbulence and icing may exist. [Figure 9-75]

The upper limit of the low-level significant weather prognostic chart is 400-millibars which is about 24,000 feet MSL.

As shown in figure 9-75, symbols used to define areas of IFR, MVFR, VFR, and moderate or greater turbulence, as well as the forecast altitude of the freezing level, are depicted in the legend between the two upper panels of the low-level significant weather prog chart.

A low-level significant weather prognostic chart depicts weather conditions forecast to exist at the specific valid time, in the future.

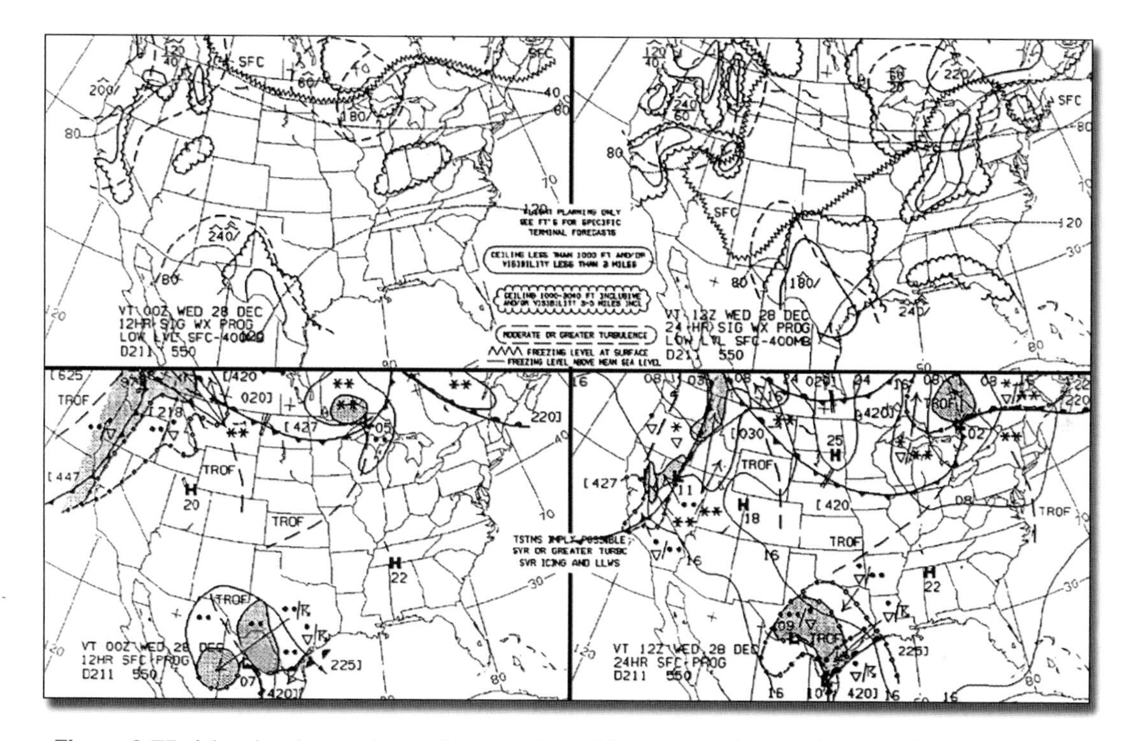

Figure 9-75. A low-level prog chart, which consists of four panels, is issued at 0000Z, 0600Z, 1200Z, and 1800Z. The two lower panels are 12- and 24-hour forecasts of surface weather conditions, while the two upper panels are 12- and 24-hour forecasts of weather between the surface and 24,000 feet. Legend information is included between the two upper panels.

Low-level prog charts are issued four times each day with the valid time printed on the lower margin of each panel. Since the two panels on the left forecast the weather 12 hours from the issue time, and the two panels on the right forecast 24 hours ahead, you can compare the two sets of panels to determine the changes expected to take place between the two time frames.

SIGNIFICANT WEATHER PANELS

The upper panels show areas of nonconvective turbulence and freezing levels as well as IFR and marginal VFR (MVFR) weather. Although the definitions of the ceiling and visibility categories for this chart are the same as the weather depiction chart, the symbolsfor portraying IFR and MVFR are different.

Areas of forecast nonconvective turbulence are shown using a dashed contour line on the significant weather panels. Concentrations of moderate or greater turbulence are enclosed in dashed lines. Numbers within these areas give the height of the turbulence in hundreds of feet MSL. Figures below a line show the expected base, while figures above a line represent the top of the turbulence. Since thunderstorms always imply moderate or greater turbulence, areas of possible thunderstorm turbulence are not outlined.

Forecast freezing levels are also depicted on the significant weather panels. Freezing level height contours for the highest freezing level are drawn at 4,000-foot intervals with dashed lines. These contours are labeled in hundreds of feet MSL. A zig-zag line labeled SFC shows the surface location of the freezing level.

SURFACE PROG PANELS

The two lower panels are the surface prog panels. They contain standard symbols for fronts and pressure centers. Direction of pressure center movement is shown by an arrow; the speed is listed in knots. In addition, areas of forecast precipitation, as well as thunderstorms, are outlined. [Figure 9-76]

Symbol	Meaning	Symbol	Meaning	Symbol	Meaning
	Showery precipitation (thunderstorms/rain-showers) covering half or more of the area.	∇	Rain Shower		Severe Turbulence
	Continuous precipitation (rain) covering half or more of the area.		Snow Shower		Moderate Icing
	Showery precipitation (snow showers) covering less than half of the area.		Thunderstorms		Severe Icing
	Intermittent precipitation (drizzle) covering less than half of the area.		Freezing Rain	•	Rain
			Tropical Storm	✱	Snow
			Hurricane (Typhoon)		Drizzle
			Moderate Turbulence		

Figure 9-76. An area which is expected to have continuous or intermittent (stable) precipitation is enclosed by a solid circle. If only showers are expected, the area is enclosed with a dot-dash pattern. A shaded area indicates that precipitation covers one-half or more of the area. Unique symbols indicate precipitation type and the manner in which it occurs. In the case of drizzle, rain, and snow, a single symbol is used to indicate intermittent conditions, while a pair of symbols denotes continuous precipitation.

Low-level significant weather prog charts graphically depict showery precipitation, thunderstorms, rain showers, and when appropriate, tropical storms. Turbulence is depicted on the low-level significant weather prog chart with a peaked symbol. Underneath the symbol, figures indicate the top and base of the turbulence. See figure 9-76.

HIGH-LEVEL SIGNIFICANT WEATHER PROG CHART

The U.S. high-level significant weather prog (prognostic) chart is derived from forecasts for both domestic and international flights and is issued for the altitudes above the 400-millibar level (24,000 feet) up to the 70-millibar level (63,000 feet). It presents a forecast of thunderstorms, tropical cyclones, severe squalls, moderate or greater turbulence, widespread duststorms and sandstorms, tropopause heights, the location of the jet streams, and volcanic activity. Since the low-level and high-level prog charts share the same symbology, reading the high-level prog is a simple process of applying the same definitions as for the low-level prog. [Figure 9-77]

In figure 9-77, the areas enclosed in scalloped lines indicate that you should expect cumulonimbus clouds (CBs), icing, and moderate or greater turbulence.

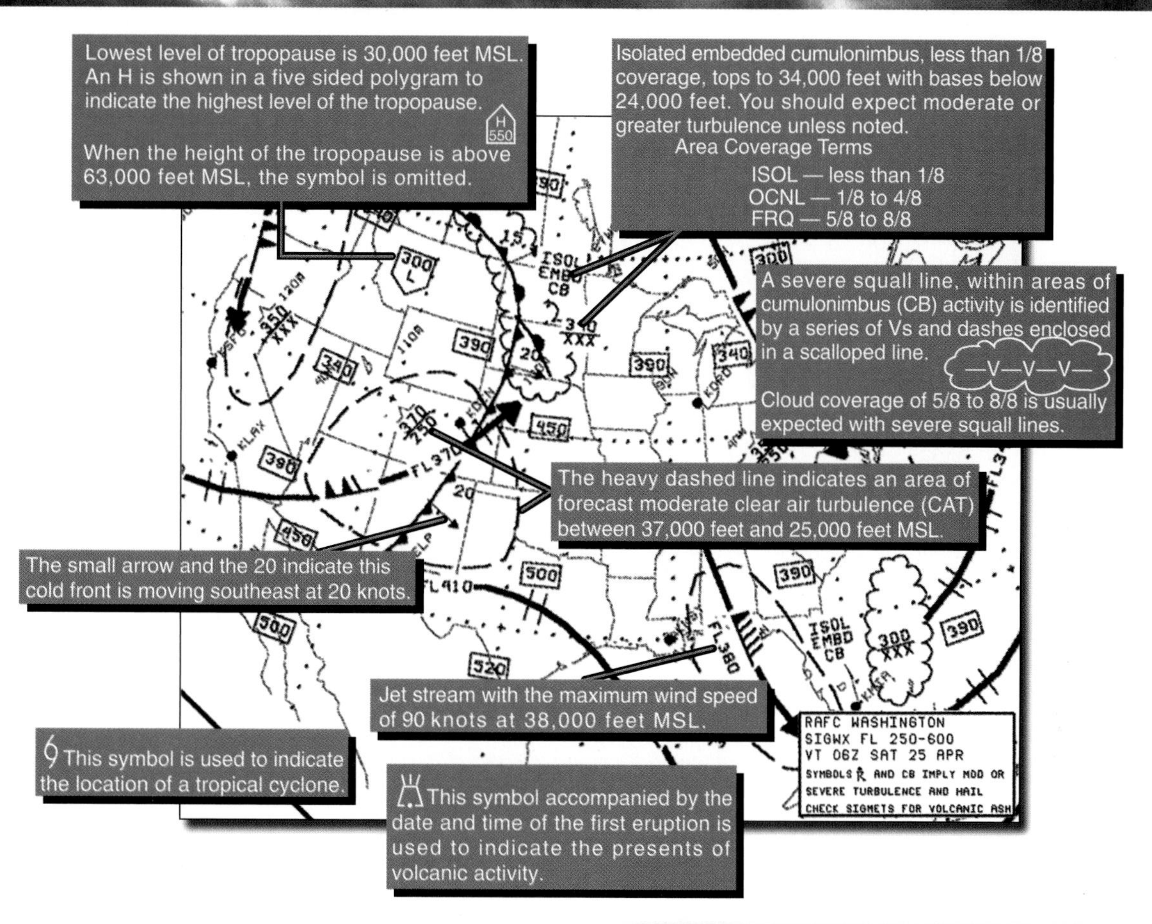

Figure 9-77. The high-level significant prog chart shown here is valid from 0600Z on Saturday the 25th. It covers the altitudes from 25,000 feet MSL (FL250) to 60,000 feet MSL (FL600). The three-digit numbers contained in the boxes represent the forecast height of the tropopause in hundreds of feet MSL. For example, over south central Nebraska the tropopause is expected to be at 45,000 feet MSL.

A high-level significant weather prognostic chart forecasts clear air turbulence, tropopause height, sky coverage, embedded thunderstorms, and jet stream velocities between 24,000 feet MSL and 63,000 feet MSL. See figure 9-77.

The severe weather outlook chart is used primarily for advanced planning and provides a 48-hour outlook for general and severe thunderstorm activity. The left panel covers the first 24 hours and the right panel, the next day.

The cross-hatched area on the severe weather outlook chart indicates areas of probable severe thunderstorm activity. Since the right panel only contains a forecast for severe thunderstorms, the letters SVR are displayed in the cross-hatched areas. General thunderstorms are indicated by a heavy line with an arrowhead. The area of activity is to the right of the line when facing in the direction of the arrow.

SEVERE WEATHER OUTLOOK CHART

The two-panel **severe weather outlook chart** is a 48-hour forecast for thunderstorm activity. The left panel depicts the outlook for general thunderstorm activity and severe thunderstorms for the first 24-hour period beginning at 1200Z. A line with an arrowhead depicts forecast general thunderstorm activity. When facing in the direction of the arrow, thunderstorm activity is expected to the right of the line. If an area is labeled APCHG, it means that the general thunderstorm activity may approach severe intensity. Hatched areas indicate possible severe thunderstorms with an associated risk factor. For example, the notation SLGT indicates that there is a slight (2 to 5 percent coverage) risk of severe thunderstorms occurring during the forecast period. The other possible risk categories include moderate (6 to 10 percent coverage) and

high (more than 10 percent coverage). The right panel of this computer-prepared chart provides a forecast for the next day beginning at 1200Z. This outlook is for the possibility of severe thunderstorms only, and does not include an associated risk factor. [Figure 9-78]

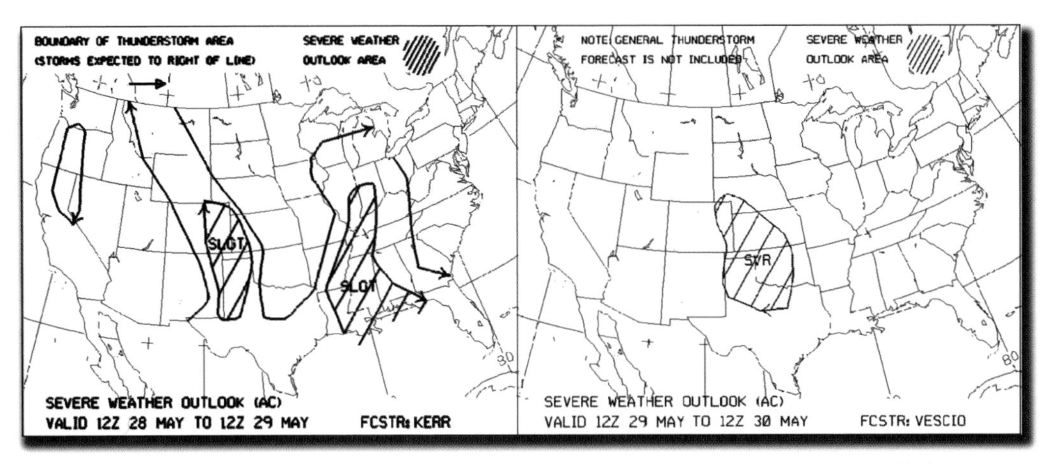

Figure 9-78. The severe weather outlook chart is issued every morning at about 0800Z. As a quick reference to identify areas of possible thunderstorm activity, the severe weather outlook chart is a very good resource for flight planning.

Forecast Winds and Temperatures Aloft Chart

The forecast winds and temperatures aloft charts (FD) are 12-hour forecasts valid at 0000Z and 1200Z daily. The charts contain eight panels, each of which corresponds to a forecast level — 6,000; 9,000; 12,000; 18,000; 24,000; 30,000; 34,000; and 39,000 feet MSL. Predicted winds are depicted using an arrow emanating from the station circle to show direction to within 10 degrees. The second digit of the wind direction is printed at the end of the arrow to help you pinpoint the forecast direction. Pennants and/or barbs at the end of the arrow depict forecast wind speed in much the same way as on the surface analysis chart. When calm or light and variable winds are expected, the arrow is eliminated and a 99 is entered below the station circle. Predicted temperatures are shown as whole degrees Celsius near the station circle. [Figure 9-79]

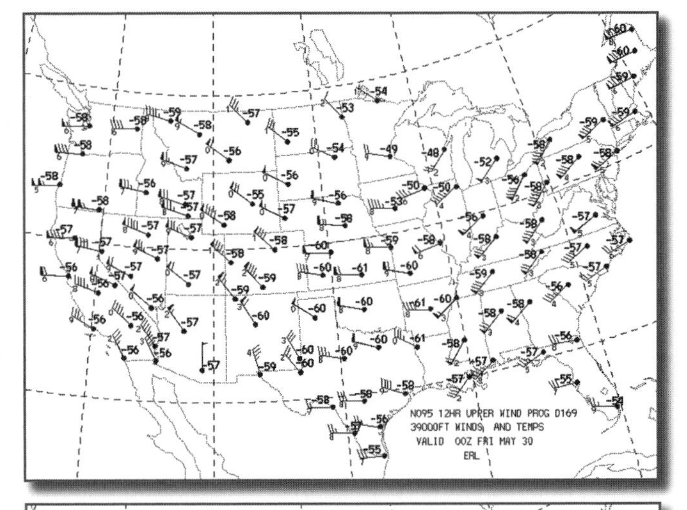

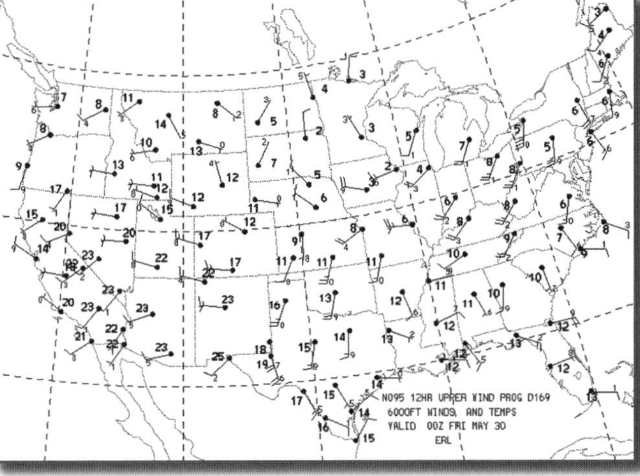

Figure 9-79. The panel on the top is for 39,000 feet. The panel on the bottom is for the same forecast period, but at 6,000 feet. All altitudes below 18,000 feet are in true altitude while those above are in pressure altitude.

TROPOPAUSE DATA CHART

The **tropopause data chart** is a two-panel chart issued once a day, which contains a maximum wind prog and a tropopause height/vertical wind shear prog. When you are planning a high-altitude flight, these charts can be effective for determining areas of both vertical and horizontal wind shear and the associated turbulence.

TROPOPAUSE WINDS PANEL

The tropopause winds prog identifies the direction of the wind by using solid lines called **streamlines**. Since streamlines are drawn so they parallel the direction of the wind it is a simple matter of following the streamlines to determine the direction of the wind. The dashed lines, called isotachs are used to show the wind speed and are labeled in 20-knot intervals. Like the 300-and 200-millibar constant pressure analysis charts, cross hatching is used to identify areas which have wind speeds from 70 knots to 110 knots and 151 knots to 190 knots. A clear area enclosed in a cross-hatched area is used to indicate wind speeds of 111 knots to 150 knots. [Figure 9-80]

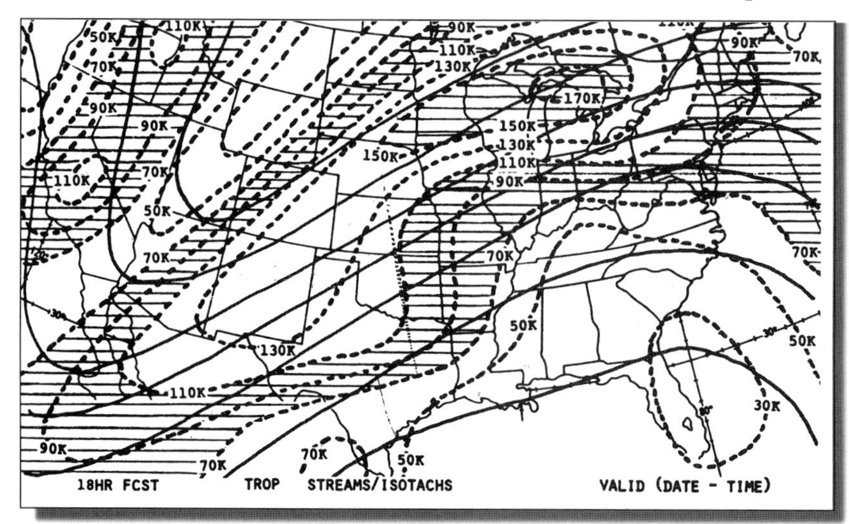

Figure 9-80. This panel is a tropopause winds prog, sometimes referred to as the maximum wind prog. The clear area over western Oklahoma bounded by the cross-hatched area indicates the winds at this pressure level of the chart are between 111 knots and 150 knots.

TROPOPAUSE HEIGHT AND VERTICAL WIND SHEAR PANEL

The tropopause height and vertical wind shear prog graphically depicts the height of the tropopause in terms of pressure altitude and vertical wind shear in knots per 1,000 feet. The tropopause height is a pressure surface shown by solid lines, which connects points where the height of the tropopause intersects standard constant pressure surfaces. These lines are labeled with the pressure altitude in hundreds of feet, preceded by the letter F. Dashed lines are used to depict the areas of vertical wind shear and are shown in 2-knot intervals. As a general rule, you should expect moderate or greater turbulence when the vertical wind shear is 6 knots or more. [Figure 9-81]

The letter F followed by three digits on the tropopause data chart indicates the height of the tropopause. Wind shear, in knots per 1,000 feet, is depicted by dashed lines at 2-knot intervals. See figure 9-81.

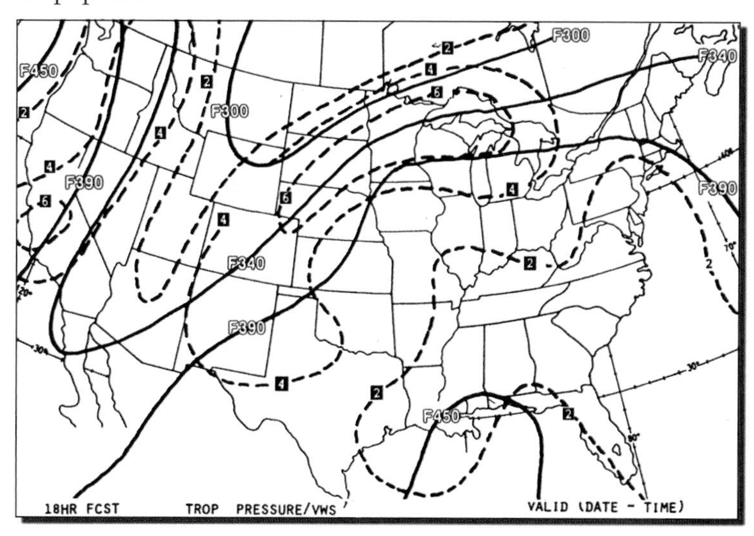

Figure 9-81. This tropopause height/vertical wind shear prog indicates the pressure altitude over New York is 39,000 feet and the wind shear is forecast to be 2 knots per 1,000 feet.

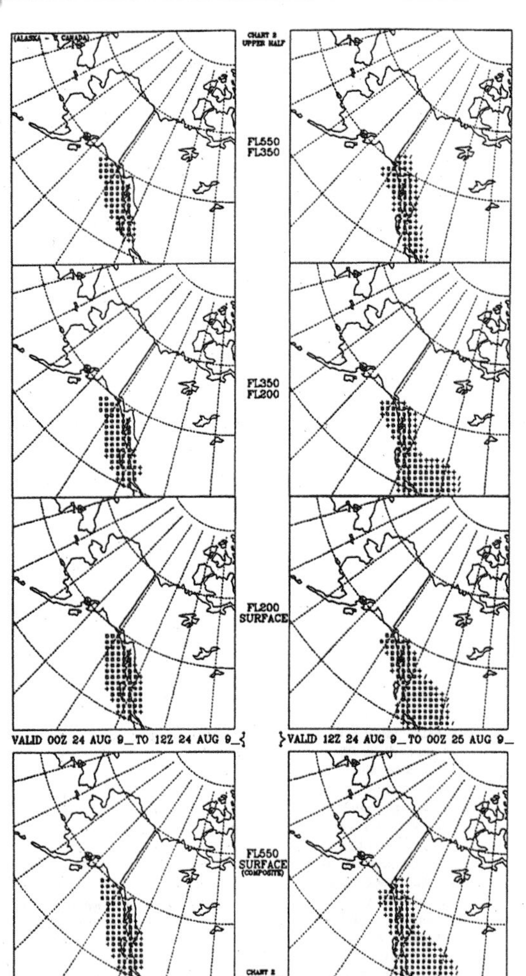

Volcanic Ash Forecast Transport and Dispersion Chart

As volcanic eruptions are reported, a **volcanic ash forecast transport and dispersion chart (VAFTAD)** is created. The chart is developed, with input from National Meteorological Center (NMC) forecasts, using a model which focuses on hazards to aircraft flight operations with emphasis on the ash cloud location. The concentration of volcanic ash is forecast over 6- and 12-hour time intervals, beginning 6 hours following the eruption. The VAFTAD uses four panels in a column for each valid time period. The top three panels in a column reflect the ash location and relative concentrations for an atmospheric layer. The bottom panel in a column shows the total ash concentrations from the surface up to 55,000 feet (FL 550). [Figure 9-82] The VAFTAD chart is designed specifically for flight planning purposes only; it is not intended to take the place of SIGMETs regarding volcanic eruptions and ash.

Figure 9-82. This VAFTAD (reduced from actual size) shows the panels associated with the 24- and 36-hour valid times. Ash concentration is depicted as low (/), moderate (+), or high (■).

Summary Checklist

✓ To get a picture of atmospheric pressure patterns at the earth's surface, you can refer to the surface analysis chart.

✓ The surface analysis chart is a good source of weather information over a wide area.

✓ The weather depiction chart is particularly useful during the preflight planning process for obtaining an overview of favorable and adverse weather conditions.

✓ The radar summary chart shows the location, size, shape, and intensity of areas of precipitation, as well as the intensity trend and direction of movement. Although the chart plots the location of lines and cells of hazardous thunderstorms, it does not show cloud formations. When single-site radar images are combined to produce a composite radar image, viewing distance is increased and ground clutter reduced.

✓ Both visible and infrared (IR) imagery are available from weather satellites. The visible picture is used generally to indicate the presence of clouds as well as the cloud shape and texture. IR photos, which depict the heat radiation emitted by the various cloud tops and the earth's surface, can be used to determine cloud height.

✓ A composite moisture stability chart can help you avoid areas of icing and thunderstorms and determine the characteristics of a particular weather system by depicting the stability and moisture of a given airmass.

- ✓ The constant pressure analysis chart depicts observed weather data at five different levels. The information includes temperature, temperature/dewpoint spread, wind direction, wind speed, height of the pressure surface in meters, the changes in height in the previous 12 hours, as well as upper air highs, lows, troughs, ridges, and fronts.

- ✓ The observed winds and temperatures aloft chart is useful for determining the optimum altitude for the most favorable winds and temperature for your flight.

- ✓ The U.S. low-level significant weather prog chart can not only help you avoid areas of significant turbulence, but it also can provide you with information to help you avoid areas where temperatures are conducive to aircraft icing. The chart is valid from the surface up to 24,000 feet.

- ✓ The upper panels of the low-level significant weather prog chart show areas of nonconvective turbulence and freezing levels, as well as areas of IFR, marginal VFR, and VFR weather from the surface to 24,000 feet. The surface prog panels, contained in the lower portion of the chart, use standard symbols for fronts and pressure centers.

- ✓ The high-level significant weather prog chart is valid for altitudes from 24,000 feet up to 63,000 feet. It presents a forecast of thunderstorms, tropical cyclones, severe squalls, moderate or greater turbulence, widespread duststorms and sandstorms, tropopause heights, the location of the jet streams, and volcanic activity.

- ✓ The severe weather outlook chart is a two-panel chart which forecasts thunderstorm activity over the next 48 hours. The left panel depicts the outlook for general thunderstorm activity and severe thunderstorms for the first 24-hour period beginning at 1200Z. The right panel provides a forecast for the next day beginning at 1200Z. Only severe thunderstorms are shown on the right panel.

- ✓ The forecast winds and temperatures aloft chart contains eight panels each of which corresponds to a forecast level — 6,000, 9,000, 12,000, 18,000, 24,000, 30,000, 34,000, and 39,000 feet MSL. The chart is issued at 1200Z and 0000Z and is valid for a 12-hour forecast period.

- ✓ The tropopause data chart is a two-panel chart which contains a maximum winds prog and the tropopause height and vertical wind shear prog. These two panels are effective for determining areas of both vertical and horizontal wind shear and the associated turbulence.

- ✓ The volcanic ash forecast transport and dispersion chart (VAFTAD) forecasts the concentration of volcanic ash over 6- and 12-hour time intervals, beginning 6 hours following a volcanic eruption. The VAFTAD chart is not intended to take the place of SIGMETs regarding volcanic eruptions; it is designed specifically for flight planning purposes.

KEY TERMS

Surface Analysis Chart

Station Model

Weather Depiction Chart

Radar Summary Chart

Composite Radar Image

Composite Moisture Stability Chart

Isopleths

Isohumes

Constant Pressure Analysis Chart

Isotherms

Isotachs

Observed Winds and Temperatures Aloft Chart

U.S. Low-Level Significant Weather Prog Chart

Freezing Level

High-Level Significant Weather Prog Chart

Severe Weather Outlook Chart

Forecast Winds and Temperatures Aloft Chart (FD)

Tropopause Data Chart

Streamlines

Volcanic Ash Forecast Transport and Dispersion Chart (VAFTAD)

QUESTIONS

Use the station model to answer questions 1 through 4.

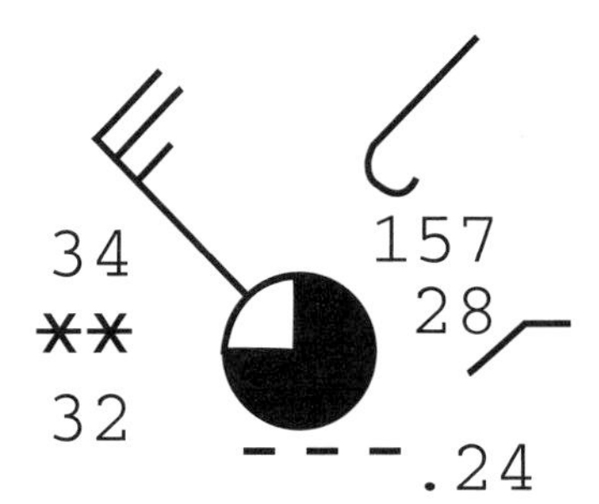

1. What is the wind direction and speed?
2. True/False. This station is reporting overcast skies.
3. What is the present weather?

   A. Light snow
   B. Continuous snow
   C. Snow mixed with fog
4. What is the sea level pressure for this station?
5. What is the significance of a bracket shown with a station model on a surface analysis chart?

Use the following weather depiction chart excerpt to answer question 6 through 8.

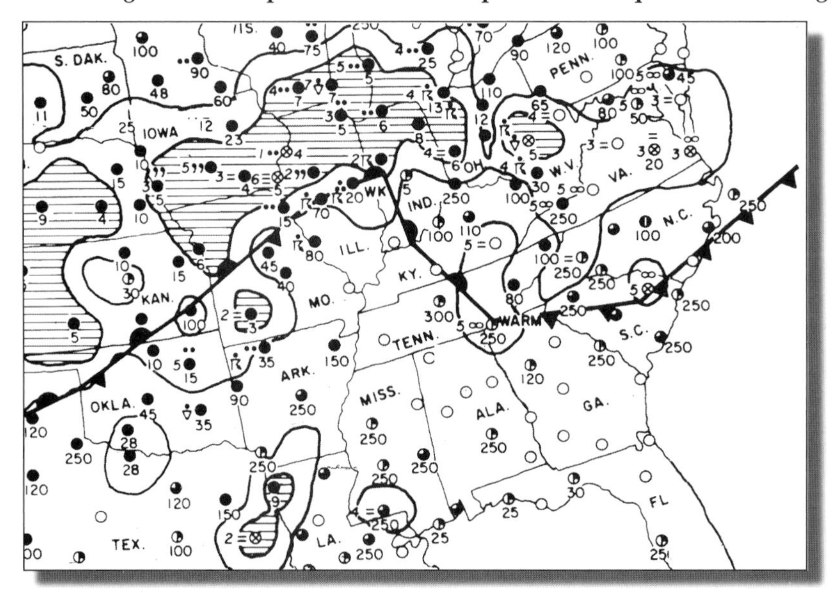

6. What type of weather condition exists over southern Iowa?

   A. IFR
   B. VFR
   C. MVFR

7. A flight from North Carolina to Georgia passes through what type of front?

   A. Cold front
   B. Warm front
   C. Stationary front

8. True/False. Thunderstorms and rain are being reported over northwestern Arkansas.

Use the radar summary chart excerpt to answer questions 9 through 12.

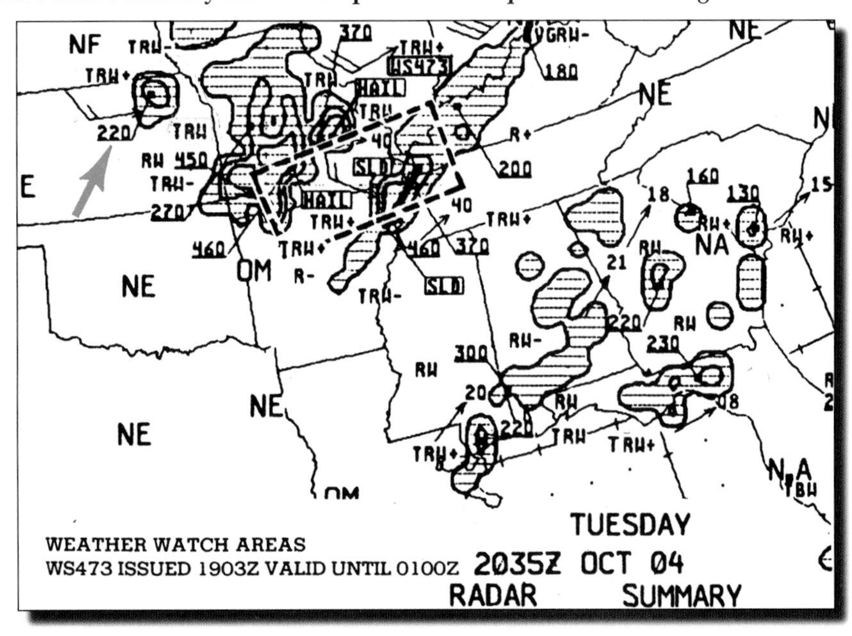

9. What is the direction of movement and speed of the thunderstorm cell in northwestern Georgia?

   A. The cell is moving northeast at 18 knots
   B. The cell is forecast to move to the northeast at 18 knots
   C. The cell is moving in a general direction of northeast at 21 knots

10. What is indicated by the 220 located in northeast Kansas?

11. What is the meaning of NE in central Oklahoma and NA in west central Florida?

12. The second level of cross-hatching indicates what type of precipitation?

    A. Light to moderate
    B. Heavy to very heavy
    C. Intense to extreme

13. How is the freezing level at the surface depicted on a composite moisture stability chart?

14. While constant pressure charts depict observed temperature, temperature/dewpoint spread, wind direction and speed, and the height of a particular pressure surface, the most useful information for general aviation pilots is what?

    A. Height of the tropopause
    B. Winds and temperatures aloft
    C. Location of fronts at the 500 millibar level

15. What is an isotherm?

    A. A line of equal wind speed
    B. A line of equal temperature
    C. International Standard Thermal Unit

Use the accompanying low-level significant prognostic chart to answer questions 16 through 19.

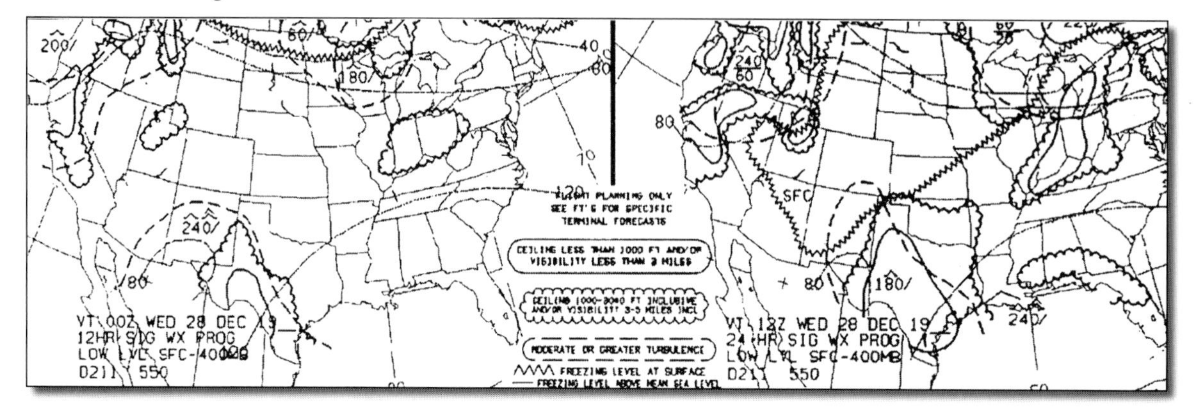

16. The low-level significant weather prognostic chart is for use from the surface up to what altitude?

    A. 10,000 feet
    B. 18,000 feet
    C. 24,000 feet

17. At 0000Z, the weather over northern Utah is forecast to be

    A. IFR
    B. VFR
    C. MVFR

18. What is the forecast weather over central Texas?

    A. IFR at 0000Z and IFR at 1200Z
    B. MVFR at 0000Z and IFR at 1200Z
    C. MVFR at 0000Z and MVFR at 1200Z

19. What does the symbol 240/ in central New Mexico indicate?

Use the following high-level significant prognostic chart to answer question 20 through 23.

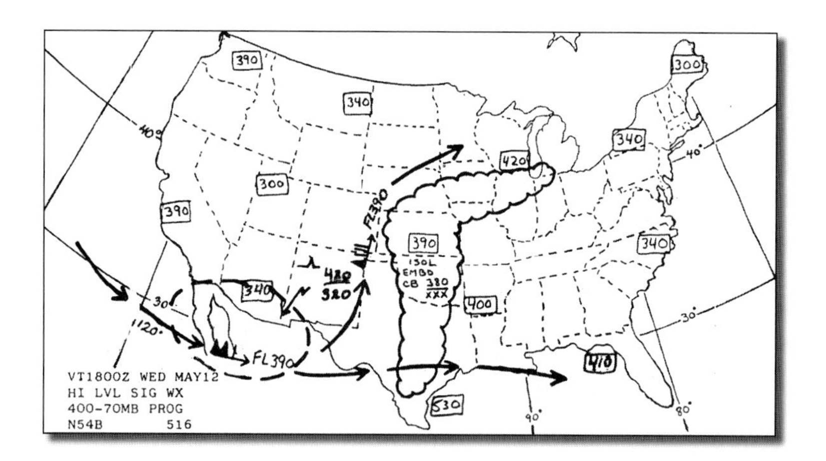

20. What does the symbol over eastern Montana indicate?

21. What is the height and speed of the jet stream over eastern Colorado?

    A. 39,000 feet AGL at 80 knots
    B. 39,000 feet MSL at 80 knots
    C. 39,000 feet MSL at 80 mph

22. True/False. The area in the scalloped line indicates isolated, embedded thunderstorms with tops at 38,000 feet and bases below 24,000 feet.

23. What conditions are forecast for the area contained within the dashed line over southern Arizona and Mexico?

24. What chart should you refer to for a forecast of general thunderstorm activity?

    A. Weather depiction chart
    B. Composite moisture stability chart
    C. Severe weather outlook chart

25. What wind speed and direction is forecast over central New Mexico in the excerpt from the forecast winds and temperatures aloft chart shown below? What is the forecast temperature?

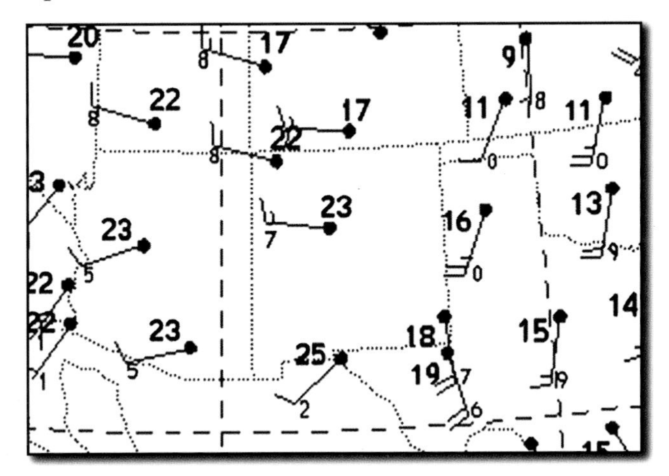

SECTION E

SOURCES OF WEATHER INFORMATION

Part of being a safe IFR pilot is keeping up to date on the latest weather developments, and maintaining the ability to adjust your plans. The weather briefing process usually begins several days before your flight, when you look at mass-disseminated weather information and form an initial opinion regarding the feasibility of your flight. As the flight gets closer, you should gather more detailed information about the weather along your route. Then, during your flight you should update your weather information using the various in-flight sources to determine what lies ahead.

PREFLIGHT WEATHER SOURCES

Federal regulations require that you obtain weather reports and forecasts any time you plan on departing IFR or when operating VFR away from an airport. Preflight weather information sources include Flight Service Station (FSS) and National Weather Service (NWS) telephone briefers, the telephone information briefing service, the Direct User Access Terminal System (DUATS), the World Wide Web, and commercial vendors.

FLIGHT SERVICE STATION

Flight service stations (FSS) are your primary sources for preflight weather information. You can obtain a preflight weather briefing from an automated FSS (AFSS) 24 hours a day by calling **1-800-WX BRIEF** almost anywhere in the U.S. In areas not served by an AFSS, National Weather Service facilities may provide pilot weather briefings. Telephone numbers for NWS facilities and additional numbers for FSSs/AFSSs can be found in the *Airport/Facility Directory* (*A/FD*) or the U.S. Government section of the telephone directory under Department of Transportation, Federal Aviation Administration, or Department of Commerce, National Weather Service.

PREFLIGHT WEATHER BRIEFING

When you request a briefing, you should identify yourself as a pilot and advise the briefer whether you plan to fly VFR or IFR. Provide your aircraft number or your name, aircraft type, departure airport, route of flight, destination, flight altitude(s), estimated time of departure (ETD), and estimated time enroute (ETE). The briefer can then provide information pertinent to your proposed flight.

While pilot weather briefers are not authorized to make original forecasts, they can translate and interpret available forecasts and reports into terms that describe the weather conditions you can expect along your route of flight and at your destination. As you know, there are three types of preflight weather briefings — standard, abbreviated, and outlook.

STANDARD BRIEFING

You should request a standard briefing when you are planning a trip and have not obtained preliminary weather or a previous briefing. This is the most complete weather briefing, and assumes you have no familiarity with the overall weather picture. When you request a standard briefing, the briefer automatically provides certain types of information in sequence, if applicable to your proposed flight. [Figure 9-83]

1. **ADVERSE CONDITIONS** — This includes the type of information that might influence you to alter your proposed route or cancel the flight altogether. Examples include such things as hazardous weather or airport closures.
2. **VFR FLIGHT NOT RECOMMENDED** — The briefer may skip this warning if your proposed flight is IFR. It is advisory in nature; the final decision whether to conduct the flight under VFR rests with you.
3. **SYNOPSIS** — The briefer will provide you with a broad overview of the major weather systems or airmasses that affect the proposed flight.
4. **CURRENT CONDITIONS** — This information is a rundown of existing conditions, including pertinent hourly, pilot, and radar weather reports. Unless you request otherwise, this item is omitted if your proposed departure time is more than two hours in the future.
5. **ENROUTE FORECAST** — The briefer will summarize the forecast conditions along your proposed route in a logical order from departure through descent for landing.
6. **DESTINATION FORECAST** — The briefer will provide the forecast for your destination at your estimated time of arrival (ETA). In addition, any significant changes predicted for an hour before or after your ETA will be included.
7. **WINDS AND TEMPERATURES ALOFT**—You will be given a summary of forecast winds for your route. If necessary, the briefer will interpolate wind direction and speed between levels and stations for your planned cruising altitude(s). Temperature information will be provided on request.
8. **NOTICES TO AIRMEN** — The briefer will supply NOTAM information pertinent to your proposed route of flight. However, information which has already been published in the Notices to Airmen publication will only be provided on request.
9. **ATC DELAYS**— You will be advised of any known air traffic control delays that might affect your proposed flight.
10. **OTHER INFORMATION** — Upon request, the briefer will provide you with other information such as approximate density altitude data, MOA and MTR activity within 100 n.m. of the flight plan area, ATC services and rules, as well as customs and immigration procedures.

Figure 9-83. The first three items of a standard briefing may be combined in any order when, in the briefer's opinion, it will help to describe conditions more clearly.

ABBREVIATED BRIEFING

When you need only one or two specific items or would like to update weather information you obtained previously, request an abbreviated briefing. Provide the briefer with the source of the prior information including the time you received it, as well as any other pertinent background information. This allows the briefer to limit the presentation to new and updated weather information. Usually, the sequence of information follows that of the standard briefing. The briefer also will advise you of adverse conditions.

OUTLOOK BRIEFING

If your proposed departure time is six or more hours away, request an outlook briefing. This helps you make an initial judgment about the feasibility of your flight, and should be updated with a standard or abbreviated briefing close to flight time.

TELEPHONE INFORMATION BRIEFING SERVICE

When you call 1-800-WX-BRIEF, a TOUCH-TONE® activated menu normally allows you to choose between a live briefer and recorded weather information. The recorded information is called the telephone information briefing service (TIBS). This service provides continuous recordings of area and/or route meteorological briefings, airspace procedures, and special aviation-oriented announcements. Other information such as METARs and TAFs also may be included. TIBS is designed to be a preliminary briefing tool; you may choose this option in place of an outlook briefing to determine the overall feasibility of your proposed flight. TIBS is not intended to replace a standard briefing from an FSS or NWS specialist. If you need more detailed information than TIBS supplies, the AFSS telephone system allows you to transfer directly to a briefer.

The telephone information briefing service (TIBS) provided by AFSSs includes continuous recordings of meteorological and/or aeronautical information available by telephone.

DIRECT USER ACCESS TERMINAL SYSTEM

The FAA-funded Direct User Access Terminal System (DUATS) allows pilots with a current medical certificate to receive weather briefings and file flight plans directly via a personal computer and modem. You can access DUATS using a toll-free number in the 48 contiguous United States. [Figure 9-84]

Companies that provide DUATS service

DTC DUATS
http://www.duat.com
Data: 1-800-245-3828
Voice: 1-800-243-3828

GTE DUATS
http://www.gtefsd.com/aviation/GTEaviation.html
Data: 1-800-767-9989
Voice: 1-800-345-3828

Figure 9-84. Check the *Aeronautical Information Manual* for the most up-to-date list of DUATS providers.

A major benefit of using DUATS is that you do not need to transcribe information that is read to you over the telephone. Plus, DUATS providers generally offer plain language translation of the abbreviations in the official FAA weather reports, as well as color weather maps that can be downloaded and printed. Flight planning computer software generally can import DUATS winds aloft data into a flight plan. Jeppesen's general aviation flight planning software can even plot SIGMETs, AIRMETs, and other advisories on a map that shows your route of flight. [Figure 9-85]

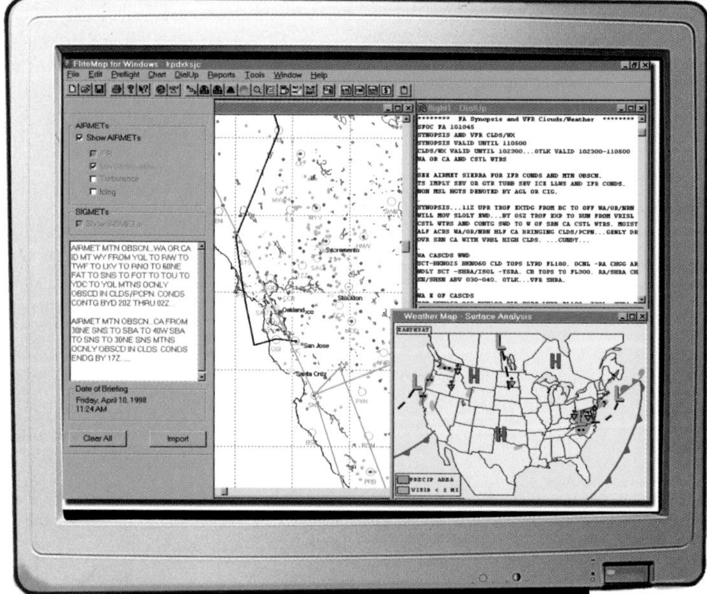

Figure 9-85. If you are using a personal computer, obtaining weather information from DUATS gives you a head start to quick, accurate flight planning.

A danger of using DUATS is that it is easy to gather many pages of weather data and throw it into your flight bag without fully understanding it. A live briefer, on the other hand, can help you interpret the raw data, and provide useful advice on how or whether to complete a flight.

PRIVATE INDUSTRY SOURCES

Prior to World War II, the U.S. Weather Bureau (now the National Weather Service) was the lone disseminator of weather data. Now, in addition to government sources, there are more than 100 companies in the 200 million-dollar-a-year business of providing weather information to the aviation industry. Jeppesen's JeppFax service is one example of a commercial source of weather information. If you have a fax machine, you can get information ranging from airport and route briefings to real-time radar maps.

THE WORLD WIDE WEB

A wealth of weather information, some of which is directed toward aviation, is available at a variety of internet web sites. You can usually locate many of the sites by choosing the search function available with your internet browser and searching for "weather." Some sites offer real-time images from TV cameras at various locations. The quality of sites providing weather information varies widely, so you should exercise caution when using this information for flight planning. A source of official information is the National Weather Service site at: www.nws.noaa.gov. [Figure 9-86]

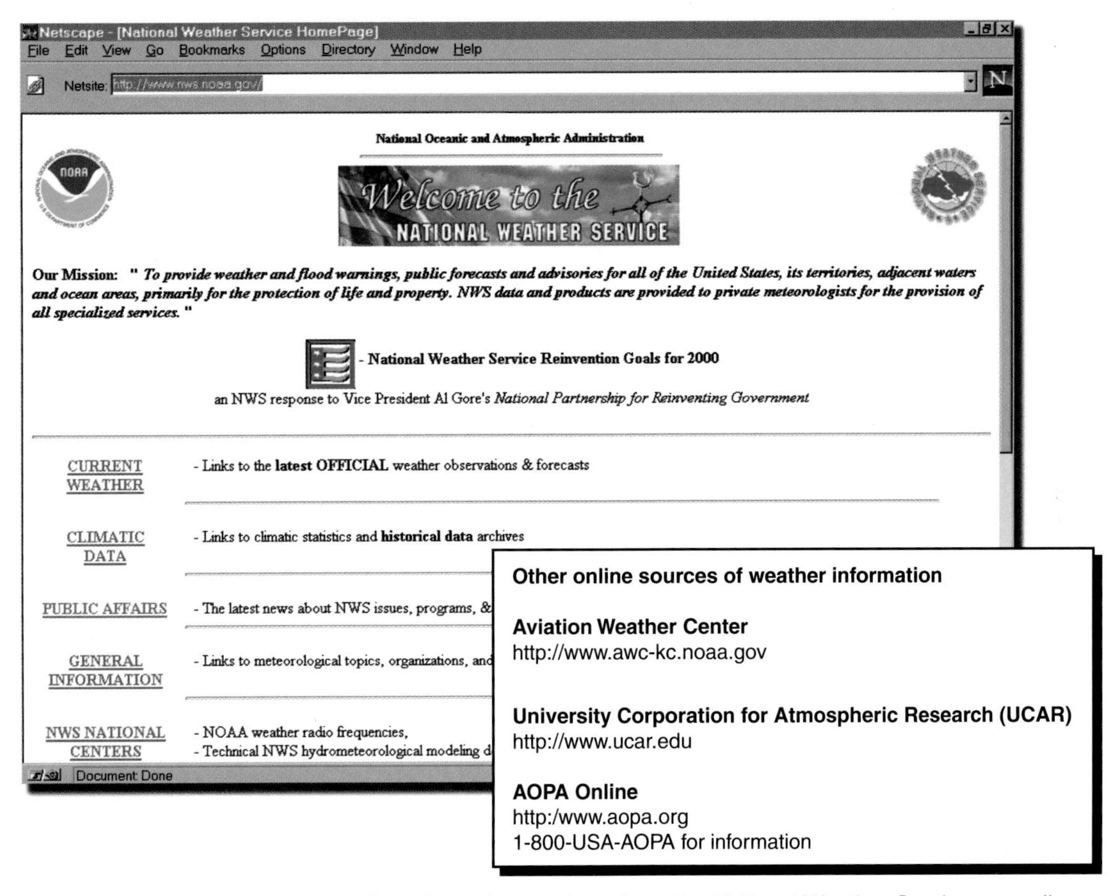

Figure 9-86. You can access a variety of weather products from the National Weather Service, as well as other weather and aviation organizations.

IN-FLIGHT WEATHER SOURCES

Since weather conditions can change rapidly between the time you obtain your preflight briefing and the time you are established enroute, it is prudent to update your weather information while in flight. You can obtain weather updates from the enroute flight advisory service, FSSs, transcribed weather broadcasts, the hazardous in-flight weather advisory service, ARTCC, and automated weather observing systems.

IN-FLIGHT WEATHER ADVISORIES

In-flight aviation weather advisories consisting of AIRMETs, SIGMETs, and convective SIGMETs are forecasts that advise enroute aircraft of the development of potentially hazardous weather. All in-flight advisories in the contiguous U.S. are issued by the National Aviation Weather Advisory Unit in Kansas City, MO. All in-flight advisories use the same location identifiers (either VORs, airports, or well-known geographic areas) to describe the hazardous weather areas.

AIRMET

AIRMET is an acronym for airmen's meteorological information. AIRMETs (WAs) are issued every six hours, with amendments issued as necessary, for weather phenomena which are of operational interest to all aircraft. These weather conditions are potentially hazardous to light aircraft or aircraft having limited capability because of lack of equipment, instrumentation, or pilot qualifications. AIRMETs are issued for moderate icing, moderate turbulence, sustained winds of 30 knots or more at the surface, ceilings less than 1,000 feet and/or visibility less than three miles affecting over 50 percent of an area at any one time, and extensive mountain obscurement. AIRMETs use the identifier Sierra for IFR conditions and mountain obscuration, Tango for turbulence, strong surface winds, and low-level wind shear, and Zulu for icing and freezing levels. After the first issuance of the day, AIRMETs are numbered sequentially for easier identification. [Figure 9-87]

The maximum forecast period for an AIRMET is 6 hours.

Figure 9-87. AIRMETs are issued for weather phenomena such as ceilings less than 1,000 feet and/or visibility less than 3 miles affecting over 50 percent of an area.

SIGMET

SIGMETs (WSs) are issued for hazardous weather (other than convective activity) which is considered significant to all aircraft. SIGMET stands for significant meteorological information and includes severe icing, severe and extreme turbulence, volcanic eruptions, and duststorms, sandstorms, or volcanic ash lowering visibility to less than three miles. [Figure 9-88]

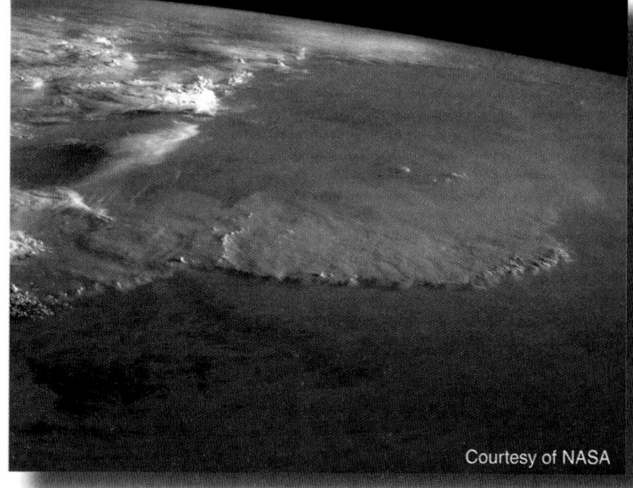

Figure 9-88. SIGMETs are issued for weather phenomena such as duststorms with visibility less than 3 miles.

SIGMETs warn of hazardous weather conditions, such as severe icing; sandstorms, duststorms, or volcanic ash lowering visibility to less than 3 miles; or volcanic eruptions, which concern all aircraft.

SIGMETs are issued whether or not the conditions are included in the area forecast. Excluding those designators reserved for scheduled AIRMETs, SIGMETs use consecutive alphanumeric designators November through Yankee. [Figure 9-89]

When used in combination with the information from PIREPs and AIRMETs, SIGMETs are your best source for information on icing conditions whether current or forecast.

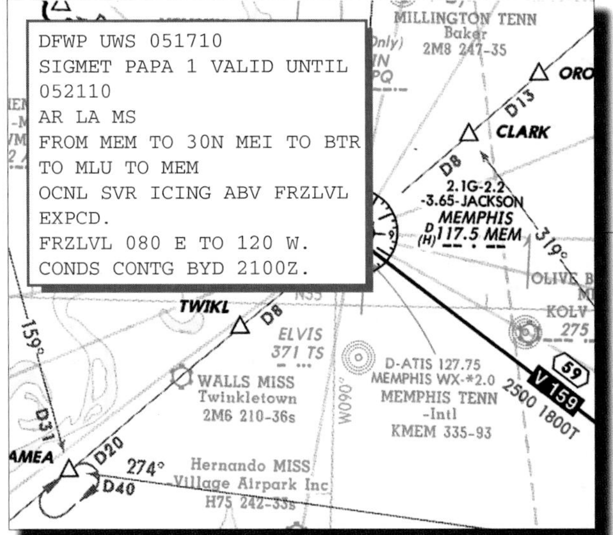

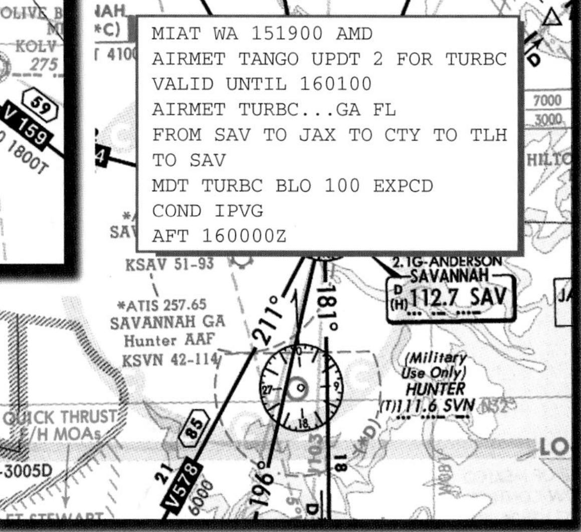

Figure 9-89. The first issuance of a SIGMET, as shown on the left, is labeled UWS (Urgent Weather SIGMET). PAPA 1 means it is the first issuance for a SIGMET phenomenon; PAPA 2 would be the second issuance for the same phenomenon. In the example on the right, AMD means this is an amended AIRMET of the phenomenon (moderate turbulence) identified as TANGO. The alphanumeric designator stays with the phenomenon even when it moves across the country.

CONVECTIVE SIGMET

Convective SIGMETs (WSTs) are issued for hazardous convective weather which is significant to the safety of all aircraft. Since they always imply severe or greater turbulence, severe icing, and low-level wind shear, these items are not specified in the advisory. WSTs include any of the following phenomena: tornadoes, lines of thunderstorms, thunderstorms over a wide area, embedded thunderstorms, hail greater than or equal to 3/4 inch in diameter, and/or wind gusts to 50 knots or greater. A WST consists of either an observation and a forecast or simply a forecast. [Figure 9-90]

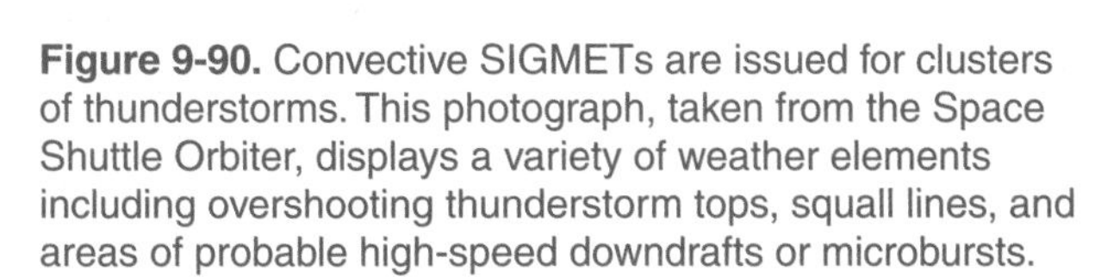

Figure 9-90. Convective SIGMETs are issued for clusters of thunderstorms. This photograph, taken from the Space Shuttle Orbiter, displays a variety of weather elements including overshooting thunderstorm tops, squall lines, and areas of probable high-speed downdrafts or microbursts.

Courtesy of NASA

Convective SIGMETs are issued for the Eastern (E), Central (C), and Western (W) United States. Individual convective SIGMETs are numbered sequentially for each area (01-99) each day. A convective SIGMET will usually be issued only for the area where the bulk of observations and forecast conditions are located. Bulletins are issued 55 minutes past each hour

Convective SIGMETs contain either an observation and a forecast, or just a forecast, for tornadoes, significant thunderstorm activity, or hail 3/4 inch or greater in diameter.

In-flight aviation weather advisories include forecasts of potentially hazardous flying conditions for enroute aircraft, including information on volcanic eruptions that are occurring or expected to occur.

unless a special update bulletin is required. SIGMET forecasts are valid for 2 hours from the time of issuance or until it is superseded by the next hourly issuance. If the criteria for a convective SIGMET is not met at the issuance time, the message, CONVECTIVE SIGMET...NONE is sent. [Figure 9-91]

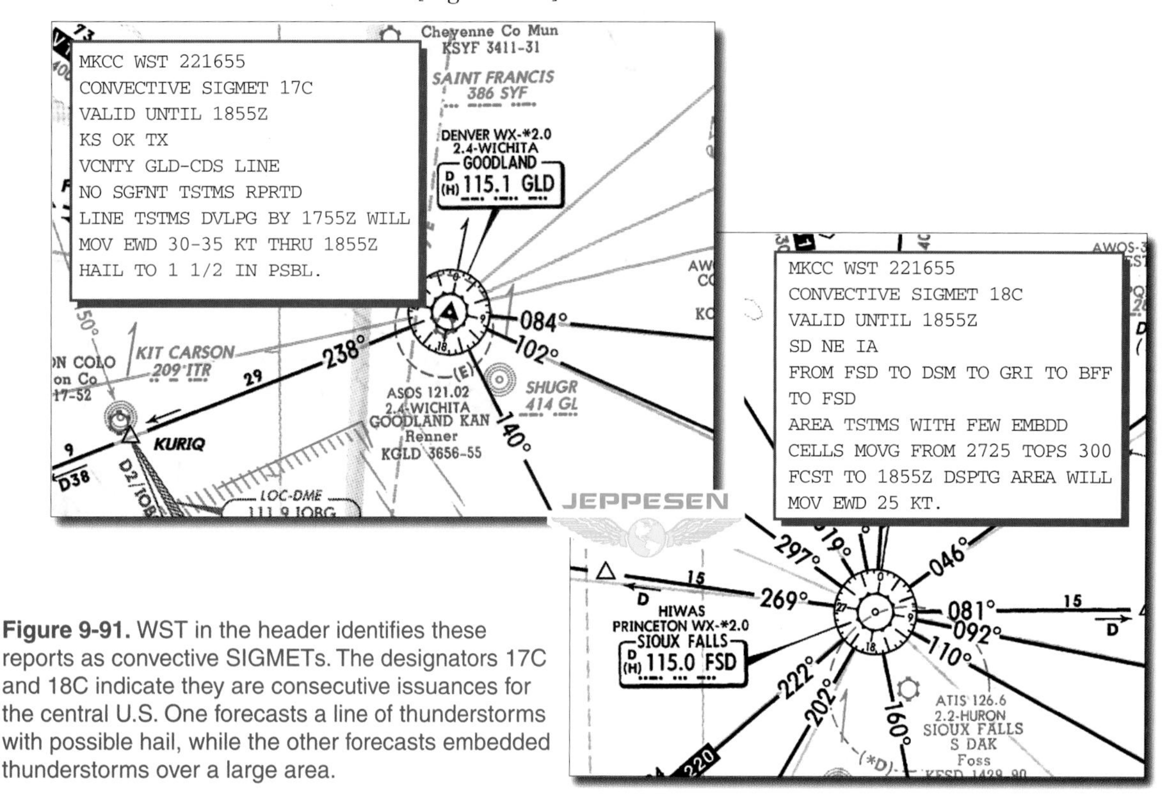

Figure 9-91. WST in the header identifies these reports as convective SIGMETs. The designators 17C and 18C indicate they are consecutive issuances for the central U.S. One forecasts a line of thunderstorms with possible hail, while the other forecasts embedded thunderstorms over a large area.

ENROUTE FLIGHT ADVISORY SERVICE

Enroute flight advisory service (EFAS), or Flight Watch, is a part of a flight service station specifically designed to provide enroute weather information upon pilot request. It acts as a central collection and distribution point for pilot reports (PIREPs), and has direct access to weather radar displays. Radar can be particularly valuable in identifying areas of possible thunderstorm activity along your route. Regular contact with Flight Watch throughout your flight gives you confidence that you can complete your flight as planned, or gives you an early warning of the need to change your plans. Other pilots benefit if you make a practice of filing a PIREP whenever you request information from Flight Watch.

Enroute flight advisory service (EFAS) provides enroute aircraft with timely and meaningful weather advisories pertinent to the type of flight intended, route, and altitude.

You can reach an EFAS specialist from 6 a.m. to 10 p.m. almost anywhere in the conterminous U.S. and Puerto Rico. If you are between 5,000 feet AGL and 17,500 feet MSL, you can use the common frequency of 122.0 MHz. At 18,000 feet MSL and above, there are different frequencies for different ARTCC regions, which you can find on a chart inside the back cover of the *Airport/Facility Directory.*

Since EFAS facilities usually serve large geographic regions through remote communications outlets (RCOs), it is helpful if you make your initial callup using the name of the ARTCC serving the area. This allows the briefer to use the RCO which will provide the best communications coverage. State your aircraft identification and the name of the VOR nearest your position, for example, "*Seattle Flight Watch, Arrow 344LC, Battleground VOR.*" If you do not know which EFAS facility you are addressing, make your initial callup as follows: "*Flight Watch, Arrow 344LC, Battleground VOR.*" The briefer will then respond with the name of the controlling facility.

Times of operation for EFAS facilities, as well as the high altitude EFAS frequencies can be found in the *Airport/Facility Directory.*

EFAS is obtained by contacting flight watch, using the name of the ARTCC facility identification in your area, your aircraft identification, and name of the nearest VOR, on 122.0 MHz below FL180.

Although EFAS normally is based at an AFSS, you should confine your EFAS requests to weather information along your route of flight. EFAS is not intended for flight plans, position reports, preflight briefings, or to obtain weather reports or forecasts unrelated to your current flight. For these items, contact FSS/AFSS personnel on other published frequencies.

FLIGHT SERVICE STATIONS

In addition to enroute flight advisory service, the FSS/AFSS also can provide other types of information during flight. Contact an FSS/AFSS on frequencies shown next to VOR communications boxes when you need to update a previous briefing, amend a flight plan, or for any service other than the specific enroute weather information you get from Flight Watch. After establishing contact, specify the type of briefing you want as well as appropriate background information similar to what you would supply when calling for a telephone briefing.

CENTER WEATHER ADVISORIES

A **center weather advisory (CWA)** is an unscheduled advisory issued by an ARTCC to alert pilots of existing or anticipated adverse weather conditions within the next two hours. A CWA may be issued prior to an AIRMET or SIGMET when PIREPs suggest AIRMET or SIGMET conditions exist. Even if adverse weather is not sufficiently intense or widespread for a SIGMET or AIRMET, a CWA may be issued if conditions are expected to affect the safe flow of air traffic within the ARTCC area of responsibility. In addition, CWAs may be issued to supplement existing in-flight advisories, as well as provide a channel through which ATC can quickly warn pilots of any pertinent weather factors not covered by existing advisories, as well as air traffic delays and other factors.

AIRMETs and center weather advisories (CWA) provide an enroute pilot with information about moderate icing, moderate turbulence, winds of 30 knots or more at the surface, and extensive mountain obscurement.

Air Route Traffic Control Centers broadcast CWAs as well as SIGMETs, convective SIGMETs, and severe weather forecast alerts (AWW) once on all but emergency frequencies when any part of the area described is within 150 miles of the airspace under the ARTCC jurisdiction. In terminal areas, local control and approach control may limit these broadcasts to weather occurring within 50 miles of the airspace under their jurisdiction. These broadcasts contain the advisory identification and a brief description of the weather activity and general area affected.

Weather advisory broadcasts, including severe weather forecast alerts (AWW), convective SIGMETs, and SIGMETs, are provided by ARTCCs on all frequencies, except emergency, when any part of the area described is within 150 miles of the airspace under their jurisdiction.

Hazardous In-Flight Weather Advisory Service

Hazardous in-flight weather advisory service (HIWAS) is a continuous broadcast of summarized AWW, SIGMETs, convective SIGMETs, CWAs, AIRMETs, and urgent PIREPs over selected VORs. When HIWAS is available, ATC generally does not broadcast the actual advisories, but instead tells pilots they exist and refers them to HIWAS or FSS. When a HIWAS is updated, ARTCC and terminal facilities broadcast an alert on all but the emergency frequencies, providing the type and number of the updated advisory and the frequencies to which you can tune for complete information. Enroute charts indicate which VORs have HIWAS capability. [Figure 9-92] In addition, a note in the *Airport/Facility Directory* indicates if a particular VOR is HIWAS-equipped.

The Hazardous In-flight Weather Advisory Service (HIWAS) is a continuous broadcast of in-flight weather advisories over selected VORs of SIGMETs, convective SIGMETs, AIRMETs, severe weather forecast alerts (AWW), and center weather advisories (CWA).

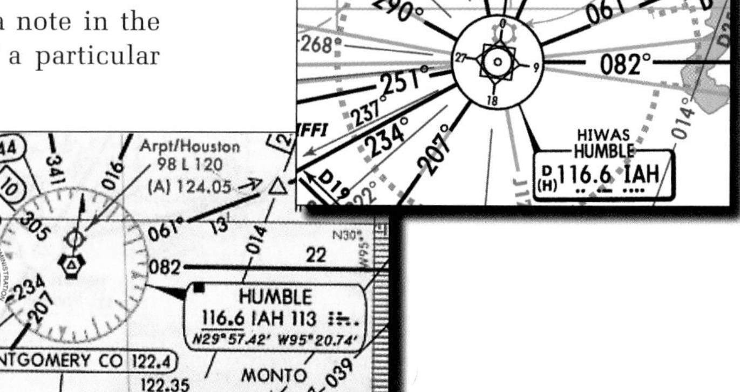

Figure 9-92. HIWAS availability is indicated on NOS enroute and VFR charts with a solid square in the navaid box. On Jeppesen charts, HIWAS is spelled out.

Transcribed Weather Broadcasts

A transcribed weather broadcast (TWEB) contains recorded weather information in a route format transmitted over selected VORs and NDBs. Generally, the broadcast includes specially prepared National Weather Service forecasts, in-flight advisories, winds aloft, and preselected information such as weather reports, NOTAMs and special notices. At some locations, the information is only broadcast over the local VOR and is limited to items such as the hourly weather for the parent station and up to five adjacent stations, local NOTAM information, the TAF for the parent station, and potentially hazardous conditions.

A transcribed weather broadcast (TWEB) provides specific information concerning expected sky cover, cloud tops, visibility, weather, and obstructions to vision in a route format. To obtain continuous transcribed information, including winds aloft and route forecasts for a cross-country flight, you could monitor a TWEB on a low-frequency radio receiver.

Typically, TWEBs are used for in-flight information purposes, however, at some locations, telephone access to the recording is available, providing an additional source of preflight information. Telephone numbers for this service, called TEL-TWEB, are listed in the *Airport/Facility Directory*.

In certain FSS regions where there has been high utilization of TWEB, the FSS puts the TWEB on their TIBS recording. NDBs and VORs which transmit TWEBs have a circled T inside the communication boxes on NOS enroute and sectional charts, and have TWEB clearly marked above the communication boxes on Jeppesen charts.

Most TWEB facilities have been replaced by HIWAS, or otherwise decommissioned. However, HIWAS does not provide route weather information, so in areas not covered

by TWEB, you may obtain that type of information prior to flight by calling 1-800-WX-BRIEF and choosing recorded information (TIBS) from a menu. During flight, your best resource for specific route information is EFAS. You also can tune in AWOS and ASOS facilities at selected airports to get up-to-the-minute weather information for those locations.

WEATHER RADAR SERVICES

The NWS operates a nationwide network of weather radar sites which provides real-time information for printed and graphic weather products. The radar network also furnishes EFAS and AFSS specialists with data they can use for in-flight advisories. Since weather radar can detect coverage, intensity, and movement of precipitation, an EFAS or AFSS specialist may be able to provide you with suggested routing around areas of hazardous weather. Since weather radar only shows heavy precipitation and does not necessarily indicate areas of clouds, you could be in IFR conditions even if you are avoiding areas of weather radar echoes.

Going in Circles Really Fast

Tornadoes carry some of the most devastating forces available in nature. The amount of energy in an F5 tornado's 250 m.p.h. wind is four times that of a 125 m.p.h. wind encountered in a hurricane. It is little wonder that tornadoes pick up houses, buildings, and large trucks and scatter them like toothpicks.

How do they measure such wind speeds? Even if a tornado should pass directly over traditional measuring equipment, it is unlikely the equipment could survive such an encounter. Professor T. Fujita of the University of Chicago, who developed the widely-used Fujita scale for measuring tornado intensity, originally determined wind speeds by analyzing film of tornadoes and calculating the velocity of the flying debris.

Now, doppler radar accurately measures wind speed inside tornadoes. Forecasters not only can determine the location, speed and direction of movement of a tornado, but also can predict how devastating it is likely to be, and warn people accordingly. The National Weather Service's WSR-88D, or NEXRAD, radar system, together with new geostationary operational environmental satellites (GOES) and the automated surface observation system (ASOS) network, is expected to save many lives by pinpointing the time and place where severe weather is likely to strike.

AUTOMATED SURFACE WEATHER REPORTING SYSTEMS

Two types of automated systems currently provide similar weather information from airports and other key locations. These systems are the automated surface observation system (ASOS) and the automated weather observation system (AWOS). Even though the names are different, both are surface observation systems, which can contain much of the same equipment. The main operational difference between ASOS and the more advanced AWOS systems is in who installs and operates the equipment. The FAA is installing and operating ASOS at AFSSs and at tower-controlled airports and the National Weather Service is installing ASOS at locations where it previously maintained fully staffed offices. AWOS typically is installed and operated at non-ASOS airports by local airport authorities or by state departments of transportation.

Many ASOS and AWOS installations broadcast over a discrete VHF frequency or over the voice portion of a local navaid. The discrete frequencies are designed to be receivable up to 25 n.m. from the site up to 10,000 feet AGL. You also can reach many ASOS facilities by telephone. The phone numbers appear in the *Airport/Facility Directory* under weather data sources for an airport. The frequencies appear in both the *A/FD* and on aeronautical charts.

AUTOMATED SURFACE OBSERVATION SYSTEM

With more than 1,000 installations operating across the U.S., the **automated surface observation system (ASOS)** is becoming the primary surface weather observing system in the United States. Many ASOS stations continuously broadcast their measurements over dedicated frequencies using computer-synthesized voices, and also can be reached by telephone. Additionally, data from ASOS facilities is sent to a central location, where it is processed and distributed over NWS networks and to FAA flight service stations.

Monitoring an ASOS frequency or listening by telephone gives you accurate, up-to-the-minute information. Although ASOS-derived data you get from ATIS or from a METAR can be up to an hour old, the ASOS software is designed to transmit a special (SPECI) report whenever it determines a significant change in conditions has occurred. For this reason, you generally can expect published ASOS observations to reflect current conditions.

ASOS equipment provides continuous minute-by-minute observations of the parameters necessary to generate an aviation routine weather report (METAR) and can provide other aviation weather information. [Figure 9-93] ASOS can measure wind shifts and peak winds. Some ASOS stations include a precipitation discriminator which can differentiate between liquid and frozen precipitation. If the station has this capability, it is designated as an AO2. Otherwise, it carries an AO1 designation. Another enhancement you are likely to see at selected ASOS installations is lightning detection equipment.

Every ASOS contains the following:

1. Cloud height indicator
2. Visibility sensor
3. Precipitation identification sensor
4. Freezing rain sensor (at select sites)
5. Pressure sensors
6. Ambient temperature and dewpoint temperature sensors
7. Anemometer (wind direction and speed sensor)
8. Rainfall accumulation sensor

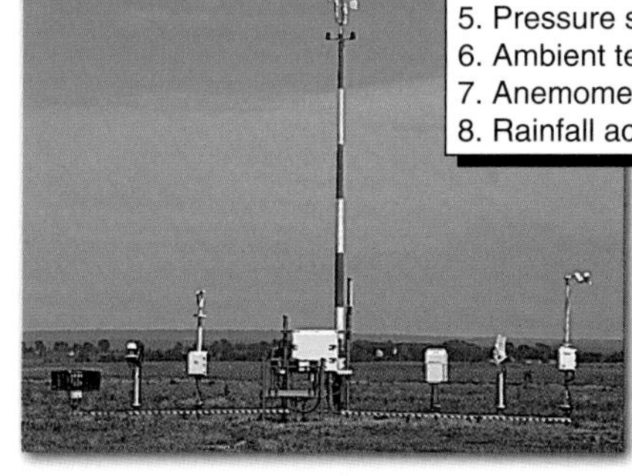

Figure 9-93. ASOS stations are becoming the primary source of U.S. surface weather observations.

ASOS is not able to distinguish between relatively harmless stratus clouds and potentially dangerous cumulonimbus. Although the technology is steadily improving, the equipment is also limited in its ability to identify restrictions to visibility. It cannot

offer intelligent judgements about prevailing, sector, or tower visibility. That is why input from a trained human observer is an integral part of the ASOS program.

ASOS is available in four levels of service, based on the amount of information added by human observers. Level A is the highest level of service which typically is available at major airports like those in or near Class B airspace. Other levels offer lesser degrees of human augmentation, with fewer types of weather phenomenon reported. Level B ASOS has human observers available 24 hours a day, while Level C is found at airports with part-time towers; human augmentation ends when the tower closes. Level D ASOS systems are found at smaller, nontowered airports that still meet the FAA or NWS criteria for installing ASOS. These systems are unattended, and always contain the AUTO designation when their observations appear in a METAR.

DOES A HUMAN PERFORM BETTER WEATHER OBSERVATIONS THAN A MACHINE?

Both human observers and automated systems have limitations when it comes to determining the complete weather picture. The measurements from AWOS/ASOS systems are completely objective, and in the strictest sense, error free. However, the AWOS/ASOS systems are limited in their ability to deduce an overall view of the weather at various locations in the airport vicinity. A human observer is able to look around the sky and quickly make a subjective judgment as to sky condition, visibility and present weather. However, depending on the complexity of the weather, and on other duties, a human observer can only make one or two observations per hour. An automated system looks at a single part of the sky over a period of time, and averages this data.

According to the FAA, the fixed time, spatial averaging technique used by human observers, and the automated system's fixed location, time averaging technique, yield remarkably similar results. Keep in mind, it is possible for an automated system to report inaccurately if there is a localized condition near the reporting equipment and little or no cloud movement.

AUTOMATED WEATHER OBSERVING SYSTEMS

First manufactured in 1979, the **automated weather observing system (AWOS)** was developed for the FAA and was the first widely installed automated weather data gathering system at U.S. airports. Installation of AWOS continues at nontowered airports not served by ASOS. Like ASOS, the AWOS uses various sensors, a voice synthesizer, and a radio transmitter to provide real-time weather data. Unlike ASOS, AWOS is available in lesser configurations that do not provide all the types of observations listed in figure 9-93. AWOS-A only reports the altimeter setting. AWOS-1 also measures and reports wind speed, direction and gusts, temperature, and dewpoint. AWOS-2 adds visibility information, and the most capable system, AWOS-3, also includes cloud and ceiling data, which essentially makes it equivalent to ASOS. Like ASOS, AWOS-3 also can include precipitation discrimination sensors, indicated by A02 in the remarks section of a METAR. Lightning detection also is a possible enhancement for selected AWOS-3 sites, and human augmentation also can be used with AWOS.

Although it can be difficult to distinguish between AWOS-3 and ASOS systems based on the information they provide, one common difference between the two systems is their ability to identify and report significant changes in surface weather. Typically, an AWOS that is connected to an FAA or NWS network transmits three reports per hour at fixed intervals. Unlike an ASOS, AWOS cannot issue a special report as needed. If a major change in weather, such as a frontal passage, occurs between AWOS reports, you may have no way of knowing about it unless you are listening to up-to-the-minute data by radio or telephone.

AIRBORNE WEATHER EQUIPMENT

No pilot, regardless of size or performance of his/her aircraft, would intentionally penetrate a thunderstorm. Ground-based radar cannot identify all hazardous areas with certainty, because the signal may be reflected back by nearer weather phenomena. To find the cells hidden from ground based weather radar, it is necessary to utilize equipment onboard the aircraft. That is why most jet aircraft and many turboprop aircraft have airborne equipment for avoiding severe weather. Certain light twin and high-performance single engine aircraft also offer this type of equipment. While having weather avoidance equipment does not necessarily make it safe to operate near embedded thunderstorms, it does offer properly trained pilots some help in avoiding the most hazardous weather.

AIRBORNE WEATHER RADAR

Airborne radar operates under the same principle as ground-based radar. The directional antenna, normally behind the airplane's fiberglass nose cone, transmits pulses of energy out ahead of the aircraft. The signal is reflected by water and ice, and is picked up by the antenna from which it was transmitted. The bearing and distance of the weather is plotted, normally on a color display. These displays are useful for indicating different degrees of precipitation severity. A radar unit might use green to indicate light showers, yellow for moderate rainfall, and red to depict heavy precipitation. [Figure 9-94]

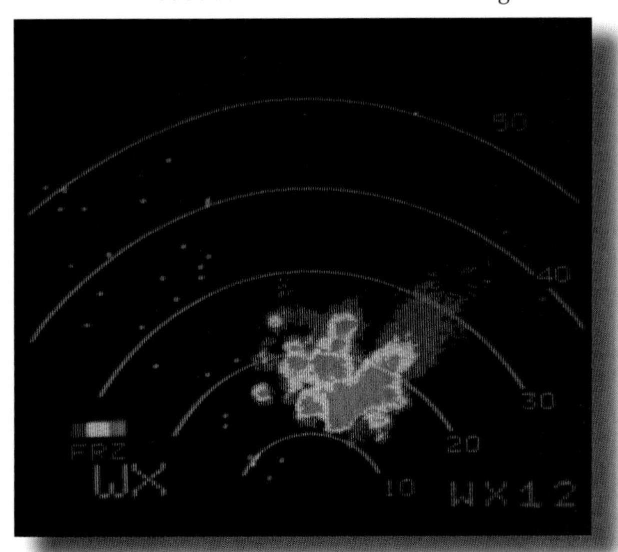

Figure 9-94. Airborne weather radar allows pilots to see and avoid many thunderstorms which do not appear on ground-based radar.

Aircraft radar generally use one of two frequency ranges — X-band and C-band. X-band systems, which are more common in general aviation aircraft, transmit on a frequency of 9,333 gigahertz (gHz), which has a wavelength of only 0.03 mm. This extremely short wave is reflected by very small amounts of precipitation. Due to the high amount of reflected energy, X-band systems provide a higher resolution and "see" farther than C-band radars. A disadvantage is that very little energy can pass through one storm to detect another which may be behind the first. The C-band frequency (5.44 gHz) can penetrate further into a storm, providing a more complete picture of the storm system. This capability makes C-band weather radar systems better for penetration into known areas of precipitation. Consequently, C-band radars are more likely to be found on large commercial aircraft.

Aircraft radar is prone to many of the same limitations as ground-based systems. It cannot detect water vapor, lightning, or wind shear. Training and experience as well as other on-board equipment and surface resources are important tools for enhancing your mental picture of the weather which lies ahead.

OTHER WEATHER DETECTION EQUIPMENT

Other, less costly, methods of thunderstorm detection have proven quite effective and have become very popular for light general aviation aircraft. Because lightning is always associated with severe thunderstorms, systems that detect lightning can reliably indicate the active parts of these storms. [Figure 9-95] Lightning detection equipment is more compact, and uses substantially less power than radar, and requires less interpretation by the pilot. The equipment is designed to help a pilot completely avoid storm cells. However, lightning detection equipment does not directly indicate areas of heavy precipitation, hail, and wind shear and may not provide as accurate information about these hazards as radar.

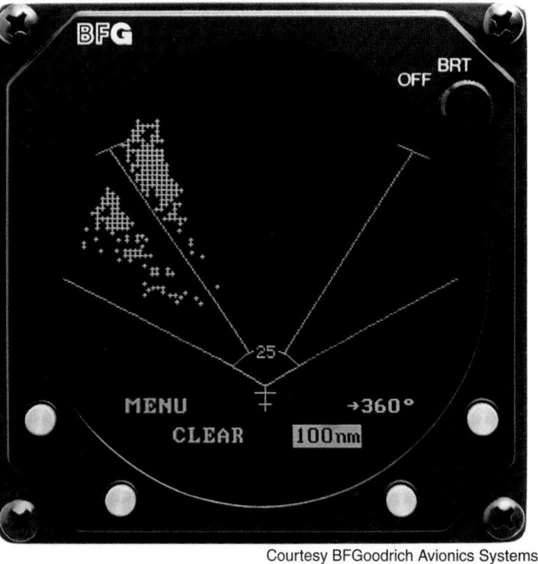

Figure 9-95. Lightning detection equipment provides many of the benefits of airborne radar at much lower cost, smaller size, and lower power consumption.

Courtesy BFGoodrich Avionics Systems
Stormscope® WX-1000

SUMMARY CHECKLIST

✓ You can obtain a preflight weather briefing from an FSS/AFSS 24 hours a day by calling the toll free number, 1-800-WX BRIEF.

✓ When connected to an AFSS, you can choose to talk to an FSS specialist, or you can use your TOUCH-TONE® phone to access the telephone information briefing service (TIBS). This recorded information typically includes area and/or route meteorological briefings, airspace procedures, and special aviation-oriented announcements.

✓ Request a standard briefing when you have received no previous weather information and an abbreviated briefing to receive an update. When your flight is more than six hours away, request an outlook briefing to make an initial judgment about the feasibility of your flight.

✓ You can receive an FAA-approved weather briefing and file a flight plan on your personal computer using the Direct User Access Terminal System (DUATS).

✓ AIRMETs are issued every six hours, with amendments issued as necessary, for weather phenomena which are potentially hazardous to light aircraft. AIRMETs are issued for moderate icing, moderate turbulence, sustained winds of 30 knots or more at the surface, ceilings less than 1,000 feet and/or visibility less than 3 miles affecting over 50 percent of an area at any one time, and extensive mountain obscurement.

✓ SIGMETs are issued for hazardous weather such as severe icing, severe and extreme turbulence, volcanic eruptions and duststorms, sandstorms, or volcanic ash lowering visibility to less than three miles.

✓ Existing or forecast hazardous convective weather which is significant to the safety of all aircraft is contained in convective SIGMETs (WSTs).

✓ Enroute Flight Advisory Service (EFAS) is one of the primary means of obtaining up-to-date weather information during flight. Below 18,000 feet, it is available throughout the conterminous United States on a frequency of 122.0 MHz. Other frequencies are available at 18,000 feet and above.

✓ In some areas, transcribed weather broadcasts (TWEB) provide recorded, route-based weather information over selected VORs and NDBs.

✓ A center weather advisory (CWA) is an unscheduled advisory issued by an ARTCC to alert pilots of existing or anticipated adverse weather conditions within the next two hours. It can precede the issuance of a SIGMET.

✓ Hazardous in-flight weather advisory service (HIWAS) is a continuous broadcast of summarized severe weather forecast alerts (AWW), SIGMETs, convective SIGMETs, CWAs, AIRMETs, and urgent PIREPs over selected VORs.

✓ The automated surface observation system (ASOS) and the automated weather observing system (AWOS) are the systems from which pilots and weather forecasters obtain most surface observations. Additional data from trained human observers supplements the data of many ASOS and AWOS systems.

✓ Airborne radar or lightning detection equipment usually can locate areas of hazardous weather ahead of your aircraft with greater reliability than ground-based radar.

KEY TERMS

1-800-WX BRIEF

Standard Briefing

Abbreviated Briefing

Outlook Briefing

Telephone Information Briefing Service (TIBS)

Direct User Access Terminal System (DUATS)

AIRMET (WA)

SIGMET (WS)

Convective SIGMET (WST)

Enroute Flight Advisory Service (EFAS)

Center Weather Advisory (CWA)

Hazardous In-Flight Weather Advisory Service (HIWAS)

Automated Surface Observation System (ASOS)

Automated Weather Observing System (AWOS)

QUESTIONS

1. What information should you provide a preflight weather briefer?

2. True/False. Calling an AFSS and listening to TIBS eliminates the need for an individual briefing from an FSS or NWS specialist.

For questions 3 through 8, match the weather source with the appropriate description.

3. Standard briefing
4. Abbreviated briefing
5. Outlook briefing.
6. DUATS
7. TIBS
8. EFAS

A. Provides FAA-approved weather briefings, weather maps, and filing of flight plans over a modem-equipped personal computer

B. Source of in-flight weather information

C. Obtained more than six hours before flight time to determine the feasibility of a proposed flight

D. Most comprehensive briefing available from an FSS briefer

E. Used to update previous weather information

F. Recorded weather information available from an AFSS.

9. True/False. Weather phenomena which are of operational interest to all aircraft are reported in an AIRMET.

10. Select the weather phenomena which can initiate the issuance of a SIGMET.

    A. Severe icing
    B. Embedded thunderstorms
    C. Hazardous convective weather

11. Convective SIGMETs include information on

    A. thunderstorms, super cells, and tornadoes.
    B. tornadoes, embedded thunderstorms, and lines of thunderstorms.
    C. embedded thunderstorms, severe thunderstorms with hail greater than or equal to 1/2 inch in diameter, and/or wind gusts 50 knots or greater.

12. What type of recorded in-flight weather advisory provides specific route forecasts and winds aloft information over selected NDBs and VORs?

    A. EFAS
    B. TWEB
    C. HIWAS

Refer to the IFR enroute chart to answer question 13.

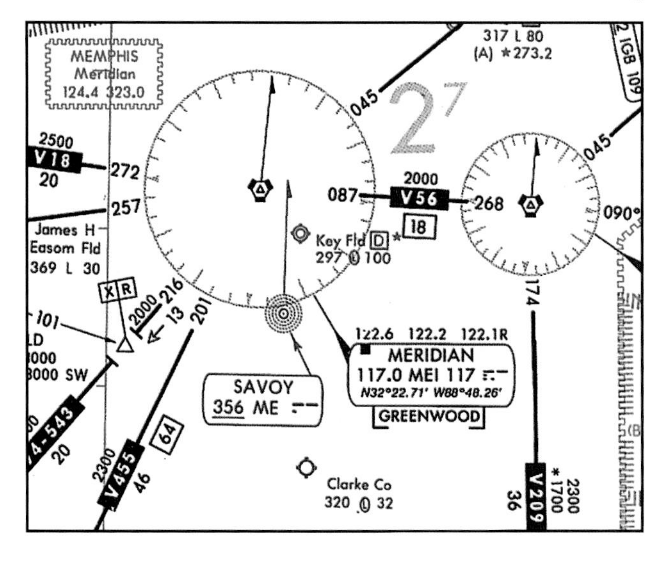

13. What type of weather information is available when monitoring Meridian VORTAC?

    A. Recorded information on expected sky cover, cloud tops, visibility, weather, and obstructions to vision in a route format
    B. Summarized SIGMETs, convective SIGMETs, AWWs, CWAs, AIRMETs, and urgent PIREPs
    C. ASOS broadcasts from nearby Key Field

14. What is the purpose of a Center Weather Advisory?

    A. To advise pilots of SIGMETs and AIRMETs so they can contact Flight Watch.
    B. To disseminate routine pilot reports so pilots do not need to leave the ATC frequency.
    C. To alert pilots of current or adverse weather conditions not covered by existing SIGMETs and AIRMETs.

CHAPTER 10

IFR Flight Considerations

Instrument/Commercial
Part III, Chapter 10 — IFR Flight Considerations

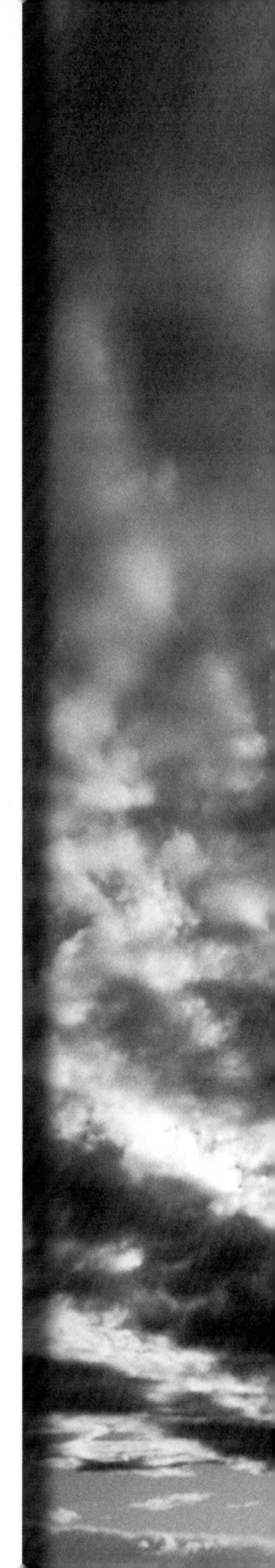

SECTION A
IFR EMERGENCIES

No one likes to think they might be required to deal with an emergency during instrument flight. However, proper planning and sound knowledge of emergency procedures can help you complete your flight safely should an emergency occur.

Naturally, you should be thoroughly familiar with the pilot's operating handbook for the aircraft you fly. The POH outlines specific emergency procedures that apply to your particular aircraft. However, the POH cannot prepare you for every event that may occur. When confronted by an emergency not covered in the aircraft's POH, or in the regulations, you are expected to exercise good judgment in responding to the situation.

The *Aeronautical Information Manual* defines an emergency as a condition of distress or urgency. Pilots in **distress** are threatened by serious and/or imminent danger and require immediate assistance. Distress conditions may include in-flight fire, mechanical failures, or structural damage. An **urgency** situation, which is not immediately dangerous, requires prompt assistance to avoid a potentially catastrophic event. Any condition that may adversely affect your flight, such as low fuel quantity or poor weather may result in an urgency condition which can develop into a distress situation if not handled in a timely manner. If you become apprehensive about your safety for any reason, you should request assistance immediately.

ATC may request a detailed report of an emergency when priority assistance has been given, even though no rules have been violated.

Other situations associated with weather, such as inadvertent entry into a thunderstorm, hail, severe turbulence, or icing, are potential emergencies. Fuel starvation or inability to maintain the MEA are undeniable emergencies. If, after considering the particular circumstances of the flight, you feel a potentially dangerous or unsafe condition exists, you should declare an emergency. During the course of the emergency, if you are given priority handling by ATC, you are required to submit a detailed report within 48 hours to the manager of that ATC facility, even though you violated no rule.

DECLARING AN EMERGENCY

As pilot in command, you are directly responsible for the safety of your flight. Under the provisions of FAR 91.3, you are allowed to deviate from any rule in FAR Part 91 and from

ATC instructions, to the extent required to meet an emergency. If you determine it is necessary to deviate from ATC instructions during an emergency, you must notify ATC as soon as possible and obtain an amended clearance.

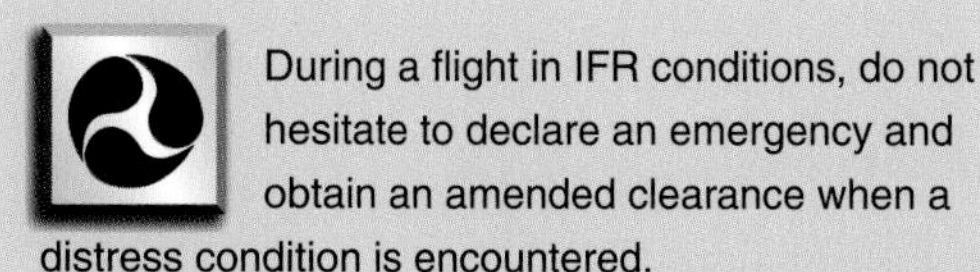

During a flight in IFR conditions, do not hesitate to declare an emergency and obtain an amended clearance when a distress condition is encountered.

To declare an emergency when operating in IFR conditions, you should contact ATC on the currently assigned frequency. If you receive no response, try calling the same facility on another available frequency. If the facility still does not respond, you should attempt to contact any other ATC facility that may be able to assist you. If you are unable to contact ATC on any of the normal frequencies, use the emergency frequency of 121.5 MHz.

In a distress situation, begin your initial call with the word **MAYDAY**, preferably repeated three times. Use **PAN-PAN** in the same manner in an urgency situation. Figure 10-1 provides an example of a distress or urgency message. You should provide ATC with only as much information as necessary under prevailing conditions. For example, when in radar contact with an ATC facility, you may not need to include your aircraft's type, position, heading, or altitude, since the controller is already aware of these facts.

INFORMATION	EXAMPLE
Distress or Urgency	*"MAYDAY, MAYDAY, MAYDAY (or PAN-PAN, PAN-PAN, PAN-PAN),*
Name of station addressed	*Seattle Center,*
Identification and type of aircraft	*1114V Cessna 182,*
Nature of distress or urgency	*severe icing,*
Weather	*IFR,*
Your intentions and request	*request immediate course reversal and lower altitude,*
Present position and heading	*JIMMY Intersection, heading 253°*
Altitude or flight level	*9,000.*
Fuel remaining in hours and minutes	*Estimate two hours fuel remaining,*
Number of people aboard	*five aboard,*
Any other useful information	*squawking 1146."*

Figure 10-1. If you are in distress, your initial communication and any subsequent transmissions should begin with the word MAYDAY, preferably repeated three times. PAN-PAN should be used in the same manner for an urgency situation. Following this, you need to provide information about your situation and the assistance that you require.

Another way to declare an emergency is to squawk **code 7700** on your transponder. This code triggers an alarm or a special indicator in radar facilities. However, if you are in radio contact with ATC, do not change your assigned transponder code from its current setting unless you are instructed to do so by that facility.

A **special emergency** is a condition of air piracy, or other hostile act by a person or persons aboard an aircraft, which threatens the safety of the aircraft or its passengers. Although these incidents rarely involve the average pilot, you should be aware of the recommended ATC procedures. In a special emergency, you should use the recommended distress or urgency procedures. When circumstances do not permit you to comply, transmit a message containing as much of the following information as possible on the frequency in use.

1. Name of the station addressed
2. Aircraft identification and present position
3. Nature of the special emergency condition and your intentions

If you are unable to provide the above information, you should alert ATC by transmitting the phrase, "*transponder seven five zero zero*" and/or squawking transponder **code 7500** Either action means, "*I am being hijacked/forced to a new destination.*"

MINIMUM FUEL

If your remaining fuel is such that you can accept little or no delay, you should advise ATC that you have **minimum fuel**. This is not an emergency, but only an advisory that any undue delays may cause an emergency situation. If your remaining usable fuel supply indicates the need for traffic priority to ensure a safe landing, do not hesitate to declare an emergency. When transmitting such a report, you should state the approximate number of minutes the flight can continue with the fuel remaining.

Declaring minimum fuel to ATC indicates an emergency situation is possible should any undue delay occur.

"BOTH FUEL CELLS WERE EMPTY..."

From the Files of the NTSB . . .
Aircraft: PA-32RT-300T
Location: Rogers, AR
Injuries: 1 Serious, 1 Minor

Narrative: According to witnesses and law enforcement personnel at the accident site, the pilot executed 3 missed approaches at the Rogers Municipal Airport while attempting to land with weather below published minimums. At 0734, the pilot reported to ATC that he was low on fuel. Two minutes later, the pilot reported that he was out of fuel. The airplane was last observed on radar descending through 1,500 feet and subsequently impacted trees approximately one mile northwest of the airport. Examination of the wreckage confirmed that both fuel cells were empty.

Had this pilot declared minimum fuel in a timely manner, ATC might have provided him the assistance he needed to complete his flight safely at an alternate airport. Waiting until just prior to fuel exhaustion can turn an urgent situation into a distress situation that may be unrecoverable.

GYROSCOPIC INSTRUMENT FAILURE

Gyroscopic instruments that stop functioning in IFR conditions due to a vacuum or electrical system failure can result in a distress situation. This type of instrument failure may develop into an emergency since your ability to immediately and accurately comply with all ATC clearances will be limited. This is particularly true in circumstances where warning indications, such as a low-vacuum warning light or low-voltage warning light fail to provide adequate warning of an impending instrument system malfunction. If you are in IFR conditions and find that one of your gyroscopic instruments has failed, you should immediately transition to partial panel and notify ATC. Since instrument failure provides little advance warning, all pilots should maintain their proficiency in partial panel instrument flying.

Detecting a problem early is important in successfully handling a gyroscopic instrument failure. A good instrument cross-check is essential for discovering a problem quickly. Your instrument cross-check should include occasional monitoring of instrument system warning indicators. If you suspect that a gyroscopic instrument has failed, verify

the problem with related flight instruments. In the following example you can confirm an attitude indicator failure by referencing your airspeed indicator, vertical speed indicator, altimeter, turn coordinator, and heading indicator. With a diligent instrument cross-check you will see a discrepancy between the failed attitude indicator and the supporting instruments. [Figure 10-2]

Figure 10-2. A good instrument cross-check allows you to quickly determine if you have an instrument failure. In this case, the attitude indicator has failed showing a climbing left turn. However, the airspeed is high, while the altimeter and VSI both show a descent. In addition, the turn coordinator displays wings-level flight.

Once you have determined an instrument has failed, cover it so you are not distracted by the incorrect information it provides. Then contact ATC and make the appropriate malfunction report. Your primary concern during an instrument failure is to fly the airplane. Once you have ensured your ability to control the airplane, your priorities are to navigate accurately and communicate your situation and intentions to ATC. The execution of good partial panel instrument flying procedures is important in maintaining attitude control of your airplane. It is also important for you to use available ATC services like radar and no-gyro approaches, to assist you in completing your flight safely. Emergency approach procedures available to you from ATC will be discussed later in this section.

COMMUNICATION FAILURE

Two-way radio communication failure procedures for IFR operations are outlined in FAR 91.185. Unless otherwise authorized by ATC, pilots operating under IFR are expected to comply with this regulation. Expanded procedures for communication failures are found in the *Aeronautical Information Manual.* In some cases, special lost communication procedures are established for certain IFR operations. [Figure 10-3]

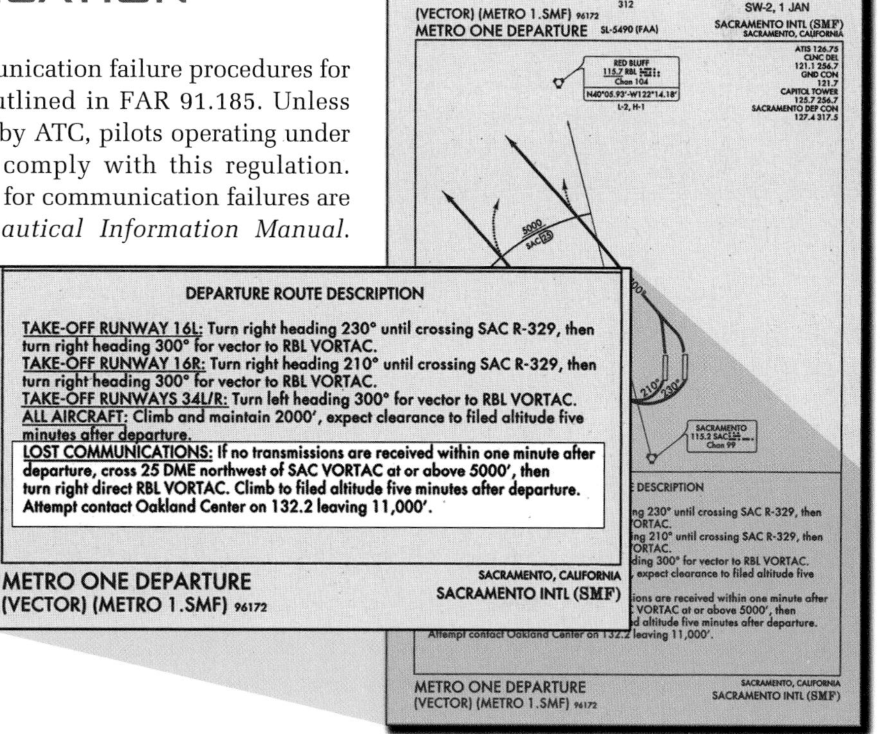

Figure 10-3. Special lost communication procedures are sometimes used with charted departures. If you are flying this departure and lose radio communication capability, you must comply with the lost communication procedures specified here instead of those in the FARs.

ALERTING ATC

You can use your transponder to alert ATC to a radio communication failure by squawking **code 7600**. [Figure 10-4] If only your transmitter is inoperative, listen for ATC instructions on any operational receiver, including your navigation receivers. It is possible ATC may try to contact you over a VOR, VORTAC, NDB, or localizer frequency. In addition to monitoring your navaid receivers, you should attempt to reestablish communications by contacting ATC on a previously assigned frequency, or calling an FSS or Aeronautical Radio/Incorporated (ARINC).

You are in IFR conditions and have two-way radio communication failure. If you do not exercise emergency authority, you should set your transponder code to 7600.

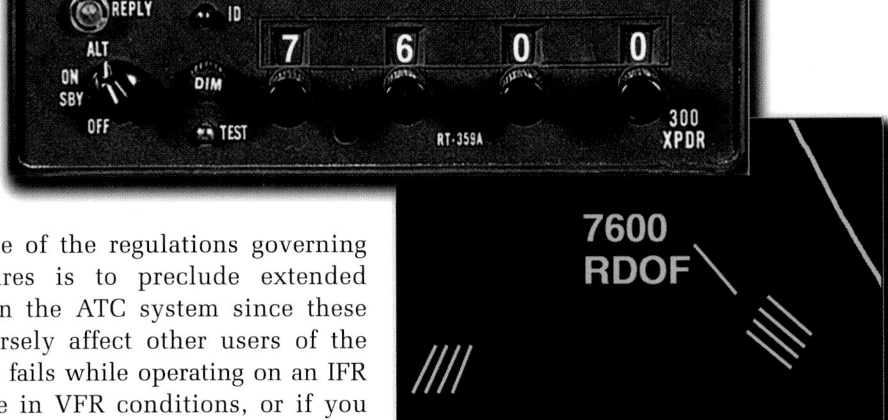

Figure 10-4. When an airplane squawks code 7600 during a two-way radio communication failure, the information block on the radar screen flashes RDOF to alert the controller.

The primary objective of the regulations governing communication failures is to preclude extended IFR operations within the ATC system since these operations may adversely affect other users of the airspace. If your radio fails while operating on an IFR clearance but you are in VFR conditions, or if you encounter VFR conditions at any time after the failure, you should continue the flight under VFR conditions, if possible, and land as soon as practicable. The requirement to land as soon as practicable should not be construed to mean as soon as possible. You retain the prerogative of exercising your best judgment and are not required to land at an unauthorized airport, at an airport unsuitable for the type of aircraft flown, or to land only minutes short of your intended destination. However, if IFR conditions prevail, you must comply with procedures designated in the FARs to ensure aircraft separation.

In the event of two-way radio communication failure while operating on an IFR clearance in VFR conditions, you should continue the flight under VFR and land as soon as practicable.

ROUTE

If you must continue your flight under IFR after experiencing two-way radio communication failure, you should fly one of the following routes in the order shown. [Figure 10-5]

1. The route assigned by ATC in your last clearance received
2. If being radar vectored, the direct route from the point of radio failure to the fix, route, or airway specified in the radar vector clearance
3. In the absence of an assigned route, the route ATC has advised you to expect in a further clearance
4. In the absence of an assigned or expected route, the route filed in your flight plan

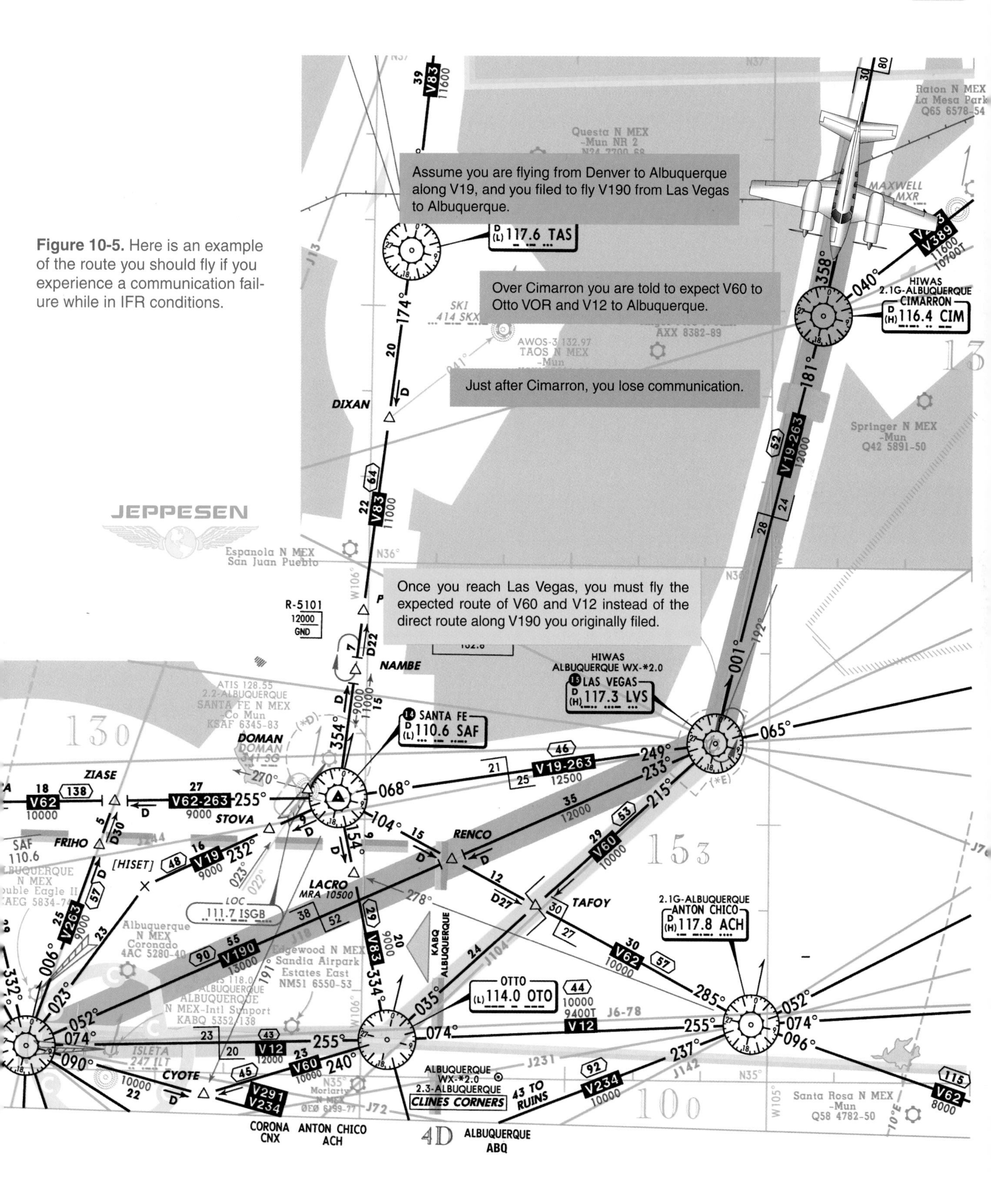

Figure 10-5. Here is an example of the route you should fly if you experience a communication failure while in IFR conditions.

ALTITUDE

It is also important for you to fly a specific altitude should you lose two-way radio communications. The altitude you fly after a communication failure can be found in FAR 91.185 and must be the highest of the following altitudes for each route segment flown.

1. The altitude (or flight level) assigned in your last ATC clearance
2. The minimum altitude (or flight level) for IFR operations
3. The altitude (or flight level) ATC has advised you to expect in a further clearance

In some cases, your assigned or expected altitude may not be as high as the MEA on the next route segment. In this situation, you normally begin a climb to the higher MEA when you reach the fix where the MEA rises. If the fix also has a published minimum crossing altitude (MCA), start your climb so you will be at or above the MCA when you reach the fix. If the next succeeding route segment has a lower MEA, descend to the applicable altitude — either the last assigned altitude or the altitude expected in a further clearance — when you reach the fix where the MEA decreases. [Figure 10-6]

The altitude and route to be used if you are flying in IFR conditions and have two-way radio communication failure are the route specified in your clearance, and the highest of the following: the altitude assigned by ATC, told to expect by ATC, or the MEA. See figures 10-5 and 10-6.

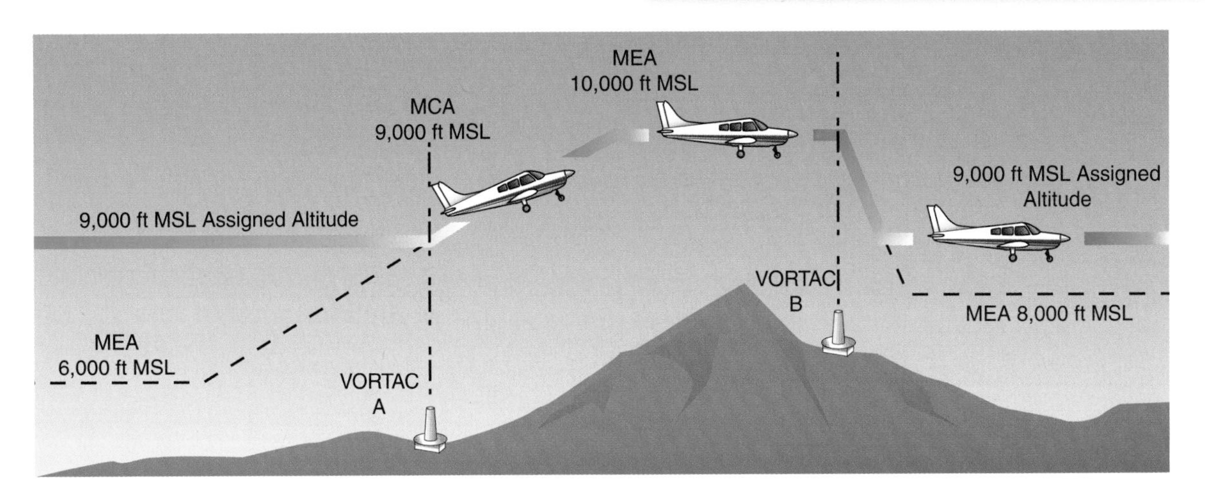

Figure 10-6. Assume your last assigned altitude is 9,000 feet MSL. At VORTAC A, the MEA increases to 10,000 feet with an MCA of 9,000 feet specified. Since your assigned altitude is 9,000 feet, you need not begin a climb before you reach the VORTAC. Upon reaching VORTAC A, climb to the higher MEA of 10,000 feet. As you arrive over VORTAC B, descend back to your assigned altitude of 9,000 feet. The MEA is lower than your assigned altitude.

LEAVING THE CLEARANCE LIMIT

When the clearance limit specified in the ATC clearance you received prior to radio failure is also a point from which an approach begins, commence your descent and approach as close as possible to the expect further clearance (EFC) time. Should you arrive at your clearance limit prior to the EFC time, you must hold at the clearance limit until that time has been reached. You may be required to adjust the holding pattern as necessary in order to begin the approach at the proper time. If an EFC time was not provided, begin your descent and approach as close as possible to the ETA calculated from your filed or amended (with ATC) time enroute. In this case, it may still be necessary to hold at the clearance limit in order to begin the approach at the proper time if you arrive at the fix early.

Upon arrival at a destination with more than one instrument approach procedure, you may fly the approach of your choice. Similarly, if more than one initial approach fix is available for the approach you choose, you may select whichever fix is appropriate. ATC provides separation for your flight, regardless of the approach selected and the initial approach fix used.

If you have an EFC time while holding at a holding fix that is not the approach fix and you experience two-way radio communication failure, depart the holding fix at the EFC time.

While holding at a DME fix for an ILS approach, ATC advises you to expect clearance for the approach at 1015. At 1000 you experience two-way radio communication failure. In this case, you should immediately squawk 7600 and plan to begin your approach at 1015.

When the clearance limit specified in your last clearance is not a fix from which an approach begins, leave the fix at the EFC time specified. After departing the fix and arriving at a point from which an approach begins, commence your descent and complete the approach. If you have not received an EFC time, continue past the clearance limit to a point at which an approach begins. You should then commence your descent and approach as close as possible to the ETA calculated from your filed or amended time enroute.

EMERGENCY APPROACH PROCEDURES

In a distress situation, such as a loss of gyroscopic instruments, radar approach procedures may be available to assist you in completing your flight safely. These procedures require a ground-based radar facility and a functioning airborne radio transmitter and receiver. They allow the controller to provide horizontal, and in some cases vertical, course guidance during the approach. However, radar guidance during the approach does not waive the prescribed weather minimums for the airport or for the particular aircraft operator concerned. Three types of radar approaches may be available to you including the airport surveillance radar approach (ASR), precision approach radar (PAR), and the no-gyro approach.

SURVEILLANCE APPROACH

An instrument procedure where a controller provides only azimuth navigational guidance is referred to as an **airport surveillance radar (ASR)** approach. The ASR approach also is referred to as a **surveillance approach**. With this type of procedure, the controller furnishes headings to fly to align your aircraft with the extended centerline of the landing runway. You are advised when to start the descent to the MDA or, if appropriate, to an intermediate stepdown fix and then to the prescribed MDA.

During the approach, you are advised of the location of the missed approach point and, while on final, of your position each mile from the runway or MAP. If you request, you are advised of the recommended altitude each mile. Normally, ATC provides navigation guidance until your aircraft reaches the MAP. At the MAP, ATC terminates guidance and instructs you to execute a missed approach unless you have reported the runway environment in sight. Also, if at any time during the approach the controller considers that safe guidance for the remainder of the approach cannot be provided, guidance is terminated and you are instructed to execute a missed approach. Radar service is automatically terminated at the completion of the approach.

In addition to headings, the information a radar controller provides without request during an ASR approach includes: when to commence descent to the MDA, the aircraft's position each mile on final from the runway, and arrival at the MAP.

The published MDA for straight-in ASR approaches is issued to you before you are instructed to begin your descent. When a surveillance approach terminates in a circle-to-land maneuver, you must furnish the aircraft approach category to the controller, who then provides you with the appropriate MDA.

The ASR approach is available only at airports for which civil radar instrument approach minimums have been published. These minimums are published on separate pages in the NOS *Terminal Procedures Publication* (TPP) and on Jeppesen Radar Approach Charts. The ASR approach is typically available as a backup procedure. Use it when you have lost the navigation equipment needed to fly the other instrument approaches available at the airport.

A surveillance approach may be used at airports for which civil radar instrument approach minimums have been published.

You've Seen One Emergency, You Haven't Seen Them All

Picture yourself flying a single-engine airplane enroute from Harbour Grace, Newfoundland, aiming for Paris, France. You have no co-pilot, no airways to follow, and no air traffic controllers to assist you. Now, imagine that you experience not one, but several emergencies with nothing but the depths of the Atlantic Ocean beneath you.

This is the same situation that Amelia Earhart faced in 1932 after she set out to fly solo across the Atlantic in a Lockheed Vega. Earhart's first crisis occurred when her altimeter failed, just a few hours after her departure. For the rest of the long flight, she had no indication of the Vega's altitude. Shortly after the altimeter failed, Earhart encountered a thunderstorm and wandered off course. About four hours into the flight, a bad weld caused the exhaust manifold seam to part and Earhart could see the glow of flames by looking through a gap under the rim of the cowling. During the night, Earhart faced increasing vibrations from the damaged manifold. As the flight progressed, the Vega began picking up ice in the clouds and then entered a spin. Earhart was able to recover just over the water with little altitude to spare. She tried to stay under the clouds, but ran into fog. Although Earhart finally found an altitude that allowed her to fly a safe distance from the water and stay clear of the icing conditions, her troubles continued.

Just prior to her transatlantic flight, Amelia Earhart poses in front of her Lockheed Vega with mechanic Edward Gorski and the pilot who test flew the Vega, Brent Balchen.

About two hours from the coast of Ireland, Earhart turned on the reserve tank and discovered that the cabin fuel gauge was leaking. She became concerned that the fumes from the leaking fuel would be ignited by the flames from the exhaust system. At this point, Earhart decided not to continue as planned to Paris, but to land as soon as she could. The Lockheed Vega touched down in a pasture near Londonderry, Northern Ireland, after a flight which lasted almost 15 hours. As Earhart climbed out of the airplane, a surprised farmhand approached the Vega. *"Where am I?"* she called out. *"In Gallagher's pasture,"* was the reply.

PRECISION APPROACH RADAR

During an approach using **precision approach radar (PAR)** the controller provides highly accurate navigational guidance in azimuth and elevation. You will be given headings to fly to align your aircraft with the runway centerline and to keep you there during the approach. The controller will advise you of glide slope intercept 10 to 30 seconds before it occurs and you will be told when to start the descent. The controller will also report your range to touchdown at least once each mile during the approach.

Course deviation and trend information will be given to you to help you make the necessary adjustments relative to the approach path. For instance, a controller may report that you are *"well above glide path, coming down rapidly."* This information allows you to adjust your airplane's flight path accordingly and keep you within the safety zone limits of the approach. Navigational guidance is provided by the controller down to the decision height. From the DH to the threshold, advisory course guidance is furnished with radar service terminated upon completion of the approach.

During a no-gyro approach and prior to being handed off to the final approach controller, the pilot should make all turns at standard rate unless otherwise advised.

Like the surveillance approach, PAR (when available) may be used when you lose the navigation instruments needed to fly the other instrument approaches available at the airport. Approach minimums can be found in the NOS TPP and the Jeppesen Radar Approach Charts.

After being handed off to the final approach controller during a no-gyro surveillance or precision approach, the pilot should make all turns at one-half standard rate.

NO-GYRO APPROACH

If your heading indicator becomes inoperative or inaccurate due to a gyroscopic instrument failure, you should advise ATC of the malfunction and request a **no-gyro vector** or **no-gyro approach**. During the vector or approach, ATC will instruct you to make turns by saying, *"turn right, stop turn,"* and *"turn left."* ATC expects these turns to be executed as soon as you receive the instructions. You should comply with ATC directions by making standard-rate turns until you have been turned onto final. Once ATC advises you that you have been turned onto final, all turns should be made at half standard rate.

MALFUNCTION REPORTS

The loss of certain equipment during an IFR flight may have little significance to you. For example, if you are flying to a destination not served by an NDB, the loss of the ADF receiver may be hardly noticeable. Regardless of your immediate need for that piece of equipment, you are required by regulations, while operating under IFR in controlled airspace, to report to ATC the loss of the ADF receiver. FAR 91.187 requires that you report the malfunction of any navigational, approach, or communication equipment.

If a VOR receiver malfunctions while operating in controlled airspace under IFR in an airplane equipped with dual VOR receivers, immediately report the malfunction to ATC.

When you make a **malfunction report**, you are expected to include the following information

1. Aircraft identification
2. Equipment affected
3. Degree to which the equipment failure will impair your ability to operate under IFR
4. Type of assistance desired from ATC

SUMMARY CHECKLIST

✓ The *Aeronautical Information Manual* defines an emergency as a condition of distress or urgency. Pilots in distress are threatened by serious and/or imminent danger and require immediate assistance. An urgency situation, such as low fuel quantity, requires timely but not immediate assistance.

✓ In an emergency, you may deviate from any rule in FAR Part 91 to the extent necessary to meet the emergency.

✓ The frequency of 121.5 MHz may be used to declare an emergency in the event you are unable to contact ATC on other frequencies.

✓ In a distress situation, begin your initial call with the word *"MAYDAY,"* preferably repeated three times. Use *"PAN-PAN"* in the same manner in an urgency situation.

✓ Your transponder may be used to declare an emergency by squawking code 7700.

✓ A special emergency is a condition of air piracy and should be indicated by squawking code 7500 on your transponder.

✓ If your remaining fuel quantity is such that you can accept little or no delay, you should alert ATC with a minimum fuel advisory.

✓ If the remaining usable fuel supply suggests the need for traffic priority to ensure a safe landing, you should declare an emergency due to low fuel and report fuel remaining in minutes.

✓ Gyroscopic instruments include the attitude indicator, heading indicator, and turn coordinator. These instruments are subject to vacuum and electrical system failures.

✓ During an instrument failure your first priority is to fly the airplane, navigate accurately, and then communicate with ATC.

✓ You can use your transponder to alert ATC to a radio communications failure by squawking code 7600.

✓ During a communication failure while operating under IFR, you are expected to follow the lost communication procedures specified in the regulations.

✓ During a communication failure in VFR conditions, remain in VFR conditions and land as soon as practicable.

✓ If you lose communication with ATC during your flight, you must fly the highest of the assigned altitude, MEA, or the altitude ATC has advised may be expected in a further clearance.

✓ If an approach is available at your clearance limit, begin the approach at the expect further clearance (EFC) time. If an approach is not available at your clearance limit, proceed from the clearance limit at your EFC to the point at which an approach begins.

✓ Radar approach procedures may be available to assist you during an emergency situation requiring an instrument approach.

✓ A radar instrument approach that provides only azimuth navigational guidance is referred to as an airport surveillance radar (ASR) approach.

✓ During a precision approach (PAR), the controller provides you with highly accurate navigational guidance in azimuth and elevation as well as trend information to help you make the proper corrections while on the approach path.

✓ A no-gyro approach may be requested when you have experienced a gyroscopic instrument failure. Controllers provide course guidance by stating *"turn right, stop turn,"* and *"turn left"* to align you with the approach path. Turns should be made at standard rate until you have been turned onto final at which point in time they should be made at half standard rate.

✓ FAR Part 91 requires that you report the malfunction of any navigational, approach, or communications equipment while operating in controlled airspace under IFR. In the malfunction report you should include the aircraft ID, equipment affected, the degree to which the flight will be impaired by the failure, and any assistance you require from ATC.

KEY TERMS

Distress

Urgency

MAYDAY

PAN-PAN

Minimum Fuel

Special Emergency

Code 7500

Code 7600

Code 7700

Airport Surveillance Radar (ASR)

Surveillance Approach

Precision Approach Radar (PAR)

No-Gyro Vector

No-Gyro Approach

Malfunction Report

QUESTIONS

1. What phrase should be used to begin a distress call to ATC?

2. Which transponder code should be used during a special emergency?

   A. 7700
   B. 7600
   C. 7500

3. True/False. Declaring minimum fuel is an emergency.

4. True/False. If you are in IFR conditions and find that one of your gyroscopic instruments has failed, you should immediately transition to the partial panel technique of instrument flying.

5. What is the primary objective of the regulations governing communication failure during IFR operations?

6. You should fly the highest of three altitudes if you lose two-way radio communication in IFR conditions. What are those altitudes?

7. List the routes, in order of importance, you are expected to fly should you lose two-way radio communication in IFR conditions.

8. If, prior to a loss of two-way radio communication, you receive an EFC time from ATC for a holding fix from which an approach begins, and you arrive at that fix early, what should you do?

9. Two specific items are required in order to use a radar approach procedure. What are they?

10. List the three types of radar approach procedures that may be available for use during an emergency.

11. Which radar approach procedure provides course deviation and trend information?

12. True/False. After being turned onto final during a no-gyro approach, all turns should be made at standard rate.

13. List the items you are expected to include in a malfunction report to ATC.

SECTION B
IFR DECISION MAKING

The only thing that keeps a man out of a storm is his own decision not to enter it, his own hands turning the airplane back to clear air, his own skill taking him back to a safe landing. Flight remains the world of the individual, where he decides to accept responsibility for his action. . . — Richard Bach, *A Gift of Wings*

Aviation differs from many pursuits since it demands so much precision from the choices that you make, especially within the IFR environment. The complexity of these decisions is one of the reasons why you must have reached a certain degree of experience before you can become an instrument-rated pilot. When you fly in poor weather, without visual references, you open yourself to situations that you would never encounter as a pilot flying VFR. You also operate in a more structured environment, and collaborate with ATC through the course of your flight.

Consider the following situations, and address each with the knowledge you have gained so far in your instrument training.

- You are on vacation and scheduled to return to work the next day for an important meeting. A slow-moving warm front sits over your entire route, and there are multiple PIREPs of moderate icing. The weather is expected to continue through the next 24 hours. Your light twin has pitot heat, but no other deicing protection. Airports along the route are at or above minimums.

- During climbout on an IFR cross-country, your attitude indicator fails. You have already picked up your clearance, and you are in solid IFR conditions.

- You are one hour from your destination, cruising between layers. You notice that the low voltage light on the panel is illuminated, but you are not sure how long it has been on.

- During your second ILS approach attempt, you see the approach lights at decision height. As you descend, they disappear behind a patch of fog.

- After a lengthy hold, you begin an approach to your destination, and the visibility reported is right at minimums. At the MDA, you cannot see the runway. You execute the missed approach and calculate that you no longer have enough fuel to fly to your filed alternate.

- The controller issues you an immediate descent to the initial approach fix altitude at your destination. You know that you are still over an area of mountainous terrain, and the descent may take you below the minimum safe altitude for the area. The frequency is congested, and you are unable to contact the controller.

The IFR environment presents unique situations in which you will have to apply the decision making process. You are simply unable to specifically prepare for every possible problem in the cockpit. However, through careful consideration of several risk elements, you can assess each situation and make an effective decision. [Figure 10-7]

You are current, having flown approaches locally in the past six months, but it has been a while since you completed an IFR cross-country flight. You are well-rested, and in good health.

Pilot — Evaluate your fitness to fly, including your competency in the airplane, instrument currency, and flight experience.

The aircraft is familiar to you, and while it is normally-aspirated, it has dual nav/coms with glideslope capability, and a GPS that is IFR enroute and approach-certified. You also have lightning detection equipment.

Aircraft — The airplane's equipment, performance, limitations, and airworthiness must be considered.

There is a cold front that you will have to cross on your route, with reports of embedded thunderstorms, moderate icing, and tops reaching 15,000 feet MSL. The ceilings are low enough to eliminate the option of flying VFR below the clouds. Circumnavigating the line would require you to extend the trip beyond your aircraft's range, necessitating a fuel stop that would cut into the time saved by flying yourself to the appointment.

Environment — Factors, such as weather, airport minimums and conditions, and air traffic control, must also be weighed.

You have a business appointment tomorrow with an important client in a city 500 miles distant.

Operation — The purpose of the flight is a factor which influences your decision to undertake or continue a flight.

After weighing the risk elements, you determine that it would not be prudent for you to attempt to fly through the front, nor would you want to divert so far out of your way. You may elect to make plans for another IFR cross country after you return, and you will fly commercially to the out-of town meeting, or choose to drive. You may also wait for the front to pass through before beginning the flight.

Situation — Overall situational awareness requires an accurate perception of how the pilot, aircraft, environment and operation interact to affect the flight.

Figure 10-7. When making a go/no-go decision, you must consider the risk elements.

APPLYING THE DECISION-MAKING PROCESS

The circumstances of flight change constantly, as you are exposed to a varying number of events. You can either react to situations as they are presented to you, or you can anticipate what is likely to occur next and plan your course of action. When you deal with events as they come, you behave in a reactive manner. If you plan for and rehearse problems before they arise, you become proactive. As a pilot, your reactive skills are important, such as when you maneuver away from other traffic, but your ability to develop proactive mental processes will allow you to make better decisions in shorter time frames.

THE IFR ACCIDENT

When you fly on an IFR flight plan, you are operating with one of the highest levels of safety available to general aviation, due to the level of support and control you receive from ATC, and the training you need in order to act as an instrument pilot. However, accidents do occur. Continued VFR flight into IFR conditions is the most hazardous of

accident scenarios. Weather-related accidents involving IFR conditions are roughly 69% fatal, and there are several different types of IFR accidents. You will acquire skills and experience to help you avoid each of these accident causes during your instrument training. The first type of accident involves controlled flight into terrain (CFIT), associated with instrument approach and departure procedures. The increased situational awareness possible through correct use of instrument procedures and visualization techniques makes CFIT less likely. A second accident cause is spatial disorientation, induced by either continued VFR flight into IFR conditions or lack of instrument proficiency. If you frequently practice flying by reference to instruments under simulated IFR conditions or in a flight training device, you are less likely to become disoriented in actual conditions. Another cause of IFR accidents is loss of control in IFR conditions, sometimes brought on by flight into convective activity, and possibly compounded by spatial disorientation. Awareness of wide-scale weather and judicious use of onboard weather-detection equipment can help you avoid flying into thunderstorms, but your best defense is the respect you will gain for adverse weather through caution and experience. Several situatuions that can lead to accidents that will be addressed throughout your instrument training, and by keeping these in mind you can become a safer pilot. [Figure 10-8]

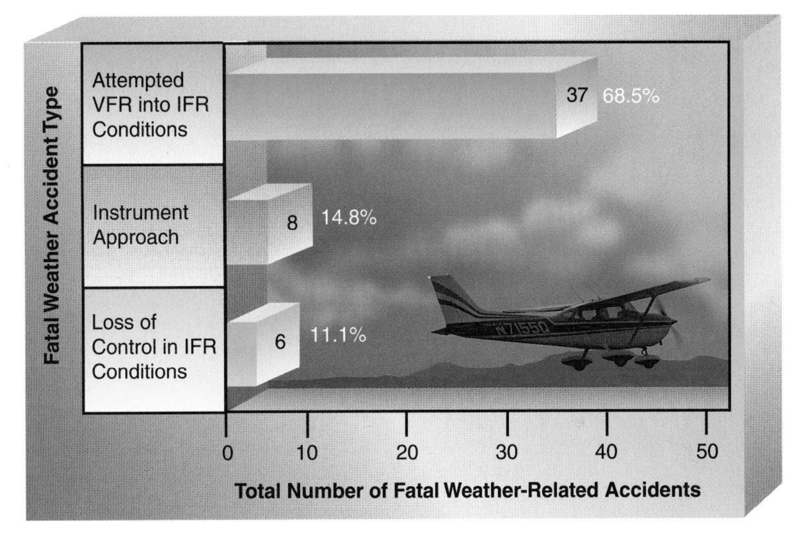

Figure 10-8. Nearly 95% of weather-related fatal accidents in single-engine aircraft in 1996 were attributed to the 3 scenarios shown here. Total percentage does not equal 100, as other accident types are not listed.

An accident, however, is rarely attributed to a single cause. Instead, a series of events, coupled with poor decisions, leads to the accident. In many cases, going back to where the first error in judgment occurred can provide a better foundation for preventing similar accidents in the future. For example, a number of accidents list fuel exhaustion as the probable cause, but many different circumstances can contribute to the accident. Many instances of fuel exhaustion are the result of poor planning, or failure to update the actual fuel burn during flight.

POOR JUDGMENT CHAIN

The string of events leading to an accident gives the pilot many opportunities to avoid disaster. One intelligent choice can mean the difference between a simple learning experience and an accident report. Often, a precautionary landing is the best option, either at an airport or other suitable landing site. Declaring an emergency and asking ATC for assistance is another viable alternative that many pilots in accident situations fail to take. Any course of action that allows you to maintain positive control of the airplane enhances your chance of bringing the flight to a successful conclusion. However, many of the judgments pilots make that lead them to an accident may have appeared plausible before the accident occurred, though in retrospect they would never choose the same course again. You are not necessarily more safety-oriented than pilots who have become involved in accidents. Sometimes pilots who may have been conscientious and skilled in the past are involved in accidents, but, during the flight in question, a checklist item was left out, a procedure was hurried, or an event happened for which they were not prepared which resulted in an accident. [Figure 10-9]

The acronym DECIDE is used by the FAA to describe the basic steps in the decision-making process.

- Detect the fact that a change has occurred.
- Estimate the need to counter or react to the change.
- Choose a desirable outcome for the success of the flight.
- Identify actions which could successfully control the change.
- Do the necessary action to adapt to the change.
- Evaluate the effect of the action.

5 Ensure That Your Decision is Producing the Desired Result

4 Implement Your Decision

3 Choose a Course of Action

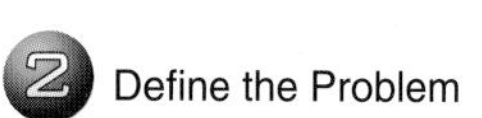

2 Define the Problem

1 Recognize a Change

While enroute on an IFR flight, you notice that ice has begun to accumulate on the leading edges of the wings.

You determine that you are experiencing light to moderate rime icing conditions. The cumulative effects of ice are a reduction in thrust and lift with an increase in drag and weight resulting in an increase in stall speed and deteriorating aircraft performance. Since your airplane does not have surface de-icing equipment, you will need to get out of the icing conditions quickly.

You remember another aircraft reported the top of the cloud layer at 10,000 feet MSL only 1,000 feet above you. You decide to climb out of the ice-producing cloud layer while you have the capability.

You advise ATC of the icing conditions, confirm the reported cloud tops, and ask for an amended clearance to climb to an altitude where you can remain clear. You climb at a slightly faster airspeed to reduce the detrimental effects of ice accumulation on the wings and tail.

Once you reach your newly assigned altitude, you will need to reevaluate the amount of ice you may have picked up during the climb, and whether the new altitude will be sufficient to keep you clear of icing conditions during the remainder of the flight.

Figure 10-9. You can use a number of methods to aid you in making good decisions. This model incorporates the FAA's DECIDE method of decision making.

ASSESSING RISK

Sources such as the NTSB, NASA, and AOPA provide accident and incident reports that will help you determine the relative safety of various IFR operations. After studying the accident reports, aviation analysts have determined that once you acquire an instrument rating, you should use it often in order to receive the greatest safety benefit. Accidents occur when instrument-rated pilots either fail to file IFR when conditions warrant, or cancel their IFR flight plans prematurely, and continue flight into poor weather. In practice, you should consider filing an IFR flight plan every time you fly cross country and closing the flight plan only when a safe landing is ensured. The practice will keep your skills sharp, and you will maintain an extra level of security that you can use if weather conditions become marginal VFR or IFR. [Figure 10-10]

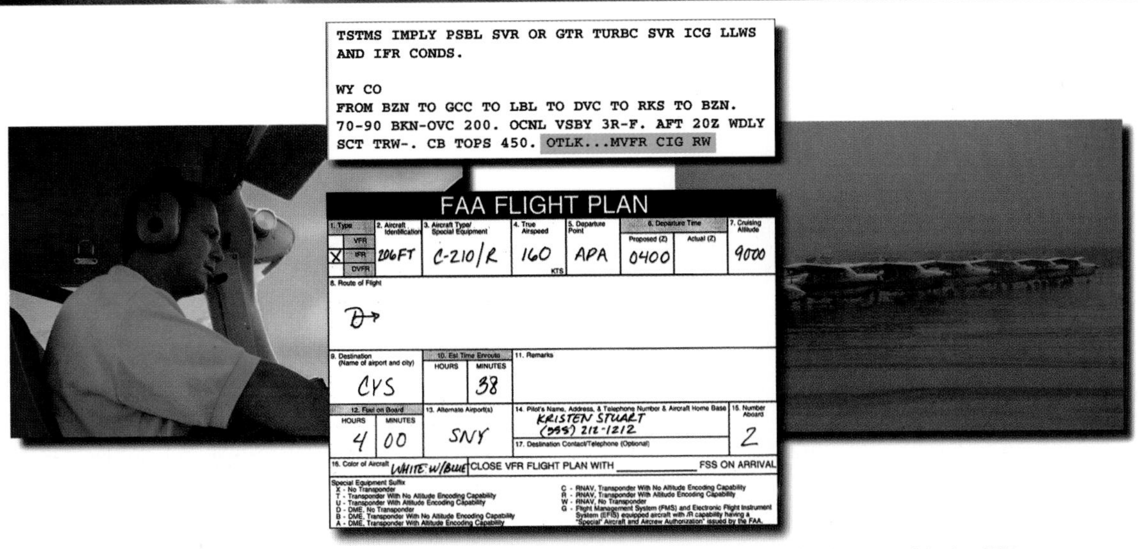

Figure 10-10. Filing an IFR flight plan each time you fly will allow you to gain familiarity with the IFR system and aid you in maintaining your IFR currency. If weather deteriorates from VFR to marginal VFR or IFR conditions, you will already have an IFR clearance and a course of action, saving you time and concern.

PILOT-IN-COMMAND RESPONSIBILITY

You are responsible for ensuring that the airplane is legal and safe for IFR flight, and that you are current and prepared to act as pilot in command in IFR conditions. Although ATC maintains a degree of control over IFR flights for separation and routing purposes, FAR 91.3 remains the ultimate rule on the pilot in command's authority within the cockpit. ATC is not prepared to assume responsibility for your flight. On the other hand, ATC must be consulted or advised concerning your decisions as PIC. You should determine a course of action and coordinate with ATC for a successful flight. [Figure 10-11]

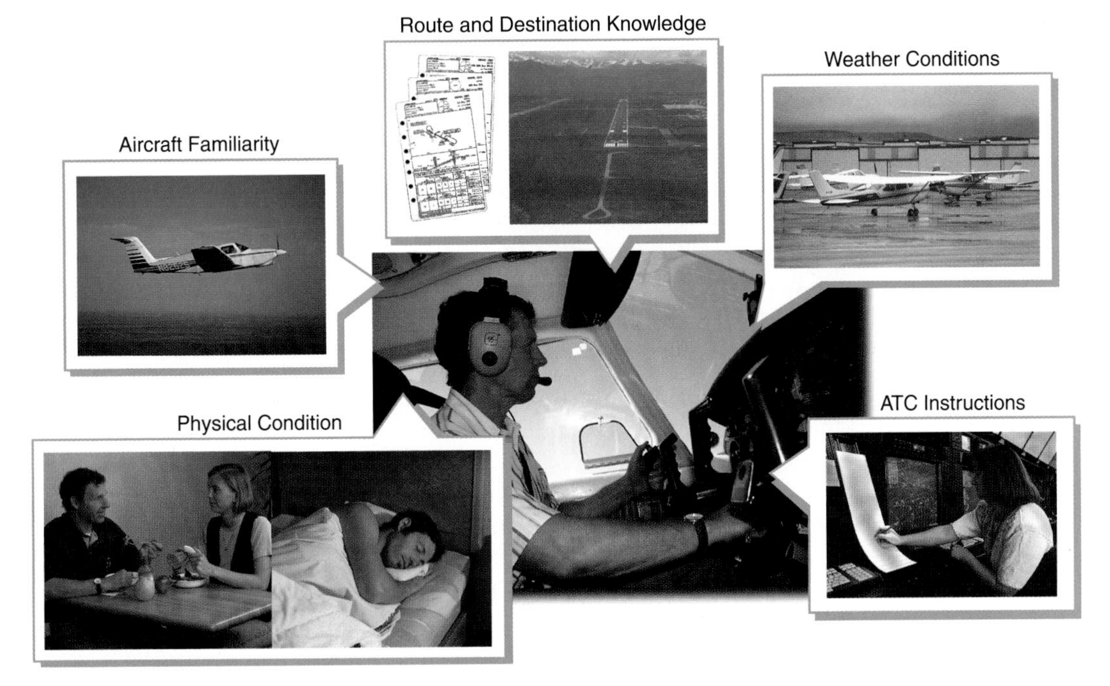

Figure 10-11. Part 91 regulations give you flexibility, but also more personal responsibility for determining what constitutes a safe flight.

In order to act proficiently as pilot in command, you will need to formulate a plan for maintaining your currency. The best course of action is to file and fly IFR on a regular basis. If circumstances prevent you from flying for a period of time, a few local flights practicing approaches with a safety pilot should be completed before you attempt a cross-country in IFR conditions. If significant time has lapsed since your last IFR flight, or you wish to increase your proficiency in actual conditions, schedule an instrument competency check with an experienced instructor. In many areas of the country, flight in IFR conditions is hard to acquire, and you may finish your training with little actual instrument time logged. Before you attempt flight as PIC in IFR weather conditions, make sure that you have tested similar weather with an experienced instrument pilot or instrument instructor.

Flight in the IFR environment demands your sharpest flying skills to begin with, so any variables, such as emotional stress, fatigue, or less-than-perfect health, should be eliminated before you take off. You will need to fly most precisely during the approach when the fatigue of conducting the flight may affect you. You may not perform very well if you begin the flight in poor physical condition.

SELF ASSESSMENT

When the pilot in command is the only rated pilot in the cockpit, the operation is referred to as **single-pilot IFR**. Flying as sole pilot presents unique challenges in the IFR environment, because of the high workload. However, it can be accomplished safely if certain standards are maintained. You should be well-rested, organized, and mentally sharp, and the route should be one that is familiar, or at least well-rehearsed in advance. Also, choose an airplane that you feel comfortable in, since you will have little time in flight to figure out unfamiliar systems.

IFR minimums are established with the intent that the approach will be flown by a current pilot at the controls of a familiar airplane. When you first receive your instrument rating, you should set far more conservative minimums than the instrument approach procedures suggest. Only after you have acquired considerable experience should you plan to fly to IFR minimums. You can create a **personal minimums checklist** to help you make go/no go decisions. [Figure 10-12]

Personal Minimums Checklist

- Ceiling/Visibility? 800 feet/ 2 miles
- Wind? 10 knots crosswind, 25 knots maximum
- Approach? ILS, VOR, NDB full panel
- Currency? Yes, with recent time in aircraft to be used
- Aircraft? Dual nav/coms, vacuum warning light, autopilot, DME or GPS/LORAN
- Enroute? No icing, stay VFR in areas of convection

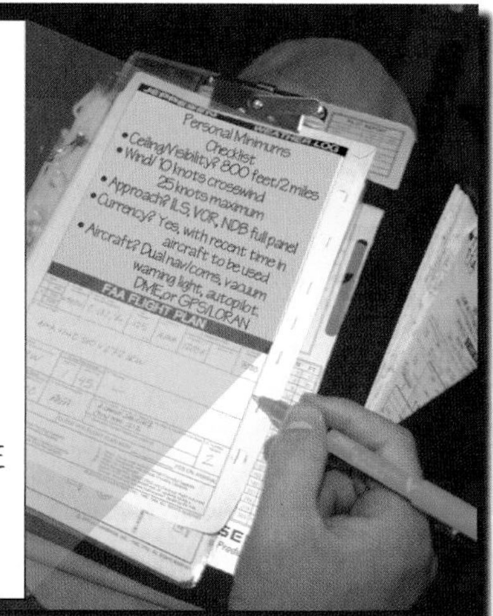

Figure 10-12. A personal minimums checklist should outline the lowest visibility, ceiling, and precipitation forecasts under which you would consider a flight. If conditions at your destination drop below your minimums while you are enroute, you may want to land instead at an alternate airport that meets your requirements.

HAZARDOUS ATTITUDES

The five hazardous attitudes that you were introduced to in Chapter 1 require your consideration during each flight. Evaluate the reasons behind the choices you make as a pilot to determine which attitudes you are most likely to exhibit. [Figure 10-13]

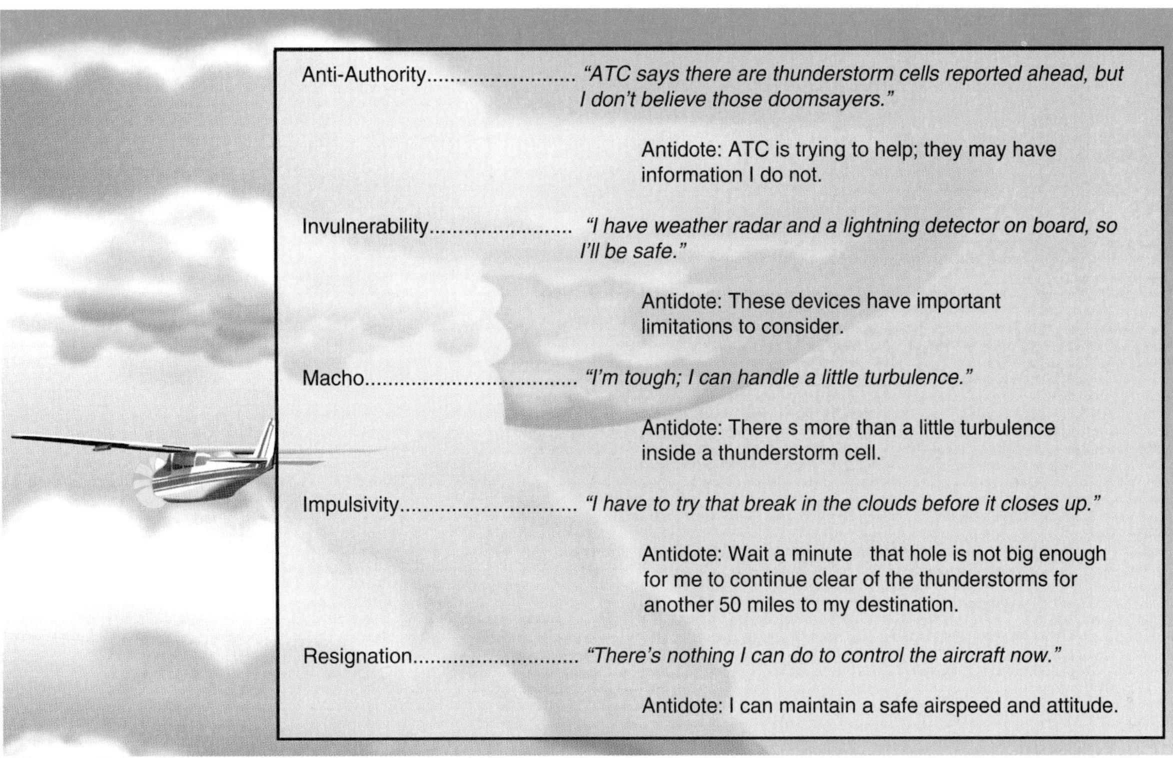

Figure 10-13. The pilot in this scenario displays each of the five hazardous attitudes that you need to avoid. A corresponding antidote follows each action, and gives you a tool for coping with similar choices that you might consider on an IFR flight.

CREW RELATIONSHIPS

Some of your flying as an instrument pilot will require you to share the cockpit. In order to maintain currency, you may need to fly with a **safety pilot**, who will be in charge of seeing and avoiding other aircraft and terrain while you practice instrument maneuvers with a view-limiting device. Anyone who holds a private pilot certificate, and is rated in the airplane you are flying, can serve as your safety pilot. Frequently, pilots take turns acting as safety pilot while the other flies with reference to the instruments. Before you take off, you should know something about the person with whom you will be flying. Does he or she have the same respect for the FARs and aircraft limitations that you do? Does he or she have more or less experience than you do in the airplane and the area in which you will be flying? What are his or her motivations and expectations concerning the flight? Even when you fly with pilots whose habits you know well, you need to establish who will act as PIC, and who will be expected to perform what duties during the flight. This also holds true when you fly with an instructor, such as during refresher training or an instrument competency check. Chapter 13, Section B — Commercial Decision Making, will explore crew relationships in more detail. [Figure 10-14]

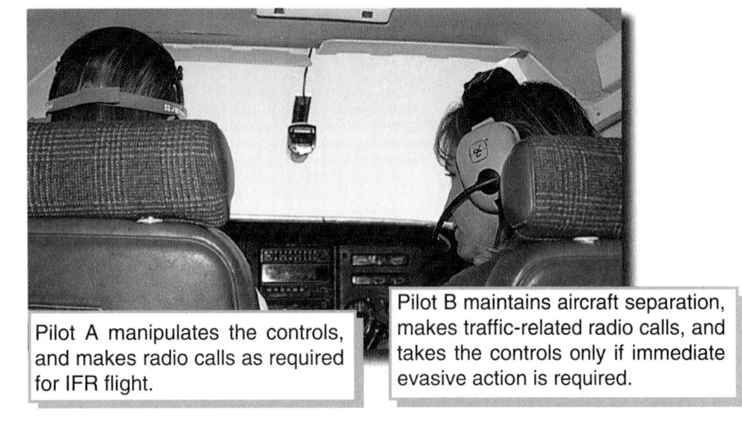

Figure 10-14. When pilots fly together, they must determine who will be responsible for completing the various duties in the cockpit.

COMMUNICATION

Though proper communication is critical to most areas of modern aviation, in IFR flight it takes on special importance. Because the coordination of your flight with others takes place over the radio, you need to pay careful attention to the manner in which you communicate your requests and intentions. The IFR system is relatively complex, and confusion occurs when you are not conscientious about how you express yourself. You should work on effective listening, by concentrating on what ATC says, and questioning any requests that you find confusing.

EFFECTIVE LISTENING

There are a variety of meanings for many common words in the English language. For example, an instruction to hold can have several different implications, depending upon the context in which it is used. You may be asked to hold before commencing an instrument approach, meaning that you need to enter a holding pattern at a fix. When used in reference to ground operations, the word hold takes on a different meaning. *"Cessna 2275C, position and hold Runway 35L"* conflicts directly with the instruction *"Cessna 2275C, hold short Runway 35L,"* even though they sound very similar. Since there is potential for confusion, you need to read back all of the key elements in a clearance. Do not hesitate to clarify with ATC if you do not understand part of a clearance, even if the controller is busy. If you proceed under the wrong interpretation of a clearance, you could make the controller's job far more difficult than if you had simply asked for clarification in the first place.

BARRIERS TO EFFECTIVE COMMUNICATION

There are few other concerns during flight as prone to human error as communication. For example, in the interest of brevity, pilots and controllers both shorten important instructions to a minimum, and at times these exchanges become too short to be effective. You should watch out for any abbreviated altitudes or headings, and, since pilots and controllers commonly drop the first three numbers of an aircraft's call sign, similar N-numbers can lead to high levels of confusion. [Figure 10-15]

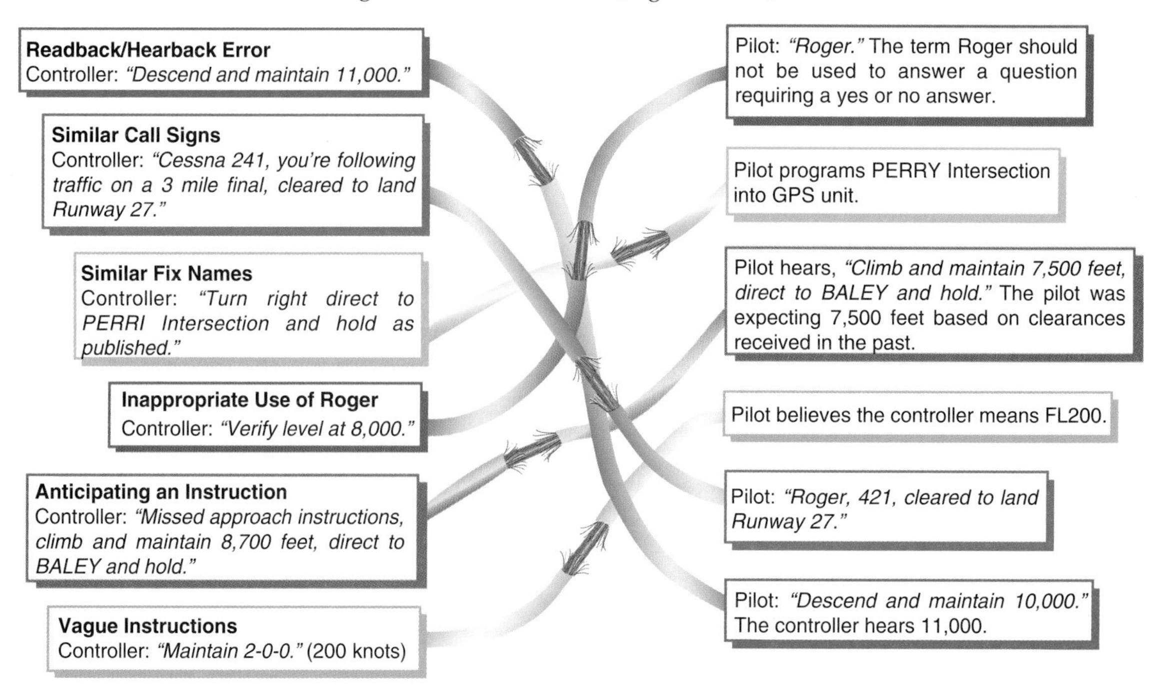

Figure 10-15. Here are some examples of commonly misunderstood phrases used by pilots when communicating with ATC.

Other barriers to communication involve preconceived notions of upcoming clearances. If you fly a particular approach to your home field many times, you may become familiar with a certain procedure normally used by ATC. However, you must remain vigilant for any changes to the normal operating practice, and listen critically to what the controller says each time you use the approach.

RESOURCE USE

Suppose you are departing from Door County-Cherryland Airport on an IFR cross-country flight back to your home base at Iowa City Municipal. Before you start the engine, you program both your filed route and an alternate route to your destination into the GPS. You also go ahead and store the first four frequencies you expect to use into the communication radios. While waiting to taxi, you write down your clearance on your kneeboard, which holds paper and several pencils ready for use in flight. As you taxi to take off, your passenger holds your kneeboard and charts. Once airborne, you call Flight Watch for updated weather and PIREPs, so you know if the weather at your destination is close to what was forecast. A short distance away from Iowa City, you use the autopilot to hold the airplane while you retrieve the proper chart for the approach. You ask ATC for a vector to the final approach course, and as you execute the approach, your passenger is able to adjust the defroster vent so that the windshield will be clear when you break out of the clouds. As you can see, proper use of available **resources** can make your flight run more smoothly.

INTERNAL RESOURCES

As more and more safety-related technology becomes affordable for general aviation pilots, the amount of resources available in the average cockpit increases steadily. Advances like GPS, airborne weather radar and TCAD are helpful tools, but only if their proper use is well understood by the pilot. Accidents have occurred when professional flight crews have entered the wrong identifier into a flight management system. Therefore, you need to fully acquaint yourself with any unfamiliar equipment on board the airplane before you take off into IFR conditions. The head-down time you spend in the airplane figuring out a piece of equipment takes away from the time you spend watching the instruments and staying abreast of the flight's progress. Program any information into navigational radios deliberately so that you are sure you are entering the proper frequencies and identifiers. If you receive a confusing signal, try to avoid fixating on that piece of equipment. Verify with another source, such as ATC, or another radio that you know for certain is working, and if the problem persists, turn off the equipment causing the confusion. The same is true in case of autopilot failure or a runaway electric trim switch. Do not hesitate to return to flying manually rather than fighting the autopilot for control of the airplane. You should never take off when an important piece of equipment is inoperative. Even if the equipment is not required by regulation, like an autopilot or a second radio, you should consider how the loss of capability will affect your flight. [Figure 10-16]

JEPPESEN NAVIGATION LOG

Aircraft Number: 5594Q | Dep: SBP | Dest: MRY | Date:

Clearance: 94Q C MRY A V 27 M 80

SQ 6374

⊛ SAN LUIS OBISPO G/S OTS

SBP/LA CENTER : 124.15

MRY/Oakland Center : 128.7

Check Points (Fixes)	Ident / Freq.	Course (Route)	Altitude	Mag Crs.	FUEL Leg / Rem.	Dist. Leg / Rem.	GS Est. / Act.	Time Off 18:46 ETE / ATE	ETA / ATA
SAN LUIS OBISPO APT.					89	112			
MORRO BAY VOR V-27	MQO	270°	6000	258°	5.0	6	165	:10	18:56
	112.4	(D→)			84.0	106			
BLANC INT (PBR)	250°	292°	MEA: 4000	278°	3.4	37		:14	19:10
	114.3				80.6	69			
BIG SUR VOR V-27	BSR	321°	MEA: 7000	304°	3.1	36		:13	19:23
	114.0				77.5	33			
MONTEREY LOC-DME V-111	IMTB	348°	MEA: 7000	330°	2.0	21		:08	19:31
	110.7				75.5	12			
MONTEREY APT. elev 254' msl		278°			4.1	12		:09	19:40
					71.4	-0-			
					17.6	112		:34	

FUEL									
Climb	5.0	Cruise	10.6	Apch.	2.0	Alt.		Res.	
Cruise Burn/Hr	14.6	Block In		Block Out		Log Time			

Figure 10-16. A flight log serves as an important resource, listing courses, altitudes, and navaid frequencies along the route.

EXTERNAL RESOURCES

When you fly under VFR, you may use ATC as a helpful tool for periodic weather and traffic updates. Now, controllers become part of your crew, and you will coordinate with them frequently, and use their knowledge and skill as an important component in each IFR flight. They are your most vital external resource. However, there are other sources of information that you should consider as well. Other airplanes in the area or ahead of you on your route can provide PIREPs, which are most helpful in cases where icing, turbulence, or thunderstorms are factors. FSSs also can provide weather information, as well as process your flight plan.

WORKLOAD MANAGEMENT

The phases of flight with the highest degree of complexity and the most critical tasks require more precise planning. Through proper workload management techniques, you can learn to plan effectively for multiple scenarios, focus your attention on the most important task at hand, and deal well with the demanding periods of flight without becoming overwhelmed.

PLANNING AND PREPARATION

FAR 91.103 provides specific requirements for preflight action before an IFR flight, and you will notice that these requirements in some instances are more extensive than those for VFR flight. Though the regulation appears thorough, mentioning weather reports, fuel requirements, and alternates available, you can increase the safety of your flight by taking your preflight planning a few steps further.

After considering weather, fuel, and alternate airports, rehearse your intended route of flight, and determine if there are any parts of the flight that make you feel uncomfortable. These could include picking up your flight plan upon departure from an unfamiliar airport, following a SID from a busy terminal area, or flying an approach at your destination for the first time. For each of these concerns, take the extra time to ask an experienced pilot about what you can expect, and to mentally prepare for this phase of the flight.

Once you have outlined the flight under ideal circumstances, direct your attention to forming an alternate plan. Go through the missed approach for each possible approach procedure at your destination. Do they differ? How do they relate to the surrounding terrain? With the proposed amount of fuel you have on board, you should determine how many approaches you could make before you would need to divert to your alternate. Then, factor in an increased headwind, such as 20 extra knots, enroute to your original destination, and recalculate the number of attempts you could make based on an updated groundspeed figure. Review the approaches available at your filed alternate, and then look at those for other airports in the vicinity. In the case of a wide-spread weather system affecting your destination, airports over a larger geographical area could fall under the same conditions, and you should formulate a plan for this possibility. [Figure 10-17]

If you follow the weather trends for several days before your flight, you will have a sense of how quickly a system is moving across the country. However, fronts are prone to changing speed, either moving more quickly than forecast, or stalling in a particular area. Determine which change in forecast conditions would most adversely affect your flight, and plan for that change to occur. Think of how the weather might affect your fuel calculations, and the approaches available. If the weather is low enough, it could be below your personal minimums, or even legal IFR minimums. You should always keep in mind where VFR weather is most likely to be found in the event you have an equipment malfunction, such as an electrical system failure or engine problems. Make sure you have the charts for that area along with you.

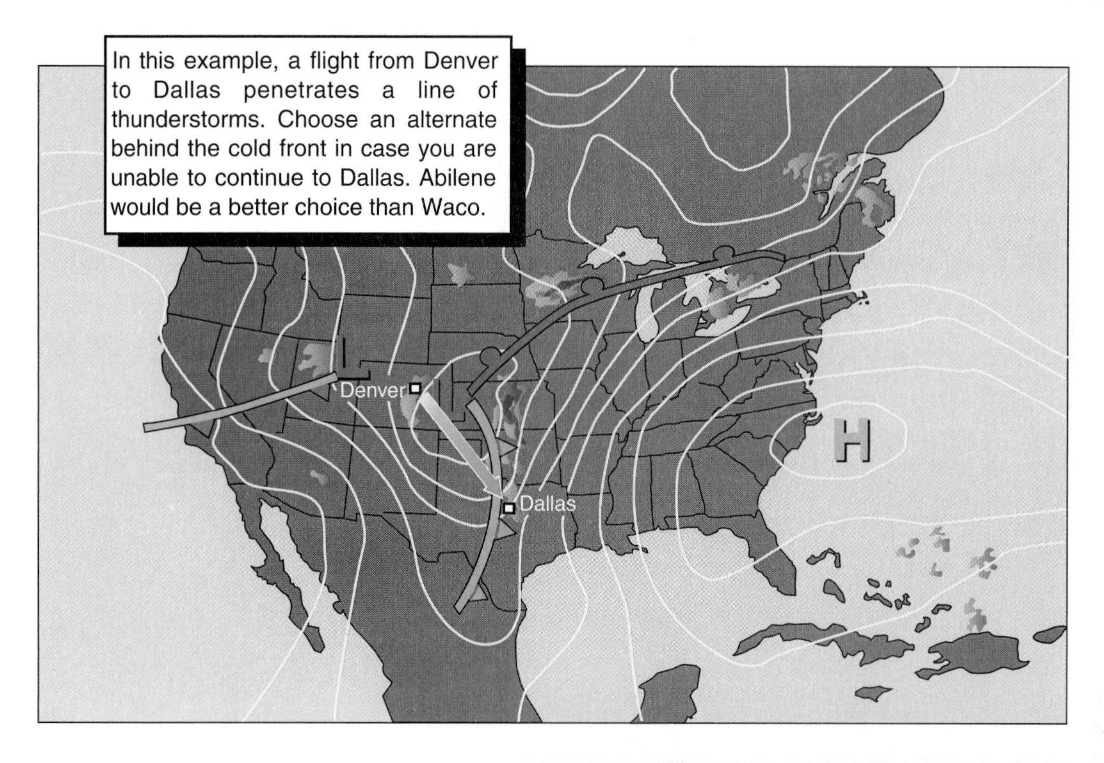

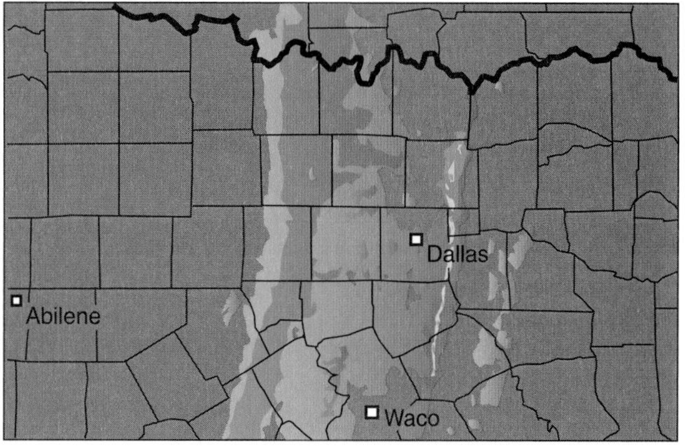

Figure 10-17. Choose your alternate(s) carefully, as weather at your destination could also affect your alternate.

If you have a GPS, or other database navigation system on board, it makes sense to program your route and alternates before you start the engine. Any head-down time you spend during flight is time away from controlling the aircraft and may detract seriously from your general awareness. In addition, if you store a couple of alternate plans into the equipment onboard, you can access them quickly.

The approach segment typically marks your highest level of workload. To help manage this, you should have rehearsed everything for the approach before you begin your descent, generally 30 miles away from the airport.

PRIORITIZING

During an IFR flight, you need to remain continually aware of what will happen next. This will assist you in **prioritizing** what items you need to accomplish, and in what order they need to occur. Prioritzing is especially important in periods of exceptionally high workload, such as during an approach in low IFR conditions.

When you encounter a high workload situation, identify your most important tasks. Usually, controlling the airplane is foremost, followed by staying aware of your position.

Equipment problems should be dealt with on an urgency basis. Obviously, emergencies such as an engine failure will require your immediate attention, but troubleshooting the problem should not interfere with positive control of the airplane. If a navigation instrument fails in the middle of the approach, such as the loss of the glideslope during an ILS, review your MDA and consider carefully whether to continue the approach or commence the missed approach. If you become distracted, proceed directly to the published missed approach to give yourself altitude and time. Communicate your intentions to ATC after you have established control of the airplane. [Figure 10-18]

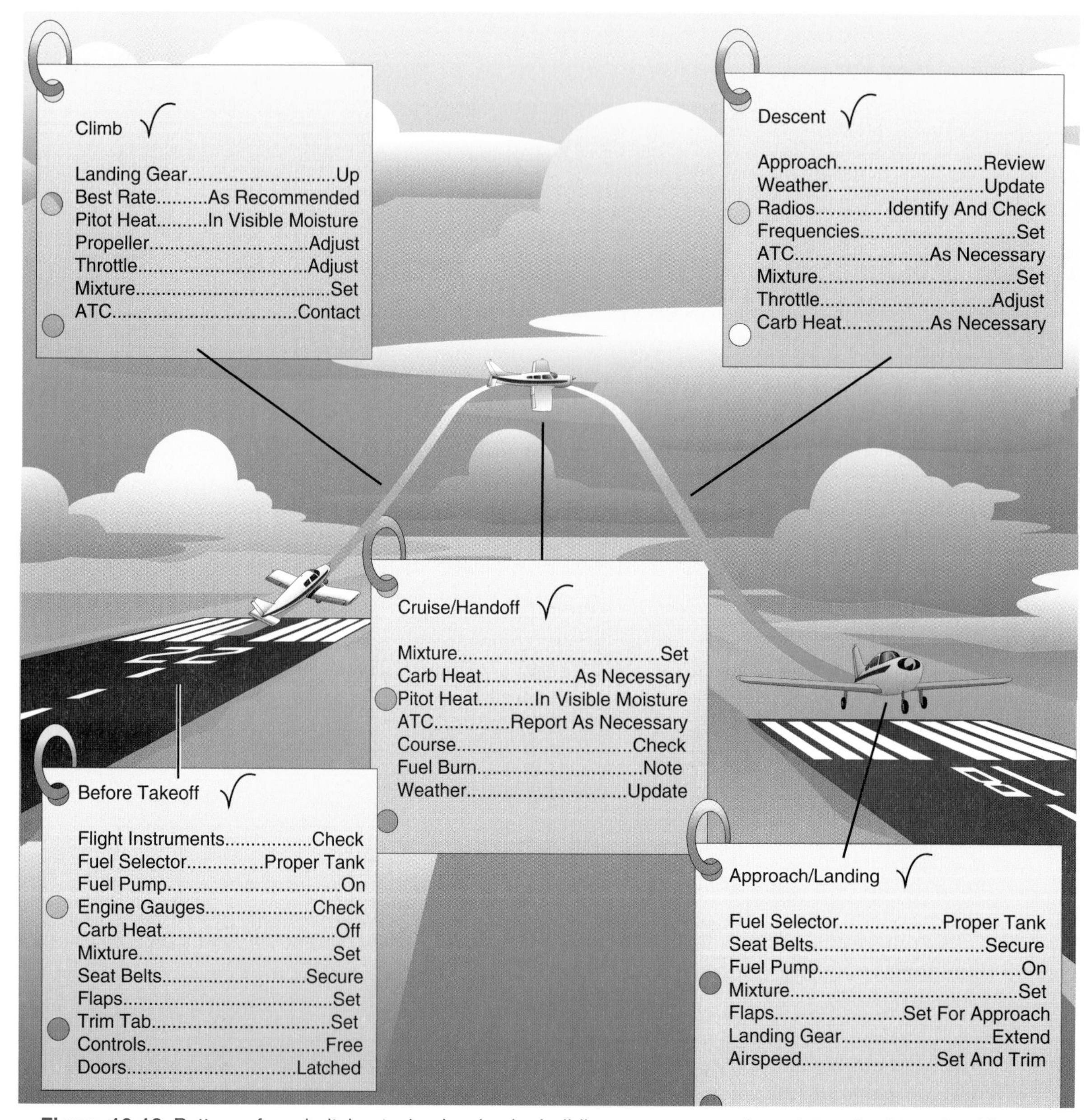

Figure 10-18. Patterns for prioritzing tasks develop by building on your use of mental and written checklists.

Work Overload

There may be times during a flight when the workload outweighs your ability to stay ahead of the airplane. When you fly a nonprecision approach, you are particularly prone to work overload, especially if the approach involves multiple step-down fixes and intermediate altitudes. The amount of numbers you must refer to increases, and your guidance information is more abstract. Any additional workload, such as an equipment failure or an

unusual request from ATC, could upset your balance in the cockpit. For this reason, you might consider setting even higher personal minimums for yourself during these approaches, and practicing them often to maintain your proficiency.

Studies have shown that most altitude deviations are the result of an improper mindset, but heading deviations are often human error due to high workload. Though altitudes may remain relatively constant from flight to flight, course headings and vectors change frequently because of varying weather and traffic situations. Some single-engine aircraft heading indicators are equipped with a heading bug that you can use to quickly record the course given to you in a clearance. [Figure 10-19]

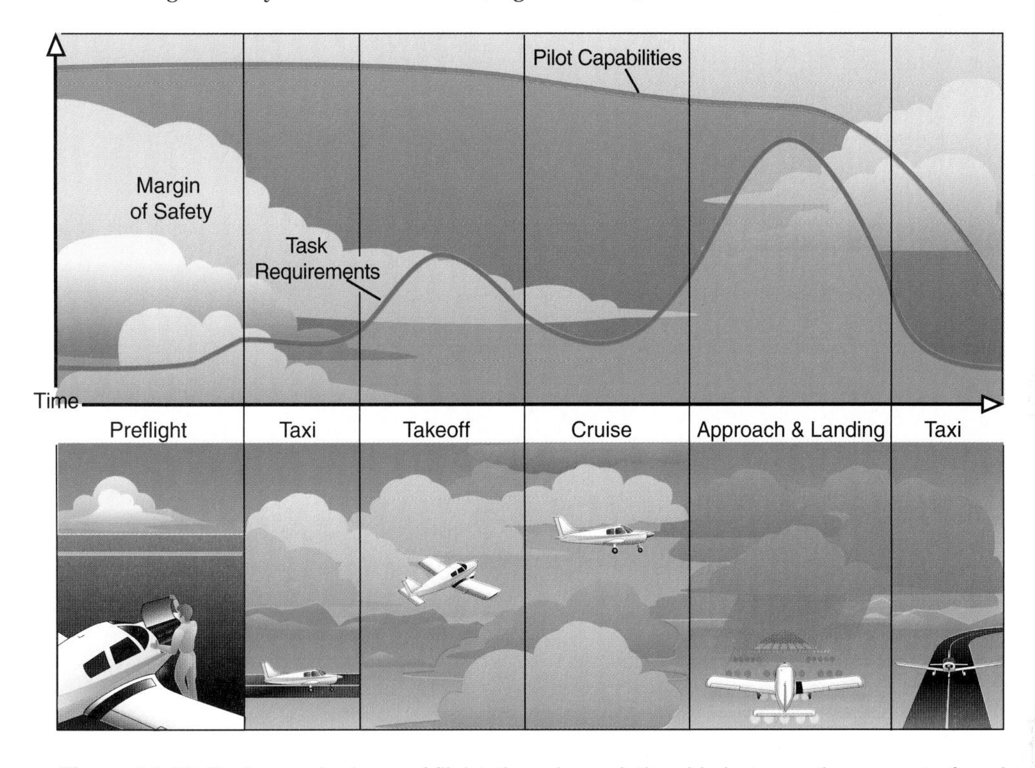

Figure 10-19. During each phase of flight, there is a relationship between the amount of work necessary and the pilot's ability to deal with the workload effectively.

SITUATIONAL AWARENESS

The skills that allow you to know where you are when you have no outside references come with your instrument training. Using these skills effectively is an important part of situational awareness, when incorporated with knowledge of the airplane and systems, the weather affecting the flight, and the requirements of ATC. Situational awareness can deteriorate when you experience problems with crew relationships, miscommunication, poor use of resources, and inadequate workload management techniques. With this in mind, your level of situational awareness is a good indicator for the overall safety of the flight.

VISUALIZATION

One way to increase your situational awareness is to use **visualization**, the ability to see in your mind an overview of the flight. Visualization has aided you throughout your pilot training. For example, you probably used detailed diagrams to help you understand airplane systems. Though you could not see an entire system at once, you could visualize how it worked through the use of the diagram. Through the use of both VFR and IFR charts, you can take this process one step further and teach yourself to picture what is going on outside the airplane when you have no visual references. You may practice visualization techniques both before and during the flight, and thereby enhance your situational awareness. [Figure 10-20]

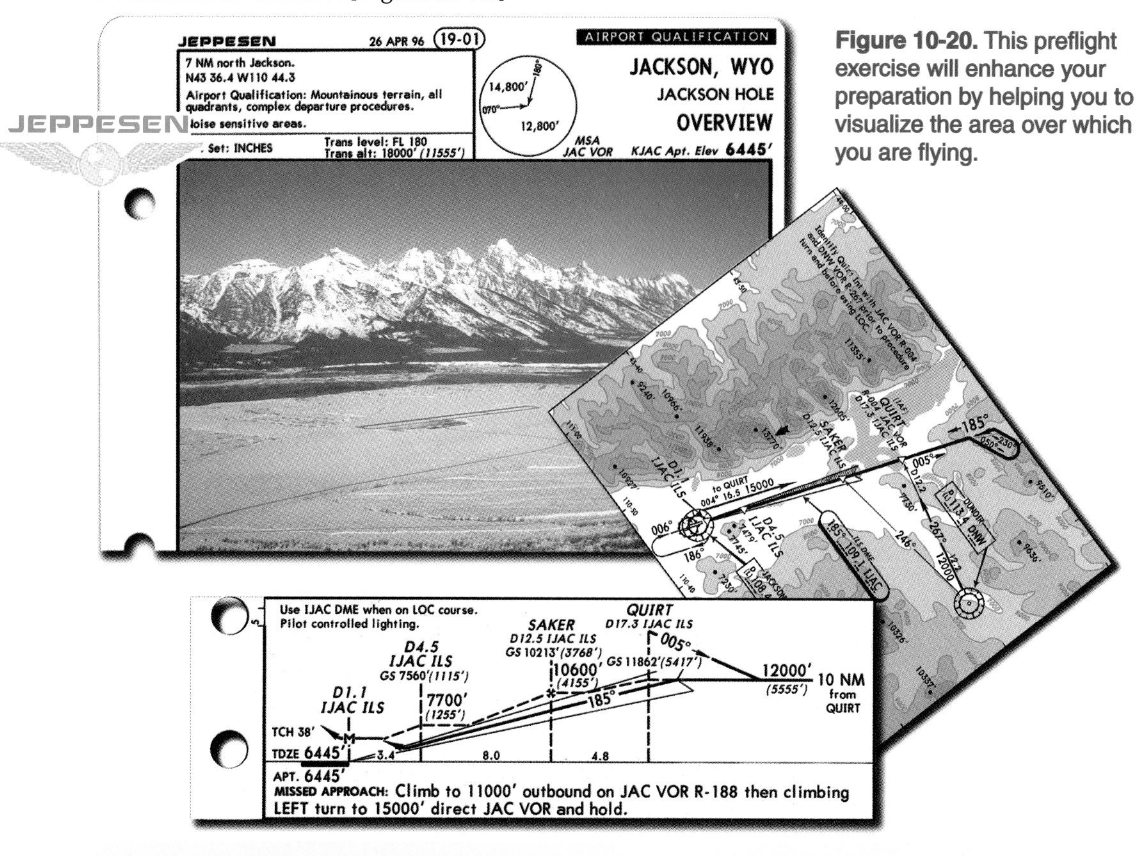

Figure 10-20. This preflight exercise will enhance your preparation by helping you to visualize the area over which you are flying.

Start with an approach you know well, over terrain with which you are familiar:

- IAP charts use two views to help you visualize the approach; create a mental three-dimensional image of the course.
- In your mind, combine the plan view of the approach with the profile view to visualize the total procedure.
- Place this three-dimensional image over your mental image of the terrain as it appears under VFR.

Using this technique, visualize an unfamiliar approach:

- Take a sectional or WAC chart of the area, and make it three-dimensional in your mind, looking at terrain features, towns and other items depicted on the chart.
- Use the plan and profile views of the approach and create a three-dimensional image of the procedure, superimposing the approach over the terrain as you have pictured it in your mind.
- Finally, fly the approach mentally, picturing where you are over the terrain in each phase of the procedure. This way you can visualize the terrain without having outside visual references.

CONTROLLED FLIGHT INTO TERRAIN

You will recall that CFIT occurs when an aircraft is flown into terrain or water with no prior awareness on the part of the crew that the crash is imminent. Through the implementation of extensive training programs and the installation of GPWS, air carriers and professional flight departments have significantly reduced the number of CFIT-related accidents in the United States. In the period between 1946 and 1972, there were 72 such accidents. From 1990 to 1997, there was only one CFIT-related accident involving a U.S.-registered Part 121 aircraft. However, CFIT-related problems remain more prevalent in general aviation due to lower technology levels in the cockpit, and lack of formal CFIT training among pilots.

You can avoid CFIT by increasing your **positional awareness**. A high level of positional awareness includes paying particular attention to your location relative to terrain, and the minimum altitudes provided on IFR charts. Visualization of your surroundings, as described previously, will aid you in maintaining positional awareness. Many general aviation CFIT accidents occur because a pilot descended below the DH or MDA without the appropriate visual references, or failed to follow a missed approach procedure properly. You should not be tempted into continuing your descent below the DH or MDA if you have not met the necessary requirements, no matter how well you know the surrounding terrain, or how much you want to land. Double-check all approach minimums, and ensure that you are following the correct procedure. Also, be aware that if there is an inoperative navigation or ground component, you will need to comply with higher approach minimums than if the full system of instruments and navigation or visual aids was available to you. Before your flight, also note what obstructions or terrain are found in the airport vicinity. [Figure 10-21]

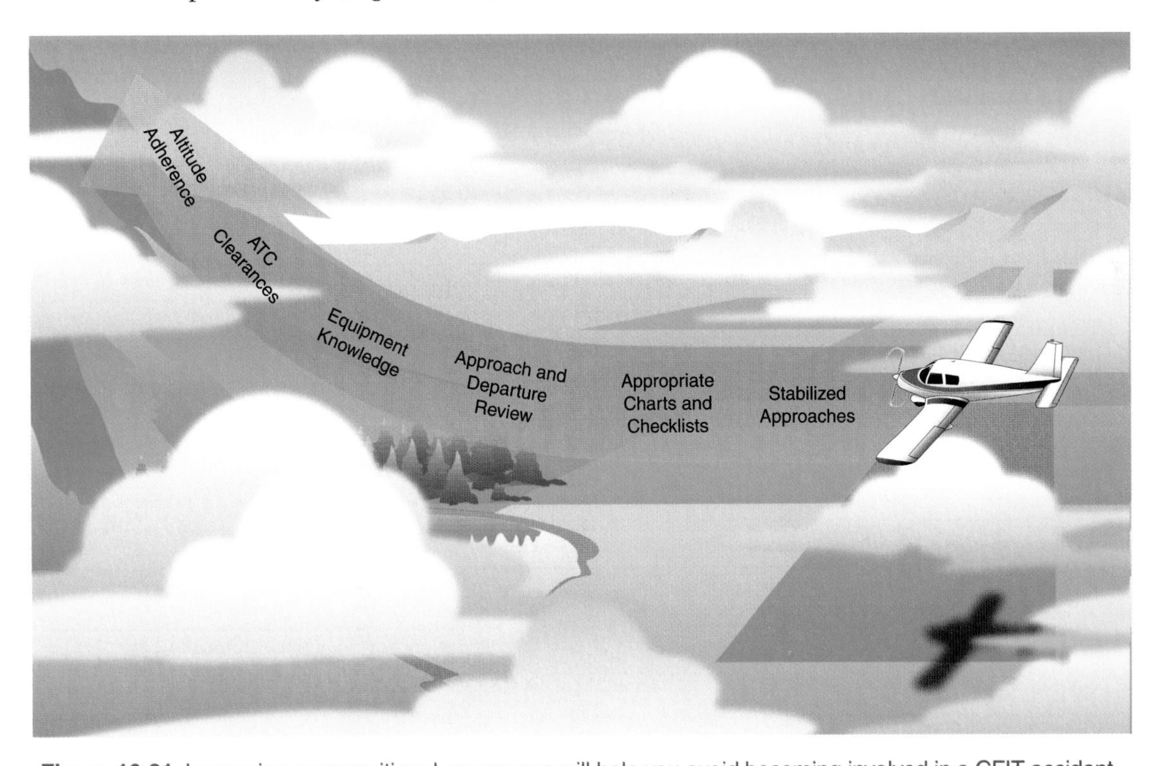

Figure 10-21. Increasing your positional awareness will help you avoid becoming involved in a CFIT accident.

As GPWS equipment becomes less expensive, you can look forward to the future when many general aviation aircraft will have the added margin of safety that it provides in high workload situations. However, as with any equipment in the cockpit, you must understand its proper use and limitations.

IFR Over the Rocks

Many pilots fly single-engine aircraft in the mountains, but most would think twice about flying over this rugged terrain in IFR conditions. On the other hand, quite a few mountain airports have published instrument approaches, and airlines and commuters use them regularly to provide service to these regions. However, if you look at an approach chart for a mountain airport such as Eagle County Regional at Eagle, Colorado, you will notice some important differences between this approach and one found over less precipitous terrain.

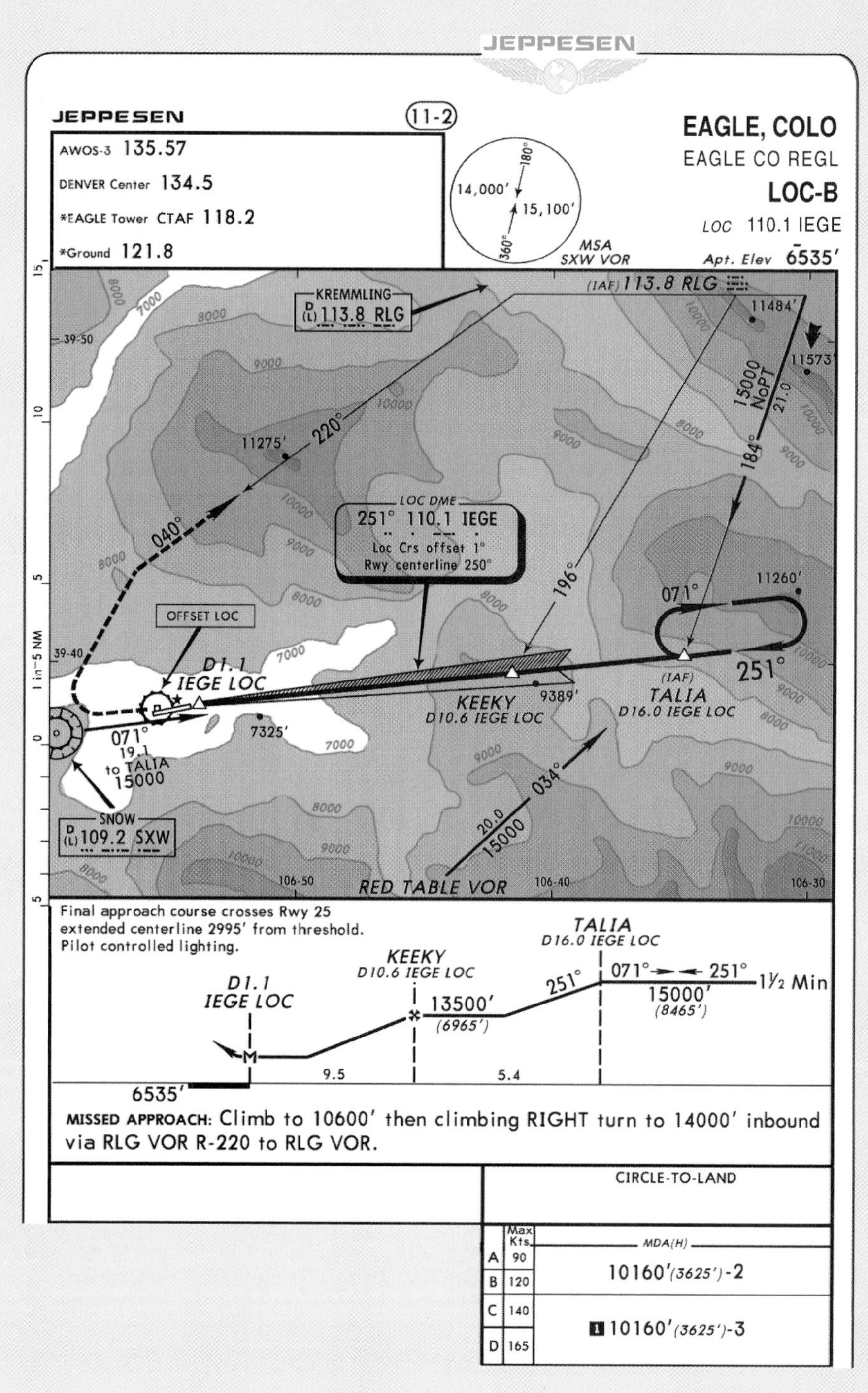

An offset localizer is necessary because of the airport's position in the valley. The elevation of Eagle is 6,535 feet MSL, and you will note that the MDA is 10,160 feet MSL, just a thousand feet below the mountain tops near the runway. The initial approach altitude of 15,000 feet MSL is a clue to the high enroute altitudes in the area. Many normally aspirated, single-engine airplanes may be unable to maintain these altitudes, and you would need to be on oxygen for much, if not all, of the flight. If you are still considering an IFR flight into a mountain airport you should also take a careful look at its established departure procedures. The departure procedure from Eagle's Runway 25 requires a minimum climb gradient of 750 feet per nautical mile to 11,200 feet, or 1,125 feet per minute at a groundspeed of 90 knots — healthy indeed for nonturbocharged aircraft at this altitude.

OBSTACLES TO SITUATIONAL AWARENESS

There are several different circumstances which can act as barriers to your situational awareness. As more automation enters the cockpit, you will become aware of its advantages and disadvantages. While it can aid you in navigation and reduce your workload, there are also times when it can have the opposite effect. When you have a system that requires complex user entries, pushing the wrong buttons can lead to a confusing readout or incorrect course. The situation becomes more serious when the equipment is used on approach. As you get closer to the ground, the time you have to ascertain and correct a mistyped entry diminishes, possibly to the point where the safety of the flight is compromised. To avoid this problem, confirm all entries with the appropriate chart, and periodically ask yourself if the readout makes sense and agrees with the other navigation equipment on board.

When you are receiving radar vectors, it may be easy to lose positional awareness. Although vectors reduce your workload, do not allow them to take you out of the decision-making loop. You should consider where the vector will take you before accepting it. Other obstacles to situational awareness include confusing clearances from ATC, a copilot that fails to tell you of a change or correction in the cockpit, or a piece of equipment that is malfunctioning and diverts your attention away from flying the airplane. The increase in workload on approach by itself can lead to a loss of situational awareness if you become fixated on a particular instrument or display. If you sense that any of these issues detract from your situational awareness, request permission to break off the approach and hold at a higher altitude until you have a better grasp of the situation.

THE APPLICATION OF THE DECISION-MAKING PROCESS

While lack of skill may lead to minor accidents, such as leaving the runway on a landing roll-out, the most serious accidents stem from poor decision making. Even if you are not the most experienced instrument pilot, you can practice sound decision making from the moment you first begin your instrument training. Talk to other pilots, fly with those who have greater experience and a record of wise judgment, and ask a lot of questions when you call for a weather briefing. Then, listen to your instincts about the flight. If there is something bothering you that you cannot place, there may be a good reason not to make the flight. The following scenarios will help you discover the difference between good IFR decision making and the poor choices that can lead to an accident.

Carelessness and overconfidence are usually more dangerous than deliberately accepted risks. — Wilbur Wright

AN ACCIDENT WAITING TO HAPPEN

The pilot of the Seneca was cleared for the ILS Runway 4 approach at North Bend, Oregon, at approximately 9 miles north of the North Bend VOR. He was told to cross the VOR at or above 3,700 feet MSL. About 14 minutes later, the pilot called North Bend FSS and reported that the localizer needle on his receiver did not seem to be moving. The airplane had a history of problems with its navigation equipment, and these difficulties continued through the rest of the flight. When the airplane was 11 miles south of the airport, Seattle Center cleared him to descend from 5,000 to 3,000 feet MSL, and to fly a heading of 360°. After a heading change to 035°, the pilot stated that the needle was centered. At the outer marker, he was at 2,000 feet MSL and he would *"try to get down."*

Approximately 2 minutes later, he reported at 800 feet. That was his last transmission. The aircraft wreckage was found 3 miles north-northeast of the airport.

The pilot began the flight with two errors in decision making. The first was that he took off in an airplane that was notorious for its navigation and communication equipment problems. Another pilot for the company said that the company sometimes put pressure on the pilots to fly aircraft with equipment problems, and this probably contributed to the pilot's decision to go ahead and make the flight. Also, the pilot had performed poorly while flying on instruments during a recent checkride. This should have been a sign that he needed to seek additional training before flying under IFR.

During the flight, the pilot never asked for assistance from ATC, though it was clear his equipment was malfunctioning. A diversion to a nearby airport with an ASR approach would have saved him from having to fly an ILS with an inoperative localizer needle. Still, he proceeded, and when the approach failed, he delayed following the published missed approach in a vain attempt to find the airport.There were several points at which the pilot could have made a wise choice, giving a successful conclusion to the flight. Even seconds before impact, adherence to the proper missed approach procedure, and a climb away from the rising terrain, might have saved him. [Figure 10-22]

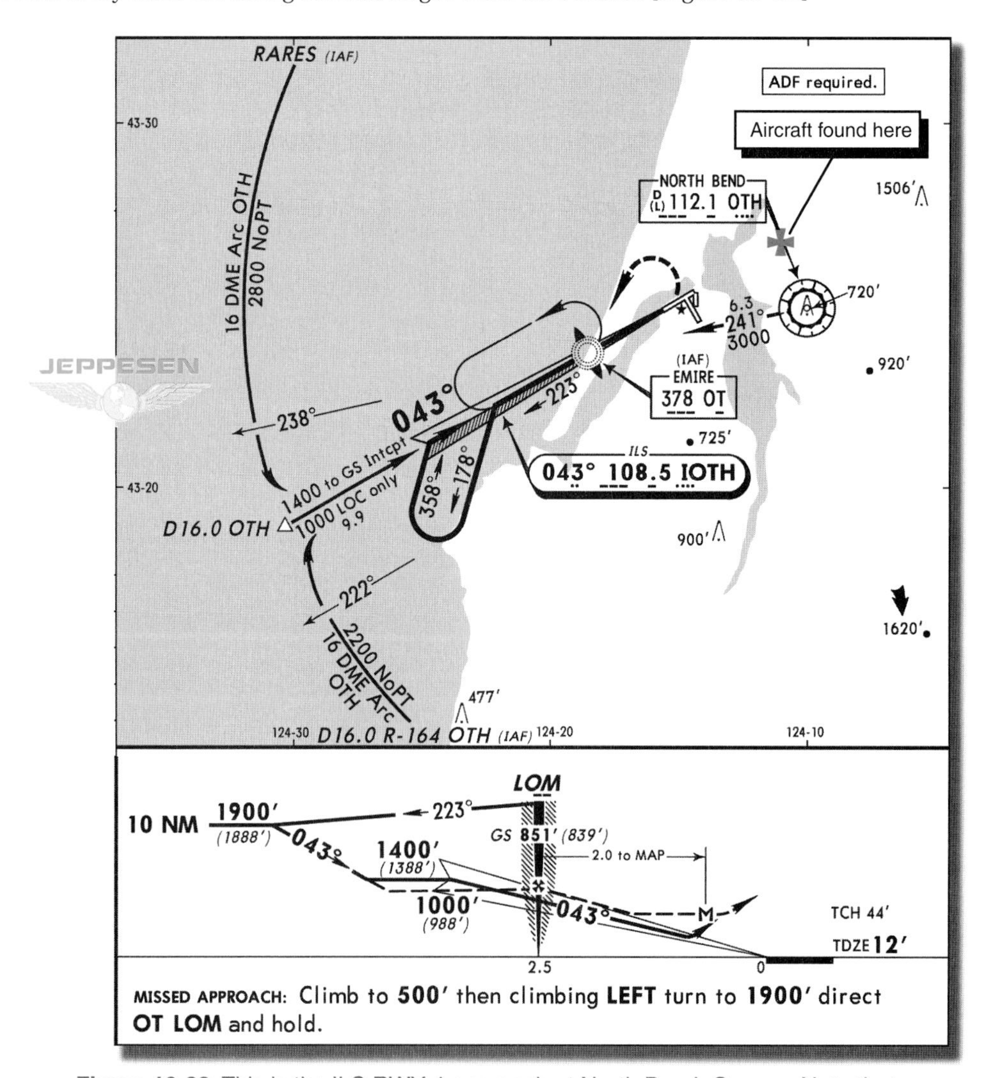

Figure 10-22. This is the ILS RWY 4 approach at North Bend, Oregon. Note that the pilot reported 2,000 feet at the outer marker, which is more than 1,000 feet above the glideslope intercept altitude.

WHAT WOULD YOU DO?

You take off into IFR conditions with the forecast that your destination will be marginal VFR and clearing by your ETA. Enroute, you are flying between layers, and you notice that the clouds ahead do not appear to be thinning at all. You call Flight Watch, only to find that the cold front has stalled, and the weather at your destination remains IFR — below your personal minimums but enough above legal minimums to be tempting. You check your fuel, and discover that headwinds stronger than forecast have cut slightly into your reserve. You are not sure you will be able to comfortably fly to your alternate if you try the approach at your original destination. On the other hand, ATC informs you that pilots ahead of you have had no trouble getting into the airport. You have been flying a lot recently, and feel your skills are pretty sharp, even though you do not have much experience in actual conditions, and this would be your first approach to minimums. What do you do?

You decide to try the approach to your original destination. ATC requires you to hold for 10 minutes before they clear you to intercept the localizer. The lower you get on the approach, the more nervous you become, and you find yourself unable to hold a steady descent down the glideslope. The steady rain beating against the windshield distracts you even more. A hundred feet above the DH, the glideslope needle becomes fully deflected — fortunately, you have erred on the high end, and you break off the approach and execute the missed approach procedure. Once you are established in the hold and calm yourself, you notice that you do not have enough fuel to safely reach your alternate. What do you do?

You collect your thoughts. When ATC clears you for another approach, you ask for assistance. They provide vectors to the localizer, and closely monitor your ground track. Knowing that someone else is paying attention makes you feel more confident. You see the approach lights, and then the runway just before you reach decision height. You promise yourself that if you make this one, you will get some more actual instrument time with an instructor on board. The landing is not your best ever, but you are simply glad to be on the ground. You know now that diverting to your alternate when you had a chance would have been the wisest plan, and carrying more fuel might have helped you avoid the tough decision in the first place. However, you broke the chain of poor judgments by asking for assistance from ATC, and you managed to land safely.

Aeronautical decision making is more of an art than a science. The more you educate yourself, through training, experience, and studying the mistakes of others, the better equipped you will be to make sound decisions as a proactive instrument pilot.

SUMMARY CHECKLIST

✓ Accidents involving IFR conditions are roughly 65% fatal. Obtaining your instrument rating and maintaining IFR currency greatly reduces your risk for these types of accidents.

✓ Accidents are rarely attributed to a single cause, but are the result of a series of poor choices.

✓ You should consider filing an IFR flight plan for every flight, and close that flight plan only when a safe landing is ensured.

✓ Though you work closely with ATC under IFR, you remain the final authority as to the safety of the flight. You may also need to coordinate responsibility with other pilots that fly with you.

✓ Flying with a safety pilot to practice instrument maneuvers will help you maintain currency and proficiency.

✓ Developing a personal minimums checklist will assist you in determining the feasibility of a particular flight. You should take into account your currency and experience when deciding which conditions you feel comfortable flying in.

✓ Five hazardous attitudes affect your decisions, and you should examine your choices to ensure that you make the proper response when one of these attitudes affects your flight.

✓ To avoid confusion, be sure to read back all important parts of a clearance, and ask for clarification when there is an instruction you do not understand.

✓ Barriers to communication include preconceived notions of upcoming clearances, abbreviated clearances, and words that have more than one meaning.

✓ Effective use of resources occurs when you understand and utilize all the people and equipment available to you during a flight.

✓ Plan for each IFR flight thoroughly before you leave the ground, including fuel requirements, alternates available, and missed approach instructions. It is also helpful to program any navigation information before engine start. The more you can rehearse ahead of time, the more prepared you will be in the event of a problem.

✓ During a high workload situation, identify the most important tasks and make those a priority. Do not allow yourself to fixate on an extraneous issue.

✓ Visualization techniques can be used to create a mental picture of the flight.

✓ You can avoid CFIT by maintaining positional awareness: staying abreast of your altitude, the proper procedures in use, and the terrain surrounding the airport.

✓ Loss of situational awareness can occur when pilots are confused by clearances, misunderstand onboard equipment, or do not communicate properly with others in the cockpit.

KEY TERMS

Decision-Making Process

Risk Elements

Reactive

Proactive

Single-Pilot IFR

Personal Minimums Checklist

Safety Pilot

Resources

Prioritizing

Visualization

Positional Awareness

QUESTIONS

1. What are the three most common causes of weather-related IFR accidents?

2. True/False. As a newly-rated instrument pilot, you should plan on making your first approach in actual conditions to IFR minimums.

For questions 3 through 6, name the hazardous attitude associated with each statement.

3. So what if the Learjet ahead of you missed the approach twice and diverted to an alternate? You're a good enough pilot to make it in.

   A. Macho
   B. Impulsivity
   C. Anti-authority

4. Ice has already started to form on the wings. There's nothing you can do about it now.

   A. Impulsivity
   B. Resignation
   C. Invulnerability

5. You hate calling ATC about every little thing you do. They always make you fly out of your way.

   A. Macho
   B. Anti-authority
   C. Invulnerability

6. The localizer needle is fully deflected to the right. You'd better make a steep turn to get back on course.

   A. Macho
   B. Impulsivity
   C. Resignation

7. What are some common sources of miscommunication with ATC during an IFR flight, and what can you do to avoid them?

8. True/False. The best time to figure out how to program a route into your GPS unit is during cruise.

9. List the preflight preparations you can make to help ensure a safe IFR flight.

10. What are some of the ways you can avoid controlled flight into terrain?

    A. Maintain altitude awareness, make a stabilized approach, and understand ATC clearances
    B. Maintain altitude awareness, contact departure control early, and ensure you have good charts
    C. Make a stabilized approach, climb out in an expeditious manner, and disregard erroneous GPS signals

SECTION C
IFR FLIGHT PLANNING

Do not let yourself be forced into doing anything before you are ready. —Wilbur Wright

Due to the complexity of IFR flight planning, Wilbur Wright's statement is sound advice. Although the **IFR flight plan** is an extension of the VFR flight planning process, it requires greater attention to detail than a VFR flight plan. IFR flight planning emphasizes route selection, the collection of communication and navigation information, knowledge of charts and flight publications, and, above all, the gathering of timely and accurate weather information.

You are required by the regulations to perform numerous preflight actions and be familiar with a variety of information. Aircraft performance, knowledge of current and forecast weather conditions, and planned alternatives are items you should pay special attention to during the IFR flight planning process. Fortunately, the resources needed to put together a comprehensive flight plan are readily accessible.

To illustrate, an IFR flight from Eppley Airfield in Omaha, Nebraska, to Rochester International Airport in Rochester, Minnesota, will be planned. The flight will depart Eppley at 2000Z and should take approximately one hour and fifteen minutes to reach Rochester. The airplane to be flown is a Cessna Turbo Skylane RG (TR182) with dual nav/coms, glide slope, ADF, Mode C transponder, DME, and GPS.

FLIGHT OVERVIEW

Before you start planning, you should take a preliminary look at factors that may prevent you from making the flight. Doing so will give you a broad **overview** of the flight and provide you with a rough idea of your ability to complete it. The preliminary overview

allows you the chance to make a go or no-go decision before you get very far into the flight planning process. If conditions look favorable to completing the flight during the overview, you can begin planning.

Initial factors to consider are weather, airplane performance and equipment, potential routes, and your instrument proficiency level. You can learn about the general weather patterns for your flight by watching The Weather Channel, the television news, reading the newspaper, contacting the National Weather Service for a long range forecast, or accessing internet weather sites several days prior to the flight. While these sources of weather information are quite good, they should not be used as substitutes for complete FSS or DUATS weather briefings. [Figure 10-23]

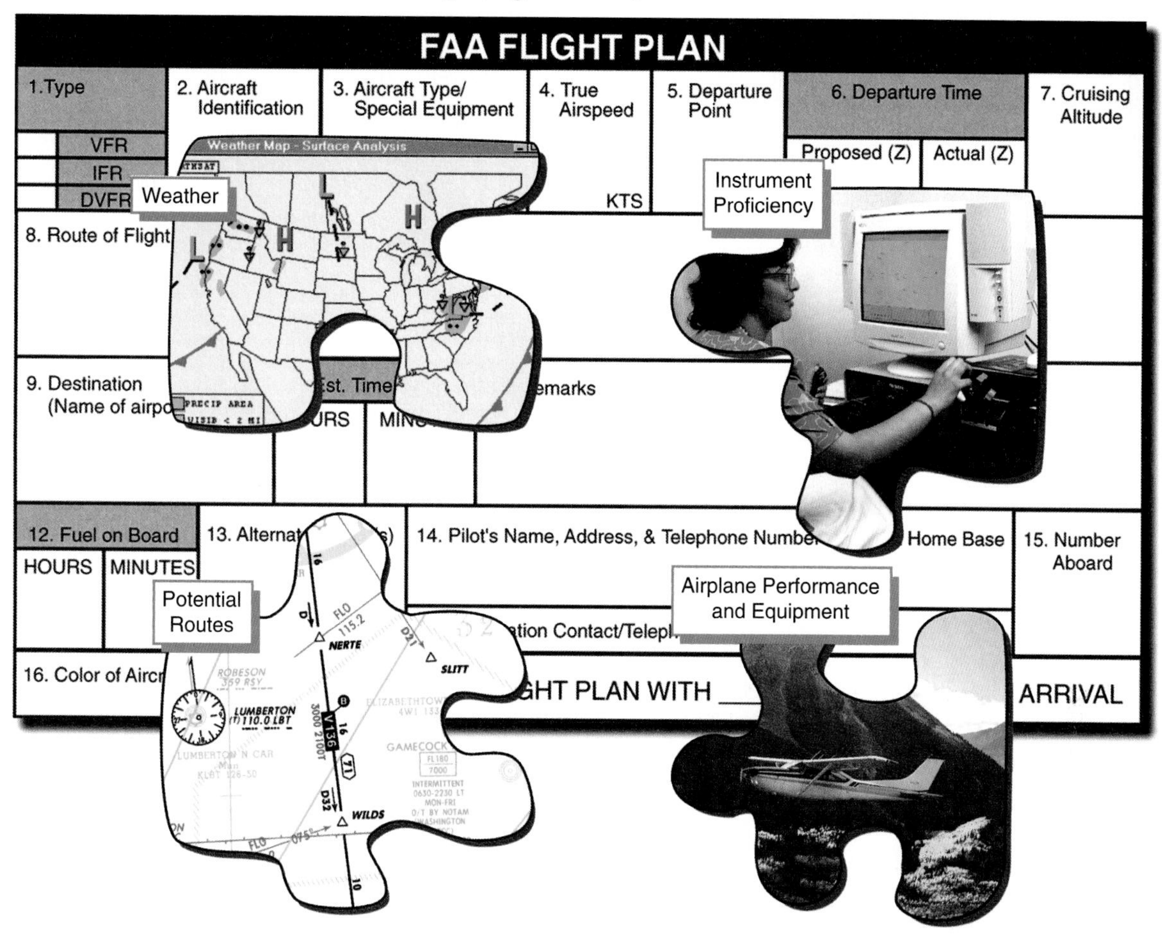

Figure 10-23. Planning an IFR flight plan is like putting together a puzzle. It requires you to fit many pieces of information together in a manner designed to give you an accurate picture of your proposed flight.

When considering airplane performance and equipment, relate them to the airports of intended use and the terrain over which you will fly. Familiarity with your airplane's performance should give you an idea of its capabilities relative to the impending flight. However, specific performance information like fuel consumption and takeoff and landing distances will need to be calculated when planning the flight. Knowledge of the general route of flight and its terrain will suggest to you whether additional equipment like supplemental oxygen is needed. Also, if your airplane is equipped with GPS or LORAN, you may want to consider making a direct flight instead of one on airways.

Finally, you should consider your instrument flying proficiency. If the rest of the overview indicates the flight may be difficult to complete, you need to decide if your current skills are up to the task. If not, you should postpone the flight until conditions improve or until you have the chance to refresh your skills with an instructor's help.

FLIGHT PLANNING

Based on factors in the overview, including the TR182's performance and equipment, as well as the relatively flat terrain over which the flight will be made, the flight from Omaha to Rochester looks like it can be completed. With a preliminary decision to go, you can plan the flight starting with the selection of your route. Once the route has been selected, you need to reference the flight publications applicable to the flight and gather detailed weather information in order to plan cruise performance and the use of alternate airports, if needed.

ROUTE

Several factors influence route selection, including the availability of route alternatives, aircraft performance considerations, and fuel economy. If a **preferred IFR route** is available, plan to fly it unless weather conditions or aircraft performance warrant a different route. Preferred routes are listed in the Enroute section of the Jeppesen Airway Manual and in the *Airport/Facility Directory*. Preferred IFR routes beginning or ending with a fix indicate that aircraft may be routed to or from these fixes via a DP, radar vector, or STAR. [Figure 10-24]

Preferred IFR routes beginning with a fix indicate that departing aircraft will normally be routed to the fix via an instrument departure procedure (DP), or radar vectors.

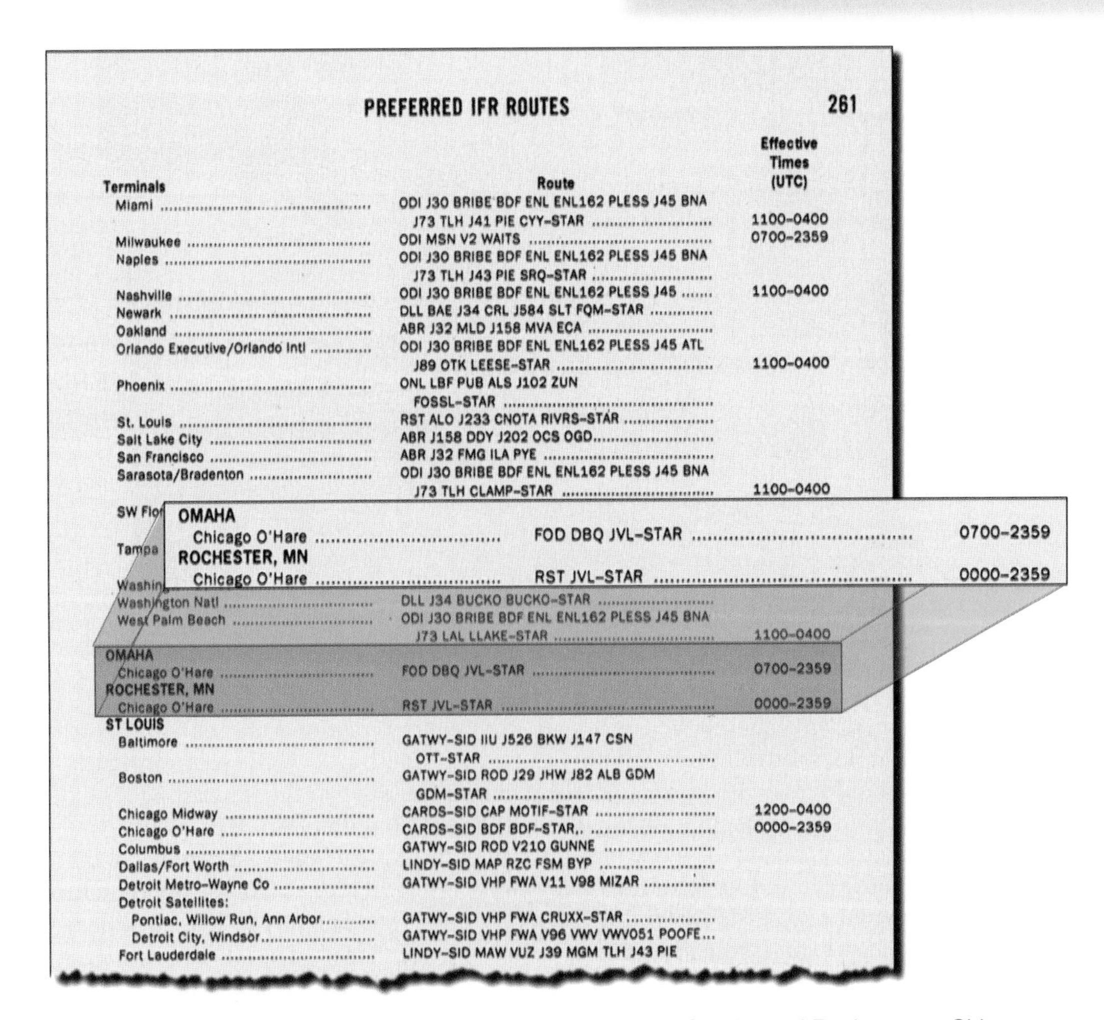

PREFERRED IFR ROUTES 261

Terminals	Route	Effective Times (UTC)
Miami	ODI J30 BRIBE BDF ENL ENL162 PLESS J45 BNA J73 TLH J41 PIE CYY-STAR	1100-0400
Milwaukee	ODI MSN V2 WAITS	0700-2359
Naples	ODI J30 BRIBE BDF ENL ENL162 PLESS J45 BNA J73 TLH J43 PIE SRQ-STAR	
Nashville	ODI J30 BRIBE BDF ENL ENL162 PLESS J45	1100-0400
Newark	DLL BAE J34 CRL J584 SLT FQM-STAR	
Oakland	ABR J32 MLD J158 MVA ECA	
Orlando Executive/Orlando Intl	ODI J30 BRIBE BDF ENL ENL162 PLESS J45 ATL J89 OTK LEESE-STAR	1100-0400
Phoenix	ONL LBF PUB ALS J102 ZUN FOSSL-STAR	
St. Louis	RST ALO J233 CNOTA RIVRS-STAR	
Salt Lake City	ABR J158 DDY J202 OCS OGD	
San Francisco	ABR J32 FMG ILA PYE	
Sarasota/Bradenton	ODI J30 BRIBE BDF ENL ENL162 PLESS J45 BNA J73 TLH CLAMP-STAR	1100-0400
Washington Natl	DLL J34 BUCKO BUCKO-STAR	
West Palm Beach	ODI J30 BRIBE BDF ENL ENL162 PLESS J45 BNA J73 LAL LLAKE-STAR	1100-0400
OMAHA		
Chicago O'Hare	FOD DBQ JVL-STAR	0700-2359
ROCHESTER, MN		
Chicago O'Hare	RST JVL-STAR	0000-2359
ST LOUIS		
Baltimore	GATWY-SID IIU J526 BKW J147 CSN OTT-STAR	
Boston	GATWY-SID ROD J29 JHW J82 ALB GDM GDM-STAR	
Chicago Midway	CARDS-SID CAP MOTIF-STAR	1200-0400
Chicago O'Hare	CARDS-SID BDF BDF-STAR	0000-2359
Columbus	GATWY-SID ROD V210 GUNNE	
Dallas/Fort Worth	LINDY-SID MAP RZC FSM BYP	
Detroit Metro-Wayne Co	GATWY-SID VHP FWA V11 V98 MIZAR	
Detroit Satellites:		
Pontiac, Willow Run, Ann Arbor	GATWY-SID VHP FWA CRUXX-STAR	
Detroit City, Windsor	GATWY-SID VHP FWA V96 VWV VWV051 POOFE...	
Fort Lauderdale	LINDY-SID MAW VUZ J39 MGM TLH J43 PIE	

Figure 10-24. While preferred IFR routes are listed from both Omaha and Rochester to Chicago O'Hare, there are no preferred routes between Omaha and Rochester.

When no preferred IFR route is available, you will have to consult the enroute chart to select the most practical route for the flight. In all cases, you should check for applicable MEAs along the route. In some parts of the country, minimum enroute altitudes may be beyond your aircraft's climb capabilities or they may require the use of oxygen. If your airplane is equipped with LORAN-C or an IFR-certified GPS, you also may consider planning a direct route. While the Turbo Skylane in this planning example does have GPS, it is not IFR-certified nor does the airplane have LORAN-C. As a result, the flight will need to be planned along airways between Omaha and Rochester. [Figure 10-25]

Figure 10-25. The most direct airway routing between Omaha and Rochester consists of V-138 from Omaha (KOMA) through Fort Dodge (KFOD) to Mason City (KMCW). The route then follows V-161 from KMCW to the Rochester VORTAC from which you can transition to several of the Rochester (KRST) approaches.

Once your route has been selected, you will need to determine if Omaha and Rochester have published departure and arrival procedures. You can find out if departure and approach procedures are published by reviewing either NOS or Jeppesen approach charts for both airports. Since the flight originates in Omaha, you should first look to see if Omaha has published departure procedures. [Figure 10-26] It may also be wise to review the approach procedures into Eppley in case you have to return shortly after takeoff.

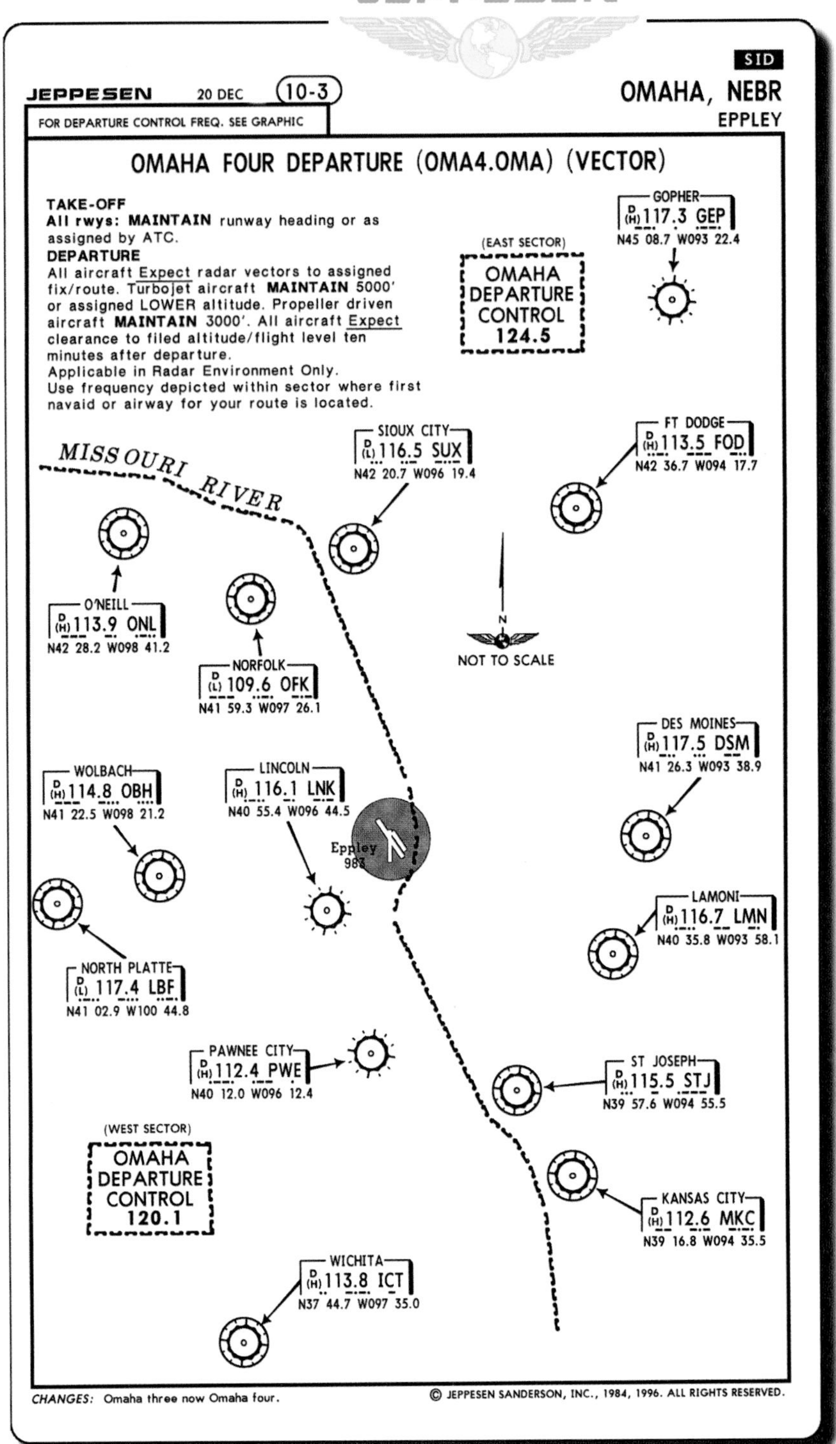

Figure 10-26. The Omaha Four Departure provides radar vectors to assigned fixes or routes for departing aircraft. For the flight to Rochester, you will be vectored directly to V-138 which will take you to Fort Dodge and Mason City.

It's a serious misunderstanding of instrument flying to think of an approach plate as a mere map for dropping out of the clouds in search of a runway, at the very least, a plate is a work of art— Paul Bertorelli, 'IFR' Magazine

In the process of looking for arrival and departure procedures, you will find there are no published arrival procedures for Rochester. Therefore, you will need to determine how to transition from the enroute to the approach segment of your flight. Reviewing the enroute charts as well as the approach plates at Rochester shows that many of the initial fixes for approaches into Rochester are based on the Rochester VORTAC, the final fix in

your enroute segment. Consequently, you will probably be able to transition directly from the enroute portion of your flight into the approach given to you by ATC. [Figure 10-27]

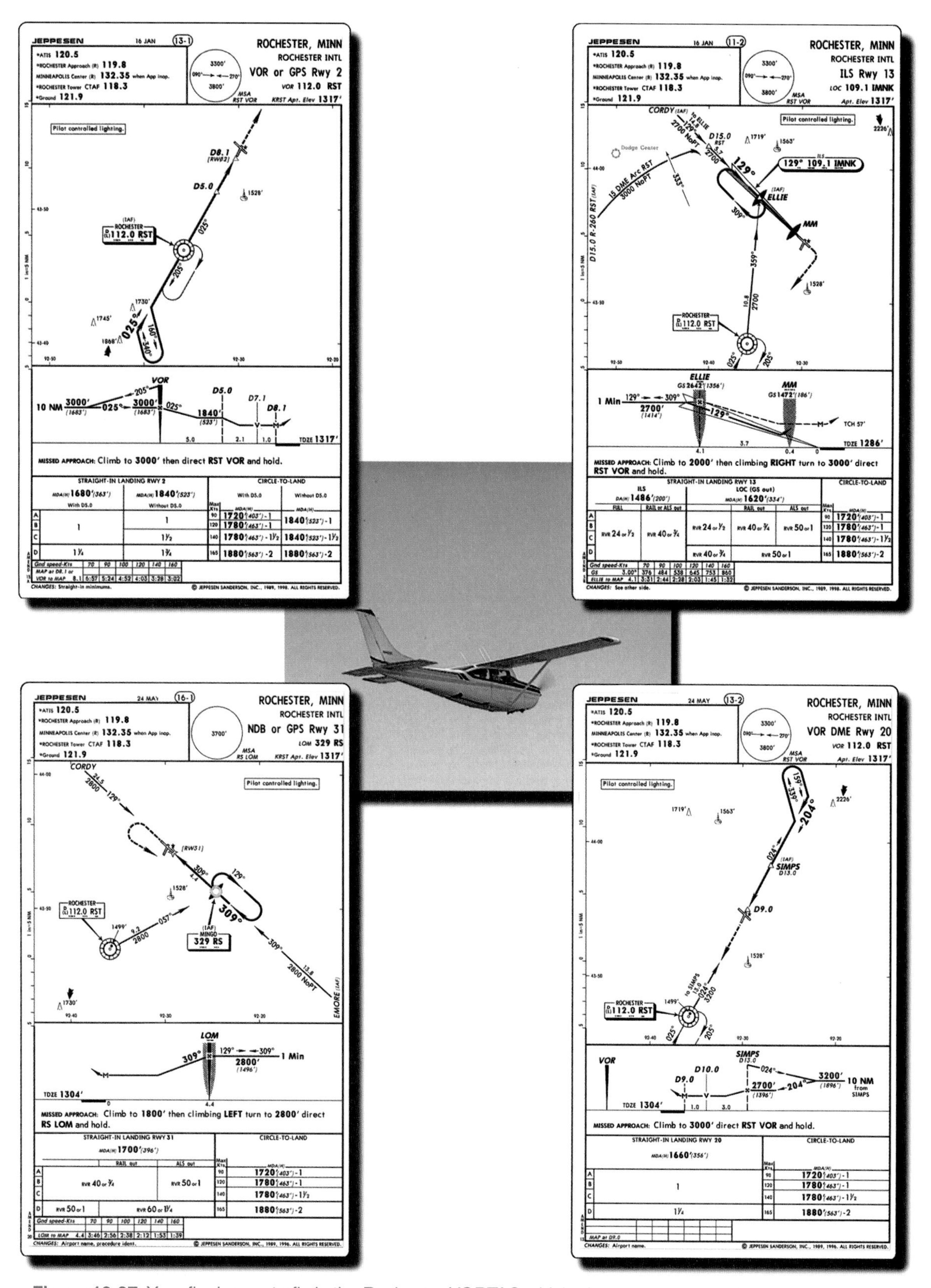

Figure 10-27. Your final enroute fix is the Rochester VORTAC which also serves as the IAF for a number of approaches into KRST.

With your primary route selected, you should consider possible **alternate airports** in case of poor weather at Rochester. Each alternate should be far enough away to avoid adverse weather affecting your destination. In addition, your alternate should have adequate communications, weather reporting, and at least one instrument approach. Other factors being equal, the more approaches an alternate has, the better. Examine the approach minimums for airports in the area and compare them to the forecast weather at your ETA. Also be sure to note whether an airport with an instrument approach may be filed as an alternate. [Figure 10-28]

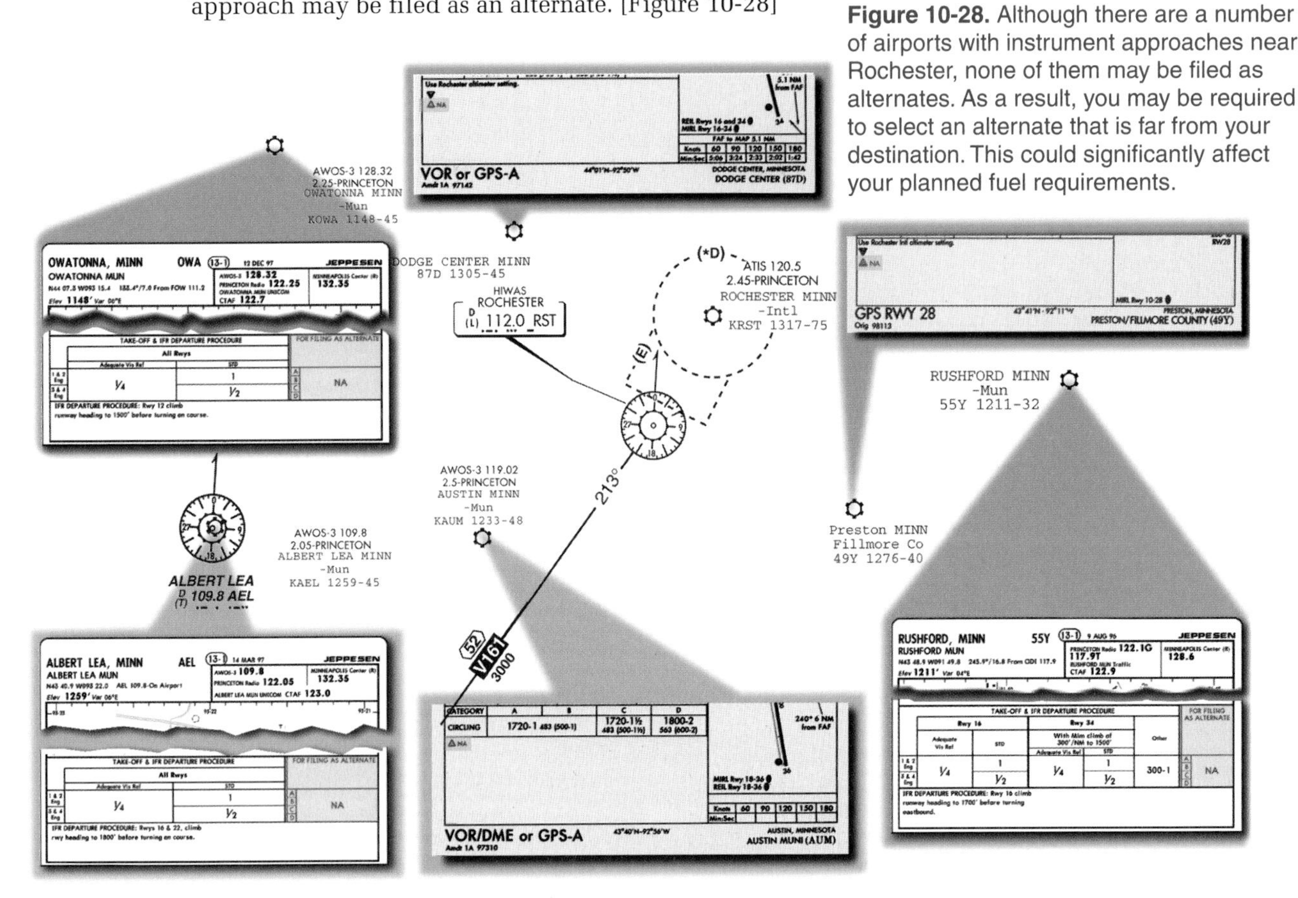

Figure 10-28. Although there are a number of airports with instrument approaches near Rochester, none of them may be filed as alternates. As a result, you may be required to select an alternate that is far from your destination. This could significantly affect your planned fuel requirements.

FLIGHT INFORMATION PUBLICATIONS

After choosing your route of flight you need to review the pertinent flight information publications. Navaid and lighting outages, runway closures, and limitations on instrument approach procedures can significantly affect your flight plan. **NOTAMs**, which are published in the Chart NOTAMs section of the Jeppesen Airway Manual and in the Notices to Airmen publications, are your best source of information regarding these factors. [Figure 10-29]

You should review the A/FD for specific information about your departure, destination, and alternate airports. Gathering data on runway lengths, fuel availability, lighting, hours of operation, and navigation, communication, and radar facilities helps to ensure that you meet the requirements of FAR 91.103 concerning preflight action. Additionally, you are required to ensure that the airplane and its equipment meet FAR inspection requirements. You also must determine whether any special navigation equipment in the airplane is IFR-certified. For instance, if your aircraft was equipped with LORAN-C you may need to reference the airplane's flight manual supplement to determine whether or not it meets IFR certification requirements.

For IFR flight, you are required to have working navigation equipment appropriate to the ground facilities to be used.

General Terminal Chart NOTAMs

KRST, ROCHESTER INTL, ROCHESTER, MN
KRST CHART NOTAMs:
Temporary, begins immediately until further notice. (13-1) VOR or GPS Rwy 2 st-in lndg with D5.0 MDA(H) 1720' (403'), VIS CAT C 1 1/4. (18-1) RADAR-1 st-in lndg ASR 2 MDA(H) 1720' (403'), VIS CAT C 1 1/4.

KMCW, MASON CITY MUN, MASON CITY, IA
KMCW CHART NOTAMs:
Temporary, begins immediately until further notice. ILS Rwy 35 (I-MCW) MM OTS.

Jeppesen Enroute Chart NOTAMs
USA - Navaids

AMES, IOWA (AMW) desmsnd. LO-27/7C
ARNOLD, TENN (AYX) VOR decmsnd. RNAV-11/2
BALLINGER, TEXAS (UBC) NDB may be shutdown. LO-11/8C, LO-16/4A
BENTON RIDGE (Findlay), OHIO (BNR) NDB may be shutdown. LO-32/2C, LO-34/2A
BETHPAGE, NY (BPA) NDB decmsnd, LO-47/11A, LO-48/6C, NE COOR-2/3D
BOYSEN RESERVOIR, WYO (BOY) VORTAC may be shutdown. No substitute routes published, LO-8/4A

Jeppesen Enroute Chart NOTAMs
USA - General Notices

Houston Approach Control Special Speed Limits. Aircraft departing from airports in the Houston Approach Control airspace and within Houston Class B airspace may be subject to removal or modification of the speed restriction contained in FAR 91.177a, at the discretion of Air Traffic Control (ATC). Houston Departure Control will be permitted to assign/authorize speed in excess of 250 knots to departing aircraft operating in the Houston Class B airspace using the phraseology "No speed limit" or "Increase speed to (number) knots".

Jeppesen Enroute Chart NOTAMs
USA Airways - Low Altitude 101-200

V115 MEA temp 3100 Choo Choo, Ten (GQO) VOR - Volunteer (VXV) VOR LO-37/8BC.
V116 MEA temp 4000 Keeler, Mich (ELX) VOR - Nepts Int (ELX R-256) and 10000 Keeler, Mich (ELX) VOR Leroy Int (ELX R-084).
V119 temp NA Bradford, PA (BFD) VOR R-232 - Midpoint. LO-45/4D.
V120 MEA 3700, Sioux Falls, S Dak (FSD) VOR - Frye Int (FSD VOR R-256). LO-27/5B

Figure 10-29. The Chart NOTAMS section of the Jeppesen Airway Manual provides descriptions of navaid, airway, and airport NOTAMs that may affect your flight.

In order to use VOR equipment for navigation under IFR, FAR 91.171 requires that it has been operationally checked within the preceding 30 days and found to be within acceptable limits. If your airplane's maintenance record indicates a VOR check is needed, you can use the information found in the VOR Receiver Check section of the A/FD to locate a suitable checkpoint for the test. After completing the VOR receiver check, you must enter in the aircraft logbook or other record the date the test was done, its location, the bearing error of the receiver, and your signature.

You may determine that a LORAN-C equipped aircraft is approved for IFR operations by checking the Airplane Flight Manual Supplement.

The national airspace system is in a continual state of change. To stay abreast of these changes and to review current procedures, you should periodically consult the AIM. It provides information on items like navigation aids, lighting and airport markings, airspace, ATC, emergency procedures, safety of flight, medical factors, and charts.

It is your responsibility as pilot in command to make sure that the VOR check has been accomplished within the past 30 days, and the transponder has been checked within the past 24 calendar months. Transponder checks must be entered in aircraft logbooks. There also must be a written record of the VOR test, which includes the date, place, bearing error, and the signature of the person performing the test.

Once you have settled on a route and consulted the appropriate publications, you can begin putting together your navigation log. You should list the routes, courses, distances, checkpoints, and necessary communication and navigation information as well as any additional information you feel is important. However, the bulk of your flight plan cannot be completed until you have collected updated weather information for your route of flight.

WEATHER CONSIDERATIONS

Computers and automated weather systems have enhanced the way weather information can be gathered. Although internet and telephone systems have made the acquisition of weather data easier and faster, the effectiveness of these systems depends on your knowledge of the system you are using and weather data in general. By the time you are ready to receive a standard briefing from Flight Service, you should already have some idea about the weather hazards along your route of flight and conditions at your destination.

You should begin gathering weather data several days before your flight in order to obtain a general overview of weather patterns. The weather section of your newspaper, the weather portion of the TV news, and The Weather Channel are good sources of general information. If you want more detailed information and have access to the internet, you can select a number of weather sites that provide current weather reports as well as radar and satellite pictures of the areas in which you plan to fly. The National Weather Service, the National Oceanic and Atmospheric Administration, the University Corporation for Atmospheric Research, the National Center for Atmospheric Research, and many other internet sites also are available to you for gathering weather information. Weather information can also be obtained through other services like AOPA Online and Jeppesen's JeppFax or GoJepp site on CompuServe.

Additional automated weather briefing sources such as Direct User Access Terminal System (DUATS) can be used for specific weather information as well. When using these services, make sure you are aware of the information they offer, as well as what they do not. For instance, many automated sources do not provide all of the necessary information for an instrument or cross-country flight. For more information on the data provided by these services, refer to the AIM.

The most current enroute and destination weather information for an instrument flight should be obtained from the FSS.

During preflight preparation, weather report forecasts which are not routinely available at the local service outlet can best be obtained by means of the request/reply service.

Once you have a general picture of the weather along your route of flight, you need to get a standard weather briefing from either an FSS or DUATS. While it is possible you might get a briefing from NWS in certain parts of the country, the NWS does not provide NOTAMs, flow control advisories, or information on MOAs and MTRs, nor do they accept flight plans. As a result, a briefing from an FSS is the best way of obtaining current enroute and destination weather.

During preflight weather briefings, you can request any reports or forecasts which are not routinely available at your service outlet. This is called request/reply service and is available at all FSS and NWS outlets. During the briefing you should use a weather log to record the information and analyze it. You want to know if the briefing is consistent with the preliminary data collected and whether or not it makes sense. If the briefing follows the general weather patterns you have already observed, the forecast weather at your destination may be fairly reliable. If the briefing and other general weather data do not coincide, then the weather may be different than forecast, and you should gather additional information.

In order to plan the Omaha to Rochester flight, you might want to begin by getting a general picture of the weather along your route of flight. Weather radar and satellite images of the region over which you will fly can provide a quick idea of the current factors you should pay attention to, as well as what you might expect to find in the forecast. [Figure 10-30]

Along with the radar and satellite images, you can use a surface analysis chart to see greater detail. The surface analysis chart is useful for determining where fronts are located, especially relative to your flight. With a front extending northeast along your route of flight, the surface analysis chart indicates you will be flying just ahead of the front during the trip. You also want to know how quickly and in what general direction the front is projected to

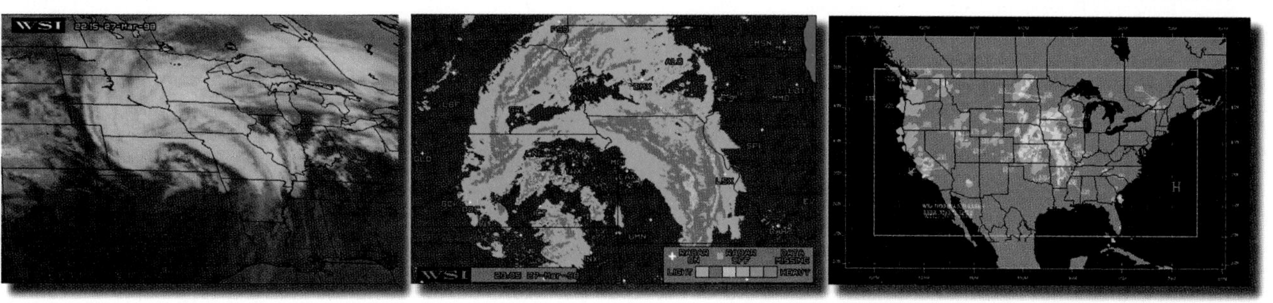

Figure 10-30. Radar and satellite images of the Nebraska, Iowa, and Minnesota region indicate a weather system with precipitation and cloud cover conditions throughout the area.

move. A review of the 12-hour surface weather prog chart can give you an idea of what the front may do during your flight, indicating where the significant weather will be as well as what you should specifically ask for during the standard briefing. [Figure 10-31]

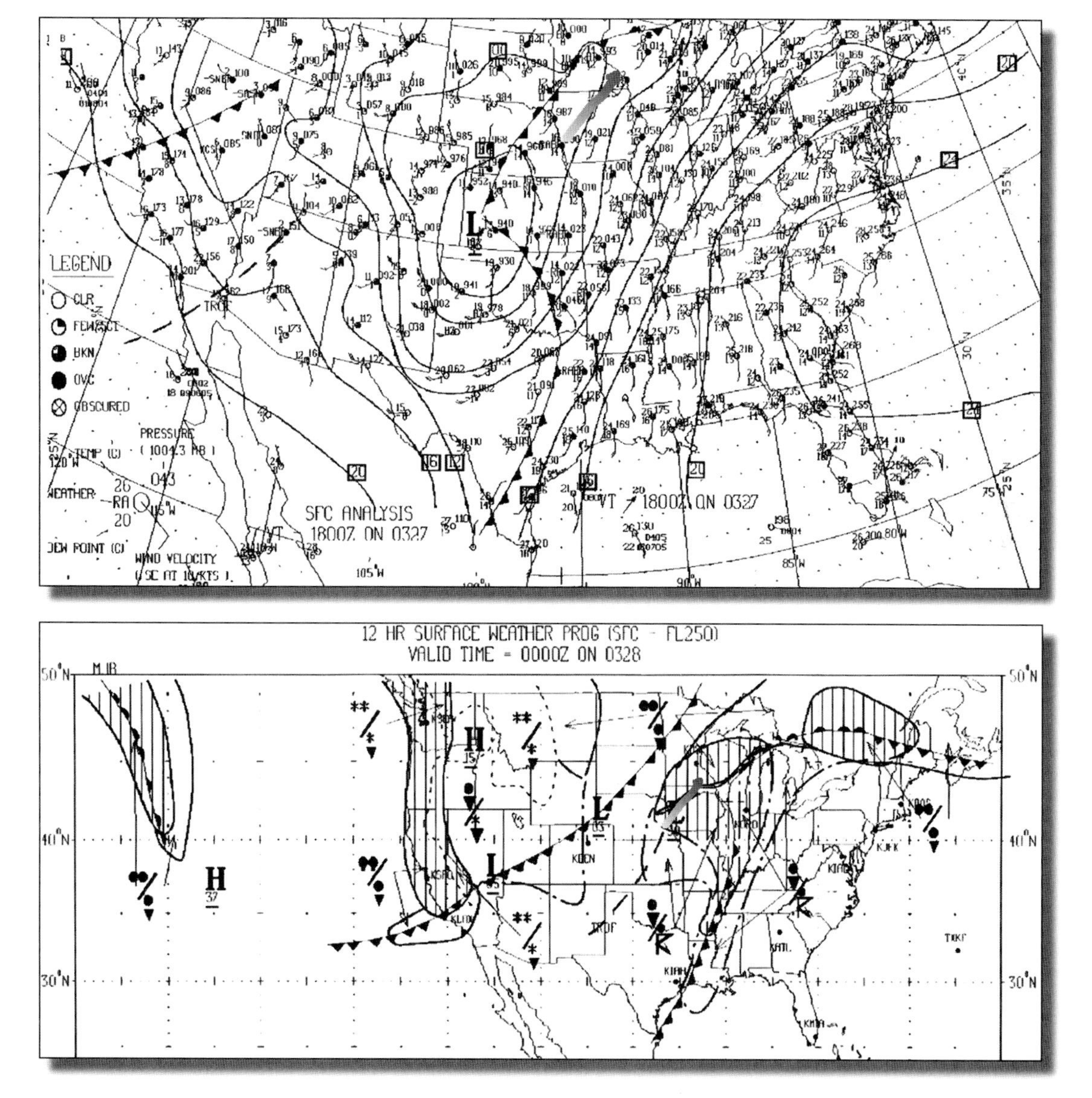

Figure 10-31. The surface analysis chart depicts a low pressure system in western Kansas with a stationary front running northeast parallel to your route of flight. However, the 12-hour surface prog chart shows the low has moved into eastern Nebraska, causing the front to shift slightly. If the front remains stationary, current reported conditions may not change much. If the front moves as predicted, however, you will need to pay close attention to the terminal forecasts for locations along your route of flight.

The picture of the weather along your route of flight should now be starting to take shape. However, the radar plots, surface analysis, and surface prog charts only tell part of the story. For instance, the radar picture indicates you can expect precipitation between Omaha and Rochester, but it may not tell you whether you will be in IFR conditions the entire time. In this case, you might review a low-level significant weather prog chart to find out where areas of IFR, marginal VFR, and VFR conditions are expected to exist. The low-level significant weather prog chart can identify areas of forecast IFR and marginal VFR weather, turbulence, and the location of the freezing level. [Figure 10-32]

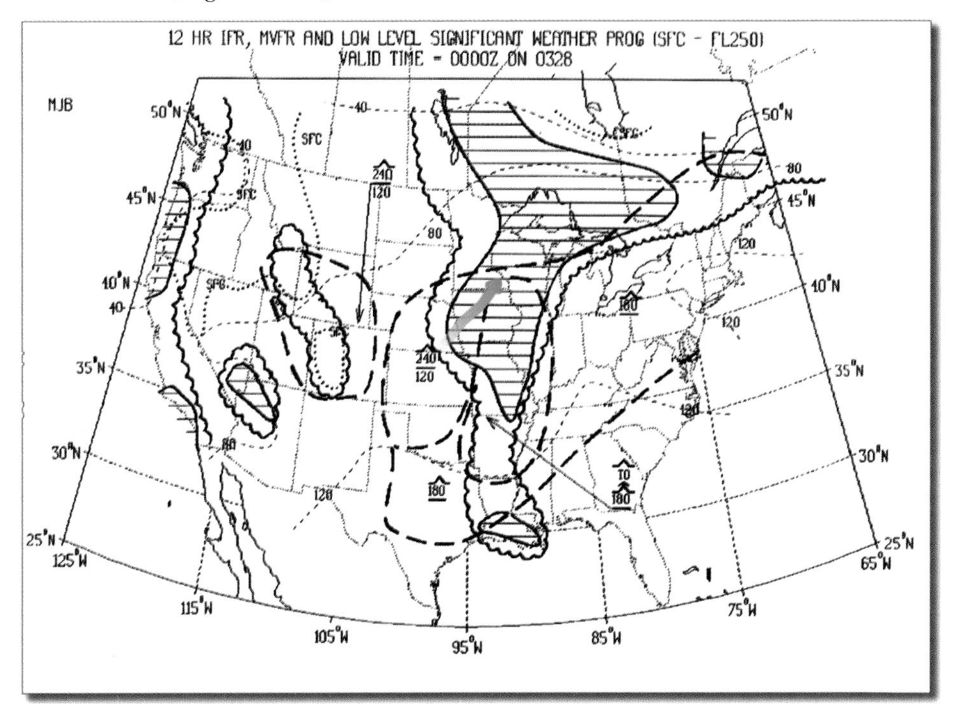

Figure 10-32. The low-level significant weather prog chart for the proposed flight time between Omaha and Rochester indicates IFR conditions and turbulence above 12,000 feet.

For greater simplicity, you may want to reference the weather depiction chart. Although it does not provide information on frontal activity or the movement of pressure systems, it does provide a clear indication of the ceiling and visibility conditions of reporting points along your route. [Figure 10-33]

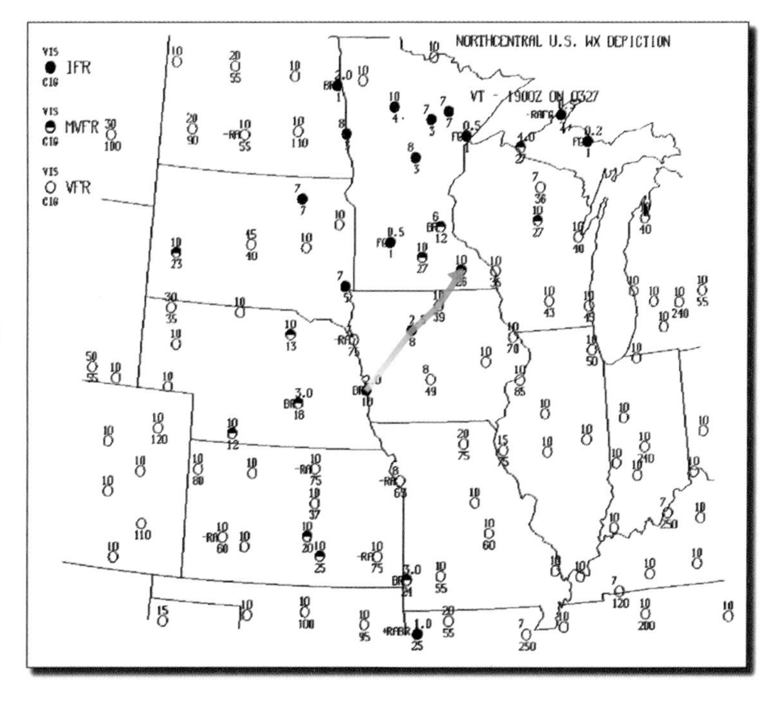

Figure 10-33. The weather depiction chart shows reporting points along your route as generally having IFR or marginal VFR conditions. These points roughly correspond with the frontal information found on the surface analysis chart.

Both the low-level significant weather prog and weather depiction charts indicate the Omaha to Rochester flight will be conducted primarily under IFR conditions. Since you will be flying in visible precipitation, you need to find out where the freezing level is along your route. The low-level significant weather prognostic chart can indicate to you whether or not to expect icing at your altitude during the trip. [Figure 10-34]

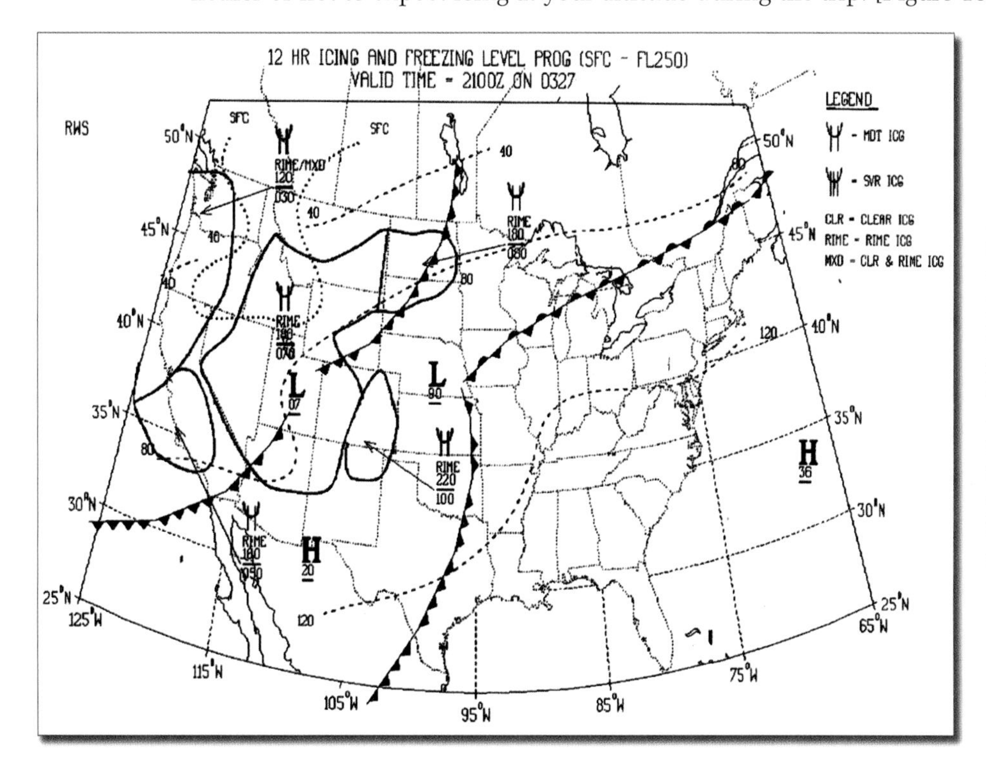

Figure 10-34. Based on the information from this chart, your route of flight lies approximately halfway between the 8,000-foot and 12,000-foot freezing level lines depicted on the chart. Since you cannot determine exactly where the freezing level is for your route, check the pilot reports for icing conditions during your briefing.

Since the radar returns indicated some heavy precipitation near your route of flight, you should check for convective activity in the area as well. The area forecast can provide you with much of the information you need to determine where you may find severe weather. You can also use it to confirm or amplify information displayed on the graphic weather charts. [Figure 10-35]

AREA FORECAST

271952
HI FA 271945 SYNOPSIS AND VFR CLDS/WX SYNOPSIS VALID UNTIL 281400 CLDS/WX VALID UNTIL 280800...OTLK VALID 280800-281400 ND SD NE KS MN IA MO WI LM LS MI LH IL IN KY. SEE AIRMET SIERRA FOR IFR CONDS AND MTN OBSCN. TS IMPLY SEV OR GTR TURB SEV ICE LLWS AND IFR CONDS. NON MSL HGTS DENOTED BY AGL OR CIG...

MN NWRN/N CNTRL...AGL BKN-SCT015 TOPS TO 060. 02Z CIG BKN020. WDLY SCT -SHRA. OTLK...MVFR CIG RA BR. SERN 1/4...CIG BKN030 TOPS TO FL250. OCNL CIG OVC010 WITH WDLY SCT -TSRA. CB TOPS TO FL350. 00Z CIG OVC010. SCT -SHRA/-TSRA. VIS 3-5SM BR. OTLK...IFR CIG RA BR. IA WRN 1/3...CIG OVC020 TOPS TO FL250. SCT -SHRA/WDLY SCT -TSRA. TS POSS SEV. CB TOPS TO FL350. 02Z CIG BKN030. WDLY SCT -SHRA.VIS 3-5SM BR. OTLK...MVFR CIG BR. ERN 2/3... CIG BKN030 OVC050 TOPS TO FL250. SCT -SHRA/WDLY SCT -TSRA CNTRL 1/3 AFT 21Z AND ERN 1/3 AFT 23Z. TS POSS SEV. CB TOPS TO FL400. OTLK...MVFR CIG BR.

NE PNHDL...AGL SCT040 SCT100. LYRS OCNL BKN. 02Z CIG BKN100 TOPS TO FL200. OTLK...VFR. ERN 1/2...CIG OVC015 TOPS TO FL250. VIS 3-5SM -RA BR. WDLY SCT EMBD -TSRA. TS POSS SEV SERN NE. CB TOPS TO FL400. 01Z CIG BKN025. WDLY SCT -SHRA/ISOL -TSRA. OTLK...MVFR CIG BECMG VFR BY 10Z.

Figure 10-35. The highlighted VFR Clouds and Weather portions of the area forecast provide a general description of the conditions to be expected for your flight. It is important to note that ceilings may be as low as 1,500 feet while cloud tops extend to FL250 and beyond. Widely scattered thunderstorms also may be possible in eastern Nebraska, Iowa, and southern Minnesota.

By now you should have a good idea of what the weather will be doing during your flight, but you still need to get specific information on the weather at your departure and destination airports as well as possible alternates near your flight path. For instance, the weather at your point of departure, (Omaha), and along your route (Fort Dodge, Mason City, Rochester) is reported as marginal VFR to IFR, which agrees with the weather depiction charts. At your estimated time of departure (2000Z), the weather at Omaha is forecast to be 2,500 feet overcast with four miles visibility in light rain and mist. This forecast is an important consideration because you need to know if it will be possible to return to Omaha should mechanical difficulties develop during departure. The forecast weather is well above the straight-in and circling minimums for all approaches into Omaha's Eppley Airfield, which means you should be able to return to the airport if necessary. [Figure 10-36]

```
TAF

KOMA  271730Z 271818 15015G24KT 4SM -RA BR OVC 025
        FM 2100 17015G25KT 5SM -RA BR OVC 040
        TEMPO 2101 1SM +TSRA GR BR OVC 008CB
        FM 0100 22012KT P6SM BKN 025 OVC 040 PROB 40 0105 2SM -SHRA BKN 008
        FM 0600 28010KT P6SM BKN 030
        FM 0900 30006KT P6SM SCT 030 SCT 250
        FM 1500 12010KT P6SM SCT 035 SCT 100
```

Figure 10-36. Except for temporary conditions, the Omaha forecast indicates that the ceiling and visibility are both expected to improve after 2100Z.

You also should check the forecast for Rochester. Remember, if the forecast weather at your destination from 1 hour prior to your arrival to 1 hour after your arrival is not expected to be 2,000 feet or higher and 3 miles or more, you need to file an alternate. As a result, you should pay close attention to the forecast weather at Rochester as well as airports in the area that may be used as alternates. The TAF for Rochester (KRST) during that time period does not indicate the need for an alternate. Since the forecast ceiling and visibility are only marginally better than the requirement for an alternate you would be wise to observe the forecasts at airports that may be used as alternates if events subsequently prove the Rochester forecast to be inaccurate. [Figure 10-37]

```
TAF

KRST    271910Z 271918 20022G32KT P6SM BKN 035
        FM 0000 19018G25KT 3SM -TSR BR OVC 008CB
        FM 0600 21015KT 1SM -SHRA BR OVC 006
        FM 1500 28016G22KT 5SM BR OVC015
```

Figure 10-37. The Rochester forecast during your estimated time of arrival (2115Z) calls for visibility to be greater than 6 miles with a 3,500-foot ceiling. However, beyond that period the forecast shows the weather deteriorating. To be safe, check the forecasts for predicted conditions at airports in the area that may be suitable for diverting to in case the weather at Rochester drops below minimums prior to your arrival.

From previous chapters you know that an alternate should have one or more approaches and weather that is forecast to be above approach minimums. For instance, the forecast weather at an alternate at the estimated time of arrival must include a ceiling of at least 600 feet and 2 miles visibility if a precision approach is available, or 800 feet and 2 miles for nonprecision approaches, in order to be filed as an alternate. Of course, should you elect to divert to an alternate and fly the approach, the landing minimums for the approach being flown apply.

The alternate minimums that must be forecast at the ETA for an airport that has a precision approach procedure are a 600-foot ceiling and 2 miles visibility.

If you elect to proceed to the selected alternate, the landing minimums used at that airport should be the minimums specified for the approach procedure selected.

By reviewing your navigation charts, graphic weather charts, and radar and satellite images, you can determine quickly which airports in the vicinity may be VFR, or at least have

better weather conditions than your destination. For instance, the radar images and surface analysis charts show VFR conditions to the west of Rochester. By checking the current and forecast conditions for airports west of Rochester, you find that Owatonna (KOWA) is above IFR minimums and conditions are forecast to improve as the weather system moves east. To the west of Owatonna, at Mankato (KMKT), current conditions are VFR and forecast to remain that way through the forecast period. If you reference the A/FD you also will find that both Owatonna and Mankato have instrument approaches. As a result, both airports make suitable alternates should conditions at Rochester preclude you from landing there. [Figure 10-38]

```
TAF

KOWA 271740 171818 17013KT 4SM OVC 018
        FM 2100 17010KT P6SM BKN 020
        FM 0000 17011KT 5SM BKN 040
        FM 0400 16008KT 5SM SCT 060
        FM 1100 16008KT P6SM SCT 100

KMKT 271750 271818 18008KT P6SM SCT 050
        FM 0000 18010KT P6SM FEW 100
        FM 1000 17010KT P6SM SCT 250
        FM 1400 17006KT P6SM SCT 250
```

Figure 10-38. The TAFs indicate that the further west you fly from Rochester, the better the conditions. This corresponds with the information gained from the radar and satellite images, as well as the surface analysis and 12-hour surface weather prog charts that indicate the weather system to be moving east.

Space Shuttle Landing Minimums

As an IFR pilot you are required to file an alternate airport if your destination airport's forecast weather does not meet certain requirements. In addition, you cannot fly an approach and land if the weather at the airport is reported as being below approach minimums for that particular approach.

Like you, the space shuttle must have certain minimum weather available in order to land. For instance, at the Kennedy Space Center, at least seven weather criteria, including the following, must be met in order for a shuttle to land at the end of a mission.

Ceiling: 8,000 feet or higher

Visibility: 5 miles or better

Winds: Crosswind – 15 knots or less
Headwind – 25 knots or less
Tailwind – 15 knots or less

Rain/Thundershowers: Not allowed within 30 nautical miles of shuttle

Turbulence: Moderate or less

With current and forecast weather in hand, your next step is to obtain the forecast winds aloft and any available pilot reports for the route of flight. In this case, the winds aloft through 12,000 feet for your route are 170° at 20 knots for the first half of the flight and 200° at 20 knots for the second half of the flight. Additionally, pilot reports indicate occasional light to moderate turbulence from 6,000 to 10,000 feet between Omaha and Fort Dodge, with light rime ice above 10,000 feet. Finally, check the NOTAMs through FSS, DUATS, or JeppView to ensure you will not encounter any problems with the airports and navigation aids you plan to use. [Figure 10-39]

Figure 10-39. NOTAMs for your flight may be obtained through FSS, DUATS, or in the Jeppesen Airway Manual. Be sure to check the NOTAMs for airports, navaids, and airways pertinent to your flight.

ALTITUDE SELECTION

Before you complete your planning, you must select a cruising altitude. With no significant terrain features between Omaha and Rochester, the highest MEA is 4,500 feet. Because ATC generally assigns altitudes which correspond with the hemispheric rule, the lowest altitude you should request for your easterly flight is 5,000 feet. And, since you are not carrying oxygen for this trip, the highest altitude you should request is 11,000 feet.

Next, consider the effects of winds aloft as well as turbulence and icing reports. From the winds and temperatures aloft forecast you know that the winds are from a southerly direction at 20 knots through 12,000 feet. Also, pilots have reported turbulence starting at 6,000 feet and light rime ice above 10,000 feet. Note that it may be possible to avoid the turbulence and stay clear of the ice while enjoying a modest tailwind by choosing to fly at 5,000 feet. This altitude also ensures that you are above the MEA and in compliance with ATC's hemispheric rule. However, if you refer back to the significant weather prognostic charts you will see that at 5,000 you will probably be in IFR conditions. You may confirm this by looking at the current observations and forecast conditions for points along your route of flight. [Figure 10-40]

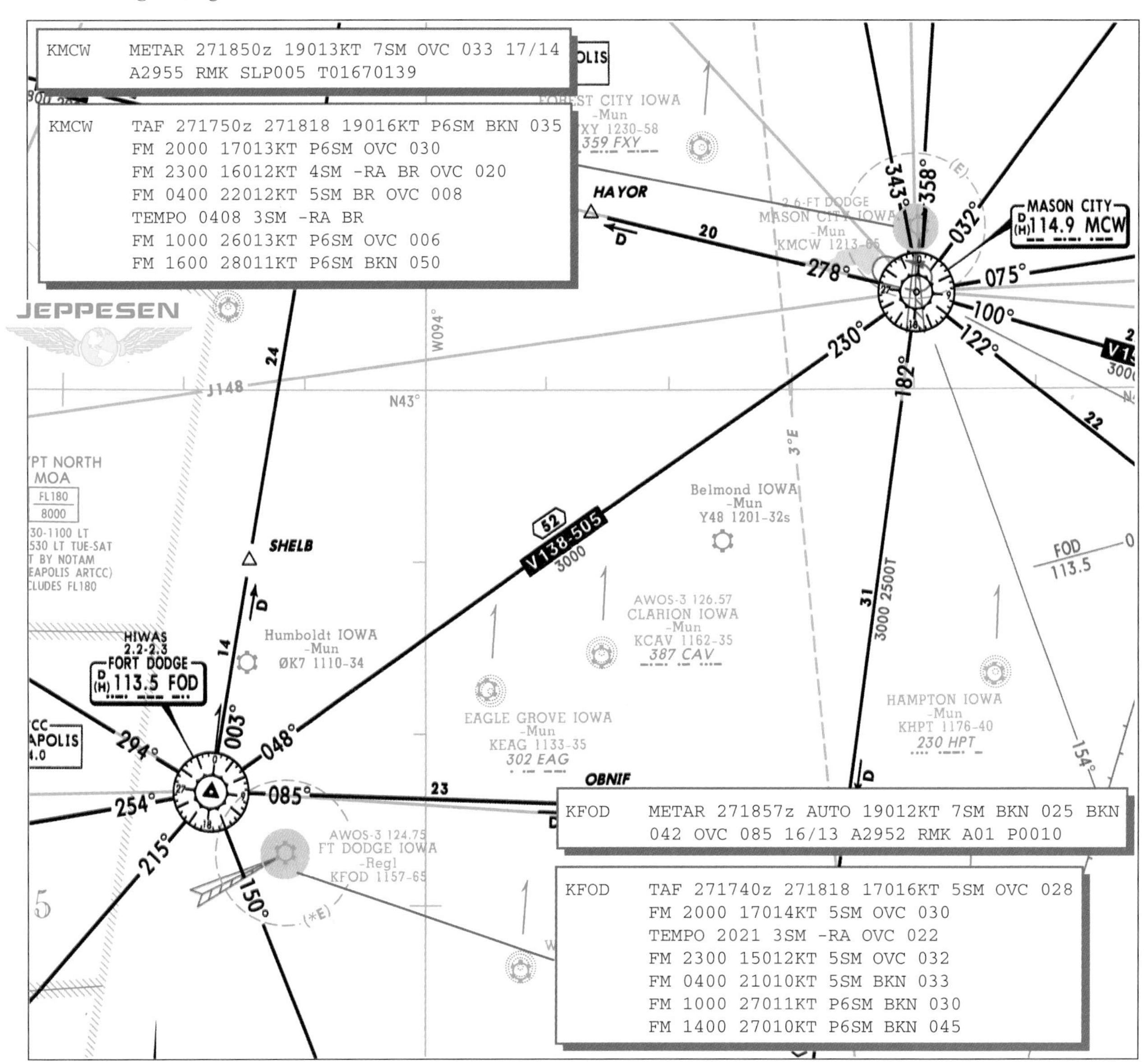

Figure 10-40. Fort Dodge is currently reporting 4,200 feet broken while Mason City has a 3,300-foot overcast. In addition, both locations are forecast to have a 3,000-foot ceiling during your proposed flight.

COMPLETING THE NAVIGATION LOG

With your weather briefing complete, you can now make your **go/no-go decision**. Your decision should be based on the weather reports and forecasts you have gathered in addition to your airplane's capabilities. You should also consider your personal limitations and level of proficiency. In considering this flight, you can see that it can be conducted below the reported turbulence and icing although it may be in IFR conditions while enroute. If you have to proceed to an alternate, there are at least two airports within an hour's flight time that are forecast to be above approach minimums or VFR. Based on this analysis, you decide the flight is within your capabilities and those of the aircraft. Your next step is to complete the navigation log. [Figure 10-41]

1

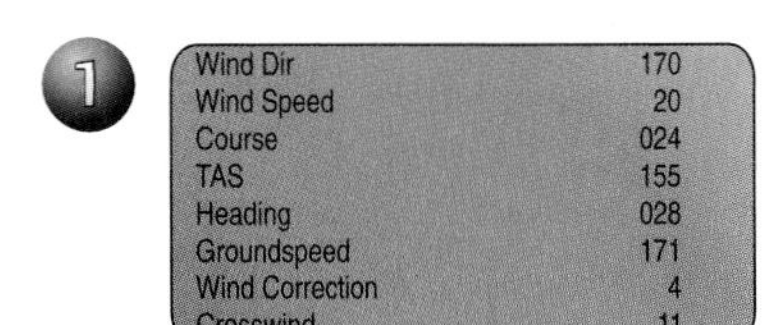

Press the WND button to get heading and groundspeed prompts. Then select heading and enter the wind direction, wind speed, a course of 024°, and a TAS of 155 knots at the corresponding prompt. The computer will provide you with a groundspeed of 171 knots and a magnetic heading of 028° which you should record on your flight log.

2

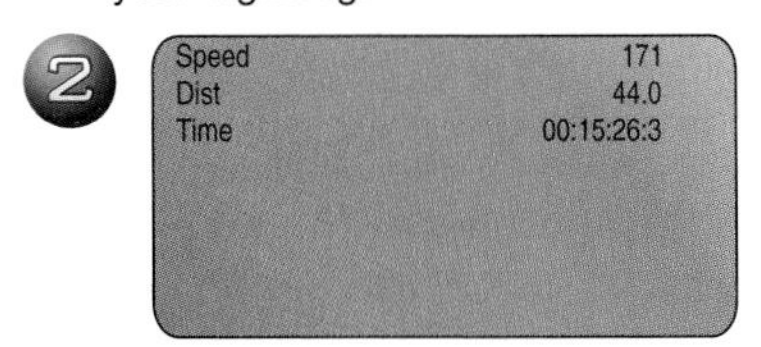

Next, press the TSD button for time, speed, and distance then choose the time function followed by the speed/distance function. Enter the groundspeed of 171 knots from ballflag one and the distance of the leg to be flown, in this case 44 miles from the top of climb to MADUP Intersection. The computer determines the flight time to be 15 minutes, 26 seconds of flight time for the leg.

3

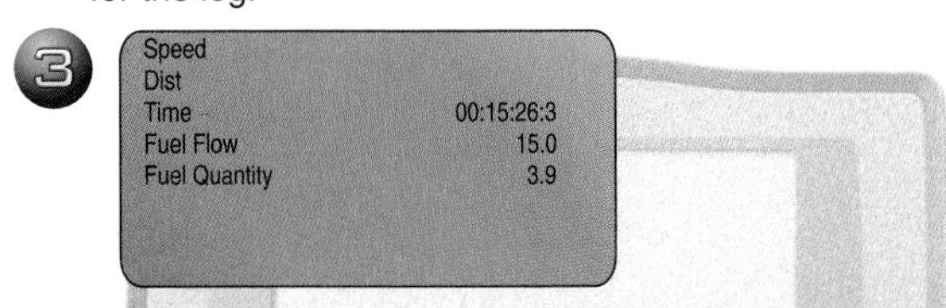

JEPPESEN NAVIGATION LOG

Aircraft Number: 3450B | Dep: KOMA | Dest: KRST | Date: 3/28

Clearance: RV THEN AS FILED 5,000

Check Points (Fixes)	Ident / Freq.	Course (Route)	Altitude	Mag Crs.	FUEL Leg / Rem.	Dist. Leg / Rem.	GS Est. / Act.	ETE / ATE	ETA / ATA
KOMA	OVR / 116.3				880	213			
TOC	OVR / 116.3	RV	↑5	024	4.3 / 83.7	11 / 202	96	7	2007
MADUP	FUD / 113.5	V138	5	035	3.9 / 79.8	44 / 158	171	15	2022
FT. DODGE	FUD / 113.5	V138	5	048	4.8 / 75.0	54 / 104	168	19	2041
MASON CITY	MCW / 114.9	V161	5	032	4.5 / 71.5	52 / 52	172	18	2059
ROCHESTER VORTAC	RST / 112.0	V161	5		4.5 / 67	52 / 0	175	18	2117
APPROACH					3.8 / 63.2			15	

FUEL									
Climb	4.3	Cruise	16.7	Apch.	3.8	Alt.		Res.	63.2
Cruise Burn/Hr	15	Block In		Block Out		Log Time			

AM436237E

Press the TSD button again, then select fuel quantity from the list. Enter the flight time of 15 minutes, 26 seconds computed for the leg as well as the fuel consumption of 15 gallons per hour found in the POH. The computer calculates a fuel burn of 3.9 gallons for that particular leg which should also be entered on the flight log.

4

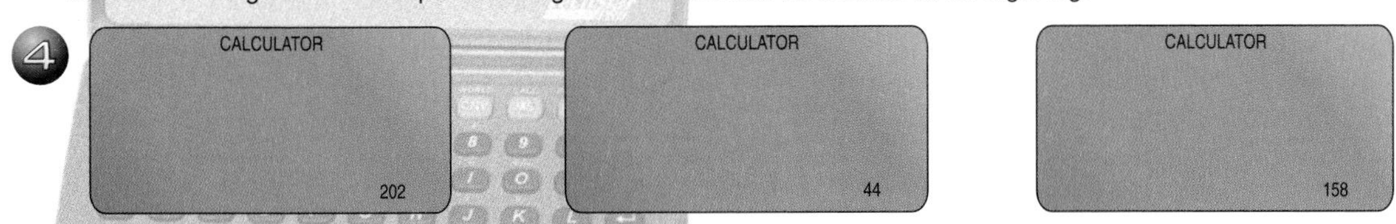

You can use the computer to determine the distance remaining for the flight by selecting the calculator function. Enter the total distance of the flight and subtract from it the distance of the leg flown to get the distance remaining. In this case, the distance from the top of climb to KRST is 202 miles. Subtract the 44 miles of the leg to MADUP Intersection from the total mileage to get 158 miles remaining. The same procedure may be used to calculate fuel remaining. The calculator function also may be used to total the time enroute and fuel used by adding the respective times and fuel consumptions per leg.

Figure 10-41. With the aid of an electronic flight computer like the TechStar Pro, you can quickly compute the information needed to complete your flight plan. If you are using a different type of computer, the exact keystrokes and information presented may be different. The computations in this figure are for the flight leg beginning at Top of Climb (TOC) and ending at MADUP Intersection.

A flight computer will aid you in computing time, speed, distance, wind, course direction, and groundspeed. See Figure 10-41.

The navigation log is a convenient way for you to complete your preflight planning and organize your flight. It provides a concise textual description of your flight and allows you to monitor your progress in order to make adjustments if conditions should change enroute. For example, if you start with 88 gallons of fuel, the fuel remaining should agree with the figures listed on the navigation log as you pass each checkpoint. By recording the time off and filling in the estimated time of arrival (ETA), actual time enroute (ATE), and actual time of arrival (ATA) boxes, you will be quickly alerted to any changes in winds aloft or in your airplane's performance.

FUEL EXHAUSTION AND PILOT CONFUSION

From the Files of the NTSB...
Aircraft: *Aronson Falco F8L*
Location: *Gainesville, FL*
Injuries: *2 Fatal*

Narrative:
The pilot received a weather briefing and filed an IFR clearance, then departed VFR. Except for a headwind, the first part of the flight was normal. In the vicinity of Macon, Georgia, the pilot requested and received an IFR clearance. Later, he elected to divert to the Gainesville Regional Airport and was vectored for an ILS approach. While diverting, he indicated some confusion concerning the approach plates. At first, he started using the Gainesville, Georgia plate, then he switched to the correct plate (Gainesville, FL). He had difficulty maintaining heading on the ILS and made a missed approach 2 miles east of the airport. He admitted having problems with the approach, declared a low fuel state and received vectors for another ILS. Again, he had difficulty maintaining heading control. On the second approach, the aircraft was observed to fly over the field and the pilot was cleared to land on any runway. However, the aircraft crashed on the approach end of Runway 6. There was evidence of little movement after impact and very little chordwise scraping of the prop. No fuel was found in the fuel tanks or fuel injector lines.

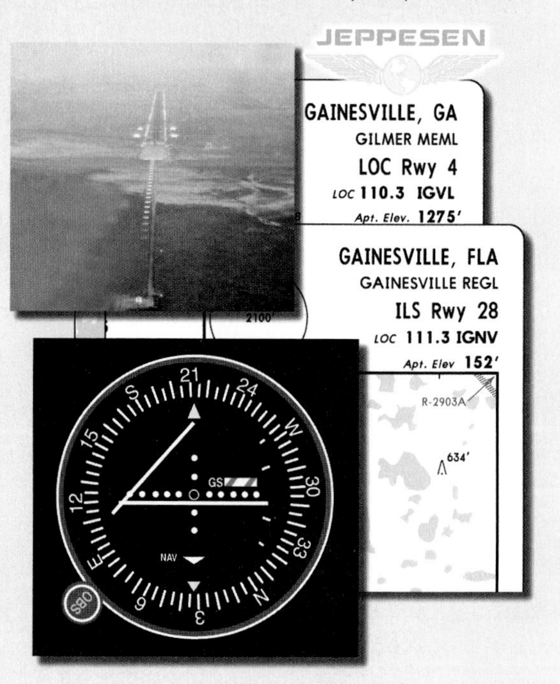

In addition to improper in-flight planning, the NTSB declared that poor IFR procedures, inadequate fuel supply, and fuel exhaustion, were causative factors in this accident. Had the pilot been better prepared for a possible diversion to an alternate airport, he may have avoided the approach chart confusion and low fuel state that ultimately resulted in a fatal accident after missing two approaches.

FILING THE FLIGHT PLAN

After you have completed the navigation log, transfer the appropriate information to your IFR flight plan. An important item to include on the IFR flight plan is the aircraft equipment code. The code is designated by a letter suffix attached to the aircraft type in block 3. The codes and their meanings are found in the *Aeronautical Information*

Manual. In this case, the /A means the aircraft is equipped with DME and a transponder with altitude encoding capability. Notice the form also lists a code indicating an airplane equipped with GPS (/G). While your Turbo Skylane RG is equipped with GPS, the receiver has not been certified for enroute, terminal, and GPS approaches which means you should not use /G to list the equipment for your airplane in box 3. [Figure 10-42]

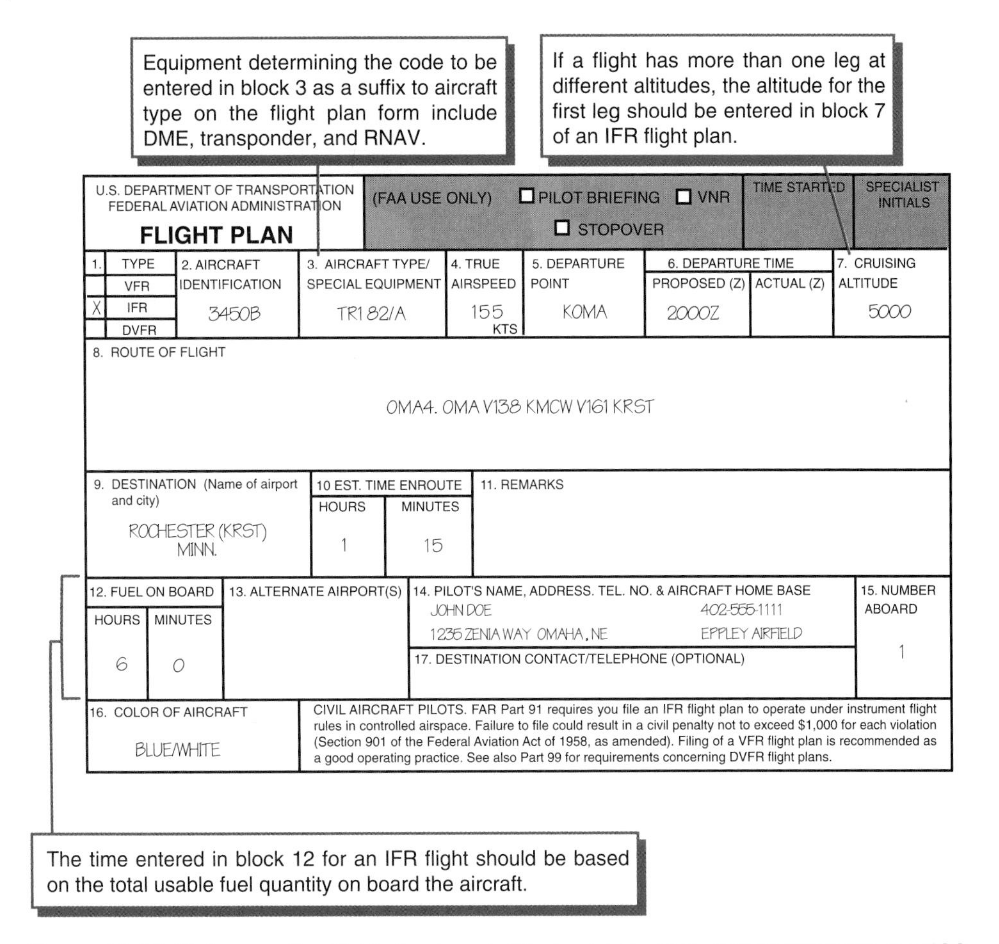

U.S. DEPARTMENT OF TRANSPORTATION
FEDERAL AVIATION ADMINISTRATION
FLIGHT PLAN

(FAA USE ONLY) ☐ PILOT BRIEFING ☐ VNR ☐ STOPOVER | TIME STARTED | SPECIALIST INITIALS

1. TYPE	2. AIRCRAFT IDENTIFICATION	3. AIRCRAFT TYPE/ SPECIAL EQUIPMENT	4. TRUE AIRSPEED	5. DEPARTURE POINT	6. DEPARTURE TIME PROPOSED (Z)	ACTUAL (Z)	7. CRUISING ALTITUDE
VFR / X IFR / DVFR	3450B	TR182/A	155 KTS	KOMA	2000Z		5000

8. ROUTE OF FLIGHT

OMA4. OMA V138 KMCW V161 KRST

9. DESTINATION (Name of airport and city)	10 EST. TIME ENROUTE HOURS	MINUTES	11. REMARKS
ROCHESTER (KRST) MINN.	1	15	

12. FUEL ON BOARD HOURS	MINUTES	13. ALTERNATE AIRPORT(S)	14. PILOT'S NAME, ADDRESS. TEL. NO. & AIRCRAFT HOME BASE	15. NUMBER ABOARD
6	0		JOHN DOE 402-555-1111 1235 ZENIA WAY OMAHA, NE EPPLEY AIRFIELD	1
			17. DESTINATION CONTACT/TELEPHONE (OPTIONAL)	

16. COLOR OF AIRCRAFT	
BLUE/WHITE	CIVIL AIRCRAFT PILOTS. FAR Part 91 requires you file an IFR flight plan to operate under instrument flight rules in controlled airspace. Failure to file could result in a civil penalty not to exceed $1,000 for each violation (Section 901 of the Federal Aviation Act of 1958, as amended). Filing of a VFR flight plan is recommended as a good operating practice. See also Part 99 for requirements concerning DVFR flight plans.

Figure 10-42. Make sure the flight plan form is filled out accurately before filing with the FSS.

You should use airport identifiers instead of the airport names to complete blocks 5, 9, and 13. Using the airport codes expedites the processing of your flight plan. If you do not know the correct code, list the name of the airport. For the route in block 8, you should use identifier codes for VORs, airways, waypoints, DPs and STARs.

In block 7 you should list your initial cruising altitude. If you want to change this altitude during the flight, you will need to direct your request to an ATC controller. Using the information from your navigation log you can compute the estimated time enroute to the point of your first intended landing. You should list this time in block 10 of the form. In block 12, record the total usable fuel on board expressed in hours and minutes.

The blocks on the flight plan form should be accurately completed with information needed by Flight Service to process your flight plan. Be sure to include the correct aircraft equipment code, route, destination, and fuel available information on the form. See Figure 10-42.

The point of first intended landing at your destination should be used to compute the estimated time enroute on an IFR flight plan.

Remember to file your IFR flight plan at least 30 minutes prior to the listed departure time. This allows sufficient time for the flight plan to be processed. [Figure 10-43]

Figure 10-43. You can file your flight plan by telephone or DUATS, or through a number of flight planning and filing software programs that are available.

Computerized Flight Planning and Filing

There are several models of flight planning and filing software available that can make the flight planning process easier. For instance, the Mentor FliteMap software allows you to access DUATS through your computer and develop a complete flight plan including the winds aloft needed to compute headings and speeds.

With a few key strokes computer-aided flight planning offers you an easy and quick way to compile a comprehensive flight plan. Some software allows you to access flexible routing capabilities, and select DPs and STARs. Computerized flight planning also allows you to quickly modify your flight plan should conditions change. In addition, the software may allow you to file your flight plan electronically through DUATS, eliminating a call to your local FSS.

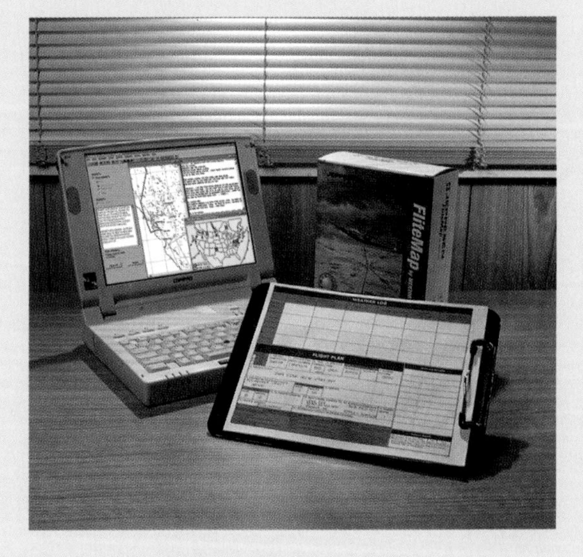

CLOSING THE IFR FLIGHT PLAN

The requirement to close your flight plan is established in FAR 91.169. You can cancel an IFR flight plan anytime you are operating in VFR conditions below 18,000 feet MSL. However, if IFR conditions are again encountered, you must receive an IFR clearance before proceeding into those conditions. If your destination airport has an operating control tower, your IFR flight plan is automatically closed when you land.

If you are flying to an airport that does not have an operating control tower, you are responsible for closing your own IFR flight plan. You can do this after landing through an FSS or by other direct communications with ATC. According to the AIM, you also can cancel in VFR conditions while you are still airborne and able to communicate with ATC by radio. This is appropriate in cases where an FSS is not available and air/ground communications with ATC are not possible at low altitudes. This saves you the time of cancelling by telephone. It also releases the airspace to other aircraft.

If your destination has IFR conditions and there is no control tower or flight service station on the field, you must close your flight plan by radio or by telephone upon landing.

SUMMARY CHECKLIST

✓ When you begin the IFR flight planning process, take a preliminary look at factors like weather, airplane performance and equipment, potential routes, and your instrument proficiency that may prevent you from making the flight.

✓ Availability of preferred IFR routes, aircraft performance considerations, and fuel economy will influence route selection.

✓ Check for published departure or arrival procedures relevant to your intended flight.

✓ NOTAMs should be reviewed for items like navaid and lighting outages or runway closures that can significantly affect your flight.

✓ Review the A/FD for specific information about departure and arrival airports as well as possible alternate airports that are pertinent to your flight.

✓ Begin gathering weather data several days before your flight in order to obtain a general overview of weather patterns.

✓ Although weather information may be obtained from numerous sources including newspapers, television, and the internet, these sources should not be considered suitable alternatives to a flight service station or DUATS standard briefing.

✓ In case the weather at your intended destination is forecast to have a ceiling less than 2,000 feet or visibility less than 3 miles, you need to file an alternate.

✓ A good alternate airport should be far enough away to be unaffected by weather at your destination, be equipped with appropriate communications and weather reporting capabilities, and have more than one approach.

✓ Once your weather briefing is complete, you can make your go/no-go decision and begin planning the flight if conditions are favorable.

✓ The navigation log is a convenient way for you to complete your preflight planning, organize your flight, and provide you with a concise textual description of your flight.

✓ Before filing your flight plan, ensure you have all of the blocks in the flight plan form filled in correctly, including the aircraft special equipment code in block 3.

✓ If you are flying to an airport that does not have an operating control tower, you are responsible for closing your own IFR flight plan by phone through FSS, or by direct communications with ATC.

KEY TERMS

IFR Flight Plan

Overview

Preferred IFR Route

Alternate Airports

NOTAMs

Go/No-go Decision

Navigation Log

Aircraft Equipment Code

QUESTIONS

1. True/False. Factors to consider during your flight overview include airplane equipment and your instrument proficiency level.

2. True/False. One factor that does not influence your route selection is fuel economy.

3. Select the item that is normally not a factor to consider when choosing an alternate airport.

   A. Adequate communications
   B. Weather reporting
   C. Courtesy car

4. True/False. The Chart NOTAMs section of the Jeppesen Airway Manual only provides descriptions of airport NOTAMs that may affect your flight.

5. True/False. It is not necessary to record the bearing error of the VOR receiver after conducting a required VOR receiver check.

6. What is the best source for obtaining current enroute and destination weather?

7. In order to file an airport with a nonprecision approach as an alternate, what minimum weather requirements must it be forecast to have at your estimated time of arrival?

8. What should your go/no-go decision be based on?

9. Select the correct information to be recorded in block 3 of the flight plan form.

   A. Departure time
   B. Aircraft type and equipment code
   C. Usable fuel available

10. How long prior to your departure time should you file your flight plan?

PART IV
COMMERCIAL PILOT OPERATIONS

"I'm challenged to do a little bit better than yesterday... each time I fly."

— Robert A. "Bob" Hoover

PART IV

As you gain experience on your journey to becoming a commercial pilot you may operate a variety of complex and high-performance airplanes. Chapter 11 will prepare you for the advanced systems you will encounter during transition to faster, more powerful airplanes. Aerodynamics and performance limitations in Chapter 12 will help you predict performance, control weight and balance, and define overall operating limitations for the airplanes you fly. Chapter 13 delves into commercial flight considerations including emergencies and human factors issues such as aeronautical decision making and crew resource management principles. Chapter 14 prepares you for the commercial pilot practical test by offering new insights when maneuvering an airplane near its maximum performance limits.

CHAPTER 11

ADVANCED SYSTEMS

Instrument/Commercial
Part IV, Chapter II—Advanced Systems

SECTION A
HIGH PERFORMANCE POWERPLANTS

Over time, aircraft engine manufacturers have improved product safety and efficiency by applying new design technology. One of the most significant improvements was a better method of delivering fuel to an engine's cylinders. This became necessary because float-type carburetors, although adequate for most lower performance aircraft, could not provide the desired performance to justify their use in higher powered aircraft. In order to overcome the carburetor's performance limitations, manufacturer's have utilized **fuel injection systems** in nearly all modern, high-powered, reciprocating engine aircraft. During your training, you will undoubtedly encounter this type of fuel system as you progress to flying complex and high performance airplanes. [Figure 11-1]

Figure 11-1. A complex airplane is one that has a retractable landing gear, flaps, and a controllable pitch propeller, whereas, a high performance airplane is one that has an engine of more than 200 horsepower. Although fuel injection systems may be installed on low performance airplanes, they are used primarily in complex or high performance designs.

To appreciate the advantages of a fuel-injected engine, first consider the performance limitations associated with the use of a carburetor. As you know, a carburetor is susceptible to the formation of venturi icing due to the venturi effect and the fuel's vaporization. If you are not constantly vigilant for the indications of carburetor ice, a partial or complete loss of engine power can occur. You also should realize that even though the use of carburetor heat can help prevent icing, the application of heat, in itself, causes the engine to suffer a reduction in power. This is due to the enriching effect caused by the warm, less dense air being mixed with the fuel.

The application of carburetor heat causes an enriching effect on the fuel/air mixture. This causes a reduction in the engine's ability to create power, causing a decrease in aircraft performance.

In another performance consideration, a carburetor is relatively inefficient because of its method of operation. For example, in a carburetor, air flowing through a venturi causes liquid fuel to be drawn out of a discharge nozzle. This action tends to be inefficient because the fuel does not readily vaporize when it is discharged into the air stream. Instead, the fuel breaks up into irregular sized droplets that are then carried through the intake manifold to the engine's cylinders. Since these droplets vary in size, it is difficult for the fuel to be proportionally distributed to each cylinder, causing some cylinders to receive a mixture that is richer than necessary for ideal combustion. An excessively rich mixture may cause spark plug fouling and a rough running engine at low r.p.m., while at higher speeds, peak engine performance cannot be achieved. In addition,

since the fuel and intake air flow through the intake manifold, the fuel/air charge is warmed by heat from the engine. As a result, the fuel/air mixture's density is decreased, causing fewer fuel and air molecules to be available for combustion. Ultimately, an additional amount of engine power is lost.

With greater efficiency and reliability, a fuel injection system significantly enhances engine performance and safety. Each of the problems associated with the use of a carburetor is greatly reduced, or eliminated, resulting in numerous benefits to the aircraft. Among these, a fuel injection system allows an engine to have a lower fuel consumption per unit of horsepower, which means that the aircraft will typically have an increased range, endurance, and fuel economy, compared to an aircraft equipped with a carbureted engine of the same horsepower. Also, an injected engine operates more smoothly, cooler, and with less weight per horsepower than a carbureted engine. However, among the most significant benefits, a fuel-injected engine increases safety of flight by eliminating the problems associated with carburetor ice. The manner in which an injection system achieves these benefits can be recognized with an understanding of its design and operation.

FUEL INJECTION SYSTEMS

Although a variety of fuel injection models are in use today, each essentially performs the same functions. Simply stated, these functions include pressurizing the fuel, proportionally metering it for a specific amount of engine power, and then atomizing the metered fuel directly into each cylinder intake port. By using this method of distribution, each cylinder receives a specific quantity of fuel, thereby providing a better-controlled mixture ratio. Additional benefits are also gained by the subsequent cooling of the fuel/air mixture as a result of the vaporization process. These benefits include a higher density fuel/air charge for combustion, as well as improved cylinder cooling. Further, since the cylinders operate at such high temperatures, there is little chance for ice to form in the induction system from fuel vaporization. [Figure 11-2]

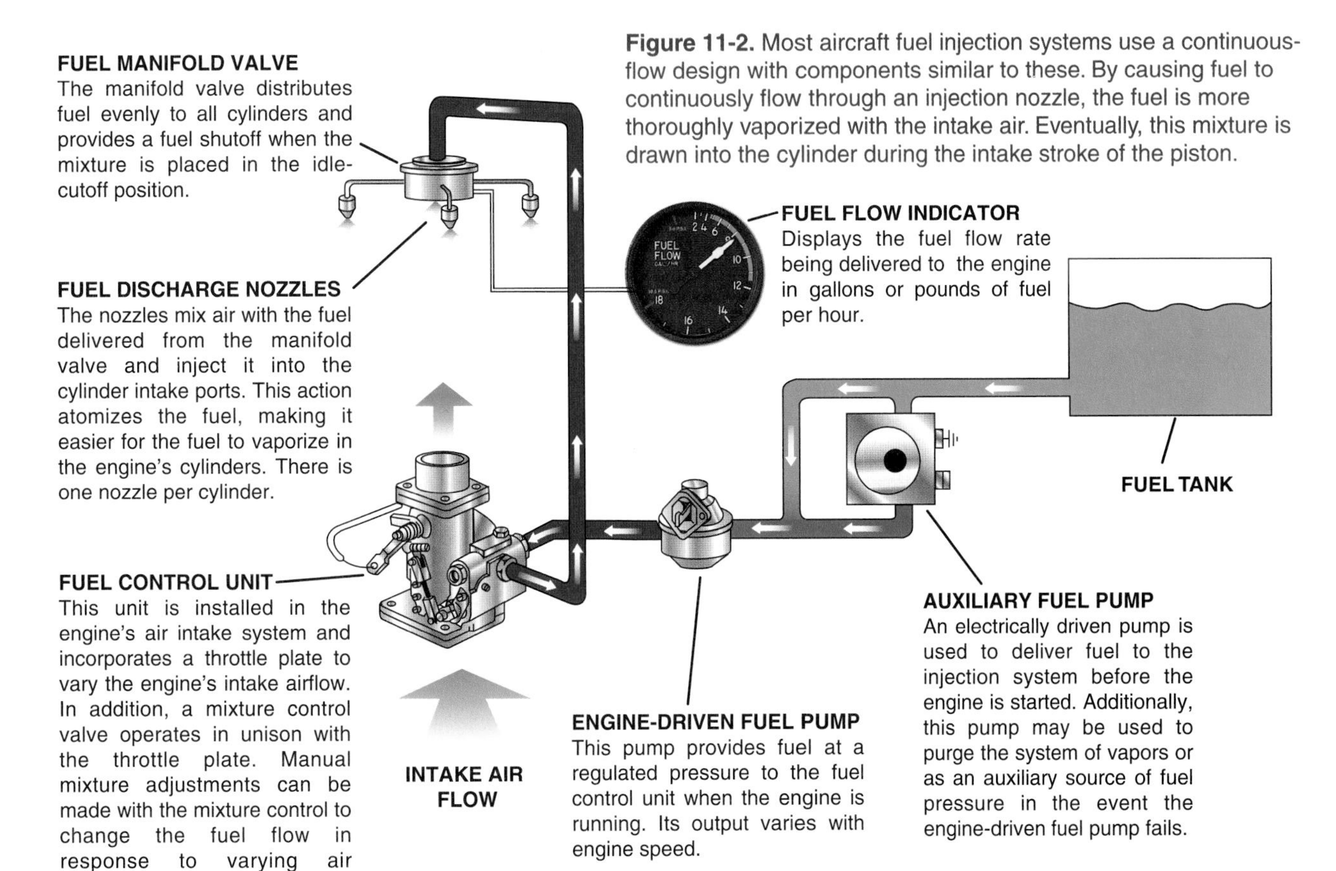

Figure 11-2. Most aircraft fuel injection systems use a continuous-flow design with components similar to these. By causing fuel to continuously flow through an injection nozzle, the fuel is more thoroughly vaporized with the intake air. Eventually, this mixture is drawn into the cylinder during the intake stroke of the piston.

The main component in the fuel injection system is the fuel control unit. Here, the fuel is metered to the engine in response to the amount of throttle opening. In order to achieve an ideal mixture ratio, the unit also includes a mixture control, which enables you to vary the amount of fuel flowing to the engine. As with a carburetor system, one purpose of this control is to allow you to lean the mixture to compensate for the lower air density encountered at higher altitudes. If the fuel flow remains constant in conjunction with an altitude increase, the resulting enriching of the fuel/air ratio may cause a reduction in power output, rough operation, and possible spark plug fouling.

A basic purpose for adjusting the mixture control, while at altitude, is to decrease the fuel flow to the engine in order to compensate for the enriching effect of the fuel/air mixture. Mixture enriching will occur if the fuel flow remains constant as the aircraft climbs into less dense air. If not corrected, spark plug fouling may occur at altitude.

Other injection system components may also have special operating considerations. For example, the **auxiliary fuel pump**, or auxiliary boost pump, as it is sometimes called, is controlled by a two-position switch on many airplanes. Although the labeling of the control switch differs between manufacturers, typically a high or low output rate can be selected. High output is primarily used in the event the engine-driven pump fails, whereas low output may be used for fuel vapor elimination or engine priming. Although intended as a safety feature, many operational problems and accidents have been attributed to inappropriate use of the auxiliary fuel pump. For example, with some injection systems, if high output is selected on the auxiliary fuel pump while the engine-driven pump is also operating, the fuel pressure developed by both pumps may cause an excessively rich mixture. This may cause the engine to lose performance, or to quit operating altogether. To preclude these problems, you should thoroughly review the auxiliary fuel pump operating procedures for any fuel-injected aircraft that you fly. [Figure 11-3]

Figure 11-3. The fuel injection system's auxiliary fuel pump is often controlled by a switch that allows for the selection of various output rates. For example, the switch pictured here produces a high output when the emergency side of the switch is placed in the HI position. Low output is provided when the start portion of the switch is selected to the ON position.

The engine-driven fuel pump may also produce special operating considerations. For instance, in some designs, the pump may deliver more fuel to the injection system's metering unit than the engine needs. In this situation, the excess fuel may be recirculated back to a reservoir tank. The tank is often located in a low part of the aircraft which may cause water and sediment to be trapped. If you neglect to drain a sample of fuel during your preflight inspection, contaminants may build up and cause the engine to fail. For this reason, you should become familiar with any special precautions regarding your aircraft's fuel system.

Another fuel injection operating consideration involves the cooling effect of the vaporizing fuel. Since the fuel vaporizes at the engine's cylinders, the cylinders' operating temperatures are reduced. Although this is usually a desirable effect, some manufacturers warn that if too much fuel is suddenly injected into a hot cylinder, shock cooling may occur. Shock cooling has been suspected of causing cylinder cracks, due to the uneven contraction of metal. To help reduce the chances of engine damage, you should perform power reductions and mixture enriching in slow or gradual increments.

As you fly more fuel-injected aircraft, you may be curious why some manufacturers use pounds per hour instead of gallons per hour as the unit of measure on the **fuel flow indicator**. You should understand that by using pounds per hour, you can obtain a more accurate indication of the fuel's performance. This is because the density of the fuel

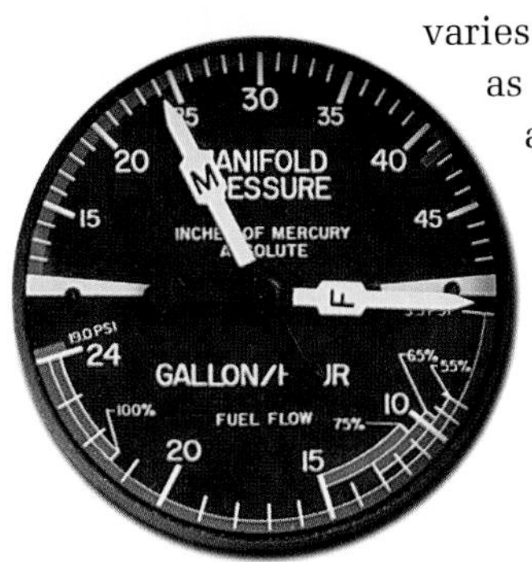

Figure 11-4. The fuel flow indicator displays the rate of fuel consumption, in gallons per hour, and may include fuel pressure indications measured in pounds per square inch. In addition, some indicators may include range marks to allow for mixture adjustments in order to obtain the best economy or best power mixture for various power settings.

varies with temperature, becoming less dense as the temperature increases. Ultimately, as fuel temperature increases, fewer molecules are available to produce power. By measuring weight instead of volume, you are provided with a better indication of the amount of energy available from the fuel. [Figure 11-4]

Technically, the fuel flow indicator measures fuel pressure. However, since a fuel injection system contains calibrated orifices, fuel flow rates can be extrapolated from the pressure measurements. In this manner, some aircraft manufacturers provide both fuel flow and pressure values on the indicator display. When using the fuel flow indicator, you should be aware of a malfunction that may occur if a fuel nozzle becomes obstructed by dirt or other foreign debris. With a blocked nozzle, the injection system's pressure will rise, causing the indicator to show a higher-than-normal fuel flow value. In response, you may be tempted to lean the mixture further in order to obtain the desired flow rate. This action may create an excessively lean mixture, which could cause substantial engine damage or a complete loss of power. To help you identify the malfunction, additional instruments are usually provided to aid you in monitoring the engine's performance. These instruments will be discussed later in this section.

OPERATING PROCEDURES

The operating procedures for fuel injection systems vary somewhat from those used with carbureted engines. Although injected engines have a reputation for being temperamental and difficult to operate, most problems are a result of not following the manufacturer's recommended procedures. The following discussion covers procedures used with various models of injected engines. However, keep in mind that these are only general guidelines. You should always refer to the pilot's operating handbook to determine the correct procedures for the specific aircraft you are flying.

STARTING

Because of the manner in which engine priming is accomplished, starting a fuel-injected engine is probably one of the more difficult procedures for most new pilots. Priming is usually accomplished with the electric, auxiliary fuel pump. By operating the pump, priming fuel is delivered through the injection system directly into the intake ports of each cylinder. This not only provides a combustible mixture for starting, but also purges any vapors, or air pockets, from the injection system. If not eliminated, fuel vapors may prevent the injection system from functioning properly. Most new pilots have difficulty developing the proper techniques for using these pumps. For example, if the pump is operated for an extended period of time with fuel flowing through the injection system, the engine may become flooded. On the other hand, by operating the pump for an insufficient length of time, a lean condition may result. Either condition can prevent the engine from starting.

Another area of difficulty involves developing the proper techniques for manipulating the engine controls during starting. Although injected engines are equipped with conventional throttle and mixture control levers, control operation during starting can be significantly different from carbureted engines. However, after a checkout with a qualified flight instructor, coupled with some operational experience, these procedures will become routine.

NORMAL STARTS

A normal starting technique can be used when mild to warm ambient temperatures exist, and the engine is cool. With these conditions, the engine should only require a small amount of priming. Normally, each aircraft manufacturer prescribes a sequence of events to follow, as detailed in the starting checklist. Although the use of a checklist is essential, you should conduct a complete review of the start checklist and then set it aside in order to make it easier to manipulate the engine controls. [Figure 11-5]

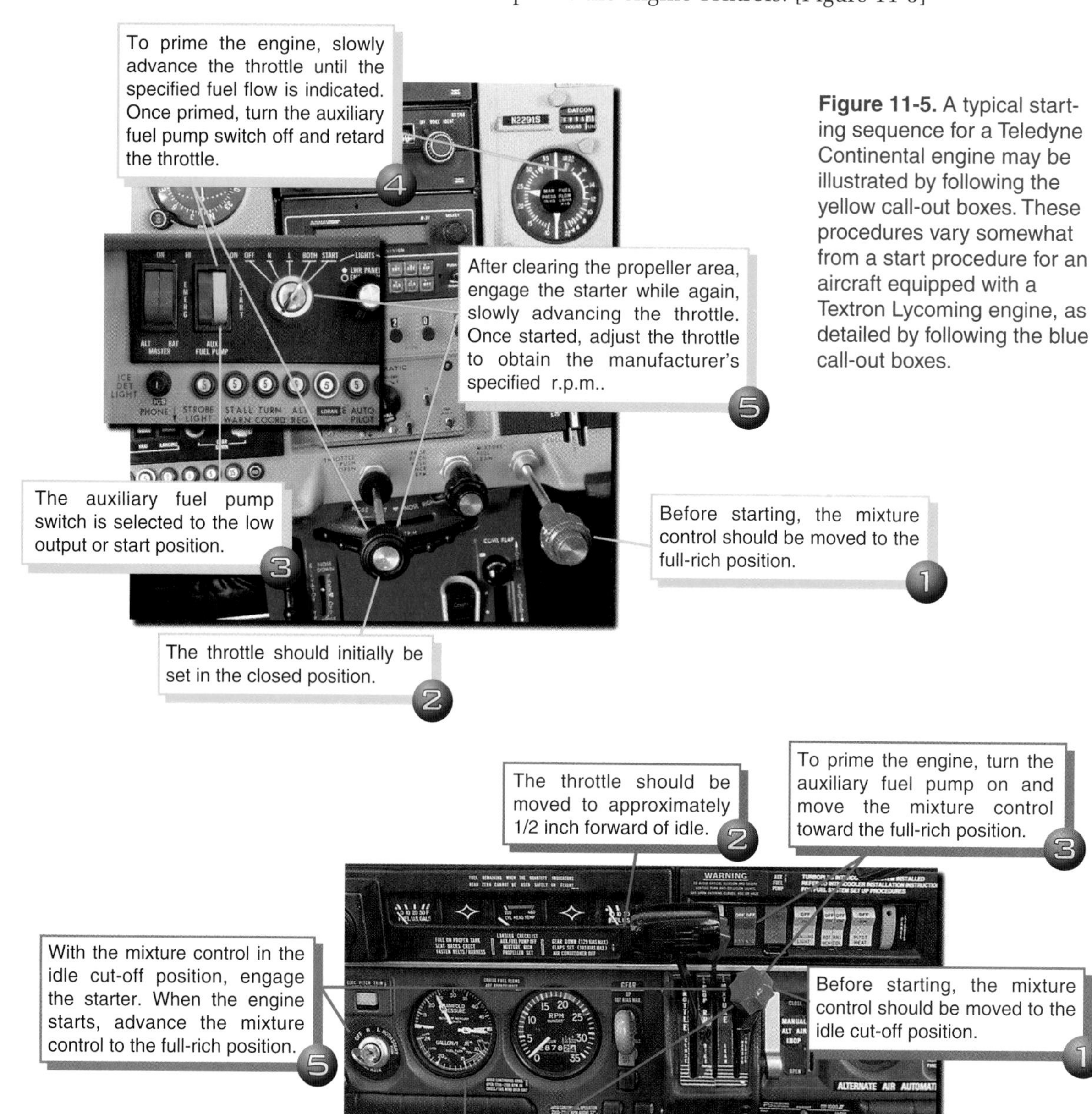

Figure 11-5. A typical starting sequence for a Teledyne Continental engine may be illustrated by following the yellow call-out boxes. These procedures vary somewhat from a start procedure for an aircraft equipped with a Textron Lycoming engine, as detailed by following the blue call-out boxes.

HOT STARTS

A fuel-injected engine tends to be difficult to restart right after shut down, particularly if it has been running for an extended period of time. This is a result of the engine's tendency to **vapor lock**, which is caused by the fuel vaporizing in the injection system's

lines and components. Without the flow of cooling air through the engine cowling, the engine's heat causes the fuel in the injection system to boil. The boiling fuel produces vapor that blocks fuel flow in the injection system. In order to restart the engine, it is necessary for you to use the auxiliary fuel pump long enough to purge fuel vapors, but not so long as to create a flooded condition. Again, as with normal starting, the engine controls and auxiliary fuel pump must be used correctly to obtain the proper fuel flow.

Once a hot engine is started, it is still possible that it will run erratically, or quit, if the auxiliary fuel pump is turned off. Again, this is caused by fuel system vapors that tend to continue forming until sufficient cooling air flows through the engine compartment. To eliminate the problem, most manufacturers require that you turn the auxiliary fuel pump back on to the low output position. You should continue operating the pump until the engine has been adequately cooled, and all fuel vapors have been purged from the injection system.

FLOODED STARTS

If too much fuel is allowed to flow to the engine during the priming procedure, engine flooding may occur. If the engine shows no sign of starting after about 30 seconds of cranking, disengage the starter to prevent it from overheating. While allowing the starter motor to cool, you should determine if the engine is flooded by checking for the presence of fuel running out onto the ground from the intake manifold drain line. If you observe fuel, you should attempt a flooded engine start procedure. [Figure 11-6]

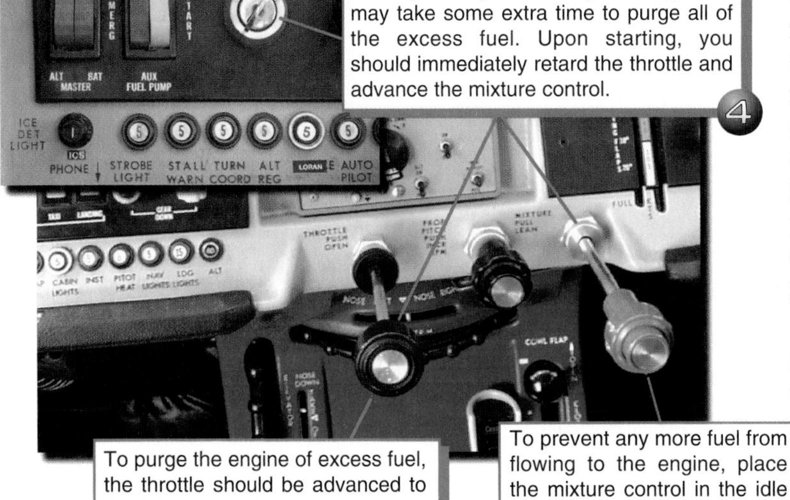

Figure 11-6. The objective of a flooded engine start procedure is to progressively purge the excess fuel from the engine until the proper fuel/air ratio is obtained. The procedures used with most fuel-injected models are essentially the same as those used with carbureted engines.

AFTER START PROCEDURES

After start procedures for fuel-injected engines are similar to those used with carbureted engines. Ground operations are generally conducted with the mixture in the full-rich position, except when operating from high elevation airports. At these airports, the mixture should be leaned in accordance with the manufacturer's recommendations. For takeoff and initial climb, the mixture control may be kept in the full-rich position, or it may be leaned slightly to an appropriate range mark on the fuel flow indicator. In either case, the fuel flow should be greater than what is normally required in cruise flight in order to provide additional fuel for engine cooling. In cruise flight, the leaning procedure is essentially the same as that for a carbureted engine, except most high performance aircraft include additional instruments to help you determine the optimum fuel/air mixture. With an understanding of the factors that affect your engine's operation, you can learn how to properly use these instruments to obtain the best performance from your aircraft.

ENGINE MONITORING

Among the instruments used to monitor performance of fuel-injected engines are the exhaust gas temperature and cylinder head temperature gauges. Although these gauges may be found on many carbureted airplanes, they are almost always found on fuel-injected airplanes. They provide you with a means to adjust the fuel/air mixture to optimize engine efficiency, reliability, and safety.

EXHAUST GAS TEMPERATURE GAUGE

The exhaust gas temperature gauge (EGT), as the name implies, measures the temperature of the exhaust gases leaving the combustion chamber. Generally speaking, the exhaust temperature increases as you lean the mixture until the optimum fuel/air ratio is achieved. Further leaning causes the temperature to begin decreasing. By combining the high reliability of a fuel injection system with the accuracy of an EGT gauge, you can perform precise adjustments to the fuel/air mixture. With proper control positioning, you will obtain the engine's best performance.

The best power mixture is the fuel/air ratio that produces the most power for a given throttle setting.

Although there are various designs and features for EGT systems, most use one or more probes, called thermocouples, to generate a voltage in conjunction with different temperatures. In response to the voltage, an indicator needle is deflected a proportionate amount. Since the actual temperature of an ideal mixture varies with the fuel and air densities, you only need to know when the mixture adjustment produces its highest, or peak temperature. To detect peak temperature, you progressively lean the mixture until the moment the EGT needle stops rising. This setting is called the best economy mixture, and produces the greatest fuel economy for the given power setting. However, most manufacturers recommend using a best power mixture at higher engine power settings. A best power mixture will produce the greatest amount of power for a particular setting while generating less heat. To obtain the setting, manufacturers recommend that you enrichen the mixture to produce a cooler EGT reading, ranging from 25°F to 125°F on the rich side of peak. [Figure 11-7]

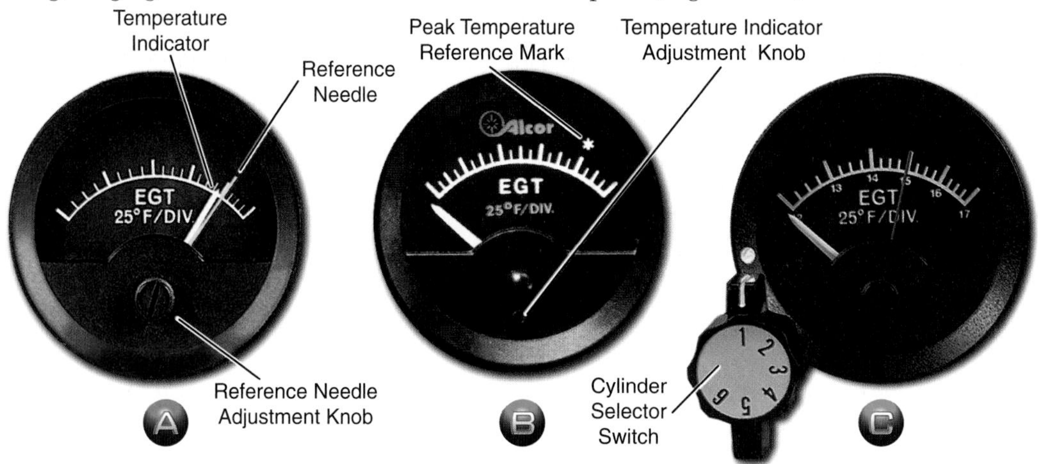

Figure 11-7. To keep track of the peak temperature indication,some EGT gauges are equipped with an adjustable reference needle (A). Once the peak EGT value is established, the reference needle is repositioned to align with the temperature indicator. On other EGT gauges, the temperature indicator can be repositioned to align with a prominent mark such as an asterisk (B). Multi-probe EGT indicators (C) allow you to select individual cylinders to monitor.

CYLINDER HEAD TEMPERATURE GAUGE

A cylinder head temperature gauge (CHT)is usually installed in most complex or high performance aircraft and, in fact, is a required instrument if the aircraft is equipped with cowl flaps. The instrument works similarly to an EGT gauge, using one or more thermocouple probes for temperature sensing. The display is marked with colored range

markings and numerical temperature values to denote the engine's normal and extreme temperature parameters. Although its operating range is fairly large, a typical procedure is to maintain a relatively constant reading toward the upper two thirds of the scale. To obtain the desired temperature, you can vary the mixture, adjust the flight attitude of the aircraft or, if installed, vary the cowl flap position to change cooling airflow. [Figure 11-8]

Figure 11-8. Regardless of the EGT value, the CHT must be maintained below the maximum limit. In some situations, you may not be able to lean to best power on the EGT, but will need to operate the engine with a significantly richer mixture.

Modern Engine Monitoring

Exhaust gas and cylinder head temperature indicating systems have been used in aircraft for many years to aid pilots with monitoring engine performance. However, the manner in which these instruments display information does not promote a rapid and easy evaluation of the engine's overall condition. To make engine monitoring easier, various aircraft equipment manufacturers have designed new engine monitoring instruments that graphically portray each cylinder's EGT and CHT indications. A typical graphic monitoring display is shown here.

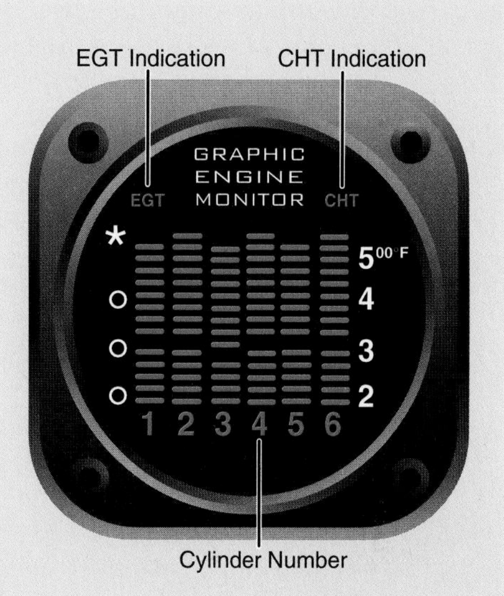

By viewing illuminated vertical bar graphs, you can determine the EGT for each cylinder. In addition, each graph incorporates a second indication for CHT by omitting a light bar at the corresponding CHT value. In this manner, both EGT and CHT indications can be easily compared between each of the cylinders. During engine leaning, the bar graph of the leanest cylinder will begin to flash when it reaches a peak value. By making the final mixture adjustments using the hottest cylinder indications, you can verify that all of the cylinders are receiving an adequate fuel/air supply. An additional benefit of a graphic engine monitoring system can be realized by observing the remaining cylinders' EGT indications. If these are significantly cooler, you can determine that a malfunction exists in the cylinder with the highest EGT. In this situation, you can specify to maintenance personnel which cylinder to check and repair, significantly reducing the aircraft's down-time and maintenance costs.

ABNORMAL COMBUSTION

The normal combustion process is accomplished when the spark plugs ignite the fuel/air mixture to produce smooth and even burning. For normal combustion, the mixture must be in the correct proportion, and the engine's ignition event must occur at the proper time. If either of these conditions are not met, abnormal combustion may result. This can be in the form of detonation or preignition, which, if not corrected, can cause substantial engine damage or a total loss of power. In order to preclude these conditions, you should be familiar with the causes, and the techniques used to avoid the onset.

Detonation is the uncontrolled, explosive combustion of fuel, which produces excessive pressures and temperatures inside the engine's combustion chambers. Detonation is usually characterized by high cylinder head temperature and is most likely to occur when operating at high power settings. Some common operational causes of detonation include using a lower fuel grade than that specified by the aircraft manufacturer, operating with extremely high manifold pressures in conjunction with low r.p.m., and operating the engine at high power settings with an excessively lean mixture. In addition, detonation can be augmented by extended ground operations or steep climbs, where cylinder cooling is reduced. Although you can employ operating procedures that help prevent its onset, various mechanical problems may also produce detonation. For this reason, you should be constantly vigilant for its indications and, if detonation does occur, be ready to take immediate actions to eliminate it. [Figure 11-9]

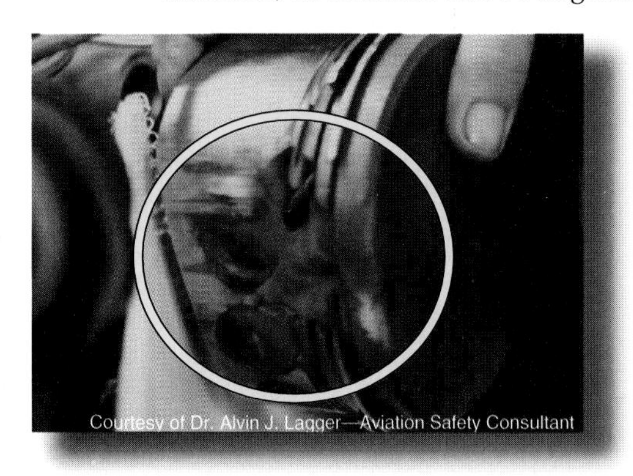

Figure 11-9. Detonation creates high cylinder temperatures and pressures that can melt or crack the piston and other cylinder parts. This piston was damaged by detonation, attributed to a restricted flow of fuel through the cylinder's fuel injection nozzle.

You can help prevent detonation by following some basic guidelines during the various phases of ground and flight operations. While on the ground, keep the cowl flaps in the full-open position to provide the maximum airflow through the cowling. During takeoff and initial climb, the onset of detonation can be reduced by using an enriched fuel mixture, as well as using a shallower climb angle to increase cylinder cooling. Extended, high power, steep climbs should be avoided. Finally, you should develop a habit of periodically monitoring the engine instruments to verify the proper operation of the engine and, also, to assist you in selecting the optimum engine control settings.

Detonation may occur at high-power settings when the unburned fuel mixture charge in the cylinders is subjected to instantaneous combustion, instead of burning progressively and evenly. This condition can be caused by a mixture set too lean for the power setting.

Preignition can occur when the fuel/air mixture ignites prior to the engine's normal ignition event. Premature burning is usually caused by a residual hot spot in the combustion chamber, often created by a small carbon deposit on a spark plug, a cracked spark plug insulator, or other damage in the cylinder that causes a part to heat sufficiently to ignite the fuel/air charge. Preignition causes the engine to lose power, and produces high operating temperatures. As with detonation, preignition may also cause severe engine damage because the expanding gases exert excessive pressure on the piston while still on its compression stroke.

The uncontrolled firing of the fuel/air charge in advance of normal ignition is known as pre-ignition.

Preignition and detonation often occur simultaneously and one may cause the other. Since either condition causes high engine temperatures accompanied by a decrease in engine performance, it is often difficult to distinguish between the two. Operating the engine within its proper temperature, pressure, and r.p.m. ranges reduces the chance of experiencing detonation or preignition.

INDUCTION ICING

Since internal icing of a fuel injection system is less likely to occur than in a carbureted system, there is no need to incorporate an induction air preheater like those used for carburetor heat. However, blockage can form over the induction air filter or intake air scoop of either system, as a result of **impact ice**. Impact icing forms when the aircraft is operated in visible moisture with the air temperature at or below freezing. Upon encountering these conditions, super-cooled water droplets are impacted, causing them to freeze on the forward facing components of the airframe. As the intake becomes blocked, the engine's performance will begin to decrease due to the restricted airflow.

To maintain engine operation, most manufacturers incorporate an **alternate air source** to provide warm, unfiltered intake air to the engine. [Figure 11-10]

Figure 11-10. Depending on the aircraft design, an alternate air source may be automatic or manual. A manual system allows you to perform a preflight run up check by observing a slight r.p.m. drop when the alternate air source is selected.

TURBOCHARGING SYSTEMS

The maximum amount of horsepower that an engine can develop is determined by engine r.p.m. and the density of intake air. In order to increase an engine's horsepower, manufacturers have developed systems that compress the intake air to increase its density. Of these systems, **turbocharging** has become the most widely used on modern, reciprocating-engine aircraft. With a turbocharger system, intake air is compressed before it enters the combustion chamber, enabling an engine to obtain much higher manifold pressure readings than with a nonturbocharged, normally aspirated engine. It also allows you to maintain sea level manifold pressure up to a much higher altitude. Ultimately, a turbocharged aircraft allows you to fly at higher altitudes where you can take advantage of higher true airspeeds and the increased ability to circumnavigate adverse weather.

TURBOCHARGING PRINCIPLES

Turbocharging is a highly innovative and efficient system, designed to increase an engine's performance. Although its components are manufactured to close tolerances, the system's theory of operation is fairly simple to understand. A reciprocating engine produces power by drawing in air and mixing it with fuel to produce a combustible mixture. When ignited, the mixture burns, causing it to release energy. If you consider a constant fuel/air ratio, the amount of power the engine develops will be directly proportional to the mass of air pumped through it. This means that the engine's horsepower is determined by the density of its intake air. A commonly used method for evaluating a high performance aircraft's power is to measure the amount of air pressure in the intake manifold. This is referred to as the **manifold absolute pressure (MAP)**. A **manifold pressure gauge** measures the MAP in the intake system in inches of mercury (in. Hg.), and often includes colored range marks to indicate the engine's operating limits.

On a standard day at sea level with the engine shut down, the manifold pressure gauge will indicate the ambient absolute air pressure of 29.92 in. Hg. Because atmospheric pressure decreases approximately 1 in. Hg. per 1,000 feet of altitude increase, the manifold pressure gauge will indicate approximately 24.92 in. Hg. while sitting at an airport that is 5,000 feet above sea level with standard day conditions. [Figure 11-11]

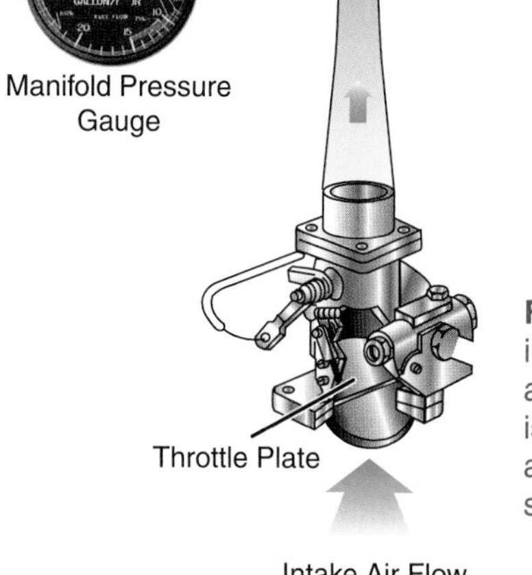

Figure 11-11. When the engine is operated at idle, the throttle plate restricts the intake airflow as the pistons are trying to draw in the manifold air. This action causes a partial vacuum to form, causing the MAP to decrease. Conversely, when the throttle is fully open, the restriction to airflow is decreased, causing the MAP to increase. On a normally aspirated engine, the maximum MAP at full power will be a value that is slightly less than the ambient atmospheric pressure.

As a normally aspirated aircraft climbs, it eventually reaches an altitude where the MAP is insufficient for further climb. A term that is used to define an altitude limit of an aircraft is called the **service ceiling**. The service ceiling is the maximum density altitude where the best rate of climb airspeed will produce a 100 foot per minute climb at gross weight while in a clean configuration with maximum continuous power. It is directly affected by the engine's ability to produce power. If the induction air entering the engine is pressurized, or boosted, the aircraft's service ceiling can be increased. This can be accomplished by utilizing a turbocharger.

A turbocharger is an engine accessory that is comprised of two main components — a compressor section, and a turbine section. The compressor section houses an impeller that turns at a high rate of speed. As induction air is drawn across the impeller blades, the impeller accelerates the air, allowing a large volume of air to be drawn into the compressor housing. The impeller's action subsequently produces high pressure, high density air, which is delivered to the engine. To turn the impeller, the engine's exhaust gases are used to drive a turbine wheel that is mounted on the opposite end of the impeller's drive shaft. By directing different amounts of exhaust gases to flow over the turbine, more energy can be extracted, causing the impeller to deliver more compressed air to the engine. [Figure 11-12]

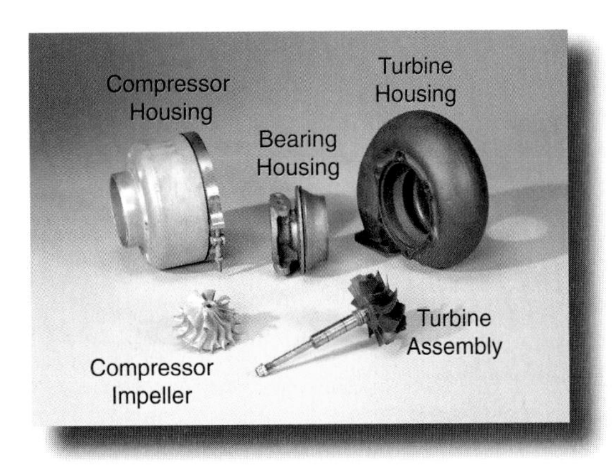

Figure 11-12. A turbocharger increases the pressure of the engine's induction air. This allows the engine to develop sea level or greater horsepower, permitting the aircraft to operate at a much higher service ceiling.

Another device, called a waste gate, is used to vary the mass of exhaust gas flowing into the turbine. A waste gate is essentially an adjustable butterfly valve that is installed in the exhaust system. When closed, most of the exhaust gases from the engine are forced to flow through the turbine. When open, the exhaust gases are allowed to bypass the turbine by flowing directly out through the engine's exhaust pipe. [Figure 11-13]

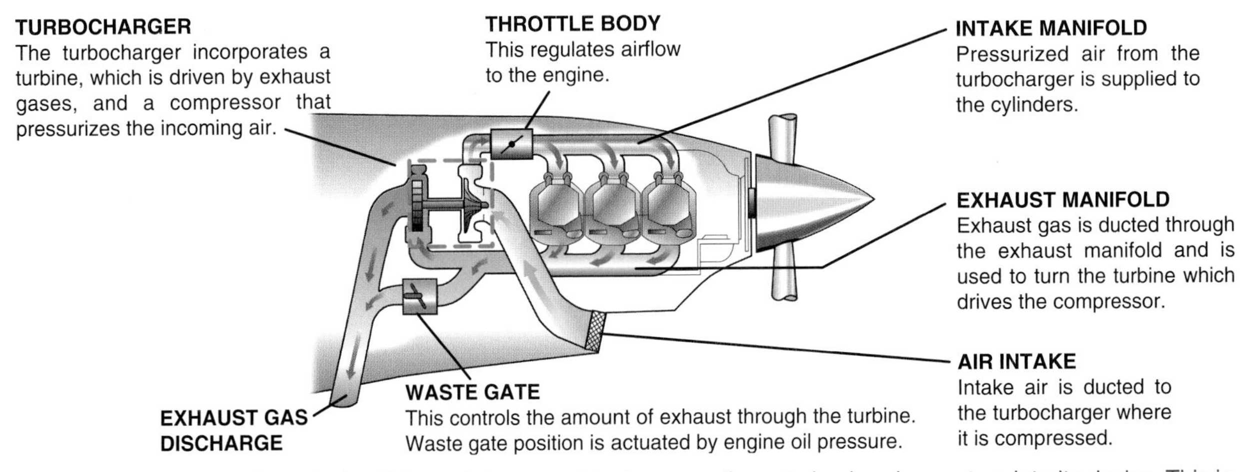

Figure 11-13. An aircraft engine's efficiency is increased by incorporating a turbocharging system into its design. This is possible because the turbocharger derives its power from the expelled exhaust gases. In this manner, the engine benefits by recapturing some of the exhaust's heat energy that is lost on a normally aspirated engine.

On some newer high performance aircraft, an intercooler may be installed to further increase the turbocharging system's efficiency. This component cools the intake air, which, due to its compression in the turbocharger, becomes hot. An intercooler functions in the same way as a radiator, allowing cool outside air to flow across a core of tubes containing the intake air. The resulting cooler intake air has a higher density which permits the engine to be operated with higher power settings.

SYSTEM OPERATION

On most modern turbocharged engines, the position of the waste gate is governed by a pressure-sensing control mechanism coupled to an actuator. Directing engine oil into or away from this actuator moves the waste gate position. On these systems, the actuator is automatically positioned to produce the desired MAP simply by changing the position of the throttle control. On the other hand, some turbocharging system designs use a separate manual control to position the waste gate. With manual control, you must closely monitor the manifold pressure gauge to determine when the desired MAP has been achieved. Manual systems are often found on aircraft that have been modified with aftermarket turbocharging systems. These systems require special operating considerations. For example, if the waste gate is left closed after descending from a high altitude, it is possible to produce a manifold pressure that exceeds the engine's limitations. This condition is referred to as an overboost, and it may produce severe detonation as a result of the leaning effect due to the increased air density.

Keep in mind that although an automatic waste gate system is less likely to experience an overboost condition, it can still occur. For example, if you try to apply takeoff power while the engine oil temperature is below its normal operating range, the cold oil may not flow out of the waste gate actuator quickly enough to prevent an overboost. To help prevent overboosting, you should advance the throttle cautiously to prevent exceeding the maximum manifold pressure limits. [Figure 11-14]

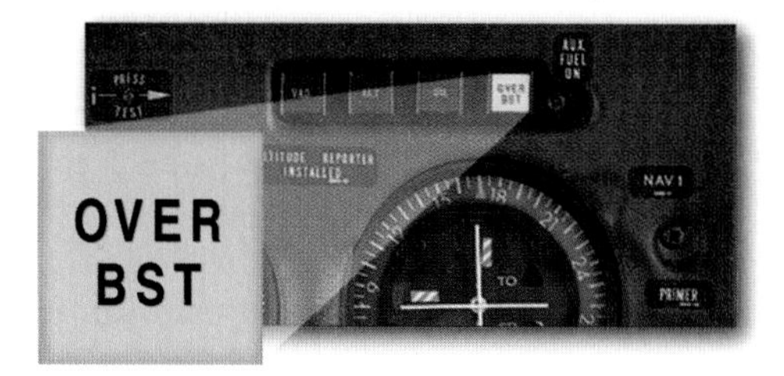

Figure 11-14. Some airplanes equipped with automatic waste gate control systems incorporate annunciator lights that warn you when an overboost condition exists. If the annunciator illuminates, you should immediately retard the throttle to an allowable manifold pressure.

There are a number of other system limitations that you should be aware of when flying an aircraft with a turbocharger. For instance, a turbocharger turbine and impeller can operate at rotational speeds in excess of 80,000 r.p.m. while at extremely high temperatures. In order to achieve high rotational speed, the turbine and impeller shaft is supported by bearings within a housing. While operating, these bearings must be constantly supplied with engine oil to reduce the frictional forces and high temperatures. To obtain adequate lubrication, the oil temperature should be in the normal operating range before high throttle settings are applied. In addition, you should allow the turbocharger to cool and the turbine to slow down before shutting the engine down. Otherwise, the oil remaining in the bearing housing will boil, causing hard carbon deposits to form on the bearings and shaft. These deposits rapidly deteriorate the turbocharger's efficiency and service life.

To help monitor the temperature of the turbine, many installations include a turbine inlet temperature (TIT) gauge. If you use an EGT to adjust the fuel/air mixture, you should verify that the TIT remains within the manufacturer's specified limits. In fact, some manufacturers may require leaning to be accomplished with the TIT instead of the EGT gauge.

HIGH ALTITUDE PERFORMANCE

As an aircraft equipped with a turbocharging system climbs, the waste gate is gradually closed to maintain the maximum allowable manifold pressure. At some point though, the waste gate will be fully closed and with further increases in altitude, the manifold pressure will begin to decrease. This altitude is specified by the aircraft manufacturer and is referred to as the aircraft's critical altitude. When evaluating the performance of the turbocharging system, if the manifold pressure begins decreasing before the specified critical altitude, the engine and turbocharging system should be inspected by a qualified maintenance technician to verify the system's proper operation.

When you are flying a turbocharged aircraft above its critical altitude, you should be aware of certain operational characteristics. One of these occurs as a result of changes to the aircraft's airspeed which causes the manifold pressure to vary. This characteristic is caused by the ram air effect that is produced in the induction system. As a result, the turbocharger receives a greater density of air, thereby causing it to produce a higher manifold pressure. In order for you to obtain a desired manifold pressure, you need to take this effect into account with any airspeed changes. For example, when you level off after a climb, set the manifold pressure to a value that is slightly less than desired, allowing it to increase as the aircraft accelerates to cruise speed.

CONSTANT-SPEED PROPELLERS

A constant-speed propeller is a system that you will usually find on either a complex or high performance airplane. The propeller's design allows you to control the angle of its blades in order to obtain the maximum amount of thrust for the engine's brake horsepower (BHP). This is an important factor when considering engine efficiency because a large amount of the engine's developed power can be lost when it is converted into thrust. In fact, a measure of a propeller's efficiency can be expressed as the ratio of the propeller's thrust horsepower to engine BHP. When comparing efficiency, a constant-speed propeller design is far better than a fixed-pitch propeller's because the constant-speed system allows you to obtain the optimum combination of engine r.p.m. and manifold pressure for the type of operation you are conducting. This means you can obtain the best overall performance by achieving a high rate of climb, increasing cruise airspeed, or extending endurance by reducing fuel consumption. In addition, a constant-speed propeller can automatically change its blade angle so the load on the engine can be varied, thereby increasing the engine's efficiency, reliability, and service life. [Figure 11-15]

Propeller efficiency is the ratio of thrust horsepower to brake horsepower.

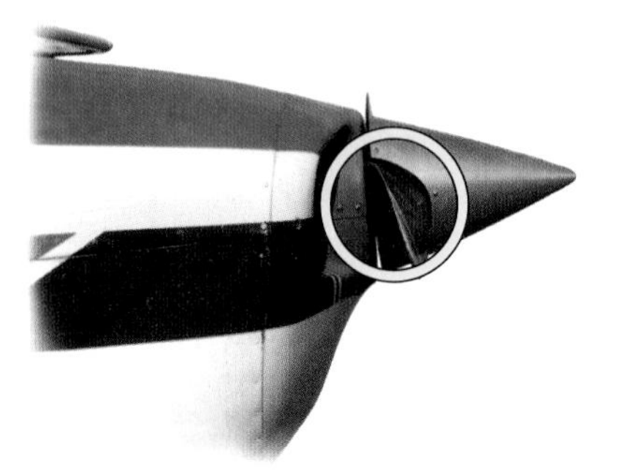

Fixed-Pitch Propeller

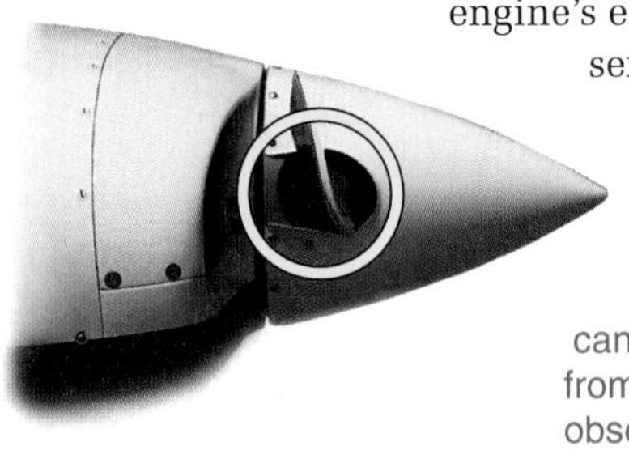

Constant-Speed Propeller

Figure 11-15. From an external view, a constant-speed propeller can be easily distinguished from a fixed-pitch propeller by observing the blade root.

PROPELLER PRINCIPLES

A propeller blade is an airfoil with its plane of rotation approximately perpendicular to the airplane's longitudinal axis. As a result, the blades of a conventional tractor (pulling) propeller cause an aerodynamic lifting force to be directed forward, thereby producing thrust. With this in mind, you can apply the same terminology to define a propeller blade cross section as you would for an airplane wing. In other words, a propeller blade has a chord, camber, and angle of attack. In addition, if either the angle of attack or rotational velocity of the blade increases, its thrust production also will increase. On the other hand, a propeller experiences an aerodynamic force that is not generally considered with an airplane wing. Dynamic airflow from the airplane's forward velocity causes a shift in the relative wind in relation to the propeller blade. This action causes the blade's angle of attack to vary with different aircraft airspeeds. [Figure 11-16]

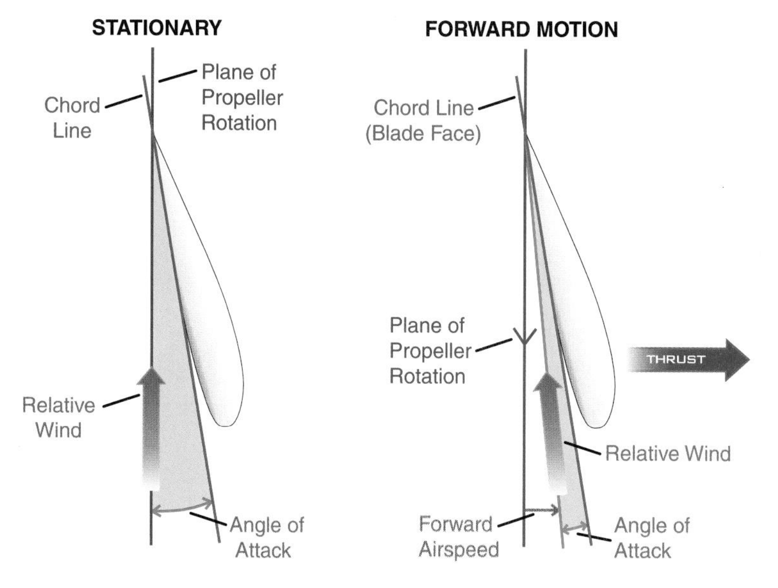

Figure 11-16. The angle of attack of a propeller is determined by the blade's chord line and the direction of the relative wind. The illustration on the left shows the effect of relative airflow over a rotating propeller on a stationary airplane. The illustration on the right shows the effects on the angle of attack caused by the shifting of the relative wind due to the forward airspeed of the airplane.

Angle of attack is defined as the acute angle formed by the chord line of an airfoil and the relative wind; the relative wind is generally considered as the airflow that is parallel and opposite to the path of the airfoil. To understand how the angle of attack is affected by the airplane's forward speed, first consider the aerodynamic forces that occur with a propeller that is operating on a stationary airplane in undisturbed air. As the propeller turns, the relative wind flows over the blade in a direction parallel to the blade's plane of rotation. This action will cause the propeller's pitch angle to be essentially the same as the blade angle. However, the terms, blade angle and pitch angle, are not the same. For clarification, a **blade angle** is formed by the chord line, or blade face, and the plane of the propeller's rotation. On the other hand, the **pitch angle** is formed by the chord line of the propeller and the relative wind. Stated another way, the pitch angle can be thought of as the blade's angle of attack. As forward speed increases, dynamic airflow is generated and enters the propeller's plane of rotation, resulting in an angular shift of the propeller's relative wind. Consequently, the pitch of the propeller changes even though the propeller's blade angle remains constant. In addition, the amount that the relative wind shifts will vary depending on the forward speed of the airplane. As the speed increases, the blade's pitch, or angle of attack, will decrease, resulting in a reduction in the amount of thrust produced by the blade. [Figure 11-17]

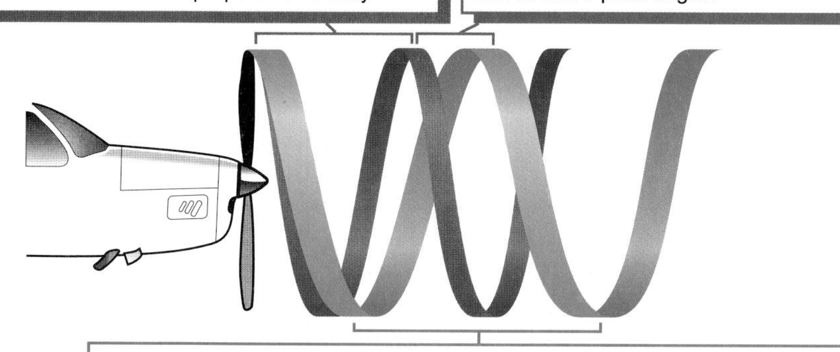

Figure 11-17. A propeller's geometric pitch can be varied by changing the blade angle in relation to the propeller's plane of rotation. When the geometric pitch (blade angle) increases, the effective pitch also increases. However, the effective pitch of a propeller also changes with other factors such as the forward speed of the airplane or the propeller r.p.m. Although a propeller may have a fixed-pitch design, its effective pitch may vary while the geometric pitch (blade angle) remains unchanged.

In the case of an airplane flying at a cruising airspeed, the blade pitch decreases as a result of the airplane's forward speed. However, if you are able to increase the blade angle, you can increase thrust by increasing the blade's angle of attack. Conversely, during a climb, the airplane's forward velocity generally is less, resulting in an increased blade angle of attack. If the propeller's blade angle is fixed, the engine will sustain an increased load, causing the engine r.p.m. to decrease. In response, the propeller's rotational speed will also decrease, resulting in less thrust being developed by the propeller due to the loss of some airflow over the blades. In order to obtain a high propeller efficiency during a climb, the blade angle should be decreased. For these reasons, fixed-pitch propellers are often designed to produce the greatest efficiency for either climbing or cruising. Unfortunately, however, a fixed-pitch propeller produces its optimum efficiency only at one particular r.p.m. and airspeed. In this case, the design is simply an attempt to achieve the maximum performance for a specific type of flight operation anticipated for the aircraft.

A fixed-pitch propeller is designed for best efficiency only at a given combination of airspeed and r.p.m.

Q-Tipped Designs for Quieter Times

Propeller designs vary significantly on different types of aircraft. One type that will probably get your immediate attention is the Q-tip design. These propellers are manufactured by Hartzell Propeller, Inc. for installation on many different models of airplanes. With this type of propeller, the tip of each blade is bent back 90 degrees to help reduce noise. When they first see a Q-tip, many pilots wonder how and where the propeller strike occurred. However, it is easy enough to distinguish a Q-tip from propeller damage because of the tip's uniform blending with the rest of the blade.

Q-tips were originally designed for use on multi-engine airplanes to increase the distance of the blade tips from the fuselage. By bending the last inch of each blade backward, the cabin interior noise level was reduced while the propeller's production of thrust remained relatively unaffected. However, since the Q-tip's original development in 1978, some single-engine airplanes have been modified with the design. In these situations, the Q-tip usually replaces a larger diameter propeller. With a smaller diameter, the airflow speed at the blade tip is decreased, resulting in less propeller tip noise. In addition to being quieter, the Q-tip's shorter blade length increases the blade-to-ground clearance. With a greater clearance, the blades tend to have less erosion and other damage caused by ground debris.

Other propeller models are being manufactured that increase efficiency while also reducing noise. One of the challenges of design engineers is to develop a propeller with a large enough diameter to produce the thrust necessary for the airplane's design while also limiting its diameter to reduce blade tip speed. Tip speed is an important factor to consider because as the airflow over the tip approaches the speed of sound, shock waves are produced. Shock waves can cause substantial losses in propeller efficiency and also produce significant noise levels.

Another relevant factor involves the different airflow velocities encountered along the blade's length. For example, even though a propeller's blade root and tip revolve at the same rate, the tip will travel through the airmass at a much faster speed. If the propeller blade angle is uniform along its length, the angle of attack at the blade tip will be greater than at the root. To distribute the thrust more evenly across the blade, the geometric pitch, or blade angle, is decreased toward the tip. By twisting the blade, the most efficient angle of attack for the entire blade length is maintained relatively constant.

The reason for variations in geometric pitch (twisting) along a propeller blade is that it permits a relatively constant angle of attack along its length in cruising flight.

CONSTANT-SPEED PROPELLER OPERATION

From the previous discussion of propeller principles, you can see that an ideal design allows the blade angle to be controlled while the engine is operating. Although there have been many types of variable-pitch propeller systems, the constant-speed propeller is the most widely used on modern, high performance aircraft. The main advantage of a constant-speed propeller is that it converts a high percentage of the engine's power into thrust over a wide range of r.p.m. and airspeed combinations. In addition, once you have selected certain r.p.m. settings for the propeller, a device called a governor can automatically adjust the blade angle to maintain the selected r.p.m. For example, during cruise flight, an increase in airspeed or reduction in propeller load will cause the blade angle to increase, thereby maintaining the selected r.p.m. On the other hand, a reduction in airspeed or increase in propeller load will cause the propeller blade angle to decrease. [Figure 11-18]

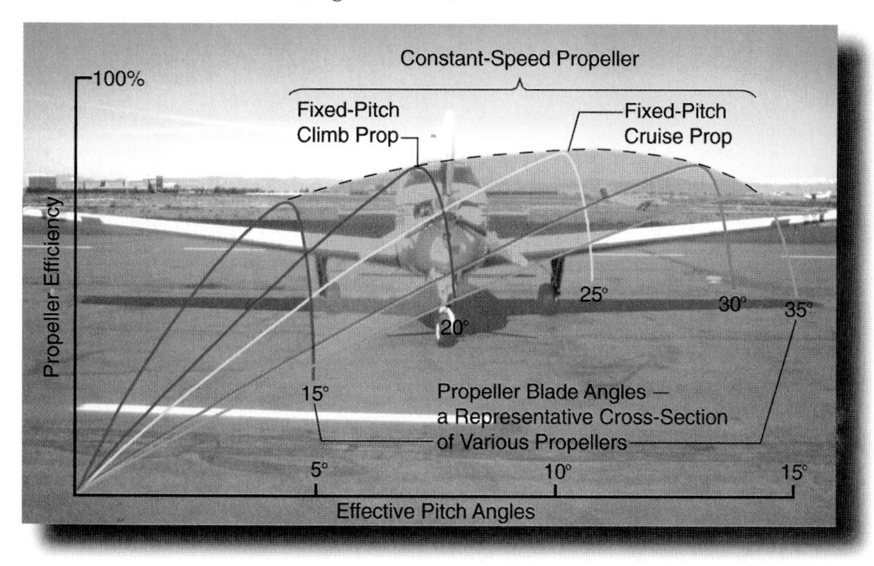

Figure 11-18. The efficiency of a fixed-pitch propeller increases with pitch angle until the optimum angle of attack is reached. Then, the efficiency of the propeller decreases rapidly. In this figure, the shaded area represents the range of operation for a constant-speed propeller. By varying the blade angle, a constant-speed propeller can maintain a high degree of efficiency over a wide range of operating conditions.

The range of possible blade angles for a constant-speed propeller is called the propeller's governing range and is defined by the limits of the blade's travel between high and low blade angle pitch stops. As long as the propeller blade angle is within the governing range and not against either pitch stop, a constant engine r.p.m. will be maintained. However, once the propeller blade reaches its pitch-stop limit, the engine r.p.m. will increase or decrease, as appropriate, with changes in airspeed and propeller load. For example, once you have selected a specific r.p.m., if the airplane's speed decreases enough, the propeller blades will reduce pitch until they contact their low-pitch stops. When these stops are reached, any further reduction in airspeed will cause the engine's r.p.m. to decrease. This action is similar to what you would encounter if a fixed-pitch propeller were installed. Conversely, when an airplane accelerates, such as in a dive, a

constant-speed propeller's blade angle will increase to maintain the selected r.p.m. until the high-pitch stop is reached. Once this occurs, the engine's r.p.m. will begin increasing. [Figure 11-19]

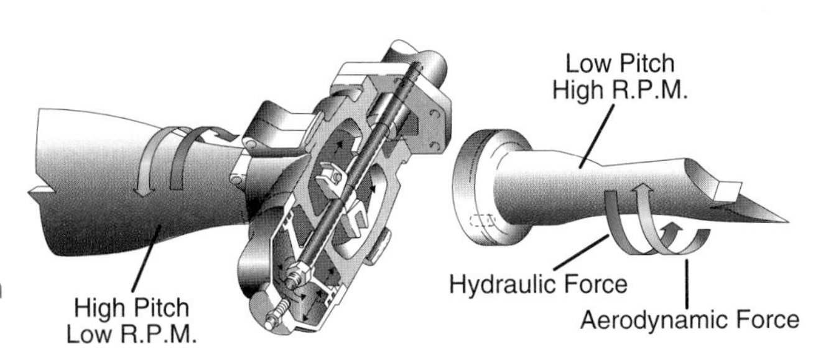

Figure 11-19. The governor senses engine speed and delivers high pressure oil to a piston located inside the propeller hub. Piston movement causes the pitch-change mechanism to alter the angle of the propeller blades. On most single-engine aircraft propellers, aerodynamic forces tend to move the blades to the low pitch, high r.p.m. position, while hydraulic pressure provides an opposing force that moves the blades to a high pitch, low r.p.m. position.

ENVIRONMENTAL IMPACT AND PROPELLER NOISE

Aircraft noise has become a significant issue at airports surrounded by highly populated residential developments as well as over National Parks, Wildlife Refuges, Wilderness Areas, and major tourist attractions. In an effort to maintain a good neighbor policy, you should be alert to these noise sensitive areas and operate your aircraft in a manner that helps to reduce its environmental impact. For example, many local airport authorities around the United States have developed noise abatement procedures that should be adhered to within manufacturer and operational safety limits. In combination with these procedures, you can further reduce your aircraft's environmental impact by following some basic operating procedures to reduce noise emissions.

From noise research studies, it has been determined that many propeller-driven airplanes produce a ground noise level as high as 88 decibels (dB) when overflying at 1,000 feet AGL. In comparison, if a person is subjected to noise levels in the range of 85 dB, hearing damage can occur. Considering this fact, you can realize that an overflying airplane can cause significant discomfort to a person on the ground. To reduce noise levels, you should avoid low altitude operations whenever possible. However, if low altitude flight is necessary, an alternative method to reduce noise emission is to decrease the r.p.m. of the propeller. For instance, after departing a noise-restricted airport, you should make an attempt to decrease the r.p.m. of the propeller as quickly as safety permits. In most cases, even a reduction of as little as 100 r.p.m. can make a significant improvement in quieting the airplane.

POWER CONTROLS

On airplanes that are equipped with a constant-speed propeller, engine power is controlled by the throttle and indicated on the manifold pressure gauge. The propeller's blade angle, on the other hand, is controlled by a **propeller control** and the resulting change in engine r.p.m., caused by a change in blade angle, is indicated on the tachometer. By providing you with a means of controlling both engine power output and propeller blade angle, the most efficient setting combinations are maintained for a variety of flight conditions. For example, during takeoff, you want the engine to develop its maximum power. Therefore, the throttle and propeller control should be positioned full forward so the engine can turn at its maximum r.p.m. while also developing the maximum amount of thrust horsepower. On the other hand, to help establish the airplane in a climb after takeoff, the throttle and propeller controls can be retarded in order to achieve more

economical fuel consumption. In addition, by retarding the propeller control, the blade pitch will increase, thereby producing more thrust with a resultant increase in airspeed. When leveling the aircraft at a particular altitude, the throttle and propeller controls can be further reduced to decrease the fuel consumption and increase thrust.

The propeller control regulates the engine's r.p.m. and in turn, the propeller's r.p.m. For takeoff, or to develop the maximum power and thrust, the control should be placed in the full forward position to provide a low pitch (low blade angle), and high r.p.m.

The order and manner in which the engine and propeller controls are operated are important considerations. With a normally aspirated engine, if the MAP is too high for a corresponding engine r.p.m., the internal pressures within the cylinder can exceed critical limits. This causes the engine to overheat, which can lead to detonation. To help prevent excessive pressures, before you decrease the engine's r.p.m. with the propeller control, you should first verify that the manifold pressure is within allowable limits. If not, the throttle should be retarded before the propeller control. For example, to establish a climb after takeoff, the throttle should first be retarded to establish the climb manifold pressure. Once adjusted, the propeller control can be reduced to obtain a lower r.p.m. Conversely, when you are increasing the throttle for high power, you should first verify that the engine's r.p.m. is high enough to permit the increased manifold pressure. This means that before the throttle can be fully advanced, the propeller control should be positioned full forward, to the low pitch, high r.p.m. position. For example, the propeller control should be full forward before landing so you can apply full throttle without causing engine damage in case a go-around is initiated.

To prevent undue stress on an engine equipped with a constant-speed propeller, apply maximum power by increasing the engine's r.p.m. before increasing the manifold pressure. For the same reason, when reducing engine power to establish a climb after takeoff, decrease the manifold pressure before decreasing the r.p.m.

As an additional consideration, you should avoid rapid throttle movement when making engine power adjustments. This is because harmonic balance weights are often installed on the engine's crankshaft. These weights are designed to position themselves by the inertia forces generated during crankshaft rotation to effectively absorb and dampen crankshaft vibration. If the throttle is abruptly moved, the balance weights may be shifted away from the optimum position, causing the crankshaft to receive a large amount of vibrational stress. This condition is referred to as detuning because of the balance weights' association with harmonic vibrations. In severe cases, rapid throttle movements can produce such significant detuning stresses that the crankshaft fails.

Detuning of an engine crankshaft is a source of overstress that may be caused by rapid opening and closing of the throttle.

SUMMARY CHECKLIST

✓ A fuel injection system increases engine efficiency by lowering its fuel consumption per unit of horsepower as compared to a carbureted engine. It also promotes safety by reducing the risk of induction system icing.

✓ Most operating difficulties of fuel-injected engines are a result of not following the manufacturer's recommended procedures.

✓ Vapor lock problems, associated with fuel-injected engines, are minimized by the proper use of the auxiliary fuel pump.

- ✓ Preignition and detonation are the result of abnormal ignition and combustion. Frequency and severity of these conditions can be greatly reduced by using the correct fuel/air mixture and by monitoring the engine instruments to maintain the proper engine operating temperatures.

- ✓ EGT and CHT gauges provide you with a reliable and accurate method for adjusting the fuel/air mixture in order to obtain optimum engine performance.

- ✓ Fuel injection systems have an alternate air source that provides unfiltered, heated intake air to the engine in the event the main air source is obstructed by impact ice.

- ✓ A turbocharger system produces increased intake air density. This allows the engine to develop more power and also provides a much higher service ceiling.

- ✓ Care must be exercised when operating an aircraft with a turbocharger system to prevent overboosting the engine. This requires slow throttle adjustments, and, with automatic systems, full throttle should not be applied until the engine's oil temperature is within normal limits.

- ✓ By allowing an extended idling time before shutting down an engine equipped with a turbocharger, you help increase the useful life and efficiency of the turbocharger.

- ✓ The critical altitude is the maximum altitude where the turbocharger waste gate is fully closed and the maximum allowable MAP is being maintained. With further increases in altitude, MAP will begin to decrease.

- ✓ A constant-speed propeller is the most efficient propeller system, commonly used on complex and high performance aircraft.

- ✓ Blade angle changes can be accomplished by the pilot but are also controlled by a governor on a constant-speed propeller.

- ✓ Before increasing MAP or reducing r.p.m. on an aircraft that is equipped with a constant-speed propeller, you should verify that the resulting MAP will not cause engine damage.

- ✓ To increase power on an engine equipped with a constant-speed propeller, first increase the engine r.p.m. by advancing the propeller control and then advance the throttle to increase MAP. To decrease power, first decrease the MAP by retarding the throttle and then decrease the engine r.p.m. by retarding the propeller control.

KEY TERMS

Fuel Injection System

Auxiliary Fuel Pump

Fuel Flow Indicator

Vapor Lock

Exhaust Gas Temperature Gauge (EGT)

Best Economy Mixture

Best Power Mixture

Cylinder Head Temperature Gauge (CHT)

Detonation

Preignition

Impact Ice

Alternate Air Source

Turbocharging

Manifold Absolute Pressure (MAP)

Manifold Pressure Gauge

Service Ceiling

Overboost

Critical Altitude

Constant-Speed Propeller

Blade Angle

Pitch Angle

Governing Range

Propeller Control

QUESTIONS

1. Where is fuel injected into an engine that is equipped with a continuous flow fuel injection system?

2. In most fuel-injected systems, if the engine-driven fuel pump fails, what should be accomplished to prevent total engine failure?

3. Why is a fuel flow indicator that measures flow rates in pounds per hour considered to be better than one that reads in gallons per hour?

4. True/False. When performing a normal start on any engine that is equipped with a fuel injection system, you should leave the mixture control in the full-rich position while operating the starter motor.

5. True/False. The best power mixture is obtained when the EGT is at its peak value, and is the condition that allows the engine to develop its maximum power for a given throttle setting.

6. What is the component that varies the flow of exhaust gases into the turbine of a turbocharger?

7. Where is a constant-speed propeller's effective pitch angle measured?

   A. Between the relative wind and the chord line of the propeller blade
   B. Between the airplane's longitudinal axis and the chord line of the propeller blade
   C. Between the propeller blade's rotational path and the blade chord line

8. True/False. A constant-speed propeller that experiences a higher engine load while operating in its governing range will automatically vary its blade angles to a lower pitch.

9. Why is the propeller control placed in the full forward position before landing an aircraft equipped with a constant-speed propeller?

10. True/False. To decrease power on an engine equipped with a constant-speed propeller, you should first reduce the engine's r.p.m. by retarding the propeller control and then reduce the MAP by retarding the throttle.

SECTION B
ENVIRONMENTAL AND ICE CONTROL SYSTEMS

Many airplanes used in private and commercial service have increased high-altitude performance capabilities which permit higher cruising speeds and allow you to avoid many adverse weather situations. However, at high altitudes, hazardous physiological conditions are likely to be encountered unless the airplane is equipped with additional advanced systems such as an oxygen or pressurization system. When these systems are available and functioning properly, you and your passengers will enjoy a safer and more comfortable flight environment. In addition to oxygen and pressurization systems, ice control systems are installed on many airplanes to further increase flight safety in the event you encounter meteorological conditions that are conducive to structural icing. Ice control systems increase time and options for maneuvering the aircraft out of hazardous icing conditions.

OXYGEN SYSTEMS

Oxygen systems for aviation use are available in three basic configurations; continuous flow, diluter demand, and pressure demand. As the name implies, a **continuous-flow** oxygen system provides continuous delivery of oxygen to the user. Because of its cost and simplicity, it is primarily installed on reciprocating-engine airplanes, but also is installed for passenger use in turboprop or jet engine aircraft. For flight crewmembers of turboprop and jet-engine aircraft, a **diluter-demand** or **pressure-demand** oxygen system is used. These systems increase the duration of the oxygen supply by providing

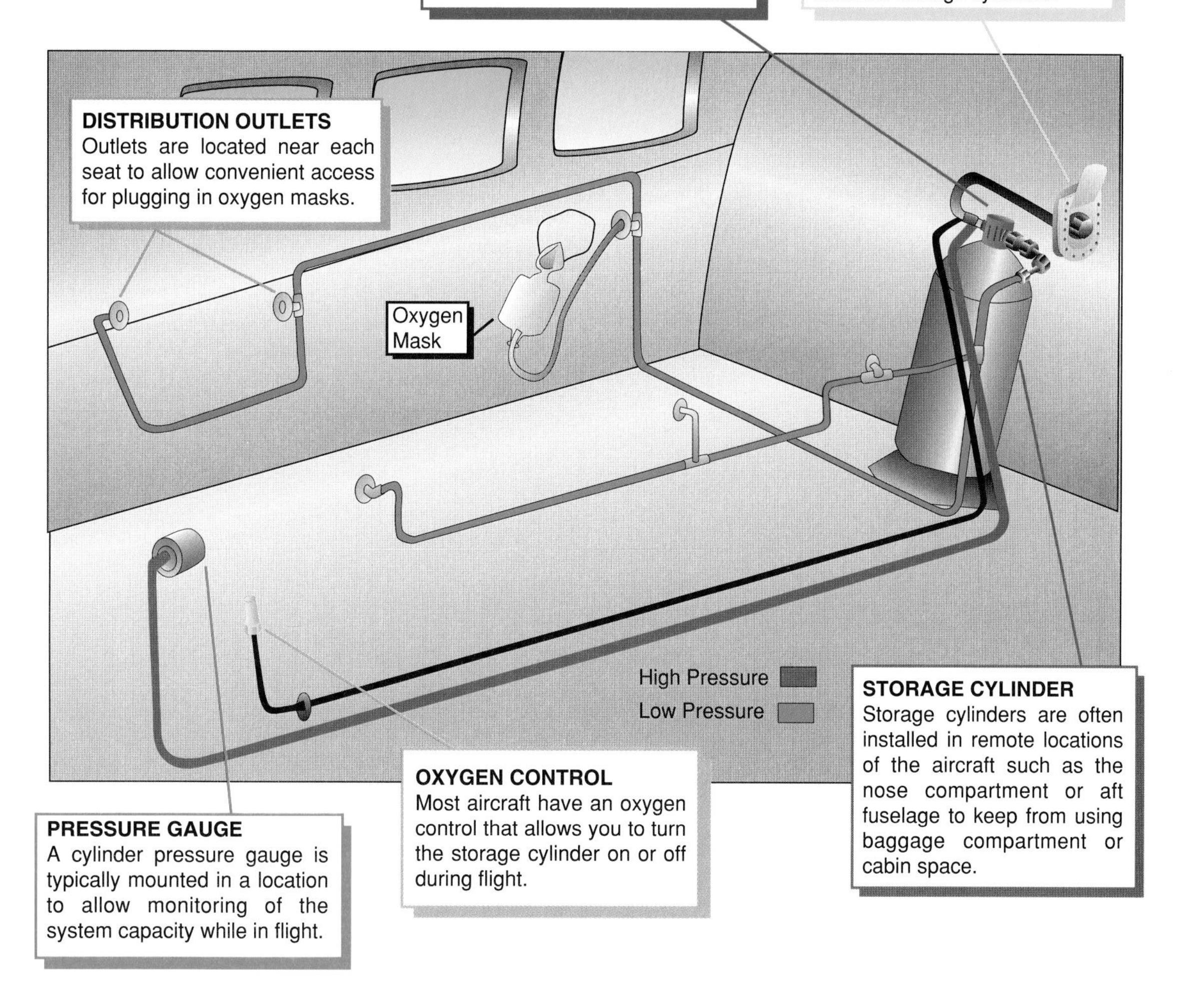

Figure 11-20. Most aviation oxygen systems contain the same basic components as those shown here. By equipping an aircraft with a built-in oxygen system, you and your passengers are provided with convenient access to supplemental breathing equipment.

oxygen flow only when inhaling, and incorporate advanced design oxygen masks and regulators that permit flight at higher altitudes than when using a continuous-flow system. [Figure 11-20]

CONTINUOUS FLOW

Continuous-flow systems provide adequate respiration up to 25,000 feet and are available in three styles: constant flow, adjustable flow, and altitude compensated. **Constant-flow** systems are used on many reciprocating-engine airplanes to provide continuous oxygen delivery at a constant-flow rate. However, at lower altitude, the constant flow of oxygen causes an excessive amount of supply usage. To increase the duration of the oxygen supply, manufacturers developed the **adjustable-flow** system to provide compensated oxygen delivery in conjunction with varying altitudes and oxygen supply pressures. For proper operation, the system requires that you manually adjust a regulator valve for different altitudes, or changes in supply pressure. An **altitude-compensated** system is a further improvement of the adjustable-flow design. The system uses a barometric control regulator that automatically adjusts the rate of oxygen flow in response to atmospheric and oxygen supply pressures. With its increased

Figure 11-21. With most types of adjustable-flow oxygen systems, you turn a regulator control valve until your aircraft flight altitude registers on a gauge.

ease of use and supply duration, the altitude-compensated system has become one of the most popular systems used in modern aircraft. [Figure 11-21]

Besides built-in designs, many continuous-flow oxygen systems are available in portable units. You can buy or rent these in the event your aircraft is not equipped with a built-in system and you are planning a flight at high altitude. However, you also can use a portable unit in combination with a built-in system to increase the duration of the oxygen supply. Another consideration might be to use a portable unit for passengers with respiration problems, which may require a higher oxygen flow rate than that required for the rest of the occupants. [Figure 11-22]

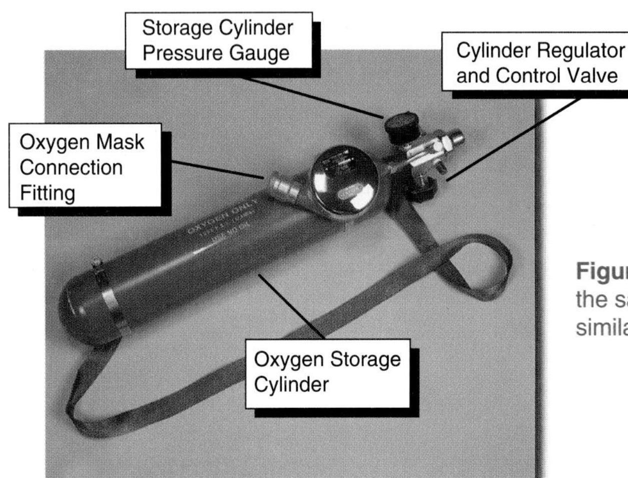

Figure 11-22. Portable continuous-flow oxygen systems use the same basic components as built-in systems, resulting in similar operating procedures.

CONTINUOUS-FLOW MASKS

The most common oxygen mask used with the continuous-flow system is an **oronasal rebreather** design. Oronasal simply means that the face mask covers both the nose and mouth, while rebreather refers to the fact that oxygen is diluted with a portion of exhaled air, and is then re-inhaled. By re-inhaling a diluted mixture, the duration of the oxygen supply is extended. [Figure 11-23]

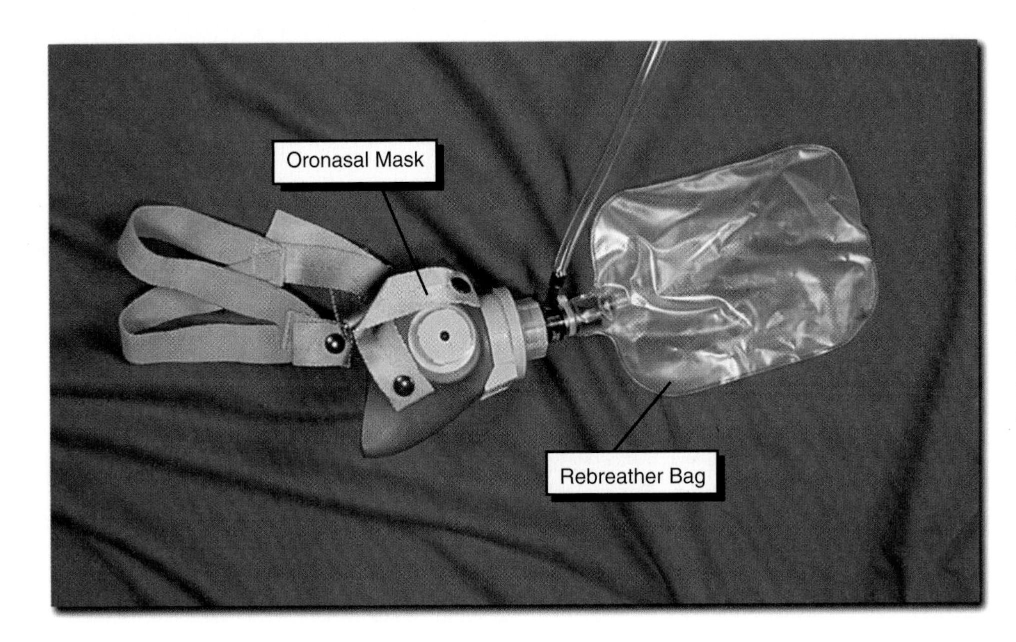

Figure 11-23. An oronasal rebreather oxygen mask uses a rebreather bag that is attached to the face mask and fills with a ratio of oxygen and exhaled air for dilution.

With adjustable-flow or altitude-compensated regulators, the amount of oxygen that flows into the rebreather bag is determined by the flow rate established by the regulator and the amount of atmospheric pressure surrounding the bag. At low altitude, the rebreather bag only partially inflates with oxygen because the oxygen flow rate is reduced and the atmospheric pressure surrounding the bag is relatively high. Upon exhaling, a large percentage of exhaled air enters the bag to combine with the oxygen. When the bag inflates fully, any additional exhaled air is diverted out the face mask through small holes or a check valve. In a similar fashion, if the bag completely deflates when you inhale, cabin air is allowed into the face mask through the small holes or another check valve. As the atmospheric pressure decreases with a gain in altitude, the bag inflates with a higher percentage of oxygen. Eventually, an altitude is reached where the bag completely inflates with oxygen, causing all exhaled air to be diverted out the mask, leaving pure oxygen for inhalation. [Figure 11-24]

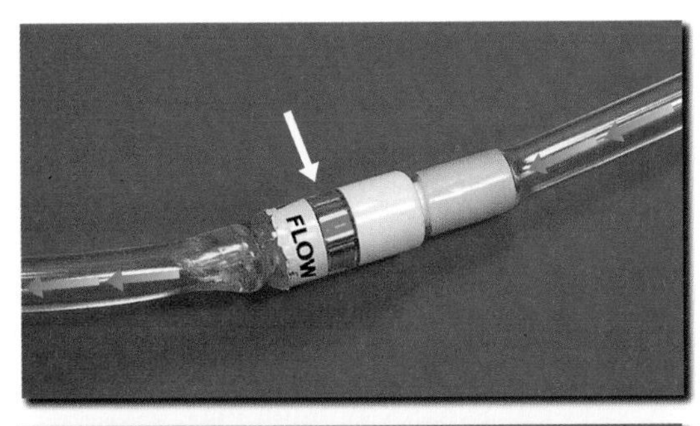

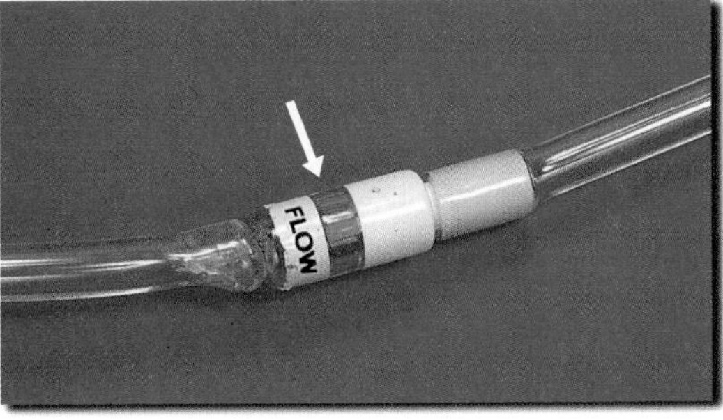

Figure 11-24. To determine if oxygen is being delivered to the mask of a continuous-flow oxygen system, most manufacturers provide a flow indicator in the delivery hose to the mask. When oxygen is flowing, the color of the indicator will be green. If oxygen flow stops, the indicator will change to red.

Upon activation of a continuous-flow system, oxygen travels to the distribution outlet where a spring-loaded valve stops the flow until a mask is connected. Once connected, the valve unseats, allowing oxygen to be dispensed to the mask. Since it is critical for the pilot to receive an ample flow of oxygen to maintain maximum mental and physical capabilities, flow restrictors are used to vary the rate of oxygen delivery between the pilot and passengers. A typical flow restrictor for a pilot provides 120 liters per hour (l.p.h.) of oxygen flow, while the passengers receive 90 l.p.h. Depending on the manufacturer, the flow restrictors may be installed inside the distribution outlet, or located in the mask connector. When installed in the mask connectors, each distribution outlet provides the same rate of oxygen flow. An advantage to this design is that you can increase the amount of oxygen flow to the passengers by providing them with a mask designed for pilot use. However, make sure you account for the increased flow rate when you determine the duration of your oxygen supply. [Figure 11-25]

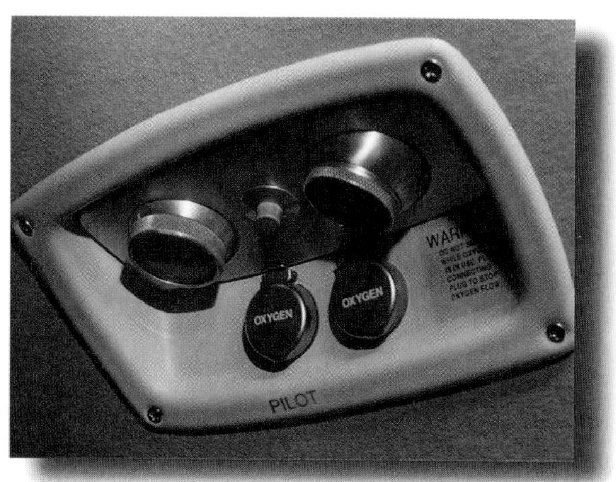

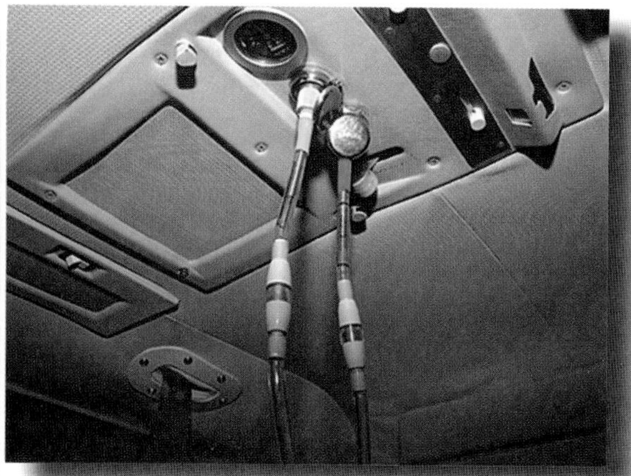

Figure 11-25. When you are using an oxygen system equipped with outlets designated for pilot or passenger use, the pilot outlet is usually identified in a manner similar to the one shown on the left. When a mask connector contains a flow restrictor, identify the proper mask by looking for a color code on or near the connector assembly, as shown on the right. The pilot's mask is red while passenger masks are yellow or gold.

Most oxygen masks use fittings that make connecting a mask to an outlet an easy task. You simply push the connector into the outlet and give it a slight turn clockwise. Although the procedure is easy, remember that passengers often will not be familiar with the task. Before conducting a high-altitude flight, you should allow each of your passengers to connect their mask to see how it is done. As an added consideration, turn the oxygen system on to let them try a few sample breaths. A couple of minutes spent instructing the passengers while on the ground, can help prevent distractions during flight. You also should explain to your passengers that smoking is not permitted while oxygen is in use. Although oxygen itself is not flammable, it causes a tremendous acceleration in the combustion rate of otherwise slow burning articles.

DILUTER DEMAND AND PRESSURE DEMAND

The components of diluter-demand and pressure-demand oxygen systems are essentially the same as those used with continuous-flow systems. Exceptions include a different style oxygen mask, and the replacement of the distribution outlet with a special regulator unit. The style of mask used with either of these systems is a quick-donning design, which simply means the mask can be put on rapidly. When compared to an oronasal rebreather mask, a **quick-donning mask** allows a tighter seal around your nose and mouth, which improves oxygen delivery when you are flying at high altitude. The mask also does not use a rebreather bag for oxygen dilution. Instead, a valve, located in the regulator unit, opens to allow cabin air into the mask to mix with oxygen when you are flying at an altitude that permits a diluted oxygen supply. As you climb, the duration and amount that the valve opens progressively decreases until above 28,000 feet, where the valve remains closed to provide 100% oxygen for inhalation. When supplying undiluted oxygen flow, the diluter-demand system is capable of supporting respiration up to 40,000 feet. At altitudes above 40,000 feet, even breathing 100% oxygen is not enough to provide adequate respiration because the atmospheric pressure is too low to support proper oxygen saturation into the bloodstream. For flight above 40,000 feet, the pressure-demand system is used to provide positive pressure oxygen, which is forced into your lungs when you inhale. Because the oxygen is delivered at a

IT'S IN THE CARDS

As the pilot in command of an airplane, you are responsible for the safety and comfort of your passengers. One requirement of FAR Part 91 is that you must brief your passengers on the use of their safety belts and, if equipped, shoulder harnesses, and advise them to fasten them before takeoff or landing. However, there are many occasions that the passengers you fly with have little experience with other important aspects of your aircraft and the flight environment. As a courtesy, and to increase flight safety, you should take time to provide your passengers with a comprehensive preflight briefing. In addition, you also may want to consider providing your passengers with a briefing card similar to the one shown here.

Briefing cards often can be purchased from the manufacturer of your aircraft or, as an option, you may want to make one of your own design. This can be especially beneficial if your aircraft is equipped with amenities that the manufacturer does not cover in their briefing card design. If you decide to make your own, consider the information that commercial operators provide to their passengers. Some of these items include the following.

- If smoking is not allowed in your aircraft, you may want to include a no smoking insignia. Otherwise, if smoking is allowed, provide information regarding when and where your passengers can smoke in the aircraft.
- Illustrate and discuss the location and means of opening the passenger door and emergency exits.
- Provide information on the location of survival and emergency equipment including fire extinguishers and flotation devices, if applicable.
- If the flight will be conducted at high altitude, provide illustrations and instructions on the normal and emergency use of oxygen equipment.
- If your passengers have portable electronic devices such as cellular telephones, hand-held video games, or laptop computers, discuss the restrictions that apply regarding the use of these items during flight.

Other items that you may want to consider will vary with your aircraft. For example, if reclining passenger seats are installed, you should include information that explains how the seats are adjusted and when it is permissible to use the reclined position. You may even want to include some basic information on how to cope with certain physiological conditions such as sinus and ear blockages, air sickness, and anxiety. By taking a few minutes to discuss briefing card information with your passengers, you will help relieve anxiety that they may have with regard to the flight.

positive pressure, you need to receive special training to develop proper respiration techniques before using this system. [Figure 11-26]

Figure 11-26. A quick-donning oxygen mask, like the one shown here, is primarily used in pressurized aircraft since it can be put on and begin delivering oxygen within 5 seconds, in case the cabin air pressure rapidly decreases.

The regulator unit of a diluter-demand or pressure-demand oxygen system also uses a diaphragm-operated demand valve, which opens by the slight suction created during inhalation. When the demand valve is open, oxygen is allowed to flow into the mask. Once you begin to exhale, the valve closes, shutting

off the flow of oxygen. The regulator unit also incorporates control switches that allow you to activate the oxygen system and to select different operating modes. One switch allows you to turn the dilution valve off to provide 100% oxygen each time you inhale, regardless of altitude. This feature might be used if you feel that the normal dilution mode is not providing an adequate supply of oxygen. Another switch allows you to select a continuous flow of oxygen for emergency purposes. For example, if smoke or fumes develop in the cabin, activation of the switch will provide 100% oxygen flow to help prevent hazardous contaminates from entering the mask. [Figure 11-27]

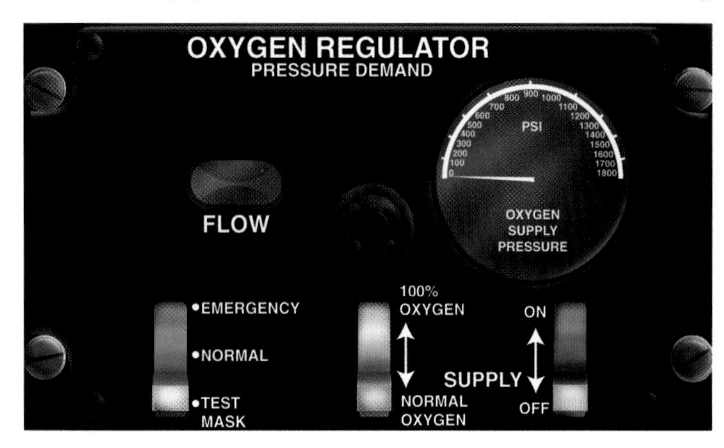

Figure 11-27. A pressure-demand regulator, similar to the one shown here, includes a test switch which allows you to check the operation of the system before flight. To perform the check, breath into the oxygen mask while holding the switch in the test position. A flow indicator flashes each time you inhale when the system is functioning properly. You can also monitor the flow indicator during normal operation to determine that oxygen is being delivered to the mask.

An O_2 Hose For The Nose?

A cannula oxygen breathing device, such as the one shown here, may look awkward and uncomfortable, but these devices are gaining popularity with many pilots. With a cannula, oxygen is delivered through a hose that is equipped with two stems that are inserted slightly into your nostrils. Although the thought of putting anything in your nose may seem offensive, most pilots have found that the device is considerably more comfortable than an oronasal style oxygen mask. Some of the advantages of a cannula include an unrestricted ability to eat, drink, or talk, and also the ability to use a standard headset and microphone in conjunction with a passenger intercom or radio transmitter.

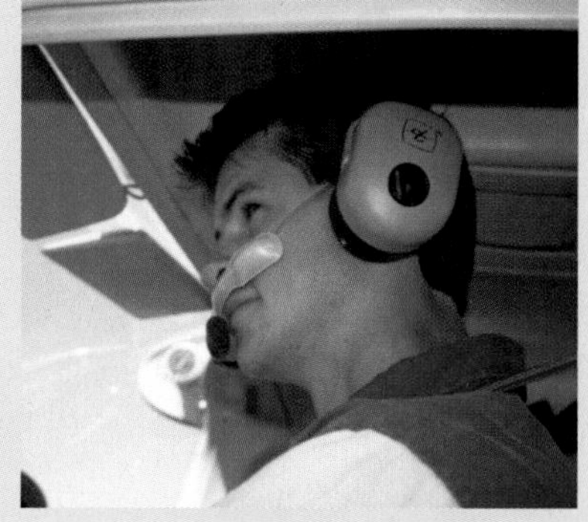

This cannula breathing device is actually one of many styles that are available for aviation applications. Some designs, similar to those used for medical equipment, are simply hoses with nasal tube attachments. The style shown here has the added feature of an oxygen reservoir that rests under your nose. When a cannula of this design is used, oxygen continuously flows into the reservoir to provide an ample supply for respiration. In fact, these cannula designs are often used with special oxygen flow regulators that allow you to significantly reduce the consumption of oxygen when compared to the amount necessary for most oronasal rebreather masks. In many cases, when a cannula is equipped with a regulator, the oxygen duration can be increased to double, or even triple your oxygen supply duration.

If you use a cannula, you should be aware that the FAA restricts its use to a maximum altitude of 18,000 feet. Above this altitude, you must use an oronasal mask that provides an adequate seal to your face. Also, you must carry an oronasal mask for each aircraft occupant in case a head cold or nasal congestion prevents proper respiration through the nose. However, even with these restrictions, you will probably prefer using the cannula instead of an oronasal rebreather oxygen mask, whenever conditions permit.

OXYGEN STORAGE

To transport and distribute oxygen, most civil aircraft use **oxygen cylinders** or **chemical oxygen generators**. High-pressure oxygen cylinders provide the most popular method of

carrying gaseous oxygen, and are available in a variety of capacities. The cylinders are usually made of steel or aluminum, while some newer airplanes use cylinders fabricated from advanced composite materials to decrease weight. Most cylinders are painted green with the words aviator's breathing oxygen stenciled on the side, along with the normal servicing pressure. When the cylinder is part of a built-in system, it is securely mounted to the aircraft structure, to prevent it from causing injuries in the event of a crash. If you are using a portable unit, you need to remember to secure the cylinder before flight. [Figure 11-28]

Figure 11-28. High-pressure cylinders permit extremely safe storage of oxygen, provided the cylinder and its fittings are kept clean from grease, oil, or hydraulic fluid. If these materials come into contact with high concentrations of oxygen, a severe fire can occur. In addition, to prevent moisture from entering an oxygen cylinder, you should never allow the supply pressure to deplete below 50 p.s.i.

Chemical oxygen generators are often used in pressurized cabin airplanes to provide an adequate oxygen supply in the event an emergency descent is necessitated by a loss of cabin pressurization. In accordance with the requirements of FAR Part 91, when you are flying a pressurized cabin airplane above FL250, you must have an oxygen supply of at least 10 minutes duration for each occupant. Chemical oxygen generators provide the necessary oxygen to meet this requirement, while also being lighter weight and taking up less space than a high-pressure oxygen cylinder. When activated, the generator burns a solid chemical to produce gaseous oxygen suitable for breathing. However, since the chemical process cannot be stopped once the generator is activated, it can only be used once. When supplemental oxygen is needed, you activate the system by pulling an oxygen mask out of a storage compartment. By pulling the mask toward you, a lanyard cord is also pulled, causing a triggering mechanism to ignite the chemical. To prevent inadvertent activation of the system, you should leave the mask stowed except when an emergency necessitates use of the oxygen system.

OXYGEN SERVICING

A typical high-pressure oxygen cylinder is charged to a pressure of 1,800 to 1,850 pounds per square inch (p.s.i.), with a maximum pressure of approximately 2,200 p.s.i. To prevent exceeding the maximum pressure, most cylinders are equipped with a pressure relief valve. When transporting a high-pressure cylinder, never leave it exposed to heat or direct sunlight for an extended period of time. Oxygen expands when heated, which could cause the cylinder pressure to increase above maximum limits. Also, since the temperature of oxygen increases as a result of compression, the cylinder pressure gauge will register high immediately after servicing, and then decrease as the oxygen cools to ambient temperature. As a result of the high pressure reading, you might inaccurately compute the duration of your oxygen supply when using an **oxygen duration chart**. [Figure 11-29]

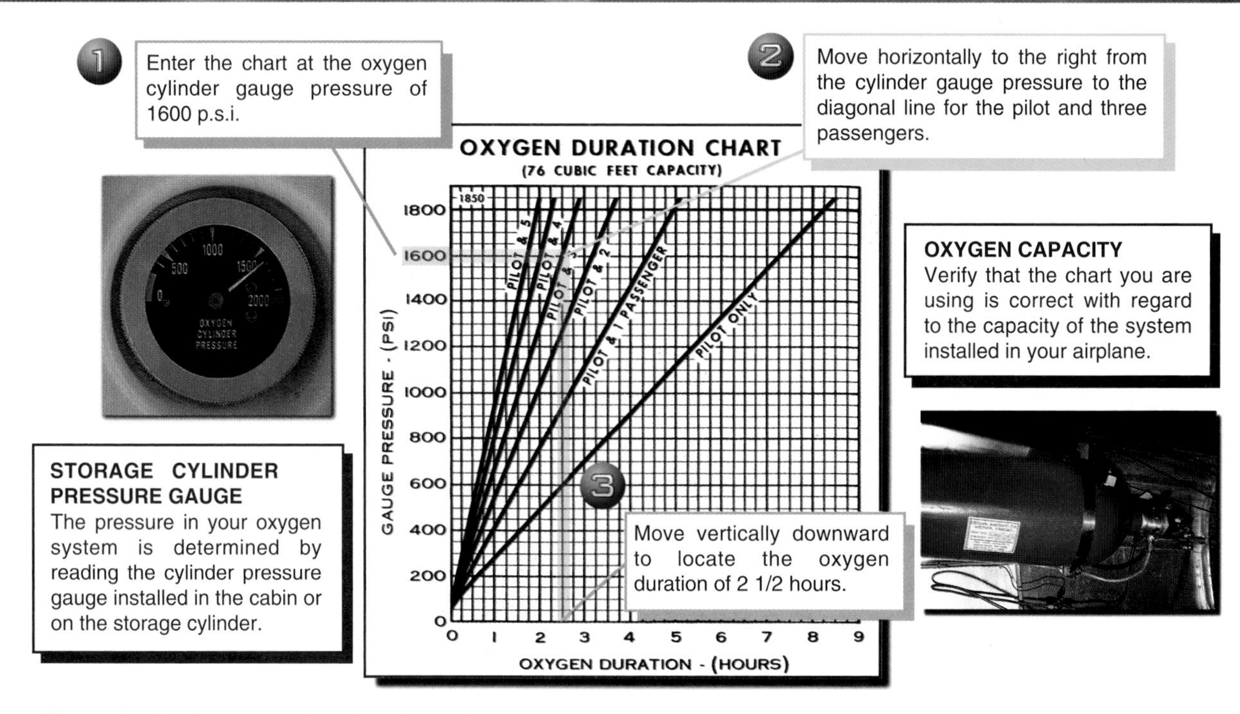

Figure 11-29. An oxygen duration chart allows you to determine the length of time that you and your passengers can remain at altitude. For the example shown here, a cylinder with a 1,600 p.s.i. pressure provides approximately 2 1/2 hours oxygen duration for a pilot and 3 passengers.

Make sure the oxygen system is filled with **aviator's breathing oxygen**. Specifications for this type of oxygen are 99.5% pure oxygen and not more than .005 milligrams of water per liter. Too much moisture in the oxygen may cause valves and lines to freeze, possibly stopping the flow of oxygen. To verify the quality of your oxygen supply, you should have the system serviced by a reputable distributor. The *Airport/Facility Directory* contains information on airports that can provide oxygen system servicing. [Figure 11-30]

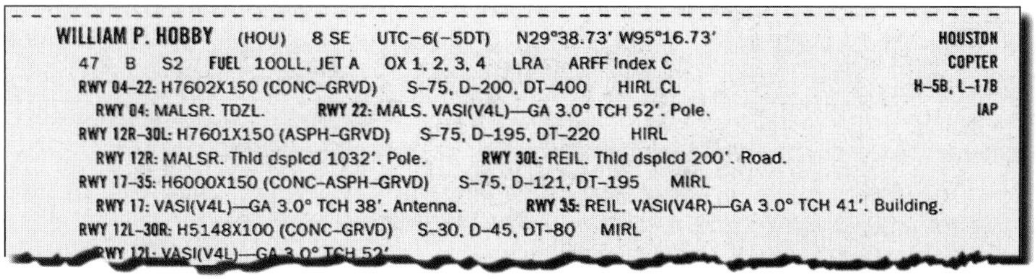

WILLIAM P. HOBBY (HOU) 8 SE UTC–6(–5DT) N29°38.73' W95°16.73' **HOUSTON**
47 B S2 **FUEL** 100LL, JET A OX 1, 2, 3, 4 LRA ARFF Index C **COPTER**
RWY 04–22: H7602X150 (CONC–GRVD) S–75, D–200, DT–400 HIRL CL **H–5B, L–17B**
RWY 04: MALSR. TDZL. **RWY 22:** MALS. VASI(V4L)—GA 3.0° TCH 52'. Pole. **IAP**
RWY 12R–30L: H7601X150 (ASPH–GRVD) S–75, D–195, DT–220 HIRL
RWY 12R: MALSR. Thld dsplcd 1032'. Pole. **RWY 30L:** REIL. Thld dsplcd 200'. Road.
RWY 17–35: H6000X150 (CONC–ASPH–GRVD) S–75, D–121, DT–195 MIRL
RWY 17: VASI(V4L)—GA 3.0° TCH 38'. Antenna. **RWY 35:** REIL. VASI(V4R)—GA 3.0° TCH 41'. Building.
RWY 12L–30R: H5148X100 (CONC–GRVD) S–30, D–45, DT–80 MIRL
RWY 12L: VASI(V4L)—GA 3.0° TCH 52'

Figure 11-30. Oxygen services can be determined by looking for the designations OX1, 2, 3, or 4 in the *Airport/Facility Directory*. OX1 indicates that high-pressure oxygen cylinders can be serviced, while 3 indicates that replacement high-pressure cylinders are available. The numbers 2 and 4 are for low-pressure oxygen systems that are sometimes installed in older aircraft.

CABIN PRESSURIZATION

Although oxygen systems provide adequate protection from hypoxia, wearing an oxygen mask for extended periods of time can become irritating. In addition, a number of other physiological effects can occur at high altitude that are not prevented by breathing supplemental oxygen. Decompression sickness and sinus or ear blockages are among some of the physiological conditions that can cause severe pain or discomfort to you or your passengers. To reduce or eliminate these problems, many airplanes in private and commercial service are equipped with a **cabin pressurization system**. With a pressurized cabin, a safer and more comfortable environment is produced, permitting high altitude flight without having to wear an oxygen mask. However, before you act as the pilot in command of a pressurized aircraft that is certified for operations above 25,000 feet MSL, you should be aware that FAR Part 61, requires you to receive and log specific training. This training

consists of ground and flight instruction to include high-altitude aerodynamics and meteorology, respiration, hypoxia, use of supplemental oxygen, and other physiological aspects of high-altitude flight.

PRESSURIZATION PRINCIPLES

Pressurization is accomplished by pumping air into an aircraft that is adequately sealed to limit the rate at which air escapes from the cabin. By limiting the rate of airflow out of the cabin, the air pressure increases to produce a cabin environment equivalent to one encountered at a lower altitude. For example, when flying at 20,000 feet, the atmosphere exerts a pressure of about 6.8 pounds per square inch (p.s.i.). If your airplane is equipped with a pressurization system that increases the cabin air pressure to 10.9 p.s.i., the **cabin pressure altitude** will be equivalent to 8,000 feet. Cabin pressure altitude is a term that describes the equivalent altitude inside the cabin and is a factor in evaluating the effectiveness of the pressurization system.

Another factor considered with a pressurization system is the amount of **cabin differential pressure** that the system produces. Cabin differential pressure is simply the difference between the outside air pressure and the cabin air pressure. In the previous example, the cabin differential pressure is equal to 4.1 pounds per square inch differential (p.s.i.d.). This is determined by subtracting the atmospheric pressure of 6.8 p.s.i. from the cabin air pressure of 10.9 p.s.i. Due to structural considerations, all pressurized aircraft have limits on the amount of differential pressure that the airframe can endure. On smaller general aviation airplanes, the maximum cabin differential pressure is typically limited to a value that ranges between 3.35 to 4.5 p.s.i.d., while most large transport category airplanes have maximum cabin differential pressures near 9.0 p.s.i.d. [Figure 11-31]

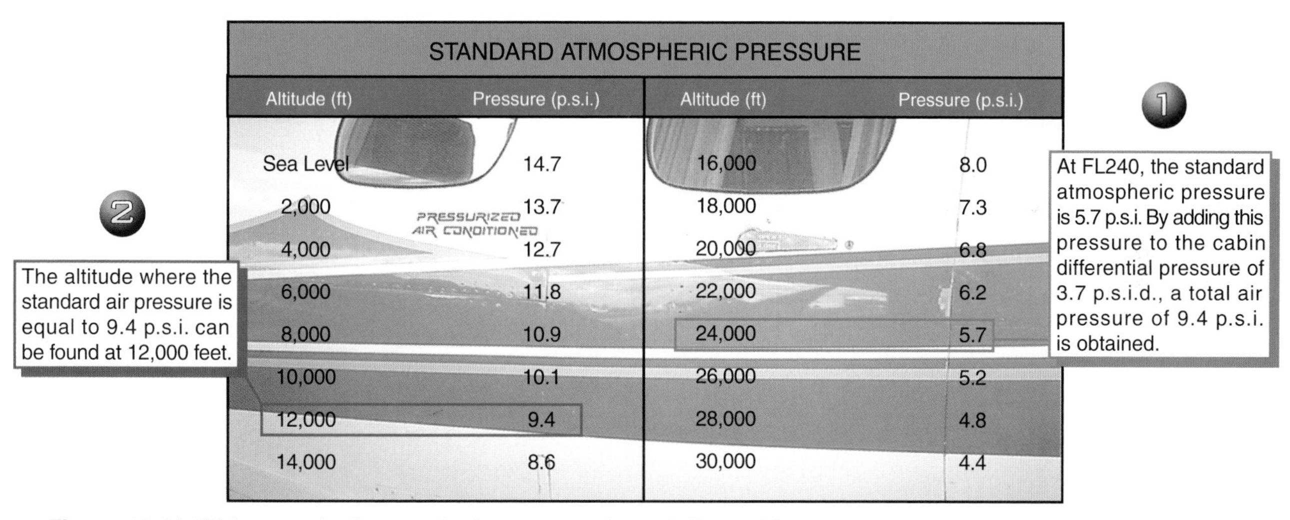

STANDARD ATMOSPHERIC PRESSURE

Altitude (ft)	Pressure (p.s.i.)	Altitude (ft)	Pressure (p.s.i.)
Sea Level	14.7	16,000	8.0
2,000	13.7	18,000	7.3
4,000	12.7	20,000	6.8
6,000	11.8	22,000	6.2
8,000	10.9	24,000	5.7
10,000	10.1	26,000	5.2
12,000	9.4	28,000	4.8
14,000	8.6	30,000	4.4

Figure 11-31. With a standard atmospheric pressure chart similar to this one, you can determine what your cabin pressure altitude will be when you know the cabin differential pressure. For example, if you are flying at FL240 with a cabin differential pressure of 3.7 p.s.i.d., your cabin pressure altitude will be approximately 12,000 feet.

PRESSURIZATION COMPONENTS

On a pressurized airplane with a turbocharged reciprocating engine, pressurization air is obtained from the compressor section of the turbocharger. As hot compressor air is discharged, it enters a sonic venturi which limits the airflow to prevent too much air from being taken away from the engine. The airflow is limited by accelerating the air to a sonic speed as it flows through the venturi, which causes a shock wave to form. The shock wave acts as a barrier, preventing a portion of the compressor discharge air from flowing to the pressurization system. From the sonic venturi, air enters a heat exchanger where it is heated, or cooled, to produce a comfortable climate in the cabin. Inside the heat exchanger, ambient air flows over hollow tubes that contain the pressurization air. By controlling the temperature of the ambient air, the amount of heat transfer between the pressurization air

and ambient air is varied. If additional cabin heat is necessary during high-altitude flight, you can adjust a cabin heat valve to direct the ambient air to become preheated by flowing through a shrouded cavity that surrounds the exhaust muffler. By varying the ambient air temperature in this manner, the pressurization air temperature can be controlled. From the heat exchanger, the pressurization air is distributed to the cabin through heating and ventilation outlets. To regulate the amount of air pressure in the cabin, an **outflow valve** modulates between open and closed, to allow the pressurization air to vent out of the cabin at a controlled rate. Another valve, referred to as a **safety/dump valve**, is used to provide a means for eliminating the pressurization air if the outflow valve fails, as well as other functions discussed later in this section. [Figure 11-32]

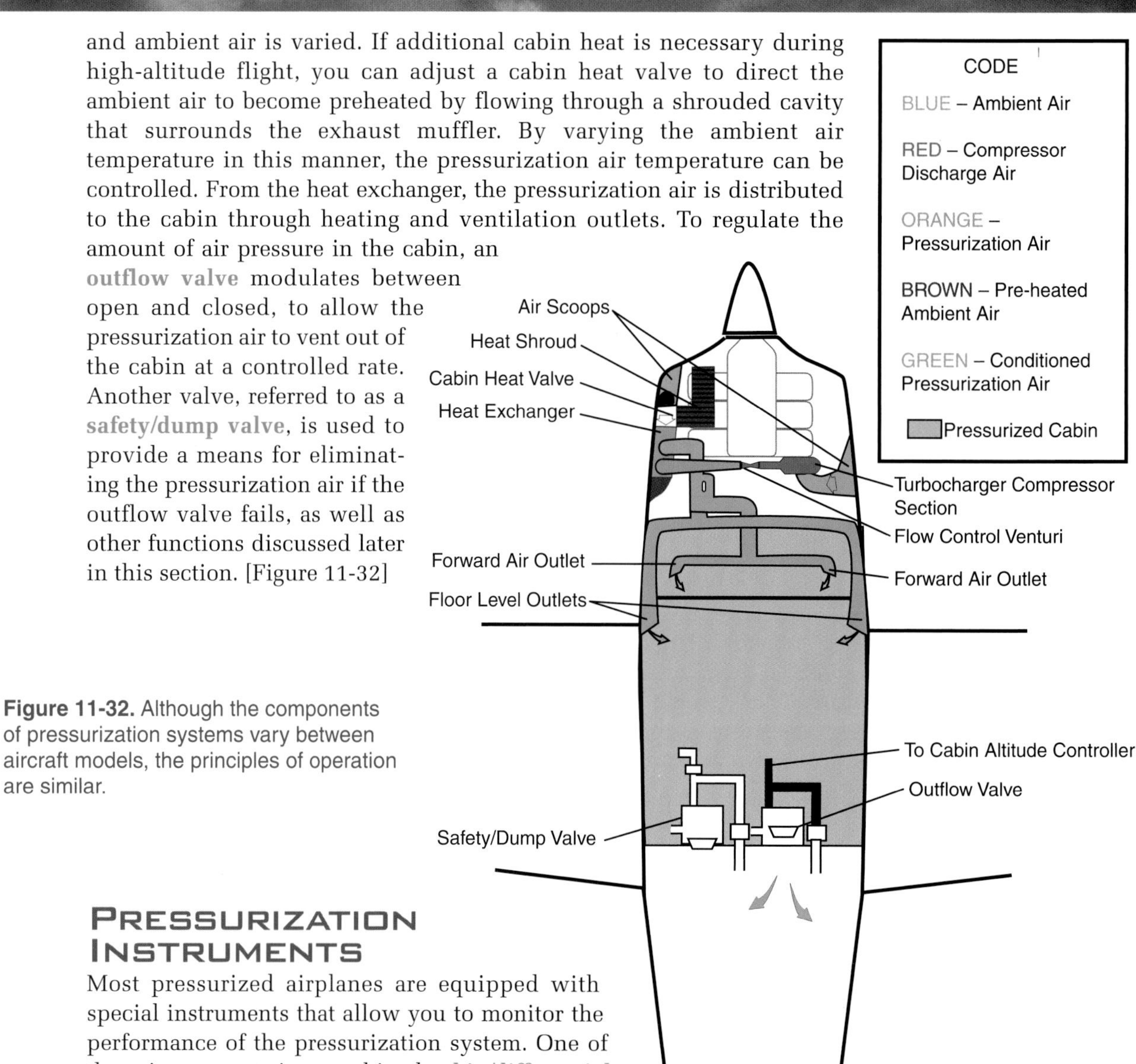

Figure 11-32. Although the components of pressurization systems vary between aircraft models, the principles of operation are similar.

PRESSURIZATION INSTRUMENTS

Most pressurized airplanes are equipped with special instruments that allow you to monitor the performance of the pressurization system. One of these instruments is a combined **cabin/differential pressure indicator**. The indicator works like an altimeter, except that it has two reference pressures: one for cabin air pressure, and another for outside air pressure. By reading the indicator, you can determine either your cabin pressure altitude or the differential pressure between the cabin and outside air. A second instrument is the **cabin rate-of-climb indicator**. This instrument is referenced to the cabin air pressure and indicates the rate of pressure change in the cabin. [Figure 11-33]

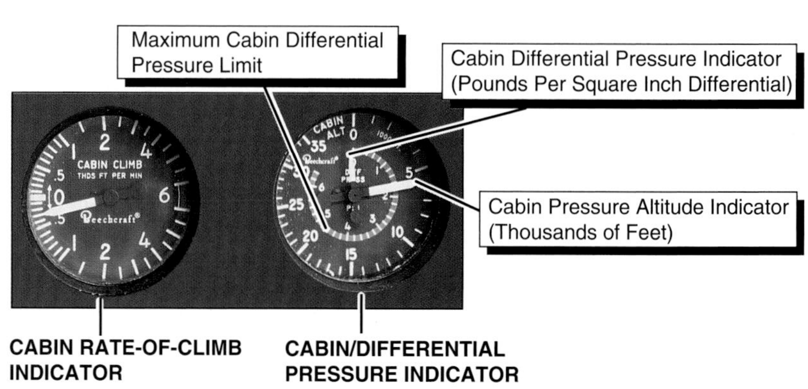

Figure 11-33. The cabin/differential pressure indicator has two indicator needles and scales. The large scale is used to determine the cabin pressure altitude in thousands of feet, while the small scale is used to determine the cabin differential pressure, measured in p.s.i.d. The cabin rate-of-climb indicator senses the rate of cabin air pressure change, and displays the rate in feet-per-minute climb or descent.

PRESSURIZATION CONTROL

There are two types of systems available to control pressurization. The first is a basic system that begins pressurizing the cabin when the airplane reaches a preset altitude. Although this altitude may vary with different manufacturers, 8,000 feet is commonly used. As the airplane ascends above the preset altitude, the outflow valve modulates between open and closed to limit the amount of air that flows out of the cabin. By limiting the outflow of air, the cabin pressure begins to increase to maintain the cabin pressure altitude at the preset level. This causes the cabin rate of climb to remain at zero, until the airplane climbs to an altitude where the maximum cabin differential pressure is obtained. Above this altitude, the pressurization system maintains the maximum cabin differential pressure, which causes the cabin pressure altitude to begin climbing. As a result, the cabin rate-of-climb indicator will show a climb rate that is slightly less than the airplane climb rate. The difference in the rate-of-climb values is because the air density inside the airplane is greater than the ambient air density. When the pressurization system is working to prevent the cabin differential pressure from exceeding maximum limits, it is operating in the **differential range**. On the other hand, when the system is working to maintain the cabin pressure altitude at the preset level, it is considered to be operating in the **isobaric range**.

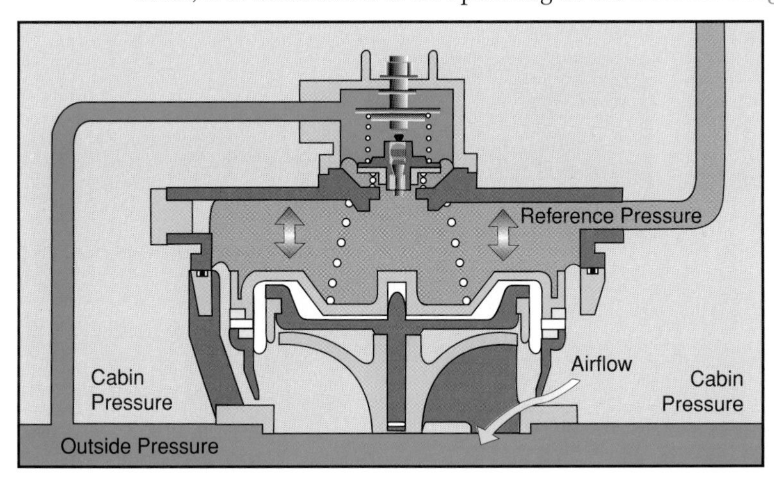

Figure 11-34. The outflow valve allows cabin air to flow from the cabin into a nonpressurized area of the airplane where it can vent to the atmosphere. If the outflow valve fails to maintain the pressurization at a value that is lower than the maximum cabin differential pressure, the airframe structure may be damaged. To prevent an over-pressurization, the safety/dump valve automatically opens when the maximum cabin differential pressure is exceeded. The safety/dump valve is similar in design to the outflow valve, but is set to open at a higher cabin differential pressure value.

A disadvantage of the basic system is that while flying below the altitude where pressurization begins, the cabin air pressure changes in response to climbs or descents. If the rate of altitude change is rapid, you and your passengers may experience physical discomfort such as sinus or ear blockage. In a similar manner, when the airplane is operating at an altitude that causes the pressurization system to function in the differential range, rapid changes in altitude may also produce physical discomfort. The second type of pressurization system alleviates these problems by using a **cabin pressure control**, which allows you to select the altitude where pressurization begins to operate. As an additional feature of most cabin pressure controls, you can also adjust the rate that the air pressure changes inside the cabin by varying the position of a rate control knob. By repositioning the knob, you are able to adjust the opening and closing of the outflow valve to increase or decrease the cabin rate of climb. To utilize the rate control, the pressurization system must be operating in the isobaric range. [Figure 11-35]

CABIN ALTITUDE CONTROLLER

Cabin Pressure Altitude Selector Scale (Thousands of Feet)

Cabin Pressure Altitude Indicator Needle

Cabin Rate-of-Climb Adjustment Knob

Aircraft Altitude Scale

Cabin Pressure Altitude Selector Knob

Figure 11-35. Although cabin pressure control styles vary, most consist of a dual-scale indicator like this one. A large scale is used to set the desired cabin pressure altitude, while a second scale, visible through a window, indicates the altitude where the maximum cabin differential pressure will be reached.

To set the cabin pressure control, you rotate an adjustment knob until your desired cabin altitude registers under the indicator needle. When adjusting the control, most manufacturers suggest that you use a value that is 500 feet higher than the desired setting, to avoid pressurization surges. These surges, or bumps as they are often called, may cause some passenger concern or physical discomfort. As the adjustment knob is rotated, the indicator needle moves as well as the altitude value showing through the window. By observing the reading in the window, you can determine the altitude where the pressurization system will begin operating in the differential range. If you will be flying at an altitude that exceeds the altitude shown in the window, you should reposition the cabin pressure control during your climb so that the altitude in the window shows 500 feet higher than your cruise altitude. By doing this, the pressurization system will continue operating in the isobaric range throughout your flight. An advantage of maintaining the control in the isobaric range is that you can adjust a rate control knob, which allows you to vary the rate that the air pressure

Figure 11-36. This illustration shows the typical operating sequence of a cabin pressure control for a direct flight from Cheyenne, Wyoming (CYS) to Fargo, North Dakota (FAR). To keep the cabin pressure control in the isobaric mode, you need to readjust the cabin pressure altitude during the climb.

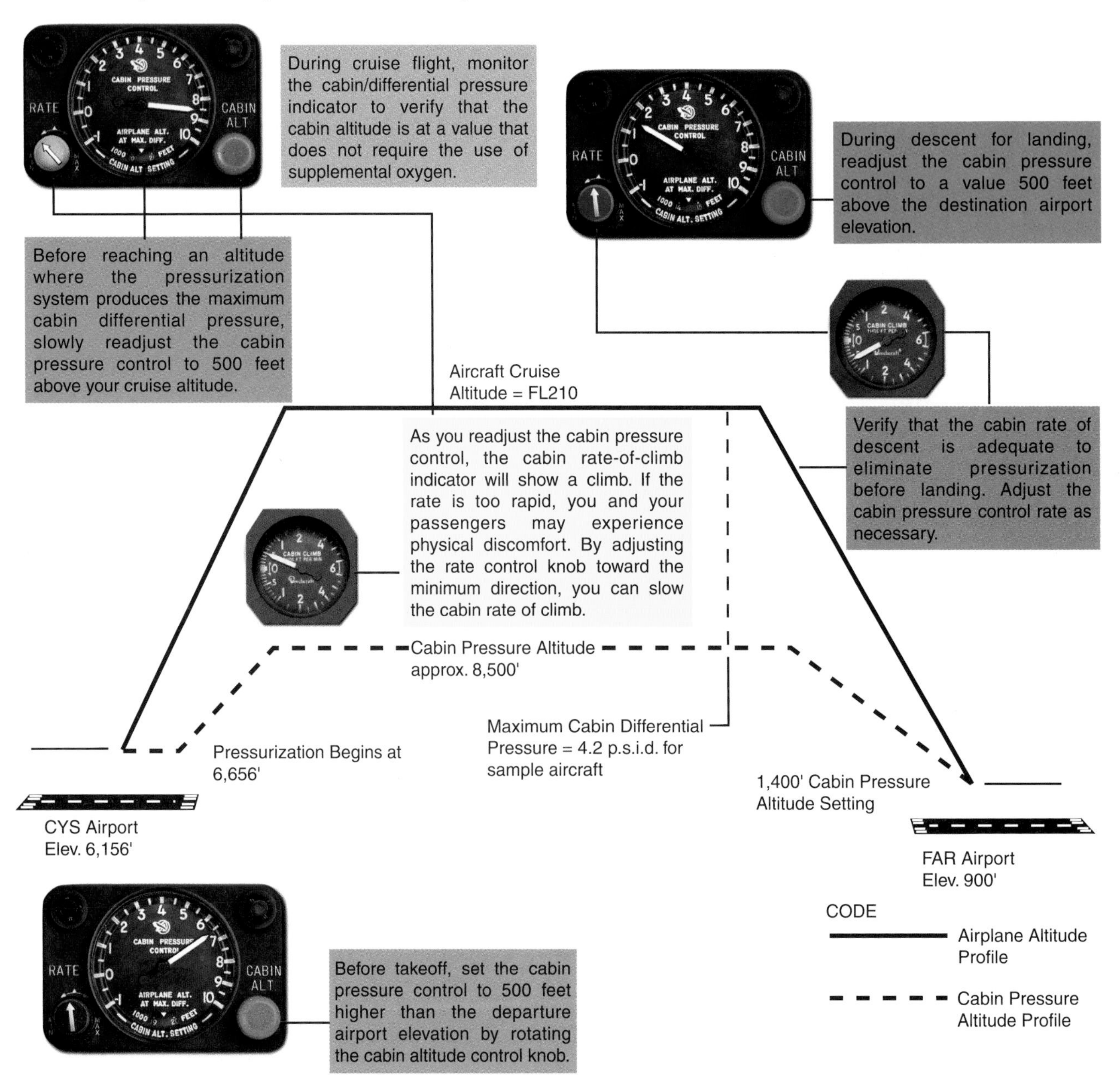

changes inside the cabin. The rate control provides a means for adjusting the rate-of-pressure change, to provide a smoother transition during altitude changes, which helps to alleviate physical discomfort. An additional fact to remember is that when the cabin pressure control is operating in the isobaric range, you need to make all cabin pressure control adjustments slowly. Failure to do so may result in extreme changes in cabin air pressure, which can cause significant physical pain to you and your passengers. [Figure 11-36]

With any pressurization system, you can de-pressurize the cabin by activating a **dump valve** in the event of certain emergencies. For example, you can use the valve to evacuate smoke if it appears in the cabin. The valve also may be designed to remain open to prevent pressurization while the aircraft is on the ground. If the cabin is pressurized when you or your passengers try to exit the aircraft, you may not be able to get the cabin door open since some doors open inward. On the other hand, if the cabin door opens outward, it can be blown open when the latch is released. While this can cause injury to an occupant inside the airplane, it is an even bigger hazard for someone opening the door from the outside. [Figure 11-37]

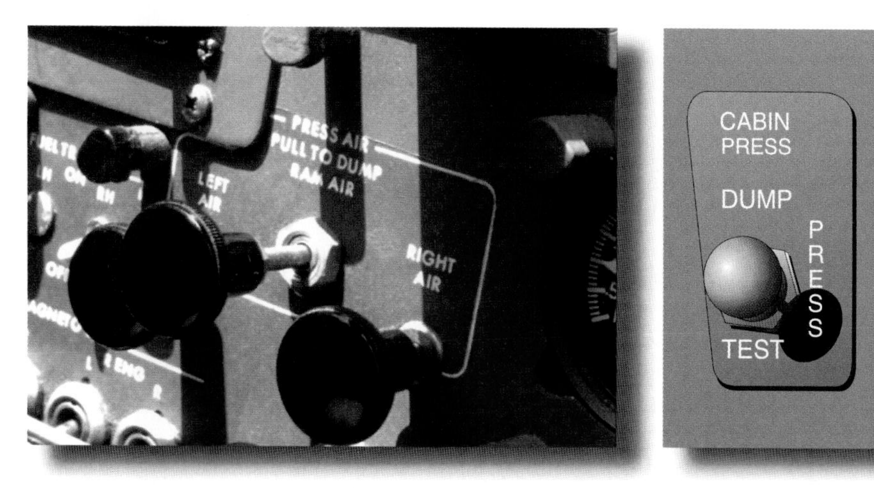

Figure 11-37. On some aircraft, pulling a handle manually activates the dump valve. With other designs, repositioning a switch from the pressurize to dump position electrically actuates the safety/dump valve to an open position to release cabin pressure.

PRESSURIZATION EMERGENCIES

Although cabin pressurization systems are highly reliable, malfunctions do occur with system components or aircraft structures. Of the malfunctions that occur, **cabin decompressions** are among the most serious. Gradual decompressions are hazardous because you may not be aware of the subtle changes in cabin pressure. As the cabin altitude increases, your ability to detect the cabin decompression may be impaired by hypoxia-induced euphoria. To aid in detecting an excessive cabin pressure altitude, the aircraft is equipped with an annunciator light that illuminates when the cabin pressure altitude exceeds a preset value. Typical settings for illumination are 10,000 feet or 12,500 feet. If the cabin altitude light illuminates, you should don an oxygen mask and activate the oxygen system. As a further safeguard, you may want to consider descending to a lower altitude, especially if your oxygen supply is limited.

In other situations, rapid decompressions can cause total loss of cabin pressurization within 1 to 10 seconds, while an explosive decompression can cause complete pressurization loss in less than 1 second. Rapid and explosive decompressions are likely to occur when a large part of the cabin structure fails. Windshield, cabin window, and door failures are typical causes of rapid or explosive decompressions. To help prevent these hazards, you should inspect the cabin structure before any pressurized flight. If cracks or crazing are apparent in the windshield or any cabin windows, have them inspected

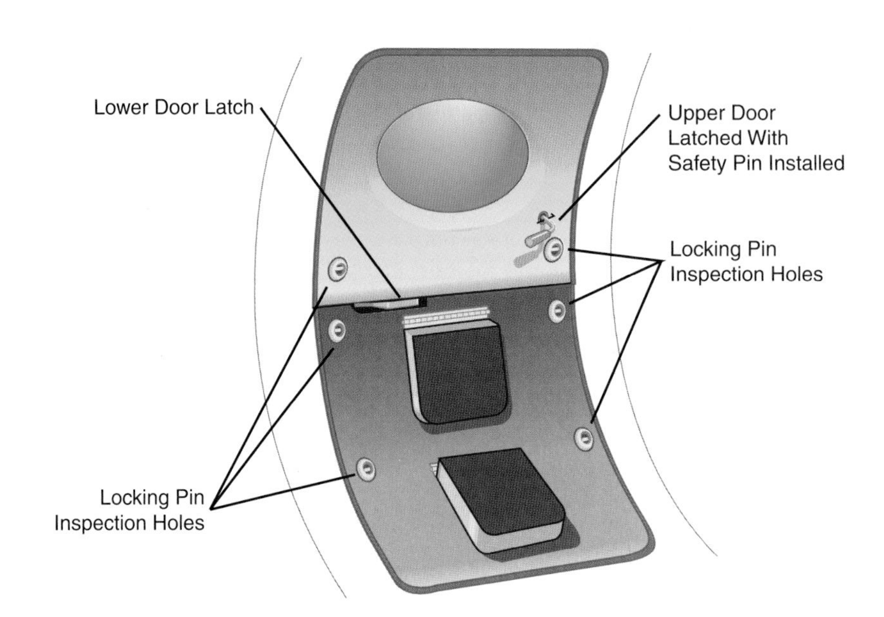

Figure 11-38. Pressurized aircraft have cabin doors that are constructed to withstand high-pressure loading. To maintain a sealed and structurally sound cabin, the doors are equipped with numerous locking pins. Most manufacturers provide viewing ports for you to determine that the door is securely closed. However, if you suspect that a door has come slightly ajar while the cabin is pressurized, do not attempt to close it in flight unless there are specific procedures to follow in the POH for the airplane.

before attempting to fly with the cabin pressurized. Also, since pressurized aircraft doors incorporate special latching mechanisms, you or another authorized flight crewmember should close and secure the cabin door. [Figure 11-38]

When rapid or explosive cabin decompressions occur, they are accompanied by a wide array of phenomena. In many cases, the cabin fills with fog due to immediate condensation of water vapor in the cabin. Also, anything that is not secured in the aircraft is pulled toward the cabin opening, causing flying dirt and debris. After the decompression, the cabin is left extremely cold since all heated air is immediately drawn out. Because the time of useful consciousness is very short, immediate actions must be taken. Your first reaction must be to don an oxygen mask and verify that the oxygen system is operating. Then, if you are on an IFR flight plan, declare an emergency to ATC and begin setting the aircraft up for an emergency descent. Most initial procedures for propeller-driven airplanes specify to retard the throttles and advance the propellers to low pitch, high r.p.m. By decreasing the pitch of the propellers, the blades create drag to slow the aircraft. Once the airspeed has decreased to allowable limits, extend the landing gear and begin a descent. When an altitude has been reached where oxygen is no longer required, you should continue the flight only to the closest usable airport. Remember, once you have experienced a cabin decompression at high altitude, you need to be aware of the onset of altitude-induced decompression sickness (DCS). To be safe, it is recommended that you seek medical attention from an aeromedical specialist, trained in DCS emergencies. DCS and other high altitude physiological effects are discussed in Chapter 1, Section B — Advanced Human Factors Concepts.

ICE CONTROL SYSTEMS

Ice control systems used on high performance or complex airplanes consist of a combination of **anti-icing** and **de-icing** equipment. Anti-icing equipment prevents the formation of ice, while de-icing equipment removes the ice once it has formed. Aircraft components that must be protected from ice accumulations include leading edges of wing and tail surfaces, pitot and static source openings, fuel tank vents, stall warning sensors, windshields, and propeller blades. In addition, engines may require alternate sources of intake air or ice protected inlets, while ice detection lighting also is required to help you determine the extent of structural icing when flying at night. Although an aircraft may be equipped with systems to protect some or all of these components, the

aircraft still may not be certified to fly in icing conditions. To determine if your airplane is certified to operate in icing conditions, you should consult the FAA-approved POH for the airplane. However, even if certified, you should keep in mind that icing conditions pose significant hazards. When encountered, always seek out regions where meteorological conditions are less conducive to ice formation. A discussion of aircraft icing causes, effects, characteristics, and avoidance procedures are found in Chapter 9, Section B — Weather Hazards. [Figure 11-39]

Figure 11-39. Static discharge wicks are required for flights in icing conditions or when operating in areas of precipitation to help prevent static electricity from arcing between the airplane and the atmosphere. If static discharges are not controlled, radio interference can become so severe that communication and navigation signals are unusable.

AIRFOIL ICE CONTROL

To protect wing and tail surface leading edges from icing, most aircraft use pneumatic devices. On many reciprocating-engine and turboprop airplanes, **de-icing boots** are pneumatically inflated to break the ice, allowing it to be carried away by the airstream. In another system, primarily used on jet airplanes, heated pneumatic air is directed through ducting in the airfoil leading edge, to thermally prevent ice from forming. This system is often referred to as a **thermal anti-ice** system.

DE-ICING BOOTS

A high-pressure de-icing boot is essentially a fabric reinforced rubber sheet that contains built-in inflation tubes and is bonded to the leading edge of the surface to be protected. For reciprocating-engine airplanes, pneumatic pressure from engine-driven pumps is supplied to the inflation tubes by suitable plumbing. The boots are installed in sections along each leading edge surface. Where boots are installed on wings and horizontal stabilizers, corresponding sections on the left and right side of the aircraft operate simultaneously. Simultaneous operation provides symmetrical airflow between the left and right side of the airplane. To limit the disruption of airflow over the surface, the inflation tubes within a boot may be alternately cycled by a sequencing timer. By cycling inflation, the load on the pneumatic pumps is reduced. [Figure 11-40]

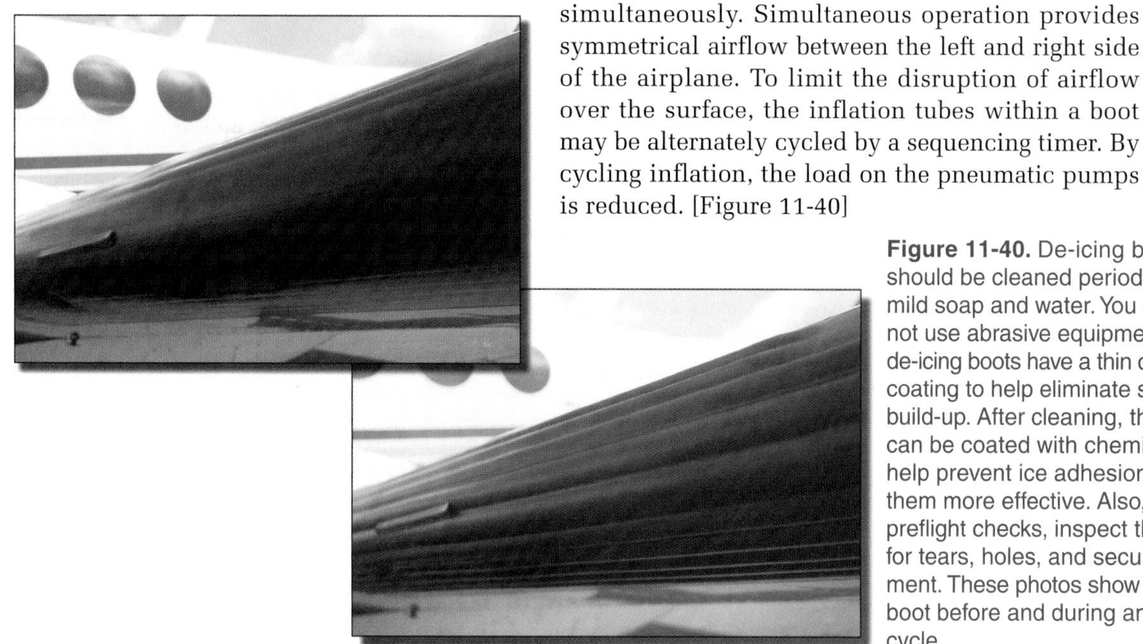

Figure 11-40. De-icing boots should be cleaned periodically with mild soap and water. You should not use abrasive equipment since de-icing boots have a thin conductive coating to help eliminate static build-up. After cleaning, the boots can be coated with chemicals that help prevent ice adhesion, making them more effective. Also, during preflight checks, inspect the boots for tears, holes, and secure attachment. These photos show a de-icing boot before and during an inflation cycle.

The pneumatic pumps that are used to operate de-icing boots are often the same pumps that power the gyroscopic flight instruments. These produce suction and pressure air, which can be directed to the inflation tubes. When the boots are not in use, suction is applied to the inflation tubes, which holds the boot in the deflated position to maintain the contour of the leading edge. Depressing a switch activates the de-icing boots, which causes pressure air to be cycled through the inflation tubes. The switch often includes two positions, one for automatic cycling and the other for a single cycle. In the automatic position, a complete cycling of the boots is accomplished at timed intervals, whereas the single-cycle position only provides one complete cycle. [Figure 11-41]

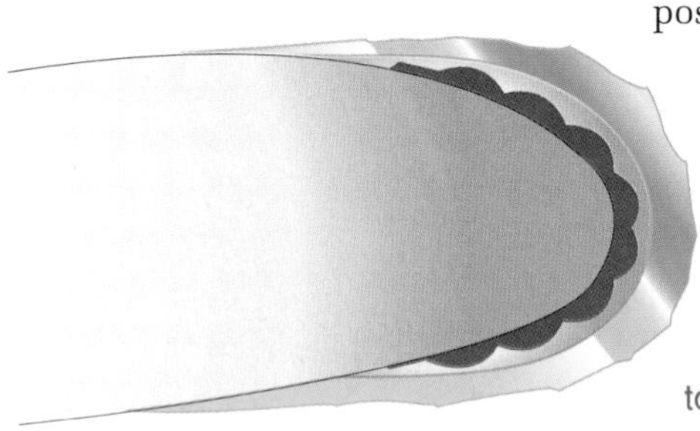

Figure 11-41. If pneumatic de-icing boots are allowed to cycle too often, ice can form to the inflated contour of the boot. This can cause a void to form under the ice, causing boot inflation to become ineffective. To prevent the condition, boots should not be activated until ice has accumulated to an amount of 1/4 to 1/2 inch thickness.

Many de-icing boot systems use the instrument system suction gauge and a pneumatic pressure gauge to indicate proper boot operation. These gauges have range marks to depict minimum and maximum operating limits for boot operation. When the boots are not in operation, the suction gauge should indicate in the normal operating range. If it is low, a boot section may have a leak, which prevents it from inflating properly. During boot cycling, the gauges fluctuate as each section inflates. While each section inflates, you should check the pressure gauge to verify that it indicates in the green, normal operating range.

Another type of indicator for de-icing boots uses annunciator lights. During boot operation, a red light illuminates until adequate pressure causes a cycling switch to actuate. When cycling occurs, a green light illuminates. If the red light remains on, the system is not operating properly and must be manually deactivated to deflate the boots.

THERMAL ANTI-ICE SYSTEMS

Pneumatic, thermal anti-ice systems are commonly installed on turbojet and some turboprop aircraft. This is because a jet engine has a ready source of hot air that can be used to heat airfoil leading edges. In this system, hot air is directed from one of the later stages of the compressor section of the engine and directed to the leading edges of the thermally protected airfoils. [Figure 11-42]

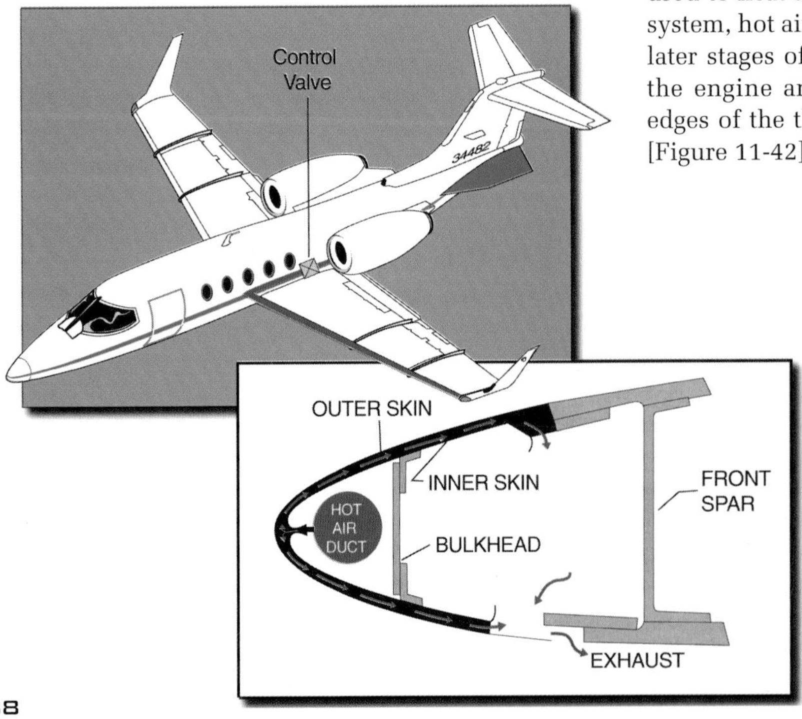

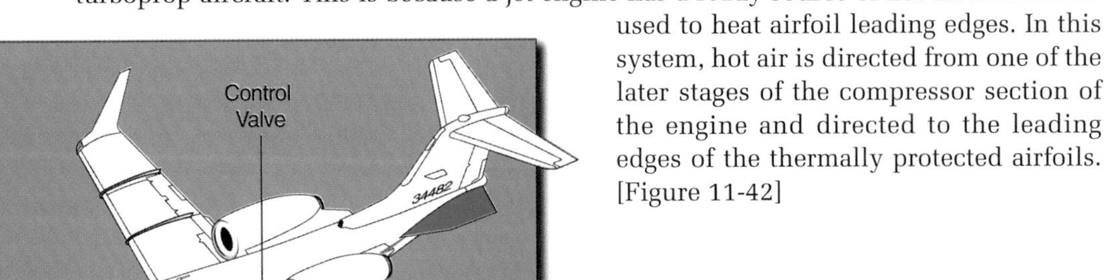

Figure 11-42. The leading edge of an airfoil that uses a thermal anti-ice system uses ducts and chambers that allow hot engine air to sufficiently heat the leading edge to prevent ice formation.

Weeping Wings Shed Ice, Not Tears

While de-icing boots and thermal anti-ice systems are the most common methods of protecting airfoil leading edges, some aircraft manufacturers are using another type of airfoil anti-ice system. A weeping wing, as it is often called, uses leading edge panels that are either laser-drilled with small holes, or fitted with a fine stainless steel mesh as shown here. By gradually pumping a mixture of ethylene glycol, isopropyl alcohol, and water through the leading edge, structural ice formations are prevented.

The anti-icing fluid is stored in a reservoir that holds enough fluid to provide two or more hours of ice protection when flying in moderate icing conditions. In addition to airfoil protection, the anti-icing fluid is used for windshield and propeller ice protection. When the system is combined with other ice control devices, many aircraft equipped with a weeping wing are certified for flight into areas of known icing conditions.

WINDSHIELD ICE CONTROL

Most aircraft are equipped with a defroster consisting of vents, which direct heated air across the windshield on the inside of the cabin. Although this system is adequate for some operations, flight in icing conditions may cause ice to adhere to the outside of the windshield, restricting your visibility. To prevent ice formation, some **windshield anti-ice** systems use a flow of alcohol to coat a small section of the windshield. Although these systems may be used for ice removal, they should be used early enough when icing conditions are encountered to prevent ice from forming. The system is easy to operate, often using a switch that allows you to vary the flow rate of the alcohol. Before a flight with this type of windshield ice protection, check to make sure the reservoir is properly serviced with the correct type of alcohol as recommended in the POH for your airplane.

Another effective way of protecting a windshield is provided by using electric heat to prevent the formation of ice. With these systems, small wires or electrically conductive materials are embedded in the windshield or in a panel of glass that is installed over the exterior of the windshield. By passing electrical current through the windshield or panel, sufficient heat is produced to prevent the formation of ice. You should be aware that when these systems are activated, the electrical current flow causes disturbances to the magnetic compass. These disturbances can generate erroneous compass readings in excess of 40 degrees when the windshield heat is operating. [Figure 11-43]

Figure 11-43. When a heated windshield or panel is installed on an aircraft, it should only be operated in flight to prevent overheating. The anti-ice windshield panel in this photo developed bubbles between the window laminations because it was left on for an extended period of time while the aircraft was on the ground.

PROPELLER ICE CONTROL

In a similar fashion to aircraft windshields, propellers can be protected from icing by using alcohol or electric heating elements. With a **propeller anti-ice** system that uses alcohol, the spinner is equipped with discharge nozzles that are pointed toward each blade root. By discharging alcohol from the nozzles, centrifugal force causes the alcohol to flow down the leading edge of the blade to prevent ice formation. The inboard portions of each propeller blade may also have rubber boots attached to the leading edges. These boots have grooves that help to direct the flow of alcohol. An electric propeller anti-ice system also uses rubber boots attached to the inboard portions of each blade. However, the boots have wires imbedded in them to carry electric current for heating. [Figure 11-44]

PROP ANTI-ICE AMMETER
When the system is operating, the prop ammeter will show in the normal operating range. As each boot section cycles, the ammeter will fluctuate.

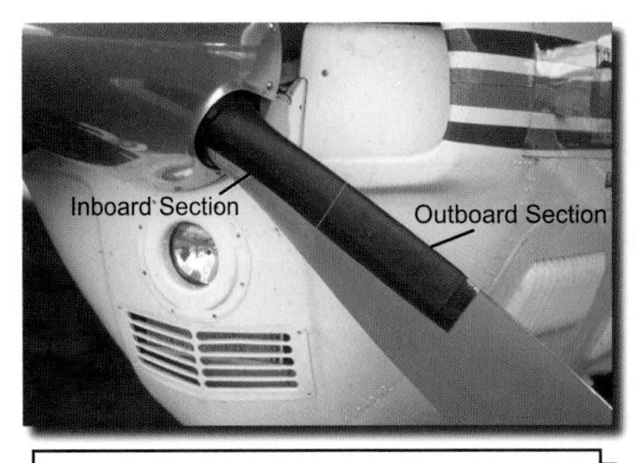

PROP ANTI-ICE BOOT
The boot is divided into two sections: inboard and outboard. When the anti-ice is operating, the inboard section heats on each blade, and then cycles to the outboard section. If a boot fails to heat properly on one blade, unequal ice loading may result, causing severe vibration.

Figure 11-44. Electrically heated propeller anti-ice boot sections are usually heated in a cycling fashion to prevent an excessive load on the electrical system. Before flight, check the system by touching each boot section to determine proper operation. In flight, you can determine that the propeller anti-ice system is operating by monitoring an ammeter that shows the amount of electrical current being supplied to the boots.

OTHER ICE CONTROL SYSTEMS

As mentioned earlier, other ice protection systems include the pitot tube and static port openings, fuel vents, and stall warning sensors. Each of these systems are typically heated by electrical elements and can be checked during a preflight inspection by touching each component when the ice protection systems are on. Use caution when checking these devices. The temperature produced by heating elements can cause severe burns. For example, never wrap your hand around a pitot tube to check heating. Instead, use a moistened finger or rag to touch the tube lightly.

Even with a thorough preflight inspection of each ice control device, you should check the operation of the devices while in flight before icing conditions are encountered. Also, since many of these systems are anti-ice devices, they should be activated before icing conditions develop. Also keep in mind that when any icing conditions are encountered, you should seek alternate altitudes or routes. With most small aircraft ice control systems, your safety is primarily increased by providing you with more time to seek meteorological conditions that are less conducive to icing.

SUMMARY CHECKLIST

✓ Built-in and portable oxygen systems are available to supplement breathing while operating at high altitude, to prevent adverse physiological conditions.

✓ There are three basic configurations of oxygen systems used on aircraft. They are the continuous-flow, diluter-demand, and pressure-demand systems.

- ✓ Constant-flow, adjustable-flow and altitude-compensated oxygen systems are all continuous-flow designs, and are capable of providing adequate respiration up to 25,000 feet.
- ✓ Oxygen masks that are used with continuous-flow oxygen systems are usually oronasal rebreather designs that dilute oxygen with a portion of exhaled air, to provide increased oxygen supply duration.
- ✓ Diluter-demand and pressure-demand oxygen systems only allow oxygen to flow when you inhale, and are capable of providing diluted or 100% oxygen.
- ✓ For flights as high as 40,000 feet, the diluter-demand or pressure-demand oxygen systems are used. For flights above 40,000 feet, a pressure-demand system must be used.
- ✓ Aviator's breathing oxygen is usually stored in a high-pressure cylinder that is serviced to between 1,800 to 1,850 p.s.i. and should never be allowed to decrease below 50 p.s.i.
- ✓ A chemical oxygen generator provides an emergency oxygen supply, and is primarily used during descents in a pressurized airplane in the event of a cabin decompression.
- ✓ To prevent a fire, never allow smoking or open flames near an oxygen system while it is in use. In addition, keep oxygen fittings free of grease, oil and hydraulic fluids.
- ✓ Pressurization systems increase the pressure inside an airplane to produce a lower cabin pressure altitude. In addition, when the pressurization system is operating, you can control the cabin rate of pressurization to maintain a comfortable environment.
- ✓ A basic pressurization system is set to begin operating at a particular altitude, whereas a system with a cabin pressure control allows you to set the altitude where pressurization begins.
- ✓ The pressurization instruments include the cabin rate-of-climb indicator and the cabin/differential pressure indicators.
- ✓ If a pressurized cabin decompresses, you must descend to an altitude where oxygen is not required. If the decompression is rapid or explosive, you should be checked by an aeromedical examiner to determine if you are suffering from altitude-induced decompression sickness (DCS).
- ✓ Aircraft ice control systems provide more time for you to maneuver your aircraft out of hazardous structural icing conditions. Even if an aircraft is certified for flight into known icing conditions, you should try to avoid icing encounters.
- ✓ Ice control systems are divided into de-ice and anti-ice systems. De-ice systems remove ice after it has formed on an aircraft component, while anti-ice systems prevent ice before it has a chance to form.

KEY TERMS

Oxygen Systems

Continuous Flow

Diluter Demand

Pressure Demand

Constant Flow

Adjustable Flow

Altitude Compensated

Oronasal Rebreather

Quick-Donning Mask

Oxygen Cylinders

Chemical Oxygen Generators

Oxygen Duration Chart

Aviator's Breathing Oxygen

Cabin Pressurization System

Cabin Pressure Altitude

Cabin Differential Pressure

Outflow Valve

Safety/Dump Valve

Cabin/Differential Pressure Indicator

Cabin Rate-of-Climb Indicator

Cabin/Differential Pressure Indicator

Cabin Rate-of-Climb Indicator

Differential Range

Isobaric Range

Cabin Pressure Control

Dump Valve

Cabin Decompressions

Ice Control Systems

Anti-Icing

De-Icing

De-Icing Boots

Thermal Anti-Ice

Windshield Anti-Ice

Propeller Anti-Ice

QUESTIONS

1. As the pilot in command of an airplane that uses color coded oxygen mask connectors, what color mask should you use?

   A. Gold
   B. Red
   C. Orange

2. What method is used to determine if oxygen is flowing to an oronasal rebreather oxygen mask when used in a continuous-flow oxygen system?

3. To prevent moisture from entering a high-pressure oxygen supply cylinder, the cylinder pressure should not deplete below

   A. 100 p.s.i.
   B. 15 p.s.i.
   C. 50 p.s.i.

4. If an outflow valve of a pressurization system fails in the closed position, what prevents the cabin pressure from exceeding the maximum cabin differential pressure?

5. True/False. When setting a cabin pressure control before takeoff, you should adjust the cabin altitude to 500 feet above your planned aircraft cruising altitude.

6. What action should you first take if you experience a rapid cabin decompression in a propeller-driven airplane, while on an IFR flight plan?

   A. Declare an emergency with ATC
   B. Increase the pitch of the propeller
   C. Don an oxygen mask and activate the oxygen system

7. Why should you allow ice to accumulate a certain amount on de-icing boots before inflating them?

8. How can you determine that an electrically powered propeller anti-ice system is operating properly while in flight?

SECTION C

RETRACTABLE LANDING GEAR

Landing gear have a simple function: to absorb the load imposed during a landing, in order to protect the airplane's fuselage and preserve the comfort of those on board. You may not give much thought to the fixed gear on an airplane, except to ensure they are aligned with the runway upon touchdown. However, when you fly an aircraft with retractable landing gear, you assume more responsibility in exchange for the advantages afforded by the more complex design.

You will find certain differences when operating an aircraft with retractable landing gear. For example, climb performance usually improves as the gear is raised after takeoff. You will fly at faster cruise speeds, in general, than in similar fixed-gear aircraft. Extending the gear during approach will help you slow down and descend more smoothly, and you may find that lowering the gear during encounters with turbulence helps to stabilize the airplane. Since there are numerous moving parts within the gear, you will need to preflight the system closely. You must always remember to lower the gear before landing, and, in case the gear will not extend or retract through normal means, there are emergency procedures to learn and follow. Ground handling and taxiing, however, are quite similar to those operations in fixed-gear aircraft.

LANDING GEAR SYSTEMS

Most aircraft with retractable landing gear use some type of mechanical aid to help the pilot raise the gear into the belly of the airplane and extend it again for landing. The gear may be operated either through a hydraulic system or an electric motor, or a hybrid of the two methods. Since systems differ between aircraft, you should familiarize yourself with the landing gear system in your airplane, using the recommendations outlined in the aircraft's POH.

On some aircraft, the main gear retract into the wings or fuselage on either side. The nose gear generally retracts into the forward fuselage, below the engine compartment. Gear doors may be installed to further streamline the fuselage after the gear has retracted.

ELECTRICAL GEAR SYSTEMS

Some aircraft employ an electrically driven motor for gear operation. The **electrical gear system** consists of a motor that is reversible to facilitate both raising and lowering the gear, and this motor drives a series of rods, levers, cables, and bellcranks that form what is essentially a motorized jack. Struts are also actuated that open and close the gear doors during the **gear cycle**. The gear cycle refers to the process that the gear goes through during extension and retraction. The electric motor moves up or down and it will continue to operate until the up or down limit switch on the motor's gearbox is tripped. [Figure 11-45]

It's All About Having the Right Gear

Different airplanes require different types of landing gear to perform their particular jobs. While you may be familiar with the tricycle and conventional gear aircraft around your local airport, there are several other gear applications of which you may not be aware. Aircraft in remote areas use extra-large tires that can roll over rough terrain, while those that routinely land on snowfields may have skis installed. Seaplanes may either use pontoons or a keel for water landings, but amphibious types have wheels that either extend out of the pontoons or the fuselage for landings on hard-surfaced runways. If you fly a seaplane, neglecting to lower the gear for landing on pavement — or to raise the gear when landing on water — can have damaging consequences.

Gliders often have a single main gear, located on the bottom of the fuselage, and a small tailwheel. Because of this, crew are required to walk the wings so they do not touch the ground during takeoff. Having one fewer gear saves on weight and parasite drag, which are even more critical in gliders than in powered aircraft.

Large transport aircraft use multiple-wheel gear to distribute the weight of the aircraft over a larger surface area on the runway. Early landing gear experiments on transport aircraft tested extremely large, single wheels, and tracked gear, like the wheels on a tank, but found that the applications for those types of gear were more limited than for multiple-wheel gear.

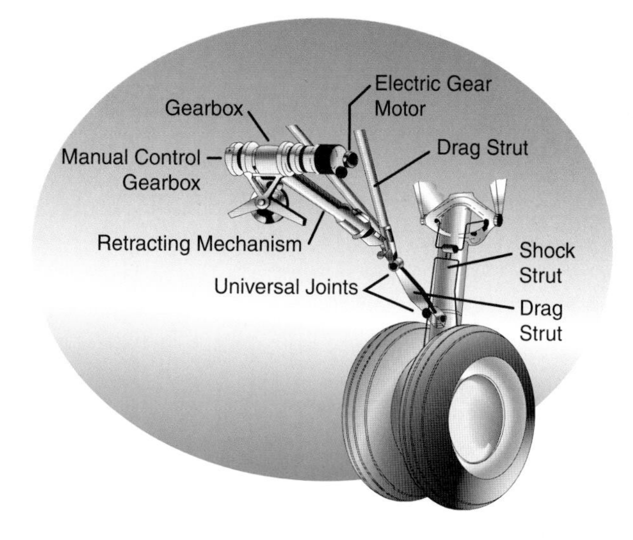

Figure 11-45. The drag struts help support the aircraft's weight, while the shock strut provides cushioning during touchdown.

HYDRAULIC GEAR SYSTEMS

Other aircraft utilize a **hydraulic gear system** to actuate the linkages that raise and lower the gear. When the gear switch is retracted, hydraulic fluid, like that used in many brake systems, is pressurized and directed into the gear up line. The fluid flows through sequence valves and downlocks to the nose gear and main gear actuating cylinders. A similar process occurs during gear extension. The pump which pressurizes the fluid in the system can be either engine driven or electrically powered. If an electrically powered pump is used to pressurize the fluid, the system is referred to as an **electrohydraulic system**. A hydraulic fluid reservoir serves to contain excess fluid, and gives you a means of determining the fluid level in the system before flight. [Figure 11-46]

The hydraulic pump, regardless of its power source, is designed to operate within a specific range. When a sensor detects excessive pressure, a relief valve within the pump opens, and hydraulic fluid is routed back to the reservoir. A second type of

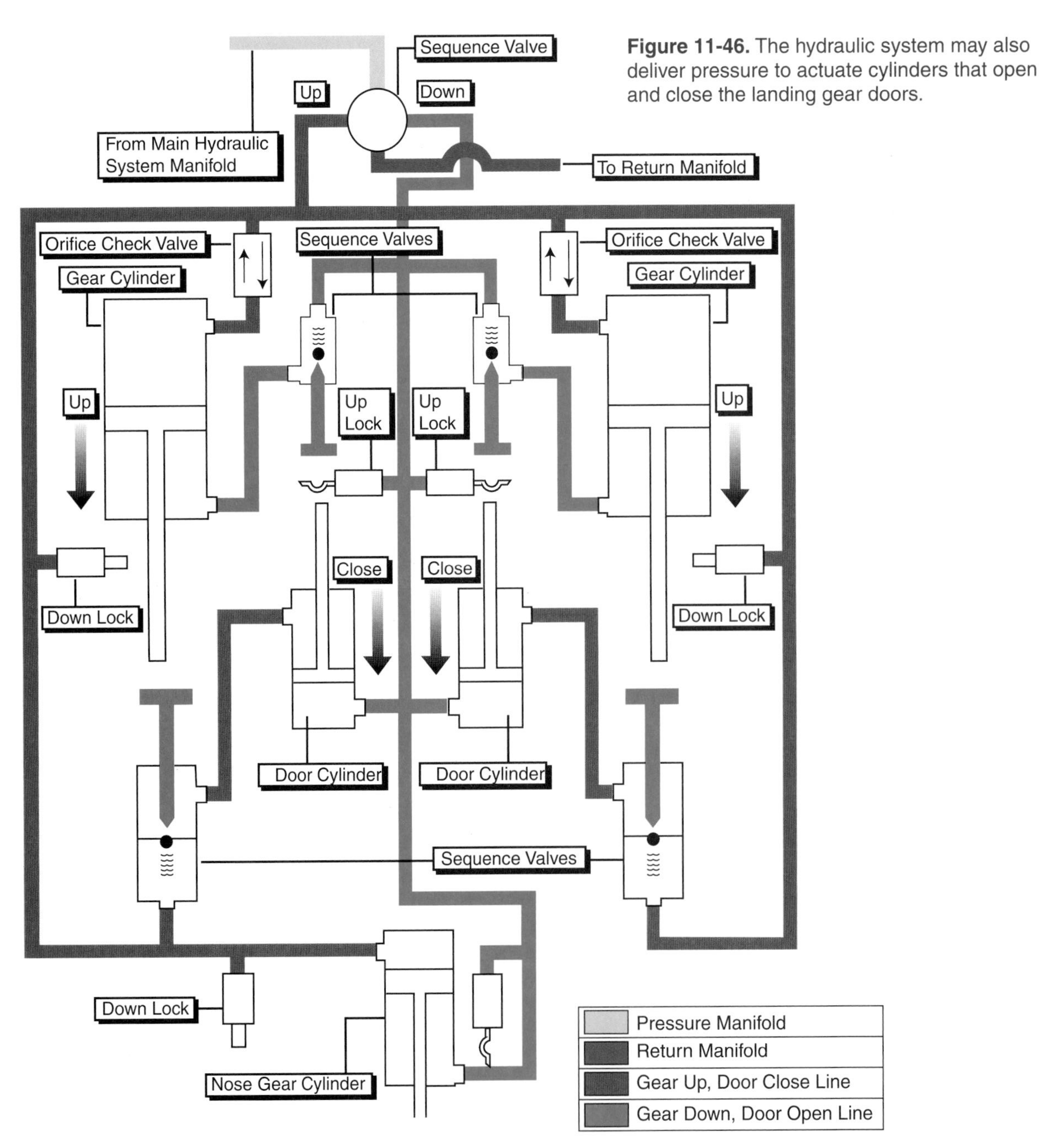

Figure 11-46. The hydraulic system may also deliver pressure to actuate cylinders that open and close the landing gear doors.

relief valve prevents the excess pressure associated with thermal expansion. The hydraulic pressure is also regulated by limit switches on each of the landing gear. Each gear has two switches, one dedicated to extension, and one for retraction. These switches deenergize the hydraulic pump after the landing gear has completed a gear cycle. In case any of these switches fail, a backup valve will relieve excess pressure in the system.

GEAR SYSTEM SAFETY

With the benefits of retractable landing gear come additional complexity and potential problems. To deal with these issues, aircraft manufacturers have installed various devices on aircraft with retractable gear to enhance safety and ease of operation.

GEAR POSITION INDICATORS

In most aircraft, not all of the landing gear are visible from the cockpit. Therefore, one of several different types of **gear position indicators** is installed to help you keep track of where the gear is in the gear cycle. Position lights are one common type of indicator. Often, the grouping will consist of one red light and one or three green lights. The red light is illuminated when the gear are in transit or unsafe for landing, and the green lights indicate when each gear is down and locked. Generally, the red light will illuminate and then go off once the gear is fully retracted or extended. These lights are often installed so that you can press on them to test for proper illumination. [Figure 11-47] Other position indicators include miniature gear icons electrically placed by the movement of the actual gear, or an arrow indicator showing gear position, and some models have separate lights that illuminate solely when the gear is in transit.

Figure 11-47. A flickering indicator light may signal either that one of the gear is not locked or that there may be a loose wire to the bulb.

GEAR WARNING HORN

As a reminder to pilots, most aircraft with retractable gear have a **gear warning horn** that will sound when the aircraft is configured for landing and the gear is still in an unsafe position. Usually, the horn is linked to the throttle or flap position, and/or the airspeed indicator, so that when the aircraft is below a certain airspeed, configuration, or power setting with the gear retracted, the warning horn will sound. The horn generally differs in sound from the stall warning horn, so that you will know which condition you are being alerted to. Some aircraft have other safety systems installed to help you avoid a gear-up landing. [Figure 11-48]

Figure 11-48. This aircraft has an automatic gear extension system that extends the gear below a certain airspeed if the pilot fails to do so. The system has an override feature for use during slow flight maneuvers.

SAFETY SWITCHES

Most retractable landing gear systems incorporate a mechanical **safety switch** that prevents the retraction of the gear when the aircraft is on the ground. The switch, sometimes referred to as a squat switch, is usually found on one of the main gear struts. When the strut is compressed, the switch opens the electrical circuit to the motor or mechanism that powers retraction. That way, if the gear switch in the cockpit is moved to the retract position when weight is on the gear, such as during taxi, the gear will remain extended. However, the gear warning horn may sound as an alert to the unsafe condition. If the gear switch remains in the retract position during takeoff, once the weight is off of the gear, the safety switch will release and the gear will retract. If the aircraft does not have a positive rate of climb, it may settle back to the ground with the gear retracted. [Figure 11-49]

DO BIRDS EVER LAND GEAR UP?

Every year, new and experienced pilots alike manage to inadvertently land aircraft with the gear retracted. Because pilots are not physically connected to the airframe, it is not intuitive to put the gear down before landing, as you would automatically extend your arms to brace a fall. You may wonder if birds and other animals that fly land on their feet by instinct, or if they have to learn to do so, as pilots do.

Just like airplanes, different species of flying creatures have "gear," or feet, that suit their environments. Belted kingfishers nest in burrows high on silt cliffs, and horned owls scavenge the nests that others leave behind. The hummingbird cannot walk — it must always use its wings to move, much like a helicopter on skids. Raptors such as the bald eagle have feet that serve double duty, being useful for snaring prey as well as landing. Bats land upside-down, which enables them to hang from cave ceilings and tree branches.

How much of an animal's ability to land safely is instinct, and how much is learned? If a bird landed simply on instinct, it would rarely suffer a bad landing. However, just as birds must learn to fly from watching their elders, so too must they perfect the skill of landing. There are moments when a young bird, struggling to touchdown, obviously forgets to put its "gear" down, and the result is nearly as embarrassing as when a pilot does the same in an airplane. The albatross, for example, is notorious for never quite perfecting the landing process: it nearly always lands with a tumble across the water, but its awkwardness on land is offset by an incredible ability to soar.

You also must work to perfect your landing skills. To help you remember critical items such as extending the gear, you should always use a prelanding checklist. Also, if you fly in a crew, work with your fellow pilot as you run through the checklist. The coordination will help you ensure that you do not land someday with perfectly good gear still up in the wheel wells.

Figure 11-49. This safety switch is located on the left main gear. While some safety switches prevent gear retraction by opening the electrical circuit, some keep the gear switch inside the cockpit from moving.

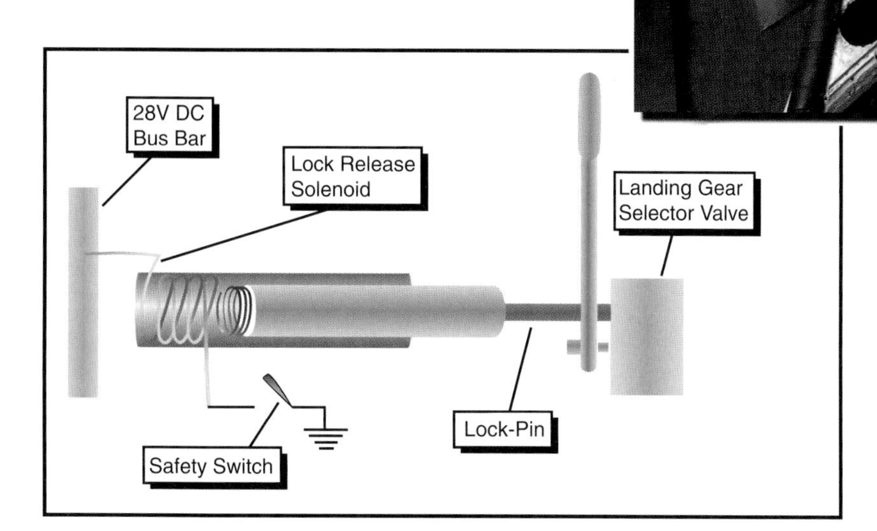

AIRSPEED LIMITATIONS

Limiting airspeeds are established for gear operation in order to protect the gear components from becoming overstressed during flight. The airspeeds for gear retraction and extension may be different in some airplanes, because of the gear cycle sequence, and the relative strength of gear doors, struts, and other parts of the mechanism. The operating loads placed on the gear at faster speeds may cause structural damage due to the forces of the airstream. Though you will not see these speeds on the airspeed indicator, they are usually found on placards in the airplane, and published in the POH. The **maximum landing gear extended speed (V_{LE})** is the maximum speed at which you can fly the aircraft with the landing gear extended. The **maximum landing gear operating speed (V_{LO})** is the highest speed at which you may operate the landing gear through its cycle. On some aircraft, V_{LO} may be slower than V_{LE}, and this means that you will have to slow the airplane down to extend the gear, after which you could fly at a slightly faster speed. [Figure 11-50]

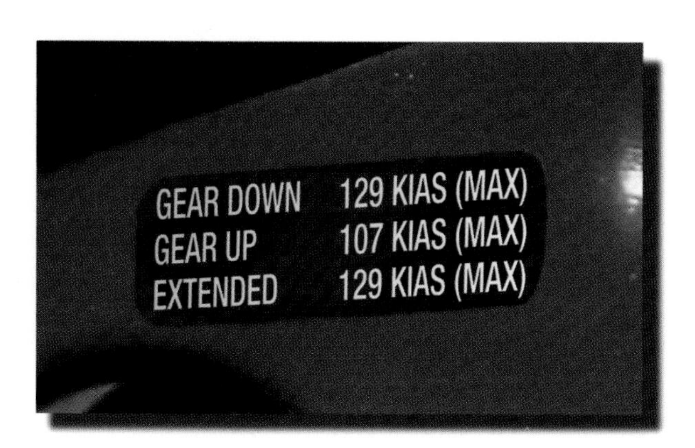

Figure 11-50. Though relevant gear speeds are posted on placards in the cockpit, you should review the procedures outlined in your aircraft's POH for specific details.

OPERATING PROCEDURES

Many potential gear system problems can be prevented if you conduct a thorough preflight. Because the system is more complex, retractable landing gear demand a close inspection before every flight. The wheel wells need to be clear of any obstructions, as a foreign object may damage the gear or cause it to become lodged inside the wells. Bent gear doors could be a sign of other problems with gear operation, and they may also become stuck during extension or retraction. Cracks, corrosion, or loose bolts in the linkage may lead to gear failure at a later time, so look carefully at each component of the system. Because of the systems' complexity, there is a good chance that airworthiness directives (ADs) exist for a given landing gear system. When you fly a particular aircraft for the first time, you should check for any ADs pertaining to the landing gear system to ensure they have been performed. The POH or a maintenance technician familiar with the type of aircraft may be able to tell you about specific items to look for on the aircraft you fly.

You control the position of the landing gear through a switch in the cockpit. When you place the switch in the DOWN position, the gear extends and locks down. Conversely, when you move the switch to the UP position, the gear retracts and locks up inside the wells, and, on many aircraft, gear doors close to seal the gear from the airstream. The nosewheel steering linkage disengages from the nose gear, and the gear slides into a straightened position so that it fits into the wheel well.

As a general rule, you will retract the gear when you no longer have usable runway left on which to land after takeoff. On a long runway, if you decide to abort the takeoff with runway remaining, you will want to have the gear down so that you do not damage the airplane's fuselage upon touchdown. Once there is no more runway left to land on, having the gear up will usually result in an increased climb rate or better glide in case of engine failure.

Some gear switches incorporate a detent over which you must lift the switch in order to raise the gear. Also, many gear switches have a wheel shape, so that you can feel the difference between the flap switch and the gear switch. However, familiarize yourself with your aircraft thoroughly, as some airplanes have different flap/gear switch locations. This is also an important reason why you may want to avoid raising the flaps while you are still rolling out after landing, unless the procedure is called for in the aircraft's POH. When your attention is divided between taxiing, looking for traffic, and reconfiguring the airplane, you may accidentally raise the gear switch instead of the flap switch. [Figure 11-51}

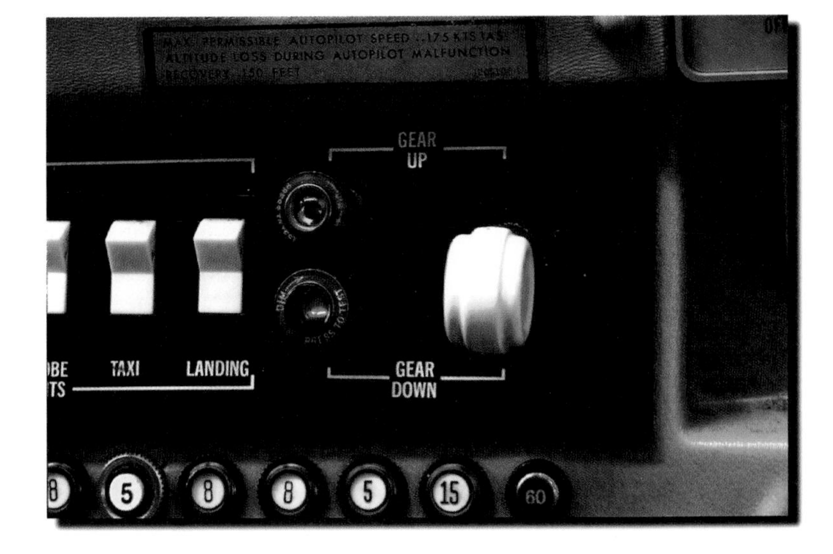

Figure 11-51. This gear switch is found on a Cessna 182RG and is typical of gear switches in many aircraft.

The gear takes several seconds to complete a cycle, so you should factor in this delay when you determine the point at which you will raise or lower the gear. The indicated airspeed will decrease when the gear is extended and increase as the gear is retracted. You also will notice that the airplane may change pitch attitude when the gear is extended, and you will need to adjust accordingly. Some aircraft demonstrate an opposite pitching moment when the flaps are extended, and operating procedures may recommend simultaneous deployment of the gear and flaps so that the aerodynamic forces offset each other and make for a relatively unchanged pitch attitude. [Figure 11-52]

Figure 11-52. The aerodynamic forces that occur with flap and gear deployment may offset each other. Nose down pitch created by the addition of 15° flaps is balanced by the nose up moment created by the extension of the landing gear.

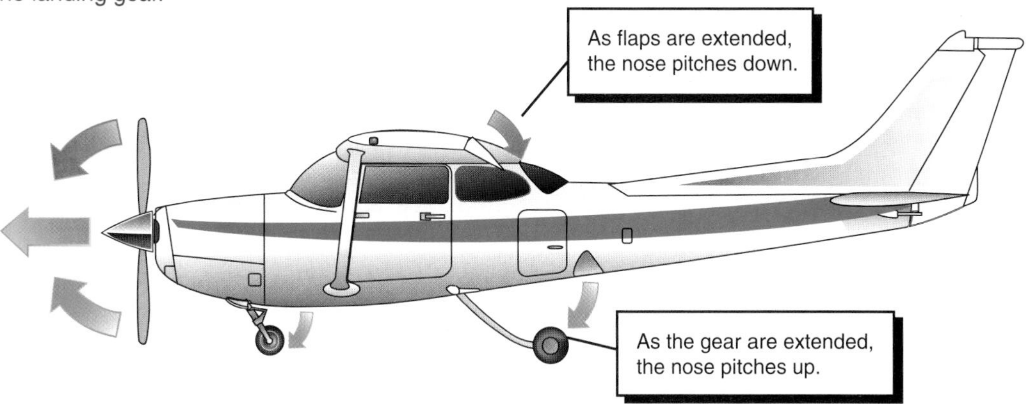

Circuit breakers protect electric landing gear circuitry from excessive loads. On some aircraft, one circuit breaker may serve to protect the landing gear circuit, while another shields the rest of the electrical system, including the gear warning system. The gear pump breaker may be pulled if the gear motor continues to operate after the landing gear has fully extended or retracted to prevent overheating. Other aircraft use a single breaker for the entire system. Because of the high amount of electrical power used in the gear cycle, the circuit breakers should be one of the first items you check in case the gear fails to operate. You should avoid pulling a circuit breaker simply to silence the gear warning horn, as you may disable an automatic extension system, or land with the gear up due to the lack of warning.

During cold weather, slush and water can splash on to the gear linkages and freeze. You should taxi slowly through snow, slush or water to prevent the gear from becoming wet, and avoid isolated areas of moisture on the ramp whenever possible. If the gear does become wet, cycle it several times after takeoff to keep the ice from adhering to and freezing moveable parts in the system. If you are unable to retract the gear because of ice accumulation, leave it down, return to the airport, and have the gear deiced.

Recycling the landing gear after takeoff from a slushy runway can help prevent the build-up of ice on the gear.

GEAR SYSTEM MALFUNCTIONS

The position indicator lights in the cockpit are your primary means of determining whether the gear has extended or retracted correctly, but the flight characteristics mentioned above are another clue as to the status of the gear. You can also listen for the gear motor, if you are able to hear it running from the cockpit on your aircraft. Some aircraft have mirrors installed on the wings or engine nacelles so that you can actually see the gear position from the cockpit. An observer on the ground may also be able to check for you, if you question whether or not the gear has extended or retracted properly.

If the position indicator lights fail to illuminate when you move the gear switch, there are several things to consider before you extend the gear manually. First of all, if you felt a change in pitch and/or airspeed, the gear may well be down, and one or more of the position indicator light bulbs is burned out. Replace the suspect bulb with a working one and check it again.

If you did not notice any other signs that the gear has operated, check the circuit breakers to see if an overload has caused a breaker to pop out. If this is the case, reduce the electrical load by turning off unnecessary equipment, reset the breaker, and try the gear switch again.

EMERGENCY GEAR EXTENSION

When you establish that the gear has failed to extend, you should be ready to use whatever type of **emergency landing gear system** is available in the aircraft. The most common kinds you will see are the hand crank, hand pump hydraulic, free fall, and carbon dioxide (CO_2) pressurized systems. Study the POH for the proper method and any caveats that apply to a particular system. Also, practice a manual gear extension with an instructor if it is feasible in your airplane. In some cases, it may be necessary to put the aircraft on jacks to reset the landing gear system after an emergency extension. If you practice an emergency extension, make sure that you stow the hand crank or pump (if used) properly, as failure to do so may keep the landing gear circuit open and leave you unable to extend or retract the gear normally.

A **hand crank system** is often found on aircraft with electric gear systems. The crank is employed to turn the gears that operate the actuators manually instead of through the electric motor. When you use this system, you will need to turn the crank a specific number of revolutions before the gear is fully extended. A good reason to practice this extension method in flight is that the position you must be in to turn the crank may be awkward, and the crank may be difficult to turn. In general, hand cranks should not be used to retract the gear in flight.

If you fly an aircraft with a hydraulic system, you may use a **hand pump hydraulic system** to extend the gear when the primary pump fails. You will supply the necessary hydraulic fluid pressure to the actuator cylinders by pumping a hand pump. With some systems, you will need to put the gear switch in the DOWN position before operating the pump. As with the hand crank system, hand pumps should not be used to retract the gear.

A **freefall system** is another method of gear extension used on aircraft that have hydraulic gear systems. A control lever is used to open a valve that releases the system pressure. Hydraulic pressure equalizes on both sides of the gear actuators and allows the landing gear to fall into position through the force of gravity and aerodynamic loads. On some aircraft, a spring assists in gear extension.

In the **carbon dioxide (CO_2) pressurized system**, compressed gas is used to apply the pressure to extend the gear if the hydraulic part of a gear system fails. When the CO_2 extension handle is pulled, pressure is released from a cylinder and directed through tubes to each landing gear actuator, which then extends the gear. Some CO_2 canisters may operate only once per flight, so consult your POH before extending the gear in this manner. You may need to be in a position where you can leave the gear down for the remainder of the flight. Other types of pressurized gas may be used in place of CO_2 for precharging the system. [Figure 11-53]

Hand Pump

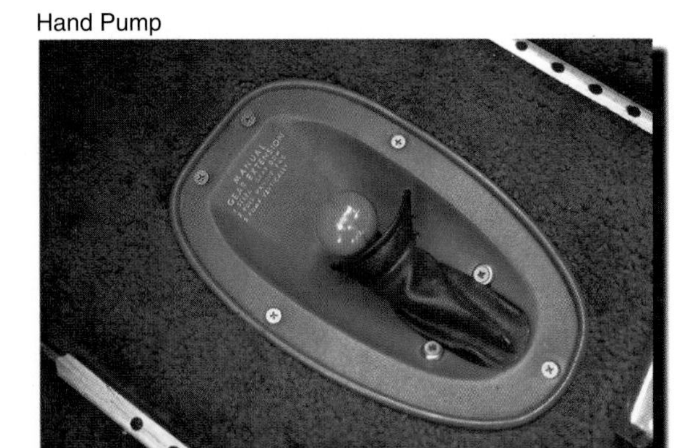

Figure 11-53. The procedures for emergency gear extension vary from aircraft to aircraft. By preparing for a manual gear extension before you need to perform one in flight, you make the operation safer, and less of an actual emergency situation.

Hand Crank

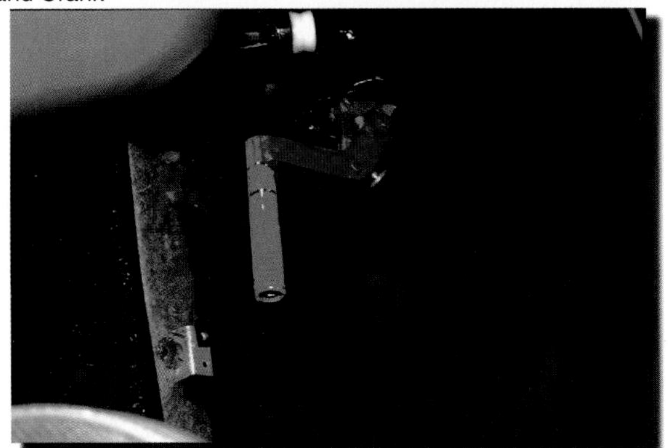

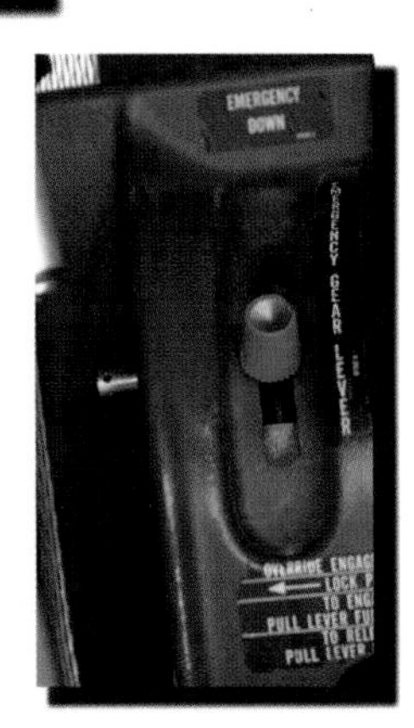

Freefall

N_2 Precharge System

SUMMARY CHECKLIST

✓ Landing gear systems may be operated through a hydraulic system, and electrically driven motor, or a hybrid of the two methods.

✓ The electrical gear system consists of a reversible motor that drives a series of rods, levers, cables, and bellcranks that raise and lower the gear.

✓ The gear cycle is the time that it takes the gear system to fully extend or retract the gear.

✓ In a hydraulic gear system, hydraulic fluid flows under pressure through valves and downlocks to the gear actuating cylinders to raise and lower the gear.

- ✓ An electrohydraulic system utilizes an electric pump to pressurize a hydraulic system for gear extension and retraction.
- ✓ Gear position indicators allow you to determine the location of the gear in the gear cycle. Position lights are one common gear position indicator, and they consist of three green lights and one red light.
- ✓ As a reminder to pilots to lower the gear before landing, a gear warning horn sounds when the aircraft is configured for landing with the gear unsafe.
- ✓ Most retractable landing gear systems incorporate a safety switch to ensure that the gear does not retract when the aircraft is sitting on the ground.
- ✓ Maximum landing gear extension speed (V_{LE}) is the fastest speed at which you can fly with the landing gear extended, and maximum landing gear operating speed (V_{LO}) is the highest speed at which you can safely operate the landing gear.
- ✓ The landing gear system should be thoroughly preflighted before each flight, including a check for loose bolts, cracks and corrosion, and foreign objects in the wheel wells.
- ✓ The gear switch is often shaped differently from the flap switch to aid you in discriminating between the two.
- ✓ Besides the gear position indicators, other flight characteristics exist which allow you to determine if the gear is operating properly. These include a pitch change, a change in airspeed, and the sound of the mechanism operating the gear.
- ✓ One or more circuit breakers protect the landing gear system from electrical overloads.
- ✓ Recycling the landing gear after takeoff from a slushy runway can help prevent the build-up of ice on the landing gear.
- ✓ The position indicator lights are your primary means of determining gear position. If one or more fail to illuminate, check the bulbs and the circuit breakers before attempting to lower the gear manually.
- ✓ The four most common kinds of emergency gear extension systems are the hand crank, the hand pump hydraulic, the freefall, and the carbon dioxide (CO_2) pressurized systems. Check the POH for your aircraft to determine the proper method for operating the system on your airplane.

KEY TERMS

Electrical Gear System

Gear Cycle

Hydraulic Gear System

Electrohydraulic System

Gear Position Indicators

Gear Warning Horn

Safety Switch

Maximum Landing Gear Extended Speed (V_{LE})

Maximum Landing Gear Operating Speed (V_{LO})

Emergency Landing Gear System

Hand Crank System

Hand Pump Hydraulic System

Freefall System

Carbon Dioxide (CO_2) Pressurized System

QUESTIONS

1. Which of these are methods of emergency gear extension?

   A. Hand crank, hand pump hydraulic, and electrohydraulic
   B. Hand pump hydraulic, freefall, and the gear warning horn
   C. Hand crank, hand pump hydraulic, and carbon dioxide system

2. True/False. An illuminated red indicator light generally means that the gear is unsafe for landing.

3. What is the symbol for the maximum speed you can fly with the landing gear extended?

   A. V_{LO}
   B. V_{LE}
   C. V_{SO}

4. How does the gear warning horn work?

5. Explain when you would retract the gear after takeoff.

6. What should you do to prevent ice from building on the wheels after takeoff from a slushy runway?

   A. Recycle the gear.
   B. Leave the gear extended for 10 minutes after takeoff.
   C. Use the hand crank or hand pump to manually retract the gear.

CHAPTER 12

AERODYNAMICS AND PERFORMANCE LIMITATIONS

Private Pilot
Part I, Chapter 3 — Aerodynamic Principles
Part IV, Chapter 8 — Predicting Performance
— Weight and Balance

SECTION A
ADVANCED AERODYNAMICS

Many of the concepts in this section will be familiar from your earlier training, but several others will present new information covering advanced aerodynamics. As a professional pilot, you will need a more detailed knowledge of the forces that govern your flight.

The four fundamental flight maneuvers — straight-and-level flight, turns, climbs, and descents — are controlled by changing the balance between the four aerodynamic forces: lift, thrust, drag, and weight. As you know, opposing aerodynamic forces are balanced in straight-and-level flight. Lift balances weight, and thrust balances drag. The airplane is in a state of dynamic equilibrium, and there is no acceleration in any direction. A change in any one of the forces will result in an acceleration until equilibrium is reestablished. For instance, you may have noticed that true airspeed increases throughout a long flight when you maintain a constant altitude and power setting. The total weight of the airplane goes down as fuel is burned, resulting in an imbalance between lift and weight. Gradually, you unconsciously reduce the angle of attack to maintain your altitude, reducing the amount of lift the wings generate, and restoring the balance between lift and weight. The decrease in lift results in a reduction in induced drag, creating an imbalance between thrust and drag. The excess thrust accelerates the airplane until increasing parasite drag again balances the two forces, and the airplane flies on at the higher airspeed. [Figure 12-1]

The four fundamentals in maneuvering an aircraft are straight-and-level flight, turns, climbs, and descents.

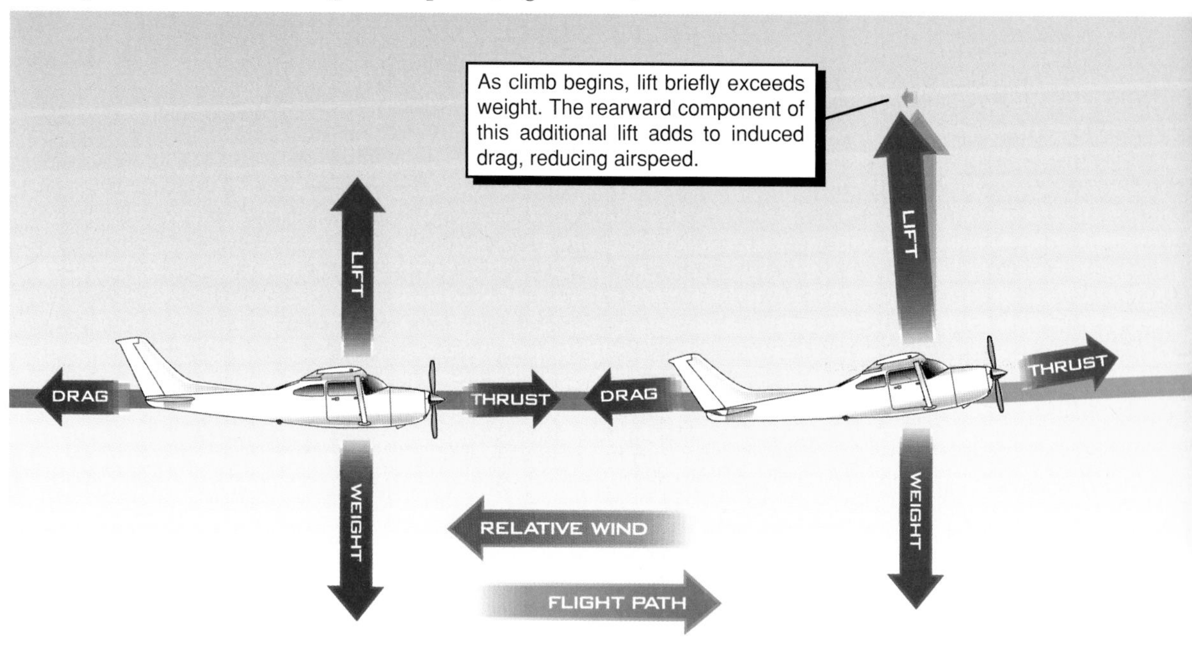

Opposing forces are equal in steady-state level flight.

The same principles apply in climbs and descents. If you create an imbalance between the forces, the airplane will accelerate until it reaches a new equilibrium. The airplane only accelerates during the initial change in direction. Once the climb or descent is established, the forces are again balanced, even though the flight path is now inclined. In a stabilized, constant airspeed climb, a portion of the thrust is acting upward, creating a component vector called the lift of thrust. In a stabilized descent, the thrust of weight becomes a factor. In both cases, upward and downward forces are balanced and forward and rearward forces are balanced. All forces are in equilibrium, as they are in straight-and-level flight.

LIFT

Formulated in the 17th century, Sir Isaac Newton's laws of motion are the foundation of classical mechanics. Two of his laws in particular are used to describe lift. Newton's second law of motion essentially says that a force results whenever a mass is accelerated. His third law states that for every action, there is an equal and opposite reaction. The wing causes the air moving past it to curve downward, creating a strong **downwash** behind the wing. As this mass of air is accelerated downward, the reaction force pushes the wing up. In level flight, this lift force is equal to the weight of the airplane. In other words, lift is simply the reaction that results from the action of forcing air downward.

Newton's laws describe what happens as air is accelerated downward, but what is the mechanism that causes the air to change direction? A flat plate could be used to deflect the air, but more efficient airfoils have been developed to maximize lift and minimize drag. In general, these airfoils are designed to smoothly and effectively bend the airmass in a curved path to create the downwash.

In normal flight, the air pressure on the bottom surface of the wing is somewhat greater than the pressure on the upper surface. When Newton's second law is applied to a fluid,

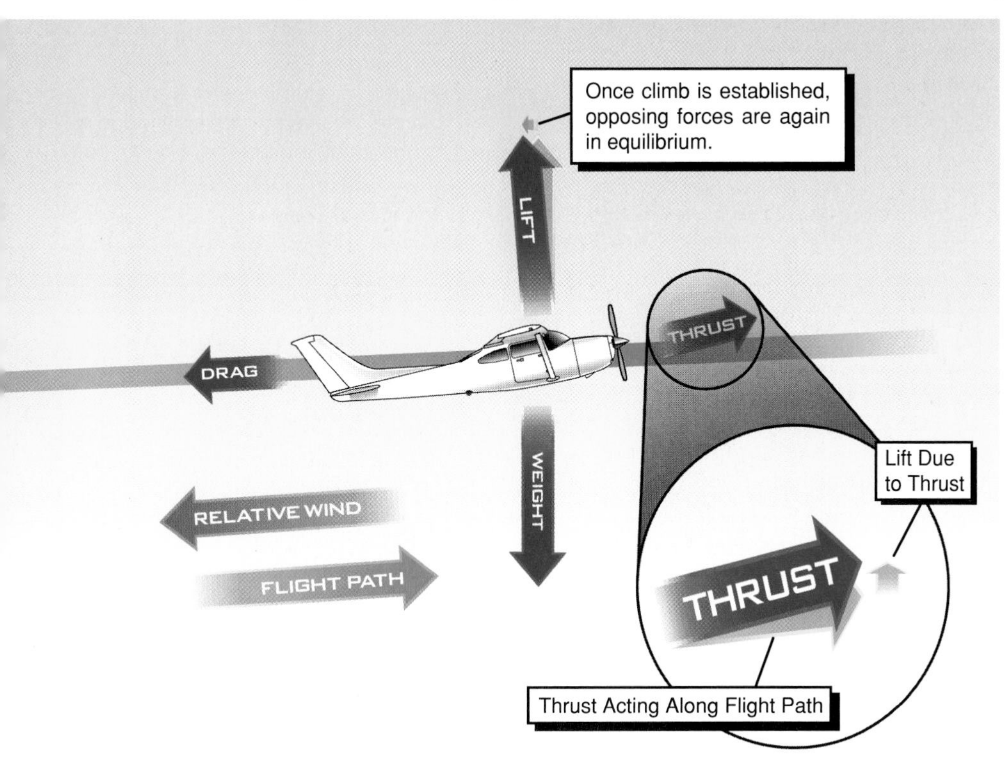

Figure 12-1. When you initiate a climb or descent, you momentarily create an imbalance in the four forces, which causes the airplane to accelerate until the forces reach equilibrium again. For example, to climb you increase the angle of attack, creating an excess of lift. If thrust remains constant, the increase in induced drag slows the airplane. This reduced airspeed decreases the amount of lift, bringing lift and weight back into balance. Because the thrust vector is now inclined upward, a portion is acting to lift the airplane.

such as air, the result is Bernoulli's equation. Bernoulli states that as the velocity of a fluid increases, its pressure decreases. The air pressure above a wing is reduced because the shape of the airfoil and the angle of attack cause air to flow faster over the upper surface of the wing. Likewise, as air moving along the lower surface of the wing slows down, its pressure increases, adding to lift. The net difference in pressure acting over the aerodynamic surfaces balances the weight of the airplane in straight-and-level flight. These pressure differences cause air to curve upward as it approaches the leading edge of the wing (upwash), and downward from the trailing edge as it leaves the wing.

Lift results from relatively high pressure below the wing's surface and lower air pressure above the wing's surface.

Relative wind is directly opposite the flight path of the airplane. Three of the four forces, lift, drag, and thrust, are associated with the relative wind. Weight, of course, always acts toward the center of the earth. The angle between the chord line of the wing and the relative wind is the **angle of attack**. Lift generally acts perpendicular to the relative wind, regardless of the wing's angle of attack. Drag acts opposite the flight path, in the same direction as the relative wind. The **angle of incidence** is the small angle formed by the chord line and the longitudinal axis of the airplane. In most airplanes, the angle of incidence is fixed and cannot be changed by the pilot. The angle of incidence generally provides the proper angle of attack for level flight at cruising speed so that the longitudinal axis of the airplane can be aligned with the relative wind, helping to minimize drag. [Figure 12-2]

Lift is defined as the force acting perpendicular to the relative wind. Drag acts parallel to the flight path, in the same direction as the relative wind.

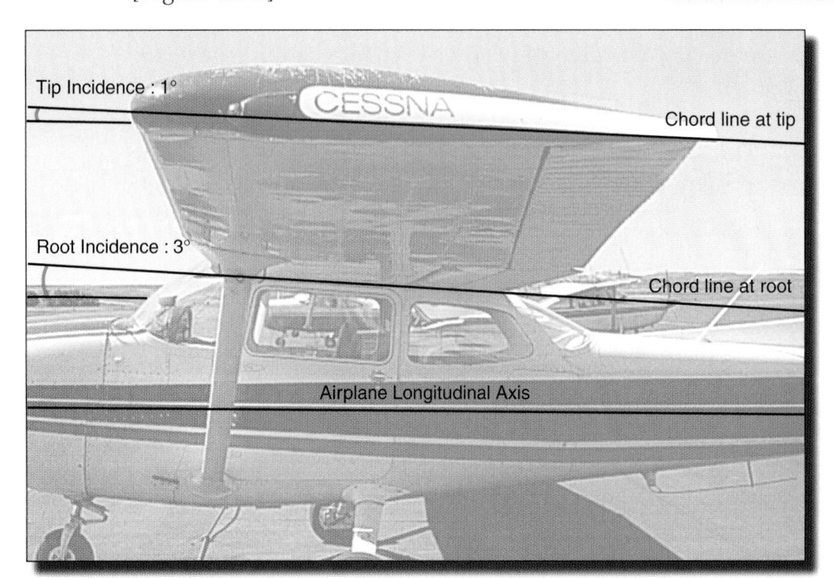

Figure 12-2. On many airplanes, the wings are built with less incidence at the tips than at the roots. The term for this is washout, and it helps ensure that the wing roots stall before the tips by placing the tips at a lower angle of attack. You can see the subtle twist from root to tip in this Cessna wing.

LIFT EQUATION

As you would expect, for any wing there is a definite mathematical relationship between lift, angle of attack, airspeed, altitude, and the size of the wing. These factors correspond to the terms coefficient of lift, velocity, air density, and wing surface area in the lift equation, and their relationship is expressed in the formula to the right.

$$L = C_L V^2 \frac{\rho}{2} S$$

L = Lift

C_L = Coefficient of lift

(This dimensionless number is the ratio of lift pressure to dynamic pressure and area. It is specific to a particular airfoil shape, and below the stall, it is proportional to angle of attack.)

V = Velocity (Feet per second)

ρ = Air density (Slugs per cubic foot)

S = Wing surface area (Square feet)

This shows that for lift to increase, one or more of the four factors on the other side of the equal sign must increase. Lift is proportional to the square of the velocity, or airspeed, so doubling your airspeed quadruples the amount of lift, if

everything else remains the same. Likewise, if the other factors remain the same while the coefficient of lift increases, lift will also increase. The coefficient of lift goes up as the angle of attack is increased, and will be discussed in more detail later. As air density increases, lift increases, but you will usually be more concerned with how lift is diminished by reductions in air density, for example, as you climb to higher altitudes or take off on a hot day. Air density varies with altitude, temperature, barometric pressure, and humidity. Finally, larger wings generate more lift, all other things being equal.

If the angle of attack and other factors remain constant and airspeed is doubled, lift will be four times greater.

There is a corresponding indicated airspeed for every angle of attack to generate sufficient lift to maintain altitude.

Another way of using the equation is to keep lift at the same value and notice how the factors on the other side of the equation affect each other. On a typical short flight, the weight of the airplane changes relatively little, so the amount of lift needed to support it stays about the same. To generate the same lift at a certain altitude with a reduced airspeed, the equation shows that you must increase the coefficient of lift, or angle of attack. Likewise, if you increase airspeed you will have to reduce the angle of attack to stay at the same altitude. You can see that for any particular amount of lift, there is a specific combination of angle of attack and airspeed. Similarly, since air density decreases with altitude, the airplane must fly with either a higher angle of attack or a higher speed to generate the same lift at a higher altitude.

To generate the same amount of lift at a higher altitude, an airplane must be flown at either a higher angle of attack or a higher true airspeed.

CONTROLLING LIFT

Four ways are commonly used to control lift. Increasing airspeed generates more lift by causing more air molecules to act on the wing, and changing the angle of attack changes the coefficient of lift. The two other methods of controlling lift consist of changing the shape of the airfoil or varying the total area of the wing. Flaps and leading edge devices are examples of how these methods are used in flight.

The most common method of controlling lift is to vary the angle of attack. This also allows you to control airspeed and drag to a certain extent. As you know, increasing the angle of attack causes the wing to generate more lift, but only up to a point. As the wing reaches a particular angle of attack, the airflow begins to separate, and the amount of lift begins to decrease. You can usually feel airframe buffeting as the airflow begins to separate just before a stall. This angle of attack is called C_{Lmax}. [Figure 12-3]

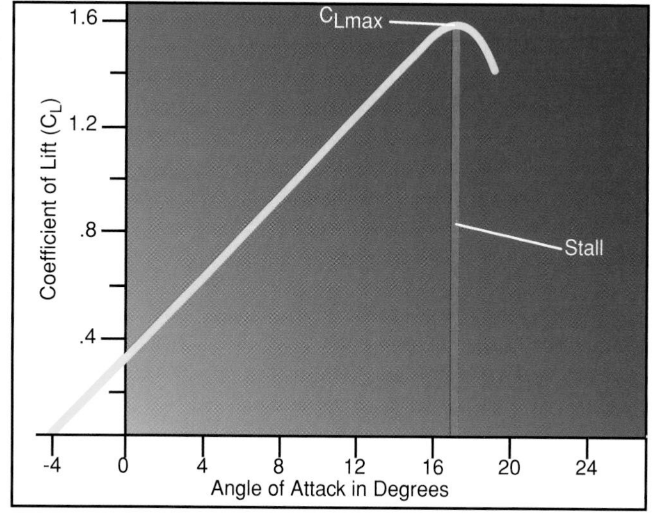

Figure 12-3. Notice that the coefficient of lift increases steadily until C_{Lmax} and then begins to drop as the wing stalls. For a given airfoil, a stall always occurs at the same angle of attack regardless of weight, dynamic pressure, bank angle, or pitch attitude. With this in mind, you can see that a wing can stall at a wide range of indicated airspeeds, and that stall speed is not a fixed value.

The angle of attack directly controls the distribution of pressures acting on a wing. By changing the angle of attack, you can control the airplane's lift, airspeed, and drag.

The increase in lift also causes an increase in induced drag, which slows the airplane unless thrust is also increased. Flight at slow airspeeds demonstrates this relationship clearly. As your airspeed decreases, the angle of attack has to increase to maintain the lift necessary for level flight. The angle of attack, however, can be increased only until C_{Lmax} is reached. At this point, you can no longer increase lift by increasing the angle of attack, and any further increase in angle of attack results in an abrupt loss of lift. In straight-and-level flight, C_{Lmax} normally is reached as the indicated airspeed approaches the published stall speed. Turbulence can cause abrupt changes in the direction of the relative wind and result in an increase in the stall speed. Stall speed also increases rapidly as G-loading increases during maneuvers such as steep turns. This is because the load factor multiplies the effective weight of the airplane, which must be balanced by a similar increase in lift from the wings, necessitating a higher angle of attack. High G-forces, such as those produced in violent or abrupt maneuvering, can cause a stall at speeds twice as high as the level flight, unaccelerated stall speed. The stall speed can be affected by a number of things such as weight, center of gravity location, load factor, and power. Finally, you should never attempt to stall an aircraft at a speed above the design maneuvering speed, because this can easily result in airframe structural damage.

The angle of attack at which a wing stalls remains constant regardless of weight, dynamic pressure, bank angle, or pitch attitude.

You usually think of changing angle of attack by changing the pitch attitude with the elevator control, but you may also change the angle of attack without changing pitch, for instance, by reducing power and holding your pitch attitude. In this situation, the airplane would slow, the wing would produce less lift due to the lower airspeed, weight would momentarily exceed lift and the airplane would begin to descend, changing the relative wind direction, and thus increasing the angle of attack.

If you were to determine the change in pressure due to the airflow over each square inch of the wing, you would find that most places have decreased pressure, but some places have increased pressure. This pressure distribution pattern changes as the angle of attack is varied. Adding the pressure vectors for the entire wing surface, top and bottom, gives a large single vector representing the net lift of the entire airfoil. This vector can be called the center of lift, or center of pressure. [Figure 12-4]

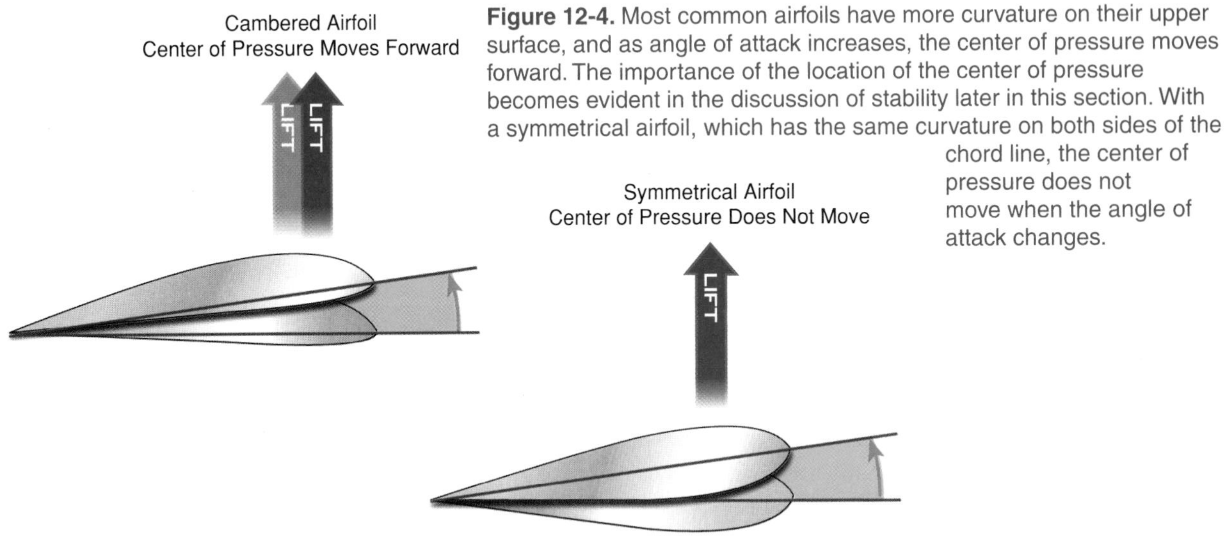

Figure 12-4. Most common airfoils have more curvature on their upper surface, and as angle of attack increases, the center of pressure moves forward. The importance of the location of the center of pressure becomes evident in the discussion of stability later in this section. With a symmetrical airfoil, which has the same curvature on both sides of the chord line, the center of pressure does not move when the angle of attack changes.

When the angle of attack of a symmetrical airfoil is increased, the center of pressure will remain unaffected.

At high angles of attack, pressure increases below the wing, and the increase in lift is accompanied by an increase in induced drag.

HIGH LIFT DEVICES

In addition to controlling lift with speed and angle of attack, wing design features may allow you to change the camber and the area of the wing through use of trailing-edge flaps and leading-edge high-lift devices. Some of these devices also delay the separation of airflow and lower the stall speed.

TRAILING EDGE FLAPS

The most common high-lift device is the trailing-edge flap, which can produce a large increase in airfoil camber, increasing both lift and drag. Using flaps can add to your rate of descent without increasing airspeed. Consequently they are often used to make steeper landing approaches. Since flaps expand the coefficient of lift, they also decrease the stall speed. On the other hand, raising the flaps increases the stall speed — something you may want to remember before you raise them at low airspeeds or high angles of attack. Four basic types of flaps are found on general aviation airplanes. [Figure 12-5]

Flaps increase lift and reduce stall speed, allowing the wing to produce the same lift at a lower airspeed. Conversely, raising the flaps increases stall speed.

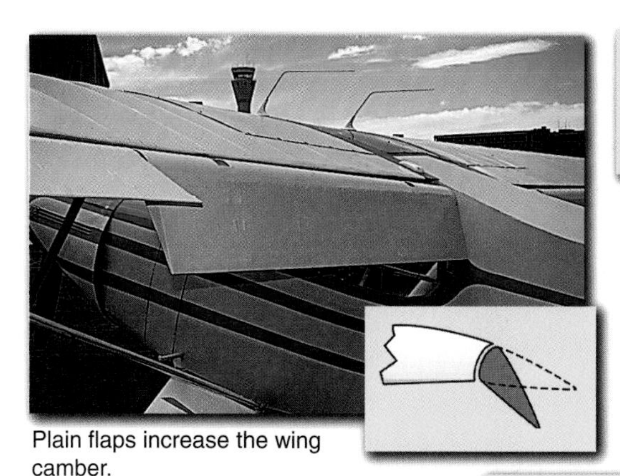

Plain flaps increase the wing camber.

Fowler flaps increase camber, increase wing area, and help prevent airflow separation.

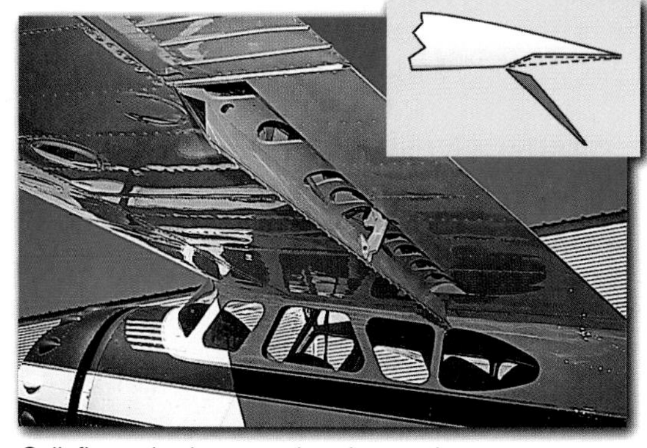

Split flaps also increase the wing camber.

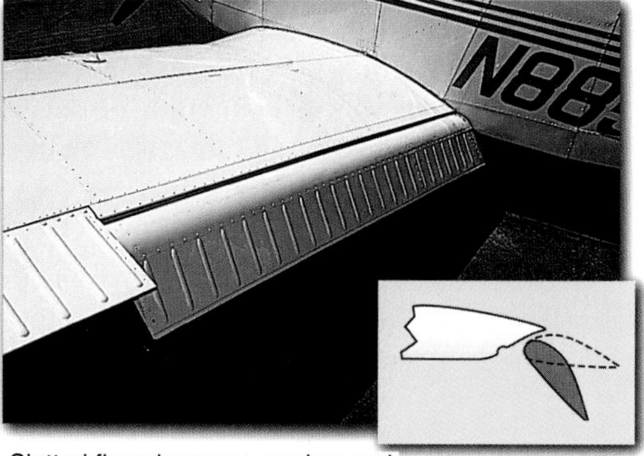

Slotted flaps increase camber and help prevent airflow separation.

Figure 12-5. Flaps increase lift (and drag) by increasing camber, which also raises the angle of attack. In some cases, flaps also enlarge the area of the wing.

As you can see, plain, split, and simple slotted flaps change the effective camber of the airfoil from a high speed, low drag configuration to a shape that is more effective at low speeds, but with a high penalty in additional drag. Moreover, flaps with slots also allow some of the high pressure air to come to the upper surface of the wing, which increases lift by accelerating the air over the top of the wing, thus reducing the pressure there. Slotted flaps also improve flow over the trailing edge, delaying separation of the airstream and reducing the stall speed. Because of the way Fowler flaps are designed, they increase the area of the wings as they move down and back. Adding to the wing area also increases lift, while imposing less of a drag penalty. Some airplanes have Fowler flaps with multiple slots, further modifying the airfoil camber, wing area, and airflow. To increase the camber of the outer portions of the wings, some airplanes are equipped with ailerons that move simultaneously downward a few degrees as the flaps are extended to further increase lift at low speeds. The mechanism allows the "drooped" ailerons to function normally for roll control. Because of the aerodynamic stresses on the flaps and their support structure, the maximum operating speed with flaps extended is typically lower than the maximum cruising speed for a particular aircraft.

Because flaps increase lift and reduce the stall speed, many airplanes use them for takeoff as well as landing. Used properly, they can reduce ground roll and increase climb performance. In general, the first few degrees of flap extension provide the greatest lift benefit with the least penalty in drag, while the last few degrees of travel add relatively little lift and a great deal of drag. Permissible use of flaps on takeoff varies from airplane to airplane, so always comply with the limitations in the POH.

LEADING EDGE DEVICES

The leading edges of the wings can also be modified in various ways to increase airfoil camber, add more wing area, or delay airflow separation at high angles of attack. The simplest leading edge device is a fixed **slot**. When slots are incorporated into the wing structure near each wingtip, they serve a purpose similar to washout, allowing the airflow to remain attached over the outer portion of the wings after the roots have stalled. When the wing is at low angles of attack, relatively little air flows from the bottom of the leading edge through the opening, so slots do not cause a substantial change in the effectiveness of the airfoil. Slots are somewhat more effective than washout, since the entire wing can be placed at the most efficient angle of attack for a given flight situation. With washout, whenever the roots are at the ideal angle of attack, the tips are at a less efficient angle.

Slats are portions of the leading edge which are moved forward and down to create a path for air similar to a slot. On many types of aircraft they are deployed automatically by aerodynamic pressures. Other aircraft use electrical or hydraulic devices to extend them mechanically. The advantage of slats is that the wing is clean when they are retracted, yet the benefits of slots are available when needed. Most slats also increase the effective wing area, providing additional lift as well as delaying airflow separation. The main disadvantages are the weight and complexity of the mechanism.

Leading edge flaps usually increase both wing camber and area. They are generally only found on jets, due to their expense, weight, and complexity, and because the airfoils of jets, which are optimized for high speeds, need them to create adequate lift at relatively low takeoff and landing speeds. [Figure 12-6]

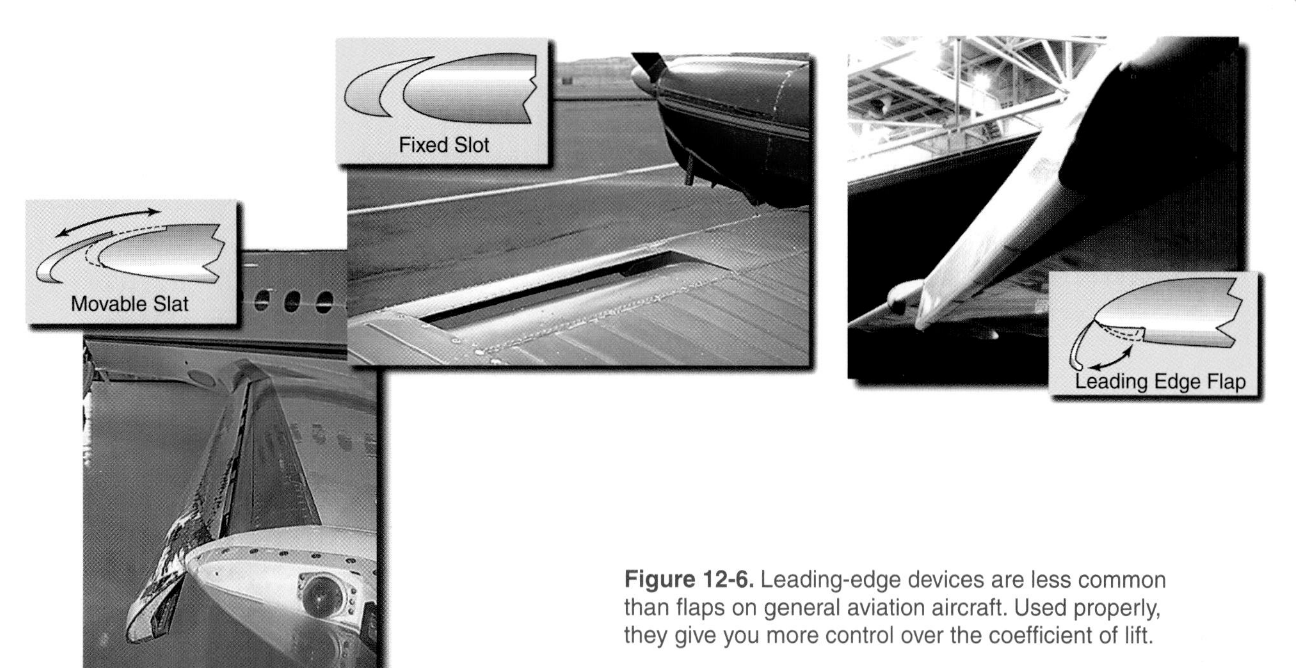

Figure 12-6. Leading-edge devices are less common than flaps on general aviation aircraft. Used properly, they give you more control over the coefficient of lift.

DRAG

Reducing the drag force means less thrust is required to fly at a given airspeed, so minimizing drag usually makes an airplane more efficient. Likewise, if drag is reduced, an airplane can go faster with the same thrust. Ordinarily, drag is separated into two types. Induced drag results from the production of lift by the wings, and decreases as the airplane goes faster. All other drag is classified as parasite drag, which increases as the square of the speed.

INDUCED DRAG

Whenever a wing is producing lift, drag is created as a by-product. The greater the angle of attack, the more drag is produced. At low speeds a wing must fly at a higher angle of attack to generate enough lift to support the airplane, and more high-pressure air from the lower surface comes around the wingtips, forming more powerful vortices, which are the primary cause of induced drag. The action of the vortices changes the local relative wind for the portion of the wing nearest the tip, which essentially reduces the angle of attack at the wingtips. At higher speeds, a lower angle of attack is sufficient to generate the necessary lift, so there are correspondingly less powerful wingtip vortices, and less induced drag. [Figure 12-7]

Induced drag is a by-product of lift and is greatly affected by changes in airspeed.

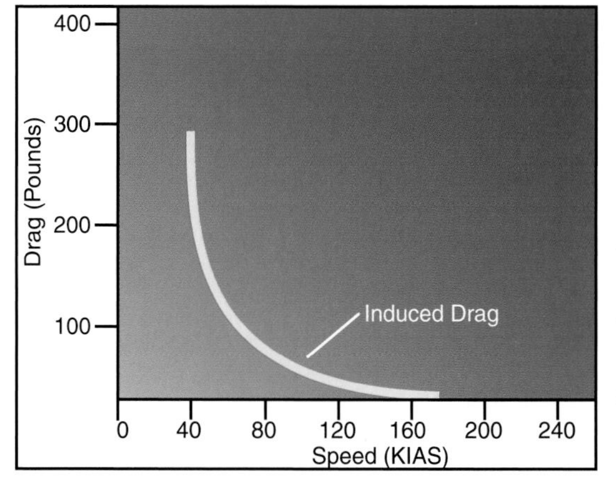

Figure 12-7. At low speeds, induced drag is at its maximum, and it diminishes as speed is increased.

WING PLANFORM

The airfoil is only one characteristic of the wing's design. Planform is the term that describes the wing's outline as seen from above. Many factors influence the final shape of the planform, including the primary purpose of the airplane, the load factors and speeds anticipated, costs of construction and maintenance, maneuverability, stability, stall and spin characteristics, and whether or not the wings

will house fuel tanks, high lift devices, landing gear, etc. The wings may be tapered, the leading or trailing edges may be straight or curved, and wingtips may be square or rounded. There are advantages and disadvantages to each configuration, and many aircraft combine the features of multiple planforms to achieve the desired flight characteristics. [Figure 12-8]

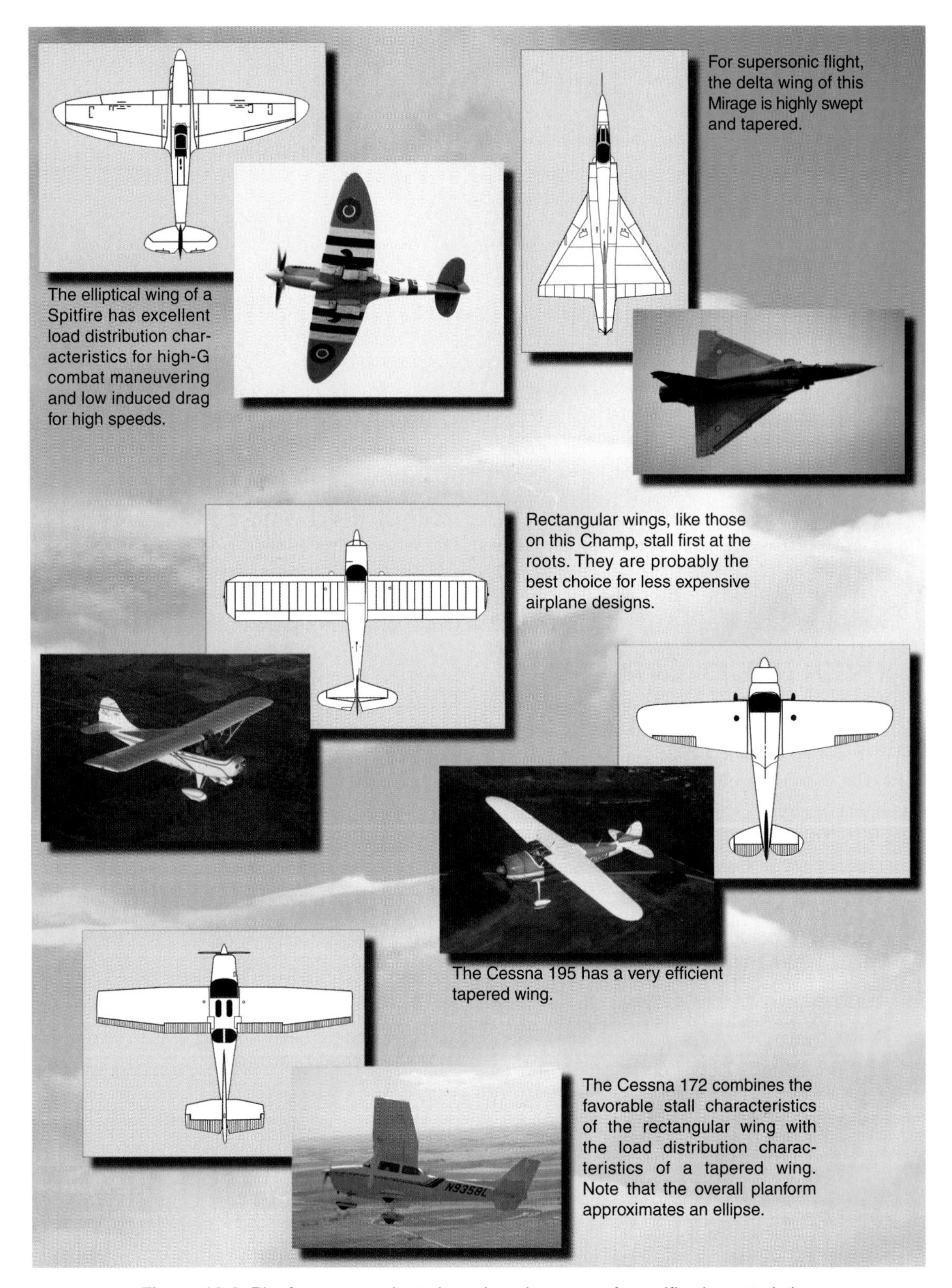

Figure 12-8. Planforms are selected to take advantage of specific characteristics.

The ratio of the root chord to the tip chord is the **taper** of the wing. Rectangular wings have a taper ratio of one, and delta wings have a taper of zero. One of the first things the early aeronautical engineers realized is that the area of a rectangular wing nearest the fuselage does most of the lifting, while the wingtips provide relatively little lift. They learned that tapering the wing saved precious weight and distributed the load more efficiently. The downside was that each wing rib needed to be a different size, adding to the time and cost of production. Rectangular wings are simpler and more economical to produce and repair, since the ribs are the same size. Another benefit of the rectangular wing is that the wing roots tend to stall first, providing more warning to the pilot and more control during stall recovery. The planform that provides the best spanwise load distribution and the lowest induced drag, at least for subsonic airplanes, is an ellipse. The price for this efficiency is that the whole wing stalls at about the same time, which is undesirable compared to wings that stall progressively from root to tip. Elliptical wings also are among the most difficult, expensive and complex to build, so they are employed primarily on very efficient airplanes such as the Supermarine Spitfire and Culver Cadet.

Rectangular wings have a tendency to stall first at the wing root, with the stall progression toward the wingtip.

Dividing the wingspan by the average chord gives the **aspect ratio**. With some wing planforms, the average chord can be difficult to determine, so the formula of wingspan squared over area is often used, but the result is the same. Aspect ratio is especially important in reducing the amount of induced drag produced by a wing. In fact, doubling the aspect ratio cuts the induced drag coefficient in half. Increasing the wingspan while keeping the wing area the same results in smaller wingtips, which generate smaller wingtip vortices. The influence of these vortices on the local relative wind affects a smaller fraction of the wing surface, further reducing induced drag. Perhaps you have noticed that airplanes renowned for their efficiency, such as sailplanes, long-range transports, high altitude research airplanes, and human-powered airplanes all have high aspect ratios. On the other hand, airplanes that require extreme maneuverability and strength, like those designed for aerobatic competition or air combat maneuvering, generally have wings with a low aspect ratio. [Figure 12-9]

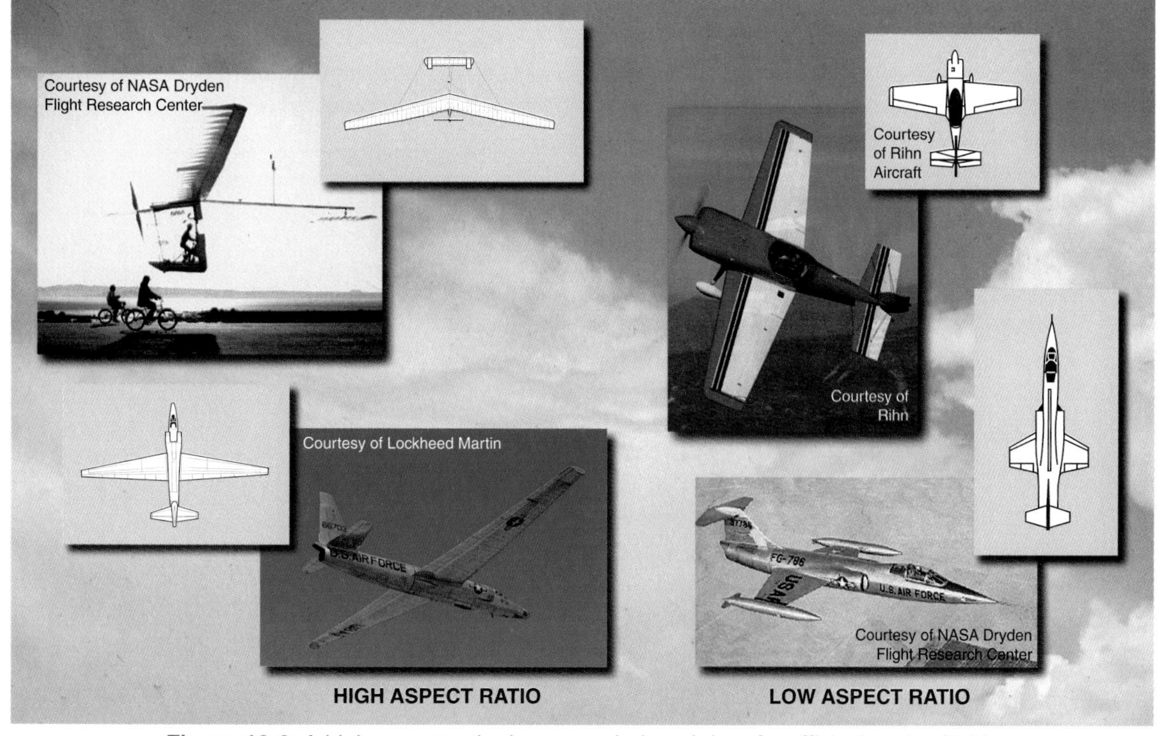

Figure 12-9. A high aspect ratio decreases induced drag for efficient cruise flight.

Another element of wing planform is **sweep**. If a line connecting the 25% chord points of all the wing ribs is not perpendicular to the longitudinal axis of the airplane, the wing is said to be swept. [Figure 12-10] The sweep can be forward, but almost all swept wings angle back from root to tip. Most swept wing designs were developed to delay the onset of compressible airflow problems and control the shockwaves that form as an airplane flies near the speed of sound and beyond, but a number of relatively low-speed airplanes employ swept wings because of their contribution to lateral stability.

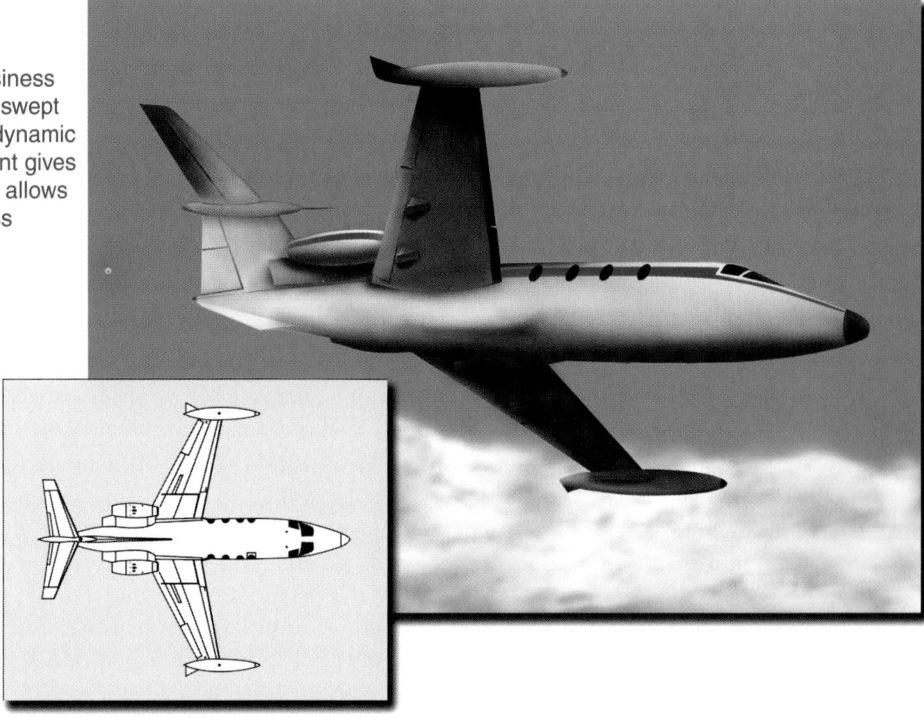

Figure 12-10. The Hansa business jet of the 1960s used forward swept wings. In addition to the aerodynamic benefits, this wing arrangement gives passengers a better view and allows the wing spar structure to pass behind the cabin.

Winglets

Wing tip vortices cost airplanes a great deal in efficiency and performance, and over the years there have been many attempts to block or diffuse them. One of the most effective devices so far seems to be the winglet developed by Richard Whitcomb of NASA. These nearly vertical extensions on the wingtips are actually carefully designed, proportioned, and positioned airfoils with their camber toward the fuselage, and with span, taper, and aspect ratio optimized to provide maximum benefit at a specific speed and angle of attack. On most jets, this is the airplane's cruise speed, but some turboprops use winglets to improve lift and reduce drag at low speeds. The winglet combines many small factors to increase performance. Downwash from the trailing edge of the winglet blocks or counteracts the vortices, and even the vortex from the tip of the winglet is positioned to affect a portion of the main wingtip vortex. The leading edges of many winglets are actually toed out about 4°, but because of the relative wind induced by the wingtip vortex, the winglet is actually at a positive angle of attack, and part of the lift generated by the winglet acts in a forward direction, adding to thrust. Many winglets are canted outward around 15°, which adds to vertical lift and increases their aerodynamic efficiency while also contributing to dihedral effect. Depending on the application, performance improvements due to winglets can increase fuel efficiency at high speeds and altitudes by as much as 16 to 26%.

Courtesy of Learjet Inc.

GROUND EFFECT

An interesting thing happens when an airplane flies within a distance from the ground (or water) surface equal to its own wingspan or less. The amount of induced drag decreases, due to changes in the upwash pattern ahead of the wing, and the downwash and wingtip vortices behind the wing. Induced drag is only about half of its usual value when the wing is at 10% of its span above the ground. With the reduction in induced drag, the amount of thrust necessary is also reduced, allowing the airplane to lift off at a lower-than-normal speed. It may feel as though you are getting something for nothing, but many pilots have gotten into trouble when using ground effect, either intentionally or unintentionally. If you try to climb out of ground effect at too low an airspeed, as normal induced drag values accumulate the airspeed may decrease, resulting in a stall. As an airplane climbs out of ground effect, more thrust will be required. Since ground effect may allow an overloaded or improperly configured airplane to lift off, the pilot may believe the airplane will climb when it may only be capable of flying in ground effect until it hits something. Ground effect is also responsible for the floating effect you may have encountered during the landing flare. Since the wing is able to create more lift at the same angle of attack in ground effect, the pitch angle may need to be reduced slightly to maintain a descent. Simultaneously, thrust will need to be reduced to continue slowing the airplane for landing. On a short field, it is possible that ground effect could cause the airplane to float so far down the runway that insufficient room would remain to stop after touchdown. [Figure 12-11]

An airplane leaving ground effect will experience an increase in induced drag and will require more thrust. If the same angle of attack is maintained, an airplane will have more lift and less induced drag when in ground effect, and, conversely, to generate the same lift within ground effect requires a lower angle of attack.

Figure 12-11. Proximity to the ground reduces upwash, downwash, and wingtip vortices, decreasing induced drag. Ground effect is greatest close to the surface and decreases rapidly with height, becoming negligible at an altitude equal to about one wingspan.

PARASITE DRAG

While induced drag diminishes with speed, parasite drag increases. The rate of increase is proportional to the square of the airspeed, so doubling your speed quadruples parasite drag. [Figure 12-12] This kind of drag is easy to understand; it is the force you feel when you put your hand out the window of a moving car. Parasite drag can be subdivided into form drag, interference drag, and drag due to skin friction. Form drag is based on the shape of the airplane, how well it is streamlined, and how much frontal area it has. For

example, an airplane with an abrupt, steep windshield will have more form drag than one with a shallow-angled, smooth-flowing windshield design, even if the frontal areas are identical. Likewise, if you had a small airplane with exactly the same overall shape as a larger airplane, the larger one would have more form drag because of its greater frontal area. Even the wing creates a certain amount of parasite drag. If you flew the wing at an angle of attack that produced zero lift, it would produce no induced drag. The form drag of the wing itself would still exist, however, and would increase with speed. Everything that sticks out into the airstream creates form drag: antennas, struts, landing gear, entry steps, door handles, and so on. Streamlining is more important than you might think. A bracing wire with a round cross section has about ten times the drag of an equal-strength wire with a streamlined, teardrop-shaped cross section. Drag reduction is the sole aerodynamic purpose of retractable landing gear, and the drag of fixed gear can be reduced considerably by enclosing the gear legs in streamlined fairings and adding wheel pants around the tires.

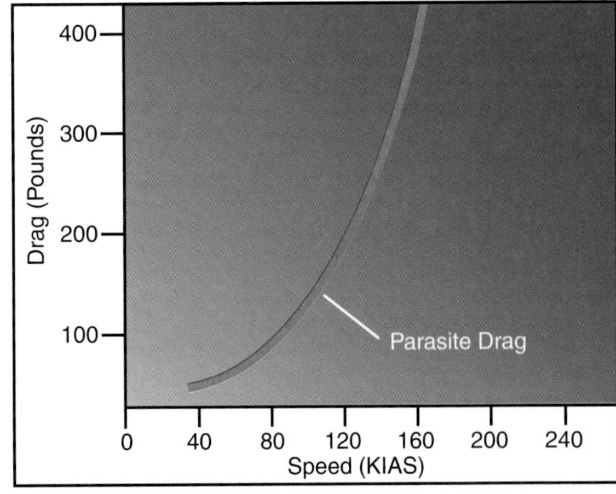

Figure 12-12. Doubling your airspeed will result in four times the parasite drag. This is the same as the relationship of speed to lift. The rapid increase in parasite drag at higher speeds is a major determinant of an airplane's top speed.

Interference drag is created when the airflow around one part of the airplane interacts with the airflow around an adjacent part. The greatest amount of interference drag usually occurs where the wing joins the fuselage of a low wing airplane, since air is traveling at a higher speed across the top of the wing than the air flowing along the fuselage. Many high performance low wing airplanes have large wing root fillets to smooth the transition and reduce interference drag. [Figure 12-13] On high wing airplanes, the high-speed air above the wings does not interfere as much with the air from around the fuselage, so they usually do not need fillets.

A

B

C

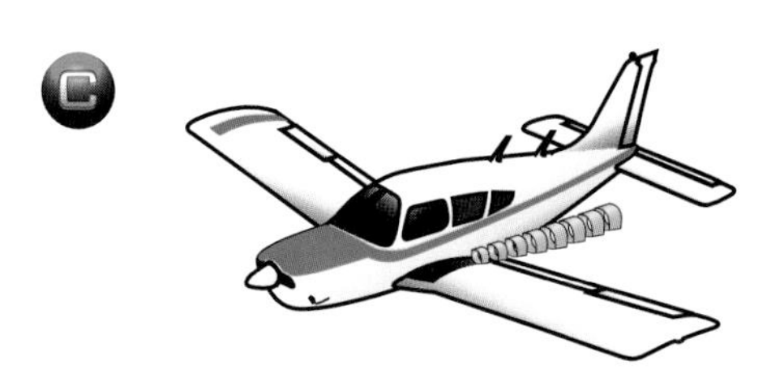

D

Figure 12-13. To understand interference drag, consider the airflow around a wing by itself [A]. Now look at the airflow around the fuselage [B]. These two separate flow patterns interact with each other when the wings are attached to the fuselage [C], the high-speed, low-pressure air flowing over the wing must mix with the relatively low-speed, higher-pressure air flowing around the fuselage. Every element of the airframe has its own flow pattern. Where these flows mix and interact, interference drag is created. Wing fillets help reduce interference drag [D]. Smaller versions can be used to blend tail surfaces and struts into the airframe.

Parasite drag increases in proportion to the square of the airspeed, thus doubling the airspeed will quadruple parasite drag.

Add One of These to Your Airplane and Decrease Drag 75%

The National Advisory Committee for Aeronautics (NACA) was created in 1915 to "to supervise and direct the scientific study of the problems of flight, with a view to their practical solution." This government-sponsored research organization has solved many major aerodynamic problems over the years. One of their developments was the NACA cowl, which reduced airframe drag by about 75% on radial engine aircraft. It allowed higher speeds and better performance on a variety of airplanes, from fighters such as the Grumman F3F to private aircraft such as the Stinson Reliant. The NACA cowl is shown here on a Cessna 195.

In 1958, NACA became the National Aeronautics and Space Administration (NASA). Although primarily known for their extraordinary successes in space exploration, they continue to perform fundamental research in aerodynamics that benefits all sectors of aviation.

Skin friction drag is due to air molecules giving up some of their kinetic energy as they contact the skin surfaces of the airplane. Both the total wetted area of the airplane and the degree of smoothness or roughness of the skin surface affect this kind of parasite drag. Wetted area is a term for the total outer surface area of the airplane. As you might guess, it comes from marine engineering, where drag on ship hulls is determined based on the part that is under water, or wetted.

TOTAL DRAG

As you would expect, adding the induced drag to the parasite drag gives **total drag**. Because induced drag decreases as parasite drag increases, the curve for the sum of both types of drag usually looks something like figure 12-14.

Total drag is lowest at the airspeed which produces the highest ratio of lift to drag (L/D_{max}). At airspeeds below L/D_{max}, total drag increases due to induced drag, and at speeds above L/D_{max}, total drag increases because of parasite drag. See figure 12-14.

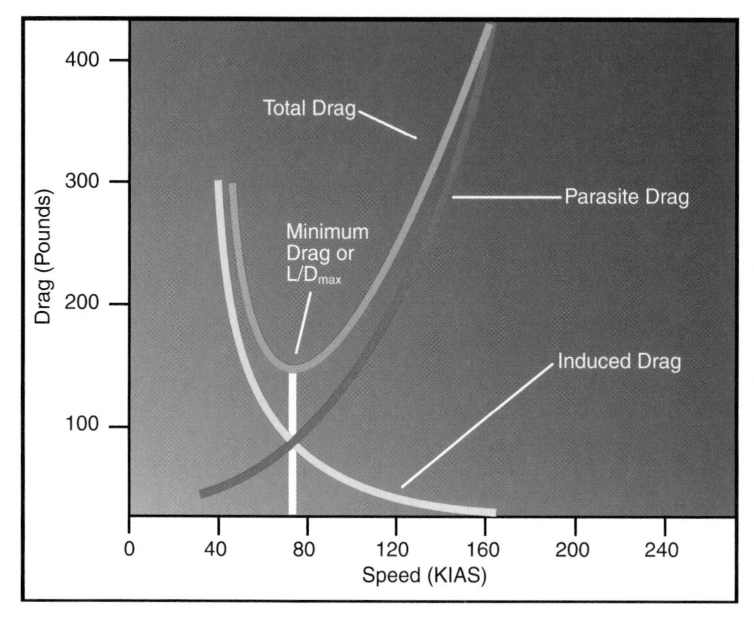

Figure 12-14. At low speeds, induced drag from the production of lift contributes most to the total drag, but at high speeds most drag is parasite drag. The point where the total drag is lowest is L/D_{max}. This is also the power-off glide speed that provides maximum gliding distance.

The Eiffel Tower and the Dimples On Golf Balls

Almost everyone recognizes the Eiffel Tower, but did you know that it was used for aerodynamic research? Alexandre Gustave Eiffel was a pioneering aerodynamicist, and the tower was utilized in 1910 to measure the air resistance of flat plates. Monsieur Eiffel also formulated many of the basic techniques of wind tunnel testing, and did research on the aerodynamic characteristics of spheres. He came up with a drag coefficient for the sphere that was considerably different from the value found by some other researchers, but both values turned out to be correct. The difference is reconciled by taking into account the turbulent flow characteristics described by another aerodynamic pioneer, Osborne Reynolds. The Reynolds number is a dimensionless ratio that relates the viscosity of a fluid with velocity and distance. It helps to predict where a smooth, laminar flow will begin to transition to a turbulent flow as air flows over a surface. It is extremely important in many aspects of aircraft design, but also helps to explain why golf balls with dimples travel more than twice as far as smooth balls of identical size and weight.

Whenever anything moves through the air, the air molecules close to the surface of the object are slowed down by friction, creating a slower-moving *boundary layer.* At the front of an object, a ball for example, the boundary layer is smooth, and the air outside the boundary layer slips over with minimal drag. But farther back, the flow separates from the ball, creating large amounts of drag. It is relatively easy to separate the laminar flow, so it breaks away near the widest part of a smooth ball. If the ball has dimples, the laminar flow is broken up near the front of the ball, but the turbulence causes the airflow around the ball to resist separation, so that it sticks to the ball further back along the trailing side, thus reducing the diameter of the drag-producing wake. It seems contradictory to say that introducing turbulence to an airflow, which causes drag, could actually reduce drag, but the amount of drag caused by disrupting the laminar flow is relatively small compared with the drag of the larger wake of a smooth ball.

Maximum Range

It stands to reason that the airspeed that produces the best power-off glide range would also be the most efficient in terms of fuel economy. If you fly at the airspeed for L/D_{max}, you will go the maximum possible distance for the amount of fuel you burn. Efficiency does not equate with speed, however. It is also important to remember that the airspeed for L/D_{max} varies with the weight of the

The airspeed that gives the lowest total drag (L/D_{max}) will provide the best power-off glide range, as well as the greatest range. See figure 12-14. As aircraft weight decreases, the airspeed for L/D_{max} also decreases.

airplane. The angle of attack for L/D_{max} does not vary significantly. Since weight determines the amount of lift that must be generated for level flight, the airspeed to produce that lift varies with weight.

HIGH DRAG DEVICES

Although engineers put a lot of effort into minimizing drag, they have also created devices solely to produce more drag. **Spoilers** are deployed from the upper surfaces of wings to spoil the smooth airflow, reducing lift and increasing drag. Spoilers are used for roll control on some aircraft, one of the advantages being the virtual elimination of adverse yaw. To turn right, for example, the spoiler on the right wing is raised, destroying some of the lift and creating more drag on the right. The right wing drops, and the airplane banks and yaws to the right. Deploying spoilers on both wings at the same time allows the aircraft to descend without gaining speed, and is the principle method of glide path control in many gliders and sailplanes. Spoilers are also deployed to help shorten ground roll after landing. By destroying lift, they transfer weight to the wheels, improving braking effectiveness. Dive bombers from the 1930s were equipped with dive brakes to permit extremely steep dives without unmanageable speed increases. During World War II, high performance airplanes were approaching the realm of transonic flight, where the effects of compressibility would cause buffeting, difficulty in recovering from dives, and loss of control. Special dive recovery flaps were created for the Lockheed P-38 Lightning and Republic P-47 Thunderbolt, among others. These flaps, located on the bottom surfaces of the wings near the leading edge, assisted in the recovery from high-speed dives.

Propeller-driven airplanes have an automatic high drag device whenever the throttle is retarded beyond the point where zero thrust is produced. At less than zero thrust, energy from the airplane's forward motion is being used to turn the propeller and engine, creating a large amount of drag. This is why most multi-engine airplanes provide a means of feathering the propeller blades on a failed engine. The advent of jet aircraft increased the need for drag-producing devices for a couple of reasons. First, aerodynamically clean jet airframes just did not slow down quickly enough when thrust was reduced. Pilots who were used to losing speed as soon as power was reduced had trouble adjusting to the slippery new jets. Also, the early jet engines took a relatively long time to go from idle to full thrust. If full thrust was needed for a go-around, for instance, several seconds could be lost waiting for the engine thrust to become effective. As a result, designers added drag-producing devices to manage airspeed so that pilots could maintain higher thrust settings during an approach. As jet engine technology progressed, spool-up time has improved, however, jet engines are still less responsive than reciprocating engines, and speed-retarding devices are useful. **Speed brakes** are found in many sizes, shapes, and locations on different airplanes, but they all have the same purpose — to increase drag and provide fairly rapid deceleration. [Figure 12-15]

Figure 12-15. Speed brakes [A] are carefully planned to avoid undesirable pitching or trim changes. Spoilers [B] can be used for roll or glide path control.

THRUST

Thrust is the force that opposes drag, and whenever there is more thrust than drag, the airplane accelerates along the flight path until the increasing drag restores the equilibrium. In conjunction with drag, thrust is the factor that limits the top speed of an airplane. Most airplanes use power generated by their engines to turn propellers, converting rotational energy into thrust. Jet engines produce thrust directly, through the expansion of burning gases. Interestingly, jets convert some of that thrust into rotational energy to turn their compressors and accessories.

The difference between thrust and power is fundamental to an understanding of aircraft performance. Thrust must equal drag in level flight. In other words, a pound of thrust must be present for each pound of drag at any airspeed. Parasite drag increases as the square of airspeed, so the thrust necessary for level flight at 150 knots will be four times the thrust needed at 75 knots, less whatever reduction there is in induced drag. On the other hand, power is the rate at which work is done, so it is tied directly to speed. It takes less power to do the same amount of work at a slower rate. If an airplane had the same amount of drag at low speed as at high speed, it would still take more power to fly at high speed, because the work of overcoming drag would be done at a faster rate. Of course, drag increases at any speed higher than L/D_{max}, so more thrust is required in addition to more power. When using our familiar units of pounds, knots, and horsepower, the power required for level flight is defined as thrust required times velocity over 325. (325 is simply a constant to convert horsepower in feet per second to knots.) If your airplane has 300 pounds of drag force at 100 knots, it requires 92.3 thrust horsepower to maintain level flight. At 125 knots, the drag force will be nearly 469 pounds, and about 180 thrust horsepower would be required. In this example, increasing speed by 25% requires a 95% increase in power to overcome a 53% increase in drag. [Figure 12-16]

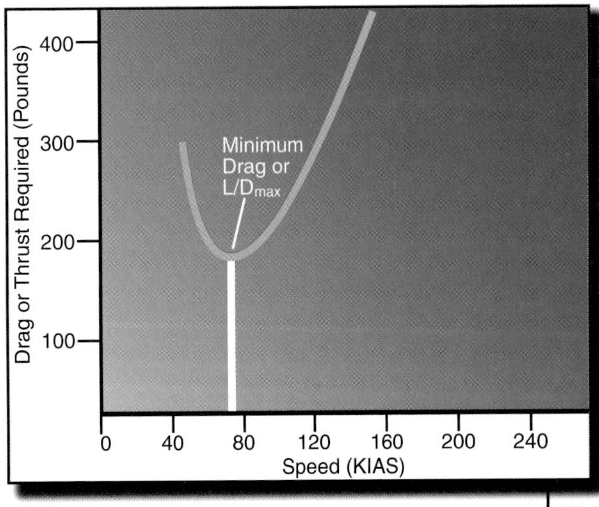

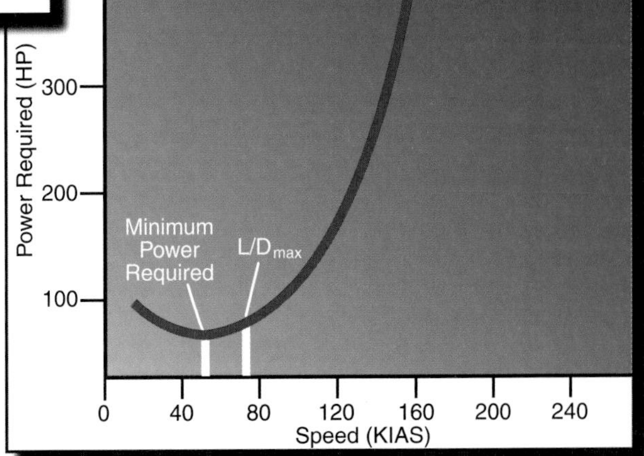

Figure 12-16. The thrust-required curve [A] looks suspiciously like the total drag curve in figure 12-14, since thrust must equal drag in level flight. The power-required curve [B] is somewhat different, indicating that less power is required at very low speeds. While the low point of the thrust-required curve defines L/D_{max}, the lowest point on the power-required curve is at a significantly lower airspeed, and marks the minimum power required for level flight. Typically, the airspeed for minimum power required is 76% of the speed for L/D_{max}.

PROPELLER EFFICIENCY

The propeller converts the engine power into thrust. In order to obtain maximum performance, the propeller must make this conversion as efficiently as possible. Because it is an airfoil, the propeller is subject to all the factors that affect airfoil efficiency, such as angle of attack and speed. There are additional factors that are unique to propellers because of their

rotation. The shape of a propeller blade reflects many of the principles you reviewed in the lift discussion. It has a low speed airfoil at a high angle of attack near the root where the local speed is low, and a high-speed airfoil at a low angle of attack at the tip where speeds are much higher. It has a high aspect ratio to minimize induced drag, and most have an elliptical planform to distribute the load. Controllable-pitch propellers allow the pilot to vary the angle of attack of the blades. As with wings, high angles of attack create high induced drag, and that is why the engine slows down when you cycle the propeller during your runup. For takeoff, the blades must be in the low-pitch, high rpm setting to make maximum use of the engine's power. In cruise flight, the forward speed of the airplane changes the relative wind that the propeller blades encounter, reducing angle of attack. When you adjust the propeller pitch in cruise, you restore the blades to an angle of attack that provides a higher coefficient of lift, increasing the thrust provided. At peak efficiency, some propellers are able to convert 85% or more of the power produced by a reciprocating engine into thrust.

When the propeller's axis of rotation is different from the airplane's relative wind, the angle of attack of each blade changes continuously through each revolution. The angle of attack is at its minimum as the blade is ascending, and reaches its maximum as it descends on the other side. This varies the amount of lift, or thrust, produced by one side of the propeller disc compared to the other, and is most noticeable during climbs. [Figure 12-17]

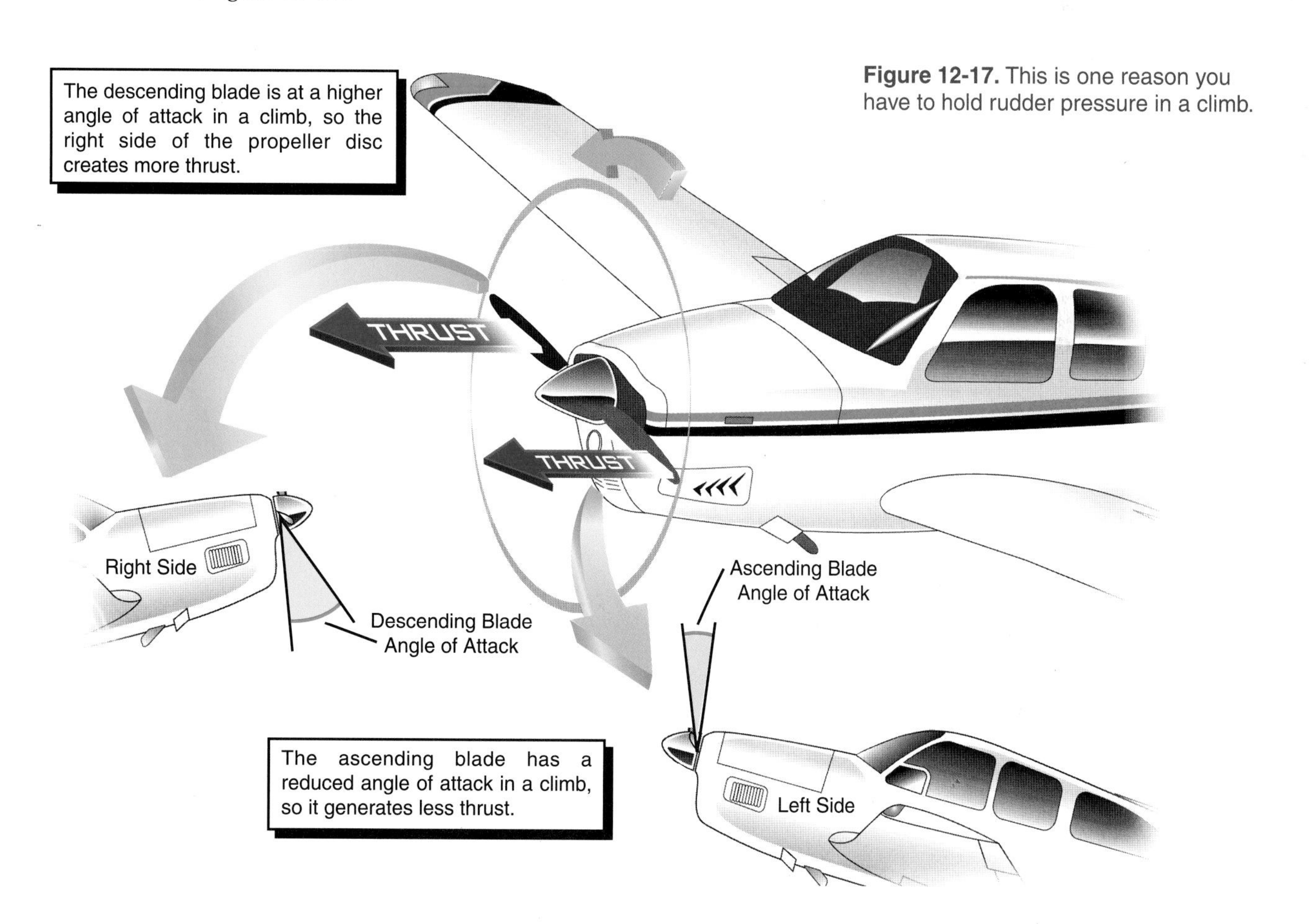

Figure 12-17. This is one reason you have to hold rudder pressure in a climb.

MAXIMUM LEVEL FLIGHT SPEED

There is an upper limit to how much thrust your engine and propeller can produce. When maximum thrust is produced, the airplane accelerates until the drag force is equal to the thrust. Power and thrust available vary with speed. [Figure 12-18]

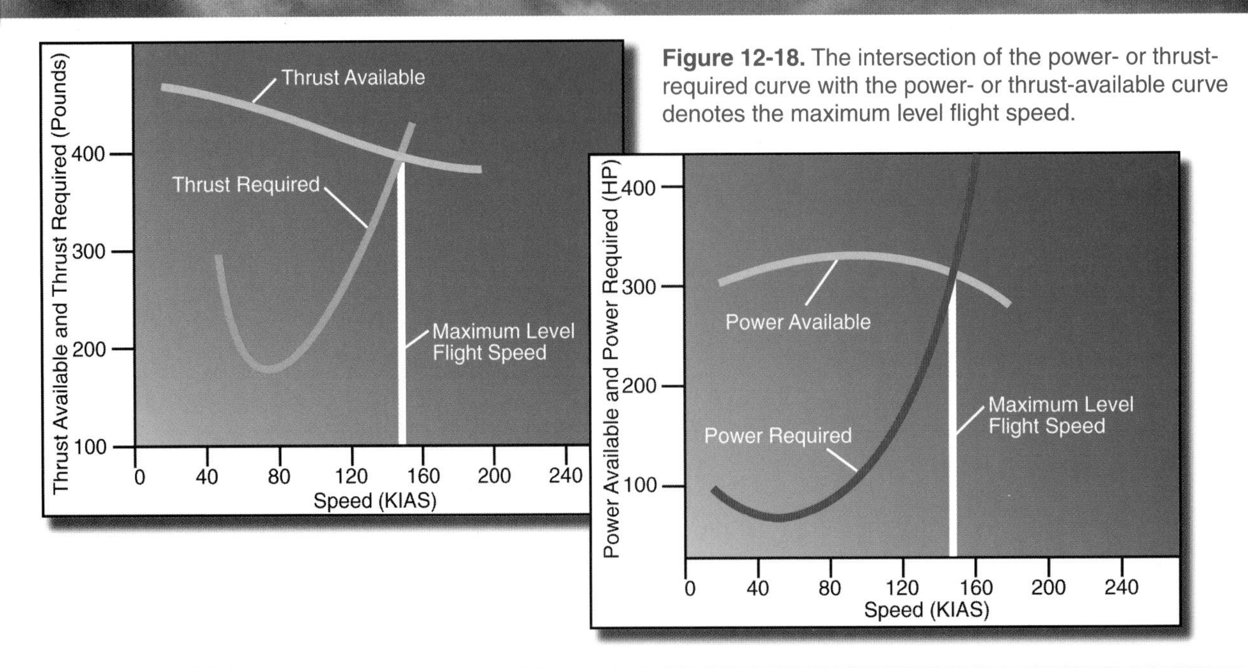

Figure 12-18. The intersection of the power- or thrust-required curve with the power- or thrust-available curve denotes the maximum level flight speed.

Richard T. Whitcomb

Many aircraft designers are associated with particular airplanes: Howard Hughes with the "Spruce Goose," Bill Lear with the Learjet, "Kelly" Johnson with most of Lockheed's advanced military projects, from the P-38 to the F-117, and Burt Rutan with innovative shapes like the Vari-Eze, Beech Starship, Voyager, and Boomerang. But some engineers have had a profound influence on the aviation industry without ever designing an airplane or working for a manufacturer. Such a person is Richard T. Whitcomb.

Courtesy of NASA Dryden Flight Research Center

Doctor Whitcomb was working for NACA, the predecessor of NASA, when Convair had a big problem with their new supersonic interceptor, the F-102 Delta Dagger. It could not exceed the speed of sound except in a dive. He had done some research a few years earlier on the way drag rises as an object nears the speed of sound. Although ridiculed at the time of publication, when his *area rule* was applied to the F-102, performance increased dramatically. Virtually every high-speed aircraft since has applied this principle. Later, he worked on developing the *supercritical wing*, an airfoil that helps to delay transonic drag rise. In the early 1970s, Dr. Whitcomb came up with the drag-reducing *winglets* seen on so many production aircraft today. What's next? Well, he has been doing some research on swept propeller blades . . . .

WEIGHT AND LOAD FACTOR

Weight is the only one of the four forces that does not depend on the flight path. Since weight is the force that results from the acceleration due to gravity, it always acts toward the center of the earth. The actual weight of the airplane in flight changes very gradually during flight as fuel is burned. However, the weight that the wings must actually support also depends on the load factor, and may be as much as several times the actual weight of the airplane.

LOAD FACTOR

The unit used to measure acceleration is the G, an abbreviation for gravity. Everything tries to accelerate toward the center of the earth. When you stand on the surface, that acceleration force is resisted by the force of the ground against the soles of your feet. You feel this as weight, even though you are not accelerating. The force of lift resists the force of gravity to keep an airplane in level flight, and the wings are bearing the weight of the airplane. In both cases, you experience an acceleration force of 1 times the force of gravity, or 1G. Even though we cannot change the force of gravity, whenever the airplane is changing speed or direction the magnitude of the acceleration can be measured in Gs. While weight always acts straight down, acceleration forces, or G-forces, can be created in any direction. For example, in a 60° banked turn, the acceleration due to gravity added to the acceleration necessary to change the flight path results in a 2G force toward the bottom of the airplane. In this situation, the wings will have to generate twice as much lift as they would with the wings level in order to maintain the same altitude. The **load factor** is defined as the load the wings are supporting divided by the total weight of the airplane. With this in mind, you can understand why airplane structures are designed to withstand greater loads than normal flight is likely to produce. When an airplane is certified in the normal category, for example, each part of the structure is designed to do its job at 3.8 positive Gs and 1.52 negative Gs. Acrobatic-category airplanes are designed for higher load factors, and airliners for lower load factors.

Load factor is the ratio between the lift generated by the wings at any given time divided by the total weight of the airplane. The design load factor takes into account the effects of acceleration on the contents of the airplane.

G, I DIDN'T KNOW THAT!

Obviously, acceleration forces act on your body as well as on the aircraft structure. Some parts of your body are more affected than others. In positive G situations, where you are being pushed down into your seat, blood is pressed down into your legs and abdomen. This normally is not harmful, except that it reduces the blood available at the top of your body — your brain. Any reduction in the blood supply to your brain is bad news. Technically, this is stagnant hypoxia, and it can lead to GILOC, or G-Induced Loss of Consciousness. The onset of G-induced hypoxia usually brings tunnel vision as the light-sensitive rods in the retina stop supplying peripheral vision. This is quickly followed by gray-out, where the remaining visual field begins to fade. At this point, sounds may also seem to fade away or become softer as hearing diminishes. Finally, blackout occurs. The onset of these symptoms may occur over a period of seconds, or may be nearly instantaneous, depending on the rate of G increase, the total acceleration, and the pilot's preparation and G tolerance. If the pilot reduces the acceleration forces at any stage before the blackout, full function returns almost instantly, and even blackouts may be momentary if Gs are reduced immediately. In World War II, dive-bomber pilots were instructed to tense their legs and abdomens and yell or grunt during pullouts from steep dives, in order to help prevent blood from sinking into their lower bodies. Most civilian aerobatic pilots use similar techniques. Modern military pilots use G suits, which fit tightly around legs and abdomen. Air bladders in the G suit are automatically inflated with compressed air during high-G maneuvers. Pressure breathing systems and whole-body G suits can help by forcing air into the lungs while squeezing the upper torso in addition to the lower body, thus helping the heart push blood into the head. These systems will probably remain too heavy and expensive for serious competition aerobatic pilots.

As with other types of flying, your physical condition for flight can have a major effect on your G tolerance. Besides the factors mentioned above, an aerobatic pilot's G tolerance also depends on proper hydration (drinking enough water), proper blood sugar levels, the oxygen-carrying capacity of the blood (no smoking), and other general fitness factors.

A heavily loaded airplane has a higher stall speed than the same airplane with a light load. This is because the heavily loaded airplane must use a higher angle of attack to generate the required lift at any given speed than when lightly loaded. Thus, it is closer to its stalling angle of attack, and will encounter C_{Lmax} at a higher airspeed. This is important to remember when flying a heavily loaded airplane at low airspeeds, since a bump, gust, or abrupt control movement could result in a stall. This property also means that at low speeds and/or high weights, abrupt use of the controls results in a stall instead of a structural overload. In this case, the stall acts as a safety valve, reducing the load factor prior to airframe damage.

At high speeds, the controls are more effective, and abrupt control movements can increase G loads so quickly that the structure could be damaged before the stalling angle of attack is reached. The maximum speed at which full or abrupt use of the controls will result in a stall rather than structural damage is the maneuvering speed (V_A). The maneuvering speed will vary with the airplane's total weight, since a lightly loaded airplane is easier to accelerate, and it will also have a larger margin between the angle of attack necessary for level flight and the stalling angle of attack. It takes less force, whether from an abrupt control movement or a from a gust or bump, to create a high-G situation, so the maneuvering speed is reduced. The maneuvering speed depicted on a cockpit placard is calculated for the maximum weight of the airplane, but V_A for other weights may be found in some POHs. The formulas used to calculate V_A are shown below.

$$V_A = V_S \sqrt{n_{limit}}$$

V_A = Maneuvering Speed (for this weight)
V_S = Stall Speed (for this weight)
n_{limit} = Limit Load Factor

$$V_{A_2} = V_A \sqrt{W_2 \div W_1}$$

V_{A_2} = Maneuvering Speed (at this weight)
V_A = Maneuvering Speed (at maximum weight)
W_2 = Actual Airplane Weight
W_1 = Maximum Weight

THE V-G DIAGRAM

Several of the factors that define an airplane's performance envelope can be combined in graphic formats. The V-g diagram is one example, relating velocity (V) to load factor (G). Each V-g diagram applies to one airplane type and the information is valid only for a specific weight, configuration, and altitude. V-g diagrams show the maximum amount of positive or negative lift the airplane is capable of generating at a given speed. They also show the safe load factor limits and the load factor, or number of Gs, the airplane can sustain at various speeds. Flying within the boundaries depicted by the diagram minimizes the risk of stalls, spins, and structural damage. [Figure 12-19]

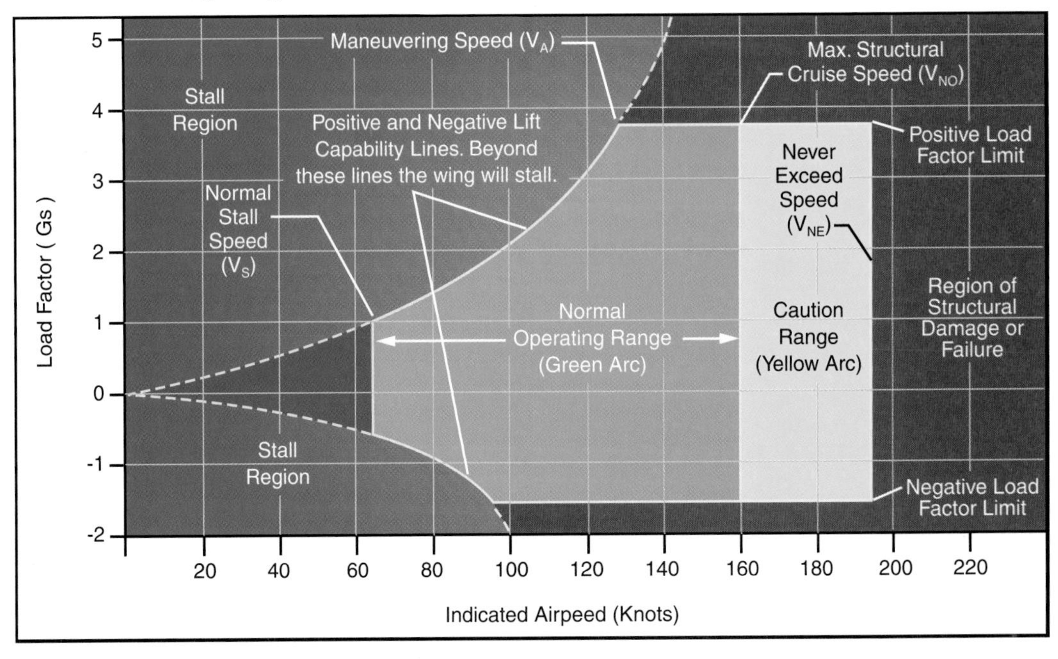

Figure 12-19. The horizontal scale indicates speed (V), and the vertical scale is load factor (G). The load factor limits shown are for a normal category airplane.

Major points of the V-g diagram include the curved lines representing positive and negative maximum lift capability. These lines portray the maximum amount of lift the airplane can generate at the specified speed. The intersection of these lines with the vertical speed lines indicates the maximum G-load capability at that speed. If you exceed the G-load limit at that speed, the airplane will stall. For example, at a speed of about 97 knots, this airplane will stall with a load factor of two Gs under the conditions represented. The horizontal positive and negative load factor limits represent the structural limitations of the airplane. Exceeding these values may cause structural damage.

Another important line on the V-g diagram is the normal stall speed, V_S. Note that at this speed, the airplane stalls at one positive G. This is the same speed shown by the lower limit of the green arc on the airspeed indicator. Above the one-G load factor, stalling speed increases. Any stall which occurs above the straight-and-level load of one G is an accelerated stall.

As discussed earlier, maneuvering speed (V_A) is the maximum speed at which you can use full and abrupt control movement without causing structural damage. V_A occurs at the point where the curved line representing maximum positive lift capability intersects the maximum positive load factor limit. At this speed, a load in excess of 3.8 Gs will result in a stall. Above this speed, G loads may cause structural damage before a stall occurs.

V_{NO} is the maximum structural cruising speed during normal operations while V_{NE} is the never-exceed speed. Above V_{NE}, design limit load factors may be exceeded if gusts are encountered, causing structural damage or failure. See figure 12-19.

The vertical line at a speed of 160 is the maximum structural cruising speed (V_{NO}). It should not be exceeded in rough air. The speed range from V_S to V_{NO} is the normal operating range and corresponds to the green arc on the airspeed indicator. The vertical line at a speed of about 195 represents the never-exceed speed, V_{NE}. This speed corresponds to the red line. If you fly faster than V_{NE}, there is a possibility of control surface flutter, airframe structural damage, or failure. The range from V_{NO} to V_{NE} is the caution range represented by the yellow arc on the airspeed indicator. Because turbulence and gusts are more likely to cause high load factors at higher speeds, this speed range should be used only in smooth air.

AIRCRAFT STABILITY

In unaccelerated flight, the airplane is in a state of equilibrium. Stability describes how the airplane reacts when that equilibrium is disturbed, either by control inputs, or by external factors such as gusts or bumps. A certain amount of stability is desirable, in most airplanes. It makes them easier to fly, especially on instruments or in turbulence, since the airplane tends to return to the trimmed attitude. It is possible to have too much stability, however. If the airplane is too resistant to changes in attitude, the pilot will have to use more effort to maneuver or change direction, so designers try to reconcile the need for stability with the contrasting needs for maneuverability and controllability. While most aspects of stability are designed into the airplane, the pilot's actions can affect stability in several ways. [Figure 12-20]

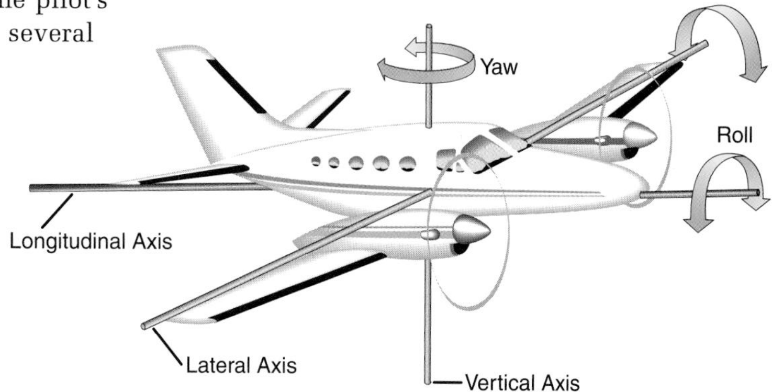

Figure 12-20. Stability characteristics can be described for each of the airplane's three axes of rotation.

STATIC STABILITY

Static stability is the most basic way of looking at stability. A system in equilibrium can only respond to a disturbance in one of three ways. If it initially tends to return toward equilibrium, it has positive static stability. If it initially tends to move away from equilibrium, it has negative static stability. If it does neither, it displays neutral static stability. [Figure 12-21]

An airplane with positive static stability tends initially to return to equilibrium if disturbed. An airplane that remains in a new attitude without returning toward equilibrium or moving farther away from equilibrium displays neutral static stability. If an airplane tends initially to move away from equilibrium when disturbed, it has negative static stability.

POSITIVE NEUTRAL NEGATIVE

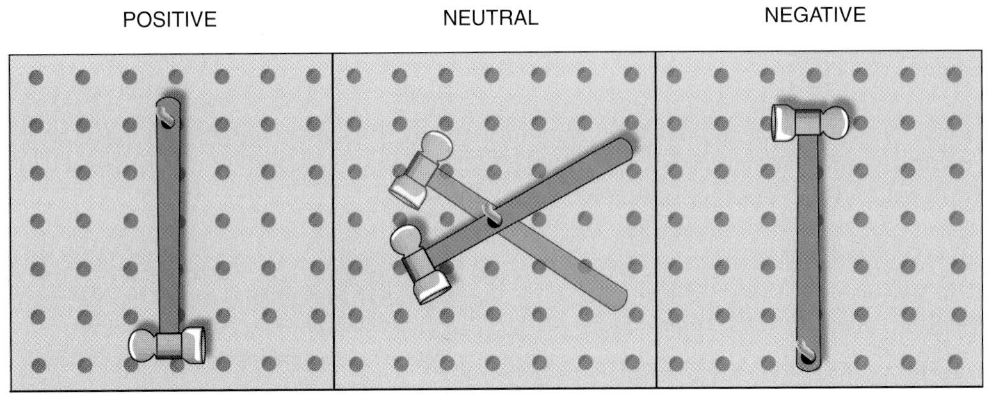

Figure 12-21. Static stability characteristics can be seen all around you. Note that in each example the system is in equilibrium for the moment. If disturbed, the systems on the left will tend to return to their former positions, those in the center will remain in a new position, and those on the right will move away from their former positions.

DYNAMIC STABILITY

While static stability refers to the initial response of a system to a disturbance, **dynamic stability** describes how the system responds over time. What we will call dynamic stability is more correctly called damping, and refers to whether the disturbed system actually returns to equilibrium or not. The degree of stability can be gauged in terms of how quickly it returns to equilibrium. Again, the response can be described as positive, neutral, or negative.

Dynamic stability can be further divided into oscillatory and non-oscillatory modes. To see the difference, think of a smooth bowl with a marble at rest on the bottom. The system is in equilibrium. If you move the marble up the side of the bowl and let it go, you disturb the equilibrium. The marble rolls down the side and up the opposite side, then back across the bottom. After a series of oscillations, the marble comes to rest on the bottom again, showing positive static and oscillatory positive dynamic stability. If you do the same thing with a cotton ball in place of the marble, you see an example of non-oscillatory positive dynamic stability, as the cotton ball simply slides to the bottom without oscillations.

Negative dynamic stability, or instability, can also be oscillatory. As a top spins, it is held by gyroscopic forces in a stable state. These forces weaken as the top slows down, and a small wobble develops, which grows until the top loses all stability, and falls over. Dynamic pitch instability in an aircraft might mean that a small disturbance in equilibrium, if uncorrected by the pilot, would result in a series of pitch oscillations, each greater than the last, until the aircraft structure failed from overload stresses. [Figure 12-22]

Longitudinal dynamic instability is usually characterized by progressively steeper pitch oscillations.

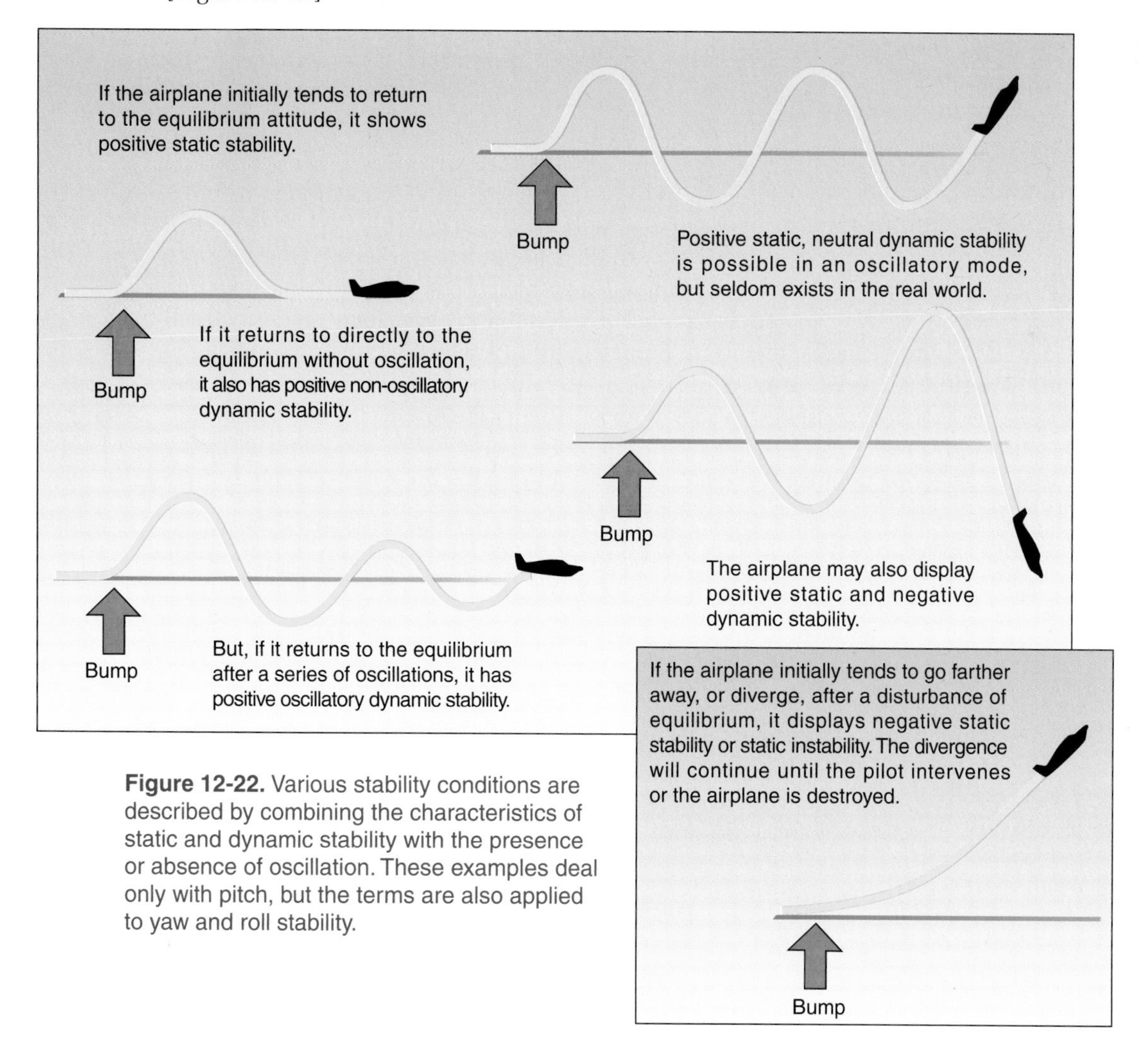

Figure 12-22. Various stability conditions are described by combining the characteristics of static and dynamic stability with the presence or absence of oscillation. These examples deal only with pitch, but the terms are also applied to yaw and roll stability.

LONGITUDINAL STABILITY

For an airplane to be stable in pitch, or longitudinally stable about the lateral axis, it must return to the trimmed pitch attitude if some force causes it to pitch up or down, whether the pitch change is caused by a bump, gust, or control input. Most training airplanes exhibit positive dynamic stability, or positive damping, in pitch.

Longitudinal stability involves the motion of the airplane controlled by the elevators.

Longitudinal stability is achieved in most airplanes by locating the center of gravity slightly ahead of the center of lift of the wings, which makes the airplane tend to pitch down. This nose-heaviness is balanced in flight by a downward force generated by the horizontal tail. On many airplanes, the horizontal stabilizer and elevator form a cambered airfoil with the greater curvature on the bottom surface to create this **tail-down force**. Others use a symmetrical airfoil so that the left and right parts are interchangeable. On some airplanes, the stabilizer is installed with a slightly negative angle of incidence; on others, pitch trim is adjusted by changing the angle of the horizontal stabilizer. In cruising flight, the wings must generate somewhat more lift than the actual weight of the airplane to compensate for the negative lift of the tail-down force. [Figure 12-23]

Figure 12-23. Tail-down force compensates for the nose-down moment that results from the center of gravity being located ahead of the center of lift.

Tail-down force contributes to pitch stability because of the way it changes in response to pitch disturbances. If a properly trimmed airplane is pitched up, the negative angle of attack of the stabilizer is reduced and increased drag causes a drop in airspeed, both of which reduce the tail-down force, allowing the airplane to pitch down. As the airplane then pitches down and accelerates, the increasing angle of attack and increasing airflow at the horizontal tail increase the tail-down force, raising the nose and causing the airspeed to decay again. After a series of progressively smaller oscillations, the airplane returns to level flight. [Figure 12-24]

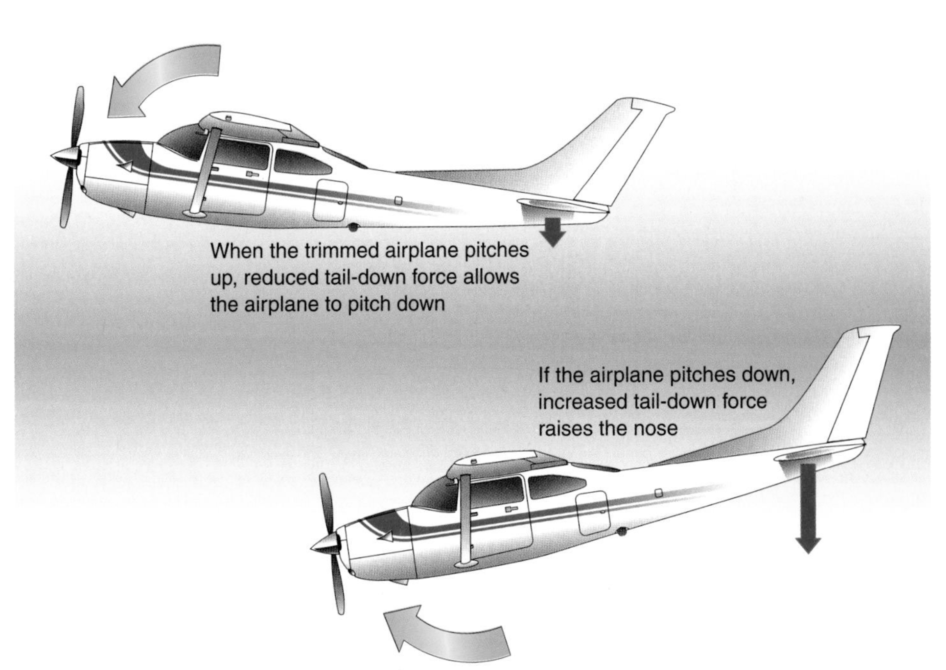

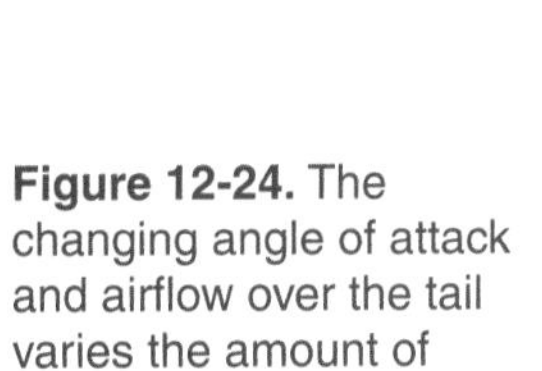
Figure 12-24. The changing angle of attack and airflow over the tail varies the amount of tail-down force.

If the airplane is loaded with the center of gravity farther forward, you can see that the tail-down force must be increased to keep the airplane in trim. This adds to longitudinal stability, since the additional nose heaviness makes it more difficult to raise the nose, while the additional tail-down force makes it more difficult to pitch the nose down. Small disturbances are opposed by more powerful forces, so they tend to damp out more quickly. If the CG is too far forward, there may not be enough elevator authority to raise the nose, either for takeoff or to flare for landing. Conversely, as the CG is located further aft, the airplane becomes less stable in pitch. If the airplane is loaded so that the CG is actually behind the center of lift, the tail must exert an upward force to prevent the nose from pitching up. If a gust pitches the nose up in this situation, the reduction of airflow over the tail will allow the nose to pitch up further, a classic example of negative stability, or instability. Any disturbance of this precarious equilibrium will result in greater divergence from the trimmed condition. This is an extremely dangerous situation. The aft CG also makes it easier to stall or spin, there is more likelihood that a spin will go flat, and recovery may be impossible. If the CG coincides with the center of lift, control forces will be at their minimum, but the airplane will be difficult to fly, requiring constant attention and continuous control input even in calm air. In most airplanes, proper loading requires the CG to be forward of the center of lift.

LATERAL STABILITY

Stability around the longitudinal axis, or **lateral stability**, is the tendency of the airplane to return to a wings-level attitude following a roll deviation. Airplanes ordinarily have little resistance to roll inputs, and if you bank a few degrees in coordinated flight and let go of the controls, most airplanes show neutral static stability in roll. Most efforts to enhance lateral stability take advantage of the fact that when roll occurs accidentally, as the result of a bump or gust, it is almost always accompanied by a sideslip. Airplane designers use the forces and relative wind effects of the sideslip to return the airplane to a wings-level attitude, and still provide comfortable roll control forces for intentional turns. Wing dihedral and sweep are two methods of increasing lateral stability without adding to roll control forces.

Dihedral works because the relative wind during a sideslip increases the angle of attack on the lower, upwind wing while the higher wing experiences a reduced angle of attack. The difference in the lift produced tends to roll the airplane out of the slip, while the increase in induced drag on the lower wing helps to yaw the airplane into the relative wind. [Figure 12-25]

Figure 12-25. Dihedral uses relative wind from the sideslip to create more lift on the upwind wing. This is how the relative wind sees the airplane in a sideslip. Notice how much more you can see of the bottom of the nearer wing. This translates to additional angle of attack, and means that the wing will develop more lift than the downwind wing, rolling the airplane back to level flight.

Walking around airport ramps, you may have noticed that low-wing airplanes generally have much more dihedral than high-wing types. The difference is due to the way the relative wind is influenced by the fuselage during a sideslip. In high-wing airplanes, the relative wind creates an upwash near the wing root at the leading edge of the upwind wing, and a slight downwash at the root of the downwind wing. This increases the local angle of attack on the upwind wing while decreasing it on the downwind wing, with the resulting difference in lift tending to roll the airplane out of the slip. The situation is reversed with a low-wing airplane; the sideslip creates a downwash for the upwind wing and an upwash for the downwind wing, which tends to increase the roll into the sideslip. Low-wing airplanes need to have a certain amount of dihedral to overcome the destabilizing effect of the fuselage, and an additional amount to provide lateral stability. [Figure 12-26]

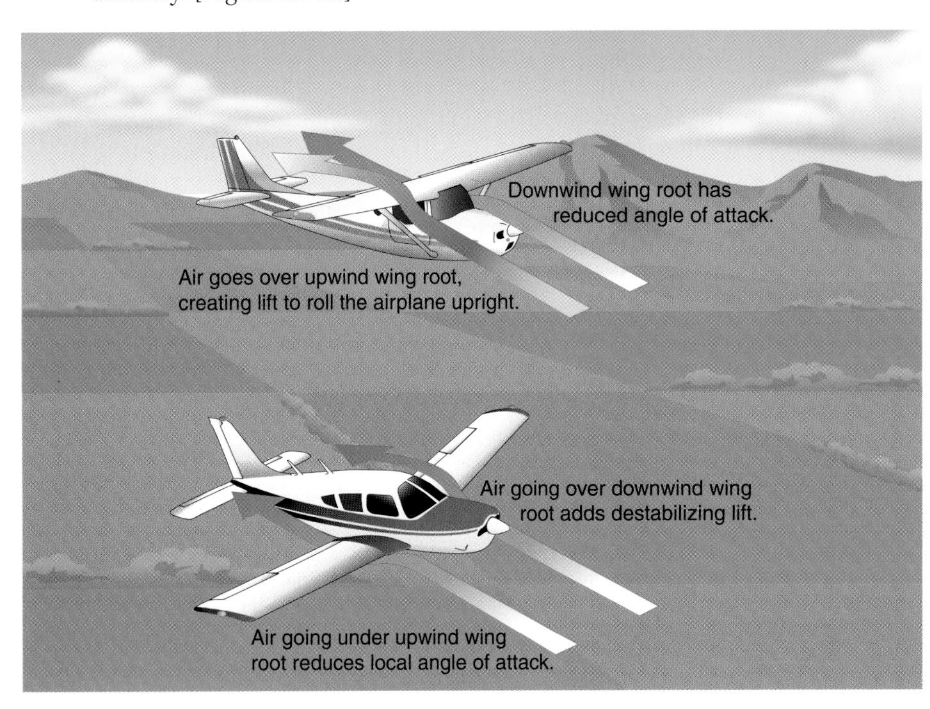

Figure 12-26. Low-wing airplanes need more dihedral because of the way the fuselage influences airflow at the wing roots. Both of these airplanes are slipping to their right. Note how the fuselage causes air to flow under the upwind wing of the low-wing airplane, and over the upwind wing of the high-wing airplane.

Long before swept wings became necessary to deal with the problems of drag rise and shockwaves at transonic speeds, designers used them to assist with lateral stability. Sweepback is especially useful when dihedral would be inappropriate, such as on an aerobatic airplane that needs lateral stability while inverted as well as when upright. [Figures 12-27 and 12-28]

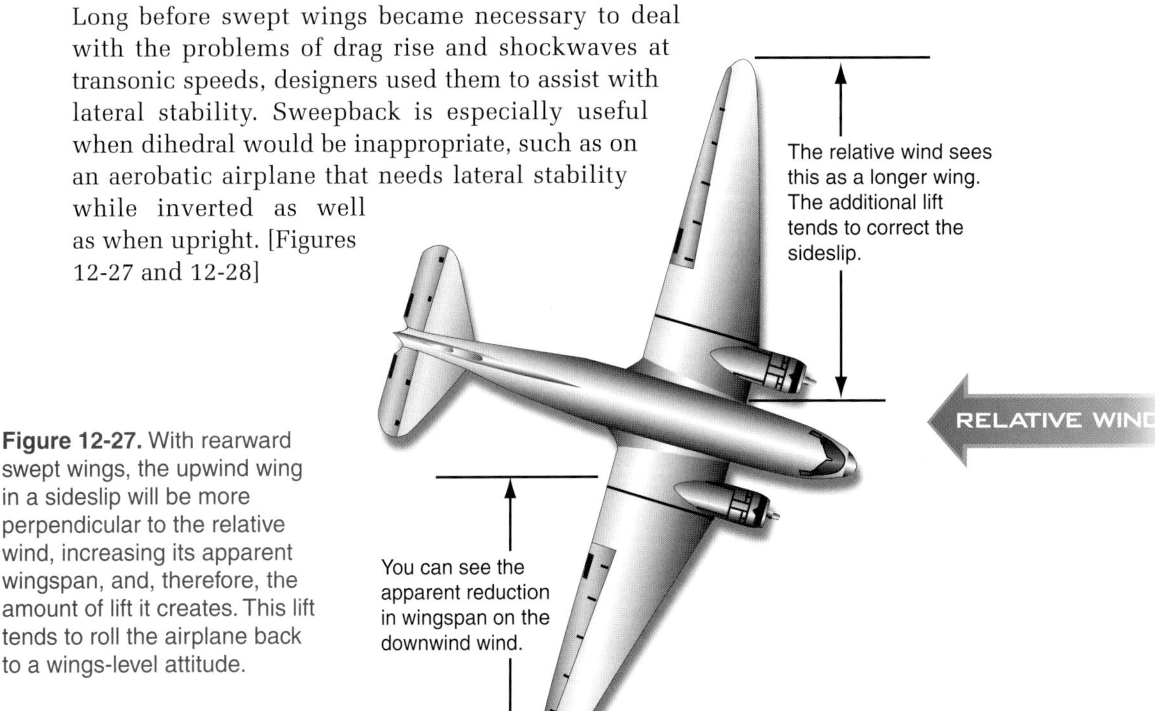

Figure 12-27. With rearward swept wings, the upwind wing in a sideslip will be more perpendicular to the relative wind, increasing its apparent wingspan, and, therefore, the amount of lift it creates. This lift tends to roll the airplane back to a wings-level attitude.

Figure 12-28. The Bucker Jungmann of 1934 used swept wings to provide lateral stability in both upright and inverted flight. This design feature has also been used in many other aerobatic airplanes.

Most airplanes have a moderate amount of lateral stability, since too much would make them difficult to control in turns, and would hamper crosswind takeoffs and landings. In fact, some airplanes are designed with negative dihedral to counteract some of the dihedral effect that results from their swept wings. [Figure 12-29]

Figure 12-29. These airplanes counteract unwanted lateral stability with negative dihedral.

DIRECTIONAL STABILITY

Stability in yaw, or resistance to undesired rotation around the vertical axis is called **directional stability**. The vertical tail and the sides of the fuselage contribute forces that help to keep the airplane's longitudinal axis aligned with the relative wind. [Figure 12-30]

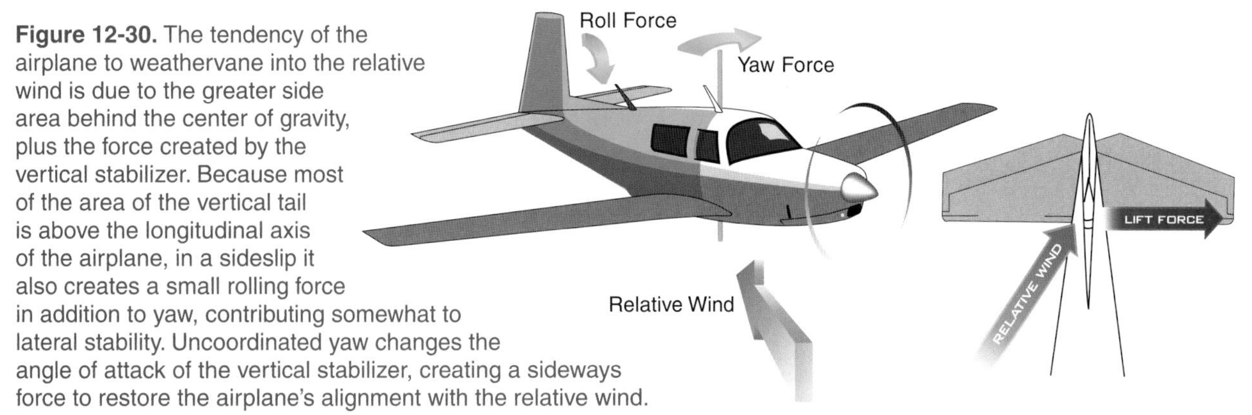

Figure 12-30. The tendency of the airplane to weathervane into the relative wind is due to the greater side area behind the center of gravity, plus the force created by the vertical stabilizer. Because most of the area of the vertical tail is above the longitudinal axis of the airplane, in a sideslip it also creates a small rolling force in addition to yaw, contributing somewhat to lateral stability. Uncoordinated yaw changes the angle of attack of the vertical stabilizer, creating a sideways force to restore the airplane's alignment with the relative wind.

EFFECTS OF LATERAL AND DIRECTIONAL STABILITY

Although the design factors that contribute to lateral and directional stability are carefully balanced to complement each other in flight, the complex manner in which they interact leaves many aircraft with a certain degree of instability in yaw and roll. Many of the design compromises are dictated by the wide speed range over which the aircraft must operate, for instance, the design elements that provide stability in a low speed configuration for takeoff and landing may lead to undesirable flight characteristics at high speed, and vice versa. Two of the most common problems are Dutch roll and spiral instability.

Dutch roll, or oscillatory lateral instability, is usually a consequence of too much dihedral effect, from the combination of actual dihedral, sweepback, a high wing configuration, relatively small vertical tail, and/or other factors. In Dutch roll, the airplane makes a continuous back-and-forth rolling and yawing motion with the roll out of phase with the yaw. Dutch roll is usually dynamically stable, that is, the oscillations tend to decrease in amplitude, but if the oscillation is weakly damped it may continue for a relatively long time, making it objectionable. In some airplanes, Dutch roll can be aggravated by the pilot's attempts to correct it, a form of pilot-induced oscillation. [Figure 12-31]

Figure 12-31. Dutch roll is a combined sequence of rolling/yawing oscillations, with the airplane always yawing away from the direction of bank.

If an airplane has strong directional stability relative to its lateral stability, it may display **spiral instability**. When a bump or gust produces a side slip in an airplane with spiral instability, its powerful directional stability tends to yaw the airplane back into alignment with the relative wind. As the outside wing travels faster it generates more lift, tending to roll the airplane in the direction of the yaw. This rolling force overcomes the comparatively weak dihedral effect, allowing the roll to progress. As the vertical component of lift decreases, the relative wind has an upward component, and as the airplane swings into alignment with the airstream, the nose begins to drop. This results in a gradually tightening spiral dive known as the graveyard spiral. Despite the name, which dates from the days when pilots attempted flight into IFR conditions without adequate instruments, spiral instability is generally

Figure 12-32. The mild divergence caused by spiral instability is usually easy to correct, but can lead to a steepening spiral if allowed to develop.

considered less objectionable than Dutch roll. A certain degree of spiral instability is considered acceptable, so designers usually try to minimize Dutch roll by placing more emphasis on directional stability. [Figure 12-32]

AERODYNAMICS AND FLIGHT MANEUVERS

Designers devote as much attention to an airplane's handling characteristics in maneuvering flight as they do to its stability. A great deal of an airplane's performance and appearance is dictated by the aerodynamic considerations of maneuvering flight.

STRAIGHT-AND-LEVEL FLIGHT

If you remain in straight-and-level flight, your maneuvering is limited to speeding up or slowing down. Maintaining straight-and-level flight while changing airspeed involves managing three of the four aerodynamic forces simultaneously. To reduce airspeed, you must create an excess of drag, usually by reducing thrust, although speed brakes, landing gear, or other devices are used in some circumstances. As the airplane slows, lift diminishes, so you increase the angle of attack to maintain altitude. When drag has dropped to a value that matches the new thrust level, equilibrium is restored. If you slow to a speed that requires an angle of attack near the stall, you may obtain more lift by changing the wing camber or area with flaps or other high lift devices. Since these also increase drag, additional thrust will be required to maintain the desired airspeed. To increase airspeed, either thrust must be added, drag reduced, or both. As the speed increases, the angle of attack will need to be reduced to keep additional lift from causing the airplane to climb.

To maintain altitude while airspeed is being reduced, the angle of attack must be increased.

CLIMBS

A climb is normally initiated by increasing the angle of attack, which momentarily increases lift. The additional induced drag slows the airplane, unless thrust is added, and the excess lift quickly returns to equilibrium. Whether thrust has been added or not, the thrust vector remains at a slight upward angle to the relative wind, and it is this vertical component of thrust that causes the airplane to climb. If you raise the nose without adjusting the power, you divert thrust from opposing drag and instead use it to create a difference between power required for level flight and power available, and it is this excess power that sustains a climb. The relationship between power and airspeed is easy to see on a graph. [Figure 12-33]

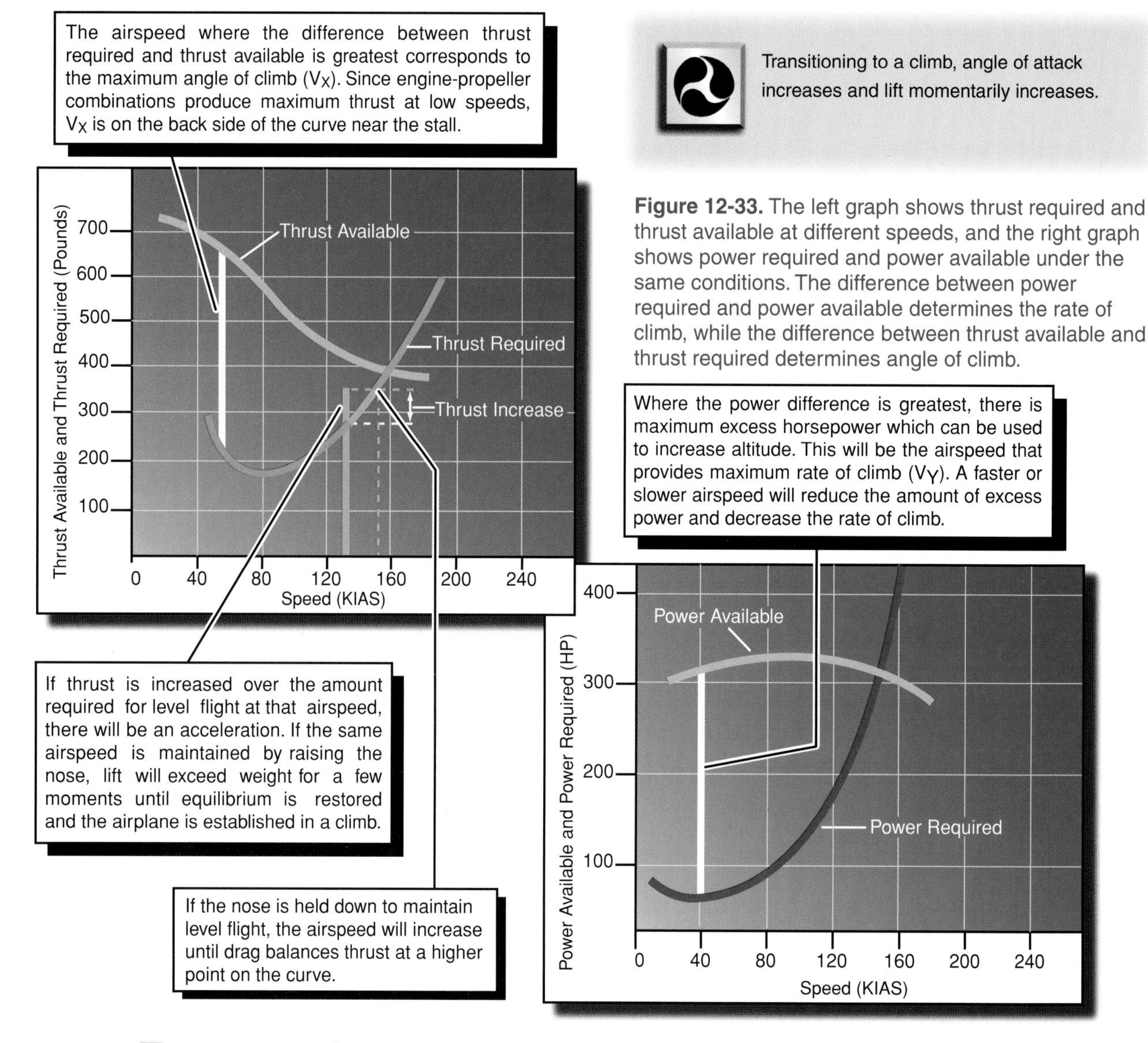

Figure 12-33. The left graph shows thrust required and thrust available at different speeds, and the right graph shows power required and power available under the same conditions. The difference between power required and power available determines the rate of climb, while the difference between thrust available and thrust required determines angle of climb.

FACTORS AFFECTING CLIMB PERFORMANCE

Airspeed affects both the angle and rate of climb. At a given airplane weight, a specific airspeed is required for maximum performance. If your airspeed is faster or slower than best climb speed, climb performance will decrease. Generally, the larger the speed variation, the larger the performance variation. Weight also affects climb performance. If you change the weight of the airplane, you also change the drag and power required. As you increase weight, you reduce the maximum rate of climb and maximum angle of climb.

As you know, climb performance decreases with altitude. This is because more power and thrust are required, but the power and thrust available decrease with altitude. The airspeed necessary to obtain the maximum angle of climb increases with altitude, but the airspeed for the best rate of climb decreases. The point where these two airspeeds converge is referred to as the airplane's **absolute ceiling**. When the best rate of climb is zero and the airplane cannot climb any higher, it has reached its absolute ceiling. Because the rate of climb deteriorates to nearly zero as the absolute ceiling is approached, a more practical value is commonly used. The **service ceiling** is the pressure altitude where the maximum rate of climb is 100 f.p.m. For multi-engine airplanes, the service ceiling with one engine inoperative is the pressure altitude where the airplane can climb at 50 f.p.m. with one propeller feathered. These ceilings vary with temperature and aircraft weight.

GLIDES

In the event of an engine failure, you will probably be interested in flying the airplane at the minimum glide angle so you can travel the maximum distance for the altitude you have available. As mentioned before, the best glide performance is at the maximum lift-to-drag ratio, or $L/D_{max.}$ This occurs at a specific value of the lift coefficient and, therefore, at a specific angle of attack. [Figure 12-34]

Approximate gliding distance can be found by multiplying your altitude AGL by the L/D ratio. Glide ratios for various angles of attack are obtained from an L/D graph. See figure 12-33.

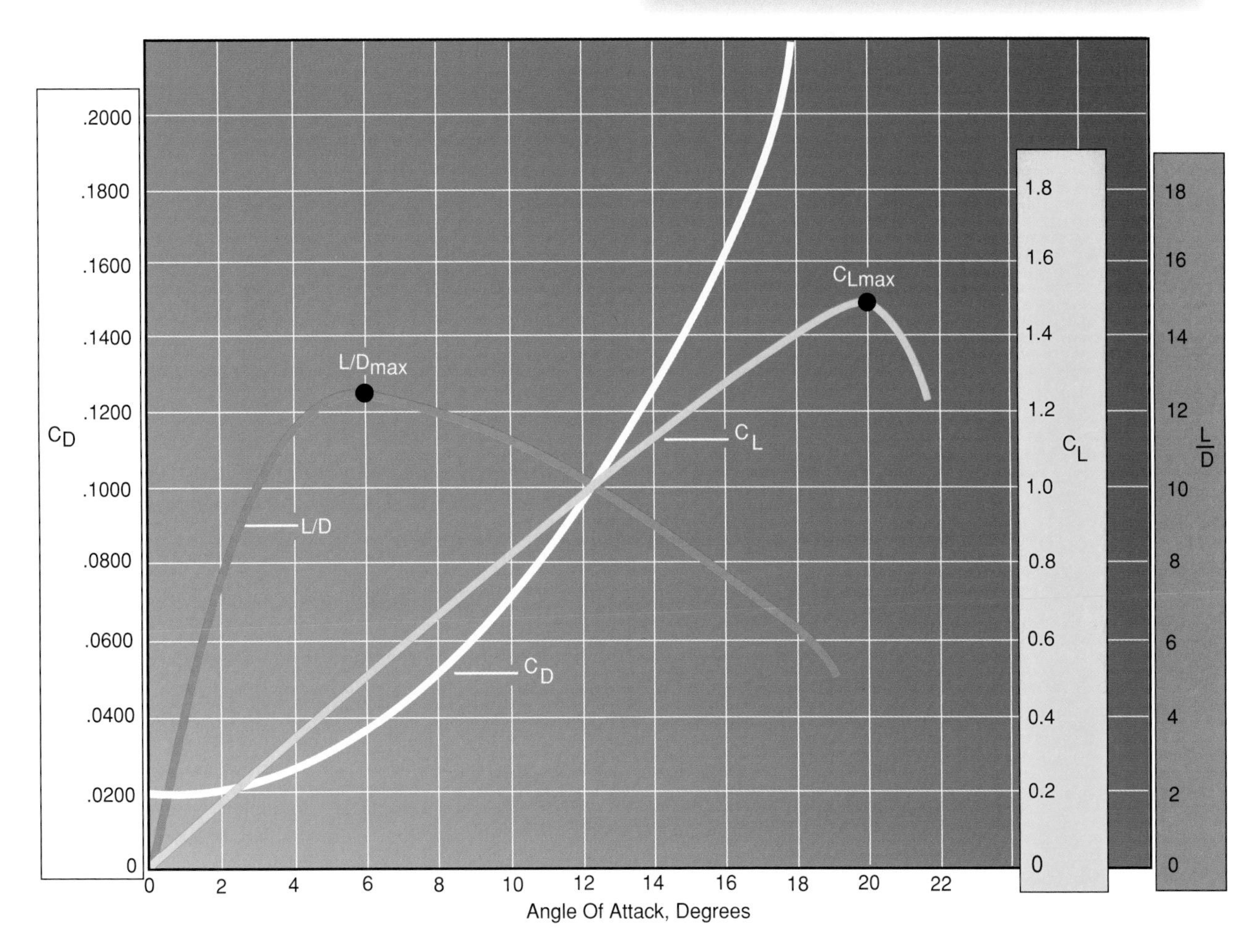

Figure 12-34. This graph shows the curves for the coefficient of drag, C_D, and coefficient of lift, C_L, as well as the curve obtained by dividing lift by drag, L/D, for a specific airfoil at positive angles of attack. In addition to the angle of attack for L/D_{max}, this graph also shows you the L/D value you could expect for other angles of attack. Since the L/D ratio is also the glide ratio, you can use this graph to determine how far this particular airplane will travel in a power-off glide at any given angle of attack, assuming there is no wind. For example, if you were gliding at an angle of attack of 10°, you would expect a glide ratio of 11 to 1, or 11 feet forward for every foot of altitude lost. If you began your glide at 3,500 feet AGL, you could glide 38,500 feet horizontally (11 × 3,500) at that angle of attack, or about 6.3 nautical miles (38,500 ÷ 6,076). Gliding from the same altitude at L/D_{max}, with a 6° angle of attack, lets you travel about 7.2 miles, or about 13% farther. Note that for most glide ratios there are two possible angles of attack. The glide ratio obtained at 10° is the same as for 3.6°.

TURNS

Newton's first law of motion says that an object in motion tends to remain in motion in a straight line unless acted on by some outside force. In order to turn an airplane, you provide that force through the flight controls, and this causes an acceleration. The acceleration continues for as long as you are preventing the airplane from following a

As the angle of bank is increased, the vertical component of lift decreases as more of the total lift is directed horizontally. To compensate for the loss of part of the vertical component of lift in a turn, you must increase the angle of attack by using elevator back pressure.

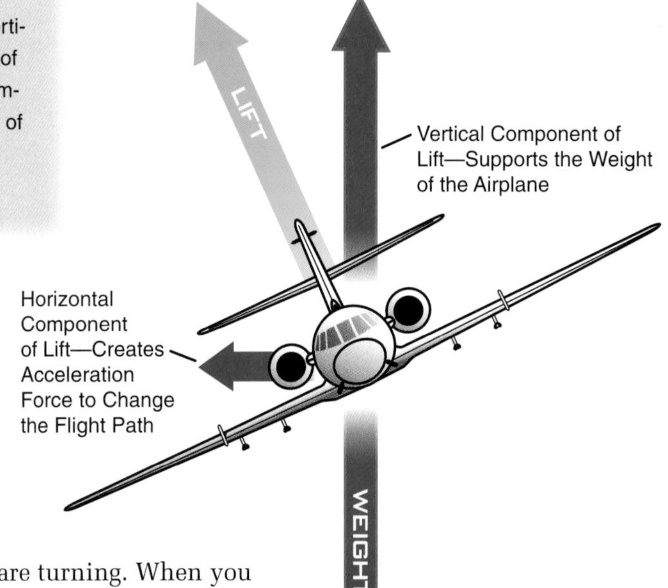

Figure 12-35. In a bank, part of the lift acts vertically to support the weight of the airplane, and part of it acts horizontally to change the flight path. As you can see, in order to balance weight, the wings must generate more lift than in straight flight, and that is why you must increase the angle of attack by adding back pressure to maintain altitude in a turn.

straight line, that is, for as long as you are turning. When you roll the airplane into a bank, some of the lift of the wings is directed to the side, and it is this **horizontal component of lift** that causes the airplane to turn. [Figure 12-35]

LOAD FACTOR IN TURNS

The increase in weight that you feel in a turn is due to the inertia of your own body as it tries to continue along a straight path rather than the curving flight path of the airplane. This is in accordance with Newton's second law of motion, which describes the relationships between force, acceleration, and mass. Briefly, a force acting on a certain mass will create a proportional acceleration. The more force you exert through the flight controls to change the airplane's path away from a straight line, the more acceleration you will feel. The lift of the wings exerts this force on every part of the airplane, increasing the load factor. A steeper bank means a greater horizontal component of lift, which corresponds to a higher load factor. Therefore, in a properly coordinated turn, the steeper the bank, the higher the flight load factor. [Figure 12-36]

1 Suppose you are flying an airplane with a wings-level stall speed of 62 KIAS. To find the stall speed in a 45° banked turn, go up to the Stall Speed Increase curve.

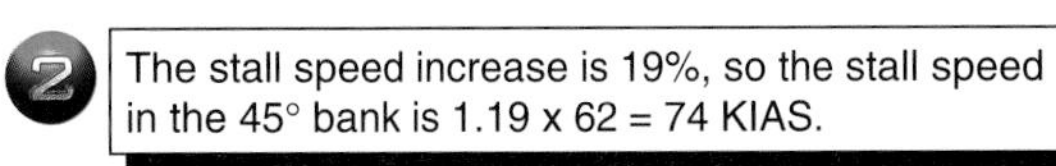

2 The stall speed increase is 19%, so the stall speed in the 45° bank is 1.19 x 62 = 74 KIAS.

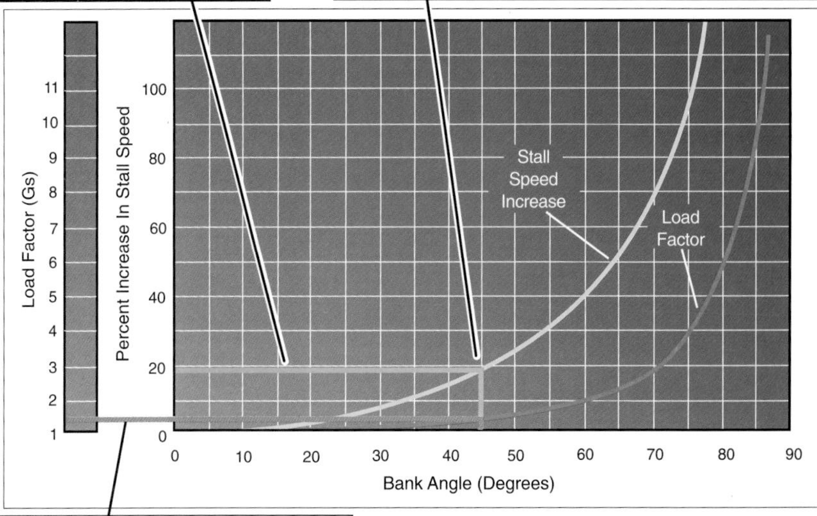

Figure 12-36. The relationship between angle of bank, load factor, and stall speed is the same for all airplanes. You can use this graph to find the load factor in turns for any airplane, as well as the percentage the stall speed will increase.

3 To find the load factor, go up to the Load Factor curve and read across to the proper scale. The load factor at 45° is less than 1.5G.

The constant load factor generated in a coordinated, level turn does not depend on the rate or radius of turn, but on the bank angle. Load factor remains constant if there is no change in bank angle. At a bank angle of 60°, the load factor is 2Gs, which means that the wings are supporting twice the weight of the aircraft.

Load factor in turns increases at steeper bank angles, as does the stall speed. In a coordinated, level turn at a given bank angle, all airplanes will experience the same load factor and the same percentage of increase in stall speed over their wings-level stall speed. See figure 12-35.

RADIUS AND RATE OF TURN

Two variables determine the rate and radius of a turn. A steeper bank reduces turn radius and increases the rate of turn, but produce higher flight load factors. Reducing airspeed does the same thing, but without increasing the load factor. The radius of turn at any given bank angle varies directly with the square of the airspeed, that is, at twice the airspeed, the radius of the turn will be quadrupled, if the bank remains the same. The rate of turn also varies with airspeed, at any given bank angle, so slower airplanes require less time and area to complete a turn than faster airplanes at the same angle of bank. [Figure 12-37]

A specific angle of bank and true airspeed will produce the same rate and radius of turn regardless of weight, CG location, or airplane type. You can also see from this chart that increasing the velocity increases the turn radius and decreases the turn rate. The load factor on the airplane remains the same because you have not changed the angle of bank. To increase the rate and decrease the radius of a turn, you should steepen the bank and/or decrease your airspeed.

A given airspeed and bank angle will produce a specific rate and radius of turn in any airplane. In a coordinated, level turn, an increase in airspeed will increase the radius and decrease the rate of turn. Load factor is directly related to bank angle, so the load factor for a given bank angle is the same at any airspeed.

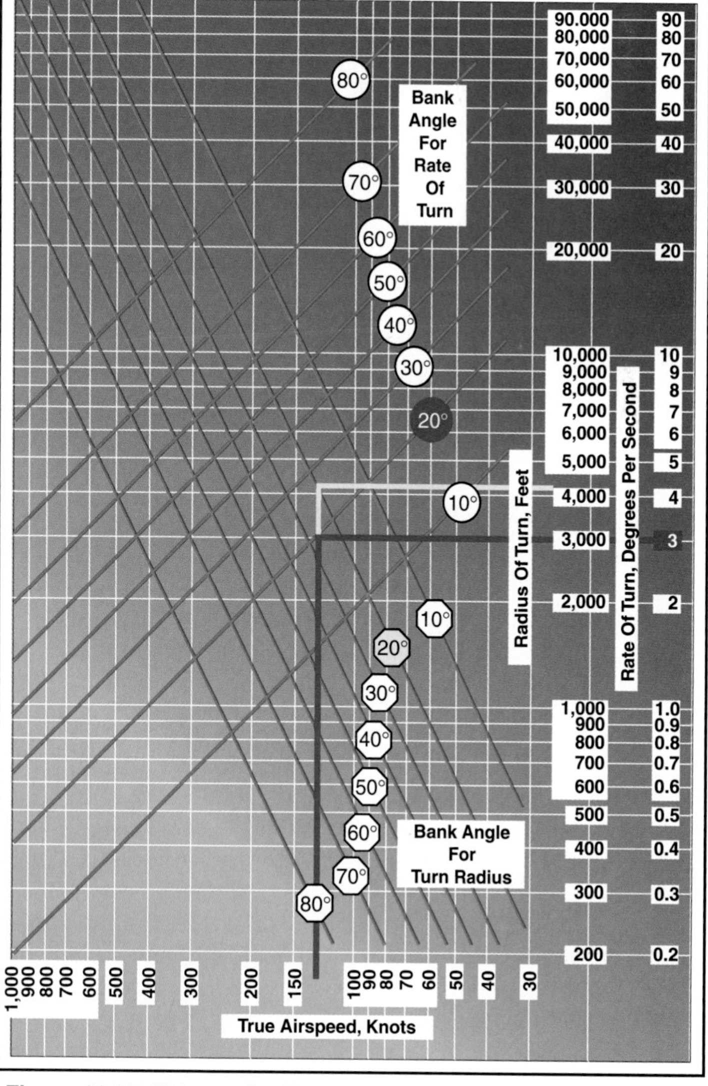

Figure 12-37. This graph will work for any airplane. The example shows that for a turn at 130 knots and a bank angle of 20°, the radius will be 4,200 feet and the rate of turn will be 3° per second.

COORDINATION IN TURNS

Before the first airplane flew, the Wright brothers experimented with control systems on their gliders. They found that they could turn by tilting the lift of the wings, which they accomplished by changing the camber on the outer portions of the wings. They used a system of wires and pulleys to twist the trailing edges of the wings down on one side and up on the other. The Wrights soon found that their gliders tended to yaw away from the direction of bank, due to the increase in induced drag on the side with the greater camber. Subsequently, they fitted a vertical rudder to assist them in keeping the aircraft aligned with the flight path, and most of the airplanes since have followed the same strategy for turning. In general, ailerons have replaced wing-warping as the method of changing the camber of the wingtips, so we refer to the tendency to yaw away from the bank as adverse aileron yaw, or simply **adverse yaw**. This effect is most noticeable at low airspeeds, where larger control movements are required for a given change in attitude, and the down-turned aileron creates proportionately more induced drag. Proper use of the rudder overcomes adverse yaw.

The ball in the inclinometer indicates whether your airplane is aligned with its flight path. If you do not use enough rudder for the bank angle, the airplane yaws to the outside of the turn, and the ball moves to the inside, indicating a slip. Conversely, using too much rudder for the amount of bank yaws the airplane to the inside of the turn, displacing the ball to the outside, indicating a skid. The same forces that cause the ball to move will try to shift your body sideways in the seat. This is one reason why maintaining coordinated flight without watching the ball is known as flying by the seat of your pants. Because slips increase drag, they are sometimes used to lose altitude without gaining speed. Some airplanes restrict intentional slips for aerodynamic or structural reasons. In addition, slipstream disturbances at the static ports may cause indicated airspeed errors during slips.

STALL AND SPIN AWARENESS

Most stall/spin accidents occur when the pilot is momentarily distracted from the primary task of flying the aircraft. Because of this, the FAA requires training that emphasizes recognition of situations that could lead to an unintentional stall and/or spin. An understanding of the aerodynamics of the stall and spin will help you avoid such situations, as well as improve your chances for a safe recovery should an unintentional stall or spin occur.

STALLS

As you know, at any angle of attack beyond C_{Lmax}, the airflow can no longer follow the upper surface of the wing, and the flow separates. The wing loses lift, and the airplane accelerates downward because weight exceeds lift. Unless you reduce the angle of attack by movement of the yoke, the airplane will remain in the stalled condition, and may enter a spin. If the stall occurs near the ground or if the controls are incapable of reducing the angle of attack, recovery may not be possible.

CAUSES OF STALLS

The stall always occurs at the same angle of attack for any particular airfoil, regardless of weight, load factor, attitude, airspeed, or thrust. However, each of these factors does influence the stall speed. An airplane can stall at any airspeed, in any attitude, and at any power setting. The airflow separation is really a progressive event. It begins as the angle of attack approaches C_{Lmax}, and the rate of increase of the C_L begins to flatten out. As the angle of attack increases to C_{Lmax} and beyond, the separation increases, lift decreases, and drag increases. There are several ways a pilot can bring the airplane to the stalling angle of attack. Slowing the airplane compels you to increase the angle of attack to generate the required lift. Likewise, to compensate for the loss of part of the vertical component of lift in a turn, the angle of attack must be increased to maintain altitude. The load factor in a turn adds to the effective weight of the airplane, and whenever you increase the weight, either through flight load factors or by loading actual weight on

board, the wings have to fly at a higher angle of attack at any given airspeed to generate a corresponding amount of lift. The location of the center of gravity also affects the load the wings must support, due to the effects of tail-down force. As the CG is moved forward, the tail-down force necessary for level flight increases. Each of these factors reduces the margin between a lift-producing angle of attack and the stalling angle of attack. If the wing is near C_{Lmax}, even a bump or gust can change the relative wind enough to produce a stall. Abrupt use of the flight controls can trigger a stall even when the airspeed is well above the published stall speed.

Total weight, load factor, power, and CG location affect stall speed.

Frost or ice on the wings can raise the stall speed in various ways. The weight of ice adds to the airplane's total weight, of course. Ice also may change the shape of the airfoil, either reducing the coefficient of lift or the stalling angle of attack, or both. Even a thin coating of ice or frost can trigger separation by disrupting the smooth flow of air. As mentioned in the discussion of stability, if the center of gravity is too far aft, the stall speed will decrease, but the controls may not have enough authority to reduce the angle of attack, making recovery impossible. The condition of the airplane itself may adversely affect its stall characteristics, for instance, if the airplane is poorly maintained, has damaged wing surfaces, or is improperly rigged.

TYPES OF STALLS

Stalls are usually classified according to whether the airplane is flying straight ahead or turning, and whether the power is on or off. You should be proficient in recognizing the most common stall situations as well as performing and recovering from the stalls associated with them. Normally, **power-on stalls** are associated with takeoff and climb configurations, and **power-off stalls** are most likely in an approach configuration. Consequently, power-on stalls are usually practiced with gear up and flaps at the takeoff setting, and power-off stalls with the gear and flaps down. Of course, stalls in either configuration can be done either straight ahead or turning. [Figure 12-38]

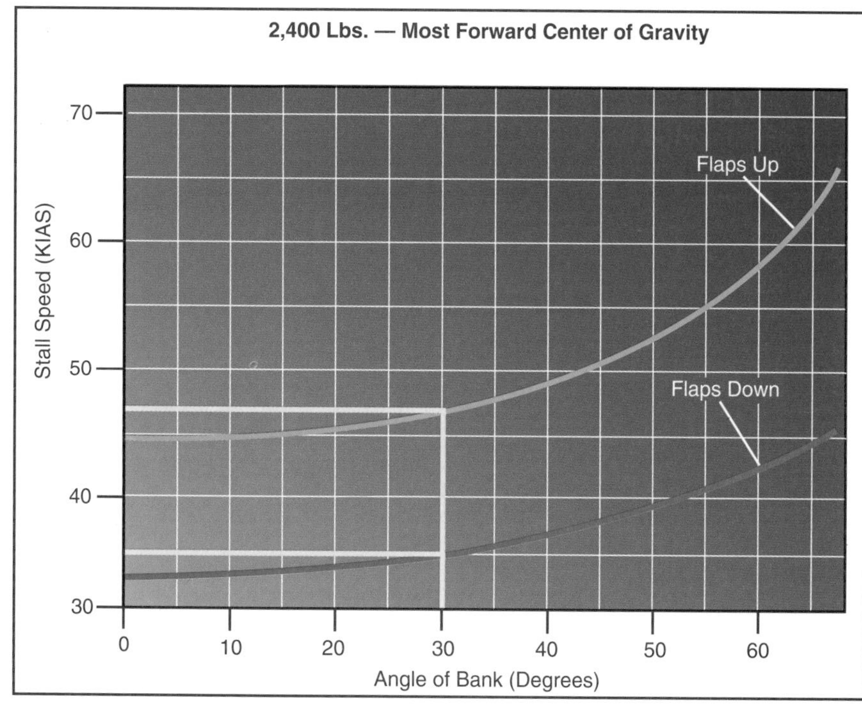

Figure 12-38. This graph shows the effect of configuration and bank angle on stall speed. For this aircraft, there is an 11 knot decrease in stall speed with the flaps down with the wings level. This difference increases with the angle of bank. At 30° bank, the stall speed is 35 KIAS with flaps down and 47 KIAS with flaps up. Remember that the speeds published in any POH apply only to the stated conditions, in this case, a weight of 2,400 pounds and the most forward CG position.

Additional descriptive terms for stalls include accelerated, secondary, crossed-control, and elevator trim. Familiarity with these terms helps you to understand descriptions you may read in the aviation literature, press reports, or accident/incident analyses.

Accelerated stalls are caused by abrupt or excessive control movement. They commonly occur during maneuvers such as steep turns or rapid dive recoveries involving a high load factor or a sudden change in the flight path. Accelerated stalls are usually more violent than unaccelerated stalls, and they are often unexpected because of the relatively high airspeed. Any time you are experiencing an increased flight load factor, that acceleration indicates that you have increased the angle of attack, and even though you may be well above the usual stall speed, you will be closer to the stalling angle of attack.

Stalling speed is most affected by load factor. In any situation involving increased load factor, such as a rapid pullout from a dive, the stall speed increases.

If you pull up too quickly during recovery from a stall, you may trigger a **secondary stall**. This usually happens as a result of an attempt to hasten a stall recovery, either by increasing the angle of attack too quickly, or by not decreasing the angle of attack enough in the first place. Relax the back pressure and allow the airplane to regain flying speed. This may be extremely difficult if the primary stall occurred near the ground. In rare instances, too much forward pressure in a stall recovery could cause a negative G stall. This type of stall is used intentionally by aerobatic pilots to enter inverted spins or outside snap-rolls. Although very unlikely in visual conditions, a primary stall on instruments coupled with some spatial disorientation could lead you to make unusual control movements.

The **crossed-control stall** occurs when the flight controls are crossed, meaning that rudder pressure is being applied in one direction while ailerons are applied in the opposite direction. When combined with excessive back pressure, conditions are ripe for a crossed-control stall. The most likely time for this is during a poorly executed turn to final approach for a landing. In a typical crossed-control stall, excessive inside rudder is used to tighten the turn, along with opposite aileron to keep the bank from becoming too steep. If too much elevator back pressure is added, or if the aircraft becomes too slow due to the added drag of the skid, the nose may pitch down and the wing on the inside of the turn may suddenly drop. Often these stalls occur with little warning and insufficient altitude for recovery; however, a displaced ball in the inclinometer or a skidding-turn feeling can prompt you to coordinate the controls and go around if necessary.

An **elevator trim stall** is most likely to occur during a go-around from a landing approach. In this case, the airplane's trim is adjusted for the configuration and slow speed of the approach with considerable nose-up trim. As power is applied for the go-around, the normal tendency is for the nose to pitch up, and if positive pressure is not used to counteract the strong trim forces, the nose will continue to pitch up. In addition, since full power is required, the aircraft is subject to left-turning forces which can easily lead to an uncoordinated flight condition. Now you have the conditions that can cause a stall — a high angle of attack, uncoordinated flight, and decreasing airspeed.

STALL RECOGNITION AND RECOVERY

A major part of stall recognition is knowing the flight characteristics of the airplane you are flying. Then, you must maintain an awareness of the flight conditions where various types of stalls are most likely to occur.

Most airplanes warn the pilot in a variety of ways. In general aviation airplanes there is usually a stall warning light or horn. This warning system is a simplified angle of attack indicator installed on the leading edge of the wing. On many airplanes it consists of a small mechanical switch attached to a vane that is activated by airflow as the wing approaches C_{Lmax}. Other systems use a reed-type horn similar to a party noisemaker connected to an opening in the wing. As the wing approaches C_{Lmax}, airflow on the wing creates suction that draws air through the horn. Most systems will trigger the stall warning horn and/or warning light about 5 to 10 knots above the stall speed. A disadvantage of this type of

system is that it is attached to only one wing. It is conceivable that an uncoordinated or asymmetric stall on the other wing might not actuate the warning signal. The systems that require electricity are subject to blown fuses, open circuits, or electrical failures, while the reed systems are sometimes rendered inoperative by bugs or debris wedged in the horn or plumbing.

Most airplanes provide additional indications of an impending stall. As the airflow begins to separate from the wing, there is often a buffeting or shaking through the entire airframe. If the tail is within the turbulent wake from the wings, you may feel a trembling through the elevator or rudder controls. At low airspeeds, the controls often feel mushy, and you may notice the sound of air rushing along the fuselage fading away. As lift is diminishing, you may notice a sinking feeling. These preliminary indications serve as a warning for you to decrease the angle of attack and increase airspeed by adding power. Ideally, you should be able to detect the signs of an imminent stall and make appropriate corrections before it actually occurs. Recovery at the first indication is quite simple, but if you allow the stalled condition to progress, recovery becomes more difficult.

The first priority in recovery from a stall is to reduce the angle of attack with forward elevator to allow the wings to regain lift. This may consist of merely releasing back pressure, or you may have to firmly move the elevator control forward, depending on the aircraft design, severity of the stall, and other factors. Excessive forward movement of the control column, however, may impose a negative load on the wings and delay the stall recovery. The next step, which should be accomplished almost simultaneously, is to smoothly apply maximum power to increase the airspeed and to minimize the loss of altitude. As airspeed increases, adjust power to return the airplane to the desired flight condition. In most training situations, this is straight-and-level flight or a climb. During the recovery, coordinated use of the controls is especially important. Uncoordinated flight control inputs can aggravate a stall and result in a spin. You also must avoid exceeding the airspeed and r.p.m. limits of the airplane.

SPINS

The spin is one of the most complex of all flight maneuvers. There are many aerodynamic and inertial factors that contribute to the spinning of an airplane. In spite of this, it is an easy maneuver to enter in most airplanes, and recovery can also be easy in many general aviation airplanes. Because of the low G-forces and the relative ease of execution, it is usually the first maneuver taught to pilots learning aerobatics. There are major differences between intentional spins performed by a well-trained pilot in a properly loaded airplane with no restrictions against spinning, and an accidental spin. The first can be enjoyable, the second deadly. [Figure 12-39]

For our purposes, a spin can be defined as an aggravated stall resulting in autorotation, which simply

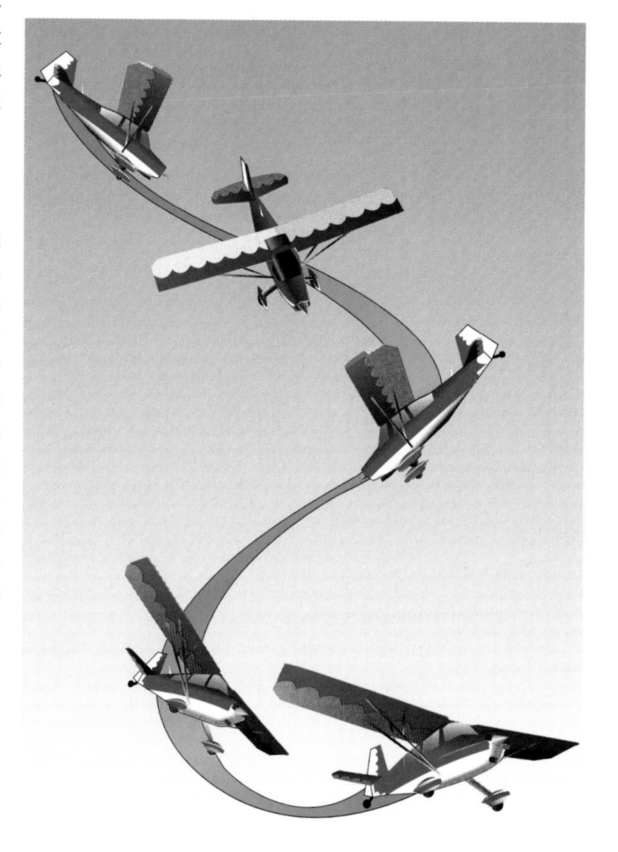

Figure 12-39. During an intentional spin in a typical aerobatic trainer, the ground seems to whirl around very fast and the nose may seem to be pointed straight down. It is hard to believe that the airspeed is near stall speed, the nose is only down 45-60°, and each turn of the spin takes three or four seconds. Altitude loss may be 500-1,000 feet or more per turn, depending on airplane type.

means that the rotation is stable and will continue due to aerodynamic forces if nothing intervenes. During the spin, the airplane descends in a helical, or corkscrew, path while the wings remain unequally stalled. Even though the nose is usually well below the horizon, the angle of attack remains greater than the stalling angle of attack. This is usually due to the pilot holding back pressure, maintaining the high angle of attack and creating high drag, combined with the large upward component of the relative wind.

PRIMARY CAUSES

A stall must occur before a spin can take place, and all of the factors that affect stalls also affect spins. The principle difference that transforms a straight-ahead stall into a spin entry is that one wing stalls more than the other. This usually happens when there is an element of yaw as the stall breaks, and occurs most often when the airplane is in uncoordinated flight. A spin also may develop if forces on the airplane are unbalanced in other ways, for example, from yaw forces due to an engine failure on a multi-engined airplane, or if the CG is laterally displaced by an unbalanced fuel load. Stalling with crossed controls is a major cause of spins, and usually happens when either too much or not enough rudder is used for existing yawing forces. If the airplane is allowed to stall with crossed controls, it is likely to enter a spin. The spin that results from crossed controls usually results in rotation in the direction of the rudder being applied, regardless of which wing is raised.

During an uncoordinated maneuver, the pitot/static instruments, especially the altimeter and airspeed indicator, may be unreliable due to the uneven distribution of air pressure over the fuselage. This error may lead you to believe that your airspeed is sufficient to maintain smooth airflow over the wing when, in fact, you are very close to stalling. Your only warnings of the stall may be aerodynamic buffeting and/or the stall warning horn.

Coordination of the flight controls is natural and easy under most conditions, but preoccupation with situations inside or outside the cockpit, maneuvering to avoid other aircraft, and maneuvering to clear obstacles during takeoffs, climbs, approaches, or landings has caused control problems for many pilots. Because of this, you will be required to learn how to recognize and cope with these distractions by practicing flight at slow airspeeds with realistic distractions in your training.

PHASES OF A SPIN

There are three phases in a complete spin maneuver. The **incipient spin** is the first phase, and exists from the time the airplane stalls and rotation starts until the spin is fully developed. A **fully developed spin** exists from the time the angular rotation rates, airspeed, and vertical descent rate are stabilized from one turn to the next. The third phase, **spin recovery**, begins when the anti-spin forces overcome the pro-spin forces.

If an airplane is near the stalling angle of attack, and if more lift is lost from one wing than from the other, that wing will drop more quickly than the other, creating a roll toward the wing with less lift. As this wing drops, its local relative wind will come more from below, further increasing the angle of attack for that wing. Likewise, as the airplane rolls around its center of gravity, the upper wing has a lower local angle of attack, and continues to develop some lift. This situation of unbalanced lift tends to increase as the airplane yaws toward the low wing, accelerating the higher outside wing while slowing the inner, lower wing still more. As with other stalls, the nose drops, and as inertial forces come into play, the spin usually stabilizes at a steady rate of rotation and descent.

It is crucial that you initiate recovery from an inadvertent spin as soon as possible, since many airplanes will not recover from a fully developed spin, and others continue for several turns before recovery control inputs become effective. The recovery from an incipient spin normally requires less altitude, and time, than the recovery from a fully developed spin. Keep in mind that although some characteristics of a spin are predictable, every airplane spins differently, and an individual airplane's spin characteristics vary depending on configuration, loading, and other factors.

How Many Ways to Spin?

The spin is perhaps the most basic of all aerobatic maneuvers. Although many early pilots were able to execute and recover from spins, it wasn't until after World War I that spin aerodynamics began to be explored in detail. The tailspin became a staple of barnstormers and stunt performers, and is still a crowd-pleaser at airshows today.

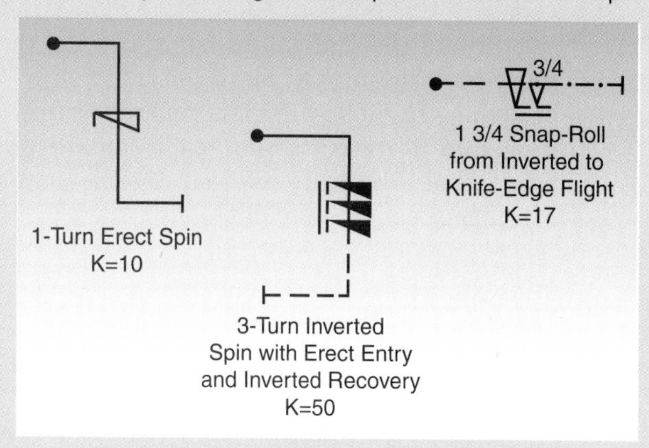

For competitive aerobatics, there is a catalog of standard aerobatic maneuvers, grouped according to the type of maneuver and assigned coefficients of difficulty (K) for judging purposes. It is named for its originator, the Spanish Count Ferdinand Aresti. There are 192 different spins listed, consisting of erect spins, inverted spins, flat spins, and inverted flat spins among others. Snap rolls are another family of maneuvers closely related to spins. In essence, a snap roll is an abrupt spin forced on the airplane at an airspeed well above the usual stall speed. The wings are unequally stalled, just as in a spin, but the stall is caused by quickly forcing the angle of attack past the critical angle of attack. Unlike spins, snap rolls can be performed horizontally or upward as well as downward. Just as spins can be erect or inverted, snap rolls can be inside or outside. If you add in all the various inside and outside snap rolls with all their possible entries and recoveries, the extended family of spins includes at least 840 different maneuvers, ranging in difficulty from the basic one-turn erect spin with right-side-up entry and recovery, with a K-factor of 10, to the 3 turn erect flat spin with inverted entry and inverted recovery with a K-factor of 74.

WEIGHT AND BALANCE

Both the total weight of the airplane and the distribution of weight influence the spin characteristics of the airplane. Higher weights generally mean slower initial spin rates, but as the spin progresses, spin rates may tend to increase. The higher angular momentum extends the time and altitude necessary for spin recovery in a heavily loaded airplane. The location of the center of gravity is even more significant, affecting the airplane's resistance to spins as well as all phases of the spin itself. Most airplanes are more stable with the CG near its forward limit. This increases control forces and makes it less likely that you will make large, abrupt control movements. It also means that when trimmed, the airplane will tend to return to level flight if you let go of the controls, but the stall speed will be higher with the CG forward. Since moving the CG aft decreases longitudinal stability and reduces pitch control forces in most airplanes, it tends to make the airplane easier to stall. Once a spin is entered, an airplane with its CG farther aft tends to have a flatter spin attitude. If the CG is outside the limits of the CG envelope, or if power is not reduced promptly, the spin will be more likely to go flat. A **flat spin** is characterized by a near level pitch and roll attitude with the spin axis near the CG of the airplane. Although the altitude lost in each turn of a flat spin may be less than in a normal spin, the extreme yaw rates (often exceeding 400° per second) result in high descent rates. The relative wind in a flat spin is nearly straight up, keeping the wings at a high angle of attack. More importantly, the upward flow over the tail may render the elevators and rudder ineffective, making recovery impossible.

Spin recovery may be difficult if the CG is too far aft and rotation is around the CG.

SPIN RECOVERY

Where the primary goal in recovery from a straight-ahead stall is to reduce the angle of attack to restore airflow over the wings, spin recoveries have the additional goal of stopping the rotation. The complex aerodynamics of spins may dictate vastly different recovery procedures for different airplanes, so no universal spin recovery procedure can exist for all airplanes. The recommended recovery procedure for some airplanes is simply to reduce power to idle and let go of the controls, and others are so resistant to spins that they can be considered spin-proof. At the other extreme, the design of some airplanes is such that recovery from a developed spin requires definite control movements, precisely timed to coincide with certain points in the rotation, for several turns. You should be intimately familiar with the procedures in the POH for the aircraft you will be flying. The following is a general recovery procedure for erect spins, but it should not be applied arbitrarily without regard for the manufacturer's recommendations.

In some airplanes, the displacement of the ball in spins will depend on where the inclinometer is installed. If the inclinometer is on the left side of the cockpit, the ball will move to the left in both left and right spins.

- Move the throttle or throttles to idle. This minimizes the altitude loss and reduces the potential for a flat spin to develop in most airplanes. It also eliminates possible assymetric thrust in multi-engine airplanes. Engine torque and precession can increase the angle of attack or the rate of rotation in single-engine airplanes, aggravating the spin.

- Neutralize the ailerons. Aileron position is often a contributing factor to flat spins, or to high rotation rates in normal spins.

- Apply full rudder against the spin. Spin direction is most reliably determined from the turn coordinator. Do not use the ball in the inclinometer. Its indications are not reliable, and may be affected by its location within the cockpit. For example, in some airplanes, if the inclinometer is on the left side of the cockpit, the ball will move to the left in both left and right spins, and if installed on the right side of the cockpit, the ball will move to the right during a spin in either direction.

- Move the elevator control briskly to approximately the neutral position. Some aircraft merely require a relaxation of back pressure, while others require full forward elevator or stabilator pressure

- As rotation stops, indicating the stall has been broken, neutralize the rudder. If you maintain rudder pressure after rotation stops, you could enter a spin in the other direction.

- Recover from the resulting dive with gradual back elevator pressure. Pulling too hard could trigger a secondary stall, or worse, could exceed the flight load factor limits and damage the aircraft structure. Conversely, recovering too slowly from the dive could let the airplane exceed airspeed limits, especially in aerodynamically clean airplanes. Avoiding excessive speed buildup in the recovery is another reason for closing the throttle during spin recovery. Add power as you resume normal flight, being careful to observe power and r.p.m. limits.

SUMMARY CHECKLIST

✓ Opposing forces are balanced in unaccelerated flight.

✓ The coefficient of lift increases with angle of attack, until the stalling angle of attack is reached.

✓ If all other factors remain constant, lift is proportional to the square of the speed.

✓ The four most common ways to increase lift are to increase angle of attack, increase airspeed, increase the camber of the wing, or increase the wing area.

- ✓ Modifying the camber of the wing with flaps and/or leading edge devices can postpone airflow separation, resulting in lower stall speeds and greater coefficients of lift.
- ✓ Induced drag increases with angle of attack, and decreases with speed. Parasite drag increases in proportion to the square of the speed.
- ✓ A wing's planform affects induced drag, stall propagation, stability, maneuverability, and other characteristics.
- ✓ At a distance of less than one wingspan above the ground, induced drag diminishes, allowing the airplane to fly with less thrust. This phenomenon is called ground effect.
- ✓ Best glide angle is achieved at the speed where the ratio of lift to drag is the greatest (L/D_{max}). This speed gives the greatest horizontal distance in a power-off glide, as well as maximum range with power.
- ✓ Maximum level flight speed is attained when maximum thrust is balanced by drag.
- ✓ Load factor is the ratio of lift being developed by the wings divided by the weight of the airplane.
- ✓ Stability characteristics can be described for each of the airplane's three axes of rotation.
- ✓ Static stability is determined by whether the airplane tends to return to equilibrium when disturbed, diverges farther from equilibrium, or maintains the same degree of displacement from equilibrium.
- ✓ Dynamic stability, or damping, describes how quickly the airplane reacts to its static stability, as well as whether it returns to equilibrium or not.
- ✓ The center of gravity is ordinarily located forward of the wings' center of lift. The resulting nose-down moment is normally balanced by the tail-down force created by the horizontal tail. This arrangement contributes to positive longitudinal stability, since pitch disturbances alter the tail-down force to return the airplane to balanced flight.
- ✓ In most airplanes, if the center of gravity is too far aft, the airplane will be unstable in pitch.
- ✓ Wing dihedral, sweepback, and mounting the wing on top of the fuselage are design factors that contribute to lateral stability.
- ✓ The vertical tail and the sides of the fuselage aft of the center of gravity tend to keep the fuselage aligned with the relative wind, contributing to directional stability.
- ✓ Too much lateral stability can lead to Dutch roll, while too much directional stability can lead to spiral instability. Spiral instability is considered less objectionable, so designers usually try to minimize Dutch roll at the expense of a little spiral instability.
- ✓ Climbs are a result of excess thrust. The maximum rate of climb (V_Y) occurs at the airspeed where the difference between power available and power required is the greatest.
- ✓ The maximum angle of climb (V_X) is achieved at the airspeed where the difference between thrust available and thrust required is the greatest.

- ✓ Higher aircraft weight reduces climb performance, glide ratio, and range.
- ✓ Load factor in coordinated, level turns is strictly a function of bank angle.
- ✓ Adverse yaw is caused by greater induced drag on the outside wing as an airplane rolls into a bank.
- ✓ Radius and rate of turn are influenced by airspeed and bank angle.
- ✓ Exceeding the stalling angle of attack (C_{Lmax}) will always result in a stall, regardless of weight, load factor, airspeed, or attitude.
- ✓ To spin, there must first be a stall. Prevent the stall and you prevent the spin.
- ✓ Spins occur when one wing stalls more fully than the other, and an autorotation develops.
- ✓ Spin recovery consists of breaking the stall by reducing angle of attack, stopping the rotation with rudder, and returning the airplane to normal flight.

KEY TERMS

Downwash
Relative Wind
Angle Of Attack
Angle Of Incidence
Slot
Slat
Leading Edge Flaps
Taper
Aspect Ratio
Sweep
Ground Effect
Form Drag
Interference Drag
Skin Friction Drag
Total Drag
Spoilers
Speed Brakes
Load Factor
Static Stability
Dynamic Stability
Longitudinal Stability
Tail-Down Force
Lateral Stability
Directional Stability
Dutch Roll
Spiral Instability
Absolute Ceiling
Service Ceiling
Horizontal Component Of Lift
Adverse Yaw
Power-On Stalls
Power-Off Stalls
Accelerated Stalls
Secondary Stall
Crossed-Control Stall
Elevator Trim Stall
Incipient Spin
Fully Developed Spin
Spin Recovery
Flat Spin

QUESTIONS

1. True/False. The angle of attack at which a wing stalls varies depending on total aircraft weight, bank angle, load factor, and airspeed.

2. Wing slats deploy at high angles of attack, reducing the stall speed. The stall speed decreases because

   A. the chord line of the wings is reduced, decreasing the local angle of attack.
   B. the slats keep the airflow attached and increase the lifting area of the wings.
   C. the airflow through the slats increases the airspeed over the wings and reduces parasite drag.

3. True/False. Airplanes with a low aspect ratio have higher fuel efficiency because they have less induced drag.

4. True/False. There is an increase in lift when flying in ground effect that may allow an airplane to lift off even when it cannot climb, and which may cause a landing airplane to balloon or float before landing.

5. L/D_{max} is the airspeed that provides

   A. maximum range with or without power.
   B. maximum endurance and maximum rate of climb.
   C. maximum angle of climb and maximum level flight speed.

6. What term is used for the airspeed at which maximum thrust is balanced by drag in level flight?

7. Which of the four aerodynamic forces does not depend on the flight path or relative wind?

8. All other things being equal, increasing the weight of an airplane causes the stall speed to

   A. remain the same.
   B. increase.
   C. decrease.

10. An airplane has positive static stability and negative dynamic stability in pitch. If it is trimmed for level flight and you disturb the equilibrium by pulling back and releasing the elevator control, how will the airplane respond after you release the controls?

    A. The airplane will continue to pitch up farther and farther without oscillating.
    B. There will be a series of increasing pitch oscillations, each steeper than the last.
    C. There will be a series of gradually decreasing pitch oscillations until the airplane is again in level flight.

11. True/False. An airplane with spiral instability has a tendency to steepen its bank and gradually nose down unless the pilot intervenes.

12. What is the term for the maximum pressure altitude that an airplane is capable of climbing 100 f.p.m.?

13. In a skidding right turn, what do you see in the inclinometer, and what should you do to coordinate the turn?

    A. The ball is to the left; add more left rudder.
    B. The ball is to the right; add more right rudder.
    C. The ball is to the right; steepen the bank with more aileron.

14. Using the adjacent chart, determine the stall speed for an airplane with a level-flight stalling speed of 42 KIAS when the airplane is in a coordinated turn at a bank angle of 55°.

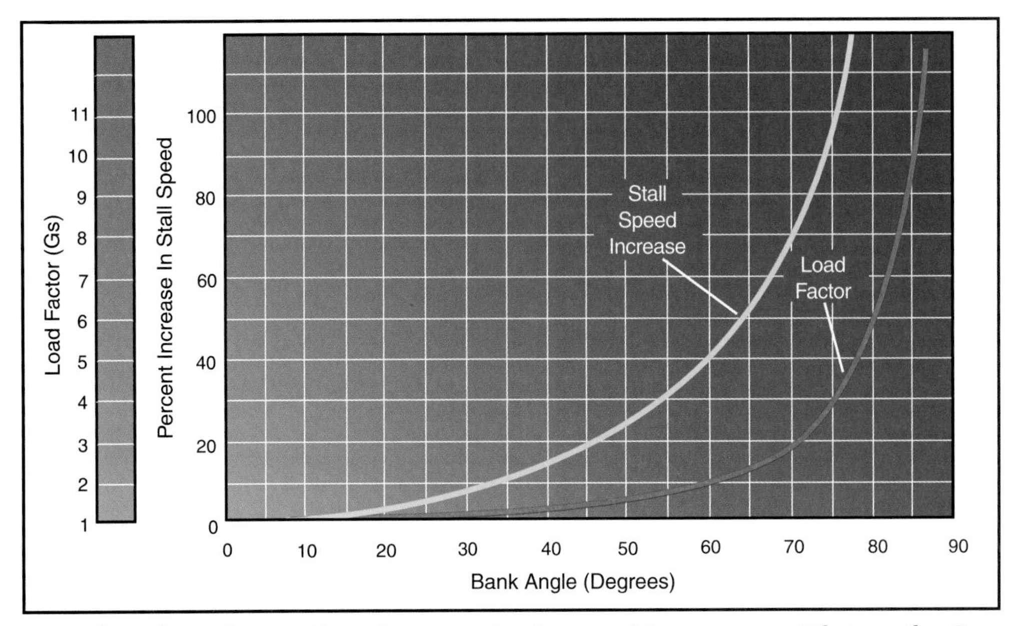

15. The radius of a coordinated turn can be decreased in two ways. What are they?

16. In most cases, stall recovery includes applying full power, but spin recovery procedures for most airplanes direct that the throttle be closed immediately. Why?

17. Why are flat spins considered more dangerous than other spins?

SECTION B
PREDICTING PERFORMANCE

As you learn to fly aircraft with higher performance, you will discover that they are flown less by sight, sound, and feel, and more by the book and by the numbers. You have seen how aircraft behavior is governed by well-established physical laws and mathematical relationships. Now, you can use these concepts to predict the airplane's performance under different conditions, as well as to stay within operating limitations. The most frequently used performance information is usually provided in the POH in the form of tables and graphs.

FACTORS AFFECTING PERFORMANCE

Airplanes operate in an environment that is constantly changing, and several physical characteristics of the atmosphere influence aircraft performance. Temperature, pressure, humidity, and the movement of the air all have definite and foreseeable effects in every phase of flight, and runway conditions affect takeoff and landing performance. Although standard atmospheric conditions are the foundation for estimating airplane performance, airplanes are rarely flown in conditions that match the standard atmosphere. Some basic performance calculations will help you to take the differences between standard and existing conditions into account.

DENSITY ALTITUDE

The density of the air directly influences the performance of your airplane by altering the lift of the airframe as well as the ability of the engine to produce power. Both reciprocating and turbine engines produce less thrust as air density decreases and/or temperature increases, since these factors reduce the mass of the air passing through the engine. In propeller-driven airplanes, propeller efficiency is reduced when there are fewer air molecules to act upon. Jet thrust is the reaction force resulting from accelerating mass backward, so a reduction in the mass going into the engine reduces thrust if other factors remain the same. **Density altitude** is pressure altitude corrected for nonstandard temperature, and is equal to pressure altitude only when standard atmospheric conditions exist at that level. Since these specific conditions are uncommon, you should understand and be able to compute density altitude. Although the performance degradation due to increased density altitude occurs throughout the airplane's operating envelope, it is especially noticeable in takeoff and climb performance. You can find density altitude using either a flight computer or a density altitude chart. [Figure 12-40]

As air temperature increases or air density decreases, engine performance decreases for both piston and gas turbine engines.

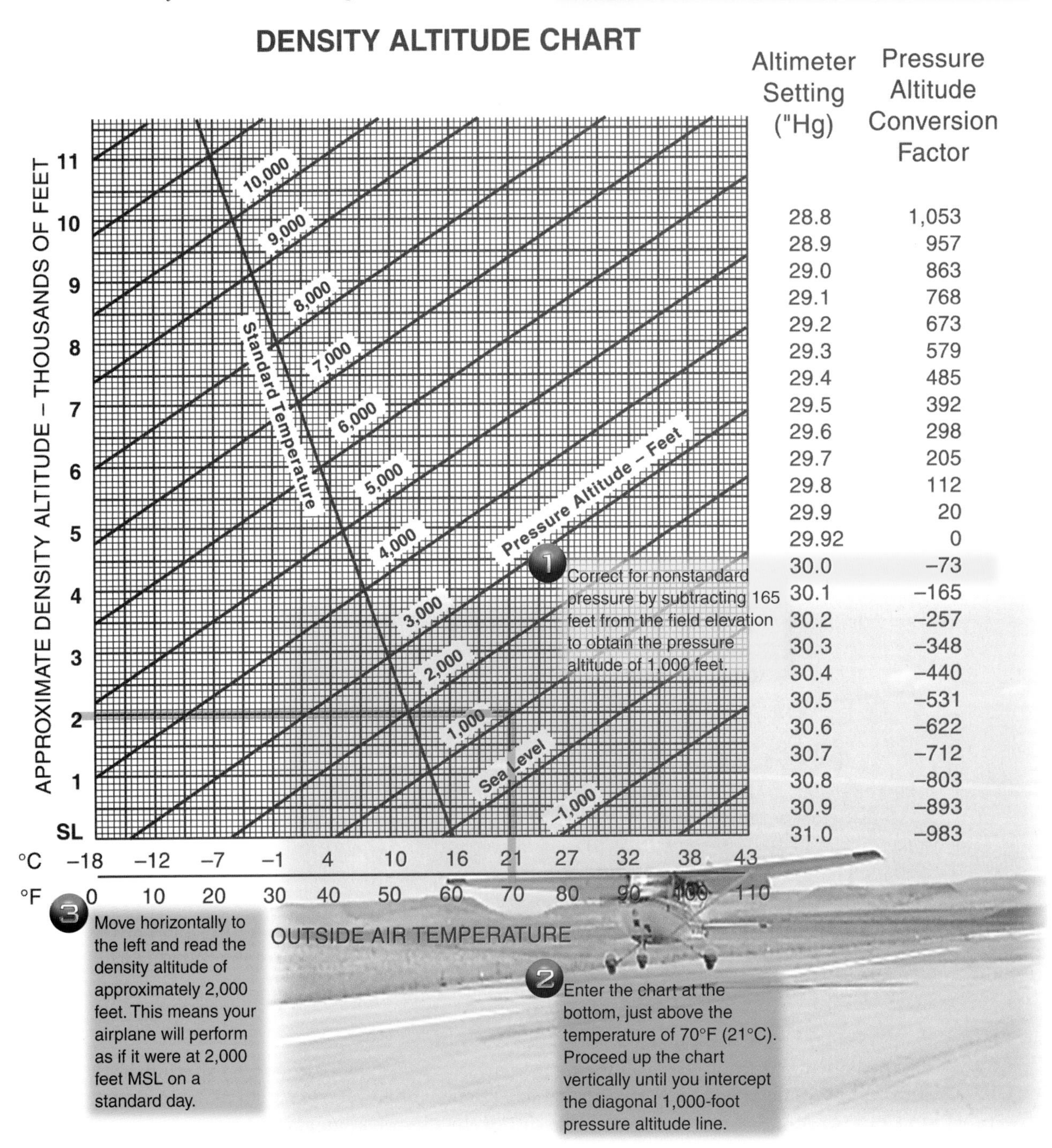

Altimeter Setting ("Hg)	Pressure Altitude Conversion Factor
28.8	1,053
28.9	957
29.0	863
29.1	768
29.2	673
29.3	579
29.4	485
29.5	392
29.6	298
29.7	205
29.8	112
29.9	20
29.92	0
30.0	–73
30.1	–165
30.2	–257
30.3	–348
30.4	–440
30.5	–531
30.6	–622
30.7	–712
30.8	–803
30.9	–893
31.0	–983

Figure 12-40. The chart combines air temperature, barometric pressure, and altitude to arrive at the density altitude. For example, at a field elevation of 1,165 feet MSL, a temperature of 70°F (21°C), and an altimeter setting of 30.10, the density altitude is almost 2,000 feet.

Records Ripe for Breaking

Getting the utmost performance out of an aircraft is a challenge that gives many pilots satisfaction. Perhaps you have thought of setting a speed, distance, or altitude record yourself. There are many aviation records that were set 30-60 years ago, so given the advances in engineering, materials and technology, it should be possible to advance some of these records today. To get your feet wet, you might take a crack at Francesco Agello's piston-engined seaplane speed record of 440.68 miles per hour. He set this record on October 23, 1934 in a bright red Macchi-Castoldi MC-72. This racing floatplane had two 1,500 HP engines in tandem, driving fixed-pitch contra-rotating propellers.

Jacqueline Cochran set more flying records than any other person, and many of them still stand. Maybe you would like to try to beat her closed-course speed records for 100 km or 1,000 km in a piston-engined airplane, set in the late 1940s. The 100 km record is 469.549 mph, and the 1,000 km record is 431.09 mph. Both records were set in a North American P-51.

Several of the speed records set in the 1950s and 1960s by Max Conrad, Sheila Scott, and Geraldine Mock still stand. These pilots did not use custom-built racing airplanes or converted fighters. They flew Cessnas and Pipers, similar to those you are using in your commercial training.

Want to build flight time? Perhaps you could try to beat the record for time aloft of 64 days, 22 hours, 19 minutes, and 5 seconds, set in a Cessna 172 from December 4, 1958 through February 7, 1959 by Robert Timm and John Cook over Nevada.

You do not need a high-tech airplane to set a new record, just careful and thorough preflight preparation, accurate navigation, and precise flying skills. Good luck!

Density altitude calculations are based on pressure altitude and air temperature. A flight computer or a density altitude chart may be used. See figure 12-40.

Most modern performance charts do not require you to compute density altitude. Instead, the computation is built into the performance chart itself. All you have to do is enter the chart with the correct pressure altitude and the temperature. Older charts, however, may require you to compute density altitude before entering them.

Water vapor also has an effect on density altitude. Humidity refers to the amount of water vapor in the atmosphere and is expressed as a percentage of the maximum amount of vapor the air can hold. This amount varies with air temperature. The warmer it is, the more water vapor the air can hold. When the air is saturated, it cannot hold any more water vapor at that temperature, and its humidity is 100%.

Although the effects of humidity are not shown on performance charts or included in density altitude computations, moist air does reduce airplane performance. For one thing, water vapor takes up volume that is normally available for combustion air, so less air enters the engine in high humidity conditions. This causes a small increase in density altitude. In conditions of very high humidity, the reduction in engine horsepower may be as much as 7%, and the airplane's overall takeoff and climb performance may be reduced by as much as 10%. Always provide yourself with an extra margin of safety by using longer runways and by expecting reduced takeoff and climb performance whenever it is hot and humid.

SURFACE WINDS

Some performance tables are based on pressure/density altitude.

Wilbur and Orville Wright insisted on a measured wind at Kitty Hawk of at least 11 miles per hour before they flew their airplane. In addition to increasing ground speed and fuel economy in cruise, favorable winds

JIMMY DOOLITTLE DISCUSSES DENSITY ALTITUDE

Aerial pioneer Jimmy Doolittle had a close encounter with the effects of density altitude during a tour in the 1930s.

"I had left the Hawk in China, so I had to borrow an aircraft to put on the show. This was almost my undoing at a field located at high altitude in the Dutch East Indies.

I borrowed a Hawk from the Dutch air force and went through my aerial routine. To conclude my act, I put the airplane into a power dive, intending to pull out as close to the ground as possible. When I started to pull out, the airplane didn't respond in the thin air as I had expected. It began to 'mush,' and the ground came up too fast. I managed to pull out, but only after my wheels hit the ground rather hard, though without damaging the Hawk. I had not calculated that the Hawk wouldn't perform in thin air as it would have at a lower altitude.

Several Dutch pilots who had seen the near accident congratulated me. The commander of the group said it was 'the most delicate piece of flying' he had ever seen. But I thought it was stupid flying and said so. He smiled and said, 'We knew; we wondered if you would lie about it.'"

That quote is from Doolittle's autobiography, *I Could Never Be So Lucky Again*. Part of maintaining situational awareness is to know what you can reasonably expect from your airplane under existing conditions.

can significantly shorten takeoff and landing distances. Since surface winds are not always aligned with the runway in use, you need to determine how much of the wind is acting along the runway and how much is acting across it. This is another instance of resolving a vector into its components.

The **headwind** or **tailwind component** of the wind vector affects the length of the ground run. Headwinds are less beneficial to takeoff distances than tailwinds are detrimental. For instance, in some airplanes a headwind of 10 knots may reduce the takeoff distance by approximately 10%, but a tailwind of the same amount increases the takeoff distance by 40%. These differentials vary from airplane to airplane, and also depend on the pilot's takeoff technique.

In your earlier training you learned that the **crosswind component** acts to push the airplane to the side, and must be compensated for to keep the airplane on the runway. Most airplanes have a maximum demonstrated crosswind component listed in the POH, although this is not an operating limitation. FARs require that all airplanes type-certificated since 1962 have safe ground handling characteristics in 90° crosswinds equal to 20% of V_{S0}. For example, if V_{S0} for an airplane is 65 knots, the manufacturer's **demonstrated crosswind component** is 13 knots. Your personal crosswind limit is based on your own skill level in taking off and landing safely in a particular airplane type.

You can easily compute headwind and crosswind components using a wind component chart. When you use the chart, remember that for takeoff or landing, surface wind is reported in magnetic direction, so it corresponds with the runway number. [Figure 12-41]

Headwind and crosswind components can be found using the wind component chart. See figure 12-41.

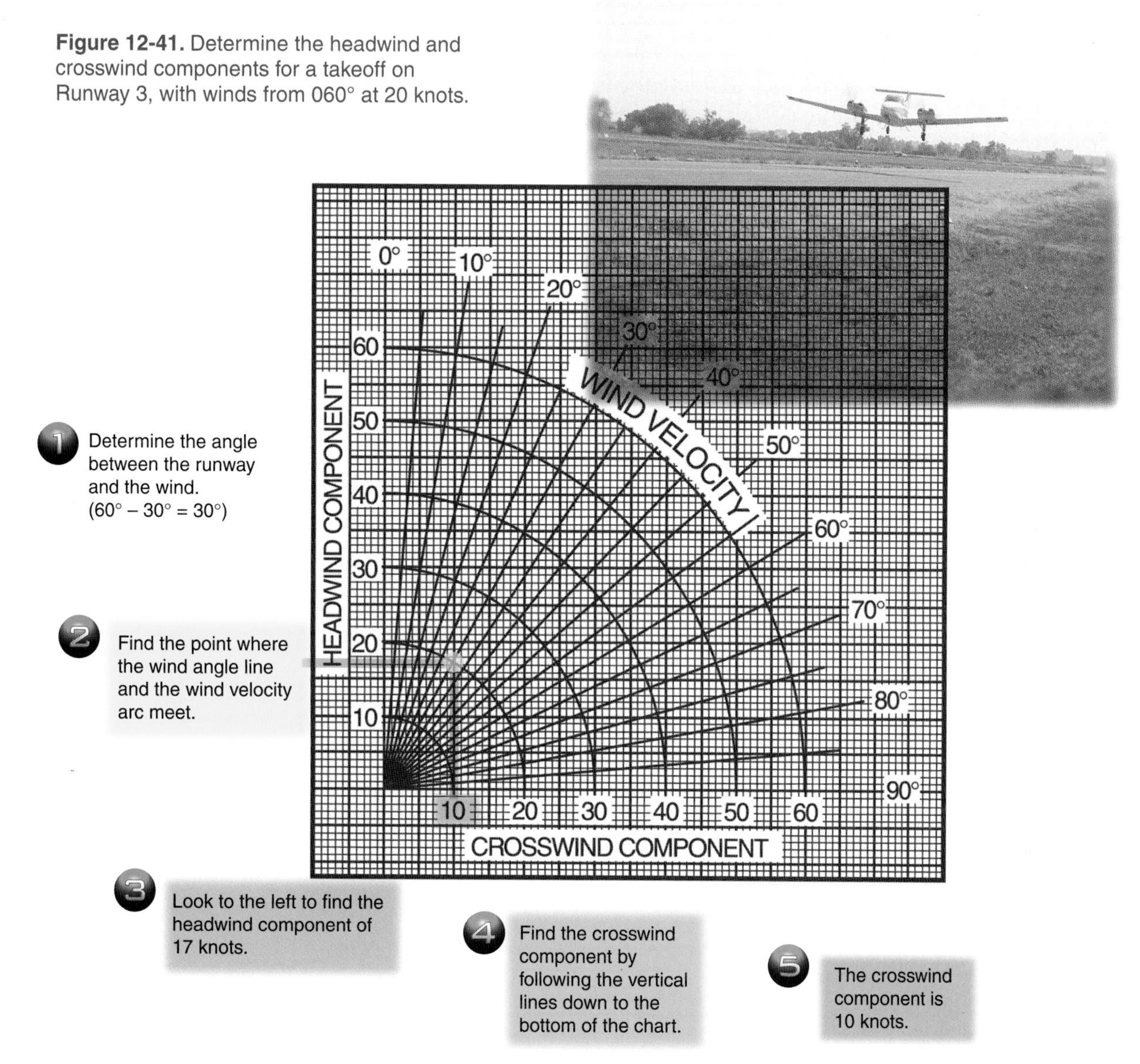

Figure 12-41. Determine the headwind and crosswind components for a takeoff on Runway 3, with winds from 060° at 20 knots.

WEIGHT

Lift must equal the effective weight of the airplane and its contents to maintain altitude during flight, thus, an increase in weight requires a proportional increase in lift. In cruise, this additional lift is usually provided by flying at a higher angle of attack, but in order to take off, a heavily loaded airplane must accelerate to a higher speed to generate the required lift. The additional weight also reduces acceleration during the takeoff roll, adding to the total takeoff distance. After liftoff, the heavily loaded airplane climbs more slowly, and its service ceiling is lower. This is because there is less difference between power required and power available. Moreover, exceeding the maximum allowable weight in any flight regime may cause the airplane to become unstable and difficult to fly. For an airplane to be safe, it must be operated within weight and balance limitations. Procedures for computing weight and CG location are covered in the next section.

RUNWAY CONDITIONS

Most published takeoff and landing distances are based on a paved, level runway with a smooth, dry surface. In the real world, completely level runways are uncommon. The gradient, or slope, of the runway is expressed as a percentage. For example, a gradient of

2 percent means the runway height changes 2 feet for each 100 feet of runway length. Positive gradients are uphill, and negative gradients are downhill. A positive gradient increases takeoff distances because the airplane must accelerate uphill. Conversely, landing uphill reduces the ground roll. A negative gradient has the opposite effects, shortening takeoff distance and lengthening the landing roll. Runways with a gradient of 0.3% or more are listed in the A/FD. Many larger aircraft incorporate runway slope into their takeoff distance calculations.

An uphill runway slope increases takeoff distance because the airplane accelerates more slowly.

The surface conditions of the runway affect both takeoff and landing distances. Rolling resistance from a rough or soft surface lengthens your takeoff roll, and the POH may include information to adjust takeoff distances for different runway surfaces. [Figure 12-42]

Figure 12-42. This takeoff distance chart suggests a 15% increase in the ground roll for takeoff from a dry, grass runway. You should be careful in applying this information, since there is wide variation between grass runways. Takeoff distance from a wild turf runway at a back-country airstrip can be much longer than from the lawn-like grass at a well-groomed airport.

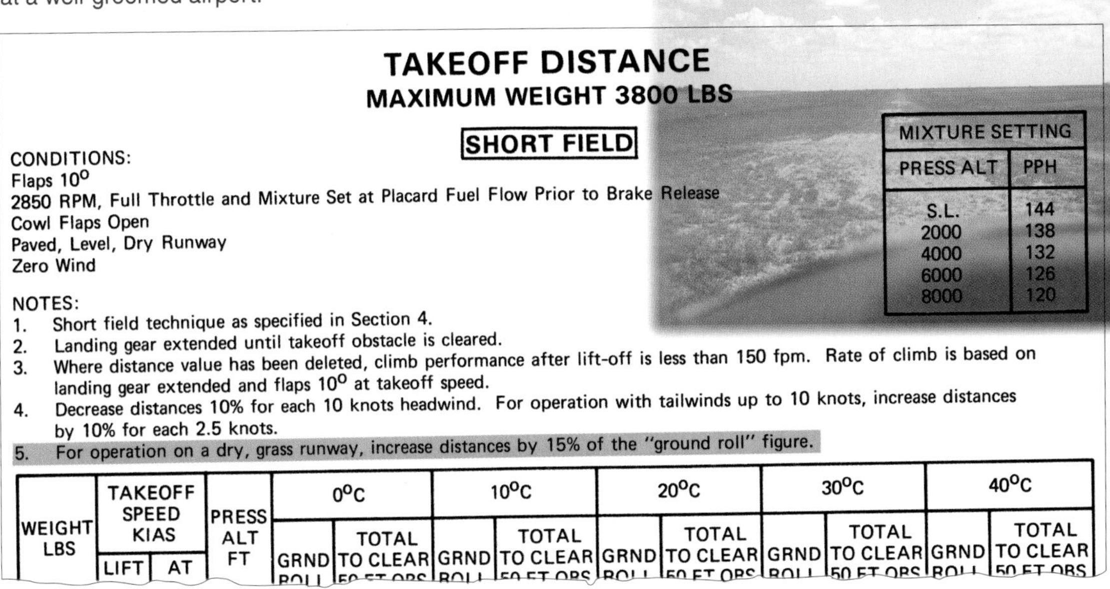

TAKEOFF DISTANCE
MAXIMUM WEIGHT 3800 LBS
SHORT FIELD

CONDITIONS:
Flaps 10°
2850 RPM, Full Throttle and Mixture Set at Placard Fuel Flow Prior to Brake Release
Cowl Flaps Open
Paved, Level, Dry Runway
Zero Wind

MIXTURE SETTING	
PRESS ALT	PPH
S.L.	144
2000	138
4000	132
6000	126
8000	120

NOTES:
1. Short field technique as specified in Section 4.
2. Landing gear extended until takeoff obstacle is cleared.
3. Where distance value has been deleted, climb performance after lift-off is less than 150 fpm. Rate of climb is based on landing gear extended and flaps 10° at takeoff speed.
4. Decrease distances 10% for each 10 knots headwind. For operation with tailwinds up to 10 knots, increase distances by 10% for each 2.5 knots.
5. For operation on a dry, grass runway, increase distances by 15% of the "ground roll" figure.

WEIGHT LBS	TAKEOFF SPEED KIAS		PRESS ALT FT	0°C		10°C		20°C		30°C		40°C	
	LIFT	AT		GRND ROLL	TOTAL TO CLEAR 50 FT OBS	GRND ROLL	TOTAL TO CLEAR 50 FT OBS	GRND ROLL	TOTAL TO CLEAR 50 FT OBS	GRND ROLL	TOTAL TO CLEAR 50 FT OBS	GRND ROLL	TOTAL TO CLEAR 50 FT OBS

During landing, **braking effectiveness** can be a major concern. A dry runway surface allows the brakes to dissipate the maximum amount of energy without skidding the tires, but when the runway is wet, the maximum coefficient of friction between the tires and runway may drop by 30-60%. The tires begin to break loose from the runway surface and slide at lower brake application forces, so you must apply the brakes more gently to maximize deceleration and maintain control. The lighter application of brakes means that energy is dissipated more slowly, increasing the length of the landing roll. Obviously, if the ground roll becomes too long, the aircraft will run off the end of the runway. An ice-covered runway may reduce friction still more, but perhaps the worst situation is to lose braking effectiveness as a result of **hydroplaning**, which happens when a thin layer of water separates the tires from the runway. Hydroplaning reduces friction even more than smooth, clear ice. [Figure 12-43] When available, ATC furnishes pilots the quality of braking action received from pilots or airport management. The quality of braking action is described by the terms good, fair, poor, nil, or a combination of these terms. When braking action approaches nil, maintaining directional control may become impossible, and a strong crosswind could push the airplane off the side of the runway.

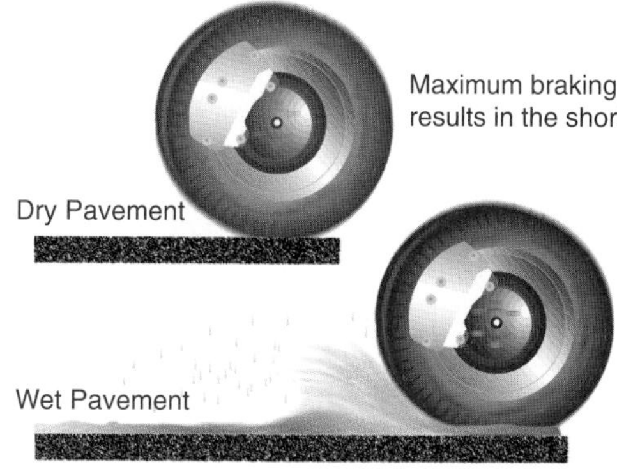
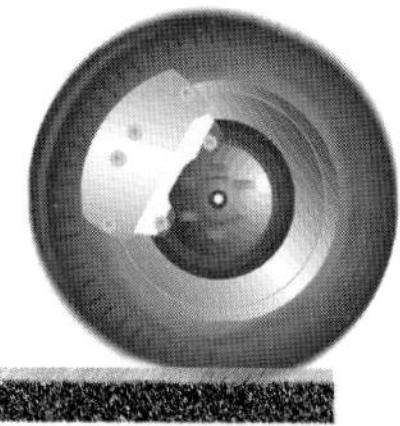
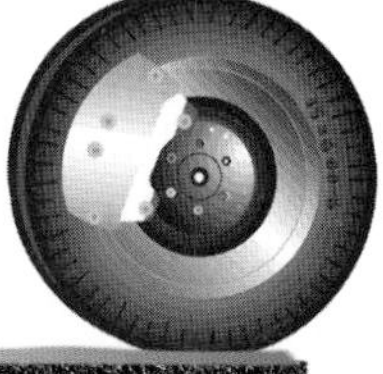
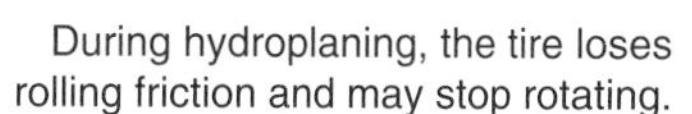

Figure 12-43. Braking effectiveness depends on the runway surface conditions.

Dynamic hydroplaning occurs when there is standing water or slush on the runway about one-tenth of an inch or more in depth. A wedge of water builds up, lifting the tires away from the runway surface. The speed of the aircraft, the depth of the water, and the air pressure in the tires are some of the factors that affect dynamic hydroplaning. There is a simple rule of thumb you can use to estimate the minimum speed at which dynamic hydroplaning will occur. Take the square root of the tire pressure and multiply by 8.6 to obtain the hydroplaning groundspeed in knots. For example, with a tire pressure of 30 p.s.i., you could expect hydroplaning at about 47 knots. Once hydroplaning has started, it can persist at lower speeds. Viscous hydroplaning can occur on just a thin film of water, not more than one-thousandth of an inch deep, if it covers a smooth surface. Reverted rubber hydroplaning is the result of a prolonged locked-wheel skid, in which reverted rubber acts as a seal between the tire and the runway. Entrapped water is heated to form steam, which supports the tire off the pavement. If you must use a runway where braking effectiveness is poor or nil, be sure the length is adequate and the surface wind is favorable.

THE PILOT'S OPERATING HANDBOOK

According to FARs, you can find your aircraft's operating limitations in the approved airplane flight manual (AFM), approved manual materials, markings and placards, or any combination of these. For most airplanes manufactured in the U.S. after March 1, 1979, the **pilot's operating handbook (POH)** is the approved flight manual, and a page in the front of the handbook will contain a statement to that effect. Prior to that date, aircraft under 6,000 pounds were not required to have AFMs, but manufacturers usually provided similar information in Owner's Manuals or Information Manuals. Format and content in these manuals varied widely. Then, the General Aviation Manufacturers Association (GAMA) adopted a standardized format so that each POH provides the same information in the same order. The GAMA framework presents performance charts in a logical order, beginning with general information and flight planning calculations and progressing through takeoff, climb, cruise, descent, and landing. If you fly older airplanes, you should keep in mind that older manuals may not use the same conventions or units of measure as those in the newer, standardized style.

PERFORMANCE CHARTS

Performance charts are compiled from information gathered during actual flight tests. Since it is impractical to test the airplane under all the conditions portrayed on the performance charts, the engineers evaluate specific flight test results and mathematically derive the remaining data. In developing performance charts, manufacturers must make assumptions about the condition of the airplane. They assume the airplane is in good condition, and that the engine is developing its full rated power. It is prudent to keep in mind that the flight tests are performed by experienced test pilots, and manufacturers often use the best results from many repetitions of each test. While the performance numbers in the POH are achievable and reliable, you should always consider your own proficiency, especially if you anticipate operating near the limits of the airplane's stated performance capabilities. If your flying skills are rusty or you have not flown for a long time, you are unlikely to be able to match the performance in the charts. Likewise, the charts were developed using factory-new airplanes, so you should consider the condition of your airplane and allow for an adequate margin of safety.

More sophisticated aircraft often have an assortment of criteria that must be considered when using performance charts. There may be optional flap settings, power settings, or correction factors for different runway conditions, so you must be careful to select the correct chart and pay attention to the conditions under which the chart is valid.

Remember that all performance charts in this section are samples and must never be used for a real airplane. For actual flight planning, you should refer only to the POH for the specific airplane you intend to fly. Performance data can vary significantly, even among aircraft of the same make and model. Although the general use of performance charts should be familiar to you, higher performance airplanes often have charts that employ more variables, more complex formats, or more conditions. This section reviews the use of some basic charts and also presents new material on higher performance equipment.

Tables provide information for a limited list of conditions, and if the conditions for your flight are not listed, you will need to interpolate to find your performance values. Interpolation is the estimation of an unknown value between two known values. For instance, if the performance table only provides information for 2,000 and 4,000 feet,

Takeoff Ground Roll: Zero

The use of rocket engines to shorten takeoff distances was pioneered in 1941. The concept was first tested using an Ercoupe, a low-wing, two-place light airplane that is still a familiar sight at airports around the country. The success of the Ercoupe experiments led to extensive military use of this technique to assist heavily-loaded airplanes in taking off from limited space. Many bombers, transports, and patrol airplanes were equipped to use a refined version of the system, usually called JATO, for jet-assisted takeoff, or RATO, for rocket-assisted takeoff.

As rocket technology developed through the 1950s, it became possible to use a single, large rocket engine to launch a fully armed fighter from a flatbed trailer and accelerate it to flying speed in a few seconds. The spent rocket was then jettisoned. Although tested successfully, the zero-length launch was never used operationally.

Courtesy USAF Museum

and you need to find a value for an altitude of 3,000 feet, you can find the midpoint between the values provided to estimate the value for 3,000 feet. Similarly, you can use simple proportions to estimate other intermediate values. If information is given for 20°C and 30°C, you can estimate the performance at 28°C by adding 80% of the difference to the value for 20°C.

TAKEOFF CHARTS

Takeoff charts usually incorporate compensation for pressure altitude, outside air temperature, aircraft weight, and headwind or tailwind component. Some manuals include separate charts for normal and obstacle takeoff distances, two- or three-bladed propellers, various flap settings, or specific pilot techniques, such as short field or maximum effort. To review, start by looking over the sample problem in figure 12-44.

To estimate takeoff ground roll and distance to clear 50 feet, use the takeoff distance charts. See figure 12-44.

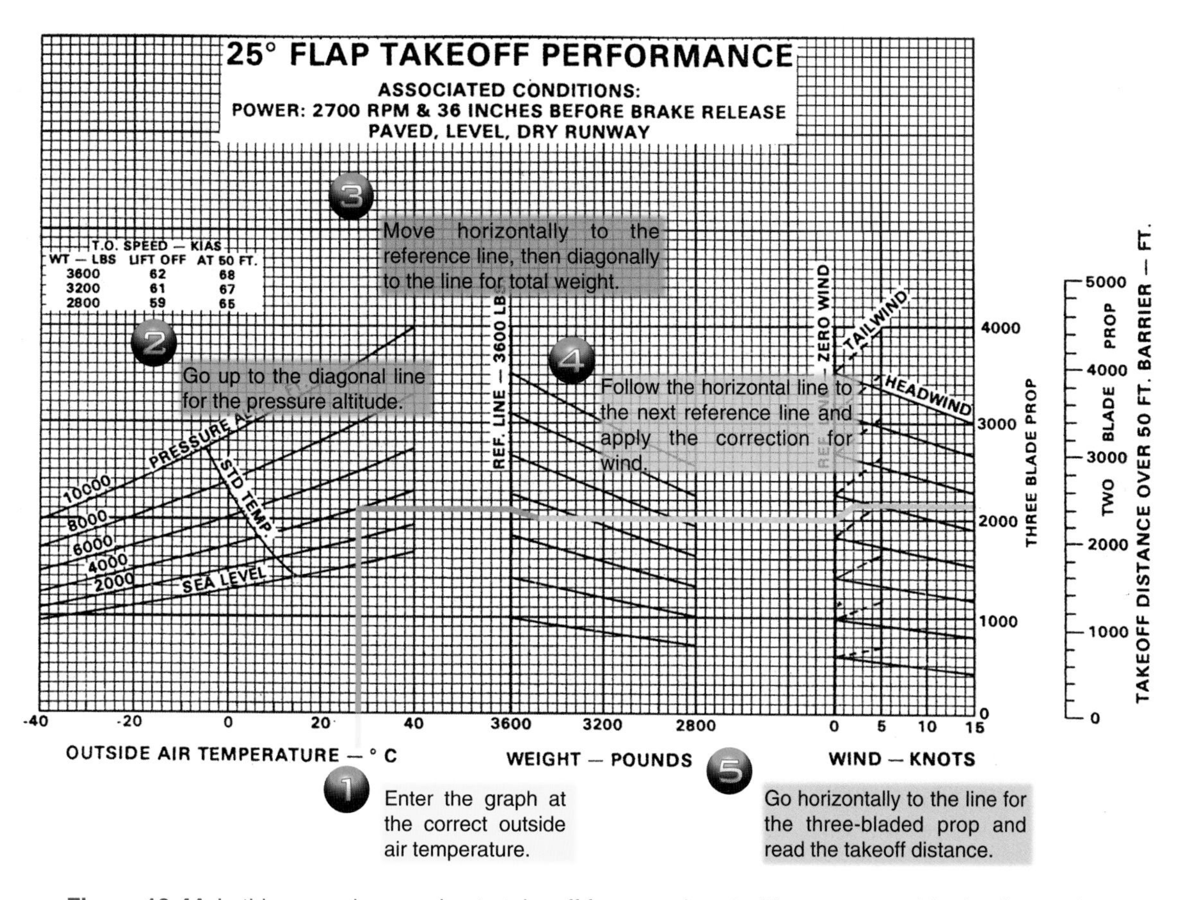

Figure 12-44. In this example, you plan to take off from an airport with a pressure altitude of 4,000 feet, with a temperature of 28°C (82°F), a total aircraft weight of 3,480 pounds, and a tailwind of 2 knots. Assume that your airplane has the three-bladed propeller and the flaps will be set at 25° for this takeoff. The graph shows that approximately 2,150 feet are required to take off and climb to 50 feet AGL.

The manuals for larger aircraft often include charts to help you determine additional takeoff and climb performance values. One example is the climb gradient chart, which can be useful when you plan takeoffs over obstacles or terrain in the first few miles after liftoff. [Figure 12-45]

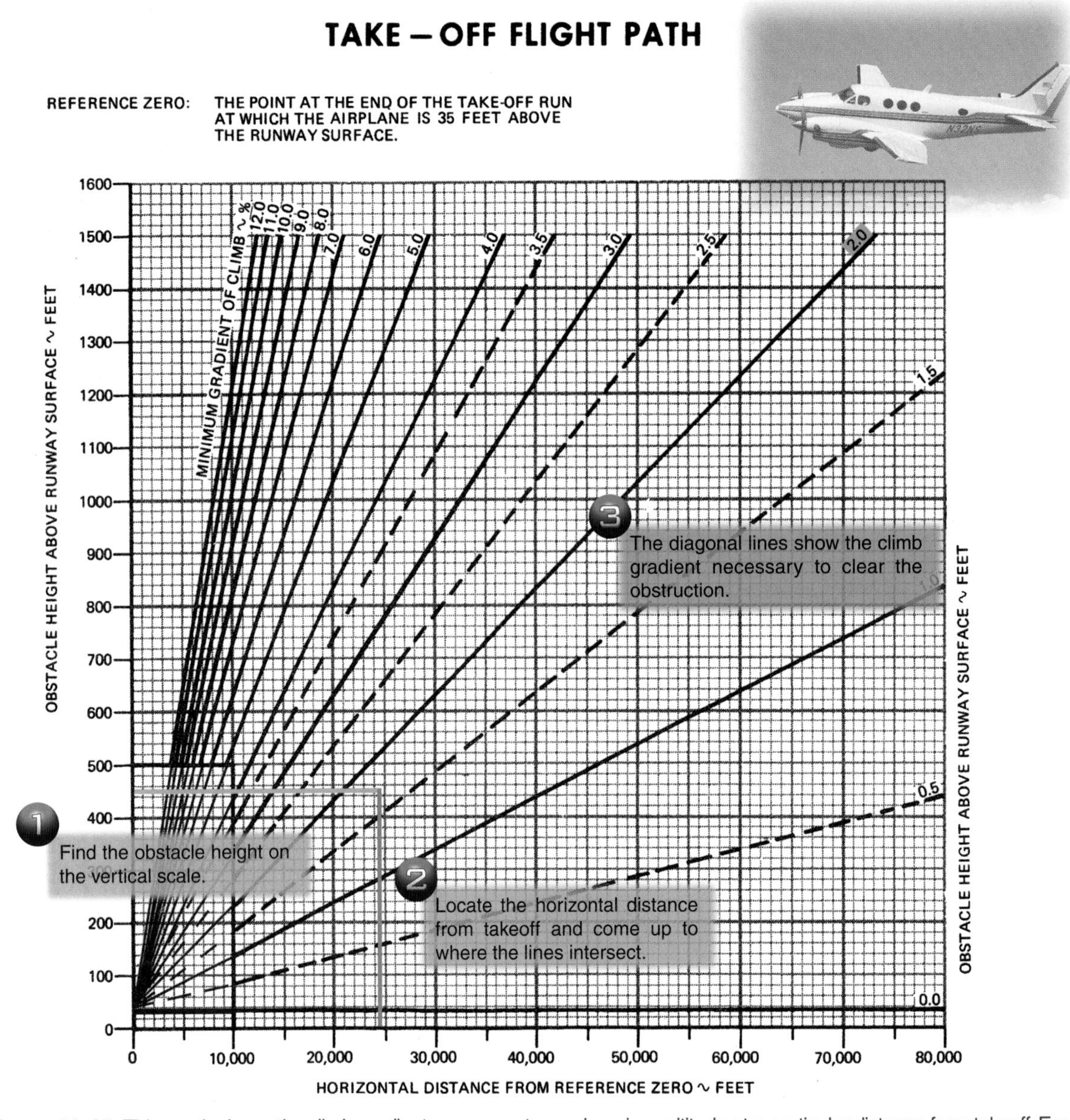

Figure 12-45. This graph shows the climb gradient necessary to reach a given altitude at a particular distance from takeoff. For example, if you know there is a 450 foot obstacle 4 n.m. away on your departure path, the graph shows that you will need a 2% climb gradient to clear that height within 24,304 feet (4 n.m.). Of course, you will want to clear all obstacles by a much greater margin.

CLIMB PERFORMANCE CHARTS

In manuals that follow the GAMA format, you will ordinarily find at least two different climb performance charts. One type shows the maximum rate of climb for various combinations of air temperature, pressure altitude, aircraft weight, and airspeed. Another chart depicts the time, fuel, and distance required to climb from one altitude to another based on temperature, aircraft weight, and the altitudes concerned.

The maximum rate-of-climb chart is commonly based on the best rate-of-climb airspeed (V_Y) for the airplane at different weights and altitudes. Some manuals offer two rate-of-climb charts, one for takeoff power, the other for maximum continuous power. Climbing at takeoff power can be hard on the engine in many high-performance airplanes. It is important to thoroughly understand the consequences of operating

Temperature, pressure altitude, and total airplane weight are used on charts to estimate maximum rate of climb. See figure 12-46.

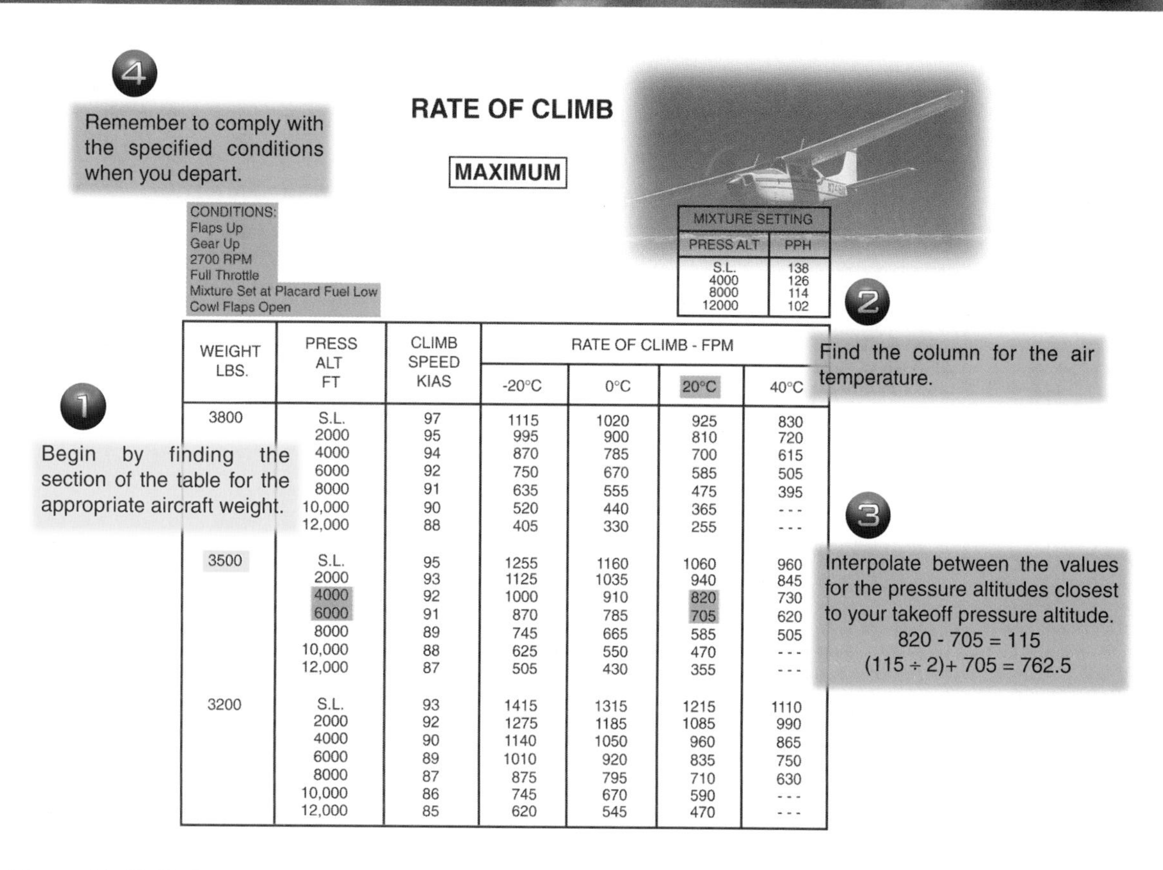

WEIGHT LBS.	PRESS ALT FT	CLIMB SPEED KIAS	RATE OF CLIMB - FPM -20°C	0°C	20°C	40°C
3800	S.L.	97	1115	1020	925	830
	2000	95	995	900	810	720
	4000	94	870	785	700	615
	6000	92	750	670	585	505
	8000	91	635	555	475	395
	10,000	90	520	440	365	- - -
	12,000	88	405	330	255	- - -
3500	S.L.	95	1255	1160	1060	960
	2000	93	1125	1035	940	845
	4000	92	1000	910	820	730
	6000	91	870	785	705	620
	8000	89	745	665	585	505
	10,000	88	625	550	470	- - -
	12,000	87	505	430	355	- - -
3200	S.L.	93	1415	1315	1215	1110
	2000	92	1275	1185	1085	990
	4000	90	1140	1050	960	865
	6000	89	1010	920	835	750
	8000	87	875	795	710	630
	10,000	86	745	670	590	- - -
	12,000	85	620	545	470	- - -

Figure 12-46. To find the maximum rate of climb for an aircraft weight of 3,500 pounds and an air temperature of 20°C at a pressure altitude of 5,000 feet, you need to interpolate between the values for 4,000 and 6,000 feet. This results in a climb rate of 763 f.p.m. As you climb higher, the indicated airspeed for V_Y decreases. This table also indicates how the mixture setting (fuel flow) should be adjusted for the pressure altitude.

your engine at high power settings for extended periods, especially at climb airspeeds where cooling airflow may be diminished. The POH and engine operating handbook provide the appropriate information. [Figure 12-46]

Using the time, fuel, and distance to climb chart is a two-step process. First, find the values for your cruising altitude, then find and subtract the values for your takeoff altitude. The difference is the time, fuel, and distance to climb from the takeoff elevation to the cruising altitude. Use the same procedure if you need to make an enroute climb from one altitude to another. You can only expect to achieve the published values by adhering to the conditions set forth in the chart, which means precise airspeed control as well as compliance with the configuration and power settings stipulated in the chart notes. Many manuals provide separate charts for different climb speeds. For example, one chart is provided for maximum rate-of-climb data, and another chart supplies information for a normal climb that uses a lower power setting. The normal climb reduces stress on the engine, provides better engine cooling, and gives you better visibility over the nose. Your passengers will appreciate the lower noise level in the cabin, and there will be less noise on the ground as you fly over. Unless there is some reason to climb at the maximum rate, such as terrain or obstructions in your departure path, you will usually prefer to use the normal climb. [Figure 12-47]

Climb performance charts usually provide information on time, fuel, and distance required to climb from one altitude to another. Remember to subtract the values for the starting altitude. See figure 12-47.

Verify that you have the proper chart and that the specified conditions are met.

TIME, FUEL, AND DISTANCE TO CLIMB

NORMAL CLIMB - 110 KIAS

CONDITIONS:
Flaps Up
Gear Up
2500 RPM
30 Inches Hg.
140 PPH Fuel Flow
Cowl Flaps Open
Standard Temperature

Add 18 pounds for engine start, taxi, and takeoff.

NOTES:
1. Add 18 pounds of fuel for engine start, taxi, and takeoff allowance.
2. Increase time, fuel and distance by 10% for each 8°C above standard temperature.
3. Distances shown are based on zero wind.

Add 20% to adjust for the higher-than-standard temperature.

WEIGHT LBS	PRESS ALT FT	RATE OF CLIMB FPM	FROM SEA LEVEL		
			TIME MIN	FUEL USED POUNDS	DISTANCE NM
4100	S.L.	760	0	0	0
	4000	740	5	12	10
	8000	700	11	25	21
	12,000	650	17	39	34
	16,000	585	23	55	48
	20,000	490	31	72	66
	24,000	360	40	94	91
3700	S.L.	900	0	0	0
	4000	880	4	10	8
	8000	840	9	21	18
	12,000	790	14	33	28
	16,000	725	19	45	40
	20,000	625	25	59	54
	24,000	485	33	76	73
3300	S.L.	1070	0	0	0
	4000	1050	4	9	7
	8000	1010	8	18	15
	12,000	955	12	27	23
	16,000	885	16	38	33
	20,000	780	21	49	45
	24,000	630	27	62	60

③ Find the time, fuel and distance to be subtracted for departing an airport at a pressure altitude of 4,000 feet.

② Read the time fuel and distance to climb to 20,000 feet at a weight of 3,700 pounds.

Subtract the values for the takeoff altitude from the values for the cruise altitude.

Time: 25 - 4 = 21
Fuel: 59 - 10 = 49
Distance: 54 - 8 = 46

Figure 12-47. In this example, you want to make a normal climb at an aircraft weight of 3,700 pounds from an airport at 4,000 feet pressure altitude to a cruising altitude of FL200, when the temperature is 16°C above standard. This results in a time of 25.2 minutes, fuel consumption of 76.8 pounds, and a distance of 55.2 nautical miles.

Since the service ceiling is the altitude where the maximum rate of climb is 100 f.p.m., you can often use a maximum rate-of-climb chart to determine the service ceiling for different aircraft weights or outside air temperatures. Start with the 100 f.p.m. rate-of-climb, and work back through the chart to the pressure altitude. [Figure 12-48]

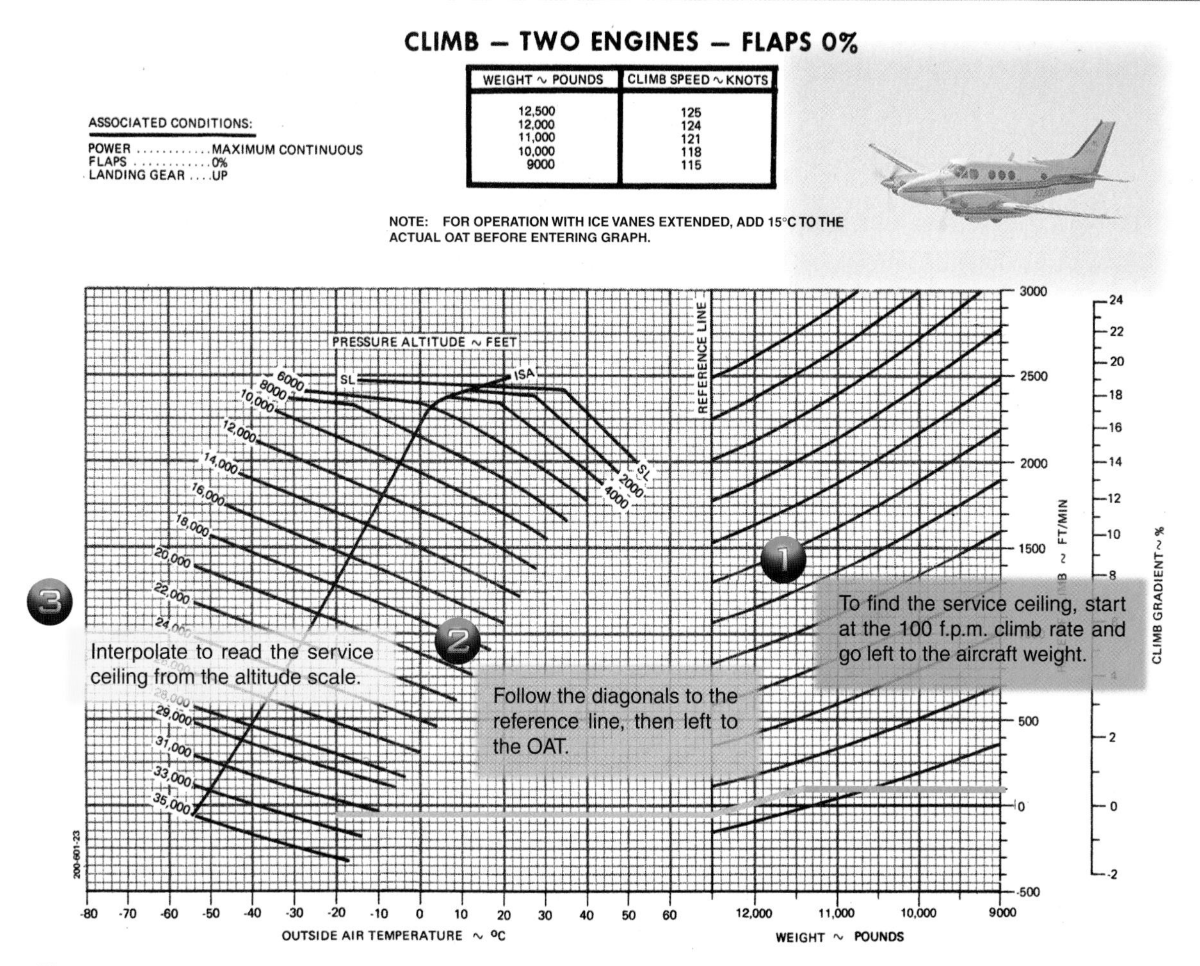

Figure 12-48. When the outside air temperature is -20°C and the aircraft weight is 11,400 pounds, the service ceiling is a pressure altitude of approximately 32,000 feet.

CRUISE PERFORMANCE CHARTS

The performance of the airplane in cruise depends on many variables. You have direct control over some of them, such as manifold pressure, propeller r.p.m., and fuel flow settings. You can also exercise indirect control over some of the other variables, for instance, selecting an altitude that provides favorable winds or temperatures. You need a thorough understanding of the effects of each variable on the airplane's overall performance, so you can decide on the best combination for each particular flight. If you increase power to add a couple of knots to your airspeed, will you cut into your fuel reserve? Do the performance penalties of flying the airplane at a lower power setting with a heavy fuel load cancel out the time savings of skipping a fuel stop? The cruise performance charts contain the information you need to answer questions like these. Besides the elements in the performance charts, you also will want to consider less tangible factors, such as the stress on the engine. The effects of high operating temperatures, high manifold pressures, excessively lean mixture settings, and other harsh operating procedures can increase maintenance costs and, in some cases, compromise engine reliability.

As you know, if you maintain the same power settings, you can generally expect higher true airspeeds at higher altitudes. This is due to the reduction in drag as air density decreases. Going faster and farther while using fuel at the same rate is an attractive idea, however, the lower airspeed and high power setting during a long climb may offset the benefit of a high cruising altitude. Always remember to consider the wind direction at altitude, since even a few degrees of wind correction angle or a few knots of headwind can reduce your groundspeed significantly, and winds are typically stronger the higher you fly.

The mixture must be adjusted to compensate for the reduced air density at higher altitudes. Two popular methods of setting the mixture involve using an exhaust gas temperature (EGT) gauge or a fuel flow meter. Always follow the manufacturer's recommendations for leaning, both to avoid possible engine damage and to obtain the fuel consumption values that you determined in your preflight planning. A seemingly minor difference in mixture setting could result in a higher fuel flow rate, or in a lower true airspeed. Either situation could lead to fuel quantity problems on a long flight. Some performance charts distinguish between the mixture settings for best power and best economy. The best economy mixture is a leaner setting that results in lower fuel consumption and higher engine temperatures. Cruise performance charts can be used to determine the true airspeed expected under a particular set of conditions. [Figure 12-49]

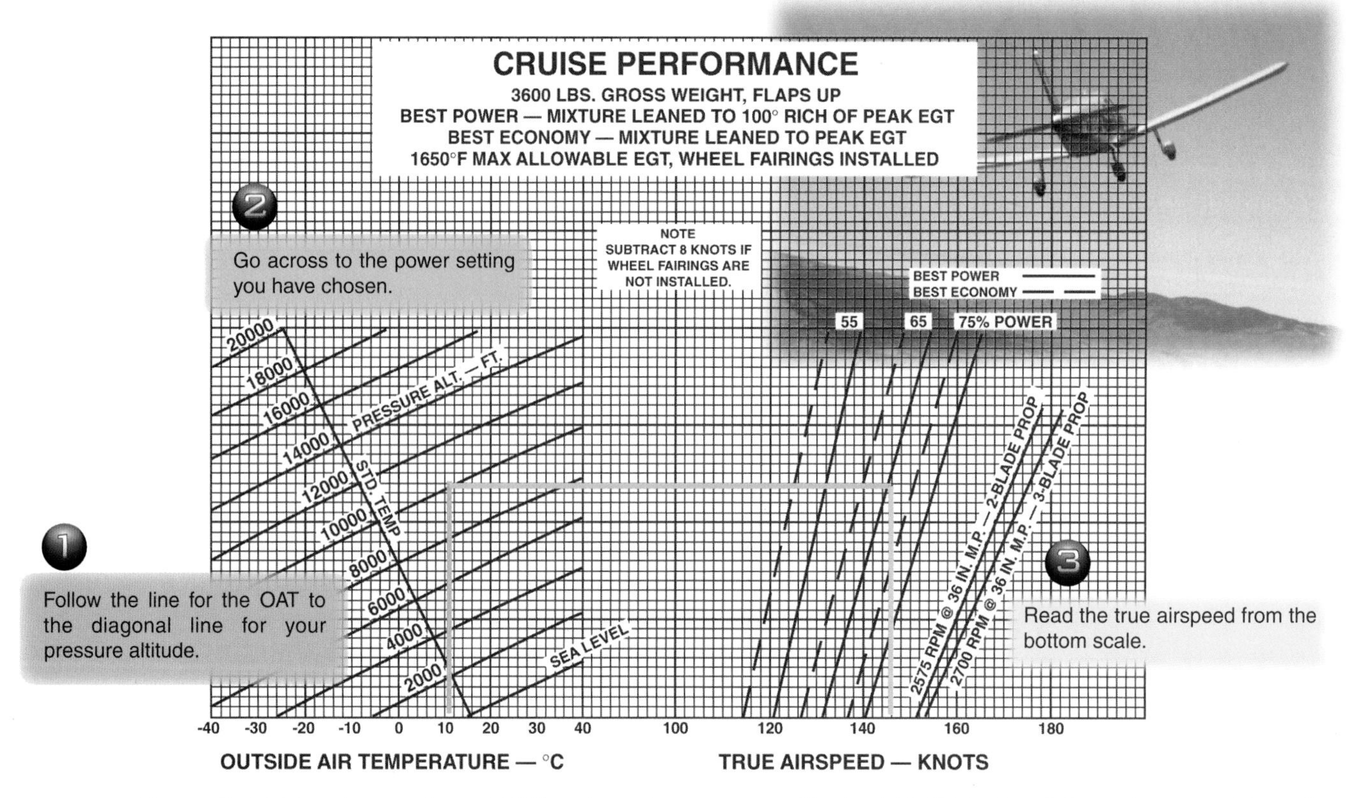

Figure 12-49. For a temperature of 10°C and a cruise pressure altitude of 10,000 feet, a 65% best power setting will result in a true airspeed of 144 knots.

Range is the distance an airplane can travel with a given amount of fuel, while **endurance** is the length of time the airplane can remain in the air. The range and endurance charts in the POH help you to estimate how far you can go and how long you can remain aloft at various cruise power settings. In most cross-country situations, range is more desirable than endurance, since you will want to maximize the distance traveled per pound of fuel. Endurance might be more important in a holding pattern, or in circumstances where you do not want to arrive at a certain point before a specific time. Examples could include a clearance limit, a landing slot assignment, or when the weather is improving at your destination. Most charts include allowances for normal climb and descent, as well as for fuel reserves. [Figure 12-50] After obtaining a preliminary range from the chart, you must still adjust the result for the effects of any headwind or tailwind.

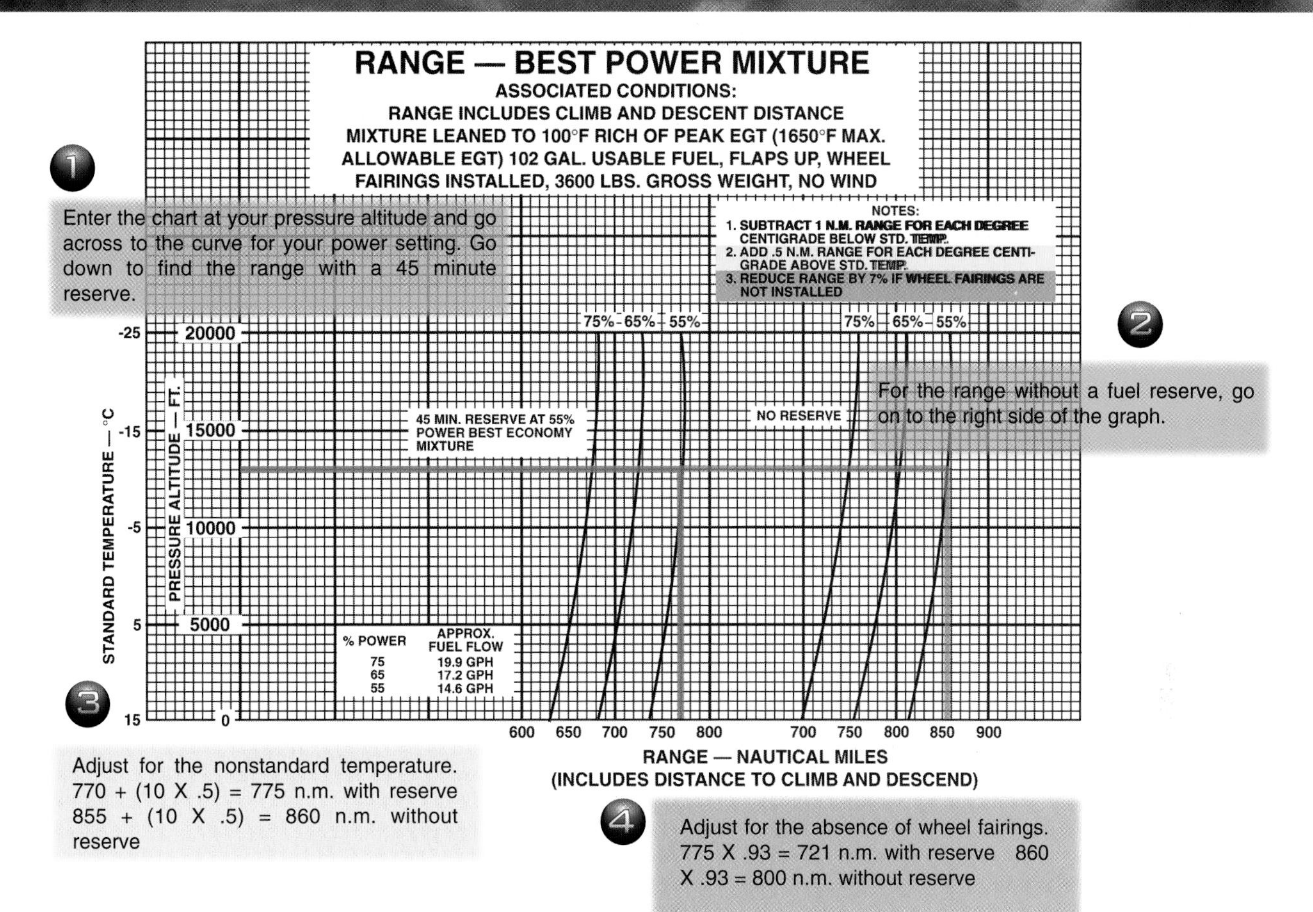

Figure 12-50. Suppose you need to determine how far you could fly with full tanks at a pressure altitude of 13,000 feet at 55% power. Assume that the temperature is 10° above standard, and that the wheel fairings are not installed. The zero-wind range with a 45 minute reserve would be 721 nautical miles, and 800 nautical miles with no reserve. The range chart shown here is similar in format to the endurance chart for the same airplane. Note that this chart specifies the best power mixture, but the 45 minute reserve is based on the best economy mixture setting. Full range information using the best economy mixture would be found on a separate chart in the POH.

Endurance: Several Months

An airplane similar to this one should be capable of what the designers call semi-perpetual flight. Operating at altitudes up to 100,000 feet, the Centurion will convert sunlight into electricity to power its 14 electric motors. During the day, electricity will also be used to break water down into hydrogen and oxygen, which will be recombined in fuel cells to provide power at night. The crewless flying wing will have a longer wingspan than a 747, yet weigh only 1,105 pounds. The Pathfinder, pictured here, is essentially a smaller version of the Centurion. It flew to 71,530 feet on solar power in 1997.

There are many possible uses for such airplanes. High altitude scientific research, reconnaissance, communications relay, environmental monitoring, and earth-resource applications are a few likely prospects.

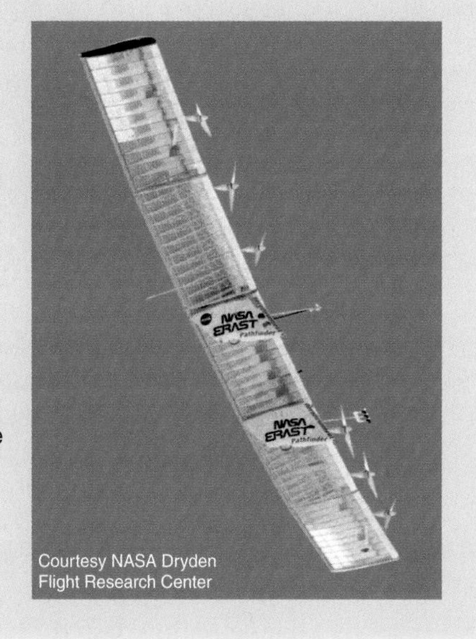

Courtesy NASA Dryden Flight Research Center

If you maintain the same engine horsepower, lower temperatures generally reduce your true airspeed, while slightly increasing your range and endurance. Higher temperatures have the reverse effect. Outside air temperatures on cruise charts may include International Standard Atmosphere (ISA) values, as well as colder and warmer temperatures. For instance, categories may specify ISA –20°C, standard day ISA, and ISA +20°C. Both Celsius and Fahrenheit values may be included. In addition, temperature figures for some manufacturers may be adjusted for frictional heating of the temperature probe at higher speeds.

Range, endurance, fuel consumption, true airspeed, and other cruise information can be obtained from cruise performance charts. See figures 12-49 and 12-50.

As discussed in the preceding section, an airplane will realize its maximum range at the airspeed with the greatest ratio of lift to drag (L/D_{max}). Maximum endurance is achieved when using the minimum power necessary to maintain level flight. [Figure 12-51].

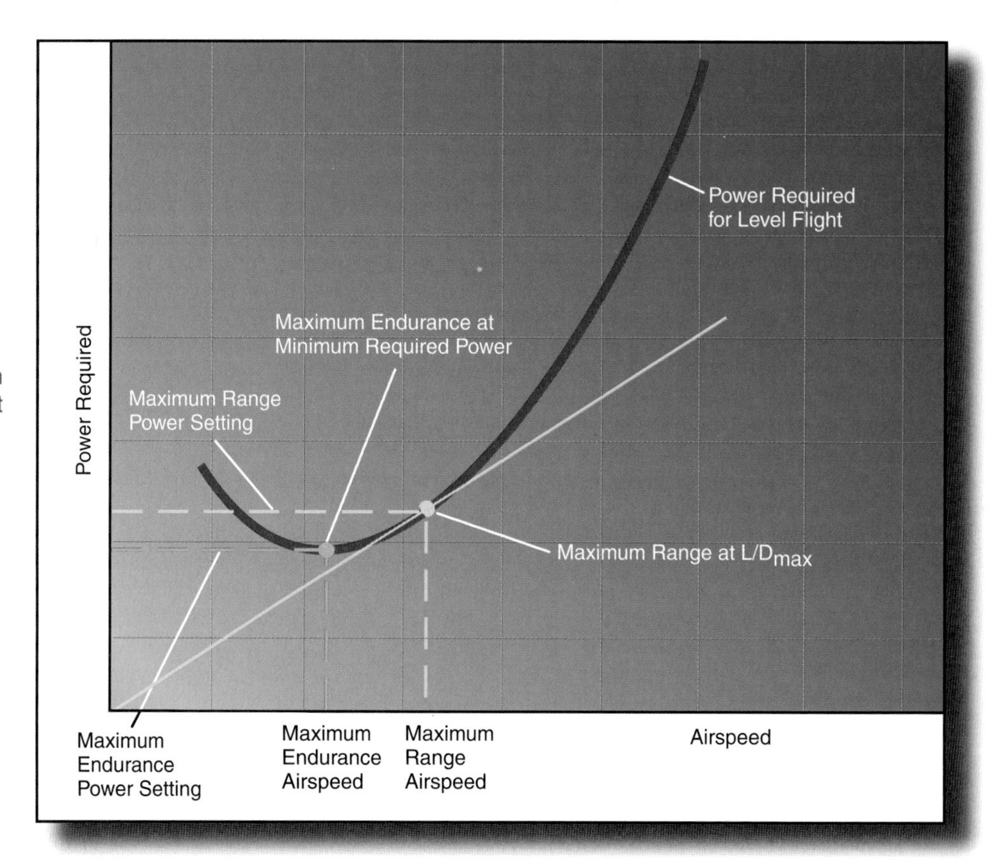

Figure 12-51. On the power-required graph, the maximum endurance power setting is at the lowest point of the curve. A line from the origin of the graph is tangent to the curve at the point corresponding to maximum range (L/D_{max}).

While the speeds for maximum range and endurance do provide high levels of efficiency, they are slow compared to normal cruise speeds. Bear in mind that most cruise performance charts do not include range or endurance information for less than 45% power, which is far more power than the minimum required for level flight. As previously noted, the airspeed for maximum range is the same as the best glide speed, and the airspeed for maximum endurance is usually about 76% of the best glide speed. In many general aviation airplanes, these speeds are very close to the stall speed.

Descent Charts

Some manufacturers supply charts to help you plan your descent. For a given airspeed and configuration, you can use them to predict the time, fuel, and distance to descend from your cruising altitude to your destination. As with other performance charts, be sure to comply with the conditions specified in the chart. [Figure 12-52]

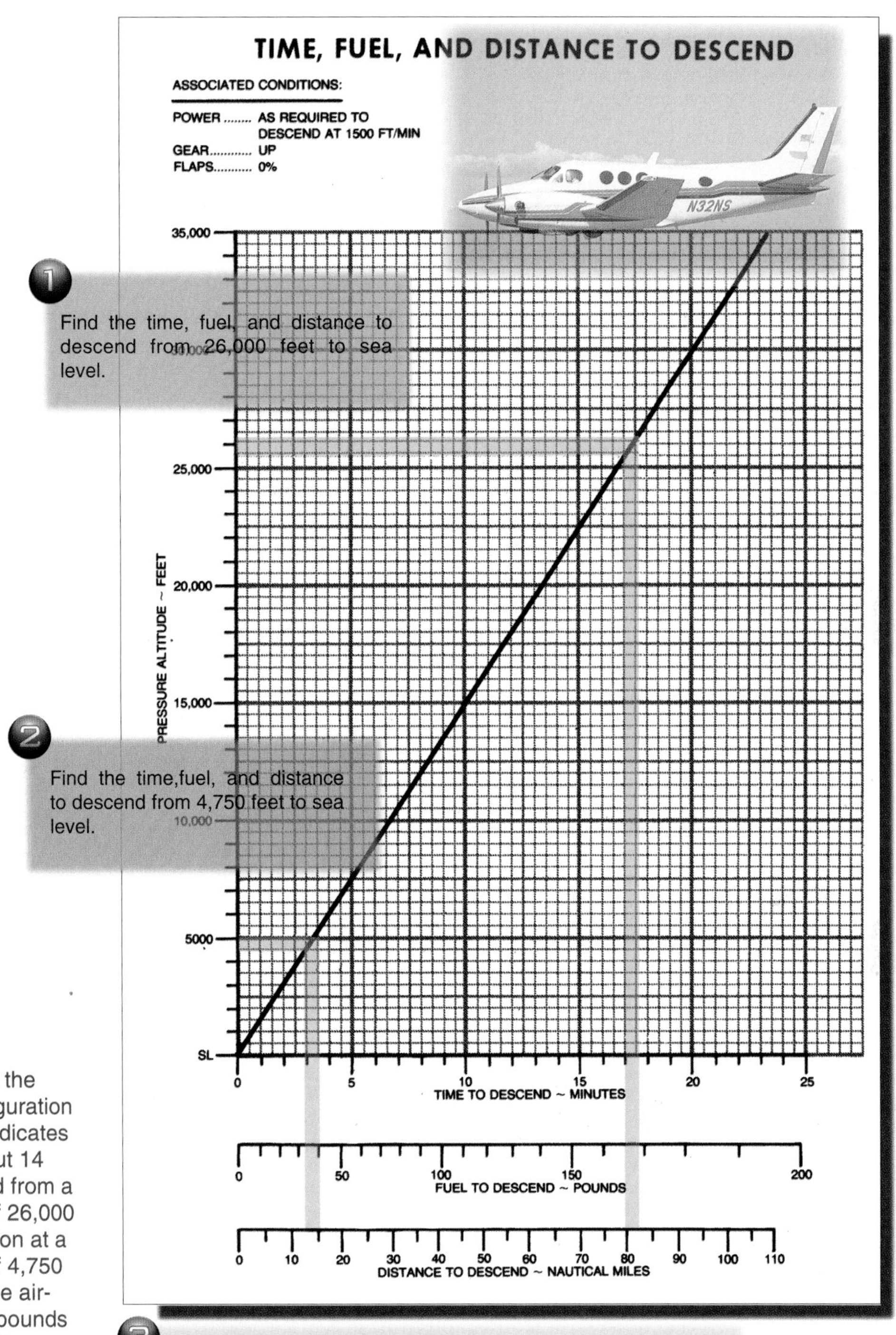

Figure 12-52. With the airspeed and configuration shown, the chart indicates that it will take about 14 minutes to descend from a pressure altitude of 26,000 feet to the destination at a pressure altitude of 4,750 feet. In that time, the airplane will use 132 pounds of fuel and cover 67 nautical miles if the winds are calm. You can use your flight computer to adjust for the effects of wind.

LANDING DISTANCE CHARTS

In many respects, landing distance charts are similar to takeoff distance charts. They normally include compensations for temperature, altitude, aircraft weight, and headwind or tailwind component, and most provide the landing distance from a height of 50 feet as well as the length of the ground roll itself. Manufacturers may incorporate other variables into landing charts, such as the configuration of flaps, touchdown speed, use of reverse propeller pitch or reverse thrust, or the amount of runway slope. Separate charts may be supplied for short field or obstacle clearance landing techniques. As with the other charts, be sure to take your own abilities and proficiency into account, and always leave yourself an adequate safety margin. [Figure 12-53]

Use landing charts to estimate landing distances and runway requirements. See figure 12-53.

LANDING DISTANCE

SHORT FIELD

CONDITIONS:
Flaps 30°
Power Off
Maximum Braking
Paved, Level, Dry Runway
Zero Wind

1. Verify that you will match the conditions on the chart, including landing technique (short field), flap setting, landing weight, and runway conditions.

2. Find the row and column for your pressure altitude and air temperature.

NOTES:
1. Short field technique as specified in Section 4.
2. Decrease distances 10% for each 10 knots headwind. For operation with tailwinds up to 10 knots, increase distances by 10% for each 2.5 knots.
3. For operation on a dry, grass runway, increase distances by 40% of the "ground roll" figure.

WEIGHT LBS	SPEED AT 50 FT KIAS	PRESS ALT FT	0°C		10°C		20°C		30°C		40°C	
			GRND ROLL	TOTAL TO CLEAR 50 FT OBS	GRND ROLL	TOTAL TO CLEAR 50 FT OBS	GRND ROLL	TOTAL TO CLEAR 50 FT OBS	GRND ROLL	TOTAL TO CLEAR 50 FT OBS	GRND ROLL	TOTAL TO CLEAR 50 FT OBS
3800	71	S.L.	725	1440	750	1480	780	1520	805	1560	830	1600
		1000	750	1480	780	1520	805	1560	835	1605	860	1645
		2000	780	1525	810	1565	835	1605	865	1650	895	1695
		3000	810	1565	840	1610	870	1660	900	1705	930	1750
		4000	840	1615	870	1660	900	1705	930	1750	965	1800
		5000	870	1660	905	1710	935	1755	965	1805	1000	1855
		6000	905	1710	940	1765	970	1810	1005	1860	1035	1910
		7000	940	1765	975	1815	1010	1870	1045	1920	1075	1970
		8000	975	1815	1010	1870	1050	1930	1085	1980	1120	2035

3. Read the ground roll and distance over 50 foot obstacle. Interpolate for intermediate values as needed.

4. Apply any necessary correction factors. In this example, subtract 10% for the 10 knots headwind.

900 x .9 = 810 feet ground roll
1,705 x .9 = 1,535 feet over 50 foot obstacle

Figure 12-53. This table indicates that at a pressure altitude of 3,000 feet and a temperature of 30°C, the minimum landing distance over a 50-foot obstacle in a 10-knot headwind is 1,535 feet. The ground roll with maximum braking is approximately 810 feet.

GLIDE DISTANCE

These charts show how far your airplane will glide at the best glide airspeed (L/D_{max}), provided there is no wind. Some manufacturers place this chart in the Performance section while others place it in the Emergency Procedures section. The simplest chart shows maximum glide distance based on height above the terrain. The chart also shows the appropriate airspeed, and provides information on the configuration to be used. The glide ratio also may be included, expressed as a distance in nautical miles per thousand feet of altitude. [Figure 12-54]

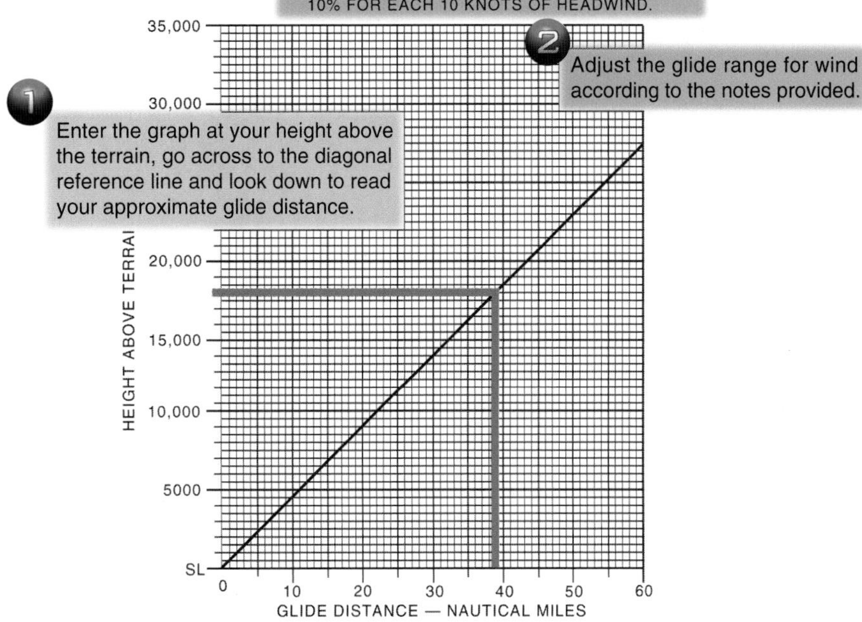

Figure 12-54. If your engine quit at 18,000 feet AGL, the chart indicates that you could glide 39 nautical miles with no wind. The chart also provides information to adjust the glide distance for the effects of wind. For example, if you were gliding from 18,000 feet AGL with a 20-knot headwind, your glide distance would be reduced by 20%, to 31.2 nautical miles.

More detailed charts incorporate air temperature and pressure altitude, and are similar in use to the time-to-climb charts. Begin by calculating the glide range from your current density altitude to sea level, and then subtract the distance that would be covered from the height of the terrain to sea level. [Figure 12-55]

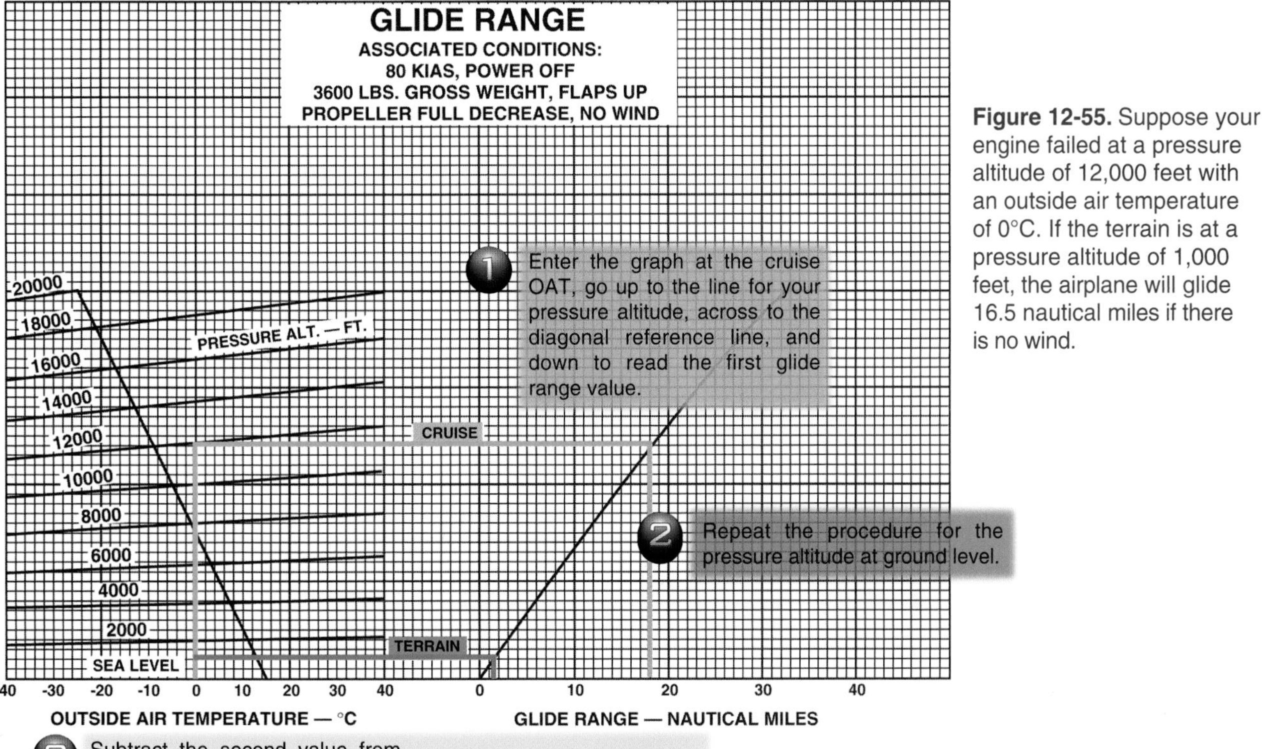

Figure 12-55. Suppose your engine failed at a pressure altitude of 12,000 feet with an outside air temperature of 0°C. If the terrain is at a pressure altitude of 1,000 feet, the airplane will glide 16.5 nautical miles if there is no wind.

STALL SPEEDS

As discussed in the previous section, stall speed varies with factors such as weight, CG location, configuration, and bank angle. Because of the change in stall speed due to weight, many airplane flight manuals specify a higher approach speed when the airplane is heavily loaded. Manufacturers provide charts to help you estimate the stall speed for various conditions. [Figure 12-56]

Stall speed charts present the stall speeds that can be expected in various configurations and at different bank angles. See figure 12-56.

STALL SPEEDS

CONDITIONS:
Power Off
Gear Up or Down

MOST REARWARD CENTER OF GRAVITY

WEIGHT LBS	FLAP DEFLECTION	ANGLE OF BANK							
		0°		30°		45°		60°	
		KIAS	KCAS	KIAS	KCAS	KIAS	KCAS	KIAS	KCAS
3800	UP	64	65	69	70	76	77	91	92
	10°	62	64	67	69	74	76	88	91
	30°	46	56	49	60	55	67	65	79

1. Find the stall speed at the most rearward CG for the conditions in question.

MOST FORWARD CENTER OF GRAVITY

WEIGHT LBS	FLAP DEFLECTION	ANGLE OF BANK							
		0°		30°		45°		60°	
		KIAS	KCAS	KIAS	KCAS	KIAS	KCAS	KIAS	KCAS
3800	UP	68	69	73	74	81	82	96	98
	10°	67	68	72	73	80	81	95	96
	30°	55	61	59	66	65	73	78	86

2. Find the corresponding stall speed for the forward CG limit.

3. Interpolate to find the stall speed for a midrange CG under those conditions.
65 – 55 = 10
10 ÷ 2 = 5
55 + 5 = 60 KIAS

Figure 12-56. To use this table, simply find the row for the appropriate flap deflection, and look across to the angle of bank in question. Note how the stall speed varies with the location of the CG. Interpolate to find stall speeds for intermediate CG locations. For example, if you know that the CG is about halfway between the limits, and want to know the stall speed with 30° of flaps and a bank angle of 45°, you can determine the stall speed to be 60 KIAS (halfway between 55 and 65 KIAS). With power on, most airplanes stall at a lower airspeed because of the airflow over the wings induced by the propeller.

SUMMARY CHECKLIST

✓ Atmospheric properties such as temperature, pressure, and humidity have predictable effects on airplane performance.

✓ Density altitude affects all aspects of airplane performance, but the effects are most noticeable during takeoff and climb.

- ✓ Your own proficiency must be taken into account when using performance charts.
- ✓ Headwind and crosswind components can be determined by using a wind component chart.
- ✓ Demonstrated crosswind capability must be at least 20% of V_{S0}.
- ✓ Increasing the weight of your airplane increases takeoff and landing distances.
- ✓ Wet or icy runways decrease braking effectiveness and increase landing distances. Braking effectiveness is reported as good, fair, poor, or nil, or some combination of these terms.
- ✓ Hydroplaning is when the tires are separated from the runway surface by a thin layer of water. Hydroplaning not only increases stopping distances, it can also lead to complete loss of airplane control.
- ✓ Operating limitations are contained in the approved flight manual (AFM), approved manual materials, markings and placards, or any combination of these. For most modern airplanes, the pilot's operating handbook (POH) is the approved flight manual.
- ✓ Even small tailwind components can increase takeoff or landing distances significantly. A tailwind increases your takeoff or landing distance much more than the same headwind reduces it. Climb and descent gradients are similarly affected.
- ✓ Range is the distance the airplane can travel on a given amount of fuel. Endurance is the amount of time an airplane can remain airborne on a given amount of fuel. The airspeeds for maximum range and endurance usually are near the airplane's stall speed.
- ✓ Because of the change in stall speed, airplanes are usually flown at a higher approach speed when they are heavily loaded.

KEY TERMS

Density Altitude

Headwind Component

Tailwind Component

Crosswind Component

Demonstrated Crosswind Component

Braking Effectiveness

Hydroplaning

Pilot's Operating Handbook (POH)

Range

Endurance

QUESTIONS

1. If the stall speed of your airplane in its landing configuration (V_{S0}) is 55 knots, what is the minimum demonstrated crosswind component?
2. True/False. All other factors being equal, an increase in density altitude causes your airplane's engine to produce more horsepower.

3. With very high relative humidity, how much of a reduction in engine output may be expected?

   A. 7%
   B. 12%
   C. 18%

Use the accompanying takeoff distance chart to answer questions 4 and 5.

TAKEOFF DISTANCE
MAXIMUM WEIGHT 4100 LBS
SHORT FIELD

CONDITIONS:
Flaps 20°
2700 RPM, 37 Inches Hg, and Mixture Set at 220 PPH Prior to Brake Release
Cowl Flaps Open
Paved, Level, Dry Runway
Zero Wind

NOTES:
1. Short field technique as specified in Section 4.
2. Decrease distances 10% for each 10 knots headwind. For operation with tailwinds up to 10 knots, increase distances by 10% for each 2.5 knots.
3. For operation on a dry, grass runway, increase distances by 15% of the "ground roll" figure.

WEIGHT LBS	TAKEOFF SPEED KIAS		PRESS ALT FT	0°C		10°C		20°C		30°C		40°C	
	LIFT OFF	AT 50 FT		GRND ROLL FT	TOTAL FT TO CLEAR 50 FT OBS	GRND ROLL FT	TOTAL FT TO CLEAR 50 FT OBS	GRND ROLL FT	TOTAL FT TO CLEAR 50 FT OBS	GRND ROLL FT	TOTAL FT TO CLEAR 50 FT OBS	GRND ROLL FT	TOTAL FT TO CLEAR 50 FT OBS
4100	69	75	S.L.	1115	1845	1215	2015	1325	2210	1450	2425	1580	2675
			1000	1185	1945	1290	2125	1410	2330	1535	2565	1680	2830
			2000	1255	2055	1370	2250	1495	2465	1635	2715	1785	2995
			3000	1335	2170	1460	2380	1595	2610	1740	2875	1905	3180
			4000	1420	2295	1555	2520	1695	2770	1855	3050	2030	3380
			5000	1515	2430	1655	2670	1810	2935	1980	3240	2165	3590
			6000	1615	2580	1765	2830	1930	3120	2110	3445	2310	3825
			7000	1725	2735	1885	3005	2060	3315	2255	3665	2470	4080
			8000	1840	2910	2015	3200	2205	3530	2410	3905	2645	4355

4. What is the minimum distance to take off and climb to an altitude of 50 feet, if the pressure altitude is 5,500 feet, the air temperature is 30°C, the total aircraft weight is 4,100 pounds, and you have a 10 knot headwind?

5. What is the minimum ground roll for takeoff from a dry, grass runway at a pressure altitude of 3,000 feet and an air temperature of 25°C, with a 5 knot tailwind? The takeoff weight is 4,100 pounds.

6. True/False. If you drained 120 pounds of fuel from the airplane in question 5, the takeoff roll would be shorter.

Use the accompanying climb performance charts to answer questions 7, 8, and 9.

7. What is the maximum rate of climb for a pressure altitude of 4,000 feet, an air temperature of 10°C, and an aircraft weight of 3,300 pounds?

   A. 1,010 f.p.m.
   B. 1,110 f.p.m.
   C. 1,190 f.p.m.

8. How much fuel would be required for a normal climb from an airport at a pressure altitude of 2,000 feet to a pressure altitude of 14,000 feet? The air temperature at ground level is 25°C, the temperature at 14,000 feet is –5°C, and the airplane weight is 3,600 pounds.

   A. 7 gallons
   B. 8 gallons
   C. 9 gallons

9. How much time would it take to climb from a pressure altitude of 4,000 feet (temperature 15°C) to an altitude of 20,000 feet (temperature –30°C)? The airplane weight is 3,600 pounds.

   A. 22 minutes
   B. 26 minutes
   C. 27 minutes

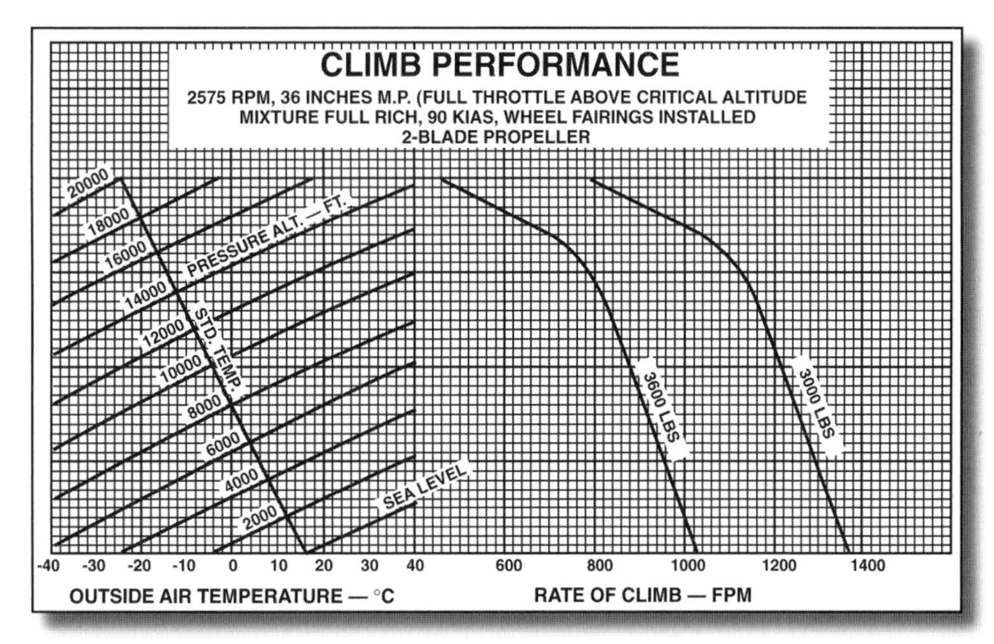

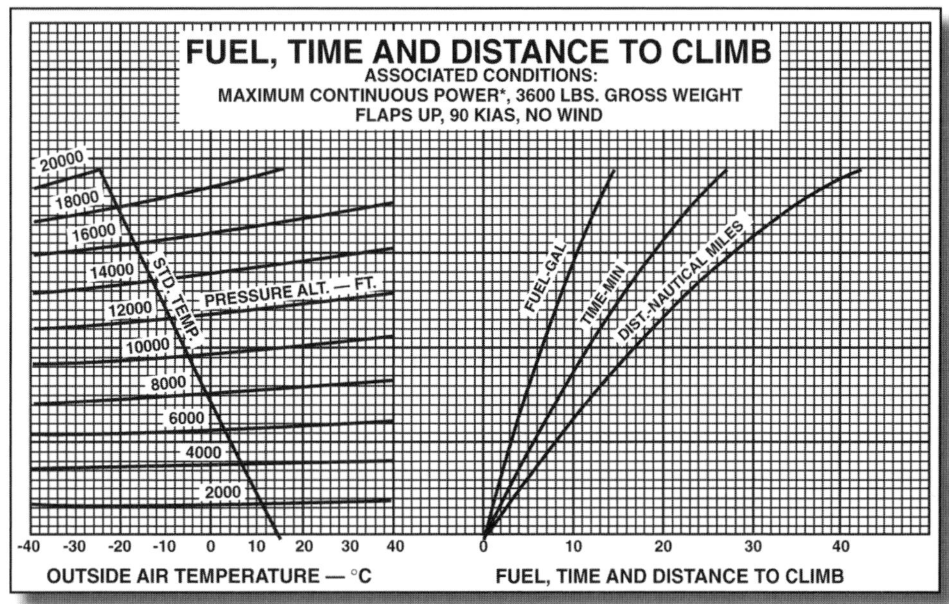

Use the accompanying cruise performance chart to answer questions 10 and 11.

CRUISE PERFORMANCE

PRESSURE ALTITUDE 18,000 FEET

CONDITIONS:
4100 Pounds
Recommended Lean Mixture
Cowl Flaps Closed

NOTE

Power settings within shaded areas may not be obtainable, but are listed to aid interpolation.

		20°C BELOW STANDARD TEMP – 41°C			STANDARD TEMPERATURE – 21°C			20°C ABOVE STANDARD TEMP – 1°C		
RPM	MP	% BHP	KTAS	PPH	% BHP	KTAS	PPH	% BHP	KTAS	PPH
2500	30	81	198	118	77	198	111	72	197	104
	28	76	193	110	72	193	104	67	192	98
	26	70	186	102	66	186	96	62	184	91
	24	64	179	93	60	178	89	56	175	86
	22	57	170	86	54	168	83	50	164	80
2400	31.5	81	198	117	77	198	111	72	197	104
	30	77	194	112	73	194	105	68	193	99
	28	72	188	104	68	188	99	64	187	93
	26	66	182	96	62	181	91	59	179	87
	24	60	174	89	57	173	85	53	169	82
2300	33	81	197	116	76	197	110	71	196	103
	31	76	192	110	71	192	103	67	191	97
	29	70	187	102	66	186	97	62	185	91
	27	65	181	95	61	180	90	58	177	86
	25	60	173	88	56	172	84	53	168	80
2200	34.5	79	196	115	75	196	108	70	195	102
	33	76	193	110	72	192	104	67	192	98
	31	71	188	104	67	187	98	63	186	92
	29	66	182	97	63	181	91	59	179	87
	27	61	176	90	58	174	86	54	171	82
	25	56	168	83	53	165	80	49	161	76
2100	34.5	75	191	108	70	191	102	66	190	96
	33	72	188	104	68	188	98	63	186	92
	31	67	183	98	63	182	92	59	180	87
	29	62	177	91	59	176	87	55	173	83
	27	57	170	85	54	168	81	51	164	77
	25	52	161	79	49	158	75	46	153	72

10. What is the rate of fuel consumption in cruise at a pressure altitude of 18,000 feet, with a manifold pressure of 28 inches, a propeller r.p.m. of 2,400, and an air temperature of –31°C? The airplane weight is 4,100 pounds.

    A. 99 pounds per hour
    B. 101.5 pounds per hour
    C. 188 pounds per hour

11. What is the true airspeed under the conditions specified in question 10?

12. True/False. If a runway has a gradient of more than .3%, it is listed in the airport information on navigation charts.

Use the accompanying landing distance chart to answer questions 13 and 14.

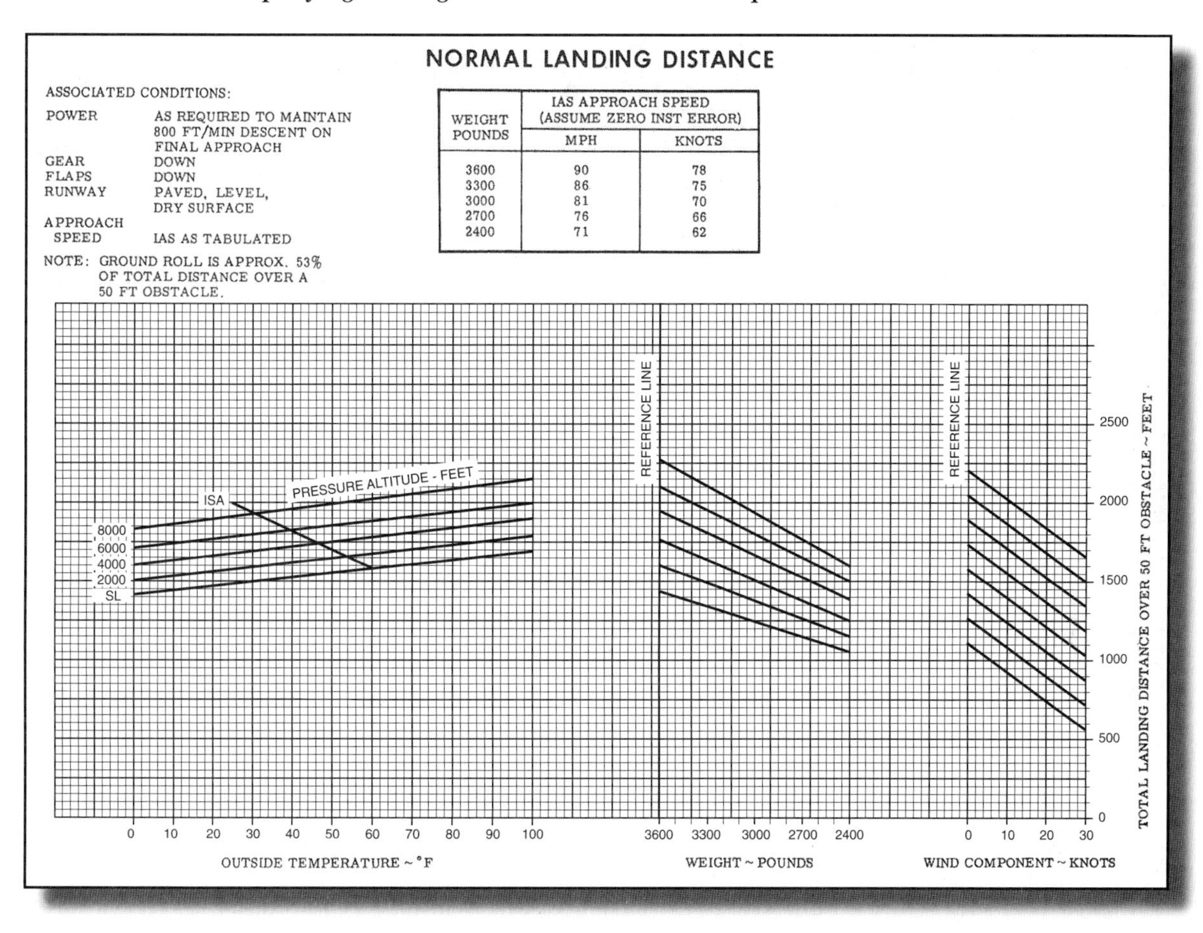

13. For a weight of 2,800 pounds, what indicated approach speeed should you use?

14. Determine the ground roll distance for this airplane at an air temperature of 42°F, a pressure altitude of 6,000 feet, an airplane weight of 3,100 pounds, and a headwind component of 6 knots?

    A. 822 feet
    B. 1,400 feet
    C. 1,550 feet

SECTION C

CONTROLLING WEIGHT AND BALANCE

As a private pilot, you became aware of the importance of weight and balance control and learned how to determine the loading conditions of an airplane by using a variety of methods. While concepts and techniques for performing weight and balance computations are similar for commercial flights, there will be occasions when last minute load changes will require you to quickly amend your original loading computations. In addition, you will probably fly airplanes that are larger and have a greater variety of loading options than the airplanes you have been flying. To conduct safe and efficient flights, you should thoroughly understand the effects of various weight and balance conditions, as well as the terminology and techniques that different manufacturers provide for weight and balance control.

WEIGHT AND BALANCE LIMITATIONS

As discussed in the previous sections of this chapter, the weight and distribution of items carried in an airplane has a tremendous effect on aerodynamic stability and control, as well as affecting overall performance. One of your responsibilities as pilot in command is to consider whether the loading conditions of your airplane are within allowable limits before beginning a flight. These limits are established by the aircraft manufacturer and have been demonstrated to the FAA to meet airworthiness certification standards. When loaded within these limits, and under normal operating conditions, the airplane will be stable and controllable throughout the flight envelope. In addition, the performance values obtained from the POH will allow you to plan your flight with sufficient accuracy to help ensure safety. On the other hand, if you fly an improperly loaded airplane, safety and performance will be compromised in many ways. [Figure 12-58]

EFFECTS OF EXCESSIVE AIRPLANE WEIGHT
— Higher takeoff speed and longer takeoff run.
— Reduced rate and angle of climb.
— Lower maximum altitude.
— Shorter range and endurance.
— Reduced cruising speed and maneuverability.
— Higher stalling speeds.
— Higher landing speed and longer landing roll.

Figure 12-58. An important part of your preflight planning responsibility includes considering the weight limitations of the airplane. If you operate an airplane in an over-loaded condition, you may encounter a number of adverse performance effects, as shown in this table.

MAXIMUM WEIGHT LIMITS

As you know, you must maintain the total, or gross, weight of an airplane at or below maximum limits to provide safe operations. These weight limits are primarily established to maintain the structural integrity of the airframe up to the limit load factors. For

low-powered airplanes, the maximum weight limit is specified as **maximum weight**, maximum certificated weight, or maximum gross weight. However, with other airplanes, the manufacturer may also designate maximum weight limits for various stages of flight and ground operations. [Figure 12-59]

Figure 12-59. In addition to the maximum weight limit established for the total weight of an airplane, manufacturers also designate weight limits for specific areas in the airplane such as baggage or cargo compartments. As seen in this photo, the areas are usually placarded to indicate the limitations. These weight limits are necessary to help prevent the airframe structure from being damaged when subjected to high G-loads.

With many high-performance airplanes, a **maximum ramp weight** is specified, which is the maximum amount the airplane can weigh while on the ramp or during taxi. This weight limit is higher than the maximum weight allowed for takeoff to provide an allowance for the amount of fuel used from engine start until the takeoff roll. Occasionally you will find the manufacturer uses the term maximum taxi weight instead of maximum ramp weight, however, these terms generally represent the same limitation. In addition to maximum ramp or taxi weight, the manufacturer may specify a **maximum takeoff weight** and **maximum landing weight**. For these limitations, the manufacturer has determined that the total weight of the airplane permitted for flight may cause structural load limits to be exceeded during landing. When maximum takeoff and landing weights are designated, the amount of fuel consumed during flight must be adequate to reduce the total weight of the airplane to the maximum landing weight before touchdown. For smaller airplanes, the difference between these weight limits is usually only a few pounds. If an emergency requires you to land immediately after takeoff, these aircraft will provide adequate structural strength for a safe landing. However, you should advise maintenance personnel of the overweight-landing occurrence, since the manufacturer generally requires special inspections to detect damage caused by the landing. [Figure 12-60]

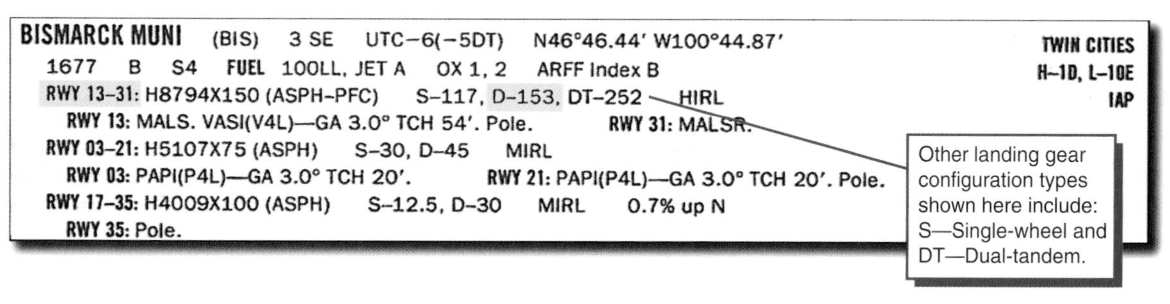

BISMARCK MUNI (BIS) 3 SE UTC–6(–5DT) N46°46.44′ W100°44.87′ **TWIN CITIES**
1677 B S4 **FUEL** 100LL, JET A OX 1, 2 ARFF Index B **H–1D, L–10E**
RWY 13–31: H8794X150 (ASPH–PFC) S–117, D–153, DT–252 HIRL **IAP**
RWY 13: MALS. VASI(V4L)—GA 3.0° TCH 54′. Pole. **RWY 31:** MALSR.
RWY 03–21: H5107X75 (ASPH) S–30, D–45 MIRL
RWY 03: PAPI(P4L)—GA 3.0° TCH 20′. **RWY 21:** PAPI(P4L)—GA 3.0° TCH 20′. Pole.
RWY 17–35: H4009X100 (ASPH) S–12.5, D–30 MIRL 0.7% up N
RWY 35: Pole.

Figure 12-60. In connection with landing weight considerations, you also are required to verify that your airplane weight will not exceed the strength capacity of runways that you plan to use. These capacities can be determined by referring to the *Airport/Facility Directory*, which indicates weight limits in relation to the type of landing gear installed on the airplane. In this example, Runway 13-31 at Bismarck Municipal (BIS), is capable of supporting 153,000 pounds when an airplane is equipped with dual-wheel-type landing gear.

For many large airplanes, the maximum ramp, takeoff, and landing weights are established for additional reasons. As with smaller airplanes, large aircraft may have maximum weight values designated for structural load limit considerations. However, manufacturers also impose maximum weight values in order for an airplane to meet various performance requirements. For example, when operating on a short runway, the maximum takeoff and landing weights may need to be reduced to provide adequate performance for the available runway length. Other performance considerations include evaluating the effects of density altitude, wind, and the mechanical condition of the airplane during takeoff and landing, as well as for climb or cruise. For example, in order

for your airplane to climb adequately, you may need to reduce the maximum amount of weight carried during takeoff. When a maximum weight value is designated for performance reasons, it is commonly referred to as an operational weight restriction. In contrast, if the maximum weight value is required for structural load considerations, it is generally referred to as a design weight limit. Operational weight restrictions can be determined by consulting the performance information in the POH. [Figure 12-61]

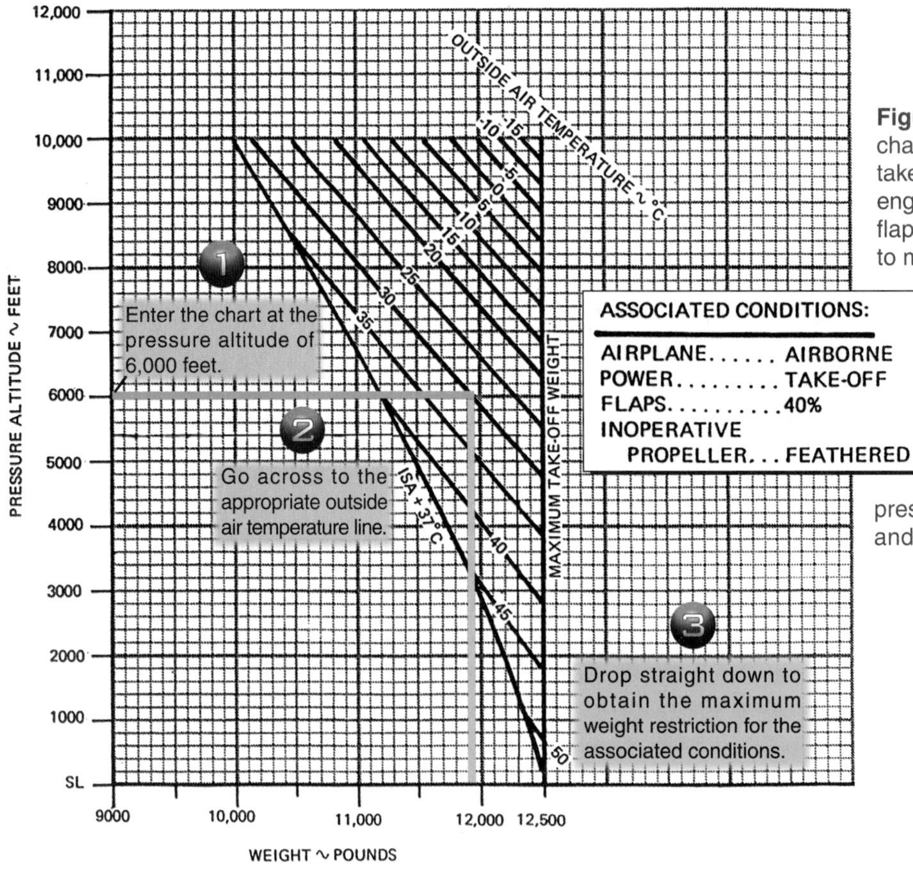

Figure 12-61. The performance chart shown here provides maximum takeoff weight restrictions for a twin-engine, turboprop airplane with 40% flaps selected for takeoff. In order to meet minimum climb performance requirements in the event an engine fails, the maximum takeoff weight may need to be adjusted. For example, a weight restriction of approximately 11,920 pounds will be required when taking off from an airport with a pressure altitude of 6,000 feet, and at a temperature of 30°C.

Weight — Before You Land

To prevent structural damage, it is important to verify that an airplane is at or below its maximum landing weight before touchdown. An example of the effects of landing a heavy airplane occurred when Scott Crossfield had to make an emergency landing with the North American X-15 rocket plane. As a result of an engine explosion and fire, Mr. Crossfield had to abort the flight and attempt to land with a heavy fuel load. Although the airplane design was adequate to handle the extra weight, as the nose gear touched down, the X-15 broke into two pieces at about the midsection. Later, it was discovered that a flaw in the nose gear design prevented it from absorbing the landing shock. When the nose gear touched down, nearly the entire nose gear landing load was transferred into the airframe structure. Once the nose gear design was corrected, the airplane was repaired and used in further research into the effects of high speed, high altitude flight.

Courtesy of NASA Dryden Flight Research Center

Some airplanes also have a **maximum zero fuel weight** limit, which is the maximum amount the airplane can weigh without usable fuel. To determine zero fuel weight, you simply subtract the weight of the usable fuel on board the aircraft from the total weight. If the zero fuel weight is greater than the maximum zero fuel weight limit, items, other than fuel, must be reduced to lighten the load. These limits are necessary to establish the maximum weight that can be carried in the fuselage. If fuselage weight is excessive, the wings may flex too far upward when generating the lift required to support the aircraft. [Figure 12-62]

Figure 12-62. By properly distributing weight between the fuselage and wings, the wings' maximum load limits will not be exceeded, provided the aircraft is operated in normal flight conditions.

CENTER OF GRAVITY LIMITS

While it is important for you to maintain an airplane within allowable weight limits, you must also ensure that items loaded in your airplane do not cause the **center of gravity** (CG) to exceed the allowable range. The manufacturer establishes the CG range and designates it by the position of the forward and aft CG limit locations. These limits are expressed as a distance measured in inches from the **reference datum**. The reference datum, as you recall, is an imaginary vertical plane established by the manufacturer from which longitudinal measurements are made. Provided you load your airplane within the CG range, and given normal flight conditions, longitudinal stability and control of the airplane will be satisfactory. However, if the load causes the CG to exceed allowable limits, airplane control, stability, and efficiency will be affected.

The effects of CG position can be better understood by reviewing the basic aerodynamic forces that act on various parts of the airplane. As previously discussed, when the wings produce lift, the resultant force is considered to act through a region called the center of pressure. In order to increase longitudinal stability and to improve stall characteristics, the center of gravity is usually located ahead of the center of pressure. For example, as the airplane approaches a stalled condition, the relationship of the CG to the center of pressure will tend to cause the airplane to pitch about the lateral axis toward a nose-down attitude. Although this is a desirable design characteristic, it causes the airplane to be in an unbalanced condition. To provide balance, the aerodynamic forces produced by the tail counteract the nose-down tendency. This means that for straight-and-level flight, the tail must exert a downward force to counteract the moment created by the relative position of the CG and the center of pressure. A **moment**, as you recall, is the product of weight, which is a force, times a distance or arm. Since the CG can be located at different distances from the center of pressure, the down force acting on the tail changes with the CG position. In reaction to the tail-down force, the airplane produces various flight and performance characteristics.

FORWARD CG EFFECTS

When the CG is located near the forward limit, the airplane becomes more stable because of the equilibrium that is set up between the forward CG and the forces produced by the tail. For example, if the nose pitches downward, the angle of attack on the tail increases to create a greater tail-down force, which causes the nose to rise. On the other hand, if the nose pitches upward, the angle of attack on the tail decreases, resulting in a reduction in tail-down force, causing the nose to drop. Through a series of damping oscillations between nose-up and nose-down, the airplane eventually returns to a level flight attitude. However, as the airplane becomes more stable, its controllability decreases. This occurs because the tail must exert a greater force to counteract the stability of the airplane. As a result, the horizontal control surface must be deflected a greater amount to cause a pitch change. In fact, if the CG is located ahead of the forward CG limit, at slow airspeed it may be impossible for the control surface to exert enough force to sufficiently raise the nose. For example, during landing, an elevator may not be able to produce enough tail-down force, even when deflected to full travel. As a result, the nose of a tricycle gear airplane may strike the runway before the main gear, resulting in heavy loads being imposed on the nose gear and its supporting structure. While stability and control are influenced by a forward CG, the overall efficiency of the airplane is also affected. [Figure 12-63]

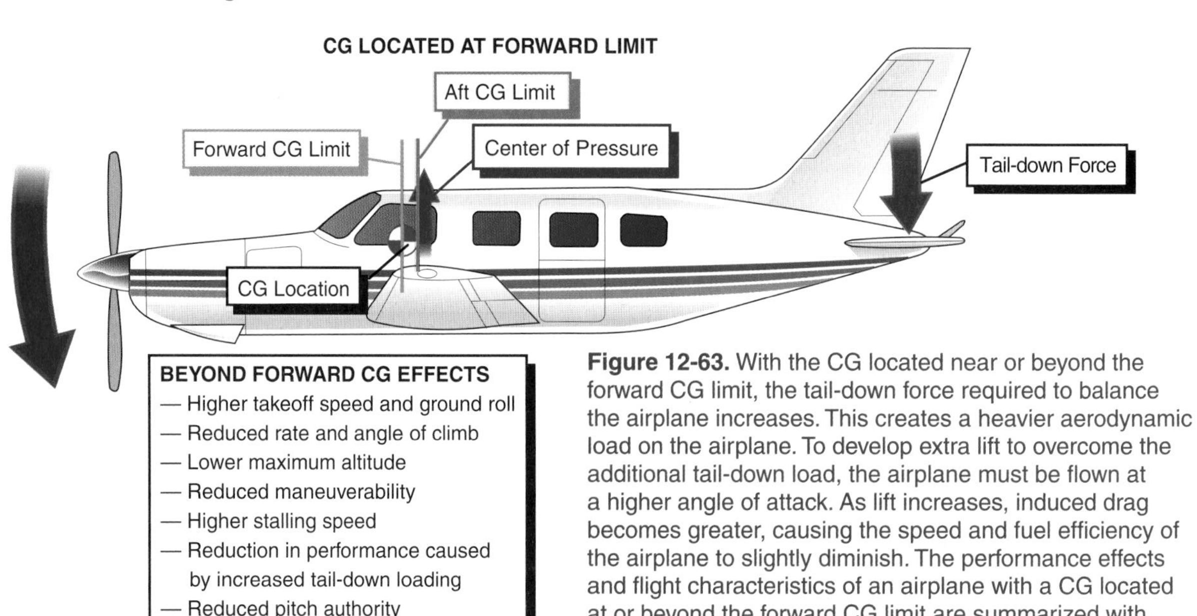

Figure 12-63. With the CG located near or beyond the forward CG limit, the tail-down force required to balance the airplane increases. This creates a heavier aerodynamic load on the airplane. To develop extra lift to overcome the additional tail-down load, the airplane must be flown at a higher angle of attack. As lift increases, induced drag becomes greater, causing the speed and fuel efficiency of the airplane to slightly diminish. The performance effects and flight characteristics of an airplane with a CG located at or beyond the forward CG limit are summarized with this diagram.

AFT CG EFFECTS

When the CG of an airplane is located at the aft limit, the amount of tail-down force required to maintain a balanced condition decreases because the CG is located closer to the center of pressure. This causes the horizontal control surface to become more effective. However, with the increased control responsiveness, it may be possible for you to over-control the airplane. In addition, the airplane will have a tendency to pitch-up due to the aft CG position, which could cause an inadvertent stall. As a further consideration, if the CG is located behind the aft limit, it may become difficult, or even impossible to recover from a stalled condition. Spins also are more critical with an aft CG, since the spin will tend to be flat, making recovery difficult or impossible. However, when the CG is located at the aft limit, there is a slight increase in TAS and fuel efficiency due to the reduction of the tail loading. [Figure 12-64]

When an airplane is loaded with the CG near the aft limit, it tends to become unstable about the lateral axis.

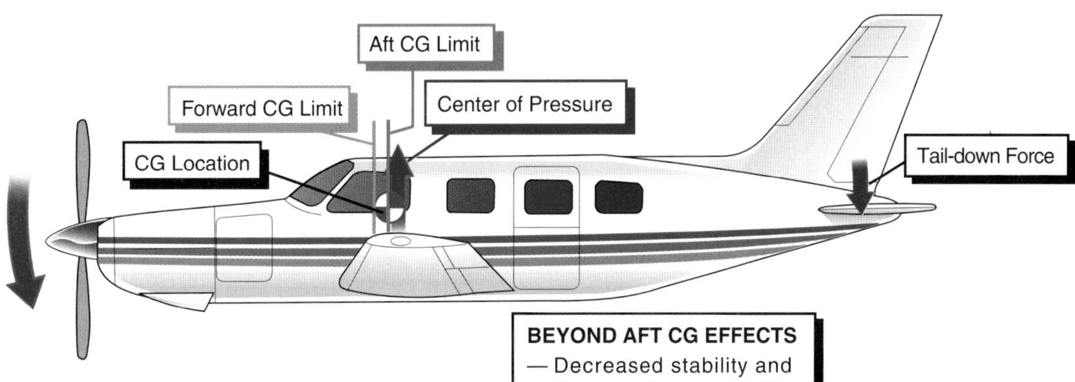

Figure 12-64. A CG that is located aft of limits creates greater hazards than a CG located beyond forward limits. For the safest condition, it is preferred to fly with the CG located near the center of the CG range. The performance effects and flight characteristics of an airplane with an aft CG are included in this diagram.

Loading the Big Rigs

The CG location in transport airplanes is usually expressed in terms of the percentage of mean aerodynamic chord (% MAC). As you know, a chord is the distance from the leading edge to the trailing edge of the wing, and on a tapered wing this distance varies from root to tip. The mean aerodynamic chord is the chord drawn through the geometric center of the planform of one wing. [Figure A] It is a convenient reference for weight and balance as well as aerodynamic purposes. The leading edge of the mean aerodynamic chord (LEMAC) is 0% MAC, and the trailing edge (TEMAC) is 100% MAC. The CG range typically falls between 15% MAC and 37% MAC.

Since transport airplanes have a relatively large CG range, control pressures can vary significantly even when the airplane is loaded within the approved CG envelope. To help compensate, the pitch trim is adjusted on the ground before takeoff according to the computed CG location. With the CG located toward the forward limit, more nose-up trim is set than when the CG is toward the aft limit. The longitudinal trim on most large airplanes is controlled by changing the angle of the horizontal stabilizer [Figure B], but at least one type of airplane moves the whole empennage.

B

STAB TRIM SETTING

CG%	Flaps		
	5	15/20	25
	Units	Airplane	Nose Up
10	6 3/4	7 1/2	8 1/4
12	6 1/2	7 1/4	8
14	6 1/4	7	7 3/4
16	6	6 3/4	7 1/2
18	5 3/4	6 1/2	7
20	5 1/2	6	6 1/2
22	5	5 3/4	6 1/4
24	4 3/4	5 1/4	5 3/4
26	4 1/2	4 3/4	5 1/4
28	4	4 1/2	4 3/4
30	3 3/4	4	4 1/4
32	3 1/2	3 3/4	4
34	3 1/4	3 1/4	3 1/2
36	2 3/4	3	3
38	2 1/2	2 1/2	2 1/2
40	2 1/2	2 1/2	2 1/2
42	2 1/2	2 1/2	2 1/2

LATERAL CG EFFECTS

Lateral weight and balance control is not considered to be as critical as longitudinal control because of the symmetry that exists between the left and right side of the airplane. An exception occurs when fuel becomes unevenly distributed between left and right side fuel tanks which causes a lateral CG shift and wing heavy condition. In this condition, the ailerons can be positioned to offset the heavy wing. However, by deflecting the ailerons, drag increases, which causes a reduction in cruise speed, range, and endurance. To prevent the lateral imbalance, you should use fuel from left and right side fuel tanks simultaneously or alternate between the tanks at regular intervals.

WEIGHT AND BALANCE DOCUMENTS

In order for you to determine the weight and balance limitations and loading conditions of an airplane, you need to be familiar with the information provided in various weight and balance documents. These documents primarily include the **weight and balance report** and **equipment list**. A weight and balance report is initially prepared by the manufacturer and is maintained by the aircraft owner or operator. In addition, the report must be carried in the airplane anytime it is flown, in order to meet standard airworthiness requirements. Equipment lists are generally considered to be part of a weight and balance report, and are used to inventory the weight and location of standard and optional equipment installed in the airplane.

WEIGHT AND BALANCE REPORT

A weight and balance report is required for all airplanes, and may be presented in a variety of formats. For airplanes that are required to have a POH, the report is retained in the handbook. For these airplanes, the POH contains charts, graphs, or tables for use in determining loading conditions. In addition, the POH includes sample problems to demonstrate methods for performing weight and balance computations. Remember that the sample problems provide weight and balance values for a hypothetical airplane. You should never use these values for performing actual load computations. [Figure 12-65]

Figure 12-65. When checking loading conditions, you must use the information provided in the specific weight and balance report developed for the airplane. For airplanes requiring a POH, the report is usually located in the weight and balance section of the handbook.

For airplanes not requiring a POH, the manufacturer compiles the weight and balance report into a single document. For reports retained in this manner, the manufacturer usually provides information similar to that contained in the weight and balance section of a POH. For some older airplanes, the weight configurations of items that will produce acceptable flight conditions are detailed in loading schedules. These schedules may be displayed inside the airplane as placards, while others are contained in the weight and balance report. Regardless of the method used to determine loading conditions, you must make sure that the weight and balance information is appropriate and current for your aircraft.

Information found in a weight and balance report includes the empty weight and **empty weight center of gravity** position of the airplane. Depending on the certification requirements in effect when the airplane was manufactured, the empty weight may be referred to as a **basic empty weight**, or a **licensed empty weight**. Airplanes manufactured after March 1, 1979 use the term basic empty weight. This weight includes the standard airplane, optional equipment, unusable fuel, and full operating fluids including full engine oil. For older airplanes, the term licensed empty weight is used, which is similar to basic empty weight, except that engine oil is not included. For these airplanes, you must remember to include the weight of the oil when you perform weight and balance computations. As you recall, engine oil weighs 7.5 pounds per gallon, or 1.875 pounds per quart. [Figure 12-66]

Figure 12-66. To determine the empty weight and empty weight center of gravity location, maintenance personnel actually weigh an airplane by placing it on scales or other weighing devices. Many airplanes that are operated under FAR Part 91 requirements are only weighed when the airplane is new. On the other hand, airplanes used for commercial operations regulated under various other FAR parts, are required to be weighed at specific intervals to maintain accurate weight and balance records.

The empty weight of an airplane includes unusable fuel, hydraulic fluid, and undrainable oil, or, in some aircraft, all of the oil.

Other information found in the weight and balance report includes maximum ramp, takeoff and landing weights, or the maximum weight only, as previously discussed. In addition, the **useful load** of the airplane will be indicated. Useful load, as you recall, includes usable fuel, pilot and crew, passengers, baggage or cargo, and engine oil, if it is not already included in the empty weight. If applicable, the maximum zero fuel weight limit will also be included in the report.

EQUIPMENT LIST

As previously mentioned, an equipment list provides an inventory of the weight and location of standard and optional equipment installed in the airplane. When a POH is required, the equipment list will usually be inserted in the weight and balance section of the handbook. On airplanes not requiring a POH, the equipment list is included in the weight and balance report. The list includes permanently installed components such as radio equipment and flight instruments, as well as some removable items such as the POH, portable fire extinguishers, tow bars, and other similar items. If you are unsure if an item is part of the empty weight, you can refer to the equipment list to determine if it is included. [Figure 12-67]

Figure 12-67. For most airplanes, the equipment list occupies a number of pages in the POH. These lists often include equipment that may be available from the manufacturer, but are not installed in your airplane. To designate installed equipment, pertinent items are marked with a check mark or an X.

When the installation or removal of equipment alters an airplane, the technician or agency conducting the alteration must revise the equipment list to reflect the change. In addition, the weight and balance report also will be revised to reflect changes to the empty weight, empty weight CG and useful load. Remember, when equipment is added to an airplane, the empty weight increases, which causes a decrease in the useful load.

When using weight information given in a typical aircraft owner's manual, the actual useful load of the aircraft may be lower than that shown in the manual because of the installation of additional equipment.

WEIGHT AND BALANCE COMPUTATIONS

To determine if an airplane is properly loaded, you must perform weight and balance computations to verify that both total weight and CG are within allowable limits. To compute total weight, you merely add the weight of useful load items to the empty weight. Although weight computations are reasonably easy to perform, balance computations require you to have an understanding of various terms and principles. Once you understand these, balance computations are easy to perform.

MOMENT COMPUTATIONS

In order to determine the effects that useful load items have on balance conditions, you must calculate the moment that each item produces. A moment, as you recall, can be found by multiplying the weight of an item by its arm. For weight and balance purposes, an **arm** is the horizontal distance in inches that an item is located away from the reference datum. To designate the position of an item as being ahead of or behind the reference datum, the arms are assigned positive or negative values. Items located ahead of the reference datum have negative values, while items located aft have positive values. To alleviate computations with negative numbers, some manufacturers locate the reference datum on the nose or at some point ahead of the airplane. For these aircraft, items cannot be loaded ahead of the reference datum, so all arm values are positive. [Figure 12-68]

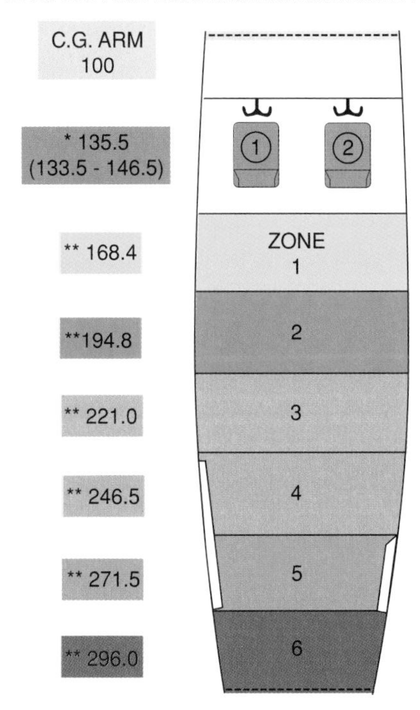

Figure 12-68. An arm may also be expressed as a fuselage station (F.S. or sta.), which identifies various longitudinal locations in the airplane. For some cabin configurations such as those used for carrying cargo, arms or F.S. values are separated into zones. Zones are often identified inside the airplane by the location of seams in floor panels, or by placards installed along the cabin walls.

The moment of an item may be regarded as a torque or twisting force that is applied about a fulcrum, and is expressed in pound-inches. Since the product of the weight and arm can result in relatively large numbers,

When the reference datum is located at or ahead of the nose, the moment index will always be positive when items are loaded in the airplane.

most manufacturers use **moment indexes**. These are merely moments that have had a reduction factor applied to help simplify weight and balance computations. For example, a moment of 130,000 pound-inches will have a moment index of 130 when a reduction factor of 1,000 is applied. [Figure 12-69]

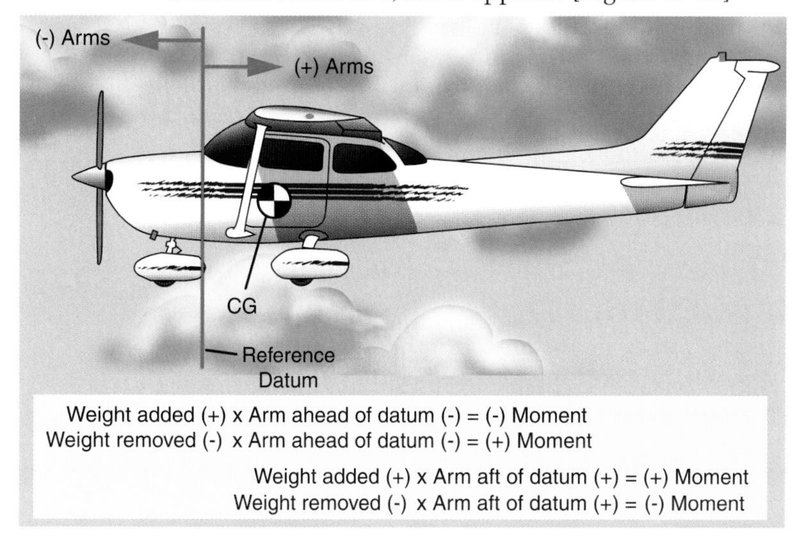

Figure 12-69. When performing moment computations, you need to remember that the moment may be positive or negative, depending on a variety of conditions. Conditions that affect moment signs are summarized in this illustration.

DETERMINING CENTER OF GRAVITY POSITION

Once you have calculated the moment of each useful load item, you can locate the CG by determining the total weight of the items and the total of the moments. By dividing total moment by total weight, you can derive the CG position in inches from the datum. An example of this procedure, using a variety of weights at different locations from a datum, is shown in figure 12-70.

To compute the combined CG of assorted weights, find the moment of each item by multiplying the item weights by the arm. To find the CG location, divide the total moment by the total weight.

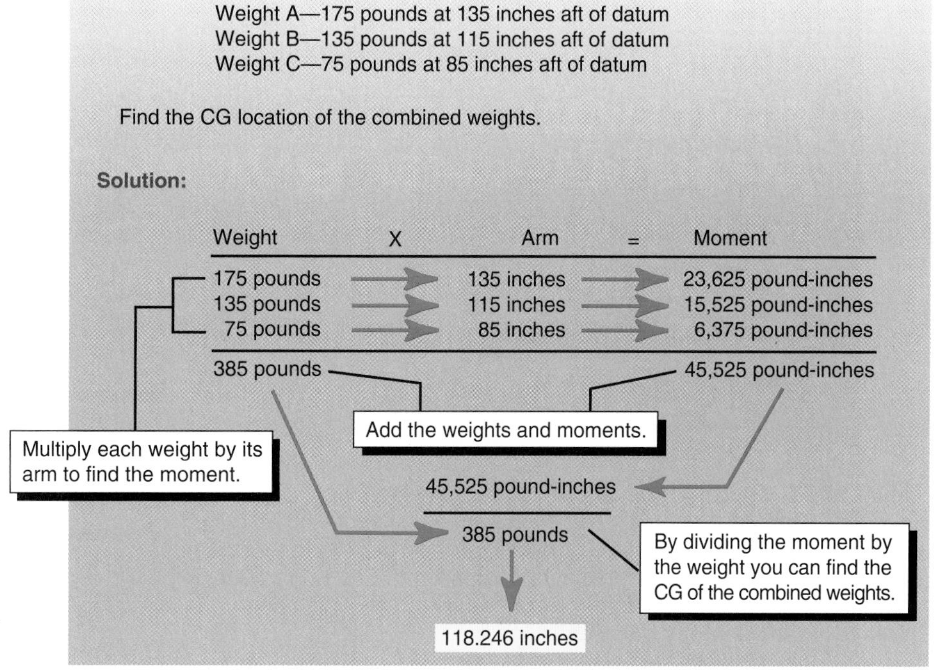

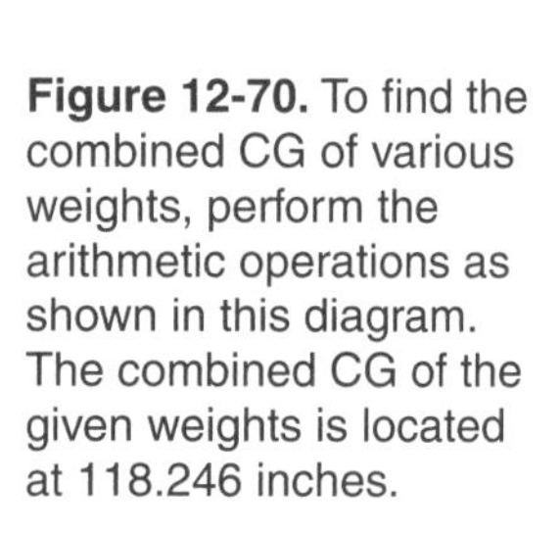

Figure 12-70. To find the combined CG of various weights, perform the arithmetic operations as shown in this diagram. The combined CG of the given weights is located at 118.246 inches.

WEIGHT AND BALANCE CONDITION CHECKS

To verify that an airplane is loaded within acceptable weight and balance limits, manufacturers prescribe various methods for checking weight and balance conditions. Among these are the computation, table and graph methods for determining the total weight and CG location of the airplane. From your previous flying experience, you are probably familiar with each of these methods. However, as you progress into flying larger and more complex aircraft, you will likely encounter a greater number of loading considerations. To help you understand some of these considerations, the following sample weight and balance condition checks represent airplanes that you will likely encounter when conducting commercial flights.

COMPUTATION METHOD

The computation method for checking weight and balance conditions can generally be used with any airplane. The first step is to locate the empty weight and empty weight CG position in the weight and balance report. In most cases, the report also includes the empty moment of the airplane. Once these are located, list them on a note pad or a copy of the weight and balance loading form supplied in the POH. Next, list the weight of all useful load items. To convert aviation fuel from gallons to pounds, you can use the standard six pounds per gallon or the weight specified by the manufacturer. By following a sample weight and balance condition check, you can see how a loading form helps you complete weight and balance computations. [Figure 12-71]

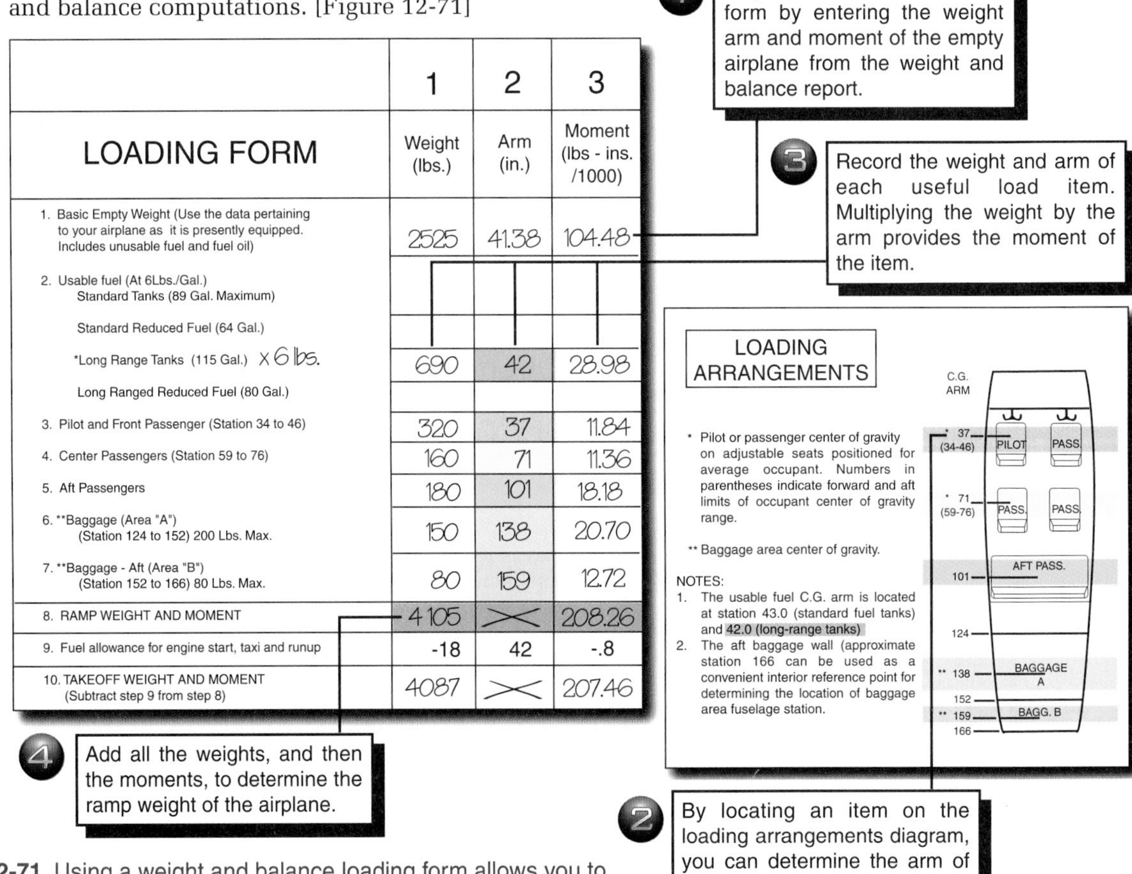

	1	2	3
LOADING FORM	Weight (lbs.)	Arm (in.)	Moment (lbs - ins. /1000)
1. Basic Empty Weight (Use the data pertaining to your airplane as it is presently equipped. Includes unusable fuel and fuel oil)	2525	41.38	104.48
2. Usable fuel (At 6Lbs./Gal.) Standard Tanks (89 Gal. Maximum)			
Standard Reduced Fuel (64 Gal.)			
*Long Range Tanks (115 Gal.) X 6 lbs.	690	42	28.98
Long Ranged Reduced Fuel (80 Gal.)			
3. Pilot and Front Passenger (Station 34 to 46)	320	37	11.84
4. Center Passengers (Station 59 to 76)	160	71	11.36
5. Aft Passengers	180	101	18.18
6. **Baggage (Area "A") (Station 124 to 152) 200 Lbs. Max.	150	138	20.70
7. **Baggage - Aft (Area "B") (Station 152 to 166) 80 Lbs. Max.	80	159	12.72
8. RAMP WEIGHT AND MOMENT	4105	✕	208.26
9. Fuel allowance for engine start, taxi and runup	-18	42	-.8
10. TAKEOFF WEIGHT AND MOMENT (Subtract step 9 from step 8)	4087	✕	207.46

Figure 12-71. Using a weight and balance loading form allows you to perform an organized and complete evaluation of the airplane's loading conditions. A sample loading form is shown here on the left. To provide a convenient method of locating arm distances to be used with the loading form, manufacturers also provide loading arrangement charts, similar to the one shown on the right. By referring to this chart, you can readily visualize the arm distance of various useful load items.

After you have entered weight and moment values on the loading form, add the weights to determine the total weight of the airplane and then add the moments to find the total moment. Before continuing, check to make sure that the total weight does not exceed the maximum ramp weight limit specified by the manufacturer. The ramp weight limit may be shown on the loading form, or you may need to determine maximum weight limits by locating them in the limitations section of the POH. By referring back to the loading form in figure 12-71, you will notice that the manufacturer has determined that 18 pounds of fuel is considered to be used during engine start, runup, and taxi. This fuel weight and moment can be deducted from the total weight and moment amounts to determine the airplane's weight and balance condition for takeoff. Once these adjustments have been made, a CG moment envelope chart can be used to determine if the adjusted weight and moment values are within allowable limits. These charts depict the forward and aft CG limits, maximum weight values and also include notes regarding special loading considerations. [Figure 12-72]

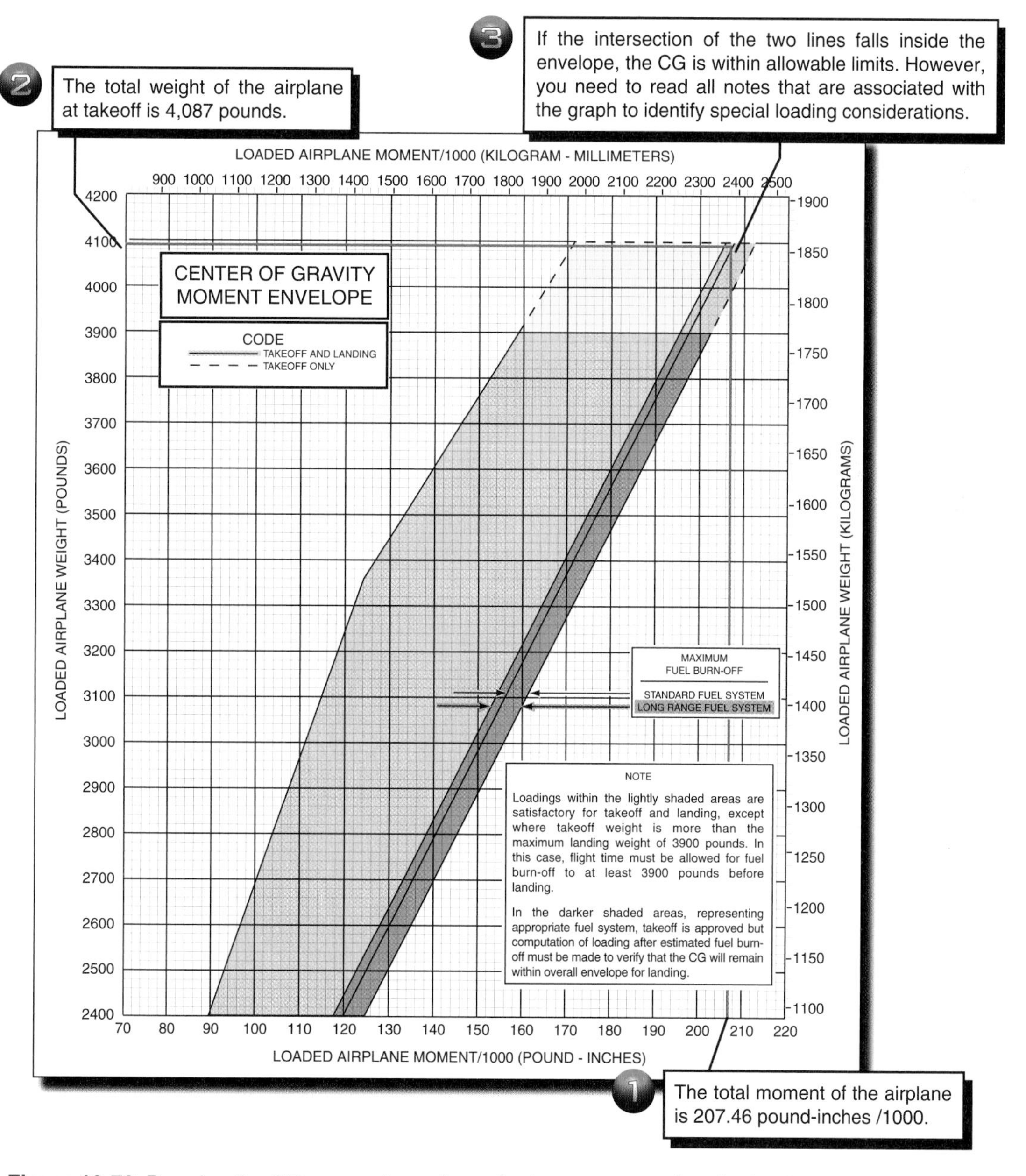

Figure 12-72. By using the CG moment envelope chart, you can see that the loading conditions of the sample airplane are within limits, but fall within a darker shaded area of the CG envelope. In this case, you need to comply with the notes contained on the chart.

To determine if the CG of the sample airplane will remain within limits, you need to compute how much fuel will be used during flight. Normally this will include the amount used for takeoff, climb, and cruise. For example, assume that the sample airplane uses all but 45 minutes of fuel during flight. If the airplane burns 13 gallons per hour (g.p.h.) during cruise, you can divide 13 g.p.h. by 60 minutes to arrive at .217 gallons per minute (g.p.m.) fuel consumption. By multiplying .217 g.p.m. by 45 minutes, you find that the fuel remaining in the aircraft will be approximately 9.75 gallons. By multiplying 9.75 gallons by 6 pounds per gallon, you can see that there will be 58.5 pounds of fuel remaining in the airplane upon landing. Since the airplane originally contained 690 pounds of fuel, the weight of the airplane will decrease by 631.5 pounds (690 pounds – 58.5 pounds). To determine the effects on the CG, follow the example in figure 12-73.

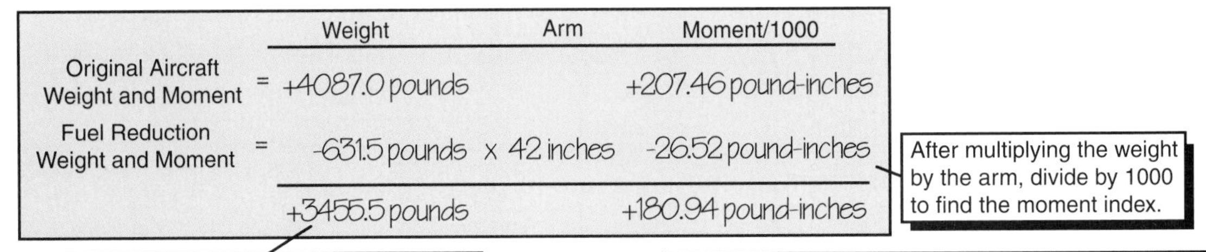

		Weight	Arm	Moment/1000
Original Aircraft Weight and Moment	=	+4087.0 pounds		+207.46 pound-inches
Fuel Reduction Weight and Moment	=	-631.5 pounds	x 42 inches	-26.52 pound-inches
		+3455.5 pounds		+180.94 pound-inches

After multiplying the weight by the arm, divide by 1000 to find the moment index.

Use the new weight and moment on the center of gravity moment envelope to determine if the CG is within allowable limits.

Since the intersection of the two lines falls outside the envelope, you will need to reposition items to maintain the CG within allowable limits. In this example, the CG is located approximately 940 pound-inches behind the aft CG limit.

The total weight is 3455.5 pounds after computing fuel usage for the flight.

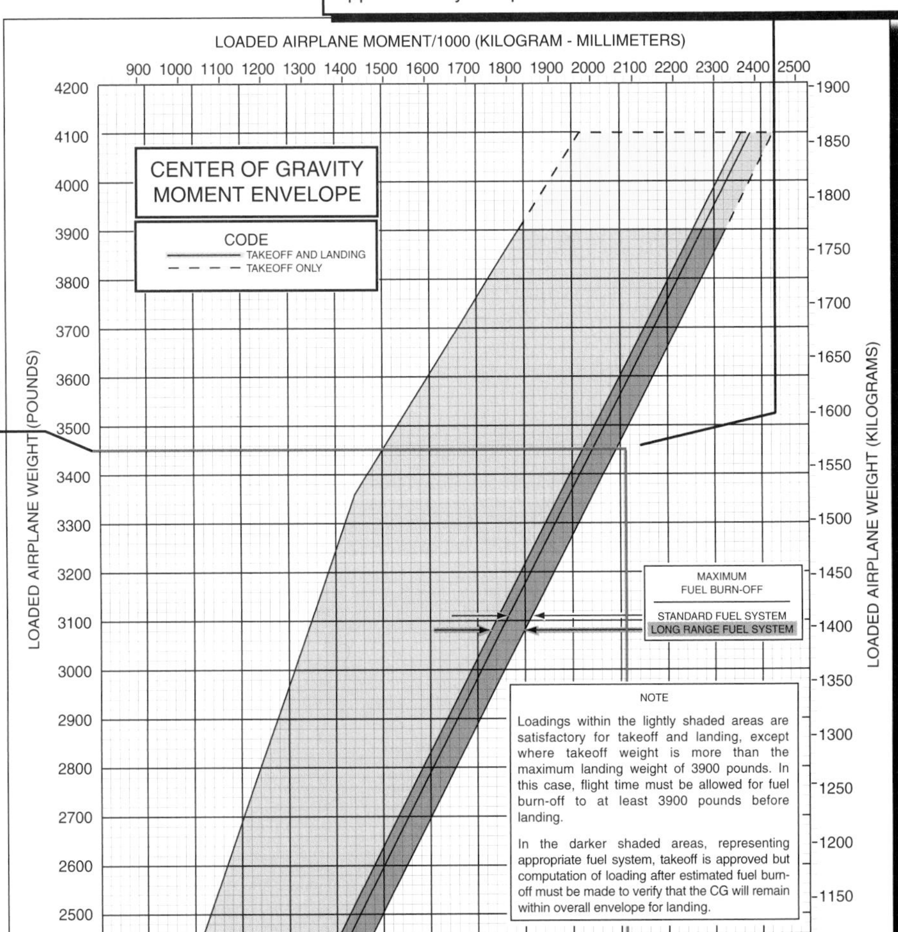

Figure 12-73. Once you have determined how much weight will be lost during flight due to fuel consumption, perform a second weight and balance computation as shown above, to determine if the airplane remains within the CG envelope. In this example, you can see that the CG falls aft of limits by approximately 940 pound-inches.

The total moment is 180.94 pound-inches/1000 after computing fuel usage for the flight.

When the CG position falls out of allowable limits, you can correct the condition by repositioning items to different locations. When possible, it is more practical to move a heavy item since you only need to reposition it a short distance. To help determine the amount of weight that must be moved, you can perform a weight shift computation. A weight shift computation example is provided later in this section, and provides a solution to the out of balance condition of this airplane.

To determine the effects of fuel consumption on weight and balance, calculate the weight of fuel consumed for the specified period of time. Once the weight has been determined, deduct it from the original fuel weight and recompute the weight and balance for the airplane.

GRAPH METHOD

To reduce the need to perform arithmetic operations when performing weight and balance condition checks, some manufacturers provide graphs to check loading conditions. With the graph method, you essentially perform the same operations as those used with the computation method. However, determining the moment of an individual item is easier when you can consult a loading graph, which provides the moment of various items at different weights. [Figure 12-74]

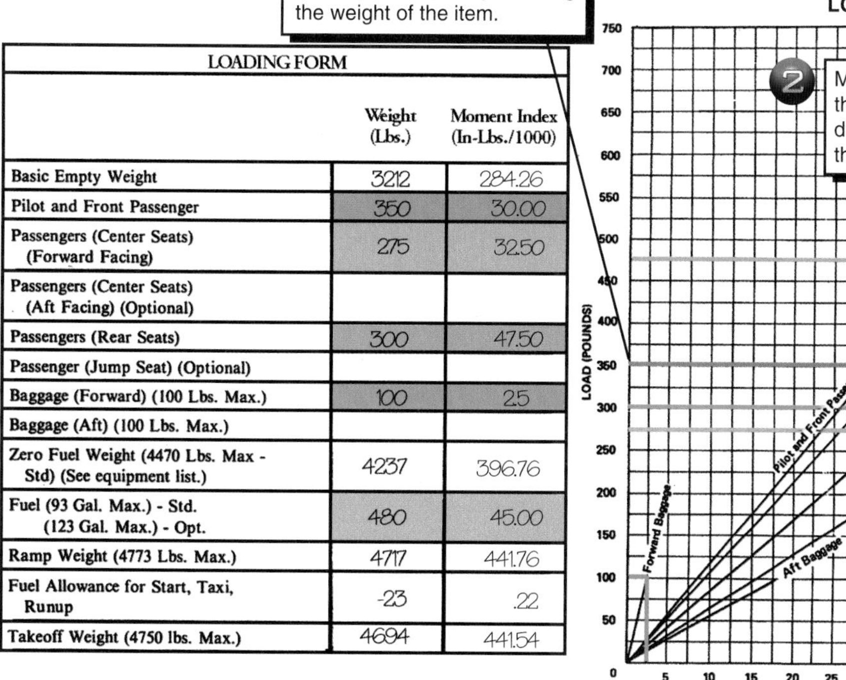

LOADING FORM

	Weight (Lbs.)	Moment Index (In-Lbs./1000)
Basic Empty Weight	3212	284.26
Pilot and Front Passenger	350	30.00
Passengers (Center Seats) (Forward Facing)	275	32.50
Passengers (Center Seats) (Aft Facing) (Optional)		
Passengers (Rear Seats)	300	47.50
Passenger (Jump Seat) (Optional)		
Baggage (Forward) (100 Lbs. Max.)	100	2.5
Baggage (Aft) (100 Lbs. Max.)		
Zero Fuel Weight (4470 Lbs. Max - Std) (See equipment list.)	4237	396.76
Fuel (93 Gal. Max.) - Std. (123 Gal. Max.) - Opt.	480	45.00
Ramp Weight (4773 Lbs. Max.)	4717	441.76
Fuel Allowance for Start, Taxi, Runup	-23	.22
Takeoff Weight (4750 lbs. Max.)	4694	441.54

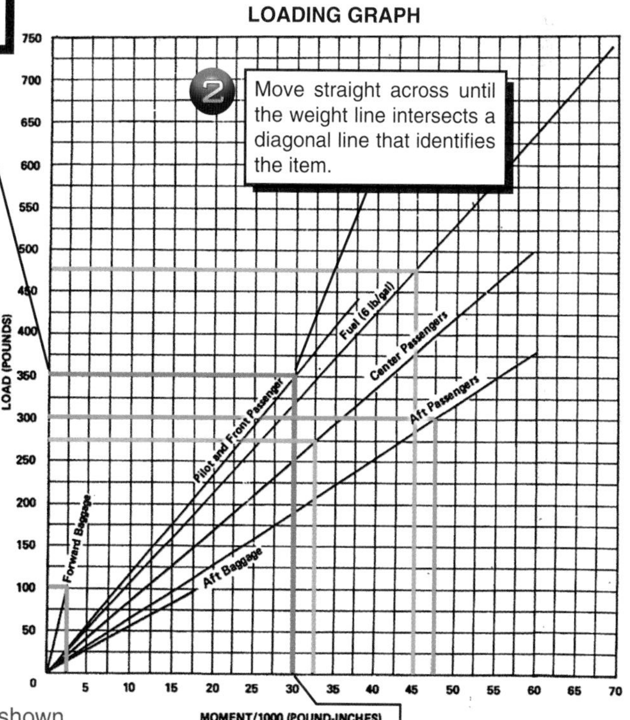

Figure 12-74. By using a loading graph such as the one shown here, you can complete a loading form with reasonable ease. In this example, you can see that the weight of the airplane without fuel is less than the maximum zero fuel weight. If the zero fuel weight were to exceed the maximum limit, the weight of useful load items, other than fuel, would need to be reduced to correct the condition.

After you have computed the total weight and total moment of the airplane, you can refer to a CG range and weight chart to determine if the airplane is loaded within the CG envelope. These charts are similar to a CG moment envelope, except that you use the arm of the CG instead of the total moment of the airplane. [Figure 12-75]

To perform weight and balance condition checks you can use a loading graph to find the moment of all useful load items. Once the total weight and total moment are computed, determine if the moment and weight are within allowable limits by using a center of gravity envelope graph.

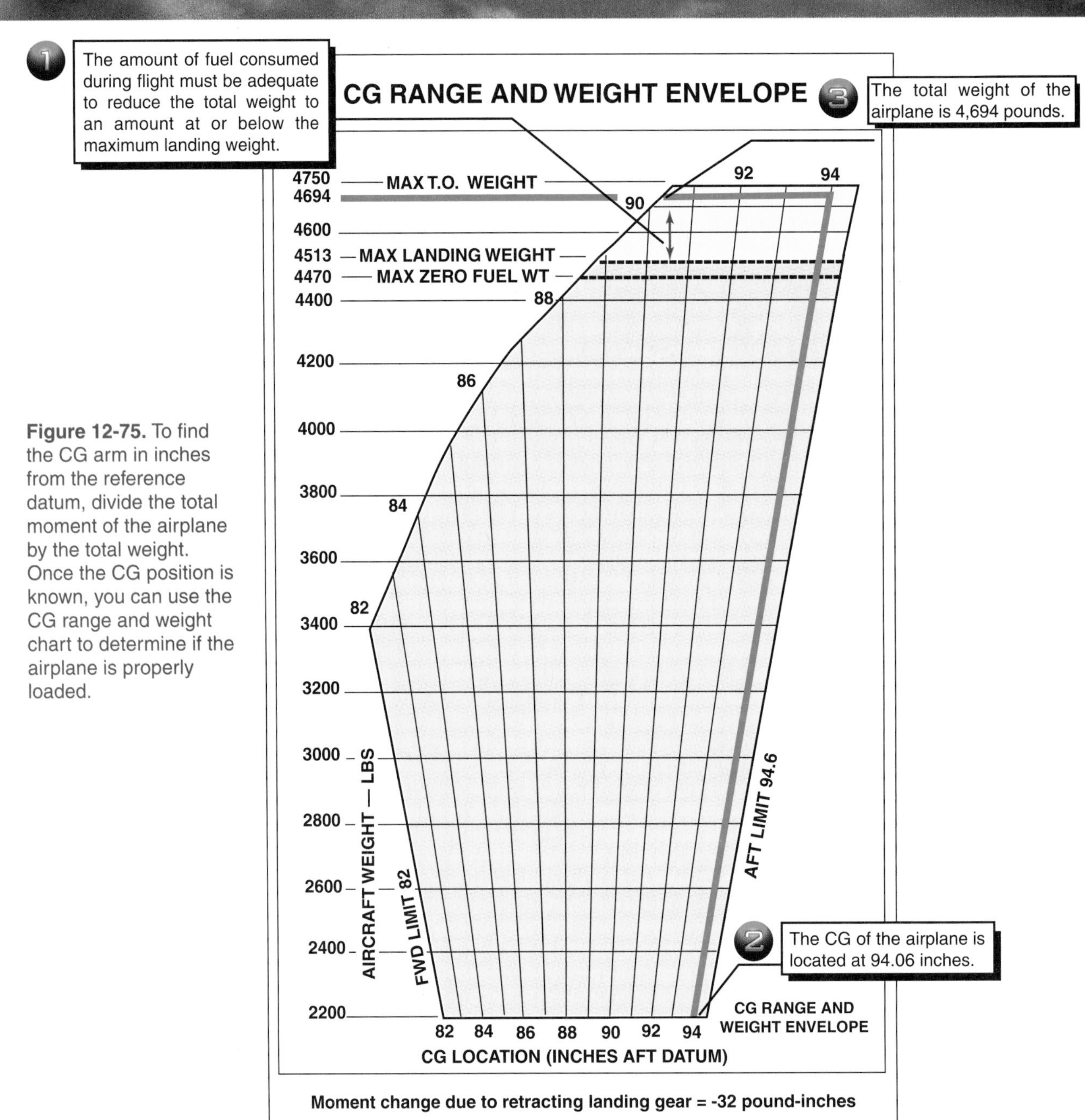

Figure 12-75. To find the CG arm in inches from the reference datum, divide the total moment of the airplane by the total weight. Once the CG position is known, you can use the CG range and weight chart to determine if the airplane is properly loaded.

TABLE METHOD

Many manufacturers of large airplanes supply tables to simplify weight and balance computations by tabulating the moments of useful load items at various weights and fuselage station locations. Because large airplanes often have a variety of possible cabin configurations, you need to first determine which tables to use. To accomplish this, manufacturers provide cabin configuration diagrams, which illustrate various positions of the cabin furnishings. By identifying the diagram that matches your airplane's configuration, you can determine which tables to use for weight and balance computations. Once the tables have been identified, use a loading form to list the weights, arms, and moments of the useful load items carried in the airplane. For large airplanes, these loading forms may include additional weight considerations such as **payload** and **basic operating weight**. Payload can be thought of as potential revenue generating items such as passengers and baggage or cargo, while basic operating weight includes the airplane and crew, but without fuel and payload items. [Figure12-76]

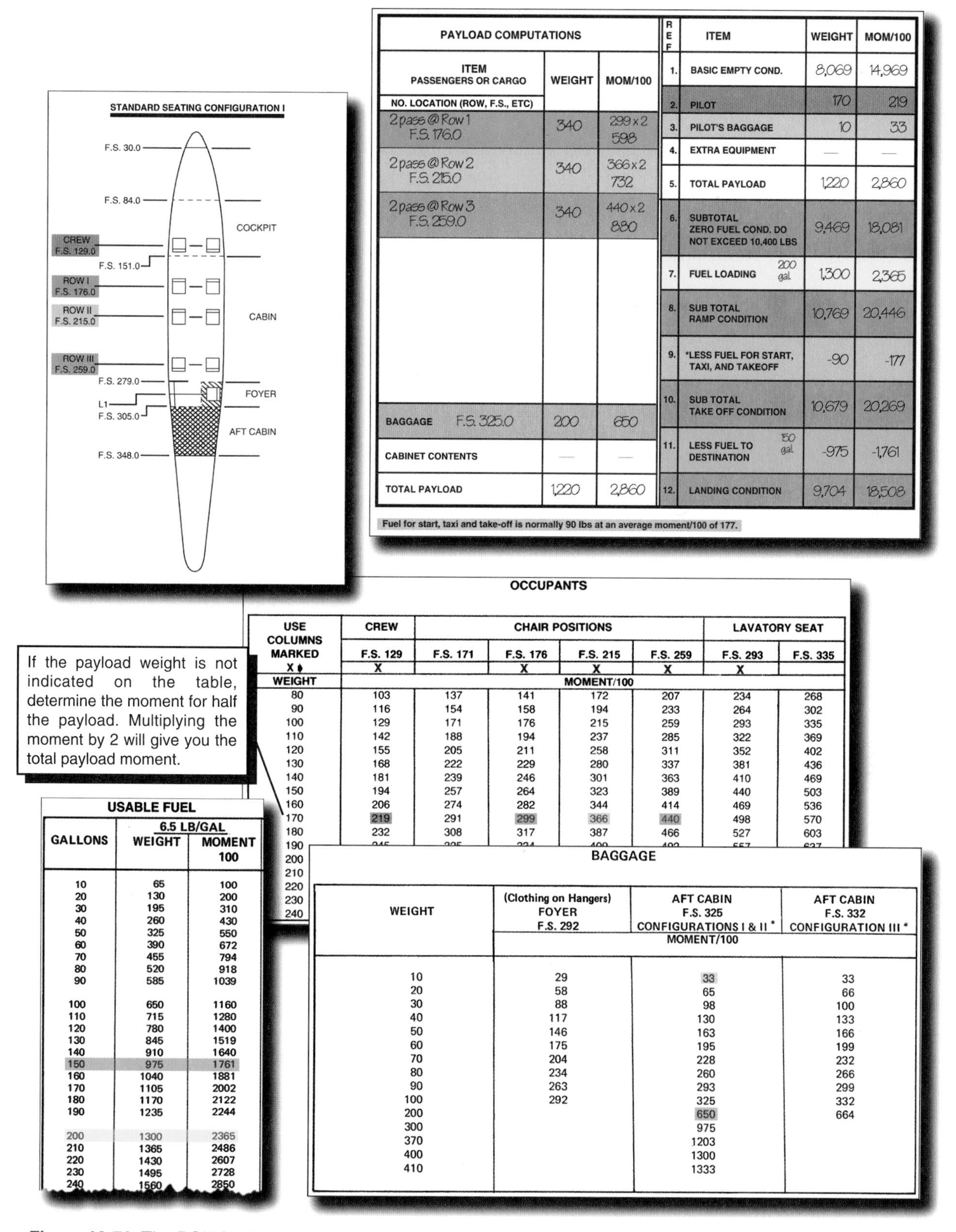

PAYLOAD COMPUTATIONS		
ITEM PASSENGERS OR CARGO NO. LOCATION (ROW, F.S., ETC)	WEIGHT	MOM/100
2pass @ Row 1 F.S. 176.0	340	299 x 2 598
2pass @ Row 2 F.S. 215.0	340	366 x 2 732
2pass @ Row 3 F.S. 259.0	340	440 x 2 880
BAGGAGE F.S. 325.0	200	650
CABINET CONTENTS	—	—
TOTAL PAYLOAD	1,220	2,860

REF	ITEM	WEIGHT	MOM/100
1.	BASIC EMPTY COND.	8,069	14,969
2.	PILOT	170	219
3.	PILOT'S BAGGAGE	10	33
4.	EXTRA EQUIPMENT	—	—
5.	TOTAL PAYLOAD	1,220	2,860
6.	SUBTOTAL ZERO FUEL COND. DO NOT EXCEED 10,400 LBS	9,469	18,081
7.	FUEL LOADING 200 gal	1,300	2,365
8.	SUB TOTAL RAMP CONDITION	10,769	20,446
9.	*LESS FUEL FOR START, TAXI, AND TAKEOFF	-90	-177
10.	SUB TOTAL TAKE OFF CONDITION	10,679	20,269
11.	LESS FUEL TO DESTINATION 150 gal	-975	-1,761
12.	LANDING CONDITION	9,704	18,508

Fuel for start, taxi and take-off is normally 90 lbs at an average moment/100 of 177.

OCCUPANTS

USE COLUMNS MARKED X ♦	CREW	CHAIR POSITIONS				LAVATORY SEAT	
	F.S. 129	F.S. 171	F.S. 176	F.S. 215	F.S. 259	F.S. 293	F.S. 335
	X		X	X	X	X	
WEIGHT	MOMENT/100						
80	103	137	141	172	207	234	268
90	116	154	158	194	233	264	302
100	129	171	176	215	259	293	335
110	142	188	194	237	285	322	369
120	155	205	211	258	311	352	402
130	168	222	229	280	337	381	436
140	181	239	246	301	363	410	469
150	194	257	264	323	389	440	503
160	206	274	282	344	414	469	536
170	219	291	299	366	440	498	570
180	232	308	317	387	466	527	603
190							
200							
210							
220							
230							
240							

USABLE FUEL

GALLONS	6.5 LB/GAL WEIGHT	MOMENT 100
10	65	100
20	130	200
30	195	310
40	260	430
50	325	550
60	390	672
70	455	794
80	520	918
90	585	1039
100	650	1160
110	715	1280
120	780	1400
130	845	1519
140	910	1640
150	975	1761
160	1040	1881
170	1105	2002
180	1170	2122
190	1235	2244
200	1300	2365
210	1365	2486
220	1430	2607
230	1495	2728
240	1560	2850

BAGGAGE

WEIGHT	(Clothing on Hangers) FOYER F.S. 292	AFT CABIN F.S. 325 CONFIGURATIONS I & II *	AFT CABIN F.S. 332 CONFIGURATION III *
	MOMENT/100		
10	29	33	33
20	58	65	66
30	88	98	100
40	117	130	133
50	146	163	166
60	175	195	199
70	204	228	232
80	234	260	266
90	263	293	299
100	292	325	332
200		650	664
300		975	
370		1203	
400		1300	
410		1333	

Figure 12-76. The POH for the airplane contains multiple cabin configuration diagrams. By using the diagram that matches your airplane configuration, you can readily identify the fuselage station of useful load items.

Once you have determined the total weight and total moment of the airplane, you can refer to a moment limit versus weight table to determine if the airplane is loaded within allowable CG limits. [Figure 12-77]

MOMENT LIMITS VS. WEIGHT			
	WEIGHT	MINIMUM MOMENT/100	MAXIMUM MOMENT/100
	7200	13032	14141
	7250	13122	14239
	7300	13213	14337
	7350	13304	14435
	7400	13394	14534
	10150	18372	19935
	10200	18462	20033
	10250	18552	20131
	10300	18643	20229
	10350	18734	20327
MAX ZERO FUEL WEIGHT	10400	18824	20426
	10450	18914	20524
	10500	19005	20622
	10550	19096	20720
	10600	19276	20818
	10650	19367	20917
	10700	19458	21015
	10750	19548	21113
	10800	19638	21211
	10850	19729	21309
	10900	19820	21408
	10950	19910	21506

By interpolating between the weight values, you can determine that the takeoff moment of 20,269 pound-inches /100 is within allowable limits.

Figure 12-77. By locating the total weight of the airplane, you can determine the range of moments that will produce an acceptable CG location for flight. In this illustration, the sample airplane's moment is found to be within allowable limits.

WEIGHT SHIFT COMPUTATION

During weight and balance computations, you may find that either the weight of the airplane or its CG location is beyond acceptable limits. Usually, decreasing weight is a fairly simple matter, especially if you accomplish it before the airplane's CG is calculated. However, if you change the load to adjust the CG, the computations are somewhat more complex. You can use the following formula to solve weight shift problems.

$$\frac{\text{Weight of cargo moved}}{\text{Airplane weight}} = \frac{\text{Distance CG moves}}{\text{Distance between arm locations}}$$

By performing a few sample weight shift problems, you can see how to apply this formula. In the first example, assume you have a loaded airplane weight of 3,200 pounds and a CG of 45.6 inches. A box weighing 120 pounds is moved from cargo area B, which has an arm of 99 inches, to cargo area A, with an arm of 70 inches. How many inches will the CG move when this weight is shifted? [Figure 12-78]

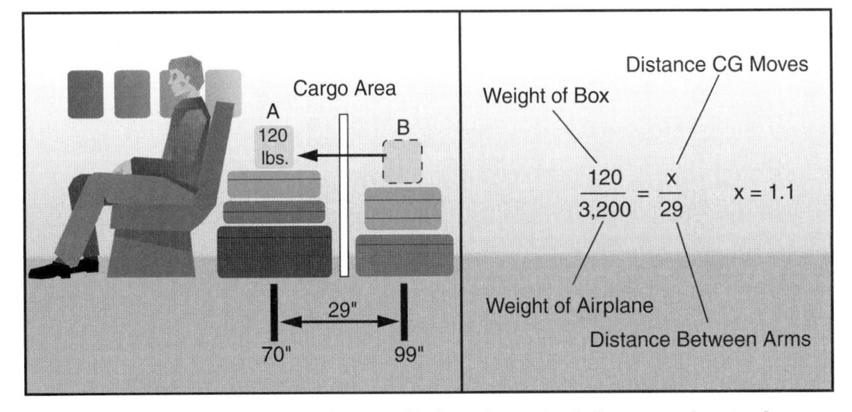

Figure 12-78. As shown in this illustration, the first step is to determine the known values. For example, the distance between cargo areas A and B is 29 inches. Next, insert the other known values into the formula and solve for the unknown variable. In this case, the unknown variable is the distance the CG moves forward when the weight is shifted. Solving this equation results in the answer of 1.1 inches.

For the next example, recall that the CG of the sample airplane used in the computation method problem exceeded aft limits when fuel was consumed during flight. To correct the aft CG condition, you need to reposition useful load items farther forward so that the CG will move approximately 940 pound-inches in order to remain within limits throughout the flight. In the sample aircraft, there are two baggage compartments; area A at 138 inches and area B at 159 inches. By moving a certain amount of weight from baggage

area B into area A, you can cause the CG position to move forward within limits. To determine how much weight to move between the compartments, you can use the weight shift formula. However, the formula uses CG positions expressed as arms, while the CG moment envelope, used in the sample problem, expresses balance limits in moment indexes. When moment indexes are used to convey balance limits, the first step to solving a weight shift problem is to convert moment values into CG locations or arms. Once these are determined, you can solve for the amount of weight to move by using the weight shift formula. [Figure 12-79]

Figure 12-79. Notice that the aft moment limit of the sample airplane is located at 180 pound-inches /1000 when the airplane weighs 3,455.5 pounds. To find the arm of the aft limit, divide the aft limit moment by the weight of the airplane. In a similar fashion, you need to find the arm or CG of the loaded airplane. This is accomplished by dividing the airplane's total moment by its total weight. Once these computations are complete, you can use the weight shift formula to determine how much weight to move between the baggage areas. The solution to the weight shift problem is shown here.

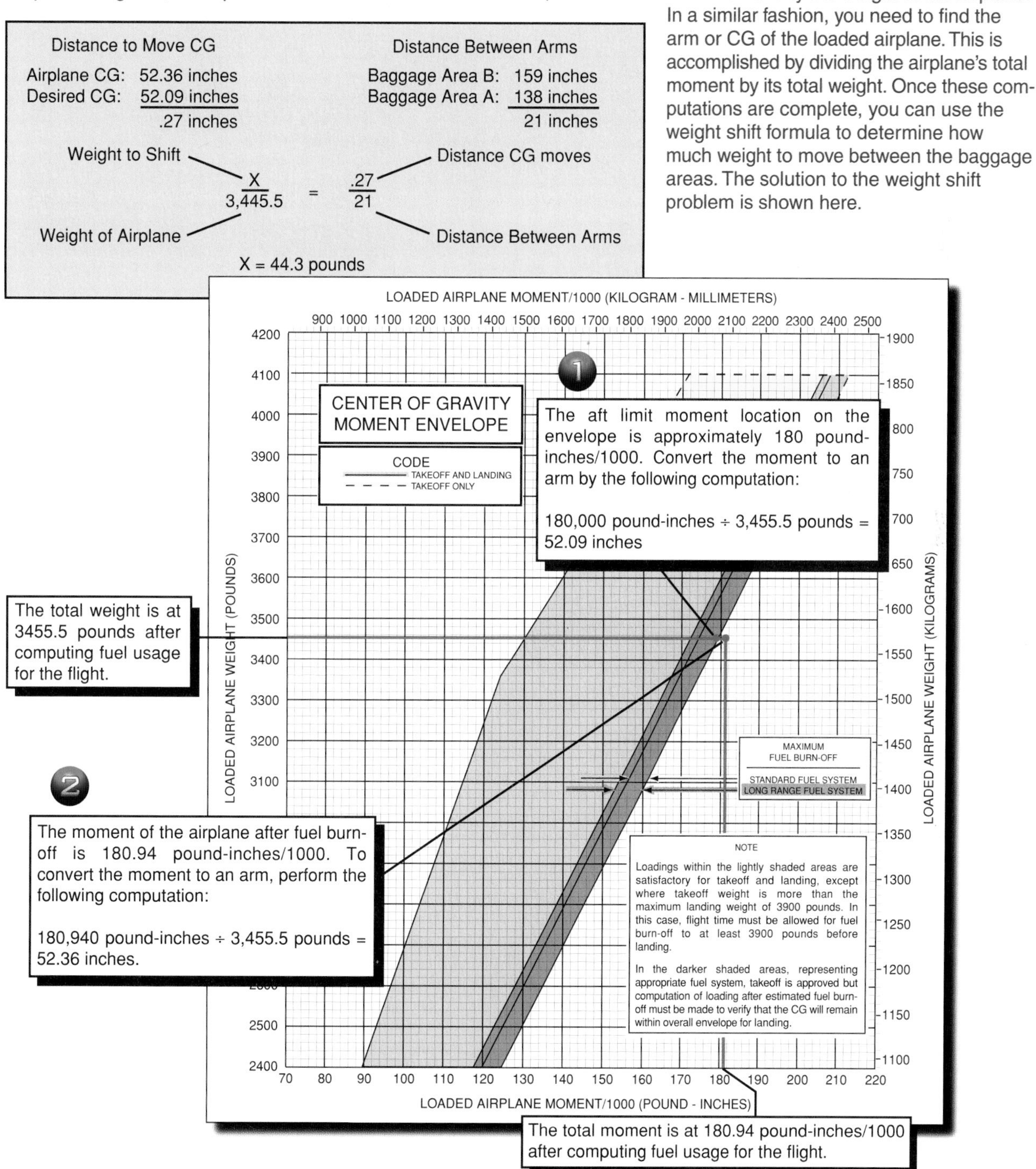

JUST SAY NO

After landing on a soft beach near a fishing camp, the pilot of the Cessna 207 was approached by eleven people that needed to be flown into a nearby town to purchase supplies. The pilot, feeling the need to satisfy his employer, an air charter operator, permitted all eleven people to board his aircraft. Although the airplane was only certified to carry seven occupants, the pilot considered that since his fuel load was light, the maximum weight of the airplane would not be exceeded. However, he failed to consider the effects on the balance condition of the airplane, which resulted in his trying to takeoff with a CG that was substantially aft of limits. As a result of the aft CG and the soft soil condition, the airplane was unable to become airborne, which caused it to collide with a boat that was moored on the beach. You can see the results of the attempted takeoff in this photo. Fortunately, none of the occupants was seriously injured, primarily because they were packed so tightly in the airplane.

Courtesy of Dr. Alvin J. Lagger — Aviation Safety Consultant

Had the pilot of this aircraft refused to carry the extra passengers, this accident probably would not have occurred. However, when feeling the pressure to perform his job, he allowed himself to compromise safety.

When you are employed as a professional pilot, you will likely encounter situations where you will be asked to fly an airplane when you know conditions are not safe or legal. Part of the responsibility of being a professional pilot is being able to respond sensibly yet emphatically to people when they try to pressure you into compromising situations. Although you may feel that you have the ability to stand your ground, there will be occasions when you will be faced with difficult decisions. When making these decisions, think of the possible consequences of your actions. If you compromise safety and are involved in an incident or accident, you will be held responsible and accountable. Remember, it is an agonizing ordeal to look back at your past decisions and wish that you had only said no.

The weight shift formula can be used to calculate the amount of weight that must be moved a specific distance or to determine the distance a specific weight would need to move to bring the CG within approved limits.

SUMMARY CHECKLIST

- ✓ Both the amount and distribution of weight significantly affect aircraft stability, control, and efficiency.
- ✓ An airplane may have a single maximum weight limit or multiple maximum weight limits for various stages of ground and flight operations.
- ✓ Maximum weight limits may be required for structural or performance considerations.
- ✓ An airplane loaded near the forward CG limit becomes more stable, but less controllable in flight, while an airplane loaded near the aft limit is less stable, and more sensitive to control input.

- ✓ Weight and balance reports and equipment lists contain essential information for weight and balance control and must be maintained current by the aircraft owner or operator.
- ✓ Weight and balance condition checks can be accomplished by the computation, graph, or table methods.
- ✓ When the CG of an airplane is outside of allowable limits, you can use the weight shift formula to determine how much weight to move or the distance that a given weight must be moved to place the CG within allowable limits.

KEY TERMS

Maximum Weight

Maximum Ramp Weight

Maximum Takeoff Weight

Maximum Landing Weight

Maximum Zero Fuel Weight

Center Of Gravity

Reference Datum

Moment

Weight and Balance Report

Equipment List

Empty Weight Center of Gravity

Basic Empty Weight

Licensed Empty Weight

Useful Load

Arm

Moment Index

Payload

Basic Operating Weight

QUESTIONS

1. Which of the following weight limits represents the heaviest total weight value?

   A. Maximum ramp weight
   B. Maximum takeoff weight
   C. Maximum landing weight

2. Which of the following terms may be used to describe the weight of an airplane without fuel or payload items?

   A. Useful load
   B. Basic operating weight
   C. Maximum zero fuel weight

3. When the CG of an airplane is near the aft limit, which of the following flight characteristics is likely to be encountered?

   A. The stability of the airplane increases and the elevator control becomes less effective.
   B. The airplane becomes less stable and the elevator control becomes less effective.
   C. The elevator controls become more effective while stability decreases.

4. True/False. Loading an airplane near the aft CG limit produces a higher true airspeed as compared to when loaded with the CG located near the forward limit.

5. The total weight of an airplane is 3,100 pounds and the total moment index is 310 pound-inches /1000. Where is the CG located in inches aft of the datum?

   A. +10 inches
   B. +100 inches
   C. +1,000 inches

6. Using the accompanying loading graph, what would be the CG of the total combined weight of the following items?

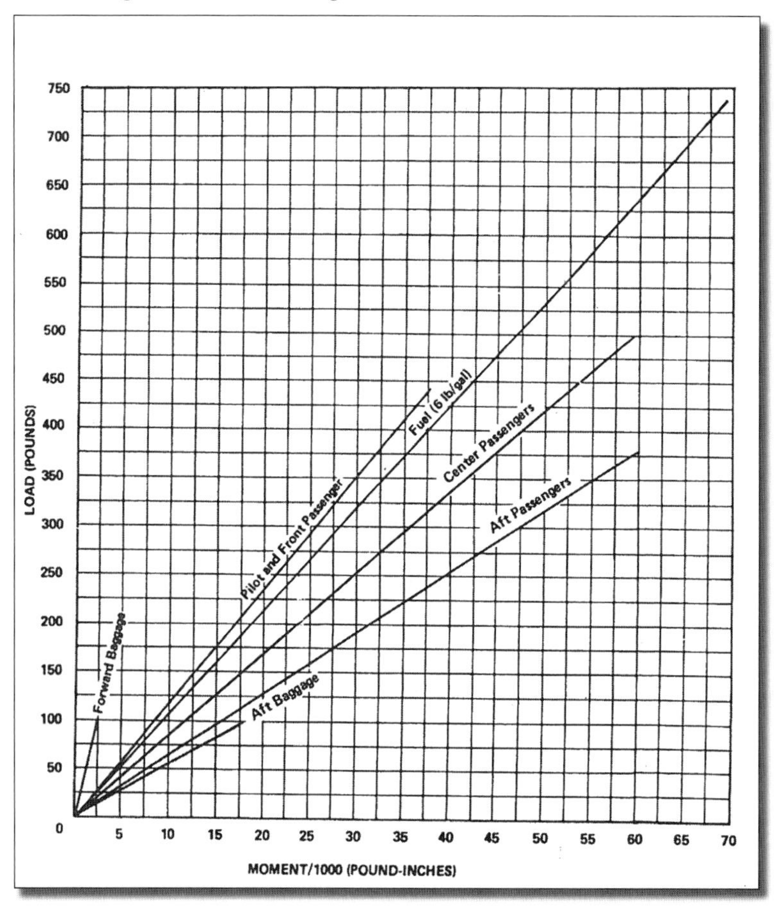

Pilot and passengers	350 pounds
Center seat passengers	300 pounds
Usable fuel	80 gallons
Aft baggage	50 pounds

   A. Approximately +101.27 inches aft of the reference datum
   B. Approximately +156.38 inches aft of the reference datum
   C. Approximately +105.13 inches aft of the reference datum

7. Use the item information from question number 6 and the CG range and weight graph shown here to determine if an airplane is loaded within the CG envelope. Assume the airplane has an empty weight of 3,200 pounds with an empty weight moment index of 2,842.6 pound-inches /100.

   A. The CG falls within the envelope and is safe for flight.
   B. The CG falls in the envelope, but exceeds the maximum zero fuel weight.
   C. The CG falls inside the envelope but the fuel burn-off in flight must be sufficient to reduce the total weight to below the maximum landing weight.

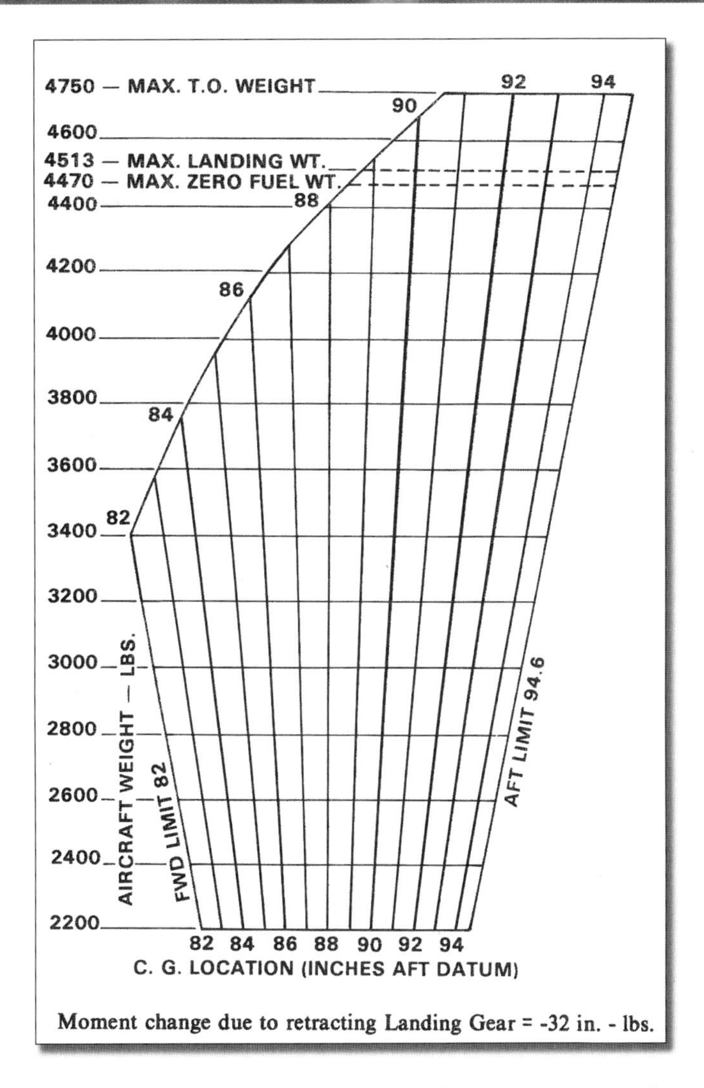

Moment change due to retracting Landing Gear = -32 in. - lbs.

8. An airplane is loaded with 100 gallons of fuel at station +38 inches which causes the airplane to have a total weight of 3,500 pounds and a CG located at +42 inches. If the airplane burns 12.6 gallons per hour, where will the CG be located after 3 hours and 20 minutes of flight? Assume the fuel weighs 6 pounds per gallon.

   A. +42.83 inches aft of datum
   B. +42.31 inches aft of datum
   C. +42.05 inches aft of datum

CHAPTER 13

COMMERCIAL FLIGHT CONSIDERATIONS

Private Pilot Maneuvers
Volume I — Emergency Landing Procedures

SECTION A
EMERGENCY PROCEDURES

Although modern aircraft are extremely reliable, the possibility of emergencies in flight due to mechanical failure, fuel problems, or pilot error has not been eliminated. Problems resulting from these factors are not common, but they are frequent enough that you should be aware of them and know how to handle them if they occur. The following discussion emphasizes VFR emergencies; you also may refer to Chapter 10, Section A — IFR Emergencies.

Unfortunately, not all in-flight emergencies will allow you sufficient time to reference your pilot's operating handbook or checklist. In fact, some emergencies may not be addressed in the POH at all. Therefore, when you fly, you need to be prepared for distress situations by knowing the emergency procedures for your airplane and by using good judgment in responding to the situation. To assist you, the FAA provides some general guidelines for dealing with various emergencies. While the specific procedures recommended by your airplane's manufacturer should be adhered to, the general emergency procedures discussed in this section may be helpful to you in those situations where time is limited and you must act quickly.

An FAA study has shown that power loss is a major contributing factor in general aviation accidents. Power loss accidents have a variety of causes. Examples include operating powerplants beyond normal limits, poor maintenance, failure of engine parts or system components, and above all, fuel starvation due to pilot error or other factors. Unexpected power loss coupled with inadequate pilot response can easily result in an accident.

A well-prepared and competent pilot can usually deal with an emergency situation in a manner resulting in a safe outcome for the flight. The most important thing to remember in an emergency situation, whether you use the POH or the following general emergency procedures, is to fly the airplane.

EMERGENCY DESCENT

An **emergency descent** is used to achieve the fastest practical rate of vertical descent to reach a safe altitude or landing during an emergency situation. You may need to perform this maneuver due to an uncontrollable fire, a sudden loss of cabin pressurization, or any other situation demanding an immediate and rapid loss of altitude. Your objective is to descend as quickly as possible without exceeding the airspeed limitations of the airplane.

To initiate the descent, reduce the throttle to idle, then roll into a bank angle of approximately 30 to 45 degrees in order to establish and maintain a positive load factor on the airplane. Doing so will help you remain within the safe operating limits of the airplane, and also will allow you to stay over, and closely observe, a potential landing area. If your airplane is equipped with a controllable propeller, set the prop control to low pitch (high RPM). This action allows the propeller to act as an aerodynamic brake preventing excessive airspeed buildup during the descent. Then, if recommended by the manufacturer, lower the landing gear and fully extend the flaps as quickly as practical. The drag produced by extending the landing gear and full flaps serves to increase the descent rate without increasing the airspeed. [Figure 13-1]

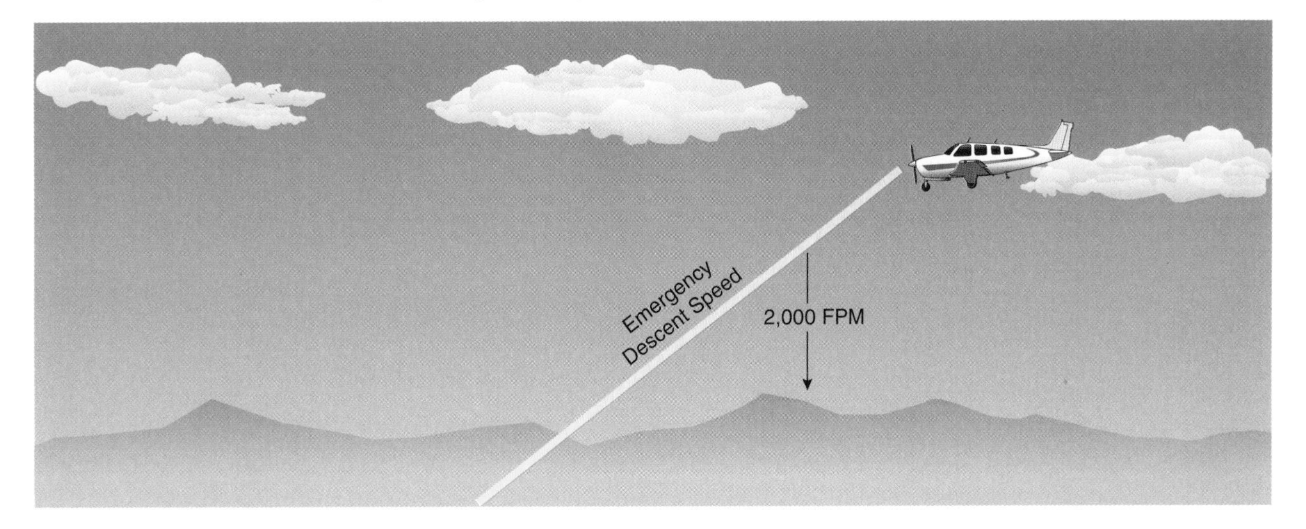

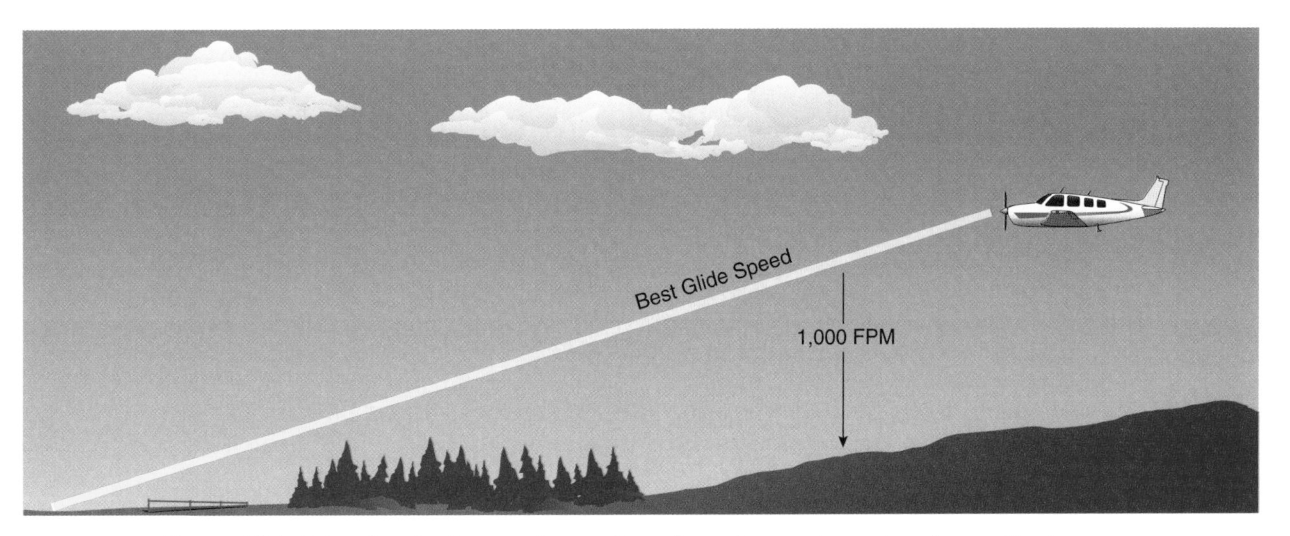

Figure 13-1. Extending the flaps and gear allows for a steeper descent angle resulting in a quicker loss of altitude than would be the case when flying at best glide airspeed. In addition, the airspeed will be higher than best glide airspeed, but remains within the airspeed limitations of the airplane.

At no time should you allow the airplane's airspeed to rise above the never exceed speed (V_{NE}), the maximum gear extended speed (V_{LE}), or the maximum flap extended speed (V_{FE}). If the descent is conducted in turbulent conditions, you also should comply with the recommended maneuvering speed (V_A) limitations. These procedures should be continued until a safe altitude has been reached or until you begin the emergency approach and landing if necessary.

EMERGENCY APPROACH AND LANDING

Your own fear of injury or of damage to the airplane may interfere with your ability to act decisively when you are faced with an emergency involving a forced landing. Records show, however, that survival rates favor those pilots who maintain their composure and know how to apply the recommended procedures and techniques that have been developed throughout the years.

EMERGENCY PARALYSIS

The NTSB has identified several factors that diminish a pilot's ability to deal with emergency landings.

1. Pilots may be reluctant to accept the emergency situation, paralyzed with the thought that the aircraft will be on the ground in a short time regardless of what they do. As a result, they delay action, fail to maintain flying speed, or attempt desperate measures at the expense of aircraft control.

2. There may be a desire on the part of the pilot to save the aircraft rather than sacrificing it to save the occupants. Such a desire may cause pilots to stretch a glide or make abrupt maneuvers at low altitude resulting in accidents.

3. The fear of injury may cause pilots to panic, inhibiting their ability to properly carry out emergency procedures resulting in the situation they wanted to avoid most.

The key to dealing with these fears is to fly the airplane. Concentrate on maintaining your glide speed, adhering to the checklists, and managing resources, and this will help you keep your mind off what might happen and allow you to continue flying the airplane. If you practice simulated emergency procedures enough, you will be more confident in your ability to deal with an emergency approach and landing.

When practicing emergency descents, make sure you clear the area below before you start. Begin recovery at an altitude high enough to ensure a safe return to level flight. In addition, emergency landings should be practiced only over favorable terrain in the event an actual emergency landing becomes necessary. Normally, in training, when all the prescribed procedures and airspeeds are established and stabilized, the maneuver should be terminated. In airplanes with piston engines, prolonged practice emergency descents should be avoided to prevent the excessive cooling of engine cylinders. Apply carburetor heat as recommended by the manufacturer.

Many factors are important in successfully executing an emergency approach and landing, the first of which is the landing field. A competent pilot normally is on alert for a suitable forced landing site. Ideally, the best option is an established airport, or hard-packed, long, smooth field with no high obstacles on the approach end and situated so you can land directly into the wind.

Since these ideal conditions rarely present themselves, you must select the next best field available. Cultivated fields are generally satisfactory since they will be fairly level. Plowed fields may be acceptable as long as you land parallel to the furrows. If the field appears to be soft or snow covered, you may even consider a gear up landing, if your airplane has retractable gear. Otherwise, you should ensure that the nose wheel is not allowed to sink in and cause the airplane to nose over.

Field size and wind direction are also important factors to consider in selecting an emergency landing site. Wind direction and speed will affect your airplane's gliding distance over the ground, the track along the ground during the approach, the groundspeed at

which the airplane touches down, and the distance required for deceleration after the landing. These factors will be important in selecting your landing field. However, do not allow yourself to be locked into landing directly into the wind. For instance, limited options in an emergency may dictate making a downwind landing to clear obstacles or because the best field may be too far upwind to reach. Furthermore, the field you select should also be wide enough to allow you to extend the base leg before turning final in the event you have misjudged your airspeed or altitude. [Figure 13-2]

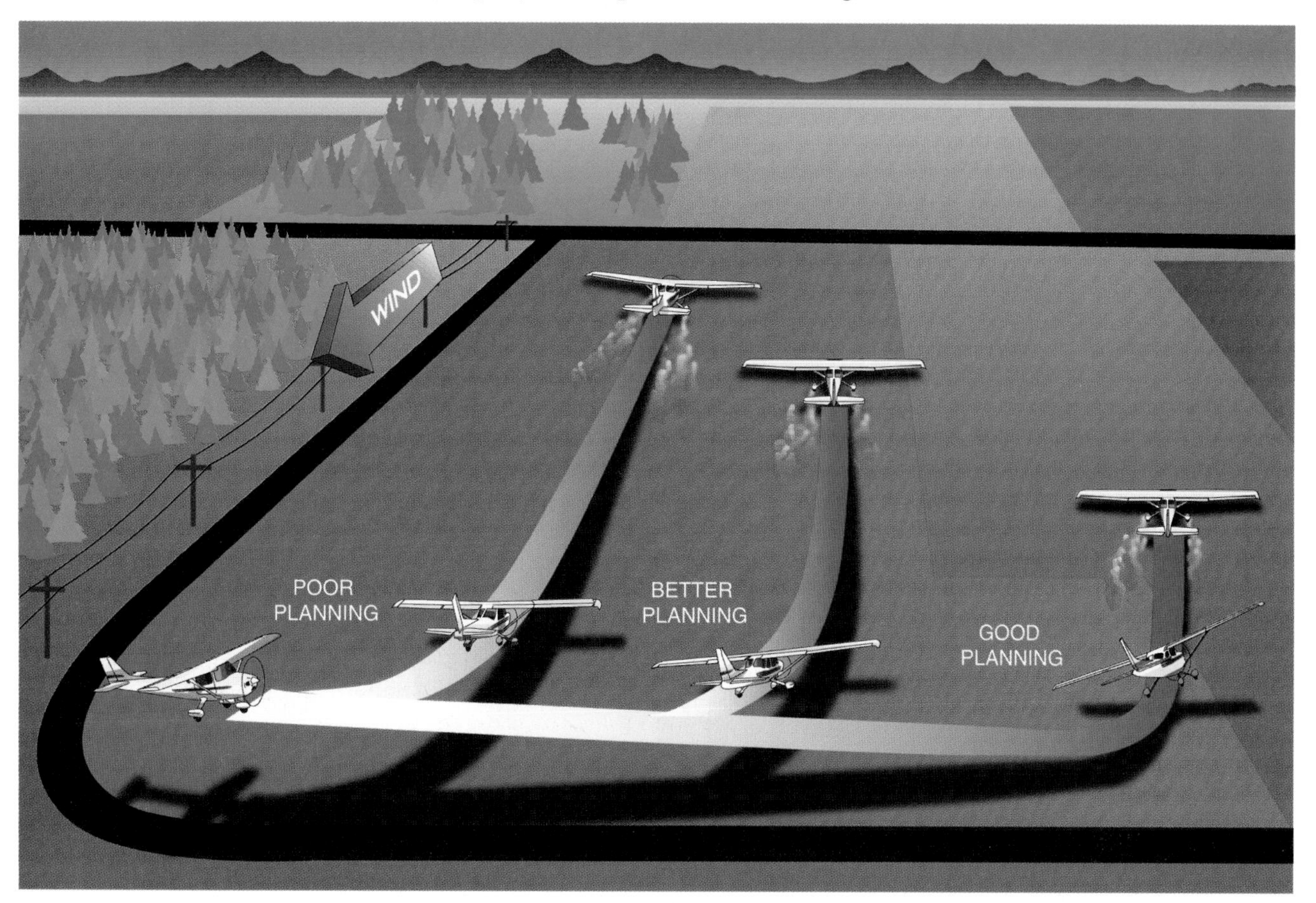

Figure 13-2. A wide field provides the opportunity to adjust your approach without making extreme maneuvers at low altitude.

In most cases, the altitude you have is the primary controlling factor in the successful accomplishment of a forced landing. The lower you are when a power loss occurs, the fewer your options. Should an engine failure happen prior to reaching a safe maneuvering altitude, your choice of actions will be limited to flying straight ahead or making shallow turns to avoid objects in your immediate path.

When a forced landing appears imminent, regardless of your altitude, the main priority is to complete a safe landing in the best field available. This involves getting the airplane on the ground in as near a normal landing attitude as possible without striking obstructions. Large, abrupt turns near the ground in an attempt to avoid a poor landing area makes attitude control difficult. If you can fly the airplane all the way to the ground and make a relatively normal landing, you increase the chances of completing the procedure without catastrophic consequences to you and your passengers.

Performing a safe approach and landing begins with airspeed control. For each particular airplane, the manufacturer recommends an airspeed and configuration that will provide the **maximum glide distance**. The **best glide airspeed**, found in the pilot's operating handbook, will determine the distance you can glide and consequently the number of

Crash Course

"The success of an emergency landing under adverse conditions is as much a matter of the mind as it is of skill!" — Mick Wilson, former Manager, FAA Aviation Safety Program, Denver Flight Standards District Office

Mick Wilson developed a Safety Seminar and text entitled *How to Crash an Airplane (and Survive!)*. His presentation includes numerous insights into basic crash safety concepts including the following observations concerning post-impact deceleration of groundspeed.

The overall severity of a deceleration process is governed by groundspeed and the stopping distance of the aircraft. For instance, doubling the groundspeed means quadrupling the total destructive energy at impact. This is why it is so important to reduce your landing speed to the lowest possible while still maintaining aircraft control.

Very little stopping distance is required if the speed can be dissipated evenly over the available distance. Assume an aircraft is exposed to deceleration forces equaling 9 times the force of gravity, the standard to which general aviation aircraft are certified. The stopping distance required while traveling at 50 mph is 9.4 feet. At 100 mph, the stopping distance is 37.6 feet, or four times the stopping distance required when traveling at 50 mph. While these are not distances you would prefer for a normal landing, it is nice to know just how little space is needed to successfully decelerate an airplane during an emergency landing.

Understanding the need for a firm but uniform deceleration process in very poor terrain provides you more landing options. As a result, you can select touchdown conditions that, while not ideal, can reduce the peak deceleration of the airplane and increase your chances of walking away from the landing. Courtesy of Mick Wilson, www.crashandsurvive.com

landing areas available to you. In some airplanes, the best glide speed may change as gross weight changes. Of course, any deviation from best glide speed may negatively affect your ability to reach a suitable landing site. [Fig. 13-3]

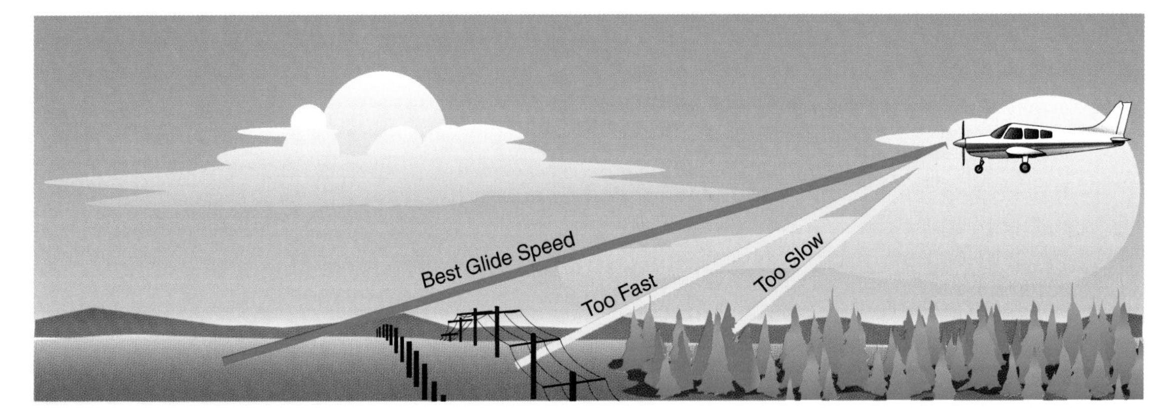

Figure 13-3. Any deviation from best glide speed will reduce the distance you can glide and may cause you to land short of a safe touchdown point.

To achieve best glide speed, the landing gear and flaps should be retracted immediately to eliminate unwanted drag. However, during a power loss immediately after takeoff you may want to keep the gear and flaps extended in preparation for touchdown. If your aircraft is equipped with a controllable propeller, set the prop control to a high pitch, low RPM setting as well. Unlike emergency descent procedures in which your focus is on a quick and controlled loss of altitude, your objective during an emergency approach is to remain airborne long enough to set up a safe landing. As a result, it is necessary to eliminate as much

180 DEGREES OF TROUBLE

From the files of the NTSB...

Aircraft: *Piper PA-18*

Injuries: *1 Minor*

Narrative: *The airplane departed a restricted landing area runway for a local flight. Shortly after departing, the airplane's engine stopped running at 300 feet above the ground. The pilot performed a 180 degree turn toward the departure runway. During the turn, the airplane stalled and subsequently collided with the ground. The airplane was destroyed by a post impact fire.*

The desire to turn back to an airport if the engine fails shortly after takeoff can be quite powerful. Unfortunately, giving in to the desire can have disastrous consequences considering the glide performance of airplanes during turns. That is why it is usually best to continue straight ahead, utilizing small heading changes, to keep the airplane under control in order to complete an emergency landing as safely as possible.

To emphasize the danger in these situations, assume you have just taken off from your local airport and climbed to 300 feet AGL when the engine fails. After a four-second reaction time, you decide to turn back to the runway. If you turn at a standard rate, it will take a minute to turn 180°. At a glide speed of 65 knots, the radius of the turn is 2,100 feet, so at the completion of the turn, you are now 4,200 feet to one side of the runway. You must turn another 45° to head the airplane toward the runway. By now, your total change of direction equals 225° equating to 75 seconds plus the four-second reaction time. If the average light airplane in a no-power gliding turn descends at approximately 1,000 feet per minute, you will have descended 1,316 feet —1,016 feet below the runway!

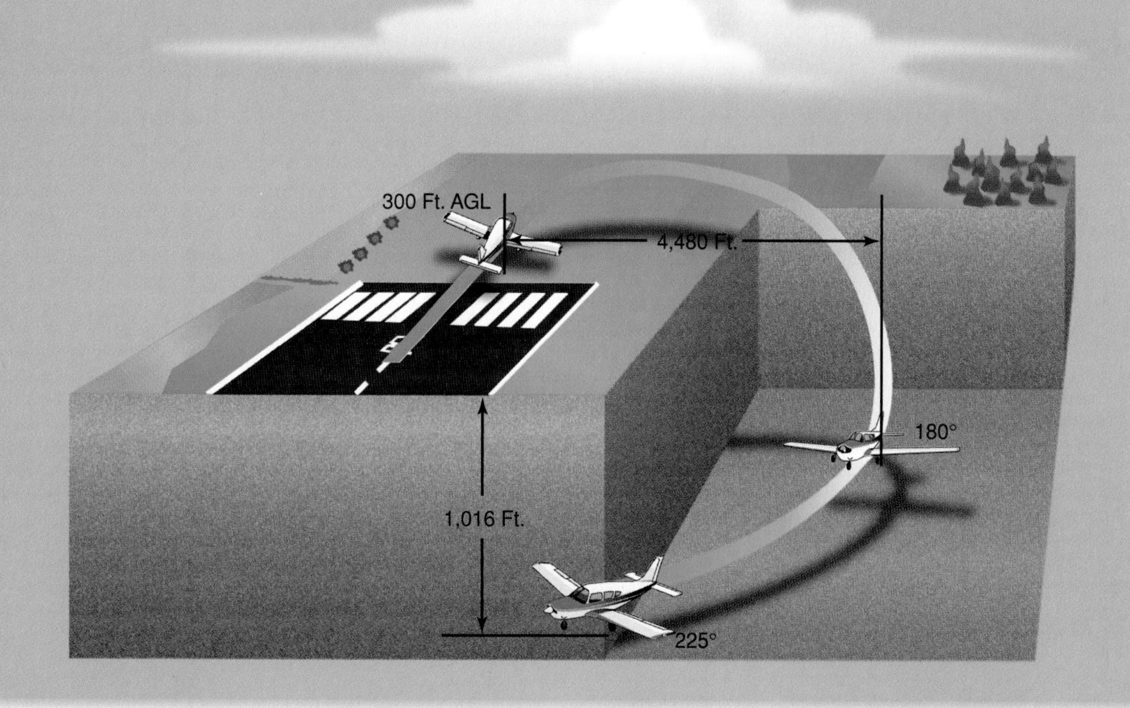

drag as possible by cleaning up the airplane. Next, adjust your airplane's pitch attitude to obtain best glide, and trim to maintain that pitch attitude while you follow the emergency checklist procedures. Remember, any deviation from best glide speed, as a result of poor pitch attitude control, will reduce your gliding distance and landing options. [Figure 13-4]

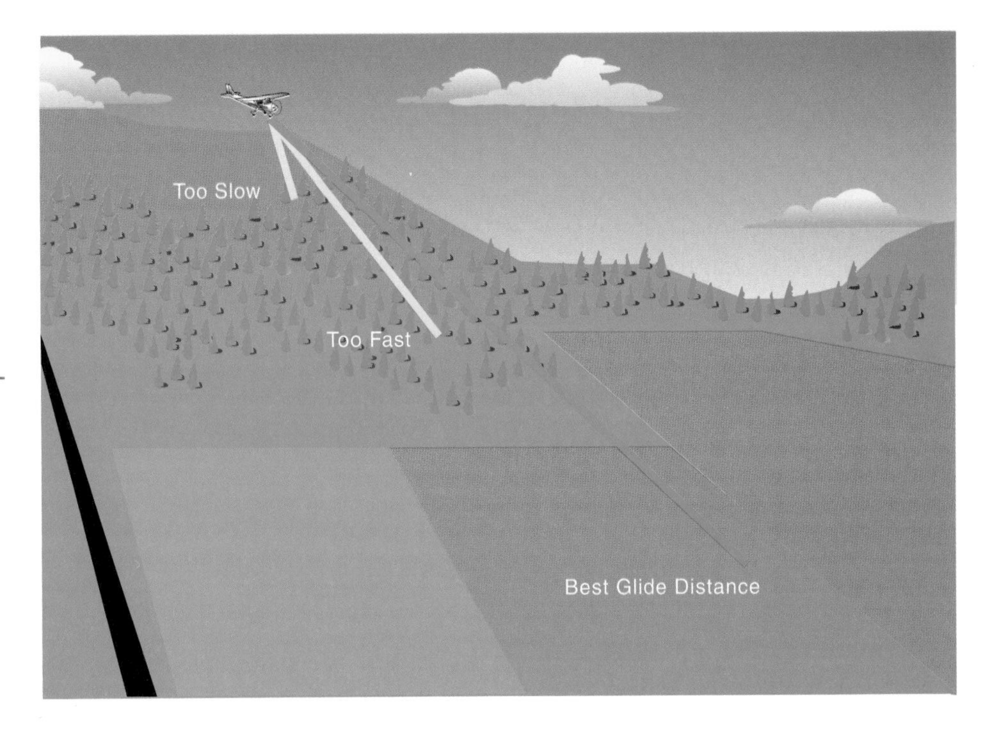

Figure 13-4. The greater your gliding distance, the more options will be available for a safe landing.

Once the proper pitch attitude and airspeed have been established, complete the appropriate emergency checklist found in the POH for that particular airplane. However, during an engine failure, a generic cockpit checklist may be suitable, particularly in situations occurring at low altitude which do not provide time to reference the POH checklist. [Figure 13-5]

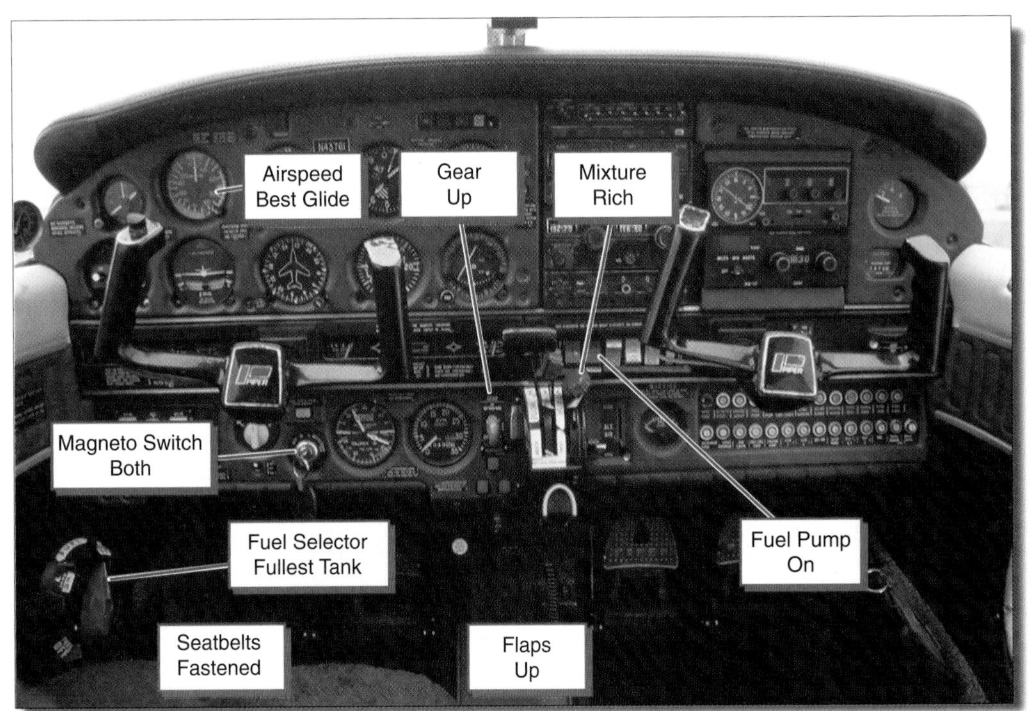

Figure 13-5. These are critical items that should be checked in order to restart the engine during an engine failure emergency.

For a successful landing to happen, it is important to set it up properly. The FAA recommends using any combination of normal gliding maneuvers from wings level to spirals to arrive at a point from which a normal landing may be made. This point, or **key position**, may be abeam your intended touchdown zone or at the turn from base to final depending on whether you are making a 180 degree or 90 degree approach. Once at the key position, follow the recommended flap and gear extension procedures in the POH and add full flaps only when you know you can reach the touchdown point. From here, any miscalculations about the glide angle can be corrected with the use of flaps, slipping, or moving the touchdown spot if feasible. [Figure 13-6]

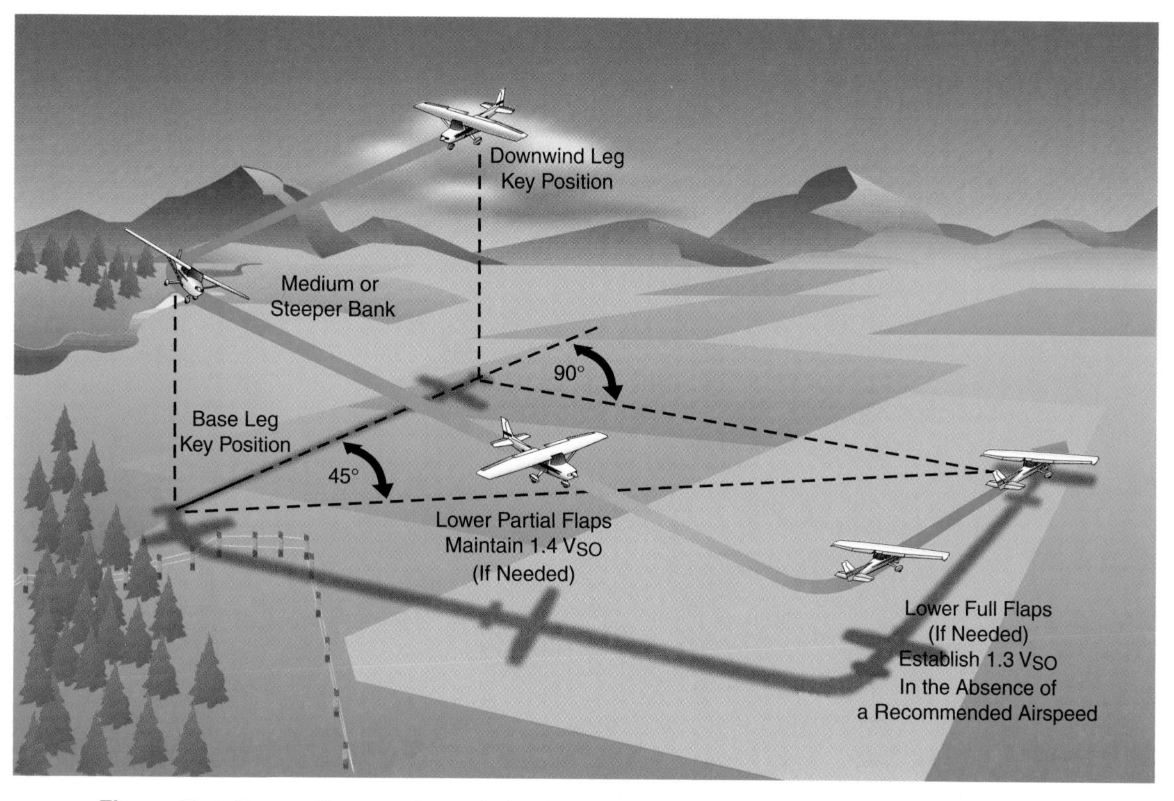

Figure 13-6. Key positions on downwind or base allow you to judge your gliding distance in the same manner you would during a normal traffic pattern.

SYSTEMS AND EQUIPMENT MALFUNCTIONS

There are some emergency conditions that do not appear in checklists or in your airplane's POH. Although these situations are not specifically addressed, the FAA provides some general recommendations for dealing with them. The procedures are not intended to be used in lieu of the particular recommendations that may be provided by a manufacturer, but rather in their absence.

IN-FLIGHT FIRE

If you experience a fire while in flight, follow the checklist procedures specified in the POH for your airplane and declare an emergency by radio. The checklist may address only one type of in-flight fire or it might include procedures for different types of fires ranging from cabin and electrical fires to engine fires. In any event, follow the appropriate procedure for the situation. In addition, a general recommendation may be of benefit to you during such an emergency. Should the fire and flames be visible outside the cabin during the emergency descent, attempt to slip away from the fire as much as possible. For example, if the fire is observed on the left side of the airplane, slip to the right. This may move the fire away from the cabin.

PARTIAL POWER LOSS

It is possible that during a flight you may experience a partial loss of engine power. Two options might be available to you, depending on the degree of power loss and the airplane's resulting decrease in performance. You may be able to continue the flight in a reduced power condition as long as you are able to hold altitude or climb. In this situation, maintain an airspeed that will provide the best airplane performance available. In most cases, the **best performance airspeed** will be approximately the best glide speed. However, it is also possible the engine will not continue to run in this condition and a forced landing will still need to be made. With an engine problem, you should continually monitor your engine instruments and update your choice of landing options.

Alternatively, your airplane's performance with partial power may not be sufficient to maintain altitude. In this case, a forced landing is imminent. Consequently, you will need to declare an emergency with ATC and begin the emergency approach and landing procedures specified in your airplane's checklist.

DOOR OPENING IN FLIGHT

A cabin or baggage compartment door opening in flight can be a disconcerting event. Although a door generally will not open very far, the sudden noise can be startling. Regardless of the noise and confusion, it is important to maintain control of the airplane, particularly during departure. Accidents have occurred on takeoff because pilots have stopped flying the airplane to concentrate on closing cabin or baggage doors.

IGNORE THE DOOR

From the files of the NTSB...

Aircraft: *Piper PA-32-300*

Injuries: *1 Fatal, 2 Serious, 2 Minor*

The forward baggage compartment door came open and began flapping during takeoff roll/initial takeoff climb. The pilot continued his takeoff and radioed that his baggage compartment door had popped open and he was returning to land. The airplane was observed flying in a steep climb attitude, at very slow airspeed, with a pronounced left crab or slip. It was in a left turn at an altitude of approximately 75 to 100 feet. The passenger in the right front seat reported hearing the stall warning horn intermittently prior to the airplane's impact with two utility poles.

A baggage or cabin door opening in flight can be extremely distracting. The natural reaction by most pilots is to immediately attempt to close the door or ask a passenger to help close it. The airstream creates a partial vacuum which makes it almost impossible to close the door. The next decision by the pilot is to slow the airplane to increase the chance of getting the door latched. This probably led to the accident described above. The best procedure is to return for landing using a normal approach speed and latch the door on the taxiway.

While an open door does not normally compromise airplane control, it is possible that control will become more difficult. If such a condition should occur it may be necessary to increase airspeed in all phases of flight, including the approach, in order to ensure that you can control the airplane. Once you have adequate airplane control, land as soon as practical and secure the door.

ASYMMETRICAL FLAP EXTENSION

An unexpected rolling motion during flap extension may be due to an asymmetrical or split flap condition. If one flap extends while the other remains in place, a differential in lift across the wing is the cause of the rolling motion. A split flap condition can be hazardous, particularly in the traffic pattern or during a turn at low altitude. [Figure 13-7]

Figure 13-7. Extending the flaps during a turn can result in a dangerous situation should an asymmetrical flap extension occur.

If the unexpected rolling motion occurs during flap extension, immediately return the flap control to the up, or the previous position, while maintaining control of the airplane. Should you be in the approach phase of the traffic pattern when an asymmetrical extension occurs, execute a go-around and adjust your airspeed for approach and landing.

EMERGENCY EQUIPMENT AND SURVIVAL GEAR

You may carry a survival kit for years and not need it. But when you do need it, you need it badly and the biggest survival kit is none to big. — Private Pilot's Survival Manual by Frank Kingston Smith

Regulations only require you to carry flotation gear and a signaling device during certain overwater commercial operations. However, it is highly recommended that you include several other basic survival items in the event of an emergency. A survival kit should be able to provide you with sustenance, shelter, medical care and a means to summon help without a great deal of effort or improvisation on your part. That is why a minimally equipped survival kit should be avoided. You can tailor the contents of your survival kit to the conditions of the flight. Any complete survival kit should contain a basic core of survival supplies around which you can assemble the additional items that are appropriate to the terrain and weather you would have to deal with in an emergency.

When making decisions about the emergency gear to take with you, consider the terrain you will be flying over, the climate or season during which the flight will be made, and what type of emergency communication equipment you may need. For instance, an emergency landing in mountainous terrain in December would require different survival gear than would a ditching in the ocean in August. Some general items you may consider as the basis on which to build a survival kit for all flight operations include: a comprehensive first aid kit and field medical guide, flashlight, supply of water, knife, matches, shelter, and a signaling device.

Survive to Fly

A wide variety of survival kits are available commercially. They may range in price up to several hundred dollars for larger kits designed to support 4 or 5 people. However, you can put together your own survival kit tailored to meet your specific flying needs.

Keeping in mind weight considerations, you may begin with the basics. Your kit should include items covering the important survival factors like water, shelter and protection, food, first aid, signals, and personal or miscellaneous supplies.

This commercially available survival kit is designed to sustain 4 to 6 people in an emergency. You may opt to purchase a kit like this, or you may elect to build your own kit.

There are many sources of information that have been published to help you determine what items are appropriate for your kit. In addition, survival courses may be available locally, particularly in mountainous parts of the western United States.

SUMMARY CHECKLIST

✓ You need to be prepared for emergency situations by knowing the procedures for your airplane.

✓ An emergency descent is used to achieve the fastest practical rate of vertical descent to reach a safe altitude or landing.

✓ NTSB records indicate accident survival rates favor pilots who maintain their composure and know how to apply the recommended procedures and techniques.

✓ Limited options during an emergency may dictate making a downwind landing.

✓ When you experience a power loss, altitude is usually the primary controlling factor in the successful accomplishment of a forced landing.

✓ The FAA recommends using any combination of normal gliding maneuvers from wings level to spirals to arrive at a point from which a normal landing may be made.

✓ Should fire be visible outside the cabin during an in-flight fire, attempt to slip away from the fire.

✓ During a complete power loss the most important thing to do is maintain control of the airplane regardless of the altitude available to you.

✓ Some general items you may consider for a survival kit include, a first aid kit, flashlight, container of water, knife, matches, and a signaling device.

KEY TERMS

Emergency Descent

Maximum Glide Distance

Best Glide Airspeed

Key Position

Best Performance Airspeed

Survival Kit

QUESTIONS

1. During an emergency descent, what airspeed limitations should you observe?

2. True/False. Practicing simulated emergency procedures helps build confidence in handling real emergencies.

3. Why should the emergency landing field you select be wide instead of narrow?

4. In most cases, what is the primary controlling factor in the successful accomplishment of a forced landing?

   A. Airspeed
   B. Altitude
   C. Wind

5. True/False. Airspeeds below best glide increase the number of landing options available to you.

6. Which of the following conditions would allow you to extend the glide range in an airplane with a controllable pitch propeller?

   A. High pitch, high RPM
   B. High pitch, low RPM
   C. Low pitch, low RPM

7. True/False. An open baggage door normally compromises airplane control.

8. An unexpected split flap extension will be most dangerous during which phase of flight?

   A. Climb
   B. Cruise
   C. Turns at low altitude

9. True/False. It is prudent to consider the type of terrain you will be flying over when selecting items for your emergency survival kit.

10. True/False. FAA regulations require flotation gear and a signalling device to be carried on board during all overwater operations.

SECTION B
COMMERCIAL DECISION MAKING

No matter what your reasons are for pursuing a commercial pilot certificate, the additional knowledge you acquire during commercial training will make you a more competent pilot. The advanced skills and experience that you gain will hone your decision-making abilities and enhance your flight safety.

Chapter 1, Section B — Advanced Human Factors Concepts introduced you to the elements involved in aeronautical decision making. This section will explore decision making from the perspective of a pilot flying for a commercial operation; however, the decision-making skills involved are essential for all pilots. Since commercial operations are often conducted under IFR, if you have already received your instrument rating and you are obtaining your commercial certificate separately, you will need to review Chapter 10, Section B — IFR Decision Making.

COMMERCIAL OPERATIONS

As a commercial pilot, decision making can be complex as you balance public perceptions of aviation, your company's schedule, and your desire to conduct safe flight operations. Previously, your motivation to fly safely was based on your own personal expectations. Now, you have passengers and company personnel to consider when you make choices regarding each flight. You also may rely upon flying for your income, and this could impinge upon your ability to make sound decisions if you are not conscientious.

Your priority as pilot in command is to stay focused on the continuous safety of the flight. You may need to make the sometimes difficult and unpopular choice to cancel a flight or divert to an alternate. You assume this responsibility when you contract to work for a commercial operator, so you should fully research such concerns before you accept a position with a particular company. A conscientious employer will factor in these issues as a part of doing business. You should never make an inflight decision based on whether it will affect your career. [Figure 13-8]

In addition to the demands of your employer, you may encounter changing personal expectations. By obtaining your commercial certificate, you have developed a higher level of skill, but along with this skill may come the temptation to take more risks. Although it is true you possess a greater level of expertise than you did earlier in your training, the increased knowledge and practiced judgment that you have gained throughout your flying career should keep you from having to exercise extraordinary flying skills. To make effective decisions, you must stay honest with yourself about your capabilities and limitations, and those of your aircraft.

Figure 13-8. When you work as a pilot for a small company, you have a lot of direct contact with your customers. You may serve as baggage handler, fuel technician and flight attendant, as well as pilot in command.

The Airmail Pilot's Oath of Office

Some of the first commercial pilots were those who flew the mail, and they were entrusted with a very serious mission — to get the mail through regardless of the weather. Below is the oath that airmail pilots swore to uphold when assuming the job. Would you commit to the same contract?

"I hereby make application for position of pilot in the Air Mail Service and hereby agree, if appointed, to fly whenever called upon and in whatever Air Mail plane that I may be directed by the superintendent of the division to which I am assigned, or his representative on the field, and in the event of my refusal to fly, such refusal shall constitute my resignation from the service, which you are hereby authorized to accept. If appointed, I pledge myself to serve the Air Mail Service of the Post Office Department for a period of one year, and to carry out its orders implicitly, unless separated for cause or in accordance with the foregoing agreement . . ."

Courtesy of Philip Martin *Pilot Wings of the United States, 1913-1995*

APPLYING THE DECISION-MAKING PROCESS

By now, you have applied the decision-making process many times as pilot in command. With experience, you become more adept at making effective choices and reducing the risks associated with each flight. Each step in the decision-making process is critical to making sound inflight judgments. [Figure 13-9]

Risk management, as a part of the ADM process, relies upon situational awareness, problem recognition, and good judgment to reduce the risks associated with each flight.

Recognize a change.

The forecast was for VFR conditions along your route of flight; however, as you near a point of interest, you notice a cloud layer obscuring terrain.

Define the problem.

You must remain VFR according to your company's operations specifications, ensure flight safety, and please your passengers to the extent possible. You contact Flight Watch and are informed that the visibility is restricted in the area you normally overfly. An updated weather briefing indicates that the marginal weather is isolated and widespread VFR conditions are still forecast to prevail.

Type of Aircraft Make/Model/Series	Class of Operation	Type of Operation En Route Flight Rule	 Day/Night Conditions	Flight Attendant or Cargo Only
C-U206F	SEL	VFR	DAY	NONE
BE-90-E90	MEL	IFR/VFR	DAY/NIGHT	NONE
BE-200-200	MEL	IFR/VFR	DAY/NIGHT	NONE

Choose a course of action.

Instead of discontinuing the flight and returning to the airport, you decide to divert to another scenic area with VFR conditions.

Implement your decision.

You request current weather for another area nearby that you know has a stunning waterfall as a point of interest. After determining that VFR conditions prevail at your new destination, you plot a course and amend your flight plan. You inform your passengers of the alternate routing and continue the flight.

Figure 13-9. You may have many more factors to consider as you make decisions in the commercial environment. For example, as a pilot flying for a scenic charter operation, you must maintain flight safety, as well as balance the demands of your employer and passengers.

Ensure that your decision is producing the desired result.

You pay close attention to the weather conditions and PIREPs in the area. Since one of your passengers is unhappy about the change in plans, you agree to fly over the original site if the weather clears soon. If the conditions do not improve, you may be able to offer a refund or discounted flight.

The acronym DECIDE is used by the FAA to describe the basic steps in the decision-making process.

Detect the fact that a change has occurred
Estimate the need to counter or react to the change.
Choose a desirable outcome for the success of the flight.
Identify actions which could successfully control the change.
Do the necessary action to adapt to the change.
Evaluate the effect of the action.

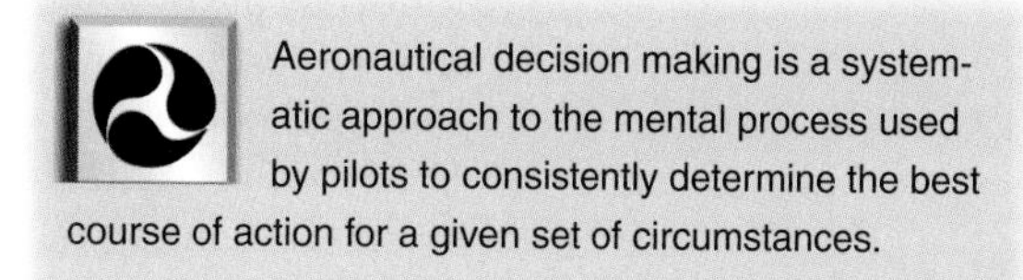

Aeronautical decision making is a systematic approach to the mental process used by pilots to consistently determine the best course of action for a given set of circumstances.

The ability to gather and interpret information becomes essential during aeronautical decision making. This is illustrated by research which separated crews flying simulators into higher and lower performing groups for comparison. During emergency situations, the higher performing crews engaged in more information gathering, monitoring, and planning. These crews requested significantly more data, such as weather conditions, from ATC. In addition, to allow more time for making decisions and accomplishing necessary tasks, higher performance crews requested ATC assistance in the form of radar vectors and clearances for holding or long final approaches.

CREW RESOURCE MANAGEMENT

Crew resource management (CRM) training is in part responsible for the marked increase in the safety of Part 121 operations. New regulatory requirements outline similar programs for certain Part 135 operators, as the FAA has recognized the success of CRM training. The investment in CRM training is no longer viewed as a luxury, but as an important part of all pilot training. If your company does not provide CRM training or you would like to prepare in anticipation of working in a crew environment, you should consider attending a CRM workshop conducted at a local college or flight school. If this is not possible, you may practice inflight scenarios with other pilots. If you have access to a simulator, flight training device, or PCATD, practice executing instrument approaches and emergency procedures in teams. [Figure 13-10]

Figure 13-10. Even if your CRM training is informal, any role-playing that you can do on the ground can aid cockpit coordination in flight.

Examining flight crew scenarios contained in ASRS accounts, NTSB reports and other safety-related publications can help increase your awareness of how crews handle situations effectively using CRM, and how the lack of CRM skills can lead to hazardous and sometimes deadly consequences. For example, pilot accounts and safety research presented in ASRS *Directline* are published to meet the needs of operators and flight crews of complex aircraft, such as commercial carriers and corporate fleets. ASRS *Directline* is distributed to operational managers, safety officers, training organizations, and publications departments. You can access *Directline* articles through the ASRS web page at http://www-afo.arc.nasa.gov/ASRS/ASRS.html.

By analyzing situations that other pilots have experienced, you can determine how the elements of pilot-in-command responsibility, communication, resource use, workload management, and situational awareness influence flight operations. After each of these topics is discussed in this section, you will be presented with scenarios which illustrate the effect of human factors principles on aeronautical decision making in the commercial environment.

PILOT-IN-COMMAND RESPONSIBILITY

You may feel a great deal of pride when you obtain your first job as a commercial pilot, but with the job comes the responsibility to fly as a professional. As pilot in command of an airplane in a commercial operation, you remain the final authority as to the safe conduct of each flight. However, you will share some of the decisions with management and crew as to where, how, and when each flight will take place. You may be required to refer to an **flight operations manual**, which outlines company procedures regarding such elements as crewmember duties and responsibilities, enroute flight, navigation and communication, weight limitations, emergencies, operating in hazardous weather, and obtaining aircraft maintenance, as well as many other company operations. The flight operations manual also includes the **operations specifications**, or ops specs, which are the conditions under which your company must operate in order to retain approval from the FAA. Ops specs include information such as the authorized areas of operations, aircraft, crew complements, types of operations (IFR, VFR, day, night, etc.), and any other pertinent information.

When you are employed as a pilot, you may feel that making decisions is easier since many choices are made for you. However, you still bear the primary responsibility for the safety of the flight. If you have a disagreement with management or passengers regarding a decision you make as PIC, be sure to complete the flight in a safe manner first. If the disagreement occurs before you take off, you may decide not to initiate the flight. You should never allow yourself to be pressured to take off if you have reservations about the flight's success. Discuss the reasons for your decision in a professional way, on the ground. State your position clearly, and provide evidence, such as weather reports, ATC recordings, or maintenance records. Telling yourself that a "real pilot" would make the flight, no matter what the situation, is an example of the macho hazardous attitude. Refer to Chapter 1, Section B — Advanced Human Factors Concepts and Chapter 10, Section B — IFR Decision Making for more information on hazardous attitudes.

Identifying hazardous attitudes through an inventory is an early part of the ADM process. There are five hazardous attitudes and corresponding antidotes.

CREW RELATIONSHIPS

If you pursue a career as a pilot, you most likely will transition from acting as sole authority in the cockpit to operating as part of a crew. While some airplanes require a second officer, or flight engineer, to operate and monitor the airplane's systems, your first experience in the crew environment normally is as the first officer, or second in command. Although the captain has the final authority as pilot in command, your experience and skill are a vital part of the cockpit resources. In certain situations, you may have information that the captain does not possess which is essential to the safety of the flight. You should voice your opinions in a respectful yet confident manner, and express your concern regarding any practices you deem unsafe. Depending on the procedures specified by your employer, you should expect to fly the airplane roughly 50% of the time that you are in the cockpit, so you can maintain proficiency and gain the experience necessary to eventually act as PIC.

When you make the transition to the left seat, you should provide leadership and exercise authority in a manner which encourages and supports an open exchange of opinions. Since it is your responsibility to provide direction for the other crewmembers, you must be assertive in your actions. However, you must be careful to avoid a domineering attitude. By allocating duties, as well as soliciting and accepting feedback from crewmembers, you will facilitate a professional and effective working environment. [Figure 13-11]

PILOT-IN-COMMAND RESPONSIBILITY

REPORT

On December 28, 1978, a DC8-61, Flight 173 departed Denver with 46,700 pounds of fuel on board. The flight plan indicated that 31,900 pounds would be required for the flight to Portland, including both FAR required fuel and company contingency fuel. On approach to Portland as the gear was lowered, the captain heard and felt an unusual sound and thought the gear went down more rapidly than normal. Neither the normal red transit light nor the gear door light illuminated. Flight attendants and passengers also reported a loud noise and severe jolt when the gear was lowered.

The crew advised Portland Approach that they had a gear problem. Approach advised that they could, "*just orbit you out there.*" The second officer reported that visual indicators on the wings confirmed that the main landing gear was down and locked. The captain discussed the problem with the first flight attendant, but not with United Maintenance Center in San Francisco until considerable time had elapsed. Eventually, he reported to Dispatch and Maintenance that they had 7,000 pounds of fuel on board and that he intended to hold for another 15 or 20 minutes.

There was some discussion about the fuel status when the captain asked the second officer to "*give us a current card on weight, figure about another fifteen minutes.*" The first officer responded "*Fifteen minutes?*" The captain answered, "*yeah, give us three or four thousand pounds on top of zero fuel weight.*" The engineer responded, "*Not enough. Fifteen minutes is really gonna run us low on fuel here.*" Later, the engineer stated, "*We got about three on the fuel and that's it.*" The captain responded "*Okay. On touchdown . . . get those boost pumps on . . .*" four minutes later, when the first officer stated "*We're going to lose an engine . . . .*" the captain asked "*Why?*" the first officer responded "*we're losing an engine.*" Again the captain asked "*Why?*" The first officer responded "*Fuel!*" Eight minutes later the airplane crashed into a wooded residential section of suburban Portland about 6 miles from the airport. There was no fire.

ANALYSIS

The NTSB determined that the probable cause of this accident was the captain's failure to properly monitor the aircraft's fuel state and to properly respond to both the low fuel state and crewmember advisories. This resulted in fuel exhaustion to all engines. The captain's inattention resulted from preoccupation with an unsafe landing gear indication and preparation for a possible emergency landing. Contributing to this accident was the failure of the other two crewmembers to successfully communicate to the captain their concern regarding the low fuel state.

Figure 13-11. As you review this accident account, think about how the crew relates to one another. What could the captain and other crewmembers have done to establish a more positive cockpit environment and prevent this accident from occurring?

COMMUNICATION

Effective cockpit communication is the cornerstone of safe commercial crew operations. When you fly as a single pilot, you are in charge of all duties during the flight, and your only communication generally is with ATC. In two-pilot crew operations, each person is responsible for completing specific tasks, and you both need to be informed of the other's actions. You also need to be able to discuss upcoming procedures and any problems that may arise.

One way to facilitate effective communication is through **departure** and **approach briefings**. Before takeoff, you should go over the proposed departure routing, either given by ATC or through published procedures. Determine who will be the **pilot flying (PF)** and who will be the **pilot not flying (PNF)**. The PF generally is in charge of manipulating the controls, while the PNF is responsible for communication with ATC and navigation. During the departure briefing you should note any obstructions or terrain which may be a factor on climbout. Depending on the situation, an approach briefing might include a discussion of terrain features, minimum altitudes to be used and an approach chart review if an IAP is necessary.

Business As Usual

While much attention during pilot training is focused on dealing with emergency situations, most of your actual time in the cockpit is spent completing routine tasks. Crew coordination during these times is just as important as it is during an emergency, as the failure to ensure that all important items have been accomplished could lead to a serious problem in flight. In light of this concern, commercial air carriers include line-oriented flight training (LOFT) within their ongoing training curricula.

LOFT was first developed in the late 1970s after a serious airline accident focused attention on CRM issues. The program is organized to evaluate how crews work together during legs of a typical trip. For example, imagine you are a first officer transitioning from domestic to international routes. You would join a small class with a captain and (if applicable) a second officer, and review the dispatch for an actual flight segment — in this case, from San Francisco to Tokyo. [Figure A] After a detailed briefing, you would rehearse the flight in the simulator, practicing all variations of normal enroute procedures. Other considerations that are addressed include operating at foreign airports, local customs, and even how to get to your hotel from the airport.

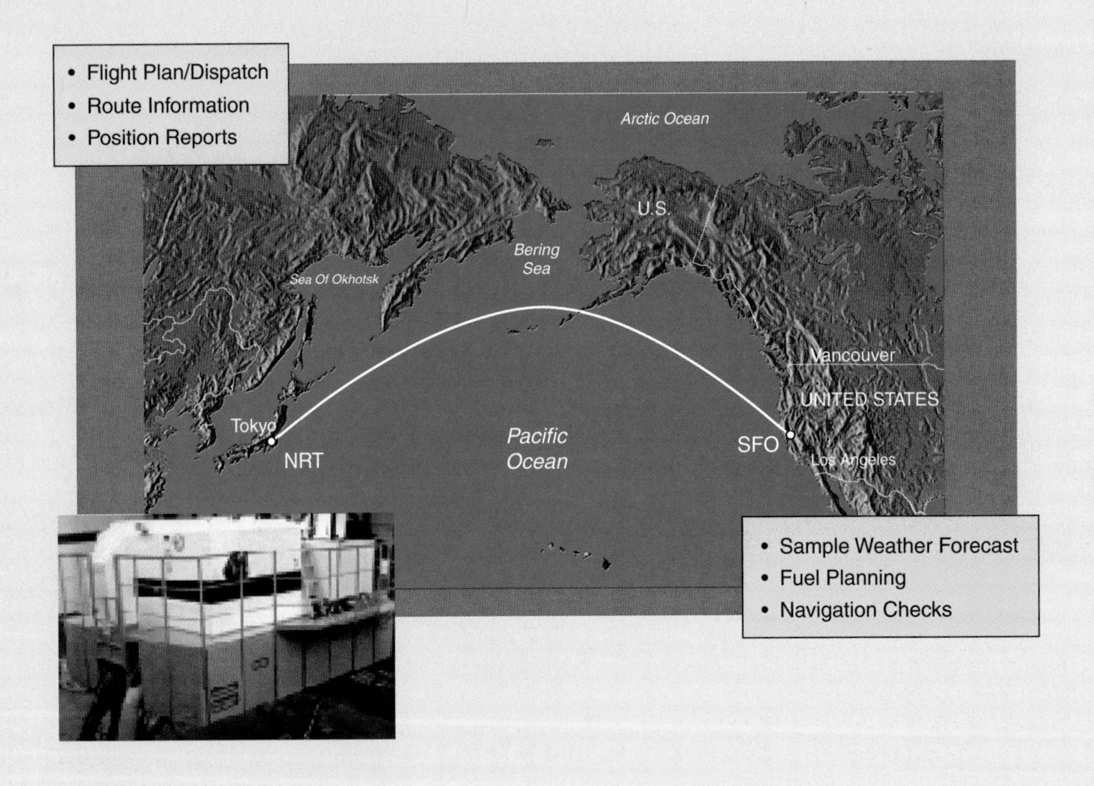

Most flight crews use the **challenge-response method** when going through checklists. This procedure requires one pilot to read the checklist item out loud, while the other crewmember completes the task and repeats the instruction verbally. Since both crewmembers are directly involved in the checklist process, the chance that a mistake or omission will occur is reduced. [Figure 13-12]

Barriers to Effective Communication

You can avoid a breakdown in communication with crewmembers by recognizing some of the barriers to effective communication that exist in the commercial environment. These problems

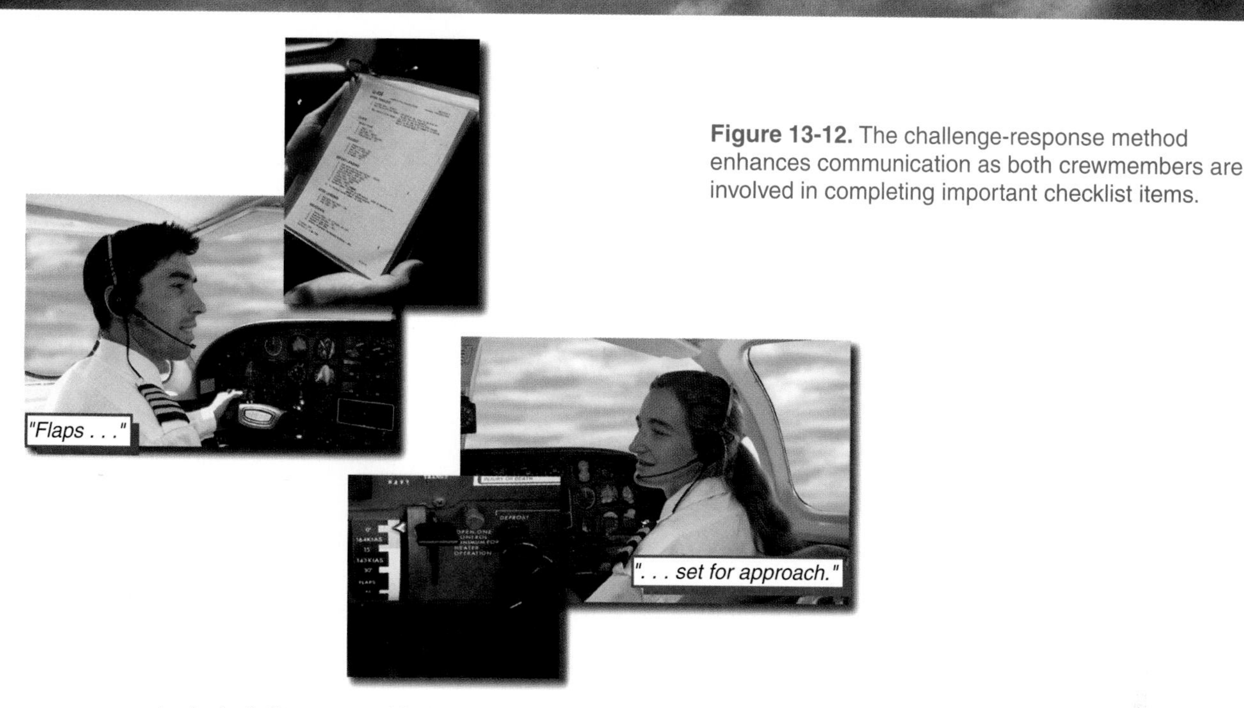

Figure 13-12. The challenge-response method enhances communication as both crewmembers are involved in completing important checklist items.

include failing to establish open communication with crewmembers, misunderstanding and conflict in the cockpit, and the distraction caused by nonessential conversation.

Of these barriers, failing to establish communication may be the easiest to correct. By simply knowing that you need to keep an open dialogue in the cockpit, you can reasonably ensure that one will exist. Both pilots should be included in critical decisions when possible, not just in emergencies, but in routine practices such as programming electronic equipment and setting autopilot functions. Anytime you change a course, or enter a new frequency or identifier into onboard equipment, you should verify the process out loud with the other crew. Both crewmembers should feel comfortable to question, scrutinize, or investigate any discrepancies which may arise during the flight and should feel an obligation to express reservations about a current or planned course of action. It is the pilot in command's responsibility to seek out viewpoints from the second in command and to listen to alternative courses of action.

You may have the opportunity to fly with wide variety of pilots who have had different training experiences than yours. To avoid misunderstandings, listen carefully to your fellow crewmembers, and clarify any procedures or instructions that you do not understand. Although you may try to maintain a positive environment, at times, conflict is inevitable. Crewmembers enter the cockpit with their own expectations and biases. Whether actual or perceived, differences in thoughts, opinions, values, or actions can lead to disagreements. Conflict can reveal unexpressed thoughts or feelings which may have been adversely affecting performance, and successful conflict resolution can strengthen team effectiveness. However, if steps are not taken to resolve conflict, the atmosphere in the cockpit can degrade to a point where effective communication is impossible and the safety of flight is at risk.

Another barrier to effective communication occurs when a flight crew's attention is diverted by conversation unrelated to the tasks at hand. A significant number of accidents have occurred when the crews were engaged in unnecessary dialogue and overlooked important procedures or checklist items, such as setting the flaps properly for takeoff or extending the landing gear. To help prevent these types of accidents, the FAA issued FAR 121.542 and FAR 135.100. Commonly referred to as the **sterile cockpit rule**, these FARs specifically prohibit crewmember performance of nonessential duties or activities while the aircraft is involved in taxi, takeoff, landing, and all other flight operations conducted below 10,000 feet MSL, except cruise flight. [Figure 13-13]

Figure 13-13. These ASRS reports illustrate how barriers to effective communication take many different forms. As you review these accounts, think about what actions you can take to prevent these types of situations during your flights.

COMMUNICATION

REPORT

The following experience was related by a pilot who admitted to not reviewing the applicable charts prior to a night flight under IFR. — "*The dim shape of the mountain came into view . . . seconds before the 'WHOOP . . . WHOOP . . . PULL UP' sounded. We both pulled back abruptly on the controls and climbed . . .*"

The controller provided additional insight regarding this flight. — "*The tapes revealed that I had told the pilot to descend to 7,000 feet (6,500 is the MEA) but he had read back 5,000. He got down to 5,700 feet, about 2 miles from a 5,687 foot mountain before I saw him.*"

ANALYSIS

The failure of the pilots to brief on the published minimum altitudes near and within the terminal area made them vulnerable to readback/hearback errors and to a CFIT accident.

REPORT

"The copilot . . . turned the radios over to me so he could do a passenger announcement. I acknowledged a frequency change. I then checked in to the new frequency, using the call sign of my previous flight. The controller, who I think was expecting me, gave me a clearance for the correct call sign. I acknowledged, apologized for the mistake, and continued without incident. In retrospect, I feel I was too quick to accept the clearance. I could have easily taken someone else's clearance . . . You lose your system of double-check when one pilot is off the air getting ATIS, doing a passenger announcement, or talking with the company."

ANALYSIS

The pilot should have verified his call sign and clearance with the controller. Although the crew environment can provide a valuable double-check of radio transmissions, there are times when the attention of the other pilot is diverted with another task. You must avoid becoming complacent and be careful not to rely on another crewmember or ATC to recognize errors.

REPORT

In this situation, the crew believed that they landed without contacting the tower and receiving landing clearance. — *"Although VMC on the approach, the new special weather was . . . [indefinite ceiling, 200 obscured, visibility 1-1/4 mile in ground fog], snow falling and some snow on the runway . . . I was flying and captain viewing PIT stadium and various sights out the window, chatting incessantly . . . Captain then reviewed procedures for short ground roll on snow-covered runways and returned to miscellaneous conversation . . . the potential for disaster scenarios should be apparent . . . The bottom line: lack of professionalism. Captain habitually rambled from push back to block-in through a four day trip. This was the first of two incidents on the same day . . . Below the line: lack of courage. F/O and F/E were not willing to ask the captain to please shut up so we could fly the airplane."*

ANALYSIS

This situation provides an excellent example of the problems that can arise when a sterile cockpit environment is not maintained. Due to the distraction of the captain's talking, the crew did not obtain a necessary clearance from ATC and crew communication suffered as the first officer and flight engineer did not voice their concerns to the captain.

RESOURCE USE

The amount of resources available to you most likely will multiply when you work as a professional pilot. Depending upon the type of operations your company conducts, you may have more advanced aircraft equipment, additional crewmembers in the cockpit or cabin, and more personnel on the ground to support each flight. The efficiency of the flight depends on your ability to effectively utilize these resources.

INTERNAL RESOURCES

Your first experience with a more complex aircraft may come during your commercial training. A thorough understanding of the systems and operation of all equipment on board the aircraft is crucial to flight safety. A number of systems are considered vital to safe commercial operations, and their use is required during certain stages of flight under FAR Parts 121, 125, or 135. For example, the use of an autopilot is required for certain aircraft to be operated under IFR with a single pilot, as stated in FAR 135.105. The autopilot frees you from holding the aircraft on a heading and altitude during cruise, and, when coupled with a navigation system, responds to course changes as programmed into the system. You can simply monitor the system to ensure that the course is correct, a task that is far less physically demanding than hand-flying the airplane.

Another resource to consider is your company's supply of aeronautical charts. For VFR operations, you should be provided with the appropriate sectional and terminal area charts for the region in which you fly. You should also consider carrying IFR charts, even if you are not approved for IFR operations. In an emergency, they could prove invaluable. If you conduct IFR operations, there should be at least one set of charts for the region on board the airplane, and though it is standard for crews to rely on one chart between two pilots, you may want to carry a second set. Whether you use Jeppesen or NOS charts, make sure that updates are completed on schedule, so that you are not caught in IFR conditions with outdated charts. [Figure 13-14]

Figure 13-14. Other crewmembers involved in commercial operations are valuable resources. For example, in addition to another pilot, a second officer and/or flight attendant may be required. If you fly for an air ambulance operation, crewmembers may include a flight nurse or medical personnel.

EXTERNAL RESOURCES

Being a professional pilot often means that others are available to take over some of the duties that you normally did as a private pilot. Your company may employ dispatch personnel, who are responsible for planning the trip and collecting weather information for your route of flight. Company maintenance technicians can answer questions about aircraft systems and, if a company frequency is established, aid you during an inflight emergency. You may have baggage handlers to load the airplane and gate personnel to assist with passengers. Though these people are there so that you can focus on flying the airplane, you still are responsible as PIC to check all weather, route, and airport information concerning the flight, as well as ensure the aircraft is airworthy. [Figure 13-15]

RESOURCE USE

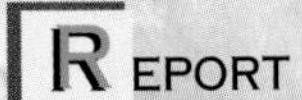

REPORT

While descending toward ABC, we were cleared to . . . intercept the localizer course for Runway 30. Center then issued a VFR traffic advisory to us — a general aviation airplane was descending into ABC. The GA airplane was also advised that we were descending. [Each aircraft] reported the other aircraft in sight. Just prior to intercepting the localizer at 12,000 feet, we received a traffic alert form our TCAS. We still had a visual on the airplane, but it was difficult to ascertain his altitude or heading due to darkness. Very quickly after that, the TCAS issued a resolution advisory to "descend, descend now!" We complied, increased our rate of descent, and turned right to avoid the target. I estimate that our aircraft passed within a half mile of each other and were separated by 100-200 feet vertically.

ANALYSIS

It is difficult to judge the altitude and distance of other aircraft at night. A TCAS is an excellent resource which can help you determine the position and closing rate of traffic. Assistance from ATC also can be enlisted to help you maintain adequate spacing from traffic. For example, you can request that ATC keep you informed of aircraft spacing.

REPORT

During cruise, we got a #1 engine overheat light . . . then it went out. [Later], the light came back on, followed by a fire loop fault light. We got clearance to divert to the nearest airport. While completing the emergency checklists, we got a #1 engine fire light and bell. We declared an emergency and fired both extinguisher bottles. We landed without further problems. The fire trucks reported no evidence of smoke or fire, and [later] the mechanics confirmed a short-circuit in the #1 engine fire detection system. I had the copilot fly while I got hold of company. We had a jumpseat pilot . . . who made an announcement to the passengers, after which he handled ATC communications. I completed checklists, kept an eye on aircraft position, and talked to the lead flight attendant. CRM can take full credit for the uneventful completion of this flight.

ANALYSIS

Through the effective use of resources, this crew was able to successfully manage their workload during an emergency situation. Each crewmember performed specific tasks and the help of a jumpseat pilot was enlisted. By declaring an emergency and requesting an amended clearance, the crew was able to utilize ATC as a valuable resource. Extensive knowledge of the aircraft's systems and the efficient use of checklists proved essential. Additional resources on the ground included an emergency crew and aviation maintenance technicians.

Figure 13-15. These ASRS reports provide examples of the effective use of resources in the commercial environment.

WORKLOAD MANAGEMENT

In a crew that is functioning efficiently, workload is divided between the pilots so that every necessary task is accomplished even during the highest workload periods. However, you need to remain vigilant to ensure that one pilot is not

Good cockpit stress management begins with good life stress management. To help you manage cockpit stress, you should avoid situations that degrade your ability to handle cockpit responsibilities.

doing too much and excluding the other pilot from sharing the workload. By managing workload so that it does not become excessive, you can avoid situations that lead to cockpit stress.

PLANNING AND PREPARATION

If you work for a large commercial operation, much of your preflight workload is assumed by other personnel, but you should not use their presence as an excuse to become complacent. Your job may shift from one of doing the actual flight planning to one of supervising and ensuring that the information presented to you is timely and correct. You can also focus on rehearsing the flight, and preparing for any contingencies that may arise. When you work for a smaller company, however, your preflight workload may increase above what you were responsible for as a private pilot. In addition to the normal required preflight duties, you may also be responsible for preparing a **load manifest**, if you fly multi-engine aircraft. The load manifest lists the number of passengers, the origin and destination of the flight, the registration number of the airplane, weight and balance information, and center of gravity limits.

Because of the high demand on your time, you need to be organized and familiar with your current aircraft and operation. FAR Parts 121 and 135 list specific flight time limitations and rest requirements for flight crews. In addition to specifying the maximum number of hours pilots can be scheduled to fly in a given time period, these regulations limit the amount of time a pilot can be on duty without a rest period. [Figure 13-16]

Figure 13-16. This example of a commercial pilot's duty schedule shows how the pilot's time is budgeted. Pilots who fly short trips may spend each evening at their home base. Pilots on longer trips often are away from home for several days and then off duty for several days. Some Part 135 and corporate operations do not have scheduled flights and require pilots to be on call.

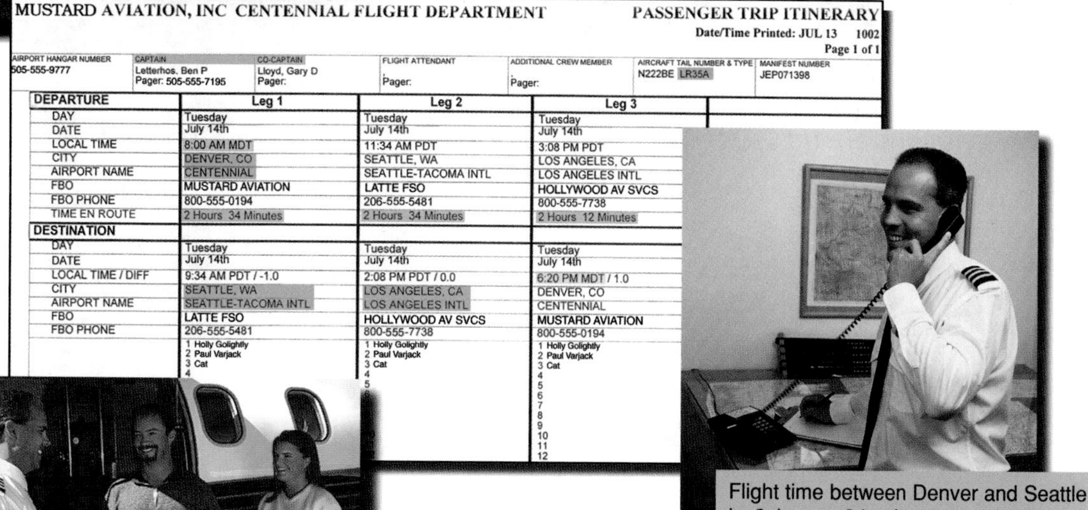

MUSTARD AVIATION, INC CENTENNIAL FLIGHT DEPARTMENT

PASSENGER TRIP ITINERARY

Date/Time Printed: JUL 13 1002

Page 1 of 1

AIRPORT HANGAR NUMBER	CAPTAIN	CO-CAPTAIN	FLIGHT ATTENDANT	ADDITIONAL CREW MEMBER	AIRCRAFT TAIL NUMBER & TYPE	MANIFEST NUMBER
505-555-9777	Letterhos, Ben P Pager: 505-555-7195	Lloyd, Gary D Pager:	, Pager:	, Pager:	N222BE LR35A	JEP071398

DEPARTURE	Leg 1	Leg 2	Leg 3
DAY	Tuesday	Tuesday	Tuesday
DATE	July 14th	July 14th	July 14th
LOCAL TIME	8:00 AM MDT	11:34 AM PDT	3:08 PM PDT
CITY	DENVER, CO	SEATTLE, WA	LOS ANGELES, CA
AIRPORT NAME	CENTENNIAL	SEATTLE-TACOMA INTL	LOS ANGELES INTL
FBO	MUSTARD AVIATION	LATTE FSO	HOLLYWOOD AV SVCS
FBO PHONE	800-555-0194	206-555-5481	800-555-7738
TIME EN ROUTE	2 Hours 34 Minutes	2 Hours 34 Minutes	2 Hours 12 Minutes
DESTINATION			
DAY	Tuesday	Tuesday	Tuesday
DATE	July 14th	July 14th	July 14th
LOCAL TIME / DIFF	9:34 AM PDT / -1.0	2:08 PM PDT / 0.0	6:20 PM MDT / 1.0
CITY	SEATTLE, WA	LOS ANGELES, CA	DENVER, CO
AIRPORT NAME	SEATTLE-TACOMA INTL	LOS ANGELES INTL	CENTENNIAL
FBO	LATTE FSO	HOLLYWOOD AV SVCS	MUSTARD AVIATION
FBO PHONE	206-555-5481	800-555-7738	800-555-0194
	1 Holly Golightly 2 Paul Varjack 3 Cat 4	1 Holly Golightly 2 Paul Varjack 3 Cat 4 5	1 Holly Golightly 2 Paul Varjack 3 Cat 4 5 6 7 8 9 10 11 12

PRIORITIZING

You are experiencing the hurry-up syndrome when your performance is degraded by a perceived or actual need to hurry or rush tasks or duties for any reason. For commercial operations, errors caused by rushing are most likely to occur in high workload operational phases, especially during preflight and taxi. Time-related pressures experienced by professional flight crews include the need of a company agent or ground personnel to open a gate for another aircraft, pressure from ATC to expedite taxi for takeoff or to meet a restriction in clearance time, the pressure to keep on schedule when delays have occurred due to maintenance or weather, or the inclination to hurry to avoid exceeding duty-time regulations.

During preflight, numerous tasks must be accomplished, but normally a logical sequence does not exist for performing these tasks as it does during other phases of flight. For example, prior to a flight you may have to handle flight planning, weather information, fuel loading, dispatch manifests, last-minute maintenance or MEL items, aircraft de-icing, and duty-time requirements. Trying to accomplish these duties without establishing priorities can lead to error. Paperwork and nonessential tasks can be relegated to low workload operational phases. Following checklists and adhering to the sterile cockpit rule also can aid in establishing priority for performing necessary tasks. [Figure 13-17]

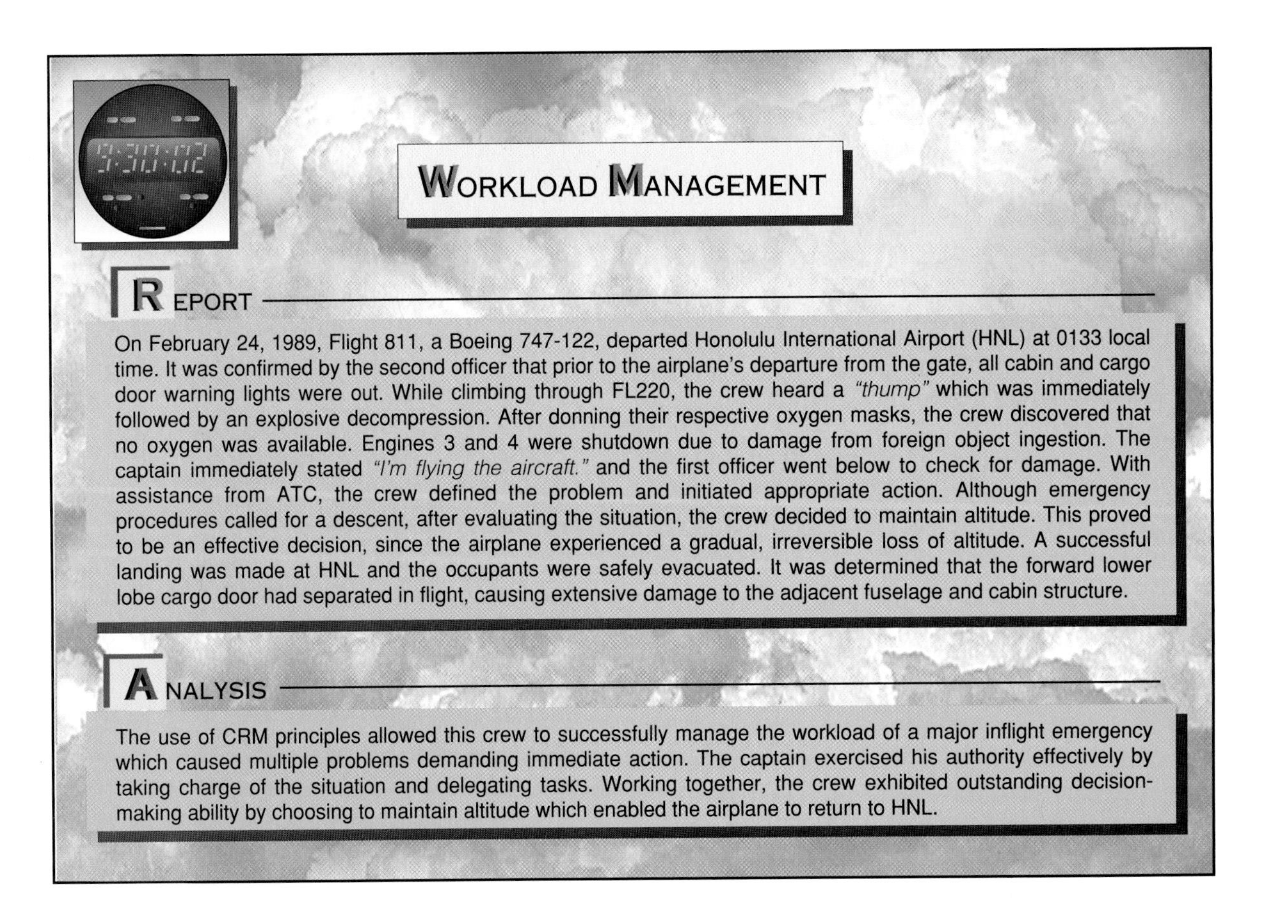

WORKLOAD MANAGEMENT

REPORT

On February 24, 1989, Flight 811, a Boeing 747-122, departed Honolulu International Airport (HNL) at 0133 local time. It was confirmed by the second officer that prior to the airplane's departure from the gate, all cabin and cargo door warning lights were out. While climbing through FL220, the crew heard a *"thump"* which was immediately followed by an explosive decompression. After donning their respective oxygen masks, the crew discovered that no oxygen was available. Engines 3 and 4 were shutdown due to damage from foreign object ingestion. The captain immediately stated *"I'm flying the aircraft."* and the first officer went below to check for damage. With assistance from ATC, the crew defined the problem and initiated appropriate action. Although emergency procedures called for a descent, after evaluating the situation, the crew decided to maintain altitude. This proved to be an effective decision, since the airplane experienced a gradual, irreversible loss of altitude. A successful landing was made at HNL and the occupants were safely evacuated. It was determined that the forward lower lobe cargo door had separated in flight, causing extensive damage to the adjacent fuselage and cabin structure.

ANALYSIS

The use of CRM principles allowed this crew to successfully manage the workload of a major inflight emergency which caused multiple problems demanding immediate action. The captain exercised his authority effectively by taking charge of the situation and delegating tasks. Working together, the crew exhibited outstanding decision-making ability by choosing to maintain altitude which enabled the airplane to return to HNL.

Figure 13-17. As you study this report, consider how the crew used planning, preparation, and prioritizing to effectively manage their workload during an emergency.

SITUATIONAL AWARENESS

Although having a second pilot will generally aid your overall situational awareness, there are issues unique to being part of a two-pilot crew that can detract from your knowledge of your position and the state of the aircraft. One concern is that you may have trouble maintaining an awareness of operations that are not under your direct control. For example, if you are not responsible for communicating with ATC, you may not be immediately aware of a radio problem. You need to ensure that you maintain an awareness of the entire flight situation by eliciting information from other crewmembers and keep the other pilot informed of the status of operations that you are responsible for. There is a flow to gathering information and subsequent actions that helps you to stay informed of the state of the aircraft, weather conditions, and ATC communication, throughout the flight. [Figure 13-18]

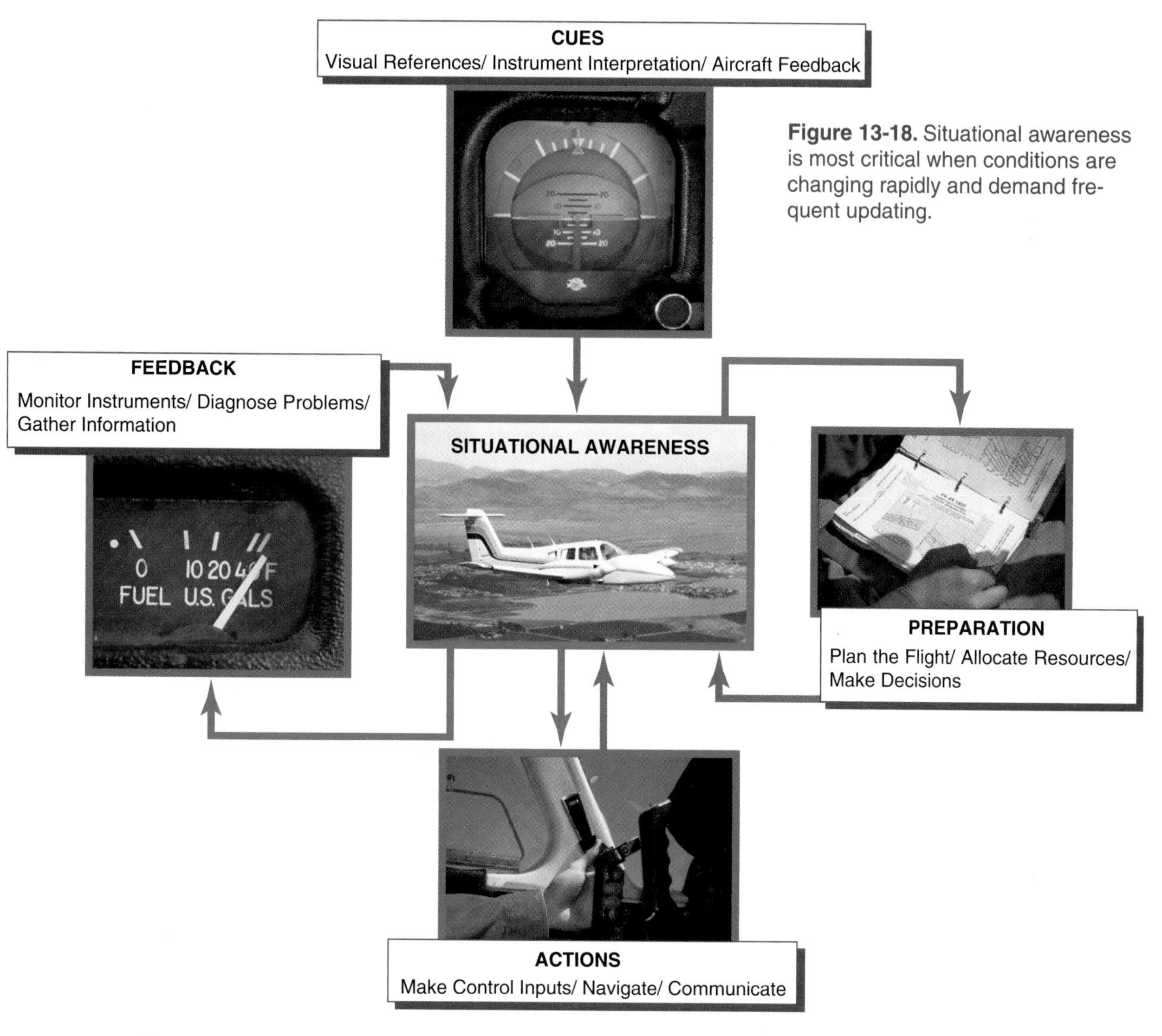

Figure 13-18. Situational awareness is most critical when conditions are changing rapidly and demand frequent updating.

CONTROLLED FLIGHT INTO TERRAIN

The special demands of some commercial operations under FAR Parts 91, 121, and 135 make them especially vulnerable to controlled flight into terrain (CFIT) accidents. When you first begin flying commercially, you may be hired by a company that does scenic flights and commuter operations in places like Alaska, Hawaii, or the Grand Canyon. The rugged terrain in these areas, and the additional duties which are required in the commercial environment can prove challenging. You can avoid becoming involved in a CFIT

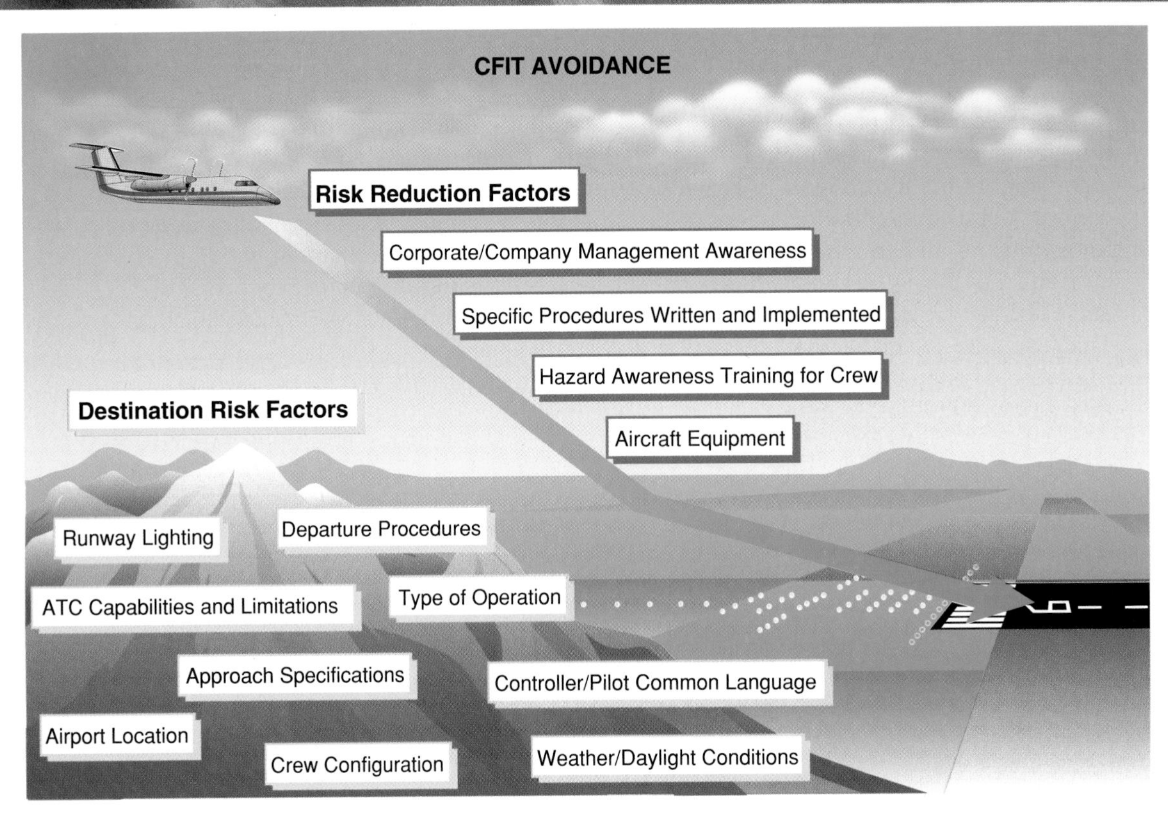

Figure 13-19. You can decrease your exposure to a CFIT accident by identifying risk factors and remedies prior to flight.

accident by strictly adhering to VFR/IFR requirements and maintaining a high degree of knowledge about the terrain over which you are flying. [Figure 13-19]

The multitude and complexity of systems on board an aircraft used for commercial operations may also add to the risk of CFIT. For example, relying too heavily on advanced navigational equipment, such as a flight management system (FMS), can lead to complacency and loss of situational awareness. The FMS is a computer system, typically used in airliners, which uses a large data base to allow routes to be preprogrammed and fed into the system by a data loader. The FMS is constantly updated with respect to position accuracy by reference to conventional navigation aids. Failure to double-check entries, actively monitor displays, and refer to conventional aeronautical charts can lead to errors and a loss of situational awareness. While you may not be using an FMS, similar problems can occur when making entries into navigation equipment such as a GPS unit. [Figure 13-20]

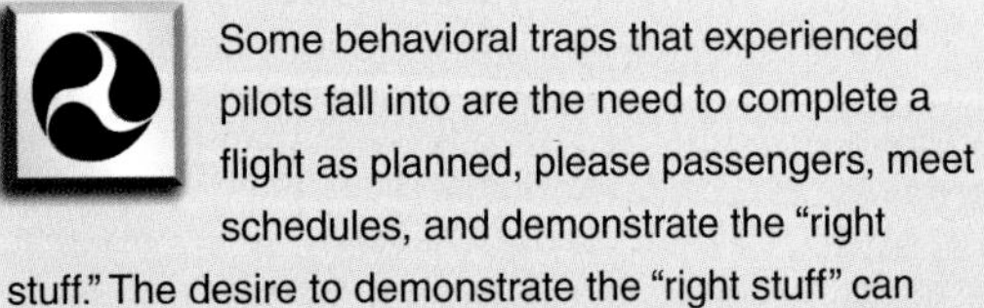

Some behavioral traps that experienced pilots fall into are the need to complete a flight as planned, please passengers, meet schedules, and demonstrate the "right stuff." The desire to demonstrate the "right stuff" can generate tendencies that lead to practices that are dangerous, often illegal, and may lead to a mishap. Commercial pilots are also just as prone to peer pressure, get-there-itis, loss of situational awareness, and operating without adequate fuel reserves.

APPLICATION OF AERONAUTICAL DECISION MAKING

By this point in your aviation career, you have reached a level of expertise that sets you apart from the majority of pilots. However, the commercial certificate is yet another step

SITUATIONAL AWARENESS

REPORT

On June 8, 1992, a Beechcraft C99 scheduled passenger flight crashed while maneuvering to land at the Anniston Metropolitan Airport in Anniston, Alabama. The NTSB determined that this CFIT accident occurred when the flight crew experienced a loss of situational awareness. When cleared for the ILS approach to Runway 5 at Anniston, the flight crew turned the airplane north away from the airport in the mistaken belief that the airplane was south of the airport. The flight crew did not fly outbound or execute the procedure turn which was required by the IAP. The C99 intercepted the back course localizer signal for the ILS approach, and the flight crew tried to fly the approach at an excessive airspeed about 2,000 feet above the specified altitude for crossing the FAF. The airplane continued a controlled descent until it impacted terrain.

0848:10
ATC: *. . . and a eight sixty one, proceed direct Bogga maintain four thousand 'til Bogga, cleared localizer run- er ILS runway five approach.*
F/O: *Direct, direct to Bogga four thousand and cleared for the ILS runway five. Eight sixty one. Thank you.*
Captain: Ask him distance from . . .
F/O: *From Bogga?*
Captain: *That's okay, I'll just . . .*
F/O: *We're ah . . . minus six point one. We're five miles from Bogga.*
F/O: *Go ahead and slow on up.*
F/O: *There you go keep the shiny side up.*
Captain: *Ah.*
F/O: *There you go. Should have moved your heading bug. Here you go I'll get you set in here.*
Captain: *Okay let's go approach flaps.*
F/O: *Speed checks coming now.*
F/O: *Didn't realize that you're going to get this much on your first day did ya.*
Captain: *Well it's all kind of ganged up here on me a little fast.*

0851:34
F/O: *Okay watch your airspeed. One fifteen on the airspeed.*
F/O: *We're inside — through twenty-two we can continue our descent on down. We're way high.*
Captain: *Okay, is the glide slope working?*
F/O: *Nope I'm not gettin' any.*
F/O: *So with no glide slope, we're down to eleven hundred.*
Captain: *You got your frequency in there?*
F/O: *Five hundred — one eleven five, double check, yup.*
Captain: *What's our missed approach point now?*
F/O: *Missed approach at the middle marker ah . . .*
F/O: *Eleven hundred but we need to add a hundred so twelve hundred.*
F/O: *Comin' up . . .*

0852:25 - Sound of impact

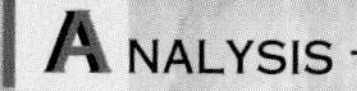

ANALYSIS

The NTSB determined that one of the causes of this accident was the failure of the flight crew to use approved instrument flight procedures, which resulted in a loss of situational awareness and terrain clearance. The crew's failure to effectively manage their cockpit duties in part stemmed from the pairing of an inadequately prepared captain with a relatively inexperienced first officer. In addition, crew coordination suffered due to a role reversal on the part of the captain and first officer.

Figure 13-20. A review of the details of this accident and an excerpt from its transcript reveals how important CRM skills are to maintaining situational awareness.

in your continuing education that will go on as long as you fly. Each flight will bring a new set of experiences that affect your future decisions. In the following example from an ASRS report, a first officer is confronted with a complex flight situation resulting from an adverse cockpit environment, and he must make a number of decisions in order to see that the flight ends safely.

A SW3 Metroliner departed from Ketchikan, Alaska, for a scheduled passenger and cargo flight to Wrangell, Petersburg, and Juneau. After climbing above the cloud deck, the DME failed, and the aircraft was then limited to VFR operations, as per the MEL. VFR conditions were initially reported at each stop, but to the north, conditions worsened significantly, and the first officer was concerned that this lowering weather would affect the flight. The captain suggested that they *"look and see,"* and the flight proceeded from Petersburg toward Juneau. Terrain in this part of Alaska is extremely rugged, with sea level fjords and mountains reaching 6,000 feet MSL in close proximity to the course of the flight. While enroute, the captain penetrated 3 rain squalls in which visibility dropped to zero, and was forced to descend to below 400 feet AGL while flying through channels and mountainous valleys. At this point, the flight became in danger of CFIT for the first time, and the crew should have requested assistance from ATC while executing a climb to a safe altitude and course.

After numerous attempts, the first officer finally convinced the captain to return to Petersburg, reminding him that they were breaking regulations and endangering themselves and their passengers. They landed at Petersburg, but the captain wanted to go on to Ketchikan, for purely personal reasons. Since Ketchikan was still reporting VFR, the first officer agreed to go, and it was his leg to fly. Once they were airborne, the first officer noted that Wrangell, located between Petersburg and Ketchikan, was obscured by clouds and rain, and he asked the captain for a weather report from Wrangell, attempting to use him as a resource. The captain refused, and the first officer then decided to return to Petersburg. As the first officer began a 180° turn, the captain took control of the airplane and turned back toward Wrangell. The aircraft entered IFR conditions, even though both pilots were aware of mountainous terrain only 3 miles in front of them. Despite the first officer's repeated attempts to make the captain turn back, he continued until the first officer called Sitka Radio to declare an emergency. The captain finally capitulated and began to turn around when he spotted the town of Wrangell from 150 feet AGL and was able to make a contact approach. They landed, and a weather observer on the field verified that IFR conditions prevailed. [Figure 13-21]

Though you may never fly in coastal Alaska, similarly hazardous conditions can form in any part of the country. The MEL limitation to VFR conditions could have been dealt with effectively had the captain practiced sound decision making. Only through the continued advocacy of the first officer, and his final attempt to declare an emergency, did the captain begin to change his mind about continuing into weather that was below VFR minimums. Still, the airplane landed at Wrangell when the better choice would have been to land at a field that was still VFR. There were several points along the chain of events where both effective and poor choices were made. The captain penetrated IFR conditions without being on an IFR flight plan, exposing the flight to a potential CFIT accident and violating FARs. The first officer voiced his concern, trying to establish communication in the cockpit. Unfortunately, the captain dismissed his concerns, and continued on, setting the stage for a hostile cockpit environment. By refusing to return to Petersburg when conditions at Wrangell were deemed unsuitable, the captain let personal issues and outside stress affect his decisions. The first officer continued to demand a return to a VFR airport, and he finally attempted to obtain outside help, in the form of Sitka Radio.

The final poor decision implemented by the crew was to land at Wrangell, again inviting a CFIT accident. Positional awareness remained high, as both pilots were intimately familiar with the terrain and airports, yet each pilot used this knowledge in a different

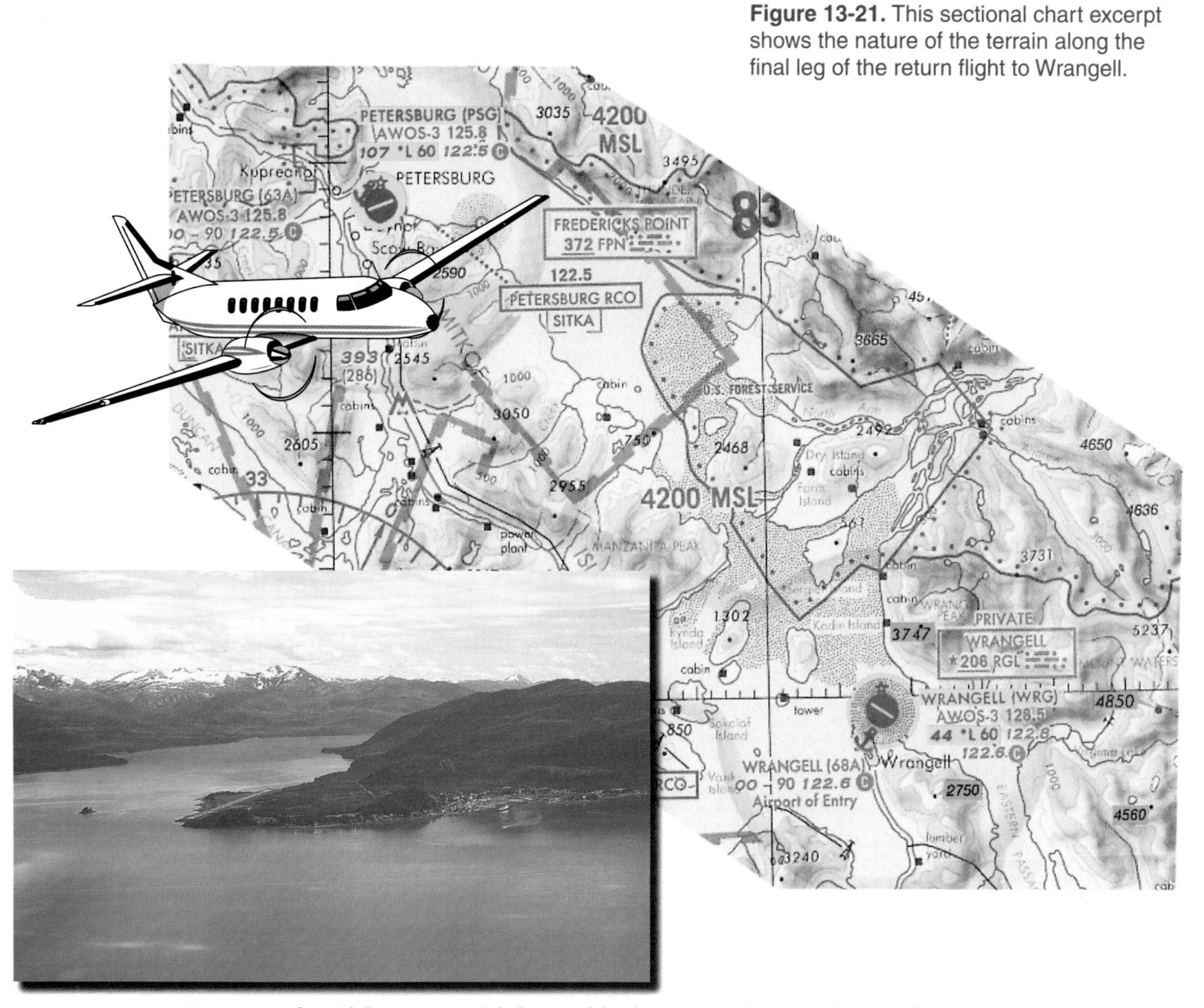

Figure 13-21. This sectional chart excerpt shows the nature of the terrain along the final leg of the return flight to Wrangell.

way. It appeared as if the captain felt he could take more risks since he was familiar with the area, while the first officer was alarmed that the captain would continue toward the mountains in IFR conditions at a high rate of speed. An effective decision may have been to turn away from the mountains and remain clear in VFR conditions while deciding on an alternate plan.

It is estimated that human factors contribute significantly to approximately three-quarters of commercial aircraft accidents. As you enter the commercial aviation environment, your understanding of basic CRM principles and the human factors elements which affect aeronautical decision making can help you avoid becoming part of this statistic.

SUMMARY CHECKLIST

✓ As a commercial pilot, decision making can be complex as you balance public perceptions of aviation, your company's schedule, and your desire to conduct safe flight operations.

✓ Aeronautical decision making is a systematic approach to the mental process used by pilots to consistently determine the best course of action for a given set of circumstances.

✓ Hazardous attitudes should be identified and corrected as an early part of the ADM process.

✓ The acronym DECIDE is used by the FAA to describe the basic steps in the decision-making process.

✓ Examining flight crew scenarios contained in ASRS accounts, NTSB reports and other safety-related publications can help increase your awareness of how CRM affects flight safety in the commercial environment.

✓ A flight operations manual outlines company procedures regarding such elements as crewmember duties and responsibilities, enroute flight, navigation and communication, weight limitations, emergencies, operating in hazardous weather, and obtaining aircraft maintenance, as well as many other company operations.

✓ Operations specifications, or ops specs, are the conditions under which your company must operate in order to retain approval from the FAA. Ops specs include information such as the authorized areas of operations, aircraft, crew complements, types of operations (IFR, VFR, day, night, etc.), and any other pertinent information.

✓ One way to facilitate effective communication and prepare for high workload phases of flight is through departure and approach briefings.

✓ Generally, the pilot flying (PF) is in charge of aircraft control, while the pilot not flying (PNF) often is responsible for radio communication and navigation.

✓ The challenge-response method requires one pilot to read the checklist item out loud, while the other crewmember completes the task and repeats the instruction verbally.

✓ Barriers to communication include failing to establish open communication with crewmembers, misunderstanding and conflict in the cockpit, and the distraction caused by nonessential conversation.

✓ Commonly referred to as the sterile cockpit rule, FAR 121.542 and FAR 135.100 prohibit crewmember performance of non-essential duties or activities while the aircraft is involved in taxi, takeoff, landing, and other flight operations conducted below 10,000 feet MSL, except during cruise flight.

✓ Resources which may be available to you during commercial operations include additional crewmembers and advanced aircraft equipment, as well as company maintenance and ground personnel.

✓ If you work for a large commercial operation, your preflight activities may shift from doing the actual flight planning to supervising and ensuring that the information presented to you is timely and correct.

✓ The load manifest lists the number of passengers, the origin and destination of the flight, the registration number of the airplane, weight and balance information, and center of gravity limits.

✓ You are experiencing the hurry-up syndrome when your performance is degraded by a perceived or actual need to hurry or rush tasks or duties for any reason.

✓ Certain commercial operations have increased exposure to situations where CFIT is a serious concern. You can increase your situational awareness by making proper use of all available resources, managing your workload effectively and staying continuously aware of your position and the state of the airplane.

KEY TERMS

Crew Resource Management (CRM)

Flight Operations Manual

Operations Specifications

Departure Briefing

Approach Briefing

Pilot Flying (PF)

Pilot Not Flying (PNF)

Challenge-Response Method

Sterile Cockpit Rule

Load Manifest

QUESTIONS

1. True/False. When you fly for a commercial operation, your employer is responsible for making all go/no-go decisions.

2 Describe methods which you can use to familiarize yourself with CRM principles.

3 What are operations specifications?

4. What is the procedure called when one pilot calls out a checklist item, and the other pilot completes the task and repeats the instructions verbally?

   A. Challenge-repeat method
   B. Challenge-response method
   C. Condition-response method

5. When do FAR 135.100 and FAR 121.542 require a sterile cockpit?

   A. During takeoff and landing only
   B. During taxi, takeoff, landing, cruise, and all other flight operations conducted below 10,000 feet MSL
   C. During taxi, takeoff, landing, and all other flight operations conducted below 10,000 feet MSL, except cruise flight

6. List some of the internal and external resources which may be available to you during commercial operations.

7. True/False. The load manifest contains information regarding center of gravity limits for the flight.

8. What are some of the time-related pressures experienced by professional flight crews during preflight and taxi?

CHAPTER 14

COMMERCIAL MANEUVERS

Instrument/Commercial
Part IV, Chapter 14— Commercial Manuevers

SECTION A

MAXIMUM PERFORMANCE TAKEOFFS AND LANDINGS

During flight training, you typically learn to fly at airports with relatively long, paved runways. However, not all airports have long runways, many have runways made of dirt, grass, or sod. It is important that you learn how to perform short-field and soft-field takeoffs, climbs, approaches, and landings. These maneuvers, also referred to as maximum performance takeoffs and landings, are designed to allow you to operate safely into and out of unimproved airports.

SOFT-FIELD TAKEOFF AND CLIMB

A soft field may be defined as any runway that measurably retards acceleration during the takeoff roll. The objective of the soft-field takeoff is to transfer the weight of the airplane from the landing gear to the wings as quickly and smoothly as possible to eliminate the drag caused by surfaces such as tall grass, soft dirt, or snow. Takeoffs and climbs from soft fields require special procedures, as well as knowledge of your aircraft's performance characteristics including, best angle-of-climb speed (V_X) and best rate-of-climb speed (V_Y). When using the FAA-approved flight manual or POH performance data, keep in mind that the figures, such as takeoff distance, apply to an airplane in good operating condition, and they are valid only for the listed conditions.

You actually begin the soft-field procedure during the taxi phase. If the taxi area surface is soft, use full back pressure on the yoke to maintain full-up elevator (or stabilator) deflection with a slight amount of power to keep the airplane moving. This technique transfers some of the airplane's weight from the nosewheel to the main wheels, resulting in lower power requirements and greater ease in taxiing.

Complete the before takeoff check on a paved or firm surface area, if practical. This helps to avoid propeller damage and the possibility of the airplane becoming stuck.

- Set the flap position as recommended by the manufacturer.
- Clear the approach and departure areas and the traffic pattern prior to taxiing onto the runway. After obtaining a clearance (at a controlled airport) or self announcing your intentions (at an uncontrolled airport), taxi into position for takeoff without stopping.

Align the airplane with the center of the runway, and while still rolling, smoothly add takeoff power. Make sure you check engine instruments as the engine reaches full power. Maintain full back pressure on the yoke to raise the nosewheel from the soft surface.

When the nosewheel is clear of the runway, nosewheel steering becomes ineffective. However, due to the increasing air flow, the rudder effectiveness becomes sufficient to maintain directional control.

- As you increase speed and the elevator (or stabilator) becomes more effective, reduce back pressure slightly. Continue to use back pressure on the yoke to hold the nose up and to reduce the amount of weight on the nosewheel.

If you do not release some back pressure, while accelerating during the takeoff roll, the airplane may assume an extremely nose-high attitude, which can cause the tail skid to come in contact with the surface.

As the airplane lifts from the runway surface, reduce back pressure to achieve a level flight attitude.

As the airspeed increases, lift increases and more of the aircraft's weight is transferred to the wings. This causes the airplane to become airborne at an airspeed slower than safe climb speed. The airplane is now flying in ground effect.

Allow the airplane to accelerate in level flight, within ground effect, to V_X or V_Y (as required) before starting a climb.

On a rough surface the airplane may skip or bounce into the air before its full weight can be supported aerodynamically. Therefore, it is important to hold the pitch attitude as constant as possible (an important application of slow flight). If you permit the nose to lower after a bounce, the nosewheel may strike the ground. On the other hand, sharply increasing the pitch attitude after a bounce may cause the airplane to stall. If obstacles are in the departure path, accelerate to V_X before climbing out of ground effect. If no obstacles are in the departure path, accelerate to V_Y and then begin to climb.

Establish the climb attitude maintaining V_Y. Once a positive rate of climb is attained, retract the landing gear and raise the flaps, then trim to relieve control pressures.

SOFT-FIELD TAKEOFF AND CLIMB

To meet the PTS standards, you should be able to:

- Exhibit knowledge of the elements related to a soft-field takeoff and climb.
- Position the flight controls and flaps for the existing conditions to maximize lift as quickly as possible.
- Clear the area; taxi onto the takeoff surface at a speed consistent with safety and align the airplane without stopping while advancing the throttle smoothly to takeoff power.
- Establish and maintain the pitch attitude that will transfer the weight of the airplane from the wheels to the wings.
- Remain in ground effect after takeoff while accelerating to V_X or V_Y, as required.
- Maintain V_Y, ±5 knots.
- Retract the landing gear and flaps after a positive rate of climb is established, or as specified by the manufacturer.
- Maintain takeoff power to a safe maneuvering altitude, then set climb power.
- Maintain directional control and proper wind-drift correction throughout the takeoff and climb.
- Complete the appropriate checklist.

SOFT-FIELD APPROACH AND LANDING

The objective of a soft-field landing is to ease the weight of the airplane from the wings to the main landing gear as gently and slowly as possible, while keeping the nosewheel off the soft surface during most of the landing roll. If executed properly, this technique will prevent the nosewheel from sinking into the soft surface and reduce the possibility of an abrupt stop or possible damage to the airplane during the landing roll. You should consult your airplane's POH for the appropriate speeds and specific procedures for performing soft-field landings. In the absence of a manufacturer's recommended approach speed, use 1.3 times the stalling speed in the landing configuration (1.3 V_{SO}). In addition a gust factor adjustment to the approach speed may be applicable.

APPROACH

Ensure that the before landing checklist is completed and that the approach and landing areas are clear. Lower the landing gear and extend approximately one-third of the available flaps.

Extend the flaps to two-thirds, while progressively reducing the airspeed. Use trim to relieve control pressures.

Unless obstacles are in the approach path, maintain the same descent angle on final as you would during a normal approach. Maintain the recommended approach speed and extend the remaining flaps.

The use of flaps during a soft-field landing is normally recommended to allow the airplane to touch down at a minimum speed. However, you must also consider the runway conditions when determining whether to use full flaps. For example, in a low-wing aircraft, the flaps may suffer damage from mud, slush, or stones thrown up from the wheels.

LANDING

Hold the airplane one to two feet above the surface as long as possible to dissipate forward speed. Maintain that attitude with power and slowly continue the descent until the airplane touches down at the lowest possible airspeed with the airplane in a nose-high attitude.

When you maintain power during the landing flare and touchdown, the slipstream flow over the empennage increases the effectiveness of the elevator (or stabilator). The amount of power required during the landing flare and touchdown varies with the weight and density altitude.

Touch down in a nose-high attitude at the slowest possible airspeed. Maintain back pressure on the yoke to hold the nosewheel off the surface as long as practical. As the airspeed decreases on the roll-out, smoothly and gently lower the nosewheel to the surface.

Adding a small amount of power after touchdown will help you to ease the nosewheel down, under control.

Increase the power slightly, if necessary, to keep the aircraft moving and prevent it from stopping suddenly on the soft surface. Avoid using the brakes since braking action may cause the nosewheel to dig into the soft surface and cause damage to the landing gear. The soft surface should provide sufficient braking action to slow the aircraft.

SOFT-FIELD APPROACH AND LANDING

To meet the PTS requirements, you should be able to:

- Exhibit knowledge of the elements related to a soft-field approach and landing.
- Consider the wind conditions, landing surface, and obstructions.
- Select the most suitable touchdown point.
- Establish the recommended approach and landing configuration, and adjust power and pitch attitude as required.
- Maintain a stabilized approach and recommended approach airspeed, or in its absence not more than 1.3 V_{SO}, with gust factor applied, ±5 knots.
- Make smooth, timely, and correct control applications during the roundout (flare) and touchdown.
- Maintain crosswind correction and directional control throughout the approach and landing.
- Touch down softly with no drift, and with the airplane's longitudinal axis aligned with the landing surface.
- Maintain the proper position of the flight controls and sufficient speed to taxi on the soft surface.
- Complete the appropriate checklist.

SHORT-FIELD TAKEOFF AND CLIMB

Short-field takeoff and climb procedures may be required when the usable runway length is short, or when the runway available for takeoff is restricted by obstructions, such as trees, powerlines, or buildings, at the departure end. During short-field practice sessions, it is usually assumed that you are departing from a short runway and that you must clear an obstacle which is 50 feet in height. To accomplish successful short-field takeoffs and climbs, you must be familiar with the best angle-of-climb speed (V_X) and the best rate-of-climb speed (V_Y) for your airplane. Many manufacturers also specify a best obstacle clearance speed. You should consult your airplane's POH for the appropriate speeds and specific procedures for performing short-field takeoffs.

Complete the before takeoff check. Ensure that the runway, as well as the approach and departure paths are clear of other aircraft. After obtaining a clearance (at a controlled airport) or self announcing your intentions (at an uncontrolled airport), taxi into position at the beginning of the runway so as to allow maximum utilization of the available runway, and align the airplane on the runway centerline.

Set the flaps as recommended by the manufacturer. The appropriate flap setting varies between airplanes and can range from no flaps to approximately one-half flaps. While holding the brakes, smoothly add takeoff power, then release the brakes to begin the takeoff roll.

Holding the brakes until you achieve full power enables you to determine that the engine is functioning properly before you take off from a field where power availability is critical and distance to abort a takeoff is limited.

Allow the airplane to accelerate with its full weight on the main wheels by holding the yoke to maintain the elevator (or stabilator) in a neutral position. Smoothly and firmly apply back pressure to the yoke to lift off at the recommended airspeed. Since the airplane accelerates quickly after lift off, you may need to apply additional back pressure to establish and maintain V_X (or best obstacle clearance speed).

Avoid raising the nose prior to the recommended liftoff speed. A premature nose-high attitude increases drag and results in a longer takeoff roll. If you attempt to lift the airplane off the runway prematurely, or to climb too steeply, the airplane may settle back to the runway. In addition, a stall may result or the airplane may impact the obstacle. Deviating from the recommended climb speed, by as little as five knots, can result in a significant reduction in climb performance in some airplanes.

Once you have cleared the obstacle and reached a safe altitude, lower the nose and accelerate to V_Y. Retract the landing gear and then retract the flaps (if applicable). If no obstacles are present during training, you should maintain V_X until you are at least 50 feet above the runway surface. Trim to relieve control pressures.

To avoid a sudden loss of lift and settling of the airplane, retract the flaps in increments.

SHORT-FIELD TAKEOFF AND CLIMB

To meet the PTS requirements, you should be able to:

- Exhibit knowledge of the elements related to a short-field takeoff and climb.
- Position the flight controls and flaps for the existing conditions.
- Clear the area; taxi into the takeoff position for maximum utilization of the available takeoff area.
- Advance the throttle smoothly to takeoff power while holding brakes, or as specified by the manufacturer.
- Rotate at the recommended airspeed.
- Climb at the manufacturer's recommended airspeed and configuration, or in the absence at V_X, +5/-0 knots until the obstacle is cleared, or until the airplane is at least 50 feet (20 meters) above the surface.
- After clearing the obstacle, accelerate to V_Y, ±5 knots.
- Retract the landing gear and flaps after a positive rate of climb is established, or as specified by the manufacturer.
- Maintain takeoff power to a safe maneuvering altitude, then set climb power.
- Maintain directional control and proper wind-drift correction throughout the takeoff and climb.
- Complete the appropriate checklist.

SHORT-FIELD APPROACH AND LANDING

A short-field landing is necessary when you have a relatively short landing area or when an approach must be made over obstacles which limit the available landing area. A short-field landing consists of a steep approach over an obstacle, using power and flaps (normally full flaps). A minimum landing speed is desired with a touchdown point as close to the threshold as possible. During short-field landing practice, assume you are making the approach and landing over a 50-foot obstacle. You should consult your airplane's POH for the appropriate speeds and specific procedures for performing short-field takeoffs. In the absence of a manufacturer's recommended speed use 1.3 times the stalling speed in the landing configuration (1.3 V_{S0}). In gusty conditions, an increase in airspeed of no more than one-half the gust factor should be added.

APPROACH

Ensure that the before landing checklist is completed and that the approach and landing area is clear. Lower the landing gear and extend approximately one-third of the available flaps.

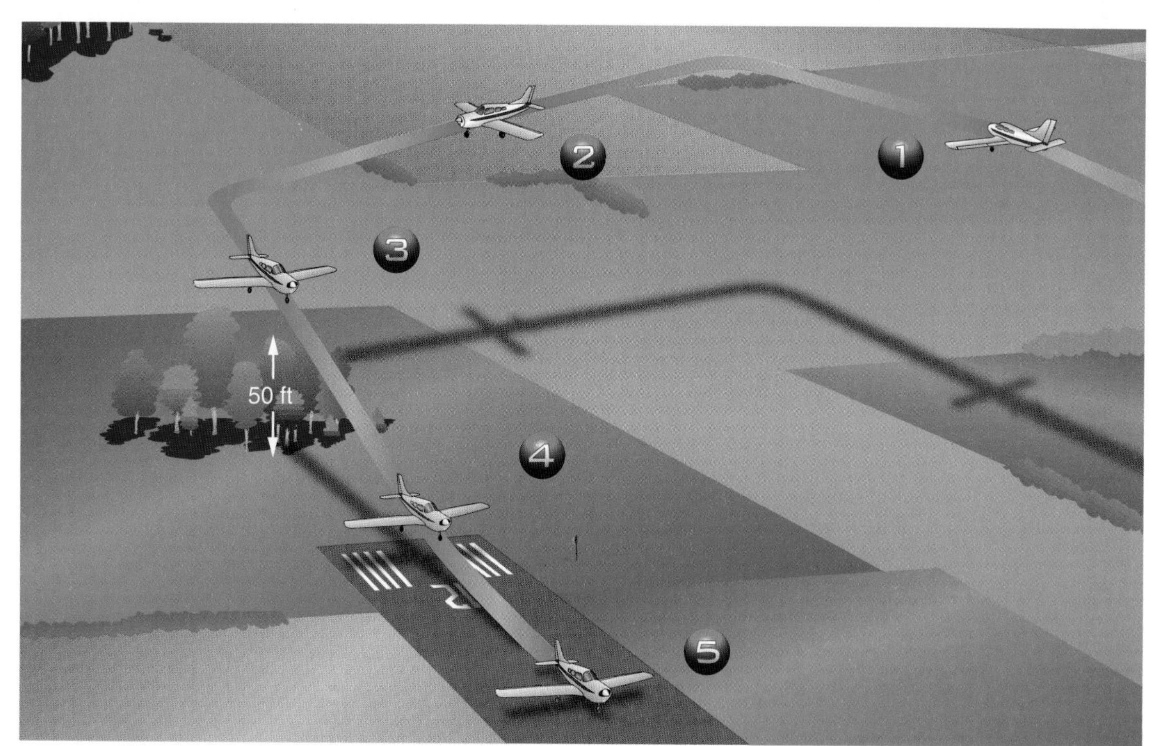

Extend the flaps to two-thirds, while progressively reducing the airspeed. Use trim to relieve control pressures.

Begin the final approach at least 500 feet higher than the touchdown area. Maintain the recommended approach speed and extend the remaining flaps.

The descent angle used for the short-field approach is steeper than that used for a normal approach. This allows you to clear an obstacle located near the approach end of the runway. Extending full flaps allows a steeper descent angle without an increase in airspeed, which results in a decrease in the distance required to bring the airplane to a full stop.

LANDING

As you begin the flare, reduce power smoothly to idle and allow the airplane to touch down in a full-stall condition. Since the short-field approach is made at a steep descent angle and close to the airplane's stalling speed, you must judge the initiation of the flare accurately to avoid flying into the ground or stalling prematurely and sinking rapidly.

Reducing power too rapidly may result in an immediate increase in the rate of descent and a hard landing. On the other hand, the airplane should touch down with little or no float. An excessive amount of airspeed may result in touchdown too far beyond the runway threshold and a roll-out which exceeds the available landing area. As your training progresses, your goal will be to touch down beyond and within 100 feet of a point specified by your instructor.

When the airplane is firmly on the runway, lower the nose, retract the flaps (if recommended) and apply the brakes, as necessary to further shorten the roll-out.

In nosewheel-type airplanes, holding the landing pitch attitude, as long as elevator authority remains effective, provides aerodynamic braking by the wings. Some manufacturers recommend retraction of flaps on the landing roll. This transfers more weight to the main gear and enhances braking.

SHORT-FIELD APPROACH AND LANDING

To meet the PTS requirements, you should be able to:

- Exhibit knowledge of the elements related to a short-field approach and landing.
- Consider the wind conditions, landing surface, and obstructions.
- Select the most suitable touchdown point.
- Establish the recommended approach and landing configuration and adjust power and pitch attitude as required.
- Maintain a stabilized approach and recommended airspeed, or in its absence not more than 1.3 V_{SO}, with gust factor applied, ±5 knots.
- Make smooth, timely, and correct control application during the roundout (flare) and touchdown.
- Remain aware of the possibility of wind shear and/or wake turbulence.
- Touch down at a specified point at or within 100 feet (30 meters) beyond the specified point, with little or no float, with no drift, and with the longitudinal axis aligned with and over the center of the landing surface.
- Maintain the crosswind correction and directional control throughout the approach and landing.
- Apply brakes, as necessary, to stop in the shortest distance consistent with safety.
- Complete the appropriate checklists.

QUESTIONS

1. True/False. The soft-field takeoff procedure begins during the taxi phase.
2. During a soft-field takeoff, liftoff normally occurs at a speed below the safe climb speed. What action should you take before starting a climb?
3. What is the correct procedure for the roll-out after a soft-field landing?

   A. Maintain power at idle and apply heavy braking.
   B. Hold forward pressure on the yoke and avoid braking.
   C. Maintain back presure on the yoke and increase power slightly, if necessary.

4. True/False. During a soft-field landing, you should lower the nosewheel to the surface as quickly as possible after touchdown.
5. Why should you hold the brakes until you achieve full power prior to beginning a short-field takeoff?
6. During short-field takeoff practice sessions, it is assumed that you must clear an obstacle which is how many feet high?
7. True/False. While executing a short-field landing, you should reduce power to idle in the flare and allow the airplane to touch down in a full stall condition.
8. Is the descent angle for a short-field approach steeper, shallower, or the same as that flown for a normal approach and landing?

SECTION B
STEEP TURNS

Steep turns are level, high-performance turning maneuvers normally performed as a series of 360° turns in opposite directions with a bank angle of approximately 50°, ±5°. The objective of the maneuver is to help develop the ability to accurately control an aircraft near its maximum performance limits. It also increases your knowledge of the associated performance factors, including load factor, angle-of-bank limitations, effect on stall speed, power required, and the overbanking tendency.

The actual turning performance of an airplane is limited by the amount of power the engine is developing, load limit (structural strength), and aerodynamic design. As you increase the bank angle, you eventually approach maximum performance or the load limit. In most light airplanes, the maximum bank angle you can maintain with full power is 50° to 60°. If you exceed the maximum performance limit while maintaining your airspeed at or below the airplane's design maneuvering speed (V_A), the airplane will either stall or will lose altitude. With airspeed above V_A, it is possible to exceed the load limit.

As is the case with all training maneuvers, you must be aware of other traffic in the area. Before you start the maneuver, make clearing turns to ensure the practice area is free of conflicting traffic. Start at an altitude which will allow you to complete the maneuver no lower than 1,500 feet AGL.

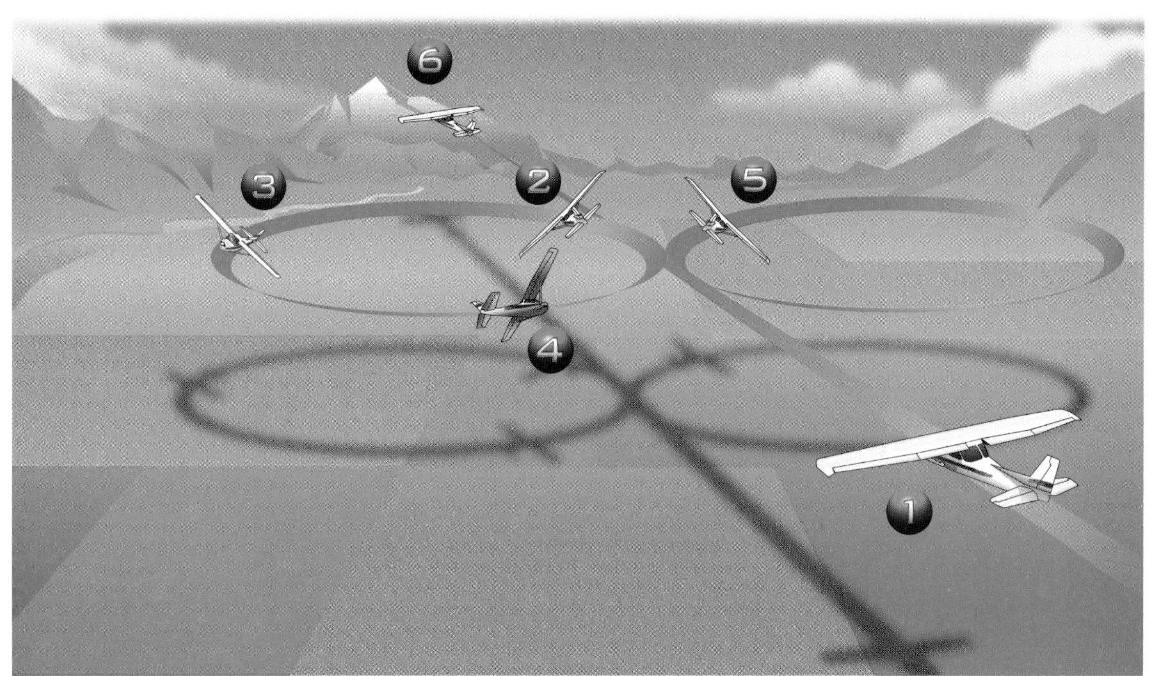

 Upon completion of your clearing turns, select a reference point on the horizon and note your heading and altitude.

 Roll into a 50° angle-of-bank turn at or below V_A. During roll-in, smoothly add power and slowly increase back pressure on the yoke to maintain altitude. Maintain coordinated flight and trim to relieve control pressures.

As you enter the turn, establish the bank at a moderate rate. If you roll the airplane too rapidly, you may have difficulty establishing the pitch attitude necessary to maintain altitude. Do not apply too much back pressure while initially entering the turn or you will gain altitude. However, as you become established in the turn, greater back pressure will be needed to maintain altitude.

 Maintain your angle of bank and altitude. Confirm your attitude by referring to both the natural horizon and attitude indicator. Use your altimeter and vertical speed indicator to determine if changes in pitch are required.

If you are losing altitude in the turn, slightly decrease the angle of bank first, then increase back pressure on the yoke to raise the nose. Once you regain your desired altitude, roll back to the desired angle of bank.

 Anticipate the change in direction of the turn by leading the roll-out heading by one-half the bank angle, approximately 25°. Roll out on the entry heading and briskly roll into a 50° banked turn in the opposite direction.

During steep turns, you will encounter an overbanking tendency which is less apparent in right turns than it is in left turns. This is because torque and P-factor tend to roll the aircraft to the left and work against the overbanking tendency during a right turn. Generally, you will need more rudder and aileron pressure during the roll-out than you needed during the roll-in. This is because the control pressures exerted during the roll-out must overcome the airplane's overbanking tendency.

 After the initial roll-in to the turn, confirm your attitude by referring to both the natural horizon and attitude indicator. Use your altimeter and vertical speed indicator to determine if changes in pitch are necessary. Anticipate the roll-out by leading the rollout heading approximately 25°.

Roll out on the entry heading and altitude. Decrease back pressure on the yoke, and reduce power to maintain altitude and airspeed. Trim to relieve control pressures.

STEEP TURNS

To meet the PTS requirements, you should be able to:

- Exhibit knowledge of the elements related to steep turns.
- Select an altitude that will allow the maneuver to be completed no lower than 1,500 feet AGL (460 meters) or the manufacturer's recommended altitude, whichever is higher.
- Establish an airspeed as recommended by the aircraft manufacturer or as specified by the examiner not to exceed V_A.
- Roll into a coordinated 360° turn; maintain a 50° angle of bank, ±5°, followed immediately by a 360° turn in the opposite direction.
- Divide your attention between airplane control and orientation.
- Roll out on the entry heading ±10°.
- Maintain entry altitude, ±100 feet (30 meters), and airspeed, ±10 knots.

QUESTIONS

1. What is the first thing you should do if you begin to lose altitude during a steep turn?

2. Why is overbanking tendency less apparent in right turns than it is in left turns?

   A. Torque and P-factor tend to roll the aircraft to the right and work against the overbanking tendency during a left turn.
   B. There is no difference between left and right turns; overbanking occurs because the angle of bank has exceeded the limits of the airplane.
   C. Torque and P-factor tend to roll the aircraft to the left and work against the overbanking tendency during a right turn.

3. True/False. The entry speed for a steep turn should be above V_A.

4. How many degrees should you lead your desired heading when you initiate the recovery from a steep turn?

SECTION C
CHANDELLES

A chandelle can be described as a maximum performance 180° climbing turn. It involves continual changes in pitch, bank, airspeed, and control pressures. During the maneuver, the airspeed gradually decreases from the entry speed to a few knots above stall speed at the completion of the 180° turn. Since you use full power (in airplanes with a fixed-pitch propeller) throughout the chandelle, you must control airspeed by adjusting the pitch attitude of the aircraft. Because of this, maintaining the proper pitch attitude is a key element of this maneuver. Due to variables, such as atmospheric density and airplane performance, altitude gain is not a criterion for successful completion of a chandelle. However, the aircraft should gain as much altitude as possible for the given bank angle and power setting without stalling. The objective of the maneuver is to help you develop good coordination habits and refine the use of aircraft controls at varying airspeeds and flight attitudes.

As is the case with all training maneuvers, you must be aware of other traffic in the area. Before you start the maneuver, make clearing turns to ensure the practice area is free of conflicting traffic. Start at an altitude recommended by the aircraft manufacturer or 1,500 feet AGL, whichever is higher.

Before entering the maneuver, configure the airplane in straight-and-level flight with the landing gear and flaps up, using the entry airspeed recommended in the pilot's operating handbook or maneuvering speed (V_A), whichever is slower.

Although the prevailing wind has little or no effect on the chandelle, you will find it is best to begin the maneuver by turning into the wind. This helps you remain within the training area. Select a prominent feature on the ground to help maintain orientation.

Establish a coordinated turn not to exceed 30° of bank. Then, simultaneously apply back elevator pressure to begin a climb and smoothly apply full power. If your airplane has a constant speed propeller, increase the RPM to the climb or takeoff setting, then advance the throttle to the climb setting.

Throughout the first 90° of the turn, the bank angle of 30° should remain the same. Pitch attitude on the other hand should gradually increase, reaching its maximum at the 90° point.

Begin a gradual reduction of bank angle, while maintaining a constant pitch attitude.

During the second 90° of turn, time the roll-out rate so you reach wings level at the 180° point. As airspeed decreases you will normally need more elevator back pressure to maintain a constant pitch attitude. You will also notice that the left-turning tendency caused by P-factor and propeller slipstream is more prevalent, and you will need to apply right rudder pressure to coordinate both right and left turns.

Reduce the pitch attitude to resume a level flight attitude, allowing your airplane to accelerate while maintaining a constant altitude.

In a chandelle to the right, the aileron on the right wing is lowered slightly during the roll-out. This causes more drag on the right wing and tends to make the airplane yaw slightly to the right. At the same time, the left-turning tendency is pulling the nose to the left. As a result, aileron drag and the left-turning tendency counteract each other, and very little left rudder pressure is required. Actually, releasing some of the right rudder pressure, which has been used to correct for left-turning tendencies, will normally have the same effect as use of left rudder pressure. In contrast, when you roll out from a chandelle to the left, two turning forces are pulling the nose of the airplane to the left. In this case, you need a significant amount of right rudder pressure.

CHANDELLES

To meet the PTS standards, you should be able to:

- Exhibit knowledge of the elements related to performance factors associated with chandelles.
- Select an altitude that will allow the maneuver to be performed no lower than 1,500 feet AGL (460 meters) or the manufacturer's recommended altitude, whichever is higher.
- Establish the entry configuration at an airspeed no greater than the maximum entry speed recommended by the manufacturer, not to exceed V_A.
- Establish appropriate bank angle, not to exceed 30°.
- Simultaneously apply specified power and pitch to maintain a smooth, coordinated climbing turn with constant bank to the 90° point.
- Begin a coordinated constant rate of rollout from the 90° point to the 180° point, maintaining specified power and a constant pitch attitude that will result in a rollout within ±10° of the desired heading and airspeed within +5 knots of power-on stall speed.
- Reduce pitch attitude to resume straight-and-level flight at the final altitude attained, ±50 feet (20 meters).

QUESTIONS

1. What is the maximum wind limit for practicing chandelles?

   A. 5 to 10 knots
   B. 15 to 20 knot surface wind
   C. Wind is not a factor when practicing chandelles

2. True/False. During the second 90° of the turn, you should always maintain the highest angle of bank and continue to increase the pitch angle.

3. What factor counteracts left-turning tendency in a chandelle to the right?

4. What maximum angle of bank should you observe during a chandelle?

   A. 10°
   B. 20°
   C. 30°

SECTION D

LAZY EIGHTS

The lazy eight is basically two 180° turns in opposite directions, with each turn including a climb and a descent. It is called a lazy eight because the longitudinal axis of the aircraft appears to scribe a flight pattern about the horizon that resembles a figure eight lying on its side. Throughout the maneuver, airspeed, altitude, bank angle, and pitch attitude, as well as control pressures, are constantly changing. Because of these constant changes, you cannot fly the lazy eight mechanically or automatically. The lazy eight requires a high degree of piloting skill and a sound understanding of the associated performance factors. The objective of this maneuver is to develop and improve your coordination, orientation, planning, division of attention, and ability to maintain precise aircraft control. A good way to visualize the lazy eight is to break each 180° turn into segments.

As is the case with all training maneuvers, you must be aware of other traffic in the area. Before you start the maneuver, make clearing turns to ensure the practice area is free of conflicting traffic. Start at an altitude that will allow you to perform the maneuver no lower than 1,500 feet AGL or an altitude recommended by the aircraft manufacturer, whichever is higher. Since you will be changing heading and altitude continually, be particularly careful to maintain vigilance throughout the maneuver. While scanning the area, look for visual reference points that you can use for orientation. You should also try to determine the direction of the wind and an entry reference point on the horizon, which will allow you to make your turns into the wind. Additional reference points at 45°, 90°, and 135° are also useful.

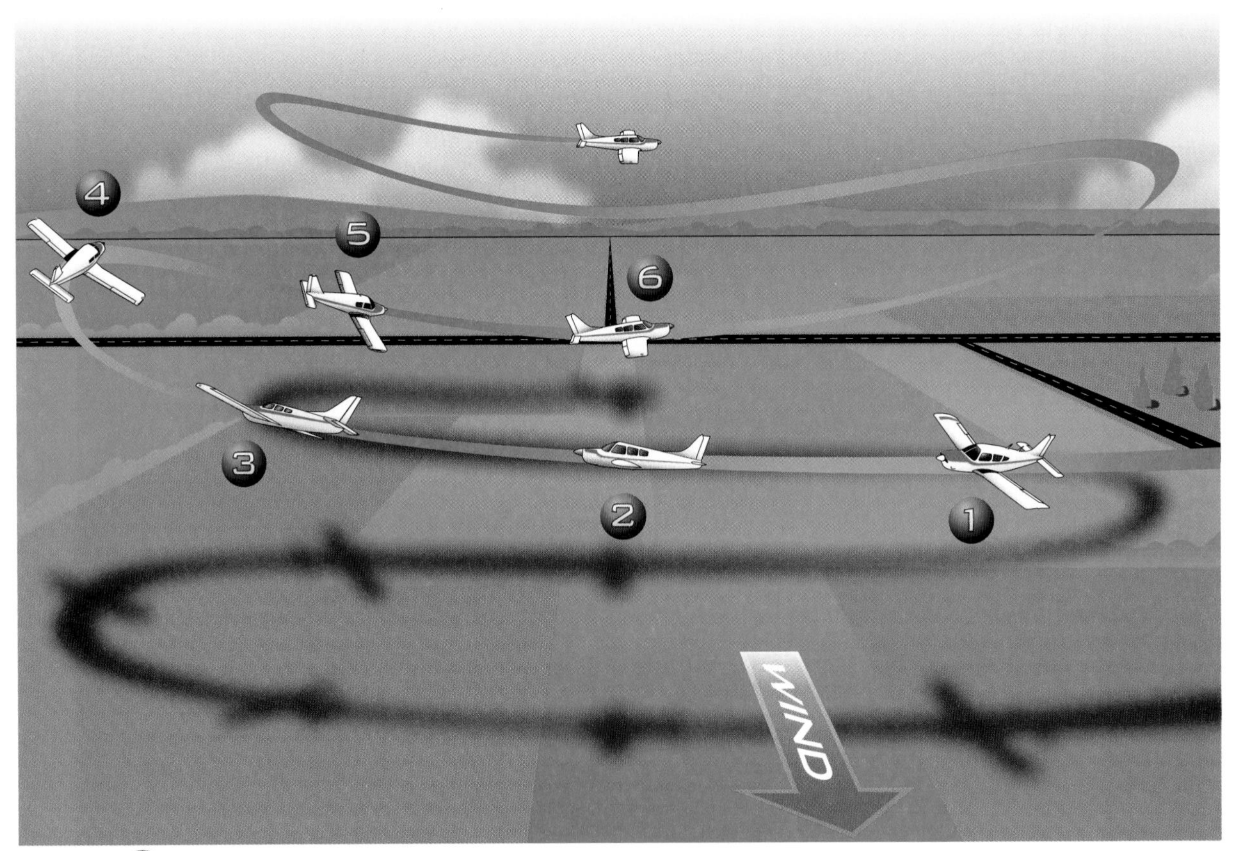

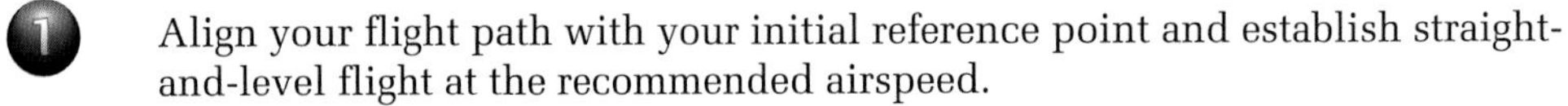

1. Align your flight path with your initial reference point and establish straight-and-level flight at the recommended airspeed.

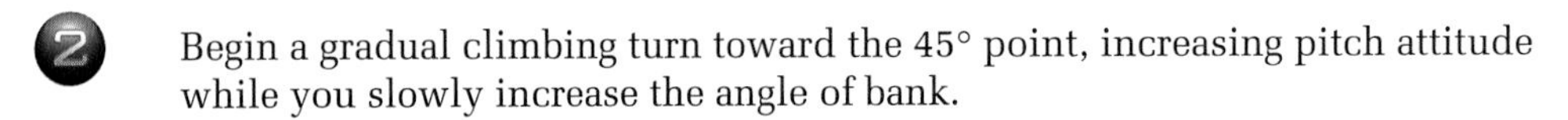

2. Begin a gradual climbing turn toward the 45° point, increasing pitch attitude while you slowly increase the angle of bank.

It is important to remember that as airspeed decreases, your rate of turn at a given angle of bank increases. If you allow the rate of turn to become too rapid, you will reach the 45° point of the turn before the maximum pitch has been attained.

3. As you pass through the 45° point of the turn, your pitch attitude should be at its maximum and your bank angle should be about 15°. From 45° to the 90° point you should begin to decrease your pitch attitude to the horizon and continue to increase the angle of bank.

A slight amount of opposite aileron pressure may be required to prevent the angle of bank from progressing beyond the maximum. To control yaw associated with the left-turning tendencies, you will need to apply right rudder pressure. More right rudder pressure is necessary during a climbing turn to the right than to the left. This is required to prevent yaw from decreasing the rate of turn. In a left turn, torque contributes to the turn so less right rudder is necessary.

4. At 90°, you should be in a level flight attitude, at the maximum angle of bank (approximately 30°), and your airspeed should be about 5 to 10 knots above the stall speed. Then, slowly begin to roll out of the 30° bank and gradually lower the nose for the descending turn as you allow the airspeed to increase.

Since the angle of bank is decreasing during the rollout, the vertical component of lift will increase. As the wings return to a level attitude, lift will continue to increase, so you will need to reduce the elevator back pressure to avoid leveling off too soon.

When you reach the 135° reference point, the nose of the airplane should be at its lowest pitch attitude. Continue a gradual rollout and allow the airspeed to continue to increase, so that you are in a level flight attitude at your entry altitude and airspeed as you reach 180° of turn.

As the airspeed increases you can gradually relax rudder and aileron pressure.

At this point you should immediately begin a climbing turn in the opposite direction toward the selected reference point to complete the second half of the eight in the same manner as the first half.

One of the key factors in making symmetrical turns is proper airspeed control. Since the power is set before you begin the maneuver, you control airspeed by varying the pitch attitude. During the first 90° of turn, which is the climbing segment, the airspeed should decrease from the entry speed to slightly above the stall speed at the 90° point. This will occur only if you constantly adjust your pitch attitude throughout the maneuver. You should pass through a level-flight pitch attitude at the 90° point, then gradually establish a nose-low pitch attitude that allows your aircraft to accelerate to the entry speed after 180° of turn.

LAZY EIGHTS

To meet the PTS standards, you should be able to:

- Exhibit knowledge of the performance factors associated with lazy eights.
- Select an altitude that will allow the maneuver to be performed no lower than 1,500 feet AGL (460 meters) or the manufacturer's recommended altitude, whichever is higher.
- Select a prominent 90° landmark in the distance.
- Establish the recommended entry power and airspeed.
- Plan and remain oriented while maneuvering the airplane with positive, accurate control, and demonstrate mastery of the airplane.
- Achieve a constant change of pitch, bank, and turn rate.
- Maintain an altitude and airspeed consistent at the 90° points, ±100 feet (30 meters) and ±10 knots, respectively.
- Using proper power settings, attain the starting altitude and airspeed at the completion of the maneuver, ±100 feet (30 meters) and ±10 knots, respectively.
- Achieve a heading tolerance of ±10° at each 180° point.
- Continue the maneuver through at least two 180° circuits and resume straight-and-level flight.

QUESTIONS

1. What is the possible cause for reaching the 45° point before the maximum pitch angle has been attained?

2. At the 90° point your airspeed should be

   A. At or above stall speed.
   B. 5 knots to 10 knots above stall speed.
   C. 10 knots to 20 knots above stall speed

3. True/False. From the 90° reference point to the 135° reference point, you should increase elevator back pressure until level flight is achieved.

4. What is the key to keeping both sides of a lazy eight symmetrical?

SECTION E
EIGHTS-ON-PYLONS

Eights-on-pylons involve flying a figure eight around two points, or pylons, on the ground. In this maneuver, you fly the airplane at an altitude and airspeed that allows you to hold a line-of-sight reference point (usually near the wingtip) on the pylon. This reference point should appear to pivot about the pylon. A complete maneuver consists of a turn in one direction around the first pylon, followed by a turn in the opposite direction around the second pylon. The objective of eights-on-pylons is to refine your ability to control the airplane at traffic pattern altitude over a varied ground track while dividing your attention between instrument indications and visual cues outside the aircraft. To accomplish this you should choose a reference point on, or near, the wing tip so your line of sight through the reference point is parallel to the lateral axis of the airplane. It is important to remember that the reference will vary considerably on different airplanes. It may be above the wingtip on a low-wing aircraft, below the wingtip on a high-wing aircraft, and ahead of the wingtip on a tapered-wing airplane.

Since you perform this maneuver at a relatively low altitude, select pylons that are in an open area and are not near hills or obstructions. Obstruction-induced turbulence, as well as updrafts and downdrafts caused by uneven terrain, increase the difficulty of the maneuver. Select pylons with approximately the same elevation to avoid the added burden of adjusting altitude for variations in terrain. The pylons also should be in a line which is perpendicular to the wind, and they should be spaced to provide three to five seconds of straight-and-level flight between the turns. Carefully select pylons that will be visible, clear of obstructions, and properly oriented in relation to the prevailing wind. The pylons you select should be in a location which will allow for a safe emergency landing.

As is the case with all training maneuvers, you must be aware of other traffic in the area. Before you start the maneuver, make clearing turns to ensure the practice area is free of conflicting traffic. Since you are continually changing direction you should keep a lookout for other aircraft throughout the maneuver.

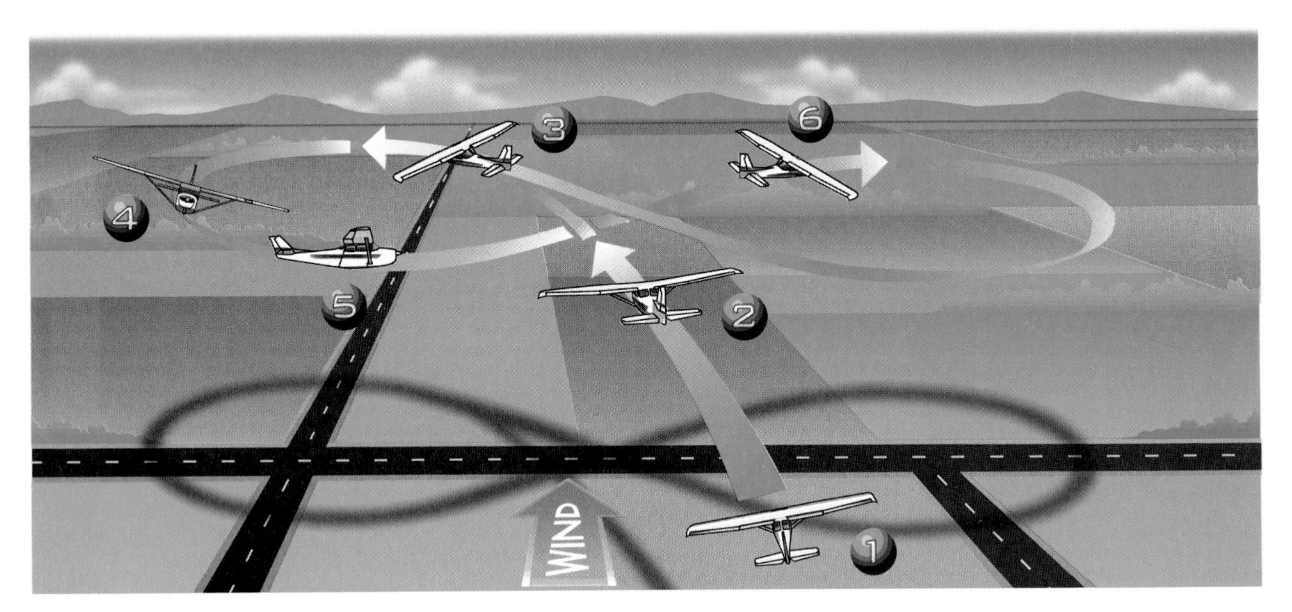

 Adjust your power to recommended entry airspeed. As you approach the pylons, the entry altitude you select should be close to the estimated pivotal altitude.

You can estimate your pivotal altitude by using a simple formula: TAS (in m.p.h.)^2 ÷ 15 = Pivotal Altitude. As an example, if your true airspeed is 115 m.p.h. (100 knots), your estimated pivotal altitude is 882 feet $(115)^2 = 13225 \div 15 = 882$. Since you will seldom know the exact TAS, groundspeed, and elevation, you will need to determine your actual pivotal altitude by experimenting while flying the maneuver.

 Enter a diagonal between the pylons from the downwind side of the pylons.

With a downwind entry you will have the highest groundspeed and the highest pivotal altitude. As groundspeed decreases so does pivotal altitude.

 Maintain straight-and-level flight until you are approximately abeam the first pylon, then roll into a 30° to 40° angle of bank.

As you proceed around the pylon, if your line-of-sight reference point moves forward of the pylons, your pivotal altitude is too low and you need to climb. Likewise, if your reference point moves aft of the pylon, you should decrease your pivotal altitude. Normally, when flying into the wind, your groundspeed decreases; therefore, you need to descend. Remember, pivotal altitude increases as groundspeed increases and decreases as groundspeed decreases.

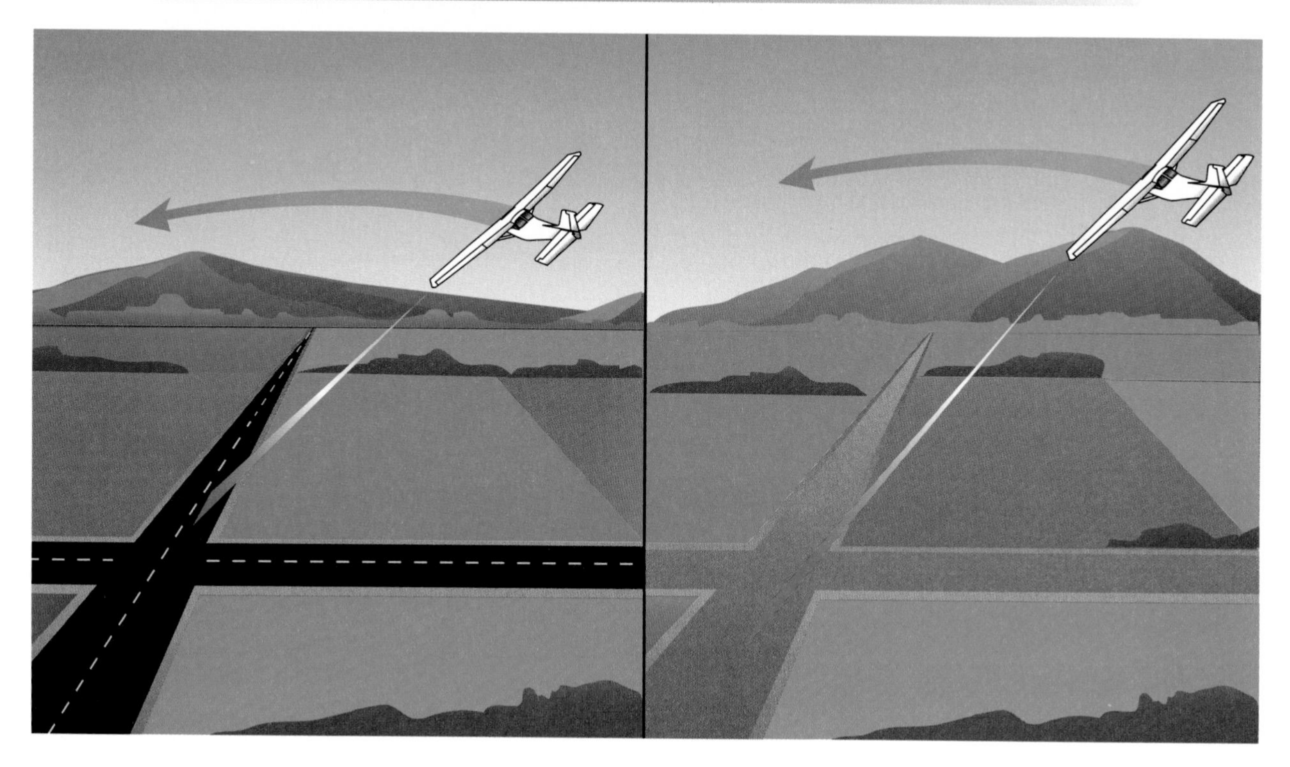

Gradually decrease your pivotal altitude and slightly reduce your angle of bank as you turn directly into the wind.

A descent has a two-fold effect on pivotal altitude. First, it provides the correction needed to hold the pylon, and second, the descent increases groundspeed. The reverse is true for climbs. It is important that you use altitude changes, rather than rudder pressure, to hold the reference point on the pylon, and you should maintain coordinated flight throughout the maneuver.

Begin the rollout to straight-and-level flight as you complete the first turn. Maintain straight-and-level flight for 3 to 5 seconds and crab into the wind, as necessary, to correct for wind draft.

Without proper wind drift correction you may fly too close to the pylon to enter the second turn and maintain a bank angle between 30° to 40°.

Initiate a turn in the opposite direction when the pylon is aligned with the wing reference point.

Repeat the same steps in the opposite direction around the second pylon. Keep in mind that your line-of-sight reference point may look different on the opposite wing.

Eights-on-Pylons

To meet the PTS standards, you should be able to:

- Exhibit knowledge of the elements related to eights-on-pylons including the relationship of groundspeed change to performance of the maneuver.
- Determine the approximate pivotal altitude.
- Select suitable pylons, considering emergency landing areas, that will permit 3 to 5 seconds of straight-and-level flight between them.
- Attain proper configuration and airspeed prior to entry.
- Apply the necessary corrections so that the line-of-sight reference remains on the pylon with minimum longitudinal movement.
- Demonstrate proper orientation, division of attention, and planning.
- Apply the necessary wind drift correction to track properly between the pylons.
- Hold pylons using pivotal altitude, avoiding slips and skids.

QUESTIONS

1. What is your pivotal altitude if your true airspeed is 120 mph?
2. Why should you begin eights-on-pylons with a downwind entry?
3. True/False. You should use altitude changes, rather than rudder pressure, to hold the reference point on the pylon.
4. You should select pylons that provide a suitable emergency landing area and allow for

   A. an immediate turn for the next pylon.
   B. 1 or 2 seconds of straight-and-level flight.
   C. 3 to 5 seconds of straight-and-level flight between them.

APPENDIX A
ANSWERS

NOTE: THERE ARE NO QUESTIONS FOR CHAPTER 1 — BUILDING PROFESSIONAL EXPERIENCE.

CHAPTER 2

SECTION A

1. B
2. The gyroscopic flight instruments are the attitude indicator, heading indicator, and turn coordinator. The two principles are rigidity in space and precession.
3. Attitude indicator
4. True
5. The turn coordinator is electrically powered, while the attitude indicator and heading indicator are vacuum powered. If either the electrical or vacuum system fails, you still have bank information.
6. 14°
7. B
8. B
9. C
10. A
11. A
12. 307°
13. Airspeed indicator, altimeter, and vertical speed indicator.
14. B
15. A
16. D
17. Indicated altitude is higher than true altitude when the temperature is lower than standard, resulting in less terrain and obstruction clearance during cold weather operations.

SECTION B

1. C
2. E
3. A
4. F
5. B
6. D
7. With fixation, the scan is completely stopped as you stare at a single instrument. With emphasis, the scan continues, but one instrument is emphasized at the expense of others.
8. Airspeed, air density, and aircraft weight
9. A, B, C, G
10. B, E, F
11. A, D, H
12. B, D, H
13. A, C, E, F, G
14. B
15. A
16. B
17. B
18. The primary pitch instrument is the attitude indicator. The primary bank instrument is the heading indicator.
19. The primary bank instrument is the attitude indicator. The primary pitch instrument is the altimeter.
20. Reduce power and level the wings using the attitude indicator or the turn coordinator. Gently raise the nose to a level pitch attitude using the attitude indicator, or by stopping the movement of the altimeter and airspeed needles.

SECTION C

1. B
2. C
3. A
4. An HSI always provides proper sensing when tuned to a VOR because the CDI is mounted on a compass card which automatically rotates to the correct heading.
5. False
6. 6 minutes; 10 nautical miles
7. 2/3 nautical mile to the right of the airway (4,000 feet)
8. The 45° index marks on the heading indicator make it easy to see when you have established the correct intercept heading; similar index marks on the ADF show when you have intercepted the bearing.
9. C
10. A
11. C
12. False
13. 130 nautical miles; 18,000 feet to 45,000 feet
14. True
15. 4°
16. False
17. Inertial navigation system (INS)
18. True
19. C
20. False

CHAPTER 3

SECTION A

1. D
2. A

3. E
4. C
5. B
6. False
7. C
8. C
9. G
10. F
11. A
12. Green, Red
13. 3,000 feet
14. It may indicate that ground visibility is less than three miles and/or the ceiling is less than 1,000 feet.
15. B
16. The base of Class A airspace is 18,000 feet MSL and extends up to and including FL600.
17. F
18. E
19. A
20. D
21. B
22. C
23. D
24. 200 knots
25. False
26. 6:00 AM — 11:00 PM (0600 —2300) local time
27. MALSR (medium intensity approach light system with runway alignment indicator lights)
28. 234 feet
29. Denver Center
30. 111.5, I-LBF
31. B
32. NOTAM(D) information is disseminated for all navigational facilities which are part of the U.S. airspace system, all public use airports, seaplane bases, and heliports listed in the A/FD. NOTAM(L)s are distributed locally only and are not attached to the hourly weather reports. Information such as taxiway closures, personnel and equipment near or crossing runways, and airport rotating beacon and lighting aid outages is included in NOTAM(L)s. FDC NOTAMs, issued by the National Flight Data Center, contain regulatory information such as temporary flight restrictions or amendments to instrument approach procedures and other current aeronautical charts.
33. *Advisory Circular Checklist* (AC-00-2)

SECTION B

1. A
2. The separation of known IFR traffic and the issuance of safety alerts
3. C
4. 30
5. False
6. B
7. C
8. When it is apparent that your aircraft is in unsafe proximity to terrain, obstructions, or other aircraft
9. Clearance delivery
10. F
11. C
12. G
13. D
14. A
15. E
16. B
17. C
18. C
19. False
20. Local airport advisory

SECTION C

1. Controlled
2. A
3. C
4. True
5. True
6. C
7. A
8. A
9. D
10. E
11. H
12. B
13. G
14. C
15. F
16. False
17. False
18. B
19. A
20. B
21. The nearest flight service station
22. A
23. True
24. "*. . . cleared to the Cheyenne Airport as filed. Climb and maintain 8,000. Departure Control frequency is 120.9. Squawk 5417.*"
25.

37R C DAL A DR BUJⓉ ↓M120 RP150
DP BUJⓉ HDG 210 RV RWY31R ILS
FAP CRS LDG RWY31R

CHAPTER 4

SECTION A

1. Runway 27
2. B
3. 275°

4. B
5. 124.35 MHz
6. B
7. 8,500 feet MSL
8. B
9. C
10. C

SECTION B

1. False
2. 1/4 statute mile
3. True
4. B
5. 550 feet per minute
6. 030°
7. At or below 9,000 feet MSL
8. C
9. At POAKE Intersection
10. False
11. NOS prints textual departure procedures in the front of each *Terminal Procedures Publication.*
12. True
13. False
14. C
15. You are responsible for your own terrain/obstruction clearance until you reach the minimum altitude for IFR operations in the area.

CHAPTER 5

SECTION A

1. False
2. D
3. A
4. E
5. B
6. C
7. B
8. True
9. C
10. C
11. False
12. False
13. B
14. C
15. False
16. C

SECTION B

1. A
2. Return to the previous frequency and ask for an alternate frequency.
3. No
4. False
5. True
6. If radar contact has been lost or radar service terminated, the FARs require you to provide ATC with position reports over compulsory reporting points. In addition, you should report the final approach fix inbound on a nonprecision approach and when you leave the outer marker inbound on a precision approach. A report also is necessary when it becomes apparent that an estimated time, that you previously submitted to ATC, will be in error in excess of 3 minutes.
7. A
8. *"Minneapolis Center, Aztec 3490R, Dickinson, 30, 9,000, ULLIN 45, Bismarck next."*
9. 17,000 feet MSL
10. B
11. False

SECTION C

1. True
2. C
3. 200 KIAS
4. C
5. A
6. B
7. B
8. C
9. True

CHAPTER 6

SECTION A

1. At WISKE Intersection.
2. C
3. 7000
4. CKB.WISKE1
5. 6,000 feet MSL
6. True
7. B
8. NOS arrival charts are located in the beginning of each booklet. Jeppesen charts are found along with the other airport charts in the appropriate binder.

SECTION B

1. You should obtain weather information as early as practical, before reaching the initial approach fix for the destination.
2. A descend via clearance authorizes you to follow the altitudes published on the STAR procedure. Descent is at your discretion; however you must adhere to the minimum crossing altitudes and airspeed restrictions printed on the chart.
3. A
4. False
5. 24 miles ahead of FREDY Intersection

CHAPTER 7

SECTION A

1. Initial, intermediate, final, and missed
2. True
3. C
4. 120.15
5. B
6. 130°, 111.5 MHz
7. True
8. B
9. B
10. False
11. B
12. A
13. A
14. 840 feet MSL
15. 2 minutes and 44 seconds
16. 1,020 feet MSL
17. C
18. False
19. C
20. True
21. The missed approach icons represent the initial pilot actions in the event of a missed approach. They provide symbolic information about the initial up and out maneuvers only. You must always refer to the missed approach instructions in the heading section and the plan view graphic for complete information about the missed approach procedure.
22. B

SECTION B

1. Procedure title, communication frequencies, primary navaid frequency, inbound course, altitude at which you will cross the FAF or FAP (nonprecision approach) or glide slope intercept altitude (precision approach), decision height (precision approach) or minimum descent altitude (nonprecision approach), airport elevation, touchdown zone elevation, missed approach instructions, special notes/procedures
2. True
3. Straight-in landing minimums normally are used when the final approach course is positioned within 30° of the runway and a minimum of maneuvering is required to align the airplane with the runway. If the final approach course is not properly aligned, or if it is desirable to land on a different runway, a circling approach may be executed and circle-to-land minimums apply. In contrast to a straight-in landing, the controller terminology, *"cleared for straight-in approach . . ."* means that you should not perform any published procedure to reverse your course, but does not reference landing minimums. A straight-in approach may be initiated from a fix closely aligned with the final approach course, may commence from the completion of a DME arc, or you may receive vectors to the final approach course.
4. 139°, 2,000 feet MSL
5. B
6. If your airplane is only equipped with single receiving equipment, a lead radial provides a cue for you to change the receiver to the localizer or other facility providing course guidance. Lead radials also ensure you are within the clearance coverage area of localizer facilities before changing frequency or accepting on-course indication.
7. False
8. C
9. MDA(H) — 880′(560′), visibility — 1 statute mile
10. 200 knots IAS
11. A
12. No
13. False
14. C
15. ATC can initiate a visual approach if the reported ceiling is at least 1,000 feet AGL and visibility is at least 3 statute miles. ATC can issue a clearance for a contact approach only upon your request when the reported ground visibility at the airport is 1 statute mile or greater. During a visual approach, you must have the airport or preceding aircraft in sight. If you report the preceding aircraft in sight, you are responsible for maintaining separation from that aircraft and avoiding the associated wake turbulence. When executing a contact approach you are responsible for your own obstruction clearance, but ATC provides separation from other IFR or special VFR traffic. Separation from normal VFR traffic is not provided.

CHAPTER 8

SECTION A

1. C
2. False
3. A
4. You should request a clearance for another approach to the same airport or a clearance to your alternate.
5. A
6. B
7. 2,600 feet MSL
8. False
9. C
10. B
11. C
12. B
13. A
14. C
15. A
16. False

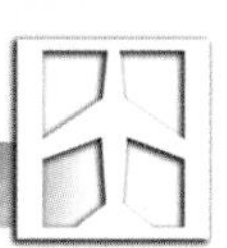

SECTION B

1. C
2. C
3. A. 1
   B. 5, 10
   C. 4, 7
   D. 2, 8
   E. 3, 6, 9
4. A. 3, 4
   B. 5
   C. 1
   D. 2, 6
   E. 2, 6
5. 710 feet to the right of the localizer centerline and 140 feet above the glide slope
6. Power
7. You can identify BEHAN using the 037° radial from PWA and the localizer, or the IOKC 11.5 DME fix.
8. A
9. You may descend to the LOC (GS out) MDA(H) of 1,620′(338′). You can identify the missed approach point using the 1.8 DME fix from IOKC or by timing from TULOO to the MAP.
10. You can identify JESKE using the 12 DME fix on the 263° radial from IRW, the 216° radial from PWA and the 263° radial from IRW, or the 156° radial from IFI and the 263° radial from IRW. You can expect to use either a teardrop or parallel entry into the published holding pattern.
11. B
12. You can identify the FAF using the 085° bearing to the LOM, Gally.
13. Left
14. False
15. An LDA has a course width of between 3° and 6°. The width of an SDF course is either 6° or 12°.

SECTION C

1. C
2. True
3. False
4. You can refer to the supplements section of the Airplane Flight Manual (AFM).
5. False
6. B
7. B
8. False
9. C
10. A
11. 2 nautical miles from BERKI inbound on the approach
12. 2.63°
13. False
14. Non-RNAV VOR indicates angular deviation from course, with each dot representing a 2° deviation. GPS indicates absolute deviation from course, with each dot representing 1 n.m. deviation during enroute operations. At 30 nautical miles, both VOR and GPS have a sensitivity of 5 nautical miles, full scale. The VOR becomes proportionately more sensitive as you get closer to the station while the GPS sensitivity increases to 1 nautical mile, then 0.3 nautical mile as you arm an approach, then approach within 2 nautical miles of the final approach fix.
15. A

CHAPTER 9

SECTION A

1. The troposphere. The change of temperature lapse rate at the tropopause acts like a lid to trap water vapor and associated weather.
2. B
3. C
4. C
5. E
6. A
7. B
8. I
9. G
10. D
11. C
12. H
13. A, C, E
14. False
15. False
16. B
17. C
18. C
19. C

SECTION B

1. Unstable air, some type of lifting action, and a relatively high moisture content
2. A
3. B
4. C
5. A
6. B
7. True
8. B
9. Slow to the airspeed recommended for rough air, such as maneuvering or penetration speed, try to maintain a level flight attitude, and accept variations in airspeed and altitude.
10. B
11. After
12. False
13. C
14. B

15. A
16. In humid climates where the bases of convective clouds tend to be low, microbursts are associated with a visible rainshaft. In drier climates, the higher thunderstorm cloud bases result in the evaporation of the rainshaft. The only visible indications under these conditions may be virga at the cloud base and a dust ring on the ground.
17. B
18. B
19. A
20. C
21. True
22. False
23. B
24. B

SECTION C

1. C
2. B
3. A
4. The actual temperature is 6.7°C and the dewpoint is 6.1°C
5. 1008.6 (hPa)
6. True
7. TEMPO indicates a temporary forecast when wind, visibility, weather, or sky conditions are expected to last less than an hour.
8. A
9. The forecast winds will be 200° true at 8 knots with visibility greater than 6 statute miles, cloud bases are 1,200 feet broken, 3,000 feet overcast. Conditions are forecast to temporarily change to 2 statute miles visibility in light rain and overcast skies with a ceiling 800 feet AGL.
10. C
11. False
12. A
13. True
14. 1,500 feet to 2,500 feet overcast with occasional ceilings of overcast conditions below 2,000 feet, visibility is expected to be below 5 miles in light rain and light snow.
15. A
16. Winds are forecast to be from 250° true at 110 knots; temperature at that altitude is forecast to be -15°C.

SECTION D

1. The wind is from the northwest at 25 knots.
2. False
3. B
4. 1015.7mb (hPa)
5. A square station model indicates an automated observation site.
6. A
7. A
8. True
9. A
10. The height of the precipitation in that cell is 22,000 feet MSL.
11. NE indicates there were no echoes detected by the radar, and NA indicates that no report was received from the radar site.
12. B
13. A dashed line is used to depict the freezing level at the surface.
14. B
15. B
16. C
17. C
18. B
19. The symbol indicates moderate to severe turbulence from the surface to 24,000 feet MSL.
20. The height of the tropopause is 34,000 feet MSL.
21. B
22. True
23. The forecast is for moderate clear air turbulence between 32,000 feet and 42,000 feet.
24. C
25. 270° at 15 knots, 23°C

SECTION E

1. When you request a briefing, identify yourself as a pilot, whether you plan to fly VFR or IFR, your aircraft number or your name, aircraft type, departure airport, route of flight, destination, flight altitude(s), estimated time of departure, and estimated time enroute.
2. False.
3. D
4. E
5. C
6. A
7. F
8. B
9. True
10. A
11. B
12. B
13. B
14. C

CHAPTER 10

SECTION A

1. MAYDAY
2. C
3. False
4. True

5. To preclude extended IFR operations by these aircraft within the ATC system since these operations may adversely affect other users of the airspace.
6. You should fly at the highest of the following altitudes: the altitude assigned in your last ATC clearance, the minimum altitude for IFR operations, or the altitude ATC has advised you to expect in a further clearance.
7. You should fly one of the following routes in the order given: the route assigned by ATC in your last clearance; if being radar vectored, the direct route from the point of radio failure to the fix route, or airway specified in the radar vector clearance; in the absence of an assigned route, the route ATC has advised you to expect in a further clearance; or the route filed in your flight plan.
8. Hold until the EFC time, then begin the approach.
9. Ground-based radar facility and a functioning airborne radio transmitter and receiver
10. Surveillance approach (ASR), precision approach radar (PAR), and no-gyro approach
11. Precision approach radar (PAR)
12. False
13. Aircraft identification, equipment affected, degree to which the equipment impairs your IFR operations, and type of assistance desired from ATC

SECTION B

1. Spatial disorientation, pilot-induced structural failure, and controlled flight into terrain.
2. False.
3. A
4. B
5. B
6. B
7. Misunderstanding ATC clearances: read back all the key elements of the clearance, especially those you do not understand; Possessing preconceived notions of ATC clearances: be prepared for a different clearance from the one you expect.
8. False.
9. Follow weather trends, determine alternate airports, calculate the fuel required, and rehearse approach procedures.
10. A

SECTION C

1. True
2. False
3. C
4. False
5. False
6. FSS Briefing
7. 800 foot ceiling and 2 miles visibility
8. Weather reports and forecasts, your airplane's capabilities, and your instrument proficiency
9. B
10. At least 30 minutes

CHAPTER 11

SECTION A

1. Fuel is injected into each cylinder's intake port.
2. You should place the auxiliary boost pump switch in the high output position.
3. Measuring the density (weight) of fuel instead of its volume gives a better indication of the fuel's available energy.
4. False
5. False
6. Waste gate
7. A
8. True
9. To prevent engine damage in the event full-power is applied during a go-around
10. False

SECTION B

1. B
2. The inline flow indicator shows green.
3. C
4. A safety valve will open when the maximum cabin differential pressure is reached.
5. False
6. C
7. To prevent the ice from forming to the shape of the boots when inflated so a void does not form under the ice
8. Check the propeller anti-ice electrical ammeter for proper indication.

SECTION C

1. C
2. True
3. B
4. When the throttle is reduced below a certain value, the flaps are extended, or the airspeed falls below a preset limit, the gear warning horn sounds to warn you that the gear is in an unsafe position for landing.
5. You should retract the gear once there is no more runway available for landing in front of the airplane.
6. A

CHAPTER 12

SECTION A

1. False
2. B
3. False
4. True
5. A
6. Maximum level flight speed

7. Weight
8. B
9. True
10. B
11. True
12. Service ceiling
13. A
14. 55 KIAS
15. Increase the angle of bank and/or reduce airspeed
16. Close the throttle to minimize altitude loss during recovery. In addition, engine rotation may cause the spin rate to increase, and thrust may cause the spin to go flat. In addition, if the spin is in a multi-engine airplane, asymmetrical thrust may aggravate the spin.
17. Because the relative wind in a flat spin is nearly straight up, the wings remain in a high angle of attack or stalled condition. In addition, the upward flow over the tail may render the elevators and rudder ineffective, making the recovery impossible.

SECTION B

1. 11 knots
2. False
3. A
4. 3,008 feet
5. 2,301 feet
6. True
7. B
8. B
9. A
10. B
11. 188 knots
12. False
13. Approximately 78 m.p.h. or 67 knots
14. A

SECTION C

1. A
2. B
3. C
4. True
5. B
6. A
7. A
8. B

CHAPTER 13

SECTION A

1. V_{NE}, V_{LE}, V_{FE}
2. True
3. To allow you to make adjustments for wind and obstacles during the approach without using extreme maneuvers at low altitude.
4. B
5. False
6. B
7. False
8. C
9. True
10. False

SECTION B

1. False
2. Attend a CRM workshop conducted at a local college or flight school; practice inflight scenarios with other pilots in a simulator, flight training device, or PCATD; examine flight crew scenarios contained in ASRS accounts, NTSB reports and other safety-related publications
3. Operations specifications (ops specs) are the conditions under which your company must operate in order to retain approval from the FAA. Ops specs include information such as the authorized areas of operations, aircraft, crew complements, types of operations (IFR, VFR, day, night, etc.), and any other pertinent information.
4. B
5. C
6. Aircraft systems and equipment, including advanced navigation equipment and autopilot; VFR and IFR aeronautical charts; crewmembers, including another pilot, second officer, or flight attendant; ground personnel including company dispatch personnel, maintenance technicians, gate agents, and baggage handlers
7. True
8. The need of a company agent or ground personnel to open a gate for another aircraft; pressure from ATC to expedite taxi for takeoff or to meet a restriction in clearance time; the pressure to keep on schedule when delays have occurred due to maintenance or weather; or the inclination to hurry to avoid exceeding duty-time regulations

CHAPTER 14

SECTION A

1. True
2. Accelerate to V_X or V_Y and then begin to climb.
3. C
4. False
5. Holding the brakes until you achieve full power enables you to determine that the engine is functioning properly before you take off from a field where power availability is critical and distance to abort a takeoff is limited.
6. 50 feet
7. True
8. Steeper

SECTION B

1. Slightly decrease the angle of bank first, then increase back pressure on the yoke to raise the nose. Once you regain your desired altitude, roll back to the desired angle of bank.
2. C
3. False
4. 25°

SECTION C

1. C
2. False
3. In a chandelle to the right, the aileron on the right wing is lowered slightly during the roll-out. This causes more drag on the right wing and tends to make the airplane yaw slightly to the right counteracting some of the left turning tendency.
4. C

SECTION D

1. A possible cause is an airspeed which is too slow causing the rate of turn to increase at a given bank angle.
2. B
3. False
4. One of the key factors in making symmetrical turns is proper airspeed control.

SECTION E

1. 960 feet
2. A downwind entry results in the highest groundspeed and the highest pivotal altitude at the start of the maneuver.
3. True
4. C

APPENDIX B
ABBREVIATIONS

AC — advisory circular
AC — convective outlook
ACARS — ARINC communications addressing and reporting system
AD — airworthiness directive
ADF — automatic direction finder
ADIZ — air defense identification zone
ADM — aeronautical decision making
A/FD — *Airport/Facility Directory*
AFM — aircraft flight manual
AFSS — automated flight service station
AGL — above ground level
AIM — *Aeronautical Information Manual*
AIREP — pirep from ACARS system
AIRMET — airman's meteorological information
ALS — approach light system
ALSF — ALS with sequenced flashers
AM — amplitude modulation
AME — aviation medical examiner
APC — approach control
ARINC — Aeronautical Radio, Incorporated
ARP — airport reference point
ARSR — air route surveillance radar
ARTCC — air route traffic control center
ASOS — automated surface observation system
ASR — airport surveillance radar
ATA — actual time of arrival
ATC — air traffic control
ATCRBS — ATC radar beacon system
ATD — along track distance
ATE — actual time enroute
ATIS — automatic terminal information service
ATP — airline transport pilot
AWOS — automated weather observation system
AWW — alert severe weather watch

BC — back course
BCN — beacon
BHP — brake horsepower
BRG — bearing

CAS — calibrated airspeed
CAT — clear air turbulence
CDI — course deviation indicator
CFI — certificated flight instructor
CFIT — controlled flight into terrain
CFR — Code of Federal Regulations
CG — center of gravity
CH — course heading
CHT — cylinder head temperature
CL — center of lift
C_L — coefficient of lift
CLC — course-line computer
C_{Lmax} — maximum coefficient of lift
CLNC — clearance
CLNC DEL — clearance delivery
CNF — computer navigation fix
CO — carbon monoxide
CO_2 — carbon dioxide
COP — changeover point
CRM — crew resource management
CTAF — common traffic advisory frequency
CVFP — charted visual flight procedures
CWA — center weather advisory

DA — density altitude
DA(H) — decision altitude (height)
DALR — dry adiabatic lapse rate
DCS — decompression sickness
DP — instrument departure procedures
DPC — departure control
DGPS — differential global positioning system
DH — decision height
DME — distance measuring equipment
DOD — Department of Defense
DUATS — direct user access terminal system
DVFR — defense visual flight rules

EAS — equivalent airspeed
EFAS — enroute flight advisory service
EFC — expect further clearance
EGT — exhaust gas temperature
ETA — estimated time of arrival
ETD — estimated time of departure
ETE — estimated time enroute

---

f.p.m. — feet per minute
FA — area forecast
FAA — Federal Aviation Administration
FAC — final approach course
FAF — final approach fix
FAP — final approach point
FARs — Federal Aviation Regulations
FAWP — final approach waypoint
FBO — fixed base operator
FD — winds and temperatures aloft forecast
FDC — Flight Data Center
FL — flight level
FLIP — flight information publication
FMS — flight management system
FSDO — Flight Standards District Office
FSS — flight service station

G — gravity; unit of measure for acceleration
GCA — ground controlled approach
GLONASS — global navigation satellite system
GNSS — global navigation satellite system
g.p.h. — gallons per hour
GPS — global positioning system
GPWS — ground proximity warning system

HAA — height above airport
HAT — height above touchdown
HF — high frequency

HIRLs — high intensity runway lights

HIWAS — hazardous in-flight weather advisory service

HMR — hazardous materials regulations

hPa — hectoPascals

HSI — horizontal situation indicator

HVOR — high altitude VOR

Hz — Hertz

IAF — initial approach fix

IAP — instrument approach procedure

IAS — indicated airspeed

ICAO — International Civil Aviation Organization

IF — intermediate fix

IFR — instrument flight rules

ILS — instrument landing system

IM — inner marker

IMC — instrument meteorological conditions

in. Hg. — inches of mercury

INS — inertial navigation system

IR — infrared

ISA — International Standard Atmosphere

IVSI — instantaneous vertical speed indicator

---

kg — kilogram

KCAS — knots calibrated airspeed

kHz — kilohertz

KIAS — knots indicated airspeed

km — kilometer

KTAS — knots true airspeed

Kts — knots

kw — kilowatt

kwh — kilowatt hour

L/MF — low/medium frequency

LAA — local airport advisory

LAHSO — land and hold short operation

LDA — localizer-type directional aid

L/D_{max} — maximum lift/drag ratio

LEMAC — leading edge mean aerodynamic chord

LI — lifted index

LIRLs — low intensity runway lights

LLT — low level turbulence

LLWAS — low level windshear alert system

LMM — middle compass locator

LOC — localizer

LOFT — line-oriented flight training

LOM — outer compass locator

LORAN — long range navigation

LVOR — low altitude VOR

M — mach

MAA — maximum authorized altitude

MAC — mean aerodynamic chord

MALS — medium intensity approach light system

MALSF — MALS with sequenced flashers

MALSR — MALS with RAIL

MAHWP — missed approach holding waypoint

MAP — missed approach point

MAP — manifold absolute pressure

MAWP — missed approach waypoint

mb — millibar

MB — magnetic bearing

MC — magnetic course

MCA — minimum crossing altitude

MDA(H) — minimum descent altitude (height)

MDH — minimum descent height

MEA — minimum enroute altitude

METAR — aviation routine weather report

MH — magnetic heading

MHA — minimum holding altitude

MHz — megahertz

MIA — minimum IFR altitude

MIRLs — medium intensity runway lights

MLS — microwave landing system

MM — middle marker

MOA — military operations area

MOCA — minimum obstruction clearance altitude

MORA — minimum off-route altitude

m.p.h. — miles per hour

MRA — minimum reception altitude

MSA — minimum safe altitude

MSAW — minimum safe altitude warning

MSL — mean sea level

MTI — moving target indicator

MTR — military training route

MVA — minimum vectoring altitude

MVFR — marginal VFR

NA — not authorized

NAR — North American Route (also refers to North Atlantic Route)

NAS — national airspace system

NASA — National Aeronautics and Space Administration

NAVAID — Navigation aid

NCAR — National Center for Atmospheric Research

NDB — nondirectional radio beacon

n.m. — nautical mile

NMC — National Meteorological Center

NOAA — National Oceanic and Atmospheric Administration

NoPT — no procedure turn

NORDO — no radio

NOS — National Ocean Service

NOTAM — Notices to Airmen

NSA — national security area

NSF — National Science Foundation

NTSB — National Transportation Safety Board

NWS — National Weather Service

OBS — omnibearing selector

OM — outer marker

OROCA — off-route obstruction clearance altitude

OTS — out of service

---

p.s.i. — pounds per square inch

p.s.i.d. — pounds per square inch differential

PAPI — precision approach path indicator

PAR — precision approach radar

PCATD — personal computer-based aviation training device

PCL — pilot controlled lighting

PF — pilot flying

PNF — pilot not flying

PIC — pilot in command

PIREP — pilot weather report

PLASI — pulsating approach slope indicator

POH — pilot's operating handbook

PPS — precise positioning service

PRM — precision runway monitor

PT — procedure turn

PTS — practical test standards

---

r.p.m. — revolutions per minute

RA — resolution advisory

RAF — Research Aviation Facility

RAIL — runway alignment indicator lights

RAIM — receiver autonomous integrity monitoring

RB — relative bearing

RCC — rescue coordination center

RCLS — runway centerline light system

RCO — remote communication outlet

REIL — runway end identifier lights

RMI — radio magnetic indicator

RMK — remarks

RNAV — area navigation

RRL — runway remaining lights

RVR — runway visual range

RVV — runway visibility value

SALR — saturated adiabatic lapse rate

SALS — short approach light system

SAR — search and rescue

SD — radar weather report

SDF — simplified directional facility

SFL — sequenced flashing lights

SIAP — standard instrument approach procedure

SID — standard instrument departure

SIGMET — significant meteorological information

SLP — sea level pressure

s.m. — satute mile

SPECI — non-routine (special) aviation weather report

SPS — standard positioning service

SSALS — simplified SALS

SSALSF — SALS with SFL

SSALSR — SALS with RAIL

SSV — standard service volume

STAR — standard terminal arrival route

SVFR — special visual flight rules

TA — traffic advisory

TACAN — tactical air navigation

TAF — terminal aerodrome forecast

TAS — true airspeed

TC — true course

TCAD — traffic alert and collision avoidance device

TCAS — traffic alert and collision avoidance system

TCH — threshold crossing height

TCU — towering cumulonimbus

TDWR — terminal Doppler weather radar

TDZE — touchdown zone elevation

TDZL — touchdown zone lighting

TEC — tower enroute control

TEMAC — trailing edge mean aerodynamic chord

TERPS — U.S. Standard for Terminal Instrument Procedures

TH — true heading

TIBS — telephone information briefing service

TIT — turbine inlet temperature

TRSA — terminal radar service area

TSO — technical standard order

TVOR — terminal VOR

TWEB — transcribed weather broadcast

UA — pilot report

UCAR — University Corporation for Atmospheric Research

UHF — ultra high frequency

UNICOM — aeronautical advisory station

UTC — Coordinated Universal Time (Zulu time)

UUA — urgent pilot report

V_1 — takeoff decision speed

V_2 — takeoff safety speed

V_A — design maneuvering speed

VAFTAD — volcanic ash forecast transport and dispersion chart

VASI — visual approach slope indicator

VDP — visual descent point

V_{FE} — maximum flap extended speed

VFR — visual flight rules

VHF — very high frequency

V_{LE} — maximum landing gear extended speed

V_{LO} — maximum landing gear operating speed

VMC — visual meteorological conditions

V_{NE} — never-exceed speed

V_{NO} — maximum structural cruising speed

VOR — VHF omnirange station

VOR/DME — collocated VOR and DME

VORTAC — collocated VOR and TACAN

VOT — VOR test facility

V_R — rotation speed

VSI — vertical speed indicator

V_{S0} — stalling speed or minimum steady flight speed in the landing configuration

V_{S1} — stalling speed or minimum steady flight speed obtained in a specified configuration

V_X — best angle of climb speed

V_Y — best rate of climb speed

WA — AIRMET

WAAS — wide area augmentation system

WAC — world aeronautical chart

WCA — wind correction angle

WFO — Weather Forecast Office

WH — hurricane advisory

WPT — waypoint

WS — SIGMET

WSP — weather systems processor

WST — Convective SIGMET

WW — severe weather watch bulletin

WX — weather

Z — Zulu time (UTC)

INDEX

Name: ______________________

CHAPTER 2
ANSWER SHEET

SECTION A

1. __________
2. ____

   ____
3. ______________________
4. __________
5. ____

   ____

   ____
6. __________
7. __________
8. __________
9. __________
10. __________
11. __________
12. __________
13. ____
14. __________
15. __________
16. __________
17. ____

    ____

    ____

SECTION B

1. __________
2. __________
3. __________
4. __________
5. __________
6. __________
7. ____

   ____

   ____
8. ____
9. ______________________
10. ______________________
11. ______________________
12. ______________________

13. ____________________
14. __________
15. __________
16. __________
17. __________
18. ____
____
19. ____
____
20. ____
____
____
____

SECTION C

1. __________
2. __________
3. __________
4. ____
____
____
5. __________
6. ____________________
7. ____
8. ____
____
____
9. __________
10. __________
11. __________
12. __________
13. ____
14. __________
15. __________
16. __________
17. ____
18. __________
19. __________
20. __________

Name: ______________________________

CHAPTER 3
ANSWER SHEET

SECTION A

1. ________ 2. ________ 3. ________

4. ________ 5. ________ 6. ________

7. ________ 8. ________ 9. ________

10. ________ 11. ________ 12. ________

13. ________

14. ________

15. ________

16. ________

17. ________ 18. ________ 19. ________

20. ________ 21. ________ 22. ________

23. ________ 24. ________ 25. ________

26. ________

27. ________

28. ________

29. ________

30. ________

31. ________

32. ________

33. ________

SECTION B

1. ________

2. ________

3. ________ 4. ________ 5. ________

6. ________ 7. ________

8. ________

9. ________

10. ________ 11. ________ 12. ________

13. ________ 14. ________ 15. ________

16. ____________ 17. ____________ 18. ____________

19. ____________ 20. ____________________________

SECTION C

1. ____________________________ 2. ____________ 3 . ____________

4. ____________ 5. ____________ 6. ____________

7. ____________ 8. ____________ 9. ____________

10. ____________ 11. ____________ 12. ____________

13. ____________ 14. ____________ 15. ____________

16. ____________ 17. ____________ 18. ____________

19. ____________ 20. ____________

21. ____

22. ____________

23. ____________

24. ____

____

25. ____

____

Name: ______________________________

CHAPTER 4
ANSWER SHEET

SECTION A

1. ______________________
2. __________
3. __________
4. __________
5. ______________________
6. __________
7. ______________________
8. __________
9. __________
10. __________

SECTION B

1. __________
2. ______________________
3. __________
4. __________
5. ______________________
6. __________
7. ______________________
8. __________
9. ______________________
10. __________
11. ____
    ____
12. __________
13. __________
14. __________
15. ____
    ____

Name: ______________________________

CHAPTER 5
ANSWER SHEET

SECTION A

1. ____________
2. ____________
3. ____________
4. ____________
5. ____________
6. ____________
7. ____________
8. ____________
9. ____________
10. ____________
11. ____________
12. ____________
13. ____________
14. ____________
15. ____________
16. ____________

SECTION B

1. ____________
2. ____
   ____
3. ____________
4. ____________
5. ____________
6. ____
   ____
   ____
   ____
   ____
7. ____________
8. ____
   ____
9. ____________________________
10. ____________
11. ____________

SECTION C

1. ____________
2. ____________
3. ____________________________
4. ____________
5. ____________
6. ____________
7. ____________
8. ____________
9. ____________

Name: ____________________

CHAPTER 6
ANSWER SHEET

SECTION A

1. ____________________
2. __________
3. ____________________
4. ____________________
5. ____________________
6. __________
7. __________
8. ____

____

____

SECTION B

1. ____

____

2. ____

____

____

____

3. __________
4. __________
5. ____

Name: ____________________

CHAPTER 7
ANSWER SHEET

SECTION A

1. ____________________
2. ________
3. ________
4. ____________
5. ________
6. ____________________
7. ________
8. ________
9. ________
10. ________
11. ________
12. ________
13. ________
14. ____________
15. ____________
16. ____________
17. ________
18. ________
19. ________
20. ________
21. ____________________

____

____

____

____

22. ________

SECTION B

1. ____________________

____

____

____

____

2. ________
3. ____________________

____

____

____

____

4. ______________________

5. __________

6. ____

____

____

____

7. __________

8. __________

9. ______________________

10. ______________________

11. __________

12. __________

13. __________

14. __________

15. ____

____

____

____

____

____

____

Name: ______________________________

CHAPTER 8
ANSWER SHEET

SECTION A

1. ____________
2. ____________
3. ____________
4. ____

   ____
5. ____________
6. ____________
7. ________________________
8. ____________
9. ____________
10. ____________
11. ____________
12. ____________
13. ____________
14. ____________
15. ____________
16. ____________

SECTION B

1. ____________
2. ____________
3. A. ____________ B. ____________ C. ____________

   D. ____________ E. ____________
4. A. ____________ B. ____________ C. ____________

   D. ____________ E. ____________
5. ____
6. ________________________
7. ____

   ____
8. ____________
9. ____

   ____

   ____
10. ____

    ____

    ____

    ____
11. ____________

12. ______________________________

13. ________

14. ________

15. ______________________________

____

SECTION C

1. ________

2. ________

3. ________

4. ______________________________

5. ________

6. ________

7. ________

8. ________

9. ________

10. ________

11. ______________________________

12. ________

13. ________

14. ______________________________

____

____

____

____

____

____

15. ________

Name: ______________________________

CHAPTER 9
ANSWER SHEET

SECTION A

1. ______________________________
____
2. __________
3. __________
4. __________
5. __________
6. __________
7. __________
8. __________
9. __________
10. __________
11. __________
12. __________
13. ____________________
14. __________
15. __________
16. __________
17. __________
18. __________
19. __________

SECTION B

1. ______________________________
2. __________
3. __________
4. __________
5. __________
6. __________
7. __________
8. __________
9. ______________________________
____
10. __________
11. __________
12. __________
13. __________
14. __________
15. __________

16. ______________________________

____

____

____

____

17. __________

18. __________

19. __________

20. __________

21. __________

22. __________

23. __________

24. __________

SECTION C

1. __________
2. __________
3. __________
4. ____________________
5. ____________________
6. __________
7. ______________________________

   ____
8. __________
9. ______________________________

   ____

   ____

   ____
10. __________
11. __________
12. __________
13. __________
14. ______________________________

    ____

    ____
15. __________
16. ______________________________

    ____

SECTION D

1. ______________________________
2. ________
3. ________
4. ________________
5. ______________________________
6. ________
7. ________
8. ________
9. ________
10. ______________________________
11. ______________________________
12. ________
13. ______________________________
14. ________
15. ________
16. ________
17. ________
18. ________
19. ______________________________
20. ______________________________
21. ________
22. ________
23. ______________________________
    ____
24. ________
25. ________________

SECTION E

1. ______________________________
   ____
   ____
   ____
2. ________
3. ________
4. ________
5. ________
6. ________
7. ________
8. ________
9. ________

10. ______________

11. ______________

12. ______________

13. ______________

14. ______________

Name: ________________________________

CHAPTER 10
ANSWER SHEET

SECTION A

1. ____________
2. ____________
3. ________
4. ________
5. ____
6. ____
7. ____
8. ____
9. ____
10. ____
11. ____________
12. ________
13. ____

SECTION B

1. ____
2. ________
3. ________
4. ________
5. ________
6. ________
7. ____
8. ________
9. ____
10. ________

SECTION C

1. ____________
2. ____________
3. ____________
4. ____________
5. ____________
6. ____________________________
7. ____
8. ____

   ____
9. ____________
10. ____________________________

Name: ______________________________

CHAPTER 11
ANSWER SHEET

SECTION A

1. ______________________________
2. ______________________________
3. ______________________________
   ____
4. __________
5. __________
6. ____________________
7. __________
8. __________
9. ______________________________
   ____
10. __________

SECTION B

1. __________
2. ______________________________
3. __________
4. ______________________________
   ____
5. __________
6. __________
7. ______________________________
   ____
8. ______________________________

SECTION C

1. __________
2. __________
3. __________
4. ______________________________
   ____
   ____
5. ______________________________
   ____
6. __________

Name: ____________________

CHAPTER 12
ANSWER SHEET

SECTION A

1. ____________
2. ____________
3. ____________
4. ____________
5. ____________
6. ____________
7. ____________
8. ____________
9. ____________
10. ____________
11. ____________
12. ____________
13. ____________
14. ____________
15. ____________
16. ____________
17. ____________

SECTION B

1. ____________
2. ____________
3. ____________
4. ____________
5. ____________
6. ____________
7. ____________
8. ____________
9. ____________
10. ____________
11. ____________
12. ____________
13. ____________
14. ____________

SECTION C

1. ______________
2. ______________
3. ______________
4. ______________
5. ______________
6. ______________
7. ______________
8. ______________

Name: ______________________

CHAPTER 13
ANSWER SHEET

SECTION A

1. ______________________
2. ____________
3. ____
____
4. ____________
5. ____________
6. ____________
7. ____________
8. ____________
9. ____________
10. ____________

SECTION B

1. ____________
2. ____
____
____
____
3. ____
____
____
____
4. ____________
5. ____________
6. ____
____
____
____
7. ____________
8. ____
____
____
____

Name: ______________________________

CHAPTER 14
ANSWER SHEET

SECTION A

1. __________
2. ______________________________
3. __________
4. __________
5. ______________________________
6. ______________
7. __________
8. ______________

SECTION B

1. ______________________________
2. __________
3. __________
4. __________

SECTION C

1. __________
2. __________
3. ______________________________
4. __________

SECTION D

1. ______________________________
2. __________
3. __________
4. ______________________________

SECTION E

1. ______________
2. ______________________________
3. __________
4. __________

NOTAMs

The NOTAMs Section is designed to inform you of recent developments that could affect your training.

Vertical Navigation (VNAV)

Jeppesen introduced some new symbology in numerous non-precision approach profiles. This symbology graphically represents coded database vertical descent information. Many modern navigation systems such as GPS and FMS use this database information to enable stabilized, constant-rate descents. The introduction of this symbology in the profile graphic is intended to support such operations.

Vertical Navigation (VNAV) descent information appears in the profile view of selected nonprecision approaches. The VNAV information appearing in the profile illustrates the geometric descent path with a descent angle from the Final Approach Fix (FAF) to the Threshold Crossing Height (TCH) at the approach end of the runway.

The VNAV descent path, depicted with a screened line, is based on the same descent angle coded into the Jeppesen NavData database. Use of this descent angle by certified VNAV-capable avionics equipment ensures a stable, constant rate of descent that clears all intervening altitude restrictions. Some approach procedures may require a delay of the start of descent beyond the FAF, until the VNAV descent path is intercepted. The profile view depicts this level segment of flight as required.

The VNAV descent angle appears in brackets along the VNAV descent path and is repeated in the conversion table. Additionally, the conversion table provides a recommended rate of descent relative to the VNAV angle and groundspeed.

The inclusion of the VNAV descent angle does not change or modify existing nonprecision approach requirements. Use of the Minimum Descent Altitude (MDA), as well as the Missed Approach Point (MAP), remains unchanged. In accordance with Federal Aviation Regulations (FARs) and ICAO PANS OPS criteria, do not descend below the MDA until attaining the required visual reference. Additionally, do not initiate the prescribed missed approach procedure prior to reaching the published missed approach point. Note: Operators may obtain permission from their controlling authority to use Decision Altitude (DA) operational techniques when making a VNAV descent. This approval is specific to the operator and to the approach.

VNAV descent is optional. Use of any VNAV approach technique is dependent on operator approval, certified VNAV-capable avionics equipment availability, and crew training. See the following VNAV profile examples.

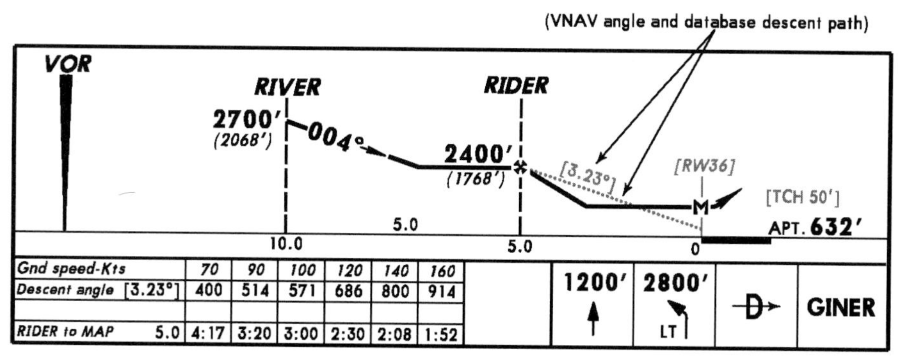

VNAV descent information from FAF to runway with TCH of 50'.

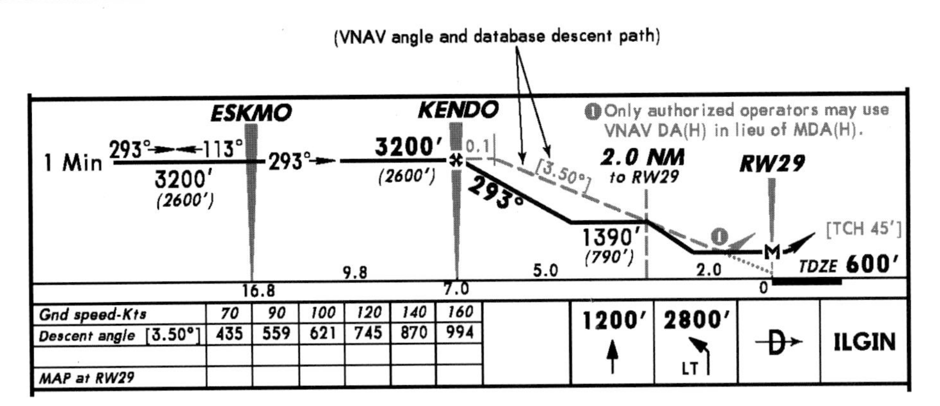

VNAV descent information from FAF to runway with TCH of 45'. Note that the VNAV descent requires level flight for 0.1 NM past the FAF, prior to intercepting the VNAV descent path of 3.50°, to cross the 2.0NM to RW29 stepdown fix at or above 1390'. For approved operators, use of DA(H) operational techniques on this approach is indicated by the ballflag note as well as by the dashed VNAV descent track in the profile view.

New NOS Chart Format

The United States Government has developed a new format for instrument approach procedure (IAP) charts. This new IAP format was found to be easier to read, and resulted in reduced head-down time caused by pilots referring to the chart during approach. The implementation plan calls for the conversion to be done airport by airport. This will result in both old and new formats being mixed in the Terminal Procedures Publications (TPPs), but with all IAP charts for any particular airport in a common format. Eventually, all U.S. Government IAP charts will be charted in the new format.

Pilot Briefing Information

The pilot briefing information format consists of three horizontal rows of boxed procedure-specific information along the top edge of the chart. Altitudes, frequency, course, and elevation values (except HATs and HAAs) are charted in bold type. The top row contains the primary procedure navigation information, final approach course, landing distance available, touchdown zone and airport elevations. The middle row contains procedure notes and limitations, icons indicating if nonstandard alternate and/or take-off minima apply, approach lighting symbology, and the full text description of the missed approach procedure. The bottom row contains air-to-ground communication facilities and frequencies in the order in which they are used during an approach.

Missed Approach Icons

In addition to the full text description of the missed approach procedure contained in the notes section of the middle briefing row, the initial (up to four) steps also are charted as boxed icons in the chart profile view. These icons provide simple-to-interpret instructions, such as direction of initial turn, next heading and/or course, next altitude, etc.

RNAV Instrument Approach Charts

Reliance on RNAV systems for instrument approach operations is becoming more commonplace as new systems such as GPS, wide area augmentation system (WAAS), and local area augmentation system (LAAS) are developed and deployed. WAAS augments the basic GPS satellite constellation with additional ground stations and enhanced position/integrity information transmitted from geostationary satellites. In order to foster and support full integration of RNAV into the National Airspace System (NAS), the FAA has developed a new charting format for RNAV IAPs. This format incorporates all types of approaches using area navigation systems, both ground and satellite based, and avoids unnecessary duplication and proliferation of charts.

The approach minima for unaugmented GPS (the present GPS approaches) and augmented GPS (WAAS and LAAS when they become operational) will be published on the same approach chart. The approach chart will be titled "RNAV RWY XX." The first RNAV approach charts may appear as stand-alone GPS procedures, prior to WAAS becoming operational. The chart may contain as many as three lines of approach minima: LNAV/VNAV (lateral navigation/vertical navigation); LNAV, and CIRCLING.

LNAV/VNAV (Lateral Navigation/Vertical Navigation)

LNAV/VNAV identifies minima developed to accommodate an RNAV IAP with vertical guidance, but with integrity limits larger than those of a precision approach. Aircraft using LNAV/VNAV minima will descend to landing via an internally generated descent path based on satellite or other VNAV systems approved for approaches. Since electronic vertical guidance is provided, the minima will be published as a DA. These minima may be used by aircraft equipped with precision approach capable WAAS equipment that has reverted to a less capable mode of operation or a WAAS lateral-only receiver integrated with an IFR approach approved BARO-VNAV. In addition, RNP-0.3 approved aircraft with an IFR approach approved BARO-VNAV system may use LNAV/VNAV minima.

LNAV (Lateral Navigation)

These minima are for lateral navigation only, and the approach minimum altitude will be published as a minimum descent altitude (MDA) because vertical guidance is not provided. LNAV provides the same level of service as the present GPS stand-alone approaches. If the quality of the WAAS navigation solution will not support vertical navigation at all, the WAAS receiver will revert to an LNAV mode. In addition, these minima may be used by aircraft with WAAS equipment approved only for nonprecision approaches and RNP-0.3 approved aircraft, as well as aircraft equipped with navigation systems using unaugmented GPS approved for approach operations in accordance with AC 20-138, and AC 20-130A.

Other Systems

Through a special authorization, aircraft equipped with other IFR approach approved RNAV systems may fly to the LNAV/VNAV and/or LNAV minima. Section A of the U.S. Terminal Procedures book provides details on the other systems which are permitted to utilize these approaches. You must obtain authorization from your local Flight Standards District Office (FSDO) for your particular equipment suite.

Required Navigation Performance (RNP)

To capitalize on the potential of RNAV systems, the FAA and the International Civil Aviation Organization (ICAO) are effecting a shift toward a new standard of navigation and airspace management called RNP. Navigation systems have typically been described as being sensor specific, such as VOR, NDB, and ILS systems. When RNP is specified, it does not matter what underlying navigation system or combination of systems is used, provided the aircraft can achieve the required navigation performance. Typically, various sensor inputs are processed by the RNAV system to arrive at a position estimate having a high statistical degree of accuracy and confidence. RNP is intended to provide a single performance standard that can be used and applied to aircraft and aircraft equipment manufacturers, airspace planners, aircraft certification and operations, pilots and controllers, and international aviation authorities. RNP can be related to obstacle clearance or aircraft separation requirements to ensure a consistent level of application.

An RNP level or type is applicable to a selected airspace, route, or procedure. The applicable RNP is expressed as a value that represents a distance in nautical miles from the intended position to the actual position of an aircraft. It is within this distance that an aircraft would normally be expected to operate. For general RNAV approach procedures, RNP-0.3 is required. You may refer to Section A of the U.S. Terminal Procedures book for aircraft approach eligibility requirements by specific RNP level requirements. Aircraft meeting RNP criteria will have an appropriate entry, including special conditions and limitations, if any, in the Aircraft Flight Manual (AFM) or its supplement.

Terminal Arrival Area (TAA)

The objective of the Terminal Arrival Area (TAA) is to provide a new transition method for arriving aircraft equipped with FMS and/or GPS navigational equipment. The TAA contains within it a "T" structure that normally provides a NoPT for aircraft using the approach. The TAA provides the pilot and air traffic controller with an efficient method for routing traffic from enroute to terminal structures. TAAs may appear on both current and new format GPS and RNAV IAP charts in the form of icons on the plan view. These icons are generally arranged on the chart in accordance with their position relative to the aircraft's arrival from the enroute structure. The TAA normally consists of three areas: the straight-in area, the left base area, and the right base area. The waypoint (WP) to which navigation is appropriate and expected within each specific TAA area will

be named and depicted on the associated TAA icon. Each depicted named WP within the icon is the IAF for arrivals within that area.

The straight-in area is defined by a semicircle with a 30 NM radius centered on and extending outward from the IF/IAF. The left and right base areas are bounded by the bottom of the straight-in TAA and the extension of the intermediate segment course. The base areas are defined by a 30 NM radius centered on the IAF on either side of the IF/IAF. The altitudes shown within the TAA icons provide minimum IFR obstacle clearance. Operating procedures for the TAA are contained in the Aeronautical Information Manual (AIM), Para 1-1-21.f.

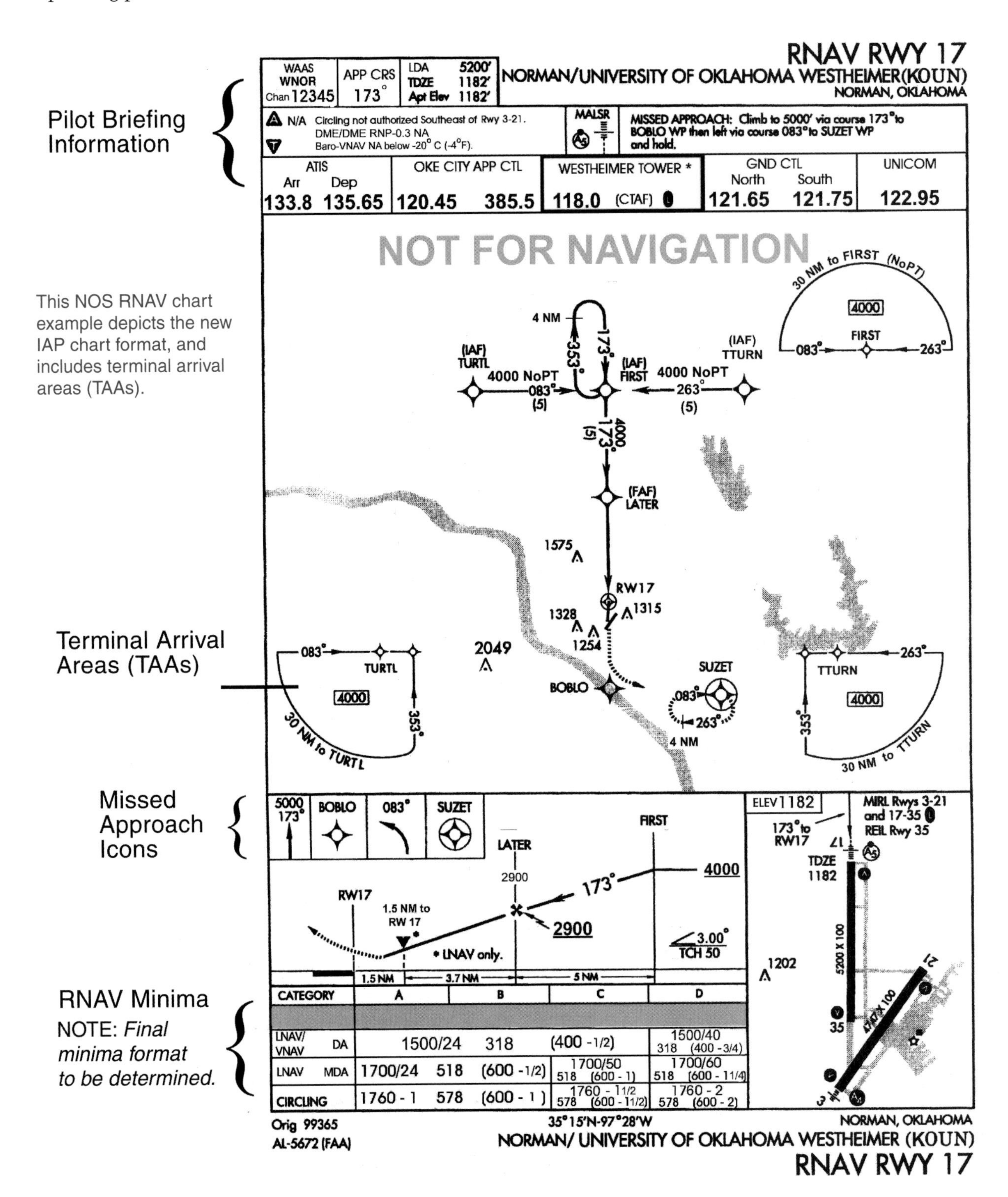

CATEGORY		A	B	C	D
LNAV/VNAV	DA	1500/24	318	(400 -1/2)	1500/40 318 (400 -3/4)
LNAV	MDA	1700/24	518 (600 -1/2)	1700/50 518 (600 - 1)	1700/60 518 (600 - 11/4)
CIRCLING		1760 - 1	578 (600 - 1)	1760 - 11/2 578 (600 - 11/2)	1760 - 2 578 (600 - 2)

Land and Hold Short Operations (LAHSO)

LAHSO is an air traffic control procedure that requires pilot participation to balance the needs for increased airport capacity and system efficiency, consistent with safety. This procedure can be done safely provided pilots and controllers are knowledgeable and understand their responsibilities.

Pilot Responsibilities and Basic Procedures

At controlled airports, air traffic may clear a pilot to land and hold short. Pilots may accept such a clearance provided that the pilot in command determines that the aircraft can safely land and stop within the available landing distance (ALD). ALD data are published in the special notices section of the *Airport/Facility Directory.* Controllers will also provide ALD data upon request. Student pilots or pilots not familiar with LAHSO should not participate in the program.

The pilot in command has the final authority to accept or decline any land and hold short clearance. The safety and operation of the aircraft remain the responsibility of the pilot. Pilots are expected to decline a LAHSO clearance if they determine it will compromise safety.

To conduct LAHSO, pilots should become familiar with all available information concerning LAHSO at their destination airport. Pilots should have, readily available, the published ALD and runway for all LAHSO runway combinations at each airport of intended landing. Additionally, knowledge about landing performance data permits the pilot to readily determine that the ALD for the assigned runway is sufficient for safe LAHSO. Pilots should determine if their destination airport has LAHSO. If so, their preflight planning should include an assessment of which LAHSO combinations would work for them given their aircraft's required landing distance. Good pilot decision making is knowing in advance whether one can accept a LAHSO clearance if offered.

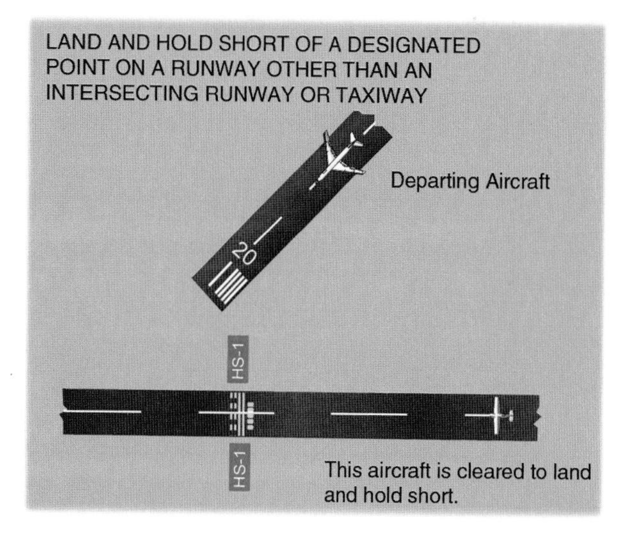

Pilots also need to have a good understanding of LAHSO markings, signage, and in-pavement lighting when installed. Examples are included in this NOTAM and Chapter 4, Section B of this manual, as well as in the chapter titled Aeronautical Lighting and Other Airport Visual Aids in the *Aeronautical Information Manual* (AIM). LAHSO visual aids consist of a three-part system of yellow hold-short markings, red and white signage, and, in some cases, in-pavement lighting.

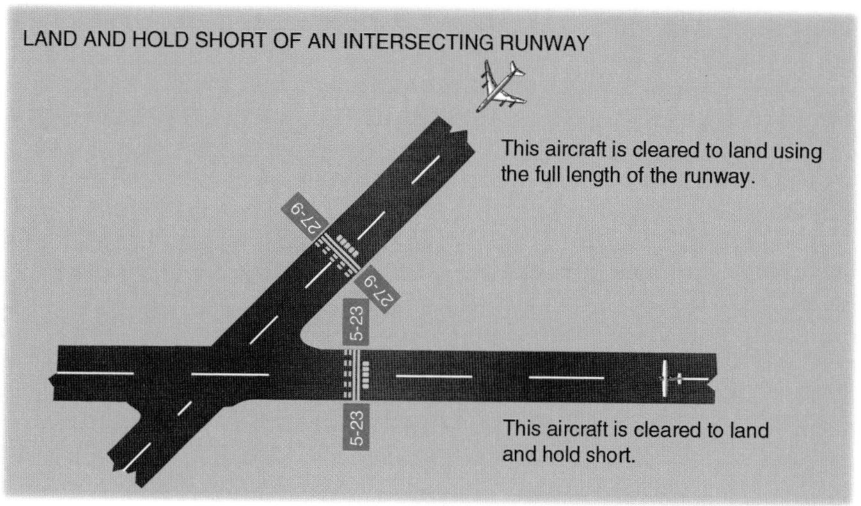

If, for any reason, such as difficulty in discerning the location of a LAHSO intersection, wind conditions, aircraft condition, etc., the pilot elects to request to land on the full length of the runway, to land on another runway, or to decline LAHSO, a pilot is expected to promptly inform air traffic, ideally even before the clearance is issued. **A LAHSO clearance, once accepted, must be adhered to, just as any other ATC clearance, unless an amended clearance is obtained**

or an emergency occurs. **A LAHSO clearance does not preclude a rejected landing.**

A pilot who accepts a LAHSO clearance should land and exit the runway at the first convenient taxiway (unless directed otherwise) before reaching the hold short point. Otherwise, the pilot must stop and hold at the hold short point. **If a rejected landing becomes necessary after accepting a LAHSO clearance, the pilot should maintain safe separation from other aircraft or vehicles, and should promptly notify the controller.**

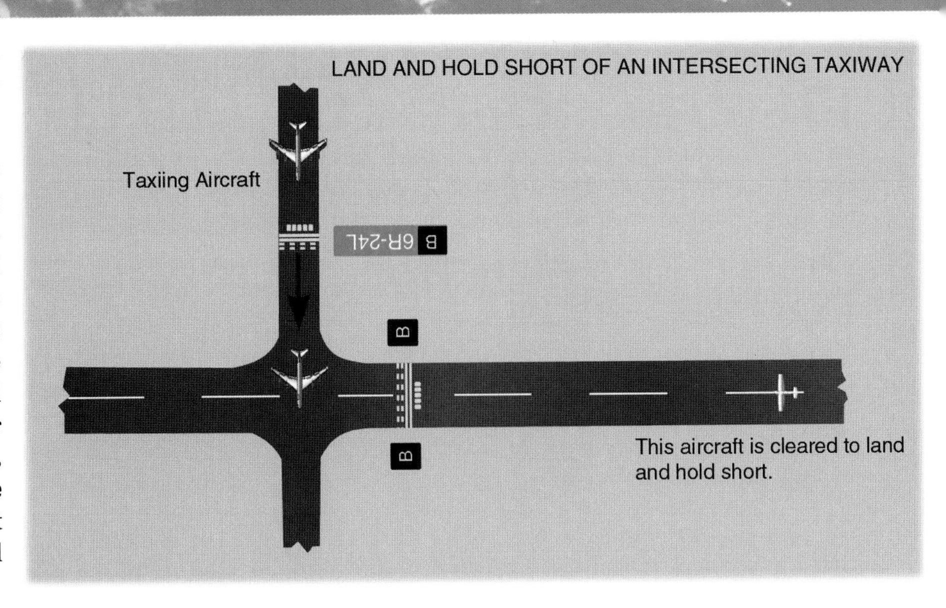

Controllers need a full readback of all LAHSO clearances. Pilots should read back their LAHSO clearance and include the words, "*Hold short of (runway / taxiway / or point)*" in their acknowledgement of all LAHSO clearances. To reduce frequency congestion, pilots are encouraged to read back the LAHSO clearance without prompting. Do not make the controller have to ask for a readback!

Situational Awareness

Situational awareness is vital to the success of LAHSO. Situational awareness starts with having current airport information in the cockpit, readily accessible to the pilot. For example, an airport diagram assists pilots in identifying their location on the airport, thus reducing requests for progressive taxi instructions from controllers. Situational awareness includes effective pilot-controller radio communication. ATC expects pilots to specifically acknowledge and read back all LAHSO clearances.

EXAMPLES OF PILOT-CONTROLLER COMMUNICATION

ATC: *"Cessna 45321, cleared to land Runway 6 Right, hold short of Taxiway Bravo for crossing traffic (type aircraft)."*

Pilot: *"Cessna 45321, wilco, cleared to land Runway 6 Right to hold short of Taxiway Bravo."*

ATC: *"Cessna 45321, cross Runway 6 Right at Taxiway Bravo, landing aircraft will hold short."*

Pilot: *"Cessna 45321, wilco, cross Runway 6 Right at Bravo, landing traffic (type aircraft) to hold."*

For those airplanes flown with two crewmembers, effective intra-cockpit communication between cockpit crewmembers is also critical. There have been several instances where the pilot working the radios accepted a LAHSO clearance but then simply forgot to tell the pilot flying the aircraft.

Situational awareness also includes a thorough understanding of the airport markings, signage, and lighting associated with LAHSO. As indicated in the accompanying illustrations, these visual aids assist the pilot in determining where to hold short. Pilots are cautioned that not all airports conducting LAHSO have installed any or all of the above markings, signage, or lighting.

Pilots should only receive a LAHSO clearance when there is a minimum ceiling of 1,000 feet and 3 statute miles visibility. The intent of having basic VFR weather conditions is to allow pilots to maintain visual contact with other aircraft and ground vehicle operations. Pilots should consider the effects of prevailing inflight visibility (such as landing into the sun) and how it may affect overall situational awareness. Additionally, surface vehicles and aircraft being taxied by maintenance personnel may also be participating in LAHSO, especially in those operations that involve crossing an active runway.

STANDARD AIRWAY MANUAL SERVICE – *THE ORIGINAL INSTRUMENT CHART SERVICE*

Designed for the pilot who requires the most up-to-date enroute and terminal charts available. This service starts you off with a complete set of current initial charts for the geographic area you request. Your initial charts are then kept up-to-date by revisions sent automatically every two weeks. With each revision, you simply replace, add or delete specific charts according to the included revision cover letter.

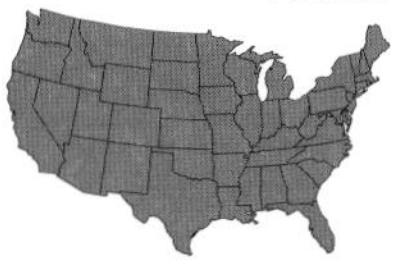

ENROUTE LOW AND TERMINAL

Region	Order Code	Initial Charts With Revision Total *without* Binders	Initial Charts With Revision Total *with* LB Binders	Annual Revision Service	Trip Kit Price	Binders Qty./Size (in inches)
FULL USA	USA05	$768	$1,316.55	$554	$285	9/2
	Trip Kit Subscriber Discount ($71)				$214	
	All 48 Contiguous States					
EAST	USB05	$461	$765.75	$339	$163	5/2
	Trip Kit Subscriber Discount ($41)				$122	
	AL, CT, DE, DC, FL, GA, IL, IN, KY, MA, MD, ME, MI, MS, NC, NH, NJ, NY, OH, PA, RI, SC, TN, VA, VT, WI, WV					
EAST & CENTRAL	USC05	$666	$1,153.60	$489	$236	8/2
	Trip Kit Subscriber Discount ($59)				$177	
	AL, AR, CT, DE, DC, FL, GA, IA, IL, IN, KS, KY, LA, MA, MD, ME, MI, MN, MO, MS, NC, ND, NE, NH, NJ, NY, OH, OK, PA, RI, SC, SD, TN, TX, VA, VT, WI, WV					
CENTRAL	USD05	$406	$710.75	$309	$129	5/2
	Trip Kit Subscriber Discount ($32)				$97	
	AR, IA, IL, IN, KS, LA, MI, MN, MO, ND, NE, OK, SD, TX, WI					
CENTRAL & WEST	USE05	$511	$876.70	$389	$163	6/2
	Trip Kit Subscriber Discount ($41)				$122	
	AR, AZ, CA, CO, IA, ID, IL, IN, KS, LA, MI, MN, MO, MT, ND, NE, NM, NV, OK, OR, SD, TX, UT, WA, WI, WY					
WEST	USF05	$251	$372.90	$181	$94	2/2
	Trip Kit Subscriber Discount ($24)				$70	
	AZ, CA, CO, ID, MT, NM, NV, OR, UT, WA, WY					
SOUTHWEST	USK05	$175	$296.90	$134	$55	2/2
	Trip Kit Subscriber Discount ($14)				$41	
	AZ, CA, CO, NM, NV, UT					
NORTHWEST	USL05	$158	$218.95	$117	$55	1/2
	Trip Kit Subscriber Discount ($14)				$41	
	ID, MT, OR, WA, WY					

ENROUTE LOW AND TERMINAL

Region	Order Code	Initial Charts With Revision Total *without* Binders	Initial Charts With Revision Total *with* LB Binders	Annual Revision Service	Trip Kit Price	Binders Qty./Size (in inches)
NORTH CENTRAL	USN05	$259	$380.90	$189	$94	2/2
	Trip Kit Subscriber Discount ($24)				$70	
	CO, IA, KS, MN, MO, MT, ND, NE, SD, WY					
GREAT LAKES	USP05	$261	$382.90	$191	$94	2/2
	Trip Kit Subscriber Discount ($24)				$70	
	IL, IN, KY, MI, OH, WI					
NORTHEAST	USR05	$262	$383.90	$192	$94	2/2
	Trip Kit Subscriber Discount ($24)				$70	
	CT, DE, DC, MA, MD, ME, NH, NJ, NY, PA, RI, VA, VT, WV					
SOUTHEAST	USS05	$259	$380.90	$189	$94	2/2
	Trip Kit Subscriber Discount ($24)				$70	
	AL, FL, GA, MS, NC, SC, TN					
SOUTH CENTRAL	UST05	$203	$324.90	$162	$55	2/2
	Trip Kit Subscriber Discount ($14)				$41	
	AR, LA, NM, OK, TX					
NORTH CENTRAL & GREAT LAKES	USU05	$396	$639.80	$299	$130	4/2
	Trip Kit Subscriber Discount ($33)				$97	
	CO, IA, IL, IN, KS, KY, MI, MN, MO, MT, ND, NE, OH, SD, WI, WY					
SOUTHEAST & SOUTH CENTRAL	USW05	$327	$570.80	$257	$94	4/2
	Trip Kit Subscriber Discount ($24)				$70	
	AL, AR, FL, GA, LA, MS, NC, NM, OK, SC, TN, TX					
GULF OF MEXICO	USY05	$146	$206.95	$120	$35	1/2
	Trip Kit Subscriber Discount ($9)				$26	
	Gulf of Mexico (Helicopter Service) Instrument approaches within 200nm of the coast at airports with weather reporting capability. Also includes Gulf of Mexico Area chart.					

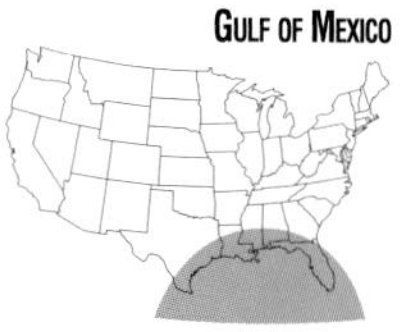

VISIT YOUR JEPPESEN DEALER OR CALL 1-800-621-5377
MAKE SURE TO CHECK OUT OUR WEB PAGE AT HTTP://WWW.JEPPESEN.COM
PRICES SUBJECT TO CHANGE.

STANDARD AIRWAY MANUAL SERVICE ON A CD-ROM

JeppView provides the same content and quality of Jeppesen's Standard Airway Manual Service in a CD-ROM format. This service provides greater convenience and reduces time spent filing revisions. Low altitude paper enroute charts are included.

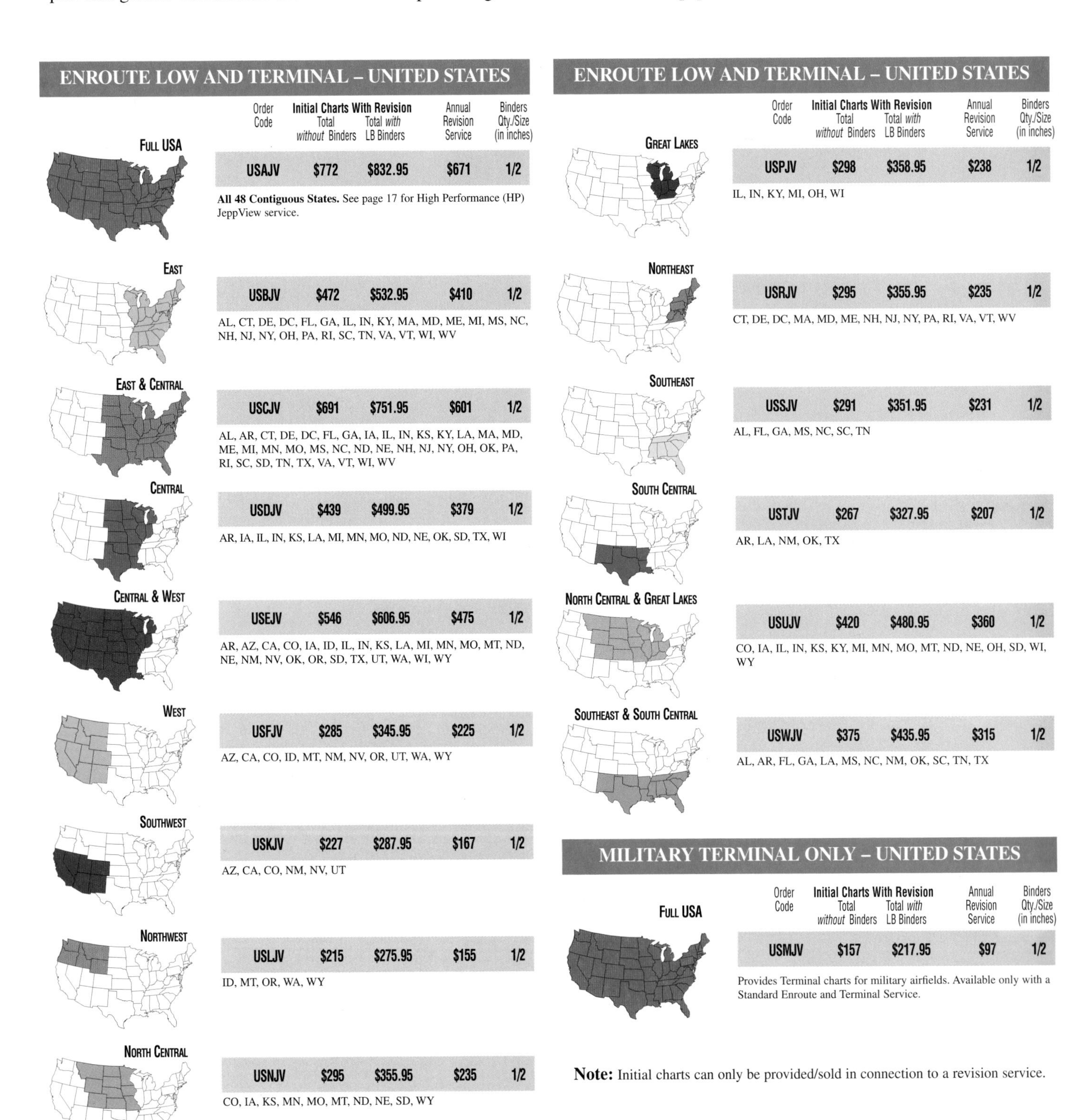

ENROUTE LOW AND TERMINAL – UNITED STATES

Region	Order Code	Initial Charts With Revision: Total *without* Binders	Initial Charts With Revision: Total *with* LB Binders	Annual Revision Service	Binders Qty./Size (in inches)	Coverage
Full USA	USAJV	$772	$832.95	$671	1/2	**All 48 Contiguous States.** See page 17 for High Performance (HP) JeppView service.
East	USBJV	$472	$532.95	$410	1/2	AL, CT, DE, DC, FL, GA, IL, IN, KY, MA, MD, ME, MI, MS, NC, NH, NJ, NY, OH, PA, RI, SC, TN, VA, VT, WI, WV
East & Central	USCJV	$691	$751.95	$601	1/2	AL, AR, CT, DE, DC, FL, GA, IA, IL, IN, KS, KY, LA, MA, MD, ME, MI, MN, MO, MS, NC, ND, NE, NH, NJ, NY, OH, OK, PA, RI, SC, SD, TN, TX, VA, VT, WI, WV
Central	USDJV	$439	$499.95	$379	1/2	AR, IA, IL, IN, KS, LA, MI, MN, MO, ND, NE, OK, SD, TX, WI
Central & West	USEJV	$546	$606.95	$475	1/2	AR, AZ, CA, CO, IA, ID, IL, IN, KS, LA, MI, MN, MO, MT, ND, NE, NM, NV, OK, OR, SD, TX, UT, WA, WI, WY
West	USFJV	$285	$345.95	$225	1/2	AZ, CA, CO, ID, MT, NM, NV, OR, UT, WA, WY
Southwest	USKJV	$227	$287.95	$167	1/2	AZ, CA, CO, NM, NV, UT
Northwest	USLJV	$215	$275.95	$155	1/2	ID, MT, OR, WA, WY
North Central	USNJV	$295	$355.95	$235	1/2	CO, IA, KS, MN, MO, MT, ND, NE, SD, WY
Great Lakes	USPJV	$298	$358.95	$238	1/2	IL, IN, KY, MI, OH, WI
Northeast	USRJV	$295	$355.95	$235	1/2	CT, DE, DC, MA, MD, ME, NH, NJ, NY, PA, RI, VA, VT, WV
Southeast	USSJV	$291	$351.95	$231	1/2	AL, FL, GA, MS, NC, SC, TN
South Central	USTJV	$267	$327.95	$207	1/2	AR, LA, NM, OK, TX
North Central & Great Lakes	USUJV	$420	$480.95	$360	1/2	CO, IA, IL, IN, KS, KY, MI, MN, MO, MT, ND, NE, OH, SD, WI, WY
Southeast & South Central	USWJV	$375	$435.95	$315	1/2	AL, AR, FL, GA, LA, MS, NC, NM, OK, SC, TN, TX

MILITARY TERMINAL ONLY – UNITED STATES

Region	Order Code	Initial Charts With Revision: Total *without* Binders	Initial Charts With Revision: Total *with* LB Binders	Annual Revision Service	Binders Qty./Size (in inches)
Full USA	USMJV	$157	$217.95	$97	1/2

Provides Terminal charts for military airfields. Available only with a Standard Enroute and Terminal Service.

Note: Initial charts can only be provided/sold in connection to a revision service.